Selected Papers
(1945–1980)
With Commentary

2005 Edition

World Scientific Series in 20th Century Physics

Published

Vol. 20 The Origin of the Third Family
edited by O. Barnabei, L. Maiani, R. A. Ricci and F. R. Monaco

Vol. 21 Spectroscopy with Coherent Radiation — Selected Papers of Norman F. Ramsey (with Commentary)
edited by N. F. Ramsey

Vol. 22 A Quest for Symmetry — Selected Works of Bunji Sakita
edited by K. Kikkawa, M. Virasoro and S. R. Wadia

Vol. 23 Selected Papers of Kun Huang (with Commentary)
edited by B.-F. Zhu

Vol. 24 Subnuclear Physics — The First 50 Years: Highlights from Erice to ELN
by A. Zichichi
edited by O. Barnabei, P. Pupillo and F. Roversi Monaco

Vol. 25 The Creation of Quantum Chromodynamics and the Effective Energy
by V. N. Gribov, G. 't Hooft, G. Veneziano and V. F. Weisskopf
edited by L. N. Lipatov

Vol. 26 A Quantum Legacy — Seminal Papers of Julian Schwinger
edited by K. A. Milton

Vol. 27 Selected Papers of Richard Feynman (with Commentary)
edited by L. M. Brown

Vol. 28 The Legacy of Léon Van Hove
edited by A. Giovannini

Vol. 29 Selected Works of Emil Wolf (with Commentary)
edited by E. Wolf

Vol. 30 Selected Papers of J. Robert Schrieffer — In Celebration of His 70th Birthday
edited by N. E. Bonesteel and L. P. Gor'kov

Vol. 31 From the Preshower to the New Technologies for Supercolliders — In Honour of Antonino Zichichi
edited by B. H. Wiik, A. Wagner and H. Wenninger

Vol. 32 In Conclusion — A Collection of Summary Talks in High Energy Physics
edited by J. D. Bjorken

Vol. 33 Formation and Evolution of Black Holes in the Galaxy — Selected Papers with Commentary
edited by H. A. Bethe, G. E. Brown and C.-H. Lee

Vol. 35 A Career in Theoretical Physics, 2nd Edition
by P. W. Anderson

Vol. 36 Selected Papers (1945–1980) with Commentary
by Chen Ning Yang

For information on Vols. 1–20, please visit http://www.worldscibooks.com/series/wsscp_series.shtml

World Scientific Series in 20th Century Physics – Vol. 36

SELECTED PAPERS (1945–1980) With Commentary

2005 Edition

Chen Ning Yang

World Scientific

NEW JERSEY • LONDON • SINGAPORE • BEIJING • SHANGHAI • HONG KONG • TAIPEI • CHENNAI

Published in 2005 by

World Scientific Publishing Co. Pte. Ltd.
5 Toh Tuck Link, Singapore 596224
USA office: 27 Warren Street, Suite 401-402, Hackensack, NJ 07601
UK office: 57 Shelton Street, Covent Garden, London WC2H 9HE

Library of Congress Cataloging-in-Publication Data
Yang, Chen Ning, 1922–
Selected papers, 1945–1980, with commentary.

Bibliography: p.
Includes index.
1. Physics—Collected works. 2. Yang, Chen Ning, 1922–
—Collected works. I. Title.
QC21.2.Y3625 2005 539.7 82-13599
ISBN 981-256-367-9

Printed in Singapore by B & JO Enterprise

Preface to the New Edition

Dr K. K. Phua proposed to reprint this book. I think it is a good idea, as it is a record of one person's journey through a very exciting period of physics research after the Second World War. I propose only to add two new items at the end of the book: one a short note written for the book *50 Years of Yang-Mills Theory* edited by G. 't Hooft (World Scientific, 2005), and another, an article *Remembering Robert Mills*, published in 2003 in *Physics Today*.

March 2005

Chen Ning Yang
Ji-Bei Hoang & Kai-Qun Lu Professor
Tsinghua University, Beijing

獻給母親

In the author's office at Stony Brook, 1981. The sculpture is by his friend Ping-Ming Hsiung, who is also the calligrapher of the Chinese title on the cover. (Courtesy of Peter Renz.)

Preface

Some time ago, several colleagues proposed that a festschrift be published for my sixtieth birthday. I thought about it and concluded that a collection of some of my papers with Commentaries might be more interesting. That is the origin of this volume.

There was no uniform criterion for deciding which papers to include. Papers that were unpublished before, or had appeared in journals that are not easily available, were given preference. Recent papers in easily available journals were, as a rule, given low priority.

The papers are generally labeled according to the year they appeared in print, but a number of exceptions have been made to this rule. The labeling of papers in one year by a, b, c, . . . was mostly by chance. A double bracket indicates that the paper is reprinted in this volume. A single bracket indicates that the paper is not reprinted here.

The Commentary is intended primarily to trace my personal development as a physicist and to sort out my interests and ideas starting from my days as a graduate student. It is not meant to be an assessment of my work. No general effort was made to follow up on the developments after the publication of the papers.

A scientist is not expected to give a balanced assessment of his papers. But he is the one who knows their stories best:

文章千古事
得失寸心知　(杜甫)

A piece of literature
Is meant for the millennium.
But its ups and downs are known
Already in the author's heart.

Tu Fu, AD *712–770*

I am grateful to a few close friends who have given me valuable advice and criticism. For her help in preparing the manuscript, I am indebted to Catherine Turpin. Special thanks are owed my friend Ping-Ming Hsiung for his calligraphy on the cover.

Chen Ning Yang
Stony Brook, New York

October 1982

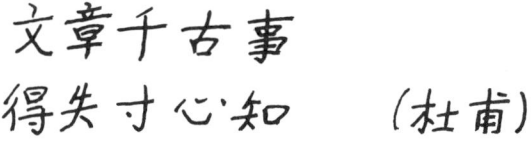

Contents

Articles

Added Items for This Edition

Commentary

A Generalization of the Quasi-Chemical Method in the Statistical Theory of Superlattices

[[45a]]
Article begins page 101

The Journal of Chemical Physics 13, 66 (1945)

C. N. Yang

Commentary After receiving a B.Sc. degree in 1942 at the Southwest Associated University in Kunming, China, I became a graduate student at the same university. For my thesis I worked on statistical mechanics. My thesis advisor was Professor J. S. Wang, who had studied with R. H. Fowler in England in the 1930s. Wang introduced me to this area of physics, which has since remained one of my fields of interest. [[45a]] was a part of my master's thesis.

The Southwest Associated University was among the best universities in China. I received from it an excellent undergraduate education and an equally excellent graduate education up to a master's degree in 1944. The material conditions of universities in China during the war were very poor. However, our university community, i.e., faculty, students, and administrators of the Southwest Associated University, made up for that by good morale and very serious attitudes toward teaching and learning: The library did not have many books; the journals usually arrived after a couple of years' delay; but I learned many, many things in that library. Our classrooms were drafty and cold in the winter; we had only limited equipment for the laboratory courses; but the lectures were generally very systematic and well prepared, going into the subject matters in great depths. I still have today the notes of the quantum mechanics course I took from Wang, which remain a useful reference. They were taken on coarse, unbleached paper that tears easily, and they serve to remind me of the difficult material conditions of those years.

The war (1937–1945) was long and devastating for China, perhaps as much as any in her long history. Indescribable catastrophe befell countless millions. There was the horror of the rape of Nanking in December 1937. There was the Japanese "three all" policy (sanko-seisaku, or kill all, burn all, destroy all), which in one area alone in North China, in 1941–1942, reduced the population from 44 million to 25 million.[1] There was the famine (what adjective should I use for this tragedy?) of Honan province in 1944.[2] There was the last Japanese push in late 1944 when they overran Kweilin and Liuchow and everyone in Kunming was wondering when Kweiyang would fall. And there were air raids and air raids. On September 30, 1940, the house that my family rented in Kunming received a direct hit, reducing most of our meager possessions to rubble. Fortunately every member of the family was in some shelter and no one was wounded. A few days later I went back with a spade and was ecstatic

[1]See Chalmers A. Johnson, *Peasant Nationalism and Communist Power* (Stanford, Calif.: Stanford University Press, 1962), chapter 2.

[2]See T. H. White, *In Search of History* (New York: Harper & Row, 1978).

when I succeeded in digging up several books, warped but quite serviceable. It is difficult now to appreciate how much a few books meant in those darkened days.

And amidst all this, there was the constant threat of inflation.[3] My father was a professor at the Southwest Associated University, and his savings were totally wiped out. At the end of the war we were literally living from hand to mouth. To feed and clothe a family of seven, my mother, a woman of great will power and self-discipline, toiled from dawn to night, year after year, with calm dignity. The family survived the war intact—lean, very lean, but healthy.

I spent the year 1944–1945 teaching mathematics at a high school in Kunming. In my spare time I studied field theory, which I had learned from Professor S. T. Ma from 1942 to 1944. I also became very much interested in the thermomechanics of deformable bodies. I developed a rather elegant formalism for this problem during April and May, 1945, and was deeply disappointed later when I found that F. D. Murnaghan had already done so in 1937.[4]

In late August, 1945, I started for the United States. There being no commercial passenger traffic between China and the United States at that time, I had to wait several months in Calcutta for a berth in a troop transport. I finally reached New York in late November and went to Chicago around Christmas. January 1946 saw me enrolled as a graduate student at the University of Chicago. I have written about this experience and my graduate student days in Chicago in a passage included in Fermi's collected works. It is reproduced as selection [[61a]] in this volume.

Thinking about my student days in China, I am moved by the memory of the atmosphere of the Southwest Associated University, which provided an opportunity for me to learn and to grow. In retrospect, my taste in physics was largely formed during the six years (1938–1944) I spent at that university. To be sure, later in Chicago I came into contact with current problems of research and was influenced especially by Professor Fermi's style. But my attitude toward what to like in physics was largely formed in those years in Kunming.

In every field of creative activity, it is one's taste, together with ability, temperament, and opportunity, that determines one's style and through it one's contribution. That taste and style have so much to do with one's contribution in physics may sound strange at first, since physics is supposed to deal *objectively* with the physical universe. But the physical universe has structure, and one's perceptions of this structure, one's partiality to some of its characteristics and aversion to others, are precisely the elements that make up one's taste. Thus it is not surprising that taste and style are so important in scientific research, as they are in literature, art, and music.

I have said that my taste in physics was largely formed during the six years from 1938 to 1944 when I was a student in Kunming. It was in those years that I learned

[3]The inflation continued after the war. In mid 1949, one U.S. dollar was worth a million Chinese yen, an inflation of 30,000,000 percent since 1937.

[4]F. D. Murnaghan, *American Journal of Mathematics* 59, 235 (1937).

to admire the work of Einstein, Dirac, and Fermi. They have, of course, very different styles. Nonetheless, they share the ability to extract the fundamentals of a physical concept, a theoretical structure, or a physical phenomenon and to zero in on the essentials. Later, when I came to know Fermi and Dirac, I realized that they spoke and thought about physics very much in the way that I had imagined them to do from studying their papers.

In contrast, I did not resonate with the style of Heisenberg. That is not to say that I did not appreciate that he was a great physicist. I did. In fact, I felt thrilled and transformed upon understanding the uncertainty principle in late 1942. But I could not appreciate his approach. My later brief encounters with him at meetings and lectures in the 1950s, 1960s, and 1970s only strengthened my earlier impression.

I did appreciate Schrodinger's approach to wave mechanics, perhaps because it was more in the tradition of classical mechanics and optics, perhaps because Schrodinger's purpose was more clearly defined. In any case, I found wave mechanics to be geometrical and appealing, to be, a priori, more easily appreciated.

On the Angular Distribution in Nuclear Reactions and Coincidence Measurements

⟦48a⟧

Article begins page 112

The Physical Review 74, 764 (1948)

C. N. Yang

Commentary In order to qualify for my bachelor's degree at the Southwest Associated University in 1942, I had to submit an undergraduate thesis. I went to Professor T. Y. Wu asking him to be my supervisor, and he gave me a copy of an article by J. E. Rosenthal and G. M. Murphy in the 1936 volume of *Reviews of Modern Physics*. It was a review paper on group theory and molecular spectra. I was thus introduced to group theory in physics. In retrospect I am deeply grateful to Wu for this introduction, since it had a profound effect on my subsequent development as a physicist.

I had actually already been exposed to the rudiments of group theory by my father when I was a high school student, and I had always been fascinated by the beautiful diagrams in the book by A. Speiser on finite groups that he had on his bookshelf. When I showed him the Rosenthal and Murphy article, he suggested that I should learn about group representations from a small book called *Modern Algebraic Theories*, by L. E. Dickson, who was his thesis advisor in the mathematics department at the University of Chicago in the 1920s. Dickson presented in a short chapter of twenty pages the essentials of the theory of characters. The elegance and potency of the chapter introduced me to the incredible beauty and power of group theory.

At Chicago, I learned more group theory both from studying it myself and from Professor E. Teller, who had an intuitive grasp of the application of group theory in atomic and molecular physics. I used what I learned to advantage in writing [[48a]], which was the result of my trying to understand why in the calculation of various angular correlations there was oftentimes surprising cancellation of terms. Teller made the suggestion to treat the relativistic case in β decay.

Teller had a large number of research associates and graduate students working with him. I was a member of that group, off and on, for about two years, during which I learned a lot of physics. I still remember the first problem Teller asked me to work on: the difference of the K-capture lifetimes of Be^7 and Be^7O crystals. He suggested that I use the Wigner–Seitz method for crystal analysis and the Thomas–Fermi–Dirac method for estimating the electron densities. I was very happy to learn about these methods. I also enjoyed the numerical calculations. But the final result involved cancellations between large terms, and I did not have confidence that it was right. So the work has remained unpublished.

Paper [[48a]] formed my Ph.D. thesis. When I arrived at Chicago I had wanted to write an experimental thesis, since I realized the deficiency in my knowledge of experimental physics. To that end I started in the fall of 1946 to work in Professor S. K. Allison's laboratory. H. M. Agnew, H. V. Argo, W. R. Arnold, G. W. Farwell, J. Hinton, L. del Rosario, H. A. Wilcox, and I, during various periods from 1946 to 1948, helped Allison construct a 400 KeV Cockcroft–Walton accelerator and did nuclear experiments with it. When my experiment to resolve the $P_{1/2}$,$P_{3/2}$ states of He^5 did not go well, Teller suggested that I abandon my plan to write an experimental thesis and said he would sponsor paper [[48a]] as my thesis. I was at first disheartened by the idea, but in a few days recovered and accepted his suggestion with relief.

The twenty months I spent in Allison's laboratory were very educational for me. I learned firsthand some of the frustrations of an experimental physicist. The accelerator always seemed to spring leaks erratically. It had the perverse habit of always starting to work well only toward the evening. In addition, I found that some of my laboratory mates had an amazing, uncanny sixth sense about where one should find the leaks and where one should kick a scaling circuit when it misbehaved. I was well received by my fellow graduate students in the laboratory, since I occasionally could help them with theory. But they told jokes about me. The one that Allison liked especially was, "Where there is a bang, there is Yang."

At the time that I was working on [[48a]], the most exciting news in field theory was renormalization. Fermi, Teller, and Wentzel went to the famous Pocono Conference in late March, 1948. They came back deeply impressed by J. Schwinger's talk on quantum electrodynamics. Fermi and Wentzel had both taken a lot of notes, and, beginning April 14, the three of them and five graduate students—G. Chew, M. L. Goldberger, M. N. Rosenbluth, J. Steinberger, and I—gathered several mornings each week in Fermi's office, trying to understand Schwinger's method. This went on for several weeks, and Goldberger wrote up the notes on our discussions. They totaled 49 pages. But we did not make much real progress.

Interaction of Mesons with Nucleons and Light Particles

The Physical Review 75, 905 (1949)
T. D. Lee, M. Rosenbluth, and C. N. Yang

⟦49a⟧ Article begins page 121

Commentary I received my Ph.D. degree in June, 1948, and spent the summer at Ann Arbor, where J. Schwinger and F. J. Dyson gave lectures. In the fall I went back to Chicago and became an Instructor in the Physics Department. I taught one course and continued my research in nuclear physics and field theory. In late 1948, T. D. Lee, M. Rosenbluth, and I studied μ–e decay and μ capture and found that these interactions and β decay have very similar strengths.

Lee had enrolled at the University of Chicago in the fall of 1946. Although we had probably met earlier in China, it was in Chicago that we really got acquainted. I found him to be exceptionally bright and hard working. We got along well and soon became close friends. Being older and several years ahead of him in my graduate studies, I tried to help him in every way. He later became Fermi's thesis student, but he turned to me for guidance and advice, and I was in effect his teacher in physics for those Chicago years.

The work on μ-decay and μ-capture was essentially finished by mid-December when the holidays started, and Rosenbluth and I took a long-distance bus ride to New York City. I remember the trip quite well, since we were stranded for many hours by snow in Pittsburg, and since I read on the bus from the newspapers that Chinese Communist troops had surrounded Peking and Tientsin. In January, 1949, back in Chicago, we were advised by Fermi to publish a short paper on the subject. That was ⟦49a⟧. The same ideas were independently pursued by several other groups[1] and formed one element in the evolving recognition that there are four basic types of interaction in nature and that there is some universality in the weak interactions.

[1]G. Puppi, *Il Nuovo Cimento* 5, 505 (1948); O. Klein, *Nature* 161, 897 (1948); J. Tiomno and J. A. Wheeler, *Review of Modern Physics* 21, 153 (1949).

Are Mesons Elementary Particles?

The Physical Review 76, 1739 (1949)
E. Fermi and C. N. Yang

⟦49b⟧ Article begins page 122

Commentary The discovery in 1947 of several kinds of mesons led to a widely held assumption that they are all elementary. Fermi suggested that we investigate an alternative, the possibility that the π-meson (not yet called a pion) is composite. We worked on this in the summer of 1949 and wrote ⟦49b⟧.

This is the only paper I wrote with Fermi. He had a very interesting way of working on a manuscript. When corrections were to be made, he would cut out portions to be deleted and scotch-tape on additions. As a result, some pages became long rolls of paper. Once he made decisions about the choice of wording and the choice of presentation, he did not easily change his mind. This is very different from my habit, since I had (and still have) an inclination to change my mind about the choice of wording.

⟦50a⟧
Article begins page 127

Selection Rules for the Dematerialization of a Particle into Two Photons

The Physical Review 77, 242 (1950)

C. N. Yang

Commentary In the years I was in Chicago there was a weekly discussion period, with the faculties of the physics and chemistry departments jointly participating. It was very informal and sometimes without a scheduled speaker. The discussion ranged over diverse topics. One week it was about carbon dating, another, about speculations on the origin of elements, and so on. With Fermi, Teller, and Urey present, the discussion was never dull. One day in 1949, it was mentioned in the discussion period that the π^0-meson had been found to disintegrate into two photons. Teller immediately argued that the observation implied the π^0 had spin zero. His argument was too simple and when challenged was demolished. I later thought about the matter and the next day worked out the correct selection rules. That resulted in ⟦50a⟧.

The first part of the paper (Section II) was simple and intuitive and gave all the results. To justify the product nature of the phase factors a field theoretical formal discussion was necessary. That was given in Section IV. The working out of this section was very useful for my future work, since it cemented my appreciation that such formulae as (8), (16), (19), and (A) are not just formally correct but have, in the right instances, experimental implications. These equations are at the very heart of the mathematical formulation of symmetry concepts in field theory.

Paper ⟦50a⟧ was my second published work on symmetry principles. It anchored my interest in this field.

A sequel to ⟦50a⟧, [50b] capitalized on the difference between the correlation of polarization planes of the two photons resulting from the decay of a scalar or a pseudoscalar meson.

In the spring of 1949, I asked Fermi and Teller to recommend me to J. R. Oppenheimer, Director of the Institute for Advanced Study in Princeton, where I was applying for a postdoctoral appointment. They kindly did that. When the appointment came, Fermi advised that I should go there for a year, but not more, since he felt the Institute's physics was generally too abstract. In fact, he, Allison, and Teller got the University of Chicago to make a commitment to hire me back in 1950.

I left the University of Chicago with the full realization that I had benefited enormously from my three-and-one-half years' association with it, from 1946 to 1949. But I also looked forward with excitement and anticipation to my year at the Institute for Advanced Study, especially since W. Pauli and S. Tomonaga were scheduled to be there, together with many bright young theorists who were known to be active in renormalization theory: K. Case, F. Dyson, R. Jost, R. Karplus, N. Kroll, J. M. Luttinger, and many others.

1949 was the year of profound change in China. The Nationalists collapsed in Shanghai on May 25, and I was deeply worried about my family, who had moved to Shanghai a few months earlier. After hesitating for a number of days, I decided that I had the right to contact my parents and inquire about their situation. So I sent them a cable and was overjoyed to receive a prompt answer the next day: "All right." This experience encouraged me to remain in contact with my parents all through the next twenty-odd years of total estrangement between China and the United States. These contacts, in retrospect, were of determining influence for much that happened to me in later years, including my decision to visit China in 1971 as soon as there were first signs of a rapproachment between the two countries.

Reflection Properties of Spin 1/2 Fields and a Universal Fermi-Type Interaction

The Physical Review 79, 495 (1950)
C. N. Yang and J. Tiomno

⟦50c⟧
Article begins page 131

Commentary Discussions with A. Wightman led J. Tiomno and me to raise the question of possible phase factors in the parity operator for a Dirac particle. That was the origin of ⟦50c⟧. The scheme proposed in the paper was later found to be irrelevant for describing experimental findings. The experience of working on the paper proved, however, to be useful to me in 1956.

Fermi was very much interested in ⟦50c⟧. At a conference at Chicago in September, 1951, he arranged a session to discuss it. The proceedings of the conference record his asking a typical question: "Fermi inquired how the different classes of particles might be experimentally distinguished."[1] Fermi's emphasis on the experimental meaning of theoretical constructs was an important element of his style. Another

[1] *Proceedings of the International Conference on Nuclear Physics and the Physics of Elementary Particles*, eds. J. Orear, A. H. Rosenfeld, and R. A. Schluter (Mimeographed notes, University of Chicago, 1951), p. 109.

example of this style in action can be found in the story told by J. J. Sakurai, which concerned the period 1953–1954:

> When Gell-Mann proposed the strangeness scheme, Fermi said to him: "I won't believe in your scheme until you have a way of telling K^0 from $\overline{K}^0$." This remark of Fermi stimulated Gell-Mann and Pais, whose subsequent investigations have led to one of the most far reaching ideas ever proposed in elementary particle physics.[2]

[2]J. J. Sakurai, *Invariance Principles and Elementary Particles* (Princeton: Princeton University Press, 1964), p. 269.

[[50d]]

Article begins page 135

The S-Matrix in the Heisenberg Representation

The Physical Review 79, 972 (1950)

C. N. Yang and David Feldman

Commentary Paper [[50d]] was the result of a collaboration with David Feldman concerning the S-matrix in the Heisenberg representation, which is the representation with the most direct physical interpretation. Our main effort, to calculate directly the S-matrix in the Heisenberg representation, was not successful, although our formalism was natural and appealing. Our efforts taught us that each Feynman diagram compactly and beautifully puts together many terms resulting from our formalism and expresses them as a single term involving the Feynman functions D_F and S_F. We did not, however, succeed in proving this statement directly in high-order processes.

Pauli was very much interested in our work, and we were warned by our fellow postdocs that this might spell trouble for us. We understood what that meant when Pauli began to drop in on us and expressed irritation if we did not make progress. I eventually learned how to deal with him: one must not be afraid of him. After that, Pauli and I were on good terms.

In the spring of 1950, Oppenheimer offered me a five-year appointment at the Institute for Advanced Study. I had a few other offers, but the real decision was whether to go back to Chicago. I remembered and appreciated Fermi's advice that one should not stay too long at the Institute for Advanced Study, where the physics is too abstract. But in the end I decided to stay on at the Institute, because I was dating Chih Li Tu, who was then studying in New York City within an hour's ride from Princeton.

Chih Li was a student in one of the classes I taught, in the year 1944–1945, at a high school in Kunming. We did not know each other very well then. One day during the Christmas holidays in 1949, while having dinner with J. M. Luttinger at the Tea Garden Restaurant on Witherspoon Street in Princeton, I ran into her. Eight months later, on August 26, 1950, we were married.

The Spontaneous Magnetization of a Two-Dimensional Ising Model

The Physical Review 85, 808 (1952)
C. N. Yang

⟦52a⟧
Article begins page 142

Commentary Following Fermi's advice, I tried to interest myself in as many aspects of physics as possible. In 1950–1951 I studied a variety of subjects, two of which were to be of great use to me later on: the question of time-reversal invariance and the theory of β decay, a subject that I had maintained an interest in since my Ph.D. thesis work (see ⟦48a⟧).

In the spring of 1951 Oppenheimer showed me a preprint that he had just received from Schwinger entitled "The Theory of Quantized Fields I," which introduced a time reversal operation later called "Schwinger time reversal." After studying this preprint I wrote a letter to Schwinger. Among other things I argued that whether or not one combines charge conjugation with time reversal was an *arbitrary* choice. In retrospect I missed discovering the CPT theorem because I was not studying the invariance properties of *all* local field theories.

In the summer of 1950, experiments were done on β–γ correlation that led to much discussion about the effect of the Coulomb field on it.[1] In October, E. Merzbacher and I studied this problem in some depth. We did not obtain any exciting results, but the study thoroughly acquainted us with M. E. Rose's article, which appeared in *The Physical Review* 51, 484 (1937). The experience proved very useful later in the study of Coulomb effects in parity nonconservation.

Starting in early 1951, I worked on the Ising model problem intensively. The Ising model is a well-known model of ferromagnetism in statistical mechanics, and I had been familiar with it because of my master's thesis. In 1944, L. Onsager produced, quite unexpectedly, an exact evaluation of the partition function of the model in two dimensions. It was a real tour de force. I had studied his paper in Chicago in the spring of 1947, but did not understand the method, which was very, very complicated, with many algebraic somersaults. One day in early November, 1949, in a ride in the station wagon that the Institute ran from Palmer Square opposite Princeton

[1]See J. M. Cork, W. C. Rutledge, A. E. Stoddard, C. E. Branyan, and J. M. LeBlanc, *The Physical Review* 79, 938 (1950); B. N. Sorensen, B. M. Dale, J. D. Kurbatove, *The Physical Review* 79, 1007 (1950).

University to the Institute, J. M. Luttinger and I happened to talk about the Ising model. Luttinger said that Bruria Kaufman had simplified Onsager's method so that the solution could be understood in terms of the representation of a system of $2n$ anticommuting hermitian matrices. I knew such representations well and understood quite readily the main points of the Onsager–Kaufman method. After arriving at the Institute, I worked out the essential steps of this approach and was very happy at finally understanding Onsager's solution. That same afternoon I proposed to Luttinger that we collaborate on generalizing the Onsager–Kaufman method to the triangular lattice. He was engaged in some other research at that time and did not want to tackle another problem. Thinking about it a little more, I also dropped the proposal myself as not challenging enough.

But I did not drop the Ising model. I kept thinking about it, and realized that Onsager and Kaufman had obtained much more information than just the partition function, which was determined by the largest eigenvalue of the transfer matrix. Their method in fact gave information about *all* eigenvalues and eigenvectors. Proceeding in this direction, I arrived in January, 1951, at the conclusion that the spontaneous magnetization is dependent on an off-diagonal matrix element between the *two* eigenvectors with the largest eigenvalues. It seemed to me that, with all the excess information latent in the Onsager–Kaufman method, I should be able to evaluate this matrix element.

I was thus led to a long calculation, the longest in my career. Full of local, tactical tricks, the calculation proceeded by twists and turns. There were many obstructions. But always, after a few days, a new trick was somehow found that pointed to a new path. The trouble was that I soon felt I was in a maze and was not sure whether in fact, after so many turns, I was anywhere nearer the goal than when I began. This kind of strategic overview was very depressing, and several times I almost gave up. But each time something drew me back, usually a new tactical trick that brightened the scene, even though only locally.

Finally, after about six months of work off and on, all the pieces suddenly fitted together, producing miraculous cancellations, and I was staring at the amazingly simple final result, equation (96) of ⟦52a⟧. Since there were some limiting procedures in my calculation that were not rigorous, I did not feel quite secure until I compared the expansion of the equation in powers of the parameter x, up to x^{12}, with the expansions of Van der Waerden and of Ashkin and Lamb, which were known to be exact to x^{12}. There was complete agreement. That was in June, 1951, about a week before my first child, Franklin, was born.

C. H. Chang generalized ⟦52a⟧ to a rectangular Ising lattice in 1952.[2] His paper also contained what was probably the earliest speculation about the "universality" of critical exponents.

The result of ⟦52a⟧ was in agreement with Onsager's computation of the long-

[2]C. H. Chang, *The Physical Review* 88, 1422 (1952).

range order. He had announced his results at a conference a few years earlier.[3] I don't think he ever published his calculations. In 1970, he gave some interesting information about his method.[4]

The Ising model is a problem that continually gave surprises. On the physics side, it was thought by some people in the 1950s that the model was a kind of interesting mathematical game, not to be taken seriously. Then in the 1960s the situation dramatically changed. The Ising model was found to be very important not only for the study of ferromagnetism but also many other kinds of phase transitions (see ⟦64e⟧). And in the 1970s it gradually became clear that the model was also very much related to field theory. On the mathematics side, several new ingenious methods have been devised to tackle the problem and to obtain many-point correlation functions,[5] and R. Baxter has found solutions to the eight-vertex model that includes the Ising model as a limiting case.

In March, 1965, M. Fisher, M. Kac, Onsager, and I were invited by W. C. de Marcus to give talks on statistical mechanics at the University of Kentucky. (It was, incidentally, a most interesting experience for all of us, staying at the sadly pretentious Carnahan House, and being commissioned Kentucky Colonels.) When we were leaving Lexington, Onsager and I had to wait for hours at the airport. I asked him how it came about that he took all those complicated algebraic steps in his paper of 1944. He said he had a lot of time during the war, so he began to diagonalize the transfer matrix, which had already been discussed by E. Montroll and by H. A. Kramers and G. H. Wannier. He started with a $2 \times \infty$, then a $3 \times \infty$, then a $4 \times \infty$ lattice. He then went on to a $5 \times \infty$ lattice, for which the transfer matrix is 32×32 in size. Such a matrix is quite large, but the experience he had gained with the smaller matrices came in handy, and he was able, after some time, to find all 32 of the eigenvalues. He proceeded then to the $6 \times \infty$ case, and eventually diagonalized the 64×64 matrix, finding that all the eigenvalues were of the form

$$\exp(\pm \gamma_1 \pm \gamma_2 \pm \gamma_3 \pm \gamma_4 \pm \gamma_5 \pm \gamma_6) \ .$$

(This is a simplification of the real regularity. See L. Onsager, *Physical Review* 65, 117 (1944), equations (97) and (98), for the exact statement. The slightly more complicated regularity of the eigenvalues presumably explains why it was not discovered earlier with smaller lattices.) That led to the concept that the algebra of the problem was a product algebra, and hence the manipulations in his paper.

[3]L. Onsager, *Il Nuovo Cimento* 6, Suppl. 261 (1949).

[4]L. Onsager, in *Critical Phenomena in Alloys, Magnets, and Superconductors,* eds. R. E. Mills, E. Ascher, and R. I. Jaffee (New York: McGraw-Hill, 1971), p. 3.

[5]See T. T. Wu, B. M. McCoy, C. A. Tracy, and E. Barouch, *Physical Review D* 13, 316 (1976).

⟦52b⟧
Article begins page 151

Statistical Theory of Equations of State and Phase Transitions. I. Theory of Condensation

The Physical Review 87, 404 (1952)
C. N. Yang and T. D. Lee

⟦52c⟧
Article begins page 157

Statistical Theory of Equations of State and Phase Transitions. II. Lattice Gas and Ising Model

The Physical Review 87, 410 (1952)
T. D. Lee and C. N. Yang

Commentary In the fall of 1951, T. D. Lee came to the Institute for Advanced Study, and we resumed our collaboration. The first problem we tackled was the susceptibility of the two-dimensional Ising model. As stated in the Commentary on ⟦52a⟧, the Onsager–Kaufman method yielded information about *all* eigenvectors of the transfer matrix. I had used some of that information to evaluate the magnetization, and I thought we might be able to use more of that information to evaluate the susceptibility by a second-order perturbation method, one order beyond that used to obtain equation (14) of ⟦52a⟧. This led to a formula that was, so to speak, an order of magnitude more difficult to evaluate than the magnetization. After a few weeks of labor we gave up and turned our attention to the lattice gas, then to J. Mayer's theory of gas–liquid transitions, and finally to the unit-circle theorem. These considerations led to papers ⟦52b⟧ and ⟦52c⟧.

The idea of the lattice gas was more or less in the minds of many authors (see reference 2 of ⟦52c⟧). We firmed it up and elaborated on it because with the result of ⟦52a⟧ we were able to construct the exact two-phase region of a simple two-dimensional lattice gas. (We were especially pleased by the "law of constant diameter," which resembled the experimental "law of rectilinear diameter".) The two-phase region consists of flat portions of the p–v diagram, bounded by the liquid and gas phases. We were thus led very naturally to the question of why Mayer's theory of condensation gave isotherms that stayed flat *into* the liquid phase, instead of becoming curves in the liquid phase.

The Mayer theory of condensation was a milestone in equilibrium statistical mechanics, for it broke away from the mean field type of approach to phase transitions. It caused quite a stir at the Van der Waals Centenary Congress on November 26, 1937 (see ⟦71b⟧). Mayer's theory led to a number of papers by Mayer himself, by B. Kahn and G. E. Uhlenbeck, and by others in succeeding years. In the early 1940s I had attended a series of lectures by J. S. Wang in Kunming on these developments and had been very much interested in the subject ever since.

Using the lattice gas model, for which we had a lot of exact information, Lee and I examined Mayer's theory as applied to this case. This led to a study of the limiting process in the evaluation of the grand partition function for infinite volume. Paper ⟦52b⟧ resulted from this study. It clarified the limiting process and made transparent the relationship between the various portions of an isotherm and the limiting process.

In late 1952, after ⟦52b⟧ had appeared in print, Einstein sent Bruria Kaufman, who was then his assistant, to ask me to see him. I went with her to his office, and he expressed great interest in the paper. That was not surprising, since thermodynamics and statistical mechanics were among his main interests. Unfortunately I did not get very much out of that conversation, the most extensive one I had with Einstein, since I had difficulty understanding him. He spoke very softly, and I found it difficult to concentrate on his words, being quite overwhelmed by the nearness of a great physicist whom I had admired for so long.

Back in the fall of 1951, Lee and I, in familiarizing ourselves with lattice gases, computed the partition function for several small lattices with 2, 3, 4, 5, etc. lattice points. We discovered to our amazement that the roots of the partition functions, which are polynomials in the fugacity, are all on the unit circle for attractive interactions. We were fascinated by this phenomenon and soon conjectured that it was a general theorem for a lattice of any size with attractive interactions. The theorem, later called the unit circle theorem, became the main element that was exploited in ⟦52c⟧ to discuss the thermodynamics of a lattice gas.

Our attempt at proving the conjecture was a struggle, which I described in a letter to M. Kac, dated September 30, 1969—when he was writing for the *Collected Papers* of George Polya. I quote now from that letter:

> *We then formulated a physicist's "proof" based on no double roots when the strength of the couplings were varied. Very soon we recognized this was incorrect; and for, I would guess, at least six weeks we were frustrated in trying to prove the conjecture. I remember our checking into Hardy's book on Inequalities, our talking to Von Neumann and Selberg. We were, of course, in constant contact with you all along (and I remember with pleasure your later help in showing us Wintner's work, which we acknowledged in our paper). Sometime in early December, I believe, you showed us the proof of the special case when all the couplings are there and are of equal strength, the case that you are now writing about in connection with Polya's collected works. The proof was fine, but we were still stuck on the general problem. Then one evening around December 20, working at home, I suddenly recognized that by making* $z_1, z_2, \ldots$ *independent variables and studying their motions relative to the unit circle one could, through an induction procedure, bring to bear a reasoning similar to the one used in your argument and produce the complete proof. Once this idea was there, it took only a few minutes to tighten up all the details of the argument.*
>
> *The next morning I drove Lee to pick up some Christmas trees, and I told him the proof in the car. Later on, we went to the Institute; and I remember telling you about the proof at a blackboard.*
>
> *I remember these quite distinctly because I'm quite proud of both the conjecture and the proof. It is not such a great contribution, but I fondly consider it a minor gem.*[1]

[1]Comments on paper 93, in *George Polya: Collected Papers,* Vol. 2, ed. R. P. Boas (Cambridge: MIT Press, 1974), p. 426.

The unit circle theorem was later generalized to very interesting additional types of interactions.[2]

With the unit circle theorem, it appeared to Lee and me in early 1952 that we could somehow figure out or guess at the root-distribution function $g(\theta)$ on the unit circle (section V of [[52c]]) for the two-dimensional Ising model. We thought that, with the exact expressions for the free energy and the magnetization already known, we had powerful handles on the structure of $g(\theta)$. Unfortunately these handles were not powerful enough, and the exact form of $g(\theta)$ remains unknown today. (The exact form of $g(\theta)$ is of course transformable into the exact partition function of the Ising model in a magnetic field.)

But our efforts in this direction did lead to two useful results. In listening to a seminar around the end of February, 1952, I learned about the new, ingenious combinatorial method of M. Kac and J. Ward for solving the Ising problem without a magnetic field.[3] It occurred to me during the seminar that, by a slight modification of the Kac–Ward method, one could find the partition function for the Ising model with an imaginary magnetic field $H = i\pi/2$. This requires the evaluation of an 8×8 matrix, which Lee and I carried out in the next couple of days, arriving at equation (48) of [[52c]] for the free energy with $H = i\pi/2$.

Comparing this expression with the known Onsager result for the same quantity for the case $H = 0$, Lee and I observed that they are very similar except for some sign changes and related alterations. Thus it seemed that the change $H = 0 \rightarrow H = i\pi/2$ is altogether minor. We therefore tried similar minor changes on the magnetization for $H = 0$ and tested the results by checking whether they were in agreement with the first few terms of a series expansion of the magnetization for $H = i\pi/2$. This was a very good method, and we soon arrived at equation (49) of [[52c]], which we knew was correct, but did not succeed in proving. It was finally proved by B. M. McCoy and T. T. Wu in 1967.[4]

---

[2]See M. Suzuki, *Progress of Theoretical Physics* 40, 1246 (1968); T. Asano, *Progress of Theoretical Physics* 40, 1328 (1968); M. Suzuki and M. E. Fischer, *Journal of Mathematical Physics* 12, 235 (1971).

[3]M. Kac and J. Ward, *The Physical Review* 88, 1332 (1952).

[4]B. M. McCoy and T. T. Wu, *The Physical Review* 155, 438 (1967).

Letter to E. Fermi dated May 5, 1952

Unpublished
C. N. Yang

⟦52d⟧
Article begins page 167

Commentary ⟦52d⟧ was written while I was visiting the University of Washington in Seattle in the spring and summer of 1952.

The academic year 1952–1953 was unproductive for me. I flirted with the strong coupling theory and the strong focusing principle in accelerator design invented by E. Courant, S. Livingston, and H. Snyder. I also kept up my interest in pion–nucleon scattering and cosmic ray experiments. My efforts did not lead to any useful results. Perhaps the best thing I did that year was to get interested in J. de Boer's lectures on liquid He, a new area of physics for me. Fortunately I was secure and confident enough not to be overly disturbed by the lack of productivity.

In mid-December 1952, I received a letter from G. B. Collins, Chairman of the Cosmotron Department at the Brookhaven National Laboratory, inviting me to visit Brookhaven for a year, 1953–1954. At the Third Rochester Conference, December 18–20, 1952, R. Serber told me more about the Laboratory and the offer, and I decided to accept.

In the summer of 1953, I moved to Brookhaven on Long Island. Brookhaven had at that time the largest accelerator in the world, the Cosmotron, operating at 3 GeV. It produced pions and "strange particles," and very interesting results were pouring out of various experimental groups working there. I made it a custom to visit the groups once every few weeks to acquaint myself with their experiments. It was a very different exposure to physics for me compared to the one in Princeton, and I recognized that each had its advantages.

During the summer there were many visitors at Brookhaven. Physics discussions, beach-going, and social activities were all at high gear. Then fall came, the visitors left, and my wife, son, and I settled into a quiet life in one of the apartments converted from old barracks on site. (The laboratory was situated at the old Upton Army Camp.) It was surrounded by woods, in which we took long walks. During weekends, we drove to explore various parts of Long Island. We grew to like Montauk Point, the Atlantic beaches, Wildwood Park, and the unpretentious islanders around Brookhaven. One snowy Sunday, quite without aim, we drove along the north shore and came upon a charming little village. We admired the atmosphere around the nice shopping center and looked in the map to find its name—Stony Brook. We did not know then that the next time we came to Stony Brook, in 1965, it was to become home.

A number of the experiments done at Brookhaven in the year 1953–1954 were about multiple meson production. R. Christian and I made an effort to compute the phase space volumes for different multiplicities. We quickly concluded that we needed to use a computer. IBM had in operation at that time, in its New York City office, the first 701—the grandfather of all IBM computers. Through Serber we arranged to use a few minutes of the machine. Christian knew how to write programs, and we wrote a simple one. That was before FORTRAN, so it was in machine language, with thousands and thousands of statements. We had to constantly estimate the magnitude of quantities and shift the registers right or left in order not to wipe out the significant figures. Christian was a master of programming, and I found it at first to be great fun. But then we began to debug, and I was disheartened to find so many careless mistakes in our statements. I gradually lost interest, but Christian persisted and pushed our project through. Years later when I learned about FORTRAN, I kicked myself for not having thought of the idea of programming language myself, first in 1952 when I was exploring with H. Goldstine the possibility of using the computer at Princeton (the predecessor of the JOHNNIAC and the MANIAC), and then in 1954 in the experience described above.

⟦54a⟧

Article begins page 169

Polarization of Nucleons Elastically Scattered from Nuclei

The Physical Review 94, 1073 (1954)

G. A. Snow, R. M. Sternheimer, and C. N. Yang

Commentary In 1953–1954, C. L. Oxley and collaborators, using the Rochester cyclotron, discovered a large polarization for scattered protons from nuclear targets. The polarization was perpendicular to the plane of scattering, because of parity conservation.[1] At Brookhaven, G. A. Snow, R. M. Sternheimer, and I got interested in this phenomenon. We made a model calculation for it, which resulted in ⟦54a⟧. This experience was later to prove useful to me in the work for ⟦56h⟧, ⟦66b⟧, and ⟦76a⟧.

[1]This was implied in L. Wolfenstein, *The Physical Review* 75, 1664 (1949).

Isotopic Spin Conservation and a Generalized Gauge Invariance

The Physical Review 95, 631 (1954)
C. N. Yang and R. L. Mills

⟦54b⟧ Article begins page 171

Conservation of Isotopic Spin and Isotopic Gauge Invariance

The Physical Review 96, 191 (1954)
C. N. Yang and R. L. Mills

⟦54c⟧ Article begins page 172

Commentary While a graduate student in Kunming and in Chicago, I had thoroughly studied Pauli's review articles on field theory.[1] I was very much impressed with the idea that charge conservation was related to the invariance of a theory under phase changes, an idea, I later found out, due originally to H. Weyl.[2] I was even more impressed with the fact that gauge invariance *determined* all electromagnetic interactions.[3] While in Chicago, I tried to generalize this to isotopic spin interactions by the procedure later written up in ⟦54c⟧, equations (1) and (2). Starting from these it was easy to get to equation (3). Then I tried to define the field strengths $F_{\mu\nu}$ by

$$F_{\mu\nu} = \frac{\partial B_\mu}{\partial x_\nu} - \frac{\partial B_\nu}{\partial x_\mu} \quad ,$$

which was a "natural" generalization of electromagnetism. This led to a mess, and I had to give up. But the basic motivation remained attractive, and I came back to it several times in the next few years, always getting stuck at the same point. This kind of repeated failure at some seemingly good idea is, of course, a common experience for all research workers. Most such ideas are eventually discarded or shelved. But some persist and may become obsessions. Occasionally an obsession does finally turn out to be something good.

As more and more mesons were discovered and all kinds of interactions were being considered,[4] the necessity to have a *principle* for writing down interactions became more obvious to me. So while at Brookhaven I returned once more to the idea of generalizing gauge invariance. My office mate was R. L. Mills, who was about to finish his Ph.D. degree at Columbia with N. Kroll. We worked on the problem and eventually produced ⟦54c⟧. We also wrote an abstract for the April 1954 meeting of the American Physical Society in Washington, which became ⟦54b⟧. Different motivations were emphasized in the two papers.

The formal aspect of our work did not take long and was essentially finished by February, 1954. But we found we were unable to conclude what the mass of the

[1]W. Pauli, *Handbuch der Physik,* 2nd ed. (Geiger and Scheel, 1933) Vol. 24 (1), p. 83; W. Pauli, *Reviews of Modern Physics,* 13, 203 (1941).

[2]H. Weyl, *Zeitschrift für Physik* 56, 330 (1929). See also Pauli, *Handbuch der Physik,* op. cit., p. 111, footnote.

[3]Gauge invariance is a misnomer. See ⟦77e⟧.

[4]See, e.g., R. E. Marshak, *Meson Physics* (New York: McGraw-Hill, 1952).

gauge particle should be. We toyed with the dimensional argument that, for a pure gauge theory, there is no quantity with the dimension of a mass to start with, and therefore the gauge particle must be massless. But we quickly rejected this line of reasoning.

Oppenheimer invited me to return to Princeton for a few days in late February to give a seminar on our work. Pauli was spending the year in Princeton, and he was deeply interested in symmetries and interactions. (He had written in German a rough outline of some thoughts, which he had sent to A. Pais. Years later F. J. Dyson translated this outline into English. It started with the remark, "Written down July 22–25, 1953, in order to see how it looks," and had the title "Meson–Nucleon Interaction and Differential Geometry.") Soon after my seminar began, when I had written down on the blackboard,

$$(\partial_\mu - i\epsilon B_\mu)\psi \quad ,$$

Pauli asked, "What is the mass of this field B_μ?" I said we did not know. Then I resumed my presentation, but soon Pauli asked the same question again. I said something to the effect that that was a very complicated problem, we had worked on it and had come to no definite conclusions. I still remember his repartee: "That is not sufficient excuse." I was so taken aback that I decided, after a few moments' hesitation, to sit down. There was general embarassment. Finally Oppenheimer said, "We should let Frank proceed." I then resumed, and Pauli did not ask any more questions during the seminar.

I don't remember what happened at the end of the seminar. But the next day I found the following message:

February 24

Dear Yang,

I regret that you made it almost impossible for me to talk with you after the seminar.

All good wishes.

Sincerely yours,

W. Pauli

I went to talk to Pauli. He said I should look up a paper by E. Schrodinger, in which there were similar mathematics.[5] After I went back to Brookhaven, I looked for the paper and finally obtained a copy. It was a discussion of space–time-dependent representations of the γ_μ matrices for a Dirac electron in a gravitational field. Equations in it were, on the one hand, related to equations in Riemannian geometry and, on the other, similar to the equations that Mills and I were working on. But it was many years later when I understood that these were all different cases of the mathematical theory of connections on fibre bundles (see Commentary on [[74c]]).

[5]E. Schrodinger, *Sitzungsberichte der Preussischen* (Akademie der Wissenschaften, 1932), p. 105.

Back at Brookhaven, G. Snow, R. M. Sternheimer, and I became interested in the newly discovered polarization phenomenon in nucleon–nuclei scattering, and we wrote [[54a]]. I then turned back to gauge fields. Mills and I wanted to eliminate the supplementary conditions. We tried to duplicate what Fermi did for electromagnetism and separate out the longitudinal field.[6] This led to very complicated computations, and we did not succeed in our attempts.

Should we publish a paper on gauge fields? That was never a real question in our minds. The idea was *beautiful* and should be published. But what is the mass of the gauge particle? We did not have firm conclusions, only frustrating experiences to show that the non-Abelian case is much more involved than electromagnetism. We tended to believe, on physical grounds, that the charged gauge particles cannot be massless. The last section of [[54c]] reveals that we leaned toward this view, without explicitly saying so. It was a section more difficult to write than all earlier sections.

Pauli was the first physicist who showed intense interest in our paper. That is not surprising, since he was familiar with Schrodinger's work[7] and had himself tried to relate interaction with geometry, as his notes to Pais, July 22–25, 1953, showed. I often wondered what he would say about the subject if he had lived into the sixties and seventies.

When I saw Oppenheimer in Princeton in February, 1954, he had appeared to me to be quite his usual self. Two months later I learned from the *New York Times* about his troubles. It was then that I realized he was in fact drafting his long letter to General Nichols of the Atomic Energy Commission when I saw him in Princeton.

[6]See E. Fermi, *Review of Modern Physics* 4, 105 (1932).
[7]Schrodinger, op. cit.

Conservation of Heavy Particles and Generalized Gauge Transformations

[[55b]]

The Physical Review 98, 1501 (1955)
T. D. Lee and C. N. Yang

Article begins page 177

Commentary I spent the summer of 1954 in Ann Arbor and moved back to Princeton in the fall. Among the ideas Lee and I tried in 1954–1955 were several variations on the gauge-field theme. [[55b]] was the result of one of these. The discussions in this paper have direct relevance to current ideas about proton decay.

Another type of variation consisted in replacing the B_μ of [[54c]] with a quadratic expression in some spinor field, but no useful combinations were found.

We also spent considerable time trying to develop a field theory based on quaternions rather than complex numbers. The idea is to write

$$\Psi = \psi_n + j\psi_p \quad , \tag{1}$$

where ψ_p and ψ_n are the usual wave functions for the proton and the neutron and are complex numbers, and j with k and i form the imaginary units of the quaternion algebra. Ψ is then regarded as *one* quaternion field. If α and β are two complex numbers satisfying

$$\alpha\alpha^* + \beta\beta^* = 1 \quad ,$$

then

$$(\alpha + j\beta)\Psi = (\alpha\psi_n - \beta^*\psi_p) + j(\beta\psi_n + \alpha^*\psi_p) \quad .$$

If we write this as

$$\psi_n' + j\psi_p' \quad ,$$

then

$$\psi_n' = \alpha\psi_n - \beta^*\psi_p$$

and

$$\psi_p' = \beta\psi_n + \alpha^*\psi_p \quad ,$$

which is an SU_2 rotation. Thus *multiplication on the left of* Ψ *by a quaternion of unit absolute value,* $\alpha + j\beta$, *generates an* SU_2 *rotation.* Furthermore, if ξ is any complex number with $\xi\xi^* = 1$, then obviously

$$\xi^{-1}\Psi\xi = \psi_n + j(\xi^2\psi_p) \quad .$$

Thus $\xi^{-1}\Psi\xi$ *generates an electromagnetic gauge transformation.* In other words, the two most important transformations on ψ_n and ψ_p are both very simply expressed in quaternion language for *one single field* Ψ .

In ordinary field theory, a complex field represents a *pair* of charged particles. In a quaternion theory, a quaternion field would represent *four* particles. The phase in complex algebra is related to electromagnetism. The phase in quaternion algebra would be related to isotopic spin gauge fields.

If quaternions are the fundamental basis of field theory, then the very *existence of isotopic spin symmetry* would be understood, in the same way that charge conjugation symmetry is understood from the complex-number algebra used in usual field theories. See my comments after Tiomno's talk in Session IX, *Proceedings of the Seventh Rochester Conference,* April 15–19, 1957 (Wiley-Interscience, 1957). Also see [[59c]] and [[72c]].

In 1954–1955 we did not succeed in developing a quaternion theory that is more than a rewriting of the usual theory in quaternion language. Nor did I succeed in many trials in later years. But I continue to believe that the basic direction is right. There must be an explanation for the existence of SU_2 symmetry: nature, we have repeatedly learned, does not do random things at the fundamental level.[1] Furthermore, the explanation is most likely in quaternion algebra: its symmetry is exactly SU_2. Besides, the quaternion algebra is a *beautiful* structure. Yes, it is noncommutative. But we have already learned that nature chose noncommutative algebra as the language of quantum mechanics. How could she resist using the only other possible nice algebra as the language to start all the complex symmetries that she built into the universe?

If all these sound good, then what is the difficulty? I do not know. Some key ideas are missing, obviously. Is it because we do not have an understanding of the theory of a function of quaternion variables? Maybe. Is it because space–time itself should be described by one quaternion variable? Maybe. Or is it something much simpler?

The quaternion was discovered by W. R. Hamilton in 1843, and he was overwhelmed by its profound beauty. He regarded it as his most important contribution and labored the rest of his life, for twenty-two years, to formulate everything in physics and astronomy in terms of quaternions. He was not successful, and his efforts have been generally regarded as misdirected. In E. T. Bell's *Men of Mathematics,* the chapter on Hamilton is entitled "An Irish Tragedy." I quote from one paragraph of this chapter:

> An honor which pleased him more than any he had ever received was the last, as he lay on his deathbed: he was elected the first foreign member of the National Academy of Sciences of the United States which was founded during the Civil War. This honor was in recognition of his work in quaternions, principally, which for some unfathomable reason stirred American mathematicians of the time (there were only one or two in existence, Benjamin Peirce of Harvard being the chief) more profoundly than had any other British mathematics since Newton's *Principia.* The early popularity of quaternions in the United States is somewhat of a mystery. Possibly the turgid eloquence of the *Lectures on Quaternions* captivated the taste of a young and vigorous nation which had yet to outgrow its morbid addiction to senatorial oratory and Fourth of July verbal fireworks.[2]

This may be Bell's sarcasm at its best, but I believe his wit may have carried him away. Nature *does* choose the most elegant and unique mathematical structures to build the universe with, and the quaternion as a number system *is* elegant and unique. Hamilton may still be vindicated.

[1]We must, of course, also seek explanations of other symmetries. See ⟦72c⟧.

[2]E. T. Bell, *Men of Mathematics* (New York: Simon & Schuster, 1937).

⟦56d⟧
Article begins page 178

Charge Conjugation, a New Quantum Number *G*, and Selection Rules Concerning a Nucleon–Antinucleon System

Il Nuovo Cimento 10 (3), 749 (1956)
T. D. Lee and C. N. Yang

Commentary In 1953, Lee moved to Columbia. To continue our collaboration, we developed a system of mutual visits. I would spend one day every week at Columbia, and he would spend one day every week at Princeton (or Brookhaven). For about six years, we maintained this routine while our interests moved between elementary particle theory and statistical mechanics. It was a very fruitful collaboration, more intensive and more extensive than any other collaboration I have participated in. Over the years we had grown to know each other so well that oftentimes it seemed we could anticipate each other's thoughts. Yet we were quite different in temperament, perception, and taste, which worked to the advantage of our collaboration.

In the fall of 1955, we turned our attention to charge conjugation because of the discovery of the antiproton at Berkeley.[1] That resulted in ⟦56d⟧. The most exciting moment of this work was when we realized that all pions, charged as well as neutral, are eigenstates of the operator G, with eigenvalue -1. This simple statement, among other things, clarified a multitude of selection rules that had been discussed by several authors before us. See footnote 3 of ⟦56d⟧.

[1] O. Chamberlain, E. Segré, C. Wiegand, and T. Ypsilantis, *The Physical Review* 100, 947 (1955).

⟦56e⟧
Article begins page 183

Introductory Talk at the 1956 Rochester Conference, Session on Theoretical Interpretation of New Particles

In *High Energy Nuclear Physics* (New York: Wiley-Interscience, 1956)
C. N. Yang

Commentary In the mid-fifties, with several large cyclotrons, the cosmotron, and the bevatron all operating, the study of elementary particles blossomed into a full-fledged field. In retrospect, the efforts of those years were mostly directed toward studying the (3, 3) resonance (now called the Δ particle) and toward identifying and classifying the new particles. The former study led to the Chew–Low theory and to dispersion relations. The latter led to A. Pais' associated production, the Gell-Mann–Nishijima strangeness scheme, and the θ–τ puzzle. Dispersion relations and the θ–τ puzzle were the theoretical foci of discussions at the Sixth Rochester Conference, which met April 3–6, 1956.

M. L. Goldberger gave the introductory talk on dispersion relations, which was just beginning to be appreciated as a major development.[1] It was to become an important theme in theoretical physics for many years to come.

The θ–τ puzzle gradually took shape from 1953 to 1955, as the multitude of particles $K_{\pi 2}$, $K_{\mu 2}$, $K_{\pi 3}$, $K_{\mu 3}$, and K_{e3} were experimentally studied. R. Dalitz contributed to sharply defining the puzzle by studying the decay products of τ ($= K_{\pi 3}$) and plotting the decay configuration in the Dalitz plot. He already concluded at the Fifth Rochester Conference, January 31–February 2, 1955, that, "if the spin of the τ meson is less than 5, it cannot decay into two π mesons" (*Proceedings of the Fifth Rochester Conference*). Such studies led to the conclusion that θ($= K_{\pi 2}$) and τ must be different particles. The puzzle associated with this conclusion was why θ and τ have masses and lifetimes that are identical within the experimental errors of measurement. On the last day of the Sixth Rochester Conference, 1956, in a session entitled Theoretical Interpretation of New Particles, I gave the introductory talk summarizing the various theoretical considerations on the strange particles. Paper ⟦56e⟧ is the text of that talk, much of which was devoted to the θ–τ puzzle.

There were a number of ideas for explaining how two different particles may exhibit very similar apparent lifetimes and masses. These included the suggestions of T. D. Lee and J. Orear and of R. Weinstein, both summarized in ⟦56e⟧. Beyond these schemes, other ideas were related to questions of symmetry. One suggestion, put forth by Lee and me, and also by M. Gell-Mann, involves a larger symmetry than usually envisaged, resulting in "parity doublets." In retrospect, that suggestion reveals how desperate we all were, trying to explain the mass and lifetime degeneracies.

Not discussed in my talk were the off-the-cuff type of discussions that always abound when physicists' attention is focused on some strange, puzzling phenomenon. Much of these discussions were related to space–time symmetries and, in particular, to parity conservation.[2]

Why were such discussions not pursued in depth? I think the answer lies in the main in three directions.

First, space–time symmetry laws have been extremely useful in atomic, molecular, and nuclear physics. Their very usefulness created the automatic assumption that they were inviolate.

Second, on the more technical side, all attempts at discussing possible parity nonconservation were usually immediately thwarted by the following problem: the

---

[1]For a short history of the developments, see M. L. Goldberger, in *Subnuclear Phenomena,* ed. A. Zichichi (New York: Academic Press, 1970), Part B.

[2]One example was the question that R. P. Feynman and M. Block asked at this session and my response, in *High Energy Nuclear Physics* (New York: Wiley-Interscience, 1956), Section VIII, p. 27:

> *Feynman* brought up a question of Block's: Could it be that the θ and τ are different parity states of the same particle which has no definite parity, i.e., that parity is not conserved. That is, does nature have a way of defining right or left-handedness uniquely? *Yang* stated that he and Lee looked into this matter without arriving at any definite conclusions . . . . Perhaps one could say that parity conservation, or else time inversion invariance, could be violated. Perhaps the weak interactions could all come from this same source, a violation of space–time symmetries . . . .

parity selection rules in nuclear physics and in β decay worked extremely well, so how could it be consistent to propose parity nonconservation? (That was in fact exactly the question raised from the audience in September, 1956, at the Seattle International Congress on Theoretical Physics, after my talk on the new particles, in which I presented the proposal Lee and I made in ⟦56h⟧, that parity may not be conserved in the weak interactions. That talk was published as ⟦57d⟧.)

Third, there was missing the idea that one should *disassociate* parity conservation in the weak interactions from parity conservation in the strong interactions. Without this idea, all discussions about parity nonconservation ran immediately into conceptual and experimental difficulties.

⟦56h⟧

Article begins page 189

Question of Parity Conservation in Weak Interactions

The Physical Review 104, 254 (1956)

T. D. Lee and C. N. Yang

Commentary The spring semester at the Institute for Advanced Study at Princeton terminated in early April, and I moved with my family to Brookhaven on April 17, 1956, to spend the summer. Lee and I continued our twice-a-week visits, now between Brookhaven and Columbia. Our interests, as always, covered various topics, but the θ–τ puzzle naturally occupied most of our attention at that time. We were especially interested in the angular distribution in the chain of reactions

$$\pi^- + p \to \Lambda^0 + \theta^0 \tag{1}$$

$$\Lambda^0 \to \pi^- + p \quad , \tag{2}$$

which R. P. Shutt, J. Steinberger, and W. D. Walker had been studying. Their results had been reported at the Rochester Conference, where there had been many confused discussions about the precise range of the variable "dihedral angle" that the three groups were using.

One day in late April or early May, I drove to Columbia for my weekly visit. I picked up Lee at his office, and we had some difficulty finding a parking space. I finally parked near the corner of Broadway and 125th Street. It was about lunchtime, but the restaurants in that neighborhood were not yet open. So we went to the White Rose Cafe nearby and pursued our discussions there. Later we had lunch at the Shanghai Restaurant (according to my memory; the Tien Tsin Restaurant according to Lee's memory). Our discussions centered on the θ–τ puzzle. At one point the idea occurred to me that one should *disassociate* the symmetry of the production process from that of the decay process: If one then assumes that only in the strong interactions is parity conserved, but not in the weak interactions, there would be no difficulty in having θ and τ the same particle with a single spin parity assignment 0^- defined by the strong interactions. This disassociation would have specific consequences for the chain of reactions (1) and (2). Lee at first resisted this idea. I tried to convince him of its attractiveness, especially since it could be tested by looking for the possible existence of an *up–down* asymmetry in reactions (1) and (2). He finally agreed with me, in the Shanghai Restaurant as I remembered it (in April, 1962), in his office later that afternoon as he remembered it.

This discussion concerns the correlation of reactions (1) and (2). That in reaction (1) the Λ produced should be polarized in a direction *perpendicular* to the production plane was entirely similar to the polarization problem in proton scattering and was familiar to me through paper ⟦54a⟧, which Snow, Sternheimer, and I had published earlier. That the decay of such polarized Λ in reaction (2) could produce an up–down asymmetry, if parity is not conserved, was familiar to me because in my thesis ⟦48a⟧ I had established the relationship between parity conservation and the absence of odd powers of $\cos\theta$ in an angular distribution. Using a superposition of s and p waves for the decay products one easily verifies the existence of an up–down asymmetry explicitly when parity is not conserved, which was how I convinced Lee of the idea.

The most-studied weak interaction at that time was β decay. Hundreds if not thousands of experiments on β decay had been performed. Were they consistent with the assumption that parity conservation is obeyed in the strong interactions but not in the weak ones? Fortunately *this question can be discussed in concrete terms.* The usual β-decay interaction was written as the sum of five terms with coefficients C. To introduce parity nonconservation one only had to *add* five more terms with coefficients C'. These latter five, in fact, had been explicitly discussed by J. Tiomno and me in ⟦50c⟧. The simultaneous presence of the C and C' types of terms leads to parity nonconservation. Later that week I realized this crucial point and decided to include all ten terms in the interaction, reexamine the known β-decay experiments (previously analyzed with only the five C-type terms), and check whether existing experimental data were consistent with the simultaneous presence of both the five C and the five C' types of interaction.

In pursuing this subject the first question that came to mind was: Wouldn't the validity of the numerous parity selection rules for β decay, which were known to work very well, imply that parity conservation must hold for β decay? In thinking about this question, I realized that, since the selection rules only concern nuclear matrix elements, including the C' terms would not change the selection rules at all. But what about the spectrum of the electrons? Here paper ⟦50c⟧ came in handy again, since Tiomno and I had investigated this subject and had shown that, for neutrino mass of zero one would not be able to differentiate, from the electron spectrum, between the C and the C' types of terms. If *both* types were present, the problem had to be reinvestigated. A short calculation showed that the spectrum *could not* yield information about whether both types or only one type was present. Such reinvestigation had to be made for all experimentally studied β decay phenomena: allowed spectrum, unique forbidden spectrum, forbidden spectrum with allowed shape, β–ν correlation, and β–γ correlation.

During the next week's visit to Columbia, while we were walking along Claremont Street near Columbia University, I outlined to Lee this list of phenomena that we had to examine.

I spent the next couple of weeks computing these processes. Lee was at that time less familiar with β decay phenomena. He became a bit anxious and advocated that we submit for publication a short paper about reactions (1) and (2). I vetoed that, because I wanted to push the β-decay calculation to the end. It took only a couple of weeks before my calculations were finished. The result was that, in all these processes, previous experiments did not yield any information about whether there was only the C type interaction or there were both the C and C′ types of interaction. In other words, *all previous β decay experiments were irrelevant as far as the question of parity conservation for β decay was concerned.* A little later, Lee made the same calculations and agreed with my conclusion.

Our reaction to these results was described a year and a half later in my Nobel lecture ⟦57s⟧:

> The fact that parity conservation in the weak interactions was believed for so long without experimental support was very startling. But what was more startling was the prospect that a space–time symmetry law which the physicists have learned so well may be violated. This prospect did not appeal to us. Rather we were, so to speak, driven to it through frustration with the various other efforts at understanding the θ–τ puzzle that had been made.

Sometime in May, I gave a seminar in the Cosmotron Department at Brookhaven on our work. At the end of my talk, Walter Selove asked for the underlying reason that none of the previous experiments had a bearing on the subject of parity conservation in β decay. I did not know. A day or two later, Lee came to visit me at Brookhaven. We pondered over this question. We wanted to prove mathematically, without computation, that previous experiments did not measure a quantity proportional to CC′. With the spins of the particles, the phases, and the plus and minus signs, the deliberation was very confusing. Furthermore, as in all symmetry arguments, intuition and logic tended to mix together in circular arguments. At the end of the day, I realized that if one introduces a *formal* transformation

$$C \to C \qquad C' \to -C'$$

then the arguments become simple and one arrives at the conclusion that the terms proportional to CC′ in the calculations must be "pseudoscalars." Since all previous experiments did not measure a pseudoscalar, they had therefore no bearing on parity conservation in β decay. I was pleased by this great clarification and explained it to Lee when we drove to my apartment for dinner.

As a corollary, we also understood the characteristic of the experiment that *would* test parity conservation in weak interactions: the experiment would have to measure a probability that contains a "pseudoscalar" term.

One possibility was the directional distribution of β decay from a polarized nucleus, but it was difficult to produce polarized nuclei and neither Lee nor I then knew that nuclei had already been polarized by low temperature techniques. In discussions later with M. Goldhaber and C. S. Wu about this problem, we learned of such techniques and settled on the suggestion that Co^{60} might be a good nucleus to study.

Toward the end of May, I got a severe attack of backache for the first time in my life. (It was diagnosed, a couple of years later, as due to a slipped disc.) I was forced to take to bed for a few days. While lying in bed I dictated a paper on our work to my wife, Chih Li, who took it down in longhand since she had had no secretarial training. The title was, "Is Parity Conserved in Weak Interactions?" I showed the manuscript to Lee, who made a few minor changes. I then signed it with our names in alphabetical order. Briefly, I considered putting my name first, but decided against it, both because I disliked name ordering and because I wanted to help Lee along in his career. The manuscript was then given to Barbara Keck of the Cosmotron Department at Brookhaven to type.

We made several mistakes in the preprint. Some were corrected during proofreading (see letter to Reinhard Oehme, quoted below in Commentary on ⟦57e⟧). One mistake was not related to parity nonconservation and was pointed out to us in 1957 by R. C. Curtis and M. Morita independently (see [57l]).

On June 5, I drove with Chih Li and our son, Franklin, to Cambridge, Massachusetts. While at Cambridge, I called Julian Schwinger to ask him about a rumor that he had some ideas on the θ–τ puzzle. I got him while he was busily packing to leave for the summer and did not succeed in getting him to elaborate on the rumor.

While in Cambridge, I gave a talk at the Massachusetts Institute of Technology (MIT) that had been postponed from the end of May because of my backache. Ed Purcell and Norman Ramsey were both in the audience and were very much interested in the suggested polarized nucleus βdecay experiment. Ramsey said he might be able to do it at the Oak Ridge National Laboratory. But apparently he did not succeed in convincing Oak Ridge to embark on the project.

Our manuscript at that point discussed β decay asymmetries and Λ^0 decay up–down asymmetry as possible tests of parity conservation in weak interactions. At a party Shirley and Al Wattenberg gave for us, in a short discussion with Al, I realized that the π–μ–e decay sequence would also provide a test. Upon returning to Brookhaven, I added this test to the manuscript, which was being typed. Later, in proofreading, we added still another test, in the Ξ–Λ–p decay sequence.

On June 22, the typed manuscript was ready and bore the Brookhaven National

Laboratory number BNL=2819.[1] We submitted it to *The Physical Review* for publication. A number of preprints were also distributed. When it appeared in print, the title had been changed to "Question of Parity Conservation in Weak Interactions" because the editor ruled that the title should not be a question. I think the original title was much more meaningful.

We were quite happy with the manuscript. We felt that we had done a good job of analyzing the physics of parity conservation and had discussed all experimental tests that appeared feasible to us at that time. We felt that the style of the work was in the good tradition of physics. Since, contrary to previous belief, parity conservation had not been tested for the weak interactions, it was important to do experiments to test it.

In a paper submitted for publication two weeks after the parity paper, we presented a table listing the experimental differences between parity doublet and parity nonconservation. That was [56i]. We were not betting on either possibility over the other. Later, in September, at the Seattle International Congress on Theoretical Physics, I again presented both possibilities (see ⟦57d⟧).

We were not alone in our hesitation to assert that parity nonconservation (that is, right–left asymmetry) in the weak interactions was indeed the solution of the θ–τ puzzle. I was told by an eminent Soviet physicist that L. Landau argued strongly against our preprint at a meeting in the USSR in October, 1956, but later changed his mind. In a famous letter to V. Weisskopf dated January 17, 1957, W. Pauli wrote:

> *I* do not *believe that the Lord is a weak left-hander, and I am ready to bet a very high sum that the experiment will give symmetric angular distribution of the electrons. I do not see any logical connection between the strength of an interaction and its mirror invariance.*

For the experimental physicists, the problem was more involved than for the theorists. There was the additional question: Was it worthwhile to launch into an experiment to test right–left asymmetry in the weak interactions? Since none of the

[1]Until now I have maintained strict public silence about the nature of my collaboration with Lee. For example, except to my immediate family and two close family friends, I have not told before the story described above of paper ⟦56h⟧, which is based on my notes of 1956 and of April 18, 1962. It would have remained untold today, had I not one day in 1979 chanced upon *Elementary Processes at High Energy, Proceedings of the 1970 Majorana School,* edited by A. Zichichi (Academic Press, 1971). In this volume I read an article by Lee, entitled "History of Weak Interactions," which included his story of the history of papers ⟦49a⟧ and ⟦56h⟧. It implied and insinuated various things, about the nature of our relationship, about parity nonconservation, and about how β decay had gotten into the θ–τ puzzle. It avoided giving a straight account of how the crucial ideas and strategies had originated and developed, and how paper ⟦56h⟧ had been written. I realized then that I must someday reveal what had actually happened. Recently (in August, 1982) I saw, again entirely by accident, A. Franklin's "The Discovery and Nondiscovery of Parity Nonconservation" in *Studies in History and Philosophy of Science,* 10 (3), 201 (1979). I am not sure I know the author. I certainly had not known about the existence of such an article. Franklin acknowledges several valuable conversations with Lee. His article quotes several passages from a manuscript by Lee bearing the title "History of Weak Interactions," which was "a talk given at Columbia University, March 26, 1971, unpublished." The manuscript is apparently largely the same as the article in the *Proceedings of the Majorana School* cited above.

experiments we suggested appeared to be simple, it was not surprising that few groups of experimental physicists took the challenge. I remember vividly a lunch conversation at Brookhaven with an experimental physicist. When I urged him to try the π–μ–e experiment, he joked he would do it when he found a very bright graduate student to be a slave.

Lee had been in contact with his Columbia colleague C. S. Wu, whose experimental contributions to the physics of β decay had been legendary. She decided to initiate a collaboration with four Bureau of Standards physicists—E. Ambler, R. W. Hayward, D. D. Hoppes, and R. P. Hudson—to test parity conservation in Co^{60} decay. Their experiment involved difficult new techniques combining the apparatus for β decay and for low temperature physics. To their courage and perception, physicists owe one of the greatest excitements in our field in recent years. Another group that decided to test parity conservation was formed in Chicago by Val Telegdi and J. Friedman. Their experiment was also difficult and met with little enthusiasm from their colleagues. Lee and I did not know of this group's activities until one day in November, 1956, when Reinhard Oehme, who was spending the academic year at the Institute in Princeton, told me that Telegdi's group had been working on the π–μ–e asymmetry, using photographic emulsions as the detector. He then swore me to secrecy to all but Lee.

Present Knowledge About the New Particles ⟦57d⟧

Lecture given at the Seattle International Conference on Theoretical Physics, September 1956
Reviews of Modern Physics 29, 231 (1957)
C. N. Yang

Article begins page 194

Commentary I have already referred, in the Commentary on ⟦56h⟧, to this report at the Seattle Conference, which took place in September, 1956.

The Seattle Conference turned out to be a very interesting one. Many ideas were discussed. There was a debate between R. P. Feynman and J. M. Blatt at the conference, which led to the beautiful experiment on the specific heat of He near the λ-point.[1] See Commentary on ⟦61c⟧.

[1] W. M. Fairbank, in *Critical Phenomena* (Washington, D. C.: Bureau of Standards, 1966), p. 71.

[[57e]]
Article begins page 199

Remarks on Possible Noninvariance Under Time Reversal and Charge Conjugation

The Physical Review 106, 340 (1957)
T. D. Lee, Reinhard Oehme, and C. N. Yang

Commentary In August, 1956, I received a letter from Reinhard Oehme of the University of Chicago that raised the question of the relationship between the violations of parity conservation, charge conjugation invariance, and time-reversal invariance in weak interactions. This was a topic that Lee and I had not considered in our preprint on parity nonconservation, and Oehme's letter was of *great importance.* We answered his letter later that month. (These two letters are printed in full at the end of this Commentary.) Oehme joined me in Princeton as a visitor in the fall of 1956. Toward the end of the year, Lee, Oehme, and I wrote [[57e]], exploring in detail the subject contained in the two letters. The paper included a section discussing, *as an example,* K^0–$\overline{K}^0$ decay modes using the Weisskopf–Wigner formalism[1]. We did not anticipate that that example was to be of practical importance a few years later, after the discovery of *CP* violation (see Commentary on [[64f]]).

Why did we not anticipate *CP* violation? Again, the reason was everybody's natural preference for more symmetry rather than less. In this case, there was furthermore the fact that there was no puzzle to compel us to venture to suggest *CP* violation. Every effort before 1964 was, in fact, directed at *saving* right–left symmetry by redefining the reflection operation to include charge conjugation.

[1]Such discussions, without symmetry violation, had been quite common in 1955–1956. See S. B. Treiman and R. G. Sachs, *The Physical Review* 103, 1545 (1956), and letter to Oehme printed below.

Letter from R. Oehme, August 7, 1956

Dear Professor Yang:

Thank you very much for sending me your very interesting paper on the possible violation of parity conservation in weak interactions. There occurred to me some questions concerning the possibilities for detecting parity violating in β *decay,* π-μ *decay, etc., and I would be very grateful if you could help me to understand why my conclusions are wrong.*

In the appendix of your paper you wrote the interaction Hamiltonian for β*-decay in a form in which the constants* c_α *and* c'_α *become real if one implies invariance with respect to time reversal. I found that invariance under charge conjugation requires that the relative phases of all* c_α *and* $c_{\tilde{\alpha}}$*, as well as all* c'_α *and* $c'_{\tilde{\alpha}}$*, must be real, but that the ratios* $c_\alpha/c'_{\tilde{\alpha}}$ *must be all imaginary. This seems to be a direct contradiction to the requirements derived from time reversal, and, if both invariance principles are taken together, one would conclude that the admixture of the two sets (*c_α *and* c'_α*) is forbidden. I understand, however, that you do not impose time reversal invariance. If one requires invariance under charge conjugation alone, the phase relations mentioned above lead to* $c'c^* + c'^*c = 0$*. This seems to* *indicate that there appear no interference terms between the parts of opposite parity in the density matrix of the final state, even if the initial nucleus is polarized (first order perturbation theory).*

At first sight it is puzzling that in a problem like β-decay charge conjugation and time reversal invariance should lead to different restrictions, but I understand that the proof for the equivalence of both invariance properties involves explicitly the assumption that parity is a good quantum number. Of course one also uses the hermitian character of the Hamiltonian and other properties.

The situation is in one respect similar for the π-μ decay. If one uses direct couplings and imposes charge conjugation and time reversal invariance, then no admixture of parity conserving and parity violating couplings seems to be allowed. But when only charge conjugation invariance is implied, I find for the density matrix ρ_f *in the rest system of the pion*

$$\rho_f = \left|T_o\right|^2 \left\{g_s^2 + g_{ps}^2 + \frac{m_\mu^2}{m_\pi^2}(f_v^2 + f_{pv}^2) + 2\frac{m_\mu}{m_\pi}(g_s f_{pv} + g_{ps} f_v)\bar{\sigma}\cdot\hat{p}\right\}$$

Here g and f are real coupling constants and $\hat{p}$ *is a unit vector in the direction of the μ-meson momentum. In this case the μ meson is polarized only if derivative and non-derivative couplings are simultaneously present. I would like to mention that Dr. Telegdi and Dr. Wright had already noticed that ps + s coupling with real constants produces no polarization.*

In view of the examples mentioned above, it looks as if under certain restrictions the simultaneous invariance under charge conjugation and time reversal would imply that parity must be conserved. But of course we do not know whether it is reasonable to impose time reversal invariance for lepton interactions.

Sincerely yours,

Reinhard Oehme

Letter to R. Oehme, August 28, 1956

Dear Dr. Oehme:

Thank you very much for your interesting letter of August 7, which prompted us to investigate in detail the question of charge conjugation and time reversal symmetry in weak interactions. We would like to make the following remarks on your observations and on the question in general:

(1) Your conclusion that C′/C = pure imaginary when $\mathfrak{C}$ *= charge conjugation is conserved is of course correct.*

(2) Formulae (A.6) and (A.7) in the preprint are not correct. (A.6) should read

$$\alpha = Re\left[C_T C_T'^{*} - C_A C_A'^{*} + i\frac{e^2 Z}{\hbar c p}(C_A C_T'^{*} + C_A' C_T^{*})\right] \times \left|M_{G.T.}\right|^2 \frac{v_e}{c}\frac{2}{\xi + (\xi b/W)}\frac{< J_z >}{J} .$$

There is a similar change for (A.7). These corrections have been made in the proofreading of the paper. The correction terms make it possible to have interference terms even though charge conjugation is conserved. This happens, however, only if (a) both axial vector and tensor type interactions are present; (the usual argument for ruling out the existence of axial vector interaction need be reexamined if one allows for the possibility of parity non-conservation) and (b) there is a coulomb distortion of the wave function of the electron. We shall return to this point in (4).

(3) The question of whether there is any experimental proof of the invariance under charge conjugation in the weak interactions was incorrectly discussed in the preprint. In proofreading, the sentence "So is the . . . and of $K^{\pm}$" in lines 7–9, p. 14 of the preprint is changed into:

"It might appear at first sight that the equality of the lifetimes of $\pi^{\pm}$ and of those of $\mu^{\pm}$ furnish proofs of the invariance under charge conjugation of the weak interactions. A closer examination of this problem reveals, however, that this is not so. In fact, the equality of the lifetimes of a particle and its charge conjugation against decay through a weak interaction (to the lowest order of the strength of the weak interaction) can be shown to follow from the invariance under proper Lorentz transformations, (i.e. Lorentz transformation with neither space nor time inversion). One has therefore at present no experimental proof of the invariance under charge conjugation of the weak interactions."

The above statement concerning the equality of the life times of a particle and its charge conjugate can be proved by utilizing a theorem due to Lüders as discussed by Pauli (Niels Bohr and the Development of Physics, *Pergamon Press, London, 1955).*

(4) Using the same theorem, one can also prove that if charge conjugate invariance holds, there can be no interference between parity conserving and parity nonconserving weak decays (to the lowest order of the strength of the weak interaction) unless there are strong interactions between the decay products to produce phase shifts which are different for opposite parities. (An example of the influence of the phase shifts is the coulomb distortion effect mentioned in (2).) It seems to us that your calculation on the π–μ decay density matrix, giving a polarization even though C is conserved, is in error.

(5) If invariance under time reversal holds, Lüder's theorem gives the very interesting result that the anti-world has always the exact opposite screw from the usual world. This seems to be of great importance if any screw exists at all, (i.e. if parity is not conserved).

(6) If charge conjugation invariance is invalid, one may ask about the $\theta_1{}^\circ$, $\theta_2{}^\circ$ situation. One sees that the long lived component would in general not be a 1:1 mixture of θ° and $\bar{\theta}^\circ$, which can be experimentally tested. However, if time reversal invariance holds, and if the 2π decay mode is the much faster one, then this ratio would still be 1:1, even though charge conjugation invariance may be violated.

We would be happy to discuss all these matters with you when you come to Princeton.

Yours sincerely,

T. D. Lee

C. N. Yang

Parity Nonconservation and a Two-Component Theory of the Neutrino

[[57f]] Article begins page 205

The Physical Review 105, 1671 (1957)

T. D. Lee and C. N. Yang

Commentary The Columbia–Bureau of Standards experiment under C. S. Wu was progressing satisfactorily during the fall of 1956. She shuttled back and forth between New York City and Washington and kept us informed of the progress of the experiment. Around Christmas she finally said they were getting an asymmetry indicating that parity was not conserved in β decay, but she warned that it was very preliminary and therefore we should not spread any rumors. We didn't, but rumors somehow were generated anyway and we answered an alarming number of phone calls asking for and/or offering information.

After additional checks verified their preliminary results, Wu and her group lifted the veil of secrecy and a shock wave swept through the physics community. That was around New Year's Day, 1957. On January 5 I sent a cable to Oppenheimer, who was vacationing in the Virgin Islands: "Wu's experiment yielding large asymmetry showing G equal to G'. Therefore neutrino is a two component wave function." He cabled back "Walked through door," referring to a talk of mine in 1956 in which I had likened the situation in 1956 of the high-energy physicists to that of a man in a dark room. He knows that in some direction there is a door. But in which direction?

In mid-January, C. S. Wu, E. Ambler, R. W. Hayward, D. D. Hoppes, and R. P. Hudson submitted their epoch-making paper. Very rapidly the R. Garwin, L. M. Lederman, M. Weinrich and the V. L. Telegdi, J. I. Friedman papers were also submitted, both showing that parity was not conserved in π–μ–e decay. See *Adventures in Experimental Physics*, gamma volume, edited by B. Maglich (Princeton: World Science Education, 1973).

The large magnitude of the asymmetry observed by the Columbia–Bureau of Standards group implied that, in β decay, invariance under charge conjugation is also violated. See Oehme's letter printed above, and [[57e]]. See also the paper of Ioffe, Okun, and Rudik referred to in [[57s]].

The discovery of parity nonconservation led Lee and me to submit [[57f]], a paper on the two-component neutrino theory, originally discussed by H. Weyl and presented by Pauli in his *Handbuch der Physik* article.[1] In November 1956, A. Salam had distributed a preprint discussing the two-component neutrino theory.[2] His preprint was written before any experimental evidence for parity nonconservation exis-

[1]See H. Weyl, *Zeitschrift fur Physik* 56, 330 (1929); W. Pauli, *Handbuch der Physik* 24(1) (1933). An important point about the two-component theory was made later by K. M. Case in *The Physical Review* 107, 85 (1957), and J. Serpe in *Nuclear Physics* 4, 183 (1957).

[2]A. Salam, *Il Nuovo Cimento* 5, 299 (1957).

ted. In January 1957, Landau also distributed a manuscript on the two-component neutrino theory.[3]

An interesting point made in section 4 of [[57f]] was that, in the two-component theory, the neutrino detection experiments should give a cross section twice as large as that in a four-component theory. We discovered this point one late evening while writing up the paper. Also included in the paper were parity nonconservation effects not previously discussed.

In early January, Lee and I wrote another paper, [57g], under the following circumstances: there were, for a short period, data from the Garwin–Lederman–Weinrich group that showed disagreement with the simple two-component neutrino theory. Lee and I rushed through a calculation to include derivative couplings á la Konopinski–Uhlenbeck and submitted a paper to explain the data. But new data were later collected, and the earlier disagreement disappeared. So we withdrew the paper.

I was greatly excited one morning in mid-January with the realization that a polarized muon moving in a magnetic field would provide a chance for an accurate measurement of its *extra* magnetic moment. I called up J. Steinberger at Brookhaven and told him about the idea. The difficulty of capturing the muon was, however, a stumbling block, and neither of us pursued the subject further. The idea occurred to other physicists independently and eventually led to beautiful experiments of great accuracy, which are known as (g-2) experiments.[4]

On January 15, the Columbia Physics Department called a press conference. J. Bernstein later called it "unprecedented."[5] I do not know whether that characterization was accurate, but I did find the idea of a press conference to announce a scientific development distasteful. I did not participate, although I was repeatedly urged to do so. The next day the *New York Times* carried a front page story about parity nonconservation.

News reached Europe a bit later, and Pauli described his reaction in a letter to V. F. Weisskopf, dated January 27:

> *Now the first shock is over and I begin to collect myself again (as one says in Munich).*
>
> *Yes, it was very dramatic. On Monday 21st at 8:15 PM I was supposed to give a talk about "past and recent history of the neutrino." At 5 PM the mail brought me three experimental papers: C. S. Wu, Lederman, and Telegdi; the latter was so kind to send them to me. The same morning I received two theoretical papers, one by Yang, Lee, and Oehme, the second by Yang and Lee about the two-component spinor theory. The latter was essentially identical with the paper by Salam, which I received as a preprint already six to eight weeks ago and to which I referred in my last short letter*

[3]L. Landau, *Nuclear Physics* 3, 127 (1957).

[4]See A. A. Schupp, R. W. Pidd, and H. R. Crane, *The Physical Review* 121, 1 (1961); G. C. Charpak, F. J. M. Farley, R. L. Garwin, T. Muller, I. C. Sens, and A. Zichichi, *Il Nuovo Cimento* 37, 1241 (1965).

[5]J. Bernstein, *New Yorker,* May 12, 1962.

> *to you. (Was this known in the USA?) (At the same time came a letter from Geneva by Villars with the* New York Times *article.)*
>
> *Now, where shall I start? It is good that I did not make a bet. It would have resulted in a heavy loss of money (which I cannot afford); I did make a fool of myself, however (which I think I can afford to do)—incidentally, only in letters or orally and not in anything that was printed. But the others now have the right to laugh at me.*
>
> *What shocks me is not the fact that "God is left-handed" but the fact that in spite of this He exhibits Himself as left/right symmetric when He expresses Himself strongly.*[6]

Lee and I received invitations to give lectures at the Brookhaven National Laboratory and at Harvard. I decided that Lee should give these lectures but that, for the annual meeting of the American Physical Society in New York, I should give the report on parity nonconservation. The annual meeting took place at the New Yorker Hotel, and the session on parity was announced for the afternoon of Saturday, February 2, 1957. It was scheduled too late to be a regular session, so it was a "post-deadline" session. News about it spread fast, resulting in a mob scene, which K. K. Darrow later described as follows: "The largest hall normally at our disposal was occupied by so immense a crowd that some of its members did everything but hang from the chandeliers."[7] There were four speakers at the session: C. S. Wu, L. M. Lederman, V. L. Telegdi, and I. The session immediately followed a *scheduled* one on astrophysics in the same hall. One of the speakers at that session, Leona Marshall, later complained that she had a huge captive audience, but no one was listening.

The up–down asymmetry in Λ^0 production and decay, which was the idea that started the examination of parity conservation in the first place (see Commentary on ⟦56h⟧), was in fact rather difficult because of the slow rate of production of Λ. Paper [57j] emphasized the usefulness of looking for this asymmetry. Several experimental groups were already working on it, but conclusive reports on its existence (which implied parity nonconservation in hyperon decay) did not come until the fall of 1957.[8]

---

[6]Reproduced and translated in *Collected Scientific Papers of Wolfgang Pauli,* Vol. 1, eds. R. Kronig and V. F. Weisskopf (New York: Wiley-Interscience, 1964).

[7]*Bulletin of the American Physical Society* 2, 249 (1957).

[8]F. S. Crawford, Jr., M. Cresti, M. L. Good, K. Gottstein, E. M. Lyman, E. T. Solmitz, M. L. Stevenson, and H. K. Ticho, *The Physical Review* 108, 1102 (1957); F. Eisler, R. Plano, A. Prodell, N. Samios, M. Schwartz, J. Steinberger, P. Bassi, V. Borelli, G. Puppi, G. Tanaka, P. Woloschek, V. Zoboli, M. Conversi, P. Franzini, I. Mannelli, R. Santangelo, V. Silvestrini, D. A. Glaser, C. Graves, and M. L. Perl, *The Physical Review* 108, 1353 (1957).

[[57h]]

Article begins page 210

Many-Body Problem in Quantum Mechanics and Quantum Statistical Mechanics

The Physical Review 105, 1119 (1957)

T. D. Lee and C. N. Yang

Commentary From July to December, 1956, Lee and I shifted our main attention to the many-body problem. Although this was partly due to the fact that we did not really expect the dramatic result of C. S. Wu's experiment, it was mainly because we happened to be making good progress in the many-body problem during that period.

In the fall of 1955, Kerson Huang became a member of the Institute for Advanced Study. He introduced me to the pseudopotential method for studying the interaction at long wavelengths between two particles with short-range interactions, a method invented by E. Fermi, G. Breit, and by J. Blatt and V. Weisskopf. Huang, Luttinger, and I were then interested in the properties of superfluid He, and we applied the pseudopotential method to the interacting many-body system. In this fashion, we obtained for a Bose system of dilute hard spheres, by a perturbation calculation, the first few terms of a series expansion in powers of (a/λ) for the fugacity series of a Bose gas, where λ is the thermal wavelength

$$\lambda = \sqrt{2\pi\hbar^2/mkT}$$

and a is the diameter of the hard sphere. We also obtained the first few terms of the ground-state energy of a Fermi system of dilute hard spheres in a series expansion in powers of $P_F a$, where P_F is the Fermi momentum of the system.

Although these results were satisfactory, a similar calculation in powers of a for the ground-state energy of a Bose system of dilute hard spheres resulted in

$$E_0 = N\frac{4\pi a(N-1)}{L^3}\left\{1 + (2.37)\frac{a}{L} + \frac{a^2}{L^2}\left[(2.37)^2 + \frac{\xi}{\pi^2}(2N-5)\right]\right\} + \ldots \quad , (1)$$

where N is the number of spheres, $L \times L \times L$ is the periodic box containing the system, and

$$\xi = \sum_{l,m,n,=-\infty}^{\infty} (l^2 + m^2 + n^2)^{-2}, \ (l, m, n) \neq (0, 0, 0) \quad .$$

We had chosen units so that $\hbar = 1$, $2m = 1$. For fixed $\rho = NL^{-3}$, as $L \to \infty$, (1) does not make sense, since the term

$$\frac{a^2\xi}{L^2\pi^2}(2N-5) \to \infty \quad .$$

We were disturbed by this and sat on the results for several months, so that the manuscripts describing our work were not submitted for publication until October 1956. They eventually appeared as [57a] and [57b].

In March and April of 1956, I began to develop a different method for calculating the fugacity expansion of a dilute system of hard spheres and for the ground-state energy of the system. Not much was done, however, because of the Rochester Conference and the intensive work on the θ–τ puzzle in April–June. In July, after manuscripts ⟦56h⟧ and [56i] were ready, I returned to the many-body problem and got Lee interested. Together we made progress developing a method of attack that we called the "binary collision expansion" method.[1] With this method, we obtained the first terms in the series expansion in powers of (a/λ) for the fugacity series of a Bose system of hard spheres and the ground-state energy for a Fermi system of hard spheres. As mentioned earlier, both results had been obtained already by Huang, Luttinger, and me using the pseudopotential method.

Lee and I then turned our attention to the ground-state energy of a Boltzmann system or a Bose system using the binary collision expansion method. I discovered a technique using a summation of most divergent terms. With this technique, we found that the troublesome term in (1) above disappeared after the summation, and obtained

$$\frac{E_o}{N} = 4\pi a\rho\left[1 + \frac{128}{15\sqrt{\pi}}\sqrt{\rho a^3} + \ldots\right] . \tag{2}$$

This result delighted us very much, especially since the form of the term $\sqrt{\rho a^3}$ was first obtained by the technique of summation of most divergent terms *without* any detailed calculations (see ⟦57q⟧).

The first term in (2), $4\pi a\rho$, was known to P. Price in 1951 and to us in 1953. W. Lenz in 1929 had given a formula for the ground-state energy of a particle moving through a collection of other fixed particles that is basically the same as this term (see footnotes 6 and 7 in [57a]). The next term in (2) was a new result.

The technique of summation over most divergent terms is highly risky from the mathematical viewpoint. But it has since been extensively used in statistical mechanics and field theory. It fails, however, in some cases.[2]

Very soon after obtaining (2), Huang, Lee, and I found that it could also be obtained through the pseudopotential method (see Commentary on ⟦57i⟧). With all these developments and with the pressure to write a paper with Oehme, Lee and I decided there was no time to write down in full the binary collision method and the derivation of (2). So on December 10, 1956, we submitted a short summary of our results, and that became ⟦57h⟧. Soon afterward, our attention became focused on parity nonconservation. But we did spare the time to write up the pseudopotential

[1]For later developments in this topic, see ⟦59b⟧ and L. D. Fadeev, *Mathematical Aspects of the Three Body Problem in the Quantum Scattering Theory,* translated by Ch. Gutfreund and I. Meroz (Israel Program for Scientific Translation, 1965).

[2]See T. T. Wu, *The Physical Review* 149, 380 (1966).

method of showing (2), and that became ⟦57i⟧. The papers describing the binary collision method were delayed and were not submitted until 1958–1959.

Formula (5) of ⟦57h⟧ was in error (see footnote 8 of ⟦58d⟧).

Another term in the expansion (2) was obtained later by T. T. Wu, by N. Hugenholtz and D. Pines, and by K. Sawada. See reference 1 of ⟦60g⟧.

⟦57i⟧

Article begins page 212

Eigenvalues and Eigenfunctions of a Bose System of Hard Spheres and Its Low-Temperature Properties

The Physical Review 106, 1135 (1957)

T. D. Lee, Kerson Huang, and C. N. Yang

Commentary Formula (2) of the Commentary on ⟦57h⟧ was obtained by the binary collision expansion method. It was natural to compare it with Kerson Huang's and my result quoted as equation (1) in that same Commentary, which contained the troublesome term $a^2\xi 2N/L^2\pi^2$. This term is divergent in the limit $NL^{-3} = \rho =$ fixed and $L \to \infty$. Using the idea of summing over most divergent terms, I realized that one should sum over an infinite series of such divergent terms, in powers of aN/L. It was hopeful that formula (2) would result.

Before this could be done, I found that in fact (1) was incorrect, since the pseudopotential was not correctly applied to this problem. A more careful examination led to a method of subtractions, which restored the correct usage of the pseudopotential. The way was now open to a fairly short calculation, which Huang, Lee, and I pushed through, resulting in ⟦57i⟧. The discussions at the end of Section 1 and in Section 4 of this paper concern the points outlined above. We also obtained the excitation spectrum near the ground state.

⟦57o⟧

Article begins page 223

General Partial Wave Analysis of the Decay of a Hyperon of Spin 1/2

The Physical Review 108, 1645 (1957)

T. D. Lee and C. N. Yang

Commentary Experiments in 1957 on parity nonconservation in Λ decay prompted Lee and me to write this paper, ⟦57o⟧, which defined the parameters α, β, γ used today in the tables of particles.[1]

[1]R. L. Kelly et al., *Reviews of Modern Physics* 52, S1 (1980).

In 1965, I found a mistake in equation (4) of ⟦57o⟧. This finding was communicated in a letter to J. W. Cronin, dated May 25, 1965.[2]

[2]Ibid., page S12, footnote 2.

Quantum Mechanical Many-Body Problem and the Low Temperature Properties of a Bose System of Hard Spheres

⟦57q⟧
Article begins page 225

Lecture given at the Stevens Conference on the Many-Body Problem, January 1957
In *The Many-Body Problem,* ed. J. K. Percus (New York: Wiley-Interscience, 1963)
Kerson Huang, T. D. Lee, and C. N. Yang

Commentary Paper ⟦57q⟧ gave an overview of the status of the binary collision method and the pseudopotential method as of January 1957. The paper was presented at the Stevens Institute Conference. Although the manuscript was submitted in time to the organizers of the Conference, the Proceedings of the Conference did not appear until 1963. For many years afterward, this paper contained the most detailed description of the method of summing the most divergent terms in the binary collision method.

The Law of Parity Conservation and Other Symmetry Laws of Physics

⟦57s⟧
Article begins page 236

In *Les Prix Nobel* (Stockholm: The Nobel Foundation, 1957); also in *Science* 127, 565 (1958)
C. N. Yang

Commentary The announcement in late October of the 1957 Nobel Physics Award to Lee and me brought great excitement. For a few days we were flooded with phone calls and cables of congratulations. It also brought personal reflection: My research interests had been almost exclusively in the areas of symmetry principles and statistical mechanics. I felt deeply grateful to T. Y. Wu and J. S. Wang for having introduced me to these areas. On October 31 I wrote to T. Y. Wu as follows:

> *At this moment of great excitement, that also calls for deep personal reflection, it is my privilege to express to you my deep gratitude for your*

having initiated me into the field of symmetry laws and group theory in the spring of 1942. A major part of my subsequent work, including the parity problem, is traceable directly or indirectly to the ideas that I learned with you that spring fifteen years ago. This is something that I have always had an urge to tell you, but today is a particularly appropriate moment.

The Nobel Committee asked each winner to write a short description of himself, which was later published in the *Les Prix Nobel en 1957*. For my piece, I emphasized that my three theses written under Professors Ta-you Wu, J. S. Wang, and E. Teller were instrumental in introducing me to my fields of interest. I also emphasized the strong influence Professor Fermi had on my development.

For our speeches in Stockholm, Lee and I agreed that I should talk about developments before the experimental discovery of parity nonconservation and he should talk about developments afterward. ⟦57s⟧ is the text of my talk. To prepare for this paper, I did some research in the history of *C, P, T* and *I* symmetries. For the *CPT* theorem, I looked into the papers of J. Schwinger, G. Lüders, W. Pauli, and R. Jost.

Today, in 1982, symmetry consideration in physics has been elevated to a new level of importance, which I described as "symmetry dictates interactions." See ⟦80b⟧.

⟦57t⟧

Article begins page 247

Speech at the Nobel Banquet, December 10, 1957

In *Les Prix Nobel* (Stockholm: The Nobel Foundation, 1957)

C. N. Yang

Commentary This talk was given in the ornate Blue Room of the City Hall of Stockholm, almost a quarter of a century ago. Much has happened since that time, but reading the talk today I know I have not changed.

⟦58a⟧

Article begins page 248

Possible Determination of the Spin of Λ^0 from Its Large Decay Angular Asymmetry

The Physical Review 109, 1755 (1958)

T. D. Lee and C. N. Yang

Commentary In April, 1957, my family and I went to Paris for our first European trip. I lectured at the École Normale. Later we moved to Geneva, and I worked for two months at the European Organization for Nuclear Research (CERN). The first buildings of CERN were just then being finished but the theorists still had offices in rented prefabs at the airport. Tatiana Fabergé was already the secretary of the theory group, as she still is today. The physicists with whom I had the most discussions that summer were L. Ferretti, O. Piccioni, and J. Steinberger.

My family and I came back to the States in late August and spent one month in Brookhaven before returning to Princeton.

With the discovery of parity nonconservation in weak interactions, nuclear physicists as well as high energy physicists rushed to study various types of weak interactions. In 1957 there were many experiments on β decay, some using parity nonconservation as a new tool, others using methods already known before 1957. A particularly important development was the measurement of the longitudinal polarization of the electrons in β decay. Lee and I had considered this type of experiment in 1956, but we thought then that longitudinal polarization was very, very difficult to detect. Thus, we did not include it in the list of proposed experiments in paper ⟦56h⟧. What we had missed was that passing the electron beam through an electric field easily turns the longitudinal polarization into a transverse one that is simple to detect.

The main push in the spring and summer of 1957 on the part of many physicists was to pin down the precise coupling for β decay. With our preoccupation with gauge concepts, Lee and I preferred it to be the vector and the axial vector type. In session IX of the Rochester Conference of April 15–19, 1957, after J. Tiomno's talk, I said, among other things, that "if the beta-decay interaction turns out to be a vector interaction and not a scalar interaction, one might ask the question if this has anything to do with the vector fields that seem always to arise out of these localized conservation law concepts".[1] Unfortunately, experimentally there was great confusion: different experiments gave different and conflicting results, and the conflict thickened as summer came. At CERN and later at Brookhaven, I decided to get to the heart of the matter by examining the details of the experimental papers and by consulting experts on β decay in the process. I finally came to the conclusion that the experiment of B. M. Rustad and S. L. Ruby on He^6 was the most reliable.

That turned out to be wrong strategy. At the end of the summer, using *theoretical* arguments, R. P. Feynman and M. Gell-Mann, E. C. G. Sudarshan and R. E. Marshak, and J. J. Sakurai had all concluded that the coupling was V and A, and that very likely the He^6 experiment was wrong. In Sakuari's paper there was the following footnote:

> It might be appropriate to remark here that the idea of fitting the β-decay experiments and the π–μ–e sequence with V and A occurred to the present author several weeks before he was informed

[1]*High Energy Nuclear Physics,* Proceedings of 1957 Rochester Conference, Session IX (New York: Wiley-Interscience, 1957).

> of the Feynman–Gell-Mann–Sudarshan–Marshak theory. However, this idea was rejected by Professor C. N. Yang on the ground that the He^6 recoil experiment is one of the more reliable experiments. C. N. Yang (private communication).[2]

He remembered correctly. I did communicate orally to him my conclusion when he visited Brookhaven that summer.

Although the proposal of V and A coupling was theoretically very attractive, especially the conserved vector-current idea of Feynman and Gell-Mann, experimental confirmation was necessary since the mistake in the He^6 experiment had not yet been found. In a devilishly ingenious experiment, M. Goldhaber, L. Grodzins, and A. W. Sunyar resolved this question by measuring the helicity of the neutrino, which they found to be -1. That was conclusive evidence of the V and A coupling, and a new chapter in our knowledge of β decay began. This experiment was finished in early December, 1957, and I remember Goldhaber calling me about the result a couple of days before Chih Li and I took off for Stockholm.

The Institute had invited Lee to visit for the academic year 1957–1958. He and his family moved to Princeton in the fall of 1957 and settled in the new housing project. Our collaboration was now more convenient, and we launched into an investigation that resulted in ⟦58a⟧. It was an elegant paper, useful for experimentally measuring the spins of hyperons.

[2] J. J. Sakurai, *Il Nuovo Cimento* 7, 649 (1958).

⟦58d⟧
Article begins page 252

Low-Temperature Behavior of a Dilute Bose System of Hard Spheres. I. Equilibrium Properties

The Physical Review 112, 1419 (1958)
T. D. Lee and C. N. Yang

Commentary This paper and its sequel, [59a], are follow-ups of ⟦57i⟧. The same method is used as in the earlier paper. Now considering incomplete zero-momentum-state occupation, we treated the degenerate phase of a dilute Bose system of hard spheres at temperatures greater than zero degrees Kelvin.

As explained in the Commentary on ⟦57h⟧, ⟦57i⟧, and ⟦57q⟧, Lee and I had spent a lot of time on the many-body problem in the fall of 1956. There were many fruitful directions to be followed, and the formulation of the binary collision expansion method was still to be written up, when news about C. S. Wu's experiment came in late 1956. We immediately turned our full attention to elementary particles, and that state of affairs lasted about one year. Later, when we could again afford to spend time thinking about topics other than elementary particles, we came back to the many-body problem and worked on ⟦58d⟧ and [59a]. These two papers yielded new results on the thermodynamical functions and transport properties of a low-temperature degenerate dilute Bose system.

Many-Body Problem in Quantum Statistical Mechanics. I. General Formulation

⟦59b⟧

Article begins page 263

The Physical Review 113, 1165 (1959)

T. D. Lee and C. N. Yang

Commentary As mentioned in the Commentary on ⟦58d⟧, Lee and I again devoted some time to statistical mechanics when pressure to work on elementary particles relaxed somewhat in 1958. We wrote five papers during 1958–1959, in a series that started with ⟦59b⟧. The work for the first three papers of the series was already completed in 1956. The last two, [60b] and [60c], were efforts to find the precise nature of the phase transition to superfluid helium.

The basic idea of [60b] was to formulate the grand partition function in terms of the average occupation numbers in momentum space. The resultant variational principle was very elegant. Lee then thought of a nice generalization of this method that enabled us to treat a degenerate system in a similar fashion. That led to the x-ensemble and [60c]. For a while, we were hopeful that we would be able to find the exact nature of the phase transition to the superfluid phase or to classify such transitions. Unfortunately, we did not succeed.

Symmetry Principles in Modern Physics

⟦59c⟧

Article begins page 276

Lecture given at the 75th Anniversary Celebration of Bryn Mawr College, Session on Symmetries, November 6, 1959
Unpublished

C. N. Yang

Commentary Bryn Mawr College celebrated its 75th Anniversary in 1959, a year ahead of time. As part of the celebration, the college organized a conference on "Symmetries." ⟦59c⟧ was my talk at the conference.

As I reread this paper today, what strikes me is not that I was obviously preoccupied with the lack of success of gauge theory, of using symmetry to generate interaction, nor that I emphasized the need to understand the relationship, and the conceptual unification, between the symmetries. I had been emphasizing these points already at the 1955, 1956, and 1957 Rochester Conferences. What strikes me today is the fact that, in 1959, although Y. Nambu, G. Chew, W. Frazer, J. Fulco, and many other theorists had already suggested that there *are* additional resonances, no one seemed to have suspected the existence of a *great number* of them, soon to be discovered.

[[60d]]

Article begins page 281

Theoretical Discussions on Possible High-Energy Neutrino Experiments

Physical Review Letters 4, 307 (1960)

T. D. Lee and C. N. Yang

Commentary In the summer of 1959, I became interested in the work of my brother, C. P. Yang, in computer simulation of phase transitions. He was a pioneer in this new field of research and made very good contributions to the theoretical analysis. He showed, for example, that after generating a collection of configurations, one should compute not only various averages, such as the average energy E and the average specific heat C, but also their standard deviations $\sigma(E)$ and $\sigma(C)$. He also analyzed the speed at which the flipping process, which is Markovian, approaches equilibrium. Later he programmed a 7090 computer at IBM's Thomas J. Watson Research Center and obtained impressive results for the two- and three-dimensional Ising lattices. His work was reported at a conference in Chicago on April 12–14, 1962, and published in 1963.[1]

In the fall of 1959, Lee got me interested in the question of how to obtain more experimental information about the weak interactions. But we did not hit upon any good methods. Lee discussed the problem with his Columbia colleagues, and one day M. Schwartz came up with the idea that one should make high-energy neutrino and antineutrino beams and study their interactions with nucleons and nuclei. That turned out to be an important suggestion, as subsequent developments clearly demonstrated. None of us knew at that time that B. Pontecorvo had already suggested such experiments several months earlier.[2]

That was the origin of [[60d]], which discussed the theoretical significance of high-energy neutrino experiments. It was published in the *Physical Review Letters*, immediately following Schwartz's article.

[1]C. P. Yang, *Proceedings of Symposia in Applied Mathematics* 15, 351 (1963).
[2]B. Pontecorvo, *JETP* 37, 1975 (1959) (in Russian).

Implications of the Intermediate Boson Basis of the Weak Interactions: Existence of a Quartet of Intermediate Bosons and Their Dual Isotopic Spin Transformation Propeterties

⟦60e⟧

Article begins page 286

The Physical Review 119, 1410 (1960)
T. D. Lee and C. N. Yang

Commentary With the establishment of the *V* and *A* couplings for β decay in 1957, theorists made many speculations, published and unpublished, on weak interactions, electromagnetic interactions, and vector mesons. The published ones included papers by J. Schwinger, S. L. Glashow, S. A. Bludman, A. Salam, J. C. Ward, J. L. Lopes, and others. Lopes' paper is particularly interesting from today's viewpoint, but it was hardly noticed at that time.[1]

The aesthetic attractiveness of non-Abelian gauge theories was also quite generally recognized by the late 1950s. Therefore, many speculations were made with the vector meson *W* for the weak interactions identified with the gauge boson.

With our generally more restrained approach, Lee and I did not want to push ideas that were too speculative, though the possible relationship between the gauge boson and *W* was something that we had always liked (see Commentary on ⟦58a⟧). We therefore concentrated on the logical and phenomenological aspects of the consequences of assuming *W* to be the transmitter of weak interactions. That was the origin of ⟦60e⟧, in which we explored a cancellation scheme that was necessitated by experiments. We called it the schizon scheme. We also spent considerable time working with Markstein in 1961 to calculate numerically the cross section for $W^{\pm}$ production by neutrino beams. The result of that calculation was [61e].

[1]J. Leite Lopes, *Nuclear Physics* 8, 234 (1958).

Imperfect Bose System

⟦60g⟧

Article begins page 296

Physica 26, S49 (1960)
C. N. Yang

Commentary This paper was my talk at the 1960 Utrecht Congress on many-particle problems.

The problem mentioned at the end of Section I has not yet been solved, to my knowledge. The discussions of Section VI have a direct relationship to a later paper, ⟦62j⟧.

⟦61a⟧
Article begins page 305

Introductory Notes to the Article "Are Mesons Elementary Particles?"

In *The Collected Papers of Enrico Fermi,* Vol. 2
(Chicago: University of Chicago Press, 1965)
C. N. Yang

Commentary E. Segré was the editor of Fermi's collected works, for which he asked me in the early 1960s to write an article as an introduction to the paper that Fermi and I wrote in 1949. I decided to describe my experience as a graduate student at Chicago under Fermi's influence. Paper ⟦61a⟧ is what I wrote.

Fermi was deeply respected by all, as a physicist and as a person. The quality about him that commands respect is, I believe, solidity. There was nothing about him that did not radiate this fundamental strength of character. One day in the early 1950s, J. R. Oppenheimer, who was the Chairman of the important General Advisory Committee (GAC) of the Atomic Energy Commission (AEC), told me that he had tried to persuade Fermi to stay on the GAC when Fermi's term was up. Fermi was reluctant. He pressed, and finally Fermi said, "You know, I don't always trust my opinions about these political matters."

⟦61b⟧
Article begins page 307

Some Considerations on Global Symmetry

The Physical Review 122, 1954 (1961)
T. D. Lee and C. N. Yang

Commentary In 1960–1961, Lee and I speculated on larger symmetry groups for the strong interactions. Paper ⟦61b⟧ was one such speculation, written after the exciting experimental discovery of several new resonances in 1960.

There were a lot of discussions at the Institute for Advanced Study in late 1960 and early 1961 on larger symmetry groups. As footnote 14 of ⟦61b⟧ indicates, D. R. Speiser and J. A. Tarski did a lot of work on the possibility of SU_3. That was before Gell-Mann's preprint on SU_3 arrived. Somehow they did not publish their paper until very much later.

The gauge theory idea was being applied in the late 1950s to strong interaction theory as well as to weak interactions. In 1960, J. Sakurai published a very enthusiastic paper proposing a non-Abelian gauge theory of strong interactions.[1] He called the theory VTSI, vector theory of strong interactions. The paper has the longest abstract of all the papers in physics I have ever seen. After the discovery of the ρ and ω, he wrote me a letter expressing his irritation that I "*seem to be detached from, and*

[1]J. Sakurai, *Annals of Physics* 11, 1 (1960).

not too sympathetic toward, the sort of program that I [Sakurai] *have been pursuing in the past two years.*" He asked why I was so "*detached*" and went on:

> *You often tell young theoreticians that one of the paramount tasks of theoreticians is to suggest a good experiment. Yet when you proposed the Yang–Mills theory in 1954, you did not encourage the experimentalists to look for the "Yang–Mills particle." Why is it?*

I remember sitting on this letter for a long time, not knowing how to answer him. I don't remember whether I eventually did answer. If I did, it must have been a polite note with no substance.

In the USSR, there was also considerable interest in gauge theories already in the late 1950s. In 1964, D. Ivanienko edited a Russian translation of fifteen articles on gauge theories. The volume was called *Elementary Particles and Gauge Fields* and included articles by J. J. Sakurai, M. Gell-Mann, S. L. Glashow, J. Schwinger, A. Salam, J. C. Ward, R. Utiyama, T. Kibble, Y. Néeman and B. d'Espagnat as well as papers ⟦54c⟧ and ⟦55b⟧.

Theoretical Considerations Concerning Quantized Magnetic Flux in Superconducting Cylinders

⟦61c⟧
Article begins page 315

Physical Review Letters 7, 46 (1961)
N. Byers and C. N. Yang

Commentary In the spring of 1961, I visited Stanford University for a couple of months. I was greatly impressed by the active research atmosphere there, as well as by the beautiful weather. Soon after my arrival, W. M. Fairbank showed me the experiment he and B. S. Deaver were doing on possible flux quantization in superconductors.

Fairbank had moved to Stanford from Duke University, where he had done a beautiful and difficult experiment in 1957, finding an infinite specific heat at the λ-transition point in liquid He.[1] The experiment was very important because of (1) the great precision achieved, (2) its settling the dispute among theorists about whether the specific heat at the λ-transition point is finite or infinite, and (3) its being the first in a series of experimental findings establishing the nature of phase transitions. These findings were to lead to the critical-indices approach to phase transitions, which was an important development in the 1960s. I had seen Fairbank's equipment at Duke and had been greatly impressed by its elegance and, of course, by his results.

[1]W. M. Fairbank, M. J. Buckingham, and C. F. Kellers, *Proceedings of the Fifth International Conference on Low Temperature Physics* (Madison, Wisconsin, 1957), p. 50.

The flux quantization experiment was again a very difficult one. After showing me what he and Deaver were doing, Fairbank asked whether, if they did succeed in finding flux quantization, that would be a new principle of physics or not. I had not thought about flux quantization before that conversation, so of course I could not answer.

N. Byers and I then began to study the subject, which had been discussed by F. London[2] and L. Onsager.[3] We gradually realized that, although their insight into possible flux quantization in superconducting rings was remarkable and might be correct, their reasoning was wrong. In particular, the most complete discussion of the subject, in London's *Superfluids,* is not correct, since it is based on the incorrect assumption that the wave function of the superconductor in the presence of a flux is proportional to that in its absence. As F. Bloch and H. E. Rorschach later remarked, if London's reasoning were right, all rings at all temperatures would have quantized flux, which cannot be the case.[4]

Byers and I were confused for a while about whether, according to the known principles of physics, there should or should not be flux quantization in superconducting rings. In the meantime, Deaver and Fairbank had some preliminary results, which they showed to F. Bloch, N. Byers, and me. To Deaver and Fairbank these results showed flux quantization. But to Bloch, Byers, and me it was not that clear. Soon afterward, I came east for a few days. When I went back to Stanford again, Deaver and Fairbank were working on a second sample, from which there resulted beautiful "steps" in the crucial diagram that they were plotting.[5] Even theorists could see that there was flux quantization. Sometime later, Byers and I finally understood that flux quantization does follow from usual physics principles á la the BCS theory of superconductivity. That is the origin of [[61c]].

The experience with the preliminary results of Deaver and Fairbank reemphasized for me the fact that experts in every field do perceive things that nonexperts, without the trained eye, cannot see. Yet, I believe the skepticism of Bloch, Byers, and myself represents a healthy attitude. Refusing to be brow-beaten is essential for progress in our field. This is, of course, equally true (maybe more so) the other way around: The experimental physicists must refuse to be brow-beaten by theorists. One hundred years ago, when Maxwell had written to Faraday about his efforts at expressing

---

[2]F. London, *The Physical Review* 74, 562 (1948); F. London, *Superfluids* (New York: Wiley, 1950), Vol. 1, p. 152.

[3]L. Onsager, *Proceedings of the 1953 International Conference on Theoretical Physics* (Tokyo: Science Council of Japan, 1954), pp. 935, 936. See also L. Onsager, *The Physical Review* 7, 50 (1961), where Onsager discusses his 1953 views. Fairbank's question to me in 1961 seemed to be prompted by Onsager's discussions of 1953.

[4]F. Bloch and H. E. Rorschach, *The Physical Review* 138, 1697 (1962); F. Bloch, *The Physical Review* 137, A787 (1965).

[5]B. S. Deaver and W. M. Fairbank, *Physical Review Letters,* 7, 43 (1961). In the same issue of *Physical Review Letters* there appeared the paper of R. Doll and M. Näbauer reporting on an independent experiment discovering flux quantization.

Faraday's physical ideas in mathematical language, Faraday wrote back on March 25, 1857:

> *My Dear Sir:*
> *I received your paper, and thank you very much for it. I do not say I venture to thank you for what you have said about "Lines of Force," because I know you have done it for the interests of philosophical truth; but you must suppose it is work grateful to me, and gives me much encouragement to think on. I was at first almost frightened when I saw such mathematical force made to bear upon the subject, and then wondered to see that the subject stood it so well.*[6]

[6]L. Campbell and W. Garrett, *Life of J. C. Maxwell* (New York: Macmillan, 1884), p. 200.

The Future of Physics

[[61f]]

Panel Discussion at the MIT Centennial Celebration, April 8, 1961
Unpublished
C. N. Yang

Article begins page 319

Commentary MIT celebrated its centennial in April, 1961, at a time when the community of physicists in the United States had a more expansive and more exuberant outlook about the future than ever before or after. The background mood of many people seemed to be either an expectation that *all* fundamental problems would be solved shortly or an assumption that physicists could overcome *any* difficulty. These feelings were hardly surprising, given the impressive developments in science and technology in the 1940s and 1950s. (As if to echo the exuberant mood, a few days after the MIT celebration, the USSR successfully launched a man into orbit around Earth.)

As part of the centennial celebration, MIT organized a panel discussion on the future of physics. R. P. Feynman, J. D. Cockcroft, R. Peierls, and I were the panelists. I decided to interject some "discordant notes" in my short talk, which is [[61f]].

Twenty years have passed. Unfortunately the cautionary remarks I made in 1961 were not wrong. There has, of course, been exciting and very important progress in the last twenty years on both the experimental and the theoretical side. But I sense today increasing difficulties for our field. On the one hand, there is the complexity and expensiveness of the present generation of experiments, each of which takes years to prepare and to execute. The cycling time in high energy experiments is now very long, and threatens, unfortunately, to become even longer in the future. On the other hand, the complexity of high-energy theory is also increasing, insulating theorists from each other and from the experiments. One result of these developments is that, more and more, our theoretical graduate students, even some experimental ones, are separated from the physical phenomena that are in the last analysis the fountainhead of our discipline. This is not anybody's fault. But it is worrisome. I am afraid that the climactic synthesis that Einstein and all of us had dreamed of will continue to elude our grasp well into the next century.

⟦62b⟧

Article begins page 322

Tests of the Single-Pion Exchange Model

Physical Review Letters 8, 140 (1962)

S. B. Treiman and C. N. Yang

Commentary The single-pion exchange model for strong interactions was very popular in the early 1960s. ⟦62b⟧ gives simple tests of how good the model is in different experiments.

⟦62g⟧

Article begins page 324

Obituary for Dr. Shih-Tsun Ma

Unpublished

T. D. Lee and C. N. Yang

Commentary S. T. Ma committed suicide in Australia in early 1962. Lee and I were shocked. We wrote ⟦62g⟧ for *Nature* magazine, but the editors refused to publish it, probably because of the critical remarks we made about the U.S. Immigration Office. These remarks were based on our personal knowledge of Ma's feelings, and on a correspondence from T. Y. Wu, who was Ma's teacher and colleague and was very close to him.

⟦62i⟧

Article begins page 325

Theory of Charged Vector Mesons Interacting with the Electromagnetic Field

The Physical Review 128, 885 (1962)

T. D. Lee and C. N. Yang

Commentary In 1960, Lee and I decided to systematically study the theory of charged vector mesons interacting with the electromagnetic field. This was prompted by our work on papers ⟦60d⟧ and [61e], which involved low-order calculations. We realized that charged vector mesons had not been systematically studied theoretically before and decided to look into this rather involved subject. Starting with the canonical formalism, we followed the usual Dyson–Wick procedure and obtained the Feynman diagram. The result was complicated. What was worse, there were very divergent terms containing $\delta^4(0)$, with coefficients that were not relativistically covariant. After some meanderings, I settled on the idea of adding to the Lagrangian a term

$$-\xi \frac{\partial \phi_\mu^*}{\partial x_\mu^*} \frac{\partial \phi_\nu}{\partial x_\nu} , \tag{1}$$

which allowed for equal treatment of the space and time indices 1, 2, 3, 4. This eliminated the bothersome $\delta^4(0)$ term. In fact, it then became possible to understand the origin of the $\delta^4(0)$ term in the limit $\xi \to 0$, as explained in Appendix E of [[62i]].

At this point, T. T. Wu made a very good suggestion that Lee and I followed, namely, that the ξ term (1) should be made electromagnetically gauge invariant.

The Feynman diagram was now covariant and free of the $\delta^4(0)$ term, but not renormalizable. Lee then proposed the introduction of a negative metric, and that rendered the theory renormalizable. (Lee's interest in the negative metric dated back at least to the discussions of the 1950s on the Lee model.)

We called the resultant theory the ξ-limiting formalism, which was described in [[62i]]. The development was later used in a different context for gauge theory.[1]

The ξ-limiting formalism has the very interesting property that, in the limit $\xi \to 0$, there is hope by the method of summation over most divergent terms that one might obtain finite results. This route was pursued by Lee in a paper in *The Physical Review* immediately following [[62i]].[2]

[[62i]] was the last paper coauthored by Lee and me.

Our relationship had started in 1946. It had been close and warm. It had been based on mutual respect, trust, and consideration. Then came 1957 and our success. Our fame, unfortunately, introduced new elements into our relationship that were not there in earlier years. Our collaboration remained fruitful, however, for five more years, but amidst slowly increasing strains. On April 18, 1962, Lee and I had a long talk in his office, in which we reviewed what had happened since 1946: our early relationship, the early 1950s, the events of 1956 that led to the parity paper, and subsequent developments. We found that we had, except for minor points, the same memory of all crucial events. Like reconciliations in family conflicts, it was an emotion-draining experience, with cathartic senses of liberation. The reconciliation, however, did not last. A few months later we parted company for good.

In the sixteen years during which Lee and I were friends, I was like an older brother to him. I was already well known in the early 1950s both in elementary particle physics and in statistical mechanics, and was the senior partner in our collaboration.

---

[1] See E. S. Abers and B. W. Lee, *Physics Reports* 9C, 1 (1973).
[2] T. D. Lee, *The Physical Review* 128, 899 (1962).

Keenly aware that he had to get out of my shadow, I bent over backward to attempt to help him in his career while maintaining strict public silence about the nature of our partnership. To the outside world, our collaboration was an extraordinarily close one. It was also extraordinarily productive for physics. It was admired and envied. And, according to Lee's own assessment, it was of determining influence on his formative years and on his career.

All in all, it was a worthwhile episode in my life. Yes, there was agony, but few things human and meaningful are entirely without pain.

⟦62j⟧

Article begins page 339

Concept of Off-Diagonal Long-Range Order and the Quantum Phases of Liquid He and of Superconductors

Reviews of Modern Physics 34, 694 (1962)

C. N. Yang

Commentary My interest in superfluid He had led me to appreciate the importance of Bose condensation in a Bose system of particles. A problem that had continued to frustrate me was that of the fraction of superfluids. It is known to be 100% at $T = 0$. How is this to be understood from first principles? I had speculated at the end of ⟦60g⟧ that it is related to the largest eigenvalues of the reduced density matrices ρ_1, ρ_2 . . . (something I no longer believe). That speculation was inspired by the earlier work of O. Penrose and L. Onsager,[1] who were the first to give a precise definition of Bose condensation in the case of interacting Bosons.

The introduction of the concept of Cooper pairs and the BCS theory was a milestone in the theory of superconductivity.[2] It clarified the earlier ideas of F. London, M. R. Schafroth, S. T. Butler, and J. M. Blatt.[3] It established that superconductivity was due to the Bose condensation of pairs of electrons. The importance of BCS's work was driven home to me in my work with N. Byers in ⟦61c⟧.

But I was not satisfied. The wave function of BCS had clearly captured the essence of superconductivity. But what *was* this essence? I wanted to know what property of the wave function of the electrons was essential for superconductivity. To put it in another way, I wanted to know the precise definition of "Bose condensation" for a system of interacting Fermions and Bosons. I explored these questions in the winter of 1961–1962. The result was ⟦62j⟧, a paper I have always been fond of, although it is clearly unfinished: The problems raised in Section 7, for example, have not been

[1]O. Penrose, *Philosophical Magazine* 42, 1373 (1951); O. Penrose and L. Onsager, *The Physical Review* 104, 576 (1956).

[2]See L. N. Cooper, *The Physical Review* 104, 1189 (1956); J. Bardeen, L. N. Cooper, and J. R. Schrieffer, *The Physical Review* 108, 1175 (1957).

[3]F. London, *Superfluids* (New York: Wiley, 1950); M. R. Schafroth, S. T. Butler, and J. M. Blatt, *Helvetica Physica Acta* 30, 93 (1957).

pursued in depth, to my knowledge. These problems are difficult, as the conjectured equation (46) clearly indicates:[4] To calculate some of the physical quantities for the superfluid phase one must go *beyond* the usual thermodynamical limit and evaluate $\ln Q$, the logarithm of the partition function, to order L and not just to order L^3.

While at CERN in the summer of 1962, I mentioned to J. S. Bell some of the conjectures in ⟦62j⟧. Within two days he produced proofs of some of them.[5]

[4]There is a misprint in equation (46). Θ should read Φ.
[5]J. S. Bell, *Physics Letters* 2, 116 (1962); see also [63b].

The Mass Formula of SU_3

⟦63e⟧
Article begins page 350

In *Some Recent Advances in Basic Sciences,* Vol. 1
(New York: Academic Press, 1966)
C. N. Yang

Commentary I spent a few months at Stanford again in the spring of 1963, where I met R. J. Oakes. We became interested in the mass formula for the decuplet of nucleons, which was very popular at the time. Studying the validity of the derivation of the formula, we made a careful analysis of whether, with large perturbations, it was justifiable to use perturbation theory, especially since the poles of the S-matrix can change Riemann sheets as the strength of interaction increases. The questions raised in our analysis were written up in [63d].

After our preprint was distributed, "Many people jumped on Yang and Oakes," in the words of F. Low. To answer their criticisms, I elaborated on our paper in a talk at the Belfer Graduate School Annual Science Conference in October, 1963. The same talk was given two months later at the Argonne User's Group Meeting, December 5, 1963. The talk was ⟦63e⟧.

Critical Point in Liquid–Gas Transitions

⟦64e⟧
Article begins page 356

Physical Review Letters 13, 303 (1964)
C. N. Yang and C. P. Yang

Commentary While I was visiting Stanford in 1963, W. Little told me about the discovery in several Russian laboratories in 1962–1963 of logarithmically divergent specific heats in liquid–gas transitions. It was clear that another chapter in the understanding of phase transitions was beginning. Later, C. P. Yang and I wrote ⟦64e⟧, where we pointed out that the lattice gas model was qualitatively a good model for liquid–gas transitions. For liquid He^4, quantum corrections would, however, have to be included, and that would decrease the peak in the specific heat. We also pointed out the necessity, in view of the newly discovered logarithmically divergent specific heat, to examine the possible corrections that had to be applied to the international standard of low-temperature calibrations.

In the spring of 1964, I became a U.S. citizen.

I had been living in the States for 19 years, 1945–1964, a period that covered most of my adult life. Yet, the decision to apply for citizenship was not an easy one. I suppose this is true of many immigrants from most countries, but it is particularly true of a person of Chinese ethnic origin. The concept of leaving China permanently, to emigrate to another country, simply did not exist in traditional Chinese culture. In fact, to emigrate was once regarded as downright treachery. Furthermore, deeply ingrained in the psyche of every Chinese is the mark of the humiliation and exploitation suffered for over a century by a once-glorious culture. It was a century that no Chinese could easily forget. My father was a professor of mathematics in Peking and Shanghai until his death in 1973. He had earned a Ph.D. degree at the University of Chicago in 1928. He was well traveled. Yet I know, in one corner of his heart, he did not forgive me to his dying day for having renounced my country of birth.

That is not all. I learned, as time went on, that the early history of the Chinese in the United States, *our* history, was drenched in unspeakable prejudice, persecution, and massacres. Betty Lee Sung summarized it as follows:

> In 1878, the entire Chinese population of Truckee was rounded up and driven from town.
>
> In 1885, the infamous massacre of 28 Chinese in Rock Springs, Wyoming, occurred. Many others were wounded and hundreds were driven from their homes.
>
> In 1886, Log Cabin, Oregon, was the scene of another brutal massacre.
>
> Professor Mary Coolidge wrote: "During the years of Kearneyism, it is a wonder that any Chinese remained alive in the U.S."[1]

Then came the Geary Act of 1892 and the Chinese Exclusion Acts of 1904, 1911, 1912, 1913, and 1924. Such legislation reduced the Chinese community in the United States to a warped, isolated, and despised pool of exploited bachelor laborers, which it still was when I came to the States in 1945.

One night in the early 1960s, I took a train ride from New York City to Patchogue on my way to Brookhaven. It was very, very late. The rickety cars were almost empty. Behind me sat an old man with whom I struck up a conversation. He was born in Chekiang around 1890 and had been in the States for some fifty years, working

[1]B. L. Sung, *Mountain of Gold* (New York: Macmillan, 1967), p. 44.

sometimes as a laundry man, sometimes as a dishwasher. He had never married. He always lived alone in a room. He had a ready, friendly smile. I kept wondering whether that meant he had no bitterness. As I watched him shuffling down the dimly lit aisle to detrain at Bayshore, bent with age, a little wobbly, I was filled with a mixture of sadness and rage.

Discrimination persists to this day, although not in the rampant form of earlier years. In late 1954, my wife and I paid a couple of hundred dollars as a deposit for a home in a new development near Princeton. A few weeks later we were told by the developer that he had to return our deposit because he was afraid that our being Chinese might affect his sales. We were furious and talked to a lawyer. He advised us not to sue, since in his opinion we had little chance of winning.

Yes, there were things that held me back. Yet I knew that America had been most generous to me. I had come very well equipped, but America had allowed me the opportunity to develop my potential. I knew there was no country in the world that was as generous to immigrants. I also realized that, almost before I recognized it, my roots here were deepening.

In January, 1961, I watched the Kennedy inauguration on television. Robert Frost was asked by Kennedy to read one of his poems. He chose "The Gift Outright." When I heard the lines:

> Possessing what we still were unpossessed by,
> Possessed by what we now no more possessed.
> Something we were withholding made us weak
> Until we found out that it was ourselves
> We were withholding from our land of living,
> And forthwith found salvation in surrender.

something seemed to go directly to my heart. I searched and found Frost's poem in an anthology. It was beautiful. It was powerful. It was to play a part in my decision to apply for U.S. citizenship.

[[64f]]
Article begins page 359

Phenomenological Analysis of Violation of *CP* Invariance in Decay of K^0 and $\overline{K}^0$

Physical Review Letters 13, 380 (1964)
Tai Tsun Wu and C. N. Yang

Commentary In the summer of 1964, when T. T. Wu and I were both visiting Brookhaven, we heard that the Princeton group was finding *CP* violation in K^0–$\overline{K}^0$ decay.[1] The result surprised all physicists, since everyone after 1957 was banking on *CP* conservation, trying to save as much symmetry as possible (see Commentary on [[57e]]).

The K^0–$\overline{K}^0$ is a remarkable system first discussed by M. Gell-Mann and A. Pais on theoretical grounds. They predicted that it should have two lifetimes, a prediction confirmed in 1956 in a beautiful experiment by K. Lande, E. T. Booth, J. Impeduglia, L. M. Lederman, and W. Chinowsky. Even before this, A. Pais and O. Piccioni had pointed out that the long-lived K^0 would in traversing through matter be partly converted into the short-lived K^0. Later, M. L. Good showed that this regeneration phenomenon has a coherent part to it, a phenomenon reminiscent of double refraction in crystals, which was brilliantly studied by W. R. Hamilton a century ago. The regeneration phenomenon was confirmed by R. H. Good, R. P. Matsen, F. Muller, O. Piccioni, W. M. Powell, H. S. White, W. B. Fowler, and R. W. Birge in 1961. Physicists were fascinated by the K^0–$\overline{K}^0$ system because of its complexity and elegance and because it provided a window on minute details of nature's interactions. But no one was prepared for the shock of 1964.

The Princeton group's experiment was an incredibly beautiful one. The workers exercised judgment to weed out all but a few dozen events from more than twenty thousand, and drew a profoundly important conclusion: that *CP* was not conserved. They further evaluated from these events the magnitude of the nonconservation. Their value has stood the test of many, many later experiments, further testimony to the reliability for which they were already famous before the experiment.

News of their discovery spread like wildfire, and within a couple of weeks it seemed that every high-energy theorist was speculating on the origin of *CP* violation. With our tendency toward restraint, Wu and I decided to make instead a phenomenological analysis of K^0–$\overline{K}^0$ decay. Although years before, in paper [[57e]], T. D. Lee, R. Oehme, and I had discussed this decay under the general assumption that *C*, *P*, and *T* conservations are all not necessarily valid, that discussion was too general to be useful for analyzing the new discovery and for suggesting additional experiments.

At first sight, there seemed to be too many parameters for a phenomenological analysis to be fruitful, but we soon found that this was not so. Our analysis was published as [[64f]].[2] It defined the parameters of experimental interest and the six

[1] J. H. Christenson, J. W. Cronin, V. L. Fitch, and R. Turlay, *Physical Review Letters* 13, 138 (1964).
[2] In equation (4) of [[64f]], A_2 and A_2^* should be switched.

equations that tie them together. It provided the framework within which subsequent experiments on K^0–$\overline{K}^0$ decay were analyzed.

This was the first paper that Wu and I coauthored. I had first met him in 1954 at Harvard, where I had gone to give a seminar. We really got to discuss physics together in the summer of 1956, when both of us were visiting Brookhaven. I found that he had a brilliant mind and would pursue questions in physics and mathematics with absolutely fearless tenacity. He was a young bachelor then, so the Yang family quickly acquired him, so to speak. We all found out soon that he had the habit of systematically understating his strength. For example, when we played bridge, he would exclaim that he had a "miserable" hand. That usually meant he did not have quite enough strength to open one no-trump, as we learned through some rather painful experiences.

Wu spent several one-year terms at the Institute for Advance Study in the late 1950s and early 1960s. We had extensive discussions, but until 1964 did not write any papers together.

Some Speculations Concerning High-Energy Large Momentum Transfer Processes

⟦65a⟧

Article begins page 365

The Physical Review 137, B708 (1965)

Tai Tsun Wu and C. N. Yang

Commentary ⟦65a⟧ was a very speculative paper.

In the early 1960s there were exciting discoveries of all kinds of resonances, which attracted the attention of many physicists. At about the same time, but much less noticed, a number of experiments with the new CERN and Brookhaven accelerators were revealing interesting gross features of collisions of hadrons at laboratory energies of more than 10 GeV. The experimental and theoretical physicists involved included G. Cocconi, G. Collins, A. D. Krisch, J. Orear, R. Serber, and L. Van Hove. There were a few others, but not many.

Wu and I were struck by these gross features. We felt that they must indicate some general structural properties of hadrons, since differential cross sections that vary by more than a factor of 10^{10} between small and large angles, for all the collisions studied, could not have done so without general reasons. We came up with the idea that these general reasons are to be found in picturing the hadrons as *extended* objects.

The idea seemed correct to us, so we expanded on it and wrote ⟦65a⟧. It was very speculative. It met with general disbelief. For example, nobody believed at that time that elastic $np \to np$ differential cross sections should be, by and large, symmetrical with respect to 90°, as our equation (3b) speculated.

Paper ⟦65a⟧ was the precursor of ⟦66b⟧ and ⟦67b⟧, which launched a geometrical model of hadronic and nuclear collisions.

Sometime in the spring of 1965, Oppenheimer told me he had decided to retire in 1966 from the Directorship of the Institute for Advanced Study. I thought the timing was right, since he had had a difficult time at the Institute for a number of years, but things were quiet just then. Also, he had received from President Johnson the Enrico Fermi Award on December 2, 1963, an award that every physicist thought the country *owed* him. To announce his retirement one-and-a-half years later seemed right.

In that conversation, Oppenheimer said he would propose to the Trustees that I be appointed his successor. I told him my instinctive reaction was that I did not want to be the Director of the Institute for Advanced Study. He asked me to think it over.

I thought it over and some days later wrote him a letter, in which I said, "It is quite uncertain that I shall make a good Director, while it is quite certain that I shall not enjoy the life of a Director." But destiny seemed to be arranging things to change my career anyway.

During 1964–1965, the legislature of the State of New York voted to establish five Einstein Professorships at universities within the State. John S. Toll, who was to take over as President of the State University at Stony Brook, T. A. Pond, the Chairman of its Physics Department, and Max Dresden, a professor of physics at Stony Brook, decided to approach me to accept an Einstein Professorship at Stony Brook that they hoped to bid for. They mobilized Sheldon Chang, a professor of electrical engineering at Stony Brook and an old classmate of mine from the Southwest Associated University in Kunming, to talk me into visiting the campus to get acquainted with its atmosphere and its faculty.

My wife and I brought our two younger children, Gilbert and Eulee, with us to visit Stony Brook in the spring of 1965. We were put up at Sunwood, the University's guest house overlooking the Long Island Sound. The first evening we were there, the window of our room framed a spectacular sunset over the sound. It captured our hearts.

But there was hesitation on my part. Stony Brook was launching a program of expansion. Toll and Pond offered to have me head an Institute of Theoretical Physics to be built up over the next few years. It would be quite small, and I wouldn't have to spend much time running it. But I was never a person whose chemistry pushes him to take charge. My first reaction to the offer was, did I know how to be in charge of a group even if it is a small one? I struggled with this question, consciously and subconsciously, and eventually decided I could learn to do the job. Around the end of April, I accepted the Stony Brook offer and told Toll I would take up my post in 1966.

πp Charge-Exchange Scattering and a "Coherent Droplet" Model of High-Energy Exchange Processes

⟦66b⟧

Article begins page 369

The Physical Review 142, 976 (1966)

N. Byers and C. N. Yang

Commentary Byers came to Princeton to spend the year 1964–1965. We resumed our collaboration and worked on several problems, among them the new experimental results on hadron–hadron collisions at multi-GeV energies. After a number of false starts, we realized that the extended object model of hadrons, which T. T. Wu and I discussed in ⟦65a⟧, offered a good starting point. The model has a large momentum transfer aspect to it, emphasized in that paper. But it also should have a small momentum transfer aspect, or coherent aspect, which can be formalized with the eikonal approach.[1] Applying this view to $\pi^- p$ charge exchange scattering, we wrote ⟦66b⟧.

We talked quite extensively to G. E. Brown, then at Princeton University (see footnote 13 of ⟦66b⟧). What we learned from him about nuclear physics convinced us that a hadron is very much similar to a nucleus and should be analyzed in that spirit. That spirit remains today the basis of the geometrical model.

A side product of my interaction with G. E. Brown was that I learned to appreciate his good physical instincts. A year later, I helped to induce him to move to Stony Brook. I have always counted that as one of my contributions to the development of the Stony Brook campus.

[1]See S. Fernbach, R. Serber, and T. B. Taylor, *The Physical Review* 75, 1352 (1949); R. J. Glauber, in *Lectures in Theoretical Physics,* eds. W. E. Brittin and L. G. Dunham (New York: Wiley-Interscience, 1959), Vol. 1, p. 315.

Remarks at the Dedication of the Einstein Stamp, March 14, 1966

⟦66c⟧

Article begins page 375

Unpublished

C. N. Yang

Commentary The U.S. Post Office issued a special stamp on March 14, 1966, bearing an engraving of a photograph of Einstein taken by the distinguished photographer Philippe Halsman. Oppenheimer was invited to speak at the dedication ceremony. A few days before the occasion he called me and asked me to speak instead, since he had the flu.

I composed a short speech, ⟦66c⟧, which I read at the ceremony. I was surprised at the end of the ceremony when Einstein's daughter Margot, his secretary Miss Dukas, and Halsman all came to me and profusely thanked me for the speech.

Halsman then said he would like to show me something, and we arranged to meet later that day in my office.

What he showed me was a draft of a letter he wanted to send to the *New York Review of Books* criticizing an article by Oppenheimer on Einstein, which was the written version of what Oppenheimer had said at a UNESCO celebration on December 13, 1965. Halsman was furious. He explained to me that he owed his life to Einstein, who had gotten him out of Nazi Germany, and he felt Einstein was like a father to him. His draft letter contained very strong language. It later appeared in the May 26 issue of the *New York Review of Books,* but had been toned down a lot.

I had not heard of either Oppenheimer's article or his speech and was totally unaware of the storm they had generated. I said nothing to Halsman, everything being such a great surprise to me.

I later got hold of a copy of Oppenheimer's article, which appeared on March 17. It was beautifully phrased. It sketched Oppenheimer's view of Einstein—the physicist and the man—in a carefully worded impressionistic way that was as revealing of the author as it was of the subject. I thought Halsman overreacted: malicious the article was not, "sophisticated"[1] and complex it was.

[1]The meaning of the word *sophistication* was one of the points of disagreement between Oppenheimer and Halsman.

⟦66e⟧
Article begins page 376

One-Dimensional Chain of Anisotropic Spin–Spin Interactions. I. Proof of Bethe's Hypothesis for Ground State in a Finite System

The Physical Review 150, 321 (1966)

C. N. Yang and C. P. Yang

Commentary Paper ⟦66e⟧ and its sequels [66f] and [66g] represented work that C. P. Yang and I carried out in 1965–1966. This work was motivated by several factors.

In early 1951, when I was beginning to calculate the spontaneous magnetization of the Ising model, I studied carefully F. Bloch's theory of spin waves. This led me to the papers of H. A. Bethe[1] and L. Hulthén,[2] which were very interesting. Unfortunately, I got sidetracked into looking for complex solutions of Hulthén's equations and finally put the problem aside.

After the work on off-diagonal long-range order in 1962 (paper ⟦62j⟧), I looked for Hamiltonian systems for which one can *prove* that such order exists. This work

[1]H. A. Bethe, *Zeitschrift für Physik* 71, 205 (1931).
[2]L. Hulthén, *Arkiv foer Matematik, Astronomi, och Fysik* 26A, No. 11 (1938).

proceeded off and on, partly in collaboration with T. T. Wu, partly with C. P. Yang. In 1963, E. H. Lieb and W. Liniger published two interesting papers on a Bose gas with δ-function interaction in one dimension,[3] and I benefited from discussing these papers with J. B. McGuire while I was visiting the University of California at Los Angeles in the summer of 1963. The method of Lieb and Liniger is the same as that of Bethe, namely, one assumes that there is no diffraction, only reflection, and verifies that the assumption is self-consistent. I tried to use similar assumptions to construct a Hamiltonian that has off-diagonal long-range order, but did not succeed.

A little later, our work on ⟦64e⟧ in 1963 led C. P. Yang and me to attempt to incorporate quantum effects into the lattice gas model, since we were convinced, first, that the lattice gas model gives an excellent approximation of the critical phenomenon in liquid–gas transitions and, second, that quantum effects are important for the light gases. We thus invented the quantum lattice gas, but learned later that it had been invented much earlier (in 1956) by T. Matsubara and H. Matsuda.[4] A paper reporting on our work, [66a], was written in December 1965. Even before we wrote that paper, we realized that these diverse attempts all pointed to the value of using Bethe's hypothesis to study a one-dimensional Heisenberg–Ising spin chain with the Hamiltonian

$$H = -\frac{1}{2}\sum\{\sigma_x\sigma_x' + \sigma_y\sigma_y' + \Delta\sigma_z\sigma_z'\} \quad . \tag{1}$$

The mathematics of this problem is the same as that of a quantum lattice gas in one dimension with nearest neighbor interaction Δ.

Special cases of equation (1) had been studied before by H. A. Bethe, L. Hulthén, R. Orbach, L. R. Walker, R. B. Griffith, J. des Cloizeau, J. J. Pearson, and others. C. P. Yang and I found that, by studying the general case, one can use *continuity* arguments with respect to the parameter Δ to gain information about the solutions of the transcendental equations that Bethe and Hulthén had already studied. Pursuing this idea, we did the work that was written up in 1966 as ⟦66e⟧, [66f], and [66g].[5]

These three papers led to a number of efforts over the next few years (1967–1971), during which C. K. Lai, Bill Sutherland, C. P. Yang, and I were active with various problems using Bethe's hypothesis. In recent years, some of the results and methods in these papers have become useful for field theory[6] and for the solution of a number of new models.[7]

---

[3]E. H. Lieb and W. Liniger, *The Physical Review* 130, 1605 (1963); E. H. Lieb, *The Physical Review* 130, 1616 (1963).

[4]See T. Matsubara and H. Matsuda, *Progress of Theoretical Physics* (Kyoto) 16, 569 (1956).

[5]By 1966 there was already a considerable number of papers using Bethe's idea in different forms. C. P. Yang and I decided to honor Bethe's insight by calling his assumption "Bethe's hypothesis." That name or, alternatively, "Bethe's ansatz," seems now to have taken hold.

[6]See Barry M. McCoy and Tai Tsun Wu, *Physics Letters* 87B, 50 (1979).

[7]See A. A. Belavin, *Physics Letters* 87B, 117 (1979); N. Andrei and J. H. Lowenstein, *Physical Review Letters* 43, 1698 (1979); N. Andrei, *Physical Review Letters* 45, 379 (1980); P. B. Wiegmann, *JETP Letters,* forthcoming.

In working on ⟦66e⟧ and [66f], C. P. Yang and I learned two useful mathematical methods. One was a topological theorem about solutions of equations and the topological concept of *index*. This theorem we learned from Hassler Whitney. It was very powerful for proving that Bethe's hypothesis does give the ground-state wave function. For this proof, the topological theorem was used in conjunction with continuity arguments with respect to Δ. Another mathematical tool we learned was the method of solving Wiener–Hopf equations. M. Kac and T. T. Wu had referred us to M. G. Krein's long review on this subject, which we found to be very beautiful.

Papers ⟦66e⟧, [66f], and [66g] were written during the period when my family and I were moving from Princeton to Stony Brook. For my family, the move to Long Island was exciting. For me the feeling was more complex. I had spent seventeen years, 1949–1966, at the Institute for Advanced Study, from age 27 to age 44. I had been productive and happy there. I liked its unpretentious Georgian buildings and its peaceful, restrained atmosphere. I liked its long, meandering walk to the little suspension bridge in the woods. It was a place out of this world. It was a place meant for contemplation, and it was populated by people who contemplated well. The permanent faculty was first class. The visitors were generally brilliant. It was an ivory tower in the best sense of the term.

I could not escape asking myself sometimes, during that period of moving, whether I was making the right decision to leave the Institute for Advanced Study. But the answer was always the same. Yes, it was the right decision: The ivory tower is not the world, and the challenge to help build a new university is exciting.

⟦66h⟧

Article begins page 383

Treatment of Overlapping Divergences in the Photon Self-Energy Function

Supplement of the Progress of Theoretical Physics 37 and 38, 507 (1966)

R. L. Mills and C. N. Yang

Commentary In 1954 at Brookhaven, R. L. Mills and I noticed that the treatment of the overlap in renormalization theory had a small gap in it. The overlap problem was most elegantly treated by Ward's method. But in applying this method to the photon propagator, we found that one must be careful in order not to encounter inconsistencies. We devised a recipe for doing this. The work was not published at that time, since we knew one need not worry about such inconsistencies until one gets to 14th-order corrections.[1] (The possible inconsistency at that order was explicitly displayed in figures 5 and 6 of a paper by T. T. Wu.[2])

[1]In recent years, Kinoshita and his associates have pushed the calculation of g-2 to the 8th order, so in thirty years physicists progressed four orders. At this rate, around the year 2020 we should come to the 14th order.

[2]T. T. Wu, *The Physical Review* 125, 1436 (1962).

In 1966, Mills and I wrote our considerations up for the festschrift for S. Tomonaga's sixtieth birthday. Our contribution to that volume became ⟦66h⟧.

The renormalization program was a great development in physics. The chief architects of the enterprise were S. Tomonaga, J. Schwinger, R. P. Feynman, and F. J. Dyson. When Tomonaga, Schwinger, and Feynman were awarded the Nobel Prize in 1965, I thought the Nobel Committee erred in not having also recognized the contribution of Dyson. I still think so today. The papers of Tomonaga, Schwinger, and Feynman did not complete the renormalization program since they confined themselves to low-order calculations. It was Dyson who dared to face the problem of high orders and brought the program to completion. In two magnificently penetrating papers, he pointed out and resolved the main problems of this very difficult analysis.[3] Renormalization is a program that converts additive subtractions into multiplicative renormalization. That it works required a highly nontrivial proof. That proof Dyson supplied. He defined the concepts of primitive divergences, skeleton graphs, and overlapping divergences. Using these concepts, he pushed through an incisive analysis and completed the proof of renormalizability of quantum electrodynamics. His perception and power were dazzling.

[3]F. J. Dyson, *The Physical Review* 75, 486 and, 1736 (1949).

Some Remarks Concerning High Energy Scattering

⟦67b⟧
Article begins page 388

In *High Energy Physics and Nuclear Structure*
(Amsterdam: North-Holland, 1967)
T. T. Chou and C. N. Yang

Commentary One of the young physicists I met on arriving at Stony Brook in 1966 was T. T. Chou. We collaborated on statistical mechanics for a while, but soon shifted our attention to high-energy collisions. We found that the geometrical idea first developed in ⟦65a⟧ for large t phenomena and later in ⟦66b⟧ for small t phenomena could be pushed further to yield a quantitative theory for elastic collisions. Paper ⟦67b⟧ was our first collaborative publication on the subject. It was to lead to a very fruitful series of work on a geometrical description of elastic and inelastic collisions (see [68b], ⟦69c⟧, and ⟦80a⟧).

Section 2 of ⟦67b⟧ was devoted to "evidences for extended structure." We argued that the experimental findings at that time, that elastic differential cross sections approach a limit at high energies, that is,

$$\frac{d\sigma}{dt} \rightarrow f(t) \quad ,$$

is a remarkable fact that is indicative of diffraction phenomena from an extended object. To us this argument was decisive. Furthermore, the beautiful experiments of R. Hofstadter had already measured the physical size of the proton. Chou and I believed that diffraction from an extended scatterer was *to be expected from general principles of wave propagation*. But during the mid-sixties, many physicists were so enthralled with the Regge roles and cuts approach that there was strong resistance to any other approach to the phenomena of high-energy scattering. To us, then as now, the two approaches are not mutually exclusive but are complementary.

Paper [[67b]] was reported at a conference that took place in Israel between February 27 and March 3, 1967. On the way there, while stopping over at Singapore, I learned that J. R. Oppenheimer had died on February 18.

I last saw Oppenheimer in late 1966, when I went back to Princeton for a short visit. By that time it was generally known that he had cancer. I called him up and went to his office to see him. I had planned to urge him to consider writing something, a kind of last testimonial, about the atomic bomb and mankind. But I found him so frail physically that I did not bring up the subject.

There has been in the history of the world few persons whose lives have been as dramatic as Oppenheimer's, or as tragic. Adding to the tragedy was the fact that in his lifetime Oppenheimer was regarded by his peers as brilliant, but not as one who has made any fundamental contributions to physics. That assessment has now changed. His pioneering work with G. Volkoff and H. Snyder on the black hole is today recognized as a great contribution. It is destined to play an increasingly important role in physics and astronomy in coming years.

[[67d]]

Article begins page 400

Some Solutions of the Classical Isotopic Gauge Field Equations

In *Properties of Matter Under Unusual Conditions*, eds. H. Mark and S. Fernbach (New York: Wiley-Interscience, 1969)

T. T. Wu and C. N. Yang

Commentary Although I did not publish anything on gauge fields from 1955 to 1967, I continued to work on the subject off and on. In the early 1960s, A. C. T. Wu and I collaborated in trying to compute gauge quantum–gauge quantum scattering and discuss the question of unitarity. We got confused by the result and put it aside. In 1967, T. T. Wu and I embarked on a different problem: an effort to find classical solutions of pure gauge fields. That resulted in [[67d]], which we contributed to the festschrift for the sixtieth birthday of E. Teller.

The strategic aim of [[67d]] was to look for *singularity-free* classical solutions and then to study perturbations around such solutions to see whether one could obtain excitations. This type of consideration was attempted already by Born and Infeld in the 1930s, but they had to adopt a rather arbitrary Lagrangian. With the gauge field we did not have to make such ad hoc assumptions. In Section III, the second variation of the Hamiltonian, $\delta^2 H$, was explicitly computed. This procedure had been pioneered by the strong coupling theory of W. Pauli, S. Dancoff, J. Schwinger, and N. Hu in the early 1940s.

Tactically, the important new point in ⟦67d⟧ was the introduction of assumption (5), which mixed isospin and space indices.

Our claim that the solution found is sourceless is, if not incorrect, certainly misleading (see ⟦75c⟧).

In 1967, Wu and I did not appreciate that our strategic aim was in fact very much related to the idea of symmetry breaking that had been discussed for field theory since the early 1960s by Y. Nambu, J. Goldstone, P. W. Anderson, P. W. Higgs, and many others. In retrospect, this was because I had an a priori argument against symmetry breaking in fundamental field theory: I felt that a symmetry that is never observable has no place in a fundamental theory. To be sure, isotropy is a symmetry not observable in a ferromagnet. But I argued that when the ferromagnet is taken apart the elementary interaction between atoms would exhibit isotropic symmetry. That is in fact the reason one approaches the magnet problem with an isotropic Hamiltonian to start with. I argued that, in a field theory with broken symmetry, the symmetry is not observable: Nothing could be taken apart.

I believe now there were several things wrong with this a priori argument: (1) Strict adherence to the philosophy that *only* observable symmetries should be considered is not good. In any case, for local gauge symmetry, the meaning of "observability" is itself unclear. (2) The symmetry may be broken only at low temperatures. At high enough temperatures, full symmetry could be restored.

It is not possible to list the many, many papers on gauge theories in the 1960s and 1970s. Of particular importance were the brilliant work on renormalization by G. t'Hooft, a student of M. Veltman, and the success of the model developed by S. L. Glashow, S. Weinberg, and A. Salam. The exciting ideas about asymptotic freedom, grand unification, and quantum chromodynamics and confinement are all manifestations of the exuberance of recent developments. But I continue to believe that fundamental new ideas are still missing. For example, the introduction of a field to break symmetry cannot be the final story, although it may be a good temporary development, perhaps not unlike Fermi's theory of β decay.

⟦67e⟧
Article begins page 406

Some Exact Results for the Many-Body Problem in One Dimension with Repulsive Delta-Function Interaction

Physical Review Letters 19, 1312 (1967)
C. N. Yang

Commentary B. Sutherland was my first graduate student at Stony Brook. He was very quiet and spoke with what appeared to me at first to be diffidence. But I soon learned that he was very solid in his knowledge of physics. Furthermore, he was very original. It was a joy to watch him develop into a full-fledged research worker in statistical mechanics.

Sutherland, C. P. Yang, and I exploited the Bethe hypothesis in several applications. ⟦67e⟧ and a later paper by Sutherland represented one of these applications.[1] The methods used in ⟦67e⟧, [66f], and ⟦69a⟧ have recently become useful in solving the Kondo problem (see Commentary on ⟦66e⟧).

Paper [68a] was a follow-up on ⟦67e⟧. [68a] is very much related to later work on S-matrices of solvable models by R. Baxter, A. A. Belavin, I. V. Cherednik, L. D. Fadeev, V. A. Fateen, M. Karowski, L. A. Takhtadzhan, H. B. Thacker, D. J. Wilkinson, A. B. Zamolodchikov, Al. B. Zamolodchikov, and others.

[1]B. Sutherland, *Physical Review Letters* 20, 98 (1968).

⟦69a⟧
Article begins page 410

Thermodynamics of a One-Dimensional System of Bosons with Repulsive Delta-Function Interaction

Journal of Mathematical Physics 10, 1115 (1969)
C. N. Yang and C. P. Yang

Commentary This paper solves the finite temperature problem for the delta-function interaction case.

The rigor with which we had established the Bethe hypothesis in ⟦66e⟧ now paid off. It allowed us to have a firm grip on the quantum numbers $I_1, I_2 \ldots$ of the present problem. The *security* generated by such an understanding allowed us to take the next jump, which led to the solution of the finite-temperature problem. It formed, in a sense, a solid platform from which to take off.

When and how to take the next jump is an important and ever-present question in research. The psychological elements involved in this process—drive, temperament, taste, and confidence—all play important roles, perhaps as important as technical skills. Our experience with ⟦69a⟧ is an illustration of the usefulness of solidifying the foundation. But excessive attention to solidification may inhibit the adventurous spirit that is also important.

The problem solved in ⟦69a⟧ is a field theory at finite temperatures. If I am not mistaken, it remains the only nontrivial example solved at finite temperatures, except for generalizations to related models. It shows the subtlety in the definitions of the vacuum, the interaction, and the excitation spectrum. See remarks at the end of [70d], and article ⟦71b⟧.

Hypothesis of Limiting Fragmentation in High-Energy Collisions

⟦69c⟧ Article begins page 418

The Physical Review 188, 2159 (1969)
J. Benecke, T. T. Chou, C. N. Yang, and E. Yen

Commentary In 1968, Chou and I wrote a formal exposition of the ideas in ⟦67b⟧ on elastic scattering, which became [68b]. Our thoughts naturally turned then to inelastic scattering. It had already been known for some time, through the experiments of the Collins group at Brookhaven, that the transverse momenta for all outgoing particles in a hadron–hadron collision are small. Thinking about this very conspicuous fact, and reviewing many earlier ideas—the "extended object" T. T. Wu and I presented in ⟦65a⟧; the geometrical model; the statistical model of Fermi; the two fireball models of G. Cocconi, of K. Niu, and of P. Coik et al.; and the diffractive dissociation idea of M. L. Good and W. D. Walker—we (Benecke, Chou, Yen, and I) gradually realized that the similarities and differences of the various models are in fact related to this experimental fact. On the basis of these *inductive* considerations, we proposed the hypothesis of limiting fragmentation, which was written up as ⟦69c⟧.

In Section 8, which presented the arguments for the hypothesis, we listed some experimental data as evidence. But these experiments had not been done for the purpose of checking the hypothesis. So we did not really, in our minds, give them very much weight. What really emboldened us to make the hypothesis was the theoretical arguments of Section 8.7, which I later elaborated on in a talk in Kiev, article ⟦70f⟧ of this volume. This theoretical argument says that, if one believes in the elastic scattering ideas of the geometrical picture, then one *must* also believe in the hypothesis of limiting fragmentation.

We were delighted, but not surprised, that from 1970 to 1973 the hypothesis received numerous experimental confirmations. (The most accurate of these was the Pisa–Stony Brook experiment.[1] The idea for the experiment probably occurred to many people. I first heard it suggested by J. Kirz.) In 1971, A.M. Baldin extended the hypothesis to nuclear collisions.[2]

Our experience in trying to describe to other physicists the idea behind the hypoth-

---

[1]Bellettini, Braccini, Bradaschia, Castaldi, del Prete, Foa, Firomini, Laurelli, Menzione, Valdata, Finocchiaro, Grannis, Green, Mustard, and Thun, *Physics Letters* 45B, 69 (1973).

[2]A. M. Baldin, *Kratk. Soobshch. Fiz* 1, 35 (1971); V. S. Stavinskii, *Fizika Elementanykh Chastis i Atomnogo Yadra* 10, 949 (1979); *Soviet Journal of Particles and Nuclei* 10, 373 (1979).

esis of limiting fragmentation was a very interesting one. On many occasions, we met first with disbelief, only to be told later that the idea was obvious and was only kinematics. Yes, kinematics did play a part in it, but the physics involved was not just kinematics. The concept of "fragmentation" and "limiting fragmentation" in the projectile and the laboratory systems of references are descriptions of dynamical properties, not of kinematics.

Our paper was reported at the Stony Brook Conference of September 1969. Also reported was a paper by R. P. Feynman with ideas in part very similar to ours but expressed in a different language and from a different viewpoint.[3] The difference between the two viewpoints stemmed from the fact that we preferred the laboratory and projectile systems of reference, whereas Feynman preferred the center of mass system. The relationship between the two was soon understood and discussed, explicitly or implicitly, in many papers. Among them was one by Chou and me, [70c].

[3]R. P. Feynman, *Physical Review Letters* 23, 1415 (1969); also in *High Energy Collisions, Third International Conference,* Stony Brook, September 1969 (New York: Gordon & Breach, 1969), p. 237.

⟦70b⟧

Article begins page 430

Some Exactly Soluble Problems in Statistical Mechanics

Lectures given at the Karpacz Winter School of Physics, February 1970
In *Proceedings of the VII Winter School of Theoretical Physics in Karpacz* (University of Wroclaw, 1970)
C. N. Yang

Commentary Paper ⟦70b⟧ sets forth a simple introduction to the essential steps in Bethe's hypothesis. It also gave a rigorous derivation of the fact that, when the density remains fixed but the system becomes infinitely large, it is legitimate to write down an integral equation for the limit.

I have already said in earlier Commentaries, those on ⟦66e⟧ and ⟦67e⟧, that Bethe's hypothesis has been used in the last few years to solve new problems. That is very good. But I do not believe that it can be generalized to higher dimensions. The assumption of no diffraction, which is the basis of Bethe's hypothesis, is not valid in any *genuinely* higher dimensional problem.

⟦70b⟧ is a collection of my lectures given at the 1970 winter school in Karpacz, Poland. I had a great time at the school and enjoyed the kind of hospitality of my hosts. I found the Poles warm and lively, with a sense of humor and a deep awareness of the turbulent nature of their national history.

High-Energy Hadron–Hadron Collisions

⟦70f⟧
Article begins page 437

Lecture given at the Kiev Conference, August 1970
In *Proceedings of the Kiev Conference—Fundamental Problems of the Elementary Particle Theory* (Academy of Sciences of the Ukranian SSR, 1970)
C. N. Yang

Commentary ⟦70f⟧ gives the physical reasons behind the hypothesis of limiting fragmentation more fully than does ⟦69c⟧.

Introductory Note on Phase Transitions and Critical Phenomena

⟦71b⟧
Article begins page 440

In Phase Transitions and Critical Phenomena, Vol. 1 eds. C. Domb and M. S. Green (New York: Academic Press, 1971)
C. N. Yang

Commentary Rereading this introductory note, I find that my admiration today for Gibbs has only increased, if that is possible. The beauty of his *Elementary Principles of Statistical Mechanics* is sheer poetry. The tenacity with which he held onto his "rational" synthesis in the face of disagreements with experiments is evidence of deep insight.

I still subscribe to the speculative remarks of the last four paragraphs of ⟦71b⟧. One remark suggested a cross fertilization of statistical mechanics and quantum field theory, which is very much in evidence today. It is bound to become more so in the next decade. See also Commentary on ⟦69a⟧.

Some Concepts in Current Elementary Particle Physics

⟦72c⟧
Article begins page 445

Lecture given at the Trieste Conference in honor of P. A. M. Dirac, September 1972
In *The Physicist's Conception of Nature,* ed. J. Mehra (Dordrecht: Reidel, 1972)
Chen Ning Yang

Commentary I have been an admirer of Dirac's for a long time and was honored to talk in 1972 at the Trieste Conference to celebrate his seventieth birthday.

⟦73b⟧
Article begins page 452

Opaqueness of *pp* Collisions from 30 to 1500 GeV/*c*

Physical Review D 8, 2063 (1973)

Alexander Wu Chao and Chen Ning Yang

Commentary At the 1972 Chicago meeting, C. Rubbia revealed the experimental result from the ISR at CERN that *pp* elastic cross section does have a conspicuous dip at $t \simeq -1.4$ (GeV/*c*).[1] This was welcome information to T. T. Chou and me, because the geometrical model had been predicting the existence of such a dip since 1968.[2] We had never doubted the general correctness of the geometrical picture from the beginning, but the excellent quantitative agreement surprised us.[3]

Also in 1972, G. B. Yodh, Y. Pal, and J. S. Trefil concluded from cosmic-ray data that at very high energies proton–nucleon total cross sections increase slowly with energy.[4] In 1973, ISR experiments showed that *pp* cross section increases with energy at very high energies. Remarkably, these results confirmed the predictions of H. Cheng and T. T. Wu in 1970.[5] With these results in mind, Alexander Chao and I made a phenomenological analysis of *pp* elastic scattering in ⟦73b⟧, using the eikonal formalism and assuming the scattering amplitude to be purely imaginary. The opaqueness function $\Omega(b)$ thus obtained for high energies was in very good agreement with the Fourier transform of $[G_E(k)]^2$, where G_E is the form factor for the proton.

In 1974, F. Hayot and U. P. Sukhatme pointed out[6] that there is one and only one natural way to accommodate increasing total cross section in the geometrical picture: to increase the coefficient K in the relation

$$\Omega = K[G_E]^2 \quad .$$

This was an excellent suggestion, and all subsequent calculations in the geometrical picture were done this way.

Chao was a graduate student of mine. He learned everything very fast and contributed in essential ways to the three papers that we wrote together in 1973–1974. I knew he would become a good theoretical high-energy physicist, but I felt if he switched fields he could have a career in accelerator design that would be more rewarding, that field being underpopulated compared with high-energy theory. So I pushed him to change fields. Happily, everything worked out well, and he is now a bright young star in accelerator theory.

---

[1]See the report of G. Giacomelli, *Proceedings of the XVI International Conference on High Energy Physics* (National Accelerator Laboratory, 1972), Vol. 3, p. 232; A. Bohm et al., *Physics Letters* 49B, 491 (1974).

[2]T. T. Chou and C. N. Yang, [68c]; L. Durand and R. Lipes, *Physical Review Letters* 20, 638 (1968).

[3]See A. W. Chao and C. N. Yang, ⟦73b⟧; M. Kac, *Nuclear Physics* B62, 402 (1973).

[4]G. B. Yodh, Y. Pal, and J. S. Trefil, *Physical Review Letters* 28, 1005 (1972).

[5]H. Cheng and T. T. Wu, *Physical Review Letters* 24, 1456 (1970).

[6]F. Hayot and U. P. Sukhatme, *Physical Review D* 10, 2183 (1974). Similar considerations for the impact picture model were made by Tai Tsun Wu and Hung Cheng in *High Energy Collisions—1973 (Stony Brook)*, ed. Chris Quigg (American Institute of Physics, 1973), p. 54.

Integral Formalism for Gauge Fields

Physical Review Letters 33, 445 (1974)

C. N. Yang

⟦74c⟧
Article begins page 457

Commentary In 1967–1968, while thinking about the gauge field concept and its possible generalizations (such as to a case when the group is discrete, or to a case when space itself is discrete), I realized that the concept of a nonintegrable phase factor is very important. Once my attention was focused on this concept, it became clear that in fact the parallel displacement concept of Levi-Civita is a special case of a nonintegrable phase factor. This led me to an understanding of the similarity between the formula in gauge field theory

$$F_{\mu\nu} = \frac{\partial B_\mu}{\partial x_\nu} - \frac{\partial B_\nu}{\partial x_\mu} + i\epsilon(B_\mu B_\nu - B_\nu B_\mu) \tag{1}$$

and the formula in Riemannian geometry

$$R^l_{ijk} = \frac{\partial}{\partial x^j}\begin{Bmatrix} l \\ ik \end{Bmatrix} - \frac{\partial}{\partial x^k}\begin{Bmatrix} l \\ ij \end{Bmatrix} + \begin{Bmatrix} m \\ ik \end{Bmatrix}\begin{Bmatrix} l \\ mj \end{Bmatrix} - \begin{Bmatrix} m \\ ij \end{Bmatrix}\begin{Bmatrix} l \\ mk \end{Bmatrix} \tag{2}$$

The formulas are similar because (2) is a special case of (1)! It is hard to describe the thrill I felt at understanding this point. I realized then that the concept of gauge fields is *deeply geometrical* from the mathematical point of view. I also realized that the similarity noticed by Pauli in 1954 between (1) above and the formulas in Schrodinger's 1932 paper was not accidental (see Commentary on ⟦54b⟧).

With an appreciation of the geometrical meaning of gauge fields, I consulted Jim Simons, a distinguished differential geometer, who was then the Chairman of the Mathematics Department at Stony Brook. He said gauge fields must be related to connections on fibre bundles. I then tried to understand fibre-bundle theory from such books as Steenrod's *The Topology of Fibre Bundles,* but learned nothing. The language of modern mathematics is too cold and abstract for a physicist. All this happened from 1967 to 1969.

I finally wrote up the idea of nonintegrable phase factors in 1974 as paper ⟦74c⟧.[1] I did not yet appreciate the global considerations necessary in gauge field concepts since I had not yet captured the spirit of the fibre bundle concept (see Commentary on ⟦75c⟧). The discussions in ⟦74c⟧ were therefore all within one "chart," or coordinate system.

The ideas summarized in ⟦74c⟧ were later discussed more fully in [75e].

From the epistemological viewpoint, the conceptual evolution in gauge theory

[1]Two years earlier, I presented the contents of ⟦74c⟧ in a seminar in Peking in July 1972. See H. Y. Kuo, Y. S. Wu, and Y. C. Chang, *Kexue Tongbao* 18, 72 (1973).

[2]It was, however, already thus understood by many physicists in the early 1960s, or even earlier. Among them were E. Lubkin, D. Finkelstein, J. M. Schiminovich, D. Speiser, J. A. Wheeler, B. S. DeWitt, A. Lichnerowicz, C. W. Misner, A. Trautman, H. G. Loos, and others. See footnote 18 of ⟦75c⟧.

originated with the replacement

$$p \to p - \frac{e}{c}A \tag{3}$$

in prequantum electrodynamics. Then quantum mechanics developed, and it became necessary to make another replacement:

$$p - \frac{e}{c}A \to -i\hbar(\partial - \frac{ie}{\hbar c}A) \quad . \tag{4}$$

The generalization to non-Abelian gauge field involved

$$\partial - \frac{ie}{\hbar c}A \to \partial - i\epsilon B \quad . \tag{5}$$

See [[77e]]. That this means that gauge potentials are mathematically "connections" was, to me at least, only apparent much later, during 1967 to 1969, as described above.[2] In this connection, the following statement by Mayer is simply not true:

> A reading of the Yang–Mills paper shows that the geometric meaning of the gauge potentials must have been clear to the authors, since they use the gauge covariant derivative and the curvature form of the connection, and indeed, the basic equations in that paper will coincide with the ones derived from a more geometric approach . . . .[3]

What Mills and I were doing in 1954 was generalizing Maxwell's theory. We knew of no geometrical meaning of Maxwell's theory, and we were not looking in that direction.

In [[74c]] I proposed that the gravitational equation should be changed to a third order differential equation. I believe today, even more than in 1974, that this is a promising idea, because the third order equation is more natural than the second order one and because quantization of Einstein's theory leads to difficulties.

Most of my physicist colleagues take a utilitarian view about mathematics. Perhaps because of my father's influence, I appreciate mathematics more. I appreciate the value judgment of the mathematicians, and I admire the beauty and power of mathematics: there are ingenuity and intricacy in tactical maneuvers, and breathtaking sweeps in strategic campaigns. And, of course, miracle of miracles, some concepts in mathematics turn out to provide the fundamental structures that govern the physical universe!

---

[3]W. Drechsler and M. E. Mayer, *Fibre Bundle Techniques in Gauge Theories* (Springer-Verlag, 1977), p. 2.

Concept of Nonintegrable Phase Factors and Global Formulation of Gauge Fields

⟦75c⟧
Article begins page 460

Physical Review D 12, 3845 (1975)
Tai Tsun Wu and Chen Ning Yang

Commentary It gradually became clear that field theorists must learn the mathematical concept of fibre bundles. In early 1975, I invited Jim Simons to give my colleagues and me a series of luncheon lectures on differential forms and fibre bundles. He kindly accepted the invitation, and we learned about the Stoke's theorem, the de Rham theorem, and so on. What we learned allowed us to understand the mathematical meaning of the Bohm–Aharonov experiment and the Dirac quantization rule of electric and magnetic monopoles. T. T. Wu and I later also understood the profound and very general Chern–Weil theorem. We appreciated that gauge fields have global geometrical connotations (not to be confused with physicists' global phase factors) that are naturally formulated in terms of fibre bundle concepts.

In ⟦75c⟧, Wu and I explored these global connotations. We showed that the gauge phase factor gives an intrinsic and complete description of electromagnetism. It neither *underdescribes* nor *overdescribes* it. Once this is accepted, the rest of the paper is an introduction of fibre bundle concepts that firm up vague ideas in the physics of gauge theory.

Hadronic Matter Current Distribution Inside a Polarized Nucleus and a Polarized Hadron

⟦76a⟧
Article begins page 473

Nuclear Physics B107, 1 (1976)
T. T. Chou and Chen Ning Yang

Commentary A nucleus with non-zero spin, when polarized, has a nucleonic current inside it. That is a well-accepted idea. A proton has inside it hadronic matter. Does it also have inside a "hadronic matter current" when it is polarized? Chou and I answered, yes, it must. From the viewpoint of the geometrical model, this is the only possible answer. We believe it is *important* to measure this hadronic-matter current distribution. (By hadronic matter we mean the constituents of hadrons, that is, quarks or stratons, or any "stuff.")

Paper ⟦76a⟧ shows that this measurement can be done with an R-type polarization parameter experiment. The crucial point here is that, with increasing total cross section, to a projectile hadron the target hadronic matter which moves toward the projectile appears blacker than that which moves away from the projectile.

The appendix was added because of a question by K. Lane. It was given the title,

"Does the spin of a Dirac particle involve motion?" It was a subject well explored in the 1920s and 1930s when people discussed Zitterbewegung. It has since been mostly forgotten.

[[76c]]

Article begins page 493

Dirac Monopole Without Strings: Monopole Harmonics

Nuclear Physics B107, 365 (1976)

Tai Tsun Wu and Chen Ning Yang

Commentary After writing [[75c]], Wu and I realized that the charting concept of the mathematicians would eliminate the forty-odd-year-old string difficulty of the theory of magnetic monopoles and bring out the true characteristics of the nontrivial bundle underlying the physics of the Dirac monopole. As our work progressed, we were delighted by the elegance of this line of development. The results were given in [[76c]] and are among the most elegant I have worked on.

The elegance also led to greater penetration. In [77b], some further properties of the monopole harmonics introduced in [[76c]] were discussed. The interaction of a Dirac electron with a Dirac monopole was explored in [77c] and [77d]. [78a] and [78g] gave generalizations to SU_2 of Dirac monopoles, which should not be confused with the t'Hooft–Polyakov monopole.

[[76d]]

Article begins page 509

Dirac's Monopole Without Strings: Classical Lagrangian Theory

Physical Review D 14, 437 (1976)

Tai Tsun Wu and Chen Ning Yang

Commentary To second-quantize a theory of electrons, monopoles, and the electromagnetic field, Wu and I tried to write down a classical Langrangian, borrowing from the ideas used in [[76c]]. The result was very satisfactory, up to a point. Among other things, it helped us to understand the conclusion Dirac made in 1948, namely, that the classical action integral was definable only modulo $4\pi eg/c$:

$$\mathfrak{A} = \mathfrak{A}_0 \pmod{4\pi eg/c} \quad .$$

This early result of Dirac's had sounded right to us for many years, but it was only in 1976 that we understood it. It is of course related via the Feynman path integral to the Dirac quantization condition $2eg/\hbar c$ = integer (see Section VI.6 of [[76d]]). This relationship seemed to us deeply satisfying, and we started to develop a second quantized theory using path integrals. This effort was, however, frustrated. See Commentary on [[78e]].

What Visits Mean to China's Scientists

⟦76g⟧
Article begins page 518

In *Reflections on Scholarly Exchanges with the People's Republic of China*, ed. A. Keatley (Committee on Scholarly Communication with the People's Republic of China, 1976)
Yang Chen-ning

Commentary The Committee on Scholarly Communication with the People's Republic of China was established in 1966 by the American Council of Learned Societies, the National Academy of Sciences, and the Social Science Research Council. With the opening of friendly relations between China and the United States in the 1970s, the Committee played a key role in establishing an important bridge between the two countries. I was asked by its able secretary, Anne Keatley, to write an article for the booklet that she was editing for the Committee. That resulted in ⟦76g⟧.

In the spring of 1971, it became clear from newspapers that the frozen state of U.S.–China relations since 1949 was showing signs of thaw. When I learned that the ban on travel to China of U.S. citizens had been lifted by the State Department, I felt this was a chance for me to again see China, my country of birth, and to visit my family, relatives, teachers, and friends. I felt an urgency, because it was not at all clear that the door that was just then being opened a crack would not be shut tight again within a few months, what with the Vietnam War and the changing geopolitical situation in Asia. So I applied for a visa in the Chinese Embassy in Paris and visited China for a month in the summer of 1971.

This is not the place to describe my experiences of that trip. But by the time of my next trip, in the summer of 1972, I had already decided that it was my responsibility as a Chinese–American scientist to help build a bridge of understanding and friendship between the two countries that are close to my heart. I also felt I should help China in her drive toward developing science and technology.

On July 1, 1972, Premier Chou En-lai invited me to a banquet in the Sinkiang Room of the Great Hall of the Peoples. He had already given a banquet for me the previous year, and this second time I felt sufficiently at ease to speak my mind more directly. I had observed that, in those years, the government's one-sided egalitarianism had ruined Chinese science. So I urged him to consider adopting a policy of increased attention to basic sciences, even if from the national viewpoint this may not seem to pay immediate dividends.

Apparently, Premier Chou discussed my suggestion with Chairman Mao Tse-tung, and both agreed that it deserved further consideration. On July 14, 1972, the Premier instructed Vice President Chou Pei-yuan of Peking University to study how to implement such a policy.[1] I have since learned that this sequence of events did produce more opportunity for basic research in China, but the main efforts were frustrated until the fall of the "gang of four" in late 1976.

[1]Chou Pei-yuan, *People's Daily*, January 13, 1977.

I did not in those years have enough understanding of what was going on in China to realize that the gang of four did not like my suggestion. In 1973 I was puzzled by the animosity exhibited toward me by Chang Chun-chaio, a member of the gang of four. He had, in fact, explicitly criticized me in one of his pronouncements, but I did not learn of that until 1977.

[[77e]]

Article begins page 519

Magnetic Monopoles, Fiber Bundles, and Gauge Fields

Annals of the New York Academy of Science 294, 86 (1977)

Chen Ning Yang

Commentary This paper was my contribution to a conference celebrating R. E. Marshak's sixtieth birthday. To prepare this paper, I looked into the early history of gauge fields. It was very interesting how the concept of phase got into physics, and how that concept has penetrated into so many areas of technology.

[[77h]]

Article begins page 531

Some Problems on the Gauge Field Theories, II

Scientia Sinica 20, 47 (1977)

Gu Chao-hao and Yang Chen-ning

Commentary In 1974, when I visited Fudan University in Shanghai, I asked whether there were mathematicians and/or physicists who might want to discuss differential geometry with me. I did this because I knew the university was strong in this field. My enquiry produced a fruitful collaboration between myself and about a dozen faculty members of Fudan that lasted several years. It was to lead to an exchange agreement between Stony Brook and Fudan, which has been beneficial to both universities.

Paper [[77h]] was one of the fruits of that collaboration. The problems discussed in this paper are basically geometrical in origin. The collaborators listed on the title page *should* all have been coauthors.

For some recent developments on a number of points covered in [[77h]], see M. A. Mostow, *Communications in Mathematical Physics* 78, 137 (1980).

Speech at the Benjamin W. Lee Memorial Session

⟦77j⟧
Article begins page 540

In Unification of Elementary Forces and Gauge Theories, ecds. D. B. Cline and F. E. Mills (Harwood Academic Publishers, 1977)
C. N. Yang

Commentary In the late spring of 1977, I spent a month at the Fermilab and saw a lot of B. W. Lee. Among other things, we discussed cyclic instanton solutions and his new interest in astrophysics. I came back to Stony Brook in mid-June. A few days later, his secretary called and stunned me with the news that Lee had just been killed in a car accident.

Lee was one of the organizers of an international conference to take place in October, 1977, at the Fermilab, on parity nonconservation, weak neutral currents, and gauge theories. After his death, the name of the conference was very appropriately changed to the Benjamin W. Lee Memorial International Conference. Selection ⟦77j⟧ was my speech at the memorial session at the conference.

I first met Lee in the fall of 1960, when he came to the Institute for Advanced Study as a young postdoc. I was impressed with his independence of mind. Interaction during the next few years convinced me that he was an excellent physicist. In the fall of 1965, when I had already decided to move to Stony Brook, I asked him to join me and was delighted when he accepted. That was very fortunate for us, since he prospered at Stony Brook. He did very important research in the succeeding years and contributed greatly to our physics department.[1] In 1973, he moved on to head the theory group at the Fermilab. But we had maintained our close contact. While his death was a shock to the physics community, for me there was in addition a deep sense of personal loss.

---

[1]See B. Lee, in *Gauge Theories and Neutrino Physics,* ed. M. Jacob (Amsterdam: North-Holland, 1978), p. 147.

Interaction of Electrons, Magnetic Monopoles, and Photons (I)

⟦78e⟧
Article begins page 541

Scientia Sinica 21, 317 (1978)
Tu Tung-sheng, Wu T'ai-tsun, and Yang Chen-ning

Commentary Although in ⟦76d⟧ T. T. Wu and I found a satisfactory classical Lagrangian for the interacting system of electrons, magnetic monopoles, and photons, we were frustrated in trying to quantize such a dynamical system using path integrals. In the summer of 1976, we worked on this problem with Tu Tung-sheng, when all three of us were visiting CERN. We found that, by generalizing the charting idea of ⟦76c⟧, it was possible to separate out the longitudinal and transverse electromagnetic fields and follow a method parallel to the canonical procedure of Fermi in usual quantum electrodynamics. That led to ⟦78e⟧, which was finally written in late 1977.

The footnote introduced by an asterisk on the title page of this paper was not in the original manuscript. It was apparently inserted by the editor. It is deleted in the reproduction in this volume.

⟦78k⟧

Article begins page 551

Pointwise SO_4 Symmetry of the BPST Pseudoparticle Solution

In *Felix Bloch and Twentieth-Century Physics,* eds. M. Chodorow et al. (Houston: Rice University, 1980)

Chen Ning Yang

Commentary The pseudoparticle solution, or the instanton, of A. A. Belavin, A. M. Polyakov, A. S. Schwartz, and Yu. S. Tyupkin is mathematically related to the Chern–Weil theorem. It has some very remarkable properties, one of which is proved in ⟦78k⟧.

⟦79b⟧

Article begins page 556

Phase Shift in a Rotating Neutron or Optical Interferometer

Physical Review D 20, 1846 (1979)

Max Dresden and Chen Ning Yang

Commentary In the spring of 1979, when I was visiting the University of Missouri, S. A. Werner showed me the beautiful neutron interferometer experiment developed there. I had always been impressed by experiments on phases and coherence, and this one was particularly interesting. Back in Stony Brook, Max Dresden and I worried about the theoretical derivation of the phase shift Werner and his colleagues had measured, and we wrote ⟦79b⟧.

Geometrical Model of Hadron Collisions

⟦80a⟧

Article begins page 559

In *Proceedings of the 1980 Guangzhou Conference on Theoretical Particle Physics* (Beijing: Science Press, 1980)
T. T. Chou and Chen Ning Yang

Commentary Selection ⟦80a⟧ is a report T. T. Chou made at the 1980 Guanghou Conference, summarizing the theoretical results from the geometrical model in comparison with experiments. Since this report was made, there have been additional experiments that confirmed the prediction of the geometrical model for the kaon charge radius[1] and the existence[2] of dips in elastic $p\bar{p}$ scattering.

[1]E. B. Dally et al., *Physical Review Letters* 45, 232 (1980).
[2]See [81a].

Einstein's Impact on Theoretical Physics

⟦80b⟧

Article begins page 563

Lecture given at the Second Marcel Grossmann Meeting held in honor of the 100th Anniversary of the birth of Albert Einstein
Physics Today 33, 42 (June 1980)
Chen-Ning Yang

Commentary In 1979, Albert Einstein would have been one hundred years old. There were meetings honoring him at many places all over the world. I participated in four: at Princeton, Bern, Jerusalem, and Trieste. Paper ⟦80b⟧ is adapted from my talk at Trieste.

Einstein in his lifetime had toiled incessantly to construct "a complete system of theoretical physics." He searched for "the concepts and fundamental principles" that would allow for a grand synthesis of the structure of the physical world. Central to this synthesis are the forces, or interactions, that hold matter together, that produce the multitude of reactions that constitute natural phenomena.

I believe we are today still very far from this grand synthesis that Einstein dreamed about. But we do have one of its key elements: the principle that *symmetry dictates interactions,* first used by Einstein himself. From the historical viewpoint, this development is particularly interesting since from ancient times philosophers have tried to relate symmetry to the structure of the universe. In ancient Greece, Timaeus and Plato associated the four "fundamental" natural substances, fire, air, water and earth, respectively with the regular tetrahedron, octahedron, icosahedron, and cube. In ancient China, the *I Ching* associated the trigram and hexagram symbols with natural phenomena. Of course, by symmetry we do not mean today the same thing that the ancient philosophers did. But that there are general conceptual relationships between these meanings is hard to deny.

Grave difficulties still lie ahead of us. Many of these difficulties are concerned with the related questions of how to deal with divergences in field theories and which theories to choose. We know that *symmetry is good for cancelling divergences*. But we do not yet comprehend the *full scope* of the concept of symmetry. It would seem that fundamental new ideas are needed, which may necessitate the introduction of new mathematical concepts.[1] What these new mathematical concepts will be we do not know today. They may be quaternions, graded Lie algebras, or complex manifolds. More likely, they will be concepts as yet unnamed or unknown. In 1873 Maxwell wrote:

> The way in which Faraday made use of his idea of lines of force in co-ordinating the phenomena of magneto–electric induction shews him to have been in reality a mathematician of a very high order— one from whom the mathematicians of the future may derive valuable and fertile methods.
>
> For the advance of the exact sciences depends upon the discovery and development of appropriate and exact ideas, by means of which we may form a mental representation of the facts, sufficiently general, on the one hand, to stand for any particular case, and sufficiently exact, on the other, to warrant the deductions we may draw from them by the application of mathematical reasoning.
>
> From the straight line of Euclid to the lines of force of Faraday this has been the character of the ideas by which science has been advanced, and by the free use of dynamical as well as geometrical ideas we may hope for a further advance. The use of mathematical calculations is to compare the results of the application of these ideas with our measurements of the quantities concerned in our experiments. Electrical science is now in the stage in which such measurements and calculations are of the greatest importance.
>
> We are probably ignorant even of the name of the science which will be developed out of the materials we are now collecting . . . .[2]

[1]For a comment, see [79d].

[2]J. C. Maxwell, *Scientific Papers,* Vol. 2, No. 61 (Cambridge: Cambridge University Press, 1890).

Photographs

The author with his parents, Ko-chuen Yang and Meng-hwa Lo Yang, in Hofei in 1923, a few days before his father sailed for the United States to pursue studies in mathematics. The author was ten months old.

The author with his parents in Amoy in 1929. His father was then a professor of mathematics at the University of Amoy.

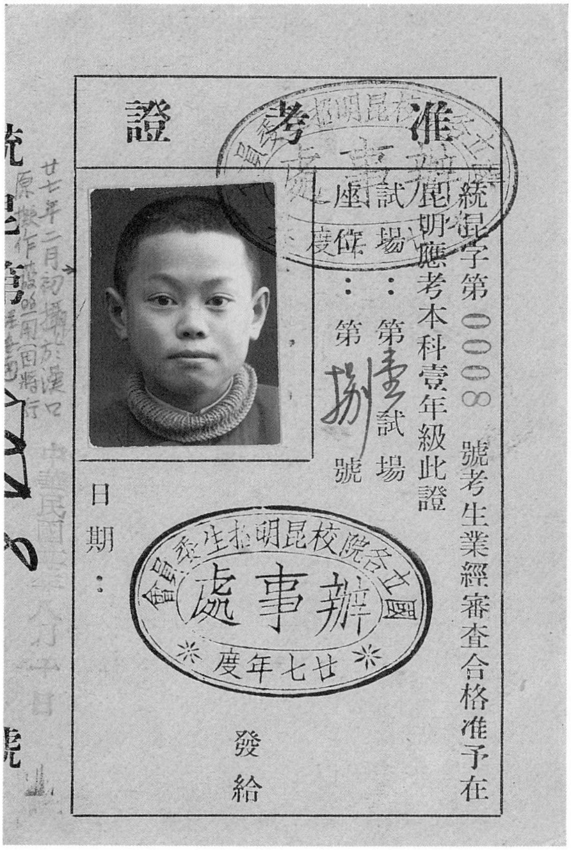

Permission Card for taking Universities Entrance Examination, Kunming, August, 1938. The photograph of the author was taken in February of that year in Hankow.

Southwest Associated University, Kunming, about 1940. Classrooms with tin roofs (left). Dormitories with thatched roofs (right).

E. Fermi at Los Alamos, 1946. (Courtesy of Los Alamos Laboratory.)

The author at the base of the Devil's Tower, Wyoming, in the summer of 1947, when he was a graduate student at the University of Chicago.

J. S. Wang in Peking, late 1950s.

Left to right: C. N. Yang, T. Y. Wu, and S. T. Ma in Wu's apartment in New York City, 1949.

The author in a park in Philadelphia in the fall of 1949, when he was a post-doctoral fellow at the Institute for Advanced Study in Princeton.

Left to right: C. N. Yang, A. Pais, T. D. Lee, G. A. Snow, S. F. Edwards, J. C. Ward, and Y. Nambu on the terrace outside Fuld Hall at the Institute for Advanced Study, Princeton, in the fall of 1952.

Portrait of Albert Einstein by Philippe Halsman in the early 1950s. This portrait was used for a United States postal stamp in 1966. See Commentary on ⟦66c⟧. (Copyright © by Philippe Halsman.)

Chih Li Yang with the King of Sweden in Stockholm, 1957.

T. D. Lee and C. N. Yang at Princeton, about 1961. (Courtesy of A. Richards.)

Left to right: A. Pais, T. D. Lee, C. N. Yang, and F. J. Dyson, outside Building E of the Institute for Advanced Study, Princeton, about 1961. (Courtesy of A. Richards.)

J. R. Oppenheimer at the European Organization of Nuclear Research, 1962. (Courtesy of CERN.)

The author in his office in Princeton, 1963. (Courtesy of Time-Life Books, photograph by Richard Kelley, © 1963, Time Inc.)

P. A. M. Dirac at Stony Brook, 1969. (Courtesy of L. Eisenbud.)

P. A. M. Dirac at Stony Brook, 1967. (Courtesy of L. Eisenbud.)

P. A. M. Dirac at Stony Brook, 1967. (Courtesy of L. Eisenbud.)

Photograph of the Physics Department at Stony Brook, spring, 1967. *Left to Right: Back row:* T. B. Sutherland, J. Cole, P. Grannis, R. B. Weinberg, W. T. Tung, A. Muriel, A. Bashian, H. Rudolph, A. Marshall, R. Jones, R. Graves, J. Marasco, P. Cooney. *Next row:* D. E. Hochman, Y. Y. Hsieh, E. Kim, S. S. Liu, E. Diener, S. F. Chen, D. E. Miller, M. Marmor, R. Loveless, S. V. K. Babu, W. Kirk, R. Kasman. *Next row:* M. T. Kane, K. K. Foo, R. Will, H. Yarger, H. Fischer, J. McFadyen, V. Hall, R. Pittman, P. Viebrock, W. T. Estler, P. Cowell, R. Orcutt, V. Nizamoff. *Next row:* R. Johnson, W. Yeh, D. Zanello, T. T. Chou, M. A. Lone, E. Barouch, D. P. Majumdar, A. Albano, F. Abbud, C. P. Fan, C. K. Lai, W. Bardeen, H. T. Nieh. *Next row:* D. Strottman, E. Yen, A. J. Bastin, D. B. Fossan, P. Paul, H. B. Silsbee, P. Kantor, L. L. Lee, D. A. Emmons, R. Vawter, S. Andrus, J. Johnson. *Next row:* C. Hansen, R. Hwa, M. A. M. Sullivan, R. Collins, J. Kahn, D. Rhome, D. DeHart, D. Fox, H. R. Muether, P. B. Kahn, C. Swartz, M. Naylor, K. Eklund. *First row:* L. Eisenbud, J. Lee-Franzini, J. S. Toll, C. N. Yang, P. A. M. Dirac, M. Dresden, N. L. Balazs, B. W. Lee, Y. H. Kao, B. J. Kayser, T. A. Pond, R. deZafra.

Premier Chou En-lai shaking the author's hand during his trip to China, 1971.

R. L. Mills at The Ohio State University, 1982. (Courtesy of The Ohio State University.)

The author with E. Teller at the Brookhaven National Laboratory, 1982. (Courtesy of Brookhaven National Laboratory.)

Articles

A Generalization of the Quasi-Chemical Method in the Statistical Theory of Superlattices

[[45a]]
Commentary begins page 3

C. N. YANG*
National Tsing Hua University, Kunming, China

(Received November 17, 1944)

The quasi-chemical method in the investigation of the equilibrium distribution of atoms in the pairs of neighboring sites in a superlattice is generalized by considering groups containing large numbers of sites. The generalized method may be used to obtain successive approximations of the free energy of the crystal. The labor of integration is avoided by the introduction of a Legendre transformation. In order to analyze the fundamental assumption underlying the method more closely, the number of arrangements of the atoms for given long-distance order is calculated and the hypothesis of the non-interference of local configurations discussed. The method is applied to the calculation of the free energy in the different approximations discussed in this paper, including Bethe's second approximation and a simple approximation for the face-centered cubic crystal Cu_3Au.

1. INTRODUCTION

IT was shown by Fowler and Guggenheim[1] that the quasi-chemical method, originally devised for the theory of regular solutions, applies equally well to the theory of superlattices with long-distance order. The method is, as they have emphasized, definitely one stage further towards an exact theory than Bragg-Williams' method.[2] Compared with Bethe's[3] or Kirkwood's[4] method it also distinguishes itself in mathematical simplicity. But in its original form the method does not lead to a consistent scheme of successive approximations and cannot be regarded as a method of approach towards the rigorous evaluation of the configurational free energy of an alloy. In the present paper a new formulation of the quasi-chemical method is developed which is capable of yielding successively higher approximations.

The free energy expression in both Bethe's method and the quasi-chemical method involves an integral. Its evaluation is very complicated and has been carried out[1,5] so far only in Bethe's approximation for simple and body-centered cubic crystals. In the new formulation of the quasi-chemical method, however, it will be shown that a Legendre transformation helps to effect the integration. (It might be noticed that a similar Legendre transformation is used to essentially the same effect in Fowler's formulation of general statistical mechanics. Compare Fowler.[6]) The free energy is obtained directly as a closed expression. Its values are given for Bethe's first and second (modified) approximations and for the face-centered alloy Cu_3Au in Sections 7 and 8.

To make sure that the quasi-chemical method may actually be used to obtain a series of successively better approximations, we must investigate the free energy in high order approximations and compare it with the partition function of the crystal. This is done in Section 5 together with a comparison of the quasi-chemical and Bethe's methods.

Except in the last section we shall be concerned only with binary alloys with atomic ratio 1:1 forming a (quadratic), simple cubic, or body-centered cubic lattice. The generalization of the method to the investigation of alloys with other atomic ratios and forming other types of lattices is easy. In fact, the superior power of the quasi-chemical method appears to be even more fully revealed when a face-centered lattice is studied. This problem is taken up at the end of the paper where an approximate free energy

* Research Fellow of the China Foundation for the Promotion of Education and Culture.

1 R. H. Fowler and E. A. Guggenheim, Proc. Roy. Soc. **A174**, 189 (1940).

2 W. L. Bragg and E. J. Williams, Proc. Roy. Soc. **A145**, 699 (1934); **151**, 540 (1935); **152**, 231 (1935).

3 H. A. Bethe, Proc. Roy. Soc. **A150**, 552 (1935).

4 J. G. Kirkwood, J. Chem. Phys. **6**, 70 (1938).

5 T. S. Chang, Proc. Camb. Phil. Soc. **35**, 265 (1939); J. G. Kirkwood, J. Chem. Phys. **8**, 623 (1940); J. S. Wang, "Free energy in the statistical theory of order-disorder transformation," Science Report of National Tsing Hua University, Series A, 30th Anniversary Memorial Number (1941), printed but failed to appear.

6 R. H. Fowler, *Statistical Mechanics*, second edition, p. 188.

□Reprinted from *The Journal of Chemical Physics* 13, 2 (February 1945), 66–76.

expression for Cu_3Au is obtained and its critical phenomena discussed.

2. REFORMULATION OF THE QUASI-CHEMICAL METHOD

Consider an alloy of type AB with altogether $N=2N_1$ atoms. Let z be the number of nearest neighbors of each atom. Those sites in the superlattice occupied by the A atoms at the absolute zero of temperature will be called the α-sites, and the rest, the β-sites. The partition function of the crystal may be written in the form

$$\sum_w P(w, T), \tag{1}$$

where $P(w, T)$ is equal to $e^{-E/kT}$ summed over all possible configurations of the crystal having N_1w α-sites occupied by B (wrong) atoms. The average energy over all these configurations is

$$\bar{E}(w, T)=kT^2\frac{\partial}{\partial T}\log P(w, T). \tag{2}$$

But evidently

$$P(w, \infty)=g(w)=\left[\frac{N_1!}{(N_1w)!(N_1-N_1w)!}\right]^2. \tag{3}$$

Hence

$$\log P(w, T)=\log g(w)+\int_\infty^T \frac{1}{kT^2}\bar{E}(w, T)dT. \tag{4}$$

The problem of finding the partition function of the crystal therefore reduces to one of finding $\bar{E}(w, T)$. An approximate solution has been obtained by Fowler and Guggenheim. Their method will now be presented in a new form better suited to generalization.

There are in the crystal zN_1 pairs of nearest neighboring sites $\alpha-\beta$. Let $[q_\alpha, q_\beta]$ of these be occupied by $q_\alpha(=0, 1)$ wrong (B) atoms on the α-site and $q_\beta(=0, 1)$ wrong (A) atoms on the β-site. For given w the following relations hold:

$$\begin{aligned}
&[0, 0]+[0, 1]+[1, 0]+[1, 1]=zN_1,\\
&[1, 0]+[1, 1]=zN_1w,\\
\text{and}\quad &[0, 1]+[1, 1]=zN_1w.
\end{aligned} \tag{5}$$

Upon the approximation of the nearest neighbor interaction the energy of the crystal may be written as

$$E(r, T)=[0, 0]V_{AB}+[0, 1]V_{AA}+[1, 0]V_{BB}+[1, 1]V_{AB}, \tag{6}$$

where the V's are the interaction energies between the different kinds of pairs of nearest neighbors.

We may give (5) and (6) a different interpretation by imagining $[0, 0]$, $[0, 1]$, $[1, 0]$, $[1, 1]$ and V_{AB}, V_{AA}, V_{BB}, V_{AB} to be, respectively, the numbers and the molecular internal energies of four different kinds of molecules, say, X, XZ, XY, XYZ of a gaseous assembly. The interpretation of (6) is that the assembly has the same internal (non-kinetic) energy as the crystal at the given value of w. Equation (5) would mean that the total numbers of X, Y, Z atoms are, respectively, zN_1, zN_1w, zN_1w.

The quasi-chemical method consists in taking the averages $\langle 0, 0\rangle_{Av}$, $\langle 0, 1\rangle_{Av}$, $\langle 1, 0\rangle_{Av}$, $\langle 1, 1\rangle_{Av}$ of the chemical assembly at any temperature as approximately representing the corresponding averages of the crystal at the same temperature.

A detailed treatment of the problem of a gaseous assembly has been given by Fowler.[7] We are only interested in our assembly of four different kinds of molecules, for which the results may be summarized as:

$$\begin{aligned}
\langle 0, 0\rangle_{Av}&=\xi \exp[-V_{AB}/kT],\\
\langle 0, 1\rangle_{Av}&=\xi\nu \exp[-V_{AA}/kT],\\
\langle 1, 0\rangle_{Av}&=\xi\mu \exp[-V_{BB}/kT],\\
\langle 1, 1\rangle_{Av}&=\xi\mu\nu \exp[-V_{AB}/kT],
\end{aligned} \tag{7}$$

where ξ, μ, and ν are to be determined from (6).

In Fowler and Guggenheim's work the starting point is the equation

$$\langle 0, 0\rangle_{Av}\langle 1, 1\rangle_{Av}/\langle 0, 1\rangle_{Av}\langle 1, 0\rangle_{Av}=\exp[(V_{AA}+V_{BB}-2V_{AB})/kT], \tag{8}$$

which can be obtained by eliminating ξ, μ, and ν from our equation (7). We shall, however, make use of (7) instead of (8) to obtain the free energy by means of a Legendre transformation which enables us to avoid the labor of integration. It will appear that this procedure is applicable to the general case studied in this paper.

In order to calculate the free energy from (4) we first write the energy of the crystal in the

[7] R. H. Fowler, *Statistical Mechanics*, second edition, pp. 162–163.

form

$$\bar{E}(w, T) = kT^2 \frac{\partial}{\partial T} \phi(\xi, \mu, \nu, T),$$

where

$$\phi(\xi, \mu, \nu, T) = \xi \exp[-V_{AB}/kT] + \xi\nu \exp[-V_{AA}/kT] + \xi\mu \exp[-V_{BB}/kT] + \xi\mu\nu \exp[-V_{AB}/kT].$$

From (5) it is evident that

$$\xi\frac{\partial \phi}{\partial \xi} = zN_1, \quad \mu\frac{\partial \phi}{\partial \mu} = zN_1 w, \quad \nu\frac{\partial \phi}{\partial \nu} = zN_1 w.$$

If these last three equations are regarded as defining ξ, μ, and ν in terms of w and T, i.e., if the Legendre transformation:

$$\log \xi, \log \mu, \log \nu, T \to zN_1, zN_1 w, zN_1 w, T$$

is made, the derivative of the function

$$\Psi(w, T) = \phi - zN_1 \log \xi - zN_1 w \log \mu - zN_1 w \log \nu,$$

with respect to T is found to be

$$\frac{\partial \Psi}{\partial T} = \frac{\partial \phi}{\partial T} = \frac{1}{kT^2}\bar{E}(w, T).$$

This shows that the integral in (4) may be expressed directly as a function of ξ, μ, ν and T:

$$\int_{\infty}^{T} \frac{1}{kT^2}\bar{E}(w, T)dT = \Psi(w, T) - \Psi(w, \infty).$$

The values of the parameters at $T = \infty$ are easily determined from (5) and (7):

$$\xi = zN_1(1-w)^2, \quad \mu = \nu = \frac{w}{1-w}.$$

Thus (4) becomes

$$\log P(w, T) = 2 \log \frac{N_1!}{(N_1 w)!\{N_1(1-w)\}!} + zN_1 \log \frac{zN_1(1-w)^2}{\xi} + zN_1 w \log \frac{w^2}{(1-w)^2 \mu\nu},$$

so that the free energy of the crystal is

$$F_0(w, T) = -kT \log P(w, T) = -kT\{zN_1 \log zN_1 + 2(z-1)N_1[w \log w + (1-w)\log(1-w)] - zN_1 \log \xi - zN_1 w \log \mu\nu\}. \quad (9)$$

We shall have occasion to return to this expression for the free energy later in Section 4.

3. GENERALIZATION TO GROUPS OF FOUR SITES

So far our attention has been fixed on the pairs of nearest neighbors in the crystal. They are classified into four different kinds and the average number of pairs in each class is obtained from chemical analogy. Now we shall generalize the whole procedure by taking into consideration all the groups of sites of a certain arbitrarily chosen form in the crystal. These groups will be classified according to the way they are occupied by atoms and the average number of groups in each class is to be obtained by chemical analogy.

To make this clear let us consider in detail groups of four sites forming the corners of squares (as shown in Fig. 1) in a quadratic lattice. We

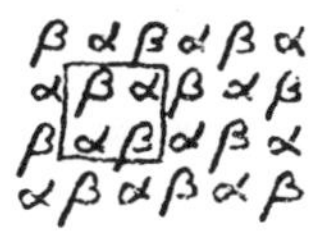

FIG. 1.

classify these groups into $2^4 = 16$ classes and denote them by (0, 0, 0, 0), (0, 0, 0, 1), $\cdots$, (1, 1, 1, 1), respectively, so that all groups in the class (q_1, q_2, q_3, q_4) have q_1 wrong atoms in their upper α-sites, q_2 wrong atoms in their lower α-sites, q_3 wrong atoms in their upper β-sites, and q_4 wrong atoms in their lower β-sites. The total number of these groups is N, hence if we denote the number of groups in the class (q_1, q_2, q_3, q_4) by $[q_1, q_2, q_3, q_4]$,

$$\sum_{q_i=0}^{1} [q_1, q_2; q_3, q_4] = N_1. \quad (10)$$

Now the number of all those groups in the crystal with a wrong atom on the upper α-site is just the number of wrong atoms on the α-sites. Similar reasoning may be applied to the other three sites, so we obtain

$$\sum_{q} q_i[q_1, q_2, q_3, q_4] = N_1 w, \quad i = 1, 2, 3, 4. \quad (11)$$

Let $\chi(q_1, q_2, q_3, q_4)$ be the energy of each group in the class (q_1, q_2, q_3, q_4). It is easy to show that

$$\chi = (2-q_1-q_2)(q_3+q_4)V_{AA} + \{(2-q_1-q_2)(2-q_3-q_4) + (q_1+q_2)(q_3+q_4)\} V_{AB}(2-q_3-q_4)(q_1+q_2)V_{BB}.$$

The total energy of the crystal is given in terms of $\chi(q_1, q_2, q_3, q_4)$ by

$$E(w, T)=\sum_q [q_1, q_2, q_3, q_4]\chi(q_1, q_2, q_3, q_4). \quad (12)$$

We may give (10), (11), and (12) interpretations similar to those given in Section 2 for Eqs. (5) and (6). The same quasi-chemical method used there to obtain (7) leads now to the following averages for given w:

$$\langle q_1, q_2, q_3, q_4\rangle_{Av}=\xi\mu_1^{q_1}\mu_2^{q_2}\mu_3^{q_3}\mu_4^{q_4} \times\exp[-\chi(q_1, q_2, q_3, q_4)/kT]. \quad (13)$$

In this expression the parameters ξ, μ_1, μ_2, μ_3, and μ_4 are to be determined from (10) and (11), which may be written in the form

$$\xi\frac{\partial\phi}{\partial\xi}=N_1, \quad \mu_i\frac{\partial\phi}{\partial\mu_i}=N_1 w, \quad (i=1, 2, 3, 4) \quad (14)$$

if we put

$$\phi(\xi, \mu_1, \mu_2, \mu_3, \mu_4)=\sum_q \xi\mu_1^{q_1}\mu_2^{q_2}\mu_3^{q_3}\mu_4^{q_4} \times\exp[-\chi(q_1, q_2, q_3, q_4)/kT]. \quad (15)$$

It can be shown[8] that ξ and μ_i are uniquely determined by (14) at given w and T. Their values at $T=\infty$ are

$$(\xi)_{T=\infty}=N_1(1-w)^4,$$

$$(\mu_i)_{T=\infty}=\frac{w}{1-w}, \quad i=1, 2, 3, 4; \quad (16)$$

as can be shown by substitution into (14).

To calculate the free energy it is necessary first to evaluate the integral in (4). By (12) and (13) the integrand may be written as

$$\frac{1}{kT^2}\bar{E}(w, T)=\frac{1}{kT^2}\sum_q \langle q_1, q_2, q_3, q_4\rangle_{Av} \times\chi(q_1, q_2, q_3, q_4)=\frac{\partial\phi}{\partial T}, \quad (17)$$

where the partial differentiation is to be taken with ξ, μ_1, μ_2, μ_3, μ_4, and T as the independent variables. If, however, we regard them as functions of w and T defined by (14) and introduce the Legendre transformation

$$\log\xi, \log\mu_1, \log\mu_2, \log\mu_3, \log\mu_4 \rightarrow N_1, N_1w, N_1w, N_1w, N_1w, \quad (18)$$

(17) reduces to

$$\frac{1}{kT^2}\bar{E}(w, T)=\frac{\partial\phi}{\partial T}=\frac{\partial}{\partial T}\Psi(w, T), \quad (19)$$

where the function Ψ is defined by

$$\Psi=\phi-N_1\log\xi-\sum_i N_1 w\log\mu_i. \quad (20)$$

Substituting (19) into (4) we get

$$\log P(w, T)=\log g(w)+\Psi(w, T)-\Psi(w, \infty), \quad (21)$$

so that the free energy may be written down directly as

$$F(w, T)=-kT\log P(w, T)$$
$$=-kT[\log g(w)+\Psi(w, T)-\Psi(w, \infty)]. \quad (22)$$

The equilibrium value $\bar{w}$ of w is obtained by minimizing F, so we have

$$0=\frac{\partial}{\partial\bar{w}}F(\bar{w}, T)=-kT\left[\frac{d}{d\bar{w}}\log g(\bar{w}) +\frac{\partial}{\partial\bar{w}}\Psi(\bar{w}, T)-\frac{\partial}{\partial\bar{w}}\Psi(\bar{w}, \infty)\right]. \quad (23)$$

But by (14) and (20)

$$\frac{\partial}{\partial w}\Psi(w, T)=-\sum_i N_1\log\mu_i,$$

and by (3)

$$\frac{d}{dw}\log g(w)=-2N_1\log\frac{w}{1-w},$$

so by (16)

$$\sum_i\log\mu_i=-2\log\frac{\bar{w}}{1-\bar{w}} +[\sum_i\log\mu_i]_{T=\infty}=2\log\frac{\bar{w}}{1-\bar{w}};$$

i.e.,

$$\prod_i\mu_i=\left(\frac{\bar{w}}{1-\bar{w}}\right)^2. \quad (24)$$

[8] The proof follows easily from Lemma 2.42 in Fowler's *Statistical Mechanics*, second edition.

It will be shown in the next section that we may put $V_{AA}=V_{BB}$, and $V_{AB}=0$ without altering the specific heat of the crystal if

$$V=\tfrac{1}{2}(V_{AA}+V_{BB})-V_{AB}$$

is left unchanged. When this is done, ϕ will be symmetrical with respect to μ_1, μ_2, μ_3, and μ_4, and we conclude that all the μ's are equal, because: (i) Eq. (14) has only one set of solutions; and (ii) if the conclusion is true, (14) becomes, with all μ_i put equal to μ,

$$\xi\frac{\partial\phi}{\partial\xi}=N_1,\quad \mu\frac{\partial\phi}{\partial\mu}=4N_1w, \tag{25}$$

which *does* have a set of solutions in ξ and μ. Now ϕ is given by

$$\phi=\xi[1+4\mu x^2+(4\mu^2x^2+2\mu^2x^4)+4\mu^3x^2+\mu^4], \tag{26}$$

where

$$x=\exp[-\tfrac{1}{2}(V_{AA}+V_{BB}-2V_{AB})/kT]. \tag{27}$$

Introducing the degree of order s by $s=1-2w$ and eliminating ξ from (25), we obtain

$$(1+s)\mu^4+(2+4s)\mu^3x^2+2sx^2(x^2+2)\mu^2 +(4s-2)x^2\mu+(s-1)=0. \tag{28}$$

The free energy is given by (16) and (22):

$$\frac{-F(w,T)}{2N_1kT}=(1-w)\log(1-w) +w\log w-2w\log\mu+\tfrac{1}{2}\log(1+4\mu x^2 +4\mu^2x^2+2\mu^2x^4+4\mu^3x^2+\mu^4), \tag{29}$$

and the condition of equilibrium by (24):

$$\mu=\left(\frac{1-s}{1+s}\right)^{\frac{1}{2}}. \tag{30}$$

To obtain the critical temperature, we expand (28) in powers of s and find, after identifying coefficients,

$$\log\mu=-\frac{1+6x^2+x^4}{2+2x^2}s+Ks^3+\cdots$$

which is the only real solution for $\log\mu$. Next we expand (30);

$$\log\mu=-s-\tfrac{1}{3}s^3-\cdots.$$

At the critical value x_c of x, these last two equations have a multiple solution at $s=0$. Hence

$$-\frac{1+6x_c^2+x_c^4}{2+2x_c^2}=-1,$$

i.e.,

$$x_c=(\sqrt{5}-2)^{\frac{1}{2}}=0.4858.$$

4. GENERAL FORM OF THE QUASI-CHEMICAL METHOD

Let us now take a group of any size and form. Let it have a α-sites, b β-sites, and γ pairs of nearest neighbors. The procedures to obtain an approximate expression for the free energy of the crystal follow exactly the same line as in the special case considered in the last section. Equations (13), (14), and (16) are essentially unchanged:

$$\langle q_1,q_2,\cdots\rangle_{Av}=\xi\mu_1^{q_1}\mu_2^{q_2}\cdots\exp[-\chi/kT], \tag{31}$$

$$\phi=\sum_q \xi\mu_1^{q_1}\mu_2^{q_2}\cdots\exp[-\chi/kT],$$

$$\xi\frac{\partial\phi}{\partial\xi}=N_1,\quad \mu_i\frac{\partial\phi}{\partial\mu_i}=N_1w, \tag{32}$$

and

$$(\mu_i)_{T=\infty}=\frac{w}{1-w}. \tag{33}$$

But now the total number of pairs in the whole set of groups is $N_1\gamma$ while the total number of pairs of nearest neighbors in the actual crystal is N_1z. Hence the energy expression (12) should be modified by multiplying the right-hand side with a factor z/γ. Thus we have, in place of (17),

$$\bar{E}=\frac{z}{\gamma}\sum_q\langle q_1,q_2,\cdots\rangle_{Av}\chi(q_1,q_2,\cdots) =\frac{z}{\gamma}kT^2\frac{\partial\phi}{\partial T}. \tag{34}$$

And (22) becomes, with the help of (3), (20), and (33),

$$F(w,T)=-\frac{z}{\gamma}N_1kT\Big[\log N_1+\left(a+b-\frac{2\gamma}{z}\right) \times\{(1-w)\log(1-w)+w\log w\} -\log\xi-w\sum_i\log\mu_i\Big]. \tag{35}$$

Differentiate (35) and we obtain

$$\frac{\partial}{\partial w}F(w,T)$$

$$=-\frac{z}{\gamma}N_1kT\log\left[\left(\prod_i \mu_i\right)\left(\frac{1-w}{w}\right)^{a+b-(2\gamma/z)}\right] \quad (36)$$

where use has been made of (32). So the condition of equilibrium is given by

$$\prod_i \mu_i=\left(\frac{w}{1-w}\right)^{a+b-(2\gamma/z)} \quad (37)$$

In actual calculations the following points may prove helpful:

(i) *The free energy is changed by a constant if V_{AA} and V_{BB} are both replaced by $\frac{1}{2}(V_{AA}+V_{BB})-V_{AB}$, and V_{AB} by 0.* To prove this let z_i be the number of sites in the group neighboring to the site i. Let χ be changed into χ' by the replacement. It is evident that $\chi'-\chi=-\gamma V_{AB}+$(No. of BB pairs)$(V_{AA}-V_{BB}/2)+$(No. of AA pairs)$\times(V_{BB}-V_{AA}/2)$, and that

$$\sum_{\alpha\text{-sites}} q_iz_i-\sum_{\beta\text{-sites}} q_iz_i$$

$$=(\text{No. of } BB \text{ pairs})-(\text{No. of } AA \text{ pairs}). \quad (38)$$

Hence

$$\langle q_1, q_2, \cdots\rangle_{N}=\xi\mu_1^{q_1}\mu_2^{q_2}\cdots\exp[-\chi/kT]$$

$$=\xi'\mu_1'^{q_1}\mu_2'^{q_2}\cdots\exp[-\chi'/kT],$$

if we put

$$\xi'=\xi\exp[-\gamma V_{AB}/kT],$$

$$\mu_i'=\mu_i\exp[\pm z_i(V_{AA}-V_{BB})/2kT]$$

where the + sign or the − sign is to be taken according as the site i is an α-site or a β-site. We can now calculate the new free energy and verify the above statement.

(ii) *Sites that are symmetrically situated in the group have μ's equal irrespective of their nature $V_{AA}=V_{BB}$ and $V_{AB}=0$.* This has already been shown in the last section for the special case considered there. With allowances for change of notation, the proof holds in general. Since the most troublesome part of the calculations is the elimination of the parameters, much might be gained by choosing a group with a large number of sites symmetrically situated.

(iii) *The free energy is a function of s^2, so that (37) is always satisfied at $w=\frac{1}{2}$ (i.e., long-distance order=0).* The proof is simple when we have already made $V_{AA}=V_{BB}$ and $V_{AB}=0$, so that an interchange of A and B atoms does not alter the energy:

$$\chi(q_1, q_2, \cdots)=\chi(1-q_1, 1-q_2, \cdots).$$

For, putting

$$\xi'=\xi\mu_1\mu_2\cdots$$

and

$$\mu_i'=1/\mu_i, \quad (39)$$

we get

$$\xi\mu_1^{q_1}\mu_2^{q_2}\cdots\exp[-\chi/kT]$$

$$=\xi'\mu_1'^{1-q_1}\mu_2'^{1-q_2}\cdots\exp[-\chi/kT].$$

Thus if (32) is satisfied

$$\sum_q(1-q_i)\xi'\mu_1'^{1-q_1}\mu_2'^{1-q_2}\cdots\exp[-\chi/kT]$$

$$=\sum_q \xi\mu_1^{q_1}\mu_2^{q_2}\cdots\exp[-\chi/kT]$$

$$-\sum_q q_i\xi\mu_1^{q_1}\mu_2^{q_2}\cdots\exp[-\chi/kT]=N(1-w);$$

i.e., ξ', μ_1', μ_2', $\cdots$ would be the solution of (32) with $(1-w)$ substituted for w. Hence b· (32) and (20)

$$\Psi(1-w,T)=N_1-N_1\log\xi'$$

$$-\sum_i N_1(1-w)\log\mu_i'=\Psi(w,T)$$

showing that

$$F(w,T)=F(1-w,T). \quad (40)$$

(iv) *The parameter for a corner site is always given by:*

$$\epsilon=\frac{1}{1+s}\{[x^2s^2+(1-s^2)]^{\frac{1}{2}}-sx\} \quad (41)$$

where x is given by (27) and $s=1-2w$, irrespective of the size of the group, if $V_{AA}=V_{BB}$, $V_{AB}=0$. By a corner site is meant a site that has only one nearest neighbor in the group. Let ϵ be the selective variable (parameter) of a corner site, and μ_1 that of its only neighbor in the group. If the corner site is dropped, a new group is obtained. We distinguish all quantities referring to this new group by a prime, and obtain at once

$$\xi'\frac{\partial\phi'}{\partial\xi'}=N_1,\quad \mu_i'\frac{\partial\phi'}{\partial\mu_i'}=N_1w,\quad i=1,2,\cdots. \tag{42}$$

The sites of the primed group are numbered in the same way as in the unprimed group. Introducing the variable x defined in (27) we may write

$$\begin{aligned}\phi&=\sum_q \xi\mu_1^{q_1}\mu_2^{q_2}\cdots\exp\left[-\chi/kT\right]\\&=\sum_{q_2\cdots}\xi(1+\epsilon x)\mu_2^{q_2}\mu_3^{q_3}\cdots\exp\left[-\chi'/kT\right]\\&\quad+\mu_1\sum_{q_2\cdots}\xi(\epsilon+x)\mu_2^{q_2}\cdots\exp\left[-\chi'/kT\right].\end{aligned} \tag{43}$$

Let these two terms be denoted by ϕ_0 and ϕ_1, respectively. Since

$$\phi=N_1,\quad \mu_1\frac{\partial\phi}{\partial\mu_1}=N_1w,$$

we have

$$\phi_0=N_1(1-w),\quad \phi_1=N_1w. \tag{44}$$

Now

$$\epsilon\frac{\partial\phi_0}{\partial\epsilon}=\frac{\epsilon x}{1+\epsilon x}\phi_0,\quad \epsilon\frac{\partial\phi_1}{\partial\epsilon}=\frac{\epsilon}{\epsilon+x}\phi_1.$$

Hence $\epsilon\dfrac{\partial\phi}{\partial\epsilon}=N_1w$ leads to

$$\frac{\epsilon x}{1+\epsilon x}N_1(1-w)+\frac{\epsilon}{\epsilon+x}N_1w=N_1w, \tag{45}$$

or

$$\frac{w}{1-w}=\frac{\epsilon(\epsilon+x)}{1+\epsilon x}, \tag{46}$$

the solution of which is (41). Thus the two parameters μ and ν in the approximation discussed in Section 2 are all equal to ϵ.

(v) *The contribution to the free energy from a corner atom is such that, in the notation of* (iv)

$$F(w,T)=\frac{\gamma-1}{\gamma}F'(w,T)+\frac{1}{\gamma}F_0(w,T), \tag{47}$$

where $F_0(w,T)$ *is the free energy obtained when* $\gamma=1$, i.e., *the free energy (9) for the approximation discussed in Section 2.* This is proved as follows. If we put

$$\xi=\xi''\frac{1}{1+\epsilon x},\quad \mu_1=\mu_1''\frac{1+\epsilon x}{\epsilon+x},\quad \mu_i=\mu_i'',\quad i\geqq 2, \tag{48}$$

it is evident from (43) that ϕ would become a function of ξ'', μ_1'', μ_2'', $\cdots$ satisfying the relations

$$\xi''\frac{\partial\phi}{\partial\xi''}\left(=\xi\frac{\partial\phi}{\partial\xi}\right)=N_1,\quad \mu_i''\frac{\partial\phi}{\partial\mu_i''}\left(=\mu_i\frac{\partial\phi}{\partial\mu_i}\right)=N_1w,$$

$$i=1,2,\cdots. \tag{49}$$

It is also evident that ϕ is the same function of ξ'', μ_1'', μ_2'', $\cdots$ as ϕ' is of ξ', μ_1', μ_2'', $\cdots$. Now (42) has only one[8] set of solutions in ξ' and μ_i'. Hence from (49) we infer that $\xi'=\xi''$, and $\mu_i'=\mu_i''$. Thus

$$\xi=\frac{\xi'}{1+\epsilon x},\quad \mu_1=\mu_1'\frac{1+\epsilon x}{\epsilon+x},\quad \mu_i=\mu_i',\quad i\geqq 2. \tag{50}$$

(41) and (50) give the parameters μ_i in terms of μ_i'. Inserting them into (35) we obtain

$$\begin{aligned}F(w,T)&=-N_1kT\frac{z}{\gamma}\Big[\log N_1+\left(a+b-\frac{2\gamma}{z}\right)\\&\quad\times\{(1-w)\log(1-w)+w\log w\}\\&\quad-\log\xi'+(1-w)\log(1+\epsilon x)\\&\quad+w\log(\epsilon+x)-w\log\epsilon-w\sum_i\log\mu_i'\Big]\\&=\frac{\gamma-1}{\gamma}F'(w,T)-N_1kT\frac{z}{\gamma}\Big[\left(1-\frac{2}{z}\right)\\&\quad\times\{(1-w)\log(1-w)+w\log w\}\\&\quad+(1-w)\log(1+\epsilon x)+w\log\frac{\epsilon+x}{\epsilon}\Big]\end{aligned} \tag{51}$$

If the original group (unprimed) is a pair of neighboring sites, we have $\gamma=1$, and (51) reduces to Eq. (9), i.e., to the free energy in the approximation discussed in Section 2:

$$\begin{aligned}F_0(w,T)&=-zN_1kT\Big[\left(1-\frac{2}{z}\right)\\&\quad\times\{(1-w)\log(1-w)+w\log w\}\\&\quad+(1-w)\log(1+\epsilon x)w\log\frac{\epsilon+x}{\epsilon}\Big].\end{aligned} \tag{52}$$

Inserting (52) back into (51) we get (47).

5. COMPARISON WITH BETHE'S METHOD

Both the quasi-chemical method and Bethe's method start from some assumption regarding the relative probabilities of finding a definite group of sites occupied in different ways by atoms. Although in the two methods the arguments leading to the assumptions are in no way similar the assumptions themselves are closely related. In fact, a comparison of Bethe's local partition function and our function ϕ shows that the quasi-chemical method would give the same probabilities of occurrence of the local configurations as Bethe's method if the free energy (35) has a minimum when

$$(\mu)_{\text{interior sites}}=1, \tag{53}$$

i.e., if (37) and (53) are mathematically equivalent. This is not true in general. But it holds approximately when the group of sites under consideration becomes very large. For then,

$$a+b-\frac{2\gamma}{z}\ll\gamma,$$

hence if (53) is true, we should have

$$\left[(\prod_i \mu_i)\left(\frac{1-w}{w}\right)^{a+b-(2\gamma/z)}\right]^{1/\gamma}$$

$$\cong[\prod(\mu)_{\text{interior sites}}]^{1/\gamma}\cong 1,$$

so that by (37)

$$\frac{\partial}{\partial w}F(w,T)\cong 0.$$

To see how the equilibrium free energy $\bar{F}(\bar{w},T)$ varies with T in high order approximations, we substitute (37) into (35) and obtain by making use of (32)

$$-\frac{F}{zN_1kT}=\frac{1}{\gamma}\log\left(\sum_q \mu_1^{q_1}\mu_2^{q_2}\cdots\exp[-\chi/kT]\right)$$

$$+\frac{1}{\gamma}\left(a+b-\frac{2\gamma}{z}\right)\log(1-\bar{w}).$$

The last term is very small for large groups, so that by (53)

$$-\frac{F}{zN_1kT}=\frac{1}{\gamma}\log\left(\sum\exp[-\chi/kT]\right).$$

This shows that in high order approximations the free energy in the quasi-chemical method reduces to the exact form demanded by statistical mechanics.

Recently a modification of Bethe's method has been developed by Wang.[9] This modified Bethe's method bears a very close relation to the quasi-chemical method. The difference between the two lies in the calculation of energy which is discussed in length in Wang's paper.

6. THE NON-INTERFERENCE OF LOCAL CONFIGURATIONS

Let us return to the fundamental assumption of the quasi-chemical method, i.e., to (31) which gives the average numbers of the different local configurations (so far called groups) in the crystal. This equation expresses the exact distribution law of an assembly of molecules (cf. the example in Section 2) which has an energy γ/z times as large as the crystal. Distinguishing all quantities referring to the assembly of molecules by a subscript m, we get from (4),

$$\frac{1}{kT}F(w,T)+\log g(w)$$

$$=\frac{z}{\gamma}\left[\frac{1}{kT}F_m(w,T)+\log g_m(w)\right].$$

But if H is the number of arrangements in the crystal lattice having the given values of $[q_1, q_2, \cdots]$,

$$F(w,T)=-kT\log(\bar{H})+\bar{E}. \tag{54}$$

Thus

$$\log\frac{\bar{H}}{g(w)}=\frac{z}{\gamma}\log\frac{\bar{H}_m}{g_m(w)}. \tag{55}$$

But*

$$H_m=\frac{N_1!}{\prod_q[q_1, q_2, \cdots]!}.$$

On substituting this into (55) and dropping the bars we get

$$H=h(w)\left\{\frac{N_1!}{\prod_q[q_1, q_2, \cdots]!}\right\}^{z/\gamma}, \tag{56}$$

---

[9] J. S. Wang, "Approximate partition function in generalized Bethe's theory of superlattices" (to be published).

* R. H. Fowler, *Statistical Mechanics*, second edition, 2.6 and 5.11.

where

$$h(w)=g(w)/\{g_m(w)\}^{z/\gamma}. \quad (57)$$

Equation (56) has been referred to in Fowler and Guggenheim's paper[1] as the mathematical expression of the "hypothesis of the non-interference of local configurations (pairs);" because when $z/\gamma=1$, the number of arrangements in the crystal consistent with the distribution law $[q_1, q_2, \cdots]$ for the groups of sites is, except for the factor $h(r)$, equal to

$$H_m=\frac{N_1!}{\prod_q [q_1, q_2, \cdots]!},$$

which would be the number of arrangements in the crystal having the given values of $[q_1, q_2, \cdots]$ if the N_1 groups in the crystal were *separated* and were filled *independently* with atoms. The term "non-interference" comes from the fact that the N_1 groups are *not separated* but are *interlocked* and *cannot be filled independently* with atoms, i.e., they "interfere" with each other.

To find the value of $g(w)$ we notice that by definition $g_m=\Sigma H_m$. But ΣH_m is the number of arrangements in the *separated* groups considered above if they are to be so filled with atoms that N_1w of them have wrong atoms on the sites i, where $i=1, 2, \cdots$. For the N_1 sites i of the N_1 groups,

$$\frac{N_1!}{\{N_1(1-w)\}!(N_1w)!}$$

different arrangements are possible. Hence

$$g_m=\sum H_m=\sum\frac{N_1!}{\prod_q [q_1, q_2, \cdots]!}$$

$$=\left[\frac{N_1!}{\{(1-w)N_1\}!(N_1w)!}\right]^{a+b}. \quad (58)$$

Thus

$$h(w)=\left[\frac{N_1!}{\{N_1(1-w)\}!(N_1w)!}\right]^{2-(z/\gamma)(a+b)}. \quad (59)$$

The free energy of the crystal may be obtained from (40), (42), and (45) as:

$$F(w,T)=\bar{E}-\frac{zN_1kT}{\gamma}\Bigg\{\left(a+b-\frac{2\gamma}{z}\right)$$

$$\times[(1-w)\log(1-w)+w\log w]+\log N_1$$

$$-\sum_q\frac{1}{N_1}\langle q_1, q_2, \cdots\rangle_{Av}\log\langle q_1, q_2, \cdots\rangle_{Av}\Bigg\},$$

which is reducible to Eq. (35) obtained above.

7. SPECIAL CONSIDERATIONS CONCERNING BETHE'S FIRST AND SECOND APPROXIMATIONS

The First Approximation

If an α-site together with its z nearest neighbors are taken as our group of interest, all the sites except the central one are corner sites. Hence their selective variables are all equal to the value of ϵ given in (41). By successive applications of (47) we see that the free energy is exactly $F_0(w, T)$, a fact which has already been pointed by Fowler and Guggenheim.[1] The selective variable of the central site is given by successive applications of (50):

$$\lambda=\frac{w}{1-w}\left(\frac{1+\epsilon x}{\epsilon+x}\right)^z. \quad (60)$$

The factor $w/(1-w)$ is the selective variable of the central site when it alone forms the group. The equilibrium condition (38) becomes

$$\lambda\epsilon^z=[\bar{w}/(1-\bar{w})]^{z-1}. \quad (61)$$

But by (60) and (45)

$$\lambda=[w/(1-w)]^{1-z}\epsilon^z.$$

Hence at equilibrium

$$\lambda=1. \quad (62)$$

Thus the approximation is completely equivalent to Bethe's first approximation, as already mentioned in Section 5.

The Second Approximation

Now consider the group of sites occurring in Bethe's second approximation. According to Section 4 (iv), the selective variables for the corner sites in the second shell are all equal to the parameter ϵ given by (41). In Bethe's original calculations, however, the selective variables for the corner sites and the medium sites are made equal, and are found to be different from ϵ. Thus if we use his original method, Eq. (32) would not be satisfied. In other words, the probabilities of occurrence of wrong atoms in the corner and medium sites would be unequal.

For simplicity we shall drop the corner sites and take as our group of interest the central site, the first shell sites and the medium sites with selective variables μ, ν, and λ, respectively. The contribution by the corner sites can be included in the free energy by simple addition, as shown in Section 4. With the notations n and g_{nm} of Bethe[3] we find

$$\phi=\xi\sum_n (x^n+\mu x^{z-n})P_n(x,\nu,\lambda), \qquad (63)$$

where

$$P_n(x,\nu,\lambda)=\nu^n\sum_m g_{nm}[(1+\lambda)x]^m \times(x^2+\lambda)^{([z/2]-1)n-(m/2)}(1+\lambda x^2)^{([z/2]-1)(z-n)-(m/2)}.$$

After the elimination of ξ and μ, (32) becomes

$$zw=\frac{\sum_{l,n}\left\{2x^{l+z-n}\left[(1-w)P_n\lambda\frac{\partial}{\partial\lambda}P_l+wP_l\lambda\frac{\partial}{\partial\lambda}P_n\right]\right\}}{(z-2)(\sum_n x^nP_n)(\sum_n x^{z-n}P_n)}$$

$$=\frac{\sum_{l,n}\left\{x^{l+z-n}\left[(1-w)P_n\nu\frac{\partial}{\partial\nu}P_l+wP_l\nu\frac{\partial}{\partial\nu}P_n\right]\right\}}{(\sum_n x^nP_n)(\sum_n x^{z-n}P_n)}. \qquad (64)$$

The free energy is obtained from (35):

$$F(w,T)=-\frac{N_1kT}{z-1}\Big[\tfrac{1}{2}(z^2-4z+4) \times\{(1-w)\log(1-w)+w\log w\} +(1-w)\log(\sum_n x^nP_n) +w\log(\sum_n x^{z-n}P_n)-zw\log\nu -w\frac{z(z-2)}{2}\log\lambda\Big]. \qquad (65)$$

8. APPLICATION TO THE CRYSTAL Cu_3Au

For the face-centered crystal Cu_3Au, we may of course follow Peierls[10] and take as our group of interest a central site together with its twelve first shell neighbors. The free energy expression would then contain seven selective variables, four of which may be easily eliminated. The resulting expression is very cumbersome and numerical calculations would be laborious. We therefore make a simpler approximation: The group is taken to be four nearest neighbors forming a tetrahedron. (See Fig. 2.) A little

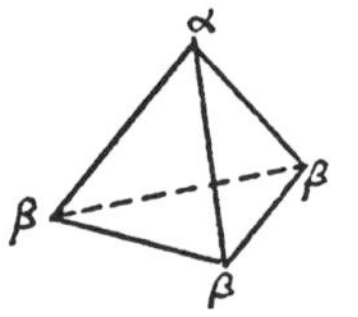

FIG. 2.

geometrical consideration assures us that all such tetrahedrons contain an α-site (for gold atoms) and three β-sites (for copper atoms), an interesting conclusion showing that the tetrahedron might be regarded as a sort of "molecular" structure in a faced-centered lattice with atomic ratio 1:3. Our approximation may thus be reasonably expected to reveal the more important features of the order-disorder transformation in such alloys.

A little reflection shows that the statement made in Section 4 (i) still holds, so that we may put $V_{AA}=V_{BB}$ and $V_{AB}=0$. Let μ and ν be the parameters (for the wrong atoms) of the β-sites and the α-sites, respectively. Let there be altogether $N=4N_1$ atoms. It is easy to see that there are $8N_1$ groups in the crystal. When N_1w atoms on the α-sites are wrong, the equations determining the parameters are

$$8N_1=\phi=\xi[x^3+3x^2\mu+3x^3\mu^2+x^6\mu^3 +\nu(x^6+3x^3\mu+3x^2\mu^2+x^3\mu^3)], \qquad (66a)$$

$$8N_1w=\nu\frac{\partial\phi}{\partial\nu}=\xi\nu(x^6+3x^3\mu+3x^2\mu^2+x^3\mu^3), \qquad (66b)$$

and

$$8N_1\left(\frac{w}{3}\right)+8N_1\left(\frac{w}{3}\right)+8N_1\left(\frac{w}{3}\right)=\mu\frac{\partial\phi}{\partial\mu} =3x^2\xi\mu[1+2x\mu+x^4\mu^2+\nu(x+2\mu+x\mu^2)], \qquad (66c)$$

where x is defined by (27). The energy of the

[10] R. Peierls, Proc. Roy. Soc. A154, 207 (1936).

crystal is [cf. (34)],

$$\bar{E}=\tfrac{1}{2}kT^2\frac{\partial\phi}{\partial T}+\text{constant}; \qquad (67)$$

so that the free energy becomes [cf. (35)]:

$$F(w, T)=-kT[\log g(w)+\tfrac{1}{2}(\phi-8N_1\log\xi$$
$$-8N_1 w\log\nu-8N_1 w\log\mu)_{T=\infty}].$$

But

$$\log g(w)=-N_1\Big[(1-w)\log(1-w)+w\log w$$
$$+w\log\frac{w}{3}+(3-w)\log\{(3-w)/3\}\Big],$$

and at $T=\infty$,

$$\nu=\frac{w}{1-w},\quad \mu=\frac{w}{3-w},\quad \xi=8N_1(1-w)\left(1-\frac{w}{3}\right)^3.$$

Hence

$$-\frac{F(w, T)}{N_1 kT}=-9\log 3+4\log(8N_1)$$
$$+6w\log w+3(1-w)\log(1-w)$$
$$+3(3-w)\cdot\log(3-w)-4\log\xi$$
$$-4w\log\mu-4w\log\nu. \qquad (68)$$

Since ξ and ν can be very easily solved from (66), numerical calculations are quite simple. The value of the free energy is plotted in the accompanying figure (Fig. 3).

The equilibrium value of w is given by [cf. (36) and (37)]:

$$0=-3\log\frac{(1-\bar{w})(3-\bar{w})}{\bar{w}^2}+4\log\mu\nu. \qquad (69)$$

This is always satisfied at $w=\frac{3}{4}$.* Actual calculation shows that the absolute minimum of the free energy is or is not at $w=\frac{3}{4}$ according as $x>0.2965$ or $x<0.2965$. From the form of the curve in Fig. 3 it is seen that the crystal has a critical temperature at which the long-distance order, and hence also the energy, is discontinuous. The critical temperature T_c and the latent heat Q are found to be given by

$$T_c=0.8228\frac{1}{k}[\tfrac{1}{2}(V_{AA}+V_{BB})-V_{AB}],$$

and

$$Q=0.8824N_1[\tfrac{1}{2}(V_{AA}+V_{BB})-V_{AB}].$$

In terms of the total energy change from $T=0$ to $T=\infty$:

$$E_0=3N_1[\tfrac{1}{2}(V_{AA}+V_{BB})-V_{AB}],$$

these quantities become, with the usual notation

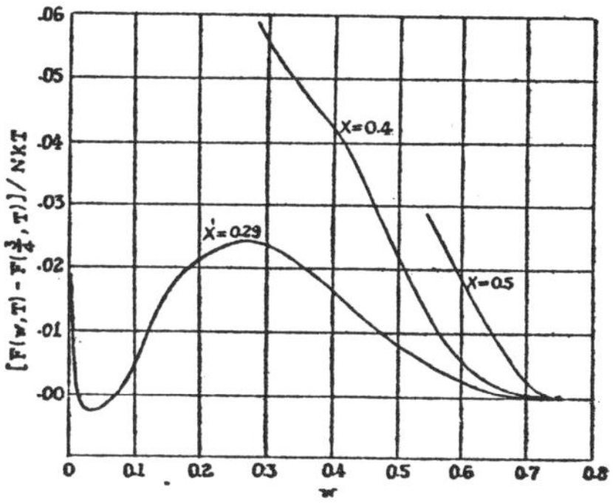

Fig. 3.

$$R=4N_1 k,$$

$$T_c=1.097E_0/R, \quad \text{and} \quad Q=0.2941E_0.$$

In Bragg and Williams' approximation, they are given by

$$T_c=2.19E_0/R, \quad \text{and} \quad Q=0.218E_0;$$

and in Peierls' approximation,[10] they are given by

$$T_c\cong 1.3E_0/R, \quad \text{and} \quad Q\cong 0.36E_0.$$

It will be noticed that because of the lack of a free energy expression Peierls did not give the exact values of these quantities.

In conclusion the author wishes to express his thanks to Professor J. S. Wang for valuable criticism and advice.

* This is not evident from (69) directly. But if we divide the whole crystal into four simple cubic sublattices and introduce a w for each sublattice so that Nw is the number of A atoms on the jth sublattice, $j=1, 2, 3, 4$, it is obvious that the free energy is symmetrical in the w's. From this we infer that (69) is satisfied at $w=\frac{3}{4}$.

⟦48a⟧ Commentary begins page 5

Commentary begins page 5

On the Angular Distribution in Nuclear Reactions and Coincidence Measurements

C. N. YANG
Department of Physics, University of Chicago, Chicago, Illinois

(Received June 9, 1948)

Theorems concerning the general form of the angular distribution of products of nuclear reactions and distintegrations are derived. These theorems are based only on the invariance properties of the physical process under space rotation and under inversion. The following examples are studied in detail: (i) angular correlation between the electron and the neutrino in β-decay; (ii) angular correlation between a β-ray and a γ-ray emitted in succession by a nucleus; and (iii) angular correlation between two γ-rays emitted in succession by a nucleus.

INTRODUCTION

IN the calculation of the angular distribution in nuclear reactions and of the angular correlation in processes involving β- and γ-decay it often happens that many terms cancel out at the end of a laborious computation. The consistency of the occurrence of such cancellation leads one to suspect that some general reasons quite independent of the particular form of interaction are at work. In this paper we shall show that this is indeed the case. In fact, the general form of the angular distribution in many cases can be obtained directly from the theorems derived in this paper.

For nuclear reactions between spinless particles the existence of a limitation on the complexity of the angular distribution for fixed orbital angular momentum of the incoming particles is well known. That the same result holds ***with the spin*** taken into consideration (for un-

 □ Reprinted from *The Physical Review* 74, 7 (October 1, 1948), 764–772.

polarized incoming beam) was first pointed out by Critchfield and Teller.[1] A proof of this statement was recently given by Eisner, Sachs, and Wolfenstein.[2] We shall in this paper formulate a new proof that lends itself easily to generalization to the case in which the particles involved have relativistic velocities.

It will be shown in general that in studying the angular correlation between two particles, as long as one of them has a wave-length long compared to the size of the nucleus, the process can be classified into different orders and for a process of given order the general form of the angular correlation is essentially known. In case both of the particles have long wave-lengths, particularly simple conclusions may be reached, as in the case of β-neutrino correlation in β-decay.

Experimentally the angular correlations β-neutrino and γ–γ have been studied by many authors. Various calculations of these correlations based on different kinds of interactions have also been made. These will be separately discussed in the different sections.

NUCLEAR REACTION

Consider the following reaction,

$$A+P\rightarrow B+Q, \tag{1}$$

and suppose both the target nucleus A and the bombarding beam of particles P are unpolarized. The complexity of the angular distribution of the outgoing particles is limited by the following theorem: *If only incoming waves of orbital angular momentum L contribute appreciably to the reaction, the angular distribution of the outgoing particles in the center of mass system is an even polynomial of cosθ with maximum exponent not higher than 2L.* Here θ is the angle between the incoming and the outgoing particles in the center-of-mass system of reference.

To prove this let us consider the collision between two particles A and P with definite ($=a$ and p) components of spin along the z axis, and definite total and z component relative orbital angular momenta L and m. (We use the center-of-mass system throughout.) The incoming wave function is, at large distances r_{AP} between A and P:

$$\frac{1}{r_{AP}}\sin(k_{AP}r_{AP}-\tfrac{1}{2}L\pi)\psi_A{}^a\psi_P{}^p Y_{Lm}(\theta_P,\,\phi_P), \tag{2}$$

where $\psi_A{}^a$, $\psi_P{}^p$ are normalized internal wave functions of particles A and P; θ_P, ϕ_P describe the direction of motion of the particle P; and $Y_{Lm}(\theta,\phi)$ is the normalized spherical harmonics of order Lm.

The asymptotic behavior of the wave function at large values of r_{BQ} is of the form

$$\frac{1}{r_{BQ}}\exp(ik_{BQ}r_{BQ})\sum_{b,q}\psi_B{}^b\psi_Q{}^q f_{bq}{}^{apm}(\theta_Q,\,\phi_Q). \tag{3}$$

In reaction (1), if we choose as the z axis the direction of motion of particle P, it is clear that when the incoming wave is expanded into partial waves with definite total and z component orbital angular momenta L and m, only terms with $m=0$ occur. Under the assumption stated in the theorem we can neglect all terms except the spherical harmonic Y_{L0}. The differential cross section of reaction (1) is, therefore,

$$d\sigma=(\text{constant})d\Omega_Q\sum_{b,q}|f_{bq}{}^{ap0}(\theta_Q,\,\phi_Q)|^2. \tag{4}$$

For unpolarized incoming particles we get

$$d\sigma=(\text{constant})d\Omega_Q\sum_{apbq}|f_{bq}{}^{ap0}(\theta,\,\phi)|^2. \tag{5}$$

The requirement of invariance under rotation will now be introduced. Consider a new coordinate system (primed system) obtained from the old by a rotation of the coordinate axis. Let $(m'/m)^{(L)}$ be the matrix element[3] of the irreducible representation $\mathfrak{D}^L$ of the three-dimensional rotation group. We have

$$\begin{aligned} Y_{Lm'}(\theta',\,\phi')&=\textstyle\sum_m(m'|m)^{(L)}Y_{Lm}(\theta,\,\phi),\\ \psi_A{}'^{a'}&=\textstyle\sum_a(a'|a)^{(S_A)}\psi_A{}^a, \end{aligned} \tag{6}$$

where $S_A=$spin of particle A, which may be an integer or a half-odd integer. By $\psi'_A{}^{a'}$ is meant the function $\psi_A{}^{a'}$ of the *primed* internal coordinates. The proof of our theorem consists in showing that

[1] C. L. Critchfield and E. Teller, Phys. Rev. **60**, 10 (1941).

[2] E. Eisner and R. G. Sachs, Phys. Rev. **72**, 680 (1947); L. Wolfenstein and R. G. Sachs, Phys. Rev. **73**, 528 (1948).

[3] E. Wigner, *Gruppentheorie und ihre Anwsndung auf die Quantenmechanik der Atomspektren* (Braunachweig, 1931), p. 180.

(a) the superposition principle requires that f be transformed according to $\mathfrak{D}^{S_A}\times\mathfrak{D}^{S_P}\times\mathfrak{D}^{L}\times\mathfrak{D}^{S_B*}\times\mathfrak{D}^{S_Q*}$ and (b) the expression

$$I^{mm'}(\theta,\phi)=\sum_{apbq}[f_{bq}^{apm}(\theta,\phi)]^* f_{bq}^{apm'}(\theta,\phi) \quad (7)$$

transforms according to

$$\mathfrak{D}^{L^*}\times\mathfrak{D}^{L}=\mathfrak{D}^{2L}+\mathfrak{D}^{2L-2}+\cdots.$$

The linear combinations of (7) that transform according to $\mathfrak{D}^{2L-1}$, $\mathfrak{D}^{2L-3}$, etc., vanish identically.

(a) Consider the following incoming wave

$$\frac{1}{r_{AP}}\sin(k_{AP}r_{AP}-\tfrac{1}{2}L\pi)\psi_{A'}{}^{a'}\psi_{P'}{}^{p'}Y_{Lm'}(\theta_{P}',\phi_{P}'). \quad (2')$$

To an observer in the primed coordinate system this has exactly the same form as (2). Hence the outgoing wave must be

$$\frac{1}{r_{BQ}}\exp(ik_{BQ}r_{BQ})\sum_{b'q'}\psi_{B'}{}^{b'}\psi_{Q'}{}^{q'}f_{b'q'}{}^{a'p'm'}(\theta_Q',\phi_Q'). \quad (3')$$

Notice that we use the *same* f instead of an f', because there is *no physically observable distinction* between the two coordinate systems. Using (6) one can express (2′) as a superposition of waves (2)

$$\sum_{apm}(a'|a)(p'|p)(m'|m)\left[\frac{1}{r_{AP}}\sin(k_{AP}r_{AP}-\tfrac{1}{2}L\pi)\times\psi_A{}^a\psi_P{}^pY_{Lm}(\theta_P,\phi_P)\right].$$

Here we have omitted the superscripts S_A, S_P, L from $(a'|a)^{(S_A)}$, $(p'|p)^{(S_P)}$, $(m'|m)^{(L)}$ for simplicity. The outgoing wave must therefore be a corresponding superposition of waves (3) with the same coefficients:

$$\sum_{apm}(a'|a)(p'|p)(m'|m)\left[\frac{1}{r_{BQ}}\exp(ik_{BQ}r_{BQ})\times\sum_{bq}\psi_B{}^b\psi_Q{}^q f_{bq}{}^{apm}(\theta_Q,\phi_Q)\right].$$

Equating this to (3′) and using (6) to express $\psi_{B'}{}^{b'}$, $\psi_{Q'}{}^{q'}$ in terms of $\psi_B{}^b$, $\psi_Q{}^q$, we get finally, by identifying the coefficient of $\psi_B{}^b\psi_Q{}^q$,

$$\sum_{apm}(a'|a)(p'|p)(m'|m)f_{bq}{}^{apm}(\theta,\phi)=\sum_{b'q'}(b'|b)(q'|q)f_{b'q'}{}^{a'p'm'}(\theta',\phi').$$

This reduces to the following form

$$f_{b'q'}{}^{a'p'm'}(\theta',\phi')=\sum_{apmbq}(a'|a)(p'|p)(m'|m)\times(b'|b)^*(q'|q)^*f_{bq}{}^{apm}(\theta,\phi) \quad (8)$$

through the orthogonality relations

$$\textstyle\sum_m(m'|m)(m''|m)^*=\delta_{m'm''}. \quad (9)$$

Equation (8) expresses the transformation property of f.

(b) To obtain the transformation property of expression (4) we investigate the behavior of expression (7) under rotation. By (8) and (9)

$$I^{m''m'}(\theta',\phi')=\sum_{abpq}\sum_{mm'''}(m''|m''')^*(m'|m)\times[f_{bq}{}^{apm'''}(\theta,\phi)]^*f_{bq}{}^{apm}(\theta,\phi)$$
$$=\sum_{mm'''}(m''|m''')^*(m'|m)I^{m'''m}(\theta,\phi). \quad (10)$$

Now the differential cross section is proportional to I^{00}. If we put in (10) $m''=m'=0$ and take the rotation from the unprimed to the primed coordinate system to be a rotation around the z axis by an angle ξ we have $\theta'=\theta$ and $\varphi'=\varphi+\xi$. Since then $(0|m)=\delta_{0m}$ it is evident that

$$I^{00}(\theta,\varphi+\xi)=I^{00}(\theta,\varphi),$$

showing that I^{00} is independent of φ. To study its dependence on θ we put in (10) $m'=m''=0$. It is well known[4] that if $\theta=\varphi=0$,

$$(0|m)=Y_{L,-m}(\theta',\varphi').$$

Hence (10) becomes

$$I^{00}(\theta',\varphi')=\sum_{mm'''}Y_{L,-m'''}{}^*(\theta',\phi')\times Y_{L,-m}(\theta',\varphi')I^{m'''m}(0,0).$$

On application of the reduction theorem[4] of products of spherical harmonics this leads directly to our theorem.

If instead of a rotation we had chosen an inversion of the coordinates, it is evident that (8)

[4] H. Bethe, *Handbuch der Physik*, (Springer, 1933) Vol. 24/1, Chapter 3, Section 65.

would become

$$f_{bq}^{apm}(\pi-\theta,\ \pi+\phi)$$
$$=P_A P_P P_B P_Q(-1)^L f_{bq}^{apm}(\theta,\ \varphi), \quad (11)$$

where P_A, P_P, etc., are the intrinsic parities of the nuclei. This shows that

$$|f(\pi-\theta,\ \pi+\varphi)|^2=|f(\theta,\ \varphi)|^2,$$

and it follows that the angular dependence must be an even function of $\cos\theta$, a fact that is already established by (10). Equation (11) further shows that any odd power of $\cos\theta$ in the angular dependence must come from an interference term between orbital wave functions of opposite parity.

The symmetry requirements of the wave function under interchanges of the nucleons do not, in general, lead to any new conclusions about the properties of f.[5] However, in the special case in which the two incoming particles or the two outgoing particles are identical, more detailed consideration is necessary. An example of such a case is the reaction

$$Li^7+H^1\rightarrow He^4+He^4.$$

Since the outgoing particles are spinless and satisfy Bose-Einstein statistics and since Li^7 has an odd parity, the value of L must be odd in order to have a balance of parity. This means that $f=0$ unless L is odd. At low energies, therefore, the effective orbital angular momentum is 1.

Another example is the D^2+D^2 reaction:

$$D^2+D^2\rightarrow n'+He^3,$$
$$D^2+D^2\rightarrow H^1+H^3.$$

This reaction has recently been considered theoretically by Konopinski and Teller.[6] Because of the symmetry nature of the deuterons it is no longer convenient to specify the spin of the two incoming particles separately. Instead we should group the nine possible incoming states into a quintet, a triplet, and a singlet. The space-wave functions for the quintet and the singlet states are symmetrical with respect to the exchange of the two deuterons, and those for the triplet states are antisymmetrical. Strictly speaking, the proof of our theorem does not apply to such a case where the space-wave function depends on the orientation of the spins of the particles. But since all the states in the same multiplet have the same *a priori* probability, it is evident that the difference of the space-wave function for the different multiplets does not affect the validity of our theorem.

The Coulomb field affects the waves of different orbital angular momenta in such a way as to favor those with higher angular momenta at low energies.[7] This accounts for the reason why at bombarding energies as low as 20 kev the angular distribution in the D+D reaction is not spherically symmetrical.[6,8] We shall not go into this point in any further detail here.

We conclude this section by stating a variation of the theorem proved above: *When contributions from incoming waves with orbital angular momenta* $>L$ *are neglected, the angular distribution in reaction* (1) *in the center-of-mass system is a polynomial of* $\cos\theta$ *with maximum exponent not higher than* $2L$. *This holds even if the contributing compound nuclear states have angular momenta* $>L$.

It will be noticed that when both even and odd values of the orbital angular momenta in the incoming beam are effective in producing the reaction, the angular distribution contains odd powers of $\cos\theta$. This, however, will not happen when either (a) the reaction goes through a *single* compound nuclear state (e.g. near a strong resonance level); or (b) symmetry requirements exclude even (or odd) L values as in the $Li^7+H^1\rightarrow He^4+He^4$ reaction discussed above.

RELATIVISTIC CASE

We shall in this section generalize the result of the last section to the case when the particle P is an electron and has relativistic velocities. (The nuclei A, B, and Q are still supposed to be non-relativistic.) No such process has been experimentally realized. We shall, however,

---

[5] To understand this it is best to introduce the idea of *channels* in the configuration space which was first discussed by G. Breit, Phys. Rev. **58**, 1068 (1940); J. A. Wheeler, Phys. Rev. **52**, 1107 (1937). An interchange of the nucleons in general results in an interchange of the channels, except for the case when either A and P or B and Q are identical.

[6] E. J. Konopinski and E. Teller, Phys. Rev. **73**, 822 (1948).

[7] H. Bethe and E. J. Konopinski, Phys. Rev. **54**, 130 (1938).

[8] E. Bretscher, A. P. French, and F. G. P. Seidl, Phys. Rev. **73**, 815 (1948).

discuss it to illustrate our method. It will be proved that *if only partial waves of orbital angular momentum L in the electron wave function contribute to the reaction, the angular distribution is a polynomial of* $\cos\theta$ *with maximum exponent not higher than* $2L+1$.

Instead of the stationary picture used in the last section, we shall here use a non-stationary description of the process. The electron wave function at time $t=0$ is a product of a spin wave function with four components and a space-wave function e^{ikz}. The spin of the electron along the z axis is a constant of motion and is denoted by $p(=\pm\frac{1}{2})$. If we expand the space-wave function into partial waves of definite orbital angular momenta L, the first term ($L=0$) would give rise to allowed transitions, the second term ($L=1$) first forbidden transitions, etc. To study the angular distribution arising from the contribution of the partial wave of orbital angular momentum L we need to decompose it again into normalized waves ψ_{LJPm} of definite L, J (total angular momentum of the electron), P (parity), and m (z component of $\mathbf{J}$).* The advantage of using these ψ's is that they have simple transformation properties under rotation. The possible values of J are $L\pm\frac{1}{2}$. Under the assumption that we are considering only the contribution from a definite L value, the wave function at $t=0$ can be replaced by

$$\sum_{P=\pm1}\sum_{J=L\pm\frac{1}{2}} \alpha_{LJPp}\psi_{LJPp}. \tag{12}$$

We have put $m=p$ because the z component of the orbital angular momentum is zero.

Let us now first study the reaction arising from the electron wave ψ_{LJPm}. Starting at $t=0$ with ψ_{LJPm} and nucleus A with a definite value a for the z component of spin, we shall denote by $f_{bq}{}^{LJPam}(\theta_Q, \varphi_Q)$ the *probability amplitude* at any later time $t>0$ for that outgoing state in which the z component of spin of the particles B and Q are b and q, and in which the momentum of Q is in the direction θ_Q, φ_Q. The absolute value of the outgoing momentum (which is not fixed because the energy is not necessarily conserved when t is small) should also enter the function f as an independent variable, but has been omitted for the sake of simplicity in writing.

Now the probability amplitudes are additive when we superpose states. Since under a rotation the different waves ψ_{LJPm} with the same LJP values combine linearly, the argument which led to (8) in the last section would now lead to

$$f_{b'q'}{}^{LJPa'm'}(\theta', \varphi') = \sum_{ambq} (a'|a)(b'|b)^* \times (m'|m)(q'|q)^* f_{bq}{}^{LJPam}(\theta, \varphi). \tag{13}$$

Returning now to the wave (12) at $t=0$, we see that the differential cross section is proportional to

$$d\Omega_Q \Big|\sum_{PJ} \alpha_{LJPp} f_{bq}{}^{LJPap}(\theta, \varphi)\Big|^2.$$

This will have to be summed over a, b, p, and q. Since the coefficients in (12′) are independent of a, b and q, the final expression is

$$d\Omega_Q \sum_{JPJ'P'}\sum_{p} \alpha_{LJPp}\alpha^*_{LJ'P'p} \times\Big[\sum_{a,b,q} \{f_{bq}{}^{LJ'P'ap}(\theta, \phi)\}^* f_{bq}{}^{LJPap}(\theta, \phi)\Big]. \tag{14}$$

By (13) the individual terms under the summation sign $\sum_{abq}$ transform under a rotation according to $\mathfrak{D}^{S_A}\times\mathfrak{D}^{S_A^*}\times\mathfrak{D}^{S_B}\times\mathfrak{D}^{S_B^*}\times\mathfrak{D}^{S_Q}\times\mathfrak{D}^{S_Q^*}\times\mathfrak{D}^{J}\times\mathfrak{D}^{J'^*}$. But after the summation over a, b, and q is carried out, the sum transforms more simply according to

$$\mathfrak{D}^{J}\times\mathfrak{D}^{J'^*} = \mathfrak{D}^{J+J'} + \mathfrak{D}^{J+J'-1} + \cdots.$$

This means that the expression in the square bracket in (14) is a sum of spherical harmonics of order $\mathfrak{L}\mathfrak{M}$ with $\mathfrak{L}\leqq J+J'$. But both J and J' are $\leqq L+\frac{1}{2}$. The theorem stated at the beginning of this section follows immediately.

If we introduce the requirement of invariance under inversion, Eq. (14) shows that those terms with $P'P=+1$ give rise to angular correlation functions that are even under the transformation $\theta\rightarrow\pi-\theta$, and those with $P'P=-1$ give rise to odd angular correlation functions. A consequence of this is the following. If the velocity v of the electron is small compared to the velocity of light c, and if the spin wave function of the electron is expanded in powers of v/c, the first term, i.e., the term that does not

* The parity can be either 1 or -1 for any given L, J, and m. However, for slow electrons the amplitude of waves with $P=-(-1)^L$ is very small. Cf. end of this section.

vanish as $v \to 0$, is invariant under an inversion. This term would therefore give rise to terms with $P = (-1)^L$. The opposite parity first appears in the next term of the expansion and is proportional to v/c. Hence those terms in (14) with $PP' = -1$ contain a factor v/c. Thus *the odd powers of $\cos\theta$ in the angular correlation have coefficients smaller than the even powers by a factor of v/c.*

β-NEUTRINO CORRELATION

In β-decay we have the particularly simple situation in which both the electron and the neutrino have wave-lengths long compared to the dimension of the nucleus. The argument of the last section can now be applied to both these particles and we can prove that *the angular correlation between the electron and the neutrino emitted in a β-decay is a polynomial of $\cos\theta$ up to a maximum exponent $K+1$, where*[9] *$K=0$ for allowed transitions, $K=1$ for first forbidden transitions, etc.*

The idea of the proof is that for first forbidden transitions one has either $L=1$ for the electron and $L_1=0$ for the neutrino or $L=0$ for the electron and $L_1=1$ for the neutrino. The waves $L=1$ and $L_1=1$ occur together only in second forbidden processes. Now the intensity produced by the $L=0$, $L_1=1$ waves has an angular correlation function that goes up to $\cos\theta$ to the first power, according to the theorem of the last section. Similarly, fixing our attention on the neutrino wave function we can draw the same conclusion about the $L=1$, $L_1=0$ waves. The interference term of the $L=1$, $L_1=0$ waves with the $L=0$, $L_1=1$ waves, however, gives an angular distribution that contains $\cos^2\theta$, which is the highest power of $\cos\theta$ possible for this case.

The proof is as follows. Consider the β-decay

$$A \to B + e + \nu.$$

Let a and b be the z components of the spin of the nuclei A and B, θ_e, ϕ_e and θ_ν, ϕ_ν the directions of motion of the electron and the neutrino, and s and s_1 the spin components of the electron and the neutrino in their respective directions of motion. Starting with the nucleus A at $t=0$, the probability amplitude at any later time t of the β-decay for given θ_e, ϕ_e, θ_ν, ϕ_ν, s, s_1, a, and b will be denoted by

$$f_{bss_1}{}^a(\theta_e, \varphi_e, \theta_\nu, \phi_\nu). \tag{15}$$

Now let the electron wave function be expanded into waves Φ_{LJPs}, as done before in (12), with the only difference that here Φ_{LJPs} represents a wave function with total angular momentum along the direction θ_e, ϕ_e (instead of along the z axis), equal to s. The coefficients α in (12) remain unchanged. Now Φ_{LJPs} can be further expanded into waves ψ_{LJPm} with definite total angular momentum along the z axis. The final result is

$$\sum_{LJPm} \alpha_{LJPs}(s|m)_e \psi_{LJPm}, \tag{16}$$

where e represents a rotation of the coordinate axes so that the z axis changes from the direction of motion of the electron (i.e., the direction specified by θ_e, ϕ_e) into the laboratory z axis. It is evident that the choice of the x and y axes perpendicular to the direction θ_e, ϕ_e affects only the phase of Φ_{LJPs} and would not in any way influence our final result. In (16) $(s|m)_e$ is the only factor that depends on θ_e, ϕ_e. A similar expansion of the neutrino wave will now be made

$$\sum_{L_1J_1P_1n} \beta_{L_1J_1P_1s_1}(s_1|n)_\nu \psi_{\nu L_1J_1P_1n}. \tag{17}$$

The wave amplitude (15) is evidently given by

$$f_{bss_1}{}^a(\theta_e, \phi_e, \theta_\nu, \phi_\nu) = \sum_{LJPL_1J_1P_1} \sum_{mn} \times \alpha_{\lambda s}{}^* \beta_{\lambda_1 s_1}{}^* (s|m)_e{}^* (s_1|n)_\nu{}^* \overset{a}{F}_{\lambda\lambda_1 bmn}, \tag{18}$$

where λ and λ_1 are abbreviations for LJP and $L_1J_1P_1$. We have taken the complex conjugates of the waves (16) and (17) because they represent final states. In (18) F represents the probability amplitude of the final state specified by b, $\psi_{\lambda m}$, and $\psi_{\nu\lambda_1 n}$, the initial state being specified by a.

The probability of the β-decay is proportional to

$$\sum_{abss_1} |\overset{a}{f}_{bss_1}(\theta_e, \varphi_e, \theta_\nu, \varphi_\nu)|^2. \tag{19}$$

Writing

$$\sum_{a,\,b} \overset{a}{F}_{\lambda\lambda_1 bmn}(\overset{a}{F}_{\bar{\lambda}\bar{\lambda}_1 b\bar{m}\bar{n}})^* = G_{\Lambda m\bar{m}n\bar{n}}, \tag{20}$$

[9] Notice that when the interaction involves derivatives of the wave function, as in the Konopinski-Uhlenbeck type of interaction, we always expand the wave function *before* taking the derivatives.

and

$$\alpha_{\lambda s}{}^*\beta_{\lambda_1 s_1}{}^*\alpha_{\bar{\lambda} s}\beta_{\bar{\lambda}_1 s_1}=\Gamma_{\Lambda s s_1}, \qquad (21)$$

where Λ is an abbreviation for λ, λ_1, $\bar{\lambda}$, $\bar{\lambda}_1$, expression (19) becomes

$$\sum_{\Lambda s s_1} \Gamma_{\Lambda s s_1} \sum_{m\bar{m}n\bar{n}} G_{\Lambda m\bar{m}n\bar{n}}(s|m)_e{}^* \times(s|\bar{m})_e(s_1|n)_\nu{}^*(s_1|\bar{n})_\nu. \qquad (22)$$

We shall show later that

$$\sum_{m\bar{m}n\bar{n}} G_{\Lambda m\bar{m}n\bar{n}}(s|m)_e{}^*(s|\bar{m})_e \times(s_1|n)_\nu{}^*(s_1|\bar{n})_\nu \qquad (23)$$

is a polynomial of $\cos\theta$ with maximum exponent $\leqq$ both $J+\bar{J}$ and $J_1+\bar{J}_1$, θ being the angle between the directions of motion of the electron and the neutrino. But $J=L\pm\frac{1}{2}$, $J_1=L_1\pm\frac{1}{2}$. Hence expression (23), which represents the (cross) term in the probability of the β-decay between waves LL_1 and $\bar{L}\bar{L}_1$, is a polynomial of $\cos\theta$ with maximum exponent $\leqq$ both $L+\bar{L}+1$ and $L_1+\bar{L}_1+1$.

The classification of β-decays into allowed, first forbidden, etc., processes consists of an expansion in powers of $r/\lambda_e \sim r/\lambda_\nu$ ($\sim\frac{1}{10}$), λ_e, λ_ν being the wave-lengths of the electron and the neutrino, and r the dimension of the nucleus. In an allowed transition only the waves $L=0$, $L_1=0$ are effective for the process. Contributions from other waves are negligible because with increasing values of L the amplitude of the wave ψ_{LJPm} inside the nucleus decreases as $(r/\lambda_e)^L$. In a first forbidden process $F^a{}_{\lambda\lambda_1 bmn}|_{L=L_1=0}$ vanishes because of selection rules and the contributing waves are the following two:[10] $L=1$, $L_1=0$ and $L=0$, $L_1=1$, In general, for a Kth forbidden transition only waves with $L+L'\leqq K$ are important.[9] This means that in the summation over Λ in (22) only $L+L_1\leqq K$, $\bar{L}+\bar{L}_1\leqq K$ terms need be retained. Hence $L_1+\bar{L}_1\leqq 2K-(L+\bar{L})$. Thus the maximum exponent of $\cos\theta$ is $\leqq$ both $L+\bar{L}+1$ and $2K-(L+\bar{L})+1$; hence it is $\leqq K+1$, which proves our theorem.

It remains to be proved that the above statement about (23) is true. This we do by noticing first that F represents the probability amplitude of the final state b, $\psi_{\lambda m}$, $\psi_{\nu\lambda_1 n}$ if the initial state is represented by a. If R is any rotation of coordinates, $\sum_a (a'/a)_R F^a{}_{\lambda\lambda_1 bmn}$ would give the probability amplitude of these *same* final states resulting from an initial state obtained by rotating nucleus A in state a' by R^{-1}. Thus

$$\sum_a (a'/a)_R \overset{a}{F}_{\lambda\lambda_1 bmn}=\sum_{b'MN} \overset{a'}{F}_{\lambda\lambda_1 b'MN} \times(b'|b)_R(M|m)_R(N|n)_R, \qquad (24)$$

which means that $F_{\lambda\lambda_1}$ is invariant under $\mathfrak{D}^{S_A}\times\mathfrak{D}^{S_B{}^*}\times\mathfrak{D}^{J^*}\times\mathfrak{D}^{J_1{}^*}$. The definition (20) therefore shows that G_Λ is invariant under $\mathfrak{D}^{J^*}\times\mathfrak{D}^{J_1{}^*}\times\mathfrak{D}^{\bar{J}}\times\mathfrak{D}^{\bar{J}_1}$. That is,

$$G_{\Lambda m\bar{m}n\bar{n}}=\sum_{M\bar{M}N\bar{N}} G_{\Lambda M\bar{M}N\bar{N}}(M|m)_R \times(\bar{M}|\bar{m})_R{}^*(N|n)_R(\bar{N}|\bar{n})_R{}^*.$$

Hence

$$\sum_{m\bar{m}}(M|m)_R{}^*(\bar{M}|\bar{m})_R G_{\Lambda m\bar{m}n\bar{n}} =\sum_{N\bar{N}} G_{\Lambda M\bar{M}N\bar{N}}(N|n)_R(\bar{N}|\bar{n})_R{}^*.$$

Putting $R=e$, $M=\bar{M}=S$, we see that (23) can be written

$$\sum_{N\bar{N}n\bar{n}} G_{\Lambda ssN\bar{N}}(N|n)_e(\bar{N}|\bar{n})_e{}^*(s_1|n)_\nu{}^*(s_1/\bar{n})_\nu$$

$$=\sum_{N\bar{N}n\bar{n}} G_{\Lambda ssN\bar{N}}(n|N)_{e^{-1}}{}^*(\bar{n}|\bar{N})_{e^{-1}} \times(s_1|n)_\nu{}^*(s_1|\bar{n})_\nu$$

$$=\sum_{N\bar{N}} G_{\Lambda ssN\bar{N}}(s_1|N)_{\nu e^{-1}}{}^*(s_1|\bar{N})_{\nu e^{-1}}. \qquad (23')$$

This is evidently independent of the choice of the laboratory coordinate system. If these be so chosen that $\theta_e=\phi_e=0$, the rotation represented by e becomes the identity and (23′) shows that (23) is a polynomial of $\cos\theta$ with maximum exponent $\leqq J_1+\bar{J}_1$. A similar argument shows that it is also $\leqq J+\bar{J}$. This completes the proof.

If we fix our attention on one end of the spectrum where the electron momentum p is $\ll$ the neutrino momentum q, the waves that contribute most in a Kth order forbidden transition are those with $L=0$, $L_1\leq K$. By the theorem proved in the last section we see that *the maximum exponent of* $\cos\theta$ *in the angular coorelation is 1*. This evidently applies also when $q\ll p$.

If $p\ll mc$ the spin function of the electron can be separated from the space-wave function.

[10] It may happen that $F^a{}_{\lambda\lambda_1 bmn}|_{L=L_1=0}$ is not zero but is $\sim F^a{}_{\lambda\lambda_1 bmn}|_{L=1,\, L_1=0}$. This happens in the usual interactions because of the presence of terms $\sim$nucleon velocity. In such cases we should include the $L=0$, $L_1=0$ wave. The conclusions are, however, unchanged as far as they concern only the complexity of the angular correlation.

Hence, after summation over the spin directions of the electron, the maximum exponent is both $\leqq L+\bar{L}$ and $J_1+\bar{J}_1 \leqq 2K-(L+\bar{L})+1$. We have $L+\bar{L}$ instead of $J+\bar{J}$, as in all non-relativistic cases. Thus *the maximum exponent is K.*

In case $p \ll q$ and $p \ll mc$, only $L=0$ wave is effective and the angular correlation is spherically symmetrical for transitions of any order. Thus when $p \to 0$ the angular correlation becomes spherically symmetrical. On the other hand, when $q \to 0$ the angular correlation becomes $1+\alpha \cos\theta$ or 1 according as the mass of the neutrino is zero or otherwise.

Actual calculations of the angular correlation between the electron and the neutrino emitted in β-decays of different orders have been carried out by Hamilton,[11] using all the five usual types of interactions. The results, of course, conform with the theorems discussed above. Experimentally,[12] information about the angular correlation has been obtained by measuring the energy spectrum of the recoil nuclei or by coincidence measurements of the electrons and the recoil nuclei. Because of the indirect nature of these experiments, the results are not as yet very quantitative.

β-γ AND γ-γ CORRELATIONS

The method used in the last three sections evidently applies also to γ-rays. The rectangular components A_x, A_y, and A_z of the vector potential of the electromagnetic field is expanded into spherical harmonics. As is well known, the term $L=0$ leads to electric dipole processes, the term $L=1$ to magnetic dipole and electric quadrupole processes, etc. For each direction of propagation of the light quantum there are two possible waves with $L=0$, corresponding to the two different polarizations. Changing the direction of propagation we obtain other waves. But altogether there are only three linearly independent waves with $L=0$, and they transform among themselves under a rotation like a vector. Hence *the angular correlation between the γ-ray and any other particle in a nuclear process is of the form $1+\alpha \cos^2\theta$ if the γ-ray process is of the electric dipole type.*

TABLE I.

		Nuclear particle	Electron or neutrino	Photon
Name for different approximations	$L=0$	*S* wave	Allowed	El. dipole
	$L=1$	*P* wave	First forbidden	Mag. dipole and el. quadrupole
	$L=2$	*D* wave	Second forbidden	Mag. quadrupole and el. octapole
Power of $\cos\theta$		Even	Even and odd	Even
Max. exponent of $\cos\theta$		$2L$	$2L+1$	$2L+2$

The odd power of $\cos\theta$ does not appear because the photon wave has a definite parity. This conclusion can be immediately generalized into *magnetic dipole and electric quadrupole processes where the angular correlation is $1+\alpha \cos^2\theta+\beta \cos^4\theta$.* This holds even when *both* the magnetic dipole and the electric quadrupole transitions are present. Similar theorems obtain in higher multipole processes.

In general, we can study a process with any number of incoming and outgoing particles. We assume that the incoming particles are unpolarized. If one of the particles (whether incoming or outgoing), say P, has a wave-length long compared to the dimension of the space-region in which it interacts with the other particles, the process can be classified according to to the effective orbital angular momentum L of P. The angular correlation between P and any other particle Q in the process would then be a polynomial of $\cos\theta$ with a maximum exponent determined by L, θ being the angle between the directions of propagation of P and Q. The presence of other particles in the process does not affect the result because a summation over the directions of motion and over the spin of these "redundant" particles must always be carried out. We may say that these particles do not produce any preferential direction in space. The general results when P is a nucleon, an electron, or a photon are summarized in Table I.

The application to the angular correlation between successive γ-rays emitted by a nucleus is straightforward. Actual calculations of this correlation for dipole-dipole, dipole-quadrupole, and quadrupole-quadrupole transitions (all electric poles) have been published.[13] They have the

[11] D. R. Hamilton, Phys. Rev. **71**, 456 (1947).

[12] J. S. Allen, Phys. Rev. **61**, 692 (1942); J. C. Jacobsen, and Kofoed-Hansen, Kgl. Danske Vid. Sels. Math.-Fys. Medd **23**, No. 12 (1945); J. S. Allen, H. R. Paneth, and A. H. Morrish, Bull. Am. Phys. Soc. **23**, No. 3 (1948); C. N. Sherwin, Phys. Rev. **73**, 216 (1948).

[13] D. R. Hamilton, Phys. Rev. **58**, 122 (1940); experimental evidence has been reported by L. Brady and M. Deutsch, Phys. Rev. **72**, 870 (1947).

form

$$1+\alpha\cos^2\theta \text{ (dipole-dipole, dipole-quadrupole)},$$
$$1+\alpha\cos^2\theta+\beta\cos^4\theta \text{ (quadrupole-quadrupole)}, \quad (25)$$

agreeing with our results. In these calculations the line width of the second γ-ray process is assumed to be large compared to the hyperfine splitting of the atom, so that the lifetime of the intermediate nucleus is small compared to the time required for the nuclear spin to precess appreciably. Also the assumption is made that there is no magnetic dipole transition mixed with the electric quadrupole. It is evident that neither of these assumptions is necessary for the validity of our theorems, and that the angular correlation is quite generally of the form (25). It should be remarked, of course, that in case either of these two assumptions is violated the coefficients α and β in (25) may not have the values tabulated by Hamilton.

Another application is found in the problem of the angular correlation between the electron and the γ-ray emitted by a nucleus in succession. Since one of the particles is a photon, only even powers of $\cos\theta$ can occur in the correlation function. Using Table I, taking the electron to be P, we conclude that *for all allowed β-transitions the correlation is spherically symmetrical.* This appears at first sight very strange because, e.g., for the Gamow-Teller type of interaction the matrix element involves the spin of the nucleus and one would expect that the emission of an electron in a definite direction would result in a preferential distribution of the spin orientation of the intermediate nucleus and hence would affect the angular distribution of the γ-rays. *For first forbidden β-transitions the correlation is $1+\alpha\cos^2\theta$.* Falkoff and Uhlenbeck have made actual calculations for the first forbidden electric dipole process, using various types of β-interactions.[14] As in the γ–γ case discussed above, we remark here that our conclusions hold independently of any assumption about the lifetime of the intermediate nucleus, and independently of the multipole nature of the γ-radiation. Also it is not necessary to neglect the term in the β-interaction that is proportional to the nucleon velocity.

[14] D. L. Falkoff and G. E. Uhlenbeck, Bull. Am. Phys. Soc. 22, No. 5 (1947).

REMARKS ABOUT OTHER PARTICLES

Table I can be extended to include mesons of spin 0 and 1. The treatment is very similar to the treatment of the electron if we use Kemmer's[15] representation of the meson wave functions. In this representation a scalar meson has a five-component and a vector meson a ten-component wave function. We shall assume that the rest mass is not zero. Let us take a plane wave

$$\psi=\phi\exp((i/\hbar)(\mathbf{p}\cdot\mathbf{x}-Et)) \quad (26)$$

and expand it into waves with definite orbital angular momentum L. Under a rotation the spin function ϕ is transformed by a matrix S. The total angular momentum can go as high as $L+1$. Notice that this is true for scalar mesons as well as vector mesons.[15] Thus if only orbital wave L contribute to the reaction the angular correlation between a meson and any other particle is a polynomial of $\cos\theta$ with maximum exponent $\leqq 2L+2$.

If further the meson has non-relativistic velocities v, as must actually be the case in order that the wave-length of the meson may be long compared to nuclear dimensions, we can expand ϕ into a power series in v/c.

$$\phi=\phi_0+\frac{|v|}{c}\phi_1+\cdots. \quad (27)$$

It can be readily proved that the following points are true:

(a) ϕ_0 has a definite parity and can be made independent of the direction of the velocity. The theorem proved in the section about nucleons can therefore be applied here and we see that to the order $(v/c)^0$ the angular correlation is an even polynomial of $\cos\theta$ with maximum exponent $\leqq 2L$.

(b) ϕ_1 has a definite parity which is the opposite of that of ϕ_0. Thus the interference term between ϕ_0 and ϕ_1 gives rise to odd powers of $\cos\theta$ only and we have the result that the terms in the angular correlation to the first order of v/c is an odd polynomial of $\cos\theta$.

The author wishes to take this opportunity to thank Professor E. Teller for invaluable discussions and advice.

[15] N. Kemmer, Proc. Roy. Soc. **A173**, 91 (1939).

Interaction of Mesons with Nucleons and Light Particles

[49a]
Commentary begins page 7

T. D. Lee, M. Rosenbluth, and C. N. Yang
Institute for Nuclear Studies, University of Chicago, Chicago, Illinois

January 7, 1949

WE have been making a phenomenological study of the various experiments which have been done in recent years on the interaction between the various types of particles. In the course of this investigation two interesting points have come to light.

First, we found that if the decay of the μ-mesons and the capture of the μ^--mesons by nuclei are described by the reactions[1]

$$\mu \to e + \nu + \nu \qquad (e = \text{electron}, \ \nu = \text{neutrino})$$
$$\mu^- + P \to N + \nu \qquad (P = \text{proton}, \ N = \text{neutron}),$$

and that the Fermi type interactions are assumed to be responsible for these processes, the coupling constants would have the values

$$g_{\mu e} \sim 3 \times 10^{-48} \text{ erg cm}^3$$

and

$$g_{\mu P} \sim 2 \times 10^{-49} \text{ erg cm}^3,$$

respectively. These values are so determined as to fit the experimental lifetime[2] of the μ-mesons and the capture probability of the μ^--mesons by nuclei.[3] It is remarkable that the three independent experiments: the β-decay of the nucleons and the μ-mesons and the interaction of the nucleons with the μ-mesons lead to coupling constants of the same order of magnitude.

One can perhaps attempt to explain the equality of these interactions in a manner analogous to that used for the Coulomb interactions, i.e. by assuming these interactions to be transmitted through an intermediate field with respect to which all particles have the same "charge." The "quanta" of such a field would have a very short lifetime and would have escaped detection.

Second, if we assume the π-mesons to have integral spin and assume direct couplings for the processes

$$\pi \to \mu + \text{anti } \nu$$
$$N \to P + \pi^-$$

with coupling constants determined from the lifetime of the π-mesons[4] and the strength of nuclear forces,[5] the interaction between the μ-mesons and the nucleons can be *quantitatively* explained as a second-order interaction through the virtual creation and annihilation of π-mesons.

After the completion of our work Mr. A. Ore has kindly informed us that similar considerations have been carried out by J. A. Wheeler and J. Tiomno.

[1] The masses of the π- and μ-mesons are taken to be

$$m_\pi = 286 m_e, \qquad m_\mu = 212 m_e.$$

[2] B. Rossi, Rev. Mod. Phys. **20**, 537 (1948).
[3] B. Rossi, Rev. Mod. Phys. **20**, 537 (1948). In the calculation for the capture process the Fermi model for the nucleus is assumed and only single particle excitations are considered. See M. Rosenbluth, Phys. Rev. **75**, 532 (1949).
[4] J. R. Richardson, Phys. Rev. **74**, 1720 (1948).
[5] H. Bethe, Phys. Rev. **57**, 390 (1940).

□ Reprinted from *The Physical Review* 75, 5 (March 1, 1949), 905.

[[49b]]
Commentary begins page 7

Are Mesons Elementary Particles?

E. FERMI AND C. N. YANG*
Institute for Nuclear Studies, University of Chicago, Chicago, Illinois
(Received August 24, 1949)

The hypothesis that π-mesons may be composite particles formed by the association of a nucleon with an anti-nucleon is discussed. From an extremely crude discussion of the model it appears that such a meson would have in most respects properties similar to those of the meson of the Yukawa theory.

I. INTRODUCTION

IN recent years several new particles have been discovered which are currently assumed to be "elementary," that is, essentially, structureless. The probability that all such particles should be really elementary becomes less and less as their number increases.

It is by no means certain that nucleons, mesons, electrons, neutrinos are all elementary particles and it could be that at least some of the failures of the present theories may be due to disregarding the possibility that some of them may have a complex structure. Unfortunately, we have no clue to decide whether this is true, much less to find out what particles are simple and what particles are complex. In what follows we will try to work out in some detail a special example more as an illustration of a possible program of the theory of particles, than in the hope that what we suggest may actually correspond to reality.

We propose to discuss the hypothesis that the π-meson may not be elementary, but may be a composite particle formed by the association of a nucleon and an anti-nucleon. The first assumption will be, therefore, that both an anti-proton and an anti-neutron exist, having the same relationship to the proton and the neutron, as the electron to the positron. Although this is an assumption that goes beyond what is known experimentally, we do not view it as a very revolutionary one. We must assume, further, that between a nucleon and an anti-nucleon strong attractive forces exist, capable of binding the two particles together. We assume that the π-meson is a pair of nucleon and anti-nucleon bound in this way. Since the mass of the π-meson is much smaller than twice the mass of a nucleon, it is necessary to assume that the binding energy is so great that its mass equivalent is equal to the difference between twice the mass of the nucleon and the mass of the meson.

According to this view the positive meson would be the association of a proton and an anti-neutron and the negative meson would be the association of an anti-proton and a neutron. As a model of a neutral meson one could take either a pair of a neutron and an anti-neutron, or of a proton and an anti-proton.

It would be difficult to set up a not too complicated scheme of forces between a nucleon and an anti-nucleon, without about equally strong forces between two ordinary nucleons. These last forces, however, would be quite different from the ordinary nuclear forces, because they would have much greater energy and much shorter range. The reason why no experimental indication of them has been observed for ordinary nucleons may be explained by the assumption that the forces could be attractive between a nucleon and an anti-nucleon and repulsive between two ordinary nucleons. If this is the case, no bound system of two ordinary nucleons would result out of this particular type of interaction. Because of the short range very little would be noticed of such forces even in scattering phenomena.

Ordinary nuclear forces from the point of view of this theory will be discussed below.

Unfortunately we have not succeeded in working out a satisfactory relativistically invariant theory of nucleons among which such attractive forces act. For this reason all the conclusion that will be presented will be

* Now at the Institute for Advanced Study, Princeton, New Jersey.

 □ Reprinted from *The Physical Review* 76, 12 (December 15, 1949), 1739–1743.

extremely tentative. It would be undesirable to assume that the attraction is due to a special field of force since in this case the quanta of this new field would be themselves new elementary particles which is just what we hope to be able to avoid. Therefore, only forces of zero range appear compatible with relativistic invariance. In Section II the attempt will be discussed to represent the interaction by a term of the fourth degree in the amplitudes of the nucleon fields. We do not know whether this attempt can be made mathematically self-consistent and we have not succeeded in finding a way to treat it, except by the most crude approximation. The main difficulty is that no stationary state exists with one pair of nucleons only, but only mixed states with one pair, two pairs and many pairs. In our simplified discussion we have neglected this important factor, and treated the problem of a nucleon and an anti-nucleon alone. Assuming hopefully that these mathematical difficulties can be overcome, we have investigated the symmetry properties of the quantum states of the system of a nucleon and an anti-nucleon, in particular for the states of total angular momentum zero, $^1S_0$ and $^3P_0$. The former of these two states corresponds to a pseudoscalar meson and the latter to a scalar meson. If the ground state of the two-nucleon system had a resultant angular momentum 1, one could get in a similar way a model of the vector meson.

A peculiarity of the wave functions of the meson is that they decrease extremely rapidly with the distance between the two nucleons, so that the dimensions of the meson appear to be of the order of magnitude of the Compton wave-length of the nucleon, which is roughly 1/10 of the classical electron radius. This feature may make the experimental detection of the complex nature of the meson extremely difficult.

In the Yukawa theory of nuclear forces it is postulated that virtual mesons are continuously created and re-absorbed in the vicinity of a nucleon. When two nucleons are close to each other, the process of absorption by one nucleon of the virtual meson originated by the other is responsible for the nuclear forces. According to the present view, the main features of this theory can be kept even when the assumption that the meson is an elementary particle is dropped.

One finds that in the vicinity of an isolated nucleon there is a tendency to pair formation of nucleons and anti-nucleons, which will be predominantly formed in the bound state, that is as π-mesons, because such bound states are energetically much lower. From this point on, the Yukawa theory can be taken over almost unchanged as a description of the mechanism of nuclear forces (see Section III).

If the program that has been outlined could be carried out in a mathematically satisfactory way, one might hope to be able eventually to establish a relationship between the strength of the ordinary nuclear forces and the meson mass. Indeed, the difference between the mass of two nucleons and the mass of the meson is the binding energy of the nucleon and the anti-nucleon system. In a consistent theory, therefore, the strength of the coupling term between a nucleon and an anti-nucleon should be adjusted to give the correct value for this binding energy. On the other hand, it is this same coupling which is responsible for the creation of virtual mesons near a nucleon and determines, therefore, the strength of the ordinary nuclear forces. In Section III an estimate of the nuclear forces, calculated as far as is possible according to this program, is given. Considering the extremely primitive mathematical means used, the agreement is not worse than what might be expected.

II. MESONS AS BOUND STATES OF A NUCLEON AND AN ANTI-NUCLEON

We proceed now to discuss the mathematical formalism needed in order to carry out the outlined program.

For this it is necessary to introduce attractive forces between a nucleon and an anti-nucleon capable of binding the two particles together into what we assume to be a meson.

As long as no requirements of relativistic invariance are introduced, this could be done merely by postulating an interaction potential of suitable depth and range. It is useful for what follows to formulate this in the language of the field theory as follows: Two types of particles, for example, protons and anti-neutrons, are described neglecting spin and relativity by two fields, P and A. It is convenient to use here these letters rather than the more usual ψ_P and ψ_A. The following Hamiltonian can be assumed in order to include the attractive potential:

$$\frac{h^2}{2M}\int \nabla P^*\nabla P d^3\mathbf{r}+\frac{h^2}{2M}\int \nabla A^*\nabla A d^3\mathbf{r}$$
$$-\int\int P^{*\prime}P^{\prime}A^{*\prime\prime}A^{\prime\prime}V(|\mathbf{r}^{\prime}-\mathbf{r}^{\prime\prime}|)d^3\mathbf{r}^{\prime}d^3\mathbf{r}^{\prime\prime}. \quad (1)$$

The first two terms are the kinetic energy of protons and anti-neutrons and the last term introduces the interaction. In this non-relativistic case, states with one proton and one anti-neutron do not mix with any other states. One can therefore confine one's attention to such states only and it is well known that the Hamiltonian (1) is then completely equivalent to that of a two-particle problem with an interaction $V(|\mathbf{r}'-\mathbf{r}''|)$.

Unfortunately no such simple situation obtains for relativistic particles in the hole theory. There are two reasons for this. One is that two-particle states mix with states in which additional pairs of particles form. The second is that only zero range forces can be used relativistically without adding an essentially new force field. For zero range forces no bound two-particle solution exists.

Since neutrons and anti-neutrons are symmetrical

particles, it is immaterial whether we call the anti-neutrons "holes" in a negative neutron sea or vice versa. Since we are interested primarily in an interaction between protons and anti-neutrons, the second alternative is preferable.

The simplest relativistically invariant interactions between these two fields are the usual[1] five types:[2]

$$\int A^*\beta A P^*\beta P d^3\mathbf{r} \qquad \text{(Scalar)}$$

$$\int \{A^*AP^*P - A^*\boldsymbol{\alpha}A\cdot P^*\boldsymbol{\alpha}P\}d^3\mathbf{r} \qquad \text{(Vector)}$$

$$\int \{A^*\beta\boldsymbol{\sigma}A\cdot P^*\beta\boldsymbol{\sigma}P + A^*\beta\boldsymbol{\alpha}A\cdot P^*\beta\boldsymbol{\alpha}P\}d^3\mathbf{r} \quad \text{(Tensor)} \qquad (2)$$

$$\int \{A^*\boldsymbol{\sigma}A\cdot P^*\boldsymbol{\sigma}P - A^*\gamma_5 A P^*\gamma_5 P\}d^3\mathbf{r} \qquad \text{(Pseudovector)}$$

$$\int A^*\beta\gamma_5 A P^*\beta\gamma_5 P d^3\mathbf{r}. \qquad \text{(Pseudoscalar)}.$$

The vector interaction in (2), like the Coulomb forces, has opposite signs for the interaction between a proton and a neutron and between a proton and an anti-neutron. It turns out that the tensor interaction also has this property while the scalar, pseudovector and pseudoscalar interactions have the same sign for a proton-neutron pair and a proton-anti-neutron pair.

As explained in the introduction, one needs an interaction that is attractive for a proton-anti-neutron pair and repulsive for a proton-neutron pair. Thus the vector and tensor interactions in (2) are the possible choices. For definiteness we shall take in what follows the vector interaction and write:

$$H^{\text{int}} = G\int \{A^*AP^*P - A^*\boldsymbol{\alpha}AP^*\boldsymbol{\alpha}P\}d^3\mathbf{r}. \qquad (3)$$

This Hamiltonian represents a δ-function interaction between a proton and an anti-neutron. Indeed, (3) may be written:

$$H^{\text{int}} = G\int\int A^{*\prime}P^{*\prime\prime} \times[\delta(\mathbf{r}'-\mathbf{r}'')(1-\boldsymbol{\alpha}_A\boldsymbol{\alpha}_P)]A'P''d^3\mathbf{r}'d^3\mathbf{r}''. \qquad (4)$$

[1] These are very similar to the interactions used in β-decay theory. See, e.g., H. A. Bethe, Rev. Mod. Phys. 8, 82 (1936). We use the same notation as Bethe's for the $\boldsymbol{\alpha}$-, β-, and γ-matrices.

[2] In the hole theory to make the vacuum expectation value of these interactions zero one needs actually to subtract from (2) certain terms. For example the correct scalar interaction to take is:

$$\int [N^*\beta N - \langle N^*\beta N\rangle_{\text{vac}}][P^*\beta P - \langle P^*\beta P\rangle_{\text{vac}}]d^3r,$$

where $\langle\ \rangle_{\text{vac}}$ means vacuum expectation value.

It has proved impossible to solve exactly the interaction problem of a proton and an anti-neutron to yield the "meson" bound state. We had to limit ourselves to the extremely crude description in terms of two-particle states only, disregarding thereby the complications due to multiple pair creation.

The following qualitative argument leads us to believe that this approximate description may be fairly good when the two particles are relatively far from each other and may break down when they are close. For a proton-anti-neutron state the unperturbed energy is larger than the actual energy by a little bit less than 2 Mc^2. For a state with an additional pair (two-pairs state), the energy difference[3] is 4 Mc^2, for an N-pairs state, $2N$ Mc^2. One might expect that an N-pair state will last a time of the order of $\hbar/(2N\ Mc^2)$ during which the particles can move away about $\hbar/(2N\ Mc)$. We expect, therefore, nucleons to be found away from the center up to about this distance. As N increases such configurations will become smaller and smaller. As a confirmation of this qualitative argument we find that actually for the two-nucleon state the wave function depends on the distance approximately as $\exp(-Mcr/\hbar)$.

We have attempted therefore to regard the effect of multiple pairs as perturbing the near parts of the single pair wave function as if the δ-function interaction were smeared over a region of dimensions about $\hbar/Mc$. This procedure is not relativistically invariant and should be substituted by a correct multiple-pairs theory. In lack of this we propose to follow up the two-particle theory assuming instead of the contact interaction one of range $\hbar/Mc$. The interaction will be modified accordingly by introducing instead of $G\delta(\mathbf{r}'-\mathbf{r}'')$ a finite range attractive potential $-V(|\mathbf{r}'-\mathbf{r}''|)$. With this the interaction term becomes

$$H^{\text{int}} = -\int\int A^{*\prime}P^{*\prime\prime}V(\mathbf{r})(1-\boldsymbol{\alpha}_A\cdot\boldsymbol{\alpha}_P)A'P''d^3\mathbf{r}'d^3\mathbf{r}''. \qquad (5)$$

For simplicity we will take for V a step function

$$\begin{aligned} V(\mathbf{r}) &= 0 && \text{for } r > \hbar/Mc \\ V(\mathbf{r}) &= V_0 = \text{constant} && \text{for } r < \hbar/Mc, \end{aligned} \qquad (6)$$

where

$$r = |\mathbf{r}'-\mathbf{r}''|, \quad \mathbf{r} = \mathbf{r}''-\mathbf{r}'.$$

We now adopt the two-particle approximation whereby the Schrödinger function will be a function of the spin and positional coordinates of the proton and the anti-neutron. The two spin indices running from 1 to 4 each yield a 16-component wave function. For states of zero total momentum each of the 16 components will depend only on the relative position $\mathbf{r}$.

The Schrödinger equation is:

$$\{-chi(\boldsymbol{\alpha}_P-\boldsymbol{\alpha}_A)\cdot\Delta + Mc^2\beta_A + Mc^2\beta_P - V(r)(1-\boldsymbol{\alpha}_A\cdot\boldsymbol{\alpha}_P)\}\psi = E\psi.$$

[3] See, however, Section III, especially footnote 5.

It is convenient to arrange the 16 components of ψ into a 4×4 matrix with the proton spin index vertical and the anti-neutron index horizontal.

For a $^1S_0$ state the rotational invariance specifies the dependence of the 16 components on the angular variables as follows:

$$\psi(^1S_0)=\begin{vmatrix} 0 & -if_1 & \frac{f_2}{r}(-x+iy) & \frac{f_2}{r}z \\ if_1 & 0 & \frac{f_2}{r}z & \frac{f_2}{r}(x+iy) \\ \frac{f_3}{r}(-x+iy) & \frac{f_3}{r}z & 0 & -if_4 \\ \frac{f_3}{r}z & \frac{f_3}{r}(x+iy) & if_4 & 0 \end{vmatrix}, \qquad (7)$$

where f_1, f_2, f_3 and f_4 are functions of the distance r only. The other state of total angular momentum 0, namely, $^3P_0$, has a wave function similar *in form* to (7) in which, however, the first and second rows are interchanged with the third and fourth rows. The $^1S_0$ state yields a particle that behaves as a pseudoscalar meson, whereas the $^3P_0$ state behaves as a scalar meson. This fact surprised us because we had thought that the opposite would be the case. The reason is connected with the different transformation properties under space reflection of the large and small components of the wave functions of a Dirac particle. No such unexpected behavior would have been found if the neutron had been treated in the sense of the hole theory as the particle and the anti-neutron as the anti-particle.

Substituting in (6) one finds for f_1, f_2, f_3, f_4 the equations:

$$2\left[r\frac{d}{dr}\left(\frac{f_2}{r}\right)+3\frac{f_2}{r}\right]=\frac{-2\,\mathrm{M}c^2+E+V}{c\hbar}f_1+\frac{3V}{c\hbar}f_4$$

$$=\frac{2\,\mathrm{M}c^2+E+V}{c\hbar}f_4+\frac{3V}{c\hbar}f_1, \qquad (8)$$

$$\frac{d}{dr}(f_1+f_4)=-\frac{E}{c\hbar}f_2,$$

$$f_2=f_3.$$

The lowest eigenvalue must be $E=\mu c^2$, the rest energy of the meson. This condition determines[4] the depth V_0 of the potential (6). Assuming the ratio 6.46 between the proton and meson masses one finds:[4]

$$V_0=26.4\ \mathrm{M}c^2=24.6\ \mathrm{Bev}. \qquad (9)$$

The corresponding normalized solution in a large volume Ω is:

$$r>r_0=\frac{\hbar}{\mathrm{M}c}\begin{cases} f_1=-\dfrac{0.236}{(r_0^3\Omega)^{\frac12}}\dfrac{1}{u}e^{-u} \\ f_2=f_3=-\dfrac{0.218}{(r_0^3\Omega)^{\frac12}}e^{-u}\left[\dfrac{1}{u}+\dfrac{1}{u^2}\right] \\ f_4=\dfrac{0.202}{(r_0^3\Omega)^{\frac12}}\dfrac{1}{u}e^{-u}, \end{cases} \qquad (10)$$

$$r<r_0\begin{cases} f_1=-\dfrac{0.0136}{(r_0^3\Omega)^{\frac12}}\dfrac{\sin v}{v} \\ f_2=f_3=\dfrac{0.370}{(r_0^3\Omega)^{\frac12}}\left[\dfrac{\cos v}{v}-\dfrac{\sin v}{v^2}\right] \\ f_4=-\dfrac{0.0147}{(r_0^3\Omega)^{\frac12}}\dfrac{\sin v}{v}, \end{cases} \qquad (11)$$

where

$$u=rc/\hbar[M^2-(\mu^2/4)]^{\frac12}, \quad v=2.03(r/r_0).$$

Notice that the wave function at large distances decreases like $\exp[-cr/\hbar(M^2-\mu^2/4)^{\frac12}]$; thus the geometrical size of the meson is of the order of $\hbar/\mathrm{M}c$ which is the Compton wave-length of the nucleon.

The inconsistencies of this representation should be emphasized. In particular we have given arguments to prove that the two-particle description breaks down at distances $\hbar/\mathrm{M}c$ and this very distance turns out to be the size of the meson. One could, therefore, state that the wave function becomes reliable only where it vanishes. Our only excuse in adopting it is that we have been unable to do better.

III. RELATIONSHIPS WITH THE YUKAWA THEORY

In spite of the differences between the Yukawa elementary particle model of the meson and the present model, most features of the Yukawa theory can be

[4] There are some undesirable solutions of (8) with energy values E that go to zero when $V_0\to 0$. These solutions are discarded because they do not adiabatically approach the state of two free particles when $V_0\to 0$. Also they would not appear at all if we had taken the neutron and the proton to be of different masses.

taken over even when the meson is pictured as a proton-anti-neutron bound pair denoted briefly as $(P+A)$.

The fundamental process of Yukawa's theory

$$P \to N+\pi^+ \tag{12}$$

now becomes

$$P \to N+(P+A). \tag{13}$$

This last process essentially is the addition to a proton P of a neutron-anti-neutron pair: N, A. Such pair formation will be induced by the postulated interaction (5). Since the energy of the bound $(P+A)$-system is much lower than that of the free particles the state (13) will be formed rather than a three free-particles state.[5] The matrix element is obtained from (5) by substituting for P the wave function of the proton that disappears, for A the wave function of the anti-neutron that disappears (neutron that appears), for $A'^*P''^*$ the complex conjugate of the wave function (7) of the bound proton-anti-neutron that appears. In order to express the wave function of the disappearing anti-neutron in terms of that of the neutron that is created, one uses the charge conjugation transformation

$$A=\gamma_2\tilde{N}^*,$$

where $\sim$ means transposed and * transposed and complex conjugate.

We calculate the matrix element for a transition from a slow proton to a slow neutron and a meson at rest. The calculation is straightforward and gives the following result:

$$\int\int V(r)N^{*\prime}Q(\mathbf{r})P''d^3\mathbf{r}'d^3\mathbf{r}'', \tag{14}$$

where Q is the matrix

$$Q=2i(f_1+f_4)\gamma_1\gamma_2\gamma_3+i(f_1-f_2)\gamma_1\gamma_2\gamma_3\gamma_4. \tag{15}$$

If the wave-length of the proton is long compared to $\hbar$/Mc (14) can be approximated by

$$\int N^*RPd^3\mathbf{r}, \tag{16}$$

where

$$R=\int V(r)Q(\mathbf{r})d^3\mathbf{r}. \tag{17}$$

[5] The contribution to the forces of the virtual creation of free-particle pairs has been discussed in Section II. It was interpreted there as modifying the interaction only at extremely short distances (Order $\hbar$/Mc). Creation of bound pairs yields inter-nucleon forces of range $\hbar/\mu c$.

Using (10), (11), and (15), and carrying out the integration one finds:

$$R=i\left(\frac{2\pi\hbar^3c^3}{\Omega\mu c^2}\right)^{\frac{1}{2}}(5.3\gamma_1\gamma_2\gamma_3+0.11\gamma_1\gamma_2\gamma_3\gamma_4). \tag{18}$$

This expression can be compared with the conventional interaction of a pseudoscalar meson with nucleons in the Yukawa theory.[6] There are two essentially independent coupling constants: f, the so-called pseudoscalar interaction, and g, the pseudovector interaction. The nucleon-meson interaction Hamiltonian is:

$$i\int N^*\left\{f\gamma_1\gamma_2\gamma_3\phi+\sum_\nu\frac{\hbar}{\mu c}g\gamma_1\gamma_2\gamma_3\gamma_\nu\frac{\partial\varphi}{\partial x_\nu}\right\}Pd^3\mathbf{r} \tag{19}$$

where ϕ is the pseudoscalar meson field.

The corresponding matrix element for the production of a meson at rest is

$$i\frac{\hbar c}{(2\Omega\mu c^2)^{\frac{1}{2}}}\int N^*(f\gamma_1\gamma_2\gamma_3+g\gamma_1\gamma_2\gamma_3\gamma_4)Pd^3\mathbf{r}.$$

Comparison with (18) gives

$$f=(4\pi\hbar c)^{\frac{1}{2}}\times 5.3,\quad g=(4\pi\hbar c)^{\frac{1}{2}}\times 0.11. \tag{20}$$

It has been proved by Case[7] that the terms f and g produce up to the second approximation nuclear forces of the same type. Indeed, their joint contribution is the same as would be obtained by putting $f=0$ and substituting g by

$$g'=g+f(\mu/2M). \tag{21}$$

We find, therefore,

$$g'=(4\pi\hbar c)^{\frac{1}{2}}\times 0.52$$

yielding for $g'^2/4\pi\hbar c$, that is for the analog of the fine structure constant, the value 0.27, which appears quite reasonable.

Naturally the similarity between the present point of view and the Yukawa theory can be carried on only up to a limited extent. The similarity breaks down on the one hand because of the finite size of the meson which introduces naturally a cut-off at short distances. On the other hand it breaks down for phenomena in which sufficiently high energies are involved to break up the meson.

[6] See for example: G. Wentzel, Rev. Mod. Phys. **19**, 1 (1947).
[7] K. M. Case, Phys. Rev. **76**, 14 (1949).

Selection Rules for the Dematerialization of a Particle into Two Photons

⟦50a⟧ Commentary begins page 8

C. N. YANG*

Institute for Nuclear Studies, University of Chicago, Chicago, Illinois

(Received August 22, 1949)

Selection rules governing the disintegration of a particle into two photons are derived from the general principle of invariance under rotation and inversion. The polarization state of the photons is completely fixed by the selection rules for initial particles with spin less than 2. These results which are independent of any specific assumption about the interactions may possibly offer a method of deciding the symmetry nature of mesons which decay into two photons.

I. INTRODUCTION

It has been pointed out[1] that a positronium in the ^3S state cannot decay through annihilation with the emission of two photons. Recent calculation[2] shows that also a vector or a pseudovector neutral meson cannot disintegrate into two photons. It is the purpose of the present paper to show that these facts are immediate consequences of certain selection rules which can be derived from the general principle of invariance under space rotation and inversion.

These selection rules also yield information on the polarization state of the two photons emitted. In particular, one concludes that the two photons resulting from the annihilation of slow positrons in matter always have their planes of polarization perpendicular to each other.[3] This has been pointed out by Wheeler who also proposed a possible experimental verification.[4]

An especially interesting consequence of these selection rules is that they could conceivably offer a means of studying the nature of particles which dematerialize into two photons. If, for example, neutral mesons are found which disintegrate into two photons,[5] one would conclude that they cannot be vector or pseudovector mesons. Besides, as will be apparent from the selection rules, the two disintegration photons from a scalar particle always have parallel planes of polarization while those from a pseudoscalar particle always have mutually perpendicular planes of polarization. An experimental determination of the relative orientation of the planes of polarization of the two disintegration photons would therefore decide whether the neutral meson is a scalar or a pseudoscalar meson.

TABLE I. Eigenvalues of the rotations $\mathcal{R}_\varphi$, $\mathcal{R}_\xi$, and the inversion $\mathcal{P}$ for the four polarization states.

	$\Psi^{RR}+\Psi^{LL}$	$\Psi^{RR}-\Psi^{LL}$	Ψ^{RL}	Ψ^{LR}
$\mathcal{R}_\phi$ (rotation around z axis through an angle φ)	1	1	$e^{2i\varphi}$	$e^{-2i\varphi}$
$\mathcal{R}_\xi$ (rotation around x axis through 180°)	1	1		
$\mathcal{P}$ (inversion)	1	-1	1	1

In Section II we shall give a simple but mathematically somewhat incomplete treatment of the symmetry nature of a state of two photons propagating in opposite directions, the detailed mathematical treatment being discussed in Section IV. The selection rules are derived in Section III and are based on the symmetry nature of the two photon states discussed in Section II. In the last section the parity of mesons and of the positronium is discussed.

II. BEHAVIOR OF THE STATE OF TWO PHOTONS UNDER ROTATION AND INVERSION

Consider two photons of equal wave-length λ_0 propagating in opposite directions along the z axis. There are four such states which we shall denote by Ψ^{RR}, Ψ^{RL}, Ψ^{LR} and Ψ^{LL}. The first index refers to the circular polarization state of the photon propagating in the $+z$ direction, the second to that of the other photon. E.g., Ψ^{RL} would represent a state with a right circularly polarized photon propagating along the $+z$ axis and a left circularly polarized photon propagating along the $-z$ axis.

In order to investigate the behavior of these four states under a space rotation or an inversion let us first write down the electric field for a right circularly polarized electromagnetic wave propagating along the z axis,

$$(E_x)_+{}^R = E_0\cos(kz-\omega t+\delta_+{}^R),\qquad (E_y)_+{}^R = E_0\sin(kz-\omega t+\delta_+{}^R). \tag{1}$$

TABLE II. Circular polarization of disintegration photons.

parity \ J	0	1	2, 4, 6 ···	3, 5, 7 ···
even	$\Psi^{RR}+\Psi^{LL}$	forbidden	$\Psi^{RR}+\Psi^{LL}$, Ψ^{RL}, Ψ^{LR}	Ψ^{RL}, Ψ^{LR}
odd	$\Psi^{RR}-\Psi^{LL}$	forbidden	$\Psi^{RR}-\Psi^{LL}$	forbidden

* Now at the Institute for Advanced Study, Princeton, New Jersey.

[1] J. A. Wheeler, Ann. N. Y. Acad. Sci. **48**, 219 (1946).

[2] S. Sakata and Y. Tanikawa, Phys. Rev. **57**, 548 (1948); R. Finkelstein, Phys. Rev. **72**, 415 (1947); J. Steinberger, Phys. Rev. **76**, 1180 (1949).

[3] They are not individually plane polarized. But if they are analyzed into plane polarized waves their planes of polarizations show the stated correlation.

[4] J. A. Wheeler, reference 1. See also M. H. L. Pryce and J. C. Ward, Nature **160**, 435 (1947), and Snyder, Pasternack and Hornbostel, Phys. Rev. **73**, 440 (1948). Experimental verification has been reported by E. Bleuer and H. L. Bradt, Phys. Rev. **73**, 1398 (1948).

[5] Bjorklund, Moyer, and York, Phys. Rev. **77**, 213 (1950).

□ Reprinted from *The Physical Review* 77, 2 (January 15, 1950), 242–245.

TABLE III. Correlation of the planes (see reference 3) of polarization of disintegration photons (⊥ = planes of polarization perpendicular, ∥ = planes of polarization parallel).

parity \ J	0	1	2, 4, 6 ···	3, 5, 7 ···
even	∥	forbidden	∥≧50% ⊥≦50%	∥ 50% ⊥ 50%
odd	⊥	forbidden	⊥	forbidden

For a right circularly polarized wave propagating in the opposite direction,

$$(E_x)_-^R = E_0 \cos(-kz-\omega t+\delta_-^R),\quad (E_y)_-^R = -E_0 \sin(-kz-\omega t+\delta_-^R). \tag{2}$$

Under a rotation through an angle φ around the z axis,

$$x = x' \cos\varphi + y' \sin\varphi,\quad y = -x' \sin\varphi + y' \cos\varphi,\quad z = z'. \tag{3}$$

We have

$$(E_x)_+^{R'} = E_0 \cos(kz-\omega t+\delta_+^R+\phi),\quad (E_y)_+^{R'} = E_0 \sin(kz-\omega t+\delta_+^R+\phi); \tag{4}$$

$$(E_x)_-^{R'} = E_0 \cos(-kz-\omega t+\delta_-^R-\varphi),\quad (E_y)_-^{R'} = -E_0 \sin(-kz-\omega t+\delta_-^R-\phi). \tag{5}$$

Thus the phase of a right circularly polarized wave along the z axis changes by $+\varphi$ while that of a right circularly polarized wave along the $-z$ axis changes by $-\varphi$ under the rotation. For the quantum state Ψ^{RR} the total phase factor is the product of the two phase factors of the two photons. (This will become evident in Section IV.) Hence we conclude that the state Ψ^{RR} is an eigenstate of the rotation (3) with the eigenvalue 1.

Mathematically the states are changed under the rotation (3) by a unitary transformation which we shall call $\mathcal{R}_\phi$. We conclude that:

$$\mathcal{R}_\varphi \Psi^{RR} = \Psi^{RR}. \tag{6}$$

Similar conclusions are reached for a rotation around the x axis through 180° and for an inversion. We summarize the results in Table I.

It is of course evident that the angular momentum along the z axis for the different states is related to the eigenvalue of $\mathcal{R}_\varphi$ in the usual way.

III. SELECTION RULES

The selection rules governing the disintegration of a particle into two photons follow immediately from Table I. We take the center-of-mass reference, system and take the z axis along the direction of one of the outgoing photons.

(i) For an odd initial state the only possible mode of disintegration is to go into the state $\Psi^{RR}-\Psi^{LL}$. For an even initial state the three final states Ψ^{RL}, Ψ^{LR}, and $\Psi^{RR}+\Psi^{LL}$ are possible.

(ii) For an initial state with total angular momentum $J=1, 3, 5\cdots$ the only possible final states are Ψ^{RL} and Ψ^{LR}. This is so because $\Psi^{RR}+\Psi^{LL}$ and $\Psi^{RR}-\Psi^{LL}$ are both simultaneous eigenstates of $\mathcal{R}_\varphi$ and $\mathcal{R}_\xi$ with eigenvalues one, while the initial state that is an eigenstate of $\mathcal{R}_\varphi$ with eigenvalue one has the rotation properties of the spherical harmonics Y_{J0} and therefore changes sign under $\mathcal{R}_\xi$ for $J=1, 3, 5\cdots$.

(iii) For an initial state with $J=0, 1$ the only possible final states are $\Psi^{RR}+\Psi^{LL}$ and $\Psi^{RR}-\Psi^{LL}$, because the states Ψ^{RL} and Ψ^{LR} have angular momentum values $\pm 2\hbar$ along the z axis, which is too big for $J=0$ or 1.

These results are summarized in Table II.

It will be shown in the next Section that

(i) $\Psi^{RR}+\Psi^{LL}$ represents two photons with their planes of polarization always[3] parallel;

(ii) $\Psi^{RR}-\Psi^{LL}$ represents two photons with their planes of polarization always perpendicular; and

(iii) Ψ^{RL} and Ψ^{LR} both represent two photons with their planes of polarization 50 percent of the time parallel and 50 percent of the time perpendicular.

These facts combined with Table II lead to the conclusions summarized in Table III concerning the correlation of the planes of polarization of the disintegration photons.

IV. SPACE ROTATION AND INVERSION IN QUANTUM ELECTRODYNAMICS

In the electromagnetic field and the meson field, the number of particles is not a constant of motion. A complete formulation of the principle of invariance can only be made with the quantized field theory. Let us first consider the electromagnetic field described by the vector potential $\mathbf{A}(xyz)$. These are operators operating on state vectors Ψ which are usually represented as functions of occupation numbers. Under a space rotation ρ defined by

$$x_i = \sum_{j=1}^{3} \rho_{ij} x_j' \quad (x_1=x,\ x_2=y,\ x_3=z), \tag{7}$$

the operators $\mathbf{A}(xyz)$ and the wave function Ψ undergo a unitary transformation $\mathcal{R}_\rho$ and the invariance under rotation requires that

$$\mathcal{R}_\rho A_i(xyz)\mathcal{R}_\rho^{-1} = \sum_{j=1}^{3} \rho_{ij} A_j(x'y'z'). \tag{8}$$

It is of course further required that the $\mathcal{R}$'s form a group isomorphic to the group of rotations.

To see in detail what this means let us expand the vector potential $\mathbf{A}$ into plane waves as usual:

$$\mathbf{A}(xyz) = \sum_{\mathbf{k}} \sum_{\lambda=1}^{2} (2\pi\hbar c/vk)^{\frac{1}{2}} \mathbf{e}_{k\lambda}(a_{k\lambda} e^{i\mathbf{k}\cdot\mathbf{r}} + a_{k\lambda}^* e^{-i\mathbf{k}\cdot\mathbf{r}}), \tag{9}$$

where $\mathbf{e}_{k1}$ and $\mathbf{e}_{k2}$ are two unit vectors forming with $\mathbf{k}/k$ a right-handed orthogonal system of unit vectors. It will be more convenient to use circular polarization for the study of rotation and we define

$$(\mathbf{e}_{k1}+i\mathbf{e}_{k2})/\sqrt{2} = \mathbf{e}_k^L,\quad (\mathbf{e}_{k1}-i\mathbf{e}_{k2})/\sqrt{2} = \mathbf{e}_k^R; \tag{10}$$

$$(a_{k1}-ia_{k2})/\sqrt{2} = a_k^L,\quad (a_{k1}+ia_{k2})/\sqrt{2} = a_k^R. \tag{11}$$

TABLE IV. The parity of particles at rest.

Scalar meson	Vector meson	Pseudovector meson	Pseudoscalar meson	Positronium in ^1S and ^3S states
even	odd	even	odd	odd

Evidently

$$\mathbf{A}(xyz)=\sum_{\mathbf{k}}(2\pi\hbar c/vk)^{\frac{1}{2}}(\mathbf{e}_k{}^L a_k{}^L e^{i\mathbf{k}\cdot\mathbf{r}}+\mathbf{e}_k{}^{L*}a_k{}^{L*}e^{-i\mathbf{k}\cdot\mathbf{r}})$$
$$+\sum_{\mathbf{k}}(2\pi\hbar c/vk)^{\frac{1}{2}}(\mathbf{e}_k{}^R a_k{}^R e^{i\mathbf{k}\cdot\mathbf{r}}+\mathbf{e}_k{}^{R*}a_k{}^{R*}e^{-i\mathbf{k}\cdot\mathbf{r}}). \quad (12)$$

The operators $a_{k\lambda}$, $a_k{}^R$ and $a_k{}^L$ all satisfy the usual commutation relations

$$a_k{}^R a_k{}^{R*}-a_k{}^{R*}a_k{}^R=1, \quad \text{etc.}$$

We are particularly interested in those modes of electromagnetic waves propagating along the $+z$ or the $-z$ direction with a definite wave-length, There are four such modes and we shall write their a operators as $a_+{}^R$, $a_+{}^L$, $a_-{}^R$ and $a_-{}^L$; $+$ and $-$ meaning the direction $+z$ or $-z$ of propagation. For definiteness we choose the phases of those modes such that the $\mathbf{e}_{k1}$, $\mathbf{e}_{k2}$ vectors in Eqs. (10) have as their xyz components:

$$\mathbf{e}_{+,1}=\mathbf{e}_{-,1}=(1,0,0), \quad \mathbf{e}_{+2}=-\mathbf{e}_{-,2}=(0,1,0). \quad (13)$$

With the operators a one can express in a very convenient form the states Ψ^{RR}, Ψ^{RL}, Ψ^{LR} and Ψ^{LL} defined in Section II.

$$\begin{aligned}\Psi^{RR}&=a_+{}^{R*}a_-{}^{R*}\Psi_{00\ldots}, & \Psi^{RL}&=a_+{}^{R*}a_-{}^{L*}\Psi_{00\ldots},\\ \Psi^{LR}&=a_-{}^{R*}a_+{}^{L*}\Psi_{00\ldots}, & \Psi^{LL}&=a_+{}^{L*}a_-{}^{L*}\Psi_{00\ldots},\end{aligned} \quad (14)$$

where $\Psi_{00\ldots}$ is defined to be the state with no photons.

We make a digression here to prove the statement at the end of the last section about the correlation in the planes of polarization of the two photons for the states $\Psi^{RR}+\Psi^{LL}$, $\Psi^{RR}-\Psi^{LL}$, Ψ^{RL} and Ψ^{LR}. By (14) and (11),

$$\begin{aligned}\Psi^{RR}+\Psi^{LL}&=(a_+{}^{R*}a_-{}^{R*}+a_+{}^{L*}a_-{}^{L*})\Psi_{00\ldots}\\ &=(a_{+,1}{}^*a_{-,1}{}^*-a_{+,2}{}^*a_{-,2}{}^*)\Psi_{00\ldots},\\ \Psi^{RR}-\Psi^{LL}&=-i(a_{+,1}{}^*a_{-,2}{}^*+a_{+,2}{}^*a_{-,1}{}^*)\Psi_{00\ldots},\\ \Psi^{RL}&=\tfrac{1}{2}(a_{+,1}{}^*a_{-,1}{}^*+a_{+,2}{}^*a_{-,2}{}^*\\ &\quad+ia_{+,1}{}^*a_{-,2}{}^*-ia_{+,2}{}^*a_{-,1}{}^*)\Psi_{00\ldots},\\ \Psi^{LR}&=\tfrac{1}{2}(a_{+,1}{}^*a_{-,1}{}^*+a_{+,2}{}^*a_{-,2}{}^*\\ &\quad-ia_{+,1}{}^*a_{-,2}{}^*+ia_{+,2}{}^*a_{-,1}{}^*)\Psi_{00\ldots}.\end{aligned} \quad (15)$$

Noticing that e.g., $a_{+,1}{}^*a_{-,1}{}^*\Psi_{00\ldots}$ represents a state with two photons with parallel planes of polarization one completes the proof with no difficulty.

Returning to the investigation of the behavior of the states Ψ^{RR}, Ψ^{RL}, etc., under rotation let us consider the rotation around the z axis through an angle φ, as defined by (3). Substitution of (13) and (12) into (8) shows[6] that

$$\begin{aligned}\mathcal{R}_\varphi a_+{}^R\mathcal{R}_\varphi{}^{-1}&=e^{-i\varphi}a_+{}^R, & \mathcal{R}_\varphi a_+{}^L\mathcal{R}_\varphi{}^{-1}&=e^{+i\varphi}a_+{}^L,\\ \mathcal{R}_\varphi a_-{}^R\mathcal{R}_\varphi{}^{-1}&=e^{+i\varphi}a_-{}^R, & \mathcal{R}_\varphi a_-{}^L\mathcal{R}_\varphi{}^{-1}&=e^{-i\varphi}a_-{}^L.\end{aligned} \quad (16)$$

These and similar equations for the other annihilation operators and for a general rotation ρ determine the operators $\mathcal{R}_\rho$ if the additional condition is imposed that $\mathcal{R}_\rho$ form a group. It is not difficult to prove that

$$\mathcal{R}_\rho\Psi_{00\ldots}=\Psi_{00\ldots}. \quad (17)$$

We can now prove Eq. (6) which asserts that Ψ^{RR} is an eigenstate of $\mathcal{R}_\varphi$ with eigenvalue 1. Take the Hermitian conjugate of (16) and multiply from the right by $\mathcal{R}_\varphi$:

$$\mathcal{R}_\varphi a_+{}^{R*}=e^{i\varphi}a_+{}^{R*}\mathcal{R}_\varphi, \quad \mathcal{R}_\varphi a_-{}^{R*}=e^{-i\varphi}a_-{}^{R*}\mathcal{R}_\varphi.$$

Hence,

$$\mathcal{R}_\varphi a_+{}^{R*}a_-{}^{R*}=a_+{}^{R*}a_-{}^{R*}\mathcal{R}_\varphi.$$

Operating on $\Psi_{00\ldots}$ with this last equation and making use of (17) and (14) one proves (6).

The other conclusions tabulated in the first row of Table I can be obtained in similar ways. For the rotation ξ:

$$x=x', \quad y=-y', \quad z=-z',$$

we have

$$\begin{aligned}\mathcal{R}_\xi a_+{}^R\mathcal{R}_\xi{}^{-1}&=a_-{}^R, & \mathcal{R}_\xi a_+{}^L\mathcal{R}_\xi{}^{-1}&=a_-{}^L,\\ \mathcal{R}_\xi a_-{}^R\mathcal{R}_\xi{}^{-1}&=a_+{}^R, & \mathcal{R}_\xi a_-{}^L\mathcal{R}_\xi{}^{-1}&=a_+{}^L.\end{aligned} \quad (18)$$

These lead to the results in the second row of Table I. Under an inversion the states are transformed by a unitary transformation $\mathcal{P}$ satisfying

$$\mathcal{P}\mathbf{A}(xyz)\mathcal{P}^{-1}=-\mathbf{A}(-x,-y,-z). \quad (19)$$

It is further required that

$$\mathcal{P}^2=1. \quad (20)$$

Expanding (19) into Fourier components we obtain

$$\begin{aligned}\mathcal{P}a_+{}^R\mathcal{P}^{-1}&=-a_-{}^L, & \mathcal{P}a_+{}^L\mathcal{P}^{-1}&=-a_-{}^R,\\ \mathcal{P}a_-{}^R\mathcal{P}^{-1}&=-a_+{}^L, & \mathcal{P}a_-{}^L\mathcal{P}^{-1}&=-a_+{}^R.\end{aligned} \quad (21)$$

It is to be noticed that (19) and (20) together do not completely determine the operator $\mathcal{P}$, as a change of sign of $\mathcal{P}$ does not affect either equation. However, changing the sign of $\mathcal{P}$ merely means a change in the name-calling of the even and odd states and is of no physical consequence. For definiteness we shall fix the sign by calling the vacuum an even state:

$$\mathcal{P}\Psi_{00\ldots}=\Psi_{00\ldots}. \quad (22)$$

Equations (21) and (22) lead to the third row of Table I.

[6] Actually since the Maxwell's equations are of the second order in $\partial/\partial t$, one should write together with (8)

$$\mathcal{R}_\rho\frac{\partial A_i(xyz)}{\partial t}\mathcal{R}_\rho{}^{-1}=\sum_j \rho_{ij}\frac{\partial A_j(x'y'z')}{\partial t}. \quad (A)$$

The operators $\partial A_i/\partial t$ are expanded into Fourier series similar to (12). Equation (A) together with (8) give Eq. (16).

V. PARITY OF MESONS AND THE POSITRONIUM

The above method for obtaining the symmetry nature of photon states can be easily extended to the meson field. In particular, if a_0 is the annihilation operator for a scalar meson at rest it is easy to see in analogy with Eq. (21) that

$$\mathcal{P}a_0\mathcal{P}^{-1}=a_0. \tag{23}$$

There is no change of sign of a_0 (see Eq. (21)) because the scalar meson field, unlike the vector potential of the electromagnetic field, retains its sign under an inversion. If we again call the state of no meson an even state, it is evident that a state with one scalar meson at rest is also even. This and similar conclusions concerning the parity of the vector, the pseudoscalar and pseudovector mesons are summarized in Table IV.

With the electron-positron field the situation is quite similar. The behavior of the field $\psi_i(xyz)$ under rotation and inversion is evidently given by

$$\mathcal{R}_\rho\psi_i(xyz)\mathcal{R}_\rho^{-1}=\sum_{j=1}^{4} S_{ij}(\rho)\psi_j(x'y'z'), \tag{24}$$

$$\mathcal{P}\psi_i(xyz)\mathcal{P}^{-1}=\sum_{j=1}^{4} \beta_{ij}\psi_j(x'y',z'), \tag{25}$$

where $S_{ij}(\rho)$ represents the spinor transformation corresponding to the rotation ρ and β_{ij} are the elements of Dirac's β-matrix.

If we expand ψ into plane waves and consider the particular mode representing an electron at rest in a positive or negative energy state it is evident that (25) shows

$$\mathcal{P}b_{0+}\mathcal{P}^{-1}=b_{0+},\quad \mathcal{P}b_{0-}\mathcal{P}^{-1}=-b_0\ , \tag{26}$$

where b_{0+} and b_{0-} are the annihilation operators for an electron at rest with positive and negative energy values, respectively. The negative sign in (26) comes from the operation on ψ with the β-matrix. It is therefore evident that an electron-positron pair, both at rest, has an odd parity. Here, as before, we adopt the convention that the state of vacuum is to be called even.

Extension of the above argument to the ^1S and ^3S states of the positronium is evident. One gets for both states an odd parity. As mentioned in the introduction, Wheeler has pointed out[1] that the annihilation photons from the ^1S state of the positronium always have mutually perpendicular[3] planes of polarization. We see from Table III that the assignment of an odd parity to the ^1S state of positronium leads directly to the same conclusion.

The author is very much indebted to Professor E. Fermi for invaluable discussions and for his kind encouragement.

Note added in proof.—Some of the results of this paper have been obtained by L. D. Landau, Dokl. Akad. Nawk., USSR **60**, 207–209 (1948). See a summary in English in Phys. Abstracts **A52**, 125 (1949).

Reflection Properties of Spin $\frac{1}{2}$ Fields and a Universal Fermi-Type Interaction

[[50c]]

Commentary begins page 9

C. N. Yang
Institute for Advanced Study, Princeton, New Jersey

AND

J. Tiomno*
Palmer Physical Laboratory, Princeton University, Princeton, New Jersey

(Received March 22, 1950)

It is pointed out that four different transformations are possible under an inversion for fields of spin $\frac{1}{2}$. The consequences are discussed, and the bearing on the possibility of a universal Fermi-type interaction is analyzed.

I. INTRODUCTION

The transformation property under an inversion of the space coordinates of the wave function of a particle of spin zero (or one) leads to the differentiation between scalar and pseudoscalar (or vector and pseudovector) particles. It is the purpose of the present note to point out that there is also a similar differentiation of the various spin $\frac{1}{2}$ particles. In contrast to the case of integral spin, however, there are for spin $\frac{1}{2}$ particles *four* different types of transformation properties under an inversion.

The types of transformation properties to which the various known spin $\frac{1}{2}$ fields belong are physical observables and could in principle be determined experimentally from their mutual interactions and their interactions with fields of integral spin. In practice this is, of course, very difficult. This problem is discussed in Section III in connection with β-decay and with the symmetry properties of the π-mesons.

In the final section, an attempt is made to see if by properly assigning the various known fields of spin $\frac{1}{2}$ to the four types one could have a universal Fermi-type interaction that would account for the experimental information accumulated in recent years about the interactions between spin $\frac{1}{2}$ fields.

II. GENERAL THEORY

We consider first the transformation property of a *single* particle under an orthochronous proper Lorentz transformation[1] (i.e., a Lorentz transformation involving neither time reversal nor space reflection). The wave function ψ undergoes[2] the following transformation

$$\psi'=S\psi. \qquad (1)$$

The matrix S is defined only up to a $(\pm)$ sign. In particular, the identity transformation is represented by

$$S_0=\pm I, \quad I=\text{unit matrix}. \qquad (2)$$

This ambiguity of sign is *necessary* because a rotation through 360° about any axis always brings about a change of sign of the wave function of a spin $\frac{1}{2}$ particle.

The space inversion P is usually represented[2] by the transformation

$$\psi'=\pm\gamma_4\psi. \qquad (3)$$

With this definition one might say that there is only *one* double-valued representation of the orthochronous Lorentz group for particles of spin $\frac{1}{2}$. This, however, leads to *two* different possibilities when the transformation properties of two different fields ψ_A and ψ_B are considered at the same time:

$$\begin{cases}\psi_A'=S\psi_A, \quad \psi_B'=S\psi_B & \text{for } L \text{ (orthochronous proper Lorentz transformation)}\\ \psi_A'=\pm\gamma_4\psi_A, \quad \psi_B'=\pm\gamma_4\psi_B & \text{for } P \text{ (space inversion)}\end{cases} \qquad (4)$$

or

$$\begin{cases}\psi_A'=S\psi_A, \quad \psi_B'=S\psi_B & \text{for } L\\ \psi_A'=\pm\gamma_4\psi_A, \quad \psi_B'=\mp\gamma_4\psi_B & \text{for } P.\end{cases} \qquad (5)$$

If case (4) holds, the fields A and B behave exactly alike under all orthochronous Lorentz transformations. This is customarily accepted as true for all spin $\frac{1}{2}$ fields. We see now that there is also possible the other case (5) in which the two fields A and B *always differ by a rotation of 360° under a space reflection.*

Racah[3] has pointed out that for the space inversion of a single particle it is also possible to have instead of Eq. (3),

$$\psi'=i\gamma_4\psi. \qquad (6)$$

This introduces two other possible transformations for a spin $\frac{1}{2}$ field under a space inversion. Altogether, we would have four kinds of fields of spin $\frac{1}{2}$ that behave differently under an inversion. Denoting by ψ_A, ψ_B, ψ_C, and ψ_D four such fields one has under a space inversion

$$\begin{bmatrix}\psi_A'\\ \psi_B'\\ \psi_C'\\ \psi_D'\end{bmatrix}=\pm\begin{bmatrix}\gamma_4 & & & 0\\ & -\gamma_4 & & \\ & & i\gamma_4 & \\ 0 & & & -i\gamma_4\end{bmatrix}\begin{bmatrix}\psi_A\\ \psi_B\\ \psi_C\\ \psi_D\end{bmatrix}. \qquad (7)$$

It is important to notice that with the simultaneous existence of the fields A, B with C, D the representation

* Rockefeller Foundation fellow of the University of São Paulo, Brazil.

[1] This nomenclature is due to H. J. Bhabha, Rev. Mod. Phys. **21**, 451 (1949).

[2] W. Pauli, *Handbuch der Physik*, Vol. 24/1, pp. 220–227.

[3] G. Racah, Nuovo Cimento **14**, 322 (1937).

becomes four-valued. For example, the identity transformation corresponds now to

$$\begin{pmatrix}\psi_A'\\ \psi_B'\\ \psi_C'\\ \psi_D'\end{pmatrix}=\pm\begin{pmatrix}\psi_A\\ \psi_B\\ \psi_C\\ \psi_D\end{pmatrix}\quad\text{or}\quad\pm\begin{pmatrix}\psi_A\\ \psi_B\\ -\psi_C\\ -\psi_D\end{pmatrix}. \tag{8}$$

The necessity of the two latter possibilities arises from the fact that if the inversion (7) is applied twice one gets a representation of the identify transformation with a relative change of sign of the fields ψ_A, ψ_B and ψ_C, ψ_D.

It is also important to notice that there is no *intrinsic* difference between the A-type fields (i.e., fields with the transformation property of ψ_A) and the B-type fields, or between the C- and D-type fields. The difference is only relative, and is there only if both types of fields exist. If, for example, only C- (or D-) type fields exist, it is impossible to tell whether it is the C-type or the D-type. Further, it would not be appropriate, for example, to call the A-type fields "fields" and the B-type fields "pseudofields." This is to be contrasted with the case of fields with integral spin, for which there is an *intrinsic* difference between fields and pseudofields.

It is easy to show that the charge conjugate field of an A-type field is B-type, and *vice versa.* On the other hand, the charge conjugate field of a C-type field (D-type) is also C-type (D-type). This fact is summarized in Table I.

TABLE I. Field type of charge conjugate field.

Field (particle)	A	B	C	D
Charge conjugate field (antiparticle)	B	A	C	D

It should be remarked that the fact that the C- and D-type fields have the same transformation properties as their charge conjugate fields has led Racah[3] to the proposal that all spin ½ fields transform like C-type fields.

With the Majorana theory, since the field and the charge conjugate field are identical, the C- and D-type transformations are the only possibilities.

III. APPLICATIONS TO ELECTRON AND MESON FIELDS

The five densities $\phi^*\gamma_4\psi$, $\phi^*\gamma_4\gamma_\mu\psi$, $\phi^*\gamma_4(\gamma_\mu\gamma_\nu-\gamma_\nu\gamma_\mu)\psi$, $\phi^*\gamma_1\gamma_2\gamma_3\gamma_\mu\psi$ and $\phi^*\gamma_1\gamma_2\gamma_3\psi$ are usually referred to as scalar, vector, tensor, pseudovector, and pseudoscalar quantities.[4] This is evidently correct if $\phi=\psi$, whatever type of field ψ is. However, if ϕ and ψ are two different fields, and if they belong to different types, it is not correct, for example, to call $\phi^*\gamma_4\psi$ a scalar quantity. In fact, if ψ is an A-type field, while ϕ is a B-type field, then $\phi^*\gamma_4\psi$ is a pseudoscalar.

[4] See, for example, H. A. Bethe and R. F. Bacher, Rev. Mod. Phys. **8**, 190 (1936). We use the same notation for the α-, β-, and γ_μ-matrices as these authors.

In the β-decay theory, the following five interactions

$$\psi_P^*\beta\psi_N\psi_e^*\beta\gamma_5\psi_\nu, \tag{9}$$

$$\psi_P^*\psi_N\psi_e^*\gamma_5\psi_\nu-\psi_P^*\alpha\psi_N\psi_e^*\sigma\psi_\nu, \tag{10}$$

$$\psi_P^*\beta\sigma\psi_N\psi_e^*\beta\alpha\psi_\nu+\psi_P^*\beta\alpha\psi_N\psi_e^*\beta\sigma\psi_\nu, \tag{11}$$

$$\psi_P^*\gamma_5\psi_N\psi_e^*\psi_\nu-\psi_P^*\sigma\psi_N\psi_e^*\alpha\psi_\nu, \tag{12}$$

$$\psi_P^*\beta\gamma_5\psi_N\psi_e^*\beta\psi_\nu, \tag{13}$$

are usually rejected on the ground that they are not invariant under a space inversion. We see now that this is not justifiable in view of the fact that the types of fields to which ψ_P, ψ_N, ψ_e, and ψ_ν belong are not known. If it turns out that protons, neutrons, and electrons are all of type A, while the neutrino is of type B, then these five interactions are the invariant ones instead of the usual five.

If the mass of the neutrino is not zero, the interactions (9) to (13) lead to β-spectra different from that predicted by the ordinary interactions, especially near the upper end of the electron spectra. Also, the angular correlation between the electron and the neutrino would be different.

On the other hand, if the mass of the neutrino is zero, it is experimentally impossible to differentiate between the five possibilities (9) to (13) and the usual five (unless one could measure the neutrino spin). This is so because when the square of the matrix element of the interaction Hamiltonian is summed over the spin of the neutrino, the result for (9), for example, is the same as the result for the ordinary scalar-type of Fermi interaction, provided the mass of the neutrino is zero.

It is interesting to notice that the original proposal of Fermi[5] is identical with the interaction (10), rather than with the usual vector interaction. Hence, Fermi's spectrum is different from that of later authors for the case in which the mass of the neutrino is not zero, a fact already pointed out by Konopinski and Uhlenbeck.[6]

The usual statements about the meson theory of nuclear forces need to be understood in a new light now that, for example, $\psi_P^*\beta\psi_N$ can be a pseudoscalar. In particular, contrary to the usually accepted concept, a *scalar* meson *can* have a "pseudoscalar" and a "pseudovector" interaction with the nucleons:

$$iF\psi_P^*\gamma_4\gamma_5\psi_N\phi+iG\psi_P^*\gamma_4\gamma_5\gamma_\mu\psi_N(\partial\phi/\partial\chi_\mu).$$

It is also to be remarked that if the process

$$P\rightarrow N+\pi$$

occurs, it is impossible that the proton field be of type A or B, while the neutron field be of type C or D, or *vice versa.* This fact will be used later in Section IV.

IV. A UNIVERSAL FERMI-TYPE INTERACTION

An analysis of the phenomena of:

$$\beta\text{-decay: } N\rightarrow P+e+\nu, \tag{14}$$

[5] E. Fermi, Zeits. f. Physik. **88**, 161 (1934).

[6] E. J. Konopinski and G. Uhlenbeck, Phys. Rev. **48**, 7 (1935).

$$\mu\text{-capture:}^{7,8}\ P+\mu^- \rightarrow N+\nu, \tag{15}$$

and

$$\mu\text{-decay:}^{8,9}\ \mu \rightarrow e+2\nu \tag{16}$$

by assuming that the charged μ-meson[10] has spin $\frac{1}{2}$, and that the interactions which lead to these processes are of the Fermi-type, gives values of the coupling constants in the three interactions of the same order of magnitude.[11] This suggests the possibility of the existence of a universal interaction of the Fermi-type between *all* particles of spin $\frac{1}{2}$,[11] provided the requirement of conservation of charge is satisfied. The difficulty immediately arises that such a universal interaction would lead to processes inconsistent with experience, such as

$$N+P \rightarrow e+\nu \quad \text{or} \quad N \rightarrow e^+ + \mu^- + \nu. \tag{17}$$

To rule out these processes, one might propose[12] that some additional conservation laws other than the conservation of charge must be fulfilled. Indeed, with the proper assignment of the spin $\frac{1}{2}$ fields to the four types discussed in Section II, one might expect to obtain such conservation laws as consequences of the principle of invariance under a space inversion.

We have attempted to carry this out but have not succeeded in finding a completely satisfactory assignment. It turns out that some additional rather arbitrary tules will still have to be introduced. In the following, a brief account of our attempt is given, together with the results obtained.

A. The Universal Interaction

To make the proposal of a universal interaction quite definite, one has to choose a definite form of Fermi interaction. We have chosen the Wigner-Critchfield interaction[13]

$$H_{rstu} = g \int d^3\mathbf{x} \sum_{\alpha\beta\gamma\delta} \epsilon_{\alpha\beta\gamma\delta} \psi_\alpha{}^r(\mathbf{x}) \psi_\beta{}^s(\mathbf{x}) \psi_\gamma{}^t(\mathbf{x}) \psi_\delta{}^u(\mathbf{x}), \tag{18}$$

where $\epsilon_{\alpha\beta\gamma\delta} = \pm 1$ is antisymmetric with respect to its four indices, and $\psi_\alpha{}^r(\mathbf{x})$ is the α-component of the q-number spinor wave function of the rth field. To avoid taking into account the explicit occurrence of the complex conjugate of the wave functions in the interaction, we consider a field and its charge conjugate field as *two separate* fields. The reason that the Wigner-Critchfield interaction is chosen is that it is the *only* interaction that is symmetrical with respect to the four fields.[14] It is also evidently invariant under an orthochronous proper Lorentz transformation.

The universal interaction is formulated thus: *Between any four fields r, s, t, u, the interaction (18) exists with the same constant g provided it is consistent with charge conservation and is invariant under a space inversion, in the sense discussed in Section I.*[14a]

This last requirement we use to eliminate the undesired interactions, such as (17). For example, the term H_{rstu} would be absent in the Hamiltonian if the fields r, s, t are of type A and u of type B, C, or D, because in any of these cases H_{rstu} is not invariant under a space inversion.

B. Restrictions on the Assignments of Spin $\frac{1}{2}$ Fields to the Different Types

We want to allow the process

$$.N \rightarrow P+e+\nu \tag{19}$$

which amounts to the same thing as

$$P \rightarrow N+\bar{e}+\bar{\nu} \tag{20}$$

where the bar over e and ν means the antiparticle. At the same time we want to forbid the process

$$N+P \rightarrow \bar{e}+\bar{\nu} \quad \text{or} \quad P \rightarrow \bar{N}+\bar{e}+\bar{\nu}. \tag{21}$$

This is only possible if N and $\bar{N}$ have different transformation properties, and from Table I we conclude that the neutron field is of type A or B. Similar reasoning, with the additional requirement that we want to forbid the process

$$N+P \rightarrow \bar{e}+\nu \quad \text{or} \quad P \rightarrow \bar{N}+\bar{e}+\nu, \tag{22}$$

leads to the conclusion that the neutrino is of type C or D. Now it is generally accepted that the π-meson has integral spin and interacts with the $N-P$ field. Therefore, the proton field, like the neutron field, must be of type A or B (see Section III). Consequently, from (19) one concludes that the electron is a C- or D-type field.

From an analogous analysis of the μ-capture, we conclude that μ also belongs to the C- or D-type field.

These assignments lead immediately to a *conservation law of heavy particles* which has been noticed by many physicists.

Another feature of this assignment is that it is consistent with the double β-decay,[15] with Majorana's theory of the neutrino,[16] and with the experiments[8] on μ-decay and π-decay.

C. Results

By writing down all the Fermi-type interactions consistent with charge conservation and continuing the

---

[7] B. Pontecorvo, Phys. Rev. **72**, 246 (1947).

[8] For more complete references to the literature see J. Tiomno and J. A. Wheeler, Rev. Mod. Phys. **21**, 144–153 (1949). See also, Taketani, Nakamura, Ono, and Sasaki, Phys. Rev. **76**, 60 (1949), and L. Michel, Nature **163**, 959 (1949). Proc. Phys. Soc. London (to be published).

[9] Leighton, Anderson, and Seriff, Phys. Rev. **76**, 159 (1949).

[10] J. Tiomno, Phys. Rev. **76**, 858 (1949).

[11] J. Tiomno and J. A. Wheeler, reference 8; Lee, Rosenbluth, and Yang, Phys. Rev. **75**, 905 (1949).

[12] This was pointed out by Professor E. Fermi in a seminar of about a year ago.

[13] C. L. Critchfield, Phys. Rev. **63**, 416 (1943).

[14] This is so because two spin $\frac{1}{2}$ fields which are not charge conjugate to each other anticommute.

[14a] One notices that the interaction Hamiltonian consistent with this proposal is Hermitian.

[15] E. L. Fireman, Phys. Rev. **75**, 323 (1949).

[16] E. Majorana, Nuovo Cimento **14**, 171 (1937).

above analysis it is found that unless some additional conditions are imposed one cannot eliminate all the undesired interactions (in particular, those indicated in footnotes 17 and 18). To be more definite, the conclusion is as follows:

The only two possible assignments are

$$(\alpha) \quad P,\, N\epsilon A, \quad \mu,\, e,\, \nu\epsilon C \tag{23}$$

where ϵ reads "belong to type." Notice that this assignment is identical, for example, with the assignment

$$P,\, N\epsilon B, \quad \mu,\, e,\, \nu\epsilon C. \tag{24}$$

(Compare Section II.) Or

$$(\beta) \quad P\epsilon A, \quad N\epsilon B, \quad \nu\epsilon C \quad \text{and} \quad \mu,\, e\epsilon D. \tag{25}$$

But the additional restrictions will have to be imposed that (a) all terms in which a field and its charge conjugate field appear,[17] and (b) all terms in which four identical fields appear[18] are to be excluded from the Hamiltonian.

If experimental results should show the existence of another neutral spin $\frac{1}{2}$ particle μ_0 such as in[8]

$$P+\mu^- \rightarrow N+\mu_0, \tag{26}$$

and[8, 9]

$$\mu \rightarrow e+\mu_0+\nu, \tag{27}$$

it would be straightforward to include μ_0 in the present scheme of considerations.

The authors wish to thank Professor J. R. Oppenheimer, Professor E. P. Wigner, and Dr. A. Wightman for helpful discussions.

[17] This is to forbid such processes as $P+\mu \rightarrow P+e$ which contradicts the experimental result that no electrons are emitted in the capture of the μ^--meson by heavy nuclei.

[18] This is to forbid such processes as $N+N \rightarrow \bar{N}+\bar{N}$ which would lead to the instability of complex nuclei.

The S-Matrix in the Heisenberg Representation

[[50d]]
Commentary begins page 10

C. N. YANG AND DAVID FELDMAN*
Institute for Advanced Study, Princeton, New Jersey
(Received May 17, 1950)

A method is described whereby the **S**-matrix can be formulated directly in the Heisenberg representation. This has the advantage over the customary formulation in the interaction representation in that the concepts of space-like surfaces and their normals need never be introduced. Quantum electrodynamics and the β formalism of charged mesons are treated as illustrative examples; in particular, it is shown that general rules for writing down the elements of the **S**-matrix for the latter case may be immediately inferred.

In the second part of this paper, a covariant procedure, independent of the canonical formalism, is carried out for making the transition from the Heisenberg to the interaction representation and is applied to several typical cases; in this way, the **S**-matrix of the Heisenberg picture is identified with that of other authors.

INTRODUCTION

IN the recent work of Tomonaga[1] and Schwinger,[2] these authors, in their successful attempts to cast quantum electrodynamics (and, in actual fact, all meson theory) into a completely covariant and practical form, found it necessary to introduce the concept of the interaction representation. The essential virtue of this representation is that it leads to an equation of motion for the state vector of the system which is covariant in all its aspects (unlike the Schrödinger representation) while, at the same time, the field variables obey free-field equations of motion and commutation relations (in marked contrast with the corresponding situation in the Heisenberg representation). Upon using this form of the theory, it becomes a simple matter to derive the **S**-matrix and, indeed, Dyson[3] has shown that the quantum electrodynamics of Tomonaga and Schwinger leads to the well-known rules of Feynman,[4] which enable one to calculate immediately the elements of the **S**-matrix.

The object of this note is to describe a method where the **S**-matrix may be formulated directly in the Heisenberg representation. Heretofore, the complexity of the commutation relations of the field quantities in this representation has been regarded as a principal deterrent to the development of a practical field theory in the Heisenberg picture. However, if one makes the basic assumption that it is valid to employ a weak-coupling approximation (which actually is already characteristic of all current relativistic field theories), then a knowledge of the complete commutation relations in the Heisenberg representation is not needed provided one can effect a separation of the motion of the system into that of a free-field part plus that of an interacting part. Tomonaga and Schwinger have done this by going over to the interaction representation. A completely equivalent procedure which we follow in this paper is simply to express each Heisenberg variable as the sum of two parts, the one being a solution of the homogeneous free-field equation (and, indeed, also satisfying the free-field commutation relations), the other being a solution of the inhomogeneous equation.

That the two methods must inevitably lead to the same final results is of course clear. Yet there are distinct advantages to the formulation in the Heisenberg representation which become readily apparent as soon as one considers a situation where one has derivative coupling (as in the pseudoscalar theory with pseudovector coupling) or where one has dynamically dependent field variables (as in the neutral vector meson theory). It is well known that for these cases one is led in the interaction representation to considerable complications involving space-like surfaces and normals thereto which however drop out at the very end;[5] on the other hand, as will be seen below, all meson theories are no more difficult to handle in the Heisenberg picture than is the case of quantum electrodynamics.

In fact, it will become apparent that one can infer the rules for writing down the elements of the **S**-matrix for all meson theories from the rules which Feynman has established for quantum electrodynamics.

The discussion which follows has been divided into two distinct sections. In the first, the **S**-matrix is defined in the Heisenberg representation for the case of quantum electrodynamics and then, as an illustration of a more complicated situation, for the case of charged scalar and vector mesons interacting with the electromagnetic field (for the sake of compactness, the β formalism is used). In the second section, it is shown by working through several typical examples that the **S**-matrix of the Heisenberg picture may be identified with the results of the interaction-representation formulation; here, the method of passing from the Heisenberg to the interaction representation is of especial interest since it is effected in a covariant manner without having to use the canonical formalism.[6]

---

* Now at the Department of Physics, University of Rochester, Rochester, New York.

1 S. Tomonaga, Prog. Theor. Phys. **1**, 27 (1946) and later papers; Phys. Rev. **74**, 224 (1948).

2 J. Schwinger, Phys. Rev. **74**, 1439 (1948); **75**, 651 (1949); **76**, 790 (1949).

3 F. J. Dyson, Phys. Rev. **75**, 486, 1736 (1949).

4 R. P. Feynman, Phys. Rev. **76**, 749, 769 (1949).

5 S. Kanesawa and S. Tomonaga, Prog. Theor. Phys. **3**, 1, 101 (1948); A. Pais and G. E. Uhlenbeck, Phys. Rev. **75**, 1321 (1949).

6 Cf. N. M. Kroll, Phys. Rev. **75**, 1321 (1949); P. T. Matthews, Phys. Rev. **76**, 1657 (1949); J. S. de Wet, Proc. Roy. Soc. **A201**, 284 (1950).

□ Reprinted from *The Physical Review* 79, 6 (September 15, 1950), 972–978.

In fact, the so-called interaction-representation Hamiltonian will appear, in a certain sense, as the density of the **S**-matrix, so that all conditions of integrability are automatically satisfied.

I. DEFINITION OF THE S-MATRIX

A. Quantum Electrodynamics

We now proceed to show how to formulate the **S**-matrix directly in the Heisenberg representation without having to make any mention of spacelike surfaces or their normals. We consider first the case of quantum electrodynamics.

The field equations for the Heisenberg variables A_ν and ψ are[7]

$$(\gamma_\nu\partial/\partial x_\nu+m)\psi=ie\gamma_\nu A_\nu\psi, \tag{1}$$

$$(\partial\bar{\psi}/\partial x_\nu)\gamma_\nu-m\bar{\psi}=-ie\bar{\psi}\gamma_\nu A_\nu, \tag{1'}$$

$$\partial^2 A_\nu/\partial x_\lambda{}^2=-j_\nu; \tag{2}$$

where

$$j_\nu=\tfrac{1}{2}ie(\bar{\psi}\gamma_\nu\psi-\psi^T\gamma_\nu{}^T\bar{\psi}^T); \tag{2'}$$

in addition, one has the supplementary condition

$$(\partial A_\nu/\partial x_\nu)\Phi=0, \tag{3}$$

where Φ is the state vector of the system.

Equation (2) can be integrated in the usual manner of classical electrodynamics where A_ν is expressed as the sum of a freely oscillating incoming field $A_\nu{}^{\text{in}}$ and a retarded potential, *viz.*,

$$A_\nu(x)=A_\nu{}^{\text{in}}(x)+\int D^{\text{ret}}(x-x')d^4x'j_\nu(x'). \tag{4}$$

The $D^{\text{ret}}(x-x')$ function is defined, in terms of Schwinger's notation,[2] as

$$D^{\text{ret}}(x-x')=\bar{D}(x-x')-\tfrac{1}{2}D(x-x'). \tag{5}$$

The corresponding function which leads to advanced potentials is

$$D^{\text{adv}}(x-x')=\bar{D}(x-x')+\tfrac{1}{2}D(x-x'). \tag{5'}$$

Upon defining $S^{\text{ret}}(x-x')$ and $S^{\text{adv}}(x-x')$ in a similar manner, it is clear that (1) and (1′) may be integrated to give

$$\psi(x)=\psi^{\text{in}}(x)-ie\int S^{\text{ret}}(x-x')d^4x'\gamma_\nu A_\nu(x')\psi(x') \tag{6}$$

and

$$\bar{\psi}(x)=\bar{\psi}^{\text{in}}(x)-ie\int \bar{\psi}(x')\gamma_\nu A_\nu(x')d^4x'S^{\text{adv}}(x'-x). \tag{6'}$$

Equations (6), (6′), and (4) should be regarded as defining the incoming fields ψ^{in}, $\bar{\psi}^{\text{in}}$ and $A_\nu{}^{\text{in}}$.[8] It is already clear that these fields satisfy the homogeneous free-field equations and that, in terms of these variables, the supplementary condition (3) is simply

$$(\partial A_\nu{}^{\text{in}}/\partial x_\nu)\Phi=0. \tag{3'}$$

It is especially important that the incoming fields also satisfy the free-field commutation relations:

$$\left.\begin{aligned}&[\psi_\alpha{}^{\text{in}}(x),\bar{\psi}_\beta{}^{\text{in}}(x')]_+=-iS_{\alpha\beta}(x-x')\\&[A_\nu{}^{\text{in}}(x),A_\lambda{}^{\text{in}}(x')]=i\delta_{\nu\lambda}D(x-x')\\&[A_\nu{}^{\text{in}}(x),\psi_\alpha{}^{\text{in}}(x')]=0\end{aligned}\right\}. \tag{7}$$

To see that Eqs. (7) hold, we notice that at $t=-\infty$ the fields ψ^{in}, $\bar{\psi}^{\text{in}}$, and $A_\nu{}^{\text{in}}$ (and their derivatives) are identical with the true fields ψ, $\bar{\psi}$, and A_ν (and their derivatives) which in turn satisfy the free-field commutation relations at $t=-\infty$. Moreover, ψ^{in}, $\bar{\psi}^{\text{in}}$, and $A_\nu{}^{\text{in}}$ develop with time according to the free-field equations so that Eqs. (7) follow as immediate consequences.

It is evident that, besides Eqs. (6), (6′), and (4), one can write down another set of solutions of the Heisenberg field equations which are expressed in terms of freely oscillating outgoing fields and advanced potentials, i.e.,

$$\left.\begin{aligned}&\psi(x)=\psi^{\text{out}}(x)-ie\int S^{\text{adv}}(x-x')d^4x'\gamma_\nu A_\nu(x')\psi(x')\\&\bar{\psi}(x)=\bar{\psi}^{\text{out}}(x)-ie\int \bar{\psi}(x')\gamma_\nu A_\nu(x')d^4x'S^{\text{ret}}(x'-x)\\&A_\nu(x)=A_\nu{}^{\text{out}}(x)+\int D^{\text{adv}}(x-x')d^4x'j_\nu(x')\end{aligned}\right\}. \tag{8}$$

Once again, the outgoing fields ψ^{out}, $\bar{\psi}^{\text{out}}$, and $A_\nu{}^{\text{out}}$ obey the interaction-free equations and the simple commutation relations

$$\left.\begin{aligned}&[\psi_\alpha{}^{\text{out}}(x),\bar{\psi}_\beta{}^{\text{out}}(x')]_+=-iS_{\alpha\beta}(x-x')\\&[A_\nu{}^{\text{out}}(x),A_\lambda{}^{\text{out}}(x')]=i\delta_{\nu\lambda}D(x-x')\\&[A_\nu{}^{\text{out}}(x),\psi_\alpha{}^{\text{out}}(x')]=0\end{aligned}\right\}. \tag{9}$$

Physically, the meaning of the incoming fields is clear. They coincide at $t=-\infty$ with the Heisenberg fields and would represent the development of the Heisenberg fields with time if the interaction were absent. A similar meaning holds for the outgoing fields except that they reduce to the true Heisenberg variables at $t=+\infty$. Since both the incoming and outgoing fields satisfy the identical commutation relations (7) and (9), we conclude that they are related by a unitary transformation in the following way:

$$\left.\begin{aligned}&\psi^{\text{out}}(x)=\mathbf{S}^{-1}\psi^{\text{in}}(x)\mathbf{S}\\&\bar{\psi}^{\text{out}}(x)=\mathbf{S}^{-1}\bar{\psi}^{\text{in}}(x)\mathbf{S}\\&A_\nu{}^{\text{out}}(x)=\mathbf{S}^{-1}A_\nu{}^{\text{in}}(x)\mathbf{S}\end{aligned}\right\}. \tag{10}$$

---

[7] We use natural units throughout with $\hbar=c=1$. The notation A^T denotes the transpose of the matrix A.

[8] These equations have also been obtained by G. Källén. The authors wish to thank Dr. Källén for sending them his manuscript before publication.

The unitary matrix **S** is the **S**-matrix of Heisenberg. It is uniquely determined by Eqs. (10) except for an arbitrary multiplicative phase factor. We shall defer until Section II the proof that the **S**-matrix thus defined is indeed identical with the **S**-matrix which is dealt with in the more customary interaction-representation formulation.

In order to write down explicitly the matrix elements of **S**, it is possible to proceed by solving Eqs. (6), (6′), and (4) by successive approximations in powers of e and then using Eqs. (8) and (10). In this way, general rules can be formulated for writing down the various terms to any order in e but, for practical computations, they are usually more complicated than Feynman's rules.

B. The Electromagnetic Properties of Charged Mesons in the β-Formalism

The **S**-matrix formalism may be readily extended to this case in the following way. The equations of motion in the Heisenberg representation now read

$$\left.\begin{aligned}[\beta_\nu(\partial/\partial x_\nu)+\mu]u&=ie\beta_\nu A_\nu u\\ (\partial u^\dagger/\partial x_\nu)\beta_\nu-\mu u^\dagger&=-ieu^\dagger\beta_\nu A_\nu\\ \partial^2 A_\nu/\partial x_\lambda^2&=-j_\nu\end{aligned}\right\}, \quad (11)$$

where

$$j_\nu=\tfrac{1}{2}ie(u^\dagger\beta_\nu u+u^T\beta_\nu{}^T u^{\dagger T}); \quad (11')$$

we follow here the notation used by Pauli.[9] Besides Eqs. (11), there is also the usual supplementary condition given by Eq. (3). As before, one may replace the differential equations (11) by corresponding integral equations with the aid of appropriate Green's functions. To determine these, we observe that

$$\left(\beta_\nu\frac{\partial}{\partial x_\nu}+\mu\right)\left[\beta_\lambda\frac{\partial}{\partial x_\lambda}-\frac{1}{\mu}\left(\beta_\lambda\frac{\partial}{\partial x_\lambda}\right)^2-\frac{1}{\mu}\left(\mu^2-\frac{\partial^2}{\partial x_\lambda{}^2}\right)\right]=\left(\frac{\partial^2}{\partial x_\nu{}^2}-\mu^2\right), \quad (12)$$

whence, defining $T^{\mathrm{ret}}(x-x')$ by

$$T^{\mathrm{ret}}(x-x')=\left[\beta_\nu\frac{\partial}{\partial x_\nu}-\frac{1}{\mu}\left(\beta_\nu\frac{\partial}{\partial x_\nu}\right)^2-\frac{1}{\mu}\left(\mu^2-\frac{\partial^2}{\partial x_\nu{}^2}\right)\right]\Delta^{\mathrm{ret}}(x-x'), \quad (13)$$

it follows that

$$[\beta_\nu(\partial/\partial x_\nu)+\mu]T^{\mathrm{ret}}(x-x')=-\delta(x-x'). \quad (14)$$

It should be noticed that, in virtue of the properties of the Δ^{ret}-function, $T^{\mathrm{ret}}(x-x')$ vanishes if x lies outside the future light cone of x'. One may also define the T^{adv} and T_F functions by replacing Δ^{ret} in (13) by Δ^{adv} and Δ_F, respectively.[10]

The solutions of (11) may now be written in the following integral form:

$$\left.\begin{aligned}u(x)&=u^{\mathrm{in}}(x)-ie\int T^{\mathrm{ret}}(x-x')d^4x'\beta_\nu A_\nu(x')u(x')\\ u^\dagger(x)&=u^{\dagger\mathrm{in}}(x)-ie\int u^\dagger(x')\beta_\nu\times A_\nu(x')d^4x'T^{\mathrm{adv}}(x'-x)\\ A_\nu(x)&=A_\nu{}^{\mathrm{in}}(x)+\int D^{\mathrm{ret}}(x-x')d^4x'j_\nu(x')\end{aligned}\right\}. \quad (15)$$

It is once again essential to note that the incoming fields u^{in}, $u^{\dagger\mathrm{in}}$ and $A_\nu{}^{\mathrm{in}}$ are to be regarded as defined by Eqs. (15). They not only satisfy the free-field equations but also, by an argument similar to that used in our earlier discussion of quantum electrodynamics, the free-field commutation relations, *viz.*,

$$\left.\begin{aligned}[u_\alpha{}^{\mathrm{in}}(x),u_\beta{}^\dagger(x')]&=-i\left[\beta_\nu\frac{\partial}{\partial x_\nu}-\frac{1}{\mu}\left(\beta_\nu\frac{\partial}{\partial x_\nu}\right)^2\right]_{\alpha\beta}\Delta(x-x')\\ [A_\nu{}^{\mathrm{in}}(x),A_\lambda{}^{\mathrm{in}}(x')]&=i\delta_{\nu\lambda}D(x-x')\\ [A_\nu{}^{\mathrm{in}}(x),u_\alpha{}^{\mathrm{in}}(x')]&=0\end{aligned}\right\}; \quad (16)$$

the supplementary condition reduces as before to Eq. (3′).

One can write down, besides Eqs. (15), another set of integral solutions of (11):

$$\left.\begin{aligned}u(x)&=u^{\mathrm{out}}(x)-ie\int T^{\mathrm{adv}}(x-x')d^4x'\beta_\nu A_\nu(x')u(x')\\ u^\dagger(x)&=u^{\dagger\mathrm{out}}(x)-ie\int u^\dagger(x')\beta_\nu\times A_\nu(x')d^4x'T^{\mathrm{ret}}(x'-x)\\ A_\nu(x)&=A_\nu{}^{\mathrm{out}}(x)+\int D^{\mathrm{adv}}(x-x')d^4x'j_\nu(x')\end{aligned}\right\}. \quad (17)$$

The outgoing fields u^{out}, $u^{\dagger\mathrm{out}}$, and $A_\nu{}^{\mathrm{out}}$ obey the free-field equations and commutation relations. It therefore follows that there exists a unitary transformation con-

[9] W. Pauli, Rev. Mod. Phys. **13**, 203 (1941).

[10] The Δ_F-function which we use is identical with Dyson's. We shall later also use the S_F function which is defined by

$$S_F(x-x')=[\gamma_\lambda(\partial/\partial x_\lambda)-m]\Delta_F(x-x').$$

necting the incoming and outgoing quantities, *viz.*,

$$\left.\begin{aligned} u^{\text{out}}(x) &= \mathbf{S}^{-1}u^{\text{in}}(x)\mathbf{S} \\ u^{\dagger\text{out}}(x) &= \mathbf{S}^{-1}u^{\dagger\text{in}}(x)\mathbf{S} \\ A_\nu^{\text{out}}(x) &= \mathbf{S}^{-1}A_\nu^{\text{in}}(x)\mathbf{S} \end{aligned}\right\}, \quad (18)$$

where **S** is once again the **S**-matrix of Heisenberg.

One can calculate the elements of the **S**-matrix from Eqs. (15) to (18) following a procedure which is identical with that described for the case of quantum electrodynamics. It is therefore clear that at no point in the evaluation of **S** is it necessary to resort to the concept of space-like surfaces and their normals. It is an essential advantage of the formulation of the **S**-matrix in the Heisenberg representation that the extraneous complications associated with space-like surfaces do not enter.[11]

A further advantage of this procedure is that it becomes possible to formulate rules for writing down the elements of the **S**-matrix which are the exact analogues of the Feynman-Dyson rules of electrodynamics. In fact, it is only necessary to replace the S_F function[10] of electrodynamics by the T_F function, and the γ_ν-matrices by β_ν-matrices; the sign of the term corresponding to any Feynman diagram differs in the β-formalism from quantum electrodynamics by a factor $(-1)^l$ where l is the number of closed meson loops.[12]

The proof is based upon the fact that, while the Tomonaga-Schwinger equations for the various cases which are encountered in meson theory assume a more or less complicated form according to whether one does or does not have derivative coupling or dynamically dependent field components, the equations defining the **S**-matrix in the Heisenberg representation are of a comparatively simple nature and, in fact, are very similar to one another in form. This last fact enables one to infer immediately the above-mentioned rules for the case of the β-formalism from the corresponding Feynman-Dyson rules for electrodynamics. The factor $(-1)^l$ is a complication which arises due to the fact that the particles with which we are dealing obey Bose statistics.[13]

We wish finally to remark that all of these considerations (*viz.*, the definition of the **S**-matrix in the Heisenberg representation and its subsequent characterization by a set of Feynman-like rules for writing down the various matrix elements) may be directly extended to the various cases of meson-nucleon couplings.

II. IDENTIFICATION OF THE S-MATRIX

A. Quantum Electrodynamics

It remains to verify that the **S**-matrix as defined in the Heisenberg representation is indeed the same as that which emerges from the Tomonaga-Schwinger theory. We shall see that, by generalizing suitably the concepts of incoming and outgoing fields which were introduced earlier, we are led directly to a product representation for the **S**-matrix which is, in fact, Dyson's representation.

In order to go to over to the interaction representation, let us introduce a set of space-like surfaces $\sigma(x)$ and denote their normals by $n_\nu(x)$. Define the function $D^\sigma(x, x')$, where x and x' are not necessarily on the surface σ, in the following way:

$$D^\sigma(x, x') = D^{\text{ret}}(x-x') \quad \text{if } x' \text{ is later than } \sigma,$$

$$D^\sigma(x, x') = D^{\text{adv}}(x-x') \quad \text{if } x' \text{ is earlier than } \sigma,$$

that is,

$$D^\sigma(x, x') = \frac{1-\epsilon(\sigma, x')}{2} D^{\text{ret}}(x-x') + \frac{1+\epsilon(\sigma, x')}{2} D^{\text{adv}}(x-x'), \quad (19)$$

where

$$\epsilon(\sigma, x') = +1 \quad \text{if } \sigma \text{ is later than } x',$$
$$\epsilon(\sigma, x') = -1 \quad \text{if } \sigma \text{ is earlier than } x'.$$

It is clear that an equivalent representation of $D^\sigma(x, x')$ is given by

$$D^\sigma(x, x') = -\tfrac{1}{2}[\epsilon(x-x') - \epsilon(\sigma, x')]D(x-x'), \quad (19')$$

where $\epsilon(x-x')$ equals $+1$ or -1 according as x is later or earlier than x'.

Let us also define

$$S^\sigma(x, x') = [\gamma_\nu(\partial/\partial x_\nu) - m]\Delta^\sigma(x, x'). \quad (19'')$$

Then, with aid of the generalized Green's functions $D^\sigma(x, x')$ and $S^\sigma(x, x')$, Eqs. (1) and (2) may be integrated to give the following:

$$\left.\begin{aligned} \psi(x) &= \psi(x, \sigma) - ie \int S^\sigma(x, x')d^4x'\gamma_\nu A_\nu(x')\psi(x') \\ A_\nu(x) &= A_\nu(x, \sigma) + \int D^\sigma(x, x')d^4x' j_\nu(x') \end{aligned}\right\}. \quad (20)$$

The quantities $\psi(x, \sigma)$ and $A_\nu(x, \sigma)$ are to be regarded as defined by Eqs. (20). For fixed σ, they satisfy the free-field equations and, when x is on σ, they reduce to the Heisenberg fields $\psi(x)$ and $A_\nu(x)$; this reduction holds also for $\partial A_\nu(x, \sigma)/\partial x_\lambda$.

[11] For a discussion of charged meson theories in the interaction representation with the β-formalism see M. Neuman and W. H. Furry, Phys. Rev. **76**, 1677 (1949); R. G. Moorhouse, Phys. Rev. **76**, 1691 (1949); D. C. Peaslee, Phys. Rev. (to be published); and T. Kinoshita, Prog. Theor. Phys. (to be published).

[12] Similar results have been noted by R. P. Feynman using his method of space-time approach to field theory. Cf. footnote 24, Phys. Rev. **76**, 769 (1949).

[13] The easiest way to see this is to derive first the rules for the case of two scalar fields φ_1 and φ_2 coupled by the term $\varphi_1^*\varphi_1\varphi_2$. This can be done by Dyson's method just as easily as in quantum electrodynamics, since there are no complications due to surfaces and their normals. One has then only to compare the equations defining the **S**-matrix in this case with Eqs. (15) to (18) to arrive at the rules for the β formalism.

It will be convenient for what follows to introduce the symbols $\psi(x/\sigma)$ and $\partial\psi(x/\sigma)/\partial x_\nu$ by

$$\psi(x/\sigma)=[\psi(x,\sigma)]_{x\text{ on }\sigma},$$
$$\partial\psi(x/\sigma)/\partial x_\nu=[\partial\psi(x,\sigma)/\partial x_\nu]_{x\text{ on }\sigma}.$$

With this notation, one then has

$$\psi(x/\sigma)=\psi(x),\quad A_\nu(x/\sigma)=A_\nu(x),$$
$$\partial A_\nu(x/\sigma)/\partial x_\lambda=\partial A_\nu(x)/\partial x_\lambda.$$

Note that, as $\sigma\to-\infty$, $D^\sigma(x,x')\to D^{\text{ret}}(x-x')$ so that $A_\nu(x,\sigma)$ goes over into the incoming field $A_\nu^{\text{in}}(x)$. Correspondingly, as $\sigma\to+\infty$, $D^\sigma(x,x')\to D^{\text{adv}}(x-x')$ so that $A_\nu(x,\sigma)$ goes over into the outgoing field $A_\nu^{\text{out}}(x)$.[14]

In contrast to Eq. (3′), the supplementary condition becomes in terms of $A_\nu(x,\sigma)$

$$\left[\frac{\partial A_\nu(x,\sigma)}{\partial x_\nu}-\int_\sigma D(x-x')j_\nu(x')d\sigma_\nu'\right]\Phi=0. \quad (21)$$

The commutation relations of the $\psi(x,\sigma)$ and $A_\nu(x,\sigma)$ are the usual free-field expressions:

$$\left.\begin{aligned}&[\psi_\alpha(x,\sigma),\bar\psi_\beta(x',\sigma)]_+=-iS_{\alpha\beta}(x-x')\\&[A_\nu(x,\sigma),A_\lambda(x',\sigma)]=i\delta_{\nu\lambda}D(x-x')\\&[A_\nu(x,\sigma),\psi_\alpha(x',\sigma)]=0\end{aligned}\right\}. \quad (22)$$

These are obviously correct for x and x' on σ and, since $\psi(x,\sigma)$ and $A_\nu(x,\sigma)$ satisfy free-field equations, they are true in general.

Now, a set of Eqs. (22) holds for every surface σ whence it follows that there exists a unitary transformation $U(\sigma,\sigma')$, such that

$$\left.\begin{aligned}\psi(x,\sigma)&=U^{-1}(\sigma,\sigma')\psi(x,\sigma')U(\sigma,\sigma')\\A_\nu(x,\sigma)&=U^{-1}(\sigma,\sigma')A_\nu(x,\sigma')U(\sigma,\sigma')\end{aligned}\right\}. \quad (23)$$

It is clear that $U(\infty,-\infty)$ is the **S**-matrix which we have defined earlier.

We proceed next to obtain an explicit representation for the **S**-matrix. To do this, we note that the following relation is valid:

$$U(\sigma,\sigma')=U(\sigma'',\sigma')U(\sigma,\sigma''). \quad (24)$$

It is therefore natural to write

$$\begin{aligned}\mathbf{S}&=U(\infty,-\infty)\\&=\cdots U(\sigma_0,\sigma_{-1})U(\sigma_1,\sigma_0)U(\sigma_2,\sigma_1)\cdots,\end{aligned} \quad (25)$$

where $\cdots\sigma_1,\sigma_0,\sigma_{-1},\cdots$ denote an infinite sequence of space-like surfaces which proceed steadily into the past. To obtain an explicit expression for **S**, we let the surfaces approach one another so that (25) expresses **S** as the product of infinitesimal unitary transformations. It is therefore sufficient to obtain $U(\sigma,\sigma')$ to the first order as $\sigma\to\sigma'$. We accordingly write

$$U(\sigma,\sigma')=1-i\int_{\sigma'}^{\sigma}H(x'/\sigma)d^4x'+\cdots, \quad (26)$$

where

$$H(x'/\sigma)=i[\delta U(\sigma,\sigma')/\delta\sigma(x')]_{\sigma=\sigma'}. \quad (26')$$

Strictly speaking, we should have used the notation $H(x',\sigma)$ in place of $H(x'/\sigma)$ in Eq. (26); however, the procedure which we have followed is not inconsistent since, in the end, σ is made to approach σ'.

Substituting (26) into (25), we get

$$\mathbf{S}=\cdots\left[1-i\int_{\sigma_{-1}}^{\sigma_0}H(x/\sigma)d^4x\right]\times\left[1-i\int_{\sigma_0}^{\sigma_1}H(x/\sigma)d^4x\right]\cdots. \quad (25')$$

To find $H(x/\sigma)$, we differentiate (23) with respect to σ and set $\sigma=\sigma'$; then,

$$\left.\begin{aligned}i\frac{\delta\psi(x,\sigma)}{\delta\sigma(x')}&=[\psi(x,\sigma),H(x'/\sigma)]\\i\frac{\delta A_\nu(x,\sigma)}{\delta\sigma(x')}&=[A_\nu(x,\sigma),H(x'/\sigma)]\end{aligned}\right\}. \quad (27)$$

Now, from (20), we have

$$\begin{aligned}\psi(x,\sigma)-\psi(x,\sigma')&=ie\int[S^\sigma(x,x')-S^{\sigma'}(x,x')]\\&\qquad\times d^4x'\gamma_\nu A_\nu(x')\psi(x')\\&=ie\int_{\sigma'}^{\sigma}S(x-x')d^4x'\gamma_\nu A_\nu(x')\psi(x'),\end{aligned} \quad (28)$$

$$A_\nu(x,\sigma)-A_\nu(x,\sigma')=-\int_{\sigma'}^{\sigma}D(x-x')d^4x'j_\nu(x'). \quad (28')$$

But, $\delta\psi(x,\sigma)/\delta\sigma(x')$ and $\delta A_\nu(x,\sigma)/\delta\sigma(x')$ can be determined directly from (28) and (28′) whence

$$\left.\begin{aligned}&[\psi(x,\sigma),H(x'/\sigma)]=-eS(x-x')\gamma_\nu\\&\qquad\qquad\times A_\nu(x'/\sigma)\psi(x'/\sigma)\\&[A_\nu(x,\sigma),H(x'/\sigma)]=-iD(x-x')j_\nu(x'/\sigma)\end{aligned}\right\}. \quad (29)$$

It is at once clear from (29) and (22) that

$$H(x'/\sigma)=-j_\nu(x'/\sigma)A_\nu(x'/\sigma), \quad (30)$$

so that the **S**-matrix assumes the form

$$\mathbf{S}=\cdots\left(1+i\int_{\sigma_{-1}}^{\sigma_0}j_\nu(x/\sigma)A_\nu(x/\sigma)d^4x\right)\times\left(1+i\int_{\sigma_0}^{\sigma_1}j_\nu(x/\sigma)A_\nu(x/\sigma)d^4x\right)\cdots. \quad (31)$$

[14] We are indebted to Dr. R. J. Glauber for pointing out this generalization of the incoming and outgoing fields and their relation to the Hamiltonian in the interaction representation. Cf. also M. Neuman and W. H. Furry, reference 11; K. V. Roberts, Phys. Rev. **77**, 146 (1950); J. S. de Wet, reference 6.

To complete the identification with Dyson's result, we note that Eqs. (23) imply

$$\left.\begin{aligned}\psi(x/\sigma)&=U^{-1}(\sigma,-\infty)\psi^{\text{in}}(x)U(\sigma,-\infty)\\ A_\nu(x/\sigma)&=U^{-1}(\sigma,-\infty)A_\nu^{\text{in}}(x)U(\sigma,-\infty)\end{aligned}\right\}\quad(32)$$

or

$$\left.\begin{aligned}U(\sigma,-\infty)\psi(x/\sigma)&=\psi^{\text{in}}(x)U(\sigma,-\infty)\\ U(\sigma,-\infty)A_\nu(x/\sigma)&=A_\nu^{\text{in}}(x)U(\sigma,-\infty)\end{aligned}\right\}.\quad(32')$$

It is clear from (32′) that pulling a factor $\psi(x/\sigma)$ [or $A_\nu(x/\sigma)$] through from the right to the left of $U(\sigma,-\infty)$ converts it into a $\psi^{\text{in}}(x)$ [or $A_\nu^{\text{in}}(x)$]. By applying this procedure successively to all the factors of Eq. (31) taken in the order of decreasing time, we find

$$\mathbf{S}=\cdots\left(1+i\int_{\sigma_0}^{\sigma_1} j_\nu^{\text{in}}(x)A_\nu^{\text{in}}(x)d^4x\right)\times\left(1+i\int_{\sigma_{-1}}^{\sigma_0} j_\nu^{\text{in}}(x)A_\nu^{\text{in}}(x)\right)\cdots,\quad(33)$$

which is identical with Dyson's expression if we take the field quantities in the interaction representation to be the incoming fields.

B. Neutral Vector Mesons with Vector Coupling

In the remainder of this paper, we shall consider briefly those complications which arise on making the transition to the interaction representation when the field components are not all dynamically independent or when one has derivative coupling. The case of neutral vector mesons interacting with nucleons through vector coupling will serve as an illustration of the former situation. The equations of motion of the field variables in the Heisenberg representation are

$$\left.\begin{aligned}[\gamma_\nu(\partial/\partial x_\nu)+M]\psi&=\tfrac{1}{2}if\gamma_\nu(A_\nu\psi+\psi A_\nu)\\ \partial A_\nu/\partial x_\nu&=0\\ (\partial^2/\partial x_\lambda^2-\mu^2)A_\nu&=-j_\nu\end{aligned}\right\},\quad(34)$$

where

$$j_\nu=\tfrac{1}{2}if(\bar\psi\gamma_\nu\psi-\psi^T\gamma_\nu{}^T\bar\psi^T).\quad(34')$$

Note that these equations have been put into a form which is invariant with respect to charge conjugation.

The solutions of (34) are

$$\left.\begin{aligned}\psi(x)&=\psi(x,\sigma)-\tfrac{1}{2}if\int S^\sigma(x,x')d^4x'\gamma_\nu\\ &\qquad\times[A_\nu(x')\psi(x')+\psi(x')A_\nu(x')]\\ A_\nu(x)&=A_\nu(x,\sigma)+\int\left(\delta_{\nu\lambda}-\frac{1}{\mu^2}\frac{\partial^2}{\partial x_\nu\partial x_\lambda}\right)\\ &\qquad\times\Delta^\sigma(x,x')d^4x'j_\lambda(x')\end{aligned}\right\}.\quad(35)$$

It is here necessary to include the term involving second derivatives in the Green's function in order to guarantee that

$$\partial A_\nu(x,\sigma)/\partial x_\nu=0.\quad(36)$$

It is important to note that, when x is on σ, $A_\nu(x,\sigma)$ and its derivatives do *not* reduce to $A_\nu(x)$ and its derivatives; in fact,

$$A_\nu(x/\sigma)=A_\nu(x)-(1/\mu^2)n_\nu n_\lambda j_\lambda(x).\quad(37)$$

On the other hand, it is the $A_\nu(x/\sigma)$ and not the $A_\nu(x)$ which obey the free-particle commutation relations; as a consequence, the following equations are valid:

$$\left.\begin{aligned}&[\psi_\alpha(x,\sigma),\bar\psi_\beta(x',\sigma)]_+=-iS_{\alpha\beta}(x-x')\\ &[A_\nu(x,\sigma),A_\lambda(x',\sigma)]=i\left(\delta_{\nu\lambda}-\frac{1}{\mu^2}\frac{\partial^2}{\partial x_\nu\partial x_\lambda}\right)\\ &\qquad\qquad\times\Delta(x-x')\\ &[A_\nu(x,\sigma),\psi_\alpha(x',\sigma)]=0\end{aligned}\right\}.\quad(38)$$

It is therefore evident from (36) and (38) that, for any two surfaces σ and σ', there exists a unitary transformation $U(\sigma,\sigma')$ such that

$$\left.\begin{aligned}\psi(x,\sigma)&=U^{-1}(\sigma,\sigma')\psi(x,\sigma')U(\sigma,\sigma')\\ A_\nu(x,\sigma)&=U^{-1}(\sigma,\sigma')A_\nu(x,\sigma')U(\sigma,\sigma')\end{aligned}\right\}.\quad(39)$$

We now take over Eqs. (26), (26′), and (25′) from the preceding case and proceed to calculate $H(x'/\sigma)$. The method is exactly the same as before; in place of Eqs. (28) and (28′), however, we have

$$\psi(x,\sigma)-\psi(x,\sigma')=\tfrac{1}{2}if\int_{\sigma'}^{\sigma}S(x-x')d^4x'\gamma_\nu\times[A_\nu(x')\psi(x')+\psi(x')A_\nu(x')],\quad(40)$$

$$A_\nu(x,\sigma)-A_\nu(x,\sigma')=-\left(\delta_{\nu\lambda}-\frac{1}{\mu^2}\frac{\partial^2}{\partial x_\nu\partial x_\lambda}\right)\times\int_{\sigma'}^{\sigma}\Delta(x-x')d^4x'j_\lambda(x').\quad(40')$$

One finds ultimately that

$$[\psi(x,\sigma),H(x'/\sigma)]=-\tfrac{1}{2}fS(x-x')\gamma_\nu\times[A_\nu(x')\psi(x')+\psi(x')A_\nu(x')],\quad(41)$$

$$[A_\nu(x,\sigma),H(x'/\sigma)]=-i\left(\delta_{\nu\lambda}-\frac{1}{\mu^2}\frac{\partial^2}{\partial x_\nu\partial x_\lambda}\right)\times\Delta(x-x')j_\lambda(x').\quad(41')$$

Equation (41) may be rewritten in the following form with aid of (37):

$$\begin{aligned}&[\psi(x,\sigma),H(x'/\sigma)]\\ &\quad=-fS(x-x')\gamma_\nu A_\nu(x'/\sigma)\psi(x'/\sigma)\\ &\qquad-(f/2\mu)n_\nu(x')n_\lambda(x')S(x-x')\gamma_\nu\\ &\qquad\times[j_\lambda(x'/\sigma)\psi(x'/\sigma)+\psi(x'/\sigma)j_\lambda(x'/\sigma)].\quad(41'')\end{aligned}$$

One must therefore have

$$H(x'/\sigma)=-j_\nu(x'/\sigma)A_\nu(x'/\sigma) -(1/2\mu^2)[n_\nu(x')j_\nu(x'/\sigma)]^2. \quad (42)$$

It is evident from the discussions of the previous case that the **S**-matrix is finally given by

$$\mathbf{S}=\cdots\left[1-i\int_{\sigma_0}^{\sigma_1} H^{\mathrm{in}}(x/\sigma)d^4x\right] \times\left[1-i\int_{\sigma_{-1}}^{\sigma_0} H^{\mathrm{in}}(x/\sigma)d^4x\right]\cdots, \quad (43)$$

where[15]

$$H^{\mathrm{in}}(x/\sigma)=-j_\nu{}^{\mathrm{in}}(x)A_\nu{}^{\mathrm{in}}(x) -(1/2\mu^2)[n_\nu(x)j_\nu{}^{\mathrm{in}}(x)]^2. \quad (44)$$

C. Pseudoscalar Mesons with Pseudovector Coupling

As a last example, let us consider a situation where we have derivative coupling, *viz.*, the case of pseudoscalar mesons interacting with nucleons through pseudovector coupling. The equations of motion have the form

$$\left.\begin{aligned}\left(\gamma_\nu\frac{\partial}{\partial x_\nu}+M\right)\psi&=\tfrac{1}{2}ig\gamma_5\gamma_\nu\left(\psi\frac{\partial\varphi}{\partial x_\nu}+\frac{\partial\varphi}{\partial x_\nu}\psi\right)\\ \left(\frac{\partial^2}{\partial x_\nu{}^2}-\mu^2\right)\varphi&=\frac{\partial j_\nu}{\partial x_\nu}\end{aligned}\right\}, \quad (45)$$

where

$$j_\nu=\tfrac{1}{2}ig(\bar\psi\gamma_5\gamma_\nu\psi-\psi^T\gamma_\nu{}^T\gamma_5{}^T\bar\psi^T). \quad (45')$$

The solutions of (45) are

$$\psi(x)=\psi(x,\sigma)-\tfrac{1}{2}ig\int S^\sigma(x,x')d^4x'\gamma_5\gamma_\nu \times\left[\psi(x')\frac{\partial\varphi(x')}{\partial x_\nu'}+\frac{\partial\varphi(x')}{\partial x_\nu'}\psi(x')\right], \quad (46)$$

$$\varphi(x)=\varphi'(x,\sigma)-\int\Delta^\sigma(x,x')d^4x'\frac{\partial j_\nu(x')}{\partial x_\nu'}. \quad (46')$$

Actually, it is more appropriate to modify the Green's function in (46′) so as to give

$$\varphi(x)=\varphi(x,\sigma)-\int\frac{\partial\Delta^\sigma(x,x')}{\partial x_\nu}d^4x'j_\nu(x'). \quad (46'')$$

By choosing the Green's function in this way, we have arranged for $\varphi(x,\sigma)$ to satisfy the following boundary conditions:

$$\left.\begin{aligned}\varphi(x/\sigma)&=\varphi(x)\\ \partial\varphi(x/\sigma)/\partial x_\nu&=\partial\varphi(x)/\partial x_\nu+n_\nu n_\lambda j_\lambda(x)\end{aligned}\right\}. \quad (47)$$

The term involving the normals is precisely what is needed in order that the commutation relations of the $\psi(x,\sigma)$ and $\varphi(x,\sigma)$ shall reduce to those of the free-field quantities, i.e.,

$$\left.\begin{aligned}[\psi_\alpha(x,\sigma),\bar\psi_\beta(x',\sigma)]_+&=-iS_{\alpha\beta}(x-x')\\ [\varphi(x,\sigma),\varphi(x',\sigma)]&=i\Delta(x-x')\\ [\varphi(x,\sigma),\psi_\alpha(x',\sigma)]&=0\end{aligned}\right\}. \quad (48)$$

From this point on, things go exactly as before. One is led to

$$[\psi(x,\sigma),H(x'/\sigma)]=-\tfrac{1}{2}gS(x-x')\gamma_5\gamma_\nu \times\left[\psi(x')\frac{\partial\varphi(x')}{\partial x_\nu'}+\frac{\partial\varphi(x')}{\partial x_\nu'}\psi(x')\right], \quad (49)$$

$$[\varphi(x,\sigma),H(x'/\sigma)]=i\frac{\partial\Delta(x-x')}{\partial x_\nu}j_\nu(x'). \quad (49')$$

We rewrite Eq. (49) using (47):

$$\begin{aligned}[\psi(x,\sigma),H(x'/\sigma)]&=-gS(x-x')\gamma_5\gamma_\nu\psi(x'/\sigma)[\partial\varphi(x'/\sigma)/\partial x_\nu']\\ &\quad+\tfrac{1}{2}gn_\nu(x')n_\lambda(x')S(x-x')\gamma_5\gamma_\nu\\ &\quad\times[\psi(x'/\sigma)j_\lambda(x'/\sigma)+j_\lambda(x'/\sigma)\psi(x'/\sigma)]. \quad (49'')\end{aligned}$$

One has therefore

$$H(x'/\sigma)=-j_\nu(x'/\sigma)\frac{\partial\varphi(x'/\sigma)}{\partial x_\nu'}+\tfrac{1}{2}[n_\nu(x')j_\nu(x'/\sigma)]^2. \quad (50)$$

Evidently Eq. (43) can be taken over for the present case with

$$H^{\mathrm{in}}(x/\sigma)=-j_\nu{}^{\mathrm{in}}[\partial\varphi^{\mathrm{in}}(x')/\partial x_\nu'] +\tfrac{1}{2}[n_\nu(x')j_\nu{}^{\mathrm{in}}(x')]^2. \quad (50')$$

We wish to thank Professors J. R. Oppenheimer and W. Pauli for helpful discussions.

[15] Cf. Y. Miyamoto, Prog. Theor. Phys. 3, 124 (1948).

[[52a]] Commentary begins page 11

The Spontaneous Magnetization of a Two-Dimensional Ising Model

C. N. YANG
Institute for Advanced Study, Princeton, New Jersey

(Received September 18, 1951)

The spontaneous magnetization of a two-dimensional Ising model is calculated exactly. The result also gives the long-range order in the lattice.

IT is the purpose of the present paper to calculate the spontaneous magnetization (i.e., the intensity of magnetization at zero external field) of a two-dimensional Ising model of a ferromagnet. Van der Waerden[1] and Ashkin and Lamb[2] had obtained a series expansion of the spontaneous magnetization that converges very rapidly at low temperatures. Near the critical temperature, however, their series expansion cannot be used. We shall here obtain a closed expression for the spontaneous magnetization by the matrix method which was introduced into the problem of the statistics of a two-dimensional Ising model by Montroll[3] and Kramers and Wannier.[4] Onsager gave in 1944 a complete solution[5] of the matrix problem. His method was subsequently greatly simplified by Kaufman,[6] and the result has been used to calculate the short-range order in the crystal lattice.[7]

The Onsager-Kaufman solution of the matrix problem will be used in the present paper to calculate the spontaneous magnetization. In Sec. I we define the specific magnetization I and express it as an off diagonal element in the matrix problem. By introducing an artificial limiting process its calculation is reduced to an eigenvalue problem in Sec. II. This is solved in the next three sections and the final result given in Sec. VI. The relation between I and the usual long-range order is discussed in Sec. I.

It will be seen that the final expression for the spontaneous magnetization is surprisingly simple, although the intermediate steps are very complicated. Attempts to find a simpler way to arrive at the same result have, however, failed.

I. SPONTANEOUS MAGNETIZATION

Using Kaufman's notation[6] we have for the two-dimensional square lattice the following expression for the partition function:

$$Z=(2\sinh 2H)^{n/2}\,\mathrm{trace}(\mathbf{V}_2\mathbf{V}_1)^m, \tag{1}$$

where

$$\mathbf{V}_1=\exp\{H^*\sum_1^n \mathbf{C}_r\}, \tag{2}$$

and

$$\mathbf{V}_2=\exp\{H\sum_1^n \mathbf{s}_r\mathbf{s}_{r+1}\}. \tag{3}$$

H^* and H are given by

$$e^{-2H}=\tanh H^*=\exp[-(1/kT)\{V_{\uparrow\downarrow}-V_{\uparrow\uparrow}\}]. \tag{4}$$

The following abbreviation will be useful:

$$x=e^{-2H}. \tag{5}$$

If a weak magnetic field is introduced the partition function becomes

$$Z_{\mathcal{H}}=(2\sinh 2H)^{n/2}\,\mathrm{trace}(\mathbf{V}_3\mathbf{V}_2\mathbf{V}_1)^m, \tag{6}$$

where

$$\mathbf{V}_3=\exp\{\mathcal{H}\sum_1^n \mathbf{s}_r\}. \tag{7}$$

For a large crystal only the eigenvector of $\mathbf{V}=\mathbf{V}_3\mathbf{V}_2\mathbf{V}_1$ with the largest eigenvalue is important. We shall be interested in the limiting form of this eigenvector as $\mathcal{H}\to 0$.

It has been shown by Onsager[5] that below the critical temperature, i.e., for

$$x<\sqrt{2}-1,$$

the largest eigenvalue of $\mathbf{V}_2\mathbf{V}_1$ is doubly degenerate. This is evidently also true of the symmetrized matrix $\mathbf{V}_1^{\frac{1}{2}}\mathbf{V}_2\mathbf{V}_1^{\frac{1}{2}}$. Let ψ_+ and ψ_- be the even and odd eigenvectors corresponding to the largest eigenvalue λ.

$$\mathbf{V}_1^{\frac{1}{2}}\mathbf{V}_2\mathbf{V}_1^{\frac{1}{2}}\psi_+=\lambda\psi_+,\quad \mathbf{V}_1^{\frac{1}{2}}\mathbf{V}_2\mathbf{V}_1^{\frac{1}{2}}\psi_-=\lambda\psi_-. \tag{8}$$

The even eigenvector remains unchanged when the spins of all atoms are reversed while the odd eigenvector changes sign. Introducing the operator

$$\mathbf{U}=\mathbf{C}_1\mathbf{C}_2\cdots\mathbf{C}_n,$$

that reverses the spins of all atoms we have

$$\mathbf{U}\psi_+=\psi_+,\quad \mathbf{U}\psi_-=-\psi_-. \tag{9}$$

With the introduction of the magnetic field $\mathcal{H}$ the degeneracy is removed. Since we are only interested in the limit as $\mathcal{H}\to 0$, we may perform a perturbation cal-

[1] B. L. van der Waerden, Z. Physik **118**, 473 (1941).
[2] J. Ashkin and W. E. Lamb, Jr., Phys. Rev. **64**, 159 (1943).
[3] E. Montroll, J. Chem. Phys. **9**, 706 (1941).
[4] H. A. Kramers and G. H. Wanner, Phys. Rev. **60**, 252, 263 (1941).
[5] L. Onsager, Phys. Rev. **65**, 117 (1944).
[6] B. Kaufman, Phys. Rev. **76**, 1232 (1949).
[7] B. Kaufman and L. Onsager, Phys. Rev. **76**, 1244 (1949).

□Reprinted from *The Physical Review* 85, 5 (March 1, 1952), 808–816.

culation and consider the largest eigenvalue of

$$\mathbf{V}_1^{\frac{1}{2}}\mathbf{V}\mathbf{V}_1^{-\frac{1}{2}}=\mathbf{V}_1^{\frac{1}{2}}\mathbf{V}_3\mathbf{V}_2\mathbf{V}_1^{\frac{1}{2}}$$
$$=\mathbf{V}_1^{\frac{1}{2}}\mathbf{V}_2\mathbf{V}_1^{\frac{1}{2}}+\mathcal{H}\mathbf{V}_1^{\frac{1}{2}}(\sum_1^n \mathbf{s}_r)\mathbf{V}_2\mathbf{V}_1^{\frac{1}{2}}. \quad (10)$$

The last term is a matrix that anticommutes with **U**. It has, therefore, no diagonal matrix element with respect to either ψ_+ or ψ_-. It is, besides, a real symmetrical matrix. Ordinary perturbation theory shows immediately that the eigenvector of (10) with the largest eigenvalue approaches, as $\mathcal{H}\to 0$

$$\psi_{\max}=(1/\sqrt{2})(\psi_++\psi_-), \quad (11)$$

if the phases of ψ_+ and ψ_- are so chosen that they are real and that[8]

$$\psi_+'\mathbf{V}_1^{\frac{1}{2}}(\sum_1^n \mathbf{s}_r)\mathbf{V}_2\mathbf{V}_1^{\frac{1}{2}}\psi_-\geqq 0. \quad (12)$$

The average magnetization per atom is, from the general definition of the matrix method,

$$I=\frac{1}{mn}\frac{m\,\mathrm{trace}(\mathbf{V}_3\mathbf{V}_2\mathbf{V}_1)^m\sum_1^n \mathbf{s}_r}{\mathrm{trace}(\mathbf{V}_3\mathbf{V}_2\mathbf{V}_1)^m}$$
$$=\frac{1}{n}\frac{\mathrm{trace}(\mathbf{V}_1^{\frac{1}{2}}\mathbf{V}_3\mathbf{V}_2\mathbf{V}_1^{\frac{1}{2}})^m(\mathbf{V}_1^{\frac{1}{2}}\sum_1^n \mathbf{s}_r\mathbf{V}_1^{-\frac{1}{2}})}{\mathrm{trace}(\mathbf{V}_1^{\frac{1}{2}}\mathbf{V}_3\mathbf{V}_2\mathbf{V}_1^{\frac{1}{2}})^m}$$
$$=\frac{1}{n}\psi_{\max}'\mathbf{V}_1^{\frac{1}{2}}(\sum_1^n \mathbf{s}_r)\mathbf{V}_1^{-\frac{1}{2}}\psi_{\max}.$$

As $\mathcal{H}\to 0$ this becomes by (11)

$$I=\frac{1}{2n}(\psi_+'+\psi_-')\mathbf{V}_1^{\frac{1}{2}}(\sum_1^n \mathbf{s}_r)\mathbf{V}_1^{-\frac{1}{2}}(\psi_++\psi_-).$$

But $\mathbf{V}_1^{\frac{1}{2}}(\sum \mathbf{s}_r)\mathbf{V}_1^{-\frac{1}{2}}$ anticommutes with **U**, and therefore has no diagonal matrix element with respect to either ψ_+ or ψ_-. Besides, by the use of (8), one shows easily that

$$\psi_-'\mathbf{V}_1^{\frac{1}{2}}(\sum_1^n \mathbf{s}_r)\mathbf{V}_1^{-\frac{1}{2}}\psi_+=\frac{1}{\lambda}\psi_-'\mathbf{V}_1^{\frac{1}{2}}(\sum_1^n \mathbf{s}_r)\mathbf{V}_2\mathbf{V}_1^{\frac{1}{2}}\psi_+,$$
$$\psi_+'\mathbf{V}_1^{\frac{1}{2}}(\sum_1^n \mathbf{s}_r)\mathbf{V}_1^{-\frac{1}{2}}\psi_-=\frac{1}{\lambda}\psi_+'\mathbf{V}_1^{\frac{1}{2}}(\sum_1^n \mathbf{s}_r)\mathbf{V}_2\mathbf{V}_1^{\frac{1}{2}}\psi_-, \quad (13)$$

which are obviously equal. Hence at zero magnetic field the spontaneous magnetization is

$$I=\frac{1}{n}\psi_-'\mathbf{V}_1^{\frac{1}{2}}(\sum_1^n \mathbf{s}_r)\mathbf{V}_1^{-\frac{1}{2}}\psi_+, \quad (14)$$

which is always positive by (13) and (12).

[8] We use the notation $A'\equiv A$ transposed.

Intuitively one would infer that the summation $\sum\mathbf{s}_r$ in (14) can be replaced by $n\mathbf{s}_1$ so that

$$I=\psi_-'\mathbf{V}_1^{\frac{1}{2}}\mathbf{s}_1\mathbf{V}_1^{-\frac{1}{2}}\psi_+. \quad (15)$$

This can also be shown in detail by introducing the orthogonal operator **L** that is equivalent to the cyclic permutation of the n spins:

$$\mathbf{L}\boldsymbol{\sigma}_i\mathbf{L}^{-1}=\boldsymbol{\sigma}_{i+1},\quad \mathbf{L}\boldsymbol{\sigma}_n\mathbf{L}^{-1}=\boldsymbol{\sigma}_1.$$

Evidently **L** commutes with $\mathbf{V}_1$, $\mathbf{V}_2$, and **U**. Therefore $\mathbf{L}\psi_+$ is also an even eigenvector of $\mathbf{V}_2\mathbf{V}_1$ with eigenvalue λ. Hence

$$\mathbf{L}\psi_+=a\psi_+.$$

L and ψ_+ are real. Therefore a is real. Since further $\mathbf{L}^n=1$, we have $a=1$, and

$$\mathbf{L}\psi_+=\psi_+. \quad (16)$$

Similarly

$$\mathbf{L}\psi_-=\psi_-.$$

Now

$$\mathbf{s}_r=\mathbf{L}^{(r-1)}\mathbf{s}_1\mathbf{L}^{-(r-1)}.$$

Substituting this into (14) and using (16) we obtain (15).

The spontaneous magnetization I per atom is exactly the usual long-range order parameter s which may be defined as the average of the absolute value of the total spin of the lattice divided by the number of atoms. That I is equal to s is easily seen from the fact that the introduction of a vanishingly weak positive magnetic field merely cuts out all states of the lattice for which the total spin is negative.

One may ask, as Zernike[9] did, what is the average value of the total spin of the lattice if it is known that at a given lattice point the spin is $+1$. We can show that the answer is NI^2 in the following way: The total spin is either $+NI$ or $-NI$. If a given lattice point has a spin $+1$, it assumes the former value more frequently than the latter in the ratio of $\frac{1}{2}(1+I):\frac{1}{2}(1-I)$. Hence the average total spin is

$$NI(1+I)/2-NI(1-I)/2=NI^2.$$

The long-distance order can also be investigated as the limit of the short-distance order which has been studied by Kaufman and Onsager. Onsager[10] has done this and obtained the correlation of the spins of two atoms in one row at an infinite distance from each other. It can be shown that the long-distance order can be obtained from this, and the result agrees with the findings of this paper.

II. REDUCTION TO EIGENVALUE PROBLEM

A.

To calculate the spontaneous magnetization as given by (15) we notice that it is the off-diagonal ele-

[9] F. Zernike, Physica **7**, 565 (1938).

[10] L. Onsager, unpublished; see also Nuovo cimento **6**, Suppl. p. 261 (1949). The author wishes to thank Bruria Kaufman for showing him her notes on Onsager's work.

ment of the matrix $\mathbf{V}_1^{\frac{1}{2}}\mathbf{s}_1\mathbf{V}_1^{-\frac{1}{2}}$ between the vectors ψ_+ and ψ_-. Onsager and Kaufman[7] have shown how to calculate diagonal elements by reducing the $2^n\times 2^n$ matrix problem to one of $2n\times 2n$. Their method, however, does not apply to off-diagonal elements. To resolve this difficulty we shall in the present section introduce an artificial limiting process and reduce the problem to an eigenvalue problem of an $n\times n$ matrix.

From Kaufman's[6] Eq. (60) we have, except for a multiplicative phase factor:

$$\psi_- = \mathbf{S}^{-1}(\mathbf{T}_-)\tau, \tag{17}$$

where

$$\tau = \mathbf{g}\begin{bmatrix}1\\0\\0\\\vdots\\0\end{bmatrix} = 2^{-n/2}\begin{bmatrix}1\\1\\\vdots\\1\end{bmatrix}. \tag{18}$$

Similarly

$$\psi_+ = \mathbf{S}^{-1}(\mathbf{T}_+)\tau.$$

Now since $\mathbf{T}_-$ is real it follows that $\mathbf{S}(\mathbf{T}_-)$ is unitary. Hence taking the complex conjugate transposed of Eq. (17) we obtain

$$\psi_-' = \tau'\mathbf{S}(\mathbf{T}_-).$$

The reality condition of ψ_- has been used. Eq. (15) therefore assumes the form

$$I = \tau'\mathbf{S}(\mathbf{T}_-)\mathbf{V}_1^{\frac{1}{2}}\mathbf{s}_1\mathbf{V}_1^{-\frac{1}{2}}\mathbf{S}^{-1}(\mathbf{T}_+)\tau. \tag{19}$$

As we have just mentioned, if the expression were of the form

$$\tau'\mathbf{S}(\mathbf{T}_-)\cdots\mathbf{S}^{-1}(\mathbf{T}_-)\tau,$$

it would have been easy to reduce because $\mathbf{S}(\mathbf{T}_-)\cdots \times\mathbf{S}^{-1}(\mathbf{T}_-)$ induces a rotation in the $2n$ dimensional space formed by the Γ's.[11] We could, however, in the present case still utilize this reduction by first writing

$$I = \mathrm{trace}\mathbf{V}_1^{\frac{1}{2}}\mathbf{s}_1\mathbf{V}_1^{-\frac{1}{2}}\mathbf{S}^{-1}(\mathbf{T}_+)\tau\tau'\mathbf{S}(\mathbf{T}_-). \tag{20}$$

Now

$$\tau\tau' = (1/2^n)(1+\mathbf{C}_1)(1+\mathbf{C}_2)\cdots(1+\mathbf{C}_n), \tag{21}$$

does not induce a rotation. But we notice that

$$\begin{aligned}1+\mathbf{C}_1 &= \lim_{a\to i\infty}(\cos a)^{-1}(\cos a - i\mathbf{C}_1\sin a)\\ &= \lim_{a\to i\infty}(\cos a)^{-1}\exp(-ia\mathbf{C}_1),\end{aligned} \tag{22}$$

and $\exp(-ia\mathbf{C}_1)$ does induce a rotation. Write

$$\mathbf{M} = \begin{bmatrix}\cos 2a & \sin 2a & & & 0\\ -\sin 2a & \cos 2a & & & \\ & & \cos 2a & \sin 2a & \\ & & -\sin 2a & \cos 2a & \\ 0 & & & & \ddots\end{bmatrix} \tag{23}$$

so that

$$\exp(-ia\sum_1^n \mathbf{C}_r)\mathbf{\Gamma}_\alpha\exp(ia\sum_1^n\mathbf{C}_r) = \sum_\beta M_{\beta\alpha}\mathbf{\Gamma}_\beta. \tag{24}$$

[11] The $\mathbf{\Gamma}$'s are defined in Kaufman's paper (see reference 6). There is a mistake of sign in her Eq. (11) which should read $\mathbf{\Gamma}_{2r}\equiv -\mathrm{C}\times\mathrm{C}\times\cdots\times is\mathrm{C}\times 1\times 1\times\cdots\equiv\mathbf{Q}_r$.

We have from (21) and (22)

$$\begin{aligned}\tau\tau' &= \lim_{a\to i\infty}(2\cos a)^{-n}\exp(-ia\sum_1^n\mathbf{C}_r)\\ &= \lim_{a\to i\infty}(2\cos a)^{-n}\mathbf{S}(\mathbf{M}).\end{aligned}$$

Substitution back into (20) gives

$$I = \lim_{a\to i\infty}(2\cos a)^{-n}\,\mathrm{trace}\mathbf{V}_1^{\frac{1}{2}}\mathbf{s}_1\mathbf{V}_1^{-\frac{1}{2}}\mathbf{S}(\mathbf{T}_+^{-1}\mathbf{M}\mathbf{T}_-). \tag{25}$$

B.

This can easily be calculated if we know the eigenvalues and eigenvectors of the $2n$-dimensional rotation $\mathbf{T}_+^{-1}\mathbf{M}\mathbf{T}_-$. The rotations $\mathbf{T}_+$ and $\mathbf{M}$ have determinants equal to 1 while $\mathbf{T}_-$ has a determinant equal to -1. Thus $\mathbf{T}_+^{-1}\mathbf{M}\mathbf{T}_-$ is an improper rotation and must have eigenvalues 1, -1, $e^{\pm i\theta_2}$, $e^{\pm i\theta_3}$, $\cdots e^{\pm i\theta_n}$. Let $\boldsymbol{\zeta}$ be an orthogonal matrix that transforms $\mathbf{T}_+^{-1}\mathbf{M}\mathbf{T}_-$ into the canonical form

$$\boldsymbol{\zeta}\mathbf{T}_+^{-1}\mathbf{M}\mathbf{T}_-\boldsymbol{\zeta}^{-1} = \begin{bmatrix}1 & & & & & & 0\\ & -1 & & & & & \\ & & \cos\theta_2 & \sin\theta_2 & & & \\ & & -\sin\theta_2 & \cos\theta_2 & & & \\ & & & & \cos\theta_3 & \sin\theta_3 & \\ & & & & -\sin\theta_3 & \cos\theta_3 & \\ 0 & & & & & & \ddots\end{bmatrix} = \mathbf{W}. \tag{26}$$

$\mathbf{W}$ is evidently orthogonal. We shall compute, first, instead of (25), the more general expression

$$\mathrm{trace}\mathbf{\Gamma}_j\mathbf{S}(\mathbf{T}_+^{-1}\mathbf{M}\mathbf{T}_-), \tag{27}$$

where $\mathbf{\Gamma}_j$ is as defined in Kaufman's paper.[11] By (26)

$$\begin{aligned}\mathrm{trace}\mathbf{\Gamma}_j\mathbf{S}(\mathbf{T}_+^{-1}\mathbf{M}\mathbf{T}_-) &= \mathrm{trace}\mathbf{\Gamma}_j\mathbf{S}(\boldsymbol{\zeta}^{-1})\mathbf{S}(\mathbf{W})\mathbf{S}(\boldsymbol{\zeta})\\ &= \mathrm{trace}\mathbf{S}(\boldsymbol{\zeta})\mathbf{\Gamma}_j\mathbf{S}(\boldsymbol{\zeta}^{-1})\mathbf{S}(\mathbf{W}).\end{aligned} \tag{28}$$

Now

$$\mathbf{S}(\boldsymbol{\zeta})\mathbf{\Gamma}_j\mathbf{S}(\boldsymbol{\zeta}^{-1}) = \sum \zeta_{\alpha j}\mathbf{\Gamma}_\alpha,$$

where $\zeta_{\alpha j}$ are the matrix elements of $\boldsymbol{\zeta}$. Moreover, the explicit form of $\mathbf{S}(\mathbf{W})$ is known:

$$\mathbf{S}(\mathbf{W}) = i\mathbf{P}_1(\mathbf{P}_2\mathbf{Q}_2)(\mathbf{P}_3\mathbf{Q}_3)\cdots(\mathbf{P}_n\mathbf{Q}_n)\exp(\tfrac{1}{2}\sum_2^n\theta_\beta\mathbf{P}_\beta\mathbf{Q}_\beta).$$

(28) therefore reduces to

$$\begin{aligned}&\mathrm{trace}\mathbf{\Gamma}_j\mathbf{S}(\mathbf{T}_+^{-1}\mathbf{M}\mathbf{T}_-)\\ &= i\,\mathrm{trace}(\sum\zeta_{\alpha j}\mathbf{\Gamma}_\alpha)\mathbf{P}_1(\prod_2^n\mathbf{P}_\alpha\mathbf{Q}_\alpha)\exp(\tfrac{1}{2}\sum\theta_\beta\mathbf{P}_\beta\mathbf{Q}_\beta)\\ &= i\zeta_{1j}\,\mathrm{trace}\prod_2^n\mathbf{P}_\alpha\mathbf{Q}_\alpha\exp(\tfrac{1}{2}\theta_\alpha\mathbf{P}_\alpha\mathbf{Q}_\alpha)\\ &= i(-1)^{n-1}2^n\zeta_{1j}\prod_2^n\sin(\theta_\alpha/2).\end{aligned} \tag{29}$$

Returning to (25) we notice that

$$\mathbf{V}_1^{\frac{1}{2}}\mathbf{s}_1\mathbf{V}_1^{-\frac{1}{2}} = \mathbf{P}_1\cosh H^* - i\mathbf{Q}_1\sinh H^*. \tag{30}$$

(25), (29), and (30) give

$$I=(\prod_2^n \lambda_\alpha)i(\xi_{11}\cosh H^*-i\xi_{12}\sinh H^*), \quad (31)$$

where

$$\lambda_\alpha=\lim_{a\to i\infty}(-\cos a)^{-1}\sin(\theta_\alpha/2), \quad (32)$$

and

$$\xi_{\alpha\beta}=\lim_{a\to i\infty}(\cos a)^{-1}\zeta_{\alpha\beta}. \quad (33)$$

C.

In this subsection we shall derive a formula for λ_α as the eigenvalue of an $n\times n$ matrix.

The matrices $\mathbf{T}_+$ and $\mathbf{T}_-$ are real, so that we can write

$$\mathbf{T}_+^{-1}\mathbf{M}\mathbf{T}_-=\tfrac{1}{2}\mathbf{G}\exp(-2ia)+\tfrac{1}{2}\mathbf{G}^*\exp(2ia), \quad (34)$$

where * means complex conjugate, and $\mathbf{G}$ is *independent* of a and is given by

$$\mathbf{G}=\mathbf{T}_+^{-1}\begin{bmatrix} 1 & i & & & 0\\ -i & 1 & & & \\ & & 1 & i & \\ & & -i & 1 & \\ 0 & & & & \ddots \end{bmatrix}\mathbf{T}_-. \quad (35)$$

Now in Eq. (34) the eigenvalues of the left-hand side are $1, -1, e^{\pm i\theta_2}, e^{\pm i\theta_3}, \cdots e^{\pm i\theta_n}$. As $a\to i\infty$, the second term of the right-hand side becomes negligible, and we see that

$$\lim_{a\to i\infty}2e^{2ia}e^{i\theta_\alpha}=l_\alpha,$$

where $l_2, l_3, \cdots l_n$ are the nonvanishing eigenvalues of $\mathbf{G}$. A relation between the l's and the λ's is found by squaring (32):

$$\lambda_\alpha^2=\lim_{a\to i\infty}(\cos a)^{-2}\sin^2(\theta_\alpha/2)=\lim_{a\to i\infty}4e^{2ia}\sin^2(\theta_\alpha/2)$$
$$=-\tfrac{1}{2}l_\alpha. \quad (36)$$

We therefore want to find the eigenvalues of the $2n\times 2n$ matrix $\mathbf{G}$ defined by (35). Now explicit matrix elements of $\mathbf{T}_+$ and $\mathbf{T}_-$ have been exhibited by Kaufman.[6] Using these matrix elements and rearranging the rows and columns of all $2n\times 2n$ matrices so that the order of the Γ's is changed into $\mathbf{P}_1, \mathbf{P}_2\cdots\mathbf{P}_n, \mathbf{Q}_1, \mathbf{Q}_2\cdots\mathbf{Q}_n$, we arrive at the following expression for $\mathbf{G}$:

$$\mathbf{G}=\begin{bmatrix}\mathbf{D}_+^{-1} & 0\\ 0 & -i\mathbf{D}_+^{-1}\end{bmatrix}\begin{bmatrix}\mathbf{p}_+^{-1}\\ \mathbf{p}_+\end{bmatrix}\begin{bmatrix}\mathbf{p}_- & \mathbf{p}_-^{-1}\end{bmatrix}\begin{bmatrix}\mathbf{D}_- & 0\\ 0 & i\mathbf{D}_-\end{bmatrix}, \quad (37)$$

where

$$\mathbf{D}_-=n^{-\frac{1}{2}}\begin{bmatrix}\epsilon^2 & \epsilon^4\cdots & \epsilon^{2n}\\ \epsilon^4 & \epsilon^8\cdots & \epsilon^{4n}\\ \cdots & \cdots & \cdots\\ \cdots & \cdots & \cdots\\ \epsilon^{2n} & \epsilon^{4n}\cdots & \epsilon^{2nn}\end{bmatrix},\quad \epsilon=\exp(\pi i/n), \quad (38)$$

$$\mathbf{D}_+=\mathbf{D}_-\begin{bmatrix}\epsilon^{-1} & & & 0\\ & \epsilon^{-2} & & \\ & & \ddots & \\ 0 & & & \epsilon^{-n}\end{bmatrix}, \quad (39)$$

$$\mathbf{p}_-=\begin{bmatrix}e^{i\delta_2'/2} & & & 0\\ & e^{i\delta_4'/2} & & \\ & & \ddots & \\ 0 & & & e^{i\delta_{2n}'/2}\end{bmatrix}, \quad (40)$$

and

$$\mathbf{p}_+=\begin{bmatrix}e^{i\delta_1'/2} & & & 0\\ & e^{i\delta_3'/2} & & \\ & & \ddots & \\ 0 & & & e^{i\delta_{2n-1}'/2}\end{bmatrix}. \quad (41)$$

The quantities δ' are defined in Kaufman's paper. Explicit expressions for them will be given later in Eq. (60). The four matrices $\mathbf{D}_-$, $\mathbf{D}_+$, $\mathbf{p}_-$, and $\mathbf{p}_+$ are all unitary. Writing the eigenvector of $\mathbf{G}$ as $\begin{bmatrix}\phi\\ \eta\end{bmatrix}$ and by the use of (37) one obtains the following eigenvalue problem

$$\mathbf{D}_+^{-1}\mathbf{p}_+^{-1}(\mathbf{p}_-\mathbf{D}_-\phi+i\mathbf{p}_-^{-1}\mathbf{D}_-\eta)=l\phi, \quad (42)$$

$$-i\mathbf{D}_+^{-1}\mathbf{p}_+(\mathbf{p}_-\mathbf{D}_-\phi+i\mathbf{p}_-^{-1}\mathbf{D}_-\eta)=l\eta. \quad (43)$$

If $l\neq 0$, this shows that

$$\mathbf{p}_+\mathbf{D}_+\phi=i\mathbf{p}_+^{-1}\mathbf{D}_+\eta.$$

With the aid of this, η could be eliminated and the eigenvalue problem is finally reduced to

$$(\mathbf{D}+\mathbf{p}_-^{-2}\mathbf{D}\mathbf{p}_+^{2})\phi_1=l(\mathbf{p}_-^{-1}\mathbf{p}_+)\phi_1, \quad (44)$$

where

$$\phi_1=\mathbf{D}_+\phi,$$

and

$$\mathbf{D}=\mathbf{D}_-\mathbf{D}_+^{-1}. \quad (44a)$$

D.

The calculation of $\xi_{1\beta}$ will be reduced in this subsection to the eigenvector problem of an $n\times n$ matrix.

From the definition of ζ in (26) we see that the column matrix

$$\zeta_1=\begin{bmatrix}\zeta_{11}\\ \zeta_{12}\\ \vdots\end{bmatrix}$$

which is the first column of ζ^{-1} is an eigenvector of $\mathbf{T}_+^{-1}\mathbf{M}\mathbf{T}_-$ with the eigenvalue $+1$:

$$(\mathbf{T}_+^{-1}\mathbf{M}\mathbf{T}_-)\zeta_1=\zeta_1. \quad (45)$$

It is easily shown that if a column matrix ξ_1 could be found such that

$$\mathbf{G}\xi_1=0, \quad (46a)$$

and

$$\mathbf{G}\xi_1^*=2\xi_1, \quad (46b)$$

then in virtue of (34)

$$\zeta_1=\tfrac{1}{2}(e^{-ai}\xi_1+e^{ai}\xi_1^*) \quad (47)$$

does satisfy (45). It is to be emphasized that the ξ_1

defined by (46) is *independent* of a so that as $a \to i\infty$ (47) shows that ζ_1 becomes proportional to ξ_1, and the first column of the matrix $\|\xi_{\alpha\beta}\|$ is exactly ξ_1.

We now tackle Eqs. (46). Equations (37) and (46a) lead to

$$\begin{Bmatrix} \mathbf{p}_-\mathbf{p}_-^{-1} \end{Bmatrix} \begin{Bmatrix} \mathbf{D}_- & 0 \\ 0 & i\mathbf{D}_- \end{Bmatrix} \xi_1 = 0,$$

showing that there exists an $n\times 1$ column matrix y such that

$$\begin{Bmatrix} \mathbf{D}_- & 0 \\ 0 & i\mathbf{D}_- \end{Bmatrix} \xi_1 = \begin{Bmatrix} \mathbf{p}_-^{-1} \\ -\mathbf{p}_- \end{Bmatrix} y, \tag{48}$$

which is both necessary and sufficient for the fulfillment of (46a). Solving (48) for ξ_1 and substituting into (46b) one obtains

$$\mathbf{D}_+^{-1}\mathbf{p}_+^{-1}(\mathbf{p}_-\mathbf{D}_-^2\mathbf{p}_- + \mathbf{p}_-^{-1}\mathbf{D}_-^2\mathbf{p}_-^{-1})y^* = 2\mathbf{D}_-^{-1}\mathbf{p}_-^{-1}y, \tag{49}$$

$$-\mathbf{D}_+^{-1}\mathbf{p}_+(\mathbf{p}_-\mathbf{D}_-^2\mathbf{p}_- + \mathbf{p}_-^{-1}\mathbf{D}_-^2\mathbf{p}_-^{-1})y^* = 2\mathbf{D}_-^{-1}\mathbf{p}_- y. \tag{50}$$

We shall show that

$$\mathbf{p}_-\mathbf{D}_-^2\mathbf{p}_- = \mathbf{p}_-^{-1}\mathbf{D}_1^2\mathbf{p}_-^{-1}. \tag{51}$$

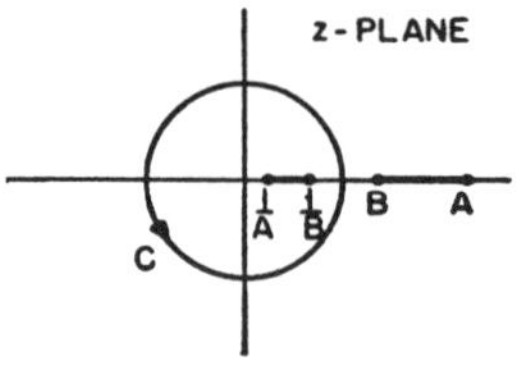

FIG. 1. Cuts in z-plane.

First, from (38)

$$\mathbf{D}_-^2 = \begin{bmatrix} 0 & & 1 & 0 \\ & & 1 & \\ & \cdot^{\cdot^{\cdot}} & & \\ 1 & & & \\ 0 & & & 1 \end{bmatrix}. \tag{51a}$$

But from Kaufman's definition of δ',

$$\delta_{2r}' = -\delta_{2n-2r}', \tag{52}$$

and

$$\exp(i\delta_{2n}') = -1, \quad T < T_C. \tag{53}$$

Hence by (40) $\mathbf{D}_-^2\mathbf{p}_-^2\mathbf{D}_-^2 = \mathbf{p}_-^{-2}$; and using $\mathbf{D}_-^4 = 1$ one immediately proves (51). (49) and (50) now simplifies to

$$\mathbf{D}\mathbf{p}_+^{-1}\mathbf{p}_-\mathbf{D}_-^2\mathbf{p}_- y^* = \mathbf{p}_-^{-1}y, \tag{54}$$

$$-\mathbf{D}\mathbf{p}_+\mathbf{p}_-\mathbf{D}_-^2\mathbf{p}_- y^* = \mathbf{p}_- y. \tag{55}$$

Elimination of y^* and simplification leads finally to

$$(\mathbf{D}^{-1} + \mathbf{p}_+^{-2}\mathbf{D}^{-1}\mathbf{p}_-^2)(\mathbf{p}_-^{-1}y) = 0. \tag{56}$$

Equations (54) and (56) together determine y, which in turn gives ξ_1 through (48).

The normalization of ξ_1 is determined by substitution of (47) into

$$\zeta_1'\zeta_1 = 1. \tag{57}$$

This results in

$$\xi_1'\xi_1 = 0, \tag{58}$$

and

$$\xi_1'\xi_1^* = 2. \tag{59}$$

(58) is automatically satisfied by virtue of (48), (51), and the fact that $\mathbf{D}_-$ and $\mathbf{p}_-$ are symmetrical matrices.

E.

To summarize the results of this section: The spontaneous magnetization I is given by (31), in which the λ's are related through Eq. (36) to the eigenvalues l of Eq. (44), and in which ξ_{11} and ξ_{12} are the first and the $(n+1)$th element of the column matrix ξ_1 calculated through (48) from the column matrix y which in turn is determined by (54) and (56). ξ_1 is to be normalized according to (59).

III. LIMIT FOR INFINITE CRYSTAL

A.

The procedure just outlined simplifies greatly when we approach the limit of an infinite crystal. To show this let us first introduce the variable

$$z = e^{i\omega} \quad (\omega = r\pi/n,\ r = 1, 2, \cdots n). \tag{59a}$$

The relationship between δ' and ω is given by Kaufman's Eq. (52). In terms of z this can be reduced to

$$e^{2i\delta'} = \frac{\tanh^2 H^*(z - \coth H \coth H^*)(z - \tanh H \coth H^*)}{(z - \coth H \tanh H^*)(z - \tanh H \tanh H^*)}. \tag{60}$$

From this we obtain $e^{i\delta'}$ and we shall write it as

$$\Theta(z) = e^{i\delta'} = (1/AB)^{\frac{1}{2}}[(z-A)(z-B)/(z-A^{-1})(z-B^{-1})]^{\frac{1}{2}} \tag{61}$$

where

$$A = \coth H \coth H^* = [(1+x)/x(1-x)], \quad B = \tanh H \coth H^* = [(1-x)/x(1+x)]. \tag{62}$$

For $T < T_C$, $A > B > 1$. $\Theta(z)$ is analytic everywhere except at the points $z = A$, B, $1/A$, or $1/B$ where it has branch points. The square root in (61) is defined to be that branch of the function that takes the value -1 at $z = 1$, in accordance with (53). (See Fig. 1.)

Consider Eq. (44). For a very large crystal

$$\mathbf{p}_- = \mathbf{p}_+ = \mathbf{p},$$

and we have

$$(\mathbf{D} + \mathbf{p}^{-2}\mathbf{D}\mathbf{p}^2)\phi_1 = l\phi_1. \tag{63}$$

By the definition of $\mathbf{D}$, Eq. (44a), the matrix elements of $\mathbf{D}$ are

$$(\mathbf{D})_{rs} = \frac{1}{n}\sum_{t=1}^{n} \epsilon^{2rt}\epsilon^t\epsilon^{-2ts} = -\frac{2}{n}\frac{1}{1-\epsilon^{2s-2r-1}}.$$

Hence $\mathbf{D}$ operating on any vector ϕ gives

$$(\mathbf{D}\phi)_r=\sum(\mathbf{D})_{rs}\phi_s=-\frac{2}{n}\sum_{s=1}^{n}\phi_s\left[1-\exp\frac{\pi i}{n}(2s-2r-1)\right]^{-1}$$

$$=-\frac{2}{n}\sum_{s=1}^{n}\frac{\phi_s}{1-z_{2s}/z_{2r}\epsilon}, \tag{64}$$

where z is the variable defined in (59a). For $s=1,2,\cdots n$ the values assumed by z_{2s} are the n nth roots of unity. As $n\to\infty$ the summation in (64) therefore becomes an integral around the unit circle:

$$(\mathbf{D}\phi)_r=-2\int_C\frac{1}{2\pi i}\frac{dz}{z}\frac{\phi(z)}{1-(z/z_r)}, \tag{65}$$

where

$$z=\exp(2\pi is/n)\quad\text{and}\quad z_r=\exp(2\pi ir/n).$$

The contour C is the unit circle. At the point $z=z_r$ the principle value of the integral is to be taken. This is necessary because of the factor ϵ in the expression (64) which prevents the denominator from assuming the value zero. Alternately, we might make a detour around the point z_r and make up the difference by adding a term to (65):

$$(\mathbf{D}\phi)_t=-\frac{1}{\pi i}\int_{C'}\frac{dz}{z}\frac{\phi(z)}{1-(z/t)}+\phi(t). \tag{66}$$

We have here used the more convenient notation t for z_r. With this definition it is evident that the point t does not have to be on the unit circle. If, however, t is inside the unit circle, it is more convenient to use the following equivalent of (66):

$$(\mathbf{D}\phi)_t=-\frac{1}{\pi i}\int_{C''}\frac{dz}{z}\frac{\phi(z)}{1-(z/t)}-\phi(t), \tag{66a}$$

where C'' is as shown in Fig. 2.

The definitions (66) and (66a) for $\mathbf{D}$ are valid when $\mathbf{D}$ operates on any function $\phi(z)$ that is analytic in a region that contains the circumference of the unit circle in its interior. It is important to notice that this region does not have to be singly connected.

We quote a few interesting properties of the operator $\mathbf{D}$:

$$\begin{aligned}\mathbf{D}z^m&=t^m\quad\text{for } m=\text{integer}\geq 1,\\ \mathbf{D}z^m&=-t^m\text{ for } m=\text{integer}\leq 0,\end{aligned} \tag{67}$$

$$\mathbf{D}^2=1. \tag{68}$$

Now return to Eq. (63). Since $\mathbf{p}^2$ is a diagonal matrix with diagonal element $\Theta(z)$ given by (61), it is evident that (63) reduces to

$$2\phi_1(t)-\frac{1}{\pi i}\int_{C'}\frac{dz}{z}\frac{\phi_1(z)}{1-(z/t)}\left(1+\frac{\Theta(z)}{\Theta(t)}\right)=l\phi_1(t). \tag{69}$$

This integral equation will be solved in the next two sections.

B.

There still remains, according to the results of the last section, the problem of solving (54) and (56). By virtue of (68), (56) reduces to the same form as (63) with $l=0$. Thus $p_-^{-1}y$ is proportional to that eigenvector $\phi_1=\Phi$ of (69) belonging to the eigenvalue $l=0$:

$$2\Phi(t)-\frac{1}{\pi i}\int_{C'}\frac{dz}{z}\frac{\Phi(z)}{1-(z/t)}\left(1+\frac{\Theta(z)}{\Theta(t)}\right)=0. \tag{70}$$

For the convenience of normalization we shall write

$$\mathbf{p}_-^{-1}y=n^{-\frac{1}{2}}\Phi. \tag{71}$$

Then by virtue of (48), Eq. (59) reduces to

$$(1/n)\Phi'\Phi^*=1,$$

or

$$\frac{1}{2\pi i}\int_C|\Phi(z)|^2\frac{dz}{z}=1. \tag{72}$$

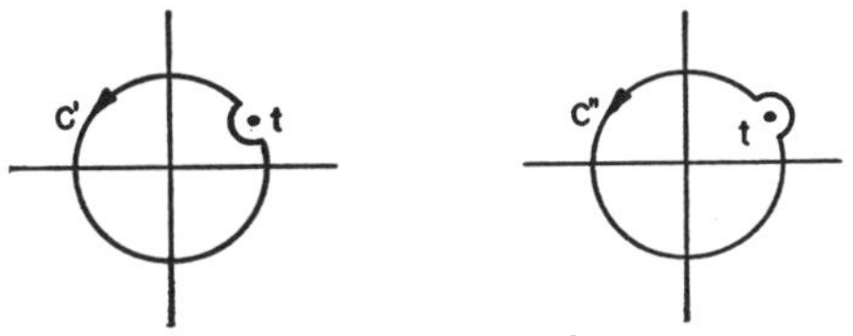

FIG. 2. Contours in z-plane.

The first and $(n+1)$th elements of ξ_1 are, according to (48) and (71):

$$\begin{aligned}\xi_{11}&=\frac{1}{n}\sum_{s=1}^{n}\epsilon^{-2s}\Phi_s=\frac{1}{2\pi i}\int_C\frac{dz}{z^2}\Phi(z),\\ \xi_{12}&=\frac{i}{n}\sum_{s=1}^{n}\epsilon^{-2s}\Theta(\epsilon^{2s})\Phi_s=\frac{1}{2\pi}\int_C\frac{dz}{z^2}\Theta(z)\Phi(z).\end{aligned} \tag{73}$$

The question of the fulfillment of Eq. (54), which now reduces to

$$\mathbf{D}\mathbf{D}_-^{2}\Phi^*=\Phi,$$

is best discussed with the aid of the introduction of the function $\Phi^\dagger(z)$ defined by

$$\Phi^\dagger(z)=[\Phi(z^*)]^*. \tag{74}$$

If Φ is analytic in a region containing the circumference of the unit circle in its interior, $\Phi^\dagger$ would be analytic in a similar region. Equation (51a) shows that

$$(\mathbf{D}_-^{2}\Phi^*)_z=[\Phi(z^*)]^*=\Phi^\dagger(z).$$

Thus (54) is fulfilled if

$$\Phi^\dagger(t)-\frac{1}{\pi i}\int_{C'}\frac{dz}{z}\frac{\Phi^\dagger(z)}{1-z/t}=\Phi(t). \tag{75}$$

C.

The integral equation (70) is easily solved by inspection:

$$\Phi(z)=Fz[(A-z)(B-z)]^{-\frac{1}{2}}, \tag{76}$$

where F is a normalization factor. F will turn out to be real, so that according to (74)

$$\Phi^{\dagger}(z)=\Phi(z).$$

It is easy to prove that (75) is satisfied. This completes the verification that (73) does indeed give the correct matrix elements of ξ_1.

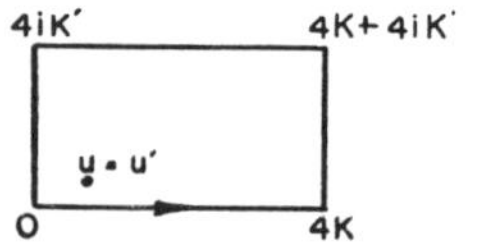

FIG. 3. Contour in u-plane.

To find F we substitute (76) into (72) and obtain

$$\frac{F^2}{2\pi i}\int_C \frac{dz}{[(A-z)(B-z)(Az-1)(Bz-1)]^{\frac{1}{2}}}=1. \tag{76a}$$

In the integrand the sign of the square root is to be so taken that at $z=1$ the integrand is positive. The integral is a complete elliptic integral and can be reduced to the standard form by a projective transformation. The result is

$$F^{-2}=\frac{4}{\pi}\frac{1}{A-B}k_{-1}^{\frac{1}{2}}K(k_{-1}),$$

where

$$k_{-1}=\left[\frac{(A^2-1)^{\frac{1}{2}}-(B^2-1)^{\frac{1}{2}}}{A(B^2-1)^{\frac{1}{2}}+B(A^2-1)^{\frac{1}{2}}}\right]^2, \tag{77}$$

and K is the complete elliptic integral of the first kind.[12] It is convenient to change the modulus and define[13]

$$k=2k_{-1}^{\frac{1}{2}}/(1+k_{-1})=4x^2/(1-x^2)^2=\sinh^{-2}2H. \tag{78}$$

Then

$$F^{-2}=2kK(k)/\pi(A-B). \tag{79}$$

The values of ξ_{11} and ξ_{12} are obtained from (73) and (76):

$$\xi_{11}=F(AB)^{-\frac{1}{2}},\quad \xi_{12}=0.$$

Substitution of these into (31), with the use of (36), leads to

$$I^4=\left[\prod_2^n(l_\alpha^2/4)\right]F^4A^{-2}B^{-2}\cosh^4H^*.$$

We have here taken the fourth power of I to eliminate the undetermined phase factor that was introduced into the expression for I as early as Eq. (17). When the explicit expressions for A, B, F, and H^* are introduced, the expression for I^4 further simplifies to

$$I^4=\left(\prod_2^n\frac{l_\alpha^2}{4}\right)\frac{\pi^2}{4}\left[\frac{1}{K(k)}\right]^2. \tag{80}$$

IV. ELLIPTIC TRANSFORMATION

It remains to find the eigenvalues l from (69) and substitute into (80). To do this we first introduce an elliptic transformation[12] that was essentially the one used in evaluating the integral in (76a):

$$z=-(\mathrm{cn}u-i[1+k]^{\frac{1}{2}}\,\mathrm{sn}u)(\mathrm{dn}u-i[k+k^2]^{\frac{1}{2}}\,\mathrm{sn}u)/(1+k\,\mathrm{sn}^2u), \tag{81}$$

the modulus k being given by (78).[13] This is the same transformation as was used by Onsager,[5] and Kaufman and Onsager[7] in their calculations. It serves to eliminate the square root in the function Θ:

$$\Theta=e^{i\delta'}=\mathrm{cn}u+i\,\mathrm{sn}u. \tag{82}$$

It is easy to verify that

$$\frac{1}{z}\frac{dz}{du}=-i\frac{1-k^2}{(1+k)^{\frac{1}{2}}}\frac{1}{\mathrm{dn}u-k^{\frac{1}{2}}\,\mathrm{cn}u}. \tag{83}$$

We shall need the following properties of the transformation (81): (A) z is doubly periodic in u with periods $4K$ and $4iK'$.

(B) z is everywhere analytic, except at $u=iK'/2$, $3iK'/2$ (mod. $4K$, $4iK'$), where $z=\infty$.

(C) In a unit cell in the complex u-plane, to every value of z there correspond exactly two values of u, except for $z=A$, B, $1/B$ or $1/A$ for which there corresponds only one value of u, namely, $u=+iK'$, $2K+iK'$, $2K-iK'$ or $-iK'$ (mod. $4K$, $4iK'$).

(D) If for a value of z there correspond in a unit cell two values of u, then at those two values Θ assume equal values but have different signs. Thus a unit cell of the u plane corresponds to both sheets of the Riemann surface in the z plane of Fig. 1 with respect to the function $\Theta(z)$.

The substitution, suggested by (76), into (69), of

$$\phi_1(z)=z[(z-A)(z-B)]^{\frac{1}{2}}\phi$$

gives, with the use of (81), (82), and (83)

$$2\phi(u')+\int_0^{4K}J(u',u)\phi(u)du=l\phi(u'), \tag{84}$$

where

$$J=\mathbf{I}\;\mathbf{II}\;\mathbf{III}\;\mathbf{IV}; \tag{85}$$

and

$$\mathbf{I}=[1-z(u)/z(u')]^{-1}, \tag{86}$$

$$\mathbf{II}=1+\Theta(z)/\Theta(z')=1+(\mathrm{cn}u+i\,\mathrm{sn}u)/(\mathrm{cn}u'+i\,\mathrm{sn}u'), \tag{87}$$

$$\mathbf{III}=\frac{z(u)}{z(u')}\left[\frac{\{z(u')-A\}\{z(u')-B\}}{\{z(u)-A\}\{z(u)-B\}}\right]^{\frac{1}{2}}, \tag{88}$$

[12] E. T. Whitaker and G. N. Watson, *Modern Analysis* (Cambridge University Press, London, 1927), fourth edition.

[13] The modulus k is the same as that used in references 5 and 7.

and

$$\mathbf{IV}=-\frac{1}{\pi}\frac{1-k^2}{(1+k)^{\frac{1}{2}}}\frac{1}{\mathrm{dn}u-k^{\frac{1}{2}}\,\mathrm{cn}u}.$$

V. SOLUTION OF INTEGRAL EQUATION (84)

We proceed by investigating the analytic behavior of $J(u', u)$ *with respect to the variable* u.

(*A*) **I**, **II**, and **IV** are all doubly periodic with periods $4K$ and $4iK'$. But **III** is doubly periodic with periods $4K$ and $8iK'$. It changes sign at periods $4iK'$:

$$\mathbf{III}(u+4iK)=-\mathbf{III}(u). \tag{89}$$

(*B*) **III** is analytic everywhere except at $z=A$ or $z=B$, i.e., $u=iK'$ or $2K+iK'$ (mod. $4K$, $4iK'$) where **III** has simple poles.

(*C*) **II** is analytic everywhere except at $u=-iK'$ or $2K-iK'$ (mod. $4K$, $4iK'$) where it has simple poles.

(*D*) **IV** is analytic everywhere except at

$$u=\pm iK'/2,\quad \pm 3iK'/2(\text{mod. } 4K,\ 4iK'),$$

where it has simple poles.

(*E*) **I** is analytic everywhere except at $z(u)=z(u')$. According to the last section in each cell there are, in general, two values of u where this exception occurs. At these two points **I** has simple poles.

However, there is *only one pole* for J in each unit cell ($4K$ by $4iK'$). This is so because of the following considerations:

(*F*) At $u=\pm iK'$, $\pm iK'+2K$ (mod. $4K$, $4iK'$), **IV** has simple zeros.

(*G*) At $u=iK'/2$, $3iK'/2$ (mod. $4K$, $4iK'$), $z(u)=\infty$, so that **I** has simple zeros.

(*H*) At $u=-iK'/2$, $-3iK'/2$ (mod. $4K$, $4iK'$), $z(u)=0$, so that **III** has simple zeros.

(*I*) According to property *D*, Sec. IV, in a unit cell ($4K$ by $4iK'$) at one of the solutions of $z(u)=z(u')$, $\Theta(u)=\Theta(u')$ so that $\mathbf{II}=2$. At the other solution, **II** has a zero.

Thus inside the rectangle in Fig. 3, J has only one pole at $u=u'$ which we assume to be inside of the rectangle. In the neighborhood of this pole $\mathbf{II}=2$, $\mathbf{III}=1$, and $\mathbf{I}\text{-}\mathbf{IV}=i\pi^{-1}(u-u')^{-1}$. Hence the residue of J at $u=u'$ is $2i/\pi$.

The solution of (84) is given by

$$\phi=\exp(im\pi u/2K),\quad m=\pm\text{integer}. \tag{90}$$

To show that this is indeed a solution we note that ϕ is periodic with period $4K$. Hence calling

$$\vartheta=\int_0^{4K} J(u', u)\phi(u)du,$$

one obtains by performing a contour integration around the rectangle of Fig. 3:

$$2\pi i(2i/\pi)\phi(u')=\vartheta[1+\exp(-2m\pi K'/K)]. \tag{91}$$

[The integration along the two vertical sides cancel each other and that along the top reduces to ϑ multiplied by a factor, in virtue of (89).] This gives

$$\vartheta=-4\phi(u')/(1+q^{2m}), \tag{92}$$

where

$$q=\exp(-\pi K'/K). \tag{93}$$

(84) is therefore satisfied with

$$l=2-4/(1+q^{2m})=2(q^{2m}-1)/(q^{2m}+1). \tag{94}$$

For $m=0$ this gives, as expected, the solution $l=0$ which was already found by inspection in Sec. IIIC.

Knowing all the nonvanishing eigenvalues we can now calculate

$$\prod_2^\infty \frac{l_\alpha^2}{4}=\prod_{\substack{m=-\infty\\ m\neq 0}}^{\infty}\left(\frac{1-q^{2m}}{1+q^{2m}}\right)^2=\prod_1^\infty\left(\frac{1-q^{2m}}{1+q^{2m}}\right)^4.$$

This infinite product can be[14] expressed in terms of the ϑ functions which are related to K. We get finally

$$\prod_2^\infty\frac{l_\alpha^2}{4}=\frac{4}{\pi^2}[K(k)]^2(1-k^2)^{\frac{1}{2}}=\frac{4}{\pi^2}K^2\frac{1+x^2}{(1-x^2)^2}(1-6x^2+x^4)^{\frac{1}{2}}. \tag{95}$$

VI. FINAL RESULTS

The spontaneous magnetization I is obtained from (95) and (80) as

$$I=\left[\frac{1+x^2}{(1-x^2)^2}(1-6x^2+x^4)^{\frac{1}{2}}\right]^{\frac{1}{8}}. \tag{96}$$

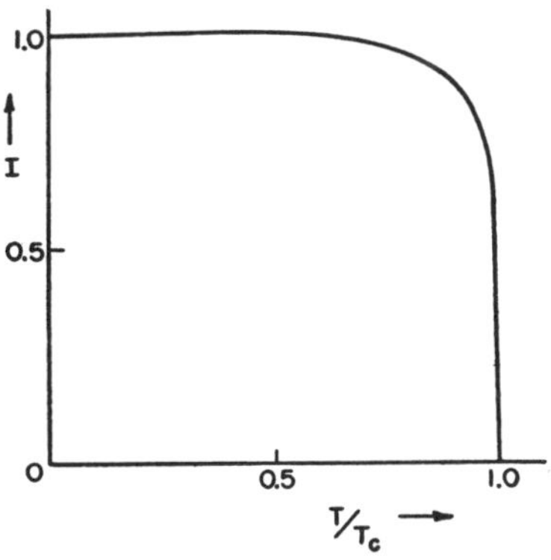

FIG. 4. Spontaneous magnetization.

At low temperatures this gives the same expansion in powers of x as obtained in previous works of Van der Waerden[1] and Ashkin and Lamb:[2]

$$I=1-2x^4-8x^6-34x^8-152x^{10}-714x^{12}-\cdots.$$

This series is convergent all the way up to the critical

[14] See reference 11, especially p. 472.

point, where

$$x = x_C = \sqrt{2} - 1.$$

Near the critical point, I has a branch point:

$$I \cong [4(\sqrt{2}+2)(x_C - x)]^{1/8}.$$

In Fig. 4, I is plotted against the temperature.

This work was completed in the summer of 1951 while the author was at the University of Illinois. He wishes to take this opportunity to thank the staff of the Department of Physics, University of Illinois for the hospitality extended him during his stay. He also wishes to thank Bruria Kaufman for many stimulating discussions.

Statistical Theory of Equations of State and Phase Transitions. I. Theory of Condensation [[52b]]

Commentary begins page 14

C. N. YANG AND T. D. LEE
Institute for Advanced Study, Princeton, New Jersey

(Received March 31, 1952)

A theory of equations of state and phase transitions is developed that describes the condensed as well as the gas phases and the transition regions. The thermodynamic properties of an infinite sample are studied rigorously and Mayer's theory is re-examined.

I. INTRODUCTION

THIS and a subsequent paper will be concerned with the problem of a statistical theory of equations of state and phase transitions. This problem has always interested physicists both from the practical viewpoint of seeking for a workable theory of properties of matter (such as a theory of liquids) and also from the more academic viewpoint of understanding the occurrence of the discontinuities associated with phase transitions in the thermodynamic functions.

The work reported in this paper is quite general and fairly abstract. We are returning in a subsequent paper to the illustration and application of the methods here outlined. In order to present the work of this present paper in its proper perspective, it may be helpful if we outline briefly the history of our own thinking on the subject.

About a year ago one of us was able to make progress[1] with the problem of the spontaneous magnetization of the Ising model, taking advantage of some special properties of this problem when treated by the Onsager-Kaufman method.[2] We then noted that the solution there obtained was also the solution of another, physically quite different, but formally identical, problem. This is the problem of a lattice gas with attractive interaction between nearest neighbors. We were thus able to follow in detail the behavior of such a lattice gas, which in many ways should reveal the features of an actual gas. In particular, we were able to study and characterize the condensation phenonenon, and to identify the liquid, gas, and transition regions in the $p-v$ diagram. The isotherms thus obtained are flat in the transition region and rise very rapidly with increasing density in the liquid phase. At this point, we were led to compare the specific solution with the well-known work[3] of Mayer on the theory of condensation of gases. In particular we were led to inquire as to why, in Mayer's theory, the isotherms stay flat beyond the condensation point and do not give the equation of state for the liquid phase. It soon became apparent that this difference lay, not in the difference of the models, but in the inadequacy of Mayer's method for dealing with a condensed phase. This led to a study of the analytical behavior of the grand partition function of an assembly of interacting atoms, and we were able, as in the special case mentioned above, to identify and characterize quite generally the condensation phenomena. These general conclusions will be presented in the present paper.

The problem is approached by allowing the fugacity to take on complex values. Although only real values of the fugacity are of any physical interest, the analytical behavior of the thermodynamic functions can only be completely revealed by going into the complex plane, whereby one is able to obtain a description of the condensed phases as well as the gas phase and the transition regions. This approach is of a very general nature and can be applied to other problems of phase transitions such as ferromagnetism, order disorder transition, etc. It will be emphasized that also this approach can lead to practical approximation methods for the description of systems undergoing transitions. These points will be discussed in paper II.

The physical conclusions of this paper derive from some mathematical results which we shall state in the form of two theorems. Due to the nature of the problem (which involves a double limiting process) it is imperative to have mathematical rigor preserved throughout. The proofs are necessarily of a mathematical nature and will be given in the appendix.

II. INTERACTION

We consider a monatomic gas with the interaction

$$U=\sum u(r_{ij}), \tag{1}$$

where r_{ij} is the distance between the ith and jth atoms. The following assumptions are made about the nature of these interactions:

(1) The atoms have a finite impenetrable core of diameter a, so that $u(r)=+\infty$ for $r\leqq a$.

(2) The interaction has a finite range b so that

$$u(r)=0 \text{ for } r\geqq b.$$

(3) $u(r)$ is nowhere minus infinity.

The theory can be easily generalized to include many body forces and forces with a weak long tail such as

[1] C. N. Yang, Phys. Rev. **85**, 808 (1952).

[2] L. Onsager, Phys. Rev. **65**, 117 (1944); B. Kaufman, Phys. Rev. **76**, 1232 (1949).

[3] J. E. Mayer, J. Chem. Phys. **5**, 67 (1937); J. E. Mayer and Ph. G. Ackermann, J. Chem. Phys. **5**, 74 (1937); J. E. Mayer and S. F. Harrison, J. Chem. Phys. **6**, 87, 101 (1938); B. Kahn and G. E. Uhlenbeck, Physica **5**, 399 (1938). M. Born and K. Fuchs, Proc. Roy. Soc. (London) **A166**, 391 (1938).

□ Reprinted from *The Physical Review* 87, 3 (August 1, 1952) 404–409.

van der Waals' force. But for clarity we shall first treat only interactions with the properties enumerated above.

Consider a box of volume V kept at a constant temperature T. If it is allowed to exchange atoms with a reservoir at a given chemical potential μ per atom, the relative probability of having N atoms in the box is

$$Q_N y^N/N!,$$

where

$$Q_N=\int\cdots\int_V d\tau_1\cdots d\tau_N \exp(-U/kT) \tag{2}$$

is the configurational part of the partition function for N atoms and

$$y=(2\pi mkT/h^2)^{\frac{3}{2}}\exp(\mu/kT). \tag{3}$$

The quantities m, k, and h have the usual meanings,

The grand partition function of the gas in the volume V is

$$\mathcal{Q}_V=\sum_{N=0}^{M}\frac{Q_N}{N!}y^N, \tag{4}$$

where M is the maximum number of atoms that can be crammed into V.

III. THE LIMIT OF INFINITE VOLUME

The average pressure and the average density of such a gas in V are calculable in terms of $\mathcal{Q}_V$ by the standard treatment of statistical mechanics, and are evidently dependent on V. In thermodynamics, however, one is only interested in an infinite sample and the thermodynamic functions are limits of these average quantities as $V\to\infty$. The pressure p and density ρ are accordingly given by

$$\frac{p}{kT}=\operatorname*{Lim}_{V\to\infty}\frac{1}{V}\log\mathcal{Q}_V, \tag{5}$$

$$\rho=\operatorname*{Lim}_{V\to\infty}\frac{\partial}{\partial\log y}\frac{1}{V}\log\mathcal{Q}_V. \tag{6}$$

The question of whether these limits do exist is usually not discussed.[4] It is, however, generally believed that in the gas phase such limits do exist and that (5) and (6) give the correct equation of state. At the point of condensation and in the liquid phase the situation has been extremely unclear. As a matter of fact doubts have been raised[5] as to whether the equation of state of both the liquid and the gas phase can be obtained from the same interaction (1) through the considerations of statistical mechanics.

We shall try to resolve these problems and prove that (5) and (6) do give a complete description of the equation of state of both the gas and condensed phases. In fact in paper II we shall give a concrete example in which it is seen how the same partition function describes both phases, and in which the two-phase-equilibrium region is exactly known.

We first state the following:

Theorem 1.—(Proved in Appendix I.) For all positive real values of y, $V^{-1}\log\mathcal{Q}_V$ approaches, as $V\to\infty$, a limit which is independent of the shape of V. Furthermore, this limit is a continuous, monotonically increasing function of y.

The assumption is made, of course, that the shape of V is not so queer that its surface area increases faster than $V^{\frac{2}{3}}$.

One might be tempted to conclude from the independence of the limit on the shape of V that the system under consideration exists only in fluid phases (i.e., gas and liquid) with no elastic resistance against shearing strain. It is to be emphasized that this is not the case. The independence of the limit on the shape of V is not due to the lack of elastic resistance against shearing strain, but rather due to the fact that for an infinite sample changing the shape of V does not produce a strain in the interior which might serve to differentiate between a fluid and a solid. This is so because the strain at the boundary only penetrates to a finite depth and is inconsequential for an infinite sample.

To study the limit of $(\partial/\partial\log y)V^{-1}\log\mathcal{Q}_V$ we notice that $\mathcal{Q}_V$ is a polynomial in y of finite degree M. This is a direct consequence of the assumed impenetrable core of the atoms. It is therefore possible to factorize $\mathcal{Q}_V$ and write

$$\mathcal{Q}_V=\prod_{i=1}^{M}\left(1-\frac{y}{y_i}\right), \tag{7}$$

where $y_1,\cdots y_M$ are the roots of the algebraic equation

$$\mathcal{Q}_V(y)=0. \tag{8}$$

Evidently none of these roots can be real and positive, since all the coefficients in the polynomial $\mathcal{Q}_V$ are positive.

As V increases these roots move about in the complex y plane and their number M increases (essentially) linearly with V. Their distribution in the limit $V\to\infty$ gives the complete analytic behavior of the thermodynamic functions in the y plane. In fact one can prove the following:

Theorem 2.—(Proved in Appendix II.) If in the complex y plane a region R containing a segment of the positive real axis is always free of roots, then in this region as $V\to\infty$ all the quantities:

$$\frac{1}{V}\log\mathcal{Q}_V,\quad\left(\frac{\partial}{\partial\log y}\right)\frac{1}{V}\log\mathcal{Q}_V,\quad\left(\frac{\partial}{\partial\log y}\right)^2\frac{1}{V}\log\mathcal{Q}_V\cdots,$$

[4] The behavior of the partition function Q_N as the volume approaches infinite was discussed by L. von Hove, Physica **15**, 951 (1949), where it is proved that $N^{-1}\log Q_N$ approaches a limit as the volume approaches infinity at constant density. His proof is similar to our proof of theorem 1.

[5] There was apparently some discussion on this point at the International Conference held in Amsterdam, 26 November 1937. The doubts can perhaps be formulated in the form of the question: "How can the gas molecules know when they have to coagulate to form a liquid or solid?" See p. 391 of reference 3 (Born and Fuchs).

approach limits which are analytic with respect to y. Furthermore the operations $(\partial/\partial \log y)$ and $\mathrm{Lim}_{V\to\infty}$ commute in R so that, e.g.,

$$\operatorname*{Lim}_{V\to\infty}\frac{\partial}{\partial \log y}\frac{1}{V}\log \mathcal{Q}_V=\frac{\partial}{\partial \log y}\operatorname*{Lim}_{V\to\infty}\frac{1}{V}\log \mathcal{Q}_V. \quad (9)$$

This gives, together with (5) and (6),

$$\rho=(\partial/\partial \log y)(p/kT). \quad (10)$$

IV. PHASE TRANSITIONS

The quantity $(\partial/\partial \log y)V^{-1}\log \mathcal{Q}_V$ does not, however, always approach a limit ρ for all values of y. Physically this must evidently be as the density of the system does not assume a single value at the point of condensation. It is clear therefore that the problem of phase transition is intrinsically related to the form of the regions R described in theorem 2. We discuss the following cases:

(1) The roots of $\mathcal{Q}_V(y)=0$ do not close in onto the positive real axis of y as $V\to\infty$, or more exactly, there exists a region R which contains the whole positive real axis and is free of roots.

In this case from the two theorems one concludes that the pressure and density of the system are analytic functions of y (along the positive real axis). They are related by Eq. (10). Furthermore, p is an increasing function of y. We shall show in Appendix III that ρ is also an increasing function of y. Consequently in the $p-\rho$ diagram, on the isotherm p increases analytically as the specific volume v decreases. The system under consideration is thus a single phase system (see Fig. 1).

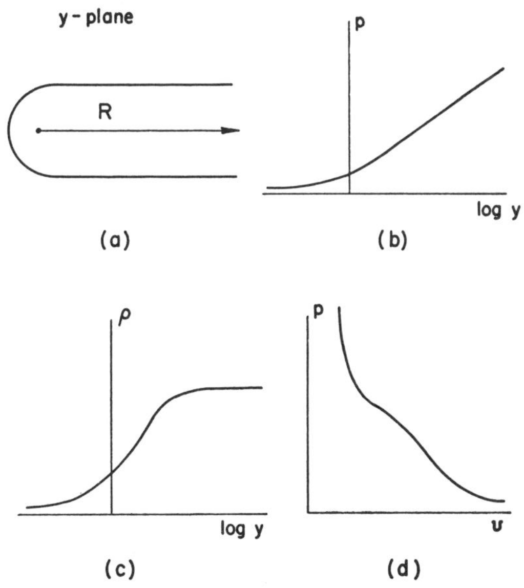

FIG. 1. Analytical behavior at a given temperature of thermodynamic functions for a single phase system. The quantity y is defined by Eq. (3) in the text. The region R is free of roots of Eq. (8). Notice that the density ρ of (c) is proportional to the slope of the $p-\log y$ curve in (b).

(2) The roots of $\mathcal{Q}_V(y)=0$ do close in onto the real axis as $V\to\infty$, say at the points $y=t_1, t_2$; and regions R_1, R_2, and R_3 free of roots enclose, respectively, the three segments of the positive real axis as in Fig. 2(a).

By the same reasoning as in the previous case one concludes, within any one of these three segments, that the system exists in a single phase, that p and ρ are analytic and increasing functions of y, that ρ is $(kT)^{-1}(\partial p/\partial \log y)$, and that on the isotherm p increases analytically as v decreases.

At the points $y=t_1, t_2$ the pressure p is continuous (by theorem 1), but its derivative ρ has in general a discontinuity. By Appendix III one shows easily that ρ increases across the discontinuity. The functions p and ρ are schematically plotted in Figs. 2(b) and 2(c) which together give the isotherm in Fig. 2(d).

As the temperature varies the points t_1 and t_2 will in general move along the y axis. If at a certain temperature T_c the roots cease to close in onto one of the points, say t_1, then T_c is the critical temperature for the transition phase $1\leftrightarrow$ phase 2. If, on the other hand, t_1 and t_2 merge together at a particular temperature T_0, we would then have a triple point at that temperature.

It may be remarked that at $y=t_1$ or t_2 the density ρ may in some cases be continuous (although its derivative will in general be discontinuous). At the critical temperature this will happen, but not at neighboring temperatures. If, however, this happens over an extended temperature range, one would have a transition of second (or higher) order.

It is clear therefore that phase transitions of the system occur only at the points on the positive real y axis onto which the roots of $\mathcal{Q}(V)=0$ close in as $V\to\infty$. For other values of the fugacity y a single phase system obtains.

As mentioned before, the theory can be easily generalized to include many body forces and forces with a weak long tail. In fact, the generalization does not lead to any alterations of the conclusions reached above.

Generalization can also be made to other kinds of phase transitions such as order-disorder phenomena and ferromagnetism, as will be discussed in paper II. The study of the equations of state and phase transitions can thus be reduced to the investigation of the distribution of roots of the grand partition function. In many cases, as will be seen in paper II, such distributions turn out to have some surprisingly simple regularities.

V. COMPARISON WITH MAYER'S THEORY

We first notice that by expanding in powers of y one obtains from (7)

$$\frac{1}{V}\log \mathcal{Q}_V=\sum_{l=1}^{\infty} b_l(V)y^l, \quad (11)$$

where

$$b_l(V) = \frac{-1}{lV}\sum_{j=1}^{M}\left(\frac{1}{y_j}\right)^l. \quad (12)$$

Combining (11) and (3) we have

$$\frac{Q_N}{N!} = \text{coefficient of } y^N \text{ in } \exp[V\sum_{l=1}^{\infty} b_l y^l]. \quad (13)$$

Comparison of this equation with Mayer's theory shows that the b_l's defined by (12) are identical with the reducible cluster integrals defined[6] by Mayer. It is interesting to notice that these reducible cluster integrals are, according to (12), closely related to the moments of the roots y_j of Eq. (8). It should be emphasized that in both (12) and in Mayer's definition the b_l's are functions of the volume V. It is evident from Mayer's definition that they approach definite limits $b_l(\infty)$ as $V\to\infty$.

In Mayer's theory the cluster integrals b_l are replaced from the very beginning by their limiting values $b_l(\infty)$. He then considers the series

$$\chi(y) = \sum_{1}^{\infty} b_l(\infty)y^l \quad (14)$$

and its analytical continuation along the positive real axis. If one calls the first singularity of $\chi(y)$ along the positive real axis t_1, one shows in Mayer's theory that

(1) for densities ρ less than

$$\rho_1 = \lim_{y\to t_1-} y\chi'(y), \quad (15)$$

the system exists in a single phase;

(2) for $\rho \geqq \rho_1$, the pressure p (at a given temperature) becomes independent of the density. Consequently, one identifies the density ρ_1 as the density of the gas at condensation.

An essential difficulty of Mayer's theory is that it does not admit of the existence of a liquid phase with finite density, since the isotherm remains horizontal for all specific volume less than ρ_1^{-1}. This is clearly due to the replacement of the volume dependent b_l's by their limiting values. The question is therefore often raised[7] as to exactly at what point on the isotherm Mayer's theory breaks down.

In the present theory by retaining the volume dependence of the partition function $\mathcal{Q}_V$ we do not encounter these difficulties. To clarify the relationship with Mayer's theory, we refer back to Fig. 2(a) and draw a small circle C within R_1 with the center at the origin. The series

$$\sum_{l=1}^{\infty} b_l(V)y^l$$

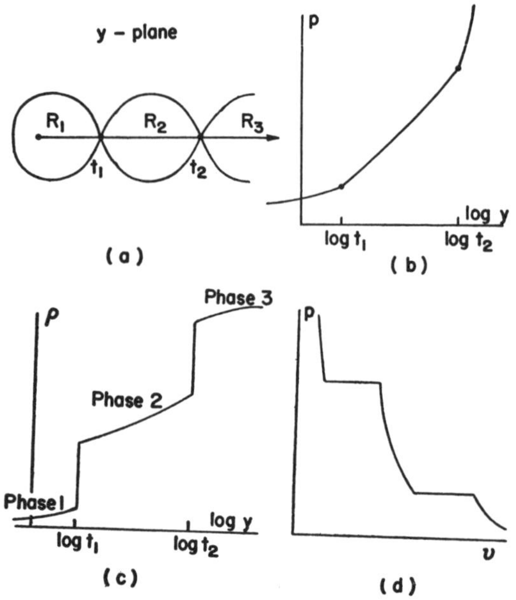

FIG. 2. Analytical behavior at a given temperature of thermodynamic functions for a system that undergoes two phase transitions. The transitions occur at t_1 and t_2 which are the points at which the roots of Eq. (8) close in onto the positive real y axis. The regions R_1, R_2, and R_3 are free of roots. The three phases 1, 2, and 3 are indicated in (c). The horizontal parts of (d) represent two-phase-equilibrium regions.

is easily shown[8] to converge uniformly in the circle C. By a well-known mathematical theorem on double limiting processes one concludes that in C

$$\lim_{V\to\infty}\sum_{1}^{\infty} b_l(V)y^l = \sum_{1}^{\infty} b_l(\infty)y^l. \quad (16)$$

The left-hand side of this equation is by definition

$$\lim_{V\to\infty} V^{-1}\log\mathcal{Q}_V,$$

and the right-hand side $\chi(y)$. Therefore within C the function $\chi(y)$ in Mayer's theory is indeed $(kT)^{-1}$ times the pressure p as defined by (5). By analytical continuation one concludes that this holds throughout the region R_1.

In the interval $0\leqq y<t_1$ it is evident that $\rho<\rho_1$ and Mayer's theory is seen to give a correct description of the system.

Beyond the point $y=t_1$ (i.e., $y\geqq t_1$) it is not possible in Mayer's theory to analytically continuate $\chi(y)$. The $p-\log y$ and $\rho-\log y$ diagrams [Figs. 2(b) and 2(c)] therefore exist in his theory only to the left of the first singularity. This explains the nonexistence of the liquid phase in Mayer's theory.

[6] See for example Eq. (13.5) in J. E. Mayer and Mayer, *Statistical Mechanics* (John Wiley and Sons, Inc., New York, 1946), p. 280.

[7] See, for example, reference 3 (Kahn and Uhlenbeck), p. 415.

[8] All roots y_j have absolute values larger than the radius σ of the circle C. By Eq. (12) we have $|b_l|\leqq(M/V)l^{-1}\sigma^{-l}$. But M/V is bounded. Hence the statement.

We thus remark:

(1) Throughout the gas phase (i.e., $\rho<\rho_1$) Mayer's theory gives correct results.

(2) For $\rho\geqq\rho_1$ Mayer's conclusion that the $p-v$ diagram becomes horizontal is, as already mentioned, incorrect for high densities due to the existence of the liquid phase. It is not even justified for densities immediately above ρ_1, as for transitions of high order the isotherm does not even have any horizontal part at all.

We are indebted to Professor J. R. Oppenheimer for criticism and comment.

APPENDIX I

To prove theorem 1 we first establish the following:

Lemma 1.—Let V and W be two cubes of linear dimensions L and $L+(b/2)$, respectively. Keeping b fixed one has as $L\to\infty$,

$$\mathrm{Lim}W^{-1}(\log\mathcal{Q}_W-\log\mathcal{Q}_V)=0. \tag{17}$$

Proof.—Put V completely inside W and write $\mathcal{Q}_W$ as the sum of contributions A_0, $A_1\cdots$ from configurations with zero, one . . . atoms outside of V. Now (a) since the interaction has finite range, each atom interacts with at most a finite and definite number of other atoms. Also (b) the available volume for the first atom outside of V is $\Delta=W-V$. If the volume of the impenetrable core of an atom is α, the available volume for the second atom outside of V is less than $\Delta-\alpha$, the third, $\Delta-2\alpha$, etc. Combining (a) and (b) one concludes that

$$A_m\leqq\beta^m[\Delta(\Delta-\alpha)(\Delta-2\alpha)\cdots(\Delta-m\alpha+\alpha)/m!]\mathcal{Q}_V, \tag{18}$$

where β is a constant. (This inequality is obtained by comparing the contributions to A_m and $\mathcal{Q}_V$ of a distribution of atoms with, say, N atoms inside of V and m atoms outside.) Adding all A's one obtains

$$\mathcal{Q}_W\leqq\mathcal{Q}_V(1+\beta\alpha)^{\Delta/\alpha}.$$

But $\Delta\sim L^2$, and clearly

$$\mathcal{Q}_V\leqq\mathcal{Q}_W.$$

Hence the lemma.

Lemma 2.—Let W_i be a cube of linear dimension 2^iL and $\mathcal{Q}_i$ an abbreviation for $\mathcal{Q}_{W_i}$. Then

$$\lim_{i\to\infty} W_i^{-1}\log\mathcal{Q}_i=K \text{ exists.}$$

Proof.—Consider W_j to be built up from 8^{j-i} smaller cubes $W_i(j>i)$. Evidently the number of atoms interacting across the boundaries of the small cubes is at most proportional to the area of such boundaries and hence is less than $8^j2^{-i}\gamma$, (γ=constant). Should one neglect these interactions, $\mathcal{Q}_j$ would become $\mathcal{Q}_i$ raised to the power 8^{j-i}. The inclusion of these terms violates this identity by not more than a constant factor β raised to the power 8^j2^{-i}, i.e.,

$$\log\mathcal{Q}_j\leqq8^{j-i}\log\mathcal{Q}_i+8^j2^{-i}\gamma\log\beta. \tag{19}$$

Next draw within each small cube W_i a concentric cube V_i with linear dimension $2^iL-(b/2)$. Since b is the range of interaction, clearly atoms in different V_i's do not interact. Hence,

$$8^{j-i}\log\mathcal{Q}_{V_i}\leqq\log\mathcal{Q}_j. \tag{20}$$

Equations (19) and (20) give

$$W_i^{-1}\log\mathcal{Q}_{V_i}\leqq W_j^{-1}\log\mathcal{Q}_j\leqq W_i^{-1}\log\mathcal{Q}_i+2^{-i}\gamma L^{-3}\log\beta. \tag{21}$$

The last term approaches zero as $i\to\infty$. Also by Lemma 1, $W_i^{-1}(\log\mathcal{Q}_i-\log\mathcal{Q}_{V_i})\to0$. Thus as $j>i\to\infty$ $\cdot W_j^{-1}\log\mathcal{Q}_j-W_i^{-1}\log\mathcal{Q}_i\to0$. Hence the lemma.

Proof of the theorem.—Given any $\epsilon>0$ there exists by Lemma 2 a large enough box W such that

$$|K-W^{-1}\log\mathcal{Q}_W|<\epsilon.$$

In fact by the same reasoning as used in proving that lemma, one easily sees that this can also be made true of any box Ω which can by adding partitions be divided into cubes of size W:

$$|K-\Omega^{-1}\log\mathcal{Q}_\Omega|<\epsilon.$$

Now consider a volume V of arbitrary shape. For sufficiently large V one can build two Ω-type boxes Ω_1 and Ω_2 such that Ω_1 is contained in V and Ω_2 contains V and that

$$|(\Omega_1/\Omega_2)-1|<\epsilon.$$

Using

$$\mathcal{Q}_{\Omega_1}\leqq\mathcal{Q}_V\leqq\mathcal{Q}_{\Omega_2},$$

one proves easily that $V^{-1}\log\mathcal{Q}_V$ also approaches K.

That this limit monotonically increases with y follows from the same property of $V^{-1}\log\mathcal{Q}_V$. That it is also continuous follows from the observation that $(\partial/\partial\log y)V^{-1}\log\mathcal{Q}_V$ has a finite and definite upper bound (equal to the density of closest packing).

APPENDIX II

We first prove the following:

Lemma 3.—Consider the series

$$\sum_{l=0}^{\infty}\mathfrak{b}_l(V)z^l=S_V(z),$$

where

$$|\mathfrak{b}_l(V)|\leqq A\sigma^{-l},$$

A and σ being positive constants. For all real z between $-\sigma$ and $+\sigma$ assume $\mathrm{Lim}S_V(z)$ to exist as $V\to\infty$. Then (a) $\mathrm{Lim}\mathfrak{b}_l(V)$ as $V\to\infty$ exists and will be denoted by $\mathfrak{b}_l(\infty)$. (b) $S_V(z)$ approaches the limit

$$\sum_{l=0}^{\infty}\mathfrak{b}_l(\infty)z^l, \tag{22}$$

as $V\to\infty$ for all $|z|<\sigma$. The series (22) is convergent for all $|z|<\sigma$.

Proof.—(a) Evidently $\mathfrak{b}_0(\infty)$ exists and is equal to $\text{Lim}S_V(0)$ as $V\to\infty$. To prove the existence of $\text{Lim}\mathfrak{b}_1(V)$: Given any real ϵ between 0 and $\sigma/2$ consider the convergence of $S_V(\epsilon)$ and $\mathfrak{b}_0(V)$ as $V\to\infty$. There exists a volume V_0 such that for any volumes V and W greater than V_0 one has

$$|S_V(\epsilon)-S_W(\epsilon)|<\epsilon^2, \tag{23}$$

$$|\mathfrak{b}_0(V)-\mathfrak{b}_0(W)|<\epsilon^2. \tag{24}$$

But

$$|\sum_{l=2}^{\infty}\mathfrak{b}_l(V)\epsilon^l|\leqq\frac{A\epsilon^2}{1-\epsilon\sigma^{-1}}\leqq 2A\epsilon^2. \tag{25}$$

The same is true if one replaces V by W. Using

$$\epsilon\mathfrak{b}_1(V)=S_V(\epsilon)-\mathfrak{b}_0(V)-\sum_{l=2}^{\infty}\mathfrak{b}_l(V)\epsilon_l,$$

one proves easily with the aid of (23), (24), and (25) that

$$\epsilon|\mathfrak{b}_1(V)-\mathfrak{b}_1(W)|<(2+4A)\epsilon^2.$$

Hence $\text{Lim}\mathfrak{b}_1(V)$ exists as $V\to\infty$. Similar proof holds for the other $\mathfrak{b}_l$'s.

(b) The series $\sum\mathfrak{b}_l(V)z^l$ evidently converges uniformly in z for $|z|<\sigma$. The lemma follows from a well-known theorem on double limits.

Proof of theorem 2.—Consider first a circle C, lying inside R, with its center at the point $y=\eta$ along the positive real axis. We shall first prove the theorem inside this circle. Making the displacement $z=y-\eta$ we express (7) in the form

$$\mathcal{Q}_V=\prod_{i=1}^{M}\left(1-\frac{z}{z_i}\right)\left(\frac{z_i}{y_i}\right), \tag{26}$$

where $z_i=y_i-\eta$ are the roots of $\mathcal{Q}_V$. Expanding $V^{-1}\log\mathcal{Q}_V$ in powers of z one obtains

$$\frac{1}{V}\log\mathcal{Q}_V=\sum_{l=0}^{\infty}\mathfrak{b}_l(V)z^l, \tag{27}$$

where

$$\mathfrak{b}_l(V)=\frac{-1}{Vl}\sum_{i=1}^{M}\left(\frac{1}{z_i}\right)^l \text{ for } l\geqq 1, \tag{28}$$

and

$$\mathfrak{b}_0(V)=\frac{1}{V}\sum_{i=1}^{M}\log\frac{z_i}{y_i}. \tag{29}$$

If σ is the radius of C, since C is free of roots we have $|z_i|\geqq\sigma$. Hence by (28)

$$|\mathfrak{b}_l(V)|\leqq(M/V)l^{-1}\sigma^{-l} \text{ for } l\geqq 1.$$

But M/V is bounded; hence we can use Lemma 3 and the theorem is proved in C.

By similar arguments we can extent the theorem into a circle C' lying inside R with its center inside C. One can easily prove the theorem in the whole region R by repeating this process.

APPENDIX III

To prove that ρ is an increasing function of y it is only necessary to show for any finite V the inequality

$$\frac{d^2}{(d\log y)^2}\log\mathcal{Q}_V>0.$$

Now $\mathcal{Q}_V$ is a polynomial in y with positive coefficients. Regarding the various terms of $\mathcal{Q}_V$ as relative probabilities we have obviously

$$\frac{d}{d\log y}\log\mathcal{Q}_V=\langle N\rangle,$$

where $\langle\ \rangle$ means "average." Also

$$\frac{d^2}{(d\log y)^2}\log\mathcal{Q}_V=\langle N^2\rangle-\langle N\rangle^2=\langle(\Delta N)^2\rangle,$$

which is always positive. Here ΔN is the deviation of N from the average

$$\Delta N=N-\langle N\rangle.$$

Statistical Theory of Equations of State and Phase Transitions. II. Lattice Gas and Ising Model

[[52c]]
Commentary begins page 14

T. D. Lee and C. N. Yang
Institute for Advanced Study, Princeton, New Jersey

(Received March 31, 1952)

The problems of an Ising model in a magnetic field and a lattice gas are proved mathematically equivalent. From this equivalence an example of a two-dimensional lattice gas is given for which the phase transition regions in the $p-v$ diagram is exactly calculated.

A theorem is proved which states that under a class of general conditions the roots of the grand partition function always lie on a circle. Consequences of this theorem and its relation with practical approximation methods are discussed. All the known exact results about the two-dimensional square Ising lattice are summarized, and some new results are quoted.

INTRODUCTION

In paper I[1] we have seen that the problem of a statistical theory of phase transitions and equations of state is closely connected with the distribution of roots of the grand partition function. It was shown there that the distribution of roots determines completely the equation of state, and in particular its behavior near the positive real axis prescribes the properties of the system in relation to phase transitions. It was also shown there that the equation of state of the condensed phases as well as the gas phase can be correctly obtained from a knowledge of the distribution of roots. While this general and abstract theory clarifies the problems underlying the statistical theory of phase transitions and condensed phases, it is natural to ask whether it also provides us with a means of obtaining practical approximation methods for calculating properties pertaining to phase transitions and condensed phases.

The problem is clearly that of seeking for the properties of the distribution of roots of the grand partition function. At first sight this appears to be a formidable problem, as the roots are in general complex and would naturally be expected to spread themselves for an infinite sample in the entire complex plane, or at least regions of the complex plane, and make it very difficult to calculate their distribution. We were quite surprised, therefore, to find that for a large class of problems of practical interest, the roots behave remarkably well in that they distribute themselves not all over the complex plane, but only on a fixed circle. This fact will be stated as a theorem in Sec. IV of the present paper and proved in the appendix. Implications of the theorem are discussed in Sec. V.

Also in this paper we shall give a proof (Sec. II) that the problem of an Ising model with a magnetic field is mathematically identical with that of a "lattice gas." From this identification we were able to trace exactly the transition region in the $p-v$ diagram of a two-dimensional lattice gas in detail. This will be presented in Sec. III and forms a clear illustration of the discussions of paper I and of Sec. V of the present paper.

At the end of Sec. V we give a summary of all the exact knowledge known to us about the two-dimensional Ising model in a magnetic field and its relationship with the distribution of roots of the partition function.

Table I. Identification of corresponding quantities in Ising model and lattice gas.[a]

Ising model		Lattice gas
No. of spins	=	volume
No. of ↓ spins	=	No. of atoms
$2/(1-I)$	=	specific volume v
$-F-H$	=	pressure p

[a] I, H, and F are respectively the intensity of magnetization, the magnetic field, and the free energy per spin in the Ising model problem.

II. ISING MODEL AND LATTICE GAS

We shall in this section show that the problems of an Ising model in a magnetic field and of a lattice gas are mathematically equivalent. In the former problem one considers a lattice of interacting spins each of which can assume two possible positions: ↑ and ↓. In the latter, one considers a corresponding lattice with each lattice point either vacant or occupied by an atom.[2] To each configuration of the lattice of spins there corresponds a configuration of the lattice gas in which a lattice point is vacant or occupied according as whether the corresponding spin is ↑ or ↓. Using this geometrical correspondence, one could establish the mathematical equivalence of the two problems.

For clarity of presentation we shall take as an example a simple cubic lattice and consider first the Ising model problem with nearest neighbor interaction. The same treatment can be applied to any lattice with arbitrary interaction between the spins. Denote by [↑↑], [↓↓], and [↑↓] the total number of nearest neighboring spins that are respectively parallel and upward, parallel and downward, and antiparallel to each other. Also denote by [↑] and [↓] the total numbers of upward and downward spins in the lattice.

[1] C. N. Yang and T. D. Lee, Phys. Rev. **87**, 404 (1952).

[2] Similar ideas have been used in the "hole theory of liquids." See, e.g., J. E. Lennard-Jones and A. F. Devonshire, Proc. Roy. Soc. (London) **A169**, 317 (1939); **A170**, 464 (1939). F. Cernuschi and H. Eyring, J. Chem. Phys. **7**, 547 (1939).

Counting the number of nearest neighbors of all the upward spins, one arrives at the following identity:

$$2[\uparrow\uparrow]+[\uparrow\downarrow]=6[\uparrow]. \tag{1}$$

Similarly

$$2[\downarrow\downarrow]+[\uparrow\downarrow]=6[\downarrow]. \tag{2}$$

Evidently

$$\mathfrak{N}=[\uparrow]+[\downarrow] \tag{3}$$

is the total number of spins in the lattice.

The interaction energy of all the spins can be written as $[\uparrow\downarrow]\epsilon$ if the interaction energy between parallel spins is taken to be zero. Here ϵ is a constant and is positive for ferromagnetic interactions and negative for antiferromagnetic interactions. In an external magnetic field H (measured in proper units), there is an additional magnetic energy so that the total energy of the Ising lattice is

$$U_I=H([\downarrow]-[\uparrow])+[\uparrow\downarrow]\epsilon. \tag{4}$$

The partition function is therefore

$$\exp(-\mathfrak{N}F/kT)=\sum \exp(-U_I/kT), \tag{5}$$

where F is the free energy per spin and the summation extends over all arrangements of the spins. F as a function of the magnetic field H and the temperature T defines completely the thermodynamic behavior of the lattice. Its derivative with respect to H gives the intensity of magnetization I,

$$\partial F/\partial H=-I, \tag{6}$$

where I is defined to be

$$I=\mathfrak{N}^{-1}([\uparrow]-[\downarrow]). \tag{7}$$

Now consider a lattice gas on the same simple cubic lattice. According to the geometrical correspondence discussed before, each downward spin $\downarrow$ corresponds to one gas atom, hence $[\downarrow]$ is equal to the number of atoms in the gas. Also the "volume" of the gas (in proper units) is simply $\mathfrak{N}$. The specific volume v per atom is, by Eqs. (3) and (7), related to the intensity of magnetization I by

$$v=2/(1-I). \tag{8}$$

To prevent more than one atom from occupying the same lattice site and to correspond to the case of nearest neighbor interaction in the Ising model problem, we consider here the following potential energy u between two atoms:

$u=+\infty$ if the two atoms occupy the same lattice site,

$u=-2\epsilon$ if the two atoms are nearest neighbors, (9)

and

$u=0$ otherwise.

It should be remarked that this interaction closely simulates the actual interaction between atoms of a atonatomic gas. For $\epsilon>0$ (ferromagnetic case), the gas moms attract each other at intermediate distances and

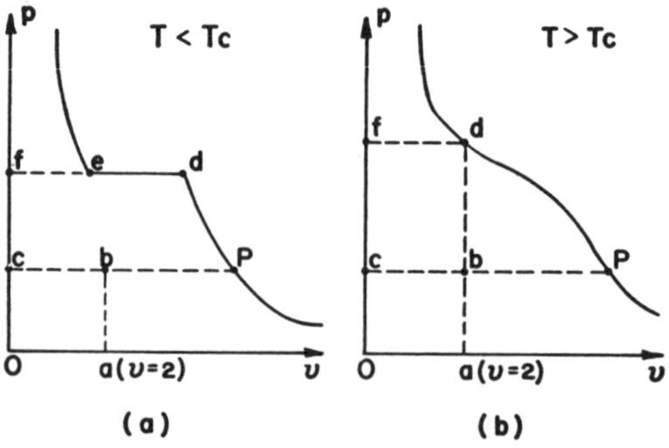

FIG. 1. Geometrical construction of the thermodynamic functions of the Ising model from the $p-v$ diagram of the corresponding lattice gas. Solid curves are isotherms. Given a point P the corresponding Ising model has a magnetic field $H=\frac{1}{2}$ (area $cPdf$), a free energy per spin $F=-\frac{1}{2}$ (area $OabPdf$).

for $\epsilon<0$ (antiferromagnetic case) they repel each other. Since $[\downarrow\downarrow]$ is by the geometrical correspondence the number of nearest neighboring pairs of atoms the energy of the gas is

$$U_G=-2[\downarrow\downarrow]\epsilon.$$

The grand partition function of the gas is

$$\exp(p\mathfrak{N}/kT)=\sum y^{[\downarrow]}\exp(2[\downarrow\downarrow]\epsilon/kT), \tag{10}$$

where p is the pressure of the gas and y the fugacity given by Eq. (3) in paper I. Now on using (2), (3), and (4), one can write Eq. (5) in the form

$$\exp\{-\mathfrak{N}(F+H)/kT\}$$
$$=\sum \exp\{(-2H[\downarrow]-6\epsilon[\downarrow]+2\epsilon[\downarrow\downarrow])/kT\}.$$

Upon comparing this with Eq. (10), one concludes that for a value of the fugacity given by

$$y=\exp\{(-2H-6\epsilon)/kT\}, \tag{11}$$

the pressure p of the lattice gas is related to the free energy F per spin of the Ising lattice by

$$p=-F-H. \tag{12}$$

The same treatment can be easily applied to *any general lattice with arbitrary interactions, yielding results identical with Eqs.* (8) *and* (12). These results are compiled in Table I which lists the corresponding quantities for the problems of an Ising model with a magnetic field and of a lattice gas. The two problems are completely equivalent; the thermodynamical properties of one system can be derived with the aid of Table I from those of the other and vice versa. In particular, the isotherms in the $I-H$ and $p-v$ diagrams bear a very close relationsip to each other. We mention specifically that a discontinuity in I corresponds to a discontinuity in v. In the case of ferromagnetism, below the Curie temperature T_C, I has a discontinuity, and this corresponds to a phase transition at the same temperature in the lattice gas. Above T_C the isotherms in

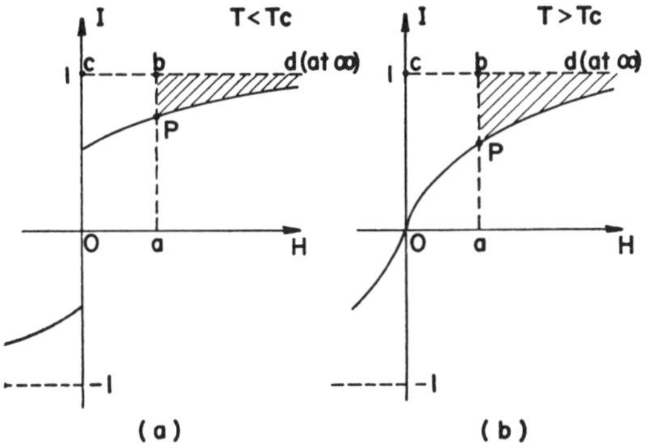

FIG. 2. Geometrical construction of the thermodynamic functions of the lattice gas from the $I-H$ diagram of the corresponding Ising model. Solid curves are isotherms. Given a point P the corresponding lattice gas had a density $\rho=\frac{1}{2}$ (length bP), a pressure $p=$area Pdb. p can be obtained from the free energy of the Ising model which is $F=-$area $OaPdc$.

both of the two diagrams become smooth, and we therefore identify the Curie temperature with the critical temperature of the lattice gas.

From the $p-v$ diagram one can construct the thermodynamic properties of the Ising model in a simple geometric manner and vice versa. These are illustrated in Figs. 1 and 2.

III. AN EXAMPLE OF A TWO-DIMENSIONAL LATTICE GAS

The question naturally arises as to the relationship between a lattice gas and a real gas in which the atoms are not confined to move on lattice points. If one replaces the configurational integral in the partition function of the real gas by a summation over lattice sites, one would obtain the partition function of the lattice gas. Theoretically speaking, by making the lattice constant smaller and smaller one could obtain successively better approximations to the partition function of the real gas.[3] In practice this is a very difficult procedure. However, referring back to the simple example discussed in detail in Sec. II one sees that if $\epsilon>0$ the interaction u there has all the characteristics of the interaction between gas atoms that are usually considered responsible for the phenomenon of condensation, namely, an attractive force with a finite range outside of a strongly repulsive core. Thus one would expect that the main features of the phenomenon of condensation should be revealed even in this simple example.

It happens fortunately that in two dimensions this example can be solved exactly in the transition region. To be more specific, the problem is that of a two-dimensional lattice gas with the interaction u specified in (9). (A square lattice is considered.) ϵ is assumed to be positive so that the gas atoms attract each other. The corresponding Ising model problem has been studied extensively. It is known that the model exhibits ferromagnetism so that according to the discussion of Sec. II the lattice gas has a phase transition. In the Ising model, for temperatures lower than the Curie temperature T_C which is given by

$$\exp(-\epsilon/kT_C)=\sqrt{2}-1, \tag{13}$$

the intensity of magnetization has a discontinuity at $H=0$. The corresponding phase transition of the lattice gas occurs at a value of the fugacity[4]

$$y=x^4,$$

where

$$x=\exp(-\epsilon/kT).$$

Since the free energy[5] and the spontaneous magnetization[6] of the Ising model problem are known exactly at zero magnetic field, the transition region in the $p-v$ diagram of the lattice gas can be mapped out completely with the aid of Table I. We list here the formulas used for the vapor pressure p and specific volumes v_g and v_l of the equilibrium gas and liquid phases:

$$\frac{p}{kT}=\log(1+x^2)+\frac{1}{2\pi}\int_0^\pi \times\log\{\tfrac{1}{2}[1+(1-k_1^2\sin^2\varphi)^{\frac{1}{2}}]\}d\varphi, \tag{14}$$

$$v_g^{-1}=\tfrac{1}{2}-\tfrac{1}{2}[(1+x^2)(1-6x^2+x^4)^{\frac{1}{2}}/(1-x^2)^2]^{\frac{1}{4}}, \tag{15}$$

and

$$v_g^{-1}+v_l^{-1}=1, \tag{16}$$

where

$$k_1=4x(1-x^2)(1+x^2)^{-2}. \tag{17}$$

In Fig. 3 the transition region in the $p-v$ diagram is plotted. The isotherms are calculated in the following way:

(1) For small values of the fugacity y, one uses Mayer's series expansion[7] in powers of y:

$$\frac{p}{kT}=y+y^2\left(\frac{2}{x^2}-\frac{5}{2}\right)+y^3\left(\frac{6}{x^4}-\frac{16}{x^2}+\frac{31}{3}\right)+y^4\left(\frac{1}{x^8}+\cdots\right)+\cdots, \tag{18}$$

---

[3] See Appendix I for a simple illustration on this point.

[4] The relation between the quantity $\exp(-2H/kT)$ of the two-dimensional Ising model in a magnetic field and the fugacity y of the corresponding lattice gas can be easily obtained by using similar reasonings as that used in deriving Eq. (11). One obtains

$$y=x^4\exp(-2H/kT). \tag{A}$$

The fourth power in x comes from the fact that each lattice point has four nearest neighbors. The corresponding equation for any general lattice with arbitrary interaction can be written as

$$y=\sigma\exp(-2H/kT), \tag{B}$$

where σ is a constant and is determined by the structure of the lattice and the interaction between the atoms.

[5] L. Onsager, Phys. Rev. **65**, 117 (1944); B. Kaufman, Phys. Rev. **76**, 1232 (1949).

[6] C. N. Yang, Phys. Rev. **85**, 808 (1952).

[7] J. E. Mayer, J. Chem. Phys. **5**, 67 (1937); J. E. Mayer and Ph. G. Ackermann, J. Chem. Phys. **5**, 74 (1937); J. E. Mayer and S. F. Harrison, J. Chem. Phys. **6**, 87, 101 (1938).

and

$$\frac{1}{v}=y\frac{\partial}{\partial y}\left(\frac{p}{kT}\right). \tag{19}$$

We shall see in the next section that (18) is convergent for all values of y less than x^4.

(2) When y is equal to x^4, the gas undergoes a phase transition at temperatures lower than the critical temperature. The vapor pressure and specific volumes of the gas and liquid at this value of y have been given before in Eqs. (14), (15), and (16).

(3) At $y=x^4$, but above the critical temperature, the pressure of the gas is still given by Eq. (14), but the specific volume v now is a constant

$$v=2. \tag{20}$$

This is so because the corresponding Ising model problem has now, according to Eq. (A),[4] zero magnetic field and consequently zero intensity of magnetization.

(4) For $y>x^4$, the Ising model problem has, according to Eq. (A),[4] a negative magnetic field. Changing the sign of the magnetic field corresponds to changing the fugacity to a value y' given by, according to Eq. (A),[4]

$$yy'=x^8. \tag{21}$$

Since the free energy F is even in H and the intensity of magnetization I is odd, one could express with the aid of Table I the pressure and density in terms of their values when the fugacity is equal to y'. But y' is less than x^4. On using (18) and Table I, one can therefore expand the pressure and density *in inverse powers of the fugacity y:*

$$\frac{p}{kT}=\log\frac{y}{x^4}+y^{-1}x^8+y^{-2}(2x^{14}-(5/2)x^{16})$$
$$+y^{-3}(6x^{20}-16x^{22}+(31/3)x^{24})$$
$$+y^{-4}(x^{24}+\cdots)+\cdots, \tag{22}$$

and

$$\frac{1}{v}=y\frac{\partial}{\partial y}\left(\frac{p}{kT}\right).$$

This series is convergent for all $y>x^4$, which holds everywhere in the liquid phase and also in part of the gas phase above the critical temperature.

It is really quite remarkable that a model with so many properties characteristic of a real gas should allow of a complete and exact solution in the transition region where the usual virial expansion does not apply. A complete solution outside of the transition region is related to the solution of an Ising model in a non-vanishing magnetic field, which still remains unknown. However, the problem can be reduced by the results of paper I, to that of the distribution of roots of the grand partition function; and although the complete distribution is not known, many of its properties have been obtained. We shall return to this problem in Sec. V.

Before closing this section we shall make the following remark. Equation (16) shows that the sum of the densities of the vapor and liquid in equilibrium is a constant independent of the temperature.[8] This closely resembles the behavior of a real gas where a law called "law of rectilinear diameter"[9] is known to hold. It states that the sum of the densities of the vapor and liquid in equilibrium increases linearly with decreasing temperature. This increment is, however, very slow, being not more than 10 percent for a temperature variation over a factor 2 for He near its critical temperature. Now our model obeys what may be called the "law of constant diameter," which provides no increment of the sum of densities with decreasing temperature and may be considered a first approximation to the law of rectilinear diameter. The difference lies, we believe, in the inadequacy of the lattice model (with a finite lattice constant) as an approximation to a true gas. One can formulate arguments which indicate that the correction to the lattice model is in the right direction (i.e., it tends to make the sum of the densities of the vapor and liquid in equilibrium increase with decreasing temperature).

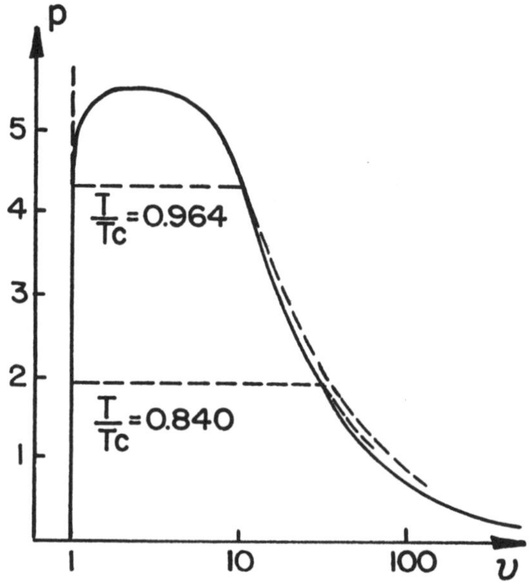

FIG. 3. $p-v$ diagrams for a two-dimensional lattice gas. The solid curve is the exact boundary of the two-phase region. The dotted curves are the isotherms.

[8] The constancy of the sum of the densities of the vapor and liquid in equilibrium is true even for any general lattice. This can be seen very easily since there exists a one to one correspondence between the configurations of the vapor phase and that of the liquid phase, which is obtained simply by replacing the lattice sites occupied by the atoms in the former by vacant sites in the latter and vice versa. Compare Cernuschi and Eyring's paper quoted in footnote 2.

[9] Mathias, Onnes, and Crommelin, Proc. Sect. Sci. Amsterdam **15**, 960 (1913).

IV. A THEOREM ON THE DISTRIBUTION OF ROOTS OF THE GRAND PARTITION FUNCTION

We shall now return to the general problem of the condensation of gases, and shall in the following apply the results of paper I to the problem of a lattice gas. There is actually no loss of generality in confining our attention to a lattice gas, as a real continuum gas can be considered as the limit of a lattice gas as the lattice constant becomes infinitesimally small.

The equivalence proved in Sec. II states that the problem of a lattice gas is identical with that of an Ising model in a magnetic field, and that the grand partition function in the former problem is proportional to the partition function in the latter problem. It is convenient to introduce in the Ising model problem the variable

$$z=\exp(-2H/kT), \tag{23}$$

which is by Eq. (B),[4] proportional to the fugacity y of the lattice gas:

$$y=\sigma z, \tag{24}$$

where σ is a constant. In terms of z the partition function $\exp(-\mathfrak{N}F/kT)$ of the Ising lattice is equal to $\exp(\mathfrak{N}H/kT)$ times a polynomial $\mathcal{P}$ in z of degree $\mathfrak{N}$:

$$\exp(-\mathfrak{N}F/kT)=\mathcal{P}\exp(\mathfrak{N}H/kT), \tag{25}$$

where

$$\mathcal{P}=\sum P_n z^n, \quad (n=0, 1, \cdots \mathfrak{N}). \tag{26}$$

The coefficients P_n are the contribution to the partition function of the Ising lattice in zero external field from configurations with the number of $\downarrow$ spins equal to n. It should be noticed that

$$P_n=P_{n'} \text{ if } n+n'=\mathfrak{N}.$$

The P_n's are, of course, real and positive.

Evidently the roots of the polynomial $\mathcal{P}$ are never on the positive real z axis, and are in general complex.

The results of paper I show that if at a given temperature as $\mathfrak{N}$ approaches infinity, the roots of the polynomial $\mathcal{P}$ do not close in onto the positive real axis in the complex z plane, the free energy F is an analytic function of the positive real variable z. Physically this means that the Ising model has a smooth isotherm in the $I-H$ diagram and that the corresponding lattice gas undergoes no phase transition at the given temperature. If, on the other hand, the roots of the polynomial $\mathcal{P}$ do close in onto the positive real z axis at the points $z=t_1, t_2\cdots$, each of these points would correspond to a discontinuity of the isotherm in the $I-H$ diagram of the Ising lattice and to a phase transition of the lattice gas.

To study the problem of phase transitions of the lattice gas (and of an Ising model), one therefore needs only to study the distribution in the complex z plane of the roots of the polynomial $\mathcal{P}$. The surprising thing is that under quite general conditions this distribution shows a remarkably simple regularity, which may be stated in the form of the following theorem:

Theorem 3. If the interaction u between two gas atoms is such that

$$u=+\infty \text{ if the two atoms occupy the same lattice and } u\leqq 0 \text{ otherwise}, \tag{27}$$

then all the roots of the polynomial $\mathcal{P}$ lie on the unit circle in the complex z-plane.

This theorem will be proved in Appendix II. It should be noticed that in the theorem no assumptions are made about (1) the range of the interaction u, (2) the dimensionality of the lattice, and (3) the size and structure of the lattice. In fact, even the periodicity property of the lattice plays no part at all in the proof.

For the Ising model problem Eq. (27) means that the interaction between all pairs of spins (not limited to pairs of nearest neighbors) are ferromagnetic.

Some immediate consequences of theorem 3 may be enumerated as follows:

(1) The lattice gas cannot undergo more than one phase transition, which must occur, if at all, at a value of the fugacity equal to σ, which according to Eq. (24), corresponds to $z=1$. The isotherms in the $I-H$ diagram of the corresponding Ising model problem is smooth everywhere except possibly at zero magnetic field (which occurs at $z=1$). This is usually believed to be true but was not proved.

(2) Mayer's series expansions[7] of the pressure p and density $1/v$ in powers of the fugacity y [see, e.g., Eq. (18)] are convergent for all values of y less than σ. This is easily proved by the argument used in footnote 9 of paper I. On using the symmetry property of the Ising model problem with respect to a reversal of sign of the magnetic field, one obtains, for values of y greater than σ, convergent series expansions of p and $1/v$ in inverse powers of the fugacity [see, e.g., Eq. (22)].

V. DISTRIBUTION OF ROOTS ON THE UNIT CIRCLE

A. Distribution Function $g(\theta)$

We have seen in the last section that for the interaction (27) the roots of $\mathcal{P}$ lie on the unit circle. Its distribution as $\mathfrak{N}\to\infty$ may be described by a density function $g(\theta)$[10] so that $\mathfrak{N}g(\theta)d\theta$ is the number of roots with z between $e^{i\theta}$ and $e^{i(\theta+d\theta)}$. Evidently one has

$$g(\theta)=g(-\theta), \tag{28}$$

---

[10] The average density of a finite lattice gas is easily seen to be

$$\sum_k z/[z-\exp(i\theta_k)],$$

where $z=\exp(i\theta_k)$ are the zeros of the grand partition function. The results of paper I show that this average density converges to an analytic function in z both inside and outside of the unit circle as the size of the lattice approaches infinity. It seems intuitively clear from this that the distribution of these roots should also approach a limiting distribution on the unit circle for an infinite lattice. This is indeed the case and a rigorous mathematical proof exists in the literature. See A. Wintner, Monatsh. Math. Phys. 4, 1 (1934). We are indebted to Professor Kac for showing us the proof.

and

$$\int_0^{\pi} g(\theta)d\theta=\tfrac{1}{2}. \qquad (29)$$

Taking the logarithm of (25) and factorizing $\mathcal{P}$ one obtains for the Ising model

$$\frac{-F}{kT}=\frac{H}{kT}+\int_0^{2\pi} g(\theta)\log(z-e^{i\theta})d\theta$$
$$=\frac{H}{kT}+\int_0^{\pi} g(\theta)\log(z^2-2z\cos\theta+1)d\theta. \qquad (30)$$

The intensity of magnetization is obtained from Eqs. (6) and (30):

$$I=1-4z\int_0^{\pi} g(\theta)\frac{z-\cos\theta}{z^2-2z\cos\theta+1}d\theta. \qquad (31)$$

For the lattice gas one has, on using Table I and Eqs. (30) and (31),

$$\frac{p}{kT}=\int_0^{\pi} g(\theta)\log(z^2-2z\cos\theta+1)d\theta, \qquad (32)$$

and

$$\frac{1}{v}=2z\int_0^{\pi} g(\theta)\frac{z-\cos\theta}{z^2-2z\cos\theta+1}d\theta. \qquad (33)$$

These equations enable one to calculate the isotherms in the $I-H$ and $p-v$ diagrams from the distribution function $g(\theta)$. The isotherms thus obtained extend to the condensed region, the two-phase region as well as the gas region. They approximate very realistically the isotherms of a real gas even for very simple distribution functions $g(\theta)$.

B. An Electrostatic Analog

A very simple analog of Eqs. (32) and (33) may be found in electrostatics in the following way: Consider a circular cylinder of unit radius perpendicular to the complex z plane discussed above, cutting it at the unit circle. Assume the cylinder to be charged with a surface charge density dependent only on the angle θ and equal to $g(\theta)$ per unit area. Denote the electrostatic potential and the field produced by this charge distribution at any point on the real axis in the z plane by ϕ and E. Since $g(\theta)=g(-\theta)$, the electric field E is evidently parallel to the real z axis. One can easily verify the following equations:

$$-2p/kT=\text{electrostatic potential } \phi, \qquad (34)$$

$$2/vz=-\partial\phi/\partial z=\text{electric field } E. \qquad (35)$$

If $g(\theta)$ vanishes in the neighborhood of $\theta=0$, the sheet of charge has the shape "C" and the electric potential and electric field are well behaved for all real positive values of z. Consequently the pressure and density of the lattice gas are analytic in the fugacity and the system undergoes no phase transition. If, on the other hand, $g(0)$ does not vanish, the electric field has a discontinuity at $z=1$ as one goes along the positive real z axis through the sheet of surface charge distribution. Evidently

$$E(z=1+)-E(z=1-)=4\pi g(0). \qquad (36)$$

This means of course that the specific volume v of the lattice gas has a discontinuity at $z=1$, showing that the system undergoes a phase transition. The specific volumes v_g and v_l of the gas and liquid phases in equilibrium are related to $g(0)$ by

$$1/v_l-1/v_g=2\pi g(0). \qquad (37)$$

Another relation between v_l and v_g is given by the "law of constant diameter" discussed in Sec. 3 (see especially footnote 8).

Equation (37) asserts that the value of the distribution function g at the transition point ($\theta=0$) is equal to $(2\pi)^{-1}$ times the difference of the densities of the liquid and gas phases in equilibrium, which is a directly physically observable quantity.

It should be emphasized that this relationship has actually a much wider range of validity and holds even when the roots are not necessarily on the unit circle.

It is evident that the variation of the pressure and density near the point $z=1$ depends very sensitively on the behavior of $g(\theta)$ near $\theta=0$. One would for example expect that the derivatives of $g(\theta)$ near $\theta=0$ should determine the derivatives of the isotherm in the $p-v$ diagram near the region of condensation. This is indeed the case, as can be seen in the example: If for small values of θ,

$$g(\theta)=g(0)+a|\theta|^n+\cdots, \quad (n>0)$$

the $g(0)$ term, (if nonvanishing), will give rise to a flat horizontal portion on the $p-v$ diagram. The next term will give rise to a discontinuity in the nth derivative of v with respect to p for integral values of n. If n is a fraction between the integers, say, m and $m-1$ then the mth derivative of v with respect to p is discontinuous.

For the Ising model problem, the relation corresponding to Eq. (37) can be obtained by using Table I. The intensity of spontaneous magnetization I is thus expressible in terms of $g(0)$:

$$I=2\pi g(0). \qquad (38)$$

C. Relationship between the Cluster Integrals and $g(\theta)$

If one expands (32) in powers of z and identifies the coefficient with Mayer's expansion,[7]

$$\frac{p}{kT}=\sum_{l=1}^{\infty} b_l y^l=\sum_{l=1}^{\infty} b_l\sigma^l z^l,$$

where σ is given by (24), one obtains the following expression for the cluster integrals b_l:

$$b_l=\frac{-1}{l\sigma^l}\int_0^{2\pi} g(\theta)\cos l\theta d\theta,\quad l\geqq 1. \tag{39}$$

In other words the b_l's are the Fourier coefficients of the distribution function $g(\theta)$. One has, of course, the inverse relation,

$$g(\theta)=\frac{1}{2\pi}-\frac{1}{\pi}\sum_{l=1}^{\infty} l\sigma^l b_l\cos l\theta, \tag{40}$$

which expresses the distribution function $g(\theta)$ in terms of the cluster integrals b_l, which in turn are themselves calculable from the virial coefficients.

Equations (40) and (37) enable one to make practical approximate calculations of the distribution function $g(\theta)$ from the virial coefficients and the change in specific volume in the phase transition, both of which are experimentally osbervable quantities.

D. $g(\theta)$ and the Analytical Behavior of the Specific Volume

The density $1/v$ of the lattice gas is evidently an analytic function of z both inside and outside of the unit circle in the complex z plane. Its value is equal to 1 at $z=\infty$ where it is also analytic. All its singularities lie on the unit circle. The values of $1/v$ on the two sides of the unit circle are related to the distribution function $g(\theta)$ by

$$\lim_{r\to 1+}\frac{1}{v}\bigg|_{z=re^{i\theta}}-\lim_{r\to 1-}\frac{1}{v}\bigg|_{z=re^{i\theta}}=2\pi g(\theta). \tag{41}$$

This equation can be proved by extending the reasoning that led to Eq. (37) in the following way: Consider the electric field at a point not necessarily on the real z axis produced by the charge distribution discussed in subsection B. Let E_x and E_y be the components of the electric field $\|$ and $\perp$ to the real z axis. Defining

$$\mathcal{E}=E_x+iE_y,\quad \mathcal{E}^*=E_x-iE_y,$$

one can easily verify the identity

$$2/vz=\mathcal{E}^* \tag{42}$$

at every point in the complex z plane. The discontinuity of $\mathcal{E}$ across the unit circle at any point $z=e^{i\theta}$ is related to the charge density $g(\theta)$ by

$$\mathcal{E}|_{\text{outside}}-\mathcal{E}|_{\text{inside}}=4\pi g(\theta)z.$$

Taking the complex conjugate of this equation and using (42) one obtains Eq. (41).

The corresponding equation for the Ising model problem is

$$\lim_{r\to 1+} I\bigg|_{z=re^{i\theta}}-\lim_{r\to 1-} I\bigg|_{z=re^{i\theta}}=-4\pi g(\theta). \tag{41a}$$

E. Example of One- and Two-Dimensional Ferromagnetic Ising Model

We shall illustrate the above discussion with the problem of a one-dimensional ferromagnetic Ising model with nearest neighbor interaction in an external magnetic field. This problem can be rigorously solved by the matrix method.[11] We quote here only the results. For a closed chain of $\mathfrak{N}$ spins denote the roots of the partition function by $e^{\pm i\theta_1}$, $e^{\pm i\theta_2}$, $\cdots$. These roots are given by

$$\cos\theta_j=-x^2+(1-x^2)\cos[\pi(2j-1)/\mathfrak{N}], \tag{43}$$

where x is defined to be $\exp(-\epsilon/kT)$ and j runs through all integers 1, $2\cdots$ less than or equal to $\frac{1}{2}(\mathfrak{N}+1)$. As $\mathfrak{N}\to\infty$ these roots distribute themselves continuously on an arc of the unit circle lying to the left of the points

$$z=(1-2x^2)\pm i2x(1-x^2)^{\frac{1}{2}}. \tag{44}$$

For all values of $x>0$, the roots therefore do not close in onto the positive real axis, confirming the well-established fact that a one-dimensional Ising model does not exhibit ferromagnetism. The density of roots is given by the distribution function

$$\begin{aligned} g(\theta)&=\frac{1}{2\pi}\frac{\sin\frac{1}{2}\theta}{(\sin^2\frac{1}{2}\theta-x^2)^{\frac{1}{2}}}, && \text{for } \cos\theta<1-2x^2,\\ g(\theta)&=0, && \text{for } \cos\theta>1-2x^2. \end{aligned} \tag{45}$$

On substituting this into Eq. (30), one obtains the correct free energy. The intensity of magnetization is given by

$$I=\left[\frac{z^2-2z+1}{z^2-2z(1-2x^2)+1}\right]^{\frac{1}{2}}.$$

This is analytic everywhere except at the two points given in Eq. (44). The cut between these two points should be made along the unit circle toward the left of the two points. One easily verifies that the discontinuity across the cut is exactly $4\pi g(\theta)$, as given by Eq. (41a).

For a two-dimensional ferromagnetic Ising square lattice the problem with a finite magnetic field, to our knowledge, has not been solved. However, the following exact knowledge is available about the problem:

(1) The free energy F at zero magnetic field (i.e., $z=1$) was obtained by Onsager[5] as

$$F(z=1)=\frac{-kT}{2\pi^2}\int_0^{\pi}\int_0^{\pi}\log[1+2x(\cos\omega+\cos\omega')+2x^2-2x^3(\cos\omega+\cos\omega')+x^4]d\omega d\omega'. \tag{46}$$

(2) The intensity of spontaneous magnetization at

[11] H. A. Kramers and G. H. Wannier, Phys. Rev. **60**, 252, 263 (1941).

zero magnetic field was obtained by one of us[6] as

$$I(z=1)=\left[\frac{1+x^2}{(1-x^2)^2}(1-6x^2+x^4)^{\frac{1}{2}}\right]^{\frac{1}{4}} \quad \text{for } x\leqq\sqrt{2}-1,$$
$$I(z=1)=0 \qquad\qquad \text{for } x>\sqrt{2}-1. \qquad (47)$$

(3) By an extension of a method due to Kac and Ward[12] for obtaining a combinatorial solution of the free energy of the Ising model problem in the absence of a magnetic field, one could obtain[13] the free energy at $z=-1$ which corresponds to a pure imaginary magnetic field equal to $i\pi/2$:

$$F(z=-1)=-\frac{i\pi}{2}-\frac{kT}{4\pi^2}\int_0^\pi\int_0^\pi \log\{(1-x^2)^2 \times[1+(6-4\cos^2\omega-4\cos^2\omega')x^2+x^4]\}d\omega d\omega'. \qquad (48)$$

(4) The intensity of magnetization at this imaginary value[13] of the magnetic field is

$$I(z=-1)=\left[\frac{(1+x^2)^2}{1-x^2}(1+6x^2+x^4)^{-\frac{1}{2}}\right]^{\frac{1}{4}}. \qquad (49)$$

This can be verified by series expansion of both sides in powers of x.

The results quoted above have direct bearing on the distribution function $g(\theta)$ of the zeros of the partition function on the unit circle. In particular (47) and (49) are precisely the value of $g(\theta)/2\pi$ at $\theta=0$ and $\theta=\pi$. It is interesting to notice that $g(0)/2\pi$ is always less than unity while $g(\pi)/2\pi$ is always greater than unity. Also $g(0)$ increases with decreasing temperature while $g(\pi)$ decreases with decreasing temperature. This shows the motion of the roots toward the right along the unit circle as the temperature decreases.

On the other hand, (46) and (48) give through Eq. (30) certain averages with respect to the distribution function $g(\theta)$. The form of these averages are extremely suggestive and we have tried to construct the distribution function from them, but without success.

VI. CONCLUDING REMARKS

The relation between the distribution of roots of a polynomial and its coefficients is mathematically a very complicated problem. It is therefore very surprising that the distribution should exhibit such simple regularities as proved in theorem 3 which appliès, as remarked before, under very general conditions. One cannot escape the feeling that there is a very simple basis underlying the theorem, with much wider application, which still has to be discovered.

It is a great pleasure to thank Professor M. Kac for many stimulating and very pleasant discussions from which we learned much in mathematics.

[12] M. Kac and J. C. Ward (to be published).

[13] We hope to publish in a future communication the details of the steps that led to (48) and (49).

APPENDIX I. ONE-DIMENSIONAL HARD RODS AND LATTICE GAS

A simple example that illustrates clearly the relationship between a continuum gas and a lattice gas is the problem of one-dimensional hard rods and the corresponding problem of a one-dimensional lattice gas, as the problem can be exactly solved in both cases. To be specific, let us consider a lattice gas of n atoms distributed on a one-dimensional lattice of total length L and lattice constant δ. The total number of lattice sites is clearly

$$\mathfrak{N}=L/\delta. \qquad (50)$$

The interaction potential $u(r)$ between the gas atoms is

$$u(r)=+\infty \text{ for } r<m\delta, \quad u(r)=0 \text{ for } r\geqq m\delta, \qquad (51)$$

where r is the distance between atoms and m is a positive integer. Evidently, r can only be 0, δ, 2δ, etc.

The evaluation of the spatial partition function Q can be reduced to that of a simple combinatorial problem of distribution $(\mathfrak{N}-mn)$ identical pieces into $(n+1)$ different bags. One easily obtains

$$Q/n!=[\mathfrak{N}-(m-1)n]!/(\mathfrak{N}-mn)!n!. \qquad (52)$$

The pressure p of the gas is related to Q through the relation

$$\frac{p}{kT}=\frac{\partial}{\partial L}\log\left(\frac{Q}{n!}\right). \qquad (53)$$

On using (50) and differentiating (52) one obtains

$$\frac{p}{kT}=\frac{1}{\delta}\log\left(1+\frac{n\delta}{L-mn\delta}\right). \qquad (54)$$

In the limit of an infinitesimal lattice constant δ but a finite value D for $m\delta$, the pressure approaches a limit given by

$$\lim_{\substack{\delta\to 0\\ m\delta=D}}\frac{p}{kT}=\frac{n}{L-nD}, \qquad (55)$$

which is precisely the expression for the pressure of a system of n hard rods each of length D, enclosed in a one-dimensional box of total length L. Thus one sees that when the lattice constant δ approaches zero the thermodynamic functions of a lattice gas indeed approach that of a continuum gas.

APPENDIX II. PROOF OF THEOREM 3

Theorem 3 is a special case of the following more general theorem:

Theorem: Let $x_{\alpha\beta}=x_{\beta\alpha}$ ($\alpha\neq\beta$, α, $\beta=1, 2, \cdots n$) be real numbers *whose absolute values are less than or equal to* 1. Divide the integers $1, 2, \cdots n$ into two groups a and b so that there are γ integers in group a and $(n-\gamma)$ in group b. Consider the product of all $x_{\alpha\beta}$ where α belongs to group a and β to group b. We shall denote by P_γ the sum of all such products over all the

$n!/[\gamma!(n-\gamma)!]$ possible ways of dividing the n integers. In other words

$$P_\gamma=\sum[\gamma!(n-\gamma)!]^{-1}\prod_{j=1}^{n-\gamma}\prod_{i=1}^{\gamma}x_{a_ib_j}, \tag{56}$$

where $a_1\cdots a_\gamma$, $b_1\cdots b_{n-\gamma}$ is any permutation of the integers 1, 2, $\cdots n$ and the summation extends over all such permutations; e.g.,

$$P_0=P_n=1,\quad P_1=P_{n-1}=x_{12}x_{13}\cdots x_{1n}+x_{21}x_{23}\cdots x_{2n}+\cdots+x_{n1}x_{n2}\cdots x_{n(n-1)}.$$

It is easy to verify that $P_\alpha=P_{n-\alpha}$. Consider the polynomial

$$\mathcal{P}(z)=1+P_1z+P_2z^2+\cdots+P_{n-1}z^{n-1}+z^n. \tag{57}$$

The theorem asserts that all the roots of the equation

$$\mathcal{P}=0$$

are on the unit circle.

To prove this theorem it is convenient to introduce the following polynomials $\mathfrak{P}$ of the variables z_1, z_2, $\cdots z_n$:

$$\mathfrak{P}(z_1,\cdots z_n)=\sum\left\{[\gamma!(n-\gamma)!]^{-1}\left(\prod_{j=1}^{n-\gamma}\prod_{i=1}^{\gamma}x_{a_ib_j}\right)\prod_{i=1}^{\gamma}z_{a_i}\right\}, \tag{58}$$

where, as in the definition of P_γ, $a_1, a_2, \cdots a_\gamma, b_1, \cdots b_{n-\gamma}$ is any permutation of the integers 1, 2, $\cdots n$ and the summation extends over all such permutations and over all γ. It is clear that

$$\mathfrak{P}(z,z,z,\cdots z)=\mathcal{P}(z). \tag{58a}$$

To prove the theorem it is only necessary to prove that there are no roots of $\mathcal{P}=0$ that have an absolute value >1. Equation (58a) therefore shows that the theorem is an immediate consequence of the following lemma:

Lemma: If $\mathfrak{P}(z_1, z_2, \cdots z_n)=0$ and none of the z's has an absolute value less than one, then

$$1=|z_1|=|z_2|=\cdots=|z_n|.$$

Proof: (1) We shall assume throughout the proof that all the x's are different from zero and ±1. The proof can then be easily generalized to include the case when one or more of the x's either vanish or are equal to ±1.

The lemma is clearly true when $n=1$, for which $\mathfrak{P}_1=1+z_1$. For $n=2$,

$$\mathfrak{P}_2=1+x_{12}(z_1+z_2)+z_1z_2.$$

$\mathfrak{P}_2=0$ therefore implies

$$z_1=-(1+x_{12}z_2)/(z_2+x_{12}).$$

It is easy to prove that for $|z_2|>1$ the right-hand side of this equation always has an absolute magnitude <1. Hence if $|z_1|\geqq1$, $|z_2|\geqq1$, they must both be equal to 1, proving the lemma for the case $n=2$.

(2) We shall prove the lemma for general values of n by induction. Assume it is true for $n=m-2$ and $n=m-1$, but not true for $n=m$. We shall call this assumption hypothesis A and prove that it is self-contradictory. Under hypothesis A there exists a set of z's equal to z_1, z_2, $\cdots z_m$ such that

$$\mathfrak{P}_m(z_1, z_2, \cdots z_m)=0 \tag{59a}$$

and

$$|z_1|>1,\quad\text{and}\quad |z_2|, |z_3|\cdots, |z_m| \text{ all } \geqq1. \tag{59b}$$

The subscript m in $\mathfrak{P}_m$ is to indicate that it has m independent variables z.

(3) We now prove in this paragraph that keeping z_3, z_4, $\cdots z_m$ fixed and regarding z_2 as a function of z_1 defined by (59a) one obtains a limit $\mathfrak{z}_2$ for z_2 as $z_1\to\infty$ and that $|\mathfrak{z}_2|<1$. To prove this we take, say, $m=3$ and notice that

$$\mathfrak{P}_3=1+z_1x_{12}x_{13}+z_2x_{21}x_{23}+z_3x_{31}x_{32}+z_1z_2x_{13}x_{23}+z_2z_3x_{21}x_{31}+z_3z_1x_{32}x_{12}+z_1z_2z_3. \tag{61}$$

As $z_1\to\infty$, $\mathfrak{P}_3=0$ gives

$$x_{12}x_{13}+\mathfrak{z}_2x_{13}x_{23}+z_3x_{32}x_{12}+\mathfrak{z}_2z_3=0, \tag{62}$$

i.e.,

$$\mathfrak{z}_2=-(x_{12}x_{13}+z_3x_{32}x_{12})/(z_3+x_{13}x_{23}). \tag{63}$$

It is easy to prove that unless the denominator vanishes this does give the correct limit for z_2. But the denominator cannot vanish, as will be seen later. Hence the limit $\mathfrak{z}_2$ is correctly given by (62). Now write

$$\mathfrak{z}_2x_{12}^{-1}=\zeta_2,\quad z_3x_{13}^{-1}=\zeta_3. \tag{64}$$

Equation (62) reduces to

$$1+\zeta_2x_{23}+\zeta_3x_{23}+\zeta_2\zeta_3=0.$$

But this is exactly an equation of the form

$$\mathfrak{P}_2(\zeta_2\zeta_3)=0,$$

where $\mathfrak{P}_2$ is a polynomial of the general form (58) with $n=2$. Now by condition (59b)

$$|\zeta_3|>|z_3|\geqq1.$$

Hence hypothesis A asserts that $|\zeta_2|<1$. Hence

$$|\mathfrak{z}_2|<|\zeta_2|<1.$$

We need now only prove that the denominator in (63) does not vanish. This is evident in the present case of $m=3$. We shall, however, give a formal proof which holds in general for any value of m. The denominator in (63) is clearly the coefficient of z_2z_1 in $\mathfrak{P}_3$ given in (61), just as the left-hand side of (62) is the coefficient of z_1 in $\mathfrak{P}_3$. We therefore make a similar transformation as (64):

$$z_3x_{13}^{-1}x_{23}^{-1}=\zeta_3',$$

and reduce the denominator in (63) into the form

$$x_{13}x_{23}\mathfrak{P}_1(\zeta_3'). \tag{65}$$

Clearly $|\zeta_3'|>1$ so that hypothesis A implies that (65) does not vanish. The above arguments evidently hold for any general value of m and this completes the proof that as $z_1\to\infty$, z_2 approaches a limit smaller than unity in absolute value.

(4) Keeping $z_3, z_4\cdots z_m$ fixed one can increase $|z_1|$ and define z_2 as a continuous function of z_1. Since by (59b), $|z_2|$ starts to be $\geqq 1$ in absolute magnitude and tends to a limit <1 in absolute magnitude as $z_1\to\infty$, there must be a value of z_1 equal to z_1' so that z_2 assumes a value z_2' equal to 1 in absolute magnitude, i.e.,

$$\mathfrak{P}_m(z_1', z_2', z_3, \cdots z_m)=0 \tag{66a}$$

and

$$|z_1'|>1, \quad |z_2'|=1, \quad |z_3|, \quad |z_4|, \quad \cdots |z_m|\geqq 1. \tag{66b}$$

We can fix z_2', z_4, z_5, $\cdots z_m$ and regard z_3 as a function of z_1' and follow the same procedure by increasing $|z_1'|$ till z_3 assumes a value equal to 1 in absolute magnitude. Continuing this way we finally get a set of values z_1'', z_2'', $\cdots z_n''$ such that

$$\mathfrak{P}_m(z_1'', z_2'', \cdots z_m'')=0, \tag{67a}$$

and

$$|z_1''|>1, \quad |z_2''|=|z_3''|=\cdots=|z_m''|=1. \tag{67b}$$

But $\mathfrak{P}_m$ is linear in z_1''. Writing $\mathfrak{P}_m=Bz_1''+C$ where B and C are independent of z_1'' one verifies easily that

$$B=z_2''z_3''\cdots z_m''\mathfrak{C},$$

where $\mathfrak{C}$ is the complex conjugate of C under the condition (67b). Hence

$$|z_1''|=|C|/|B|=1, \tag{68}$$

which contradicts (67b). (It is easy to show that B does not vanish by making a transformation similar to (64) and reduce B to products of some x's with $\mathfrak{P}_{m-1}$.)

This completes the proof by induction.

Letter to E. Fermi dated May 5, 1952

C. N. Yang

[52d]

Commentary begins page 17

May 5, 1952

Professor E. Fermi
Institute for Nuclear Studies
University of Chicago
Chicago 37, Illinois

Dear Professor Fermi:

Many thanks to you and Herb for sending me your manuscripts.

I have spent some time looking into the question of the uniqueness of the phase shifts. It is found that the two sets of phase shifts:

$$\text{Set } A: \quad \alpha_3 = 26°,\ \alpha_1 = -1°,\ \alpha_{33} = -37°,$$
$$\alpha_{31} = -11°,\ \alpha_{13} = 2°,\ \alpha_{11} = 20°;$$
$$\text{Set } B: \quad \alpha_3 = 26°,\ \alpha_1 = -1°,\ \alpha_{33} = -20°,$$
$$\alpha_{31} = -46°,\ \alpha_{13} = 15°,\ \alpha_{11} = -3°;$$

(set A is the set you gave in your paper) lead to identical cross sections at all angles for the $\pi^+ + p \to \pi^+ + p$ scattering. For the two other scatterings the differential cross sections at 180° and 0° are also identical for the two sets A and B. At 90°, however, they give slightly different cross sections:

$$\frac{[\sigma_{90°}(\pi^- + p \to \pi^- + p)]_A}{[\sigma_{90°}(\pi^- + p \to \pi^- + p)]_B} = 1.1$$

$$\frac{[\sigma_{90°}(\pi^- + p \to \pi^° + n)]_A}{[\sigma_{90°}(\pi^- + p \to \pi^° + n)]_B} = 0.91$$

Within the present experimental errors, the two sets of phase shifts A and B are, I believe, both possible.

Whether there are other possible sets it is very difficult to decide. I tend to believe there are none. If this is correct the phase shifts for the S waves (which are identical for sets A and B) are determinate, although not those for the P waves.

The set B is constructed from the set A by the formulae:

$$\begin{aligned}(\alpha_{33})_B &= \eta_3 - (\alpha_{33})_A,\ (\alpha_{31})_B = \eta_3 - (\alpha_{31})_A \\ (\alpha_{13})_B &= \eta_1 - (\alpha_{13})_A,\ (\alpha_{11})_B = \eta_1 - (\alpha_{11})_A\end{aligned} \tag{1}$$

where

$$\eta_3 = \text{angle of } (2e^{2i\alpha_{33}} + e^{2i\alpha_{31}}) = -57°$$
$$\eta_1 = \text{angle of } (2e^{2i\alpha_{13}} + e^{2i\alpha_{11}}) = 17°$$

The reason for the possibility of two sets A and B of phase shifts that give approximately identical cross sections is that (a) they give identical scattered waves for non-spin–flip scattering and (b) for spin–flip scattering they give scattered waves of identical magnitude but different phases in the $I = 3/2$ state; and also in the $I = 1/2$ state. The only possible difference therefore lies in the interference between scattered spin–flip $I = 3/2$ and $I = 1/2$ states, which only shows up in the scattering at 90° for $\pi^- + p \to \pi^- + p$ and $\pi^- + p \to \pi^° + n$.

It is evident that if it happens that the interference term between the scattered spin–flip waves in the $I = 3/2$ and $I = 1/2$ states is zero, the two sets A and B of phase shifts defined by (1) would give exactly identical cross sections at all angles and for all three reactions. This would be the case if

$$(A) \quad \alpha_{33} = \alpha_{31}$$

or

$$(B) \quad \alpha_{13} = \alpha_{11}$$

If neither (A) nor (B) is fulfilled, but

$$(C) \quad \alpha_{33} - \alpha_{31} = \alpha_{13} - \alpha_{11}$$

then the two sets of phase shifts would still give

identical cross sections. In your experiment this is almost the case, so that the difference in the cross sections is only $\sim 10\%$.

At 110 MeV a similar set *B* of phase shifts can be found:

Set *A*: $\alpha_3 = 15°$ $\alpha_{33} = -25°$ $\alpha_{31} = 0°$

Set *B*: $\alpha_3 = 15°$ $\alpha_{33} = -8°$ $\alpha_{31} = -33°$

There is of course no a priori reason why condition (C) should still be approximately valid, so that experimental results at this energy might differentiate between the two sets *A* and *B*. If that is the case, by arguments of continuity with respect to the energy one would also be able to differentiate between the two possibilities in the 135 MeV case.

Our trip to Seattle was very pleasant although slow. The flood was ahead of us and did not give us any trouble. We like Seattle very much and are having a nice time here.

With best regards.

Sincerely,

C. N. Yang

Polarization of Nucleons Elastically Scattered from Nuclei*

[[54a]]
Commentary begins page 18

G. A. Snow, R. M. Sternheimer, and C. N. Yang†
Brookhaven National Laboratory, Upton, New York
(Received March 30, 1954)

SEVERAL recent experiments[1–3] have shown that protons of energy 200–350 Mev scattered from nuclei are polarized. The elastically scattered protons have a polarization (~60 percent) that is somewhat larger than the inelastically scattered ones. Fermi[4] has proposed an explanation of the polarization for *elastic* scattering in terms of a nuclear spin-orbit interaction potential similar to that assumed in the nuclear shell model. He used the Born approximation in his estimates. The purpose of this note is to investigate the polarization effects of a nuclear spin-orbit potential using the transparent nuclear model of Serber.[5] That is, we add to the nuclear complex potential which is constant over a sphere of radius R, a term

$$-\hbar^{-1}U(r)\mathbf{L}\cdot\boldsymbol{\sigma}, \tag{1}$$

where $\mathbf{L}$ and $\hbar\boldsymbol{\sigma}/2$ are the orbital and spin angular momenta of the proton. U is taken to be real since the absorption cross section of protons in nuclear matter is independent of its spin. The radial and energy dependence of U are unknown, as well as its variation with atomic number. We have assumed two specific forms for U:

$$U_1(r)=u_1R\delta(r-R), \quad U_2(r)\cong u_2(R/r)^2. \tag{2}$$

$U_1(r)$ is the one considered by Fermi and is suggested by the Thomas precession of a particle with spin under acceleration, which, in the transparent nuclear model, is concentrated at the boundary of the potential hole. By contrast, $U_2(r)$ is concentrated near the origin. [The exact form of U_2, which was chosen for convenience in numerical computation, is implied in Eq. (8).]

The calculations are performed by a partial-wave analysis. The degree of polarization (as defined by Oxley *et al.*[1]) is

$$P(\theta)=(|f-ig|^2-|f+ig|^2)/(|f-ig|^2+|f+ig|^2), \tag{3}$$

where

$$f(\theta)=\sum_{l=0}^{L}[(l+1)A_l^+ + lA_l^-]P_l^0(\cos\theta), \tag{4}$$

$$g(\theta)=-\sum_{l=0}^{L}[A_l^+ - A_l^-]P_l^1(\cos\theta), \tag{5}$$

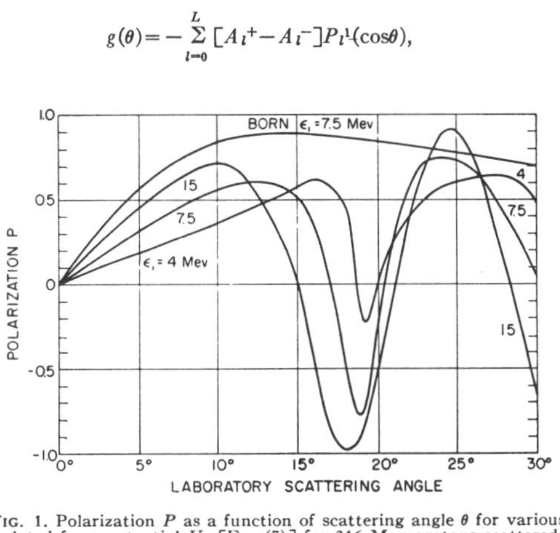

Fig. 1. Polarization P as a function of scattering angle θ for various ϵ_1 calculated from potential U_1 [Eq. (7)] for 316-Mev protons scattered by Be. The top curve gives the Born approximation values for $\epsilon_1=7.5$ Mev.

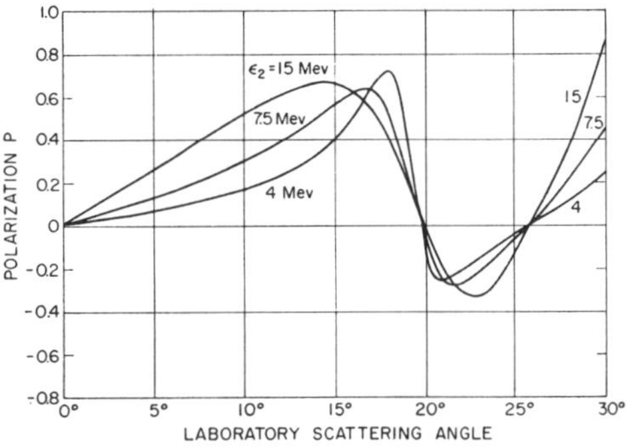

Fig. 2. Polarization P as a function of scattering angle θ for various ϵ_2 calculated from potential U_2 [Eq. (8)] for 316-Mev protons scattered by Be.

with

$$A_l^\pm=[\exp(2i\delta_l^\pm)-1]/2i. \tag{6}$$

$\delta_l^\pm$ are the phase shifts for the partial waves with $J=l\pm\frac{1}{2}$, P_l^m = associated Legendre polynomial, and L = largest integer $<kR$. Using the WKB approximation to evaluate $\delta_l^\pm$, we obtain

$$(\delta_l^\pm)_1=[(iK/2)+k_1][R^2-(l+\tfrac{1}{2})^2/k^2]^{\frac{1}{2}} \pm\tfrac{1}{2}\epsilon_1 lT^{-1}[1-(l+\tfrac{1}{2})^2/(L+\tfrac{1}{2})^2]^{-\frac{1}{2}}, \tag{7}$$

and

$$(\delta_l^\pm)_2=[(iK/2)+k_1\pm(k\epsilon_2/2T)][R^2-(l+\tfrac{1}{2})^2/k^2]^{\frac{1}{2}}, \tag{8}$$

for the two forms of spin-orbit interaction respectively. Here T = kinetic energy, k_1 and K are the usual optical parameters. $\epsilon_{1,2}$ are energies that characterize the depth of the spin-orbit coupling. $\epsilon_1=Lu_1$ and $\epsilon_2=Lu_2$. The two potential differ in that U_1 emphasizes the high, U_2 the low angular momenta phase shifts.

For definiteness, we have calculated P, using Eqs. (3)–(8), foi 316-Mev nucleons scattered by Be. This energy corresponds to the experiments of Marshall *et al.*[2] We take the optical parameters from a recent paper by Taylor,[6] i.e., $R=3.2\times10^{-13}$ cm, $k_1=0.86\times10^{12}$ cm^{-1}, and $K=1.7\times10^{12}$ cm^{-1}, corresponding to a complex nuclear potential $V=(-13+25.6i)$ Mev, and $L=13$. Figure 1 shows $P(\theta)$ for the spin-orbit potential $U_1(r)$ with $\epsilon_1=4$, 7.5, and 15 Mev. For comparison, the top curve in this figure shows the Born approximation result with $\epsilon_1=7.5$ Mev and the same nuclear parameters. Figure 2 shows $P(\theta)$ for $U_2(r)$ with $\epsilon_2=4$, 7.5, and 15 Mev and Fig. 3 shows the differential cross section averaged over spin directions,

$$\langle d\sigma/d\Omega\rangle=\tfrac{1}{2}k^{-2}[|f-ig|^2+|f+ig|^2], \tag{9}$$

for $\epsilon_2=0$–15 Mev.

It is clear from the curves that one can fit the result of Marshall *et al.*[2] that $P(14°)\cong0.6$ with $\epsilon_1=7.5$ Mev (i.e., $u_1=0.58$ Mev) or $\epsilon_2=12$ Mev (i.e., $u_2=0.92$ Mev). These represent spin-orbit interactions that agree roughly in order of magnitude with that in the shell model theory provided one assumes a linear dependence of the spin-orbit interaction with the momentum of the nucleon.

□ Reprinted from *The Physical Review* 94, 4 (May 15, 1954), 1073–1074.

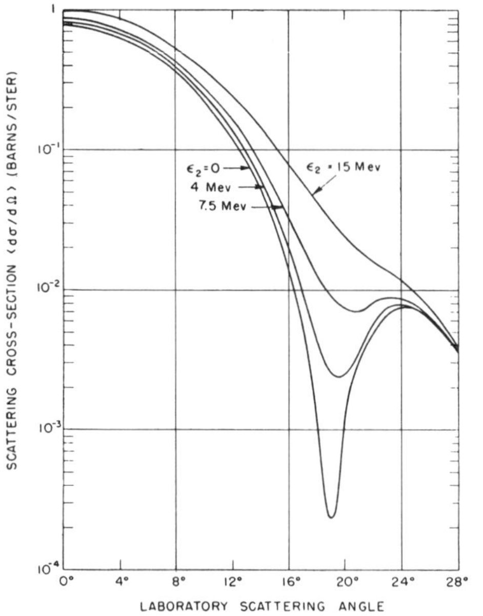

FIG. 3. Differential scattering cross section $\langle d\sigma/d\Omega\rangle$ averaged over spin directions for 316-Mev protons scattered by Be. The curves were calculated from potential U_2 for various ϵ_2.

The curves show that for small θ, P is proportional to ϵ. For a given ϵ, P is larger for interaction U_1 than U_2. This results from the fact already mentioned that U_1 emphasizes the larger and more heavily weighted l values more than U_2.

Perhaps the most interesting result of these calculations is the oscillatory nature of $P(\theta)$. $P(\theta)$ first becomes negative in the region of the first diffraction minimum ($\sim 20°$) due to a change of sign of $f(\theta)$. It becomes positive again when $g(\theta)$ also changes sign. The fact that the Born approximation applied to U_1 does not show this, is fortuitously due to f and g having identical angular dependences. The search for negative $P(\theta)$ may be hampered of course by the low intensity of the diffracted beam beyond the first minimum.

Finally it is interesting to note in Fig. 3 how the spin-orbit potential has the effect of washing out the deep minima in $\langle d\sigma(\theta)/d\Omega\rangle$. There are some experimental indications of such an effect.[7] It is also seen that the total diffraction cross section σ_d increases with ϵ. The values of σ_d as obtained from integration of $\langle d\sigma(\theta)/d\Omega\rangle$ are 61, 67, 75, and 105 mb for $\epsilon_2=0$, 4, 7.5, and 15 Mev, respectively.

Perhaps detailed investigations of $P(\theta)$ for various elements and at lower energies will be able to determine the radial and velocity dependence of $U(r)$, although it must be borne in mind that $P(\theta)$ is a sensitive function of all the optical parameters R, k_1, and K as well as $U(r)$. One can also conjecture that the polarization of elastically scattered nucleons will still be present in the Bev range. For example, a calculation was carried out for 1.4-Bev neutrons on Be with $\epsilon_2=15$ Mev, and one obtains $P(\theta=5°)=0.46$ and $P(\theta=7°)=0.81$.

We wish to thank Professor Fermi and Professor Marshall for sending us copies of their papers in advance of publication.

* Work performed under the auspices of the U. S. Atomic Energy Commission.

† On leave of absence from the Institute for Advanced Study, Princeton, New Jersey.

[1] Oxley, Cartwright, and Rouvina, Phys. Rev. **93**, 806 (1954).
[2] Marshall, Marshall, and de Carvalho, Phys. Rev. **93**, 1431 (1954).
[3] Chamberlain, Segrè, Tripp, Wiegand, and Ypsilantis, Phys. Rev. **93**, 1430 (1954).
[4] E. Fermi, Nuovo cimento **11**, 407 (1954).
[5] Fernbach, Serber, and Taylor, Phys. Rev. **75**, 1352 (1949).
[6] T. B. Taylor, Phys. Rev. **92**, 831 (1953).
[7] Richardson, Ball, Leith, and Moyer, Phys. Rev. **83**, 859 (1951).

Isotopic Spin Conservation and a Generalized Gauge Invariance*

[[54b]]

Commentary begins page 19

C. N. Yang† *and R. Mills*
Brookhaven National Laboratory

The conservation of isotopic spin points to the existence of a fundamental invariance law similar to the conservation of electric charge. In the latter case, the electric charge serves as a source of electromagnetic field; an important concept in this case is gauge invariance which is closely connected with (1) the equation of motion of the electromagnetic field, (2) the existence of a current density, and (3) the possible interactions between a charged field and the electromagnetic field. We have tried to generalize this concept of gauge invariance to apply to isotopic spin conservation. It turns out that a very natural generalization is possible. The field that plays the role of the electromagnetic field is here a vector field that satisfies a nonlinear equation even in the absence of other fields. (This is because unlike the electromagnetic field this field has an isotopic spin and consequently acts as a source of itself.) The existence of a current density is automatic, and the interaction of this field with any fields of arbitrary isotopic spin is of definite form (except for possible terms similar to the anomalous magnetic moment interaction terms in electrodynamics).

---

* Work performed under the auspices of the U. S. Atomic Energy Commission.

† On leave of absence from the Institute for Advanced Study, Princeton, New Jersey.

□ Reprinted from *The Physical Review* 95 (1954), 631.

[54c]
Commentary begins page 19

Conservation of Isotopic Spin and Isotopic Gauge Invariance*

C. N. YANG † AND R. L. MILLS
Brookhaven National Laboratory, Upton, New York
(Received June 28, 1954)

It is pointed out that the usual principle of invariance under isotopic spin rotation is not consistant with the concept of localized fields. The possibility is explored of having invariance under local isotopic spin rotations. This leads to formulating a principle of isotopic gauge invariance and the existence of a **b** field which has the same relation to the isotopic spin that the electromagnetic field has to the electric charge. The **b** field satisfies nonlinear differential equations. The quanta of the **b** field are particles with spin unity, isotopic spin unity, and electric charge $\pm e$ or zero.

INTRODUCTION

THE conservation of isotopic spin is a much discussed concept in recent years. Historically an isotopic spin parameter was first introduced by Heisenberg[1] in 1932 to describe the two charge states (namely neutron and proton) of a nucleon. The idea that the neutron and proton correspond to two states of the same particle was suggested at that time by the fact that their masses are nearly equal, and that the light stable even nuclei contain equal numbers of them. Then in 1937 Breit, Condon, and Present pointed out the approximate equality of $p-p$ and $n-p$ interactions in the ${}^1S$ state.[2] It seemed natural to assume that this equality holds also in the other states available to both the $n-p$ and $p-p$ systems. Under such an assumption one arrives at the concept of a total isotopic spin[3] which is conserved in nucleon-nucleon interactions. Experi-

* Work performed under the auspices of the U. S. Atomic Energy Commission.

† On leave of absence from the Institute for Advanced Study, Princeton, New Jersey.

1 W. Heisenberg, Z. Physik **77**, 1 (1932).

2 Breit, Condon, and Present, Phys. Rev. **50**, 825 (1936). J. Schwinger pointed out that the small difference may be attributed to magnetic interactions [Phys. Rev. **78**, 135 (1950)].

3 The total isotopic spin **T** was first introduced by E. Wigner, Phys. Rev. **51**, 106 (1937); B. Cassen and E. U. Condon, Phys. Rev. **50**, 846 (1936).

□ Reprinted from *The Physical Review* 96, 1 (October 1, 1954), 191–195.

ments in recent years[4] on the energy levels of light nuclei strongly suggest that this assumption is indeed correct, An implication of this is that all strong interactions such as the pion-nucleon interaction, must also satisfy the same conservation law. This and the knowledge that there are three charge states of the pion, and that pions can be coupled to the nucleon field *singly*, lead to the conclusion that pions have isotopic spin unity. A direct verification of this conclusion was found in the experiment of Hildebrand[5] which compares the differential cross section of the process $n+p\rightarrow\pi^0+d$ with that of the previously measured process $p+p\rightarrow\pi^++d$.

The conservation of isotopic spin is identical with the requirement of invariance of all interactions under isotopic spin rotation. This means that when electromagnetic interactions can be neglected, as we shall hereafter assume to be the case, the orientation of the isotopic spin is of no physical significance. The differentiation between a neutron and a proton is then a purely arbitrary process. As usually conceived, however, this arbitrariness is subject to the following limitation: once one chooses what to call a proton, what a neutron, at one space-time point, one is then not free to make any choices at other space-time points.

It seems that this is not consistent with the localized field concept that underlies the usual physical theories. In the present paper we wish to explore the possibility of requiring all interactions to be invariant under *independent* rotations of the isotopic spin at all space-time points, so that the relative orientation of the isotopic spin at two space-time points becomes a physically meaningless quantity (the electromagnetic field being neglected).

We wish to point out that an entirely similar situation arises with respect to the ordinary gauge invariance of a charged field which is described by a complex wave function ψ. A change of gauge[6] means a change of phase factor $\psi\rightarrow\psi'$, $\psi'=(\exp i\alpha)\psi$, a change that is devoid of any physical consequences. Since ψ may depend on x, y, z, and t, the relative phase factor of ψ at two different space-time points is therefore completely arbitrary. In other words, the arbitrariness in choosing the phase factor is local in character.

We define *isotopic gauge* as an arbitrary way of choosing the orientation of the isotopic spin axes at all space-time points, in analogy with the electromagnetic gauge which represents an arbitrary way of choosing the complex phase factor of a charged field at all space-time points. We then propose that all physical processes (not involving the electromagnetic field) be invariant under an isotopic gauge transformation, $\psi\rightarrow\psi'$, $\psi'=S^{-1}\psi$, where S represents a space-time dependent isotopic spin rotation.

To preserve invariance one notices that in electrodynamics it is necessary to counteract the variation of α with x, y, z, and t by introducing the electromagnetic field A_μ which changes under a gauge transformation as

$$A_\mu'=A_\mu+\frac{1}{e}\frac{\partial\alpha}{\partial x_\mu}.$$

In an entirely similar manner we introduce a B field in the case of the isotopic gauge transformation to counteract the dependence of S on x, y, z, and t. It will be seen that this natural generalization allows for very little arbitrariness. The field equations satisfied by the twelve independent components of the B field, which we shall call the **b** field, and their interaction with any field having an isotopic spin are essentially fixed, in much the same way that the free electromagnetic field and its interaction with charged fields are essentially determined by the requirement of gauge invariance.

In the following two sections we put down the mathematical formulation of the idea of isotopic gauge invariance discussed above. We then proceed to the quantization of the field equations for the **b** field. In the last section the properties of the quanta of the **b** field are discussed.

ISOTOPIC GAUGE TRANSFORMATION

Let ψ be a two-component wave function describing a field with isotopic spin $\frac{1}{2}$. Under an isotopic gauge transformation it transforms by

$$\psi=S\psi', \tag{1}$$

where S is a 2×2 unitary matrix with determinant unity. In accordance with the discussion in the previous section, we require, in analogy with the electromagnetic case, that all derivatives of ψ appear in the following combination:

$$(\partial_\mu-i\epsilon B_\mu)\psi.$$

B_μ are 2×2 matrices such that[7] for $\mu=1$, 2, and 3, B_μ is Hermitian and B_4 is anti-Hermitian. Invariance requires that

$$S(\partial_\mu-i\epsilon B_\mu')\psi'=(\partial_\mu-i\epsilon B_\mu)\psi. \tag{2}$$

Combining (1) and (2), we obtain the isotopic gauge transformation on B_μ:

$$B_\mu'=S^{-1}B_\mu S+\frac{i}{\epsilon}S^{-1}\frac{\partial S}{\partial x_\mu}. \tag{3}$$

The last term is similar to the gradiant term in the gauge transformation of electromagnetic potentials. In analogy to the procedure of obtaining gauge invariant field strengths in the electromagnetic case, we

---

[4] T. Lauritsen, Ann. Rev. Nuclear Sci. **1**, 67 (1952); D. R. Inglis, Revs. Modern Phys. **25**, 390 (1953).
[5] R. H. Hildebrand, Phys. Rev. **89**, 1090 (1953).
[6] W. Pauli, Revs. Modern Phys. **13**, 203 (1941).

[7] We use the conventions $\hbar=c=1$, and $x_4=it$. Bold-face type refers to vectors in isotopic space, not in space-time.

define now

$$F_{\mu\nu}=\frac{\partial B_\mu}{\partial x_\nu}-\frac{\partial B_\nu}{\partial x_\mu}+i\epsilon(B_\mu B_\nu-B_\nu B_\mu). \qquad (4)$$

One easily shows from (3) that

$$F_{\mu\nu}'=S^{-1}F_{\mu\nu}S \qquad (5)$$

under an isotopic gauge transformation.‡ Other simple functions of B than (4) do not lead to such a simple transformation property.

The above lines of thought can be applied to any field ψ with arbitrary isotopic spin. One need only use other representations S of rotations in three-dimensional space. It is reasonable to assume that different fields with the same total isotopic spin, hence belonging to the same representation S, interact with the same matrix field B_μ. (This is analogous to the fact that the electromagnetic field interacts in the same way with any charged particle, regardless of the nature of the particle. If different fields interact with different and independent B fields, there would be more conservation laws than simply the conservation of total isotopic spin.) To find a more explicit form for the B fields and to relate the B_μ's corresponding to different representations S, we proceed as follows.

Equation (3) is valid for any S and its corresponding B_μ. Now the matrix $S^{-1}\partial S/\partial x_\mu$ appearing in (3) is a linear combination of the isotopic spin "angular momentum" matrices T^i $(i=1, 2, 3)$ corresponding to the isotopic spin of the ψ field we are considering. So B_μ itself must also contain a linear combination of the matrices T^i. But any part of B_μ in addition to this, $\bar{B}_\mu$, say, is a scalar or tensor combination of the T's, and must transform by the homogeneous part of (3), $\bar{B}_\mu'=S^{-1}\bar{B}_\mu S$. Such a field is extraneous; it was allowed by the very general form we assumed for the B field, but is irrelevant to the question of isotopic gauge. Thus the relevant part of the B field is of the form

$$B_\mu=2\mathbf{b}_\mu\cdot\mathbf{T}. \qquad (6)$$

(Bold-face letters denote three-component vectors in isotopic space.) To relate the $\mathbf{b}_\mu$'s corresponding to different representations S we now consider the product representation $S=S^{(a)}S^{(b)}$. The B field for the combination transforms, according to (3), by

$$B_\mu'=[S^{(b)}]^{-1}[S^{(a)}]^{-1}BS^{(a)}S^{(b)}$$
$$+\frac{i}{\epsilon}[S^{(a)}]^{-1}\frac{\partial S^{(a)}}{\partial x_\mu}+\frac{i}{\epsilon}[S^{(b)}]^{-1}\frac{\partial S^{(b)}}{\partial x_\mu}.$$

But the sum of $B_\mu^{(a)}$ and $B_\mu^{(b)}$, the B fields corresponding to $S^{(a)}$ and $S^{(b)}$, transforms in exactly the same way, so that

$$B_\mu=B_\mu^{(a)}+B_\mu^{(b)}$$

(plus possible terms which transform homogeneously, and hence are irrelevant and will not be included). Decomposing $S^{(a)}S^{(b)}$ into irreducible representations, we see that the twelve-component field $\mathbf{b}_\mu$ in Eq. (6) is the same for all representations.

To obtain the interaction between any field ψ of arbitrary isotopic spin with the $\mathbf{b}$ field one therefore simply replaces the gradiant of ψ by

$$(\partial_\mu-2i\epsilon\mathbf{b}_\mu\cdot\mathbf{T})\psi, \qquad (7)$$

where T^i $(i=1, 2, 3)$, as defined above, are the isotopic spin "angular momentum" matrices for the field ψ.

We remark that the nine components of $\mathbf{b}_\mu$, $\mu=1, 2, 3$ are real and the three of $\mathbf{b}_4$ are pure imaginary. The isotopic-gauge covariant field quantities $F_{\mu\nu}$ are expressible in terms of $\mathbf{b}_\mu$:

$$F_{\mu\nu}=2\mathbf{f}_{\mu\nu}\cdot\mathbf{T}, \qquad (8)$$

where

$$\mathbf{f}_{\mu\nu}=\frac{\partial\mathbf{b}_\mu}{\partial x_\nu}-\frac{\partial\mathbf{b}_\nu}{\partial x_\mu}-2\epsilon\mathbf{b}_\mu\times\mathbf{b}_\nu. \qquad (9)$$

$\mathbf{f}_{\mu\nu}$ transforms like a vector under an isotopic gauge transformation. Obviously the same $\mathbf{f}_{\mu\nu}$ interact with all fields ψ irrespective of the representation S that ψ belongs to.

The corresponding transformation of $\mathbf{b}_\mu$ is cumbersome. One need, however, study only the infinitesimal isotopic gauge transformations,

$$S=1-2i\mathbf{T}\cdot\delta\boldsymbol{\omega}.$$

Then

$$\mathbf{b}_\mu'=\mathbf{b}_\mu+2\mathbf{b}_\mu\times\delta\boldsymbol{\omega}+\frac{1}{\epsilon}\frac{\partial}{\partial x_\mu}\delta\boldsymbol{\omega}. \qquad (10)$$

FIELD EQUATIONS

To write down the field equations for the $\mathbf{b}$ field we clearly only want to use isotopic gauge invariant quantities. In analogy with the electromagnetic case we therefore write down the following Lagrangian density:[8]

$$-\tfrac{1}{4}\mathbf{f}_{\mu\nu}\cdot\mathbf{f}_{\mu\nu}.$$

Since the inclusion of a field with isotopic spin $\frac{1}{2}$ is illustrative, and does not complicate matters very much, we shall use the following total Lagrangian density:

$$\mathcal{L}=-\tfrac{1}{4}\mathbf{f}_{\mu\nu}\cdot\mathbf{f}_{\mu\nu}-\bar{\psi}\gamma_\mu(\partial_\mu-i\epsilon\boldsymbol{\tau}\cdot\mathbf{b}_\mu)\psi-m\bar{\psi}\psi. \qquad (11)$$

One obtains from this the following equations of motion:

$$\partial\mathbf{f}_{\mu\nu}/\partial x_\nu+2\epsilon(\mathbf{b}_\nu\times\mathbf{f}_{\mu\nu})+\mathbf{J}_\mu=0,$$
$$\gamma_\mu(\partial_\mu-i\epsilon\boldsymbol{\tau}\cdot\mathbf{b}_\mu)\psi+m\psi=0, \qquad (12)$$

‡ *Note added in proof.*—It may appear that B_μ' could be introduced as an auxiliary quantity to accomplish invariance, but need not be regarded as a field variable by itself. It is to be emphasized that such a procedure violates the principle of invariance. Every quantity that is not a pure numeral (like 2, or M, or any definite representation of the γ matrices) should be regarded as a dynamical variable, and should be varied in the Lagrangian to yield an equation of motion. Thus the quantities B_μ must be regarded as independent fields.

[8] Repeated indices are summed over, except where explicitly stated otherwise. Latin indices are summed from 1 to 3, Greek ones from 1 to 4.

where

$$\mathbf{J}_\mu = i\epsilon\bar{\psi}\gamma_\mu \boldsymbol{\tau}\psi. \quad (13)$$

The divergence of $\mathbf{J}_\mu$ does not vanish. Instead it can easily be shown from (13) that

$$\partial \mathbf{J}_\mu/\partial x_\mu = -2\epsilon \mathbf{b}_\mu \times \mathbf{J}_\mu. \quad (14)$$

If we define, however,

$$\mathfrak{J}_\mu = \mathbf{J}_\mu + 2\epsilon \mathbf{b}_\nu \times \mathbf{f}_{\mu\nu}, \quad (15)$$

then (12) leads to the equation of continuity,

$$\partial \mathfrak{J}_\mu/\partial x_\mu = 0. \quad (16)$$

$\mathfrak{J}_{1,2,3}$ and $\mathfrak{J}_4$ are respectively the isotopic spin current density and isotopic spin density of the system. The equation of continuity guarantees that the total isotopic spin

$$\mathbf{T} = \int \mathfrak{J}_4 d^3x$$

is independent of time and independent of a Lorentz transformation. It is important to notice that $\mathfrak{J}_\mu$, like $\mathbf{b}_\mu$, does not transform exactly like vectors under isotopic space rotations. But the total isotopic spin,

$$\mathbf{T} = -\int \frac{\partial \mathbf{f}_{4i}}{\partial x_i} d^3x,$$

is the integral of the divergence of $\mathbf{f}_{4i}$, which transforms like a true vector under isotopic spin space rotations. Hence, under a general isotopic gauge transformation, if $S \to S_0$ on an infinitely large sphere, $\mathbf{T}$ would transform like an isotopic spin vector.

Equation (15) shows that the isotopic spin arises both from the spin-$\frac{1}{2}$ field ($\mathbf{J}_\mu$) and from the $\mathbf{b}_\mu$ field itself. Inasmuch as the isotopic spin is the source of the $\mathbf{b}$ field, this fact makes the field equations for the $\mathbf{b}$ field nonlinear, even in the absence of the spin-$\frac{1}{2}$ field. This is different from the case of the electromagnetic field, which is itself chargeless, and consequently satisfies linear equations in the absence of a charged field.

The Hamiltonian derived from (11) is easily demonstrated to be positive definite in the absence of the field of isotopic spin $\frac{1}{2}$. The demonstration is completely identical with the similar one in electrodynamics.

We must complete the set of equations of motion (12) and (13) by the supplementary condition,

$$\partial \mathbf{b}_\mu/\partial x_\mu = 0, \quad (17)$$

which serves to eliminate the scalar part of the field in $\mathbf{b}_\mu$. This clearly imposes a condition on the possible isotopic gauge transformations. That is, the infinitesimal isotopic gauge transformation $S = 1 - i\boldsymbol{\tau}\cdot\delta\boldsymbol{\omega}$ must satisfy the following condition:

$$2\mathbf{b}_\mu \times \frac{\partial}{\partial x_\mu}\delta\boldsymbol{\omega} + \frac{1}{\epsilon}\frac{\partial^2}{\partial x_\mu{}^2}\delta\boldsymbol{\omega} = 0. \quad (18)$$

This is the analog of the equation $\partial^2\alpha/\partial x_\mu{}^2 = 0$ that must be satisfied by the gauge transformation $A_\mu' = A_\mu + e^{-1}(\partial\alpha/\partial x_\mu)$ of the electromagnetic field.

QUANTIZATION

To quantize, it is not convenient to use the isotopic gauge invariant Lagrangian density (11). This is quite similar to the corresponding situation in electrodynamics and we adopt the customary procedure of using a Lagrangian density which is not obviously gauge invariant:

$$\mathfrak{L} = -\frac{1}{2}\frac{\partial \mathbf{b}_\mu}{\partial x_\nu}\cdot\frac{\partial \mathbf{b}_\mu}{\partial x_\nu} + 2\epsilon(\mathbf{b}_\mu \times \mathbf{b}_\nu)\frac{\partial \mathbf{b}_\mu}{\partial x_\nu} - \epsilon^2(\mathbf{b}_\mu \times \mathbf{b}_\nu)^2 + \mathbf{J}_\mu\cdot\mathbf{b}_\mu - \bar{\psi}(\gamma_\mu\partial_\mu + m)\psi. \quad (19)$$

The equations of motion that result from this Lagrangian density can be easily shown to imply that

$$\frac{\partial^2}{\partial x_\nu{}^2}\mathbf{a} + 2\epsilon \mathbf{b}_\nu \times \frac{\partial}{\partial x_\nu}\mathbf{a} = 0,$$

where

$$\mathbf{a} = \partial \mathbf{b}_\mu/\partial x_\mu.$$

Thus if, consistent with (17), we put on one space-like surface $\mathbf{a} = 0$ together with $\partial\mathbf{a}/\partial t = 0$, it follows that $\mathbf{a} = 0$ at all times. Using this supplementary condition one can easily prove that the field equations resulting from the Lagrangian densities (19) and (11) are identical.

One can follow the canonical method of quantization with the Lagrangian density (19). Defining

$$\boldsymbol{\Pi}_\mu = -\partial \mathbf{b}_\mu/\partial x_4 + 2\epsilon(\mathbf{b}_\mu \times \mathbf{b}_4),$$

one obtains the equal-time commutation rule

$$[b_\mu{}^i(x), \Pi_\nu{}^j(x')]_{t=t'} = -\delta_{ij}\delta_{\mu\nu}\delta^3(x-x'), \quad (20)$$

where $b_\mu{}^i$, $i = 1, 2, 3$, are the three components of $\mathbf{b}_\mu$. The relativistic invariance of these commutation rules follows from the general proof for canonical methods of quantization given by Heisenberg and Pauli.[9]

The Hamiltonian derived from (19) is identical with the one from (11), in virtue of the supplementary condition. Its density is

$$H = H_0 + H_{\text{int}},$$

$$H_0 = -\tfrac{1}{2}\boldsymbol{\Pi}_\mu\cdot\boldsymbol{\Pi}_\mu + \frac{1}{2}\frac{\partial \mathbf{b}_\mu}{\partial x_j}\cdot\frac{\partial \mathbf{b}_\mu}{\partial x_j} + \bar{\psi}(\gamma_j\partial_j + m)\psi, \quad (21)$$

$$H_{\text{int}} = 2\epsilon(\mathbf{b}_i \times \mathbf{b}_4)\cdot\boldsymbol{\Pi}_i - 2\epsilon(\mathbf{b}_\mu \times \mathbf{b}_j)\cdot(\partial \mathbf{b}_\mu/\partial x_j) + \epsilon^2(\mathbf{b}_i \times \mathbf{b}_j)^2 - \mathbf{J}_\mu\cdot\mathbf{b}_\mu.$$

The quantized form of the supplementary condition is the same as in quantum electrodynamics.

---

[9] W. Heisenberg and W. Pauli, Z. Physik **56**, 1 (1929).

PROPERTIES OF THE *b* QUANTA

The quanta of the **b** field clearly have spin unity and isotopic spin unity. We know their electric charge too because all the interactions that we proposed must satisfy the law of conservation of electric charge, which is exact. The two states of the nucleon, namely proton and neutron, differ by charge unity. Since they can transform into each other through the emission or absorption of a **b** quantum, the latter must have three charge states with charges $\pm e$ and 0. Any measurement of electric charges of course involves the electromagnetic field, which necessarily introduces a preferential direction in isotopic space at all space-time points. Choosing the isotopic gauge such that this preferential direction is along the z axis in isotopic space, one sees that for the nucleons

$$Q = \text{electric charge} = e(\tfrac{1}{2} + \epsilon^{-1}T^z),$$

and for the **b** quanta

$$Q = (e/\epsilon)T^z.$$

The interaction (7) then fixes the electric charge up to an additive constant for all fields with any isotopic spin:

$$Q = e(\epsilon^{-1}T^z + R). \tag{22}$$

The constants R for two charge conjugate fields must be equal but have opposite signs.[10]

FIG. 1. Elementary vertices for **b** fields and nucleon fields. Dotted lines refer to **b** field, solid lines with arrow refer to nucleon field.

We next come to the question of the mass of the **b** quantum, to which we do not have a satisfactory answer. One may argue that without a nucleon field the Lagrangian would contain no quantity of the dimension of a mass, and that therefore the mass of the **b** quantum in such a case is zero. This argument is however subject to the criticism that, like all field theories, the **b** field is beset with divergences, and dimensional arguments are not satisfactory.

One may of course try to apply to the **b** field the methods for handling infinities developed for quantum electrodynamics. Dyson's approach[11] is best suited for the present case. One first transforms into the interaction representation in which the state vector Ψ satisfies

$$i\partial\Psi/\partial t = H_{\text{int}}\Psi,$$

where H_{int} was defined in Eq. (21). The matrix elements of the scattering matrix are then formulated in terms of contributions from Feynman diagrams. These diagrams have three elementary types of vertices illustrated in Fig. 1, instead of only one type as in quantum electrodynamics. The "primitive divergences" are still finite in number and are listed in Fig. 2. Of these, the one labeled a is the one that effects the propagation function of the **b** quantum, and whose singularity determines the mass of the **b** quantum. In electrodynamics, by the requirement of electric charge conservation,[12] it is argued that the mass of the photon vanishes. Corresponding arguments in the **b** field case do not exist[13] even though the conservation of isotopic spin still holds. We have therefore not been able to conclude anything about the mass of the **b** quantum.

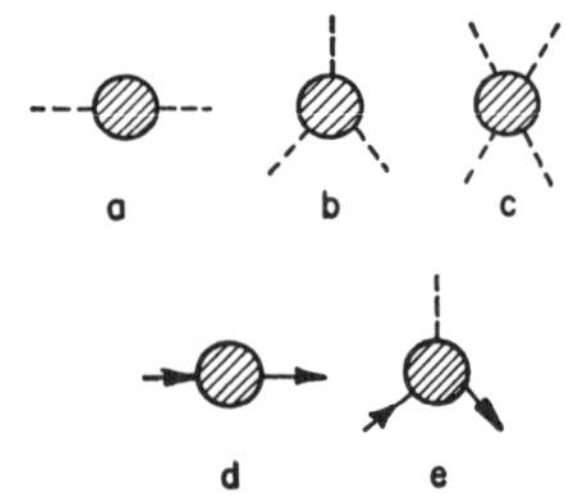

FIG. 2. Primitive divergences.

A conclusion about the mass of the **b** quantum is of course very important in deciding whether the proposal of the existence of the **b** field is consistent with experimental information. For example, it is inconsistent with present experiments to have their mass less than that of the pions, because among other reasons they would then be created abundantly at high energies and the charged ones should live long enough to be seen. If they have a mass greater than that of the pions, on the other hand, they would have a short lifetime (say, less than 10^{-20} sec) for decay into pions and photons and would so far have escaped detection.

[10] See M. Gell-Mann, Phys. Rev. **92**, 833 (1953).

[11] F. J. Dyson, Phys. Rev. **75**, 486, 1736 (1949).

[12] J. Schwinger, Phys. Rev. **76**, 790 (1949).

[13] In electrodynamics one can formally prove that $G_{\mu\nu}k_\nu = 0$, where $G_{\mu\nu}$ is defined by Schwinger's Eq. (A12). ($G_{\mu\nu}A_\nu$ is the current generated through virtual processes by the arbitrary external field A_ν.) No corresponding proof has been found for the present case. This is due to the fact that in electrodynamics the conservation of charge is a consequence of the equation of motion of the electron field alone, quite independently of the electromagnetic field itself. In the present case the **b** field carries an isotopic spin and destroys such general conservation laws.

Conservation of Heavy Particles and Generalized Gauge Transformations

T. D. LEE, *Columbia University, New York, New York*

AND

C. N. YANG, *Institute for Advanced Study, Princeton, New Jersey*

(Received March 2, 1955)

Commentary begins page 21

The possibility of a heavy-particle gauge transformation is discussed.

THE conservation laws of nature fall into two distinct categories: those that are related to invariance under space-time displacements and rotations, and those that are not. In the former category there are the conservation laws of momentum, energy, and angular momentum. In the latter category we find the conservation laws of electric charge, of heavy particles, and the approximate conservation laws of isotopic spin, and perhaps others.[1] We notice that the best known within this second category, the conservation of electric charge, is related to invariance under gauge transformations,[2] which expresses the nonmeasurability of the phase of the complex wave function of a charged particle.

We want to ask here whether similar gauge invariances should be related to all conservation laws of the second category. This question has been discussed in connection with the conservation of isotopic spin by Yang and Mills.[3] We wish here to discuss the problem in connection with the conservation of heavy particles.

If we take the conservation of heavy particles to mean invariance under the transformation

$$\psi_N \to e^{i\alpha}\psi_N, \quad \psi_P \to e^{i\alpha}\psi_P, \tag{1}$$

for the wave function of the heavy particles (neutrons and protons), a general gauge transformation (heavy-particle gauge transformation) is a transformation like (1) with the phase α an arbitrary function of space-time. Invariance under such a transformation means that the relative phase of the wave function of a heavy particle at two different space-time points is not measurable.

Such a gauge transformation is formally completely identical with the electromagnetic gauge transformation. Invariance under such a transformation therefore necessitates the existence of a neutral vector massless field coupled to all heavy particles. A nucleon would have a "heavy-particle charge" of $+\eta$ in such a field and an antinucleon would have a "heavy-particle charge" of $-\eta$. The force between two massive bodies therefore would contain a contribution from the Coulomb-like repulsion between such "heavy-particle charges." The total force including the gravitational attraction is:

$$\text{Force} = -G(M_1M_2/R^2)+\eta^2(A_1A_2/R^2). \tag{2}$$

Here M_1, M_2, A_1, and A_2 are the inertia masses and mass numbers of the two bodies. There should also be a magnetic-dipole-like interaction between individual nuclei because the nucleons are in constant motion in a nucleus. But in a macroscopic object the nuclear spins average out so that (2) is correct unless the two bodies are spinning at high speeds.

Now the packing fraction of various atoms differ so that M/A varies fractional-wise from substance to substance by $\sim 10^{-3}$. This means that the observed gravitation mass [which contains a contribution from the η^2 term in (2)] divided by the inertia mass would vary fractional-wise from substance to substance by $10^{-3}\eta^2/G(M_P)^2$, where M_P is the mass of the proton[4] Very careful measurements by Eötvös and co-workers. have shown this variation to be $<10^{-8}$. Therefore

$$\eta^2/G(M_P)^2<10^{-5}.$$

It may be remarked that since the packing fraction differs most between hydrogen and, say, carbon, Eötvös' experiment could yield a more sensitive detection of η^2 by a factor of 10 if repeated with a comparison of hydrogen and carbon.

The assumption that leads to the above line of reasoning and the force expression (2) is that the phase factor α in (1) should be space-time-dependent. It should be noticed that in addition the assumption has also been made that the transformation that generates the conservation of heavy particles is of the specific form (1).

We wish to thank Dr. J. Robert Oppenheimer for an interesting discussion.

[1] See M. Gell-Mann and A. Pais, Proceedings of the Glasgow Conference, July, 1954 (to be published).
[2] W. Pauli, Revs. Modern Phys. **13**, 203 (1941).
[3] C. N. Yang and R. L. Mills, Phys. Rev. **96**, 191 (1954).
[4] Eötvös, Pekár, and Fekete, Ann. Physik **68**, 11 (1922).

⟦56d⟧
Commentary begins page 24

Charge Conjugation, a New Quantum Number G, and Selection Rules Concerning a Nucleon-Antinucleon System.

T. D. LEE
Columbia University - New York, N.Y.

C. N. YANG
Institute for Advanced Study - Princeton, N.J.

(ricevuto il 30 Gennaio 1956)

Summary. (*). — A new quantum number is introduced which leads to selection rules concerning transitions between states with heavy particle number = 0. They are applied to the system nucleon-antinucleon; it is also shown that they can be extended to include the conservation of strangeness.

(*) *Editor's care.*

Whenever several conservation laws operate for the same system it is oftentimes possible to obtain new quantum numbers and new selection rules. For the pion-nucleon-antinucleon system in strong interaction one has besides the usual spin parity conservation laws the additional conservation laws of isotopic spin, charge conjugation and heavy particle number. We shall see that these conservation laws together do lead to a new quantum number for all systems with heavy particle number = 0. Application of this result to a nucleon-antinucleon system gives the selection rules tabulated in Tables I and II.

We shall be concerned with these operators:

$$I_1, I_2, I_3 = \text{isotopic spin operators},$$

$$G = C \exp[i\pi I_2],$$

$$\text{and} \quad N = \text{number of heavy particles}$$

 □Reprinted from *Il Nuovo Cimento* 10, 3 (April 1956), 749–753.

TABLE I. – *Selection Rules for* $\bar{p} + n \to m\pi$.

State	Spin parity	I	G	$\pi^- + \pi^0$	$2\pi^- + \pi^+$	$\pi^- + 2\pi^0$	$2\pi^- + \pi^+ + \pi^0$	$\pi^- + 3\pi^0$	$3\pi^- + 2\pi^+$	$2\pi^- + \pi^+ + 2\pi^0$	$\pi^- + 4\pi^0$
$^1S_0$	0^-	1	—	×			—	—			
$^3S_1$	1^-	1	+		—	—			—	—	—
$^1P_1$	1^+	1	+	×	—	—			—	—	—
$^3P_0$	0^+	1	—	—	×	×	—	—			
$^3P_1$	1^+	1	—	×			—	—			
$^3P_2$	2^+	1	—	—			—	—			

× means strictly forbidden and — means forbidden so far as the isotopic spin is a good quantum number.

Here C is the charge conjugation operator. The advantage of using the operator G instead of C is that G commutes with I_1, I_2, and I_3, while C does not.

To see this one need only write down in the rest system the explicit form for these operators between the four states (1) with the same z component of spatial spin:

$$\begin{pmatrix} p \\ n \\ \bar{n} \\ -\bar{p} \end{pmatrix}.$$

They are

$$I_1 = \frac{1}{2}\begin{pmatrix} 0 & 1 & 0 & 0 \\ 1 & 0 & 0 & 0 \\ 0 & 0 & 0 & 1 \\ 0 & 0 & 1 & 0 \end{pmatrix}, \quad I_2 = \frac{1}{2}\begin{pmatrix} 0 & -i & 0 & 0 \\ i & 0 & 0 & 0 \\ 0 & 0 & 0 & -i \\ 0 & 0 & i & 0 \end{pmatrix}, \quad I_3 = \frac{1}{2}\begin{pmatrix} 1 & 0 & 0 & 0 \\ 0 & -1 & 0 & 0 \\ 0 & 0 & 1 & 0 \\ 0 & 0 & 0 & -1 \end{pmatrix},$$

$$G = \begin{pmatrix} 0 & 0 & 1 & 0 \\ 0 & 0 & 0 & 1 \\ -1 & 0 & 0 & 0 \\ 0 & -1 & 0 & 0 \end{pmatrix}, \quad N = \begin{pmatrix} 1 & 0 & 0 & 0 \\ 0 & 1 & 0 & 0 \\ 0 & 0 & -1 & 0 \\ 0 & 0 & 0 & -1 \end{pmatrix}.$$

(1) Notice the minus sign in front of the $\bar{p}$ state in the convention here. The definition of $\bar{n}$ and $\bar{p}$ as the antineutron and antiproton states is here so chosen that they are identically related to the neutron and proton states. In other words, charge conjugation is here defined in the same way for the neutron and the proton.

TABLE II. – *Selection Rules for* $\overline{p}+p \to m\pi$ *or* $\overline{n}+n \to m\pi$.

State	Spin parity	C	I	G	$2\pi^0$	$\pi^++\pi^-$	$3\pi^0$	$\pi^++\pi^-+\pi^0$	$4\pi^0$	$\pi^++\pi^-+2\pi^0$	$2\pi^++2\pi^-$	$5\pi^0$	$\pi^++\pi^-+3\pi^0$	$2\pi^++2\pi^-+\pi^0$
$^1S_0$	0^-	+	0	+	×	×	—	—				—	—	—
			1	—	×	×			—	—	—			
$^3S_1$	1^-	—	0	—	×	—	×		×	—	—	×		
			1	+	×		×	—	×			×	—	—
$^1P_1$	1^+	—	0	—	×	×	×		×	—	—	×		
			1	+	×	×	×	—	×			×	—	—
$^3P_0$	0^+	+	0	+			×	×				—	—	—
			1	—	—	—	×	×	—	—	—			
$^3P_1$	1^+	+	0	+	×	×	—	—				—	—	—
			1	—	×	×			—	—	—			
$^3P_2$	2^+	+	0	+			—	—				—	—	—
			1	—	—	—			—	—	—			

× means strictly forbidden and — means forbidden so far as the isotopic spin is a good quantum number.

For a multiple particle system, I and N are additive and G multiplicative. One therefore has *in general*

(1) $$[I_i, G] = [I_i, N] = 0 \, ,$$

(2) $$GN + NG = 0 \, ,$$

(3) $$G^2 = (-1)^N \, .$$

Besides G is unitary by definition and N has integral eigenvalues. Solving the relations (1) to (3) one obtains the following:

(1) For a state with a definite value $\neq 0$ for N, the quantum numbers are N, I^2 and I_3. The operation of G on the system gives a state with the same I^2, I_3 but changes the sign of N.

(2) For a state with $N = 0$ the quantum numbers are $N = 0$, I^2, I_3 and $G = \pm 1$. Besides, *all components of the same I-multiplet have the same value for G*. The pions obviously belong to this category, with $I^2 = 1(1+1)$ and $I_3 = \pm 1$ or 0. To find their value for G we notice that a π^0 has matrix elements connecting it to the $^1S_0$ state of a proton-antiproton system. Such a system has $C = +1$ (2). Now π^0 has a total I-spin equal to 1. Therefore $\exp[i\pi I_2]$ for π^0 is -1. Hence by definition $G = -1$ for π^0. *Therefore* $G = -1$ *for all pions, charged or neutral.*

The existence of the quantum number G leads to interesting selection rules (3) concerning transitions between states with $N = 0$. In particular one sees that *an even number of pions cannot by strong interaction go into an odd number of pions*. When applied to a system consisting of a nucleon and an antinucleon one obtains the selection rules tabulated in Tables I and II.

The above consideration can easily be extended to include the conservation (4) of strangeness S. (S is defined to be $2Q - 2I_3 - N$). The commutation relations between S and the other operators are

$$[N, S] = [S, I_i] = 0 \, ,$$

$$GS + SG = 0 \, ,$$

---

(2) L. Wolfenstein and D. G. Ravenhall: *Phys. Rev.*, **88**, 279 (1952); A. Pais and R. Jost: *Phys. Rev.*, **87**, 871 (1952).

(3) Most of these results have been stated in the literature in various forms. See A. Pais and R. Jost: *Phys. Rev.*, **87**, 871 (1952); L. Michel: *Nuovo Cimento*, **10**, 319 (1953); D. Amati and B. Vitale: *Nuovo Cimento*, **2**, 719 (1955).

(4) M. Gell-Mann: *Phys. Rev.* (in press); T. Nakano and K. Nishijima: *Prog. Theor. Phys.*, **10**, 581 (1954).

and

$$G^2 = (-1)^{N+S}.$$

Solving these commutation relations one obtains results similar to (1) and (2) discussed above, except that S is now an additional quantum number. G is a good quantum number $= \pm 1$ in this case only if both N and S are zero.

RIASSUNTO (*)

Si introduce un nuovo numero quantico G che porta a regole di selezione riguardanti le transizioni fra stati con un numero N di particelle pesanti nullo. Tali regole sono poi applicate al sistema nucleone-antinucleone e viene anche mostrato che esse possono essere estese in modo da comprendere la conservazione della « stranezza » S.

(*) *A cura della Redazione.*

Introductory Talk at the 1956 Rochester Conference, Session on Theoretical Interpretation of New Particles

C. N. Yang

[[56e]]
Commentary begins page 24

1. After being introduced to the zoo last Wednesday we have taken excursions in it for two days. This morning, before we leave the zoo, we want to ask, "What have we learned?" I am supposed to present to you the theoretical arguments in this direction. What I shall tell you will not form a clear picture: a clear picture does not exist. But I do hope I can present to you an exciting and challenging picture that provokes further experiments and further speculations.

The past year has witnessed very interesting developments in our knowledge of the strange particles. Perhaps the most important of these is the firm establishment of the "strangeness" quantum number. The starting point of these considerations was, as you remember, the puzzle that while the strange particles are produced quite abundantly (say 5 per cent of the pions) at BeV energies and up, their decays into pions and nucleons are rather slow ($\sim 10^{-10}$ sec). Since the time scale of pion–nucleon interactions is of the order of 10^{-23} sec, it was very puzzling how to reconcile the abundance of these objects with their longevity (10^{13} units of time scale). In 1952 Pais proposed that a way out of this difficulty is to assume that a strange particle is always produced in association with other strange particles. This proposition was very soon supported by direct experimental evidence.

A natural way to explain the associated production phenomenon is to say that there are some selection rules which prevent the strong interactions from being operative in the decay mechanism. A glance at the many observed production, reaction and decay schemes shows indeed that one could assign to each strange particle a strangeness quantum number and stipulate that in all fast interactions the strangeness is additively conserved, and that in all observed decays it is not. This was first discussed by Gell-Mann and by Nishijima in 1953. The assignment is:

$S = 0$: ordinary particles (π, N, P, and γ)

$S = +1$: θ°, K^+

$S = -1$: Λ°, $\Sigma^\pm$, Σ°(?), K^-, $\bar{\theta}^\circ$

$S = -2$: Ξ^-

One remark is proper here: two charge conjugate particles must have equal and opposite values of S. This is because, in a fast reaction, a particle can always be moved to the other side of the reaction and become its antiparticle, keeping the reaciton fast.

The use of the concept of conservation of strangeness is as follows: A reaction is fast (time scale 10^{-23} sec) if it satisfies all conservation laws, namely, of energy, momentum, angular momentum, parity, charge, heavy particle number and strangeness, and if it does not contain a γ-ray. It is weaker by a factor of 1/137 if it involves a γ-ray. If it violates the strangeness selection rule it is weaker by a factor of, say, 10^{-12}. μ, e and ν are not assigned any strangeness, but except for electromagnetic interactions they are supposed to interact with a strength 10^{-12} weaker than the strong interactions. We shall return to this point later.

This conservation of strangeness was proposed by Gell-Mann and by Nishijima in 1953, and during the past year it was given very strong experimental support. These supports are:

a. Associated production seems to be the general rule. But the associated production $N + N \rightarrow \Lambda^\circ + \Lambda^\circ$, although of the lowest threshold, is definitely of much lower probability. The significance of this is that this makes any multi-

□ Reprinted from *High Energy Nuclear Physics, 1956*. New York: Wiley-Interscience, 1956, VIII 1–18.

plicative selection rule modeled after parity impossible.

b. To stabilize the cascade particle Ξ^- in the strangeness scheme it was proposed that its strangeness is -2. The observation of a reaction producing $\Xi^- + K^\circ + K^\circ$ is in conformity with this proposal.
c. K^+ and K^- behave very differently in matter: K^+ scatters, but does not cause big stars, K^- causes big stars and oftentimes changes into a Λ° or $\Sigma^\pm$. In the strangeness scheme, this is evident because K^+ is of the lowest excitation in the family $S = 1$, while K^- is very vulnerable in that in its family $S = -1$ there are many particles (Σ°, Λ°) which have much lower excitation.
d. K^+/K^- is very large (>50) at cosmotron energies. This was predicted on the strangeness scheme as due to the fact that K^- with $S = -1$ must be produced with a partner with $S = +1$, i.e., a partner with at least the excitation of K^+ mass ($= 965\ m_e$) while a K^+ with $S = +1$ is produced with a partner with $S = -1$, such as Λ° with an excitation of only 340 m_e. The large ratio K^+/K^- is therefore a threshold effect.
e. There are no known violations of the strangeness selection rule.

On the strength of these experimental findings it seems that the conservation of strangeness gives such a consistent picture of the interaction of strange particles that it is certainly part of the truth in this subject.

Actually, the Gell-Mann–Nishijima scheme finds experimental support from another set of experimental results. This concerns the charge degeneracy of the strange particle states, and its relationship to the isotopic spin, a concept familiar in nuclear and pion physics.

We recall that in pion physics there was the relationship

$$Q = I_3 + \frac{N}{2} \tag{1}$$

between the charge, the third component I_3 of the I-spin and the number of nucleons N. Suppose, for strange particles, the conservation law of I holds in strong interactions, but this relation breaks down. Then the balance

$$\left(Q - I_3 - \frac{N}{2}\right)$$

would, for strange particles, not be zero. However, it still must be a quantity which is additively conserved in any strong interactions, since Q, I_3 and N all are. This in fact was the starting point of the Gell-Mann–Nishijima scheme; namely, to ask whether the new quantum number, defined to be

$$S = 2\left(Q - I_3 - \frac{N}{2}\right) \tag{2}$$

could stabilize the strange particles.

This connection between the strangeness and isotopic spin provides us with three more kinds of results that can be directly checked with experiments. They are all related to the conservation of the total I-spin, not only of I_3:

a. The assignment of S leads immediately to a value for I_3, which, if non-vanishing, in turn would imply the existence of other particles of approximately the same mass but different charges. E.g., the particle Σ^- has $S = -1$, $Q = -1$, and $N = 1$. Therefore, $I_3 = -1$. Therefore, it must have at least two partners of approximately the same mass with $I_3 = 0$, 1 respectively, and charge 0 and +1. The latter is indeed found, experimentally called Σ^+. The other one, Σ°, is perhaps on the way to be found. Applied to the particle Λ° one gets the result that $S = -1$ is consistent with an $I = 0$ assignment for Λ°. This agrees with the experimental picture of no observed charged particles degenerate with Λ°. One may ask, of course, in this connection, whether different particles of the same multiplet may not be separated by several hundred MeV in mass values. It seems, however, that such a separation would in itself indicate that the interactions that violate conservation of I are large and would render I meaningless.

b. The light hyperfragments would form isotopic spin multiplets, such as $_{\Lambda}He^4$ and $_{\Lambda}H^4$. This was discussed by Dalitz. Experimental evidence of such multiplets is expected to be found.

c. The existence of certain relationships between different reaction rates. E.g.,

$$K^- + d \to \Sigma^- + p$$

$$K^- + d \to \Sigma^0 + n$$

would be in the ratio of 2:1. No direct experimental evidence of this kind yet exists to my knowledge.

Before I conclude this discussion of the strangeness scheme, two remarks are in order:

a. There was, in the preceding discussion, the implicit assumption that all particles of the same multiplet have the same strangeness quantum number. It is tortuous not to have this assumption. But it is important to recognize that this is one of the fundamental points that we do not understand at all.

b. As we just said, the origin of the strangeness quantum number is the question whether the un-understood empirical relation (1) could be violated. Now there exists another un-understood empirical relation which has not been found violated so far, and that is that all particles with $N = 1$ (i.e., conserved together with the nucleons) have half integral spin and all particles participating in fast interactions with $N = 0$ (e.g., mesons) have integral spin. Violation of this rule would immediately lead to new kinds of quantum numbers. It is perhaps useful to bear this in mind when puzzling new stabilities occur.

We might say that our knowledge of the strange particles at the moment consists of a convergent part and a divergent part. The convergent part I have just given a discription of and constitutes an essentially closed chapter. More will be added to it, to be sure; but until an over-all understanding is gained, the pattern of this chapter will most probably remain as it is.

There have been many theoretical papers written about schemes that do not differ essentially in their conclusions from the Gell-Mann–Nishijima scheme. The authors include M. Goldhaber, Sachs, Salam, d'Espagnat and others. We shall not have time to go into them in detail.

2. Now we come to the divergent part of the subject. This consists of the many loose ends that need to be tied together. Foremost among these is the question of the K-mesons ($K^+_{\pi 2}$, $K^+_{\mu 2}$, $K^+_{\pi 3}$, $K^+_{\mu 3}$, K^+_{e3}). The puzzle, about which we have heard so much in the last two days, is that they have very approximately the same masses and the same life times. The latter were measured both at rest and in flight. Of course, if they are all different decay modes of the same particle, the puzzlement would vanish. (Even then one still needs to understand how this multitude of decay modes comes about.) But the situation is that Dalitz's argument strongly suggests that it is not likely that $K^+_{\pi 3}(\equiv\tau^+)$ and $K^+_{\pi 2}(\equiv\theta^+)$ are the same particle. I hope we may have some discussion of this point this morning.

If θ and τ do not have the same spin and parity, one needs to explain (a) why the masses are so close to each other and (b) why the measured life times are so close to each other.

A few months ago, Lee and Orear suggested that the life time identity may be due to a genetic relationship between the two particles. E.g., the decay diagram may be like:

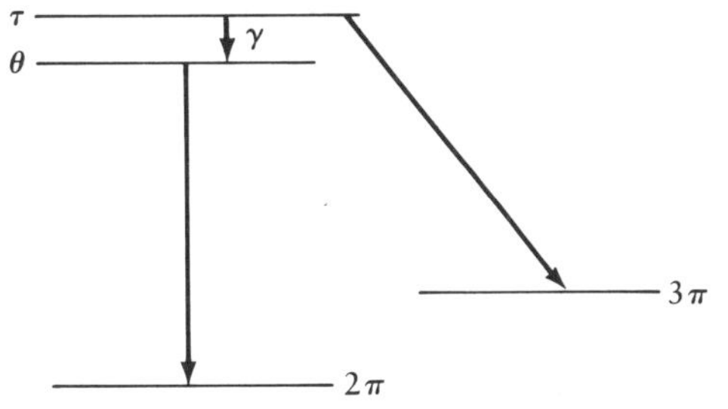

If the true life time of θ is short compared to 10^{-9} seconds, but that of the τ is the observed 10^{-8} sec, then a few meters from the source, the θ's, if originally produced, would all have decayed and one would observe the well-known single-life-time phenomena familiar in radioactivity. This suggestion has been extensively discussed by many people. At the moment, the experimental conclusion seems to be against the existence of γ-rays > 1 MeV, as we heard from Alvarez. Theoretically, if τ and θ are 0^- and 0^+ particles, and if their masses do get within, say, 1 MeV of each other, the electromagnetic transition (double γ emission) would become so slow that this explanation would not be tenable. If, on the other hand, they are 0^- and 2^+ particles, single γ-ray transition would be possible (magnetic quadrupole) and a mass difference of, say, 1 MeV would be appropriate.

A relevant suggestion has recently been made by Weinstein that when the mass difference is small, say $<10^{-5}$ eV, the two states may get mixed in passing through matter, so that the measured life times become identical. For illustration let us take the spin parity assignment to be 2^- and 2^+. There would then be, in general, a static electric dipole strength between the two states. It arises, for example, from processes like this:

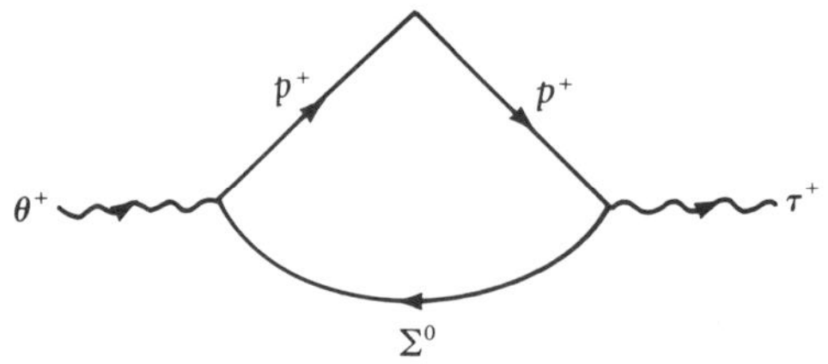

This strength of such electric dipoles is expected to be $\sim e\hbar/mc$ on dimensional grounds. We take m to be the mass of the θ. This causes an energy split $\sim(e\hbar/mc)E \sim 10^{-3}$ eV in an atomic electric field. this is bigger than the mass difference, so the two split states are complete mixtures of the two states θ and τ. Their relative phase would change with time as $(10^{-3}\text{ eV})t/\hbar \sim t/10^{-12}$ sec. If the field were uniform, the two particles would then be completely mixed in 10^{-12}sec. Actually the problem is very involved as **E** is a vector and is not uniform. It is even more involved if a mixed magnetic dipole moment is the cause of the transitions. This problem should be examined in closer detail. Also the problem of the two different methods of life time measurement, namely at rest and in flight, should be examined if the Weinstein suggestion is operative.

If one takes the two most likely assignments, $(0^-, 0^+)$ and $(0^-, 2^+)$, however, the coupling with the electric and magnetic fields would be so weak that no mixing occurs in 10^{-8} sec, and this explanation for equal life times would not work.

Concerning the mass degeneracy of the τ and θ, assuming that they are 0^- and 0^+ particles, Lee and I had discussed the following point: If this degeneracy is *not* accidental, then it follows that all particles whose strangeness is odd must exist in two states of opposite parity. In particular, there would then have to be two Λ°'s: Λ_1° and Λ_2° of opposite relative parity such that

$$\pi^+ + n \rightarrow \Lambda_1^\circ + \theta^+$$

and

$$\pi^+ + n \rightarrow \Lambda_2^\circ + \tau^+$$

occur with equal amplitude. In fact, the symmetry must extend to all fast interactions so that one can define the simultaneous switching of θ and τ, and of Λ_1° and Λ_2°, etc. as an operation that commutes with the strong part of the Hamiltonian. We shall call this operation parity conjugation and shall denote it by C_p. All ordinary particles are eigenstates of C_p with $C_p = +1$. All particles with odd strangeness would exist as a parity multiplet, i.e., two particles with opposite parity that switch to each other under C_p.

The following remarks are in order:

a. If θ and τ have different spins, the whole concept does not work.
b. Particles with $S = 0$ change into themselves under C_p. They may therefore have $C_p = \pm 1$. The possibility of particles with $S = 0$, $C_p = -1$ therefore offers itself as a selection rule to stabilize particles.

c. The reactions

$$\pi^+ + n \to \Lambda_2^\circ + \theta^+$$

$$\pi^+ + n \to \Lambda_1^\circ + \tau^+$$

have equal amplitudes. They may occur together with the two reactions listed before. Their relative rates are not fixed by the invariance requirements. However, it is evident that θ^+ and τ^+ are always produced with equal abundance.

d. The symmetry elements *may* be much more numerous than C_p alone. But C_p represents the minimum symmetry to have equal masses.

e. The electromagnetic interaction may or may not be conserved under parity conjugation. If it is, the only interactions that violate C_p conservation would be the weak interactions. The mass difference between two elements of a parity multiplet would then be exceedingly small (say $< 10^{-5}$ eV). In such a case the Lee–Orear scheme would not be tenable. To illustrate that the electromagnetic interaction may not conserve C_p, we may mention, for example, that Λ_1° and Λ_2° may have different magnetic moments.

Let me summarize the situation now as I visualize it in the following table:

3. I shall now call your attention to the following problems:

a. The $\theta\bar{\theta}$ problem. This was discussed in the published work of Gell-Mann and Pais, and Pais and Piccioni.

b. Possible ways to measure the spin of strange particles. Adair, Treiman, and others have discussed this problem. We heard, Thursday, Karplus and Primakoff's suggestion in this connection.

c. What are the weak interactions? Concerning the parts that are responsible for the decay of the strange particles it is evident that, to facilitate the discussion of selection rules, it is best to split the weak coupling constants into many additive parts, each of which, if thought of as carrying an isotopic spin, would leave the weak interactions invariant. The simplest possibility would then be to have only one such constant with $I = 1/2$, so that $\Delta I = \pm 1/2$, $\Delta I_3 = \pm 1/2$. This has been discussed by Gell-Mann in some unpublished preprints and more recently by Wentzel, by Gatto and by Nishijima. I believe Professor Wentzel will discuss this later.

d. There are, in addition, the well-known weak interactions involving π-decay, μ-decay, β-decay and μ-nucleon interactions. It is very

Spin and parity: τ	Spin and parity: θ	Fit Dalitz plot?	Mass degeneracy could be due to	Life time identity could be due to	Remarks
same particle		??			
0^-	0^+	easily	C_p conservation	?	
	odd				$\theta^\circ \mapsto \pi^\circ + \pi^\circ$
0^-	2^+	easily	?	Cascade ($E_\gamma \sim$ MeV)	Alvarez says no $\gamma \sim 1$ MeV
2^-	0^+	O.K.	?	Same	"
2^-	2^+	O.K.	C_p conservation	Cascade ($E\gamma \sim$ KeV)	
				Weinstein's idea(?)	
Others					

remarkable that the strange particle decays and these interactions all have comparable strength, i.e.,

Strong interactions	strength ~ 1
Electromagnetic interactions	strength $\sim 10^{-2}$
Strange particle decay π, μ, β-decays; μ-nucleon interactions	strength $\sim 10^{-12}$–10^{-14}

One should notice that the bunching of the interaction strength near these widely separated regions can *not* be explained as due to the time scale of available experimental techniques.

4. I shall conclude with a discussion of the over-all view of all the known conservation laws. We first list the conserved quantities other than those due to space time invariance: (C_p is not included here).

1. For all interactions: N, Q and $\mathscr{C}$ (charge conjugation).
2. For all but the weak interactions: N, Q, $\mathscr{C}$, and S.
3. For strong interactions only: N, (Q), $\mathscr{C}$, S and $\vec{\mathbf{I}}$.

We put Q in parentheses because its conservation in case (3) follows from those of $\mathbf{I}$, N, and S.

It is straight-forward to write down all the commutation rules between these quantities. It is then found to be more convenient in case (3) to use a quantity G defined by

$$G = \mathscr{C} e^{i\pi I_2}$$

in place of $\mathscr{C}$. This is because G commutes with $\mathbf{I}$ while $\mathscr{C}$ does not. The commutation rules can be realized as follows:

Case 1: Two independent axes of rotation, the "angular momenta" around which are Q and N: $\underset{Q}{\circlearrowright}\ \underset{N}{\circlearrowright}$.
$\mathscr{C}$ is an operation that turns both axes simulataneously by 180°.

Case 2: Three axes: $\underset{Q}{\circlearrowright}\ \underset{N}{\circlearrowright}\ \underset{S}{\circlearrowright}$, with $\mathscr{C}$ simultaneously turning all three axes by 180°.

Case 3: Two axes $\underset{S}{\circlearrowright}\ \underset{N}{\circlearrowright}$ and a spherical symmetry $\mathbf{I}$.

G turns the S and N axes simultaneously by 180°.

In algebraic language, the mass degeneracies are related to the irreducible representations of the group of symmetries named above. Additional symmetries, such as C_p, would enlarge such irreducible representations and consequently increase the degeneracy.

The purpose of this visualization of the symmetries is to see whether a general, integrated pattern would emerge. E.g., one of Pais' schemes is in this picture equivalent to proposing that in cases (2) and (3) the axis for S is really one of three, forming a spherical by symmetry. The difficulty with such a scheme is that the increased symmetry gives rise to a greater degeneracy than that which is observed. (E.g., it would imply the existence of a Λ^+.) In contrast, experimentally we are peculiarly beset with rather strange degeneracies. More symmetries appear to be called for. It is very interesting that these further symmetries seem to get entangled with space–time concepts. Let us hope that this entanglement would lead rapidly to a resolution of the present situation which is characterized by a directionless growth of more and more quantum numbers.

Question of Parity Conservation in Weak Interactions*

[[56h]]

Commentary begins page 26

T. D. LEE, *Columbia University, New York, New York*

AND

C. N. YANG,† *Brookhaven National Laboratory, Upton, New York*

(Received June 22, 1956)

The question of parity conservation in β decays and in hyperon and meson decays is examined. Possible experiments are suggested which might test parity conservation in these interactions.

RECENT experimental data indicate closely identical masses[1] and lifetimes[2] of the $\theta^+(\equiv K_{\pi 2}^+)$ and the $\tau^+(\equiv K_{\pi 3}^+)$ mesons. On the other hand, analyses[3] of the decay products of τ^+ strongly suggest on the grounds of angular momentum and parity conservation that the τ^+ and θ^+ are not the same particle. This poses a rather puzzling situation that has been extensively discussed.[4]

One way out of the difficulty is to assume that parity is not strictly conserved, so that θ^+ and τ^+ are two different decay modes of the same particle, which necessarily has a single mass value and a single lifetime. We wish to analyze this possibility in the present paper against the background of the existing experimental evidence of parity conservation. It will become clear that existing experiments do indicate parity conservation in strong and electromagnetic interactions to a high degree of accuracy, but that for the weak interactions (i.e., decay interactions for the mesons and hyperons, and various Fermi interactions) parity conservation is so far only an extrapolated hypothesis unsupported by experimental evidence. (One might even say that the present $\theta-\tau$ puzzle may be taken as an indication that parity conservation is violated in weak interactions. This argument is, however, not to be taken seriously because of the paucity of our present knowledge concerning the nature of the strange particles. It supplies rather an incentive for an examination of the question of parity conservation.) To decide unequivocally whether parity is conserved in weak interactions, one must perform an experiment to determine whether weak interactions differentiate the right from the left. Some such possible experiments will be discussed.

PRESENT EXPERIMENTAL LIMIT ON PARITY NONCONSERVATION

If parity is not strictly conserved, all atomic and nuclear states become mixtures consisting mainly of the state they are usually assigned, together with small percentages of states possessing the opposite parity. The fractional weight of the latter will be called $\mathfrak{F}^2$. It is a quantity that characterizes the degree of violation of parity conservation.

The existence of parity selection rules which work well in atomic and nuclear physics is a clear indication that the degree of mixing, $\mathfrak{F}^2$, cannot be large. From such considerations one can impose the limit $\mathfrak{F}^2 \lesssim (r/\lambda)^2$, which for atomic spectroscopy is, in most cases, $\sim 10^{-6}$. In general a less accurate limit obtains for nuclear spectroscopy.

Parity nonconservation implies the existence of interactions which mix parities. The strength of such interactions compared to the usual interactions will in general be characterized by $\mathfrak{F}$, so that the mixing will be of the order $\mathfrak{F}^2$. The presence of such interactions would affect angular distributions in nuclear reactions. As we shall see, however, the accuracy of these experiments is not good. The limit on $\mathfrak{F}^2$ obtained is not better than $\mathfrak{F}^2 < 10^{-4}$.

To give an illustration, let us examine the polarization experiments, since they are closely analogous to some experiments to be discussed later. A proton beam polarized in a direction z perpendicular to its momentum was scattered by nuclei. The scattered intensities were compared[5] in two directions A and B related to each other by a reflection in the x–y plane, and were found to be identical to within $\sim 1\%$. If the scattering originates from an ordinary parity-conserving interaction plus a parity-nonconserving interaction (e.g., $\boldsymbol{\sigma}\cdot\mathbf{r}$), then the scattering amplitudes in the directions A and B are in the proportion $(1+\mathfrak{F})/(1-\mathfrak{F})$, where $\mathfrak{F}$ represents the ratio of the strengths of the two kinds of interactions in the scattering. The experimental result therefore requires $\mathfrak{F} < 10^{-2}$, or $\mathfrak{F}^2 < 10^{-4}$.

The violation of parity conservation would lead to an electric dipole moment for all systems. The magnitude of the moment is

$$\text{moment} \sim e\mathfrak{F} \times (\text{dimension of system}). \qquad (1)$$

* Work supported in part by the U. S. Atomic Energy Commission.

† Permanent address: Institute for Advanced Study, Princeton, New Jersey.

[1] Whitehead, Stork, Perkins, Peterson, and Birge, Bull. Am. Phys. Soc. Ser. II, **1**, 184 (1956); Barkas, Heckman, and Smith, Bull. Am. Phys. Soc. Ser. II, **1**, 184 (1956).

[2] Harris, Orear, and Taylor, Phys. Rev. **100**, 932 (1955); V. Fitch and K. Motley, Phys. Rev. **101**, 496 (1956); Alvarez, Crawford, Good, and Stevenson, Phys. Rev. **101**, 503 (1956).

[3] R. Dalitz, Phil. Mag. **44**, 1068 (1953); E. Fabri, Nuovo cimento **11**, 479 (1954). See Orear, Harris, and Taylor [Phys. Rev. **102**, 1676 (1956)] for recent experimental results.

[4] See, e.g., *Report of the Sixth Annual Rochester Conference on High Energy Physics* (Interscience Publishers, Inc., New York, to be published).

[5] See, e.g., Chamberlain, Segrè, Tripp, and Ypsilantis, Phys. Rev. **93**, 1430 (1954).

The presence of such electric dipole moments would have interesting consequences. For example, if the proton has an electric dipole moment $\cong e\times(10^{-16}$ cm), the perturbation caused by the presence of the neighboring $2p$ state of the hydrogen atom would shift the energy of the $2s$ state by about 1 Mc/sec. This would be inconsistent with the present theoretical interpretations of the Lamb shift. Another example is found in the electron-neutron interaction. An electric dipole moment for the neutron $\cong e\times(10^{-18}$ cm) is the upper limit allowable by the present experiments.

By far the most accurate measurement of the electric dipole moment was made by Purcell, Ramsey, and Smith. They gave[6] an upper limit for the electric dipole moment of the neutron of $e\times(5\times10^{-20}$ cm). This value sets the upper limit for $\mathfrak{F}^2$ as $\mathfrak{F}^2<3\times10^{-13}$, which is also the most accurate verification of the conservation of parity in strong and electromagnetic interactions. We shall see, however, that even this high degree of accuracy is not sufficient to supply an experimental proof of parity conservation in the weak interactions. For such a proof an accuracy of $\mathfrak{F}^2<10^{-24}$ is necessary.

QUESTION OF PARITY CONSERVATION IN β DECAY

At first sight it might appear that the numerous experiments related to β decay would provide a verification that the weak β interaction does conserve parity. We have examined this question in detail and found this to be not so. (See Appendix.) We start by writing down the five usual types of couplings. In addition to these we introduce the five types of couplings that conserve angular momentum but do not conserve parity. It is then apparent that the classification of β decays into allowed transitions, first forbidden, etc., proceeds exactly as usual. (The mixing of parity of the *nuclear states* would not measurably affect these selection rules. This phenomenon belongs to the discussions of the last section.) The following phenomena are then examined: allowed spectra, unique forbidden spectra, forbidden spectra with allowed shape, β-neutrino correlation, and $\beta-\gamma$ correlation. It is found that these experiments have no bearing on the question of parity conservation of the β-decay interactions. This comes about because in all of these phenomena no interference terms exist between the parity-conserving and parity-nonconserving interactions. In other words, the calculations always result in terms proportional to $|C|^2$ plus terms proportional to $|C'|^2$. Here C and C' are, respectively, the coupling constants for the usual parity-conserving interactions (a sum of five terms) and the parity-nonconserving interactions (also a sum of five terms.) Furthermore, it is well known[7] that without measuring the spin of the neutrino we cannot distinguish the couplings C from the couplings C' (provided the mass of the neutrino is zero). The experimental results concerning the above-named phenomena, which constitute the bulk of our present knowledge about β decay, therefore cannot decide the degree of mixing of the C' type interactions with the usual type.

The reason for the absence of interference terms CC' is actually quite obvious. Such terms can only occur as a pseudoscalar formed out of the experimentally measured quantities. For example, if three momenta $\mathbf{p}_1$, $\mathbf{p}_2$, $\mathbf{p}_3$ are measured, the term $CC'\mathbf{p}_1\cdot(\mathbf{p}_2\times\mathbf{p}_3)$ may occur. Or if a momentum $\mathbf{p}$ and a spin $\boldsymbol{\sigma}$ are measured, the term $CC'\mathbf{p}\cdot\boldsymbol{\sigma}$ may occur. In all the β-decay phenomena mentioned above, no such pseudoscalars can be formed out of the measured quantities.

POSSIBLE EXPERIMENTAL TESTS OF PARITY CONSERVATION IN β DECAYS

The above discussion also suggests the kind of experiments that could detect the possible interference between C and C' and consequently could establish whether parity conservation is violated in β decay. A relatively simple possibility is to measure the angular distribution of the electrons coming from β decays of oriented nuclei. If θ is the angle between the orientation of the parent nucleus and the momentum of the electron, an asymmetry of distribution between θ and $180°-\theta$ constitutes an unequivocal proof that parity is not conserved in β decay.

To be more specific, let us consider the allowed β transition of any oriented nucleus, say Co^{60}. The angular distribution of the β radiation is of the form (see Appendix):

$$I(\theta)d\theta=(\text{constant})(1+\alpha\cos\theta)\sin\theta d\theta, \qquad (2)$$

where α is proportional to the interference term CC'. If $\alpha\neq0$, one would then have a positive proof of parity nonconservation in β decay. The quantity α can be obtained by measuring the fractional asymmetry between $\theta<90°$ and $\theta>90°$; i.e.,

$$\alpha=2\left[\int_0^{\pi/2}I(\theta)d\theta-\int_{\pi/2}^{\pi}I(\theta)d\theta\right]\Big/\int_0^{\pi}I(\theta)d\theta.$$

It is noteworthy that in this case the presence of the magnetic field used for orienting the nuclei would automatically cause a spatial separation between the electrons emitted with $\theta<90°$ and those with $\theta>90°$. Thus, this experiment may prove to be quite feasible.

It appears at first sight that in the study of γ-radiation distribution from β-decay products of oriented nuclei one can form a pseudoscalar from the spin of the oriented nucleus and the γ-ray momentum $\mathbf{p}_\gamma$. Thus it may seem to offer another possible experimental test of parity conservation. Unfortunately, the nuclear levels have definite parities, and electromagnetic inter-

[6] E. M. Purcell and N. F. Ramsey, Phys. Rev. 78, 807 (1950); Smith *et al.* as quoted in N. F. Ramsey, *Molecular Beams* (Oxford University Press, London, 1956).

[7] C. N. Yang and J. Tiomno, Phys. Rev. 79, 495 (1950).

actions conserve parity. (Any small mixing of parities characterized by $\mathcal{F}^2 < 3\times 10^{-15}$ would not affect the arguments here.) Consequently the γ rays carry away definite parities. Thus the observed probability function must be an even function of $\mathbf{p}_\gamma$. This property eliminates the possibility of forming a pseudoscalar quantity. It is therefore not possible to use such experiments as a test of parity conservation.

In β-γ-γ' triple correlation experiments one can, by some rather similar but more complicated reasoning, prove that a measurement of the three momenta cannot supply any information on the question of parity conservation in β decay.

In β-γ correlation experiments the nature of the polarization of the γ ray could provide a test. To be more specific, let us consider the polarization state of γ rays emitted parallel to the β ray. If parity conservation holds for β decay, the γ ray will be unpolarized. On the other hand, if parity conservation is violated in β decay, the γ ray will in general be polarized. However, this polarization must be circular in nature and therefore may not lend itself to easy experimental detection. (The usual ways of measuring polarization through Compton effect, photoelectric effect, and photodisintegration of the deuteron are all incapable of detecting circular polarization. This is because circular polarization is specified by an axial vector parallel to the direction of propagation. From the observed momenta in these detection techniques such an axial vector cannot be formed.) For other directions of γ-ray propagation, elliptical polarization will result if parity is not conserved. This effect will thus be more difficult to detect.

QUESTION OF PARITY CONSERVATION IN MESON AND HYPERON DECAYS

If the weak interactions, such as the β-decay interactions or the decay interactions of mesons and hyperons, do not conserve parity, parity mixing will occur in all interactions by means of second-order processes. To examine this effect let us consider, for example, the decay of the Λ^0:

$$\Lambda^0 \to p + \pi^-.$$

The assumption that parity is not conserved in this decay implies that the Λ^0 exists virtually in states of opposite parities. It could therefore possess an electric dipole moment of a magnitude

$$\text{moment} \sim e\mathcal{G}^2 \times (\text{dimension of } \Lambda^0), \tag{3}$$

where $\mathcal{G}$ is the coupling strength of the decay interaction of the Λ^0. ($\mathcal{G}^2 \lesssim 10^{-12}$.) The electric dipole moment of the Λ^0 is therefore $\lesssim e\times(10^{-25}\text{ cm})$.

Clearly the proton would have an electric dipole moment of the same order of magnitude. The existence of such a small electric dipole moment is, as we have seen, completely consistent with the present experimental information. Another way of putting this is to observe that by comparing Eq. (3) with Eq. (1), one has

$$\mathcal{F} \sim \mathcal{G}^2.$$

Since all the weak interaction including β interactions are characterized by coupling strengths $\mathcal{G}^2 < 10^{-12}$, a violation of parity in weak interactions would introduce a parity mixing characterized by an $\mathcal{F}^2 < 10^{-24}$. This is outside the present limit of experimental knowledge, as we have discussed before.

If the weak interactions violate parity conservation, parity would be defined and measured in strong and electromagnetic interactions only, just as strangeness is. Furthermore it is important to notice that with the conservation of strangeness, as with every conservation law, there is an element of arbitrariness introduced into the parity of all systems. The parity of all strange particles would be defined only up to a factor of $(-1)^S$, where S is the strangeness. The parity of the Λ^0 (relative to the nucleons) is therefore a matter of definition. But once this is defined, the parity of other strange particles would be measurable from the strong interactions.

POSSIBLE EXPERIMENTAL TESTS OF PARITY CONSERVATION IN MESON AND HYPERON DECAYS

To have a sensitive unequivocal test of whether parity is conserved in weak interactions, one must decide whether the weak interactions differentiate between the right and the left. This is possible only if one produces interference between states of opposite parities. The mere observation of two decay products of opposite parities originating from a "particle" cannot provide conclusive evidence that parity is not conserved. Such indeed is the state of affairs of the present $\theta-\tau$ puzzle.

As we have discussed before, these interference terms are possible only if the observed quantities can form a pseudoscalar such as $\mathbf{p}_1\cdot(\mathbf{p}_2\times\mathbf{p}_3)$. The observation of Λ^0 decays in association with their production does provide such a possible pseudoscalar and hence a possible test of whether parity is conserved in the Λ^0 decay interaction. Let us consider the experiment

$$\pi^- + p \to \Lambda^0 + \theta^0, \quad \Lambda^0 \to p + \pi^-. \tag{4}$$

Let $\mathbf{p}_{\text{in}}$, $\mathbf{p}_\Lambda$, and $\mathbf{p}_{\text{out}}$ be, respectively, the momenta in the laboratory system of the incoming pion, the Λ^0, and the decay pion. We define a parameter R as the projection of $\mathbf{p}_{\text{out}}$ in the direction of $\mathbf{p}_{\text{in}}\times\mathbf{p}_\Lambda$. The value of R ranges from approximately -100 Mev/c to approximately $+100$ Mev/c. Switching from a right-handed convention for vector products (which we use) to a left-handed convention means a switch of the sign of R. Parity conservation in the weak decay interaction of Λ^0 can therefore be experimentally checked by investigating whether $+R$ and $-R$ have equal probabilities of occurrence.

To see more clearly the meaning of the parameter R, one transforms $\mathbf{p}_{out}(\rightarrow\mathbf{p}')$ into the center-of-mass system of Λ^0. The new vector $\mathbf{p}'$ has a constant magnitude $\cong 100$ Mev/c. The frequency distribution of this vector $\mathbf{p}'$ can then be plotted on a spherical surface. Taking the z axis for this sphere to be in the direction of $\mathbf{p}_{in}\times\mathbf{p}_{\Lambda}$, one can prove the following two symmetries:

(*a*) The frequency distribution on the sphere remains unchanged under a rotation through 180° around the z axis. This symmetry follows from parity conservation in the strong reaction producing the Λ^0. It does not depend on the nature of the weak interaction.

(*b*) If parity is conserved in the decay interaction of Λ^0, the frequency distribution on the sphere is unchanged under a reflection with respect to the production plane of Λ^0.

To prove statement (*a*), one need only consider the invariance of the production process under a reflection with respect to the production plane defined by $\mathbf{p}_{in}$ and $\mathbf{p}_{\Lambda}$. This reflection is the resultant of an inversion and a rotation through 180° around the z axis (which is normal to the production plane). The state of polarization of Λ^0 is thus invariant under a 180° rotation around the z axis, leading to the stated symmetry.[8]

Statement (*b*) follows[8] directly from the assumption that the weak interaction as well as the strong interaction conserves parity. A reflection with respect to the production plane must then leave the whole process invariant.

The frequency distribution of R is just the projection of the distribution on the sphere onto the z axis. An asymmetry between $+R$ and $-R$ therefore implies parity nonconservation in Λ^0 decay. However, if the spin of Λ^0 is unpolarized, no asymmetry[9] can obtain even if parity is not conserved in Λ^0 decay. To obtain a polarized Λ^0 beam, the experiment is therefore best done at a definite nonforward angle of production of Λ^0 and at a definite incoming energy.

The above discussions apply also to any other strange particle decay if (1) the particle has a nonvanishing spin and (2) it decays into two particles at least one of which has a nonvanishing spin, or it decays into three or more particles. Thus the above considerations can be applied also to the decays of $\Sigma^\pm$ and *maybe* also to $K_{\mu 2}^\pm$, $K_{\mu 3}^\pm$ and $K_{\pi 3}^\pm$ $(\equiv\tau^\pm)$.

In the decay processes

$$\pi\rightarrow\mu+\nu, \tag{5}$$

$$\mu\rightarrow e+\nu+\nu, \tag{6}$$

starting from a π meson at rest, one could study the distribution of the angle θ between the μ-meson momentum and the electron momentum, the latter being in the center-of-mass system of the μ meson. If parity is conserved in neither (5) nor (6), the distribution will not in general be identical for θ and $\pi-\theta$. To understand this, consider first the orientation of the muon spin. If (5) violates parity conservation, the muon would be in general polarized in its direction of motion. In the subsequent decay (6), the angular distribution problem with respect to θ is therefore closely similar to the angular distribution problem of β rays from oriented nuclei, which we have discussed before. (Entirely similar considerations can be applied to $\Xi^-\rightarrow\Lambda^0+\pi^-$ and $\Lambda^0\rightarrow p+\pi^-$.)

REMARKS

If parity conservation is violated in hyperon decay, the decay products will have mixed parities. This, however, does not affect the arguments of Adair[10] and of Treiman[11] concerning the relationship between the spin of the hyperons and the angular distribution of their decay products in certain special cases.[12]

One may question whether the other conservation laws of physics could also be violated in the weak interactions. Upon examining this question, one finds that the conservations of the number of heavy particles, of electric charge, of energy, and of momentum all appear to be inviolate in the weak interactions. The same cannot be said of the conservation of angular momentum, and of parity. Nor can it be said of the invariance under time reversal. It might appear at first sight that the equality of the life times of $\pi^\pm$ and of those of $\mu^\pm$ furnish proofs of the invariance under charge conjugation of the weak interactions. A closer examination of this problem reveals, however, that this is not so. In fact, the equality of the life times of a charged particle and its charge conjugate against decay through a weak interaction (to the lowest order of the strength of the weak interaction) can be shown to follow from the invariance under proper Lorentz transformations (i.e., Lorentz transformation with neither space nor time inversion). One has therefore at present no experimental proof of the invariance under charge conjugation of the weak interactions. In the present paper, only the question of parity nonconservation is discussed.

[8] This proof for statement (*a*) is correct only if Λ^0 exists as a single particle with a definite parity in the strong interactions, (as discussed in the last section); i.e. if Λ^0 does not exist as two degenerate states Λ_1^0 and Λ_2^0 of opposite parity, as has been suggested [T. D. Lee and C. N. Yang, Phys. Rev. **102**, 290 (1956)]. [It is to be emphasized, that if parity is indeed not conserved in the weak interactions, there would be (at present) no necessity to introduce the complication of two degenerate states of opposite parity at all.] On the other hand, statement (*b*) is correct even if Λ^0 exists as two degenerate states Λ_1^0 and Λ_2^0 of opposite parity. *To summarize, violation of the symmetry stated in (a) implies the existence of the parity doublets Λ_1^0 and Λ_2^0 with a mass difference less than their widths. Violation of the symmetry stated in (b) implies the nonconservation of parity in Λ decay.* See also footnote 12 and T. D. Lee and C. N. Yang, Phys. Rev. (to be published).

[9] Also the interference may accidentally be absent if the relative phase between the two parities in the decay product is 90°. This, however, cannot be the case if time-reversal invariance is preserved in the decay process.

[10] R. K. Adair, Phys. Rev. **100**, 1540 (1955).

[11] S. B. Treiman, Phys. Rev. **101**, 1216 (1956).

[12] The existence of Λ_1^0 and Λ_2^0 of opposite parity may affect these relationships. This is similar to the violation of symmetry (*a*) discussed in footnote 8. See T. D. Lee and C. N. Yang, Phys. Rev. (to be published).

The conservation of parity is usually accepted without questions concerning its possible limit of validity being asked. There is actually no *a priori* reason why its violation is undesirable. As is well known, its violation implies the existence of a right-left asymmetry. We have seen in the above some possible experimental tests of this asymmetry. These experiments test whether the present elementary particles exhibit asymmetrical behavior with respect to the right and the left. If such asymmetry is indeed found, the question could still be raised whether there could not exist corresponding elementary particles exhibiting opposite asymmetry such that in the broader sense there will still be over-all right-left symmetry. If this is the case, it should be pointed out, there must exist two kinds of protons p_R and p_L, the right-handed one and the left-handed one. Furthermore, at the present time the protons in the laboratory must be predominantly of one kind in order to produce the supposedly observed asymmetry, and also to give rise to the observed Fermi-Dirac statistical character of the proton. This means that the free oscillation period between them must be longer than the age of the universe. They could therefore both be regarded as stable particles. Furthermore, the numbers of p_R and p_L must be separately conserved. However, the interaction between them is not necessarily weak. For example, p_R and p_L could interact with the same electromagnetic field and perhaps the same pion field. They could then be separately pair-produced, giving rise to interesting observational possibilities.

In such a picture the supposedly observed right-and-left asymmetry is therefore ascribed not to a basic non-invariance under inversion, but to a cosmologically local preponderance of, say, p_R over p_L, a situation not unlike that of the preponderance of the positive proton over the negative. Speculations along these lines are extremely interesting, but are quite beyond the scope of this note.

The authors wish to thank M. Goldhaber, J. R. Oppenheimer, J. Steinberger, and C. S. Wu for interesting discussions and comments. They also wish to thank R. Oehme for an interesting communication.

APPENDIX

If parity is not conserved in β decay, the most general form of Hamiltonian can be written as

$$\begin{aligned}H_{\text{int}}= &(\psi_p^\dagger\gamma_4\psi_n)(C_S\psi_e^\dagger\gamma_4\psi_\nu+C_S'\psi_e^\dagger\gamma_4\gamma_5\psi_\nu)\\ &+(\psi_p^\dagger\gamma_4\gamma_\mu\psi_n)(C_V\psi_e^\dagger\gamma_4\gamma_\mu\psi_\nu+C_V'\psi_e^\dagger\gamma_4\gamma_\mu\gamma_5\psi_\nu)\\ &+\tfrac{1}{2}(\psi_p^\dagger\gamma_4\sigma_{\lambda\mu}\psi_n)(C_T\psi_e^\dagger\gamma_4\sigma_{\lambda\mu}\psi_\nu\\ &+C_T'\psi_e^\dagger\gamma_4\sigma_{\lambda\mu}\gamma_5\psi_\nu)+(\psi_p^\dagger\gamma_4\gamma_\mu\gamma_5\psi_n)\\ &\times(-C_A\psi_e^\dagger\gamma_4\gamma_\mu\gamma_5\psi_\nu-C_A'\psi_e^\dagger\gamma_4\gamma_\mu\psi_\nu)\\ &+(\psi_p^\dagger\gamma_4\gamma_5\psi_n)(C_P\psi_e^\dagger\gamma_4\gamma_5\psi_\nu+C_P'\psi_e^\dagger\gamma_4\psi_\nu),\end{aligned} \quad \text{(A.1)}$$

where $\sigma_{\lambda\mu}=-\frac{1}{2}i(\gamma_\lambda\gamma_\mu-\gamma_\mu\gamma_\lambda)$ and $\gamma_5=\gamma_1\gamma_2\gamma_3\gamma_4$. The ten constants C and C' are all real if time-reversal invariance is preserved in β decay. This however, will not be assumed in the following.

Calculation with this interaction proceeds exactly as usual. One obtains, e.g., for the energy and angle distribution of the electron in an allowed transition

$$N(W,\theta)dW\,\sin\theta d\theta=\frac{\xi}{4\pi^3}F(Z,W)pW(W_0-W)^2 \times\left(1+\frac{ap}{W}\cos\theta+\frac{b}{W}\right)dW\,\sin\theta d\theta, \quad \text{(A.2)}$$

where

$$\begin{aligned}\xi=&(|C_S|^2+|C_V|^2+|C_S'|^2+|C_V'|^2)|M_{\text{F.}}|^2\\ &+(|C_T|^2+|C_A|^2+|C_T'|^2+|C_A'|^2)|M_{\text{G.T.}}|^2,\end{aligned} \quad \text{(A.3)}$$

$$\begin{aligned}a\xi=&\tfrac{1}{3}(|C_T|^2-|C_A|^2+|C_T'|^2-|C_A'|^2)|M_{\text{G.T.}}|^2\\ &-(|C_S|^2-|C_V|^2+|C_S'|^2-|C_V'|^2)|M_{\text{F.}}|^2,\end{aligned} \quad \text{(A.4)}$$

$$\begin{aligned}b\xi=&\gamma[(C_S^*C_V+C_SC_V^*)+(C_S'^*C_V'+C_S'C_V'^*)]|M_{\text{F.}}|^2\\ &+\gamma[(C_T^*C_A+C_A^*C_T)+(C_T'^*C_A'+C_A'^*C_T')]\\ &\times|M_{\text{G.T.}}|^2.\end{aligned} \quad \text{(A.5)}$$

In the above expression all unexplained notations are identical with the standard notations. (See, e.g., the article by Rose.[13])

The above expression does not contain any interference terms between the parity-conserving part of the interactions and the parity-nonconserving ones. It is in fact directly obtainable by replacing in the usual expression the quantity $|C_S|^2$ by $|C_S|^2+|C_S'|^2$, and $C_SC_V^*$ by $C_SC_V^*+C_S'C_V'^*$, etc. This rule also holds in general, except for the cases where a pseudoscalar can be formed out of the measured quantities, as discussed in the text.

When a pseudoscalar can be formed, for example, in the β decay of oriented nuclei, interference terms would be present, as explicitly displayed in Eq. (2). In an allowed transition $J\rightarrow J-1$ (no), the quantity α is given by

$$\alpha=\beta\langle J_z\rangle/J,$$

$$\beta=\text{Re}\left[C_TC_T'^*-C_AC_A'^*+i\frac{Ze^2}{\hbar cp}(C_AC_T'^*+C_A'C_T^*)\right] \times|M_{\text{G.T.}}|^2\frac{v_e}{c}\frac{2}{\xi+(\xi b/W)}, \quad \text{(A.6)}$$

where $M_{\text{G.T.}}$, ξ, and b are defined in Eqs. (A.3)–(A.5), v_e is the velocity of the electron, and $\langle J_z\rangle$ is the average spin component of the initial nucleus. For an allowed transition $J\rightarrow J+1$ (no), α is given by

$$\alpha=-\beta\langle J_z\rangle/(J+1). \quad \text{(A.7)}$$

The effect of the Coulomb field is included in all the above considerations.

[13] M. E. Rose, in *Beta- and Gamma-Ray Spectroscopy* (Interscience Publishers, Inc., New York, 1955), pp. 271–291.

[[57d]]
Commentary
begins
page 31

Present Knowledge About the New Particles

C. N. YANG
Institute for Advanced Study, Princeton, New Jersey

THIS discussion is divided into the following four sections: First an introduction to the subject, dealing mostly with generalities; then, a brief description of what seems at present to be established knowledge concerning the strange particles; thirdly, some topics currently under discussion by theoretical physicists; lastly, a question which has in the past year occupied a great deal of attention, namely, the identity of the K particles.

The discovery of the strange particles dates back to 1947, when Rochester and Butler[1] found, in a magnetic cloud chamber, V-shaped events in a penetrating shower induced by cosmic rays. This was followed in 1949 by the observation and identification[2] by Powell's group in England of the decay of the τ meson in electron sensitive emulsions. Experimental work on these strange particles, and others found later, rapidly expanded in the last few years. With the completion in 1953 of the cosmotron and in 1955 of the bevatron, experiments entered into a new era in which one can work with controlled beams of strange particles and in which one is able to make relatively accurate measurements on their interactions and decay. The scope and intensity of the experimental activities on this subject may be gauged from the fact that at the Brookhaven National Laboratory in the United States, 60% of the cosmotron time at present is devoted to the study of strange particles, the remaining time being divided between pion physics, radio chemistry, machine design, and other subjects.

What have we learned about these particles? We know the identity of quite a number of them; that is to say their masses, their charges, their lifetimes and decay modes. We know something about the processes by which they are produced. We could say that we know enough, so that there has emerged a common language among us in talking about them. We know enough to be able to plan controlled experiments which will tell us more about them. But the sum total of our knowledge goes very little beyond this. We are still confused by the identity of some of them, and we know very little about their spins and their parities. Above all, we do not have any idea or even any hint about the reason for their existence. Their relationship to each other and to the familiar particles do not seem to emerge in a simple pattern. Unlike Yukawa's π mesons, which fit into the physicist's picture of the world, the strange particles seem to be uncalled for, and unwanted. Do they presage the physicist's future understanding of small distances? Of the mass spectrum of particles? Or do they presage the breakdown of the concepts and laws that we have learned in the physics of nonstrange particles? These questions remain completely unanswered.

In view of our profound ignorance of what the whole subject is about, we must admit that we are very fortunate indeed to find in the experimental data many regularities, and many irregularities, which we hope may provide keys to a future understanding. We cherish such a hope mainly because of two facts: (a) The experiments consistently confirm a concept which we have had before, namely, that there is a natural classification of interactions into strong and weak interactions with very widely different orders of magnitude in their strengths. One senses that perhaps herein lies a very fundamental order of nature. (b) We recall that in the past, the concept of symmetry has played a very important role in physics. To quote a few examples: The theories of special and general relativity are founded upon such concepts of symmetry. Again, we could say that in essence the structure of the periodic table and therefore the properties of matter are largely determined by the isotropy of space, or in other words, by the symmetry of nature under space rotations. The prediction of the existence of the positron and of the antiproton were based upon the symmetry of nature under Lorentz transformations. Now it is observed that symmetries, and the related concepts of quantum numbers and selection rules are very much involved in the regularities and irregularities in the behavior of the new particles. We may therefore hope that the study of them will lead to a further understanding of nature.

To go on to the second topic, let us briefly outline our knowledge about the new particles. The well-established law of conservation of the number of nucleons requires that all particles have a "heavy particle quantum number" N. The nucleons have $N=1$. The pions, photons, electrons, neutrinos, and μ mesons have $N=0$. The antinucleons have $N=-1$. It is customary to call the strange particles with $N=1$ hyperons, those with $N=0$ mesons. Table I lists the experimental knowledge about the hyperons. Table II lists that about the positive and neutral mesons. Negative mesons are expected to behave like the charge conjugates of the positive ones, and they seem to do just that. They are therefore omitted from the table. In these tables, I have given numbers in round figures. More complete and detailed information with

[1] G. D. Rochester and C. C. Butler, Nature **160**, 855 (1947).
[2] Brown *et al.* Nature **163**, 82 (1949).

□ Reprinted from *Reviews of Modern Physics* 29, 2 (April 1957), 231–235.

TABLE I. Experimental knowledge about hyperons.

Hyperon	Decay product	Q	Mean life
Λ^0	$p+\pi^-$	37 Mev	3.7×10^{-10} sec
Σ^+	$n+\pi^+$	110	$\sim1\ \times10^{-10}$
	$p+\pi^0$		
Σ^0	$\Lambda^0+\gamma$		short
Σ^-	$n+\pi^-$	110	$\sim2\ \times10^{-10}$
Ξ^-	$\Lambda^0+\pi^-$	~67	$\sim1\ \times10^{-10}$

references can be found in the various review compilations.[3]

In addition to tabulating the strange particles, we must say something about their interactions. There seem to be three kinds of interactions, the strong interactions, the electromagnetic interactions and the weak interactions. By the weak interactions we mean those responsible for the decay of the strange particles, for beta decay, for μ-meson decay, for π-meson decay, and for the μ-meson nucleon interaction. The reason why one feels some confidence in such a classification is that if one makes a rough dimensional analysis of the coupling strengths, one finds that the π-meson interactions and the production processes for the strange particles have intensities of around 1 to 10^{-1}; the electromagnetic interaction, as is well known, has strength 10^{-2}; and the weak interactions have intensities ranging from 10^{-12} to 10^{-14}. The wide gap separating the weak interactions from the other two forms such a striking and consistent pattern that one cannot but feel that there is some fundamental reason for it. A characteristic feature of the strange particles is that their production seems to fall into the class of strong interactions and their decays into the class of weak interactions. The practical manifestation of the weakness of the decay interaction is the fact that the strange particles all have lifetimes of the order of 10^{-10} sec or longer; i.e., more than 10^{+13} cy of nuclear time.

Let us write down the various invariance properties of these interactions besides invariance with respect to space time transformations. For all interactions it is found that the number of nucleons is conserved, and that charge is conserved. If we however limit ourselves to the strong and the electromagnetic interactions, we find that there is another conservation law, which one refers to as the conservation of "strangeness." If one makes the assignments of strangeness as follows:

$$
\begin{array}{ll}
n,\ p,\ \pi^+,\ \pi^0,\ \pi^- & S=0 \\
K^-,\ \bar{K}^0,\ \Lambda^0,\ \Sigma^+,\ \Sigma^0,\ \Sigma^- & S=-1 \\
\Xi^- & S=-2 \\
K^+,\ K^0 & S=+1
\end{array}
$$

then the experimental production data all conform to the statement that the strong interactions conserve strangeness. Thus, for example, in

$$\pi^-+p\rightarrow K^0+\Lambda^0,$$
$$n+n\rightarrow\Lambda^0+\Lambda^0,$$

the first reaction is expected to be a strong interaction, and the second one should be forbidden as a strong interaction. These indeed have been borne out experimentally. Finally, if we confine ourselves to the strong interactions only, the isotopic spin seems to be a good quantum number, just as in pion physics and in nuclear physics for light nuclei. One of the components, I_3, is related to the charge Q, the heavy particle number N, and the strangeness quantum number S in the following way:

$$I_3=Q-\frac{N}{2}-\frac{S}{2}.$$

The concept of strangeness has proved to be extremely fruitful in the past year or two, and although we do not understand the origin of this conservation law, nor do we know what new additional selection rules and quantum numbers will turn up in the future, it seems fair to say that strangeness conservation is here to stay.

As to theoretical problems of current interest: first there is the question of $\theta^0-\bar{\theta}^0$, (or $K^0-\bar{K}^0$). It was pointed out a few years ago that since the θ^0 particle has a strangeness of $+1$, its antiparticle $\bar{\theta}^0$ must have strangeness -1 and must therefore be a different particle. The decay of these particles was then expected to exhibit rather queer properties. This is of great interest, because these properties of the decay were predicted purely on the basis of symmetry principles. An experimental test of these predictions is in progress. Another problem which has been occupying a number of physicists is the question of the possible methods for the determination of the spins of the various strange particles Then there is the question: just how do the weak interactions violate strangeness? Also there is the study of hyperfragments, which are nuclei with Λ^0 particles bound in them It has been pointed out that an examination of the binding energies of light hyperfragments will yield some information about the Λ^0-nucleon force such as its spin dependence Lastly there

TABLE II. Experimental knowledge about neutral and positive mesons.

Meson	Decay product	Mass	Mean life	Fraction of all K^+
θ°	$\pi^++\pi^-$	See below	1.7×10^{-10} sec	
$\theta_2{}^\circ$	?+?+?			
$K_{\pi3}{}^+\equiv\tau^+$	$2\pi^++\pi^-$	All K^+ and K° have mass	All K^+ about	6%
$\tau^{+\prime}$	$2\pi^\circ+\pi^+$	$\sim966m_e$	$1\ \times10^{-8}$ sec	2%
$K_{\pi2}{}^+$	$\pi^++\pi^\circ$			25%
$K_{\mu2}{}^+$	$\mu^++\nu$			60%
$K_{\mu3}{}^+$	$\mu^++\nu+\pi^\circ$			~3%
$K_{e3}{}^+$	$e^++?+?$			~5%

[3] See, e. g., A. M. Shapiro, Revs. Modern Phys. 28, 164 (1956).

is the subject of the so-called "freak" particles, which have turned up periodically, and which do not seem to fit into any established categories These and other questions have been discussed in the literature

I now turn to the last topic, namely, the question of the identity of the K particles Interest in this subject started when it was pointed out that by studying the decay of the $K_{\pi 3}^+$, also known as the τ^+ meson, into three charged π mesons. it is possible to conclude something about its spin and parity Let me give you a simple example of the kind of arguments involved. Suppose one of the π mesons resulting from the decay of a τ meson is very slow, so that one may assume that it comes out in an s state relative to the center of mass of the other two π mesons. What can one conclude under these circumstances? Well the remaining two pions will be in one of the states $0^+, 1^-, 2^+, \cdots$, because the pions are spinless, and their intrinsic parities add up to 'even', so that the spin and parity of the two pion system are identical with those of the orbital part of their wave function. The extra π meson is odd, and contributes no angular momentum; we can therefore conclude that the spin and parity of the 3 pions together can only be 0^- or 1^+ or $2^- \cdots$. Since spin and parity are conserved in any process, the τ meson spin parity also must be $0^-, 1^+, 2^-$, etc. The $K_{\pi 2}$ by the same argument can only be $0^+, 1^-, 2^+, \cdots$. Therefore we may conclude that if ever the τ meson is found to decay into three pions one of which is slow, the $K_{\pi 2}$ and τ cannot be the same particle. Of course, the analysis is not quite so simple, because there is always the question: how slow is "slow" for the decay pion? To investigate this problem, the distribution of the momenta of the decay products of the τ has been plotted in a triangular plot and the full distribution studied against the various possible spin parity assignments for τ. With the one-thousand or so points available at the present moment the general opinion is that the τ is not the same particle as the $K_{\pi 2}^+ \equiv \theta^+$.

However it will not do to jump to hasty conclusions. This is because experimentally the K mesons seem all to have the same masses and the same lifetimes. The masses are known to an accuracy of say from 2 to 10 electron masses, or a fraction of a percent, and the lifetimes are known to an accuracy of say 20%. Since particles which have different spin and parity values, and which have strong interactions with the nucleons and pions, are not expected to have identical masses and lifetimes, one is forced to keep the question open whether the inference mentioned above that the τ^+ and θ^+ are not the same particle is conclusive. Parenthetically, I might add that the inference would certainly have been regarded as conclusive, and in fact more well founded than many inferences in physics, had it not been for the anomaly of mass and lifetime degeneracies.

There has been in the past year a number of speculations concerning the reason for this anomaly. There was the suggestion that the equality of masses of two particles, say τ and θ, which have the same spin but opposite parity might come about in the following way: consider the process $\theta^+ \to p + \bar{\Lambda}^0 \to \theta^+$: the θ^+ particle virtually dissociates into a proton and an anti-Λ^0, which then recombine. Such a process contributes to the mass of the θ^+ partic.e, and does so appreciably, because all the interactions involved are strong interactions. Consider the same process for the τ^+ particle: $\tau^+ \to p + \bar{\Lambda}^0 \to \tau^+$. Because of its different parity the contribution to the mass of τ^+ from this virtual dissociation will be different (as the virtual particles will be in different orbital states relative to one another). After looking at this for a while, it appears that the only way for the contribution to be equal is that there exist two Λ^0 particles: Λ_1^0 and Λ_2^0, which have opposite parities, so that the process $\theta^+ \to p + \bar{\Lambda}_1^0$ has its counterpart in $\tau^+ \to p + \bar{\Lambda}_2^0$ with the same orbital motion. The proposal has therefore been made that there may exist another symmetry operation called "parity conjugation," C_p, which leaves the strong interactions invariant, and which has the property of interchanging θ and τ. Also it is assumed that there are two kinds of Λ^0's, and two kinds of Σ's, and that they go into one another under the operation of C_p.

It can easily be shown that the weak interactions which violate strangeness conservation cannot be invariant under C_p. In fact since Λ_1^0 and Λ_2^0 do not have the same parity, they decay in quite different ways into their final products. For example, the final product of the decay

$$\Lambda^0 \to \pi^- + p$$

would in one case be in even relative orbital states, and in the other case be in odd relative orbital states, and therefore there is no reason why the two Λ^0's should have the same lifetimes. This is also true for the θ and the τ. Thus the parity conjugation concept does not explain the equality of lifetimes.

An obvious test whether this speculation is correct is to see whether there do indeed exist such "parity doub.ets." One could for example look for evidences for two lifetimes for the Λ^0's or the Σ's. Another way of detecting parity doublets is through the interference of the two decay modes of the two members of a parity doublet This will work if the mass difference between the two members of the doublet is very small, say of the order of 10^{-5} ev. In that case, let us consider an experiment currently being done in Berkeley: the capture of K^- particles in a hydrogen bubble chamber, and single out the process

$$K^- + p \to \Sigma^- + \pi^+.$$

What is seen in the experiment is a K^- coming to rest in the chamber, and a π^+ and a Σ^- coming out. The Σ^- is relatively short-lived, and most of them decay before coming to rest, as follows

$$\Sigma^- \to \pi^- + n.$$

Let us now analyze the angular distribution of the decay particles (say the pion) under the assumption that the spin of Σ is $\frac{1}{2}$. The angle in question is the angle θ between the decay π^- and the line of flight of Σ^- in the rest system of Σ^-. We can analyze the spin into "up" and "down" components along the direction of motion of the Σ particle, and after doing this, let us make a reflection in the plane containing the lines of flight of the Σ^- and the π^- coming from its decay. Since under this reflection the spin flips (to see this picture the spin as a rotating top, so that under the reflection the sense of rotation changes, and so does the spin), and it is assumed that under a reflection the interactions are invariant, the Σ^- must have equal populations of up and down spins. In other words it is unpolarized. The angular distribution with respect to θ must then be spherically symmetric for ordinary Σ^-. However if the Σ^- does not have a definite parity, i.e. if it consists of a Σ_1^- and a Σ_2^- having opposite parities, then under an inversion they still both flip the spin. But due to their different parities they acquire a 180° phase difference between them, and therefore they would interfere, giving a forward-backward asymmetry (i.e., an asymmetry between θ and $\pi-\theta$,) a conclusion that can be experimentally tested.

With the existence of parity doublets of nearly equal masses one expects some very interesting phenomena. A beam of spin $\frac{1}{2}$ particles is usually described by four real parameters: its intensity, and a polarization vector which is the average spin of the beam. For a beam of spin $\frac{1}{2}$ particles with parity doublet structure the description is vastly more complex, because you now need four real intensities, one for the intensity of the particles of one parity, one for that of the other parity, and two for the real and the imaginary parts of the interference intensity. Also, whereas only one space vector is needed to describe the polarization in the usual case, now four space vectors are needed to describe the polarization: namely one describing the average spin of one parity one for that of the other, and two describing the interference polarization. The interference intensities are pseudoscalars instead of scalars, and the interference polarization vectors are vectors instead of axial vectors. The recognition of these transformation properties leads immediately to a number of conclusions about interference effects of parity doublets. These can all be subjected to experimental tests, and we may hope to have a clarification of the situation in the next few months.

A much more drastic thought has been advanced regarding this problem, namely the observation that both the mass identity and the lifetime identity can be understood if parity is not conserved in the decay. There need then be only one K particle, which has a definite parity upon production, but which can decay into various parities. This again is a proposal which can be subjected to experimental tests. The first question which comes to mind is: is the present evidence for parity conservation good enough to rule out such a proposal? One finds that the most accurate experiment to date, which tests and confirms the conservation of parity in strong and electromagnetic interactions, is the measurement of the electric dipole moment of the neutron which was found to be accurately zero. This experiment, though extremely accurate, is however not sufficient to yield a conclusive statement about whether the weak interactions also conserve parity, because the weak interactions are just too weak to influence the experiment. Furthermore, an examination of the tremendous bulk of knowledge about the beta interaction shows that to date there is no evidence there either for or against parity conservation in the weak interactions.

Fortunately, one can propose experiments which can test parity conservation in weak interactions, experiments which essentially test whether the weak interactions show a preference of one-sensed screw to the other. To give a typical example, consider the process

$$\pi^- + p \rightarrow \Lambda^0 + K^0$$

and examine the π^- coming from the decay of the Λ^0,

$$\Lambda^0 \rightarrow p + \pi^-.$$

Let us now ask whether the decay product π^- has an average component of momentum perpendicular to the plane of the incident π^- and the Λ^0, this plane being defined in a particular way, say, by having the Λ^0 direction always to the right of the incoming π^-. If there is such an average component, it would define a sense of screw, and its existence would indicate a non-conservation of parity in one of the processes involved, which would have to be the decay process, since the strong interaction is invariant under inversion. Another way of putting it is this: if the momentum of the incident π^- is $\mathbf{k}_1$, that of the Λ^0 is $\mathbf{k}_2$, and that of the outgoing π^- is $\mathbf{k}_3$, what we are looking for is the distribution with respect to the sign of $(\mathbf{k}_1 \times \mathbf{k}_2) \cdot \mathbf{k}_3$. This expression changes sign under space inversion. Unequal distribution with respect to its sign therefore means violation of invariance under inversion.

Another experiment, which is perhaps more easily done, is to study the angular distribution of the beta decay of an oriented nucleus. What is involved is to see whether there is an asymmetry with respect to the sign of $\boldsymbol{\sigma} \cdot \mathbf{K}$, where $\mathbf{K}$ is the electron momentum, and $\boldsymbol{\sigma}$ the spin of the nucleus. I understand that this experiment is currently being done at Columbia, and at the National Bureau of Standards.

When one opens the question of whether parity is or is not conserved for the weak interactions, one must also raise the question of the validity of the other conservation laws. Let us list the various conservation laws: they are the conservation of energy, momentum, angular momentum, parity, the invariance under time reversal (T) and the invariance under charge conjugation (C). Energy and momentum seem to be conserved in the weak interactions. The analysis of the conservation of angular momentum in weak interactions is more complex than the corresponding analysis of parity conservation, because parity conservation can be violated in only one way, whereas angular momentum conservation can be violated in a large number of ways. To my knowledge this analysis has not been carried out. Let us now consider T and C. A closer look at these conservation laws shows that there is no proof that these are conserved in weak interactions. One might at first think that the equality of the lifetimes of the π^+ and π^-, or the μ^+ and μ^-, which are charge conjugate particles, says something about this matter. But an examination of this problem shows that the lifetimes will be equal to the lowest order in the weak coupling constant (which is to an order of accuracy of 10^{-12}, and therefore essentially infinite accuracy) merely on the assumption of invariance under orthochronous Lorentz transformations (i.e., Lorentz transformations in which there are neither space nor time inversions). Because of this, the absolute invariance under C must be regarded as experimentally not proved.

Let me conclude with the point that if parity is indeed not strictly conserved, there would be a preference for either right handedness or left handedness in the universe, and this would appear to be unaesthetic. One may however observe that if one's definition of invariance is generalized, the question may appear in a quite different light. Let me give a simple example which T. D. Lee and I have speculated about: suppose that parity is not conserved, i.e., that there is a difference on going from a left-handed to a right-handed coordinate system. It may turn out, however, that by simultaneously going from one coordinate system to the other *and* switching to the anti-world, i.e., replacing π^+ by π^-, protons by antiprotons, etc., symmetry is regained. This merely shows that there is a richness in the structure of these symmetry laws which we are quite far from comprehending.

Remarks on Possible Noninvariance under Time Reversal and Charge Conjugation*

⟦57e⟧ Commentary begins page 32

T. D. LEE, *Columbia University, New York, New York*

AND

REINHARD OEHME AND C. N. YANG, *Institute for Advanced Study, Princeton, New Jersey*

(Received January 7, 1957)

Interrelations between the nonconservation properties of parity, time reversal, and charge conjugation are discussed. The results are stated in two theorems. The experimental implications for the K-$\bar{K}$ complex are discussed in the last section.

IN a recent paper[1] the question has been raised as to whether the weak interactions are invariant under a space inversion. It was also pointed out there that similar to the situation for space inversion there exists at present no experimental proof that weak interactions are invariant under charge conjugation. Consequently the absolute invariance under charge conjugation is also an open question.

The present note is devoted to a study of questions concerning the invariance under charge conjugation C, and under time reversal T (which is defined to be the Wigner time[2] reversal. It *does not* switch a particle into its antiparticle; nor does it change the sign of the spatial coordinates).

1. *CPT* THEOREM

For the discussion of the experimental consequences of possible nonconservation of P, C, and/or T, a theorem[3] which we shall call the *CPT* theorem proves very important.

To understand the meaning of the theorem one recalls first that the operations P and C in any many-particle system (with possibilities of creation and annihilation) are represented by unitary operators that operate on the state vectors. The operation T, on the other hand, is represented[2] by the *operator* of complex conjugation *multiplied* by a unitary operator. In the Schrödinger representation the transformation of a second quantized spin 0 field described by $\varphi(r)$ and $\pi(r)$ under these operations[4] can be brought into the following form:

$$\begin{aligned} P\phi(r)P^{-1}&=\eta_P\phi(-r), & P\pi(r)P^{-1}&=\eta_P^*\pi(-r),\\ C\phi(r)C^{-1}&=\eta_C\phi^\dagger(r), & C\pi(r)C^{-1}&=\eta_C^*\pi^\dagger(r),\\ T\phi(r)T^{-1}&=\eta_T\phi(r), & T\pi(r)T^{-1}&=-\eta_T^*\pi(r), \end{aligned} \tag{1}$$

where † means Hermitian conjugate and the phases η_P, η_C, and η_T have absolute values equal to 1. For the spin $\frac{1}{2}$ field $\psi(r)$, the transformations are

$$\begin{aligned} P\psi(r)P^{-1}&=\eta_P\gamma_4\psi(-r),\\ C\psi(r)C^{-1}&=\eta_C\psi^\dagger(r),\\ T\psi(r)T^{-1}&=\eta_T\gamma_1\gamma_2\gamma_3\psi(r), \end{aligned} \tag{2}$$

where the γ matrices are so chosen that γ_1, γ_2, and γ_3 are real and γ_4 is pure imaginary. The phases η_P, η_C, and η_T also have absolute values equal to unity. The transformation properties of fields of higher spin are similar.

From the *CPT* theorem one concludes that for any local Hermitian Hamiltonian H which is invariant under proper Lorentz transformations (i.e., Lorentz transformations that involve neither space nor time inversions), there always exists a choice of the phases η_C, η_P, and η_T for the various fields (usually in more than one way) with the following properties: (a) H commutes with the product of the operators P, C, and T taken in any order; and (b) if this choice of phases does not make H commute with P, then no other choice does, and the theory is not invariant under space inversion. (Of course, if this choice of phases makes H commute with P, then the theory is invariant under space inversion.) The same holds for C and T.

We shall illustrate this theorem by an example where H is invariant under proper Lorentz transformations. Let ψ_p, ψ_n, ψ_e, and ψ_ν be the fields describing the proton, the neutron, the electron, and the neutrino. The neutrino is assumed to be a non-Majorana particle with a nonvanishing mass. Consider

$$\begin{aligned} H=H_{\text{free}}+\int\{&g_1(\psi_p^\dagger\gamma_4\psi_n)(\psi_e^\dagger\gamma_4\psi_\nu)\\ &+g_2(\psi_p^\dagger\gamma_4\psi_n)(\psi_e^\dagger\gamma_4\gamma_5\psi_\nu)\\ &+g_3(\psi_p^\dagger\gamma_4\gamma_5\psi_n)(\psi_e^\dagger\gamma_4\gamma_5\psi_\nu)\\ &+g_4(\psi_p^\dagger\gamma_4\gamma_5\psi_n)(\psi_e^\dagger\gamma_4\psi_\nu)\\ &+\text{Hermitian conjugate}\}d^3(r), \end{aligned} \tag{3}$$

* *Note added in proof.*—This paper was written in December, 1956, before parity nonconservation was experimentally established.

[1] T. D. Lee and C. N. Yang, Phys. Rev. **104**, 254 (1956).

[2] E. P. Wigner, Gött. Nachr., Math. Naturw. Kl. (1932), p. 546.

[3] See W. Pauli's article in *Niels Bohr and the Development of Physics* (Pergamon Press, London, 1955). G. Lüders, Kgl. Danske Videnskab. Selskab, Mat.-fys. Medd. **28**, No. 5 (1954). J. Schwinger, Phys. Rev. **91**, 720, 723 (1953); **94**, 1366, formula (54) and p. 1576, discussions after formula (208). We are indebted to Professor Pauli for informing us of the work of Schwinger.

[4] We discuss here only the usual "type" of fields. The possibility of the existence of unusual "types" has been pointed out by Wigner. [An account of these unusual types has been given by L. Michel and A. S. Wightman, Princeton University lecture notes (unpublished).] An examination of these unusual "types" would be an important task if space-time conservation laws should indeed be found to break down for the weak interactions.

□ Reprinted from *The Physical Review* 106, 2 (April 15, 1957), 340–345.

where $\gamma_5=\gamma_1\gamma_2\gamma_3\gamma_4$. This example is a special case of an example considered by Pauli.[3]

Writing H as

$$H=H(g_1,g_2,g_3,g_4),$$

one easily proves that

$$PHP^{-1}=H(g_1\eta_P,\ -g_2\eta_P,\ g_3\eta_P,\ -g_4\eta_P), \tag{4}$$

$$CHC^{-1}=H(g_1{}^*\eta_C{}^*,\ -g_2{}^*\eta_C{}^*,\ g_3{}^*\eta_C{}^*,\ -g_4{}^*\eta_C{}^*), \tag{5}$$

$$THT^{-1}=H(g_1{}^*\eta_T,\ g_2{}^*\eta_T,\ g_3{}^*\eta_T,\ g_4{}^*\eta_T). \tag{6}$$

In deriving these formulas, use has been made of the fact that $TgT^{-1}=g^*$ and $T\gamma_iT^{-1}=\gamma_i{}^*$. The phases η_P, η_C, and η_T are products of the respective phases of the four interacting fields. They are given by

$$\begin{aligned}\eta_P&=\eta_P{}^*(p)\eta_P(n)\eta_P{}^*(e)\eta_P(\nu),\\ \eta_C&=\eta_C{}^*(p)\eta_C(n)\eta_C{}^*(e)\eta_C(\nu),\\ \eta_T&=\eta_T{}^*(p)\eta_T(n)\eta_T{}^*(e)\eta_T(\nu).\end{aligned} \tag{7}$$

Using (4), (5), and (6), one can calculate the commutation relation between H and the six operators TCP, TPC, $\cdots$, PCT. It is found that with suitable choices of the phases η, the Hamiltonian H commutes with all of the six, as required by the CPT theorem. In fact the conditions on the phases η are simply

$$\eta_P=\eta_C\eta_T=\pm 1. \tag{8}$$

It follows from the CPT theorem that, if one of the three operators P, C, and T is not conserved, at least one other must also be not conserved. It is of course also possible that all three are separately not conserved. In the example above, by assigning suitable values to the coupling constants g, one can construct examples for all the five possibilities of conservation or nonconservation of P, C, and T. These examples are displayed in Table I.

2. LIFETIME OF CHARGE CONJUGATE PARTICLES AGAINST WEAK DECAY

Consider now a Hamiltonian

$$H=H_{\text{strong}}+H_{\text{weak}},$$

where both terms are invariant under a proper Lorentz transformation. In all subsequent discussions we shall assume that H_{strong} is invariant under C, P, and T. The phases η of the fields are defined (up to, possibly, some arbitrary factors) by this invariance; i.e., by the requirements that

$$CH_{\text{strong}}C^{-1}=H_{\text{strong}},\ \text{etc.}$$

On the other hand, the weak interactions may violate the invariance of C, P, and T. One can prove the following theorem.

Theorem 1.—If a particle A decays through the interaction H_{weak}, and if the particle and its antiparticle $\bar{A}$ do not decay into the same final products (as e.g. when A is charged), then to the lowest order of H_{weak} the lifetimes of A and $\bar{A}$ are the same, even if H_{weak} is *not* invariant under charge conjugation.

Proof.—Consider the case that particle A has spin zero. [The proof for the general case follows along the same lines.] Then the final states B and $\bar{B}$ in the decays

$$A\rightarrow B,\quad \bar{A}\rightarrow\bar{B}$$

also have spin zero. Using the identity

$$\langle\psi_1|\psi_2\rangle^*=\langle T\psi_1|T\psi_2\rangle, \tag{9}$$

one obtains

$$\begin{aligned}\langle B|H_{\text{weak}}|A\rangle^*&=\langle TB|TH_{\text{weak}}T^{-1}|TA\rangle\\ &=\langle TB|C^{-1}P^{-1}H_{\text{weak}}PC|TA\rangle,\end{aligned}$$

by the CPT theorem. Consider first the case that H_{weak} commutes (or anticommutes) with P. Then

$$\langle B|H_{\text{weak}}|A\rangle^*=\pm\langle TB|C^{-1}H_{\text{weak}}C|TA\rangle.$$

For a spinless system,

$$|TA\rangle=|A\rangle,\quad |TB\rangle=|B\rangle. \tag{10}$$

Hence

$$\begin{aligned}\langle B|H_{\text{weak}}|A\rangle^*&=\pm\langle B|C^{-1}H_{\text{weak}}C|A\rangle\\ &=\pm\langle CB|H_{\text{weak}}|CA\rangle=\pm\langle\bar{B}|H_{\text{weak}}|\bar{A}\rangle.\end{aligned} \tag{11}$$

This shows that the lifetimes of A and $\bar{A}$ are the same. If H_{weak} does not commute with P, we write

$$H_{\text{weak}}=H_1+H_2, \tag{12}$$

where

$$\begin{aligned}H_1&=\tfrac{1}{2}[H_{\text{weak}}+PH_{\text{weak}}P^{-1}],\\ H_2&=\tfrac{1}{2}[H_{\text{weak}}-PH_{\text{weak}}P^{-1}].\end{aligned} \tag{13}$$

Then

$$PH_1P^{-1}=H_1, \tag{14}$$

$$PH_2P^{-1}=-H_2. \tag{15}$$

The decays of A through H_1 and through H_2 lead to states B_1 and B_2 with opposite parities. They are orthogonal to the order considered, and hence they contribute independently without interference to the decay rate of A. The lifetimes of A and $\bar{A}$ are therefore again the same.

A consequence of this theorem has already been mentioned in a previous paper[1]: The identity of the experimental lifetimes of $\pi^\pm$ and of $\mu^\pm$ does not con-

TABLE I. Examples of theories with various possible nonconservation properties.

Value of coupling constants	Conserved operators	Nonconserved operators
g_1=real, g_3=real, $g_2=g_4=0$	P, C, T	...
g_1=real, g_3=complex, $g_2=g_4=0$	P, CT, TC	C, T
g_1=real, g_2=imaginary, $g_3=g_4=0$	C, PT, TP	P, T
g_1=real, g_2=real, $g_3=g_4=0$	T, CP, PC	C, P
g_1=real, g_2=complex, $g_3=g_4=0$	PCT, *and* permutations	P, C, T

stitute a proof that charge conjugation invariance holds for the weak interactions.

For a discussion of a case where A and $\bar{A}$ may decay into the same final channels, see Sec. 4.

3. DEPENDENCE OF INTERFERENCE EFFECTS ON CONSERVATION OF C AND T

One would like to ask what are the experimentally detectable manifestations of a weak nonconservation of P, C, or T? For the nonconservation of parity, the answer is clearly to be sought in experiments to differentiate the right-handed screw from the left-handed. Some such experiments have been discussed before.[1]

If parity is indeed not strictly conserved, some of these experiments could *also reveal whether C and/or T are or are not conserved.* To illustrate this let us consider the experiment of the angular distribution of β decay from oriented nuclei. The degree of asymmetry was given in the appendix of reference 1 as proportional to

$$\mathrm{Re}\left[C_T C_T'^* - C_A C_A'^* + i\frac{Ze^2}{\hbar cp}(C_A C_T'^* + C_A' C_T^*)\right]. \quad (16)$$

Applying the arguments of Sec. 1 we recognize that the two terms in (16) are present or absent depending on whether C or T are not conserved. To be more specific: The first term vanishes if C is strictly conserved, the second term vanishes if T is strictly conserved. If this experiment shows any asymmetry, the p dependence and the Z dependence of the asymmetry could therefore reveal whether C and/or T are nonconserved. (The existence of any asymmetry rules out the possibility that both C and T are conserved, a conclusion we already drew on general grounds in Sec. 1.)

We notice that if C is strictly conserved, the asymmetry discussed above vanishes in the absence of the Coulomb distortion of the electron wave function. In fact, when C is conserved the asymmetry is directly dependent on the existence of a difference of the Coulomb phase shifts for opposite parities. It turns out that this is a consequence of a general theorem which we state and prove below:

Theorem 2.—If, in addition to the assumptions stated in Sec. 2 concerning H_{strong} and H_{weak}, we assume that H is strictly invariant under charge conjugation, (i.e., $[H,C]=0$) and if the decay products in the final state B are *free* particles, then to the lowest order of H_{weak} there is no interference between the parity-conserving and the parity-nonconserving parts of H in the decay of A, provided the interference is sought for in experiments measuring a term of the form $\boldsymbol{\sigma}\cdot\mathbf{p}^1$.

Proof.—We again illustrate the proof by considering the case that A is spinless. The general proof follows along the same lines. We perform the decomposition of H_{weak} as in Eqs. (12)–(15). The final state B consists of two states B_1 and B_2 of opposite parities reached from A through H_1 and H_2, respectively. Clearly H_1 commutes with C, and also, by the CPT theorem, commutes with CPT. Hence using identity (9) and Eq. (10) one obtains

$$\begin{aligned}\langle B_1|H_1|A\rangle^* &= \langle TB_1|TH_1T^{-1}|TA\rangle \\ &= \langle B_1|TH_1T^{-1}|A\rangle \\ &= \langle B_1|P^{-1}C^{-1}H_1CP|A\rangle \\ &= \langle B_1|P^{-1}H_1P|A\rangle = \langle B_1|H_1|A\rangle.\end{aligned}$$

Thus $\langle B_1|H_1|A\rangle$ is real. Similarly one easily proves that $\langle B_2|H_2|A\rangle$ is pure imaginary.

In the above the states B_1 and B_2 are taken as stationary states of H_{strong} consisting of standing waves. [Otherwise Eq. (10) does not hold.] Transition amplitudes into them have a relative phase factor which is, according to the above, pure imaginary. The observed final states are equal to these amplitudes multiplied by the *outgoing* part of the stationary states B_1 and B_2. Such outgoing parts always have real relative amplitudes if the stationary states B_1 and B_2 represent free particles. The theorem now follows immediately.

Using this theorem, one concludes that if any left-right asymmetry of the form $\boldsymbol{\sigma}\cdot\mathbf{p}$ is found, the part of this asymmetry that is independent of the distortion of the final-state wave functions can arise only if charge conjugation symmetry breaks down for the weak interactions. In particular, in decays where there is no strong final-state interactions, as, e.g., in $\pi\to\mu+\nu$ and $\mu\to e+\nu+\nu$ decays, the detection[1] of parity nonconservation through the observation of $\boldsymbol{\sigma}\cdot\mathbf{p}$ becomes impossible if C is strictly conserved.

4. K^0, $\bar{K}^0$ DECAY MODES

The existence of the particle $\bar{K}^0$ and some properties of its decay were predicted[5] and discussed under the assumption that charge conjugation is strictly conserved. We wish to discuss in this section the decay of K^0 and $\bar{K}^0$ under the assumption that C, P, and T are conserved for the strong interactions, but are not necessarily conserved in the weak decay interactions.

In the first place, the conservation of strangeness with respect to the strong interactions still requires that two particles K^0 and $\bar{K}^0$ with opposite strangeness exist. To understand their decay processes it is interesting to consider the charge conjugation symmetrical and antisymmetrical combinations introduced in reference 5 (compare, however, footnote 11):

$$K_1 = \frac{1}{\sqrt{2}}(K^0 + \bar{K}^0), \quad K_2 = \frac{1}{\sqrt{2}}(K^0 - \bar{K}^0). \quad (17)$$

Unlike the situation in reference 5, if C is now not invariant in the decay process, K_1 and K_2 can decay into the same final states:

$$\begin{aligned}&K_1\to\pi^+ + \pi^-,\\ &K_2\to\pi^+ + \pi^-,\\ &K_1\to\pi^\pm + e^\mp + \nu,\\ &K_2\to\pi^\pm + e^\mp + \nu,\\ &K_1\to\pi^+ + \pi^- + \pi^0,\\ &K_2\to\pi^+ + \pi^- + \pi^0.\end{aligned}$$

[5] M. Gell-Mann and A. Pais, Phys. Rev. 97, 1387 (1955).

Interference effects would therefore set in in these decay processes. The questions that one would like to ask are then: what the the lifetimes of the particles K^0 and $\bar{K}^0$? What are the branching ratios into various decay modes?

These questions can be answered by using a Weisskopf-Wigner type of treatment[6] of the time-dependent Schrödinger equation. We write the time-dependent amplitudes of the particles[7] K^0 and $\bar{K}^0$ as $a(t)$ and $b(t)$. The various channels of decay are denoted by j. $F_j(\omega)e^{-i\omega t}$ represents the amplitude of the decay product in the channel j with the energy ω.[8] [We choose units such that $\hbar=1$.] The zero of energy is taken to be the rest energy of K. The Schrödinger equations are then

$$i\frac{da}{dt}=\sum_{j,\omega} H_{aj}(\omega)F_j(\omega)e^{-i\omega t}, \tag{18}$$

$$i\frac{db}{dt}=\sum_{j,\omega} H_{bj}(\omega)F_j(\omega)e^{-i\omega t}, \tag{19}$$

$$i\frac{\partial F_j(\omega)}{\partial t}=e^{i\omega t}[H_{ja}(\omega)a+H_{jb}(\omega)b], \tag{20}$$

where $H_{ja}=H_{aj}^*$ are the matrix elements. The Weisskopf-Wigner treatment consists of first assuming an exponential time dependence for a and b, and then in sums over ω neglecting the variation of the matrix elements with ω in the interval $|\omega|\lesssim$uncertainty of energy of the original state. Using this treatment, one obtains

$$\begin{pmatrix} a \\ b \end{pmatrix}=\psi e^{-\frac{1}{2}\lambda t}. \tag{21}$$

The amplitude ψ and the decay constant λ are given by the eigenequation

$$\Gamma\psi=\lambda\psi. \tag{22}$$

Γ is a 2×2 Hermitian matrix with matrix elements given by

$$\begin{aligned}\Gamma_{11}=\Gamma_{22}&=\sum_j \Gamma_{aj}=\sum_j \Gamma_{bj},\\ \Gamma_{12}&=\sum_j (\Gamma_{aj}\Gamma_{bj})^{\frac{1}{2}}e^{i\delta_j}=\Gamma_{21}^*,\end{aligned} \tag{23}$$

where

$$\begin{aligned}\Gamma_{aj}&=2\pi|H_{aj}|^2 \text{ (density of states per unit } d\omega)_{\omega=0},\\ \Gamma_{bj}&=2\pi|H_{bj}|^2 \text{ (density of states per unit } d\omega)_{\omega=0},\end{aligned} \tag{24}$$

and

$$e^{i\delta_j}=(\text{phase of } H_{aj}H_{bj}^{-1})_{\omega=0}. \tag{25}$$

In the foregoing derivation, use has been made of Eq. (11) which leads to

$$H_{aj}^*=\pm H_{bj'},$$

where j' is the charge conjugate channel of j (which may or may not be the same as j). It is important to notice that this equation is a consequence of the CPT theorem.

The two eigenvalues λ_+, λ_- of (22) correspond to the two decay lifetimes. The general solution is a linear superposition of two solutions $\psi_\pm$ of the form (21), each of which is characterized by a pure exponential decay. Since the 2×2 matrix Γ is Hermitian, the two solutions represent linear orthogonal combinations of the states K and $\bar{K}$.

In writing down Eqs. (18) and (19) we did not include a slight difference of mass in the form of a mass operation M for the states of the K particle. This restriction can be easily removed by adding to the right-hand sides of (18) and (19) the terms $\frac{1}{2}(M_{11}a+M_{12}b)$ and $\frac{1}{2}(M_{21}a+M_{22}b)$, respectively. The mathematical treatment is very similar to the above simple case except that we have now

$$-d\psi/dt=(\Gamma+iM)\psi, \tag{26}$$

where Γ is the same Hermitian matrix given by Eq. (23). (iM) is an anti-Hermitian matrix representing the effects of the mass shifts. By using Eq. (11) one can show that, similar to Eq. (23), M is a Hermitian matrix with

$$M_{11}=M_{22}. \tag{27}$$

Equation (26) can now be readily solved. Its eigenstates, defined by

$$(\Gamma+iM)\psi_\pm=\lambda_\pm\psi_\pm,$$

are

$$\psi_\pm=\begin{pmatrix} p \\ \pm q \end{pmatrix}(|p|^2+|q|^2)^{-\frac{1}{2}}, \tag{28}$$

with the corresponding time constants

$$\lambda_\pm=\Gamma_{11}+iM_{11}\pm(pq); \tag{29}$$

where p and q are two complex numbers given by

$$p^2=\Gamma_{12}+iM_{12},\quad q^2=\Gamma_{21}+iM_{21}=\Gamma_{12}^*+iM_{12}^*. \tag{30}$$

If at $t=0$ a K particle is produced, then at a later time the state function ψ can be expressed in terms of these two eigenstates $\psi_\pm$ as

$$\psi(t)=\begin{pmatrix} 1 \\ \overline{2p} \end{pmatrix}(|p|^2+|q|^2)^{\frac{1}{2}}[\psi_+e^{-\frac{1}{2}\lambda_+t}+\psi_-e^{-\frac{1}{2}\lambda_-t}]. \tag{31}$$

It is convenient to separate the real and imaginary parts of $\lambda_\pm$. Without loss of generality, we may write

$$\lambda_+=\gamma_+,\quad \lambda_-=\gamma_-+2i\Delta, \tag{32}$$

---

[6] V. F. Weisskopf and E. P. Wigner, Z. Physik **63**, 54 (1930); **65**, 18 (1930).

[7] In this paper we assume that the K^0 particle (strangeness $=+1$) is a single state.

[8] Each channel j represents a possible decay state that is an eigenstate of H_{strong}. Thus it has a definite spin, charge, and parity.

where γ_+, γ_- are two real numbers representing the reciprocal lifetimes of the short-lived ones and the long-lived ones respectively and Δ is the mass difference between these two eigenstates. One notices that *these two eigenstates* ψ_+ *and* ψ_- *do not in general represent the states* K_1 *and* K_2 introduced in (17). *In fact they even may not be orthogonal to each other* (see footnote 11).

A general discussion of the decay processes is rather involved. We shall make only the following remarks:

(*A*) The fractional number of K mesons that decay at time t after its production is given by

$$N(t)dt=-d[\psi^\dagger\psi]. \tag{33}$$

Using (26), one easily shows that

$$-\frac{d}{dt}[\psi^\dagger\psi]=\psi^\dagger\Gamma\psi.$$

By using (28)–(31), Eq. (33) becomes

$$N(t)=\tfrac{1}{2}(1+\alpha)^{-1}\{\gamma_+e^{-\gamma_+t}+\gamma_-e^{-\gamma_-t}+\alpha e^{-\frac{1}{2}(\gamma_++\gamma_-)t}\times[(\gamma_++\gamma_-)\cos\Delta t-2\Delta\sin\Delta t]\}, \tag{34}$$

where

$$\alpha=\psi_+{}^\dagger\psi_-=[|p|^2-|q|^2][|p|^2+|q|^2]^{-1} \tag{35}$$

is a real number representing the nonorthogonality of these two eigenstates. The four real numbers γ_+, γ_-, Δ, and α characterize the decay of the K particle. They satisfy the inequalities

$$\gamma_\pm\geqq0,\quad \alpha^2\leqq\frac{4\gamma_+\gamma_-}{(\gamma_++\gamma_-)^2+4\Delta^2}, \tag{36}$$

which follow from the fact that Γ is a positive Hermitian matrix. These conditions also insure that $N(t)\geqq0$ for all t.

Experimentally $N(t)$ is measurable. From $N(t)$ one can in principle determine all four constants γ_+, γ_-, Δ, and α. Indications from presently existing experiments[9] show that probably $\gamma_+/\gamma_->100$. Equation (36) then shows that $\alpha^2<4\gamma_-/\gamma_+<0.04$.

(*B*) The above discussion also leads easily to a determination of the branching ratio of the long-lived component (and the short-lived component) into the various decay modes. If charge conjugation is conserved, the long-lived component is an eigenstate of charge conjugation.[5] Consequently its decay into charge conjugate channels such as $\pi^+e^-\nu$ and $\pi^-e^+\bar{\nu}$ must be equally probable, as is well known. If charge conjugation is not strictly conserved, decays into $\pi^+e^-\nu$ and $\pi^-e^+\nu$ may have different probabilities for the long-lived component.

A more complete discussion of the charge asymmetry of the decay of the long-lived K^0 will be given in the appendix. We mention here only that Lederman[10] has kindly informed us that experimental work in this direction is in progress. It is important to notice that if the experiments should yield a large asymmetry, and a small α (as mentioned above), Eq. (A7) would impose very strict conditions on the relative magnitudes of the amplitudes f_1, g_1, f_2 and g_2. (To see this roughly we need only examine the limiting case $\alpha=0$ discussed below.)

(*C*) If $\alpha=0$, the two eigenstates are orthogonal.[11] Also $|p|=|q|$. [See (35).] This is the case if the mass matrix is negligible. In this case $\psi_\pm$ are both 1:1 superpositions of the particle K^0 and $\bar{K}^0$. The fraction of particles decaying in dt, namely $N(t)dt$, becomes the sum of two pure exponentials by (34). Furthermore (A7) shows that the decays of the long-lived component into charge conjugate channels such as $\pi^++e^-+\nu$ and $\pi^-+e^++\bar{\nu}$ are equally probable.

One of us (Reinhard Oehme) would like to express his gratitude to Professor Robert Oppenheimer for his kind hospitality at the Institute for Advanced Study.

APPENDIX

In this appendix we shall show the interrelationship between the parameters p, q and the branching ratio for the decay of, say, the long-lived component of K particle into various charge conjugate states.

Consider first the following decay channel of the K particle

$$K\rightarrow e^-+\pi^++\nu. \tag{A1}$$

The final product may be in states with either parity $=+1$ or parity $=-1$. Let us denote the matrix elements for the decay process into these two types of states by f_1 and f_2. Similarly, we denote the matrix elements for

$$K\rightarrow e^++\pi^-+\bar{\nu}, \tag{A2}$$

with the final state having parity $=+1$ and parity $=-1$, by g_1 and g_2.

By using the *CPT* theorem and Eq. (11), the corresponding matrix elements for the decay of $\bar{K}$,

$$\bar{K}\rightarrow e^++\pi^-+\bar{\nu}, \tag{A3}$$

are related to that of (A1). These elements are f_1^* and $-f_2^*$. Similarly the matrix elements for

$$\bar{K}\rightarrow e^-+\pi^++\nu \tag{A4}$$

are g_1^* and $-g_2^*$. Let ψ_+ represent the long-lived component K_+ of the K particle. The matrix elements for

[9] K. Lande *et al.*, Phys. Rev. **103**, 1901 (1956).

[10] L. Lederman (private communication).

[11] We recall that since the strangeness S is conserved in the strong interaction, the phase η_c of a K particle ($S=+1$) under charge conjugation is not fixed by the strong interactions. If the weak interaction is not invariant under charge conjugation, the phase η_c is defined only up to a factor $e^{iS\theta}$. If ψ_+ is orthogonal to ψ_-, there exists however a most convenient choice which makes $\psi_\pm$ identical with the K_1, K_2 defined in (17).

the decay of K_+,

$$K_+ \to e^- + \pi^+ + \nu, \quad \text{(A5)}$$

into the two different final parity states are proportional to $pf_1 + qg_1^*$ and $pf_2 - qg_2^*$, respectively, while the corresponding elements for

$$K_+ \to e^+ + \pi^- + \bar{\nu} \quad \text{(A6)}$$

are proportional to $pg_1 + qf_1^*$ and $pg_2 - qf_2^*$. The branching ratio r for the decay of K_+ into $e^- + \pi^+ + \nu$ and $e^+ + \pi^- + \bar{\nu}$ is, therefore,

$$r = \frac{|pf_1 + qg_1^*|^2 + |pf_2 - qg_2^*|^2}{|pg_1 + qf_1^*|^2 + |pg_2 - qf_2^*|^2}. \quad \text{(A7)}$$

Parity Nonconservation and a Two-Component Theory of the Neutrino

[[57f]]

Commentary begins page 35

T. D. LEE, *Columbia University, New York, New York*

AND

C. N. YANG, *Institute for Advanced Study, Princeton, New Jersey*

(Received January 10, 1957; revised manuscript received January 17, 1957)

A two-component theory of the neutrino is discussed. The theory is possible only if parity is not conserved in interactions involving the neutrino. Various experimental implications are analyzed. Some general remarks concerning nonconservation are made.

RECENTLY the question has been raised[1,2] as to whether the weak interactions are invariant under space inversion, charge conjugation, and time reversal. It was pointed out that although these invariances are generally held to be valid for all interactions, experimental proof has so far only extended to cover the strong interactions. (We group here the electromagnetic interactions with the strong interactions.) To test the possible violation of these invariance laws in the weak interactions, a number of experiments were proposed. One of these is to study the angular distribution of the β ray coming from the decay of oriented nuclei. We have been informed by Wu[3] that such an experiment is in progress. The preliminary results indicate a large asymmetry with respect to the spin direction of the oriented nuclei. Since the spin is an axial vector, its observed correlation with the β-ray momentum (which is a polar vector) can be understood only in terms of a violation of the law of space inversion invariance in β decay.

In view of this information and especially in view of the large asymmetry found, we wish to examine here a possible theory of the neutrino different from the conventionally accepted one. In this theory for a given momentum p the neutrino has only *one* spin state, the spin being always parallel to p. The spin and momentum of the neutrino together therefore automatically define the sense of the screw.

In this theory the mass of the neutrino must be zero, and its wave function need only have two components instead of the usual four. That such a relativistic theory is possible is well known.[4] It was, however, always rejected because of its intrinsic violation of space inversion invariance, a reason which is now no longer valid. (In fact, as we shall see later, in such a theory the violation of space inversion invariance attains a maximum.)

In Sec. 1 we describe this two-component theory of the neutrino. It is then shown in Sec. 2 that this theory is mathematically equivalent to a familiar four-component neutrino formalism for which all parity-conserving and parity-nonconserving Fermi couplings C and C' (as defined in the appendix of reference 1) are always related in the following manner: $C_S=C_S'$, $C_V=C_V'$, etc. or $C_S=-C_S'$, $C_V=-C_V'$, etc. Sections 3 to 8 are devoted to the physical consequences of the theory that can be put to experimental test. In the last section some general remarks about nonconservation are made.

[1] T. D. Lee and C. N. Yang, Phys. Rev. **104**, 254 (1956).

[2] Lee, Oehme, and Yang, Phys. Rev. (to be published).

[3] Wu, Ambler, Hayward, Hoppes, and Hudson. We wish to thank Professor C. S. Wu for informing us of the progress of the experiment.

[4] See, e.g., W. Pauli, *Handbuch der Physik* (Verlag Julius Springer, Berlin, 1933), Vol. 24, 226–227.

I. NEUTRINO FIELD

1. Consider first the Dirac equation for a free spin-$\frac{1}{2}$ particle with zero mass. Because of the absence of the mass term, one needs only three anticommuting Hermitian matrices. Thus the neutrino can be represented by a spinor function φ_ν which has only two components.[4] The Dirac equation for φ_ν can be written as ($\hbar=c=1$)

$$\boldsymbol{\sigma}\cdot\mathbf{p}\varphi_\nu=i\partial\varphi_\nu/\partial t, \tag{1}$$

where σ_1, σ_2, σ_3 are the usual 2×2 Pauli matrices. The relativistic invariance of this equation for proper Lorentz transformations (i.e., Lorentz transformations without space inversion and time inversion) is well known. In particular, for the space rotations through an angle θ around, say, the z axis, the wave function transforms in the following way:

$$\varphi\rightarrow\exp(-i\sigma_3\theta/2)\phi. \tag{2}$$

The σ matrices are therefore the spin matrices for the neutrino. For a state with a definite momentum $\mathbf{p}$, the energy and the spin along $\mathbf{p}$ are given, respectively, by

$$H=(\boldsymbol{\sigma}\cdot\mathbf{p}),$$

$$\sigma_p=(\boldsymbol{\sigma}\cdot\mathbf{p})/|\mathbf{p}|.$$

They are therefore related by

$$H=|\mathbf{p}|\sigma_p. \tag{3}$$

In the c-number theory, for a given momentum, the particle has therefore two states: a state with positive energy, and with $\frac{1}{2}$ as the spin component along $\mathbf{p}$, and a state with negative energy and with $-\frac{1}{2}$ as the spin component along $\mathbf{p}$.

It is easy to see that in a hole theory of such particles, *the spin of a neutrino* (defined to be a particle in the

□ Reprinted from *The Physical Review* 105, 5 (March 1, 1957), 1671–1675.

positive-energy state) *is always parallel to its momentum while the spin of an antineutrino* (defined to be a hole in the negative-energy state) *is always antiparallel to its momentum* (i.e., the momentum of the antineutrino). Many of the experimental implications discussed in later sections are direct consequences of this correlation between the spin and the momentum of a neutrino. We have remarked in the introduction that such a correlation defines automatically the sense of a screw. With the usual (right-handed) conventions which we adopt throughout this paper, the spin and the velocity of the neutrino represent the spiral motion of a right-handed screw while the spin and the velocity of the antineutrino represent the spiral motion of a left-handed screw.

We shall now discuss some general properties[5] of this neutrino field:

(A) In this theory it is clear that the neutrino state and the antineutrino state cannot be the same. A Majorana theory for such a neutrino is therefore impossible.

(B) The mass of the neutrino and the antineutrino in this theory is necessarily zero. This is true for the physical mass even with the inclusion of all interactions. To see this, one need only observe that all the one-particle *physical* states consisting of one neutrino (or one antineutrino) must belong to a representation of the inhomogeneous proper Lorentz group identical with the representation to which the free neutrino states discussed above belong. For such a representation to exist at all, the mass must be zero.

(C) That the theory does not conserve parity is well known. We see it also in the following way: Under a space inversion P, one inverts the momentum of a neutrino but not its spin direction. Since in this theory the two are always parallel, the operator P applied to a neutrino state leads to a nonexisting state. Consequently the theory is not invariant under space inversion.

(D) By the same reasoning one concludes that the theory is also not invariant under charge conjugation C which changes a particle into its antiparticle but does not change its spin direction or momentum.

(E) It is possible, however, for the theory to be invariant under the operation CP, as this operation changes a neutrino into an antineutrino and simultaneously reverses its momentum while keeping the spin direction fixed. By the Lüders-Pauli theorem[6] it follows that the theory can be invariant under time reversal T.

For the free neutrino field, as described by (1), one can prove that the theory is indeed invariant under time reversal and under CP.

2. We shall in this section indicate how one can use the conventional four-component formalism of the neutrino (with violation of parity conservation) and obtain the same results as the present theory.

We start from Eq. (1) and enlarge the matrices by the following definitions (1 represents a 2×2 unit matrix):

$$\boldsymbol{\alpha}\equiv\begin{pmatrix}\boldsymbol{\sigma} & 0\\ 0 & -\boldsymbol{\sigma}\end{pmatrix},\quad \beta\equiv\begin{pmatrix}0 & 1\\ 1 & 0\end{pmatrix}, \tag{4}$$

$$\psi_\nu\equiv\begin{pmatrix}\phi\\ 0\end{pmatrix}, \tag{5a}$$

$$\boldsymbol{\gamma}\equiv -i\beta\boldsymbol{\alpha},\quad \gamma_4\equiv\beta,\quad \gamma_5\equiv\gamma_1\gamma_2\gamma_3\gamma_4=\begin{pmatrix}-1 & 0\\ 0 & 1\end{pmatrix}. \tag{6}$$

An immediate consequence of these definitions is

$$\gamma_5\psi_\nu=-\psi_\nu. \tag{7a}$$

The free neutrino part of the Lagrangian is, as usual,

$$L_\nu=\psi_\nu^\dagger\gamma_4\left(\gamma_\mu\frac{\partial}{\partial x_\mu}\right)\psi_\nu, \tag{8}$$

where $\psi_\nu^\dagger$ = Hermitian conjugate of ψ_ν. The most general interaction Lagrangian not containing derivatives for the process

$$n\to p+e+\bar{\nu} \tag{9a}$$

is exactly as usual; namely, it is the sum of the usual S, V, T, A, and P couplings:

$$+L_{\text{int}}=-H_{\text{int}}=\sum[-2C_i(\psi_p^\dagger O_i\psi_n)(\psi_e^\dagger O_i\psi_\nu)], \tag{10a}$$

where i runs over S, V, T, A, and P and

$$\begin{aligned} O_S&=\gamma_4,\\ O_V&=\gamma_4\gamma_\mu,\\ O_T&=-\frac{1}{2\sqrt{2}}i\gamma_4(\gamma_\lambda\gamma_\mu-\gamma_\mu\gamma_\lambda),\\ O_A&=-i\gamma_4\gamma_\mu\gamma_5,\\ O_P&=\gamma_4\gamma_5. \end{aligned} \tag{11}$$

It is not difficult to prove that Eqs. (5a) and (7a) are consistent with a relativistic theory even in the presence of the interaction (10a). Another way of proving this is to start from the conventional theory of the neutrino with the interaction Hamiltonian given in (A.1) of reference 1 and observe that when

$$C_S=-C_S',\quad C_V=-C_V',\ \text{etc.} \tag{12a}$$

the neutrino field ψ_ν there always appears in interactions in the combination $(1-\gamma_5)\psi_\nu$. In the explicit representa-

[5] We have received a manuscript from Professor A. Salam on a theory of the neutrino similar to the one discussed in the present paper. He specifically discussed points (A) and (B) that we discuss here. He also gave the Michel parameter for the μ decay that agrees with the ones obtained below in Sec. 6.

[6] G. Lüders, Kgl. Dansk Videnskab. Selskab, Mat.-fys. Medd. **28**, No. 5 (1954); W. Pauli, *Niels Bohr and the Development of Physics* (Pergamon Press, London, 1955).

tion that we have adopted above, this means that only the first two components of ψ_ν contribute to the interaction. *All calculations using the conventional theory of the neutrino with the Hamiltonian* (A.1) *of reference* 1 *concerning* β *decay therefore gives the same result as the present theory if we take the choice of constants* (12a). There exists, however, the possibility that in the decay of the neutron a neutrino[7] is emitted:

$$n \rightarrow p+e+\nu. \tag{9b}$$

The corresponding general form (not including derivatives of the fields) of the Hamiltonian is

$$H_{\text{int}}=\sum[2C_i(\psi_p^\dagger O_i \psi_n)(\psi_e^\dagger O_i \psi_\nu')], \tag{10b}$$

where O_i has been defined in Eq. (11). The field ψ_ν' is a four-component spinor defined in terms of the two-component neutrino field ϕ by

$$\psi_\nu'=\begin{pmatrix} 0 \\ \sigma_2\varphi^\dagger \end{pmatrix}. \tag{5b}$$

From Eq. (6), we see that

$$\gamma_5\psi_\nu'=+\psi_\nu'. \tag{7b}$$

It can be shown that (5b) and (7b) are consistent with a relativistic theory even in the presence of interaction (10b). *It can also be proved that one can use again the Hamiltonian* (A.1) *of reference* 1 *for the conventional theory of the neutrino with the choice of the coupling constants*

$$C_S=C_S', \quad C_V=C_V', \text{ etc.} \tag{12b}$$

and obtain the same result as the present theory.

The two possible choices (12a) and (12b) depend on whether, in the β decay of the neutron, process (9a) or (9b) prevails, i.e., whether a neutrino[7] or an antineutrino is emitted. We shall see in Sec. 3 that experimentally it will be easy to decide which of the two choices is appropriate (if the theory is correct). [We do not consider the possibility here of the simultaneous presence of (9a) and (9b), since the double beta decay process does not seem to be observed experimentally.]

II. EXPERIMENTAL IMPLICATIONS

3. We consider in this section the experiment of the β decay of oriented nuclei already discussed in reference 1, and currently being carried out.[3] For the present theory, according to Eqs. (12a) or (12b), Eq. (A.6) reduces to

$$\beta=\mp\left(\frac{v_e}{c}\right)\frac{|C_T|^2-|C_A|^2-(2Ze^2/\hbar cp)\,\text{Im}(C_AC_T^*)}{|C_T|^2+|C_A|^2}. \tag{13}$$

The choice of the $\mp$ sign depends on whether

$$\begin{aligned} &n\rightarrow p+e+\bar{\nu} \quad (\bar{\nu}=\text{left-handed screw}),\\ \text{or}\quad &\\ &n\rightarrow p+e+\nu \quad (\nu=\text{right-handed screw}). \end{aligned} \tag{14}$$

In writing down (13) the Fierz interference terms has been set equal to zero, which is in conformity with the experimental results,[8] and which implies [see Eq. (A.5) of reference 1]:

$$\text{Real part of } C_AC_T^*=0. \tag{15}$$

By measuring the momentum dependence of the asymmetry parameter β, one can test whether the present theory is correct.

It is interesting to notice that for a positron emitter the asymmetry parameter has the opposite sign. This is a direct consequence of the fact that in positron and electron emission, the neutrino and antineutrino emitted have opposite spirality.

4. An experiment such as the one being carried out by Cowan and collaborators[9] measures the cross section for neutrino absorption, which can be calculated in both the present theory and the usual theory. Now one determines the magnitude of the β-coupling constants to give the observed lifetimes of nuclei against β decay. The calculated value of the cross section turns out then to be *twice as great* in the present theory as in the usual theory. This follows from the following simple reasoning: The neutrino flux is an experimental quantity independent of the theory. If the neutrinos in a given direction have only one spin state instead of the usual two, by a detailed balancing argument they must have twice the cross section for absorption as the usual ones.

5. In the decay of $\pi^\mp$ mesons at rest, let us consider the component of angular momentum along the direction of $\mathbf{p}_\mu$, the momentum of the μ meson. The orbital angular momentum contributes nothing to this component. The μ spin component is therefore completely determined (irrespective of its total spin) by the spin component of the ν or $\bar{\nu}$. There are then two possibilities:

$$\text{(A)}\quad \begin{aligned} &\pi^+\rightarrow\mu^++\nu, \quad (\mu^+ \text{ spin along } \mathbf{p}_\mu)=+\tfrac{1}{2},\\ &\pi^-\rightarrow\mu^-+\bar{\nu}, \quad (\mu^- \text{ spin along } \mathbf{p}_\mu)=-\tfrac{1}{2}; \end{aligned} \tag{16}$$

or

$$\text{(B)}\quad \begin{aligned} &\pi^+\rightarrow\mu^++\bar{\nu}, \quad (\mu^+ \text{ spin along } \mathbf{p}_\mu)=-\tfrac{1}{2},\\ &\pi^-\rightarrow\mu^-+\nu, \quad (\mu^- \text{ spin along } \mathbf{p}_\mu)=+\tfrac{1}{2}. \end{aligned} \tag{17}$$

In each case the μ mesons with fixed $\mathbf{p}_\mu$ form a polarized beam. (It was pointed out in reference 1 that if parity is not conserved in the decay of π mesons, the μ mesons would in general be polarized.) Furthermore, the polarization is now complete (i.e., in a pure state). If this theory of the neutrino is correct, then the $\pi-\mu$ decay is a perfect polarizer of the μ meson, offering a

---

[7] The neutrino as defined in Sec. 1 is a particle with spin parallel to its momentum representing a right-hand screw. Similarly, the antineutrino as defined there is a particle with spin antiparallel to its momentum representing a left-handed screw. We use this definition throughout the present paper.

[8] See, e.g., R. Sherr and R. H. Miller, Phys. Rev. **93**, 1076 (1954).

[9] See C. L. Cowan, Jr. *et al.*, Science **124**, 103 (1956).

natural way to measure the spin and the magnetic moment of the μ meson. (It turns out that the $\mu-e$ decay may serve as a good analyzer, as we shall discuss in the next section.)

The choice of the two possibilities (16) and (17) will be further discussed in Sec. 7.

6. For the $\mu^- - e^-$ decay the process can be

$$\mu^- \to e^- + \nu + \bar{\nu}, \tag{18}$$

or

$$\mu^- \to e^- + 2\nu, \tag{19}$$

or

$$\mu^- \to e^- + 2\bar{\nu}. \tag{20}$$

Consider process (18) first. The decay coupling can be written with the notations defined in Eq. (11). (We assume no derivitive coupling.)

$$H_{\text{int}} = \sum_{i=V,A} f_i(\psi_e^\dagger O_i \psi_\mu)(\psi_\nu^\dagger O_i \psi_\nu). \tag{21}$$

It is easy to see that in the present theory, where ψ_ν satisfies (7a), the S-, T-, and P-type couplings do not exist. We have assumed in writing down (21) that the spin of the μ meson is $\frac{1}{2}$. For a μ^- at rest with spin completely polarized, the normalized electron distribution is given by

$$dN = 2x^2[(3-2x) + \xi \cos\theta(1-2x)]dx d\Omega_e(4\pi)^{-1}, \tag{22}$$

where p = electron momentum, $x = p$/maximum electron momentum, θ = angle between electron momentum and the spin direction of the μ, Ω_e = solid angle of electron momentum, and

$$\xi = [|f_V|^2 + |f_A|^2]^{-1}[f_V f_A^* + f_A f_V^*]. \tag{23}$$

The mass of the electron is neglected in this calculation. The decay probability per unit time is ($\hbar = c = 1$):

$$\lambda = M^5[|f_A|^2 + |f_V|^2]/(3\times 2^8 \pi^3), \tag{24}$$

where M is the mass of the μ meson. The spectrum (22) for a nonpolarized μ meson,

$$dN = 2x^2[3-2x]dx d\Omega_e(4\pi)^{-1}, \tag{25}$$

is characterized[5] by a Michel[10] parameter $\rho = \frac{3}{4}$, which is consistent with known[11] experimental results.

One sees that for not too small values of ξ, the spectrum (22) is sensitive to $\cos\theta$, especially in the region of large momentum for the electrons. Therefore the $\mu - e$ decay may turn out to be a very good analyzer of the μ-meson spin.

An analysis of the so-called universality of the Fermi couplings is easier in this theory because there are fewer coupling constants, and also because $\pi-\mu-e$ decay measurements would supply information concerning the parameter ξ of (23).

If process (19) or (20) prevails, the spectrum becomes

$$dN = 12x^2(1-x)dx[1+\eta \cos\theta]d\Omega_e(4\pi)^{-1}. \tag{26}$$

This is characterized[5] by a Michel parameter[10] $\rho = 0$ which is not consistent with experiments.[11] One therefore concludes that (18) is the correct process.

A general theorem concerning the relationship between μ^+ and μ^- decays will be stated in Sec. 9.

7. If experiments should show that in the decay of the π meson, process (16) prevails, and in the β-decay process (9a) prevails, then one would say that the ν (the right-handed screw), the μ^-, and the e^- are light particles, and there is a conservation of light particles. If processes (17) and (9b) prevail, one would say that the $\bar{\nu}$ (the left-handed screw), the μ^-, and the e^- are light particles, and there is a conservation of light particles. Similar concepts have been discussed before.[12]

We have already seen in Sec. 3 that the sign of β in Eq. (13) determines whether

$$n \to p + e + \bar{\nu} \tag{9a}$$

or

$$n \to p + e + \nu \tag{9b}$$

is the process for β decay. To decide whether

$$\pi^+ \to \mu^+ + \nu, \quad (\mu^+ \text{ spin along } \mathbf{p}_\mu) = \tfrac{1}{2} \tag{16}$$

or

$$\pi^+ \to \mu^+ + \bar{\nu}, \quad (\mu^+ \text{ spin along } \mathbf{p}_\mu) = -\tfrac{1}{2} \tag{17}$$

one will have to determine the spin of μ^+ along its direction of motion.

8. The $\pi-\mu-e$ type experiment discussed in Secs. 6 and 7 can be done with the $K_{\mu 2}-\mu-e$ decays. The analysis is dependent on the spin of $K_{\mu 2}$. If this spin is not zero, the polarization of the μ meson is not necessarily complete. The degree of polarization can be experimentally found by a comparison of the angular distribution of the electrons in $\pi-\mu-e$ decay and in $K-\mu-e$ decay.

Another interesting experiment is to measure the momentum and polarization of the electron emitted in a β decay. A polarization of the electron results only if parity is not conserved; a measurement of this polarization is a measurement of a quantity similar to the parameter β in Eq. (13). The polarization in such a case will be along the direction of the momentum of the electron. Polarization along other directions can result if the momentum of the recoil nucleus is also determined. Theoretical considerations of such possibilities are being made by Dr. R. R. Lewis.

GENERAL REMARKS

9. Some general remarks concerning the conservation and nonconservation of the parity P, the charge conjugation C, and the time reversal T will be made in this section. Except for the last paragraph, no assump-

[10] L. Michel, Proc. Phys. Soc. (London) **A63**, 514 (1950).
[11] See, e.g., Sargent *et al.*, Phys. Rev. **99**, 885 (1955).
[12] E. J. Konopinski and H. M. Mahmoud, Phys. Rev. **92**, 1045 (1953).

tion that the neutrino is a two-component wave is made.

Since the preliminary result of the oriented nucleus experiment that there is a strong asymmetry, Eq. (A.6) of reference 1 shows that not only parity, but also charge conjugation is not conserved[2] in β decay. A measurement of the velocity dependence of the asymmetry parameter could supply[2] some information concerning time reversal invariance or noninvariance. If the $\pi-\mu-e$ decay should show any forward-backward asymmetry (as discussed in reference 1, and further analyzed above in Sec. 6 for the two-component neutrino theory), it can be shown from theorem 2 of reference 2 that charge conjugation invariance must be violated in both the $\pi-\mu$ and $\mu-e$ decays.

It is, however, easy to show from the Lüders-Pauli theorem[6] that even if C, T, and P are all not conserved, a stable particle ($e^{\pm}$ or $p^{\pm}$, or a deuteron, etc.) must have *exactly* the same mass as its antiparticle.

One can also prove that even if C, T, and P are all not conserved, the e^+ angular distribution in $\pi^+-\mu^+-e^+$ decay is exactly the same as the e^- angular distribution in $\pi^--\mu^--e^-$ decay. The only difference in the two cases is that the average spin of μ^+ along $\mathbf{p}_\mu$ is the opposite of that of μ^- along $\mathbf{p}_\mu$. (The decays are here assumed to occur in free space from $\pi^{\pm}$ at rest.)

It is further obvious from the Lüders-Pauli theorem[6] that if time reversal invariance is not violated, the operation CP is conserved. This means that the left-right asymmetry that is found in a laboratory is always exactly opposite to that found in the antilaboratory.

Should it further turn out that the two-component theory of the neutrino described above is correct, one would have a natural understanding of the violation of parity conservation in processes involving the neutrino. An understanding of the $\theta-\tau$ puzzle presents now a problem on a new level because no neutrinos are involved in the decay of $K_{\pi 2}$ and $K_{\pi 3}$. Perhaps this means that a more fundamental theoretical question should be investigated: the origin of all weak interactions. Perhaps the strange particles belong to strange representations of the Lorentz group. (Nature seems to make use of simple but odd representations.) It is also interesting to note that the massless electromagnetic field is the cause of the breakdown of the conservation of isotopic spin. The similarity to the massless two-component neutrino field that introduces the nonconservation of parity may not be accidental.

⟦57h⟧

Commentary begins page 38

Many-Body Problem in Quantum Mechanics and Quantum Statistical Mechanics

T. D. LEE, *Columbia University, New York, New York*

AND

C. N. YANG, *Institute for Advanced Study, Princeton, New Jersey*

(Received December 10, 1956)

THIS is a progress report on some work[1] concerning the quantum mechanical calculation of the fugacity coefficients b_l (which correspond to the classical cluster integrals) of a Bose, a Fermi, and a Boltzmann gas at low temperatures. A "binary collision expansion" method is developed which allows for the systematic calculation of b_l as expansions in powers of a/λ, where a represents the parameters of the dimensions of length that characterize the low-energy two-body collision and λ is the thermal wavelength. To any power of (a/λ) the calculation of any specific b_l is reduced to a finite number of quadratures. The method, therefore, is the low-temperature counterpart of the high-temperature expansion of b_l.

By going to the limit $T\to 0$, the binary collision expansion method can also yield the ground-state energy and ground-state wave function in a systematic expansion. It also supplies information concerning the density of energy levels near the ground state.

The method is applied to the case where the interaction is a hard-sphere interaction with diameter a. The particles are assumed to have spin J, where J is taken to be zero for the Bose and Boltzmann gases, and left arbitrary for the Fermi gas. We use the following notations and units: $\hbar=1$, $m=$ mass of particles $=\frac{1}{2}$, $N=$ number of particles, $V=$ volume of box, $\rho=N/V$, $P_F=$ maximum Fermi momentum $=[6\pi^2\rho/(2J+1)]^{1/3}$, and $\lambda=(4\pi/kT)^{1/2}$. The results are tabulated below:

(A) For the Fermi gas, the fugacity expansion is

$$\begin{aligned}\lambda^3 p/kT=\lambda^3\sum_1^\infty b_l z^l &= -(2J+1)g_{5/2}(-z)\\ &\quad -2J(2J+1)[g_{3/2}(-z)]^2(a/\lambda)\\ &\quad -8J^2(2J+1)g_{1/2}(-z)[g_{3/2}(-z)]^2(a/\lambda)^2\\ &\quad +8J(2J+1)F(-z)(a/\lambda)^2+O(a^3/\lambda^3), \qquad (1)\end{aligned}$$

where

$$g_n(z)=\sum_{l=1}^\infty l^{-n}z^l, \qquad (2)$$

$$F=\sum_{r,s,t=1}^\infty (rst)^{-1/2}(r+s)^{-1}(r+t)^{-1}z^{r+s+t}. \qquad (3)$$

(B) For the Bose gas, the fugacity expansion is

$$\begin{aligned}\lambda^3 p/kT &= g_{5/2}(z)-2[g_{3/2}(z)]^2(a/\lambda)\\ &\quad +8g_{1/2}(z)[g_{3/2}(z)]^2(a/\lambda)^2+8F(z)(a/\lambda)^2+O(a^3/\lambda^3). \qquad (4)\end{aligned}$$

(C) For the Bose gas, the pressure and density at the transition point are given by

$$\begin{aligned}\lambda^3 p/kT &= 1.34-2(2.61)^2(a/\lambda)+O[(a/\lambda)^{3/2}],\\ \lambda^3\rho &= 2.61-4(2.61\pi)^{1/2}(a/\lambda)^{1/2}+O[a/\lambda]. \qquad (5)\end{aligned}$$

To obtain this expression it was necessary to sum the dominant terms in the fugacity expansion to all orders of (a/λ) near $z=1$.

(D) The ground-state energy per particle for a Fermi gas at a finite density ρ and infinite volume is given by

$$\begin{aligned}E/N &= (3P_F^2/5)+8\pi a\rho J(2J+1)^{-1}\\ &\quad\times[1+6(11-2\log_e 2)P_F a/35\pi+O(P_F^2a^2)]. \qquad (6)\end{aligned}$$

(E) The ground-state energy per particle for a Boltzmann gas and for a Bose gas at a finite density and infinite volume is

$$E/N=4\pi a\rho[1+128(\rho a^3)^{1/2}/15\pi^{1/2}+O(\rho a^3)]. \qquad (7)$$

The parameters of expansion in Eqs. (6) and (7) are determinable by a simple argument without explicit calculation.

The ground-state wave function and the thermodynamical behavior in cases (D) and (E) near $T=0$ were also obtainable in these computations. Details of the binary collision expansion method and the above calculations will appear in a later publication.

□ Reprinted from *The Physical Review* 105, 3 (February 1, 1957), 1119–1120.

The binary collision expansion method is being applied to a more realistic interaction. Calculation with the Lennard-Jones potential is feasible for b_3 at low T. Work in this direction is under contemplation.

Equations (4) and (6) have been obtained before[2] by the method of pseudopotentials. By using the same method it is not difficult to obtain also Eq. (1). It was emphasized in reference 1 that the method of pseudopotentials is not applicable to all orders of a. With the binary collision expansion method the full range of applicability of the pseudopotential becomes clear. One concludes that it should be possible to obtain also Eqs. (5) and (7) by the pseudopotential method. This problem was looked into in collaboration with Dr. Kerson Huang. It turns out that it is extremely simple to obtain (7) by the pseudopotential method. A detailed description of this computation will be published shortly.

It should be emphasized that the expansions quoted above are probably all asymptotic expansions. One is led to this conclusion by the following argument. A small and negative value of a corresponds to the case where the force is purely attractive with the scattering length $-a$, for which the gas (for any statistics) collapses. The formula should therefore become meaningless for negative a. This would result if, for example, the physical quantities contain such terms as exp. $[-a^{-3}\rho^{-1}]$.

[1] A description of the binary collision expansion method, together with formulas (4) and (5) below, had previously been given at the International Conference on Theoretical Physics at Seattle, 1956 (unpublished).

[2] K. Huang and C. N. Yang, Phys. Rev. **105**, 767 (1957); Huang, Yang, and Luttinger, Phys. Rev. **105**, 776 (1957).

[57i]
Commentary begins page 40

Eigenvalues and Eigenfunctions of a Bose System of Hard Spheres and Its Low-Temperature Properties

T. D. LEE, *Columbia University, New York, New York*

AND

KERSON HUANG AND C. N. YANG, *Institute for Advanced Study, Princeton, New Jersey*

(Received March 19, 1957)

It is shown that the pseudopotential method can be used for an explicit calculation of the first few terms in an expansion in power of $(\rho a^3)^{\frac{1}{2}}$ of the eigenvalues and the corresponding eigenfunctions of a system of Bose particles with hard-sphere interaction. The low-temperature properties of the system are discussed.

THIS paper is concerned with the low-temperature properties of a dilute system of Bose particles with hard-sphere interactions, at a low but finite density. An explicit mathematical calculation is made of the energies and wave functions of the ground state and the low-lying excited states. The results confirm the usual notion of phonon waves as the only low-lying excitation, and the idea of momentum space ordering. One concludes from the calculation that such a system does show superfluidity and exhibit the two-fluid behavior at low temperatures.

It may be appropriate here to describe the motivation underlying the study of a system of hard spheres. One would like, of course, to study the general many-body problem with any potential of interaction between the particles. Such a program can be formalistically carried out. It is, however, generally recognized that to draw any definite physical conclusions from such a general program is very difficult. If one makes approximations on the general problem in order to arrive at concrete results, one usually encounters the great difficulty of defining and justifying the validity of the approximation made. We therefore start instead from the concrete model of hard-sphere interactions, which is sufficiently simple so that one might hope to be able to discuss the validity of the method of approach.

The interaction between real He atoms contains besides a hard repulsive core, also an attractive interaction outside of the core. This attractive interaction is responsible for many properties of the He liquid. For example, the ground state of a system of He atoms is known to have a negative energy corresponding to a binding energy per He atom of $(k\times 7°)$, as determined from the experimental vapor pressure curve near the absolute zero of temperature. Such a bound system owes its origin, of course, to the attractive force. The strength of the attractive force also determines the density of the He atoms in the ground state. Now at this density the total attractive potential that a He atom experiences from its neighbors is expected not to fluctuate very much. This fact suggests the following approximate picture: One replaces the attractive interparticle forces by a constant uniform negative external potential that acts on the individual particles, the repulsive core is retained, and the system is kept by an external pressure at a density equal to that of the ground state of He. Many qualitative features of the behavior of this hypothetical model may then be expected to resemble those of real He. Since the uniform external potential does not influence the system except to give it a negative total energy, one may consider simply a system of hard spheres at a given density and in the end add the external potential separately. This kind of reasoning is essentially contained in the work

□Reprinted from *The Physical Review* 106, 6 (June 15, 1957), 1135–1145.

of London[1] on the density and the energy of liquid He in the ground state.

In Secs. 1 and 2 the method of the pseudopotential[2,3] is applied to the problem. It is seen that the energy per particle in the ground state and the energy level spectrum near the ground state can be very easily obtained as power series expansions in the parameter $(\rho a^3)^{\frac{1}{2}}$, where ρ is the particle density and a the hard-sphere diameter. That the expansion parameter should be $(\rho a^3)^{\frac{1}{2}}$ was already pointed out before.[4] The ground state energy per particle calculated with the present method agrees with that given in reference 4. The excited levels immediately above the ground state represent "phonon" states. The excitation spectrum is the same as that of Bogoliubov's.[5]

In Sec. 3 the same method is used to calculate the wave functions for the ground state, and the pair distribution function for the ground state. The results are compared with the work of Feynman[6] and of Penrose and Onsager.[7] It emerges from these results that one can define a "correlation length" which characterizes the spatial extension of the correlation introduced by the hard-sphere interactions.

Section 4 is devoted to a critical discussion of the validity of the method of the pseudopotential in the present problem. The order of magnitude of the expected corrections to the present calculation is analyzed.

In Sec. 5 the physical properties of a dilute system of a gas of hard spheres are discussed briefly on the basis of the energy spectrum obtained in Sec. 2. The energy spectrum near the ground state is shown to be that of a collection of "phonons." The properties of the system, such as the existence of a normal fluid and a superfluid component, can therefore be inferred immediately from the work of Landau,[8] Kramers,[9] and others.[9]

In Sec. 6 the concept of a "correlation length" introduced in Sec. 3 is further emphasized, and related to London's idea[10] of an order in momentum space. The question of the flow of the superfluid is discussed by the method of Sec. 1. It is indicated that the superfluid flow is irrotational, as was pointed out by Onsager and Feynman.[11]

---

[1] F. London, *Superfluids* (John Wiley and Sons, Inc., New York, 1954), Chap. B.
[2] K. Huang and C. N. Yang, Phys. Rev. **105**, 767 (1957).
[3] Huang, Yang, and Luttinger, Phys. Rev. **105**, 776 (1957).
[4] T. D. Lee and C. N. Yang, Phys. Rev. **105**, 1119 (1957).
[5] N. N. Bogoliubov, J. Phys. U.S.S.R. **II**, 23 (1947).
[6] R. P. Feynman, Phys. Rev. **94**, 262 (1954).
[7] O. Penrose and L. Onsager, Phys. Rev. **104**, 576 (1956).
[8] L. D. Landau, J. Phys. U.S.S.R. **5**, 71 (1940).
[9] H. A. Kramers, Physica **18**, 653 (1952). R. B. Dingle, *Advances in Physics* (Taylor and Francis, Ltd., London, 1952), Vol. 1, p. 112.
[10] F. London, *Superfluids* (John Wiley and Sons, Inc., New York, 1954), pp. 142–144 and pp. 199–201.
[11] L. Onsager, Suppl. Nuovo cimento **6**, 249 (1949). R. P. Feynman, in *Progress in Low Temperature Physics*, edited by C. J. Gorter (North Holland Publishing Company, Amsterdam, 1955), Vol. 1, p. 17.

1. GROUND STATE ENERGY

We use mostly the same notation as that of reference 2 but choose units so that $\hbar=1$, $2m=1$, and recall that the Hamiltonian of a system of hard spheres can be replaced in certain approximations by the pseudopotential Hamiltonian [see Eqs. (32) and (33) of reference 2]:

$$H=-\sum_{i=1}^{N}\nabla_i^2+V,\qquad V=8\pi a\sum_{i<j}\delta(\mathbf{r}_i-\mathbf{r}_j)\frac{\partial}{\partial r_{ij}}r_{ij}. \tag{1}$$

By using the language of quantized fields, the pseudopotential V can be recast in the form [see Eq. (38) of reference 2]:

$$V=4\pi a\int d^3\mathbf{r}_1d^3\mathbf{r}_2\psi^*(\mathbf{r}_1)\psi^*(\mathbf{r}_2)\delta(\mathbf{r}_1-\mathbf{r}_2)\frac{\partial}{\partial r_{12}}\times[r_{12}\psi(\mathbf{r}_1)\psi(\mathbf{r}_2)]. \tag{2}$$

We shall not enter here into a discussion of the region of validity of the use of the pseudopotential, a subject that we shall come back to in Sec. 4. In the present section and the next section it will be shown that the pseudopotential (2) leads directly and simply to an expression of the ground state energy per particle of the Bose gas and to the energy spectrum near the ground state.

It was already observed and emphasized in reference 2 that the pseudopotential V, when operating on a wave function that is not singular at $r_{ij}=0$, is equivalent to the operator

$$V'=4\pi a\int d^3\mathbf{r}_1d^3\mathbf{r}_2\psi^*(\mathbf{r}_1)\psi^*(\mathbf{r}_2)\delta(\mathbf{r}_1-\mathbf{r}_2)\psi(\mathbf{r}_1)\psi(\mathbf{r}_2). \tag{3}$$

It was further observed that using the potential (3) leads to divergences which arise from the singularities of the correct wave function. The use of the correct pseudopotential V, however, does not lead to any divergencies. For clarity we shall adopt the following procedure in the present paper. The potential V' will first be used to compute the ground state energy per particle. It will be found that the expression obtained is divergent, as expected. It will then be easy to see that substituting the correct pseudopotential V, [Eq. (2)], for the potential V', [Eq. (3)], in the calculation leads very simply to a subtraction procedure which yields a correct finite result.

By expanding ψ into free-particle waves as was done in reference 2, we obtain

$$V'=\Omega^{-1}4\pi a\sum_{\alpha,\beta,\mu,\nu}a_\alpha^*a_\beta^*a_\mu a_\nu\delta(\mathbf{k}_\alpha+\mathbf{k}_\beta-\mathbf{k}_\mu-\mathbf{k}_\nu), \tag{4}$$

where a_α^* and a_α are, respectively, the creation and annihilation operators of the free-particle states with momentum $\mathbf{k}_\alpha$, and $\Omega = L^3$ is the volume of the cube in which the N particles move. The delta symbol $\delta(\mathbf{k}_\alpha+\mathbf{k}_\beta-\mathbf{k}_\mu-\mathbf{k}_\nu)$ appearing in (4) is a Kronecker delta function. It is essential that the boundary condition at the edge of the box be taken to be the usual periodicity condition [compare reference 17]. The diagonal elements of (4) are

$$\langle n|V'|n\rangle = \Omega^{-1}4\pi a(2N^2-N-\textstyle\sum_\alpha n_\alpha^2), \quad (5)$$

where n_α is the occupation number $a_\alpha^* a_\alpha$. Equation (5) has already been obtained in reference 2. Subtracting a constant term $4\pi a\rho(N-1)$ from expression (5), one obtains

$$\langle n|V'|n\rangle - 4\pi a\rho(N-1)$$

$$= 8\pi a\rho \sum_{\alpha\neq 0} n_\alpha - \frac{4\pi a}{\Omega}\Big(\sum_{\alpha\neq 0} n_\alpha\Big)^2 - \frac{4\pi a}{\Omega}\sum_{\alpha\neq 0} n_\alpha, \quad (6)$$

with $\rho = N/\Omega$. If one takes a system for which the density ρ is fixed and for which N and Ω both approach infinity, Eq. (6) reduces to[12]

$$\langle n|V'|n\rangle - 4\pi a\rho N = 8\pi a\rho \sum_{\alpha\neq 0} n_\alpha. \quad (7)$$

The off-diagonal matrix elements of the potential V' cause transitions in which two particles of momenta $\mathbf{k}_\alpha$ and $\mathbf{k}_{\beta'}$ collide and go into the states $\mathbf{k}_\mu$ and $\mathbf{k}_\nu$. The periodicity boundary condition that we took insures that the matrix element is nonvanishing only if momentum is conserved: $\mathbf{k}_\alpha+\mathbf{k}_\beta=\mathbf{k}_\mu+\mathbf{k}_\nu$. The value of such an off-diagonal matrix element is equal to

$$(4\pi a/\Omega)[n_\alpha n_\beta(n_\mu+1)(n_\nu+1)]^{\frac{1}{2}}. \quad (8)$$

The crucial point is now to observe that as the total number of particles N approaches infinity, each of the n_α's is finite except n_0, which is $N-\sum_{\alpha\neq 0} n_\alpha$. For large values of N, the off-diagonal matrix elements fall into three categories in magnitude:

(1) Those in which two of the four momenta $\mathbf{k}_\alpha$, $\mathbf{k}_\beta$, $\mathbf{k}_\mu$, $\mathbf{k}_\nu$ are equal to 0. Such matrix elements are proportional to $8\pi a\rho$.

(2) Those for which only one of the four momenta $\mathbf{k}_\alpha$, $\mathbf{k}_\beta$, $\mathbf{k}_\mu$, $\mathbf{k}_\nu$ is equal to 0. Such matrix elements are smaller than those of the first category by a factor $N^{-\frac{1}{2}}$.

(3) Those for which none of the four momenta $\mathbf{k}_\alpha$, $\mathbf{k}_\beta$, $\mathbf{k}_\mu$, $\mathbf{k}_\nu$ is 0. Such matrix elements are smaller than those of the category (1) by a factor N^{-1}.

[12] The neglect of the second and third terms of the right hand side of (6) as compared to the first term is consistent with the power series expansion of the energy in the parameter $(\rho a^3)^{\frac{1}{2}}$. It is shown later [see (41)] that

$$N^{-1}\langle \sum_{\alpha\neq 0} n_\alpha\rangle \sim (\rho a^3)^{\frac{1}{2}},$$

where the expectation value is taken with respect to the *perturbed* ground state of the total system.

Starting from the free-particle ground state, by first considering only matrix elements of category (1), we would obtain the dominant term of the energy of the system. The matrix elements of categories (2) and (3) will later be shown in Sec. 4 to give rise to higher order corrections. To calculate the dominant term of energy, we thus need *only consider those free-particle states S which are connected to the free-particle ground state, directly or indirectly, through off-diagonal matrix elements of category* (1), i.e., matrix elements that represent the scattering of two particles of momenta $\mathbf{k}$ and $-\mathbf{k}$ into the ground state or vice versa. Evidently a state in S is specified by l_1 pairs of particles each with momenta $\mathbf{k}_1$ and $-\mathbf{k}_1$, l_2 pairs of particles each with momenta $\mathbf{k}_2$ and $-\mathbf{k}_2$, etc., and $N-2\sum l_i$ particles with momentum zero. We denote such a state by

$$|l_1, l_2, \cdots\rangle. \quad (9)$$

In terms of the annihilation operators $a_\mathbf{k}$, where $\mathbf{k}\neq 0$ ranges over *half* of the momentum space, we can write down the diagonal matrix elements (7) for the pseudopotential V' between the states of S:

$$4\pi a\rho N + 16\pi a\rho \textstyle\sum' a_\mathbf{k}^* a_\mathbf{k}, \quad (10)$$

where $\sum'$ represents a summation over *half* of the $\mathbf{k}$ space with $\mathbf{k}\neq 0$. The off-diagonal matrix elements of V' are given by those of

$$8\pi a\rho \textstyle\sum' B_0(k), \quad (11)$$

where

$$B_0(\mathbf{k}) = \begin{bmatrix} 0 & 1 & 0 & 0 & \\ 1 & 0 & 2 & 0 & \\ 0 & 2 & 0 & 3 & \\ 0 & 0 & 3 & 0 & \\ & & & & \cdots \end{bmatrix}, \quad (12)$$

in the standard representation in which $a_\mathbf{k}^* a_\mathbf{k}$ is diagonal:

$$a_\mathbf{k}^* a_\mathbf{k} = \begin{bmatrix} 0 & 0 & 0 & 0 & \\ 0 & 1 & 0 & 0 & \\ 0 & 0 & 2 & 0 & \\ 0 & 0 & 0 & 3 & \\ & & & & \cdots \end{bmatrix}. \quad (13)$$

One has evidently the commutation relations

$$0 = [B_0(\mathbf{k}), a_{\mathbf{k}'}] = [B_0(\mathbf{k}), a_{\mathbf{k}'}^*] \quad \text{if} \quad \mathbf{k}\neq\mathbf{k}'. \quad (14)$$

The Hamiltonian $H' = -\sum \nabla_i^2 + V'$ between the states of S is then

$$H' = 4\pi a\rho N + 2\textstyle\sum'(k^2+k_0^2)[a_\mathbf{k}^* a_\mathbf{k} + y_\mathbf{k} B_0(\mathbf{k})], \quad (15)$$

where

$$k_0^2 = 8\pi a\rho, \quad (16)$$

$$y_\mathbf{k} = \tfrac{1}{2}k_0^2(k^2+k_0^2)^{-1}. \quad (17)$$

The summation $\sum'$ in (15) is a sum of mutually commuting operators. Its lowest eigenvalue is therefore the sum of the lowest eigenvalues of the individual terms. It will be shown in Appendix I that the eigenvalues of

$$a^*a+yB_0$$

are

$$\lambda_m=-\tfrac{1}{2}+(m+\tfrac{1}{2})(1-4y^2)^{\frac{1}{2}}, \tag{18}$$

with $m=0, 1, 2, \cdots$. One thus obtains the lowest eigenvalue of the Hamiltonian (15):

$$\begin{aligned} E_0'&=4\pi a\rho N+\sum'(k^2+k_0^2)[-1+(1-4y_{\mathbf{k}}^2)^{\frac{1}{2}}] \\ &=4\pi a\rho N-\sum'[k^2+k_0^2-k(k^2+2k_0^2)^{\frac{1}{2}}]. \end{aligned} \tag{19}$$

The above expression contains a spurious term which makes the sum divergent. This is because we have used V' instead of the correct pseudopotential V. The situation is easily remedied by identifying the spurious term and subtracting it.

The correct interaction V, Eq. (2), expressed in momentum space, reads

$$V=\lim_{r\to 0}4\pi a\Omega^{-1}\frac{\partial}{\partial r}\{r\sum_{\mu,\nu}\exp[\tfrac{1}{2}i(\mathbf{k}_\mu-\mathbf{k}_\nu)\cdot\mathbf{r}] \times\sum_{\alpha,\beta}a_\alpha^*a_\beta^*a_\mu a_\nu\delta(\mathbf{k}_\alpha+\mathbf{k}_\beta-\mathbf{k}_\mu-\mathbf{k}_\nu)\}. \tag{20}$$

It can be seen that the replacement of (4) by (20) does not affect in any essential way the general arguments that led to the Hamiltonian (15), which is now replaced by

$$H=4\pi a\rho N+2\sum'(k^2+k_0^2)a_{\mathbf{k}}^*a_{\mathbf{k}} +\tfrac{1}{2}k_0^2\operatorname{Lim}_{r\to 0}\frac{\partial}{\partial r}\{r\sum_{\mathbf{k}\neq 0}e^{i\mathbf{k}\cdot\mathbf{r}}B_0(\mathbf{k})\}. \tag{21}$$

Using this Hamiltonian, the calculation of E_0 proceeds in the same way as before except that in the final expression (19), the simple sum over $\mathbf{k}$ is replaced by a limiting process, namely

$$E_0=4\pi a\rho N-\tfrac{1}{2}\operatorname{Lim}_{r\to 0}\frac{\partial}{\partial r}\{r\sum_{\mathbf{k}\neq 0}e^{i\mathbf{k}\cdot\mathbf{r}}[k^2+k_0^2 -k(k^2+2k_0^2)^{\frac{1}{2}}]\}. \tag{22}$$

The mathematical problem of evaluating this expression is similar to the corresponding problems encountered in reference 2. It can be shown without difficulty that

$$E_0=4\pi a\rho N-\sum'\left[k^2+k_0^2-k(k^2+2k_0^2)^{\frac{1}{2}}-\frac{k_0^4}{2k^2}\right]. \tag{23}$$

The sum can easily be evaluated in the limit $\Omega\to\infty$:

$$E_0=4\pi a\rho N+\frac{\Omega k_0^5}{4\pi^2}\int_0^\infty dy y^2\left[-1-y^2 +y(y+2)^{\frac{1}{2}}+\frac{1}{2y^2}\right], \tag{24}$$

or

$$E_0=4\pi aN\rho\left[1+\frac{128}{15\sqrt{\pi}}(a^3\rho)^{\frac{1}{2}}\right], \tag{25}$$

a result which was first obtained in reference 4 by the "binary collision expansion method."

Another way of proving that the correct pseudopotential V of Eq. (1) leads to the convergent expression (23) while V' leads to the divergent one [Eq. (19)] is the following: Treating the pseudopotential V or V' as a perturbation, one can calculate the ground state energy E_0 as a power series expansion in a. This was the procedure followed in reference 2. In the order a^2, using the potential V', one obtains a divergent expression. Using the correct pseudopotential V, however, one obtains zero in the order a^2. [See Eq. (53) below. Notice that $a/L=0$ in the limit $L\to\infty$.] Except for the order a^2, V and V' give the same results. [We stay here within the approximation of neglecting small off-diagonal matrix elements. As will be discussed in Sec. 4, this approximation is equivalent to retaining the maximum power of N to each order of a.] To obtain the energy expression when V is used, one therefore need only take the divergent expression (19) for the case of V' and expand it in powers of a and strike out the term a^2. Now

$$\sum'[k^2+k_0^2-k(k^2+2k_0^2)^{\frac{1}{2}}]=\sum'\left[\frac{k_0^4}{2k^2}-\frac{k_0^6}{2k^4}+\cdots\right],$$

$$k_0^2=8\pi a\rho.$$

Striking out the term a^2 therefore means subtracting from the summand $k_0^4/2k^2$, leading immediately to (23). We shall return to this discussion in Sec. 4.

2. ENERGY LEVELS NEAR THE GROUND STATE; PHONON SPECTRUM

The method of the last section can also be applied to discuss the energy of a state with a nonvanishing momentum. We start from an unperturbed state $|\mathbf{q}\rangle$ in which all particles have momentum zero except one, which has momentum $\mathbf{q}$. The set of unperturbed states, denoted by S', connected to $|\mathbf{q}\rangle$ by large off diagonal matrix elements are all of the form

$$|\mathbf{q}; l_{\mathbf{q}}; l_1, l_2, \cdots\rangle, \tag{26}$$

which means that there is a particle of momentum $\mathbf{q}$, and in addition, there are $l_{\mathbf{q}}$ pairs of particles $\mathbf{q}, -\mathbf{q}$; l_1 pairs $\mathbf{k}_1, -\mathbf{k}_1$; l_2 pairs $\mathbf{k}_2, -\mathbf{k}_2$; etc., with $\mathbf{k}_i\neq\mathbf{q}$. The rest of the particles, $(N-2\sum l_i-1)$ in number, have momentum $\mathbf{k}=0$. The total momentum of every state in S' is $\mathbf{q}$. The Hamiltonian H' for the states S' is very similar to that for the states S given before by Eq. (15). It is

$$H'=4\pi a\rho N+2\sum_{\mathbf{k}\neq\mathbf{q}}'(k^2+k_0^2)[a_{\mathbf{k}}^*a_{\mathbf{k}}+y_{\mathbf{k}}B_0(\mathbf{k})] +2(q^2+k_0^2)[N_{\mathbf{q}}+y_{\mathbf{q}}B_1(\mathbf{q})]+8\pi a\rho+q^2, \tag{27}$$

where

$$N_{\mathbf{q}}=\begin{bmatrix} 0 & 0 & 0 & 0 & \\ 0 & 1 & 0 & 0 & \\ 0 & 0 & 2 & 0 & \\ 0 & 0 & 0 & 3 & \\ & & & & \cdots \end{bmatrix} \tag{28}$$

has diagonal values equal to $l_{\mathbf{q}}$, and

$$B_1(\mathbf{q})=\begin{bmatrix} 0 & (1\times 2)^{\frac{1}{2}} & 0 & \\ (1\times 2)^{\frac{1}{2}} & 0 & (2\times 3)^{\frac{1}{2}} & \\ 0 & (2\times 3)^{\frac{1}{2}} & 0 & \\ & & & \cdots \end{bmatrix}. \tag{29}$$

The matrix $B_0(\mathbf{k})$ is given by Eq. (12). The eigenvalue of $N+yB_1$ is discussed in Appendix I. The lowest eigenvalue is

$$-1+[1-4y^2]^{\frac{1}{2}}. \tag{30}$$

The difference of the lowest eigenvalue of (27) and that of (15) is the energy of excitation into a state of momentum $\mathbf{q}$. From (30) and (18) it is evidently equal to[13]

$$E_{\mathbf{q}}-E_0=q(q^2+2k_0^2)^{\frac{1}{2}}=q(q^2+16\pi a\rho)^{\frac{1}{2}}. \tag{31}$$

It will be shown in Appendix II that the wave function in coordinate space for the state we just discussed, i.e., for the lowest excited state with momentum $\mathbf{q}$, is to the order of approximation considered equal to

$$\sum_{j=1}^{N} e^{i\mathbf{q}\cdot\mathbf{r}_j}\Psi_0,$$

where Ψ_0 is the wave function of the ground state. This means that these excitations are density fluctuations (i.e., sound waves, or phonons), as has been discussed by Bijl[14] and Feynman.[6]

The velocity v of sound waves of infinite wavelength is directly related to the macroscopic compressibility, which can in turn be computed from the energy expression Eq. (25) for the ground state. In fact, remembering that in our units $m=\frac{1}{2}$, one has

$$v=\left(2\frac{dp}{d\rho}\right)^{\frac{1}{2}}, \quad \text{and} \quad p=\rho^2\frac{d}{d\rho}(E_0/N). \tag{32}$$

Equations (25) and (32) together give

$$v=(16\pi a\rho)^{\frac{1}{2}}[1+16\pi^{-\frac{1}{2}}(a^3\rho)^{\frac{1}{2}}]. \tag{33}$$

The first term of (33) agrees with the velocity that one computes from (31) for the sound waves with momentum $\mathbf{k}=0$, as it should. The second term in (33) represents a correction term that is beyond the accuracy of (31).

In an entirely similar way, one can solve other eigenvalues and eigenstates of the Hamiltonian (1), by considering the excited states of (15) and (27) and also considering the states connected to an unperturbed state that contains more than one particle having nonvanishing momentum. This is discussed in detail in Appendix I. The eigenvalues for these states can be shown to be

$$E=E_0+\sum_{\mathbf{k}\neq 0} m_{\mathbf{k}}k(k^2+16\pi a\rho)^{\frac{1}{2}}, \tag{34}$$

with the corresponding total momentum

$$\mathbf{P}=\sum m_{\mathbf{k}}\mathbf{k}, \quad m_{\mathbf{k}}=0,\ 1,\ 2,\ \cdots. \tag{35}$$

They represent therefore states with $m_{\mathbf{k}}$ phonons of momentum $\mathbf{k}$.

3. WAVE FUNCTIONS AND THE PAIR DISTRIBUTION FUNCTION

The ground state wave function Ψ_0 of the Hamiltonian (15) can be written in terms of the free-particle states $|l_1,l_2,\cdots\rangle$ [Eq. (9)] as

$$\Psi_0=\sum_{l_i=0}^{\infty} A(l_1,l_2,\cdots)|l_1,l_2,\cdots\rangle, \tag{36}$$

with $A(l_1,l_2,\cdots)$ representing the probability amplitudes. The value of $A(l_1,l_2,\cdots)$ is found to be (see Appendix I)

$$A(l_1,l_2,\cdots)=C\prod_i{}'[-\alpha(\mathbf{k}_i)]^{l_i}, \tag{37}$$

where

$$\alpha(\mathbf{k})=(8\pi a\rho)^{-1}[k^2+8\pi a\rho-k(k^2+16\pi a\rho)^{\frac{1}{2}}], \tag{38}$$

and C is a normalization constant given by

$$C=\prod_i{}'[1-\alpha^2(\mathbf{k}_i)]^{\frac{1}{2}}. \tag{39}$$

In Eqs. (37) and (39) the product $\prod_i{}'$ extends over half of the $\mathbf{k}$ space with $\mathbf{k}_i\neq 0$.

Upon using Eq. (37), it is easy to compute the average occupation number $\langle n_{\mathbf{k}}\rangle$ of the free-particle states with momentum $\mathbf{k}$ for the ground state wave function Ψ_0. One finds

$$\langle n_{\mathbf{k}}\rangle=\frac{\alpha^2(\mathbf{k})}{1-\alpha^2(\mathbf{k})} \quad \text{for} \quad \mathbf{k}\neq 0, \tag{40a}$$

and

$$\langle n_{\mathbf{k}=0}\rangle=N\left[1-\frac{8}{3\sqrt{\pi}}(a^3\rho)^{\frac{1}{2}}\right], \tag{40b}$$

where N is the total number of particles and $\langle\ \rangle$ means taking the average over the ground state of the system. For an ideal Bose system the ground state of the system is characterized by the fact that all particles are in the free-particle ground state. In the present case, owing to the interactions, particles are excited from the state, $\mathbf{k}=0$, into various free-particle states

[13] To calculate the excitation energy $(E_{\mathbf{q}}-E_0)$, the identical result is obtained by using either V' [Eq. (3)] or the correct pseudopotential V [Eq. (2)].

[14] A. Bijl, Physica 7, 869 (1940).

with $\mathbf{k}\neq 0$. Let f be the total fractional number of particles excited. We find for the ground state of the entire system, this fraction is

$$f\equiv N^{-1}\sum_{\mathbf{k}\neq 0}\langle n_{\mathbf{k}}\rangle=\frac{8}{3\sqrt{\pi}}(\rho a^3)^{\frac{1}{2}}. \tag{41}$$

It is important to note that the occupation number of the free-particle ground state $\langle n_{\mathbf{k}=0}\rangle$ is proportional to N while all the other free-particle states have finite occupation numbers as $N\to\infty$. The significance of these free-particle state occupation numbers in the discussion of a Bose system with interactions has recently been pointed out and emphasized by Penrose and Onsager.[7]

Another important quantity is the pair distribution function $D(r_{12})$, defined by

$$D(r_{12})\equiv\rho^{-2}\langle\psi^*(\mathbf{r}_1)\psi^*(\mathbf{r}_2)\psi(\mathbf{r}_2)\psi(\mathbf{r}_1)\rangle. \tag{42}$$

The pair distribution function $D(r)$ describes the relative probability for finding two particles at a distance r apart. The normalization of the function is so chosen that $D(r)\to 1$ as $r\to\infty$. By using Eqs. (36)–(39), the function $D(r)$ can be readily evaluated. It is

$$D(r)=[1+G(r)]^2+[1+F(r)]^2-1 -4f[G(r)+F(r)], \tag{43}$$

where

$$F(r)=\frac{1}{8\pi^3\rho}\int\frac{\alpha^2(\mathbf{k})}{1-\alpha^2(\mathbf{k})}e^{i\mathbf{k}\cdot\mathbf{r}}d^3\mathbf{k},$$
$$G(r)=-\frac{1}{8\pi^3\rho}\int\frac{\alpha(\mathbf{k})}{1-\alpha^2(\mathbf{k})}e^{i\mathbf{k}\cdot\mathbf{r}}d^3\mathbf{k}, \tag{44}$$

with f and $\alpha(\mathbf{k})$ given by Eq. (41) and Eq. (38). To study the behavior of these two functions F and G, it is convenient to introduce a "correlation length" r_0, defined as

$$r_0\equiv(8\pi a\rho)^{-\frac{1}{2}}. \tag{45}$$

r_0 is the inverse of k_0 introduced in Eq. (16). For $r\gg r_0$, the functions F and G approach, respectively,

$$F(r)\to+\frac{1}{\pi^2\rho r_0 r^2}$$

and

$$G(r)\to-\frac{1}{\pi^2\rho r_0 r^2}, \tag{46}$$

while for small distances $r\ll r_0$,

$$F(r)\to f=\frac{8}{3\sqrt{\pi}}(\rho a^3)^{\frac{1}{2}}$$

and

$$G(r)\to-\frac{a}{r}+\frac{8}{\sqrt{\pi}}(\rho a^3)^{\frac{1}{2}}. \tag{47}$$

Correspondingly, we see that for $r\ll r_0$,

$$D(r)\to\left(1-\frac{a}{r}\right)^2+O\left(\frac{a}{r_0}\right)$$

and for $r\gg r_0$,

$$D(r)\to 1+O\left(\frac{1}{r^4}\right). \tag{48}$$

Thus the correlation length r_0 characterizes the extension of the correlation between particles introduced by the hard-sphere interaction. Qualitative discussion of the physical implications of this correlation length will be given in Sec. 6.

It is of interest to compare the present result with the work of Feynman.[6] The function $S(\mathbf{k})$ in Feynman's paper can be defined in terms of the Fourier transform of $D(r)$ as

$$S(\mathbf{k})\equiv 1+\rho\int D(r)e^{i\mathbf{k}\cdot\mathbf{r}}d^3\mathbf{k}. \tag{49}$$

From Eq. (44), one finds

$$S(\mathbf{k})=k(k^2+16\pi a\rho)^{-\frac{1}{2}}[1+O(\rho a^3)^{\frac{1}{2}}],\quad (\mathbf{k}\neq 0). \tag{50}$$

Substitution into the Feynman-Bijl relation[6,14] for the phonon energy,

$$E_{\mathbf{k}}-E_0=k^2/S(\mathbf{k}) \tag{51}$$

leads to

$$E_{\mathbf{k}}-E_0=k(k^2+16\pi a\rho)^{-\frac{1}{2}}, \tag{52}$$

in agreement with Eq. (31). This is not surprising since we shall see in Appendix II that the wave functions of the excited states have the form used by Feynman and Bijl from which Eq. (51) was derived.[6]

4. CRITICAL DISCUSSION OF THE VALIDITY OF THE PSEUDOPOTENTIAL METHOD FOR THE PRESENT PROBLEM

The method used in the present paper evokes many questions concerning its validity. In particular the following points need be analyzed:

(1) It has been emphasized in reference 2 that the pseudopotential (1) is in general accurate only to the order a^2, and that as applied to the ground state energy it is only accurate to the order a^3. The approximations involved include the neglect of the D-wave scattering and the genuine triple collisions as explained in Fig. 2 of reference 2. In the present paper we have used the pseudopotential (1) to calculate quantities which certainly involve contributions from infinitely high powers of a. How could one then be sure that such use of the pseudopotential is justified? Also, in reference 2 the energy per particle for the ground state was calculated

up to a^3. The result was

$$\frac{E_0}{N}=\frac{4\pi a(N-1)}{L^3}\left\{1+2.37\frac{a}{L}+\frac{a^2}{L^2}\left[(2.37)^2+\frac{\xi}{\pi^2}(2N-5)\right]\right\}, \quad (53)$$

$$\xi=\sum_{l,m,n=-\infty}^{\infty}\frac{1}{(l^2+m^2+n^2)^2}; \quad (l,m,n)\neq(0,0,0).$$

If one keeps $\rho=N/\Omega$ constant and allows $\Omega=L^3$ to approach ∞, expression (53) diverges as $N^{\frac{1}{3}}$. How does one reconcile this divergence with the finite result obtained in Sec. 1 of the present paper?

(2) Even assuming the validity of the use of the pseudopotential (1), how can one justify the neglect of the small off-diagonal matrix elements (8)?

(3) What is the nature of the series expansion of which (25) gives the first two terms? What is the limit of validity of the phonon spectrum (31)?

We start with a discussion of point (1) by examining the divergence of formula (53). If the expansion is carried out to higher orders of a, one can express the energy per particle E_0/N as a power series in a/L. The coefficient of $(a/L)^m$, $m\geqq 3$, is a polynomial in N:

$$\frac{1}{NL^2}\left(\frac{a}{L}\right)^m[AN^\nu+BN^{\nu-1}+\cdots+Z],$$

where A, B, $\cdots Z$ are numerical constants independent of a, L, or N, and ν is an integer depending on m, giving the maximum power of N that occurs in the coefficient of $(a/L)^m$. Of the terms in the polynomial, the most divergent one in the limit $N\rightarrow\infty$ at constant ρ is

$$\frac{1}{NL^2}\left(\frac{a}{L}\right)^m AN^\nu. \quad (A,\ \nu=\text{functions of } m). \quad (54)$$

Now in the discussion of Sec. 1 the guiding principle was that to each order of a, only the term with the maximum power for N be retained. The calculation that leads to (25) is therefore a calculation of the sum of the terms (54). This calculation shows that for the order $(a/L)^m$, the maximum exponent of N is

$$\nu=m \quad (m\geqq 3),$$

as one verifies immediately by expanding (23) in powers of a. The power series for E_0/N can therefore be written in the following way:

$$\frac{E_0}{N}-4\pi a\rho=\frac{1}{NL^2}\left[A\left(\frac{aN}{L}\right)^3+A'\left(\frac{aN}{L}\right)^4+A''\left(\frac{aN}{L}\right)^5+\cdots+B\left(\frac{aN}{L}\right)^3\frac{1}{N}+B'\left(\frac{aN}{L}\right)^4\frac{1}{N}+B''\left(\frac{aN}{L}\right)^5\frac{1}{N}+\cdots+C\left(\frac{aN}{L}\right)^3\frac{1}{N^2}+\cdots+\cdots\right], \quad (55)$$

where terms of the form (54) are written in the first line. The calculation that leads to (25) consists of summing the first line of the foregoing expression, and the result shows that this series, namely

$$\frac{1}{NL^2}\left[A\left(\frac{aN}{L}\right)^3+A'\left(\frac{aN}{L}\right)^4+\cdots\right],$$

approaches the finite limit

$$\frac{1}{NL^2}\frac{4\pi\times 128}{15\sqrt{\pi}}\left(\frac{Na}{L}\right)^{\frac{5}{2}}=4\pi a\rho\frac{128}{15\sqrt{\pi}}(a^3\rho)^{\frac{1}{2}},$$

as $Na/L\rightarrow\infty$.

It is clear that D-wave scattering introduces terms that contain higher powers of a for given powers of N. Triple collisions give rise to terms also of such nature. Therefore, their inclusion does not affect the first line of (55), but only subsequent lines.

It seems reasonable to expect that the sum of the terms in the second line of (55), i.e.,

$$\frac{1}{NL^2}\frac{1}{N}\left[B\left(\frac{aN}{L}\right)^3+B'\left(\frac{aN}{L}\right)^4+\cdots\right]$$

would also converge to a finite number of the limit $aN/L\rightarrow\infty$. This can happen only if the series in the square bracket approaches $(aN/L)^4$ as $aN/L\rightarrow\infty$. In that case the second line of (55) reduces to an expression of the form

$$(\text{constant})\rho^2a^4,$$

indicating that the expansion (25) is in powers of $(a^3\rho)^{\frac{1}{2}}$.

One arrives at the same conclusion in discussing question (2) mentioned at the beginning of this section. If one attempts to include the next dominant off-diagonal matrix elements, the additional perturbation energy is of the form

$$\Delta E=\sum(\text{matrix element})^2/(\text{energy difference}).$$

The matrix elements are of the order $N^{-\frac{1}{2}}a\rho$ and connects the ground state with states in which three phonons $\mathbf{k}_1$, $\mathbf{k}_2$, $\mathbf{k}_3$ are present, where $\mathbf{k}_1+\mathbf{k}_2+\mathbf{k}_3=0$. One therefore has a sum of the form

$$\Delta E=\sum\delta(\mathbf{k}_1+\mathbf{k}_2+\mathbf{k}_3)\frac{(a\rho)^2N^{-1}}{E(\mathbf{k}_1,\mathbf{k}_2,\mathbf{k}_3)}.$$

Using the energy spectrum for the phonons calculated in Sec. 2, one obtains

$$\Delta E=(a\rho)^2N^{-1}L^6\int d\mathbf{k}_1d\mathbf{k}_2F((a\rho)^{\frac{1}{2}},\mathbf{k}_1,\mathbf{k}_2).$$

By a dimensional argument one obtains

$$\Delta E=(a\rho)^2N^{-1}L^6(a\rho)^2(\text{constant})=(\text{constant})Na^4\rho^2,$$

indicating again that the expansion (25) is in powers of $(a^3\rho)^{\frac{1}{2}}$.

The surmise that the expansion (25) is in powers of $(a^3\rho)^{\frac{1}{2}}$ is in agreement with a conclusion already drawn[4] from the "binary collision expansion method."

We now come to the third point raised at the beginning of this section: the limit of validity of the formulas (25) and (31). The above discussions indicate that they represent the first terms of expansions in $(a^3\rho)^{\frac{1}{2}}$. As has been pointed out before,[4] such expansions are probably asymptotic expansions which even may not converge. For the phonon spectrum (31) the limit of validity,

$$ka \ll 1, \tag{56}$$

has to be imposed in addition to the condition

$$(a^3\rho)^{\frac{1}{2}} \ll 1.$$

Condition (56) is necessary for the validity of the pseudopotential (1).

We conclude this section by stating that to develop a systematic expansion method starting from the pseudopotential method of the present paper seems difficult, because the inclusion of triple collision terms presents grave obstacles. On the other hand, in the "binary collision expansion method"[4] triple and higher order collision terms can be automatically included. A systematic approach starting from the "binary collision expansion method" appears hopeful.

5. "TWO-FLUID MODEL" AND THE LOW-TEMPERATURE PROPERTIES OF THE HARD-SPHERE SYSTEM

In Sec. 2 we obtained the low-lying energy levels of a Bose system of hard spheres. The levels can be described as those of a collection of phonons with a spectrum given by (34). If one examines, by a method similar to the one already used, the low-lying energy levels of a corresponding Fermi-Dirac system, one finds that the energy level density near the ground state is infinitely greater than in the Bose case. The scarcity of low-lying energy levels in the Bose case has long been recognized[15] as the reason for the superfluid behavior of liquid helium. Feynman[15] has given arguments to show that for a Bose system of interacting particles such scarcity is to be expected. The results of Secs. 1 and 2 of the present paper confirms this conclusion in the case of a dilute hard sphere gas by an explicit mathematical treatment.

Knowing the spectrum of the phonons (i.e., of the low-lying states), one can easily obtain the specific heat of the system at low temperatures. Furthermore, by the reasoning developed by Landau,[8] Kramers,[9] and others[9] one can conclude that the system shows a two-fluid[16] behavior. According to these authors the ground state of the system is looked upon as a pure "superfluid." The low-lying excited states are looked upon as a mixture of "superfluid" and "normal fluid" components, with the collection of phonons constituting the "normal fluid" component. The "normal fluid" thus can be said to be moving against a "background superfluid." With such an identification of the two fluids, one can use all the formulas which the previously mentioned authors have established for the two-fluid model, and one can compute the density of the normal fluid, the velocity of second sound, and the magnitude of the fountain effect at very low temperatures. We shall not go into these discussions in detail as we have nothing new to add to the reasonings already developed in the literature quoted. It is to be noticed, however, that the present explicit mathematical treatment of a definite model allows one to visualize very clearly the fact that a phonon does carry a momentum equal to $\hbar\mathbf{k}$, where $\mathbf{k}$ is its wave number, and that by a superposition of phonon waves one does obtain a mass transport of the Bose particles.

6. MOMENTUM SPACE ORDER, CORRELATION LENGTH, AND SUPERFLUID FLOW

The method of Secs. 1 and 2 can be applied easily to the case where one starts from an unperturbed state in which almost all particles are in a given state of momentum $\mathbf{k}_0 \neq 0$. The lowest perturbed eigenstate there describes a background superfluid flow with velocity $2\mathbf{k}_0$ (notice that the mass per particle is $\frac{1}{2}$). The excited states represent various phonon states in such a background superfluid.

Is it possible to start from an unperturbed state in which a finite fraction of the particles occupy each of two different momentum states? In other words, is it possible to have an interpenetration of two superfluid velocities? The answer is no, because the method of Sec. 1 leads in this case to very large perturbations, indicating[17] that the unperturbed state is very far from an eigenstate.

The condensation of nearly all particles into a single free-particle momentum state is what London[10] called momentum space ordering. The foregoing discussion and the wave function and eigenvalues found in Secs. 1, 2, and 3 give explicit demonstrations of this concept for the special model of a dilute Bose system of hard spheres.

The influence of the order in momentum space does not, however, extend over infinite spatial distances. If it did, there would not be the possibility of superfluid flow, but only uniform motion of the superfluid as a whole. We shall in the following give a qualitative

[15] See, e.g., R. P. Feynman, in *Progress in Low Temperature Physics*, edited by C. J. Gorter (North Holland Publishing Company, Amsterdam, 1955), Vol. 1, p. 17.

[16] L. Tisza, J. phys. radium 1, 164 (1940).

[17] For the same reason it is important to take periodic boundary conditions, as we remarked in Sec. 1. If one had chosen, e.g., the boundary condition $\Psi=0$ on the surface of the box, the unperturbed ground state would have an unphysical density variation across the box, so that the hard-sphere interaction would not be a small perturbation.

discussion[18] of the superfluid flow in the present model and of the stability of the flow. *The discussion is to be regarded as suggestive, rather than mathematically conclusive.*

We first notice that the number of particles within one correlation distance $r_0 = k_0^{-1} = (8\pi a\rho)^{-\frac{1}{2}}$ is

$$\sim \rho r_0^3 \sim (\rho a^3)^{-\frac{1}{2}} \gg 1.$$

The number of excited particles among these is computable from the fraction (41), and is a finite number of the order of 1. The correlation distance is therefore the distance within which the momentum space ordering is strongly effective.

In order to allow for a variation of the superfluid velocity, we divide the system into small boxes each of which is of the dimension of the correlation length, within which the ordering in momentum space forces practically all the particles to have the same momentum. The correlation between two different boxes is, however, not so strong, with the result that the superfluid velocity may vary from one small box to the other. This suggests that one makes use of the method of Secs. 1, 2, and 3, but takes the individual particle wave functions to be

$$e^{i\varphi + i\mathbf{k}\cdot\mathbf{r}}, \tag{57}$$

which form a complete set. Here, φ is a function of $\mathbf{r}$ (independent of $\mathbf{k}$) and $\nabla\varphi$ varies little within each small box. Expanding the second quantized wave function into these individual particle waves,

$$\psi(\mathbf{r}) = \sum_{\mathbf{k}} a_{\mathbf{k}} e^{i\varphi + i\mathbf{k}\cdot\mathbf{r}},$$

one can calculate the matrix elements of the kinetic energy and the pseudopotential for the various eigenstates of the occupation numbers $a_{\mathbf{k}}^* a_{\mathbf{k}}$. It is then seen that the pseudopotential has the same matrix elements as in Sec. 1, and that the diagonal matrix elements of the kinetic energy is also the same as in Sec. 1 except for a uniform increment of the amount

$$\rho \int (\nabla\phi)^2 d\tau. \tag{58}$$

To give a physical meaning to $\nabla\varphi$, we notice that in each small box $\nabla\varphi$ may be taken as a constant vector. It is then evident that for the ground state in each small box the momentum of the superfluid is equal to $\nabla\varphi$ per particle. In other words,

$$\mathbf{v}_s = 2\nabla\phi. \tag{59}$$

The expression (58) then gives simply the kinetic energy of the superfluid flow, which according to (59) is irrotational.

Neglecting the off-diagonal matrix elements of the kinetic energy, one could solve for the excited states too. The excited states are again describable as the states of phonon waves. The off-diagonal matrix elements of the kinetic energy then give rise to a possible transfer of momentum and energy from the superfluid background flow into the phonon waves.

The above discussion leads to the conclusion that the superfluid flow is described by a condensation of almost all particles [i.e., other than a fraction $\sim(\rho a^3)^{\frac{1}{2}}$] into the single-particle state (57). This is clearly exactly what London[10] meant by a macroscopic quantum state. It is clear that from the single-valuedness of φ one would obtain a quantization of the vortices, an interesting conclusion that has been discussed in detail by Onsager and by Feynman.[11]

One of us (K. Huang) would like to thank Dr. J. Robert Oppenheimer for the hospitality extended him during his stay at the Institute for Advanced Study.

APPENDIX I

In this Appendix, we discuss the eigenvalues and eigenfunctions of the matrix

$$M_s = N + yB_s, \tag{A1}$$

where

$$N = \begin{bmatrix} 0 & 0 & 0 & 0 & \cdot \\ 0 & 1 & 0 & 0 & \cdot \\ 0 & 0 & 2 & 0 & \cdot \\ 0 & 0 & 0 & 3 & \cdot \\ \cdot & \cdot & \cdot & \cdot & \cdots \end{bmatrix}, \tag{A2}$$

and

$$B_s = \begin{bmatrix} 0 & [1\times(s+1)]^{\frac{1}{2}} & 0 & \cdot \\ [1\times(s+1)]^{\frac{1}{2}} & 0 & [2(s+2)]^{\frac{1}{2}} & \cdot \\ 0 & [2(s+2)]^{\frac{1}{2}} & 0 & \cdot \\ \cdot & \cdot & \cdot & \cdots \end{bmatrix}. \tag{A3}$$

Let ψ be an eigenstate, with

$$M_s\psi = \lambda\psi, \tag{A4}$$

and

$$\psi = \begin{bmatrix} A_0 \\ A_1 \\ A_2 \\ \cdot \\ \cdot \\ \cdot \end{bmatrix}. \tag{A5}$$

By substituting ψ into (A4), we have

$$nA_n + y\{A_{n-1}[n(n+s)]^{\frac{1}{2}} + A_{n+1}[(n+1)(n+s+1)]^{\frac{1}{2}}\} = \lambda A_n.$$

It is convenient to introduce A_n' defined by

$$A_n' = \left[\frac{n!}{(n+s)!}\right]^{\frac{1}{2}} A_n. \tag{A6}$$

The difference equation for the A_n' becomes

$$(n-\lambda)A_n' + y[nA_{n-1}' + (n+s+1)A_{n+1}'] = 0, \tag{A7}$$

[18] See similar discussions by Onsager and Feynman, reference 11.

which can be readily solved by defining a generating function

$$H(z)\equiv\sum_{n=0}^{\infty}A_n'z^n. \tag{A8}$$

From (A7), we obtain the differential equation for H as

$$\frac{dH}{dz}[z+yz^2+y]=H\left[\lambda-yz-\frac{sy}{z}\right]. \tag{A9}$$

In order that ψ be normalizable we must have

$$\sum_{n=0}^{\infty}|A_n|^2=\text{finite},$$

which in turn implies that in the complex z plane except for $z=0$, $H(z)$ has no singularity inside the unit circle $|z|<1$. Thus, the eigenvalues of M_s are immediately determined. They are

$$\lambda_m=-\tfrac{1}{2}(1+s)+(\tfrac{1}{2}+m+\tfrac{1}{2}s)(1-4y^2)^{\frac{1}{2}}, \tag{A10}$$

with $m=0, 1, 2, \cdots$.

The corresponding eigenstates are given by Eqs. (A5) and (A6), with

$$A_n'=\text{coefficient of } z^n \text{ in } H_m(z),\quad (n\geq 0).$$

The generating function $H_m(z)$ is

$$H_m(z)=z^{-s}(z+\alpha)^{m+s}(1+\alpha z)^{-(m+1)}, \tag{A11}$$

with

$$\alpha=(2y)^{-1}[1-(1-4y^2)^{\frac{1}{2}}]. \tag{A12}$$

In particular, for $s=0$ and $m=0$

$$\lambda=-\tfrac{1}{2}+\tfrac{1}{2}(1-4y^2)^{\frac{1}{2}}, \tag{A13}$$

and the corresponding unnormalized A_n are

$$A_n=(-\alpha)^n,\quad (n=0, 1, 2\cdots), \tag{A14}$$

which yields Eq. (37).

The Hamiltonians (15) and (27) are related to the matrices M_s with $s=0$ and $s=1$. Consider now the more general case of starting with any unperturbed state which has $s_{\mathbf{k}}$ free particles with momentum $\mathbf{k}$. (Without loss of generality we can restrict the momentum $\mathbf{k}$ to range over only half of the $\mathbf{k}$ space.) Using the same arguments as that of Sec. 1, it is easy to see that the dominant part of the Hamiltonian H, Eq. (1), connects this state with other states which has in addition to these $s_{\mathbf{k}}$ particles also $l_{\mathbf{k}}$ pairs of particles each of momentum $\mathbf{k}$ and $-\mathbf{k}$, etc. Thus the Hamiltonian reduces to

$$H'=4\pi a\rho N+2\sum_{\mathbf{k}\neq 0}{}'(k^2+k_0^2)[N_{\mathbf{k}}+y_{\mathbf{k}}B_s(\mathbf{k})+\tfrac{1}{2}s_{\mathbf{k}}], \tag{A15}$$

where k_0^2 and $y_{\mathbf{k}}$ are given by Eqs. (16) and (17). The sum $\sum'$ extends over half of the $\mathbf{k}$ space with $\mathbf{k}\neq 0$. From the solution (A.10), we obtain immediately the complete phonon spectrums which are listed in Eqs. (34) and (35).

APPENDIX II

In this Appendix, we discuss the properties of the wave functions in the configuration space. From Eqs. (36) and (37), the ground state wave function Ψ_0 can be written in the configuration space as

$$\Psi_0=C\sum_{n=0}^{N/2}\chi_n, \tag{A16}$$

where

$$\chi_n=\Omega^{-N/2}\left[\frac{(N-2n)!}{N!}\right]^{\frac{1}{2}}N^n\sum f(r_{12})f(r_{34})\cdots,\quad (n\neq 0) \tag{A17}$$

and C is the normalization constant. The functions χ_n represent the part in which n pairs of particles are excited. In (A17), the sum extends over all different combinations of selecting n pairs made of $2n$ different particles among a total of N particles. Each term in the sum is a product of n functions $f(r_{ij})$ with the distances between these n pairs as arguments. Altogether there are

$$\frac{N!}{(N-2n)!n!2^n}$$

terms in the sum. The function $f(r)$ is

$$f(r)=-\frac{1}{8\pi^3\rho}\int\alpha_{\mathbf{k}}e^{i\mathbf{k}\cdot\mathbf{r}}d^3\mathbf{k}, \tag{A18}$$

with

$$\alpha_{\mathbf{k}}=(8\pi a\rho)^{-1}[k^2+8\pi a\rho-k(k^2+16\pi a\rho)^{\frac{1}{2}}].$$

Its behaviors at large and small distances are as follows:

$$f(r)\to -a/r\quad\text{as}\quad r\to 0, \tag{A19}$$

and

$$f(r)\to -(2\pi^{\frac{3}{2}}a^{\frac{1}{2}}\rho^{\frac{3}{2}}r^4)^{-1}\quad\text{as}\quad r\to\infty.$$

Using the ground state wave function Ψ_0 in the configuration space, it is also possible to obtain directly the pair distribution function $D(r_{12})$ [Eq. (43)] by integrating over the remaining spatial coordinates $\mathbf{r}_3, \cdots, \mathbf{r}_n$.

Our ground state wave function Ψ_0 satisfies the boundary condition,

$$\Psi_0=0\quad\text{at}\quad r_{ij}=a, \tag{A20}$$

only *approximately*. Its violation of this boundary condition, however, has an effect on the energy spectrum only in higher orders of $(\rho a^3)^{\frac{1}{2}}$. To see this more clearly, let us consider the wave function

$$\Psi_0'=C'\Omega^{-N/2}\prod_{i<j}[1+f(r_{ij})], \tag{A21}$$

which satisfies the required boundary conditions. We

can obtain Ψ_0 from the above wave function by expanding the above product in powers of f and then omitting all terms in which the coordinate of any particle, say $\mathbf{r}_i$, occurs more than once. For example, a term like

$$f(r_{12})f(r_{13}) \tag{A22}$$

must be omitted. The difference between Ψ_0' and Ψ_0, therefore, consists of terms like (A22), which expresses a correlation among more than two particles. Such terms belong to a higher order of $(a^3\rho)^{\frac{1}{2}}$ than we have considered. For example, upon Fourier-analyzing (A22), we find that it is of the form of a sum over three momenta $\mathbf{k}_1$, $\mathbf{k}_2$, $\mathbf{k}_3$, subject to $\mathbf{k}_1+\mathbf{k}_2+\mathbf{k}_3=0$. Such terms arise from a calculation of order a^4, as shown in Sec. 4.

The wave function for the one-phonon state can be obtained directly from (A11). By an argument similar to the above one, it can be shown that upon neglecting terms of higher orders in $(\rho a^3)^{\frac{1}{2}}$ the wave function $\Psi_\mathbf{q}$ of one phonon with momentum $\mathbf{q}$ in the configuration space is

$$\Psi_\mathbf{q}=\sum_{i=1}^{N} e^{i\mathbf{q}\cdot\mathbf{r}_i}\Psi_0, \tag{A23}$$

where Ψ_0 is the ground state wave function [Eq. (A16) or Eq. (A21)]. Thus it is to be expected that the Feynman-Bijl relations [Eq. (52)] correlating the excitation energy of a phonon with the pair distribution function is satisfied for a dilute system of hard spheres with Bose statistics.

General Partial Wave Analysis of the Decay of a Hyperon of Spin $\frac{1}{2}$

[[570]]
Commentary begins page 40

T. D. LEE* AND C. N. YANG
Institute for Advanced Study, Princeton, New Jersey
(Received October 22, 1957)

THIS note is to consider the general problem of the decay of a hyperon of spin $\frac{1}{2}$ into a pion and a nucleon under the general assumption of possible violations of parity conservation, charge-conjugation invariance, and time-reversal invariance. The discussion is in essence a partial wave analysis of the decay phenomena and is independent of the dynamics of the decay. *Nonrelativistic approximations are not made on either of the decay products.*

In the reference system in which the hyperon is at rest there are two possible final states of the pion-nucleon system: $s_{\frac{1}{2}}$ and $p_{\frac{1}{2}}$. Denoting the amplitudes of these two states by A and B, one observes that the decay is physically characterized by *three* real constants specifying the magnitudes and the relative phase between these amplitudes. One of these constants can be taken to be $|A|^2+|B|^2$, and is evidently proportional to the decay probability per unit time. The other two constants are best defined in terms of experimentally measurable quantities. We discuss three types of experiments:

(a) The angular distribution of the decay pion from a completely polarized hyperon at rest.

It has been pointed out before[1] that the distribution is proportional to

$$[1+\alpha \cos\chi]d\Omega, \tag{1}$$

where $d\Omega$ is the solid angle of the pion momentum vector $\mathbf{p}_\pi$ and χ is the angle between $\mathbf{p}_\pi$ and the spin of the hyperon. The constant α is given by

$$\alpha=2\,\mathrm{Re}(A^*B)/(|A|^2+|B|^2), \tag{2}$$

and characterizes the degree of mixing of parities in the decay.

That the distribution is of the form (1) follows immediately from the assumption that the spin of the hyperon is $\frac{1}{2}$. One easily proves (2) by considering the decay probabilities for the cases $\chi=0$ and $\chi=\pi$. In the former case the amplitude of decay is $(A+B)$ and in the latter $(A-B)$. One therefore obtains $(1+\alpha)/(1-\alpha)=|A+B|^2/|A-B|^2$, which results in (2).

Recent experiments[2] have indicated that the absolute value of α is quite large for Λ^0 decay. With improved statistics these experiments can establish beyond doubt that parity nonconservation is not peculiar to neutrino processes. It is, however, not possible to determine the sign of the parameter α through the experiments quoted above, as the sign of the polarization of the Λ^0 in the production process is unknown. Further, it appears that these experiments cannot give an accurate measurement of the magnitude of α because of the difficulty in determining the degree of polarization of the Λ^0 produced. Another method of measuring α which does not depend on the degree and sign of a polarized hyperon beam is found in the following type of experiment:

(b) The longitudinal polarization of the nucleon emitted in the decay of unpolarized hyperons at rest.

The degree of longitudinal polarization (i.e., average spin along the direction of motion divided by $\frac{1}{2}\hbar$) is easily shown to be $-\alpha$. To see this, we consider the decay of a hyperon at rest, with spin$=+\frac{1}{2}\hbar$ along the $+z$ axis, into a nucleon traveling along the $+z$ axis and a meson along the $-z$ axis; and then the decay of a hyperon, with spin$=-\frac{1}{2}\hbar$ along the $+z$ axis, into the same final states. An incoherent mixture of the two cases gives a description of the decay of unpolarized hyperons. By the conservation of the z component of angular momentum the spin of the nucleon is respectively equal to $+\frac{1}{2}\hbar$ and $-\frac{1}{2}\hbar$ for the two cases. Furthermore the two cases correspond to the experiment discussed under (a) for $\chi=\pi$ and $\chi=0$, respectively. The probabilities for the two cases are, according to previous discussions, proportional to $|A-B|^2$ and $|A+B|^2$. The incoherent mixture of the two cases therefore gives a longitudinal polarization for the nucleon equal to

$$(|A-B|^2-|A+B|^2)/(|A-B|^2+|A+B|^2)=-\alpha.$$

It is well known[3] that the scattering of high-energy protons by nuclei offers a good method of analyzing the polarization of high-energy protons as well as determinating the sign of the polarization. A measurement of the parameter α, together with its sign, through such methods may not be impossible. We remark that once the value of α is determined, the experiments of type (a) could be used to determine the degree and sign of the polarization of the hyperons.

(c) Transverse polarization of the nucleon emitted in a given direction in the decay of a polarized hyperon.

The remaining parameter describing the decay process can be determined only through a measurement of the transverse polarization of the nucleon emitted from a polarized hyperon decaying at rest. Let $\frac{1}{2}\hbar\hat{s}$ be

□ Reprinted from *The Physical Review* 108, 6 (December 15, 1957), 1645–1647.

the spin of the hyperon, where $\hat{s}$ is a unit vector. The probability for the emission of a nucleon in the direction $\hat{p}$ (=unit vector) is, according to (1), proportional to $1+\alpha\cos\chi$, where $\cos\chi\equiv-\hat{p}\cdot\hat{s}$. The spin of the nucleon in its own rest system can be shown to be $\frac{1}{2}\hbar$ times the unit vector

$$(1+\alpha\cos\chi)^{-1}[(-\alpha-\cos\chi)\hat{p}+\beta\hat{p}\times\hat{s}+\gamma(\hat{p}\times\hat{s})\times\hat{p}], \quad (3)$$

where

$$\beta=-\mathrm{Im}(A^*B)/(|A|^2+|B|^2), \quad (4)$$

$$\gamma=(|A|^2-|B|^2)/(|A|^2+|B|^2). \quad (5)$$

The three parameters α, β, and γ are related through the identity

$$\alpha^2+\beta^2+\gamma^2=1. \quad (6)$$

One could therefore write also

$$\beta=(1-\alpha^2)^{\frac{1}{2}}\cos\phi, \quad \gamma=(1-\alpha^2)^{\frac{1}{2}}\sin\phi. \quad (7)$$

The geometrical description of the polarization vector (3) is quite simple. It is a unit vector, in agreement with the fact that the nucleon moving in a given direction $\hat{p}$ is in a pure state. Its longitudinal component (i.e., projection along $\hat{p}$) has the value

$$(-\alpha-\cos\chi)/(1+\alpha\cos\chi). \quad (8)$$

Its transverse component has the polar angle ϕ in the plane determined by $\hat{p}\times\hat{s}$ and $(\hat{p}\times\hat{s})\times\hat{p}$ if one chooses the former vector to be along the $+x$ axis and the latter along the $+y$ axis.

The two parameters α and ϕ together with the decay probability determine completely the kinematical aspects of the decay of the hyperon. One easily verifies from (7), (4), (5), and (2) that knowing α and ϕ one can compute A and B up to a common multiplicative factor.

The ranges of the parameters α and ϕ are given by

$$-1\leqq\alpha\leqq1, \quad -\pi\leqq\phi<\pi. \quad (9)$$

The sign of ϕ has a physical meaning: positive values of ϕ imply positive values for γ, and consequently a preponderance of $s_{\frac{1}{2}}$ over $p_{\frac{1}{2}}$ in the final states. Negative values of ϕ imply the reverse situation. Geometrically, a positive ϕ implies an acute angle between the transverse polarization and the spin of the hyperon, therefore a preference for non-spin-flip decays, i.e., a preference for $s_{\frac{1}{2}}$ final states.

Additional requirements are imposed on the parameters if time-reversal invariance is assumed to hold. For Λ^0 decay the conclusion[1] is essentially that A and B are real, relative to each other, implying that $\beta\cong0$, or in other words $\phi\cong\pm\frac{1}{2}\pi$. A measurement of ϕ in Λ^0 decay therefore gives a test of time-reversal invariance in Λ^0 decay.

In the case of a hyperon decay with two final channels, such as $\Lambda^0\to p+\pi^-$ and $\Lambda^0\to n+\pi^0$, there would appear six parameters describing the transition, three for each channel. In principle there exists another real parameter which describes the relative phase of the transition amplitudes into the two channels. Such a parameter is, however, extremely hard to measure experimentally, and is at present only of academic interest. If the Λ^0 decay interaction is invariant under time reversal, then the number of real parameters is reduced from seven to four.

We conclude with the remark that the large asymmetry observed in the experiments of reference 2 shows that the production process $\pi^-+p\to\Lambda^0+K^0$ is a surprisingly good polarizer of Λ^0 spin, and that the decay $\Lambda^0\to\pi^-+p$ is a good convenient analyzer. These facts open the way to a possible measurement[4] of the magnitude and the *sign* of the gyromagnetic ratio of Λ^0 which does not seem completely hopeless. For example, in a magnetic field of 200 000 gauss, the spin of Λ^0 would precess through an angle of 33° in 3×10^{-10} sec if its magnetic moment is one nuclear Bohr magneton.

* On leave of absence from Columbia University, New York, New York.

[1] Lee, Steinberger, Feinberg, Kabir, and Yang, Phys. Rev. **106**, 1367 (1957).

[2] F. S. Crawford *et al.*, Phys. Rev. **108**, 1102 (1957); F. Eisler *et al.*, Phys. Rev. **108**, 1353 (1957); L. Leipuner and R. Adair, Phys. Rev. (to be published).

[3] See, e.g., the review article by L. Wolfenstein, *Annual Review of Nuclear Science* (Annual Reviews, Inc., Stanford, 1956), Vol. 6, p. 43.

[4] It has been pointed out before that the magnetic moment of a hyperon may be measured by using the angular asymmetries in the hyperon decay as an analyzer. M. Goldhaber, Phys. Rev. **101**, 1828 (1956).

Quantum Mechanical Many-Body Problem and the Low Temperature Properties of a Bose System of Hard Spheres†

[[57q]]
Commentary begins page 41

KERSON HUANG, T. D. LEE, and C. N. YANG

I wish to discuss some recent work with Drs. Kerson Huang and T. D. Lee on the quantum statistical many-body problem.

1. *Binary collision expansion method*

First we shall describe a binary collision expansion method [1] developed by Lee and myself. We recall that Ursell [2] and Kahn and Uhlenbeck [3] had discussed a long time ago a method of decomposition of the propagation function W_N^s for an N-particle Bose system into functions U^s:

$$
\begin{aligned}
&W_N^s(\mathbf{x}_1', \mathbf{x}_2' \ldots; \mathbf{x}_1, \mathbf{x}_2 \ldots) \equiv N! \sum_{\text{Sym}\,\psi} \psi_i(\mathbf{x}_1', \mathbf{x}_2', \ldots)\, \psi_i^*(\mathbf{x}_1, \mathbf{x}_2, \ldots)\, e^{-\beta E_i} \\
&W_1^s(1', 1) \equiv U_1^s(1', 1) \\
&W_2^s(1', 2'; 1, 2) \equiv U_1^s(1', 1) U_1^s(2', 2) + U_2^s(1', 2'; 1, 2) \qquad (1) \\
&W_2^s(1', 2', 3'; 1, 2, 3) \equiv U_1^s(1', 1) U_1^s(2', 2) U_1^s(3', 3) \\
&\quad + U_1^s(1', 1) U_2^s(2', 3'; 2, 3) + U_1^s(2', 2) U_2^s(1', 3'; 1, 3) \\
&\quad + U_1^s(3', 3) U_2^s(1', 2'; 1, 2) + U_3^s(1', 2', 3'; 1, 2, 3) \qquad \text{etc.}
\end{aligned}
$$

where $1' \equiv \mathbf{x}_1'$, $2' \equiv \mathbf{x}_2'$, etc., and $\beta \equiv 1/kT$. The superscript s in the propagation functions W^s, and in the U^s functions in these equations, all refer to the symmetry of the Bose system. If one had treated a system with Boltzmann statistics, one could easily write down the corresponding decompositions

† As presented by C. N. Yang.

□ Lecture at the Stevens Conference on the Many-Body Problem, January 1957. Reprinted from *The Many-Body Problem*, J. K. Percus, ed. New York: Wiley-Interscience, 1963, 165–175.

$$
\begin{aligned}
&W_1(1', 1) \equiv U_1(1', 1)\\
&W_2(1', 2'; 1, 2) \equiv U_1(1', 1)U_1(2', 2) + U_2(1', 2'; 1, 2)\\
&W_3(1', 2', 3'; 1, 2, 3) \equiv U_1(1', 1)U_1(2', 2)U_1(3', 3) \qquad (2)\\
&\qquad + U_1(1', 1)U_2(2', 3'; 2, 3) + U_1(2', 2)U_2(1', 3'; 1, 3)\\
&\qquad + U_1(3', 3)U_2(1', 2'; 1, 2) + U_3(1', 2', 3'; 1, 2, 3) \qquad \text{etc.}\\
&W_N(\mathbf{x}_1', \mathbf{x}_2', \ldots; \mathbf{x}_1, \mathbf{x}_2, \ldots) \equiv \sum_{\text{all } \psi} \psi_i(\mathbf{x}_1', \mathbf{x}_2', \ldots)\psi_i^*(\mathbf{x}_1, \mathbf{x}_2, \ldots)e^{-\beta E_i}
\end{aligned}
$$

The reason for the introduction of the U functions and the U^s functions is that they are related simply to the fugacity expansion [4] coefficients b_l:

$$
\begin{aligned}
\Omega b_l &= (1/l!)\int_{\Omega^l} d^{3l}x\, U_l(\mathbf{x}, \mathbf{x}) \qquad (\mathbf{x} \equiv \mathbf{x}_1, \ldots, \mathbf{x}_l)\\
\Omega b_l^s &= (1/l!)\int_{\Omega^l} d^{3l}x\, U_l^s(\mathbf{x}, \mathbf{x})
\end{aligned} \qquad (3)
$$

where Ω is the volume of the box. One recalls [5] that the equations of state are expressible in terms of b_l by

$$
\begin{aligned}
p/kT &= \sum_1^\infty b_l z^l\\
\rho \equiv N/\Omega &= \sum_1^\infty l b_l z^l
\end{aligned} \qquad (4)
$$

The first step in the binary collision expansion method is to express the propagation function with Bose statistics in terms of the U functions for Boltzmann statistics. (This is desirable because, as one recalls, for non-interacting particles $U_2 = U_3 = \ldots = 0$. To express everything in terms of this Boltzmann U therefore allows an easier treatment for expansions near the non-interacting system.) To complete this step, one needs the following equation, which can be easily proved:

$$
W_N^s(\mathbf{x}', \mathbf{x}) = \sum_{P'} P'\{W_N(\mathbf{x}', \mathbf{x})\} \qquad (5)
$$

where the sum extends over the $N!$ permutations P' of the coordinates $\mathbf{x}_i'$. Starting from the U functions, one constructs W from (2) and then W^s from (5). Substituting this W^s into eq. (1), one can solve for the functions U^s. The fugacity coefficients b_l^s can then be easily computed from (3).

The second step in the binary collision expansion method is to express the functions U_3, U_4, . . . in terms of U_2. This is accomplished by an expansion method that can be derived in the following way. One writes

$$W_N(\mathbf{x}', \mathbf{x}) = \langle \mathbf{x}' | e^{-\beta H} | \mathbf{x} \rangle \equiv \langle \mathbf{x}' | e^{-\beta(T+V)} | \mathbf{x} \rangle \qquad V = \sum_{i \neq j} V_{ij}$$

Now [6]

$$e^{-\beta(T+V)} = e^{-\beta T} + \int_0^\beta e^{-\beta' T}(-V)\, e^{-(\beta-\beta')T}\, d\beta' + \iint_0^{\beta'+\beta''\leq\beta} e^{-\beta' T}(-V)\, e^{-\beta'' T}(-V)\, e^{-(\beta-\beta'-\beta'')T} d\beta'\, d\beta'' + \ldots \tag{6}$$

The individual terms in this sum can be graphically represented as follows: (for $N = 3$)

$$\langle \mathbf{x}' | e^{-\beta T} | \mathbf{x} \rangle =$$

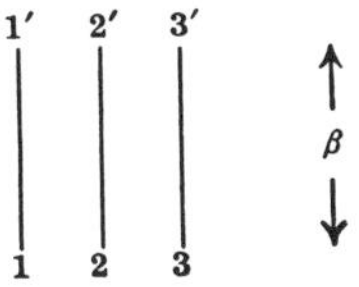

$$\langle \mathbf{x}' | \int_0^\beta e^{-\beta' T}(-V)\, e^{-(\beta-\beta')T}\, d\beta' | \mathbf{x} \rangle =$$

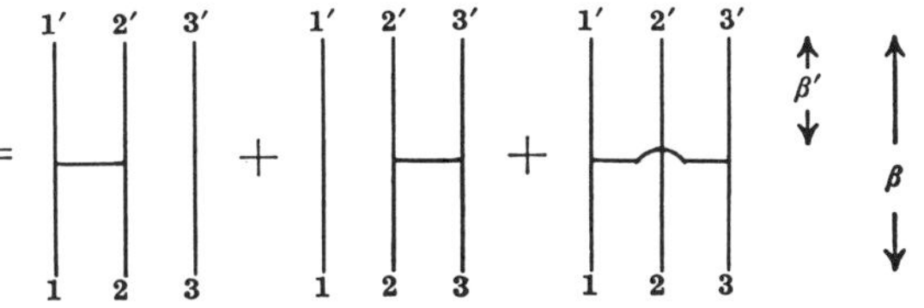

etc., where vertical lines represent $\langle \mathbf{x}' | e^{-\beta T} | \mathbf{x} \rangle$ for individual particles and horizontal lines represent the potential $(-V)$ between two particles. In terms of these graphs, the first two equations of (2) yield immediately, on solving for the U_l,

$$U_1 =$$

$$U_2 =$$

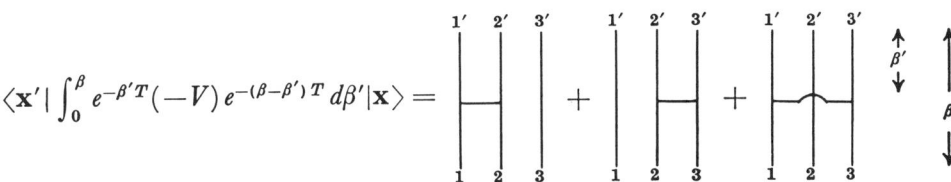

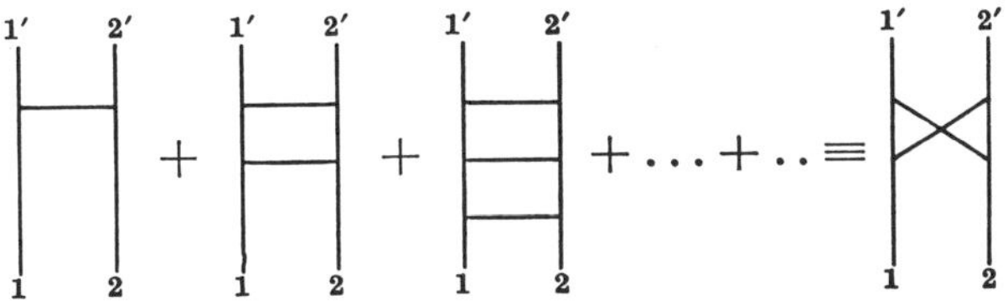

If one applies the idea of these graphs to the third equation of (2), one sees that the decomposition of W into U is just a sorting out of graphs with different connectedness. One obtains

$$U_3 = \begin{array}{c}\text{[graph]}\end{array} + \begin{array}{c}\text{[graph]}\end{array} + \begin{array}{c}\text{[graph]}\end{array} + \begin{array}{c}\text{[graph]}\end{array} + \begin{array}{c}\text{[graph]}\end{array} + \begin{array}{c}\text{[graph]}\end{array}$$

$$\text{+ terms with 3 or more horizontal lines}$$

Graphs in which the points 1, 2, or 3 are isolated from each other are not included in the sum.

It is obvious that if one introduces the sum

$$X \equiv \begin{array}{c}\text{[graph]}\end{array} + \begin{array}{c}\text{[graph]}\end{array} + \begin{array}{c}\text{[graph]}\end{array} + \ldots + \ldots \equiv \begin{array}{c}\text{[graph]}\end{array} = V_{12}\, W_2 \quad (7)$$

one can express U_3 in the following way

$$U_3 = \begin{array}{c}\text{[graph]}\end{array} \quad \begin{array}{l}\text{+ five other terms obtained by switching the crosses}\\ \text{around}\end{array}$$

$$\qquad\qquad\qquad\qquad\qquad\qquad\qquad\qquad\qquad\qquad (8)$$

$$+ \begin{array}{c}\text{[graph]}\end{array} + \text{ etc.}$$

To be explicit we write out the algebraic meaning of the first graph:

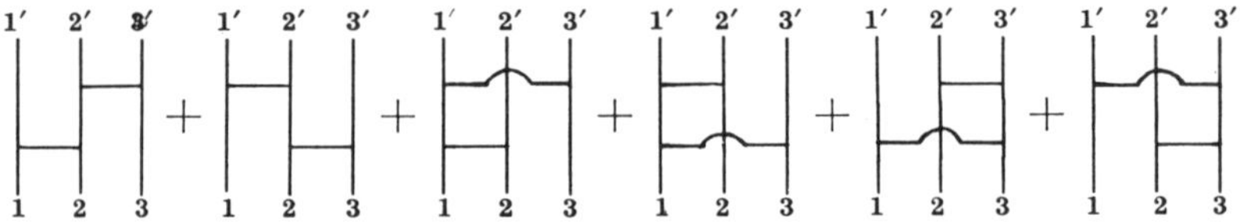

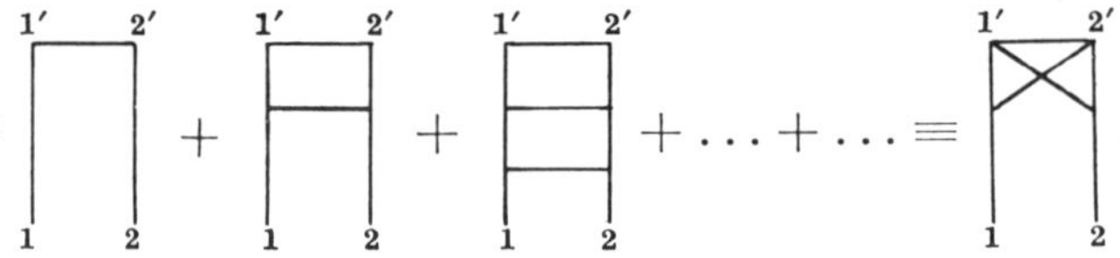

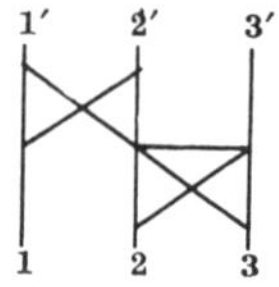

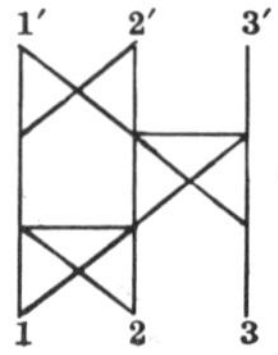

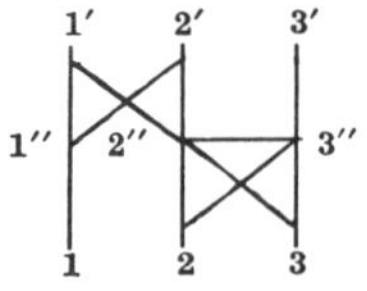

$$\begin{array}{c}\text{[graph]}\end{array} = \int_0^\beta \left\{ \int [U_2(1', 2'; 1'', 2'')\, U_1(3', 3'')|_{\beta'} \right.$$

$$\left. U_1(1'', 1')\, X(2'', 3''; 2, 3)]|_{\beta-\beta'}\, d^9x'' \right\} d\beta' \quad (8')$$

It is easy to prove that X is related to U_2 through the equation

$$X(1', 2'; 1, 2) = -\partial U_2(1', 2'; 1, 2)/\partial\beta + (\nabla_{1'}^2 + \nabla_{2'}^2)U_2 \quad (9)$$

in units defined by $2m = 1$, $\hbar = 1$. Equations (8), (8′) and (9) together allow one to calculate U_3 in terms of U_2. [It can be seen without difficulty that for hard sphere interactions of diameter a, the function U_2 is of first order in a. By (9), X is also of the same order. Equation (8) therefore expresses U_3 in powers of a, beginning with the order a^2.] For the calculation of U_4, U_5, . . . the procedure is exactly similar to that for U_3; in each case, an expansion in X is possible, except for the final interaction, which requires U_2.

Using the two steps described above, one can compute the fugacity coefficients b_l^s starting from the function U_2, which can be obtained because the solution of the two-body problem is known. The procedure described above can obviously also be applied to a system with Fermi statistics. Only minor changes of sign are necessary in the formulae.

The introduction of the sum X, it appears, corresponds to Professor Bethe's summation of diagrams in his discussion [7] of the Brueckner method.

2. *Application*

Lee and I have applied [1] this binary collision expansion method to a system of hard spheres with diameter a. Two types of singularities may be expected in such a problem. The first, because of the severe singularity of the hard sphere potential, has been eliminated by our expansion in X rather than in the potential V. The second is associated with the non-convergence of typical expansions at transition temperatures. To illustrate the treatment that is required in such cases, we consider the ground state ($T = 0°$ K) energy per particle for a hard sphere Bose gas at finite density and infinite volume. The following notation and units will be used:

$\hbar = 1$, m = mass of particles = 1/2, N = number of particles, Ω = volume of box, $\rho = N/\Omega$, $\lambda = (4\pi/kT)^{\frac{1}{2}}$.

Now since Bose and Boltzmann systems have the same ground state, we confine our attention to a Boltzmann system. The method described above leads to expressions for b_l of the following form:

$$b_l = (1/\lambda^3)(a/\lambda)^{l-1}\{\alpha_l + (a/\lambda)\,\beta_l + O[(a/\lambda)^2]\} \tag{10}$$

The coefficients α_l and β_l in this equation can be explicitly computed, but the form of the equation is obvious without a detailed calculation (U_l contains an integral over a product of one U_2 and at least $l-2$ X's. Hence b_l is of order a^{l-1}). We now try to approach the limit $T \to 0$, i.e., $\lambda = \infty$. Writing

$$\sum_1^\infty \alpha_l x^l \equiv f(x) \qquad \sum_1^\infty l\alpha_l x^l \equiv f_1(x) = df(x)/d\ln(x)$$
$$\sum_1^\infty \beta_l x^l \equiv g(x) \qquad \sum_1^\infty l\beta_l x^l \equiv g_1(x) = dg(x)/d\ln(x) \tag{11}$$

and

$$x \equiv az/\lambda \tag{12}$$

one therefore has from (4):

$$p/4\pi = (1/a\lambda^4)[f(x) + (a/\lambda)\, g(x) + \ldots]$$
$$\rho = (1/a\lambda^2)[f_1(x) + (a/\lambda)\, g_1(x) + \ldots] \tag{13}$$

Taking only the first term on the right-hand side of (13), one sees that at a finite ρ, as $\lambda \to \infty$, one obtains a finite pressure p if and only if

$$f_1(x) \to \text{constant } (\lambda^2) \qquad \text{and } f(x) \to \text{constant } [f_1(x)]^2$$

This is possible only if

$$f(x) \to \text{constant } (\ln x)^2 \qquad \text{and } x \to \infty$$

A detailed computation of the coefficients α_l explicitly bears out this conjecture.

One obtains consequently, as $T \to 0$

$$p \propto (\ln x)^2/a\lambda^4 \qquad \rho \propto \ln x/a\lambda^2 \tag{14}$$

Hence

$$p \propto a\rho^2 \tag{15}$$

Now the pressure at $T \to 0$ is related to the ground state energy E_0 in the following way

$$p = \rho^2\, d(E_0/N)/d\rho \tag{16}$$

Equations (15) and (16) together yield

$$E_0/N \propto a\rho \tag{17}$$

The coefficient in (17), on detailed computation, turns out to be 4π.

Now we can include the second term in the right-hand side of (13). Remembering that

$$\lambda \to (\ln x)^{1/2} (a\rho)^{-1/2}$$

one easily sees that these terms contribute a finite correction to p if and only if

$$g(x) \to (\ln x)^{5/2} \tag{18}$$

which is also borne out by a detailed computation of the coefficients. The correction term $(a/\lambda)\, g(x)$ compared with $f(x)$ is thus of order $(a/\lambda)(\ln x)^{1/2} \propto (\rho a^3)^{1/2}$.

We indicate briefly the considerations involved in determining the explicit coefficients α_l. To first order in $a \ll \lambda$, U_2 and X are very easily obtained in momentum space:

$$\begin{aligned} \langle \mathbf{p}' | U_1 | \mathbf{p} \rangle &= \delta^3(\mathbf{p}-\mathbf{p}') e^{-\beta p^2} \\ \langle \mathbf{p}_1' \mathbf{p}_2' | U_2 | \mathbf{p}_1 \mathbf{p}_2 \rangle &= -(a\beta/\pi^2)\, \delta^3(\mathbf{p}_1+\mathbf{p}_2-\mathbf{p}_1'-\mathbf{p}_2')\, e^{-\beta(p_1^2+p_2^2)} \\ \langle \mathbf{p}_1' \mathbf{p}_2' | X | \mathbf{p}_1 \mathbf{p}_2 \rangle &= -(1/\beta) \langle \mathbf{p}_1' \mathbf{p}_2' | U_2 | \mathbf{p}_1 \mathbf{p}_2 \rangle \end{aligned} \tag{19}$$

Now all diagrams representing U_l must be connected and contain one U_2; the lowest order, required for f, will then have one U_2 and $l-1$ X's. The integrations, as in (3) and (6), are readily performed and, given l, are independent of the diagram chosen. Hence, we need merely the number of $(l-1)$-branch, connected diagrams with l vertices (i.e., trees), and this may be determined by use of a generating function.

The higher order terms, representing β_l, etc., are found by a similar process, but here several types of diagram must be distinguished. For the ground state energy of a Bose (or Boltzmann) gas, one finds [8]

a. $$E/N = 4\pi a\rho[1+128(\rho a^3)^{1/2}/15\sqrt{\pi}+O(\rho a^3)] \tag{20}$$

in agreement with the prediction of (17) and (18). Further, the following results [8] are similarly obtained (particles of spin J are considered, where $J = 0$ for Bose and Boltzmann gases, and is arbitrary for a Fermi gas; P_F denotes the maximum Fermi momentum $[6\pi^2\rho/(2J+1)]^{1/3}$):

b. The ground state energy per particle for a Fermi gas at a finite density ρ and infinite volume is given by

$$E/N = (3P_F^2/5)+8\pi a\rho J(2J+1)^{-1}[1+6(11-2\ln 2)P_F a/35\pi+O(P_F^2 a^2)] \tag{21}$$

c. For the Fermi gas the fugacity expansion is

$$\begin{aligned} \lambda^3 p/kT &= \lambda^3 \sum_1^\infty b_l z^l \\ &= -(2J+1)g_{5/2}(-z)-2J(2J+1)[g_{3/2}(-z)]^2(a/\lambda) \\ &\quad -8J^2(2J+1)\,g_{1/2}(-z)[g_{3/2}(-z)]^2(a/\lambda)^2 \\ &\quad +8J(2J+1)\,F(-z)(a/\lambda)^2+O(a^3/\lambda^3) \end{aligned} \tag{22}$$

where

$$g_n(z) = \sum_1^\infty l^{-n} z^l \tag{23}$$

and

$$F(z) = \sum_{r,s,t=1}^\infty (rst)^{-1/2}(r+s)^{-1}(r+t)^{-1}z^{r+s+t} \tag{24}$$

d. For the Bose gas the fugacity expansion is

$$\begin{aligned} \lambda^3 p/kT &= g_{5/2}(z)-2[g_{3/2}(z)]^2(a/\lambda)+8g_{1/2}(z)[g_{3/2}(z)]^2(a/\lambda)^2 \\ &\quad +8F(z)(a/\lambda)^2+O(a^3/\lambda^3) \end{aligned} \tag{25}$$

e. For the Bose gas the pressure and density at the transition point are given by

$$\begin{aligned} \lambda^3 p/kT &= 1.34+(2.61)^2 2(a/\lambda)+O[(a/\lambda)_{3/2}] \\ \lambda^3 \rho &= 2.61+O(a/\lambda) \end{aligned} \tag{26}$$

To obtain this expression it was necessary to sum the dominant terms in the fugacity expansion to all orders of a/λ near $z = 1$.

3. *Pseudopotential method*

The pseudopotential method was discussed long ago by many people. It was very clearly discussed in the book of Blatt and Weisskopf [9]. Using the pseudopotential to simulate the hard sphere potential, the energy per particle, E/N, of a system of hard sphere Bosons has been previously computed [10] in powers of a up to a^3. Because in this computation the coefficient of a^2 was 0; that of a^3 being ∞, a comparison with the result we just discussed, which contains fractional powers of a, was very illuminating. Huang, Lee, and I were led [11] by such a comparison to a new use of the pseudopotential method consisting essentially of calculating to all integral powers of a the terms that have a maximum dependence on N. The sum of such terms yields exactly eq. (20), which is discussed above.

A natural generalization is to a calculation of the energies and wave functions of the low-lying excited states [11]. The results confirm the usual notion of phonon waves [with velocity corresponding to that calculated from the energy expression (20)] as the *only* low-lying excitations, and the idea of *momentum-space ordering*. Due to the low energy-level density associated with phonon states, one concludes [12] from the calculation that a dilute system of hard sphere Bose particles shows superfluidity and exhibits the two-fluid behavior at low temperatures. A similar computation for a Fermi-Dirac system reveals an infinitely greater energy-level density, and so superfluidity, e.g., in He^3, is not to be expected. One also concludes that a correlation length $(8\pi a\rho)^{-1/2}$ can be defined within which London's idea [13] of an ordering in momentum space is effective; residual spatial correlations vary as $1/r$ well within this length, and as $1/r^4$ outside it. It is interesting to observe that if the volume is restricted to this region of superfluidity and high correlation, then even the unmodified approach of Huang and Yang [10] yields the correct form for the mean particle energy. It is also suggestive that the superfluid flow for the hard sphere system is irrotational, as was pointed out by Onsager [14] and Feynman [12].

4. *Further remarks*

I shall make two more remarks. The first has to do with what happens when the system becomes very dense. One can conclude from a different kind of reasoning that when the system is so dense as to be near "jamming", the qualitative properties of the energy-level density near the ground state are very different from the case of low density — they resemble rather a collection of Fermions. What happens between the two extremes of high and low density is very interesting, but seems difficult to study.

The other remark has to do with the motivation underlying the study of a system of hard spheres. One would like, of course, to study the general many-body problem with any interaction potential between the particles. Such a program can be formalistically carried out. It is, however, generally recognized that to draw any definite physica. conclusions from such a general program is very difficult. If one makes approximations on the general problem to arrive at concrete results, one usually encounters the great difficulty of defining and justifying the validity of the approximation made. We therefore start instead from the concrete model of hard sphere interactions, which is sufficientlly simple so that one can discuss the validity of the method of approach. The interaction between real He atoms contains, besides a hard repulsive core, an attractive interaction outside the core. This attractive interaction is responsible for many properties of the He liquid. For example, the ground state of a system of He atoms is known to have a negative energy corresponding to a binding energy per He atom of ($k \times 7°$K), as determined from the experimental vapor pressure curve near the absolute zero of temperature. Such a bound system owes its origin, of course, to the attractive force. The strength of the attractive force also determines the density of the He atoms in the ground state. Now at this density the total attractive potential that a He atom experiences from its neighbors is expected not to fluctuate very much. This fact suggests the following approximate picture: One replaces the attractive interparticle forces by a constant uniform negative external potential that acts on the individual particles, the repulsive core is retained, and the system is kept by an external pressure at a density equal to that of the ground state of He. Many qualitative characteristics of this hypothetical model may then be expected to resemble those of real He. Since the uniform external potential does not influence the system except to give it a negative total energy, *one may consider simply a system of hard spheres at a given density* and in the end add the external potential separately. I should add that this kind of reasoning is essentially contained in the work of London on the density and the energy of liquid He in the ground state.

NOTE ADDED IN PROOF (Jan. 1962). The considerations outlined in this report and further developments were subsequently reported in full in the following publications: *Phys. Rev.* **106,** 1135 (1957); **112,** 1419 (1958); **113,** 1406 (1959); **113,** 1165 (1959); **116,** 25 (1959); **117,** 12, 22, 897 (1959). See also *Physica Suppl.* **26,** S49 (1960).

Notes and references

1. Reported at the International Conference on Theoretical Physics, Seattle, Wash., Sept. 1956, see also T. D. Lee and C. N. Yang, *Phys. Rev.* **105**, 1119 (1957).
2. H. D. Ursell, *Proc. Cambridge Phil. Soc.* **23**, 685 (1927).
3. B. Kahn and G. E. Uhlenbeck, *Physica* **5**, 399 (1938).
4. See, e.g., ter Haar, *Elements of Statistical Mechanics*, (Rinehart, New York, 1954, Chap. 8.
5. J. E. Mayer and M. G. Mayer, *Statistical Mechanics*, Wiley, New York, 1940, Chap. 13.
6. See, e.g., R. P. Feynman, *Phys. Rev.* **84**, 108 (1951).
7. H. A. Bethe, Chap. II, this volume.
8. T. D. Lee and C. N. Yang, *Phys. Rev.* **105**, 1119 (1957).
9. J. M. Blatt and V. F. Weisskopf, *Theoretical Nuclear Physics*, Wiley, New York, 1952, p. 76.
10. K. Huang and C. N. Yang, *Phys. Rev.* **105**, 767 (1957).
11. Lee, Huang, and Yang, *Phys. Rev.* **106**, 1135 (1957).
12. R. P. Feynman, in *Progress in Low Temperature Physics*, C. J. Gorter, ed., Interscience, New York, 1955, Vol. I.
13. F. London, *Superfluids*, Wiley, New York, 1954, Vol. II, Chap. B.
14. L. Onsager, *Nuovo cimento* **6**; *Suppl.*, 249 (1949).

[[57s]]
Commentary begins page 41

THE LAW OF PARITY CONSERVATION AND OTHER SYMMETRY LAWS OF PHYSICS

C. N. YANG

Nobel Lecture, December 11., 1957.

It is a pleasure and a great privilege to have this opportunity to discuss with you the question of parity conservation and other symmetry laws. We shall be concerned first with the general aspects of the role of the symmetry laws in physics, second, with the development that led to the disproof of parity conservation, and last, with a discussion of some other symmetry laws which physicists have learned through experience, but which do not yet together form an integral and conceptually simple pattern. The interesting and very exciting developments since parity conservation was disproved will be covered by Dr. LEE in his lecture (1).

I.

The existence of symmetry laws is in full accordance with our daily experience. The simplest of these symmetries, the isotropy and homogeneity of space, are concepts that date back to the early history of human thought. The invariance of physical laws under a coordinate transformation of uniform velocity, also known as the invariance under Galilean transformations, is a more sophisticated symmetry that was early recognized and formed one of the corner stones of Newtonian mechanics. Consequences of these symmetry principles were greatly exploited by physicists of the past centuries and gave rise to many important results. A good example in this direction is the theorem that in an isotropic solid there are only two elastic constants.

Another type of consequences of the symmetry laws relates to the conservation laws. It is common knowledge today that in general a symmetry principle (or equivalently an invariance principle) generates a conservation law. For example, the invariance of physical laws under space displacement has as a consequence the conservation of momentum, the invariance under space rotation has as a consequence the conservation of angular momentum. While the importance of these conservation laws was fully understood, their

□Nobel Lecture (December 11, 1957). Reprinted from *Les Prix Nobel*. Stockholm: The Nobel Foundation, 1957, 95–105.

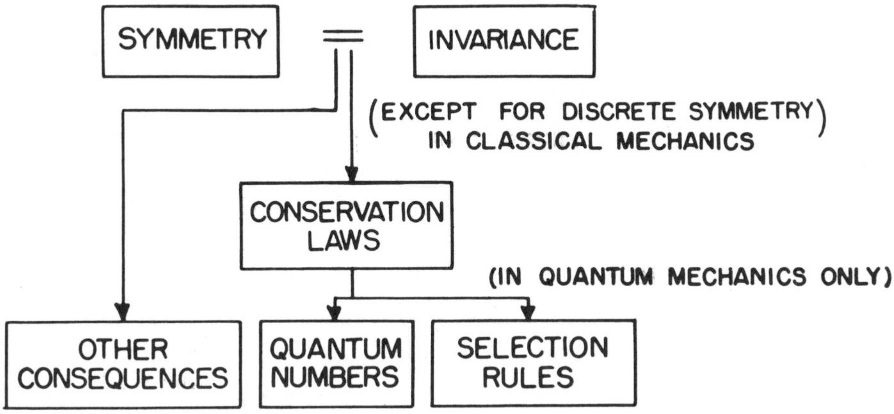

Fig. 1.

close relationship with the symmetry laws seemed not to have been clearly recognized until the beginning of the twentieth century (2). [Compare Fig. 1.]

With the advent of special and general relativity the symmetry laws gained new importance: Their connection with the dynamic laws of physics takes on a much more integrated and interdependent relationship than in classical mechanics, where logically the symmetry laws were only consequences of the dynamical laws that by chance possess the symmetries. Also in the relativity theories the realm of the symmetry laws was greatly enriched to include invariances that were by no means apparent from daily experience. Their validity rather were deduced from or were later confirmed by complicated experimentation. Let me emphasize that the conceptual simplicity and intrinsic beauty of the symmetries that so evolve from complex experiments are for the physicists great sources of encouragement. One learns to hope that nature possesses an order that one may aspire to comprehend.

It was, however, not until the development of quantum mechanics that the use of the symmetry principles began to permeate into the very language of physics. The quantum numbers that designate the states of a system are often identical with those that represent the symmetries of the system. It indeed is scarcely possible to overemphasize the role played by the symmetry principles in quantum mechanics. To quote two examples: The general structure of the periodic table is essentially a direct consequence of the isotropy of Coulomb's law. The existence of the antiparticles — namely the positron, the antiproton and the antineutron, were theoretically anticipated as consequences of the symmetry of physical laws with respect to Lorentz transformations. In both cases nature seems to take advantage of the simple mathematical representations of the symmetry laws. When one pauses to consider the elegance and the beautiful perfection of the mathematical reasoning involved and contrast it with the complex and far reaching physical consequences, a deep sense of respect for the power of the symmetry laws never fails to develop.

One of the symmetry principles, the symmetry between the left and the right, is as old as human civilization. The question whether nature exhibits such symmetry was debated at length by philosophers of the past (3). Of course, in daily life, left and right are quite distinct from each other. Our hearts, for example, are on our left sides. The language that people use both in the orient and the occident, carries even a connotation that right is good and left is evil. However, the laws of physics have always shown complete symmetry between the left and the right, the asymmetry in daily life being attributed to the accidental asymmetry of the environment, or initial conditions in organic life. To illustrate the point, we mention that if there existed a mirror image man with his heart on his right side, his internal organs reversed compared to ours, and in fact his body molecules, for example sugar molecules, the mirror image of ours, and if he ate the mirror image of the food that we eat, then according to the laws of physics, he should function as well as we do.

The law of right left symmetry was used in classical physics, but was not of any great practical importance there. One reason for this derives from the fact that right left symmetry is a discrete symmetry, unlike rotational symmetry which is continuous. Whereas the continuous symmetries always lead to conservation laws in classical mechanics, a discrete symmetry does not. With the introduction of quantum mechanics, however, this difference between the discrete and continuous symmetries disappears. The law of right left symmetry then leads also to a conservation law: the conservation of parity.

The discovery of this conservation law dates back to 1924 when LAPORTE (4) found that energy levels in complex atoms can be classified into "gestrichene" and "ungestrichene" types, or in more recent language, even and odd levels. In transitions between these levels during which one photon is emitted or absorbed, LAPORTE found that the level always changes from even to odd or vice versa. Anticipating later developments, we remark that the evenness or oddness of the levels was later referred to as the parity of the levels. Even levels are defined to have parity $+1$, odd levels parity -1. One also defines the photon emitted or absorbed in the usual atomic transitions to have odd parity. LAPORTE's rule can then be formulated as the statement that in an atomic transition with the emission of a photon, the parity of the initial state is equal to the total parity of the final state, *i. e.* the product of the parities of the final atomic state and the photon emitted. In other words, parity is con served, or unchanged, in the transition.

In 1927 WIGNER (5) took the critical and profound step to prove that the empirical rule of LAPORTE is a consequence of the reflection invariance, or right left symmetry, of the electromagnetic forces in the atom. This fundamental idea was rapidly absorbed into the language of physics. Since right left symmetry was unquestioned also in other interactions, the idea was further taken over into new domains as the subject matter of physics extended into nuclear reactions, β-decay, meson interactions, and strange particle physics. One became accustomed to the idea of nuclear parities as well as atomic parities, and one discusses and measures the intrinsic parities of the mesons. Throughout these developments the concept of parity and the law of parity conservation proved to be extremely fruitful, and the success had in turn been taken as a support for the validity of right left symmetry.

II.

Against such a background the so-called ϑ—τ puzzle developed in the last few years. Before explaining the meaning of this puzzle it is best to go a little bit into a classification of the forces that act between subatomic particles, a classification which the physicists have learned through experience to use in the last 50 years. We list the four classes of interactions below. The strength of these interactions is indicated in the column at right. The strongest interactions are the nuclear interactions

1. NUCLEAR FORCES	1
2. ELECTROMAGNETIC FORCES	10^{-2}
3. WEAK FORCES (Decay Interactions)	10^{-13}
4. GRAVITATIONAL FORCES	10^{-38}

which include the forces that bind nuclei together and the interaction between the nuclei and the π mesons. It also includes the interactions that give rise to the observed strange particle production. The second class of interactions are the electromagnetic interactions of which physicists know a great deal. In fact, the crowning achievement of the physicists of the 19th century was a detailed understanding of the electromagnetic forces. With the advent of quantum mechanics, this understanding of electromagnetic forces gives in principle an accurate, integral and detailed description of practically all the physical and chemical phenomena of our daily experience. The third class of forces, the weak interactions, was first discovered around the beginning of this century in the β-radioactivity of nuclei, a phenomena which especially in the last 25 years has been extensively studied experimentally. With the discovery of π-μ, μ-e decays and μ capture it was (6) noticed independently by KLEIN, by TIOMNO and WHEELER and by LEE, ROSENBLUTH and me that these interactions have roughly the same strengths as β-interactions. They are called weak interactions, and in the last few years their rank has been constantly added to through the discovery of many other weak interactions responsible for the decay of the strange particles. The consistent and striking pattern of their almost uniform strength remains today one of the most tantalizing phenomena — a topic which we shall come back to later. About the last class of forces, the gravitational forces, we need only mention that in atomic and nuclear interactions they are so weak as to be completely negligible in all the observations with existing techniques.

Now to return to the $\vartheta-\tau$ puzzle. In 1953 DALITZ and FABRI (7) pointed out that in the decay of the ϑ and τ mesons

$$\vartheta \to \pi + \pi$$
$$\tau \to \pi + \pi + \pi$$

some information about the spins and parities of the τ and ϑ mesons can be obtained. The argument is very roughly as follows. It has previously been determined that the parity of a π meson is odd, (*i. e.* $= -1$). Let us first neglect the effects due to the relative motion of the π-mesons. To conserve parity in the decays the ϑ meson must have the total parity, or in other words, the product parity, of two π mesons, which is even (*i. e.* $= +1$). Similarly, the τ meson must have the total parity of three π mesons, which is odd. Actually because of the relative motion of the π mesons the argument was not as simple and unambiguous as we just discussed. To render the argument conclusive and definitive it was necessary to study experimentally the momentum and angular distribution of the π mesons. Such studies were made in many laboratories and by the spring of 1956 the accumulated experimental data seemed to unambiguously indicate, along the lines of reasoning discussed above, that ϑ and τ do not have the same parity, and consequently are not the same particle. This conclusion, however, was in marked contradiction with other experimental results which also became definite at about the same time. The contradiction was known as the $\vartheta-\tau$ puzzle and was widely discussed. To recapture the atmosphere of that time allow me to quote a paragraph concerning the conclusion that ϑ and τ are not the same particle from a report entitled "Present Knowledge about the New Particles" which I gave at the International Conference on Theoretical Physics in Seattle in September 1956: (8)

"However it will not do to jump to hasty conclusions. This is because experimentally the K mesons (*i. e.* τ and ϑ) seem all to have the same masses and the same lifetimes. The masses are known to an accuracy of, say, from 2 to 10 electron masses, or a fraction of a percent, and the lifetimes are known to an accuracy of, say, 20 percent. Since particles which have different spin and parity values, and which have strong interactions with the nucleons and pions, are not expected to have identical masses and lifetimes, one is forced to keep the question open whether the inference mentioned above that the τ^+ and ϑ^+ are not the same particle is conclusive. *Parenthetically, I might add that the inference would certainly have been regarded as conclusive, and in fact more well founded than many inferences in physics, had it not been for the anomaly of mass and lifetime degeneracies.*"

The situation that the physicist found himself in at that time has been likened to a man in a dark room groping for an outlet. He is aware of the fact that in some direction there must be a door which would lead him out of his predicament. But in which direction?

That direction turned out to lie in the faultiness of the law of parity conservation for the weak interactions. But to uproot an accepted concept one must first demonstrate why the previous evidences in its favor were insufficient. Dr. LEE and I (9) examined this question in detail, and in May 1956 we came to the following conclusions: (A) Past experiments on the weak interactions had actually no bearing on the question of parity conservation. (B) In the strong interactions, *i. e.* interactions of classes 1 and 2 discussed above, there were indeed many experiments that established parity conservation to a high degree of accuracy, but not to a sufficiently high degree to be able to reveal the effects of a lack of parity conservation in the weak interactions.

The fact that parity conservation in the weak interactions was believed for so long without experimental support was very startling. But what was more startling was the prospect that a space time symmetry law which the physicists have learned so well may be violated. This prospect did not appeal to us. Rather we were, so to speak, driven to it through frustration with the various other efforts at understanding the ϑ—τ puzzle that had been made (10).

As we shall mention later there is known in physics a conservation law — the conservation of isotopic spin — that holds for interactions of class 1 but breaks down when weaker interactions are introduced. Such a possibility of an approximate symmetry law was, however, not expected of the symmetries related to space and time. In fact one is tempted to speculate, now that parity conservation is found to be violated in the weak interactions, whether in the description of such phenomena the usual concept of space and time is adequate. At the end of our discussion we shall have the occasion to come back to a closely related topic.

Why was it so that among the multitude of experiments on β-decay, the most exhaustively studied of all the weak interactions, there was no information on the conservation of parity in the weak interactions? The answer derives from a combination of two reasons. First, the fact that the neutrino does not have a measurable mass introduces an ambiguity that rules out (11) indirect information on parity conservation from such simple experiments as the spectrum of β-decay. Second, to study directly parity conservation in β-decay it is not enough to discuss nuclear parities, as one had always done. One must study parity conservation of the *whole* decay process. In other words, one must design an experiment that tests right left symmetry in the decay. Such experiments were not done before.

Once these points were understood it was easy to point out what were the experiments that would unambiguously test the previously untested assumption of parity conservation in the weak interactions. Dr. LEE and I proposed (9) in the summer of 1956 a number of these tests concerning β-decay, π-μ, μ—e and strange particle decays. The basic principles involved in these experiments are all the same: *One constructs two sets of experimental arrangements which are mirror images of each other, and which contain weak interactions. One then examines whether the two arrangements always give the same results in terms of the readings of their meters* (or counters). If the results are not the same, one would have an unequivocal proof that right left symmetry, as we usually understand it, breaks down. The idea is illustrated in Fig. 2 which shows the experiment proposed to test parity conservation in β-decay.

This experiment was first performed in the latter half of 1956 and finished early this year by WU, AMBLER, HAYWARD, HOPPES and HUDSON (12). The actual experimental setup was very involved, because to eliminate disturbing outside influences the experiment had to be done at very low temperatures. The technique of combining β-decay measurement with low temperature apparatus was unknown before and constituted a major difficulty which was successfully solved by these authors. To their courage and their skill physicists owe the exciting and clarifying developments concerning parity conservation in the past year.

The results of Drs. WU, AMBLER and their collaborators was that there is a very large difference in the readings of the two meters of Fig. 2. Since the

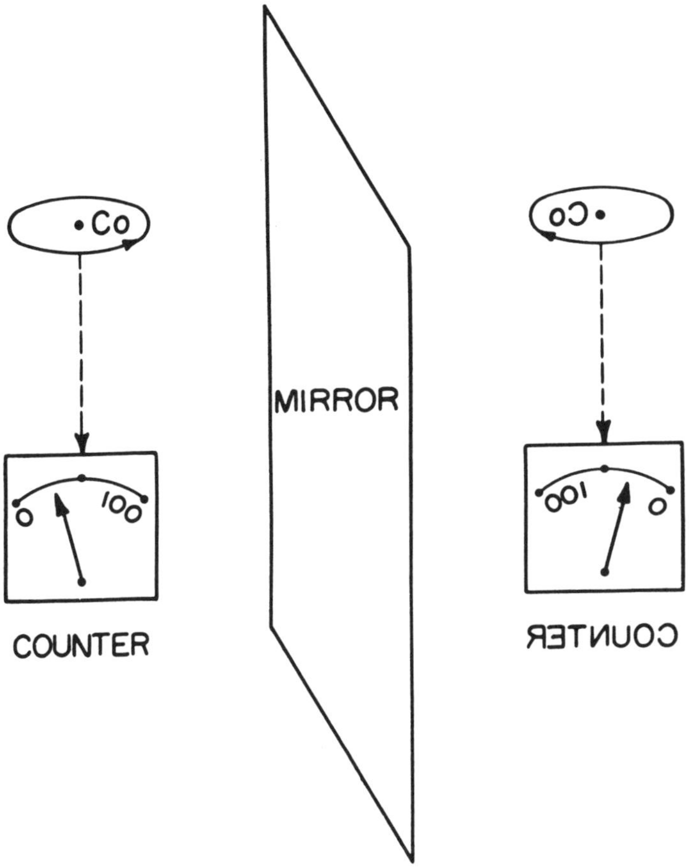

Fig. 2.

behavior of the other parts of their apparatus observes right left symmetry, the asymmetry that was found must be attributed to the β-decay of cobalt. Very rapidly after these results were made known, many experiments were performed which further demonstrated the violation of parity conservation in various weak interactions. In his lecture (1) Dr. LEE will discuss these interesting and important developments.

III.

The breakdown of parity conservation brings into focus a number of questions concerning symmetry laws in physics which we shall now briefly discuss in general terms:

(A) As Dr. LEE (1) will discuss, the experiment of WU, AMBLER and their collaborators also proves (13,14) that charge conjugation invariance (15) is violated for β-decay. Another symmetry called time reversal invariance (16) is at the present moment still being experimentally studied for the weak interactions.

The three discrete invariances — reflection invariance, charge conjugation invariance and time reversal invariance — are connected by an important theorem (17) called the CPT theorem. Through the use of this theorem one can prove (13) a number of general results concerning the experimental manifestations of the possible violations of the three symmetries in the weak interactions.

Of particular interest is the *possibility* that time reversal invariance in the weak interactions may turn out to be intact. If this is the case, it follows from the CPT theorem that although parity conservation breaks down, right left symmetry will still hold if (18) one switches all particles into antiparticles in taking a mirror image. In terms of Fig. 2 this means that if one changes *all* the matter that composes the apparatus at the right into anti-matter, the meter reading would become the same for the two sides if time reversal invariance holds. It is important to notice that in the usual definition of reflection, the electric field is a vector and the magnetic field a pseudovector while in this changed definition their transformation properties are switched. The transformation properties of the electric charge and the magnetic charge are also interchanged. It would be interesting to speculate on the possible relationship between the nonconservation of parity and the symmetrical or unsymmetrical role played by the electric and magnetic fields.

The question of the validity of the continuous space time symmetry laws has been discussed to some extent in the past year. There is good evidence that these symmetry laws do not break down in the weak interactions.

(B) Another symmetry law that has been widely discussed is that giving rise to the conservation of isotopic spin (19). In recent years the use of this symmetry law has produced a remarkable empirical order among the phenomena concerning the strange particles (20). It is however certainly the least understood of all the symmetry laws. Unlike Lorentz invariance or reflection invariance, it is not a "geometrical" symmetry law relating to space time invariance properties. Unlike charge conjugation invariance (21) it does not seem to originate from the algebraic property of the complex numbers that occurs in quantum mechanics. In these respects it resembles the conservation laws of charge and heavy particles. These latter laws, however, are exact while the conservation of isotopic spin is violated upon the introduction of electromagnetic interactions and weak interactions. An understanding of the origin of the conservation of isotopic spin and how to integrate it with the other symmetry laws is undoubtedly one of the outstanding problems in high energy physics today.

(C) We have mentioned before that all the different varieties of weak interactions share the property of having very closely identical strengths. The experimental work on parity nonconservation in the past year reveals that they very likely also share the property of not respecting parity conservation and charge conjugation invariance. They therefore serve to differentiate between right and left once one fixes one's definition of matter versus anti-matter. One could also use the weak interactions to differentiate between matter and anti-matter once one chooses a definition of right versus left. If time reversal invariance is violated, the weak interactions may even serve to differentiate simultaneously right from left and matter from anti-matter. One senses herein that maybe the origin of the weak interactions is intimately tied in with the question of the differentiability of left from right and of matter from anti-matter.

REFERENCES.

1. T. D. LEE, Nobel Lecture, 1957.
2. For references to these developments see E. P. WIGNER, Proc. Am. Phil. Soc. *93*, 521 (1949).
3. Cf. the interesting discussion on bilateral symmetry by H. WEYL. *Symmetry*, Princeton University Press (1952).
4. O. LAPORTE, Z. Physik *23*, 135 (1924).
5. E. P. WIGNER, Z. Physik *43*, 624 (1927).
6. O. KLEIN, Nature *161*, 897 (1948); J. TIOMNO and J. A. WHEELER, Rev. Mod. Phys. *21*, 144 (1949); T. D. LEE, M. ROSENBLUTH and C. N. YANG, Phys. Rev. *75*, 905 (1949).
7. R. DALITZ, Phil. Mag. *44*, 1068 (1953); E. FABRI, Nuovo Cimento *11*, 479 (1954).
8. C. N. YANG, Rev. Mod. Phys. *29*, 231 (1957).
9. T. D. LEE and C. N. YANG, Phys. Rev. *104*, 254 (1956).
10. T. D. LEE and J. OREAR, Phys. Rev. *100*, 932 (1955); T. D. LEE and C. N. YANG, Phys. Rev. *102*, 290 (1956); M. GELL-MANN, unpublished; R. WEINSTEIN, private communication; a general discussion of these ideas can be found in the *Proceedings of the Rochester Conference* (April 1956, Interscience Publishers), Session VIII.
11. C. N. YANG and J. TIOMNO, Phys. Rev. *79*, 495 (1950).
12. C. S. WU, E. AMBLER, R. W. HAYWARD, D. D. HOPPES, and R. P. HUDSON, Phys. Rev. *105*, 1413 (1957).
13. T. D. LEE, R. OEHME and C. N. YANG, Phys. Rev. *106*, 340 (1957).
14. B. L. IOFFE, L. B. OKUN′ and A. P. RUDIK, J. E. T. P. (U. S. S. R.) *32*, 396 (1957). English translation in Soviet Phys. JETP, *5*, 328 (1957).
15. Charge conjugation invariance is very intimately tied with the hole theory interpretation of Dirac's equation. The development of the latter originated with P. A. M. DIRAC, Proc. Roy. Soc. *A126*, 360 (1930); J. R. OPPENHEIMER, Phys. Rev. *35*, 562 (1930) and H. WEYL, *Gruppentheorie und Quantenmechanik*, 2nd edition, p. 234 (1931). An account of these developments is found in P. A. M. DIRAC, Proc. Roy. Soc. *A133*, 60 (1931). Detailed formalism and application of

charge conjugation invariance started with H. A. KRAMERS, Proc. Acad. Amst. *40*, 814 (1937) and W. FURRY, Phys. Rev. *51*, 125 (1937).
16. E. P. WIGNER, Nachr. Akad. Wiss. Göttingen, Math.-physik. 1932, p. 546. This paper explains in terms of time reversal invariance the earlier work of H. KRAMERS, Proc. Amsterdam, *33*, 959 (1930).
17. J. SCHWINGER, Phys. Rev. *91*, 720, 723 (1953); G. LÜDERS, Kgl. Danske Videnskab. Selskab, Mat.-fys. Medd. *28*, No. 5 (1954); W. Pauli's article in *Niels Bohr and the Development of Physics* (Pergamon Press, London, 1955). See also reference 21.
18. This possibility was discussed by T. D. LEE and C. N. YANG and reported by C. N. YANG at the international Conference on Theoretical Physics in Seattle in September 1956. (See reference 8.) Its relation with the CPT theorem was also reported in the same conference in one of the discussion sessions. The speculation was later published in T. D. LEE and C. N. YANG, Phys. Rev. *105*, 1671 (1957). Independently the possibility has been advanced as the correct one by L. LANDAU, J. E. T. P. (U. S. S. R.) *32*, 405 (1957). An English translation of Landau's article appeared in Soviet Phys. JETP, *5*, 336 (1957).
19. The concept of a total isotopic spin quantum number was first discussed by B. CASSEN and E. U. CONDON, Phys. Rev. *50*, 846 (1936) and E. P. WIGNER, Phys. Rev. *51*, 106 (1937). The physical basis derived from the equivalence of p-p and n-p forces, pointed out by G. BREIT, E. U. CONDON and R. D. PRESENT, Phys. Rev. *50*, 825 (1936). The isotopic spin was introduced earlier as a formal mathematical parameter by W. HEISENBERG, Z. Physik *77*, 1 (1932).
20. A. PAIS, Phys. Rev. *86*, 663 (1952) introduced the idea of associated production of strange particles. An explanation of this phenomenon in terms of isotopic spin conservation was pointed out by M. GELL-MANN, Phys. Rev. *92*, 833 (1953) and by K. NISHIJIMA, Progr. Theoret. Phys. Japan, *12*, 107 (1954). These latter authors also showed that isotopic spin conservation leads to a convenient quantum number called strangeness.
21. R. JOST, Helv. Phys. Acta *30*, 409 (1957).

Stockholm 1958. Kungl. Boktr. P. A. Norstedt & Söner

Speech at the Nobel Banquet, December 10, 1957

C. N. Yang

[[57t]]

Commentary begins page 42

Your Majesties, Your Royal Highnesses, Ladies and Gentlemen:

First of all allow me to thank the Nobel Foundation and the Swedish Academy of Sciences for the kind hospitality that Mrs. Yang and I have so much enjoyed. I also wish to thank especially Professor Karlgren for his quotation and his passage in Chinese, to hear which is to warm my heart.

The institution of the awarding of Nobel prizes started in the year 1901. In that same year another momentous event took place of great historical importance. It was, incidentally, to have a decisive influence on the course of my personal life and was to be instrumental in relation to my present participation in the Nobel festival of 1957. With your kind indulgence I shall take a few minutes to go a little bit into this matter.

In the latter half of the last century the impact of the expanding influence of Western culture and economic system brought to China a severe conflict. The question was heatedly debated of how much Western culture should be brought into China. However, before a resolution was reached reasons gave way to emotions, and there arose in the eighteen nineties groups of people called I Ho Tuan in Chinese, or Boxers in English, who claimed to be able to withstand in bare flesh the attack of modern weapons. Their stupid and ignorant action against the Westerners in China brought in 1900 the armies of many European countries and of the U.S. into Peking. The incident is called the Boxer War and was characterized on both sides by barbarous killings and shameful lootings. In the final analysis, the incident is seen as originating from an emotional expression of the frustration and anger of the proud people of China who had been subject to ever increasing oppression from without and decadent corruption from within. It is also seen in history as settling, once and for all, the debate as to how much Western culture should be introduced into China.

The war ended in 1901 when a treaty was signed. Among other things the treaty stipulated that China was to pay the powers the sum of approximately 500 million ounces of silver, a staggering amount in those days. About ten years later, in a typically American gesture, the U.S. decided to return to China her share of the sum. The money was used to set up a Fund which financed a University, the Tsinghua University, and a fellowship program for students to study in the U.S. I was a direct beneficiary of both of these two projects. I grew up in the secluded and academically inclined atmosphere of the campus of this University where my father was a professor and enjoyed a tranquil childhood that was unfortunately denied most of the Chinese of my generation. I was later to receive an excellent first two years' graduate education in the same University and then again was able to pursue my studies in the U.S. on a fellowship from the aforementioned fund.

As I stand here today and tell you about this, I am heavy with an awareness of the fact that I am in more than one sense a product of both the Chinese and Western cultures, in harmony and in conflict. I should like to say that I am as proud of my Chinese heritage and background as I am devoted to modern science—a part of human civilization of Western origin—to which I have dedicated and I shall continue to dedicate my work.

□Reprinted from *Les Prix Nobel*. Stockholm: The Nobel Foundation, 1957, 53–54.

[[58a]] Commentary begins page 42

Possible Determination of the Spin of Λ^0 from Its Large Decay Angular Asymmetry

T. D. Lee* and C. N. Yang
Institute for Advanced Study, Princeton, New Jersey

(Received November 7, 1957)

General consideration of the angular distribution of the decay products of a hyperon into a pion and a nucleon is carried out for arbitrary values J of the hyperon spin. Limitations on the magnitude of the asymmetry in the angular distribution are found. These limitations are formulated in terms of certain test functions which when applied to experimental results may lead to an unambiguous determination of the value of the hyperon spin. These considerations and the large "up-down" asymmetry in Λ^0 decay reported in recent literature suggest that the spin of Λ^0 is $\frac{1}{2}$.

I

RECENT experiments[1] have shown that in the decay of (partially) polarized Λ^0 there is a strong asymmetry in the distribution of the momentum of the decay π^-. We wish to point out in the present note that such a strong asymmetry may also serve to rule out high values of spin for Λ^0. No nonrelativistic approximation on either of the decay products is made in this note.

We consider a sample $\mathcal{S}$ of Λ^0's in their rest system, produced in any process or collection of processes, selected in any manner as long as *the selection process does not involve the angular distribution of the decay products of* Λ^0. It is well known that the angular distribution $\mathcal{g}$ of the decay pion from such a sample $\mathcal{S}$ when expanded into spherical harmonics Y_{LM} involves only L values up to $2J$, a conclusion that follows from the law of invariance under space rotation. Such conclusions on the maximum complexity of an angular distribution have been widely used to yield information on the spins of various systems. An additional type of conclusion, which we shall discuss below, is that the coefficients of such an expansion in Y_{LM} satisfy certain inequalities, which in the case of Λ^0 decay can lead also to useful information about the spin of Λ^0. To be more specific, let

$$\tfrac{1}{2}\mathcal{g}(\xi)d\xi, \quad (1\geqq\xi\geqq-1), \tag{1}$$

where

$$\xi=\cos\zeta, \tag{2}$$

be the distribution with respect to $\cos\zeta$. Here ζ is the angle between the decay proton momentum and the z axis, which is any direction fixed in any manner (e.g., with respect to the production process), but independent of the angular distribution of the decay products. As examples, one can quote the following two theorems (proved in the appendix):

Theorem 1

$$\frac{-1}{2J+2}\leqq\langle\xi\rangle\leqq\frac{1}{2J+2}, \tag{3}$$

where

$$J\equiv\text{spin of }\Lambda^0$$

and $\langle\ \rangle\equiv$average over the distribution $\mathcal{g}$.

Theorem 2

If one assumes that $\mathcal{g}$ is linear in ξ;

$$\mathcal{g}(\xi)=1+\alpha\xi, \tag{4}$$

then

$$\frac{-1}{6J}\leqq\langle\xi\rangle\leqq\frac{1}{6J}. \tag{5}$$

The experimental results[1] so far, with a total of $N\sim500$ cases indicate that [with the $+z$ axis chosen along $\mathbf{p}_\Lambda\times\mathbf{p}_{in}$]

$$\langle\xi\rangle\sim0.17. \tag{6}$$

This large value of the average of ξ compared to the maxima given by (3) and (4) lends hope that perhaps one can narrow down in an unambiguous way, *free from assumptions* (such as K spin$=0$), the value of the spin of Λ^0.

In the next section the most general conditions on the distribution function for $J=\frac{3}{2}$ and $J=\frac{5}{2}$ are given. These lead to certain test functions which when applied to the experimental data may give a determination of J. The limit of confidence of such tests is also discussed. In the appendix we give the general conditions on the distribution function for arbitrary J.

II

Since we do not discuss the azimuthal distribution of the decay protons, the sample $\mathcal{S}$ can[2] be considered as an *incoherent* mixture of states each with a definite angular momentum $m(=J, J-1, \cdots \text{ or } -J)$ in units of $\hbar$ along the z axis. We denote by I_m the statistical weight of the

* On leave of absence from Columbia University, New York, New York.

[1] F. S. Crawford *et al.*, Phys. Rev. **108**, 1102 (1957); F. Eisler *et al.*, Phys. Rev. **108**, 1353 (1957); L. Leipuner and R. Adair, Phys. Rev. **109**, 1358 (1958).

[2] Any state function ψ can be decomposed into a coherent mixture of such states, but the interference term between the states characterized by m and $m'\neq m$ always have the azimuthal distribution (constant)$\times e^{i(m-m')\phi}+$complex conjugate, which contributes zero when integrated over the azimuthal angle ϕ. If azimuthal distribution is also studied, more inequalities can be derived than those discussed in the present paper. They are however of more complicated forms, involving quadratic or higher rank forms in the averages of the spherical harmonics.

□ Reprinted from *The Physical Review* 109, 5 (March 1, 1958), 1755–1758.

state characterized by m. By definition

$$\sum_{m=-J}^{J} I_m = 1, \tag{7}$$

and

$$I_m \geqq 0. \tag{8}$$

The inequalities that will be derived follow from (8). To illustrate the reasoning involved, let us take as an example the case $J=\frac{3}{2}$. The decay proton and π^- must then be in a mixture of $p_{\frac{3}{2}}$ and $d_{\frac{3}{2}}$ states. If the amplitudes of the p and d waves are respectively A and B where $|A|^2+|B|^2=1$, the final state wave function resulting from a Λ^0 with spin component m is

$$Ap_{\frac{3}{2},m}+Bd_{\frac{3}{2},m}.$$

One therefore obtains for the incoherent mixture that makes up $\mathcal{S}$,

$$\mathcal{I}=\sum_{m=-J}^{J} I_m(Ap_{\frac{3}{2},m}+Bd_{\frac{3}{2},m})^\dagger(Ap_{\frac{3}{2},m}+Bd_{\frac{3}{2},m}). \tag{9}$$

We use the notation $(\ )^\dagger(\ )$ to indicate a matrix multiplication with respect to the proton spin coordinate only. $\mathcal{I}$ is therefore a function of the proton momentum direction.

It is easy to verify that

$$F_{\frac{3}{2},m}\equiv p_{\frac{3}{2},m}{}^\dagger p_{\frac{3}{2},m}=d_{\frac{3}{2},m}{}^\dagger d_{\frac{3}{2},m}, \tag{10}$$

$$G_{\frac{3}{2},m}\equiv d_{\frac{3}{2},m}{}^\dagger p_{\frac{3}{2},m}=p_{\frac{3}{2},m}{}^\dagger d_{\frac{3}{2},m}, \tag{11}$$

$$F_{\frac{3}{2},m}=F_{\frac{3}{2},-m}, \tag{12}$$

$$G_{\frac{3}{2},m}=-G_{\frac{3}{2},-m}. \tag{13}$$

One can therefore write (9) in the form

$$\mathcal{I}=(I_{\frac{3}{2}}+I_{-\frac{3}{2}})F_{\frac{3}{2},\frac{3}{2}}+\alpha(I_{\frac{3}{2}}-I_{-\frac{3}{2}})G_{\frac{3}{2},\frac{3}{2}} + (I_{\frac{1}{2}}+I_{-\frac{1}{2}})F_{\frac{3}{2},\frac{1}{2}}+\alpha(I_{\frac{1}{2}}-I_{-\frac{1}{2}})G_{\frac{3}{2},\frac{1}{2}}, \tag{14}$$

where

$$\alpha=2\,\mathrm{Re}(A^*B)/(|A|^2+|B|^2) \tag{15}$$

is a real constant between -1 and 1 that characterizes an interference between the two final states of different parities. The functions F and G can be easily calculated from their definitions (10) and (11);

$$\begin{aligned} F_{\frac{3}{2},\frac{3}{2}}&=\tfrac{3}{2}(1-\xi^2), & F_{\frac{3}{2},\frac{1}{2}}&=\tfrac{1}{2}(1+3\xi^2),\\ G_{\frac{3}{2},\frac{3}{2}}&=\tfrac{3}{2}(-\xi+\xi^3), & G_{\frac{3}{2},\frac{1}{2}}&=\tfrac{1}{2}(5\xi-9\xi^3), \end{aligned} \tag{16}$$

where $\xi=\cos\zeta$ and ζ is the angle between decay proton momentum and the z axis.

From the intensity formula (14) one can compute the various averages (over $\mathcal{I}$) of the Legendre polynomials P_L;

$$P_L(\xi)=\frac{1}{2^L L!}\frac{d^L}{d\xi^L}(\xi^2-1)^L,$$

$$\begin{aligned} \langle P_1\rangle&=-(3/15)(I_{\frac{3}{2}}-I_{-\frac{3}{2}})\alpha-(1/15)(I_{\frac{1}{2}}-I_{-\frac{1}{2}})\alpha,\\ \langle P_2\rangle&=-\tfrac{1}{5}(I_{\frac{3}{2}}+I_{-\frac{3}{2}})+\tfrac{1}{5}(I_{\frac{1}{2}}+I_{-\frac{1}{2}}),\\ \langle P_3\rangle&=(3/35)(I_{\frac{3}{2}}-I_{-\frac{3}{2}})\alpha-(9/35)(I_{\frac{1}{2}}-I_{-\frac{1}{2}})\alpha. \end{aligned} \tag{17}$$

These equations, together with (7), yield the following identities:

$$\begin{aligned} I_{\frac{3}{2}}+I_{-\frac{3}{2}}&=\tfrac{1}{2}-\tfrac{5}{2}\langle P_2\rangle,\\ I_{\frac{1}{2}}+I_{-\frac{1}{2}}&=\tfrac{1}{2}+\tfrac{5}{2}\langle P_2\rangle,\\ \alpha(I_{\frac{3}{2}}-I_{-\frac{3}{2}})&=-(9/2)\langle P_1\rangle+(7/6)\langle P_3\rangle,\\ \alpha(I_{\frac{1}{2}}-I_{-\frac{1}{2}})&=-\tfrac{3}{2}\langle P_1\rangle-\tfrac{7}{2}\langle P_3\rangle. \end{aligned} \tag{18}$$

The I's are all positive. Since $|\alpha|\leqq 1$, one must have

$$(I_m+I_{-m})\geqq|\alpha(I_m-I_{-m})|.$$

Hence

$$\begin{aligned} [\tfrac{1}{2}-\tfrac{5}{2}\langle P_2\rangle]&\geqq|-(9/2)\langle P_1\rangle+(7/6)\langle P_3\rangle|,\\ [\tfrac{1}{2}+\tfrac{5}{2}\langle P_2\rangle]&\geqq|-\tfrac{3}{2}\langle P_1\rangle-\tfrac{7}{2}\langle P_3\rangle|. \end{aligned} \tag{19}$$

Clearly one has also from (14) and (16);

$$\langle P_L\rangle=0 \quad \text{for} \quad L\geqq 4. \tag{20}$$

Equations (19) and (20) are necessary conditions for $J=\frac{3}{2}$, for an infinite sample $\mathcal{S}$ (for which the determination of the averages $\langle P_L\rangle$ is exact). *In the absence of additional knowledge* concerning the sample $\mathcal{S}$ and the quantity α, these equations also insure the possibility of mathematically constructing a sample $\mathcal{S}$ of a kind of particle of spin $\frac{3}{2}$ with a suitable mixing of parities in its decay so that the decay product has the distribution $\mathcal{I}$. Mathematically speaking (19) and (20) therefore give a necessary and sufficient condition for $J=\frac{3}{2}$, provided the azimuthal dependence is not considered.

For higher values of J the generalization of (20) is immediate and is well known; $\langle P_L\rangle=0$ for $L\geqq 2J+1$. The generalization of (19) is given in the appendix. The explicit form for the case $J=\frac{5}{2}$ is listed below

$$\begin{aligned} &[\tfrac{1}{3}-(25/12)\langle P_2\rangle+\tfrac{3}{2}\langle P_4\rangle]\\ &\qquad\geqq|-5\langle P_1\rangle+(35/12)\langle P_3\rangle-(11/30)\langle P_5\rangle|,\\ &[\tfrac{1}{3}+(5/12)\langle P_2\rangle-(9/2)\langle P_4\rangle]\\ &\qquad\geqq|-3\langle P_1\rangle-(49/12)\langle P_3\rangle+(11/6)\langle P_5\rangle|,\\ &[\tfrac{1}{3}+(5/3)\langle P_2\rangle+3\langle P_4\rangle]\\ &\qquad\geqq|-\langle P_1\rangle-(7/3)\langle P_3\rangle-(11/3)\langle P_5\rangle|. \end{aligned} \tag{21}$$

III

To test the inequalities (19) it is convenient to define the following four test functions:

$$\begin{aligned} T_{\frac{3}{2},\frac{3}{2}}&\equiv 9P_1+5P_2-(7/3)P_3,\\ T_{\frac{3}{2},-\frac{3}{2}}&\equiv -9P_1+5P_2+(7/3)P_3,\\ T_{\frac{3}{2},\frac{1}{2}}&\equiv 3P_1-5P_2+7P_3,\\ T_{\frac{3}{2},-\frac{1}{2}}&\equiv -3P_1-5P_2-7P_3. \end{aligned} \tag{22}$$

The inequalities (19) then reduce to

$$\langle T_{\frac{3}{2},m}\rangle\leqq 1, \quad m=\tfrac{3}{2},\cdots,-\tfrac{3}{2}. \tag{23}$$

Similarly, we define

$$T_{\frac{5}{2},\frac{5}{2}} \equiv 15P_1+(25/4)P_2-(35/4)P_3 -(9/2)P_4+(11/10)P_5,$$
$$T_{\frac{5}{2},-\frac{5}{2}} \equiv -15P_1+(25/4)P_2+(35/4)P_3 -(9/2)P_4-(11/10)P_5,$$
$$T_{\frac{5}{2},\frac{3}{2}} \equiv 9P_1-(5/4)P_2+(49/4)P_3 +(27/2)P_4-(11/2)P_5, \quad (24)$$
$$T_{\frac{5}{2},-\frac{3}{2}} \equiv -9P_1-(5/4)P_2-(49/4)P_3 +(27/2)P_4+(11/2)P_5,$$
$$T_{\frac{5}{2},\frac{1}{2}} \equiv 3P_1-5P_2+7P_3-9P_4+11P_5,$$
$$T_{\frac{5}{2},-\frac{1}{2}} \equiv -3P_1-5P_2-7P_3-9P_4-11P_5.$$

Eq. (21) now becomes

$$\langle T_{\frac{5}{2},m}\rangle \leqq 1, \quad m=\tfrac{5}{2}, \cdots, -\tfrac{5}{2}. \quad (25)$$

The inequalities (23) and (25), for $J=\frac{3}{2}$ and $\frac{5}{2}$, apply to infinite samples. For any finite sample $\mathcal{S}$ of N cases the determination of the average $\langle T\rangle$ of any function T has the standard statistical uncertainty;

$$\langle T\rangle = \left(\frac{1}{N}\sum T\right) \pm N^{-\frac{1}{2}}\left[\left(\frac{1}{N}\sum T^2\right)-\left(\frac{1}{N}\sum T\right)\right], \quad (26)$$

where the sums are extended over the finite sample $\mathcal{S}$. By using the various test functions $T_{J,m}$ in (26), (23), and (25) one obtains tests for various values of J together with confidence limits.

The large experimental[1] value of $\langle P_1\rangle=\langle\xi\rangle\cong 0.17$, together with the large positive coefficients of P_1 in $T_{\frac{3}{2},\frac{3}{2}}$ and $T_{\frac{5}{2},\frac{5}{2}}$ make these test functions the most sensitive ones. Lacking the detailed experimental information, we make the following estimates which, it may be hoped, are not too different from the experimental data. We assume *in calculating the right-hand side* of (26) that the experimental distribution of $N=500$ cases follows a linear law like (4). We assume also $\langle P_1\rangle=0.17$. The test

$$\langle T_{\frac{5}{2},\frac{5}{2}}\rangle \leqq 1$$

then becomes

$$2.55\pm 0.45 \leqq 1,$$

showing that $J=\frac{5}{2}$ is very improbable. For the test

$$\langle T_{\frac{3}{2},\frac{3}{2}}\rangle \leqq 1$$

one obtains

$$1.53\pm 0.27 \leqq 1,$$

which indicates that $J=\frac{3}{2}$ may also become unlikely.

The discussion in the last paragraph is meant only for orientation purposes. One sees from the estimate that if the sample has a slightly larger value of $\langle\xi\rangle=\langle P_1\rangle$, say $\langle P_1\rangle=0.20$, one can conclude that $J=\frac{5}{2}$ and $J=\frac{3}{2}$ are both quite impossible. It may therefore be worthwhile to select from the experimental sample those Λ^0 produced at an angle of production θ between, say, 30° and 150° for which the average polarization is likely to be greater than the over-all average, and thereby obtain a sample with a larger value for $\langle\xi\rangle$.

APPENDIX

We give here the formulas for general values of J. Derivation of these formulas follows the same line as for the case $J=\frac{3}{2}$. The distribution function $\mathcal{I}$ is expressible in terms of the diagonal elements I_m of the density matrix of Λ^0;

$$\mathcal{I}=\sum_{m>0} F_{J,m}(I_m+I_{-m})+\sum_{m>0} G_{J,m}(I_m-I_{-m})\alpha, \quad (A1)$$

where α, as before, is a parity mixing parameter[3] with its value between -1 and 1. The functions F and G are given by

$$F_{J,m}=\sum_L D(J-\tfrac{1}{2}, J-\tfrac{1}{2}; J, m, m; L, 0)P_L, \quad (A2)$$
$$G_{J,m}=\sum_L D(J+\tfrac{1}{2}, J-\tfrac{1}{2}; J, m, m; L, 0)P_L, \quad (A3)$$

where

$$D(l',l; J,m',m; L,M) \equiv (-1)^{m'+(1/2)+M+L}(2J+1)(2L+1)(2l+1)^{\frac{1}{2}}(2l'+1)^{\frac{1}{2}} \times\begin{pmatrix} l & l' & L \\ 0 & 0 & 0\end{pmatrix}\begin{pmatrix} L & J & J \\ M & m' & -m\end{pmatrix}\begin{Bmatrix} L & J & J \\ \frac{1}{2} & l & l'\end{Bmatrix}. \quad (A4)$$

The ( ) and { } symbols are the $3j$ and $6j$ symbols which are symmetrical versions of the Clebsch-Gordan-Wigner coefficients and the Racah coefficients.[4]

The averages $\langle P_L\rangle$ are given by

$$\langle P_L\rangle = \sum_{m>0} S_J(L,m)(I_m+I_{-m}), \text{ (even } L), \quad (A5)$$

$$\langle P_L\rangle = \sum_{m>0} S_J(L,m)(I_m-I_{-m})\alpha, \text{ (odd } L), \quad (A6)$$

where

$$S_J(L,m)=2(-1)^{m-J-\frac{1}{2}L} \times(J, m; J, -m|J, J; L, 0)/U_{J,L}, \quad \text{(even } L), \quad (A7)$$

$$S_J(L,m)=2(-1)^{m-J-\frac{1}{2}L-\frac{1}{2}} \times(J, m; J, -m|J, J; L, 0)/V_{J,L}, \quad \text{(odd } L), \quad (A8)$$

---

[3] It is of interest to note that for arbitrary spin value J of Λ^0 the longitudinal polarization of the decay proton from unpolarized Λ^0 is always $-\alpha$. The special case of $J=\frac{1}{2}$ has been discussed recently. T. D. Lee and C. N. Yang, Phys. Rev. **108**, 1645 (1957). See also R. Gatto, University of California Radiation Laboratory Report UCRL-3795 (unpublished).

[4] See, e.g., A. R. Edmonds, *Angular Momentum in Quantum Mechanics* (Princeton University Press, Princeton, 1957).

and

$$U_{J,L}=\left[\frac{(2J+L+1)!(2L+1)}{(2J-L)!}\right]^{\frac{1}{2}}\times\frac{(\frac{1}{2}L)!(\frac{1}{2}L)!(J-\frac{1}{2}L-\frac{1}{2})!}{L!(J+\frac{1}{2}L+\frac{1}{2})!}, \quad (A9)$$

$$V_{J,L}=\left[\frac{(2J+L+1)!(2L+1)}{(2J-L)!}\right]^{\frac{1}{2}}\times\frac{(\frac{1}{2}L-\frac{1}{2})!(\frac{1}{2}L-\frac{1}{2})!(J-\frac{1}{2}L)!}{L!(J+\frac{1}{2}L)!}. \quad (A10)$$

The symbols $(J, m; J, -m|J, J; L, 0)$ are the standard[4] Clebsch-Gordan-Wigner coefficients. The inverse of the Eqs. (A5) and (A6) are given by

$$I_m+I_{-m}=\sum_{\text{even } L} Q_J(m,L)\langle P_L\rangle \quad (A11)$$

and

$$\alpha(I_m-I_{-m})=\sum_{\text{odd } L} Q_J(m,L)\langle P_L\rangle, \quad (A12)$$

where

$$Q_J(m,L)=(-1)^{m-J-\frac{1}{2}L}\times(J, m; J, -m|J, J; L, 0)U_{J,L}, \quad (\text{even } L), \quad (A13)$$

$$Q_J(m,L)=(-1)^{m-J-\frac{1}{2}L-\frac{1}{2}}\times(J, m; J, -m|J, J; L, 0)V_{J,L}, \quad (\text{odd } L). \quad (A14)$$

The test functions are

$$T_{J,m}=-\sum_{L=1}^{2J}(J+\tfrac{1}{2})Q_J(m,L)P_L, \quad (m=J, J-1, \cdots, -J). \quad (A15)$$

The inequalities are

$$[\langle T_{J,m}\rangle]\leqq(J+\tfrac{1}{2})Q_J(m,0)=1, \quad (m=J, \cdots, -J). \quad (A16)$$

Equation (A16) together with

$$\langle P_L\rangle=0 \quad \text{for} \quad L\geqq 2J+1 \quad (A17)$$

give a necessary and sufficient condition for the case of spin J, in the mathematical sense discussed in Sec. II.

To prove Theorem 1, one computes from (A8) that

$$S_J(1,m)=-m/[2J(J+1)].$$

Thus (A6) shows that

$$\langle P_1\rangle=-\sum_{m>0}\alpha m(I_m-I_{-m})/[2J(J+1)]$$
$$=-\sum_m\frac{\alpha m I_m}{2J(J+1)}=-\left[\sum_m\frac{\alpha m I_m}{2J(J+1)}\right]\bigg/[\sum_m I_m].$$

I.e., $\langle P_1\rangle$ is equal to a weighed average of $[-\alpha m/2J(J+1)]$ with positive weights. Theorem (1) then follows immediately.

To prove Theorem 2, one computes from (A14) that

$$Q_J(m,1)=-6m/(J+\tfrac{1}{2}).$$

Thus if $\langle P_L\rangle=0$ for $L\geqq 2$, (A15) shows that

$$\langle T_{J,J}\rangle=6J\langle P_1\rangle.$$

(A16) then gives directly Theorem 2.

We remark that the functions $F_{J,\frac{1}{2}}$ and $G_{J,\frac{1}{2}}$ have been used before in recent literature.[5]

[5] R. K. Adair, Phys. Rev. **100**, 1540 (1955); S. B. Treiman, Phys. Rev. **101**, 1216 (1956); T. D. Lee and C. N. Yang, Phys. Rev. **104**, 822 (1956).

[[58d]]
Commentary begins page 44

Low-Temperature Behavior of a Dilute Bose System of Hard Spheres. I. Equilibrium Properties*

T. D. LEE† AND C. N. YANG‡
Brookhaven National Laboratory, Upton, New York

(Received August 13, 1958)

By a generalization of the method used in a previous paper the distribution of energy levels of a dilute Bose system of hard spheres is found. These energy levels are then used to compute the statistical properties for the system. A phase transition is found, and the transition point is calculated to the lowest order of a, the diameter of the hard spheres. Furthermore, the thermodynamical functions of the system in both the gas phase and the degenerate phase are obtained, provided $(a/\lambda)\ll 1$ and $\rho a^3 \ll 1$, where λ is the thermal wavelength and ρ is the particle density.

1. INTRODUCTION

IN a recent paper[1] the energy levels of a dilute Bose system of hard spheres near the ground state were calculated using the pseudopotential method.[2] From the distribution of these energy levels, it is possible to calculate the thermodynamic behavior near $T=0$. To extend these considerations to higher excitations and higher temperatures, and especially to investigate the phase transition of the system corresponding to the Bose-Einstein condensation of a free Bose gas is the aim of the present paper.

For the free Bose gas it is well known that between the transition temperature and $T=0$ there is macroscopic but incomplete occupation of the free-particle ground state. It turns out that this possibility of a macroscopic but incomplete occupation of a microscopic state can be incorporated into the method of reference 1. One obtains in this way a calculation which furnishes a natural connection between the concept of the degenerate phase in the sense of London's work[3] and the concept of phonon excitations in the sense of Landau's treatment.[4] The latter is particularly useful in the $T\cong 0$ region, the former at higher temperatures. The dual applicability of both concepts in the present calculation therefore makes possible a discussion of the thermodynamical properties of a dilute (i.e., $\rho a^3 \ll 1$) hard-sphere Bose system at all temperatures for which $a/\lambda \ll 1$, where a is the diameter of the hard spheres, ρ is the particle density, and λ is the thermal wavelength.

The system is found to exist in two thermodynamical phases, which correspond to the two phases of the free Bose gas. In each phase, corrections to the thermodynamical functions to the lowest order of a are obtained. The change in the transition point pressure and density is also obtained. In Sec. 8 a comparison is made between the calculated change of the transition point and the results of the binary collision method.

The method of the present paper for obtaining the excited energy levels can be easily generalized by a Galilean transformation. This will be discussed in a subsequent paper (paper II) and leads to the two-fluid concept[5] and to superfluidity.

If one continues along the line of approximation used in reference 1 and in this paper, one arrives at calculations concerning phonon scatterings and phonon lifetimes, which will be described in detail in future communications.

2. REVIEW

The method used in reference 1 for computing the ground-state energy and the distribution of energy

* Work performed under the auspices of the U. S. Atomic Energy Commission.
† Permanent address: Columbia University, New York, New York.
‡ Permanent address: The institute for Advanced Study, Princeton, New Jersey.
[1] Lee, Huang, and Yang, Phys. Rev. **106**, 1135 (1957). We adopt the notation of this paper in the present work.
[2] Kerson Huang and C. N. Yang, Phys. Rev. **105**, 767 (1957).
[3] See, e.g., F. London, *Superfluids II* (John Wiley and Sons, Inc., New York, 1954).
[4] L. D. Landau, J. Phys. U.S.S.R. **5**, 71 (1940).
[5] L. Tisza, J. phys. radium **1**, 164 (1940).

 □Reprinted from *The Physical Review* 112, 5 (December 1, 1958), 1419–1429.

levels near the ground state are based on the following points:

(i) The use of the pseudopotential in higher approximations. This was discussed in detail in reference 2. The crucial point is that the correct pseudopotential

$$V = 8\pi a \sum_{i<j} \delta^3(\mathbf{r}_i - \mathbf{r}_j) \frac{\partial}{\partial r_{ij}}(r_{ij}) \tag{1}$$

must be used. If one uses instead

$$V' = 8\pi a \sum_{i<j} \delta^3(\mathbf{r}_i - \mathbf{r}_j), \tag{1'}$$

then in higher approximations spurious infinities are encountered. In most computations, however, it is a great practical convenience to use V' first and to switch to V in the end.

(ii) In considering the Hamiltonians (in units $2m = \hbar = 1$),

$$H = -\sum \nabla_i^2 + V$$

and

$$H' = -\sum \nabla_i^2 + V', \tag{2}$$

one chooses the free-particle representation with periodic boundary conditions. (The importance of choosing the periodic boundary condition was emphasized in reference 1, footnote 17.) An off-diagonal matrix element of

$$V' = \Omega^{-1} 4\pi a \sum a_\alpha^* a_\beta^* a_\mu a_\nu \delta^3(\mathbf{k}_\alpha + \mathbf{k}_\beta - \mathbf{k}_\mu - \mathbf{k}_\nu) \tag{3}$$

then characterizes the collision of two particles of momenta $\mathbf{k}_\mu$, $\mathbf{k}_\nu$ going into the individual particle states with momenta $\mathbf{k}_\alpha$, $\mathbf{k}_\beta$, while all the other particles preserve their momenta. The value of such a matrix element is

$$\Omega^{-1} 4\pi a [n_\alpha n_\beta (n_\mu + 1)(n_\nu + 1)]^{\frac{1}{2}}. \tag{4}$$

Through such matrix elements one mixes states with different free particle occupation numbers. However, we saw in reference 1 that *the states that need be mixed to form an eigenstate of the Hamiltonian have occupation numbers which are the same to the order of N, if the parameter $(\rho a^3)^{\frac{1}{2}}$ is regarded as small.*

We recall in fact that in reference 1 one starts from the free-particle ground state and combines it with other states S that can be reached from the free-particle ground state through a series of off-diagonal elements (4). The assumption is made that all such states S have as the occupation number for the free-particle ground state $n_0 \sim N$. This assumption is later justified because to deplete the ground state of $2m$ particles one must go through the off-diagonal elements (4) at least m times. Since a is a small parameter, this gives a small probability for those states S for which n_0 is much different from N. In fact, it was shown that the probability falls off in a geometrical series with increasing m, making it a good approximation to take $n_0 \sim N$ for all states S of importance.

(iii) A further simplification occurs when one first drops all the off-diagonal elements (4) except those for which two of the four occupation numbers n_α, n_β, n_μ, n_ν are of the order of a finite fraction of N. The dropped elements are smaller than those kept by a factor of at least $\sim N^{\frac{1}{2}}$. However, there are more of these dropped elements than of those kept by a factor of at least N. It is not immediately obvious that these more numerous smaller matrix elements contribute less than the large off-diagonal elements. This question was discussed in reference 1, where it was shown that by treating these small matrix elements as a perturbation one obtains correction terms in the form of an expansion in powers of the parameter $(\rho a^3)^{\frac{1}{2}}$.

Based on the points reviewed above, calculations were made in reference 1 for those states for which *the occupations of the free particle ground state is essentially complete.* The ground-state energy was found to be

$$E_0 = 4\pi a \rho N [1 + (128/15\sqrt{\pi})(\rho a^3)^{\frac{1}{2}} + O(\rho a^3)]. \tag{5}$$

The excited states near the ground state were shown to have energies and momenta

$$E(m_\mathbf{k}) = E_0 + \sum_{\mathbf{k}\neq 0} m_\mathbf{k} (k^4 + 16\pi a \rho k^2)^{\frac{1}{2}},$$
$$\mathbf{P}(m_\mathbf{k}) = \sum m_\mathbf{k} \mathbf{k}, \tag{6}$$

where $m_\mathbf{k}$ are any positive integers that satisfy the additional condition

$$N^{-1} \sum_{\mathbf{k}\neq 0} m_\mathbf{k} = O(N^{-1}). \tag{7}$$

Furthermore, these states were shown to correspond to states with $m_\mathbf{k}$ quanta of compressional waves with wave number $k/2\pi$.

3. MACROSCOPIC BUT INCOMPLETE OCCUPATION OF THE FREE-PARTICLE GROUND STATE

We shall show now that the method of approximation reviewed in the last section is applicable to states for which there is macroscopic but incomplete occupation of the free-particle ground state, i.e., states for which the free-particle ground state occupation number n_0 is essentially $N\xi$ where ξ is a finite fraction, and for which no other state is occupied by a finite fraction of the N particles.

To discuss such states, we start from a state with the following occupation numbers:

$$n_0 = n_0^0 = N\xi,$$
$$n_\mathbf{k} = n_\mathbf{k}^0, \quad N^{-1} n_\mathbf{k}^0 = O(N^{-1}) \quad (\mathbf{k}\neq 0), \tag{8}$$
$$\sum_{\mathbf{k}\neq 0} n_\mathbf{k}^0 = (1-\xi)N.$$

Similarly to point (ii) of the last section, we need only consider states for which, to the order of N, the occupation number $n_{\mathbf{k}}$ does not drastically differ from $n_{\mathbf{k}}^0$. Writing

$$n_{\mathbf{k}}=n_{\mathbf{k}}^0+\delta_{\mathbf{k}} \quad (\mathbf{k}\neq 0), \tag{9}$$

it is possible to show in the same *a posteriori* manner, as done in Sec. 3 of reference 1 for the problem when $\xi=1$, that

$$\langle\delta_{\mathbf{k}}\rangle=O(1) \quad \text{for} \quad \mathbf{k}\neq 0,$$

and

$$N^{-1}\sum_{\mathbf{k}\neq 0}\langle\delta_{\mathbf{k}}\rangle=O[(\rho a^3)^{\frac{1}{2}}]. \tag{10}$$

The diagonal elements of V' were computed in reference 1:

$$\langle n|V'|n\rangle=\Omega^{-1}4\pi a(2N^2-N-\sum_{\mathbf{k}}n_{\mathbf{k}}^2).$$

Using (8) and (9), one can write this as

$$\langle n|V'|n\rangle$$
$$=\Omega^{-1}4\pi a[N^2+(1-\xi)^2N^2+2\xi N\sum_{\mathbf{k}\neq 0}n_{\mathbf{k}}]+V_1, \tag{11}$$

where

$$V_1=-\Omega^{-1}4\pi a[N+\sum_{\mathbf{k}\neq 0}n_{\mathbf{k}}^2+(\sum_{\mathbf{k}\neq 0}\delta_{\mathbf{k}})^2] \tag{12}$$

will be neglected because (10) shows that its contribution is of higher orders in $(\rho a^3)^{\frac{1}{2}}$. One then has

$$\langle n|V'|n\rangle=4\pi a\rho N[1+(1-\xi)^2]+8\pi a_n\xi\sum_{\mathbf{k}\neq 0}a_{\mathbf{k}}^*a_{\mathbf{k}}$$
$$+\text{negligible terms.} \tag{13}$$

To discuss the off-diagonal matrix elements of V' we take advantage of the simplification discussed in point (iii) of the last section. We keep, therefore, only those off-diagonal elements that correspond to a collision of two particles with momenta 0, going into momenta states $\mathbf{k}$ and $-\mathbf{k}$ or vice versa. All states that can be reached from the state (8) by such off-diagonal elements have the same value for $(n_{\mathbf{k}}-n_{-\mathbf{k}})=(n_{\mathbf{k}}^0-n_{-\mathbf{k}}^0)$.

It is not difficult to see that for these states the Hamiltonian assumes a form very similar to that discussed in reference 1 (which in fact is a special case of the present problem corresponding to $\xi=1$).

$$H'=4\pi a\rho N[1+(1-\xi)^2]+2\sum{}'a_{\mathbf{k}}^*a_{\mathbf{k}}(k^2+8\pi a\xi\rho)$$
$$+\sum{}'8\pi a\xi\rho B_{S(\mathbf{k})}(\mathbf{k}), \tag{14}$$

where the $\sum'$ extends over half of the $\mathbf{k}$ space excluding $\mathbf{k}=0$, $S(\mathbf{k})=|n_{\mathbf{k}}^0-n_{-\mathbf{k}}^0|$, and $B_S(\mathbf{k})$ is defined in Eq. (A3) of reference 1. It is important to notice that $B_S(\mathbf{k})$ and $B_S(\mathbf{k}')$ for different $\mathbf{k}$ and $\mathbf{k}'$ operate on different occupation numbers, and therefore commute with each other.

Equation (14) can be written as

$$H'=4\pi a\rho N[1+(1-\xi)^2]$$
$$+2\sum{}'(k^2+8\pi a\xi\rho)[a_{\mathbf{k}}^*a_{\mathbf{k}}+y_{\mathbf{k}}'B_{S(\mathbf{k})}(\mathbf{k})], \tag{15}$$

where

$$y_{\mathbf{k}}'=(k^2+8\pi a\xi\rho)^{-1}4\pi a\xi\rho.$$

The individual terms in the summation in (15) commute and can be simultaneously diagonalized.

The eigenfunctions and the eigenvalues of these terms have been evaluated in the Appendix of reference 1. As remarked in Sec. 2, calculation of the energy levels using V' lead to spurious terms which are divergent. The substitution of the true pseudopotential V for the V' lead to the correct final answer for the energy spectrum. Following arguments which run exactly parallel to that of reference 1, we find the energy spectrum to be

$$E(\xi,m_{\mathbf{k}})=E_0(\xi)+E_{\text{phonon}}(\xi), \tag{16}$$

where

$$E_0(\xi)=4\pi a\rho N[1+(1-\xi)^2$$
$$+(128/15\sqrt{\pi})(\rho a^3\xi^3)^{\frac{1}{2}}+O(\rho a^3)], \tag{17}$$

and

$$E_{\text{phonon}}(\xi)=\sum_{\mathbf{k}\neq 0}m_{\mathbf{k}}k(k^2+16\pi a\xi\rho)^{\frac{1}{2}}. \tag{18}$$

Here $m_{\mathbf{k}}$ are positive integers satisfying

$$N^{-1}\sum_{\mathbf{k}\neq 0}m_{\mathbf{k}}=(1-\xi)+O(N^{-1}). \tag{19}$$

The momentum of the system in the state specified by ξ, $m_{\mathbf{k}}$ is

$$\mathbf{P}(\xi,m_{\mathbf{k}})=\sum m_{\mathbf{k}}\mathbf{k}. \tag{20}$$

We remark that for $\xi=1$, these results reduce to those obtained in reference 1. It is furthermore important to remember that these equations do not give correct results to the order a, unless $ka\ll 1$, which is a condition for the applicability of the pseudopotential.

The physical interpretation of (16), (17), and (18) is exactly the same as before: $m_{\mathbf{k}}$ is the number of elementary excitons with wave number $(k/2\pi)$. One notices, however, that the energy of the excitation is now $(k^2+16\pi a\xi\rho)^{\frac{1}{2}}k$ which depends on ξ. For $\xi\neq 0$, this excitation energy varies linearly with k for small k and represents phonons. For $\xi=0$, the excitations assume the character of individual particle excitations.

The parameter ξ is, according to the spirit of the calculation above, the fraction of particles that are in the unperturbed ground state [neglecting terms of $O(\rho a^3)^{\frac{1}{2}}$]. Also the fraction $1-\xi$ is, according to (19), the ratio of the total number of excitations to the total number of particles. That the two are related is a natural consequence of the method of approach adopted above. It is perhaps worth emphasizing that the present calculation thus provides, in the model under discussion, a link between the two different viewpoints that have greatly influenced the development of our understanding of He II: the concept of the degenerate occupation of a single state, which led to London's proposal[3] that the λ transition in He is a consequence of Bose-Einstein condensation, and the concept of elementary phonon excitations, which led to Landau's theory[4] of liquid He. The dual applicability of both concepts in the present calculation also enables us to carry out calculations at all temperatures, including both the transition region and the region near $T=0$. In a subsequent paper where transport phenomena are discussed it will become clear that the parameter ξ corresponds to the fraction of superfluid, a concept first phenomenologically introduced by Tisza.[5]

The two terms $E_0(\xi)$ and $E_{\text{phonon}}(\xi)$ in (16) are accurate to the order of N (and not 1), which is sufficient for the calculation of the thermodynamical properties of the system. However, from the derivation it is clear that (16) can also be used to compute the energy difference (which is of the order of 1) between two states for which there is only a difference in a finite number of m_k's. For example, two states with a difference of one phonon belong in this computation to the *same* value of ξ and consequently their energy difference is

$$k(k^2+16\pi a\rho\xi)^{\frac{1}{2}},$$

which is of the order of 1.

4. PARTITION FUNCTION

Using the energy spectrum, Eqs. (16)–(19), we proceed to calculate the partition function Q for a dilute system of hard spheres obeying Bose statistics. We first define

$$Q(\xi)=\sum_{m_k}\exp[-\beta E(\xi,m_k)], \tag{21}$$

where β is related to the Boltzmann constant κ and the temperature T by

$$\beta=(\kappa T)^{-1}, \tag{22}$$

and the sum over m_k extends over all positive integral values provided

$$N^{-1}\sum_{k\neq 0} m_k=(1-\xi). \tag{23}$$

The partition function Q is a sum of $Q(\xi)$ over all allowable ξ's. However, for an infinite sytem ($N\rightarrow\infty$, but keeping ρ finite) the logarithm of Q is replaceable by the maximum of $\ln Q(\xi)$ with respect to ξ.

$$\ln Q(\bar{\xi})=\ln Q, \tag{24}$$

where the most probable value $\bar{\xi}$ is determined by the condition

$$\ln Q(\bar{\xi})=\text{maximum}[\ln Q(\xi)], \tag{25}$$

for all values of ξ between 0 and 1.

The quantity $\ln Q(\xi)$ can be readily evaluated by means of the method of steepest descent. We have

$$-\kappa T\Omega^{-1}\ln Q(\xi)=f(\xi,\zeta),$$

where

$$f(\xi,\zeta)=4\pi a\rho^2[1+(1-\xi)^2]+(1-\xi)\kappa T\rho\ln\zeta +(8\pi^3)^{-1}\kappa T\int\ln(1-\zeta e^{-\beta\omega})d^3k. \tag{26}$$

The parameter ζ is related to ξ by

$$(\partial f/\partial\zeta)_\xi=0,$$

i.e.,

$$(1-\xi)\rho=(8\pi^3)^{-1}\int\left(\frac{\zeta e^{-\beta\omega}}{1-\zeta e^{-\beta\omega}}\right)d^3k,\quad(0\leqq\zeta\leqq1). \tag{27}$$

The quantity ω is given by

$$\omega=k(k^2+16\pi a\xi\rho)^{\frac{1}{2}}. \tag{28}$$

Equation (27) is the condition for steepest descent. We notice that the free energy per unit volume takes the *maximum* value of the function $f(\xi,\zeta)$ with respect to ζ. In Eq. (26), as in all subsequent calculations, we neglect terms of higher powers in $(\rho a^3)^{\frac{1}{2}}$.

It is convenient to introduce a $\xi-\zeta$ plot in which, for fixed T and fixed ρ, ξ is plotted against ζ as a graphical representation of (27). Along each curve one locates the most probable (i.e., equilibrium) ξ by *minimizing* f. For orientation purposes one first investigates the case of free Bose particles (i.e., $a=0$). At fixed T the locus of the equilibrium points in the graph is easily seen to be the curve $OBCDE$ in Fig. 1.

To investigate the change in these equilibrium points brought about by the introduction of the hard-sphere interaction, we notice that *along* the $\xi-\zeta$ curves, one has

$$\frac{df}{d\xi}=\left(\frac{\partial f}{\partial\xi}\right)_\zeta=-\kappa T\rho\ln\zeta -a\rho\pi^{-2}\int\frac{\zeta e^{-\beta\omega}}{1-\zeta e^{-\beta\omega}}(1-k^2\omega^{-1})d^3k. \tag{29}$$

We take a point P along the curve $OBCDE$ and study the locus of this point at fixed T and ρ for infinitesimal values of a. [It is important to notice that the extension of this neighborhood is nonuniform in P.] It is convenient to divide the curve $OBCDE$ into three sections:

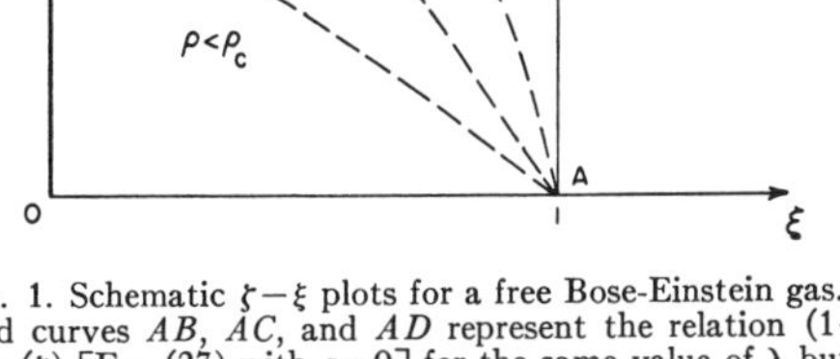

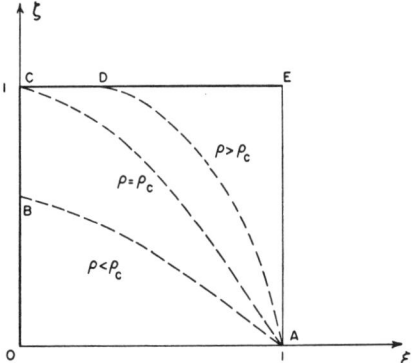

FIG. 1. Schematic $\zeta-\xi$ plots for a free Bose-Einstein gas. The dashed curves AB, AC, and AD represent the relation $(1-\xi)\rho=\lambda^{-3}g_{\frac{3}{2}}(\zeta)$ [Eq. (27) with $a=0$] for the same value of λ but different ρ. $\rho_c=\lambda^{-3}(2.612)$. Curve $OBCDE$ represents the locus of the equilibrium points for the free Bose gas, determined by Eq. (25).

(i) P is in the section OC open at C. For sufficiently small values of a the general shape of the $\xi-\zeta$ curves follow their counterparts in the case of $a=0$. Furthermore, for $a=0$, by (29), the value of $df/d\xi$ along any curve AB in Fig. 1 is always bigger than $-\rho\kappa T\ln(\zeta)_B$ which is positive. For sufficiently small values of a, therefore, $df/d\xi$ remains positive and the minimum of f occurs always at $\xi=0$. Thus the locus of equilibrium points remains along OC (open at C). In algebraic language this means that

$$\bar{\xi}=0. \tag{30}$$

(ii) P is in the section CE open at C. To find the equilibrium value of ξ we must solve the equations $df/d\xi=0$ and (27) for ξ and ζ. It follows from (29) that for $df/d\xi=0$,

$$\zeta=1-O[(\rho a^3)^{\frac{1}{2}}]. \tag{31}$$

Substituting this into (27), one obtains

$$1-\bar{\xi}=(2.612)(\rho\lambda^3)^{-1}-8\left(\frac{a\bar{\xi}}{\pi\rho\lambda^4}\right)^{\frac{1}{2}}+O(a\rho\lambda^2), \tag{32}$$

where

$$\lambda=(4\pi/\kappa T)^{\frac{1}{2}}.$$

The equilibrium point $\bar{\xi}-\zeta$ thus follows a curve slightly below the line CE in Fig. 1.

(iii) P is at C. This is the region near the transition point in which the behavior of a model satisfying strictly the spectrum (16)–(19) is very complicated. It will be discussed in detail in the Appendix.

In the gas region (i) and the degenerate region (ii) it is straightforward to substitute the equilibrium conditions (30), (31), and (32) into (26) and obtain the free energy of the system. The results are summarized in the next two sections.

5. GASEOUS PHASE

This is the phase satisfying $\bar{\xi}=0$. The Helmholtz free energy F_g for the system is given by

$$\Omega^{-1}F_g=-\lambda^{-3}\kappa T g_{\frac{5}{2}}(\zeta)+\rho\kappa T\ln\zeta +8\pi a\rho^2+O[a^{\frac{5}{2}}\lambda^{-13/2}], \tag{33}$$

where the parameter ζ is related to ρ through

$$\rho=\lambda^{-3}g_{\frac{3}{2}}(\zeta), \tag{34}$$

and

$$g_n(\zeta)=\sum_{l=1}^{\infty} l^{-n}\zeta^l. \tag{35}$$

Correspondingly, the energy E_g, entropy S_g, pressure p_g, and specific heat $(C_v)_g$ and the compressibility $(\partial p_g/\partial\rho)^{-1}$ are given by

$$\Omega^{-1}E_g=\tfrac{3}{2}\lambda^{-3}\kappa T g_{\frac{5}{2}}(\zeta)+8\pi a\rho^2, \tag{36}$$

$$\Omega^{-1}S_g=\tfrac{5}{2}\lambda^{-3}\kappa g_{\frac{5}{2}}(\zeta)-\rho\kappa\ln\zeta, \tag{37}$$

$$p_g=\lambda^{-3}\kappa T g_{\frac{5}{2}}(\zeta)+8\pi a\rho^2, \tag{38}$$

$$(\Omega^{-1}C_v)_g=(15/4)\lambda^{-3}\kappa g_{\frac{5}{2}}(\zeta)-(9/4)\kappa\rho g_{\frac{3}{2}}(\zeta)[g_{\frac{1}{2}}(\zeta)]^{-1}, \tag{39}$$

and

$$(\partial p_g/\partial\rho)_T=4\pi\rho\lambda[g_{\frac{1}{2}}(\zeta)]^{-1}+16\pi a\rho. \tag{40}$$

6. DEGENERATE PHASE

In this phase, (31) and (32) obtain. The thermodynamical functions can again be calculated by using the results obtained in Sec. 4 provided

$$\rho a^3\ll 1 \quad\text{and}\quad (a/\lambda)\ll 1. \tag{41}$$

In the following, we list the explicit forms of the thermodynamical functions for this dilute system at moderate temperature $a\rho\lambda^2\ll 1$, [case (i)], at fairly low temperature $a\rho\lambda^2\sim 1$ [case (ii)], and at very low temperature $a\rho\lambda^2\gg 1$, [case (iii)].

(i) The Helmholtz free energy F_d in the degenerate phase for $(\rho a^3)\ll 1$ and $a\rho\lambda^2\ll 1$ is given (neglecting terms proportional to $a^{\frac{5}{2}}$) by

$$\Omega^{-1}F_d=-(1.342)\lambda^{-3}\kappa T+4\pi a(\rho^2+2\rho\rho_c-\rho_c^2), \tag{42}$$

where

$$\rho_c=(2.612)\lambda^{-3}. \tag{43}$$

The other thermodynamical functions are given by

$$\Omega^{-1}E_d=\tfrac{3}{2}(1.342)\lambda^{-3}\kappa T+4\pi a(\rho^2-\rho\rho_c+2\rho_c^2), \tag{44}$$

$$\Omega^{-1}S_d=\tfrac{5}{2}(1.342)\lambda^{-3}\kappa+3\kappa a\lambda^2(\rho_c^2-\rho\rho_c), \tag{45}$$

$$p_d=(1.342)\lambda^{-3}\kappa T+4\pi a(\rho^2+\rho_c^2), \tag{46}$$

$$\Omega^{-1}(C_v)_d=(15/4)(1.342)\lambda^{-3}\kappa+\tfrac{3}{2}\kappa a\lambda^2(4\rho_c^2-\rho\rho_c) \tag{47}$$

and

$$(\partial p_d/\partial\rho)_T=8\pi a\rho. \tag{48}$$

(ii) At much lower temperature $a\rho\lambda^2\sim 1$ but $\rho a^3\ll 1$, the Helmholtz free energy becomes

$$\Omega^{-1}F_d=4\pi a\rho^2[1-4(2\rho a^3)^{\frac{1}{2}}t^{-1}\mathcal{G}(t)+O(\rho a^3)], \tag{49}$$

where

$$t=2a\rho\lambda^2, \tag{50}$$

and

$$\mathcal{G}(t)=-\frac{2}{\sqrt{\pi}}\int_0^\infty x\ln(1-e^{-tx})\left[\frac{(x^2+1)^{\frac{1}{2}}-1}{x^2+1}\right]^{\frac{1}{2}}dx. \tag{51}$$

The series expansion of $\mathcal{G}(t)$ for small and large values of t are given by

$$\mathcal{G}(t)=1.342t^{-\frac{3}{2}}-2.612t^{-\frac{1}{2}} +\tfrac{4}{3}(2\pi)^{\frac{1}{2}}-\tfrac{3}{2}(1.460)t^{\frac{1}{2}}+\cdots, \tag{52}$$

and

$$\mathcal{G}(t)=(2/\pi)^{\frac{1}{2}}\left[\frac{\pi^4}{45t^3}-\frac{\pi^6}{63t^5}+\cdots\right], \tag{53}$$

respectively. For intermediate values of t, $\mathcal{G}(t)$ can be evaluated numerically.

The other thermodynamical functions are given

[neglecting terms of higher orders in (ρa^3)] by

$$\Omega^{-1}E_d=4\pi a\rho^2[1-4(2\rho a^3)^{\frac{1}{2}}d\mathcal{G}/dt], \quad (54)$$

$$\Omega^{-1}S_d=\kappa(2a\rho)^{\frac{3}{2}}[\mathcal{G}-td\mathcal{G}/dt], \quad (55)$$

$$p_d=4\pi a\rho^2[1-2(2\rho a^3)^{\frac{1}{2}}(t^{-1}\mathcal{G}+2d\mathcal{G}/dt)], \quad (56)$$

and

$$\Omega^{-1}(C_v)_d=\kappa(2a\rho)^{\frac{3}{2}}t^2d^2\mathcal{G}/dt^2. \quad (57)$$

In this low-temperature range, Eq. (32) is not useful; instead one obtains from (27)

$$(1-\bar{\xi})\rho=(2a\rho)^{\frac{3}{2}}\mathfrak{F}(t), \quad (58)$$

where

$$\mathfrak{F}(t)=\frac{2}{\sqrt{\pi}}\int_0^\infty \frac{x}{e^{tx}-1}\left[\frac{(x^2+1)^{\frac{1}{2}}-1}{x^2+1}\right]^{\frac{1}{2}}dx. \quad (59)$$

For small values of t

$$\mathfrak{F}(t)=2.612t^{-\frac{3}{2}}-4(2/\pi)^{\frac{1}{2}}t^{-1}+1.460t^{-\frac{1}{2}}-\cdots, \quad (60)$$

and for large values of t

$$\mathfrak{F}(t)=(2/\pi)^{\frac{1}{2}}[(2.404)t^{-3}-(15.5535)t^{-5}+\cdots]. \quad (61)$$

(iii) In the extremely low temperature region $a\rho\lambda^2\gg1$ (but $\rho a^3\ll1$), the Helmholtz free energy becomes

$$\Omega^{-1}F_d=4\pi a\rho^2\left[1-\frac{1}{90}\left(\frac{a}{\lambda}\right)\left(\frac{\pi}{a\rho\lambda^2}\right)^{7/2}\right]. \quad (62)$$

The other thermodynamical functions are

$$\Omega^{-1}E_d=4\pi a\rho^2\left[1+\frac{1}{30}\left(\frac{a}{\lambda}\right)\left(\frac{\pi}{a\rho\lambda^2}\right)^{7/2}\right], \quad (63)$$

$$\Omega^{-1}S_d=\frac{2}{45}\pi\rho\kappa\left(\frac{a}{\lambda}\right)\left(\frac{\pi}{a\rho\lambda^2}\right)^{5/2}, \quad (64)$$

$$p_d=4\pi a\rho^2\left[1+\frac{1}{36}\left(\frac{a}{\lambda}\right)\left(\frac{\pi}{a\rho\lambda^2}\right)^{7/2}\right], \quad (65)$$

and

$$\Omega^{-1}(C_v)_d=\frac{2\pi}{15}\rho\kappa\left(\frac{a}{\lambda}\right)\left(\frac{\pi}{a\rho\lambda^2}\right)^{5/2}. \quad (66)$$

At these very low temperatures, $\bar{\xi}$ is given by

$$1-\bar{\xi}=\frac{1.202}{\rho\lambda^3\sqrt{\pi}}\left(\frac{1}{a\rho\lambda^2}\right)^{\frac{1}{2}}. \quad (67)$$

It is interesting to notice that in both the gas region and the degenerate region at moderate temperatures, (i.e., $a\rho\lambda^2\ll1$) the thermodynamical functions listed above, to the order included, are the same as the results of a calculation by Huang, Yang, and Luttinger.[6] This is not really surprising because the relevant excitations in these regions have momenta $k\sim\lambda^{-1}\gg(a\rho)^{\frac{1}{2}}$. The spectrum (18) used in the present paper than reduces to

$$E_{\text{phonon}}(\xi)\cong\sum_{\mathbf{k}\neq0} m_{\mathbf{k}}k^2+\sum_{\mathbf{k}\neq0} m_{\mathbf{k}}8\pi a\xi\rho$$
$$=\sum_{\mathbf{k}\neq0} m_{\mathbf{k}}k^2+N8\pi a\rho\xi(1-\xi), \quad (68)$$

which, when added to (17), gives

$$E(\xi,m_{\mathbf{k}})\cong4\pi aN\rho(2-\xi^2)+\sum_{\mathbf{k}\neq0} m_{\mathbf{k}}k^2. \quad (69)$$

This formula is exactly identical with the spectrum from which reference 6 starts. It differs from the spectrum (16)–(19) of the present paper because it does not take into account the off-diagonal elements of the potential in Eq. (1). One concludes that for the temperature region under discussion in this paragraph, these off-diagonal elements do not lead to important contributions.

The situation is completely different, however, for the very low temperature region [case (ii) and case (iii) above] which is characterized by phonon excitation with long wavelengths. One observes, for example, that in the degenerate phase C_v varies at moderate temperatures as $T^{\frac{3}{2}}$, but at extremely low temperatures as T^3. This difference of temperature dependence stems from the fact that for long wavelengths the phonon energy is linear instead of quadratic in the momentum. We shall see in a subsequent paper that this change of the energy spectrum is also responsible for the superfluidity of the system.

7. PHASE TRANSITION

By comparing the thermodynamical functions for the two phases, one finds that the phase transition occurs at

$$\lambda^3p_c/\kappa T=1.342+2(2.612)^2(a/\lambda)+O[(a/\lambda)^{\frac{3}{2}}], \quad (70)$$

and

$$\lambda^3\rho_c=2.612+O[a/\lambda], \quad (71)$$

where p_c and ρ_c are the pressure and density or densities at the transition point. It is important to notice that to the order $(a/\lambda)^{\frac{1}{2}}$ there is no discontinuity in density at the transition point. However, *this does not mean that in higher orders there will be no discontinuity in the densities of the two phases.* Similarly one sees that to the order (a/λ) there is no discontinuity in entropy, and to the order that we have calculated there is a discontinuity in the specific heat and a discontinuity in the compressibility. These discontinuities[7] are given by

[6] Huang, Yang, and Luttinger, Phys. Rev. **105**, 776 (1957).

[7] The exact meaning of (72) is as follows: Consider $(C_v)_d$ and $(C_v)_g$ as functions of T, ρ, and a. Then

$$\lim_{\rho\to\rho_c+}\lim_{a\to0+}(C_v)_d-\lim_{\rho\to\rho_c-}\lim_{a\to0+}(C_v)_g=0,$$

and

$$\lim_{\rho\to\rho_c+}\lim_{a\to0+}\frac{\partial(C_v)_d}{\partial a}-\lim_{\rho\to\rho_c-}\lim_{a\to0+}\frac{\partial(C_v)_g}{\partial a}=\frac{9}{2}(2.612)\lambda^{-1}N\kappa.$$

Similarly one can write down the exact meanings of (73). In these equations the order of the two limits $\rho\to\rho_c$ and $a\to0$ may not be exchanged. Consequently it is not possible to state the exact order of the transition for a finite a.

$$(C_v)_d-(C_v)_g=(9/2)(2.612)(a/\lambda)N\kappa, \tag{72}$$

and

$$\left(\frac{\partial p}{\partial \rho}\right)_d-\left(\frac{\partial p}{\partial \rho}\right)_g=-8\pi a\rho_c \tag{73}$$

at the transition point.

These properties are illustrated by Fig. 2 and Fig. 3.

8. COMPARISON WITH BINARY COLLISION METHOD

The thermodynamical functions in the gaseous phase and the transition point have previously been calculated by a different approach using the binary collision method.[8] The results, of course, agree with that obtained by the present method. It is of interest to compare these two different approaches.

First we compare the present results with the virial type expansion for the gaseous phase displayed in reference 8:

$$\lambda^3 p/\kappa T=g_{5/2}(z)-2(a/\lambda)[g_{3/2}(z)]^2+8(a/\lambda)^2[g_{3/2}(z)]^2 g_{1/2}(z)+8(a/\lambda)^2F(z)+O[(a/\lambda)^3], \tag{74}$$

where

$$\rho=\frac{\partial}{\partial \ln z}\left(\frac{p}{\kappa T}\right), \tag{75}$$

z is the fugacity, and $F(z)$ is

$$F(z)=\sum_{r,s,t=1}^{\infty}\frac{z^{r+s+t}}{(rst)^{1/2}(r+s)(r+t)}. \tag{76}$$

We remark that if we set

$$\ln z=\ln\zeta+4(a/\lambda)g_{3/2}(\zeta)+O[(a/\lambda)^2], \tag{77}$$

then, neglecting $O[(a/\lambda)^2]$, (74) and (75) are identical with (38) and (34). That (77) holds is to be anticipated since the parameter ζ was introduced in (27) in a steepest-descent calculation while the fugacity z is defined by

$$(\partial/\partial N)[\ln Q+N\ln z]=0. \tag{78}$$

A comparison of (78) and (27) leads immediately to (77).

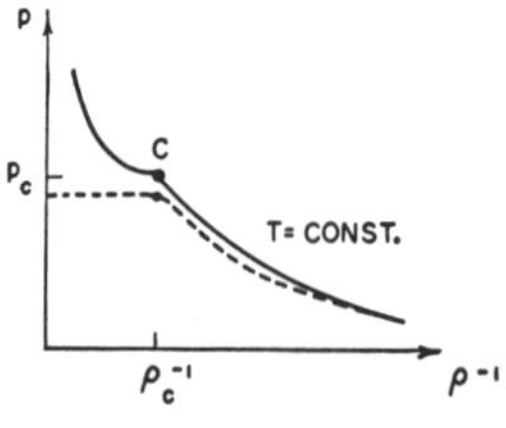

FIG. 2. Schematic $p-V$ diagram for a dilute Bose gas with hard-sphere interactions. The dotted line is the corresponding isotherm for the free Bose gas at the same T.

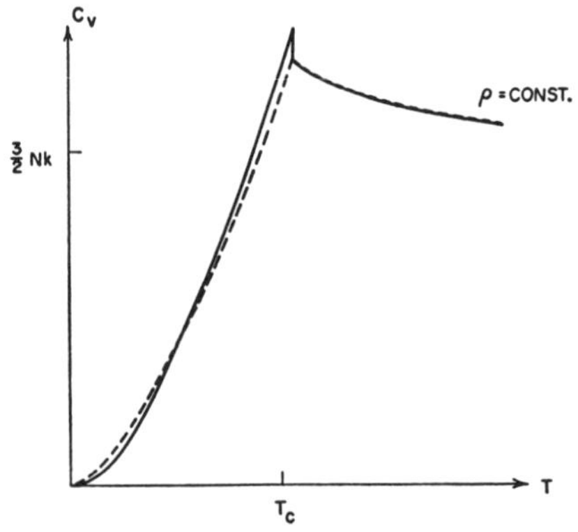

FIG. 3. Schematic C_v-T diagram for a dilute Bose gas with hard-sphere interactions. The dotted line is the corresponding curve for the free Bose gas at the same ρ.

In the treatment of the binary collision method it is pointed out that if we write the virial type expansion in the form

$$\lambda^3 p/\kappa T=\sum_{l=0}^{\infty}(a/\lambda)^l f_l(z), \tag{79}$$

then as $z\to 1_-$ the function $f_l(z)$ becomes singular for all values of $l\geq 2$. Furthermore, it is proved that if we set

$$z=e^{-\epsilon},$$

then as $\epsilon\to 0_+$, f_l can be expanded as a series. For $l\geq 2$, we have

$$f_l(z)=\epsilon^{-l}[A_l\epsilon^{3/2}+B_l\epsilon^2+C_l\epsilon^{5/2}+\cdots], \tag{80}$$

where A_l, B_l, etc., are independent of ϵ. Thus we may regroup the series in (79) and write

$$\lambda^3 p/\kappa T=1.342-(2.612)\epsilon-2(2.612)^2(a/\lambda)+\epsilon^{3/2}\mathbf{F}_{3/2}(a/\epsilon\lambda)+\epsilon^2\mathbf{F}_2(a/\epsilon\lambda)+\cdots, \tag{81}$$

where

$$\mathbf{F}_{3/2}(x)=\sum_{l=0}^{\infty}A_l x^l,\qquad \mathbf{F}_2(x)=\sum_{l=0}^{\infty}B_l x^l,\ \text{etc.} \tag{82}$$

In (82), A_0 and B_0 are coefficients of $\epsilon^{3/2}$ and ϵ^2 in the series expansion for f_0 for small ϵ; A_1 and B_1 are the corresponding coefficients of $\epsilon^{1/2}$ and ϵ in the series expansion for f_1. The other A_l and B_l are defined by (80). These functions can be explicitly calculated by using the binary collision method. They are found to be

$$\mathbf{F}_{3/2}(a/\lambda\epsilon)=\tfrac{4}{3}\pi^{1/2}[1+4(2.612)a/\lambda\epsilon]^{3/2},$$

and

$$\mathbf{F}_2(a/\lambda\epsilon)=B_0+B_1(a/\lambda\epsilon)+B_2(a/\lambda\epsilon)^2, \tag{83}$$

[8] Some results obtained with the binary collision method have been summarized in T. D. Lee and C. N. Yang, Phys. Rev. **105**, 1119 (1957). Notice that Eq. (5) of this reference was incorrect. It should read like Eqs. (70) and (71) of the present paper.

where

$$B_0 = -0.730,$$
$$B_1 = -2[4\pi + 2(1.460)(2.612)],$$
$$B_2 = -32\pi(2.612) - 8(1.460)(2.612)^2 + 8F(z=1),$$
$$B_l = 0, \quad \text{for} \quad l \geqq 3.$$

Thus, we find that $\lambda^3 p/\kappa T$ is *regular* at $\epsilon=0$ and that the singularity occurs at

$$\epsilon = -4(2.612)(a/\lambda) + O[(a/\lambda)^{3/2}], \tag{84}$$

which is identified as the transition point. The pressure and density at the transition point are given by (81):

$$\lambda^3(p_c/\kappa T) = 1.342 + 2(2.612)^2(a/\lambda) + O[(a/\lambda)^{3/2}],$$

and

$$\lambda^3 \rho_c = 2.612 + O(a/\lambda),$$

which are identical with those obtained in Sec. 7. We remark that it is indeed gratifying to find the results obtained in Sec. 7 not only agree with the virial-type expansion to order (a/λ) for all values of $z<1$, but also agree with the sum of the most singular part of $(a/\lambda)^l \times f_l(z)$ (including all values of l) as z is near 1.

9. DISCUSSIONS

We make a few remarks in this section about the present paper.

(i) The general line of reasoning described in Sec. 2 and developed in detail in the subsequent sections is also applicable to those states for which there is an incomplete but macroscopic occupation of a state of momentum $\mathbf{k}\neq 0$. Such states lead to superfluid flows and will be discussed in detail in a subsequent paper. For equilibrium thermodynamics, however, they are of no importance.

(ii) From the energy spectrum (18), we find the velocity of phonons of very long wavelength to be $(16\pi a\rho\xi)^{1/2}$. Thus, if the system is in thermodynamical equilibrium, the velocity of long-wavelength phonons has a statistical value $\bar{v}_{\text{phonon}}$, given by

$$\bar{v}_{\text{phonon}} = (16\pi a\rho\bar{\xi})^{1/2}, \tag{85}$$

which varies from $(16\pi a\rho)^{1/2}$ at $T=0$ to zero at the critical temperature. The interesting relationship between $\bar{v}_{\text{phonon}}$ and the macroscopic sound velocities will also be discussed in detail in the subsequent paper.

(iii) In Sec. 2, it was pointed out that for a dilute system of hard spheres one need only include those off-diagonal matrix elements of V for which two of the four occupation numbers $n_\alpha\ n_\beta\ n_\mu\ n_\nu$ are of the order of a finite fraction of N. The inclusion of the other off-diagonal matrix elements of V contribute to higher order corrections in energy. They also give rise to the scattering of the phonons and to the decay of a single phonon into two or three phonons of longer wavelengths. Let $\mathbf{k}$ be the momentum of the initial phonon and $\mathbf{k}_1$, $\mathbf{k}_2$ that of the final two. These momenta are related by

$$\mathbf{k} = \mathbf{k}_1 + \mathbf{k}_2,$$

and (86)

$$k(k + 16\pi a\xi\rho)^{1/2} = k_1(k_1^2 + 16\pi a\xi\rho)^{1/2} + k_2(k_2^2 + 16\pi a\xi\rho)^{1/2}.$$

Summing over all final states of $\mathbf{k}_1$, $\mathbf{k}_2$ we find the mean life τ of a phonon to be[9] (at $T\cong 0$)

$$\tau = (160\pi/3)\rho k^{-5} \quad \text{for} \quad k \ll (8\pi a\rho)^{1/2},$$

and (87)

$$\tau = (16\pi a^2\rho k)^{-1} \quad \text{for} \quad k \gg (8\pi a\rho)^{1/2}.$$

As is expected, the smaller the value of a the longer is the mean life.

The existence of these lifetimes throws new light on the nature of the "eigenstates" that were obtained in Sec. 3. Strictly speaking, these states are not eigenstates of the entire Hamiltonian. Transitions between them take place because of the decay, recombination and scattering of the phonons. The fact that one obtains states which are not eigenstates of the Hamiltonian is not really surprising. In fact, the exact eigenstates of the system are boundary-sensitive and are of no great physical interest. Furthermore, to the lowest order in a the approximate eigenstates give the equilibrium thermodynamical properties of the system. The situation is quite similar to a system of one hydrogen atom and the electromagnetic field: The excited states, $2s$, $2p$, etc., of an isolated hydrogen atom do not correspond to any real eigenstates of the system. However, if one neglects high-order terms in the fine structure constant, the eigenstate of the system becomes the product states of the free electromagnetic field and the isolated hydrogen atom. One can then compute the approximate thermodynamical behavior of the system using these metastable states.

(iv) We have seen that the presence of hard-sphere interactions for a dilute Bose gas gives rise to, on the one hand, phonon excitations which are superposable collective modes (i.e., that behave approximately as independent free modes), and also on the other hand, interactions between these modes which cause their scattering and their instability. It seems extremely plausible that the existence of certain superposable excitations that scatter and decay is a general characteristic of any nonlinear quantum mechanical system with a large number of degrees of freedom. In such a system, then, the problem of the interactions and the instability of the excitations and the problem of the very existence of these excitations must both go back to the nonlinearity of the original system, and cannot be separately understood.

ACKNOWLEDGMENT

The authors wish to thank the Brookhaven National Laboratory for the hospitality extended them during

[9] Details will be published in a subsequent communication.

their stay at Brookhaven in the Spring of 1958, when this paper was written.

APPENDIX

In this Appendix we shall study the behavior of a *model* whose energy spectrum is given *exactly* as

$$E=4\pi a\rho N[1+(1-\xi)^2]+\sum m_{\mathbf{k}}\omega_k, \tag{A1}$$

where

$$N^{-1}\sum_{\mathbf{k}\neq 0} m_{\mathbf{k}}=(1-\xi)+O(N^{-1}),$$

and

$$\omega_k=k(k^2+16\pi a\xi\rho)^{\frac{1}{2}}.$$

To the first order in (a/λ) [for $\rho a^3\ll 1$] the thermodynamical functions E, S, p, etc., of this system have been calculated in the text. These first-order results are the same as that of a dilute system of hard spheres obeying Bose statistics. To higher orders in (a/λ) the thermodynamical functions of a physical dilute system of hard spheres are not expected to be the same as that of this model. In fact, we shall show that if we take this model seriously and evaluate the partition function to higher orders in (a/λ) the resulting $p-V$ diagram exhibits unphysical behavior of the Van der Waals type. In particular, the results of the *partition function* of this model show that as one follows the isotherm from the gaseous phase to the liquid phase there is a sudden drop of pressure. This discontinuity of pressure occurs at a critical density

$$\rho_c=\lambda^{-3}(2.612)-(9.63)\lambda^{-3}(a/\lambda)+O[a^2/\lambda^5], \tag{A2}$$

with an amount

$$p_d-p_g=-(29.3)\kappa T\rho_c(a/\lambda)^2, \tag{A3}$$

where p_d and p_g are, respectively, the pressures in the degenerate phase and in the gaseous phase as the density approaches ρ_c.

On the other hand, one may use the *grand partitilon function*, instead of the partition function, then the corresponding $p-V$ diagram would exhibit a flat portion. There will not be any discontinuity in pressure, but instead a discontinuity of density with the density ρ_α of the degenerate phase different from the gas density ρ_g. Both ρ_d and ρ_g differ from $\lambda^{-3}(2.612)$ by an amount $O(a/\lambda^4)$. This discontinuity of density occurs at a pressure

$$\lambda^3 p_c/\kappa T=1.342+(2.612)^2(2a/\lambda)+O[(a/\lambda)^2]. \tag{A4}$$

To prove (A2), (A3), and (A4), let us first consider the partition function. It is necessary to investigate in the (ξ,ζ) plane the detailed behavior of the curves $\xi=\xi(\rho,\zeta)$ [Eq. (27)] and to find the maximum value of $\Omega^{-1}\ln Q(\xi)$ along these curves for fixed values of ρ. The functional form of $\Omega^{-1}\ln Q(\xi)$ is given by Eq. (26). We notice that at $\xi=0$ and $\zeta<1$

$$\frac{\partial}{\partial\xi}[\Omega^{-1}\ln Q(\xi)]<0. \tag{A5}$$

Since ξ cannot be less than 0, the points with $\xi=0$ and $\zeta<1$ always correspond to local maxima of $\Omega^{-1}\ln Q(\xi)$. It is easy to show that for low density,

$$[\rho\lambda^3-(2.612)]<O(a/\lambda), \tag{A6}$$

these local maxima are also absolute maxima of $\Omega^{-1}\ln Q(\xi)$. The above condition [Eq. (A6)] is satisfied in the gaseous region. The most probable value of ξ in the gaseous region is, then,

$$\bar{\xi}=0. \tag{A7}$$

However, as we increase density such that

$$[\rho\lambda^3-(2.612)]\sim O(a/\lambda), \tag{A8}$$

then these local maxima may not be the absolute maxima of $\Omega^{-1}\ln Q(\xi)$.

To investigate the detail change of $\bar{\xi}$ in the liquid phase near transition, it is necessary to calculate the integrals $\Omega^{-1}\ln Q(\xi)$ and $(1-\xi)\rho$ [Eqs. (26) and (27)] to very high orders in (a/λ). Let us write

$$\zeta=e^{-\epsilon}. \tag{A9}$$

It is convenient to define a parameter x,

$$x\equiv\epsilon/(2a\xi\rho\lambda^2). \tag{A10}$$

Equations (26) and (27) can then be written as

$$\begin{aligned}\Omega^{-1}\ln Q(\xi)=&\lambda^{-3}g_{\frac{5}{2}}(\zeta)-4\pi a\rho^2\beta[1+(1-\xi)^2]+(1-\xi)\rho\epsilon\\&-2a\xi\rho\lambda^{-1}g_{\frac{3}{2}}(\zeta)+(2a\xi\rho)^{\frac{3}{2}}B(x)\\&+\tfrac{3}{2}(2a\xi\rho)^2\lambda[g_{\frac{1}{2}}(\zeta)-(\pi/\epsilon)^{\frac{1}{2}}]+O[(a\xi)^{\frac{5}{2}}],\end{aligned} \tag{A11}$$

and

$$\begin{aligned}(1-\xi)\rho=&\lambda^{-3}g_{\frac{3}{2}}(\zeta)+\lambda^{-2}(2a\xi\rho)^{\frac{1}{2}}A(x)\\&-2a\xi\rho\lambda^{-1}[g_{\frac{1}{2}}(\zeta)-(\pi/\epsilon)^{\frac{1}{2}}]+O[(a\xi)^2],\end{aligned} \tag{A12}$$

where $A(x)$ and $B(x)$ are defined as

$$A(x)\equiv\frac{2}{\sqrt{\pi}}\int_0^\infty\frac{t}{x+t}\left\{\left[\frac{(t^2+1)^{\frac{1}{2}}-1}{t^2+1}\right]^{\frac{1}{2}}-t^{-\frac{1}{2}}\right\}dt. \tag{A13}$$

and

$$B(x)\equiv\frac{2}{\sqrt{\pi}}\int_0^\infty\frac{1}{x+t}\{\tfrac{2}{3}[(t^2+1)^{\frac{1}{2}}-1]^{\frac{3}{2}}-\tfrac{2}{3}t^{\frac{3}{2}}+t^{\frac{1}{2}}\}dt. \tag{A14}$$

The functions $A(x)$ and $B(x)$ satisfy the relation

$$dB/dx=-A(x)-(\pi/x)^{\frac{1}{2}}. \tag{A15}$$

The integral $A(x)$ can be evaluated explicitly. It is

$$\begin{aligned}A(x)=&-4(2/\pi)^{\frac{1}{2}}+2(\pi x)^{\frac{1}{2}}-4x[\pi(x^2+1)]^{-\frac{1}{2}}\\&\times[1+(x^2+1)^{\frac{1}{2}}]^{\frac{1}{2}}\left\{\tan^{-1}\left(\frac{1-x+(x^2+1)^{\frac{1}{2}}}{\sqrt{2}[1+(x^2+1)^{\frac{1}{2}}]^{\frac{1}{2}}}\right)\right.\\&+\tan^{-1}\left(\frac{x}{\sqrt{2}[1+(x^2+1)^{\frac{1}{2}}]^{\frac{1}{2}}}\right)\\&-\tfrac{1}{2}x[1+(x^2+1)^{\frac{1}{2}}]^{-\frac{1}{2}}\ln\left(x^{-1}[1+(x^2+1)^{\frac{1}{2}}]\right.\\&\left.\left.+\sqrt{2}x^{-1}[1+(x^2+1)^{\frac{1}{2}}]^{\frac{1}{2}}\right)\right\}.\end{aligned} \tag{A16}$$

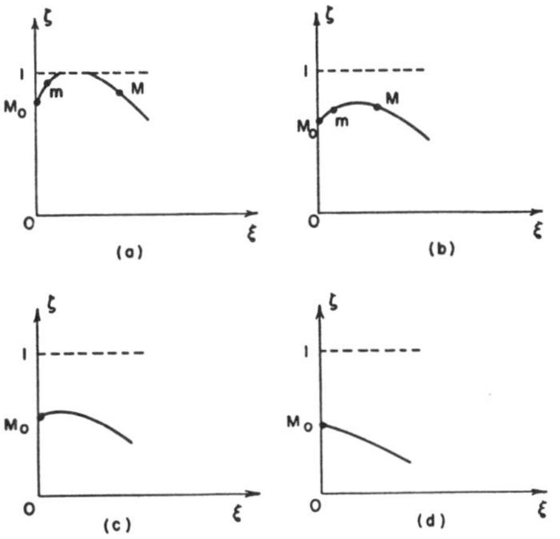

FIG. 4. Schematic plots of $\xi=\xi(\zeta,\rho)$ [Eq. (A19)] for the model (discussed in the Appendix). Figures 4(a), 4(b), 4(c), 4(d), are for the density ranges (a) $0<-\delta\rho<(16/\pi)(a/\lambda^4)$, (b) $(16/\pi)\times(a/\lambda^4)<-\delta\rho<(10.4)(a/\lambda^4)$, (c) $(10.4)(a/\lambda^4)<-\delta\rho<4\pi(a/\lambda^4)$, and (d) $-\delta\rho>4\pi(a/\lambda^4)$, respectively. The points m and M denote the minimum and maximum of $\Omega^{-1}\ln Q(\xi)$ along these curves.

Let us define

$$\delta\rho=\rho-\lambda^{-3}(2.612). \quad \text{(A17)}$$

In the following we shall consider only the region near the transition point, i.e.,

$$\lambda^3\delta\rho\sim O(a/\lambda),$$

and (A18)

$$\xi\sim O(a/\lambda).$$

For these ranges it is easy to see that Eq. (A12) becomes

$$-\xi\rho+\delta\rho=-\lambda^{-3}2(\pi\epsilon)^{\frac{1}{2}} +\lambda^{-2}(2a\xi\rho)^{\frac{1}{2}}A(x)+O(a^2/\lambda^5). \quad \text{(A19)}$$

In Fig. 4 we plot the curves $\xi=\xi(\rho,\zeta)$ in the (ξ,ζ) plane for the various values of $\delta\rho$. In the following we discuss the different physical conditions for these four figures: 4(a), 4(b), 4(c), and 4(d).

1. In Fig. 4(a), we plot schematically a typical curve $\xi=\xi(\zeta,\rho)$ [Eq. (A19)] for a ρ value which satisfies

$$0<-\delta\rho<16\pi^{-1}(a/\lambda^4). \quad \text{(A20)}$$

We see that since

$$\zeta\leqq 1,$$

this curve is broken into two pieces. The maximum value of $\Omega^{-1}\ln Q(\xi)$ along this curve can be found by setting the derivative of $\Omega^{-1}\ln Q(\xi)$ with respect to ξ equal to zero,

$$\frac{\partial}{\partial\xi}[\Omega^{-1}\ln Q(\xi)]=0. \quad \text{(A21)}$$

Upon using Eq. (A11), (A21) becomes

$$\epsilon=(2a)^{\frac{1}{2}}(\xi\rho)^{\frac{1}{2}}[\tfrac{3}{2}B+(1+x)A+(\pi x)^{\frac{1}{2}}]. \quad \text{(A22)}$$

Combining with (A19), (A22) yields two roots ξ_m and ξ_M corresponding to a minimum and a maximum of $\Omega^{-1}\ln Q(\xi)$, respectively. In Fig. 4(a) we denote these two points as m and M. As explained before, the point M_0 which denotes $\xi=0$ corresponds to a local maximum. However, by using the above expressions, it can be proved that

$$\Omega^{-1}\ln Q(\xi_M)>\Omega^{-1}\ln Q(\xi=0). \quad \text{(A23)}$$

Thus the most probable value is

$$\bar{\xi}=\xi_M.$$

Figure 4(a), then, corresponds to the degenerate phase.

2. In Figs. 4(c) and 4(d) we plot the curves $\xi=\xi(\rho,\zeta)$ for

$$10.4(a/\lambda^4)<-\delta\rho<4\pi(a/\lambda^4), \quad \text{(A24)}$$

and

$$-\delta\rho>4\pi(a/\lambda^4), \quad \text{(A25)}$$

respectively. In both cases, (A21) and (A19) yield only complex solutions for ξ. Consequently, along these curves

$$\frac{\partial}{\partial\xi}[\Omega^{-1}\ln Q(\xi)]<0. \quad \text{(A26)}$$

Thus we have

$$\bar{\xi}=0, \quad \text{(A27)}$$

provided

$$-\delta\rho>(10.4)(a/\lambda^4).$$

Figures 4(c) and 4(d), then, correspond to the gas phase.

3. In Fig. 4(b) we plot the curve $\xi=\xi(\zeta,\rho)$ for a density

$$16\pi^{-1}(a/\lambda^4)<-\delta\rho<10.4(a/\lambda^4). \quad \text{(A28)}$$

In this case (A21) and (A19) still determine two real roots ξ_m and ξ_M correspond to a minimum value and a maximum value of $\Omega^{-1}\ln Q(\xi)$. However, if

$$-\delta\rho<9.63(a/\lambda^4),$$

then we have

$$\Omega^{-1}\ln Q(\xi_M)>\Omega^{-1}\ln Q(\xi=0).$$

For densities with

$$-\delta\rho>9.63(a/\lambda^4),$$

we have

$$\Omega^{-1}\ln Q(\xi_M)<\Omega^{-1}\ln Q(\xi=0).$$

Consequently, we have for the most probable value of ξ,

$$\bar{\xi}=\xi_M \quad \text{for} \quad -\lambda^3\delta\rho<(9.63)(a/\lambda),$$

and

$$\bar{\xi}=0 \quad \text{for} \quad -\lambda^3\delta\rho>(9.63)(a/\lambda).$$

Thus at the critical density value,

$$\rho_c=\lambda^{-3}(2.612)-9.63(a/\lambda^4)+O(a^2/\lambda^5),$$

there is a discontinuity of $\bar{\xi}$ and a discontinuity of pressure.

(i) On the degenerate side, as the density $\rho\to\rho_c+$, the parameters ξ and x approach

$$\xi \to 4.23(a/\lambda), \tag{A29}$$

and

$$x \to 0.66. \tag{A30}$$

Correspondingly, $\Omega^{-1}\ln Q$ and the pressure p_d approach

$$\Omega^{-1}\ln Q \to [\lambda^{-3}(1.342) - 8\pi a\rho_c^2\beta - 23.6\lambda^{-3}(a/\lambda)^3], \tag{A31}$$

$$p_d \to (p_d)_c = \lambda^{-3}\kappa T(1.342) + 8\pi a\rho_c^2 - (36.7)(a/\lambda)^2\rho_c\kappa T. \tag{A32}$$

(ii) On the gaseous side, the parameter ξ is always given by

$$\xi = 0. \tag{A33}$$

As the density ρ approaches the critical value ρ_c, the pressure p_g and $\Omega^{-1}\ln Q$ approach

$$\Omega^{-1}\ln Q \to [\lambda^{-3}(1.342) - 8\pi a\rho_c^2\beta - (23.6)\lambda^{-3}(a/\lambda)^3], \tag{A34}$$

and

$$p_g \to (p_g)_c = \lambda^{-3}\kappa T(1.342) + 8\pi a\rho_c^2 - (7.4)(a/\lambda)^2\rho_c\kappa T.$$

Thus we find that as we change from the gas phase to the degenerate phase there is a sudden unphysical drop of the pressure of the order of $(a/\lambda)^2$,

$$(p_d)_c - (p_g)_c = -29.3(a/\lambda)^2\rho_c\kappa T. \tag{A35}$$

This shows that although to first order of (a/λ) the thermodynamical functions are correctly evaluated by using the energy spectrum [Eq. (A1)], the higher order terms of this model do not correspond to any real physical system.

On the other hand, one may use the grand partition function instead of the partition function, to calculate the thermodynamical functions for this model. It is easy to show that the use of the grand partition function leads to the well-known application of Maxwell's rule of equal area on the Van der Waals type isotherm obtained from the partition function. From the previous results for the partition function one finds that the resulting isotherm by using the grand partition function has no discontinuity in pressure, but, instead, a discontinuity of density

$$(\rho_d - \rho_g) \sim O(a/\lambda^4),$$

which occurs at a pressure p_c given by Eq. (A4).

Many-Body Problem in Quantum Statistical Mechanics. I. General Formulation*

[[59b]]

Commentary begins page 45

T. D. LEE, *Columbia University, New York, New York*

AND

C. N. YANG, *Institute for Advanced Study, Princeton, New Jersey*

(Received October 2, 1958)

A formulation is given whereby the grand partition function of a many-body system satisfying Bose-Einstein or Fermi-Dirac statistics is expressed in terms of certain U functions defined for the same system with Boltzmann statistics. It is then shown that these U functions can be evaluated in successive approximations in terms of a binary kernel B which can be computed from a solution of the two-body problem. The approach to the limit of infinite volume is studied. The example of a hard sphere interaction is discussed in some detail.

1. INTRODUCTION

THE present series of papers is devoted to a method of treating the many-body problem in quantum-statistical mechanics. Much of this work was performed in the summer and the fall of 1956 and has been reported[1] in abbreviated versions before.

From the general formulations of statistical mechanics it is known that the thermodynamical properties of a system can be obtained from its partition function. However, the actual task of evaluating the partition function from the atomic or molecular interactions is both complicated and difficult. In classical statistical mechanics, for a system in the gaseous phase, the problem has been reduced to a series of quadratures through the work of Mayer and others.[2] In quantum mechanics, the enormous difficulty of solving the N-body eigenvalue problem ($N\geq 3$), allows so far only for a systematic method[3] of computing the second virial coefficient, and for computations[4] of the quantum corrections to the classical results. For a discussion of phenomena at very low temperatures (e.g., the problem of Bose-Einstein transition and the problem of the many-body ground-state energy) where quantum effects are dominant, these known methods[5] are not applicable.

The purpose of this and the subsequent papers is to develop a systematic method that is suitable to treat problems in which quantum effects are important. The general procedure followed is to first separate out the effect of the statistics (i.e., Bose-Einstein statistics or Fermi-Dirac statistics) of the quantum-mechanical problem and to express the grand partition function in terms of certain U functions defined in terms of the quantum-mechanical problem with Boltzmann statistics. Such a separation of the effect of statistics is formulated in general in this paper and will be further developed in later papers. It is particularly useful in treating the phenomena of Bose-Einstein condensation.

The second step is to formulate a method whereby these U functions can be computed from a solution of the two-body problem. In effect, the computation of U is through an expansion, loosely speaking, in powers of a function, called the binary kernel, which is obtainable from a solution of the two-body problem. The method is applicable in cases where the two-body interaction may contain a singular repulsive core.

For the gaseous phase the formulation in this paper yields a recipe, much like Mayer's method in classical statistical mechanics, for computing the equation of state through a series of quadratures. The method can also be used to calculate the ground-state energy, to obtain the limiting forms of thermodynamical functions at very low temperatures and to study the problem of Bose-Einstein transition. These topics will be discussed in later papers.

2. SOME DEFINITIONS

We consider an N-particle Hamiltonian

$$H_N=-\sum_{i=1}^{N}\nabla_i^2+\sum_{i>j}V(\mathbf{r}_i-\mathbf{r}_j), \qquad (I.1)$$

where for simplicity units are chosen so that $\hbar=1$ and mass of the particles $=\frac{1}{2}$. Three- and more-particle interactions are not considered, although their inclusion would not introduce real complications in much of the following discussions.

To be specific, the N particles are considered to move in a cubic box of dimensions $L\times L\times L$ with *periodic boundary conditions*. We use the symbol $\Omega=L^3$ for the volume of the box.

We now discuss separately the cases when the

* Work supported in part by the U. S. Atomic Energy Commission.

[1] International Conference on Theoretical Physics at Seattle, September, 1956 (unpublished); *Conference on the Many-Body Problem, Stevens Institute, Hoboken, New Jersey, January, 1957* (to be published). Some preliminary results have been summarized in T. D. Lee and C. N. Yang, Phys. Rev. **105**, 1119 (1957).

[2] H. D. Ursell, Proc. Cambridge Phil. Soc. **23**, 685 (1927). J. E. Mayer, J. Chem. Phys. **5**, 67 (1937); J. E. Mayer *et al.*, J. Chem. Phys. **5**, 74 (1937); **6**, 87 (1938); **6**, 101 (1938).

[3] G. E. Uhlenbeck and E. Beth, Physica **3**, 729 (1936); **4**, 915 (1937). For other references, see, e.g., D. ter Haar, *Elements of Statistical Mechanics* (Rinehart and Company, New York, 1954).

[4] See, e.g., J. de Boer's review article in *Reports on Progress in Physics* (The Physical Society, London, 1949), Vol. 12, p. 305, Sec. 6(III).

[5] See some recent developments: W. B. Riesenfeld and K. M. Watson, Phys. Rev. **104**, 492 (1956); **108**, 518 (1957); A. J. F. Siegert and Ei Teramoto, Phys. Rev. **110**, 1232 (1958); E. W. Montroll and J. C. Ward, Phys. Fluids **1**, 55 (1958).

particles satisfy Boltzmann, Bose-Einstein, and Fermi-Dirac statistics.

A. Boltzmann Statistics

We follow the standard treatment and introduce the operator

$$W_N \equiv \exp(-\beta H_N), \tag{I.2}$$

where

$$\beta = (\kappa T)^{-1}. \tag{I.3}$$

We shall also use the thermal wavelength

$$\lambda = (4\pi\beta)^{\frac{1}{2}}.$$

The matrix elements of W_N in coordinate representation is

$$\langle 1',2'\cdots N'|W_N|1,2\cdots N\rangle = \sum_i \psi_i(1',2'\cdots N')\psi_i^*(1,2\cdots N)\exp(-\beta E_i). \tag{I.4}$$

Here

$$1 \equiv \mathbf{r}_1 \equiv (x_1,y_1,z_1), \text{ etc.,} \quad 1' \equiv \mathbf{r}_1' \equiv (x_1',y_1'z_1'), \text{ etc.,}$$

and $\psi_i(\mathbf{r}_\alpha)$, E_i are the normalized eigenfunctions and eigenvalues of H_N with *periodic boundary conditions* in a cubic box of volume Ω. The summation in (I.4) extends over all eigenfunctions ψ_i. It is useful to notice that the exchange of any pair $\mathbf{r}_i'$, $\mathbf{r}_i$ with $\mathbf{r}_j'$, $\mathbf{r}_j$ leaves W_N unchanged.

The partition function is

$$Q_0 \equiv 1,$$

$$Q_N \equiv \sum_i \exp(-\beta E_i) = \int_\Omega \langle 1,2\cdots N|W_N|1,2\cdots N\rangle d^{3N}r. \tag{I.5}$$

To obtain the logarithm of the grand partition function in a simple form we follow a procedure first introduced by Ursell[2] and by Mayer[2] for classical statistical mechanics and by[6] Kahn and Uhlenbeck for quantum-statistical mechanics. One defines U_l functions by

$$\begin{aligned}
\langle 1'|W_1|1\rangle &\equiv \langle 1'|U_1|1\rangle,\\
\langle 1',2'|W_2|1,2\rangle &\equiv \langle 1'|U_1|1\rangle\langle 2'|U_1|2\rangle + \langle 1',2'|U_2|1,2\rangle,\\
\langle 1',2',3'|W_3|1,2,3\rangle &\equiv \langle 1'|U_1|1\rangle\langle 2'|U_1|2\rangle\langle 3'|U_1|3\rangle\\
&\quad + \langle 1'|U_1|1\rangle\langle 2',3'|U_2|2,3\rangle\\
&\quad + \langle 2'|U_1|2\rangle\langle 1',3'|U_2|1,3\rangle\\
&\quad + \langle 3'|U_1|3\rangle\langle 1',2'|U_2|1,2\rangle\\
&\quad + \langle 1',2'3'|U_3|1,2,3\rangle, \text{ etc.}
\end{aligned} \tag{I.6}$$

Putting $\mathbf{r}_1 = \mathbf{r}_1'$, $\mathbf{r}_2 = \mathbf{r}_2'$ in these equations and integrating over $\mathbf{r}_1$, $\mathbf{r}_2$, $\cdots$, one can show[2,6] that

$$\mathcal{Q}_\Omega \equiv \sum_{N=0}^{\infty} (N!)^{-1} Q_N z^N = \exp\left\{\sum_{l=1}^{\infty} z^l (l!)^{-1} \int_\Omega \langle 1,\cdots,l|U_l|1,\cdots,l\rangle d^{3l}r\right\}. \tag{I.7}$$

For the sake of completeness we give a proof of this formula in Appendix A.

According to the principles of statistical mechanics, the equilibrium pressure p and density ρ of the system are given by

$$\frac{p}{\kappa T} = \lim_{\Omega\to\infty} \Omega^{-1} \ln \mathcal{Q}_\Omega,$$

and

$$\rho = \lim_{\Omega\to\infty} \Omega^{-1} (\partial \ln \mathcal{Q}_\Omega / \partial \ln z). \tag{I.8}$$

Using (I.7) one obtains

$$\frac{p}{\kappa T} = \lim_{\Omega\to\infty} \sum_{l=1}^{\infty} b_l(\Omega) z^l,$$

$$\rho = \lim_{\Omega\to\infty} \sum_{l=1}^{\infty} l b_l(\Omega) z^l, \tag{I.9}$$

where

$$b_l(\Omega) = (l!\Omega)^{-1} \int_\Omega \langle 1,\cdots,l|U_l|1,\cdots,l\rangle d^{3l}r. \tag{I.10}$$

It is important to remember that the W and U functions are defined for fixed Ω. The question of whether $b_l(\Omega)$ approaches a limit as $\Omega \to \infty$ will be discussed in Sec. 5.

B. Bose-Einstein (i.e., Symmetrical) Statistics

For symmetrical statistics, the corresponding function $W_N{}^S$ is

$$\langle 1',2',\cdots,N'|W_N{}^S|1,2,\cdots,N\rangle \equiv N! \sum_{\text{sym. }\psi} \psi_i(1',2',\cdots,N') \times \psi_i^*(1,2,\cdots,N)\exp(-\beta E_i). \tag{I.11}$$

We define

$$Q_0{}^S \equiv 1, \quad Q_N{}^S \equiv \int_\Omega \langle 1,2,\cdots,N|W_N{}^S|1,2,\cdots,N\rangle d^{3N}r.$$

We also define $U_l{}^S$ functions in complete analogy with (I.6):

$$\begin{aligned}
\langle 1'|W_1{}^S|1\rangle &\equiv \langle 1'|U_1{}^S|1\rangle,\\
\langle 1',2'|W_2{}^S|1,2\rangle &\equiv \langle 1'|U_1{}^S|1\rangle\langle 2'|U_1{}^S|2\rangle + \langle 1',2'|U_2{}^S|1,2\rangle, \text{ etc.}
\end{aligned} \tag{I.12}$$

---

[6] B. Kahn and G. E. Uhlenbeck, Physica 5, 399 (1938).

The grand partition function is

$$\mathcal{Q}_\Omega{}^S \equiv \sum (N!)^{-1} Q_N{}^S z^N$$
$$= \exp\left\{\sum_{l=1}^{\infty} (l!)^{-1} z^l \int_\Omega \langle 1,\cdots,l|U_l{}^S|1,\cdots,l\rangle d^{3l}r\right\}, \quad (I.13)$$

which is proved in the same way that (I.7) was proved. One obtains then again (I.9) with b_l replaced by $b_l{}^S$, where

$$b_l{}^S(\Omega) = (l!\Omega)^{-1} \int \langle 1,\cdots,l|U_l{}^S|1,\cdots,l\rangle d^{3l}r. \quad (I.14)$$

C. Fermi-Dirac (i.e., Antisymmetrical) Statistics

For antisymmetrical statistics one has

$$\langle 1',2',\cdots,N'|W_N{}^A|1,2,\cdots,N\rangle = \sum_{\text{antisym. }\psi} \psi_i(1',2',\cdots,N') \times \psi_i{}^*(1,2,\cdots,N) \exp(-\beta E_i). \quad (I.15)$$

Equations (I.12), (I.13), and (I.14) remain the same in form if one replaces all superscripts S (for symmetrical statistics) by A (for antisymmetrical statistics).

In the following we list some simple examples to illustrate these definitions:

Example 1:

$$\langle 1'|W_1|1\rangle = \langle 1'|W_1{}^S|1\rangle = \langle 1'|W_1{}^A|1\rangle = \langle 1'|U_1|1\rangle = \sum_{\mathbf{k}} \Omega^{-1} \exp[i\mathbf{k}\cdot(\mathbf{r}_1-\mathbf{r}_1') - \beta \mathbf{k}^2], \quad (I.16)$$

where the summation extends over $\mathbf{k}=2\pi L^{-1}(l,m,n)$ with l, m, $n=0$ and $\pm$ integers. In the limit $\Omega \to \infty$,

$$\langle 1'|U_1|1\rangle \to (8\pi^3)^{-1} \int d^3k \exp[i\mathbf{k}\cdot(\mathbf{r}_1-\mathbf{r}_1') - \beta\mathbf{k}^2]$$
$$= \lambda^{-3} \exp[-(\mathbf{r}_1-\mathbf{r}_1')^2/(4\beta)]. \quad (I.17)$$

One easily computes b_1:

$$b_1(\Omega) = b_1{}^S(\Omega) = b_1{}^A(\Omega) = \sum_{\mathbf{k}} \Omega^{-1} \exp(-\beta \mathbf{k}^2). \quad (I.18)$$

In the limit $\Omega \to \infty$, one obtains

$$b_1(\Omega) \to \lambda^{-3}. \quad (I.19)$$

Example 2:

For free particles it is clear from the definition (I.4) that

$$\langle 1',\cdots,N'|W_N|1,\cdots,N\rangle = \langle 1'|W_1|1\rangle\langle 2'|W_1|2\rangle\cdots\langle N'|W_1|N\rangle.$$

Hence one obtains from (I.6)

$$U_2=0, \quad U_3=0, \quad \text{etc.} \quad (I.20)$$

Example 3:

For Bose-Einstein statistics, the wave functions for two free particles are

$$\psi = \Omega^{-1} \exp[i\mathbf{k}\cdot\mathbf{r}_1 + i\mathbf{k}\cdot\mathbf{r}_2],$$

and

$$\psi = 2^{-\frac{1}{2}}\Omega^{-1}[\exp(i\mathbf{k}_a\cdot\mathbf{r}_1 + i\mathbf{k}_b\cdot\mathbf{r}_2) + \exp(i\mathbf{k}_b\cdot\mathbf{r}_1 + i\mathbf{k}_a\cdot\mathbf{r}_2)], \quad (\mathbf{k}_a \neq \mathbf{k}_b).$$

Substituting these into (I.11), one obtains

$$\langle 1',2'|W_2{}^S|1,2\rangle = \langle 1'|W_1|1\rangle\langle 2'|W_1|2\rangle + \langle 2'|W_1|1\rangle\langle 1'|W_1|2\rangle.$$

Comparison with (I.12) shows therefore that for free particles

$$\langle 1',2'|U_2{}^S|1,2\rangle = \langle 2'|W_1|1\rangle\langle 1'|W_1|2\rangle. \quad (I.21)$$

Similarly one finds

$$\langle 1',2'|U_2{}^A|1,2\rangle = -\langle 2'|W_1|1\rangle\langle 1'|W_1|2\rangle. \quad (I.22)$$

3. $U_N{}^S$ AND $U_N{}^A$ IN TERMS OF U_N

In the last section we wrote down the main formulas in the Ursell-Mayer-Kahn-Uhlenbeck treatment of the equations of state. To calculate the coefficients b_l, $b_l{}^S$, and $b_l{}^A$ one first calculates the functions U_l, $U_l{}^S$, and $U_l{}^A$. Now the functions $U_l{}^S$ and $U_l{}^A$ are considerably more complicated than U_l. [We saw, e.g., in (I.20) and (I.21) that for free particles, U_2, $U_3\cdots$ vanish, but not $U_2{}^S$ and $U_2{}^A$.] In this section we shall formulate explicit rules by which $U_N{}^S$ and $U_N{}^A$ can be computed once the functions U_N are known.

Such rules exist because the U's are defined in terms of the W's, and the W's as defined in (I.4), (I.11), and (I.15) are related through the equations

$$\langle 1',\cdots,N'|W_N{}^S|1,\cdots,N\rangle = \sum_{P'} P'\langle 1',\cdots,N'|W_N|1,\cdots,N\rangle, \quad (I.23)$$

and

$$\langle 1',\cdots,N'|W_N{}^A|1,\cdots,N\rangle = \sum_{P'} \mathcal{C}_{P'} P'\langle 1',\cdots,N'|W_N|1,\cdots,N\rangle, \quad (I.24)$$

where

P' = any one of $N!$ operators that permute the variables $\mathbf{r}_1'$, $\mathbf{r}_2'$, $\cdots$, $\mathbf{r}_N'$, (I.25)

and

$$\mathcal{C}_{P'}=1 \quad \text{for even permutations } P', \qquad \mathcal{C}_{P'}=-1 \quad \text{for odd permutations } P'. \quad (I.26)$$

Equations (I.23) and (I.24) are proved in Appendix B.

Starting from U_l one can construct W_l from (I.6). Equation (I.23) then enables one to compute $W_l{}^S$ which in turn leads to a computation of $U_l{}^S$ through (I.12). The details of these procedures appear in Appendix C. Here we only state the results as:

Rule A.—To calculate $U_l{}^S$ we first distribute the l integers 1, 2, $\cdots l$ into m_α groups each containing α integers, with $\sum_\alpha m_\alpha \alpha = l$. Such a grouping may be represented as follows:

$$\{(a)(b)\cdots\}\{(cd)(ef)\cdots\}\{(ghi)\cdots\}\cdots, \quad (I.27)$$

where a, b, c, $\cdots$ are the various integers. In the first curly bracket there are m_1 round brackets with one integer in each ($m_1=0, 1, 2\cdots$), and in the second curly bracket there are m_2 round brackets with two integers in each ($m_2=0, 1, 2\cdots$), etc. Within each round bracket the integers are arranged in ascending order. Within each curly bracket the round brackets are arranged such that their first integers follow an ascending sequence.

We then form the sum

$$\sum\{\langle\mathfrak{a}'|U_1|a\rangle\langle\mathfrak{b}'|U_1|b\rangle\cdots\}\times\{\langle\mathfrak{c}',\mathfrak{d}'|U_2|c,d\rangle\langle\mathfrak{e}',\mathfrak{f}'|U_2|e,f\rangle\cdots\}\cdots, \quad (I.28)$$

where $\mathfrak{a}'$, $\mathfrak{b}'$, $\cdots\mathfrak{c}'$, $\mathfrak{d}'$, $\mathfrak{e}'$, $\mathfrak{f}'$, $\cdots$ is any permutation of the coordinates $1'$, $2'\cdots l'$. The summation in (I.28) extends over all such permutations of $1'$, $2'\cdots l'$ which satisfy the condition [see examples (1) and (3) below] that upon putting $\mathbf{r}_i=\mathbf{r}_i'$ (all i), the summand cannot be written as a product of two factors, one of which depends only on some, but not all, of the coordinates $\mathbf{r}_1$, $\mathbf{r}_2$, $\cdots\mathbf{r}_l$, while the other depends only on the rest of these coordinates. We then sum up all expressions (I.28) over the different groupings (I.27). This total sum is equal to U_l^S.

Rule B.—To calculate U_l^A we proceed in exactly the same way as rule A, with only the following change: we replace (I.28) by

$$\sum\mathfrak{C}\{\langle\mathfrak{a}'|U_1|a\rangle\langle\mathfrak{b}'|U_1|b\rangle\cdots\}\times\{\langle\mathfrak{c}',\mathfrak{d}'|U_2|c,d\rangle\cdots\}\cdots, \quad (I.29)$$

where

$$\mathfrak{C}=+1 \quad \text{if the permutation} \begin{pmatrix} a & b & c & d & \cdots \\ \mathfrak{a} & \mathfrak{b} & \mathfrak{c} & \mathfrak{d} & \cdots \end{pmatrix} \text{is even,}$$

and

$$\mathfrak{C}=-1 \quad \text{if it is odd.}$$

The meaning of these rules is actually quite simple even though their statements appear to be so long. We quote a few examples.

Example 1:

$$U_2^S=\langle 2'|U_1|1\rangle\langle 1'|U_1|2\rangle+\langle 1',2'|U_2|1,2\rangle+\langle 2',1'|U_2|1,2\rangle.$$

The term $\langle 1'|U_1|1\rangle\langle 2'|U_1|2\rangle$ is not included because when

$$\mathbf{r}_1=\mathbf{r}_1', \quad \mathbf{r}_2=\mathbf{r}_2'$$

it splits into two factors, one of which depends only on $\mathbf{r}_1$, the other only on $\mathbf{r}_2$.

Example 2:

$$U_2^A=-\langle 2'|U_1|1\rangle\langle 1'|U_1|2\rangle+\langle 1',2'|U_2|1,2\rangle-\langle 2',1'|U_2|1,2\rangle.$$

Example 3:

A term like

$$\langle 2'|U_1|5\rangle\langle 1',6'|U_2|1,8\rangle\langle 5',3'|U_2|2,3\rangle\langle 8',7',4'|U_3|4,6,7\rangle$$

is not to be included in U_8^S because upon putting $\mathbf{r}_i=\mathbf{r}_i'$ (all i) the first and third factors together depend only on $\mathbf{r}_2$, $\mathbf{r}_3$, and $\mathbf{r}_5$ while the other two together only on $\mathbf{r}_1$, $\mathbf{r}_4$, $\mathbf{r}_6$, $\mathbf{r}_7$, and $\mathbf{r}_8$.

Example 4:

It is easy to see that for the case of Bose-Einstein statistics the combination

$$\langle 1',\cdots,l'|\Upsilon_l^S|1,\cdots,l\rangle \equiv\sum_{P'} P'\langle 1',\cdots,l'|U_l|1,\cdots,l\rangle \quad (I.30)$$

always occurs instead of U_l alone. By introducing (I.30) one could simplify the formulas for U_l^S. Developments along these lines have been pursued and led to calculations of the transition point in a Bose-Einstein gas with interactions. These developments will be presented in a later paper.

Example 5:

Similarly, for the case of Fermi-Dirac statistics, the combination

$$\langle 1',\cdots,l'|\Upsilon_l^A|1,\cdots,l\rangle \equiv\sum_{P'} \mathfrak{C}_{P'}P'\langle 1',\cdots,l'|U_l|1,\cdots,l\rangle \quad (I.31)$$

always occurs instead of U_l alone.

Example 6:

For free particles, using (I.20), one sees that U_l^S is equal to a sum of products of l U_1 functions. It is easy to prove that there are $(l-1)!$ terms in the sum. Equation (I.14) therefore leads to

$$b_l^S(\Omega) =l^{-1}\Omega^{-1}\int\langle 1|U_1|2\rangle\langle 2|U_1|3\rangle\langle 3|U_1|4\rangle\cdots\langle l|U_1|1\rangle d^{3l}r =l^{-1}\Omega^{-1}\,\text{trace}\,(U_1)^l.$$

Now by (I.16) the momentum representation of U_1 is

$$\langle\mathbf{k}'|U_1|\mathbf{k}\rangle=\delta_{\mathbf{k}\mathbf{k}'}\exp(-\beta\mathbf{k}^2). \quad (I.32)$$

Hence

$$b_l^S(\Omega)=l^{-1}\Omega^{-1}\sum_{\mathbf{k}}\exp(-l\beta\mathbf{k}^2).$$

As $\Omega\to\infty$,

$$b_l^S(\Omega)\to l^{-1}(8\pi^3)^{-1}\int\exp(-l\beta\mathbf{k}^2)d^3k=\lambda^{-3}l^{-\frac{5}{2}}, \quad (I.33)$$

which agrees with well-known results. One obtains similarly

$$b_l^A(\Omega)\to(-1)^{l-1}\lambda^{-3}l^{-\frac{5}{2}}. \quad (I.34)$$

4. U_l IN TERMS OF THE BINARY KERNEL

The functions U_l will be expressed in this section in terms of a function B, called the binary kernel, which is calculable from a solution of the two-particle problem.

We treat W_N, U_N as *operators* and write

$$W_N(\beta)=\exp(-\beta H_N). \quad (I.35)$$

Here we have explicitly indicated the β dependence of W_N. Writing $H_N=T_N+V_N$, where T_N and V_N are, respectively, the operators for the kinetic and potential energies, one notices that

$$W_N^0(\beta)\equiv\exp(-\beta T_N)=\prod_{i=1}^{N} w(\beta;i),$$

where

$$w(\beta;i)\equiv\exp(\beta\nabla_i^2). \tag{I.36}$$

The explicit form of $w(\beta;1)=\langle 1'|W_1|1\rangle$ is given by (I.16). Thus $W_N^0(\beta)$ is a product of N operators each of which operates on the coordinates of one particle. If V is finite, one can expand W_N into an exponential series in powers of V:

$$W_N(\beta)=W_N^0(\beta)+\int_0^\beta W_N^0(\beta-\beta')(-V_N)W_N^0(\beta')d\beta'$$
$$+\int_0^\beta d\beta'\int_0^{\beta'} d\beta''\, W_N^0(\beta-\beta')(-V_N)$$
$$\times W_N^0(\beta'-\beta'')(-V_N)W_N^0(\beta'')+\cdots. \tag{I.37}$$

If V is $+\infty$ for some configurations, this series ceases to be meaningful. For the time being we regard V as finite everywhere, and shall allow for the possibility of V going to $+\infty$ *after* some rearrangements of terms to be discussed later.

It is of great convenience to represent the sum in (I.37) by diagrams. We shall represent $W_N(\beta)$ as a sum of operators, each corresponding to a different diagram which consists of N vertical lines ii' connecting the points i with i' $(i=1,\cdots,N)$. The points i are all on the same horizontal base level. All the vertical lines ii' are of the same length β and are linked by some horizontal links, *no two of which are at the same height* (i.e., the vertical distances between the horizontal base level and different links are different).

In Fig. 1 we give some examples of the diagrams for W_1, W_2, and W_3. To specify a diagram, one does not specify the exact heights of the horizontal links (since they are to be integrated over), only the vertical sequence in which they are drawn. Thus, for example, in Fig. 1 the sixth and the seventh diagrams for $W_3(\beta)$ are counted as different diagrams.

To obtain the operator that corresponds to a diagram, one proceeds as follows:

A line segment of length γ along the vertical line ii' stands for the operator $w(\gamma;i)=\exp(\gamma\nabla_i^2)$. A horizontal link between ii' and jj' represents the operator $-V(\mathbf{r}_i-\mathbf{r}_j)d\beta'$ where β' is the height of the link above the base line. Multiplying all the operators represented by all the vertical line segments and by the horizontal links, and integrating over the heights β' of the various horizontal links, one obtains the operator represented by the diagram. Two important further rules must be followed:

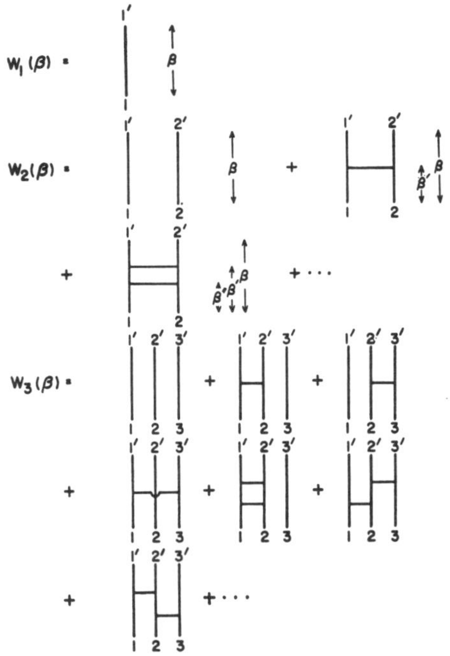

FIG. 1. Diagram representation of W_N as a power series in V. W_3 has altogether 9 diagrams with 2 horizontal links, 27 diagrams with 3 horizontal links, etc. The operators that correspond to these diagrams are given by Eqs. (I.38), (I.39), and (I.40).

(a) The order in which the operators stand must be such that those representing line elements lower down in the diagram stand to the right of those representing line elements higher up in the diagram.

(b) The limits of integration of the heights β' of the horizontal links are defined by the conditions that $\beta\geqq\beta'\geqq 0$, and that the relative height of any two links remain of the same sign within the limits of integration as in the diagram.

Explicitly, the diagram for $W_1(\beta)$ in Fig. 1 corresponds to the operator

$$W_1(\beta)=w(\beta;1). \tag{I.38}$$

The first three diagrams for $W_2(\beta)$ in Fig. 1 correspond to, respectively, the first three terms in the sum

$$W_2(\beta)=w(\beta;1)w(\beta;2)+\int_0^\beta w(\beta-\beta';1)w(\beta-\beta';2)$$
$$\times(-V_{12})w(\beta';1)w(\beta';2)d\beta'+\int_0^\beta d\beta'\int_0^{\beta'} d\beta''$$
$$\times w(\beta-\beta';1)w(\beta-\beta';2)(-V_{12})w(\beta'-\beta'';1)$$
$$\times w(\beta'-\beta'';2)(-V_{12})w(\beta'';1)w(\beta'';2)+\cdots. \tag{I.39}$$

Similarly, the first three diagrams for $W_3(\beta)$ in Fig. 1 correspond, respectively, to the first three terms in the

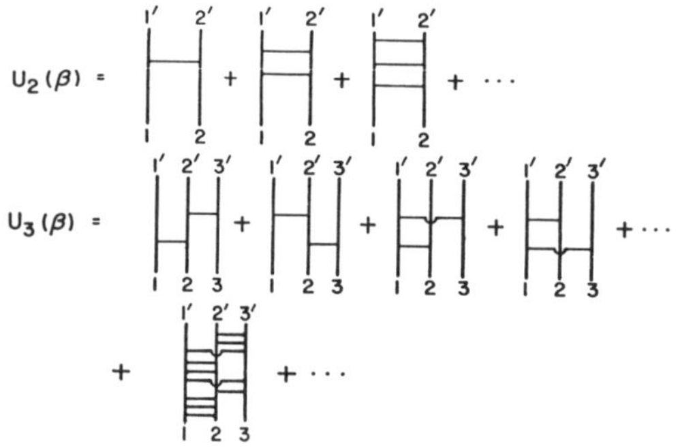

FIG. 2. Diagram representation of U_N as a power series in V. U_3 has altogether 6 diagrams with 2 horizontal links and 24 diagrams with 3 horizontal links, etc. [See Eq. (I.42).]

following sum:

$$W_3(\beta)=w(\beta;1)w(\beta;2)w(\beta;3)+\left[\int_0^\beta w(\beta-\beta';1)\right.$$

$$\left.\times w(\beta-\beta';2)(-V_{12})w(\beta';1)w(\beta';2)d\beta'\right]w(\beta;3)$$

$$+w(\beta;1)\left[\int_0^\beta w(\beta-\beta';2)w(\beta-\beta';3)(-V_{23})\right.$$

$$\left.\times w(\beta';2)w(\beta';3)d\beta'\right]+\cdots. \quad \text{(I.40)}$$

In terms of these diagrams, Eq. (I.37) becomes

$$W_N(\beta)=\sum(\text{all different diagrams with the parameter } \beta \text{ and with } N \text{ particles}). \quad \text{(I.41)}$$

As is clear from the examples illustrated above, some of these diagrams may have unconnected parts. The unconnected parts of a diagram represent commuting operators whose product is the operator corresponding to the whole diagram.

Let us define a "connected diagram" to be one in which all parts are connected through the vertical lines and the horizontal links. By comparing Eq. (I.6) with Eqs. (I.39) and (I.40) (or with their corresponding diagrams in Fig. 1), one finds that U_2 is the sum of all "connected diagrams" in $W_2(\beta)$ and U_3 is the sum of all "connected diagrams" in $W_3(\beta)$. This property is illustrated in Fig. 2.

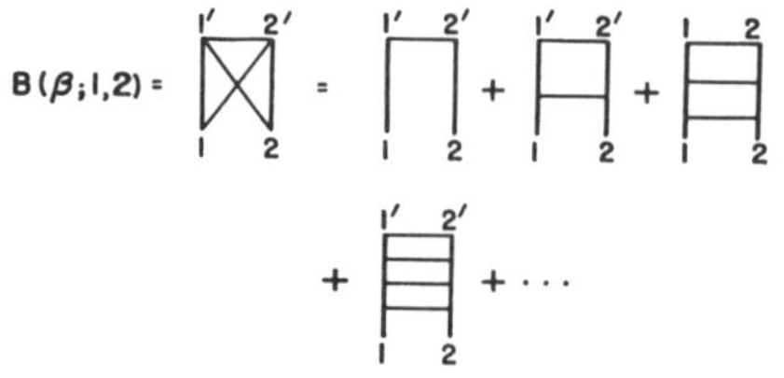

FIG. 3. Diagram representation of B.

In fact, one sees in general that

$$U_N(\beta)=\sum(\text{all different "connected diagrams" with the parameter } \beta \text{ and with } N \text{ particles}). \quad \text{(I.42)}$$

A comparison of (I.41) and (I.42) shows that the diagrams that contribute to $W_N(\beta)$ but not to $U_N(\beta)$ are the unconnected ones which are grouped in the Ursell expansion (I.6) into products of U_l with values of $l<N$. We shall see in the next section and in Appendix E that the connectedness of the diagrams in $U_N(\beta)$ also determines the behavior of $U_N(\beta)$ as the positions of the particles become far distant from each other.

It is convenient to give an explicit operator form for U_2. Using (I.6) and (I.35), we find

$$U_2(\beta)=\exp(-\beta H_2)-\exp(+\beta\nabla_1^2)\exp(+\beta\nabla_2^2). \quad \text{(I.43)}$$

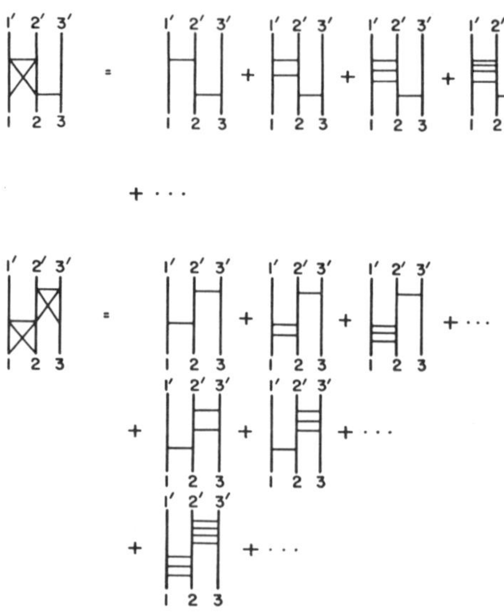

FIG. 4. Examples of diagrams which contain B as part of their structure.

We shall now define the binary kernel $B(\beta;1,2)$:

$$B(\beta;1,2)\equiv(-V_{12})W_2(\beta)=(-V_{12})\exp(-\beta H_2). \quad \text{(I.44)}$$

From (I.43) one obtains, by differentiating with respect to β,

$$B(\beta;1,2)=\frac{\partial U_2(\beta)}{\partial\beta}-(\nabla_1^2+\nabla_2^2)U_2(\beta). \quad \text{(I.45)}$$

It is important to notice that by using the solutions of the two-body problem (in a box Ω) one can compute $\exp(-\beta H_2)$. Equations (I.43) and (I.45) then lead to a computation of $U_2(\beta)$ and $B(\beta;1,2)$. Furthermore, these two equations do not contain V explicitly. Therefore, the explicit form of $B(\beta;1,2)$ can be evaluated even if $V=+\infty$ for some spatial configurations of $\mathbf{r}_1$ and $\mathbf{r}_2$. The example of hard spheres will be given in a later section.

Substituting the diagrams for W_2 in Fig. 1 into (I.44), one is led to the representation of B in terms of the diagrams of Fig. 3. The top horizontal links in the last four diagrams represent factors $(-V_{12})$. With this definition it is clear that any group of graphs with a part that has the same form as these diagrams can be summed to yield a factor B. In Fig. 4 we give two such examples.

We have seen before that U_N is equal to a sum of connected diagrams. The sum can be rearranged and grouped together in the same manner as the two examples in Fig. 4. One then obtains U_N as a sum of diagrams in which only B appears with no isolated horizontal links. In Fig. 5 we express $U_2(\beta)$, $U_3(\beta)$, etc. in terms of sums of such diagrams.

From these diagrams the explicit form of U_l in terms of the binary kernel B can be readily written down:

$$U_2(\beta)=\int_0^\beta d\beta'\, w(\beta-\beta';1)w(\beta-\beta';2)B(\beta';1,2), \quad \text{(I.46)}$$

$$\begin{aligned}U_3(\beta)=&\int_0^\beta d\beta'\int_0^{\beta'} d\beta''\, w(\beta-\beta'';1)w(\beta-\beta';2)\\ &\times w(\beta-\beta';3)B(\beta'-\beta'';2,3)B(\beta'';1,2)w(\beta'';3)\\ &+\int_0^\beta d\beta'\int_0^{\beta'} d\beta''\, w(\beta-\beta';1)w(\beta-\beta';2)\\ &\times w(\beta-\beta'';3)B(\beta'-\beta'';1,2)B(\beta'';2,3)w(\beta'';1)\\ &+\text{four other terms of order } B^2\\ &+\text{terms of higher orders in } B. \quad \text{(I.47)}\end{aligned}$$

$$\begin{aligned}U_4(\beta)=&\int_0^\beta d\beta'\int_0^{\beta'} d\beta''\int_0^{\beta''} d\beta'''\, w(\beta-\beta';1)\\ &\times w(\beta-\beta''';2)w(\beta-\beta'';3)w(\beta-\beta';4)\\ &\times B(\beta'-\beta'';1,4)B(\beta''-\beta''';3,4)\\ &\times B(\beta''';2,3)w(\beta'';1)w(\beta''';4)\\ &+95 \text{ other terms of order } B^3\\ &+\text{terms of higher order in } B. \quad \text{(I.48)}\end{aligned}$$

$$U_5(\beta)=\cdots. \quad \text{(I.49)}$$

In (I.47) the first and the second operator on the right-hand side represent, respectively, the first and the second diagram in the corresponding sum for U_3 in Fig. 5, and the first operator on the right-hand side of (I.48) represents the corresponding first diagram for U_4 in Fig. 5.

These equations express U_l for $l\geq 3$ as a sum of integrals of products of w and B. Now B vanishes for free particles. It characterizes the perturbation on W_2 due to the interactions. The expansion in Fig. 5 and in (I.47), (I.48), (I.49) for U_l are thus expansions in terms of increasingly higher order effects of such a perturbation. The convergence of such expansions is not clearly understood by the authors. It is, however, hoped that for interactions for which three-particle bound states do not exist these expansions do converge.

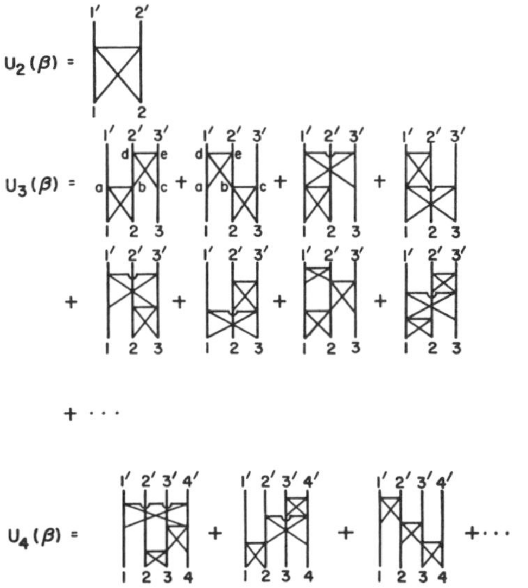

FIG. 5. Diagram representation of U_N in terms of the binary kernel B. In these diagrams there is no horizontal link (which corresponds to V). The vertical lines are all connected through structures representing B. The operators that correspond to these diagrams are given by Eqs. (I.46)–(I.48).

Equations (I.46)–(I.48) are operator equations. For illustration we give the explicit matrix element of, say, $U_3(\beta)$ between the states $\langle 1',2',3'|$ and $|1,2,3\rangle$ in the coordinate representation. From (I.47) we obtain

$$\begin{aligned}&\langle \mathbf{r}_1',\mathbf{r}_2',\mathbf{r}_3'|U_3|\mathbf{r}_1,\mathbf{r}_2,\mathbf{r}_3\rangle\\ &=\int_0^\beta d\beta'\int_0^{\beta'} d\beta''\int d^3\mathbf{r}_a d^3\mathbf{r}_b d^3\mathbf{r}_c d^3\mathbf{r}_d d^3\mathbf{r}_e\\ &\quad\times\langle\mathbf{r}_1'|W_1(\beta-\beta'')|\mathbf{r}_a\rangle\langle\mathbf{r}_2'|W_1(\beta-\beta')|\mathbf{r}_d\rangle\\ &\quad\times\langle\mathbf{r}_3'|W_1(\beta-\beta')|\mathbf{r}_e\rangle\langle\mathbf{r}_d,\mathbf{r}_e|B(\beta'-\beta'')|\mathbf{r}_b,\mathbf{r}_c\rangle\\ &\quad\times\langle\mathbf{r}_a,\mathbf{r}_b|B(\beta'')|\mathbf{r}_1,\mathbf{r}_2\rangle\langle\mathbf{r}_c|W_1(\beta'')|\mathbf{r}_3\rangle\\ &\quad+\int_0^\beta d\beta'\int_0^{\beta'} d\beta''\int d^3\mathbf{r}_a\cdots\\ &\quad\times d^3\mathbf{r}_e\langle\mathbf{r}_1'|W_1(\beta-\beta')|\mathbf{r}_d\rangle\langle\mathbf{r}_2'|W_1(\beta-\beta')|\mathbf{r}_e\rangle\\ &\quad\times\langle\mathbf{r}_3'|W_1(\beta-\beta'')|\mathbf{r}_c\rangle\langle\mathbf{r}_d,\mathbf{r}_e|B(\beta'-\beta'')|\mathbf{r}_a,\mathbf{r}_b\rangle\\ &\quad\times\langle\mathbf{r}_b,\mathbf{r}_c|B(\beta'')|\mathbf{r}_2,\mathbf{r}_3\rangle\langle\mathbf{r}_a|W_1(\beta'')|\mathbf{r}_1\rangle+\cdots. \quad \text{(I.50)}\end{aligned}$$

where the first two terms again correspond to the first two diagrams for U_3 in Fig. 5. The five corners in each of these two diagrams are denoted by a, b, c, d, and e.

In (I.50) we denote the spatial coordinates of these five corners by $\mathbf{r}_a, \mathbf{r}_b, \cdots \mathbf{r}_e$ and the heights of the horizontal lines ab, de by β'' and β', respectively.

5. THE LIMIT $\Omega\to\infty$. MOMENTUM REPRESENTATION

The results of the last three sections hold for arbitrary but finite values of Ω. If one starts with the Hamiltonian (I.1), but without the periodicity boundary condition, one is dealing[7] with the case $\Omega=\infty$. One can then still define the functions W_N, $W_N{}^S$, and $W_N{}^A$ by (I.4), (I.11), and (I.15). Also the definitions (I.6), (I.12) for the U_l, $U_l{}^S$, and $U_l{}^A$ functions are unchanged. All the discussions and results of Secs. 3 and 4 apply to the case $\Omega=\infty$ as well as to the case of finite Ω. However, the discussions in Sec. 2 concerning the partition function, the grand partition function, and the thermodynamical behavior of the system cannot apply to a system with $\Omega=\infty$ (for which the partition function is clearly ∞).

To emphasize the dependence on Ω we shall in the rest of this section add the inferior indices Ω and ∞ to indicate the cases of finite Ω and infinite Ω, respectively.

One can rewrite (I.10) in the form [to be proved in Appendix D]:

$$b_l(\Omega)=(l!)^{-1}\int_\Omega \langle 0,\mathbf{r}_2,\mathbf{r}_3,\cdots\mathbf{r}_l|U_{l\Omega}|0,\mathbf{r}_2,\mathbf{r}_3,\cdots\mathbf{r}_l\rangle \times d^3r_2d^3r_3\cdots d^3r_l. \quad \text{(I.51)}$$

Similarly one has identical equations for $b_l{}^S$ and $b_l{}^A$ with $U_{l\Omega}{}^S$ and $U_{l\Omega}{}^A$ replacing $U_{l\Omega}$. In (I.51) the region of integration of $\mathbf{r}_2, \mathbf{r}_3, \cdots \mathbf{r}_l$ is the box

$$\tfrac{1}{2}L\geqq x_i\geqq -\tfrac{1}{2}L,\quad \tfrac{1}{2}L\geqq y_i\geqq -\tfrac{1}{2}L,$$
$$\tfrac{1}{2}L\geqq z_i\geqq -\tfrac{1}{2}L;\quad i\geqq 2;\quad (\Omega=L^3). \quad \text{(I.52)}$$

It will be demonstrated in Appendix E that as $\Omega\to\infty$, the matrix elements of all the $W_{N\Omega}$, $U_{N\Omega}$, $W_{N\Omega}{}^S$, etc., operators for fixed $\mathbf{r}_1, \mathbf{r}_2, \cdots \mathbf{r}_N$ and $\mathbf{r}_1', \mathbf{r}_2', \cdots \mathbf{r}_N'$ approach the matrix elements of the corresponding operators at $\Omega=\infty$. Furthermore, it will also be demonstrated there that from (I.51) one obtains, as $\Omega\to\infty$,

$$b_l(\Omega)\to b_l(\infty)$$
$$\equiv (l!)^{-1}\int \langle 0,\mathbf{r}_2,\mathbf{r}_3,\cdots\mathbf{r}_l|U_{l\infty}|0,\mathbf{r}_2,\cdots\mathbf{r}_l\rangle \times d^3r_2d^3r_3\cdots d^3r_l,$$

$$b_l{}^S(\Omega)\to b_l{}^S(\infty)$$
$$\equiv (l!)^{-1}\int \langle 0,\mathbf{r}_2,\mathbf{r}_3,\cdots\mathbf{r}_l|U_{l\infty}{}^S|0,\mathbf{r}_2,\mathbf{r}_3,\cdots\mathbf{r}_l\rangle \times d^3r_2d^3r_3\cdots d^3r_l, \quad \text{(I.53)}$$

$$b_l{}^A(\Omega)\to b_l{}^A(\infty)$$
$$\equiv (l!)^{-1}\int \langle 0,\mathbf{r}_2,\mathbf{r}_3,\cdots\mathbf{r}_l|U_{l\infty}{}^A|0,\mathbf{r}_2,\cdots\mathbf{r}_l\rangle \times d^3r_2d^3r_3\cdots d^3r_l,$$

the limits of integration being $-\infty$ to ∞. The U_l functions for $\Omega=\infty$ are therefore useful for calculating the limits of b_l, $b_l{}^S$, and $b_l{}^A$ as $\Omega\to\infty$.

To summarize the results of these sections: One has a program[8] of computing the limits $b_l(\infty)$, $b_l{}^S(\infty)$, and $b_l{}^A(\infty)$ using exclusively quantities defined for $\Omega=\infty$. The method is usable even if the two-body interaction $V(r)$ is equal to $+\infty$ for some values of $\mathbf{r}$. The quantities to be computed in successive steps in the program are: first, $W_{2\infty}$ by solving the two-body problem; second, B_∞ through (I.43) and (I.45); third, $U_{l\infty}$ through (I.46)–(I.49); fourth, $U_{l\infty}{}^S$ and $U_{l\infty}{}^A$ through the rules A and B of Sec. 3; and last, $b_l(\infty)$, $b_l{}^S(\infty)$, and $b_l{}^A(\infty)$ through (I.53).

In carrying out this program it is sometimes convenient to use the momentum representation. One notices first that (I.43), (I.45), and (I.46)–(I.49) as operator equations are of course valid in any representation. As to rules A and B in Sec. 3, if one understands 1, 2, $\cdots$ to mean $\mathbf{k}_1, \mathbf{k}_2, \cdots$ the rules remain valid and in fact give the momentum representation of $U_l{}^S$ and $U_l{}^A$ in terms of those of U_l. Consequently (I.43), (I.45)–(I.49), and rules A and B are applicable to both coordinate and momentum representations, and to the case of $\Omega=$finite and $\Omega=\infty$.

To compute $b_l(\infty)$ etc. from the matrix elements of U_l in momentum representation, one writes

$$\langle \mathbf{k}_1',\mathbf{k}_2'\cdots\mathbf{k}_l'|U_{l\infty}|\mathbf{k}_1,\mathbf{k}_2\cdots\mathbf{k}_l\rangle \equiv \delta^3(\textstyle\sum \mathbf{k}_\alpha-\sum \mathbf{k}_\alpha')\langle \mathbf{k}_1'\cdots\mathbf{k}_l'|u_l|\mathbf{k}_1\cdots\mathbf{k}_l\rangle,$$
$$\langle \mathbf{k}_1',\mathbf{k}_2'\cdots\mathbf{k}_l'|U_{l\infty}{}^S|\mathbf{k}_1,\mathbf{k}_2\cdots\mathbf{k}_l\rangle \equiv \delta^3(\textstyle\sum \mathbf{k}_\alpha-\sum \mathbf{k}_\alpha')\langle \mathbf{k}_1'\cdots\mathbf{k}_l'|u_l{}^S|\mathbf{k}_1\cdots\mathbf{k}_l\rangle, \quad \text{(I.54)}$$
$$\langle \mathbf{k}_1',\mathbf{k}_2'\cdots\mathbf{k}_l'|U_{l\infty}{}^A|\mathbf{k}_1,\mathbf{k}_2\cdots\mathbf{k}_l\rangle \equiv \delta^3(\textstyle\sum \mathbf{k}_\alpha-\sum \mathbf{k}_\alpha')\langle \mathbf{k}_1'\cdots\mathbf{k}_l'|u_l{}^A|\mathbf{k}_1,\cdots\mathbf{k}_l\rangle.$$

The presence of the δ function is a consequence of the conservation of momentum for the Hamiltonian (I.1). The functions u_l etc. are defined through (I.54) only for those values of $\mathbf{k}_\alpha'$ and $\mathbf{k}_\alpha$ which satisfy $\sum \mathbf{k}_\alpha' = \sum \mathbf{k}_\alpha$. Now for $\Omega=\infty$

$$\langle \mathbf{k}_1\cdots\mathbf{k}_l|\mathbf{r}_1\cdots\mathbf{r}_l\rangle=(8\pi^3)^{-l/2}\exp\left[-\sum_1^l i\mathbf{k}_\alpha\cdot\mathbf{r}_\alpha\right].$$

Hence

$$\langle \mathbf{r}_1'\cdots\mathbf{r}_l'|U_{l\infty}|\mathbf{r}_1\cdots\mathbf{r}_l\rangle=\int\langle \mathbf{k}_1'\cdots\mathbf{k}_l'|U_{l\infty}|\mathbf{k}_1\cdots\mathbf{k}_l\rangle \times d^{3l}k'd^{3l}k(8\pi^3)^{-l}\exp(i\textstyle\sum \mathbf{k}_\alpha'\cdot\mathbf{r}_\alpha'-i\sum \mathbf{k}_\alpha\cdot\mathbf{r}_\alpha).$$

[7] This approach to the problem at $\Omega=\infty$ is along the lines of Dirac's representation theory [P. A. M. Dirac, *The Principles of Quantum Mechanics* (Oxford University Press, Oxford, 1947)]. An alternative way of defining the W_N functions for $\Omega=\infty$ appears in Appendix E.

[8] The question of whether

$$\lim_{\Omega\to\infty}\sum b_l(\Omega)z^l=\sum b_l(\infty)z^l$$

is not discussed here. However, see C. N. Yang and T. D. Lee, Phys. Rev. 87, 404 (1952).

Using (I.53) and (I.54), one obtains

$$b_l(\infty)=(l!8\pi^3)^{-1}\int\langle\mathbf{k}_1\cdots\mathbf{k}_l|u_l|\mathbf{k}_1\cdots\mathbf{k}_l\rangle d^{3l}k. \quad \text{(I.55)}$$

Similarly,

$$b_l^S(\infty)=(l!8\pi^3)^{-1}\int\langle\mathbf{k}_1\cdots\mathbf{k}_l|u_l^S|\mathbf{k}_1\cdots\mathbf{k}_l\rangle d^{3l}k,$$

and

$$b_l^A(\infty)=(l!8\pi^3)^{-1}\int\langle\mathbf{k}_1\cdots\mathbf{k}_l|u_l^A|\mathbf{k}_1\cdots\mathbf{k}_l\rangle d^{3l}k. \quad \text{(I.56)}$$

6. BINARY KERNEL *B* FOR HARD-SPHERE INTERACTION

To calculate B for $\Omega=\infty$ and for a central potential V, one first introduces center-of-mass and relative coordinates:

$$\mathbf{R}=\tfrac{1}{2}(\mathbf{r}_1+\mathbf{r}_2),\quad \mathbf{r}=\mathbf{r}_1-\mathbf{r}_2. \quad \text{(I.57)}$$

The Jacobian of the transformation is equal to unity. Spherical coordinates r,θ,ϕ will be introduced to replace the vector $\mathbf{r}$. Now

$$H_2=-\nabla_1^2-\nabla_2^2+V(r)=H_R+H_r,$$

where

$$H_R=-\tfrac{1}{2}\nabla_R^2\quad\text{and}\quad H_r=-2\nabla_r^2+V(r). \quad \text{(I.58)}$$

Let the normalized bound state solution in the center-of-mass system of the two particles be $\psi_i(\mathbf{r})$ $(i=1,2,\cdots)$ so that

$$H_r\psi_i(\mathbf{r})=E_i\psi_i(\mathbf{r})$$

and

$$\int|\psi_i(\mathbf{r})|^2d^3r=1. \quad \text{(I.59)}$$

Also let the continuum solutions be ψ_{klm} which satisfy

$$\psi_{klm}=(2\pi^{-1})^{\frac{1}{2}}r^{-1}\mathcal{R}_{kl}(r)Y_{lm}(\theta,\phi),$$

$$H\psi_{klm}=2k^2\psi_{klm},$$

and

$$\mathcal{R}_{kl}\rightarrow\sin(kr-\tfrac{1}{2}l\pi+\delta_{kl})\quad\text{as}\quad r\rightarrow\infty. \quad \text{(I.60)}$$

The spherical harmonics Y_{lm} are here defined so that

$$\int_0^\pi\sin\theta d\theta\int_0^{2\pi}|Y_{lm}|^2d\phi=1.$$

The normalization condition (I.60) insures[9] that

$$\int\psi_{k'l'm'}{}^*(\mathbf{r})\psi_{klm}(\mathbf{r})d^3\mathbf{r}=\delta_{l',l}\delta_{m',m}\delta(k'-k). \quad \text{(I.61)}$$

Therefore

$$\langle\mathbf{r}'|\exp(-\beta H)|\mathbf{r}\rangle=\sum_i\psi_i(\mathbf{r}')\psi_i^*(\mathbf{r})\exp(-\beta E_i)$$
$$+\sum_{lm}\int_0^\infty dk\,\psi_{klm}(\mathbf{r}')\psi_{klm}^*(\mathbf{r})\exp(-2\beta k^2). \quad \text{(I.62)}$$

[9] In the sense of the transformation theory of Dirac (see reference 7).

The last term can be written as

$$\sum_l 2(\pi)^{-1}\int_0^\infty dk(rr')^{-1}\mathcal{R}_{kl}(r')\mathcal{R}_{kl}^*(r)$$
$$\times\exp(-2\beta k^2)\sum_m Y_{lm}(\theta',\phi')Y_{lm}^*(\theta,\phi)$$
$$=\sum_l(2l+1)(2\pi^2rr')^{-1}P_l(\cos\Theta)$$
$$\times\int_0^\infty dk\,\mathcal{R}_{kl}(r')\mathcal{R}_{kl}^*(r)\exp(-2\beta k^2), \quad \text{(I.63)}$$

where

$$P_l(\xi)=\frac{1}{2^ll!}\frac{d^l}{d\xi^l}(\xi^2-1)^l,$$

and

$$\cos\Theta=(rr')^{-1}\mathbf{r}\cdot\mathbf{r}'.$$

To complete the calculation of W_2 it is also necessary to compute the operator $\exp(-\beta H_R)=\exp(+\frac{1}{2}\beta\nabla_R^2)$. The eigenvalues and eigenfunctions of ∇_R^2 are $-K^2$ and $(8\pi^3)^{-\frac{1}{2}}\exp(i\mathbf{K}\cdot\mathbf{R})$. Hence

$$\langle\mathbf{R}'|\exp(\tfrac{1}{2}\beta\nabla_R^2)|\mathbf{R}\rangle$$
$$=(8\pi^3)^{-1}\int d^3K\exp[-\tfrac{1}{2}\beta\mathbf{K}^2+i\mathbf{K}\cdot(\mathbf{R}'-\mathbf{R})]$$
$$=8^{\frac{1}{2}}\lambda^{-3}\exp[-(\mathbf{R}'-\mathbf{R})^2/(2\beta)].$$

Collecting all terms we thus obtain a general formula for the coordinate representation of W_2 for the case $\Omega=\infty$ and $V=$ central potential:

$$\langle\mathbf{r}_1',\mathbf{r}_2'|\exp(-\beta H_2)|\mathbf{r}_1,\mathbf{r}_2\rangle$$
$$=8^{\frac{1}{2}}\lambda^{-3}\exp[-(\mathbf{r}_1'+\mathbf{r}_2'-\mathbf{r}_1-\mathbf{r}_2)^2/(8\beta)]$$
$$\times\Big\{\sum_i\psi_i(\mathbf{r}')\psi_i^*(\mathbf{r})\exp(-\beta E_i)$$
$$+\sum_l(2l+1)(2\pi^2rr')^{-1}P_l(\cos\Theta)$$
$$\times\int_0^\infty dk\mathcal{R}_{kl}(r')\mathcal{R}_{kl}^*(r)\exp(-2\beta k^2)\Big\}. \quad \text{(I.64)}$$

To obtain U_2 one subtracts from this the corresponding expression for free particles. In other words, if one replaces $\mathcal{R}_{kl}(r')\mathcal{R}_{kl}^*(r)$ in (I.64) by

$$\mathcal{R}_{kl}(r')\mathcal{R}_{kl}^*(r)-[\mathcal{R}_{kl}(r')\mathcal{R}_{kl}^*(r)]_{\text{free}}, \quad \text{(I.65)}$$

one obtains the coordinate representation of U_2.

For the hard-sphere interaction,

$$V(\mathbf{r}_1-\mathbf{r}_2)=+\infty\quad\text{for}\quad|\mathbf{r}_1-\mathbf{r}_2|\leqq a,$$
$$V(\mathbf{r}_1-\mathbf{r}_2)=0\quad\text{for}\quad|\mathbf{r}_1-\mathbf{r}_2|>a;$$

there are no bound states. Now in the integral in (I.64) the exponential factor limits the important values of k to $\lesssim\lambda^{-1}$. If $(a/\lambda)\ll1$, for these small wave numbers the effect of the hard sphere is masked by the centrifugal force in all states $l>0$. In fact, e.g., the phase shift for $l=1$ is $\delta_{kl}\cong-\frac{1}{3}(ka)^3\approx-\frac{1}{3}(a/\lambda)^3$. Neglecting such small contributions of the order of $(a/\lambda)^3$ one obtains only the contribution from the S state (i.e., $l=0$ state), for which (I.65) becomes

$$\sin(kr'-ka)\sin(kr-ka)-\sin kr'\sin kr$$
$$=\tfrac{1}{2}\cos(kr'+kr)-\tfrac{1}{2}\cos(kr'+kr-2ka)\quad\text{for } r'>a \text{ and } r>a,$$

and

$$-\sin kr'\sin kr=\tfrac{1}{2}\cos(kr'+kr)-\tfrac{1}{2}\cos(kr'-kr)\quad\text{for } r'<a \text{ or } r<a.$$

The integration over k is straightforward. One obtains

$$\langle \mathbf{r}_1',\mathbf{r}_2'|U_2|\mathbf{r}_1,\mathbf{r}_2\rangle=(2\pi\lambda^4rr')^{-1}\exp[-(\mathbf{r}_1'+\mathbf{r}_2'-\mathbf{r}_1-\mathbf{r}_2)^2/(8\beta)]\times\begin{cases}\exp[-(r+r')^2/(8\beta)]-\exp[-(r+r'-2a)^2/(8\beta)] & \text{for } r>a,\ r'>a,\\ \exp[-(r+r')^2/(8\beta)]-\exp[-(r-r')^2/(8\beta)] & \text{otherwise.}\end{cases}\quad\text{(I.66)}$$

We recall that $r=|\mathbf{r}_1-\mathbf{r}_2|$, $r'=|\mathbf{r}_1'-\mathbf{r}_2'|$. To go into the momentum representation, we use

$$\langle\mathbf{k}_1',\mathbf{k}_2'|U_2|\mathbf{k}_1,\mathbf{k}_2\rangle=\int\langle\mathbf{k}_1'|\mathbf{r}_1'\rangle\langle\mathbf{k}_2'|\mathbf{r}_2'\rangle\langle\mathbf{r}_1|\mathbf{k}_1\rangle\langle\mathbf{r}_2|\mathbf{k}_2\rangle\times\langle\mathbf{r}_1',\mathbf{r}_2'|U_2|\mathbf{r}_1,\mathbf{r}_2\rangle d^3r_1d^3r_2d^3r_1'd^3r_2',$$

and

$$\langle\mathbf{r}|\mathbf{k}\rangle=(8\pi^3)^{-\frac{1}{2}}\exp(i\mathbf{k}\cdot\mathbf{r}).$$

The computation is tedious but straightforward. One first integrates over the center-of-mass coordinates $\mathbf{R}$ and $\mathbf{R}'$. Then one integrates over the angles of $\mathbf{r}$ and $\mathbf{r}'$. After the transformation

$$r+r'=\xi,\quad r-r'=\eta,$$

one integrates over ξ or over η and obtains finally

$$\begin{aligned}\langle\mathbf{k}_1',\mathbf{k}_2'|U_2|\mathbf{k}_1,\mathbf{k}_2\rangle&=[4\pi^2kk'(k^2-k'^2)]^{-1}\delta^3(\mathbf{k}_1'+\mathbf{k}_2'-\mathbf{k}_1-\mathbf{k}_2)\\&\times\{[\sin(k+k')a][k\exp(-\beta E)-k'\exp(-\beta E')]\\&-[\sin(k-k')a][k\exp(-\beta E)+k'\exp(-\beta E')]\\&+\pi^{-\frac{1}{2}}2[\cos(k+k')a-\cos(k-k')a]\\&\times[kM(\sqrt{2}\beta^{\frac{1}{2}}k)\exp(-\beta E)\\&-k'M(\sqrt{2}\beta^{\frac{1}{2}}k')\exp(-\beta E')]\},\end{aligned}\quad\text{(I.67)}$$

where

$$k=\tfrac{1}{2}|\mathbf{k}_1-\mathbf{k}_2|,\qquad E=k_1^2+k_2^2,$$
$$k'=\tfrac{1}{2}|\mathbf{k}_1'-\mathbf{k}_2'|,\quad E'=k_1'^2+k_2'^2,$$

and

$$M(y)=\int_0^y\exp(x^2)dx.\quad\text{(I.68)}$$

For large y,

$$M(y)=[\exp(y^2)][(2y)^{-1}+(4y^3)^{-1}+\cdots]\quad\text{(I.69)}$$

asymptotically. One notices that the factor $[kk'(k^2-k'^2)]^{-1}$ does not introduce any singularities in (I.67) because the other factors vanish at $k=k'$, and at $k=0$ and $k'=0$. We can now use (I.45) to compute the binary function B. In momentum representation (I.45) assumes the form

$$\langle\mathbf{k}_1',\mathbf{k}_2'|B|\mathbf{k}_1,\mathbf{k}_2\rangle=\frac{\partial}{\partial\beta}\langle\mathbf{k}_1',\mathbf{k}_2'|U_2|\mathbf{k}_1,\mathbf{k}_2\rangle+E'\langle\mathbf{k}_1',\mathbf{k}_2'|U_2|\mathbf{k}_1,\mathbf{k}_2\rangle.$$

Using (I.67), one obtains (contributions due to S states only)

$$\begin{aligned}\langle\mathbf{k}_1',\mathbf{k}_2'|B|\mathbf{k}_1,\mathbf{k}_2\rangle&=-(\pi^2kk')^{-1}\delta^3(\mathbf{k}_1'+\mathbf{k}_2'-\mathbf{k}_1-\mathbf{k}_2)\\&\times[\sin(k'a)][\exp(-\beta E)]\{k\cos ka-\pi^{-\frac{1}{2}}\sin ka\\&\times[2kM(\sqrt{2}\beta^{\frac{1}{2}}k)-(2\beta)^{-\frac{1}{2}}\exp(2\beta k^2)]\}\\&-(4\pi^2kk')^{-1}\delta(\beta)\delta^3(\mathbf{k}_1+\mathbf{k}_2-\mathbf{k}_1'-\mathbf{k}_2')\\&\times[(k-k')^{-1}\sin(k-k')a\\&-(k+k')^{-1}\sin(k+k')a].\end{aligned}\quad\text{(I.70)}$$

The first term in (I.70) is obtained by a straightforward differentiation of (I.67) with respect to β. The presence of the second term is due to the condition that at $\beta=0$, by definition (I.43), $U_2=0$. The explicit expressions (I.66) and (I.67) do not approach zero as $\beta\to0+$. Hence they are valid only for $\beta>0$, and at $B=0+$ a step function in β must be added. Taking the derivative of such a step function with respect to β gives rise to a $\delta(\beta)$ function which constitutes the second term in (I.70). At low temperatures, $k\lesssim(\lambda)^{-1}$, the contribution of the second term is $\sim(a/\lambda)^3$.

We notice that as $k\to\infty$, $\exp(-\beta E)$ varies as $\exp(-2\beta k^2)$. Using (I.69) one sees that, for $\beta>0$, $B\sim k^{-3}\sin ka$ which damps down very rapidly for large k.

For $a=0$, $B=0$ as it should. Expanding B according to powers of a, one obtains $B=B_1+B_2+\cdots$, where

$$B_1=-a\pi^{-2}\delta^3(\mathbf{k}_1+\mathbf{k}_2-\mathbf{k}_1'-\mathbf{k}_2')\exp(-\beta E),$$
$$B_2=\pi^{-\frac{5}{2}}a^2\delta^3(\mathbf{k}+\mathbf{k}_2-\mathbf{k}_1'-\mathbf{k}_2')[\exp(-\beta E)]\times[2kM(\sqrt{2}\beta^{\frac{1}{2}}k)-(2\beta)^{-\frac{1}{2}}\exp(2\beta k^2)].\quad\text{(I.71)}$$

The first order term, B_1, may be put in the form

$$B_1=-V_{12}'\exp(+\beta\nabla_1^2+\beta\nabla_2^2),$$

where

$$V_{12}'=8\pi a\delta^3(\mathbf{r}_1-\mathbf{r}_2).$$

In this form, it is closely related to the pseudopotential[10] discussed in the literature.

The binary kernel in coordinate representation can be obtained from (I.45) and (I.66). It is

$$\begin{aligned}\langle\mathbf{r}_1',\mathbf{r}_2'|B|\mathbf{r}_1,\mathbf{r}_2\rangle&=-2\lambda^{-6}\{\exp[-(\mathbf{r}_1'+\mathbf{r}_2'-\mathbf{r}_1-\mathbf{r}_2)^2/(8\beta)]\}\\&\times\{a^{-1}\delta(r'-a)\}(1-ar^{-1})\exp[-(r-a)^2/(8\beta)]\quad\text{for } r>a,\\&=\delta(\beta)[-4\pi rr']^{-1}\delta(r-r')\\&\times\delta^3[\tfrac{1}{2}(\mathbf{r}_1'+\mathbf{r}_2'-\mathbf{r}_1-\mathbf{r}_2)]\quad\text{for } r\leqq a.\end{aligned}\quad\text{(I.72)}$$

where $r=|\mathbf{r}_1-\mathbf{r}_2|$, $r'=|\mathbf{r}_1'-\mathbf{r}_2'|$.

[10] K. Huang and C. N. Yang, Phys. Rev. **105**, 767 (1957); Lee, Huang, and Yang, Phys. Rev. **106**, 1135 (1957).

ACKNOWLEDGMENT

We wish to thank the Brookhaven National Laboratory for the hospitality extended us during the summer of 1956 when much of this work was done and to thank the University of Wisconsin and the University of California Radiation Laboratory for their hospitality during the summer of 1958 when this paper was written.

APPENDIX A

To prove (I.7) we first observe that a general term in Eq. (I.6) for W_N is a product of m_1 U_1 functions, m_2 U_2 functions, etc., where $\sum lm_l=N$. Such a term gives a contribution

$$\prod_l [l!\Omega b_l(\Omega)]^{m_l}$$

to Q_N. Now in W_N there are, for fixed m_1, m_2, $\cdots$, $N![\prod_l (l!)^{m_l} m_l!]^{-1}$ such terms. Their total contribution to $Q_N(N!)^{-1}$ is thus

$$\prod_l (m_l!)^{-1}[\Omega b_l(\Omega)]^{m_l}.$$

One thus obtains

$$\mathcal{Q}_\Omega=\sum_N z^N \sum_{m_l=0,1,\cdots}{}' \prod_l \frac{1}{m_l!}[\Omega b_l(\Omega)]^{m_l}$$

$$=\sum_N \sum_{m_l}{}' \prod_l \frac{1}{m_l!}[\Omega z^l b_l(\Omega)]^{m_l},$$

where the summation, $\sum_{m_l}{}'$, over m_l is subject to the condition that $\sum lm_l=N$. The subsequent sum $\sum_N$, over N, is equivalent to a removal of this condition on the summation over m_l. Thus

$$\mathcal{Q}_\Omega=\sum_{m_l=0,1,\cdots} \prod_l \frac{1}{m_l!}[\Omega z^l b_l(\Omega)]^{m_l},$$

which leads directly to (I.7).

One may question the mathematical rigor of the above derivation. To make it rigorous we observe that for a fixed Ω,

(a) all Q_N are positive, and

(b) Q_N vanishes for sufficiently large values of N, if the interaction V has a hard repulsive core. $\mathcal{Q}_\Omega$ is thus a polynomial in z with no zeros on the positive real axis. Its logarithm is therefore an analytic function near the origin and all along the positive real axis. Near the origin this logarithm can be expanded as a Taylor's series. It is then easy to see that this Taylor's series is exactly the curly bracket in (I.7). Furthermore, since $\log\mathcal{Q}_\Omega$ is analytic along the positive real axis in the complex z plane, (I.7) is valid for all positive values of z, if one understands the curly bracket to mean the analytic continuation of the power series within.

APPENDIX B

To prove (I.23) and (I.24): The eigenfunctions of H_N can be classified according to the irreducible representations of the permutation group of N objects. If $\psi_i(1',2',\cdots N')$ belongs to an irreducible representation D, then $P'\psi_i(1'\cdots N')$ also belongs to the same representation D. Hence $\sum_{P'} P'\psi_i(1'\cdots N')$ belongs to D. But $\sum_{P'} P'\psi_i(1'\cdots N')$ is symmetrical. Hence if D is not the symmetrical representation, $\sum_{P'} P'\psi_i(1'\cdots N')=0$. On the other hand, if D is the symmetrical representation, then

$$\sum_{P'} P'\psi_i(1'\cdots N')=N!\psi_i(1'\cdots N').$$

Using the definitions (I.4) and (I.11), one obtains immediately (I.23). The proof of (I.24) is similar.

Formula (I.23) was first used by Feynman[11] in his treatment of the Bose condensation problem by path integrals.

APPENDIX C

To prove rule A: Each term of the right-hand side of (I.6) is characterized by a grouping of the form (I.27) of the coordinates 1, 2, $\cdots l$. Application of the operation $\sum_{P'} P'$ to both sides of (I.6) therefore naturally leads on the right-hand side to a summation S of the form (I.28), but without the condition stated in the paper for (I.28). The left-hand side is, by (I.23), equal to $\langle 1',\cdots l'|W_l^S|1,\cdots l\rangle$. One thus proves that $\langle 1',\cdots l'|W_l^S|1,\cdots l\rangle$ is equal to the sum of all S over the different groupings (I.27).

One then substitutes the above result for the left-hand side of (I.12). Solving the resultant equations for U_1^S, U_2^S, U_3^S, $\cdots$ in succession, one obtains rule A by induction.

APPENDIX D

Equation (I.51) is intuitively quite obvious. It is a consequence of the fact that for fixed $\mathbf{r}_1$, integration of U_l over the other $\mathbf{r}$'s gives a result independent of $\mathbf{r}_1$. To fill in the logical steps, we consider the equation $H_N\psi=E\psi$ as an eigenvalue problem in $3N$-dimensional space, within the basic cube

$$\begin{aligned}&\tfrac{1}{2}L\geqq x_i\geqq -\tfrac{1}{2}L,\quad \tfrac{1}{2}L\geqq z_i\geqq -\tfrac{1}{2}L,\\ &\tfrac{1}{2}L\geqq y_i\geqq -\tfrac{1}{2}L,\quad (i=1,2,\cdots N),\end{aligned}\tag{I.73}$$

with periodic boundary conditions in each of the $3N$ dimensions. An eigenfunction ψ can be continued outside of the cube to all space as a periodic function. Extending the definition of the operator H_N in a periodic way to all space, one has $H_{N\Omega}\psi=E\psi$ *everywhere*. The explicit form of $H_{N\Omega}$ is

$$H_{N\Omega}=-\sum_1^N \nabla_\alpha^2+\sum_{\alpha>\beta}\sum_{\mathbf{m}} V(\mathbf{r}_\alpha-\mathbf{r}_\beta+\mathbf{m}L),\tag{I.74}$$

where $\mathbf{m}=(m_x,m_y,m_z)$ has $\pm$ integers as components. [For simplicity we assume here that $V(\mathbf{r})=0$ for $|\mathbf{r}|\geqq r_0$ where $r_0=$range and is $\ll\frac{1}{2}L$.] Within the basic

[11] R. P. Feynman, Phys. Rev. 91, 1291 (1953).

cube (I.73), (I.74) is the same as (I.1) except for regions near such edges as, e.g.,

$$x_1=\frac{L}{2},\quad x_2=-\frac{L}{2},\quad y_1=y_2,\quad z_1=z_2.$$

Definition (I.1) *should be amended* to become everywhere identical to (I.74) inside the basic cube. Otherwise the discussion would become more clumsy.

Equation (I.4) can then be regarded as defining $W_{N\Omega}$ for arbitrary values of $\mathbf{r}_1', \cdots \mathbf{r}_N', \mathbf{r}_1, \cdots \mathbf{r}_N$. *It is a function periodic in each of the 6N linear dimensions with a period L.* Furthermore, since the potential energy in (I.74) depends only on the relative coordinates, and not on the coordinates of the center of mass of the N particles, all eigenfunctions can be chosen to be eigenstates of the total momentum so that

$$\psi_\alpha(\mathbf{r}_1+\xi;\mathbf{r}_2+\xi;\cdots\mathbf{r}_N+\xi)=\psi_\alpha(\mathbf{r}_1,\mathbf{r}_2,\cdots\mathbf{r}_N)\exp(i\mathbf{K}_\alpha\cdot\xi).$$

[It is important to realize that the periodicity condition does not invalidate this statement.] One then concludes from (I.4) that

$$\langle\mathbf{r}_1'+\xi,\mathbf{r}_2'+\xi,\cdots|W_{N\Omega}|\mathbf{r}_1+\xi,\mathbf{r}_2+\xi,\cdots\rangle \quad (I.75)$$

is independent of ξ.

It is clear from the structure of (I.6) that, like $W_{N\Omega}$, $U_{N\Omega}$ also has a $6N$-dimensional periodicity in coordinate space, and also is invariant under a simultaneous displacement of all $2N$ vector coordinates $\mathbf{r}_\alpha'$ and $\mathbf{r}_\alpha$ by the same displacement ξ. From these two properties of $U_{l\Omega}$ it is easy to see that

$$\int\langle\mathbf{r}_1,\mathbf{r}_2,\cdots\mathbf{r}_l|U_{l\Omega}|\mathbf{r}_1,\mathbf{r}_2,\cdots\mathbf{r}_l\rangle d^3r_2d^3r_3,\cdots d^3r_l$$

over the box (I.52) is independent of $\mathbf{r}_1$. Equation (I.51) then follows from (I 10).

APPENDIX E

To establish (I.53) we notice that $W_{l\infty}$, when regarded as a function of β and $\mathbf{r}_1', \mathbf{r}_2', \cdots\mathbf{r}_l'$, with $\mathbf{r}_1, \mathbf{r}_2, \cdots\mathbf{r}_l$ as parameters, satisfies

$$\frac{\partial}{\partial\beta}W_{l\infty}=[-\sum\nabla_\alpha'^2+\sum_{\alpha>\beta}V(\mathbf{r}_\alpha'-\mathbf{r}_\beta')]W_{l\infty}, \quad (I.76)$$

and

$$W_{l\infty}|_{\beta=0}=\delta^3(\mathbf{r}_1'-\mathbf{r}_1)\delta^3(\mathbf{r}_2'-\mathbf{r}_2)\cdots\delta^3(\mathbf{r}_l'-\mathbf{r}_l). \quad (I.77)$$

These two equations can also be used to define $W_{l\infty}$. Now we adopt the definition of $W_{l\Omega}$ *over all space* introduced in Appendix D. $W_{l\Omega}$ satisfies

$$\frac{\partial}{\partial\beta}W_{l\Omega}=[-\sum\nabla_\alpha'^2+\sum_{\alpha>\beta}\sum_{\mathbf{m}}V(\mathbf{r}_\alpha'-\mathbf{r}_\beta'+\mathbf{m}L)]W_{l\Omega}, \quad (I.78)$$

and

$$W_{l\Omega}|_{\beta=0}=[\sum_{\mathbf{m}}\delta^3(\mathbf{r}_1'-\mathbf{r}_1+\mathbf{m}L)]\cdots\times[\sum_{\mathbf{m}}\delta^3(\mathbf{r}_l'-\mathbf{r}_l+\mathbf{m}L)]. \quad (I.79)$$

These two equations can also be used to define $W_{l\Omega}$.

Equations (I.76) and (I.78) may be regarded as "diffusion equations" (with β=time) while (I.77) and (I.79) serve as the initial conditions. One then sees that $W_{l\infty}$ and $W_{l\Omega}$ differ for two reasons:

(i) The potential energy in (I.78) is different from that in (I.76). The former includes a sum

$$\sum_{\mathbf{m}}V(\mathbf{r}_\alpha'-\mathbf{r}_\beta'+\mathbf{m}L),$$

while the latter contains only the term $\mathbf{m}=0$. However, the terms with $\mathbf{m}\neq0$ may be regarded as the "images" of this term, and are ineffective except for separations $|\mathbf{r}_\alpha'-\mathbf{r}_\beta'|\gtrsim L$.

(ii) The initial condition (I.79) gives $W_{l\Omega}$ at $\beta=0$ as a sum of many δ^{3l} functions in the $3l$-dimensional space $\mathbf{r}_1', \mathbf{r}_2', \cdots\mathbf{r}_l'$, forming a periodic lattice with period L in each linear dimension. On the other hand, (I.77) gives W_l at $\beta=0$ as a *single* δ^{3l} function.

While diffusion is a process that goes with arbitrary speed, the larger speeds have progressively smaller probability. For finite and fixed values of $\mathbf{r}_1, \mathbf{r}_2, \cdots\mathbf{r}_l$ (which determine the positions of the δ^{3l} functions in the initial conditions), and at a fixed β and fixed $\mathbf{r}_1', \mathbf{r}_2', \cdots\mathbf{r}_l'$, the two differences between $W_{l\Omega}$ and $W_{l\infty}$ described above become unimportant as $L\to\infty$. Therefore $W_{l\Omega}\to W_{l\infty}$. By (I.23) and (I.24), the same holds for $W_{l\Omega}{}^S$ and $W_{l\Omega}{}^A$. From (I.6) and (I.12) one easily proves that also $U_{l\Omega}\to U_{l\infty}$, $U_{l\Omega}{}^S\to U_{l\infty}{}^S$, and $U_{l\Omega}{}^A\to U_{l\infty}{}^A$.

In fact, for a given potential V it is possible to evaluate the difference between $W_{l\Omega}$ and $W_{l\infty}$ in terms of B_∞. We shall illustrate such a computation for the case of hard sphere interactions.

For simplicity let us first consider the case $l=2$. In this case, B_∞ and $U_{2\infty}$ (consequently, also $W_{2\infty}$) are given explicitly by (I.72) and (I.66). To express $W_{2\Omega}$ and $U_{2\Omega}$ in terms of these functions, we introduce a W_2' function which satisfies the differential equation

$$\frac{\partial}{\partial\beta}W_2'=\left[-\sum_{\alpha=1}^{2}\nabla_\alpha'^2+\sum_{\mathbf{m}}V(\mathbf{r}_1'-\mathbf{r}_2'+\mathbf{m}L)\right]W_2', \quad (I.80)$$

but with the initial condition

$$W_2'|_{\beta=0}=\delta^3(\mathbf{r}_1'-\mathbf{r}_1)\delta^3(\mathbf{r}_2'-\mathbf{r}_2). \quad (I.81)$$

It is easy to see that $W_{2\Omega}$ is related to W_2' by

$$\langle\mathbf{r}_1',\mathbf{r}_2'|W_{2\Omega}|\mathbf{r}_1,\mathbf{r}_2\rangle=\sum_{\mathbf{m}_1,\mathbf{m}_2}\langle\mathbf{r}_1'+\mathbf{m}_1L,\mathbf{r}_2'+\mathbf{m}_2L|W_2'|\mathbf{r}_1,\mathbf{r}_2\rangle. \quad (I.82)$$

Although W_2' satisfies a different "differential equation" from $W_{2\infty}$, they both satisfy the same initial condition. By going through a series of arguments similar to those used in Sec. 4, we can express W_2' in terms of $W_{2\infty}$ and B_∞:

$$\begin{aligned}
\langle \mathbf{r}_1',\mathbf{r}_2'|W_2'|\mathbf{r}_1,\mathbf{r}_2\rangle &= \langle \mathbf{r}_1',\mathbf{r}_2'|W_{2\infty}|\mathbf{r}_1,\mathbf{r}_2\rangle \\
&+ \sum_{\mathbf{m}\neq 0} \langle \mathbf{r}_1'-\mathbf{m}L, \mathbf{r}_2'|U_{2\infty}|\mathbf{r}_1-\mathbf{m}L, \mathbf{r}_2\rangle \\
&+ \sum_{\mathbf{m}\neq\mathbf{m}'}\sum_{\mathbf{m}'}\int_0^\beta d\beta' \int d^3\mathbf{r}_1'' d^3\mathbf{r}_2'' \\
&\times \langle \mathbf{r}_1'-\mathbf{m}L, \mathbf{r}_2'|U_{2\infty}(\beta-\beta')|\mathbf{r}_1''-\mathbf{m}L, \mathbf{r}_2''\rangle \\
&\times \langle \mathbf{r}_1''-\mathbf{m}'L, \mathbf{r}_2''|B_\infty(\beta')|\mathbf{r}_1-\mathbf{m}'L, \mathbf{r}_2\rangle \\
&+ \sum_{\mathbf{m}\neq\mathbf{m}'}\sum_{\mathbf{m}'\neq\mathbf{m}''}\sum_{\mathbf{m}''}\int_0^\beta d\beta' \int_0^{\beta'} d\beta'' \int d^3\mathbf{r}_1'' d^3\mathbf{r}_2'' d^3\mathbf{r}_1''' \\
&\times d^3\mathbf{r}_2''' \langle \mathbf{r}_1'-\mathbf{m}L, \mathbf{r}_2'|U_{2\infty}(\beta-\beta')|\mathbf{r}_1''-\mathbf{m}L, \mathbf{r}_2''\rangle \\
&\times \langle \mathbf{r}_1''-\mathbf{m}'L, \mathbf{r}_2''|B_\infty(\beta'-\beta'')|\mathbf{r}_1'''-\mathbf{m}'L, \mathbf{r}_2'''\rangle \\
&\times \langle \mathbf{r}_1'''-\mathbf{m}''L, \mathbf{r}_2'''|B_\infty(\beta'')|\mathbf{r}_1-\mathbf{m}''L, \mathbf{r}_2\rangle + \cdots .
\end{aligned} \tag{I.83}$$

Equations (I.82) and (I.83) express precisely the physical effects discussed in (i) and (ii) above. Together these two equations give the explicit form of $W_{2\Omega}$ in terms of the binary function B_∞.

By using (I.72) and (I.66) it can be shown that for fixed positions of $\mathbf{r}_1$, $\mathbf{r}_2$ and $\mathbf{r}_1'$, $\mathbf{r}_2'$ as $L\to\infty$, (I.82) becomes

$$W_{2\Omega} = W_2' + O[\exp(-L^2/4\beta)], \tag{I.84}$$

and (I.83) becomes

$$W_2' = W_{2\Omega} + O[\exp(-L^2/2\beta)]. \tag{I.85}$$

Thus we have

$$W_{2\Omega} \to W_{2\infty} \quad \text{as} \quad \Omega\to\infty. \tag{I.86}$$

By using (I.82) and (I.83) we can also express $b_2(\Omega)$ explicitly in terms of B_∞. It is easy to see that $\lim_{\Omega\to\infty} b_2(\Omega)$ is given by (I.53).

We remark that

$$W_{2\Omega} = W_{2\infty} + O[\exp(-L^2/4\beta)] \tag{I.87}$$

is due to the familiar property of diffusion equations: i.e., the probability for a particle to travel a distance L in a time "β" is proportional to $\exp[-L^2/(4\beta)]$. Consequently, (I.87) is not limited to the case of hard spheres. It is valid for any potential with a finite range.

In an entirely similar manner one can express the difference between $W_{l\Omega}$ and $W_{l\infty}$ in terms of B_∞ and show explicitly that (I.53) is correct for any $l\geqq 2$. The main points in the proof are the following two facts. First, the integrand in (I.51) approaches that in (I.53) as $L\to\infty$. Second, in the region of integration defined by (I.52), the integrand in (I.51) is peaked at the center. It becomes exponentially small far away from the center, as can be seen by arguments similar to the above arguments for the case $l=2$. The limit as $\Omega\to\infty$ of the integral (I.51) is thus equal to the integral of the limit of the integrand over all space.

[59c]
Commentary begins page 45

Symmetry Principles in Modern Physics

C. N. Yang

We have heard this morning discussions of the concept of symmetry in mathematics and its applications in chemistry. I should like now to discuss with you some aspects of the role that the symmetry principles play in the physics of the twentieth century.

In classical physics various forms of symmetry laws were recognized and were used, but it is only in quantum mechanics that symmetry principles assume an essential role. There is an intrinsic reason for this: the superposition principle of the description of the state of a system in quantum mechanics. E.g. because of this principle, the elliptical orbits as well as circular orbits of an electron in a hydrogen atom possess symmetry properties that allow for a very elegant and far reaching mathematical formulation. It is upon such a formulation that one builds a most beautiful understanding of the periodic table. The general relationship between the symmetry principle and its physical consequences is illustrated in Figure 1. We notice first that a direct consequence of a symmetry principle, such as rotational symmetry, or invariance under rotation, is a conservation law. In this particular case, the conservation of angular momentum. Conservation laws were important in classical mechanics and the relationship between conservation laws and the symmetry principles were recognized in classical mechanics. But it is the relationship of the symmetry principles to quantum numbers and to selection rules that makes these concepts inseparable from the language of the physics of today. We in fact identify and distinguish between various dynamical systems through a classification of them by quantum numbers. This was done in atomic and molecular physics. This is again done today in the study of the building block of the atoms—the so-called elementary particles. In identifying these particles we specify their quantum numbers. Some of these quantum numbers have a meaning that is easy to grasp: the mass, the electric charge, the spin, or intrinsic angular momentum. Some of the rest, such as parity, are understandable only in the quantum language. Then there are other quantum numbers such as strangeness, the charge conjugation operator and the isotopic spin, the existence of which follow from rather unusual and previously unsuspected symmetries.

Let us first have a look at the varieties of particles that are being studied in the laboratories of physics. In Figure 2 we show the particles whose masses are less than the mass of the proton. These are again divided into particles whose spins are integral multiples of the Planck constant divided by 2π and those whose spins are half integral multiples of the Planck constant divided by 2π. We notice that the spin must be integral multiples of one-half of the Planck constant divided by 2π: a fact that follows from the quantum formulation of rotational symmetry. The ordinate here is schematically the mass of the particle in energy units, and the abcissa the electric charge.

In this figure the positron, the electron, and the gamma ray are relatively familiar objects. The mu meson and pi meson were discovered, respectively, immediately before and after the Second World War. The *K* mesons belong to a new class of particles to which the name *strange particles* is attached, betraying the ignorance of the physicists about their properties. The neutrino and the anti-neutrino are two of the queerest particles of the collection. The quantum numbers of the particles in this figure are relatively well identified.

In Figure 3, particles heavier than the proton are listed. All of these have half integral spin values. Why this is so remains a mystery. In this figure the neutron and the proton are the familiar objects which form the building blocks of the atomic nuclei. The heavier particles belong to the category of strange particles about which intensive studies

□ Lecture given at the 75th anniversary celebration of Bryn Mawr College, Session on Symmetries, November 6, 1959.

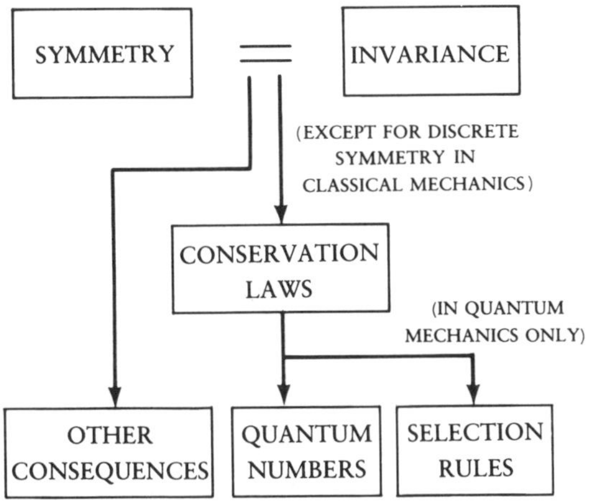

Figure 1

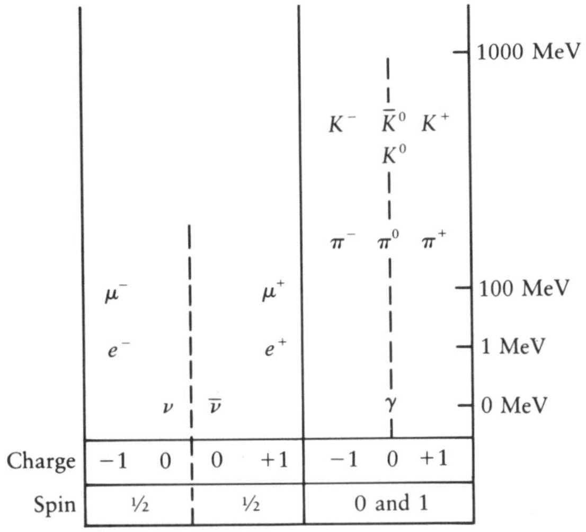

Figure 2

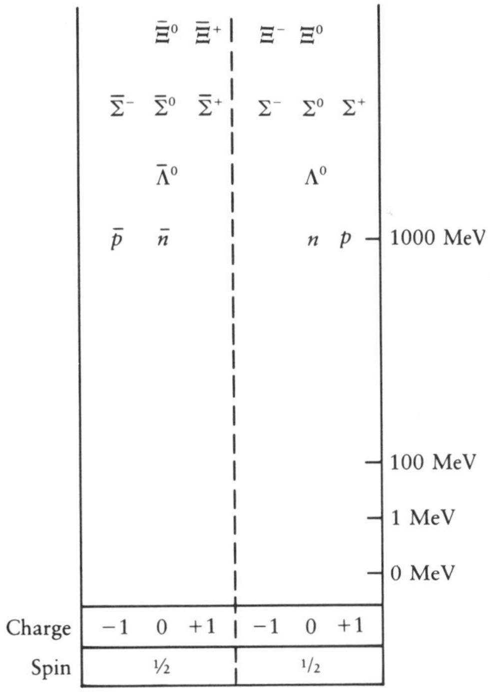

Figure 3

have been made in recent years. Their spins and parities, are, however, still not determined. On the right-hand side of this diagram we find the antiparticles to those on the left. We should add that in the same language a positron is an antielectron.

The study of the properties of these particles has led to a classification of the types of forces that govern their mutual interactions. A consistent and well defined pattern emerges as represented in the next figure. We see here that there are four types of forces or interaction between these particles, with drastically different strengths. Among these types the electromagnetic forces are relatively well understood. The nuclear forces have been intensively studied, but their great strength has so far frustrated a simple mathematical description. Under the heading of weak forces there are a great variety of different types of interactions. The surprising thing is that despite their dissimilarity, they all share approximately the same strength, as illustrated here. We shall have occasion to come back to this point later. The gravitational forces are very, very weak and have so far escaped detection on an atomic scale. Although ultimately it cannot fail to affect the influence of the other forces, its study has so far been completely outside of the scope of particle physics. We shall omit them in our discussions.

In studying the particles and the forces between them one finds that almost all of the macroscopic symmetry principles, which actually basically represent the symmetries observed by the electromagnetic interactions, hold intact. Some, however, are found to be violated in the weak interactions, and some new ones are found that govern the complex pattern of the mutual interaction between the two dozen odd particles. All these symmetries fall into one of four classes.

In Class A are the space–time symmetries, which include symmetries resulting from relativistic invariance and symmetries resulting from reflection invariance and time-reversal invariance. These symmetries are relatively well understood and possess direct geometrical significance. They played an important role already in classical physics and they are thoroughly incorporated into the theoretical language that describes the elementary particles. In a sense this incorporation may be even considered too intimate, since it handicaps our quest into the possibility of deviations from the space–time continuum structure, a possibility that may not be purely academic.

In Class B there is one and only one symmetry, and that is charge conjugation symmetry, which states roughly that to every particle there is a charge-conjugate particle with the opposite charge. The charge-conjugate particle is also called the anti-particle. We have seen illustrations of this in the list of elementary particles. The concept of charge conjugate symmetry is a purely quantum mechanical concept and is not related to any geometrical concepts such as rotational invariance. It derives its origin from the Dirac theory of the electron, which in turn is, viewed today, a logical consequence of the fusion of the quantum theory with the requirement of relativistic invariance. To first postulate the charge conjugation concept, as Dirac did about thirty years ago, was, however, a most daring and profound step, not unlike the first introduction of the negative numbers. The later experimental verification of the existence of the antiparticles constituted not only one of the most beautiful and forceful demonstrations of the practical consequences of the symmetry principles, it represented actually one of the most gratifying and far reaching triumphs of theoretical reasoning.

In Class C we put four conservation laws: The conservation of the number of heavy particles, the conservation of charge, the conservation of light fermions—that is, of e^-, μ^-, and ν—and the conservation of isotopic spin. These are conservation laws, and yet we put them together as representing a type of symmetry principle. The reason for this is, of course, the fact that symmetry principles and conservation laws are intrinsically related.

In the last class of symmetries we group the symmetry between identical particles and a type of symmetry called crossing symmetry which relates reaction rates of the two processes

$$A + B \to C + D$$

$$A + \bar{D} \to C + \bar{B}$$

This latter type of symmetry is of course related to the concept of charge conjugation. Its detailed implications are, however, very different from those of charge conjugation, and are still only vaguely understood.

It is most interesting to observe that the discovery of these symmetry principles follows in almost every case an inductive–deductive process. That is to say, the discovery starts with some experimental observation of some consequences of the symmetry principle involved. This then leads to the formulation of the invariance principle, from which, in turn, many other experimental consequences were derived and tested. This inductive–deductive process was involved in the discovery of relativistic invariance, of reflection invariance, of time-reversal invariance, and of the conservation laws of Class C. It is true that the dominant role played by invariance principles has led to many attempts at the reversal of this process. In other words, it has been attempted to formulate invariance principles *ad hoc* to start with, and to derive properties of the dynamical system therefrom. Such attempts—with the exception of Einstein's theory of general relativity—have invariably failed. We seem to be unable to anticipate nature in the types of symmetry to be used in a new domain of phenomena. The depth and penetration of Einstein's general relativity theory on the one hand, and the consistent failure at guessing at invariance principles on the other, indicate but one thing: that the pattern of symmetries is a complex and highly nontrivial system. We get a glimpse of this fact also from a look at the growth of our knowledge of the symmetries. We notice that as the domain of physics expands, while symmetry principles increase in number, the increase does not seem to follow independent and unrelated lines. To understand the relationship and the conceptual unification between them is a major challenge.

At the present moment, it seems to me that three related questions in this pattern of symmetry laws are particularly worth studying. The first is that of the symmetry principle that gives rise to isotopic spin conservation. There is a tendency today to forget that such a question exists. One has a tendency to say that isotopic spin conservation is a consequence of the invariance of an isotopic spin space rotation. But this is surely no explanation whatsoever, because the isotopic spin space is a fictitious concept imagined in order to "explain" isotopic spin conservation. We cannot later forget this origin and take the isotopic spin space as more real and primary than the conservation law itself. We may illustrate this with the following example: Suppose before the discovery of quantum mechanics, or even after the discovery of quantum mechanics but before Dirac's theory of the electron—suppose it was then discovered that positrons and anti-protons exist. It would have become obvious that there exists a certain charge-conjugate symmetry principle. While that would have given rise to an empirically correct description, let us compare it with our present viewpoint. We have mentioned before that according to our present understanding, charge conjugation symmetry is a consequence of quantum theory and the principle of relativistic invariance. We have also mentioned before that historically a dramatic demonstration of this origin of charge conjugation symmetry is the theoretical prediction of the existence of the symmetry before experimental verification. It need not then be emphasized that charge conjugation symmetry today is not just a shallow empirical working rule, it is rooted very deeply in the fundamentals of physical principles. The corresponding question I would like to raise about isotopic spin conservation is, that this conservation law is so far exclusively only an empirical description reaching no farther than the experimental conservation law of isotopic spin.

Another question that is most engaging is the question of the violation of some of the symmetry laws in the weaker interactions. It was found in recent years that violation of parity conservation, which represents a space–time symmetry concept, occurs rather generally among the weak interactions. Of course it does not make sense to have less symmetry for the strong interactions and more symmetry for the weak interactions—since the lack of symmetry in the strong interactions would mask the symmetry effects in the weak interactions so as to make the latter unobservable and meaningless. However, this does not explain why weak interactions *are* less symmetrical than the stronger ones. A possibility to consider here is that the symmetries satisfied by the various interactions may be tabulated as:

Forces (1):	*CP*, *T*, *C*, I
Forces (1) + (2):	*CP*, *T*, *C*,
Forces (1) + (2) + (3):	*CP*, *T*

where the force classifications are as shown in Figure 4 and the symbols, *C, P, T,* and I mean charge conjugation, space inversion, time reversal and isotopic spin summetries. We have omitted

Force (1)	Nuclear Interaction
Force (2)	Electromagnetic Interaction
Force (3)	Weak Interaction
Force (4)	Gravitational Interaction

Figure 4

here the proper Lorentz invariance and the conservation of hyperons, of leptons and of the electric charge, which are observed by all interactions. Reinterpreting *CP* as the operation that corresponds to space inversion, we thus conclude that all interactions satisfy all space time symmetry properties, i.e. proper Lorentz invariance, space inversion and time inversion invariance. This would then also emphasize the similarity between C and I as symmetries that are violated by the inclusion of weaker forces.

A third question concerns the strength of the weak interactions. It has been suggested upon good evidence that at least part of the source of the weak interactions bears the same relation to the source of the electromagnetic field as the various components of the isotopic spin. This suggestion, if verified, is most interesting. It raises the question of the structure of the electromagnetic field in relation to the light fermions. It raises the question of the behavior of the electromagnetic field in distances of the order of 10^{-16} cm. It further raises the question of the structure of the weak interactions themselves within such small distances.

The study of the symmetry laws which are related to each other and other principles of physics in puzzling and intricate ways, is one of the characteristics of elementary particle physics today. It is the hope of all that such studies will lead to a unification and consequent conceptual simplification and integration of the role of these symmetry principles in physics.

THEORETICAL DISCUSSIONS ON POSSIBLE HIGH-ENERGY NEUTRINO EXPERIMENTS*

[[60d]]

Commentary begins page 46

T. D. Lee
Columbia University, New York, New York
and
C. N. Yang
Institute for Advanced Study, Princeton, New Jersey

(Received February 23, 1960)

The weak interaction so far most extensively studied is β decay, in which the momentum transfer is of the order of a few Mev. In μ decay and μ capture, momentum transfers of the order of 100 Mev are involved. In the theory of these processes, because of the limited region of momentum transfer studied, the phenomena can be described by a few parameters usually called coupling constants. For larger momentum transfers it is obvious that the weak interactions cannot continue to be described by these constants, because of the clothed structure of the nucleons due to the strong interaction, and also because of the reasonable expectation that the weak interactions, even without the interference of the strong interactions, may not be of the simple four-spinor product form in Fermi's theory.

In the preceding Letter,[1] Schwartz points out that the neutrinos from the decay of high-energy mesons can be used to study weak interactions. We have investigated the theoretical implication of such possible experiments. Efforts are made to separate and dissociate the inferences that can be drawn from different assumptions concerning the weak interactions. In this Letter we report briefly on this work.

1. The identity of the neutrinos. In the processes

$$\pi^+ \to \mu^+ + \nu_1, \qquad (\pi \text{ decay}) \qquad (1)$$

$$\mu^- + p \to n + \nu_2, \qquad (\mu \text{ capture}) \qquad (2)$$

$$Z \to (Z-1) + e^+ + \nu_3, \qquad (\beta^+ \text{ decay}) \qquad (3)$$

it is easy to see that ν_1 and ν_2 are the same par-

□Reprinted from *Physical Review Letters* 4, 6 (March 15, 1960), 307–311.

ticle. Experimentally it is known that ν_1 and ν_3 both have helicity -1. It is simplest to assume that ν_1 and ν_3 are also the same particle. However, a test of this assumption is clearly desirable. To obtain such a test it is necessary to do some kind of capture experiment on the neutrinos or antineutrinos. For example, if ν_1 and ν_3 are different particles, then the reaction

$$n+\nu_1 \to p+e^- \tag{4}$$

does not occur.

2. Conservation of leptons. The conservation of leptons can be studied with neutrino capture experiments. For example, if both

$$\nu_1+p \to \Lambda^0+e^+ \tag{5}$$

and

$$\bar{\nu}_1+p \to \Lambda^0+e^+ \tag{6}$$

occur, there would be a violation of lepton conservation. While it is possible to study lepton conservation by helicity measurements, neutrino capture experiments seem to be the most direct and clean cut for such purposes.

In the rest of this Letter we shall assume that $\nu_1=\nu_2=\nu_3\equiv\nu$ and that the conservation of leptons holds.

3. Possible existence of a neutral lepton current. By using high-energy neutrinos it becomes possible to study whether reactions such as

$$\nu+p\to\nu+p \quad \text{and} \quad \nu+n\to\nu+n$$

exist or not, and if they exist, whether there is any similarity between these "neutral lepton currents" and the electromagnetic field. [See also the discussion in Sec. 8.]

4. Point structure of the lepton current. In the present theory of β decay, μ capture, etc., one assumes that all the weak reactions that contain both leptons and heavy particles can be represented by an effective Lagrangian of the type

$$-\mathcal{L}_{\text{eff}}=\sum_{\lambda=1}^{4}[J_\lambda(x)j_\lambda(x)+J_\lambda'(x)j_\lambda'(x)], \tag{7}$$

where

$$j_\lambda(x)=-i[\psi_l^\dagger\gamma_4\gamma_\lambda(1+\gamma_5)\psi_\nu], \tag{8}$$

$$j_\lambda'(x)=-i[\psi_\nu^\dagger\gamma_4\gamma_\lambda(1+\gamma_5)\psi_l], \tag{9}$$

l stands for either e^- or μ^-, ψ_ν and ψ_l are the field operators for ν and l, and $J_\lambda(x)$ and $J_\lambda'(x)$ are operators that act on heavy particles (including the pions, K mesons) only. Because of the Hermiticity of $\mathcal{L}_{\text{eff}}$, we have

$$J_\lambda'=\eta_\lambda J_\lambda^\dagger, \tag{10}$$

where

$$\eta_\lambda=+1 \quad \text{for} \quad \lambda=1,2,3,$$

$$\eta_\lambda=-1 \quad \text{for} \quad \lambda=4. \tag{11}$$

The nature of J_λ and J_λ' is known so far only in the nonrelativistic region. In the low-energy limit, the matrix elements of these heavy-particle current operators in the case of β decay are of the form

$$\langle p|J_\lambda|n\rangle=(i/\sqrt{2})u_p^\dagger\gamma_4\gamma_\lambda(G_V-G_A\gamma_5)u_n, \tag{12}$$

$$\langle n|J_\lambda'|p\rangle=(i/\sqrt{2})u_n^\dagger\gamma_4\gamma_\lambda(G_V^*-G_A^*\gamma_5)u_p, \tag{13}$$

where u_n and u_p are the spinor solutions of the free Dirac equations with the same 4-momenta as the physical neutron and proton; G_V and G_A are the Fermi and Gamow-Teller coupling constants. Due to the presence of strong interactions it is expected that in the high-energy region (12) and (13) do not hold. But one expects (7), which represents a "point interaction" for the leptons, to have a wider range of applicability.

The mere assumption that in the effective Lagrangian (7) the lepton current acts only at a single space-time point introduces rather strong restrictions on the forms of the cross sections for all neutrino and antineutrino reactions. For example, in either

$$\nu+n\to p+l^-+\text{pions} \tag{14}$$

or

$$\bar{\nu}+p\to n+l^++\text{pions}, \tag{15}$$

suppose one measures in the laboratory system the incoming momentum $\vec{k}_\nu$ and the outgoing lepton momentum $\vec{k}_l$ and does not measure the other kinematic quantities describing the reaction. The experimental cross section is then in a general case a function of the three real variables k_ν, k_l, and θ (= the angle between $\vec{k}_l$ and $\vec{k}_\nu$). Independently of the form of J_λ, assumption (7) restricts this function to a sum of three structure functions each of which has an unknown dependence on only two real variables: $E=k_\nu-k_l$ and $P=|\vec{k}_\nu-\vec{k}_l|$, which represent the energy transfer and the magnitude of the momentum transfer between the leptons and the strongly interacting particles. More explicitly, assumption (7) implies[2] that the cross section for (14) [also for (15)] is of the form

$$d\sigma = dk_l d(\cos\theta)(4\pi k_\nu)^{-1} k_l[(k_l+k_\nu)^2 - P^2] \times [xA_+ + x^{-1}A_- + B], \quad (16)$$

where $x = (k_l + k_\nu - P)(k_l + k_\nu + P)^{-1}$ and A_+, A_-, and B are functions of E and P only.

To test the validity of (16) it is not necessary to perform a detailed experiment for specific values of E and P. One could perform a capture experiment with a neutrino beam with a known spectrum $I(k_\nu)dk_\nu$ and measure for each event the values of x, P, and E. If $N(x, P, E)dxdPdE$ is the number of events, then (16) implies that

$$I^{-1}k_\nu^2 N(x,P,E) = [A_1 + A_2 x + A_3 x^2](1-x)^{-4}, \quad (17)$$

where A_i $(i = 1, 2, 3)$ are functions of P and E. Integrating (17) over P and E, one obtains

$$\sum I^{-1}k_\nu^2 = (a_1 + a_2 x + a_3 x^2)(1-x)^{-4}, \quad (18)$$

where a_1, a_2, a_3 are numerical constants, and the sum extends over all events with fixed x. To test the validity of (18), less than a thousand events could be enough provided that they do not cluster around one value of x.

5. Universality of weak interactions involving $e^\pm$ and $\mu^\pm$. Neglecting the mass of the μ meson, the assumption of the universality of the weak interactions implies equal differential cross sections for μ^+ and e^+ production and for μ^- and e^- production. If the mass of the μ meson is not neglected, comparison should be made between μ and e production processes in which the energy transfer E and the magnitude P of the momentum transfer from the leptons to the strongly interacting particles are fixed, provided the point structure of the lepton current discussed in Sec. 4 is valid.

6. S-symmetry. In (16) the structure functions determined by using ν are related to the appropriate matrix elements of J_λ in Eq. (7) while those determined by using $\bar{\nu}$ are related to that of J_λ'. Thus, unless J_λ and J_λ' obey some further symmetry property the reaction rates of, e.g., (14) and (15) in general are not related to each other in any simple way. A symmetry which will be called S-symmetry is of a type so as to link J and J'.

To explain the meaning of S-symmetry we observe that by using (12) and (13) at the low-energy limit it is readily verified that J_λ and J_λ' satisfy the following relation:

$$J_\lambda' = S J_\lambda S^{-1}, \quad (19)$$

where S is the product of a 180° rotation along the y axis $[\exp(i\pi I_y)]$ in the isotopic spin space multiplied by the time-reversal operator T, i.e.,

$$S = [\exp(i\pi I_y)]T. \quad (20)$$

Condition (20) will be defined as the condition for S-symmetry.[3] If it is satisfied then there should exist identities among the matrix elements of J and J' and consequently relations between reaction rates caused by ν and $\bar{\nu}$. For example, if the mass of the lepton is neglected, the structure functions [defined in (16)] for ν and $\bar{\nu}$ processes are related to each other by

$$(A_+)_\nu = (A_-)_{\bar{\nu}}, \quad (A_-)_\nu = (A_+)_{\bar{\nu}}, \quad (B)_\nu = (B)_{\bar{\nu}}. \quad (21)$$

Experimentally it is not known at present whether S-symmetry is satisfied at energies of the order of 100 Mev and up. Theoretically it has been customary[4] to assume bare particle universal Fermi interactions for all Fermions. If such an assumption is made, S-symmetry follows.

7. Conserved vector current and proportionality with the electromagnetic current. The heavy-particle current J_λ can be written as a sum,

$$J_\lambda = V_\lambda + A_\lambda, \quad (22)$$

where V_λ and A_λ are, respectively, its vector part and axial-vector part. Recently, Feynman and Gell-Mann[5] proposed that the vector part V_λ satisfies the conservation law

$$\partial V_\lambda / \partial x_\lambda = 0. \quad (23)$$

Furthermore it is proposed that V_λ is equal to the corresponding isotopic vector part of the electromagnetic current times a constant (hereafter called the proportional vector current hypothesis). A sensitive test of the proposal can be given by studying reactions such as

$$\nu + n \to e^- + p, \quad (24)$$

$$\bar{\nu} + p \to e^+ + n, \quad (25)$$

and comparing them with existing data on electron scattering by nucleons[6] at the same momentum transfer to the nucleons.

Making the proportional vector current hypothesis, the electron scattering experiments[6] show that V_λ is proportional to

$$F(q^2) = [1 + \frac{1}{12} q^2 \alpha^2]^{-2}, \quad \alpha = 0.8 \times 10^{-13} \text{ cm},$$

where q^2 is the invariant four-momentum transfer squared. We determine the constant of proportionality by the condition $G_V = 10^{-5}/M^2$. For orientation purposes we assumed that A_λ is given by

$$\langle p | A_\lambda | n \rangle = 1.2\ G_V F(i/\sqrt{2}) u_p^{\dagger} \gamma_4 \gamma_\lambda \gamma_5 u_n,$$

and calculated the cross sections for (24) and (25). The results are exhibited in Fig. 1.

8. Possible existence of a weakly coupled Boson $W^\pm$. The question whether the weak interactions are "transmitted" by a Boson field was already discussed in Yukawa's original work on the meson. If such a Boson field $W^\pm$ exists, it must have spin 1, and one can also conclude that its mass m_W is $\gtrsim m_K$. The nonlocality of the weak interactions implied by the finite mass of the transmitting field W has been discussed before.[7] For μ decay, the change in the Michel parameter ρ is given by

$$\rho - 0.75 \cong \tfrac{1}{3}(m_\mu / m_W)^2.$$

A value of $m_W > m_K$ is thus consistent with existing experiments.

The coupling of the $W^\pm$ to the leptons is characterized by a coupling constant,

$$g^2/4\pi = (\pi\sqrt{2})^{-1} G_V m_W^{\ 2} < 6.4 \times 10^{-7}.$$

The decay rate of $W \to e + \nu$ is found to be

$$\lambda_{W \to e + \nu} = G_V m_W^{\ 3} (6\pi\sqrt{2})^{-1}, \qquad (26)$$

and the ratio of the decay rate of $W \to \mu + \nu$ to that of $W \to e + \nu$ is

$$2 v_\mu^{\ 2} (3 + v_\mu)/(1 + v_\mu)^3, \qquad (27)$$

where v_μ is the velocity of the muon in the rest system of W. Decays of W into pions are also possible. The lifetime of W is $< 10^{-17}$ sec.

While the existence of a $W^\pm$ as a virtual particle does not change in any essential way the considerations of the previous sections, its production makes possible a much higher cross section for neutrino reactions through the pair creation of W^+ and l^- in the Coulomb field of a target nucleus:

$$\nu + Z \to W^+ + l^- + Z. \qquad (28)$$

The cross section for (28) is large compared with those without W production. At values of $k_\nu >> m_W^{\ 2}/2q_0 \approx 2$ Bev, where $q_0 = \hbar/$nucleon radius, the cross section is given by

$$\sigma \cong (6\pi\sqrt{2})^{-1} (137)^{-2} Z^2 G_V [\ln(2k_\nu q_0 / m_W^{\ 2})]^3, \qquad (29)$$

which, for $Z = 26$, is of the order of 10^{-35} cm^2. The $W^\pm$ produced are easily identifiable through their decay products $e + \nu$ or $\mu + \nu$. If experimentally no $W^\pm$ is found, it would be possible to set a lower limit on the value of m_W.

The existence of the intermediate Boson $W^\pm$ has[8] been discussed in connection with the question of the absence of $\mu^\pm \to e^\pm + \gamma$. On reasonably sure grounds one could conclude that the existence of $W^\pm$ requires that in the notation of (1), (2), (3), $\nu_1 \neq \nu_3$. In other words, the existence of W would imply that reaction (4) does not occur.

For processes without W production, the discussions of Sec. 4 above concerning a point structure of the lepton current remain unchanged, but the current J_λ now includes the effect of the propagator of W. Furthermore a corresponding change in the proportional vector current hypothesis is necessary.

The question of a neutral W^0 will not be examined here.

9. Interactions with extremely large momentum transfers. For momentum transfers of the order of $(G_V)^{-1/2} \sim 300$ Bev/c, quantities other

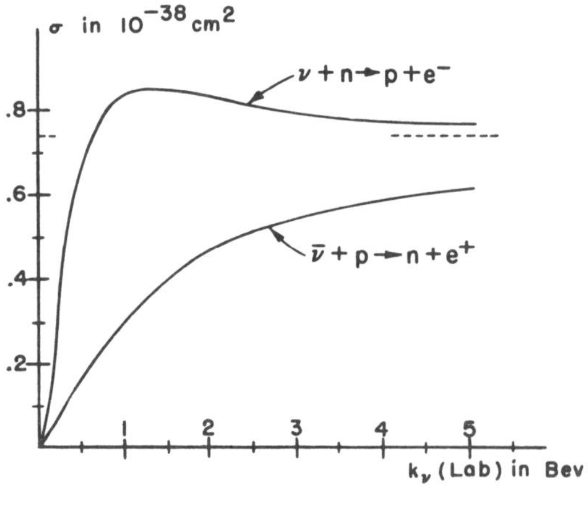

FIG. 1. "Elastic" neutrino cross sections. The dashed line represents the limit of σ as $k_\nu \to \infty$.

than the lowest power of G_V become important. For example, for $e^- + \nu \to \mu^- + \nu$, the cross section predicted by the lowest perturbation formula approaches the limit set by the unitary condition as the neutrino momentum in the center-of-mass system, $p_\nu \to 300$ Bev. For processes involving pion clouds, high-order terms in G_V presumably also become important as $p_\nu \to 300$ Bev. The description of the weak interaction at such high energies would seem to need ideas radically different from the current picture.

We wish to thank Professor M. Schwartz for many discussions.

---

*Work supported in part by the U. S. Atomic Energy Commission.

[1]M. Schwartz, preceding paper [Phys. Rev. Letters, 4, 306 (1960)].

[2]In Eq. (16), we neglect the mass of the lepton. For reactions in which the mass of the lepton may not be neglected, equations similar to but slightly more complicated than (16) result. For reactions in which strange particles are produced, (16) remains valid.

[3]By using the *CPT* theorem it is easy to see that the *S*-symmetry is closely related to the classification of weak interactions given by S. Weinberg, Phys. Rev. 112, 1375 (1958).

[4]See, e.g., M. L. Goldberger and S. B. Treiman, Phys. Rev. 111, 354 (1958).

[5]R. P. Feynman and M. Gell-Mann, Phys. Rev. 109, 193 (1958).

[6]See, e.g., R. Hofstadter, F. Bumiller, and M. R. Yearian, Revs. Modern Phys. 30, 482 (1958).

[7]T. D. Lee and C. N. Yang, Phys. Rev. 108, 1611 (1957).

[8]G. Feinberg, Phys. Rev. 110, 1482 (1958); J. Schwinger, Ann. Phys. 2, 407 (1957).

[[60e]]

Commentary begins page 47

Implications of the Intermediate Boson Basis of the Weak Interactions: Existence of a Quartet of Intermediate Bosons and Their Dual Isotopic Spin Transformation Properties

T. D. Lee
Columbia University, New York, New York

AND

C. N. Yang
Institute for Advanced Study, Princeton, New Jersey

(Received April 11, 1960)

Assuming that all weak interactions are transmitted through an intermediate boson field W, it is shown that the observed $|\Delta \mathbf{I}|=\frac{1}{2}$ rule and the small observed mass difference between K_1 and K_2 lead to the conclusion that there exist four W particles: $W^{\pm}$, W^0, and $\bar{W}^0$. Furthermore, a natural assignment of the isotopic spin transformation property of these W particles follows a dual scheme in which the W's behave sometimes as $I=\frac{1}{2}$ and sometimes as $I=1$ particles. Various experimental implications are discussed, including neutrino capture experiments, strong collisions exhibiting apparent nonconservation of strangeness, and strong collisions with apparent lepton production.

I. INTRODUCTION

It is the purpose of this paper to study the consequences of the following three propositions:

(i) All weak interactions are transmitted through an intermediate boson field W.

(ii) The mass difference between K_1 and K_2 is of the order of $\sim 10^{-5}$ ev and not ~ 10 ev. This implies[1] that $\Delta S=\pm 2$ interactions are absent in the usual weak interactions.

(iii) The $|\Delta \mathbf{I}|=\frac{1}{2}$ rule holds for the strangeness nonconserving decays of particles, where $\mathbf{I}$ is the total isotopic spin of the strongly interacting particles (i.e., baryons and the K and π mesons).

Of these propositions, (iii) has had quite impressive experimental support.[2] Evidence for (ii) has been re-

[1] L. B. Okun and B. M. Pontecorvo, J. Exptl. Theoret. Phys. (U.S.S.R.) **32**, 1587 (1957) [translation: Soviet Phys.-JETP **5**, 1297 (1957)].

[2] See the review article by R. Dalitz, Revs. Modern Phys. **31**, 823 (1959). See also F. Crawford et al., Phys. Rev. Letters **2**, 266 (1959); J. L. Brown et al., Phys. Rev. Letters **3**, 563 (1959).

□ Reprinted from *The Physical Review* 119, 4 (August 15, 1960), 1410–1419.

ported[3] recently. (i) is so far a purely theoretical speculation.

The main conclusions of this paper are: (a) that there must exist at least two neutral W fields, and (b) that the three propositions (i), (ii), and (iii) lead naturally to a quite definite interaction scheme between the W's and the strongly interacting particles which seems to put the $|\Delta \mathbf{I}|=\frac{1}{2}$ rule on a less *ad hoc* basis than in various previous discussions.[4] This scheme is first deduced in Secs. IV and V for a specific model from propositions (i), (ii), and (iii). It is then discussed for the general case in the next three sections. The W particles behave in this scheme sometimes as $I=\frac{1}{2}$ and sometimes as $I=1$ particles. For this reason they are referred to as schizons. The usual $|\Delta \mathbf{I}|=\frac{1}{2}$ rule is shown to consist of two different *types* of selection rules: one originating from the $I=\frac{1}{2}$ aspect of the schizon, the other from the extent of the *difference* of the $I=\frac{1}{2}$ and $I=1$ aspects of the schizons. It also follows that there are decays and reactions which show a $|\Delta \mathbf{I}|=1$ rule originating from the $I=1$ aspect of the schizon. [A possible variation of the scheme is discussed in Sec. VII which allows for an $I=0$ component of the schizons.]

Various experimental implications and therefore tests of the schizon basis of the weak interactions are discussed, especially in Secs. VI, X, XI, and XII.

II. SOME PROPERTIES OF $W^{\pm}$

We first summarize here some immediate consequences of (i). The spin of W is 1 in order to transmit the V and A type of weak interactions. Its mass m_W is $>m_K$ in order to prevent a fast decay $K^{\pm}\rightarrow W^{\pm}+\gamma$.

To reconcile[5] the absence of $\mu^{\pm}\rightarrow e^{\pm}+\gamma$, it seems necessary to have two sets of two-component neutrino fields ψ_ν and $\psi_{\nu'}$ coupled, respectively, to the e^- and μ^- fields. Both ψ_ν and $\psi_{\nu'}$ represent left-handed ν particles and right-handed $\bar{\nu}$ particles. The charged $W^{\pm}$ particles are coupled to the leptons through the interaction

$$ig_{e\nu}\psi_e^{\dagger}\gamma_4(1+\gamma_5)\psi_\nu\phi_\lambda^{\star}+ig_{\mu\nu}\psi_\mu^{\dagger}\gamma_4(1+\gamma_5)\psi_{\nu'}\phi_\lambda^{\star} +\text{Hermitian conjugate}, \quad (1)$$

where ψ_e, ψ_μ, ψ_ν, $\psi_{\nu'}$, and ϕ_λ denote the fields describing e^-, μ^-, ν, ν' and W_λ^+. The operator $\phi_\lambda^{\star}$ is related to the Hermitian conjugate field $\varphi_\lambda^{\dagger}$ by[6]

$$\phi_\lambda^{\star}=\eta_\lambda\phi_\lambda^{\dagger},$$

where $\eta_\lambda=+1$ for $\lambda=1, 2, 3$, and $\eta_\lambda=-1$ for $\lambda=4$. The coupling of $W^{\pm}$ to the proton and neutron fields p and n is given by

$$J_\lambda\phi_\lambda^{\star}+\text{Hermitian conjugate}. \quad (2)$$

The low momentum transfer matrix element of J_λ is related to the transition amplitudes of β decay. Let us write the matrix element between the physical states of a neutron and a proton at rest:

$$\langle p|J_\lambda^{\star}|n\rangle=ig_{np}{}^{*}u_p^{\dagger}\gamma_4\gamma_\lambda(1+a\gamma_5)u_n, \quad (3)$$

where u_p and u_n are the spinor solutions of the free Dirac equations for the proton and the neutron. By suitably choosing the phases of ϕ_λ, ψ_μ and ψ_e we shall make g_{np}, $g_{e\nu}$, and $g_{\mu\nu}$ all real and positive. The β-decay coupling constants G_V and G_A are then given by

$$G_V=\sqrt{2}g_{e\nu}g_{np}(m_W)^{-2}, \quad (4)$$

and

$$G_A=-aG_V. \quad (5)$$

Comparison of the μ-decay rate and the experimental magnitude of $G_V=10^{-5}\ M^{-2}$ where M=nucleon mass shows that[7]

$$g_{np}=g_{\mu\nu}. \quad (6)$$

The ratio of the experimental decay rates[8] $\pi^+\rightarrow e^++\nu$ and $\pi^+\rightarrow\mu^++\nu'$ leads to the conclusion[7]

$$g_{e\nu}=g_{\mu\nu}. \quad (7)$$

Combining (4), (6), and (7) one obtains

$$g_{e\nu}=g_{\mu\nu}=g_{np}=m_W G_V{}^{\frac{1}{2}}2^{-\frac{1}{4}}. \quad (8)$$

The strength of the lepton-W coupling is measured by

$$(2g_{e\nu})^2/4\pi=(\pi\sqrt{2})^{-1}G_V m_W{}^2>6.4\times10^{-7}. \quad (9)$$

The $W^{\pm}$ particles are unstable against decays into $e^{\pm}+\nu$, $\mu^{\pm}+\nu$, and 2π, 3π, etc., modes. The decay rates[9] into leptons are given by

$$\lambda_{W\rightarrow\mu+\nu}\cong\lambda_{W\rightarrow e+\nu}=G_V m_W{}^3(6\pi\sqrt{2})^{-1}>8\times10^{16}\ \text{sec}^{-1}. \quad (10)$$

The existence of W implies a "nonlocality" of a size $\sim m_W{}^{-1}$ for the presently observed weak interactions. For μ decay the Michel parameter ρ is given by[10]

$$\rho-0.75\cong\tfrac{1}{3}(m_\mu/m_W)^2,$$

which is consistent with the present experimental results.[11]

Furthermore (i) implies that $W^{\pm}$ is also coupled to a strangeness-nonconserving current $\mathcal{S}$ (generated by the strongly interacting particles):

$$\mathcal{S}_\lambda\phi_\lambda^{\star}+\text{Hermitian conjugate} \quad (11)$$

[3] F. Muller et al., Phys. Rev. Letters **4**, 418 (1960).

[4] See, e.g., S. B. Treiman, Nuovo cimento **15**, 916 (1960), B. d'Espagnat, J. Prentki, and A. Salam, Nuclear Phys. **5**, 447 (1958).

[5] G. Feinberg, Phys. Rev. **110**, 1482 (1958); J. Schwinger, Ann. Phys. **2**, 407 (1957); M. Gell-Mann, Revs. Modern Phys. **31**, 834 (1959). See also B. Pontecorvo, J. Exptl. Theoret. Phys. (U.S.S.R.) **37**, 1751 (1959) [translation: Soviet Phys.-JETP (to be published)], for possible neutrino experiments to test the existence of ν and ν'. See also S. A. Bludman, Bull. Am. Phys. Soc. **4**, 80 (1959), and Gatlinburg Conference on Weak Interactions, 1958 (unpublished).

[6] Throughout this paper, we use the superscript $\star$ to indicate the product of η_λ times the Hermitian conjugation operator.

[7] See R. P. Feynman and M. Gell-Mann, Phys. Rev. **109**, 193 (1958).

[8] T. Fazzini et al., Phys. Rev. Letters **1**, 247 (1958). G. Impeduglia et al., Phys. Rev. Letters **1**, 249 (1958).

[9] T. D. Lee and C. N. Yang, Phys. Rev. Letters **4**, 307 (1960).

[10] See T. D. Lee and C. N. Yang, Phys. Rev. **108**, 1611 (1957).

[11] R. J. Plano and A. Lecourtois, Bull. Am. Phys. Soc. **4**, 82 (1959).

to make possible the observed decays,

$$\Lambda \to p + e^- + \bar{\nu}, \tag{12}$$

and

$$K^+ \to \pi^0 + \mu^+ + \nu', \text{ etc.} \tag{13}$$

Such couplings introduce further decay modes of the $W^\pm$ particles such as $W^\pm \to K^\pm + \gamma$, $W^\pm \to K^\pm + \pi^0$, etc.

III. CONSEQUENCES OF PROPOSITIONS (i) AND (ii)

Reaction (12) implies the existence of the transitions

$$\Lambda \rightleftharpoons p + W^-,$$

and therefore also of

$$\bar{K}^0 \rightleftharpoons \pi^+ + W^-.$$

Proposition (ii) then implies the absence of

$$K^0 \rightleftharpoons \pi^+ + W^-.$$

In other words the current $\mathcal{S}_\lambda$ associated with the annihilation of a W^- must not[12] increase the strangeness of a state by $+1$. One easily concludes that this implies

$$\mathcal{S}_\lambda S = (S+1)\mathcal{S}_\lambda, \tag{14}$$

where S is the strangeness operator. A well-known consequence[12] is that

$$\Sigma^+ \not\to n + e^+ + \nu.$$

Another consequence is, e.g., that

$$\nu' + \text{nucleon} \not\to \mu^- + (\text{system with strangeness} -1). \tag{15}$$

Thus,

$$\nu' + n \not\to \Sigma^+ + \mu^-,$$

$$\nu' + n \not\to \Lambda + \mu^- + \pi^+.$$

Still another consequence is, e.g.,

$$K^+ \not\to \pi^+ + \pi^+ + \mu^- + \bar{\nu}'. \tag{16}$$

IV. A SIMPLE MODEL

We shall in this and the following section demonstrate conclusions (a) and (b) stated in the introduction. For the sake of clarity of presentation let us consider first a specific model in which J_λ and $\mathcal{S}_\lambda$ each consists of only one term[13]

$$J_\lambda \equiv (\bar{n}p) f_1,$$

and

$$\mathcal{S}_\lambda \equiv (\bar{\Lambda}p) f_2, \tag{17}$$

where f_1 and f_2 are real numerical constants. [The phase of f_2 can be arbitrarily chosen because of strangeness conservation in the strong and electromagnetic interactions, which leaves arbitrary the choice of the phase of $(\bar{\Lambda}p)$.] Under an isotopic rotation J_λ forms a vector together with $(1/\sqrt{2})[(\bar{p}p)-(\bar{n}n)]f_1$ and $(\bar{p}n)f_1$, while $\mathcal{S}_\lambda$ forms a doublet with $(\bar{\Lambda}n)f_2$. For a strangeness nonconserving decay such as $\Lambda \to p + \pi^-$, after the elimination of the virtual W field the effective matrix element is that of $J_\lambda \mathcal{S}_\lambda^\star$. To satisfy the $|\Delta\mathbf{I}| = \frac{1}{2}$ rule it is clear that the other isotopic partners of J_λ and $\mathcal{S}_\lambda$ will have to enter the picture. Neutral currents[14] and neutral W's will therefore have to be introduced.

We now examine W-J couplings and W-$\mathcal{S}$ couplings so as to generate the $|\Delta\mathbf{I}| = \frac{1}{2}$ rule. There seems at this point to be two possibilities:

(A) $\mathbf{I}$ conservation is preserved in the W-J couplings, while $|\Delta\mathbf{I}| = \frac{1}{2}$ is caused by the W-$\mathcal{S}$ couplings. In other words under an isotopic spin rotation the W-J coupling is a scalar while the W-$\mathcal{S}$ coupling is one component of a doublet. Since J behaves like a vector, this arrangement requires that W^+, W^0, and W^- form a triplet (like the pions) to which one assigns the isotopic spin $I=1$. The W-J coupling is then[15]

$$f_1\{(\bar{n}p)W^\star + (1/\sqrt{2})[(\bar{p}p)-(\bar{n}n)]W^0 + (\bar{p}n)W\}, \tag{18}$$

and the W-$\mathcal{S}$ coupling,

$$f_2\{(\bar{\Lambda}p)W^\star - (1/\sqrt{2})(\bar{\Lambda}n)W^0\} + \text{Hermitian conjugate}. \tag{19}$$

The W^0 term in (19) implies the existence of

$$n \rightleftharpoons \Lambda + W^0; \tag{20}$$

its Hermitian conjugate that of

$$\Lambda \rightleftharpoons n + W^0. \tag{21}$$

Together they give rise to the transition

$$n + n \rightleftharpoons \Lambda + W^0 + n \rightleftharpoons \Lambda + \Lambda \tag{22}$$

in contradiction to proposition (ii).

This possibility therefore does not work out in the simple form described above.[16]

(B) $\mathbf{I}$ is conserved in W-$\mathcal{S}$ couplings, while $|\Delta\mathbf{I}| = \frac{1}{2}$ is caused by the W-J couplings. To satisfy $\mathbf{I}$ conservation in W-$\mathcal{S}$ couplings it is necessary to have W^+ and W^0 form an isotopic doublet. Therefore $\bar{W}^0$ and W^- also form a doublet. Since W^0 and $\bar{W}^0$ have different isotopic rotation properties, they cannot be the same particle. The four W's thus form a quartet very similar

[12] Such possibilities have been discussed extensively in the literature in connection with the proposal that all weak interactions originate from couplings of the form (current)$\times$(current). See, in particular, R. P. Feynman and M. Gell-Mann, Phys. Rev. **109**, 193 (1958).

[13] From here on we drop the index λ. Notice that $(\bar{n}p) \equiv i\bar{\psi}_n\gamma_\lambda(1+\gamma_5)\psi_p$ and $(\bar{n}p)^\star = (\bar{p}n) = i\bar{\psi}_p\gamma_\lambda(1+\gamma_5)\psi_n$ where $\bar{\psi} \equiv \psi^\dagger\gamma_4$.

[14] The possible existence of neutral currents has also been discussed in the literature. See, in particular, S. Treiman, Nuovo cimento **15**, 916 (1960).

[15] From here on, we use for convenience W to represent ϕ_λ which annihilates a W^+ particle. Thus, for example, $(\bar{n}p)W^\star$ represents $i\bar{\psi}_p\gamma_\lambda(1+\gamma_5)\psi_n\phi_\lambda^\star$. Similarly [see (25) and (26)], $(\bar{p}p)W_a{}^0$ and $(\bar{p}p)W_b{}^0$ represent, respectively, $i\bar{\psi}_p\gamma_\lambda(1+\gamma_5)\psi_p(\phi_a{}^0)_\lambda$ and $i\bar{\psi}_p\gamma_\lambda(1+\gamma_5)\psi_p(\phi_b{}^0)_\lambda$ where $(\phi_a{}^0)_\lambda = -(\phi_\lambda{}^0 + \phi_\lambda{}^{0\star})/\sqrt{2}$ and $(\phi_b{}^0)_\lambda = i(\phi_\lambda{}^0 - \phi_\lambda{}^{0\star})/\sqrt{2}$. As another example, in (29) and (30) $JW^\star$ and $\mathcal{S}^0 W^{0\star}$ represent, respectively, $J_\lambda\phi_\lambda{}^\star$ and $\mathcal{S}_\lambda{}^0(\phi_\lambda{}^0)^\star$.

[16] As we shall see in the next section, actually possibility (B) can lead to a final result which is expressible as possibility (A) plus an additional neutral W field whose effect is to cancel out the process (22).

to the quartet of K particles. The W-$\mathcal{S}$ coupling is

$$f_2\{(\bar{\Lambda}p)W^{\star}+(\bar{\Lambda}n)W^{0\star}\}+f_2\{(\bar{n}\Lambda)W^0+(\bar{p}\Lambda)W\}. \quad (23)$$

The W-J coupling is now one component of an isotopic doublet. Thus it is

$$f_1\{(\bar{n}p)W^{\star}-\tfrac{1}{2}[(\bar{p}p)-(\bar{n}n)]W^{0\star}\} + \text{Hermitian conjugate}. \quad (24)$$

The interactions (1), (23), (24) taken together with the strong and electromagnetic interactions clearly are consistent with propositions (i), (ii), and (iii).

V. A SIMPLE MODEL (CONTINUED)

We have seen in the last section that propositions (i), (ii), and (iii) lead to the existence of W^0 and $\bar{W}^0$ forming with $W^{\pm}$ a quartet of two isotopic doublets. We shall now write the W-J interaction (24) in the following form[15]

$$f_1\{(\bar{n}p)W^{\star}+(1/\sqrt{2})[(\bar{p}p)-(\bar{n}n)]W_a^0+(\bar{p}n)W\}, \quad (25)$$

where

$$W_a^0=(-W^0-W^{0\star})/\sqrt{2}. \quad (26)$$

In this form it closely resembles the rejected expression (18), and demonstrates the following fact:

If one regards W^+, W_a^0, and W^- as forming an isotopic vector then the W-J interaction conserves **I**. [The difficulty discussed under (A) does not now arise because the field

$$W_b^0\equiv i(W^0-W^{0\star})/\sqrt{2} \quad (27)$$

describes another neutral particle W_b^0 and the process

$$n+n\rightleftharpoons\Lambda+W_b^0+n\rightleftharpoons\Lambda+\Lambda$$

exactly cancels $n+n\rightleftharpoons\Lambda+W_a^0+n\rightleftharpoons\Lambda+\Lambda$. See reference 16.]

The picture that emerges is as follows:

The four W fields are coupled to the strongly interacting particles by W-J and W-$\mathcal{S}$ interactions which are roughly comparable in strength. Each of these interactions taken separately with the strong interactions satisfy **I** conservation. For the W-J interaction, **I** conservation is satisfied with the assignment that W^+, W_a^0, W^- form an isotopic triplet. For the W-$\mathcal{S}$ interaction, **I** conservation is satisfied with the assignment that W^+, W^0 and $\bar{W}^0$, W^- form two isotopic doublets. Violation of **I** conservation only occurs when the mixed effects of W-J and W-$\mathcal{S}$ interactions are observed. In such cases, to the order of the strength of the usual weak interactions (i.e., amplitude $\propto G_V$) the violation of **I** conservation satisfies $|\Delta\mathbf{I}|=\frac{1}{2}$ since that represents the extent of the difference between the two isotopic spin transformation properties of the W particles.

VI. THE *W* PARTICLES AS SCHIZONS

The reasonings and conclusions of the last two sections are obviously not restricted to the specific model discussed. One can conclude in general that propositions (i), (ii), and (iii) lead[17] to the existence of $W^{\pm}$, W^0, and $\bar{W}^0$ as transmitters of weak interactions. The W's are generated by charge-current densities formed by the leptons, and by the strongly interacting particles in strangeness conserving motions and in strangeness nonconserving motions. A natural possibility is that these charge-current densities have the same transformation properties as those discussed in the model above. We shall now discuss these properties explicitly.

One may write the interaction Lagrangian density in the following form:

$$\mathcal{L}_{\text{strong}}+\mathcal{L}_{\gamma}+\mathcal{L}_{Wl}+\mathcal{L}_{WJ}+\mathcal{L}_{W\mathcal{S}}, \quad (28)$$

where $\mathcal{L}_{\gamma}$ denotes the electromagnetic interactions, $\mathcal{L}_{Wl}$ denotes the W-lepton interaction (1) [neutral lepton currents will be discussed in Sec. VIII], and $\mathcal{L}_{WJ}$ and $\mathcal{L}_{W\mathcal{S}}$ are given by

$$\mathcal{L}_{WJ}=JW^{\star}+J_a^0W_a^0+J^{\star}W, \quad (29)$$

and

$$\mathcal{L}_{W\mathcal{S}}=\{\mathcal{S}W^{\star}+\mathcal{S}^0W^{0\star}\}+\text{Hermitian conjugate}. \quad (30)$$

Here W and W^0 represent the fields for the W particles,[15] W_a^0 is defined in (26). $\mathcal{S}$ and $\mathcal{S}^0$ represent currents for which[18] $\Delta N=0$, $\Delta S=-1$, where N = number of baryons. Thus both satisfy (14) and

$$\mathcal{S}N-N\mathcal{S}=0. \quad (31)$$

J, J_a^0, and $J^{\star}$ represents currents for which $\Delta N=0$, $\Delta S=0$. Under an isotopic rotation, J, J_a^0, $J^{\star}$ transform[19] like an isotopic vector and $\mathcal{S}$, $\mathcal{S}^0$ an isotopic doublet. One also has the additional condition[6,13]

$$(J^0)^{\star}=J^0. \quad (32)$$

Under an isotopic rotation, $\mathcal{L}_{\text{strong}}+\mathcal{L}_{WJ}$ is invariant if W^+, W_a^0, and W^- transform like an isotopic vector[20] (and are therefore considered to have $S=0$), while $\mathcal{L}_{\text{strong}}+\mathcal{L}_{W\mathcal{S}}$ is invariant if W^+, W^0 and $\bar{W}^0$, W^- transform[21] like two isotopic doublets (and are therefore considered to have strangenesses 1, 1, -1, and -1, respectively).

The dual isotopic spin transformation property of the W particles gives rise to an integrated view of many interesting characteristics of the weak interactions, such

---

[17] It is possible to have more W fields than these four. E.g., it may be that the neutral lepton currents $(\bar{e}e)$, etc., generate additional neutral W fields. To have more W fields, however, is contrary to the spirit of proposition (i). For reasons of economy of the number of fields we shall not further discuss such possibilities.

[18] We use the following convention: $\Delta S=j$ if $\langle b|\mathcal{S}|a\rangle\neq 0$ only for $S_b-S_a=j$.

[19] To be specific, we adopt the convention that the fields $\psi_p^{\dagger}$ and $\psi_n^{\dagger}$ transform under an isotopic rotation like $|\frac{1}{2},\frac{1}{2}\rangle$ and $|\frac{1}{2},-\frac{1}{2}\rangle$ where we use the notations of A. Edmonds, *Angular Momentum in Quantum Mechanics* (Princeton University Press, Princeton, New Jersey, 1957). Then $J^{\star}$, $-J_a^0$, and $-J$ transform like $|1,1\rangle$, $|1,0\rangle$ and $|1,-1\rangle$, and $\mathcal{S}^0$ and $-\mathcal{S}$ like $\frac{1}{2},\frac{1}{2}\rangle$ and $|\frac{1}{2},-\frac{1}{2}\rangle$.

[20] To be more precise, in the convention of footnote 19, $\phi^{\star}$, $-\phi_a^0$ and $-\phi$ transform like $|1,1\rangle$, $1,0\rangle$ and $|1,-1\rangle$.

[21] To be more precise, in the convention of footnote 19, $\phi^{\star}$, $\phi^{0\star}$ transform like $|\frac{1}{2},\frac{1}{2}\rangle$ and $|\frac{1}{2},-\frac{1}{2}\rangle$. So do ϕ^0, $-\phi$.

as the $|\Delta \mathbf{I}|=\frac{1}{2}$ rule, and the so-called $\Delta Q=\Delta S$ rule discussed in Sec. III. Because of this dual property the W particles will be called schizons. [One may mention that in fact the transformation property of W under a space inversion (*without* charge conjugate) also manifests a dual character, because J and $\mathcal{S}$ both contain vector and axial vector parts.]

The reactions that are caused by the W interactions are classifiable into the following classes (cases where the electromagnetic processes are important will not be considered here):

(α) Those in which one real (not virtual) W particle is involved, e.g.,

$$\pi^- + p \rightarrow \Lambda^0 + W^0,$$

and

$$\pi^- + p \rightarrow p + W^-.$$

These involve transition amplitudes of the first order of either $\mathcal{L}_{WJ}$ or $\mathcal{L}_{W\mathcal{S}}$. This class of reactions is characterized by the strength $\sim g^2/4\pi \sim 10^{-6}$. In these reactions $\mathbf{I}$ and S are conserved ($\mathcal{L}_{WJ}$ and $\mathcal{L}_{W\mathcal{S}}$ terms do not interfere with each other) provided the W particles receive the proper $\mathbf{I}$ and S assignments stated above. However, because of the short lifetimes of the W particles, "apparent" violation of $\mathbf{I}$ conservation and S conservation may occur. This will be discussed more in detail in Sec. X.

(β) Those in which four leptons and no (real particles) W are involved, e.g., μ decay. This and the subsequent classes are characterized by the strength $(g^2/4\pi)^2 \sim 10^{-13}$.

(γ) Those in which two leptons and no (real particles) W are involved, and in which there is no change in strangeness among the strongly interacting particles, e.g., β decay. For this class, the leptons interact through a W particle. The interaction of this W with the baryons and bosons is described by $\mathcal{L}_{WJ}$ and therefore conserves $\mathbf{I}$ and S with the proper assignments.

Examples of this class of reactions are the decays[22]

$$\Sigma^+ \rightarrow \Lambda^0 + e^+ + \nu, \tag{33}$$

$$\Sigma^- \rightarrow \Lambda^0 + e^- + \bar{\nu}. \tag{34}$$

It is easy to prove that they have the same rate except for the phase space factor due to the difference between $\Sigma^\pm$ masses. The identity of their rates is a consequence of the requirement that J, $J_a{}^0$ and $J^\star$ form an isotopic vector, which in turn is an essential feature of the present schizon interpretation of the weak interactions. Intensity rules such as these can be described (in analogy with the usual $|\Delta \mathbf{I}|=\frac{1}{2}$ rule) as given by $|\Delta \mathbf{I}|=1$.

Still another type of reactions of this class are found in the neutrino capture reactions.[23,9] These will be discussed later in Sec. XI.

(δ) Those in which two leptons and no (real particle) W are involved, and in which there is a change of strangeness $\Delta S=\pm 1$ among the strongly interacting particles. This is similar to the above case except that the interaction between the strongly interacting particles and the virtual W is described by $\mathcal{L}_{W\mathcal{S}}$.

One example of this class of reactions is the leptonic decay[24] mode of K. The $\mathbf{I}$ conservation property of $\mathcal{L}_{W\mathcal{S}}$ implies here that for the strongly interacting particles $|\Delta \mathbf{I}|=\frac{1}{2}$. Consequences of this rule have been explored before.[24] Another consequence is, e.g., (16). (It is important to remember here that $\mathcal{L}_{Wl}$ does not seem to involve W^0. See Sec. VIII.) Further examples will be discussed in Sec. XI.

(ϵ) Those in which no leptons and no (real particles) W are involved. The transition amplitudes are proportional to some elements of $\mathcal{L}_{WJ}{}^2$, $\mathcal{L}_{W\mathcal{S}}{}^2$ or $(\mathcal{L}_{WJ}\mathcal{L}_{W\mathcal{S}})$. Those proportional to $\mathcal{L}_{WJ}{}^2$ and $\mathcal{L}_{W\mathcal{S}}{}^2$ observe $\mathbf{I}$ and S conservations, and are therefore of no experimental interest since they are thoroughly masked by the strong interactions. Those proportional to $(\mathcal{L}_{WJ}\mathcal{L}_{W\mathcal{S}})$ satisfy $|\Delta \mathbf{I}|=\frac{1}{2}$, and therefore $\Delta S=\pm 1$. This is so, because (29) can also be written as [in analogy with (24)]

$$\mathcal{L}_{WJ}=\{JW^\star-(1/\sqrt{2})J^0W^{0\star}\}+\text{Hermitian conjugate},$$

showing that if $W^\star$ and $W^{0\star}$ are taken to be a doublet, $\mathcal{L}_{W\mathcal{S}}$ causes $\Delta\mathbf{I}=0$ and $\mathcal{L}_{WJ}$ causes $|\Delta \mathbf{I}|=\frac{1}{2}$.

The $\Delta S=\pm 1$ rule leads directly to proposition (ii). (See Sec. VII about electromagnetic corrections.) The nonleptonic decay modes of K and of hyperons are examples of this class of reactions. The $|\Delta \mathbf{I}|=\frac{1}{2}$ rule for these reactions is due to the dual aspects of the isotopic spin properties of the W particles (just as in the model discussed in Secs. IV and V). In contrast, the $|\Delta \mathbf{I}|=\frac{1}{2}$ rule for reactions of class (δ) is due to the fact that for those reactions W behaves like a particle with $I=\frac{1}{2}$.

VII. REMARKS

We make a few general remarks here about the latitude allowed in the interaction scheme described in the last section.

1. In (29) a $W_b{}^0$ interaction was not included. It is clear that it may be included if it involves a neutral current $J_b{}^0$ that is Hermitian and is an isotopic scalar:

$$\mathcal{L}_{WJ}=JW^\star+J_a{}^0W_a{}^0+J^\star W+J_b{}^0W_b{}^0. \tag{35}$$

Also $J_b{}^0$ must satisfy[18]

$$\Delta N=0,\ \Delta S=0.$$

Inclusion of this term does not change any of the considerations of the last section. A possible form for $J_b{}^0$ is,

$$(\bar{p}p)+(\bar{n}n), \tag{36}$$

or

$$2(\bar{\Lambda}\Lambda)-(\bar{p}p)-(\bar{n}n). \tag{37}$$

[22] See Appendix for a more detailed analysis of these $\Sigma^\pm$ decays. See also S. Treiman, Nuovo cimento **15**, 916 (1960).

[23] M. Schwartz, Phys. Rev. Letters **4**, 306 (1960).

[24] S. Okubo, R. E. Marshak, E. C. G. Sudershan, W. B. Teutsch, and S. Weinberg, Phys. Rev. **112**, 665 (1958).

However, the introduction of (36) or (37) or both would lead to the violation of time reversal invariance. (It is of course, possible to construct more complicated form for $J_b{}^0$ which satisfies time reversal invariance.)

A remark about time reversal invariance is in order here. It has been pointed out by Dalitz[2] that in a theory in which $|\Delta\mathbf{I}|=\frac{1}{2}$ is satisfied, there is little existing experimental verification of time reversal invariance other than that contained in neutron decay measurements which, of course, is completely unrelated to the couplings of the neutral $W_b{}^0$.

2. In the scheme discussed in Sec. VI the electromagnetic interactions introduce corrections to the selection rules and intensity rules. However, since $\mathfrak{L}_\gamma$ commutes with I_z, the strangeness selection rule holds intact. An important consequence is the following: The amplitude for the transiton $K^0\rightarrow\bar{K}^0$ (for which $\Delta S=-2$), as discussed under class (ϵ) of the last section, vanishes in the order $(\mathfrak{L}_{WJ}+\mathfrak{L}_{WS})^2$ because of the strangeness selection rule $\Delta S=\pm 1$. Electromagnetic correction to this therefore also vanishes to all orders of $(e^2/\hbar c)$. The matrix element for $K^0\rightarrow\bar{K}^0$ only becomes nonvanishing in the order $\mathfrak{L}_{WJ}{}^2\mathfrak{L}_{WS}{}^2\sim g^4\sim 10^{-13}$. This is consistent with proposition (ii).

$|\Delta\mathbf{I}|=\frac{1}{2}$ selection rules are, however, corrected by the electromagnetic interaction. The correction introduces $|\Delta\mathbf{I}|=\frac{3}{2}$ and $|\Delta\mathbf{I}|=\frac{5}{2}$ components with comparable strengths, and higher $|\Delta\mathbf{I}|$ values only in higher orders of $e^2/\hbar c$. If experiments on the branching ratio of $K_1{}^0$ decays become more accurate, it may be possible to obtain a lower limit to the amplitude of the $|\Delta\mathbf{I}|=\frac{5}{2}$ component in K decay.

3. The conserved current hypothesis[7] is consistent with the schizon interactions discussed in the last section. It is equivalent to the statement that the vector part of the interaction $\mathfrak{L}_{WJ}$ describes the vector field W as originating from a source J which is the isotopic spin density-current of the strongly interacting particles, in complete analogy with the generation of the vector electromagnetic field A_λ from the electric charge density-current. If the conserved vector current hypothesis is correct, a pertinent question would be the interpretation of the generation of W through the term $\mathfrak{L}_{WS}$.

VIII. LEPTON COUPLINGS OF W^0

The lepton coupling $\mathfrak{L}_{Wl}$ in (28) should in general include, in addition to (1) which represents $W^\pm$ couplings to the leptons, also lepton couplings with W^0 and $\bar{W}^0$. We write these neutral couplings as

$$[g_{\mu\mu}(\bar{\mu}\mu)+g_{ee}(\bar{e}e)+g_{\nu\nu}(\bar{\nu}\nu)+g_{\nu'\nu'}(\bar{\nu}'\nu')]W^0 + \text{Hermitian conjugate.} \quad (38)$$

Comparison of (38) with (1) shows that the ratio of the rates of $K^+\rightarrow\pi^++e^++e^-$, $K^+\rightarrow\pi^++$neutrinos and $K^+\rightarrow\pi^0+e^++\nu$ are

$$R(K^+\rightarrow\pi^++e^++e^-)/R(K^+\rightarrow\pi^0+e^++\nu) \cong 2|g_{ee}|^2/|g_{e\nu}|^2, \quad (39)$$

$$R(K^+\rightarrow\pi^++\text{neutrinos})/R(K^+\rightarrow\pi^0+e^++\nu) \cong 2\{|g_{\nu\nu}|^2+|g_{\nu'\nu'}|^2\}/|g_{e\nu}|^2. \quad (40)$$

A cursory survey of the experimental limits[25,26] on the absence of $K^+\rightarrow\pi^++e^++e^-$ and $K^+\rightarrow\pi^++$neutrinos indicates

$$[|g_{ee}|^2/|g_{e\nu}|^2]<[\sim\tfrac{1}{2}\times 10^{-2}], \quad (41)$$

$$[(|g_{\nu\nu}|^2+|g_{\nu'\nu'}|^2)/|g_{e\nu}|^2]<[\sim\tfrac{1}{6}]. \quad (42)$$

To set an experimental upper limit on $g_{\mu\mu}$ let us first consider the absence of $K_2{}^0\rightarrow\mu^++\mu^-$. The state of $\mu^++\mu^-$ in $K^0\rightarrow\mu^++\mu^-$ is an eigenstate of CP with eigenvalue -1. If time reversal invariance holds for W interactions, this state is also the decay product of $K_2{}^0\rightarrow\mu^++\mu^-$. The rate of this last process is then

$$\frac{R(K_2{}^0\rightarrow\mu^++\mu^-)}{R(K^+\rightarrow\mu^++\nu')}=4\frac{|g_{\mu\mu}|^2m_K{}^3(m_K{}^2-4m_\mu{}^2)^{\frac{1}{2}}}{|g_{\mu\nu}|^2(m_K{}^2-m_\mu{}^2)^2}, \quad (43)$$

where m_K and m_μ are the masses of K and μ, respectively. Experimentally[27] this ratio is $<10^{-8}$. Thus

$$|g_{\mu\mu}|^2/|g_{\mu\nu}|^2<(2.5)\times 10^{-4}. \quad (44)$$

If time reversal invariance is not assumed, an upper limit can be set on $g_{\mu\mu}$ by considering the absence of $K^+\rightarrow\pi^++\mu^++\mu^-$. This process is theoretically similar to $K^+\rightarrow\pi^0+\mu^++\nu'$, with an amplitude ratio of $\sqrt{2}g_{\mu\mu}:g_{\mu\nu}$, except for kinematical differences. The Q values for the two processes are 143 Mev and 241 Mev, respectively. A conservative estimate then gives

$$\frac{R(K^+\rightarrow\pi^++\mu^++\mu^-)}{R(K^+\rightarrow\pi^0+\mu^++\nu')}>[\sim\tfrac{1}{2}|g_{\mu\mu}|^2/|g_{\mu\nu}|^2].$$

Experimentally $K^+\rightarrow\pi^++\mu^++\mu^-$ resembles a τ decay which has been extensively analyzed. It is safe to conclude that the ratio is less than 10^{-3}, giving

$$|g_{\mu\mu}|^2/|g_{\mu\nu}|^2\lesssim 2\times 10^{-3}.$$

The absence of W^0 and $\bar{W}^0$ couplings to the leptons makes it difficult to understand (8) in terms of a "universal" W interaction. It is to be emphasized, however, that this particular difficulty is not a consequence of the schizon theory, but rather is inherent in the experimental absence of neutral leptonic decay modes and the experimental rule $|\Delta\mathbf{I}|=\frac{1}{2}$.

One may also set an upper limit on the strength $g_{e\mu}$ of the W^0 coupling to $(e\mu)$. One has

$$\frac{R(K^+\rightarrow\pi^++\mu^++e^-)}{R(K^+\rightarrow\pi^0+\mu^++\nu')}\cong\frac{2|g_{e\mu}|^2}{|g_{\mu\nu}|^2}.$$

This is experimentally $\lesssim 10^{-3}$.

[25] F. Anderson, G. Lawlor, and T. E. Nevin, Nuovo cimento 2, 608 (1955). See also R. Dalitz, Proc. Phys. Soc. (London) A69, 527 (1956).
[26] R. W. Birge et al., Nuovo cimento 4, 834 (1956).
[27] M. Bardon et al., Ann. Phys. 5, 156 (1958).

IX. DECAY OF THE W PARTICLES

The leptonic decay modes of the $W^{\pm}$ were mentioned in Sec. II. Those of W^0 and $\bar{W}^0$ are absent as discussed in the last section. It is important to notice that decay modes such as

$$W \to \mu+\nu'+\text{pions}$$

occur with an amplitude smaller than $\sim ge^2$, and are therefore negligible.

The nonleptonic modes of decay include various channels: 2π, 3π, $K+\pi$, $\pi+\gamma$, $K+\gamma$, etc. To discuss the selection and intensity rules we shall neglect electromagnetic correction terms, but shall include the $J_b{}^0W_b{}^0$ term of (35).

The decay of W^0 and $\bar{W}^0$ resembles the corresponding situation in the decay of K^0 and $\bar{K}^0$. In the present schizon interaction scheme, through $\mathfrak{L}_{WJ}$ the particle $W_a{}^0$ and $W_b{}^0$ can make transitions into pion channels. These channels have, however, isotopic spins 1 and 0 for $W_a{}^0$ decay and for $W_b{}^0$ decay, respectively. [See (35).] There is therefore no interference between them.

Using the notations of Lee et al.,[28] contributions to the decay matrix $\Gamma+iM$ from $\mathfrak{L}_{W\mathfrak{S}}$ are proportional to the unit matrix. It follows from these considerations that $W_a{}^0$ and $W_b{}^0$ are the eigenstates ψ_+ and ψ_-, so that each follows a single exponential decay law with respect to the time. Their mass difference is ~ 10 ev. These conclusions are independent of CP invariance.

The nonleptonic decays of W^+, $W_a{}^0$, $W_b{}^0$, and W^- into particles with total strangeness $S=0$ thus obeys $\mathbf{I}$ conservation, with the assignment $I=1$ for W^+, $W_a{}^0$, and W^-, and the assignment $I=0$ for $W_b{}^0$. The nonleptonic decays of these particles into particles with total strangeness $S=+1$ is not possible for W^-. It is possible for W^+ and for the W^0 part of $W_a{}^0$ and $W_b{}^0$. Furthermore, $\mathbf{I}$ conservation is observed for W^+ and W^0 decay, with W^+, W^0 forming an isotopic doublet. Similar conclusions hold for decays into particles with total $S=-1$. Some detailed examples of these intensity and selection rules will now be given.

For the 2π modes we have the following equalities

$$\begin{aligned} R(W^+\to\pi^++\pi^0)&=R(W_a{}^0\to\pi^++\pi^-)\\ &=R(W^-\to\pi^-+\pi^0),\\ R(W_a{}^0\to 2\pi^0)&=R(W_b{}^0\to 2\pi^0)\\ &=R(W_b{}^0\to\pi^++\pi^-)=0. \end{aligned} \qquad (45)$$

For the 3π modes, if barrier penetration factors play an important role,

$$\begin{aligned} R(W_a{}^0\to 3\pi^0)&\approx 0,\\ R(W_b{}^0\to 3\pi)&\approx 0,\\ R(W^+\to\pi^++\pi^++\pi^-)&\cong R(W^+\to\pi^0+\pi^0+\pi^+)\\ &\cong\tfrac{1}{2}R(W_a{}^0\to\pi^++\pi^-+\pi^0). \end{aligned} \qquad (46)$$

Furthermore the density distribution in a Dalitz plot[29] for the last three processes are the same and are proportional to p^2 where p is the momentum of the π^-, π^+, and π^0 in the three cases, respectively.

For the $K+\pi$ modes one has the following relations:

$$\begin{aligned} &R(W^+\to K^++\pi^0)\\ &=R(W^-\to K^-+\pi^0)=\tfrac{1}{2}R(W^+\to K^0+\pi^+)\\ &=\tfrac{1}{2}R(W^-\to\bar{K}^0+\pi^-)=2R(W_\alpha{}^0\to K^0+\pi^0)\\ &=2R(W_\alpha{}^0\to\bar{K}^0+\pi^0)=R(W_\alpha{}^0\to K^++\pi^-)\\ &=R(W_\alpha{}^0\to K^-+\pi^+), \end{aligned} \qquad (47)$$

where the subscript $\alpha=a$ or b.

The decay of $W_b{}^0$ into 2π is forbidden and into 3π is hindered by barrier penetration factors, as shown by (45) and (46). If, therefore, $m_W<m_K+m_\pi$, the decay modes

$$\begin{aligned} &W_b{}^0\to\pi^0+\gamma,\\ &W_b{}^0\to\pi^++\pi^-+\gamma,\ \text{etc.} \end{aligned} \qquad (48)$$

become important. If further the $J_b{}^0W_b{}^0$ coupling is absent in $\mathfrak{L}_{WJ}$, the decay modes

$$W_b{}^0\to K^0+\gamma \text{ and } W_b{}^0\to\bar{K}^0+\gamma, \qquad (49)$$

which have equal rate, become important.

It is important to notice that the W particles in general are polarized when produced through either neutrino capture experiments (see Sec. XI) or collisions between strongly interacting particles. The spin states of W can be easily analysed by measuring the angular distributions of its decay products.

X. "APPARENT" NONCONSERVATION OF STRANGENESS

In the usual theory in a collision between pions and nucleons the probability of a reaction exhibiting a strangeness change $\Delta S=\pm1$ is $\sim 10^{-12}$ compared with that of the strong processes. That of a reaction showing a strangeness change $\Delta S=\pm2$ is, by proposition (ii), $\sim 10^{-24}$ compared with that of the strong processes. In the present theory these conclusions remain true. However, in a process in which a real W particle is emitted, its short lifetime causes its immediate disintegration, and the disintegration products would exhibit apparent strangeness changes $\Delta S=0$, ±1 for the charged $W^{\pm}$ particles, and $\Delta S=0$, ±1, ±2 for the neutral W's. For collisions with enough energy to produce a real W, the probability of such processes is $\sim 10^{-6}$ of the strong processes.

We give some examples below:

1.
$$\begin{aligned} &\pi^++p\to W^++p,\\ &W^+\to K^++\pi^0. \end{aligned} \qquad (50)$$

Apparent process:

$$\pi^++p\to p+K^++\pi^0 \quad (\Delta S=1) \qquad (51)$$

2.
$$K^++Z\to W^0+\text{nucleons and pions},$$
$$W^0\to \text{all decay products of } W_a{}^0 \text{ and } W_b{}^0.$$

[28] T. D. Lee, R. Oehme, and C. N. Yang, Phys. Rev. **106**, 340 (1957).

[29] R. Dalitz, Phil. Mag. **44**, 1068 (1953); Phys. Rev. **94**, 1046 (1954); E. Fabri, Nuovo cimento **11**, 479 (1954).

For the decay mode $W_\alpha^0 \to K^- + \pi^+ (\alpha = a,b)$ the apparent process becomes

$$K^+ + Z \to K^- + \pi^+ + \text{(nucleons and pions)} \quad (\Delta S = -2) \quad (52)$$

Detection and positive identification of such phenomena, which occur with a cross section 10^{-6} times that of the strong processes, is of course very difficult. If one thinks in terms of counter experiments, a source of difficulty is the competing apparent change of strangeness involved in the decay of the $K^0 - \bar{K}^0$ complex. One way to avoid this difficulty is to do an experiment below the threshold of strange particle production, such as (51) at a pion energy above the threshold for W^+ production but below the threshold of

$$\pi^+ + p \to K^+ + \Sigma^+.$$

This is feasible only if $m_K + 135$ Mev $< m_W < m_K + 250$ Mev. If $m_W < m_K + 135$ Mev, the apparent process

$$\pi^+ + p \to p + K^+ + \gamma$$

can occur, but with a probability only $\sim 10^{-8}$ times that of the strong processes. It seems worthwhile to explore these and other possibilities for a detection of an apparent strangeness violation. In any case it is desirable to improve the present experimental limit of strangeness nonconservation in a collision process involving only strongly interacting particles.

XI. NEUTRINO CAPTURE EXPERIMENTS

It has already been pointed out[9] that the creation of the pair of particles $\mu^- + W^+$ in the Coulomb field of a nucleus by a neutrino has a relatively high cross section:

$$\nu' + Z \to Z + \mu^- + W^+. \quad (53)$$

It seems[23] that high-energy neutrino experiments may be quite feasible in the near future. We shall in this section discuss some implications of the schizon interaction scheme for those neutrino capture reactions in which no W particle is emitted.

1. Some implications were already mentioned in Sec. III. [See especially (15).] Some others result from the fact that in $\mathcal{L}_{WS}$ S transforms like an isotopic doublet. Thus, e.g. the cross sections for

$$\bar{\nu}' + n \to \mu^+ + \Sigma^-,$$

and

$$\bar{\nu}' + p \to \mu^+ + \Sigma^0 \quad (54)$$

are in the ratio of 2 to 1 and have the same angular distribution. The same holds for the pair

$$\bar{\nu}' + n \to \mu^+ + \Lambda^0 + \pi^-,$$

and

$$\bar{\nu}' + p \to \mu^+ + \Lambda^0 + \pi^0. \quad (55)$$

These implications can all be summarized by the rule that $|\Delta \mathbf{I}| = \frac{1}{2}$ for the strongly interacting particles.

2. Another type of implication can be summarized by the rule that $|\Delta \mathbf{I}| = 1$ for the strongly interacting particles. These result from the fact that in $\mathcal{L}_{WJ}$ the current J transforms like an isotopic vector. One consequence is, e.g., that if the differential cross sections for

$$\begin{aligned} &\nu' + n \to \mu^- + n + \pi^+, \\ &\nu' + n \to \mu^- + p + \pi^0, \\ &\nu' + p \to \mu^- + p + \pi^+ \end{aligned} \quad (56)$$

are denoted by $\sigma_1, \sigma_2, \sigma_3$, respectively, then $(\sigma_1)^{\frac{1}{2}}$, $(2\sigma_2)^{\frac{1}{2}}$ and $(\sigma_3)^{\frac{1}{2}}$ satisfy the triangular inequalities

$$\begin{aligned} &(\sigma_1)^{\frac{1}{2}} + (2\sigma_2)^{\frac{1}{2}} \geqq (\sigma_3)^{\frac{1}{2}}, \\ &(2\sigma_2)^{\frac{1}{2}} + (\sigma_3)^{\frac{1}{2}} \geqq (\sigma_1)^{\frac{1}{2}}, \\ &(\sigma_3)^{\frac{1}{2}} + (\sigma_1)^{\frac{1}{2}} \geqq (2\sigma_2)^{\frac{1}{2}}. \end{aligned} \quad (57)$$

Another consequence is, e.g., found in the reactions

$$\nu' + n \to \mu^- + \Gamma, \quad (58)$$

$$\bar{\nu}' + p \to \mu^+ + \Gamma', \quad (59)$$

where Γ and Γ' are complexes of strongly interacting particles with total strangeness $= 0$. The strongly interacting particles contribute factors $\langle \Gamma | J^\star | n \rangle$ and $\langle \Gamma' | J | p \rangle$ to the matrix elements for the transitions. The fact that J and $J^\star$ transform into each other under an $\mathbf{I}$ rotation means that these two factors are identical for pairs of states Γ and Γ' which are isotopic spin partners of each other. The contribution of $\mathcal{L}_{Wl}$ to the matrix element consists of factors that can be explicitly computed in terms of the momenta and spins of the leptons in the reactions (58) and (59). The result of such an analysis is that the differential cross sections for (58) and (59) can both be expressed[9] in terms of certain structure functions, and that the structure functions for (58) and (59) are related to each other. More explicitly, the differential cross section for (58), is of the form

$$\begin{aligned} d\sigma(\nu' \to \mu_L^- + \Gamma) &= dk_\mu d(\cos\theta)(4\pi k_\nu)^{-1} \\ &\quad \times k_\mu[(k_\mu + k_\nu)^2 - P^2]^{\frac{1}{2}}(1 + v_\mu) D \\ &\quad \times [xA_+ + x^{-1}A_- + yB_+ + y^{-1}B_- + C], \end{aligned} \quad (60)$$

$$\begin{aligned} d\sigma(\nu' \to \mu_R^- + \Gamma) &= dk_\mu d(\cos\theta)(4\pi k_\nu)^{-1} \\ &\quad \times k_\mu[(k_\mu + k_\nu)^2 - P^2]^{\frac{1}{2}}(1 - v_\mu) D \\ &\quad \times [xB_+ + x^{-1}B_- + yA_+ + y^{-1}A_- - C], \end{aligned} \quad (61)$$

and that for (59) is of the form

$$\begin{aligned} d\sigma(\bar{\nu}' \to \mu_R^+ + \Gamma') &= dk_\mu d(\cos\theta)(4\pi k_\nu)^{-1} \\ &\quad \times k_\mu[(k_\mu + k_\nu)^2 - P^2]^{\frac{1}{2}}(1 + v_\mu) D \\ &\quad \times [xA_- + x^{-1}A_+ + yB_+ + y^{-1}B_- + C], \end{aligned} \quad (62)$$

$$\begin{aligned} d\sigma(\bar{\nu}' \to \mu_L^+ + \Gamma') &= dk_\mu d(\cos\theta)(4\pi k_\nu)^{-1} \\ &\quad \times k_\mu[(k_\mu + k_\nu)^2 - P^2]^{\frac{1}{2}}(1 - v_\mu) D \\ &\quad \times [xB_+ + x^{-1}B_- + yA_- + y^{-1}A_+ - C]. \end{aligned} \quad (63)$$

The notations in these formulas are defined as follows: $\mu_L^- = \mu^-$ with left-handed helicity, etc., $\mathbf{k}_\mu$, $\mathbf{k}_\nu$ = momenta of μ and ν' (or $\bar{\nu}'$) in the laboratory system, k_μ, $k_\nu = |\mathbf{k}_\mu|$, $|\mathbf{k}_\nu|$, θ = angle between $\mathbf{k}_\mu$ and $\mathbf{k}_\nu$, $\mathbf{P} = \mathbf{k}_\nu - \mathbf{k}_\mu$, v_μ = velocity of $\mu^\pm$, $x = (k_\mu + k_\nu + P)^{-1}(k_\mu + k_\nu - P)$, $y = (P + k_\nu - k_\mu)^{-1}(P - k_\nu + k_\mu)$, $E_\mu = (m_\mu^2 + k_\mu^2)^{\frac{1}{2}}$ = total energy of $\mu^\pm$ in the laboratory system, $D = [P^2 - (k_\nu - E_\mu)^2]^{-1}[P^2 - (k_\nu - k_\mu)^2]$, and A_+, A_-, B_+, B_-, and C are structure functions depending only on the state Γ and on the magnitude of the momentum transfer P and the energy transfer $k_\nu - E_\mu$ from the leptons to the strongly interacting particles.

One notices that in the forward direction, though $y^{-1} = \infty$, Dy^{-1} = finite. If the mass of μ is negligible, $D = 1$, $v_\mu = 1$, y = functions of P and $k_\nu - E_\mu$ and (60)–(63) reduce to Eqs. (16) and (21) of reference 9.

XII. CONCLUDING REMARKS

It is seen above that from propositions (i), (ii), and (iii) stated in the Introduction one is quite naturally led to the schizon interaction scheme. This scheme gives rise to a rather integrated picture of the various $|\Delta \mathbf{I}| = \frac{1}{2}$ rules, and of the so-called $\Delta Q = \Delta S$ rule.

To test the existence of the W particles and the validity of such a scheme four types of experiment seem worth considering:

(a) Neutrino capture with the production of $\mu^- + W^+$. This was touched upon in reference 9 and in Sec. XI above.

(b) Chamber type experiment of W production in pion-nucleon or nucleon-nucleon collisions. The main difficulty here is of course the fact that one can only have one W production event in millions of interactions.

(c) Counter experiment on apparent nonconservation of strangeness. This was discussed in Sec. X.

(d) Counter experiment on apparent lepton production in pion-nucleon or nucleon-nucleon collisions, such as

$$\pi^+ + p \to W^+ + p \to \begin{Bmatrix} \mu^+ + \nu' \\ e^+ + \nu \end{Bmatrix} + p. \tag{64}$$

Such processes occur with a probability of $\sim 10^{-6}$ of the strong interactions, provided the threshold of W production is exceeded. The difficulty here is to separate these events from the background of μ^+ and e^+ produced in $\pi^+ \to \mu^+ + \nu'$, $\pi^0 \to e^+ + e^- + \gamma$, and K decays. To achieve this separation suppose one measures the momenta $\mathbf{P}_\pi$, $\mathbf{P}_p$, and $\mathbf{P}_l$ of the incoming π, the final p and the outgoing lepton. One then describes the observed process as

$$\pi^+ + p \to l^+ + p + X, \tag{65}$$

where X is not detected, and is in general a complex of particles. By energy and momentum conservation one easily computes the energy m_X of X in its center-of-mass system, and the energy m_{X+l} of the complex $X+l$ in its center-of-mass system:

$$m_X^2 = (E_\pi + m_p - E_l - E_p)^2 - (\mathbf{P}_\pi - \mathbf{P}_l - \mathbf{P}_p)^2, \tag{66}$$

and

$$m_{X+l}^2 = (E_\pi + m_p - E_p)^2 - (\mathbf{P}_\pi - \mathbf{P}_p)^2, \tag{67}$$

where E_π, E_l, and E_p are the energies of the incoming pion, the lepton and the final proton. Process (64) is uniquely determined by the specifications

$$m_X = 0, \tag{68}$$

$$m_{X+l} = m_W \text{ (which is } > m_K). \tag{69}$$

To discuss the sensitivity of such a separation let us take, say, the example of a π^+ beam with good momentum resolution on a target of liquid hydrogen and detect μ^+ and p in coincidence. The background μ mesons ln this case come mainly from

$$\pi^+ + p \to \pi^+ + p \to \mu^+ + \nu' + p, \tag{70}$$

and

$$\pi^+ + p \to n\pi + \pi^+ + p \to n\pi + \mu^+ + \nu' + p, \quad (n \geq 1), \tag{71}$$

and, e.g.,

$$\pi^+ + p \to K^+ + \Sigma^+ \to \mu^+ + \nu' + \pi^0 + p. \tag{72}$$

Reaction (70) is identifiable by the conditions,

$$m_{X+l} = m_\pi, \quad m_X = 0, \tag{73}$$

and (71) and (72) by the condition that m_X and m_{X+l} both have continuous spectra with the lower limits:

$$m_X > m_\pi, \quad m_{X+l} \geq 2m_\pi. \tag{74}$$

The residual background is then due to the imperfect separation of processes (70)–(72), and due to chance coincidence. Amidst such background the desired events (64) constitute a peak in both m_X and m_{X+l} at the values of 0 and m_W, respectively. The identification and separation of (64) from the background in a counter experiment may thus be feasible.

APPENDIX

In this appendix we study the decays

$$\Sigma^- \to \Lambda^0 + e^- + \bar{\nu}, \tag{A.1}$$

and

$$\Sigma^+ \to \Lambda^0 + e^+ + \nu. \tag{A.2}$$

Their rates are unfortunately very small (see below). For completeness, however, we analyze in some detail these decays as an illustration of the $|\Delta \mathbf{I}| = 1$ rule in a decay process.

Throughout the Appendix we shall neglect the mass of the electron and consider only the decays of unpolarized $\Sigma^\pm$. Let $\mathbf{k}$ and $\mathbf{q}$ be, respectively, the momenta of $e^\pm$ and Λ^0 in the rest system of $\Sigma^\pm$. The Λ^0 particle would, in general, be longitudinally polarized. We define $P_L^-(q,k)dqdk$ and $P_R^-(q,k)dqdk$ to be the rates for the decay (A.1) of Σ^- in which the final Λ^0 has a helicity (i.e., spin component along its direction of motion) $= -\frac{1}{2}$

and $+\frac{1}{2}$, respectively. Similarly, let $P_R^+(q,k)dqdk$ and $P_R^+(q,k)dqdk$ be the corresponding rates for the decay (A.2) of Σ^+.

By using the Lagrangian (28), the dependence of $P_L^\pm$ and $P_R^\pm$ on k can be calculated explicitly. The following theorem can be readily established:

Theorem

$$P_L^\pm(q,k)=A_L[q\pm(Q-2k)]^2+B_L[q^2-(2k-Q)^2], \quad (A.3)$$

and

$$P_R^\pm(q,k)=A_R[q\mp(Q-2k)]^2+B_R[q^2-(2k-Q)^2], \quad (A.4)$$

where

$$Q=m_\Sigma-(m_\Lambda^2+q^2)^{\frac{1}{2}}, \quad (A.5)$$

and A_L, A_R, B_L, B_R are *functions of q only*. In (A.5) m_Λ is the mass of Λ^0 and m_Σ is the appropriate mass of Σ^+ or Σ^-.

It is important to notice that the explicit dependence of $P_\alpha^\pm$ $(\alpha=L,R)$ on k follows from the special form of lepton current in $\mathfrak{L}_{Wl}$ [Eq. (1)]. In $\mathfrak{L}_{WJ}$, $J_\mu^\star$ and J_μ belong to the same isotopic spin multiplet. Consequently, $J_\mu^\star$ and J_μ are related by a 180° rotation along the y-axis in the isotopic spin space

$$J_\mu^\star=-e^{-i\pi I_y}J_\mu{}^{i\pi I_y}, \quad (A.6)$$

which leads to the result that in (A.4) and (A.5) the same structure functions A_L, A_R, B_L, B_R occur in both Σ^+ decay and in Σ^- decay. In terms of the matrix elements of J_μ these structure functions are given by

$$A_L=(8\pi^3q)^{-1}(Q^2-q^2)|\langle\Lambda_\downarrow|J_z|\Sigma_\uparrow\rangle|^2\Delta,$$
$$A_R=(8\pi^3q)^{-1}(Q^2-q^2)|\langle\Lambda_\uparrow|J_z|\Sigma_\downarrow\rangle|^2\Delta,$$
$$B_L=(8\pi^3q)^{-1}|Q\langle\Lambda_\downarrow|J_z|Z_\downarrow\rangle-iq\langle\Lambda_\downarrow|J_4|\Sigma_\downarrow\rangle|^2\Delta,$$

and

$$B_R=(8\pi^3q)^{-1}|Q\langle\Lambda_\uparrow|J_z|\Sigma_\uparrow\rangle-iq\langle\Lambda_\uparrow|J_4|\Sigma_\uparrow\rangle|^2\Delta, \quad (A.7)$$

where the z axis is parallel to $\mathbf{q}$ and $\uparrow$, $\downarrow$ indicate the appropriate spin states of Σ and Λ with respect to the z axis, and Δ is related to the coupling constant $g_{e\nu}$ in $\mathfrak{L}_{W-l}$ and the propagator of the $W^\pm$ particle by

$$\Delta=|g_{e\nu}|^2[q^2+m_W^2-Q^2]^{-2}. \quad (A.8)$$

We may expand Δ and the matrix elements of J_μ in powers of q and neglect terms that are proportional to either $(q/m_W)^2$ or (q/m_Λ). Similar to the case of neutron decay, we find that in such a nonrelativistic limit A_α and B_α $(\alpha=L,R)$ depend only on two constans C_1 and C_2:

$$A_L=A_R=(16\pi^3q)^{-1}(Q^2-q^2)|C_2|^2, \quad (A.9)$$

$$B_L=(16\pi^3q)^{-1}|C_2Q-C_1q|^2, \quad (A.10)$$

and

$$B_R=(16\pi^3q)^{-1}|C_2Q+C_1q|^2. \quad (A.11)$$

It is interesting to notice that in this nonrelativistic approximation, if we sum over the helicity of Λ^0, the spectrum $P_L^+(q,k)+P_R^+(q,k)$ for Σ^+ decay is the same as that for Σ^- decay except for the mass difference between Σ^+ and Σ^-. Using the known masses of $\Sigma^\pm$ we find that the total rates R for these decays are given by

$$\frac{R(\Sigma^-\to\Lambda^0+e^-+\bar{\nu})}{R(\Sigma^+\to\Lambda^0+e^++\nu)}\cong 1.57, \quad (A.12)$$

and

$$\frac{R(\Sigma^-\to\Lambda^0+e^-+\bar{\nu})}{R(\Sigma^-\to n+\pi^-)}\cong(2\times10^{-4})\eta, \quad (A.13)$$

where

$$\eta=\frac{|C_1|^2+3|C_2|^2}{|G_V|^2+3|G_A|^2},$$

and G_V, G_A are the Fermi and Gamow-Teller coupling constants in neutron decay.

[[60g]]
Commentary begins page 47

IMPERFECT BOSE SYSTEM

by C. N. YANG

Institute for Advanced Study, Princeton, New Jersey

Introduction. This report consists of a short summary of some published works on the dilute hard sphere Bose system, and on a method of treating Bose condensation in a general interacting system. Also included will be remarks on some problems and difficulties in the field, together with preliminary ideas on how some of them may perhaps be resolved.

We shall take units such that

$$\hbar = 1, \quad 2m = 1,$$

where m is the mass of the particles.

I. *Ground State Energy of a Bose System of Hard Spheres.* The ground state energy has been computed by many authors using different methods. The results are in agreement with each other. While none of the computations is mathematically rigorous, the result [1] is clearly correct:

$$\frac{\text{Energy}}{N} = 4\pi a\rho \; [1 + \frac{128}{15\sqrt{\pi}} \sqrt{\rho a^3} + 8\left(\frac{4\pi}{3} - \sqrt{3}\right) \rho a^3 \ln(\rho a^3) + \kappa\rho a^3 + \text{higher order terms in } \rho a^3], \tag{1}$$

here a = diameter of hard sphere,

$\rho = N/\Omega$ = density of particles,

and N = number of particles.

The nature of the expansion in (1) is not understood. It has been argued [2] that it represents an asvmptotic expansion.

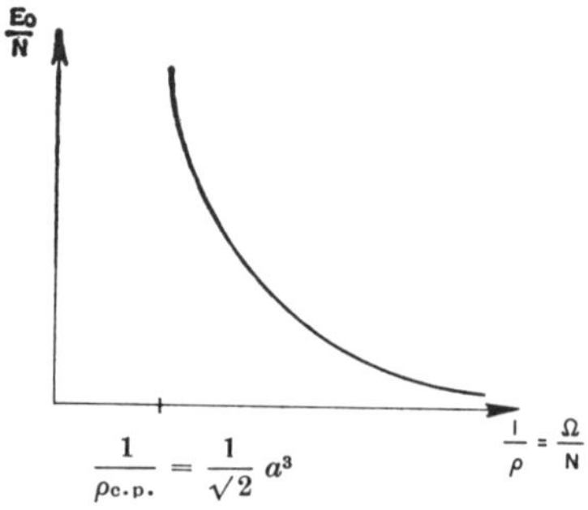

Fig. 1. Energy per particle vs. volume per particle diagram.

□Reprinted from *Physica* 26 (1960), S49–S57.

The low density expansion (1) is useless for high densities. The density $\rho_{\text{C.P.}}$ for closest packing is given by

$$\frac{1}{\rho_{\text{C.P.}}} = \frac{a^3}{\sqrt{2}}. \tag{2}$$

It can be shown that near this closest packing density the ground state energy per particle is given by

$$\frac{E_0}{N} = \frac{\alpha}{(l-l_0)^2}[1 + \text{terms that vanish as } l \to l_0], \tag{3}$$

where

$$l = \rho^{-\frac{1}{3}}, \quad l_0 = (\rho_0)^{-\frac{1}{3}}$$

and ρ_0 = density of the jammed (or closest packing) configuration. The value of the constant α is not yet known.

A "polyhedron" method is probably useful for an attempt to evaluate α. The basic idea is as follows: For N spheres in a crowded box, the allowed region in the $3N$ dimensional space $r_1, r_2, \ldots, r_N$ consists of pockets connected by long and narrow channels, each pocket representing essentially the N particles in a particular configuration of closest packing. The different pockets represent different arrangements of closest packing, such as face centered cubic, hexagonal closest packing or irregular closest packing, or, for the same arrangement, different permutations of the N spheres, such as

For a very crowded box, the channels connecting the pockets are very long and narrow since to go from one pocket to another necessitates the moving around of at least $\sim a/(l-l_0)$ particles. To the lowest order, one can therefore neglect the channels and discuss the eigenvalue problem in a single pocket. Now each pocket is bounded by curved hypersurfaces. (The hyper-surfaces are curved because the spheres have curved three dimensional surfaces.) For large values of $a/(l-l_0)$, however, only a very small portion of each hypersurface serves as the boundary of the allowed region. One can then replace these hypersurfaces by their tangent hyperplanes. The eigenvalue problem thus becomes that of a resonance cavity with the shape of a *polyhedron* in $3N$ dimensions.

The polyhedron method described above gives

$$\alpha = \pi^2 \tag{4}$$

for a simple cubic packing of the spheres. This result is different from that given by a simple cell theory for which

$$\alpha = 3\pi^2. \tag{5}$$

(4) is obtained by an application of the method of images. Unfortunately the cavity problem corresponding to a face centered cubic or hexagonal closest packing configuration is not solvable by a simple method of images. Maybe a multiplevalued image method would help. It seems to me that this is a problem that deserves further examination.

II. *Excited States just above the Ground State.* For a dilute gas, ($\rho a^3 \ll 1$) the excitations above the ground state have [3] the energy

$$\begin{aligned} &\sum m_k \omega_k, \\ &\omega_k = k\sqrt{k^2 + 16\pi a\rho}, \end{aligned} \tag{6}$$

where $\boldsymbol{k}$ = momentum of the excitation. The next order correction to this has also been computed. [4] These excitations are unstable and have mean lives which have been calculated [4] [5].

It must be emphasized that the exact meaning of the position of the excited states and their life time is not completely clear. This problem becomes especially bothersome when one wishes to break away from the low density region.

For a dense system near closest packing, the polyhedron method described in Sec. I also leads to a calculation of the excited states. For the "pocket" near a simple cubic arrangement the density of excited states resembles that of a Fermi gas. It is reasonable to expect that the same holds for the "pocket" near a face centered cubic arrangement also. Comparison of such a spectrum with that represented by the phonon excitations shows that even the *qualitative* features of the excitation spectrum change when one goes from a dilute to a dense system.

III. *Higher Excited States.* For a dilute gas the spectrum at higher energies [5] is given by

$$E = E_0(\xi) + E_{\text{phonon}}\ (\xi),$$

$$E_0(\xi) = 4\pi a\rho N\left[1 + (1-\xi)^2 + \frac{128}{15\sqrt{\pi}}\sqrt{\rho a^3}\sqrt{\xi^5} + O(\rho a^3)\right], \tag{7}$$

$$E_{\text{phonon}} = \sum m_k\, k\sqrt{k^2 + 16\pi a\rho\xi} \tag{8}$$

and m_k are integers satisfying

$$N^{-1}\sum m_k = 1 - \xi + O(N^{-1}). \tag{9}$$

The parameter ξ in these formulae, as is evident from (9), represents a quantity that is useful in describing the system when the number of particles approaches infinity. Using this parameter is analogous to the approximation of cutting off the long narrow channels discussed earlier in sec. I for a very crowded hard sphere system. Two states with different parameters ξ have their analogy there in wave functions localized in different pockets. The analogy seems perfect, but leaves unanswered the question of how many parameters there should be in each particular case, e.g. in a case of medium density for which $\rho a^3 \sim 1$. That other useful and physically important parameters should exist is evident in both the low and high density cases: In the former, the possibility of superfluid flow shows the necessity of introducing additional parameters that describe the macroscopic system. In the latter, the existence of pockets describing various "jammed" positions, (such as a simple cubic structure) also points to the necessity of additional parameters. The general question of the nature of these parameters and the associated quasistationary states for a system in the limit of $N \to \infty$ seems to be a most important but difficult subject.

IV. *Equilibrium Properties*. Using the spectrum (7)–(9) the equilibrium thermodynamical properties of a dilute hard sphere Bose gas have been computed [5]. The computation is straightforward and yields the equation of state of the system. There is found to be a phase transition at a pressure p_c and density ρ_c given by

$$\frac{\lambda^3 p_c}{kT} = 1{\cdot}342 + 2(2{\cdot}612)^2 \frac{a}{\lambda} + O\left[\left(\frac{a}{\lambda}\right)^{\frac{3}{2}}\right], \tag{10}$$

$$\lambda^3 \rho_c = 2{\cdot}612 + O\left(\frac{a}{\lambda}\right), \tag{11}$$

where

$$\lambda = \sqrt{\frac{4\pi}{kT}}\,.$$

To the order $\sqrt{a/\lambda}$ there is no discontinuity of density at the transition point, the discontinuity in $(\partial p/\partial \rho)_T$ is equal to $8\pi a \rho_c$, and the specific heat in the condensed side exceeds that on the gas side by $\frac{9}{2}(2{\cdot}612)(a/\lambda)R$.

It was explicitly discussed in reference [5]) that although the above quoted

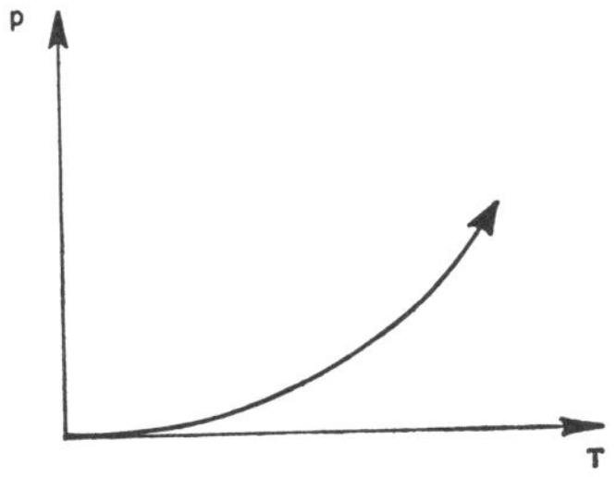

FIG. 2

Fig. 2. p-T diagram for a dilute hard sphere system.

results are very plausible, the derivation is subject to the criticism that all computations are done by first fixing the density and calculating the limit as $a \to 0$, and then making $\rho \to \rho_c + 0$ or $\rho \to \rho_c - 0$. Since one does not know how to switch these limits, the conclusions cannot be claimed as being completely firmly established.

The p-T phase diagram for the transition follows closely the free Bose-Einstein transition: $p \propto T^{5/2}$. This is of course quite different from the same diagram for real He. This qualitative difference mainly originates from the fact that the real He atoms have an attractive force for each other, strong enough for the system to have a binding energy per particle at zero temperature of order $B \sim (k \times 7°)$. The p-T diagram for real He thus looks like that indicated in fig. 3, where the dotted line represents a gas-liquid transition. It is represented by

$$p \propto T^{5/2} e^{-B/kT} \tag{12}$$

for low T. It is easy to show, however, that from the hard sphere calculation, by adding a shallow square well attractive interaction of depth Δ and radius

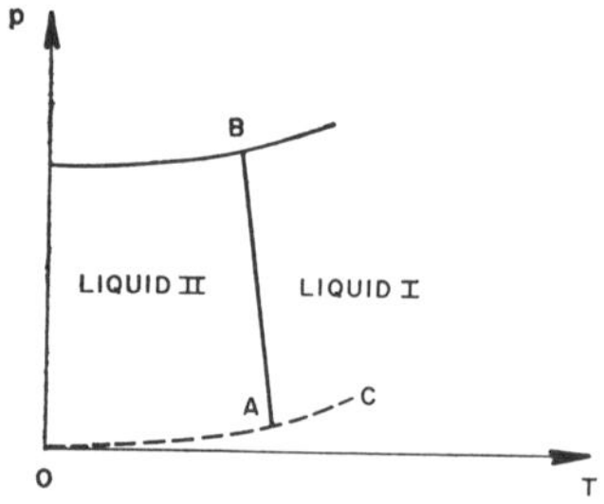

FIG. 3

Fig. 3. p-T diagram for He.

R, one could obtain the phase diagram in the limit that $A = (2\pi/3)R^3\Delta$ remains fixed but $\Delta \to 0$. To see this we notice that for a very shallow but extensive attractive square well, the attractive force merely produces a negative energy per particle of $-(4\pi/3)R^3\Delta \cdot (\rho/2) = -A\rho$. Changes in the wave function do not effect this contribution as long as $R^3\rho \gg 1$. Thus the net effect is to add the term $-NA\rho$ to the free energy of the system.

For suitable values of A the system has a binding energy for the ground state and the p-T diagram then assumes the form illustrated in fig. 4.

One sees that there is a liquid-gas transition (dotted line) as well as a Bose-Einstein transition (solid line). The former obeys the rule (12) for such transitions.

K. Huang has considered a more general attractive envelope and obtained similar results.

A different approach [6]) to the equilibrium problem is to consider the density matrix $e^{-\beta H + N \ln z}$ for the grand canonical ensemble and calculate the average occupation number $\langle n_k \rangle$ for a single particle state with momentum k. In this way one can completely and rigorously separate the effects of Bose statistics from the dynamical effects. The former is formulated in terms of a variational principle in which the pressure of the system is given by the maximum of a functional of $\langle n_k \rangle$ with respect to variations of $\langle n_k \rangle$. The explicit quantitative form of the functional is dependent on the dynamical effects in the problem. But the general characteristic of the functional is determined by the Bose statistical character of the particles alone.

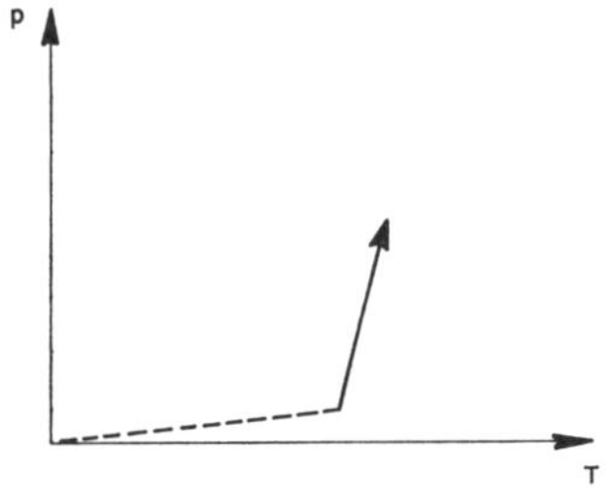

Fig. 4

This formulation is quite general, and is independent of the assumption of a dilute hard sphere system. Furthermore, the dependent and independent variables that appear in the formalism are physical variables independent of any approximation scheme. The conclusion [6]) that one can draw is that in the phase diagram Fig. 3 for He, in the limit that the volume $\Omega \to \infty$,

along AB, $\langle n_k \rangle|_{k=0} = \infty$ on both sides,
along OA, $\langle n_k \rangle|_{k=0} = \infty$ on the liquid side,
$= $ finite on the gas side,

and along AC, $\langle n_k \rangle|_{k=0} =$ finite on both sides, but is not continuous in crossing AC.

It is hoped that through such considerations one may be able to draw general conclusions about the order of the λ-transition and the nature of the singularity at the transition point for the various thermodynamical quantities. But this project has not yet been successfully carried out.

V. *Non-Equilibrium Properties of a Dilute Hard Sphere Bose System, Quasi Equilibrium States.* The spectrum (7)–(9) not only leads to the equilibrium properties, but also forms the basis of a detailed microscopic understanding [7]) of the two fluid concept first discussed by Tisza [8]) and by Landau [9]). The idea is that the state whose energy is given by (7) and (8) has a momentum

$$\boldsymbol{P} = \sum \boldsymbol{k} m_k.$$

This state has a macroscopic occupation for the single particle state $k = 0$.

Of course a similar calculation can be made for a state with a macroscopic occupation for a single particle state with $k_s \neq 0$. The result is clearly identical with that of a Galilean transformation on the state described by (7)–(9), which can be trivially performed. For such a state

$$\boldsymbol{P} = \sum \boldsymbol{k} m_k + N\xi \boldsymbol{k}_s \tag{13}$$

The states are thus described by the "quantum numbers" k_s and m_k. Their energies and momenta are explicit functions of these quantum numbers, with the parameter ξ given by (9). Since the excitations have finite lifetimes and scatter on each other, the numbers m_k reach statistical equilibrium in microscopic time scales. Not so for the "quantum number" k_s. For a given total momentum P, and a given value of k_s the system therefore reaches a *quasi equilibrium state* with an equilibrium statistical distribution for the number of excitations m_k and a free energy $F(N, T, \Omega, P, k_s)$. This free energy can be easily computed from (7)–(9). Because of Galilean invariance it is easily shown that

$$F - \frac{1}{2Nm} \boldsymbol{P}^2$$

is independent of the transformation

$$\begin{cases} \boldsymbol{P} \to \boldsymbol{P} + N\boldsymbol{p} \\ \boldsymbol{k}_s \to \boldsymbol{k}_s + \boldsymbol{p} \end{cases}$$

Choosing $\boldsymbol{p} = -(1/N)\boldsymbol{P}$ one obtains

$$F - \frac{1}{2Nm} \boldsymbol{P}^2 = F\left(N, T, \Omega, 0, \boldsymbol{k}_s - \frac{\boldsymbol{P}}{N}\right) \equiv F_{\text{internal}} \tag{14}$$

which is a function of the usual N, T, Ω, but also of the additional variable $\boldsymbol{k}_s - \boldsymbol{P}/N$. The explicit expression for F_{int} has been computed in reference 7.

It is clear that the entropy of the system resides in the distribution of m_k. The system of phonons (described by m_k), if in motion, would thus cause an entropy flux. (13) shows, however, that the total momentum $\boldsymbol{P}$ for such a system is not uniquely determined by that of the phonons. One can have, for example, a system with $\boldsymbol{P} = 0$, but $\sum m_k \boldsymbol{k} \neq 0$. Such a system would thus manifest a momentum-less isothermal entropy transfer, a phenomenon analogous to the so called infinite heat conductivity property of He II. It represents a reversible heat flow.

The spectrum (7)–(9) also shows [10]) that the system possesses superfluid properties.

A very difficult problem concerning a quasi equilibrium state specified by N, T, Ω, $\boldsymbol{P} = 0$ and $\boldsymbol{k}_s \neq 0$ (i.e., a momentumless isothermal entropy transfer state) is its possible relaxation time. It is thoroughly unclear how this problem may be approached, let alone solved.

To obtain hydrodynamical equations, a quantum transport theory has to be first worked out. Lacking that, it seems that the best way is to follow Lin [11]) who has developed a macroscopic variational principle for a system with an additional internal degree of freedom, [such as given in (14)] representing a reversible heat flow.

VI. *Superfluid Flow and the Density Matrix.* The discussions of the last section do not throw any light on the nature of superfluid flow from a molecular viewpoint .This is because the molecular theory has only been carried out so far up to the spectrum (7)–(9) and the Galilean transformation of these states. One obtains in this way only states with uniform superfluid flow. The difficult question of a nonuniform superfluid flow has so far only been discussed from a heuristic viewpoint.

Perhaps an approach to the nonuniform superfluid flow problem can be made by looking at the large eigenvalues of the contracted density matrix. Penrose and Onsager [12]) have discussed the eigenvalues of the $(N-1)$-fold contracted density matrix ρ_1 of an N-particle system. One can extend these considerations and investigate the eigenvalues of the $(N-2)$-fold contracted density matrix ρ_2, etc. One can prove that if ρ_1 has a large eigenvalue Nx, then ρ_2 has one between N^2 and N^2x^2. One can extend these considerations to ρ_3, ρ_4, ... A nonuniform superfluid flow is perhaps related to these large eigenvalues of ρ_1, ρ_2, ... Also for a general system (not necessarily dilute) the fractions of superfluid is perhaps related to these large eigenvalues of ρ_1, ρ_2, ...

Discussion remark by C. Domb

For the classical problem of hard sphere interactions, although an exact solution is not available, Monte Carlo calculations have indicated a phase transition at a particular volume which is associated with the disappearance of long range order. This is presumably a geometrical property and should also affect the quantum mechanical calculation of the ground state energy. One might thus perhaps expect a singularity of some kind in the curve of energy as a function of density, and not a smooth curve as drawn by Prof. Yang.

Answer by C. N. Yang

I would like to emphasize that to prove the existence of a singularity in the classical p-v diagram by a Monte Carlo calculation necessitates much more extensive calculation then exists in the literature.

The E_0/N vs. $1/\rho$ curve for $N \to \infty$ may indeed have singularities, but whether a singularity exists or not it is difficult to have intuitive arguments pro or con.

REFERENCES

1) The first two terms were first obtained by Lee, T. D., and Yang, C. N., Phys. Rev. **105** (1957) 1119; Lee, T. D., Huang, K. and Yang, C. N., Phys. Rev. **106** (1957) 1135. The next term was obtained independently by Wu, T. T., Phys. Rev. **115** (1959), 1390, Hugenholtz, N. and Pines, D., Phys. Rev. **116** (1959) 489 and Sawada, K., Phys. Rev. **116** (1959) 1344.
2) Lee, T. D. and Yang, C. N., Phys. Rev. **105** (1957) 1119.
3) Bogolubov, N. N., J. Phys. U.S.S.R. **2** (1947) 23; Lee, T. D., Huang, K. and Yang, C. N., Phys. Rev. **106** (1957) 1135.
4) Beliaev, Soviet Physics, JETP, **7** (1958) 299.
5) Lee, T. D. and Yang, C. N., Phys. Rev. **112** (1958) 1419.
6) Lee, T. D. and Yang, C. N., Phys. Rev. **113** (1959) 1165; **117** (1960) 22, 897.
7) Lee, T. D. and Yang, C. N., Phys. Rev. **113** (1959) 1406.
8) Tisza, L., J. Phys. Radium **1** (1940) 164.
9) Landau, L. D., J. Phys. U.S.S.R. **5** (1940) 71.
10) The argument was originally due to Landau, ref. [9]). In the present case, see ref. [7]).
11) Lin, C. C., to be published.
12) Penrose, O. and Onsager, L., Phys. Rev. **104** (1956) 576.

Introductory Notes to the Article "Are Mesons Elementary Particles?"

[[61a]]

C. N. Yang

Commentary begins page 48

At the end of the Second World War Fermi joined the University of Chicago, in the Physics Department and in the newly established Institute for Nuclear Studies (which now bears his name). That was the time when academic research work and graduate teaching were being resumed in the universities, and students, delayed by the war, thronged back to the campus. The University of Chicago had a particularly large enrollment of graduate students of physics. How many of them were attracted to Chicago by the name of Fermi we probably shall never know. In the case of myself, who was one of them, it had been my determination, in coming to the U.S. from China in November 1945, to study with Fermi or with Wigner. But I knew that war work had taken them from their universities. I remember that one day, soon after my arrival in New York, I trudged uptown and went up to the eighth floor of Pupin to inquire whether Professor Fermi would be giving courses soon. The secretaries met me with totally blank faces. I then went to Princeton, and found to my deep despair that Wigner would be mostly unavailable to students for the next year. But in Princeton I learned through W. Y. Chang that there were rumors of a new Institute to be established at Chicago and that Fermi would join the Institute. I went to Chicago, registered at the University, but did not feel completely secure until I saw Fermi with my own eyes, when he began his lectures in January 1946.

As is well known, Fermi gave extremely lucid lectures. In a fashion that is characteristic of him, for each topic he always started from the beginning, treated simple examples and avoided as much as possible "formalisms." (He used to joke that complicated formalism was for the "high priests".) The very simplicity of his reasoning conveyed the impression of effortlessness. But this impression is false: The simplicity was the result of careful preparation and of deliberate weighing of different alternatives of presentation. In the spring of 1949 when Fermi was giving a course on Nuclear Physics (which was later written up by Orear, Rosenfeld and Schluter and published as a book), he had to be away from Chicago for a few days. He asked me to take over for one lecture and gave me a small notebook in which he had carefully prepared each lecture in great detail. He went over the lecture with me before going away, explaining the reasons behind each particular line of presentation.

It was Fermi's habit to give, once or twice a week, informal unprepared lectures to a small group of graduate students. The group gathered in his office and someone, either Fermi himself or one of the students, would propose a specific topic for discussion. Fermi would search through his carefully indexed notebooks to find his notes on the topic and would then present it to us. I still have the notes I took of his evening lectures during October 1946–July 1947. It covered the following topics in the original order: theory of the internal constitution and the evolution of stars, structure of the white dwarfs, Gamow–Schönberg's idea about supernovae (neutrino cooling due to electron capture by nuclei), Riemannian geometry, general relativity and cosmology, Thomas–Fermi model, the state of matter at very high temperatures and densities, Thomas factor of 2, scattering of neutrons by para and ortho hydrogen, synchrotron radiation, Zeeman effect, "Johnson effect" of noise in circuits, Bose–Einstein condensation, multiple periodic system and Bohr's quantum condition, Born–Infeld theory of elementary particles, brief descirption of the foundation of statistical mechanics, slowing down of mesons in matter, slowing down of neutrons in matter. The discussions were kept at an elementary level. The emphasis was always on the essential and the practical part of the topic; the approach was almost always intuitive and geometrical, rather than analytic.

The fact that Fermi had kept over the years detailed notes on diverse subjects in physics, ranging from the purely theoretical to the purely experimental, from such simple problems as the best coordinates to use for the three-body problem to such deep subjects as general relativity, was an important lesson to all of us. We learned that *that* was physics. We learned that physics should not be a specialist's subject, physics is to be built from the ground up, brick by brick, layer by layer. We learned that abstractions come *after* detailed foundation work, not before. We also learned in these lectures of Fermi's delight in, rather than aversion to, simple numerical computations with a desk computer.

Besides the formal and informal classes Fermi also devoted almost all of his lunch hours to the graduate students (at least that was the state of affairs before 1950). The conversations in these lunch hours naturally covered a wide range of subjects. We observed Fermi as a somewhat conservative man with a very independent mind. We observed his dislike of pretension of whatever kind. Sometimes he wuld give general advice to us about our research work. I remember his emphasizing that as a young man one should devote most of one's time to attacking simple practical problems rather than deep fundamental ones.

Paper N° 239 was written by Fermi and me in the summer of 1949. As explicitly stated in the paper, we did not really have any illusions that what we suggested may actually correspond to reality. In fact, I was inclined to bury the work in notebooks and not publish it at all. Fermi said, however, that as a student one solves problems, but as a research worker one raises questions; and he considered the question we raised as worthy of publication. I may add here that the question remains unsolved today (1963).

As remarked by Segrè in his introduction to this collected works of Fermi, a very important question which Fermi had helped to raise was the spin orbit interaction in the shell model of nuclei; [See M. G. Mayer, *Phys. Rev.* 75, 1969 (1949), acknowledgement at end of paper]. Another question that Fermi was the first to raise was the concept of the conservation of nucleons. [See C. N. Yang and J. Tiomno, *Phys. Rev.* 79, 495 (1950), footnote 12]. I may also mention that Fermi was always very much interested in the question of parity conservation. [See the *Proceedings of the International Conference on Nuclear Physics and the Physics of Elementary Particles*, edited by J. Orear, A. H. Rosenfeld and R. A. Schluter, The Institute for Nuclear Studies, The University of Chicago, 1951, p. 2 and p. 109.] (See paper N° 245.)

Fermi fell critically ill in the fall of 1954. Murray Gell-Mann, who was then at Columbia University, and I went to Chicago to see him in Billings Hospital. As we entered his room he was reading a book which was a collection of stories about men who by their will power succeeded in overcoming fantastic natural obstacles and misfortune. He was very thin, but only a little sad. He told us very calmly about his conditions. The doctors had said that in a few days he may go home, but he would not have more than months to live. He then showed us the notebook by his bedside, and said that it was his own notes on nuclear physics. He planned, when he left the hospital, in the two months' time left, to revise it for publication. Gell-Mann and I were so overwhelmed by his simple determination and his devotion to physics that we were afraid for a few moments to look into his face. (Fermi died within three weeks of our visit.)

It has been said that the length of a man's life should not be measured in years, but in the different careers that he successfully goes through. Enrico Fermi, in one of his many careers, as a teacher in Chicago, had directly and indirectly influenced so many physicists of my generation that the record speaks for itself. The following is a list of the names of some of the physicists who received their graduate education in Chicago in the years 1946–1949 (I left Chicago in 1949 and am not familiar with Fermi's later students): H. M. Agnew, H. V. Argo, O. Chamberlain, G. F. Chew, G. W. Farwell, R. L. Garwin, M. L. Goldberger, D. Lazarus, T. D. Lee, A. Morrish, J. R. Reitz, M. N. Rosenbluth, W. Selove, J. Steinberger, R. M. Sternheimer, S. Warshaw, A. Wattenberg, L. Wolfenstein, H. A. Wilcox, C. N. Yang.

Some Considerations on Global Symmetry

⟦61b⟧

Commentary begins page 48

T. D. LEE AND C. N. YANG
Institute for Advanced Study, Princeton, New Jersey

(Received February 3, 1961)

If the recently discovered Y^* state is related to the $T=\frac{3}{2}$, $J=\frac{3}{2}$ resonance in πp scattering, global symmetry considerations should become relevant. In this paper, global symmetry is discussed with a view to understanding its group structure. Also discussed is a possibility of reconciling the conflict, pointed out by Pais, between certain experimental results and global symmetry. The partial widths of the Y^* state are calculated and also those of the companion excited states Z^* and Ξ^*. A generalization of the quantum number G is discussed.

1. INTRODUCTION

RECENT experiments[1,2] have established the existence of an excited state $Y^{*\pm}$ in the $\Lambda+\pi^\pm$ system. The spin and parity of the state are not yet measured. As discussed in reference 1, the state shows certain resemblance to the $J=\frac{3}{2}$, $T=\frac{3}{2}$, p-state resonance N^* of the $p+\pi$ system, and the resemblance is reminiscent of the concept of global symmetry.

In this paper we proceed along this line of thinking and assume that indeed the $Y^*=\Lambda+\pi^\pm$ resonance is in the $J=\frac{3}{2}$, p state, and that the resonance is related to the $J=\frac{3}{2}$, $T=\frac{3}{2}$, p-state resonance N^* of the $p+\pi$ system by global symmetry. To analyze this relation it is necessary to know the quantum numbers of various states with respect to the global symmetry operations. It is therefore important to know the structure of the global symmetry group. Now global symmetry[3–6] means some symmetry, larger than isotopic spin invariance, that describes an approximate analogy between the various baryons. But in the literature its group property has not been fully discussed. We shall in this paper formulate in mathematical terms the requirements that global symmetry must satisfy. It appears that the simplest group $\mathcal{G}_0$ satisfying these requirements can be generated by three independent 2-dimensional unitary unimodular transformations together with a discrete transformation. Adopting this group as the global symmetry group we then try to assign quantum numbers to the various particles and the resonance states Y^*, N^*.

Certain approximate relations are then written down between the widths of Y^* and N^*, and also for the various partial widths of Y^*. Companion resonance states Z^* and Ξ^* are also discussed.

It has been pointed out by Pais[5] that any kind of global symmetry is in conflict with certain experimental facts. We suggest in Sec. 6 that if global symmetry is needed to understand the resonant state Y^*, a way to resolve Pais' conflict is to have the K mesons as a mixture of states which have different quantum numbers under the global symmetry transformations. The interactions between each of these states and other particles could still predominantly satisfy global symmetry. Such a picture, while not completely satisfactory, does offer a possible consistent scheme incorporating global symmetry that leads to useful experimental information.

It should be emphasized that much of our results about widths have already been discussed in the literature from the viewpoint of symmetry considerations.[7,8] Furthermore a detailed calculation using a specific dynamical model has been performed by Amati, Stanghellini, and Vitale[7] for the states Y^* and Z^*. Various discussions on the global symmetry group properties have also existed in the literature.[3–6] The present paper is written not in the spirit of presenting something entirely new, nor even in that of presenting something which we believe to be necessarily relevant[9] to physical facts. But if a similarity between Y^* and N^* exists, an analysis along the present line would be useful.

For completeness we include in Sec. 8 a discussion of charge conjugation invariance, together with a generalization of the quantum number G. Also included are some remarks in Sec. 9 concerning a global symmetry that does not put Ξ and nucleons in the same multiplet.

2. REQUIREMENTS ON THE GLOBAL SYMMETRY GROUP

The global symmetry group must by definition contain the isotopic spin group and the strangeness group [defined by the operators $\exp(iS\theta)$ or $\exp(i(S+N)\theta)$, where S=strangeness and N=baryon number; the strangeness group commutes with the isotopic spin group]. It has an 8×8 unitary representation to which the 8 baryons belong. [Cf. Sec. 9 for the case of a symmetry between N, Σ, and Λ only.] In order that the

[1] M. Alston, L. W. Alvarez, P. Eberhard, M. L. Good, W. Graziano, H. K. Ticho, and S. G. Wojcicki, Phys. Rev. Letters **5**, 520 (1960).
[2] M. Ferro-Luzzi, J. P. Berge, J. Kirz, J. J. Murray, A. H. Rosenfeld, and M. Watson, Bull. Am. Phys. Soc. **5**, 509 (1960); H. J. Martin, W. Chinowsky, L. B. Leipuner, F. T. Shively, and R. K. Adair, Bull. Am. Phys. Soc. **6**, 40 (1961).
[3] J. Schwinger, Phys. Rev. **104**, 1164 (1956); Ann. Phys. **2**, 407 (1957).
[4] M. Gell-Mann, Phys. Rev. **106**, 1296 (1957).
[5] A. Pais, Phys. Rev. **110**, 574 (1958); **110**, 1480 (1958); **112**, 624 (1958); **122**, 317 (1961). See also A. Pais, *Proceedings of the Fifth Annual Rochester Conference on High-Energy Physics* (Interscience Publishers, Inc., New York, 1955).
[6] J. Tiomno, Nuovo cimento **6**, 69 (1957); R. E. Behrends, Nuovo cimento **11**, 424 (1959). See also G. Feinberg and F. Gürsey, Phys. Rev. **114**, 1153 (1959).
[7] D. Amati, A. Stanghellini, and B. Vitale, Nuovo cimento **13**, 1143 (1959); Phys. Rev. Letters **5**, 524 (1960).
[8] Ph. Meyer, J. Prentki, and Y. Yamaguchi, Phys. Rev. Letters **5**, 442 (1960).
[9] See the discussion of R. H. Dalitz and S. Tuan, Phys. Rev. Letters **2**, 425 (1959), Ann. Phys. **10**, 307 (1960). See also M. Ross and G. Shaw, Phys. Rev. Letters **5**, 578 (1960).

□ Reprinted from *The Physical Review* 122, 6 (June 15, 1961), 1954–1961.

8 baryons be analogous to each other under the global symmetry group, the representation must be irreducible. For the isotopic spin rotation subgroup the 8×8 representation breaks up into two doublets (N and Ξ), one triplet (Σ), and one singlet (Λ). For the strangeness subgroup this breakup must conform with the usual assignments of $S+N=+1, -1, 0, 0$ for N, Ξ, Σ, and Λ, respectively.

To state the above requirements explicitly we introduce as usual the states

$$\begin{aligned} Y^+ &= \Sigma^+, \\ Y^0 &= \frac{1}{\sqrt{2}}(\Sigma^0+\Lambda^0), \\ Z^0 &= \frac{1}{\sqrt{2}}(\Sigma^0-\Lambda^0), \\ Z^- &= \Sigma^-, \end{aligned} \tag{1}$$

and write

$$N\equiv\left\|\begin{matrix} p \\ n \end{matrix}\right\|,\quad Y\equiv\left\|\begin{matrix} Y^+ \\ Y^0 \end{matrix}\right\|,\quad Z\equiv\left\|\begin{matrix} Z^0 \\ Z^- \end{matrix}\right\|,\quad \Xi=\left\|\begin{matrix} \Xi^0 \\ \Xi^- \end{matrix}\right\|, \tag{2}$$

$$B\equiv\left\|\begin{matrix} N \\ \Xi \\ Y \\ Z \end{matrix}\right\|\equiv\left\|\begin{matrix} p \\ n \\ \Xi^0 \\ \Xi^- \\ Y^+ \\ Y^0 \\ Z^0 \\ Z^- \end{matrix}\right\|. \tag{3}$$

It is now useful to introduce (operating on the column matrix B) the following three sets of operators, each set satisfying the commutation relations for angular momenta:

$$(\mathcal{L}_1,\mathcal{L}_2,\mathcal{L}_3):\quad \mathcal{L}_i\equiv\tfrac{1}{2}\left\|\begin{matrix} \sigma_i & 0 & 0 & 0 \\ 0 & \sigma_i & 0 & 0 \\ 0 & 0 & \sigma_i & 0 \\ 0 & 0 & 0 & \sigma_i \end{matrix}\right\|,\quad (i=1, 2, 3), \tag{4}$$

$$(\mathfrak{M}_1,\mathfrak{M}_2,\mathfrak{M}_3):\quad \mathfrak{M}_1\equiv\tfrac{1}{2}\left\|\begin{matrix} 0 & 0 & 0 & 0 \\ 0 & 0 & 0 & 0 \\ 0 & 0 & 0 & I \\ 0 & 0 & I & 0 \end{matrix}\right\|,\quad \mathfrak{M}_2\equiv\tfrac{1}{2}i\left\|\begin{matrix} 0 & 0 & 0 & 0 \\ 0 & 0 & 0 & 0 \\ 0 & 0 & 0 & -I \\ 0 & 0 & I & 0 \end{matrix}\right\|,\quad \mathfrak{M}_3\equiv\tfrac{1}{2}\left\|\begin{matrix} 0 & 0 & 0 & 0 \\ 0 & 0 & 0 & 0 \\ 0 & 0 & I & 0 \\ 0 & 0 & 0 & -I \end{matrix}\right\|, \tag{5}$$

$$(\mathfrak{N}_1,\mathfrak{N}_2,\mathfrak{N}_3):\quad \mathfrak{N}_1\equiv\tfrac{1}{2}\left\|\begin{matrix} 0 & I & 0 & 0 \\ I & 0 & 0 & 0 \\ 0 & 0 & 0 & 0 \\ 0 & 0 & 0 & 0 \end{matrix}\right\|,\quad \mathfrak{N}_2\equiv\tfrac{1}{2}i\left\|\begin{matrix} 0 & -I & 0 & 0 \\ I & 0 & 0 & 0 \\ 0 & 0 & 0 & 0 \\ 0 & 0 & 0 & 0 \end{matrix}\right\|,\quad \mathfrak{N}_3\equiv\tfrac{1}{2}\left\|\begin{matrix} I & 0 & 0 & 0 \\ 0 & -I & 0 & 0 \\ 0 & 0 & 0 & 0 \\ 0 & 0 & 0 & 0 \end{matrix}\right\|, \tag{6}$$

where

$$\sigma_1\equiv\left\|\begin{matrix} 0 & 1 \\ 1 & 0 \end{matrix}\right\|,\quad \sigma_2\equiv\left\|\begin{matrix} 0 & -i \\ i & 0 \end{matrix}\right\|,\quad \sigma_3\equiv\left\|\begin{matrix} 1 & 0 \\ 0 & -1 \end{matrix}\right\|,$$

and

$$I=\left\|\begin{matrix} 1 & 0 \\ 0 & 1 \end{matrix}\right\|.$$

Clearly $\mathcal{L}_i$, $\mathfrak{M}_j$, $\mathfrak{N}_k$ all commute for any i, j, k.

The physical observables, Q=charge, S=strangeness, N=baryon number, T_1, T_2, T_3=isotopic spin are related to $\mathcal{L}$, $\mathfrak{M}$, and $\mathfrak{N}$ by

$$T_i=\mathcal{L}_i+\mathfrak{M}_i, \tag{7}$$

$$\tfrac{1}{2}(S+N)=\mathfrak{N}_3, \tag{8}$$

$$Q=T_3+\mathfrak{N}_3=\mathcal{L}_3+\mathfrak{M}_3+\mathfrak{N}_3. \tag{9}$$

The 8×8 representation must (i) contain $\exp(iT_1\theta_1)$, $\exp(iT_2\theta_2)$, $\exp(iT_3\theta_3)$, $\exp(i\mathfrak{N}_3\theta_4)$. In other words, T_1, T_2, T_3, $\mathfrak{N}_3$ are among the infinitesimal generators of the representation. Furthermore (ii) the 8×8 representation is irreducible.

Many possible 8×8 representations of groups can be found that satisfy (i) and (ii). Since we are not interested in unnecessary additional symmetries, we take the group to be *isomorphic* with the representation, and shall identify a group element with its 8×8 representation. The simplest such possibility, $\mathcal{G}_0$, is discussed in the following sections. Other possibilities are discussed in Appendix A.

3. THE GROUP $\mathcal{G}_0$

The group contains arbitrary $\mathcal{L}$ transformations [i.e., $U_{\mathcal{L}}\equiv\exp i(l_1\mathcal{L}_1+l_2\mathcal{L}_2+l_3\mathcal{L}_3)$, where $l_1l_2l_3$ are real numbers], arbitrary $\mathfrak{M}$ transformations [$\exp i(m_1\mathfrak{M}_1+m_2\mathfrak{M}_2+m_3\mathfrak{M}_3)$, which mixes Y and Z], and arbitrary $\mathfrak{N}$ transformations [$\exp i(n_1\mathfrak{N}_1+n_2\mathfrak{N}_2+n_3\mathfrak{N}_3)$, which mixes N and Ξ]. The product of all these is reducible since no mixing of N, Ξ with Y, Z has been introduced. To effect such a mixing we introduce the discrete element[10]

$$R\equiv\left\|\begin{matrix} 0 & 0 & I & 0 \\ 0 & 0 & 0 & I \\ I & 0 & 0 & 0 \\ 0 & I & 0 & 0 \end{matrix}\right\|, \tag{10}$$

[10] This discrete operator has been discussed by various authors. See reference 5.

TABLE I. The irreducible representations of the group $\mathcal{G}_0$. The quantity $\chi_{\mathcal{M}}(U)$ is the character of the 2×2 matrix U for the irreducible representations of dimension $2\mathcal{M}+1$.

Representation	Dimension	Character for elements (11)	Character for elements (12)
$(\mathcal{L}\mathcal{M}\mathcal{M}+1)$	$(2\mathcal{L}+1)(2\mathcal{M}+1)^2$	$\chi_{\mathcal{L}}(U_{\mathcal{L}})\chi_{\mathcal{M}}(U)\chi_{\mathcal{M}}(U')$	$\chi_{\mathcal{L}}(U_{\mathcal{L}})\chi_{\mathcal{M}}(UU')$
$(\mathcal{L}\mathcal{M}\mathcal{M}-1)$	$(2\mathcal{L}+1)(2\mathcal{M}+1)^2$	$\chi_{\mathcal{L}}(U_{\mathcal{L}})\chi_{\mathcal{M}}(U)\chi_{\mathcal{M}}(U')$	$-\chi_{\mathcal{L}}(U_{\mathcal{L}})\chi_{\mathcal{M}}(UU')$
$(\mathcal{L}\mathcal{M}\mathcal{N})$ $\mathcal{M}>\mathcal{N}$	$2(2\mathcal{L}+1)(2\mathcal{M}+1)(2\mathcal{N}+1)$	$\chi_{\mathcal{L}}(U_{\mathcal{L}})[\chi_{\mathcal{M}}(U)\chi_{\mathcal{N}}(U')$ $+\chi_{\mathcal{M}}(U')\chi_{\mathcal{N}}(U)]$	0

and other necessary elements to form a group. The elements of the group are then the 8×8 matrices of the form

$$U_{\mathcal{L}}\begin{Vmatrix} aI & bI & 0 & 0 \\ -b^*I & a^*I & 0 & 0 \\ 0 & 0 & a'I & b'I \\ 0 & 0 & -b'^*I & a'^*I \end{Vmatrix}, \tag{11}$$

and those of the form

$$U_{\mathcal{L}}\begin{Vmatrix} 0 & 0 & aI & bI \\ 0 & 0 & -b^*I & a^*I \\ a'I & b'I & 0 & 0 \\ -b'^*I & a'^*I & 0 & 0 \end{Vmatrix}. \tag{12}$$

Here

$$\begin{Vmatrix} a & b \\ -b^* & a^* \end{Vmatrix} \equiv U = \text{arbitrary } 2\times 2 \text{ unimodular unitary matrix,}$$

and (13)

$$\begin{Vmatrix} a' & b' \\ -b'^* & a'^* \end{Vmatrix} \equiv U' = \text{arbitrary } 2\times 2 \text{ unimodular unitary matrix.}$$

This group will be called $\mathcal{G}_0$. It has an invariant subgroup (11) which is the direct product[11] of three SU_2, and the quotient of the group by this invariant subgroup is the two-element group. [It is, however, not the direct product of the two-element group with the invariant subgroup.]

In Appendix A we shall give a few other possible groups satisfying conditions (i) and (ii).

[11] We use SU_n to denote the group of unitary $n\times n$ matrices with determinant unity. Strictly speaking, the invariant subgroup (11) is only locally identical with $SU_2\times SU_2\times SU_2$. If one changes the signs of $U_{\mathcal{L}}$ and a, b, a', b' at the same time, (11) is unchanged. Hence (11) has a 1 to 2 homomorphism with $SU_2\times SU_2\times SU_2$. In other words, $SU_2\times SU_2\times SU_2$ is the covering group of (11). Similarly $\mathcal{G}_0$, which is defined to be the group of matrices of the form (11) and (12), has a covering group which we shall call $\mathcal{G}_0'$. $\mathcal{G}_0'$ has a 2 to 1 homomorphism with $\mathcal{G}_0$. It has an invariant subgroup $SU_2\times SU_2\times SU_2$. The relationship between $\mathcal{G}_0'$ and $\mathcal{G}_0$ is entirely similar to the familiar relationship between SU_2 and O_3, the group of 3×3 rotations. To simplify matters we shall not dwell on the difference between $\mathcal{G}_0$ and $\mathcal{G}_0'$ in the text. The irreducible representations in Table I are actually representations of $\mathcal{G}_0'$. Those representations $(\mathcal{L}\mathcal{M}\mathcal{M}\lambda)$ with $\mathcal{L}+2\mathcal{M}=$integer, and those representations $(\mathcal{L}\mathcal{M}\mathcal{N})$, $\mathcal{M}>\mathcal{N}$, with $\mathcal{L}+\mathcal{M}+\mathcal{N}=$integer are also representations of $\mathcal{G}_0$. (The others are not single-valued representations of $\mathcal{G}_0$.) Since the baryons belong to a representation of this type, and all known particles have transition elements to a collection of baryons and antibaryons, only representations of the same type (i.e., single-valued representations of $\mathcal{G}_0$) enter into the discussion of known particles.

4. IRREDUCIBLE REPRESENTATION OF $\mathcal{G}_0$

For any representation of $\mathcal{G}_0$, the infinitesimal operators $\mathcal{L}_1$, $\mathcal{L}_2$, $\mathcal{L}_3$, $\mathcal{M}_1$, $\mathcal{M}_2$, $\mathcal{M}_3$, $\mathcal{N}_1$, $\mathcal{N}_2$, $\mathcal{N}_3$ form three sets of commuting angular momenta. One can diagonalize $\mathcal{L}^2$, $\mathcal{L}_3$, $\mathcal{M}^2$, $\mathcal{M}_3$, $\mathcal{N}^2$, $\mathcal{N}_3$ simultaneously. Now

$$\begin{aligned} R\mathcal{L}_iR^{-1}&=\mathcal{L}_i,\\ R\mathcal{M}_iR^{-1}&=\mathcal{N}_i,\\ R\mathcal{N}_iR^{-1}&=\mathcal{M}_i. \end{aligned} \tag{14}$$

Hence in any representation the set of eigenvalues of $\mathcal{M}^2$ must be the same as those of $\mathcal{N}^2$. One has therefore two kinds of irreducible representations:

(α) $\mathcal{L}^2$, $\mathcal{M}^2$, $\mathcal{N}^2$ have unique eigenvalues $\mathfrak{L}(\mathfrak{L}+1)$, $\mathfrak{M}(\mathfrak{M}+1)$, $\mathfrak{M}(\mathfrak{M}+1)$, respectively. Since R commutes with $\mathcal{M}_3+\mathcal{N}_3$, the state with $\mathcal{M}_3=\mathcal{N}_3=\mathfrak{M}$ is an eigenstate of R. The eigenvalue λ can be ± 1. We denote this representation by the symbol $(\mathfrak{L}\mathfrak{M}\mathfrak{M}\lambda)$, where $2\mathfrak{L}=\text{integer}\geq 0$, $2\mathfrak{M}=\text{integer}\geq 0$, $\lambda=\pm 1$.

The states of this representation are designated by $\mathcal{L}_3$, $\mathcal{M}_3$, $\mathcal{N}_3$, each running in integral steps between and including $\pm\mathfrak{L}$, $\pm\mathfrak{M}$, $\pm\mathfrak{M}$ respectively. The operator R switches the indices $\mathcal{M}_3$ and $\mathcal{N}_3$ for a state:

$$R|\mathcal{M}_3=a,\quad \mathcal{N}_3=b\rangle=\lambda|\mathcal{M}_3=b,\quad \mathcal{N}_3=a\rangle. \tag{15}$$

(β) $\mathcal{L}^2$ has a unique eigenvalue $\mathfrak{L}(\mathfrak{L}+1)$. $\mathcal{M}^2$ and $\mathcal{N}^2$ each has two eigenvalues $\mathfrak{M}(\mathfrak{M}+1)$ and $\mathfrak{N}(\mathfrak{N}+1)$ where $\mathfrak{M}\neq\mathfrak{N}$. We denote this representation by the symbol $(\mathfrak{L}\mathfrak{M}\mathfrak{N})$, where $2\mathfrak{L}$, $2\mathfrak{M}$, and $2\mathfrak{N}$ are integers ≥ 0 and $\mathfrak{M}>\mathfrak{N}$.

The states of this representation are states for which

$$\{\mathcal{M}^2=\mathfrak{M}(\mathfrak{M}+1),\quad \mathcal{M}_3=-\mathfrak{M},-\mathfrak{M}+1,\cdots+\mathfrak{M},$$
$$\text{while } \mathcal{N}^2=\mathfrak{N}(\mathfrak{N}+1),\quad \mathcal{N}_3=-\mathfrak{N},-\mathfrak{N}+1,\cdots+\mathfrak{N}\};$$

and

$$\{\mathcal{M}^2=\mathfrak{N}(\mathfrak{N}+1),\quad \mathcal{M}_3=-\mathfrak{N},-\mathfrak{N}+1,\cdots+\mathfrak{N},$$
$$\text{while } \mathcal{N}^2=\mathfrak{M}(\mathfrak{M}+1),\quad \mathcal{N}_3=-\mathfrak{M},-\mathfrak{M}+1,\cdots+\mathfrak{M}\}.$$

The operator R switches the states between these two sets. In a suitable representation, R satisfies

$$R|\mathfrak{M}\mathfrak{M}_3\mathfrak{N}\mathfrak{N}_3\rangle=|\mathfrak{N}\mathfrak{N}_3\mathfrak{M}\mathfrak{M}_3\rangle.$$

The dimensions of the irreducible representations are tabulated in Table I. Also given are the characters of the representations. From the characters the decomposition of the direct product of two representations can be easily

TABLE II. Quantum number assignments for particles and excited states. $T=\mathfrak{L}+\mathfrak{M}$=isotopic spin. P_K=parity of K meson. The quantum number G_1 is explained in Sec. 8. In this table only the $(0,\frac{1}{2},\frac{1}{2},\lambda_K)$ part is listed for the K mesons.

Particle	Representation	$\mathfrak{L}$	$\mathfrak{L}_3$	$\mathfrak{M}$	$\mathfrak{M}_3$	$\mathfrak{N}$	$\mathfrak{N}_3$	G_1	T	R
p, n	$(\frac{1}{2},\frac{1}{2},0)$	$\frac{1}{2}$	$\pm\frac{1}{2}$	0	0	$\frac{1}{2}$	$\frac{1}{2}$		$\frac{1}{2}$	
Ξ^0, Ξ^-		$\frac{1}{2}$	$\pm\frac{1}{2}$	0	0	$\frac{1}{2}$	$-\frac{1}{2}$		$\frac{1}{2}$	
Y^+, Y^0		$\frac{1}{2}$	$\pm\frac{1}{2}$	$\frac{1}{2}$	$\frac{1}{2}$	0	0		1, 0	
Z^0, Z^-		$\frac{1}{2}$	$\pm\frac{1}{2}$	$\frac{1}{2}$	$-\frac{1}{2}$	0	0		1, 0	
π^+	$(1,0,0,\lambda_\pi)$	1	1	0	0	0	0	-1	1	λ_π
π^0	$\lambda_\pi=\pm1$	1	0	0	0	0	0	-1	1	λ_π
π^-		1	-1	0	0	0	0	-1	1	λ_π
K^+	$(0,\frac{1}{2},\frac{1}{2},\lambda_K)$	0	0	$\frac{1}{2}$	$\frac{1}{2}$	$\frac{1}{2}$	$\frac{1}{2}$	$-\lambda_K$	$\frac{1}{2}$	λ_K
K^0	$\lambda_K=\pm1$	0	0	$\frac{1}{2}$	$-\frac{1}{2}$	$\frac{1}{2}$	$\frac{1}{2}$	$-\lambda_K$	$\frac{1}{2}$	$\{K_1^0: \lambda_K P_K$
$\bar{K}^0$		0	0	$\frac{1}{2}$	$\frac{1}{2}$	$\frac{1}{2}$	$-\frac{1}{2}$	$-\lambda_K$	$\frac{1}{2}$	$K_2^0: -\lambda_K P_K$
K^-		0	0	$\frac{1}{2}$	$-\frac{1}{2}$	$\frac{1}{2}$	$-\frac{1}{2}$	$-\lambda_K$	$\frac{1}{2}$	λ_K
N^*	$(\frac{3}{2},\frac{1}{2},0)$	$\frac{3}{2}$	···	0	0	$\frac{1}{2}$	$\frac{1}{2}$		$\frac{3}{2}$	
Ξ^*		$\frac{3}{2}$	···	0	0	$\frac{1}{2}$	$-\frac{1}{2}$		$\frac{3}{2}$	
Y^*		$\frac{3}{2}$	···	$\frac{1}{2}$	···	0	0		1	
Z^*		$\frac{3}{2}$	···	$\frac{1}{2}$	···	0	0		2	

found in the standard way. (Except for the quantum number λ, it can also be found by the usual vector sum rule for $\mathfrak{L}$, $\mathfrak{M}$, and $\mathfrak{N}$ separately.)

5. QUANTUM NUMBERS

To assign quantum numbers to the states we first notice that Eqs. (7)–(9) give the isotopic spin, the strangeness, and the charge in terms of these quantum numbers.

The 8 baryons clearly belong to the representation $(\frac{1}{2},\frac{1}{2},0)$. It seems natural that the pions should be assigned to the 3×3 representation $(1,0,0,\lambda_\pi)$. The two possibilities $\lambda_\pi=\pm1$ are, of course, physically different, and differentiable. It seems natural to assign the K mesons to the representation $(0,\frac{1}{2},\frac{1}{2},\lambda_K)$ with again the two possibilities $\lambda_K=\pm1$. These assignments are tabulated in Table II.

For the state $N^*=\pi+p$ we notice that $\pi+p$ always belongs to either $(\frac{3}{2},\frac{1}{2},0)$ $T=\frac{3}{2}$, or $(\frac{1}{2},\frac{1}{2},0)$ $T=\frac{1}{2}$. But N^* has a total $T=\frac{3}{2}$. Hence it belongs to $(\frac{3}{2},\frac{1}{2},0)$. The natural assumption is therefore that Y^* is in the same multiplet structure $(\frac{3}{2},\frac{1}{2},0)$, as indicated in Table II. Since Y^* can go into $\Lambda+\pi$, its isotopic spin $T=1$. The multiplet $(\frac{3}{2},\frac{1}{2},0)$ also contains a $T=2$ state[1,4,7] which will be called Z^*. In addition to the 3 Y^* states and 5 Z^* states there should also be 4 Ξ^* states with $T=\frac{3}{2}$.

6. BREAKDOWN OF GLOBAL SYMMETRY

Even if global symmetry has any valid basis, there must be relatively strong interactions that violate it. One manifestation of this violation lies in the mass difference between the hyperons. Another manifestation was first pointed out by Pais,[5] who showed that the following reactions

$$\pi^+ + p \to \Sigma^+ + K^+, \tag{16}$$

$$K^+ + n \to K^0 + p, \tag{17}$$

and many others violate global symmetry. For the group $\mathcal{G}_0$ discussed above, the conservation of $\mathfrak{L}_3$ is violated by both (16) and (17).

In face of these difficulties, does global symmetry have any validity at all? And if it has, does it ever produce useful physical information?

It would be difficult to answer these questions. But if the answers to the above questions are affirmative, presumably the baryons and the states N^*, Y^* allow more directly the application of global symmetry than reactions (16) and (17). For example, if the global-symmetry-destroying interactions produce relatively little mixing for the baryons, pions and Y^*, N^*, but produce large mixing for the K mesons, then apparent violation of global symmetry is not unnatural for (16) and (17). While the mixing may be the multiplet $(0,\frac{1}{2},\frac{1}{2})$ with any multiplet possessing a $T=\frac{1}{2}$, $\mathfrak{N}=\frac{1}{2}$ component, it seems that the mixing of $(0,\frac{1}{2},\frac{1}{2})$ with $(1,\frac{1}{2},\frac{1}{2})$ is the simplest possibility.

In this view, then, the usual interactions (baryons, pions, and K mesons) are regarded as predominantly globally symmetrical. The globally unsymmetrical interactions give rise to, among others, two effects: (a) mass splitting of the states within each multiplet. (b) a strong mixing for the K meson of a $(1,\frac{1}{2},\frac{1}{2})$ $T=\frac{1}{2}$ component with the $(0,\frac{1}{2},\frac{1}{2})$ state. The globally unsymmetrical interactions may, for example, have a very small range, so that the two effects (a) and (b) are the only ones that one need consider as causing global unsymmetry in the zeroth approximation. The influence of global unsymmetry is then quite limited in scope, though not in magnitude, and one can derive consequences that can be checked with experimental information. [This is true only insofar as one does not probe into the *very small* range where the strong unsymmetric force is assumed to be effective. It may be instructive to recall the well-known symmetry between $e^\pm$ and $\mu^\pm$. In that case, the asymmetry between these two particles

TABLE III. Phase-space factor Ω, projection weight w, and relative partial widths of resonance levels. The weights w are calculated from the quantum number assignments. The phase space factor Ω is computed from experimental resonance energies for N^* and Y^*, and from an assumed energy spectrum for Z^* and Ξ^*. The partial widths of other disintegration processes, such as $Z^{*+}\to\pi^+ +\Sigma^0$ etc., can be inferred from the table through a simple isotopic spin rotation.

Particle	Total energy (Mev)	Disintegration products	Ω (Mev³)	w	Computed relative partial width
$(N^*)^{++}$	1237	π^++p	9.7×10^6	1	1
Y^{*+}	1385	$\pi^++\Lambda$	7.3×10^6	$\frac{2}{3}$	0.5
		$\Sigma^++\pi^0$	1.6×10^6	$\frac{1}{6}$	0.03
		$\Sigma^0+\pi^+$	1.6×10^6	$\frac{1}{6}$	0.03
Z^{*++}	∼1539(?)	$\pi^++\Sigma^+$	$\sim18(?)\times10^6$	1	∼1.9(?)
Ξ^{*+}	∼1637(?)	$\pi^++\Xi^0$	$\sim15(?)\times10^6$	1	∼1.5(?)

seems to be completely characterized by their large mass difference which, presumably, is also generated by some unsymmetrical forces, strong in magnitude but remarkably limited in its symmetry destroying effects.]

If one asks how does it happen that only the K particles have a strong mixing, a possible answer could be that without the globally unsymmetrical interaction two multiplets $(1,\frac{1}{2},\frac{1}{2})$ and $(0,\frac{1}{2},\frac{1}{2})$ happen to lie relatively close together and, therefore, result in large mixings for the states with $T=\frac{1}{2}$. It would then be reasonable to expect the existence of other excited states K^*.

In such a picture the K meson is a mixture of $(0,\frac{1}{2},\frac{1}{2})$ and $(1,\frac{1}{2},\frac{1}{2})$, each of which, in interacting with the other particles, still predominantly *satisfies* global symmetry. Thus, e.g., in reactions such as K^-+baryon with multiple pion productions, one can apply the global symmetry arguments and obtain equalities and inequalities between the various related processes.

7. POSITIONS AND WIDTHS OF N^*, Y^*, Z^*, AND Ξ^*

The discussion of Sec. 6 suggests a zeroth order calculation of the partial widths of $N^*Y^*Z^*$ and Ξ^*, for the processes tabulated in Table III. These processes represent a transition from a $(\frac{3}{2},\frac{1}{2},0)$ multiplet to a product of a $(\frac{1}{2},\frac{1}{2},0)$ multiplet and a $(1,0,0,\lambda_\pi)$ multiplet. The decomposition therefore yields unique weights w which are related to the squares of the appropriate transition amplitude. The calculation of these weights from the usual tables of Clebsch-Gordan coefficients is straightforward and the result is tabulated in Table III. Besides these weights due to the projection of the initial state on final states, there is also a phase-space-potential-barrier factor Ω for the p-wave state. We take it to be given by

$$\Omega=q^3E_B/(E_B+E_\pi), \tag{18}$$

where $q=$ momentum of pion in the rest system of the resonance state, $E_B=$ total energy of the final baryon, and $E_\pi=$ total energy of the final pion. In the approximation that other effects due to global asymmetrical interactions are neglected, the partial widths of each resonance level are proportional to the appropriate products of w and Ω. To calculate Ω one needs the excitation energy of the resonance states. For N^* and Y^* we take the values in reference 1. To guess at the energies of Z^* and Ξ^* we write the total energy E of the excited state in the multiplet $(\frac{3}{2},\frac{1}{2},0)$ in the form

$$E(\equiv E_B+E_\pi)=E_{N^*}+\alpha'\mathcal{L}\cdot\mathfrak{M}+\beta'(\mathfrak{N}^2-\tfrac{3}{4})+\gamma'(\mathfrak{N}_3-\tfrac{1}{2}), \tag{19}$$

where α', β' and γ' are constants. Similarly, for the 8 baryons in the multiplet $(\frac{1}{2},\frac{1}{2},0)$ we have an analogous expression

$$E=E_N+\alpha\mathcal{L}\cdot\mathfrak{M}+\beta(\mathfrak{N}^2-\tfrac{3}{4})+\gamma(\mathfrak{N}_3-\tfrac{1}{2}), \tag{20}$$

where

$$\begin{aligned}\alpha&=E_\Sigma-E_\Lambda\cong77\text{ Mev},\\ \beta&=\tfrac{4}{3}[\tfrac{1}{2}(E_\Xi+E_N)-\tfrac{1}{4}(3E_\Sigma+E_\Lambda)]\cong-59\text{ Mev},\\ \gamma&=-(E_\Xi-E_N)\cong-380\text{ Mev}.\end{aligned} \tag{21}$$

One sees that by taking[12]

$$\alpha'=\alpha,\quad \beta'=\beta,\quad \gamma'\cong-400\text{ Mev}, \tag{22}$$

one obtains the experimental resonance energy for Y^*. With this choice the resonance energies for Z^* and Ξ^* can be computed and are tabulated in Table III, with the corresponding phase space factors Ω.

The last column of Table III shows a smaller total width for Y^* than N^* [in the ratio of approximately 0.56:1], and shows a very small branching ratio of $Y^*\to\Sigma+\pi$. Both of these are in general agreement with experimental information.[1]

8. CHARGE CONJUGATION INVARIANCE

With the inclusion of the unitary operator C, representing charge conjugation, the symmetry group is enlarged. The irreducible representations become larger in general, corresponding to, e.g., the fact that a particle and its antiparticle have the same mass. To study the combined group generated from $\mathcal{G}_0$, C and the baryon number gauge transformation $\exp(i\theta N)$, we start from their commutation relations. For the sake of clarity we shall formulate this discussion in theorems. We shall also only deal with particles and states that have transition matrix elements into n baryons and antibaryons, $n=1, 2, 3, \cdots$.

Theorem 1.

$$G_1\equiv C\exp[i\pi(\mathcal{L}_2+\mathfrak{M}_2+\mathfrak{N}_2)] \tag{23}$$

commutes with all elements of the group $\mathcal{G}_0$.

Proof: For a single baryon-antibaryon the explicit representations of C, $\mathcal{L}$, $\mathfrak{M}$, $\mathfrak{N}$, R are given in Appendix B. The theorem follows from a straightforward explicit

[12] A similar guess on the masses of excited levels has been made by A. Pais (private communication). See also reference 7.

computation of the commutators. For other states the theorem follows because G_1, C, $\exp(i\pi\mathcal{L}_2)$, $\exp(i\pi\mathfrak{M}_2)$, and $\exp(i\pi\mathfrak{N}_2)$ are all multiplicative for a collection of particles.

Theorem 2.

$$G_1N+NG_1=0, \quad [N,\mathcal{G}_0]_-=0, \quad G_1^2=1. \tag{24}$$

Proof: This theorem follows directly from the explicit representation of Appendix B.

Theorem 3. The full group generated by $\mathcal{G}_0$, G_1 and $\exp(iN\theta)$ is the direct product group $\mathcal{G}_0\times O_2^{\pm}$, where $O_2^{\pm}$ is the group of all 2×2 real orthogonal matrices with determinant$=\pm1$.

The proof of this theorem is again straightforward.

The irreducible representations of $O_2^{\pm}$ are either (a) 2×2 in size, in which $N=+\alpha$ and $N=-\alpha$ [α=integer] each occurs once, representing physically a pair of particles, and G_1 switches the two states; or (b) the representation is of dimension 1×1 in which $N=0$ and either $G_1=+1$ or $G_1=-1$.

For a state with $N=0$, the operator G_1 is *one and the same numerical constant* ($=\pm1$) *for all states in a multiplet of* $\mathcal{G}_0$. By (23), C brings one state into another in the same multiplet of $\mathcal{G}_0$.

Theorem 4. For the pions, $G_1=-1$.

Proof: The pions are eigenfunctions of $\mathfrak{N}$ with eigenvalue $\mathfrak{N}=0$. Hence[13] $G_1=G=-1$.

Theorem 5. For the $(0,\frac{1}{2},\frac{1}{2},\lambda)$, $N=0$ representation, if the total angular momentum $J=0$, and the system has transition matrix elements into a baryon-antibaryon pair, then

$$G_1\lambda=-1. \tag{25}$$

For the $(1,\frac{1}{2},\frac{1}{2},\lambda)$, $N=0$ representation under the same assumption,

$$G_1\lambda=1. \tag{26}$$

Proof: In the notation of Appendix B, we have four possible states for the baryon-antibaryon pair that belong to $(0,\frac{1}{2},\frac{1}{2})$, with $\mathfrak{M}_3=\frac{1}{2}$, $\mathfrak{N}_3=\frac{1}{2}$:

$$\{p\bar{Z}^0\}+\{n\bar{Z}^+\}, \tag{27}$$

$$\{\bar{Z}^0p\}+\{\bar{Z}^+n\}, \tag{28}$$

$$\{Y^+\bar{\Xi}^0\}+\{Y^0\bar{\Xi}^+\}, \tag{29}$$

and

$$\{\bar{\Xi}^0Y^+\}+\{\bar{\Xi}^+Y^0\}, \tag{30}$$

where each curly bracket represents a state, for which the first symbol inside specifies the state of particle a and the second symbol that of particle b. Consider a [c number] product wave function of an orbital part, a spin part, and a charge part [depending on the other quantum numbers, $\mathcal{L}$, $\mathcal{L}_3$, $\mathfrak{M}$, $\mathfrak{M}_3$, etc.] of the two particles. For a state $J=0$, the product of the first two parts is antisymmetry in the interchange $a\leftrightarrow b$. Hence the charge part is symmetric since the entire wave function must be antisymmetric. Now under $a\leftrightarrow b$,

$$(27)\leftrightarrow(28), \quad (29)\leftrightarrow(30).$$

Hence the states are either

$$(27)+(28) \quad \text{or} \quad (29)+(30), \tag{31}$$

or superpositions of these two. By using the explicit matrices listed in Appendix B it can be directly verified that under

$$R\colon\quad (27)\leftrightarrow(29), \qquad (28)\leftrightarrow(30),$$

$$G_1\colon\quad (27)\leftrightarrow-(30), \qquad (28)\leftrightarrow-(29).$$

Hence under RG_1 both wave functions in (31) remain themselves but change sign. Thus (25) is proved. A similar proof holds for (26).

Applied to the K mesons, if the K meson admixture $(1,\frac{1}{2},\frac{1}{2})$ can be neglected, Theorem 5 states that

$$(G_1)_K=-\lambda_K.$$

While all the four states $\mathfrak{M}_3=\pm\frac{1}{2}$ and $\mathfrak{N}_3=\pm\frac{1}{2}$ are eigenstates of G_1 with eigenvalues $-\lambda_K$, only two: $\mathfrak{M}_3=\mathfrak{N}_3=\pm\frac{1}{2}$ are eigenstates of R with eigenvalues λ_K.

If, further, one assumes that time reversal invariance holds, then the two states K_1^0 and K_2^0 have simple behaviors under R. To see this, we notice that $(G_1)_K$ is a numerical constant. Hence (23) shows that

$$C|\mathfrak{M}_2=\tfrac{1}{2},\ \mathfrak{N}_2=-\tfrac{1}{2}\rangle=-G_1|\mathfrak{M}_2=-\tfrac{1}{2},\ \mathfrak{N}_2=\tfrac{1}{2}\rangle.$$

If P is the parity operator, we have

$$CP|\mathfrak{M}_2=\tfrac{1}{2},\ \mathfrak{N}_2=-\tfrac{1}{2}\rangle=-G_1P|\mathfrak{M}_2=-\tfrac{1}{2},\ \mathfrak{N}_2=\tfrac{1}{2}\rangle.$$

Now K_1^0, (K_2^0) is an eigenstate of CP with eigenvalue $+1$, (-1). Hence

$$|K_{1,2}^0\rangle=|\mathfrak{M}_2=\tfrac{1}{2},\ \mathfrak{N}_2=-\tfrac{1}{2}\rangle\mp(G_1P)|\mathfrak{M}_2=-\tfrac{1}{2},\ \mathfrak{N}_2=\tfrac{1}{2}\rangle.$$

One obtains with the use of (15):

$$R|K_{1,2}^0\rangle=\mp(G_1P)|K_{1,2}^0\rangle=\pm(\lambda_KP_K)|K_{1,2}^0\rangle,$$

where P_K is the parity of K^0 [with respect to, say, $\bar{\Lambda}n$]. Thus we obtain the entries in Table II for K_1^0, K_2^0 under R.

9. REMARKS

For a symmetry to exist between Y^* and N^* it is not necessary that all 8 baryons be brought into global symmetry. For example, one could have a symmetry between N, Y, and Z without Ξ. A simple symmetry group in such a case[11] is $SU_2\times SU_3$ which has more parameters than $\mathcal{G}_0$. The irreducible representations in such a case can be written down and an analysis like the above for $\mathcal{G}_0$ can be made. There also would be a companion Z^* $T=2$ state together with N^* and Y^*. The weight factors w for N^*, Y^*, Z^* remain the same as in Table III.

APPENDIX A

We give here several examples other than $\mathcal{G}_0$ satisfying conditions (i) and (ii) of Sec. 2.

(A) The group $\mathcal{G}_1$. The group is isomorphic with $SU_2\times SU_4$ where SU_2 consists of the $\mathcal{L}$ transformations

[13] T. D. Lee and C. N. Yang, Nuovo cimento 3, 749 (1956).

$U_{\mathcal{L}}=\exp i(l_1\mathcal{L}_1+l_2\mathcal{L}_2+l_3\mathcal{L}_3)$ and SU_4 consists of all the unimodular unitary transformations between N, Y, Z, Ξ. Clearly R is an element of SU_4. Hence $\mathcal{G}_0$ is a subgroup of $\mathcal{G}_1$. The usual $g_1=g_2=g_3=g_4$ case[4,5] [where g_1, g_2, g_3, g_4 are, respectively, the coupling constants between pion and $\bar{N}N$, $\bar{Y}Y$, $\bar{Z}Z$, $\bar{\Xi}\Xi$] has this larger symmetry $\mathcal{G}_1$ rather than $\mathcal{G}_0$. Notice that if the coupling constants $g_1=-g_2=-g_3=g_4$, the symmetry group is the smaller $\mathcal{G}_0$, with the pion assignment $(1,0,0,\lambda_\pi)$ where $\lambda_\pi=-1$.

(B) The group O_7'. The group is generated by the invariant subgroup $SU_2\times SU_2\times SU_2$ of $\mathcal{G}_0$, together with the elements

$$\exp(i\theta)\begin{Vmatrix}0&0&\sigma_1&0\\0&0&0&\sigma_1\\\sigma_1&0&0&0\\0&\sigma_1&0&0\end{Vmatrix}. \tag{32}$$

To find the group we introduce the matrices

$$\rho_1=\begin{Vmatrix}0&0&I&0\\0&0&0&I\\I&0&0&0\\0&I&0&0\end{Vmatrix},\quad \rho_2=i\begin{Vmatrix}0&0&-I&0\\0&0&0&-I\\I&0&0&0\\0&I&0&0\end{Vmatrix},\quad \rho_3=\begin{Vmatrix}I&0&0&0\\0&I&0&0\\0&0&-I&0\\0&0&0&-I\end{Vmatrix},$$

$$\tau_1=\begin{Vmatrix}0&I&0&0\\I&0&0&0\\0&0&0&I\\0&0&I&0\end{Vmatrix},\quad \tau_2=i\begin{Vmatrix}0&-I&0&0\\I&0&0&0\\0&0&0&-I\\0&0&I&0\end{Vmatrix},\quad \tau_3=\begin{Vmatrix}I&0&0&0\\0&-I&0&0\\0&0&I&0\\0&0&0&-I\end{Vmatrix}.$$

The infinitesimal generators of the invariant subgroup $SU_2\times SU_2\times SU_2$ of $\mathcal{G}_0$ are

$$\sigma_1,\ \sigma_2,\ \sigma_3,\quad (1+\rho_3)\tau_1,\quad (1+\rho_3)\tau_2,\quad (1+\rho_3)\tau_3,$$
$$(1-\rho_3)\tau_1,\quad (1-\rho_3)\tau_2,\quad (1-\rho_3)\tau_3, \tag{33}$$

or

$$\sigma_i,\ \tau_i,\ \rho_3\tau_i.$$

The infinitesimal generator for (32) is

$$\rho_1\sigma_1.$$

Taking the commutator of this generator with those listed in (33) one obtains additional generators. Altogether by repeatedly taking commutators one obtains the following 21 infinitesimal generators:

$$\sigma_i,\ \tau_i,\ \rho_1\sigma_i,\ \rho_3\tau_i,\ \rho_2\sigma_i\tau_j. \tag{34}$$

Taking further commutators gives rise to no new independent generators. The group obtained from these 21 generators is O_7' which has a 2-to-1 homomorphism with the group of 7×7 proper real orthogonal matrices O_7, as already discussed by various authors[6,3] in the literature. To see this we define seven anticommuting Hermitian matrices

$$\gamma_1=\rho_3\sigma_1,\quad \gamma_2=\rho_3\sigma_2,\quad \gamma_3=\rho_3\sigma_3,\quad \gamma_4=\rho_1\tau_1,$$
$$\gamma_5=\rho_1\tau_2,\quad \gamma_6=\rho_1\tau_3,\quad \gamma_7=\rho_2. \tag{35}$$

Then

$$\gamma_\mu\gamma_\nu+\gamma_\nu\gamma_\mu=2\delta_{\mu\nu}, \tag{36}$$

$$\gamma_1\gamma_2\gamma_3\gamma_4\gamma_5\gamma_6\gamma_7=-i. \tag{37}$$

The 21 infinitesimal generators (34) are then $i\gamma_\mu\gamma_\nu$ $(\mu\neq\nu)$. The group generated by (34) is therefore of the form

$$\exp(\textstyle\sum a_{\mu\nu}\gamma_\mu\gamma_\nu),$$

where $a_{\mu\nu}$ are real numbers. Now

$$[\exp(\textstyle\sum a_{\mu\nu}\gamma_\mu\gamma_\nu)]\gamma_i[\exp(-\sum a_{\mu\nu}\gamma_\mu\gamma_\nu)]=\sum_{j=1}^{7} b_{ij}\gamma_j, \tag{38}$$

where $\|b_{ij}\|$ is a 7×7 real orthogonal matrix with determinant$=1$. It is easy to prove that, conversely, for every such $\|b_{ij}\|$, there exist two sets of real $a_{\mu\nu}$'s satisfying (38). The group has thus a 2-to-1 homomorphism with O_7. [T. A. Tarski has pointed out to us that O_7' is called spin (7) in the standard language.]

In terms of the γ's, the element R of $\mathcal{G}_0$ is

$$R=\rho_1=-i\gamma_4\gamma_5\gamma_6=\gamma_1\gamma_2\gamma_3\gamma_7$$
$$=[\exp(\pi\gamma_1\gamma_2/2)][\exp(\pi\gamma_7\gamma_3/2)]. \tag{39}$$

Thus R is an element of O_7', hence $\mathcal{G}_0$ is a subgroup of O_7'.

Both of the above two groups contain $\mathcal{G}_0$ as a subgroup. There exist also groups that satisfy conditions (i) and (ii) but do not contain $\mathcal{G}_0$ as a subgroup.

(C) The group SU_3. It was pointed out to us by Speiser and Tarski[14] that the group[11] SU_3 has an irreducible 8×8 representation which satisfies both conditions (i) and (ii). However, in this case it seems impossible to incorporate π mesons and K mesons without introducing more new bosons. It is clear that $\mathcal{G}_0$ is not a subgroup of SU_3. The full implications and consequences of such possibilities still need to be investigated.

APPENDIX B

We give in this Appendix explicit matrices for G_1, R, $\mathcal{L}$, $\mathcal{M}$, $\mathcal{N}$, and N between the 16 states that describe a

[14] D. R. Speiser and J. A. Tarski (private communication).

single baryon or antibaryon:

$$\begin{Vmatrix} B \\ B' \end{Vmatrix},$$

where

$$B=\begin{Vmatrix} p \\ n \\ \Xi^0 \\ \Xi^- \\ Y^+ \\ Y^0 \\ Z^0 \\ Z^- \end{Vmatrix} \quad \text{and} \quad B'=\begin{Vmatrix} \bar{\Xi}^+ \\ -\bar{\Xi}^0 \\ -\bar{n} \\ \bar{p} \\ \bar{Z}^+ \\ -\bar{Z}^0 \\ -\bar{Y}^0 \\ \bar{Y}^- \end{Vmatrix}.$$

The antibaryon states[15] are defined such that under the charge conjugation operation all baryon states p, n, Ξ^0, Ξ^-, Y^+, Y^0, Z^0, Z^- are transformed in an *identical* way into their respective antibaryon states $\bar{p}$, $\bar{n}$, $\bar{\Xi}^0$, $\bar{\Xi}^+$, $\bar{Y}^-$, $\bar{Y}^0$, $\bar{Z}^0$, and $\bar{Z}^+$. The minus signs in B' are so chosen that the matrices G_1, R, $\mathcal{L}$, $\mathfrak{M}$, $\mathfrak{N}$, and N are given by

$$G_1=\left\|\begin{array}{cccccccc} & & & & I & 0 & 0 & 0 \\ & & & & 0 & I & 0 & 0 \\ & & & & 0 & 0 & I & 0 \\ & & & & 0 & 0 & 0 & I \\ I & 0 & 0 & 0 & & & & \\ 0 & I & 0 & 0 & & & & \\ 0 & 0 & I & 0 & & & & \\ 0 & 0 & 0 & I & & & & \end{array}\right\|,$$

$$R=\left\|\begin{array}{cccccccc} 0 & 0 & I & 0 & & & & \\ 0 & 0 & 0 & I & & & & \\ I & 0 & 0 & 0 & & & & \\ 0 & I & 0 & 0 & & & & \\ & & & & 0 & 0 & I & 0 \\ & & & & 0 & 0 & 0 & I \\ & & & & I & 0 & 0 & 0 \\ & & & & 0 & I & 0 & 0 \end{array}\right\|,$$

$$N=\left\|\begin{array}{cccccccc} I & 0 & 0 & 0 & & & & \\ 0 & I & 0 & 0 & & & & \\ 0 & 0 & I & 0 & & & & \\ 0 & 0 & 0 & I & & & & \\ & & & & -I & 0 & 0 & 0 \\ & & & & 0 & -I & 0 & 0 \\ & & & & 0 & 0 & -I & 0 \\ & & & & 0 & 0 & 0 & -I \end{array}\right\|,$$

$$\mathcal{L}_i=\tfrac{1}{2}\left\|\begin{array}{cccccccc} \sigma_i & 0 & 0 & 0 & & & & \\ 0 & \sigma_i & 0 & 0 & & & & \\ 0 & 0 & \sigma_i & 0 & & & & \\ 0 & 0 & 0 & \sigma_i & & & & \\ & & & & \sigma_i & 0 & 0 & 0 \\ & & & & 0 & \sigma_i & 0 & 0 \\ & & & & 0 & 0 & \sigma_i & 0 \\ & & & & 0 & 0 & 0 & \sigma_i \end{array}\right\|, \quad (i=1, 2, 3)$$

$$\mathfrak{M}_1=\tfrac{1}{2}\left\|\begin{array}{cccccccc} 0 & 0 & 0 & 0 & & & & \\ 0 & 0 & 0 & 0 & & & & \\ 0 & 0 & 0 & I & & & & \\ 0 & 0 & I & 0 & & & & \\ & & & & 0 & 0 & 0 & 0 \\ & & & & 0 & 0 & 0 & 0 \\ & & & & 0 & 0 & 0 & I \\ & & & & 0 & 0 & I & 0 \end{array}\right\|,$$

$$\mathfrak{M}_2=\tfrac{1}{2}i\left\|\begin{array}{cccccccc} 0 & 0 & 0 & 0 & & & & \\ 0 & 0 & 0 & 0 & & & & \\ 0 & 0 & 0 & -I & & & & \\ 0 & 0 & I & 0 & & & & \\ & & & & 0 & 0 & 0 & 0 \\ & & & & 0 & 0 & 0 & 0 \\ & & & & 0 & 0 & 0 & -I \\ & & & & 0 & 0 & I & 0 \end{array}\right\|,$$

$$\mathfrak{M}_3=\tfrac{1}{2}\left\|\begin{array}{cccccccc} 0 & 0 & 0 & 0 & & & & \\ 0 & 0 & 0 & 0 & & & & \\ 0 & 0 & I & 0 & & & & \\ 0 & 0 & 0 & -I & & & & \\ & & & & 0 & 0 & 0 & 0 \\ & & & & 0 & 0 & 0 & 0 \\ & & & & 0 & 0 & I & 0 \\ & & & & 0 & 0 & 0 & -I \end{array}\right\|,$$

$$\mathfrak{N}_1=\tfrac{1}{2}\left\|\begin{array}{cccccccc} 0 & I & 0 & 0 & & & & \\ I & 0 & 0 & 0 & & & & \\ 0 & 0 & 0 & 0 & & & & \\ 0 & 0 & 0 & 0 & & & & \\ & & & & 0 & I & 0 & 0 \\ & & & & I & 0 & 0 & 0 \\ & & & & 0 & 0 & 0 & 0 \\ & & & & 0 & 0 & 0 & 0 \end{array}\right\|,$$

$$\mathfrak{N}_2=\tfrac{1}{2}i\left\|\begin{array}{cccccccc} 0 & -I & 0 & 0 & & & & \\ I & 0 & 0 & 0 & & & & \\ 0 & 0 & 0 & 0 & & & & \\ 0 & 0 & 0 & 0 & & & & \\ & & & & 0 & -I & 0 & 0 \\ & & & & I & 0 & 0 & 0 \\ & & & & 0 & 0 & 0 & 0 \\ & & & & 0 & 0 & 0 & 0 \end{array}\right\|,$$

and

$$\mathfrak{N}_3=\tfrac{1}{2}\left\|\begin{array}{cccccccc} I & 0 & 0 & 0 & & & & \\ 0 & -I & 0 & 0 & & & & \\ 0 & 0 & 0 & 0 & & & & \\ 0 & 0 & 0 & 0 & & & & \\ & & & & I & 0 & 0 & 0 \\ & & & & 0 & -I & 0 & 0 \\ & & & & 0 & 0 & 0 & 0 \\ & & & & 0 & 0 & 0 & 0 \end{array}\right\|,$$

where $\sigma_i (i=1, 2, 3)$ are the 2×2 Pauli spin matrices and I is the 2×2 unit matrix. All empty places in the above matrices are zeroes.

[15] We use the notation that, e.g., $\bar{Y}^-$ is the antiparticle of Y^+ and is negatively charged.

THEORETICAL CONSIDERATIONS CONCERNING QUANTIZED MAGNETIC FLUX IN SUPERCONDUCTING CYLINDERS*

[61c]

Commentary begins page 49

N. Byers and C. N. Yang†

Institute of Theoretical Physics, Department of Physics, Stanford University, Stanford, California

(Received June 16, 1961)

In a recent experiment,[1] the magnetic flux through a superconducting ring has been found to be quantized in units of $ch/2e$. Quantization in twice this unit has been briefly discussed by London[2] and by Onsager.[3] Onsager[4] has also considered the possibility of quantization in units $ch/2e$ due to pairs of electrons forming quasi-bosons.

The previous discussions[3] leave unresolved the question whether quantization of the flux is a new physical principle or not. Furthermore, sometimes the discussions seem[2] to be based on the assumption that the wave function of the superconductor in the presence of the flux is proportional to that in its absence, an assumption which is not correct. We shall show in this Letter that (i) no new physical principle is involved in the requirement of the quantization of magnetic flux through a superconducting ring, (ii) the Meissner effect is closely related to the requirement that the flux through any area with a boundary lying entirely in superconductors is quantized, and (iii) the quantization of flux is an indication of the pairing of the electrons in the superconductor.

Macroscopic discussion. Consider a multiply connected superconducting body P with a tunnel O (Fig. 1). We shall only discuss macroscopic

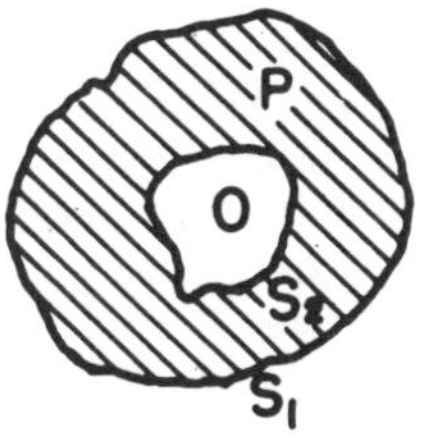

FIG. 1. Multiply connected superconductor.

□ Reprinted from *Physical Review Letters* 7, 2 (July 15, 1961), 46–49.

dimensions much larger than the penetration depth. The Meissner effect then states that inside the superconductor P the magnetic field is zero, and the current is zero. Surface currents, however, do exist and persist on the surfaces S_1 and S_2. The surface currents and the external sources of magnetic fields together produce no magnetic flux in the interior P of the superconductor. They in general, however, produce a net magnetic flux through O, to be denoted by Φ.

The energy eigenfunction ψ of the electrons in the superconductor satisfies

$$\sum_j \frac{1}{2m}[-i\hbar\vec{\nabla}_j + \frac{e}{c}\vec{A}(\vec{r}_j)]^2\psi + V\psi = E\psi, \tag{1}$$

where $\vec{A}$ is the vector potential due to the surface currents and external magnetic sources. Inside P,

$$\vec{\nabla}\times\vec{A} = 0.$$

Hence $\vec{A} = \vec{\nabla}\chi$, where χ is not single valued in P but increases by

$$\Delta\chi = \oint\vec{A}\cdot d\vec{l} = \iint\vec{H}\cdot d\vec{\sigma} = \Phi, \tag{2}$$

whenever one goes around the tunnel O once. Defining

$$\psi' = \psi \exp\sum_j i\frac{e}{c\hbar}\chi(\vec{r}_j), \tag{3}$$

we see that (1) reduces to

$$\sum_j \frac{1}{2m}(-i\hbar\,\vec{\nabla}_j)^2\psi' + V\psi' = E\psi'. \tag{4}$$

The vector potential $\vec{A}$ is eliminated from this equation. However, the boundary condition for ψ' is that when all electron coordinates are fixed, except for one, $\vec{r}_l$, and $\vec{r}_l$ is brought around O once, ψ' changes by a constant factor

$$\psi' \to \psi' e^{i(e/c\hbar)\Phi}. \tag{5}$$

[To prove (5) we use (3) and (2) and the fact that ψ is single valued.]

The eigenvalues E are determined by the differential equation (4) and the boundary condition (5). It is thus obvious that we have:

Theorem 1. The energy levels are periodic in the magnetic flux Φ with a period ch/e.

If the surfaces S_1 and S_2 are concentric cylinders on which $\psi = 0$, and V is put equal to zero, the energy levels E can be explicitly solved for, illustrating this theorem. One notices that ψ' is not simply proportional to $\psi(\Phi = 0)$, as is sometimes[2] assumed in the literature.

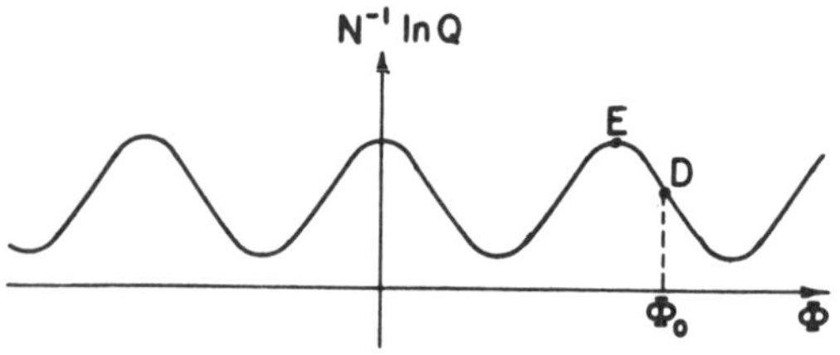

FIG. 2. Periodic variations in $N^{-1}\ln Q$ as a function of trapped flux Φ.

If V is a real function of r_j, by taking the complex conjugate of (4) and (5), we have:

Theorem 2. The energy levels are even functions of Φ.

It is clear that theorems 1 and 2 remain valid if we introduce the lattice coordinates of the metal and if we introduce the spin of the electrons. The operation of complex conjugation in the proof of theorem 2 has then to be replaced by the time reversal operation and the proof depends on the time reversal invariance of the interactions.

From these theorems it follows trivially that:

Theorem 3. The partition function Q of the system is an even periodic function of Φ with period ch/e.

At $\Phi = (ch/2e)\times$integer, this theorem shows that

$$\partial\ln Q/\partial\Phi = 0, \tag{6}$$

and that $\ln Q$ has the general form shown in Fig. 2.

Now the body current in the superconductor around O is

$$I = kTc\,\partial\ln Q/\partial\Phi. \tag{7}$$

(c = velocity of light.) In the differentiation we keep the temperature T constant.

The Meissner effect requires that $I = 0$. Thus the equilibrium states are given by the maxima and minima on the curve in Fig. 2. We shall now give an argument to show that the maxima, not the minima, are the equilibrium states realized. A point D in Fig. 2 is not an equilibrium state so the calculation of the partition function at that point is strictly speaking meaningless. But the slope of the curve at that point indicates that if a flux Φ_0 is made to pass through O, a body current would be induced, the sense of the current being negative according to (7). The additional flux due to this body current causes the flux through O to decrease. The equilibrium state reached would therefore be E

which is a maximum of the curve. We state this as:

Theorem 4. The superconducting state is given by the maxima of $\ln Q$ as a function of Φ.

If the external flux does not assume a value for which $\ln Q$ is a maximum, surface currents will flow on S_1 and S_2 to make up a total flux Φ for which $\ln Q$ is a maximum.

The experiment of reference 1 and theorem 4 together prove that $\ln Q$ has maxima at integral values of $\Phi/(ch/2e)$. Whether a microscopic theory yields these maxima will be discussed in the next section. If it does, then theorem 4 shows the following: The flux through any surface whose boundary loop lies entirely in superconductors is quantized in units $ch/2e$. The requirement of this quantization in turn clearly implies that the flux through any small area in a superconductor is zero; hence it implies the Meissner effect.

In a loose sense the above argument can be used to "derive" the Meissner effect itself: If the magnetic flux in a superconductor is not zero, body currents will flow around all loops through which the flux is not quantized. The system cannot reach a steady state until all magnetic flux is expelled from the interior of the superconductor.

Microscopic considerations. We now want to see whether a microscopic calculation does or does not lead to maxima of $\ln Q$ at $\Phi/(ch/2e)$ = integer.

To investigate this point we first take a collection of noninteracting spinless electrons between two concentric cylinders S_1 and S_2. The electrons at a point at a distance r from the axis have momenta p_r, p_θ, and p_z in the radial, azimuthal, and z directions. Clearly

$$p_\theta r = n\hbar. \quad (n = \text{integer})$$

The energy of the electron is

$$\frac{1}{2m}\left[p_r^{\,2} + p_z^{\,2} + \frac{\hbar^2}{r^2}\left(n + \frac{e}{ch}\Phi\right)^2\right]. \tag{8}$$

The partition function Q can be computed from such an energy spectrum. The resultant $N^{-1}\ln Q$ [to the order N^0] does not depend on Φ. Thus, according to theorem 4, for a collection of noninteracting electrons, the flux in the tunnel does not have to be quantized.

It is not difficult to understand why for such a model $N^{-1}\ln Q$ does not depend on Φ. To see this we suppress the p_r and p_z degrees of freedom and take the temperature $T = 0$. The one-dimensional Fermi sea problem,

$$\frac{\hbar^2}{2mr^2}\sum\left(n + \frac{e}{ch}\Phi\right)^2,$$

gives an average energy per particle of

$$\frac{E}{N} = \text{constant} + \frac{\hbar^2}{2mr^2}\frac{e^2}{c^2h^2}\Phi^2, \quad \left|\frac{e\Phi}{ch}\right| \leqslant \tfrac{1}{2}, \tag{9}$$

if $N \equiv$ the number of particles is odd. But if N is even, then

$$\frac{E}{N} = \text{constant} + \frac{\hbar^2}{2mr^2}\left(\frac{e\Phi}{ch} - \tfrac{1}{2}\right)^2, \quad \tfrac{1}{2} \geqslant \frac{e\Phi}{hc} \geqslant 0 \tag{10}$$

$$= \text{constant} + \frac{\hbar^2}{2mr^2}\left(\frac{e\Phi}{ch} + \tfrac{1}{2}\right)^2, \quad 0 \geqslant \frac{e\Phi}{hc} \geqslant -\tfrac{1}{2}. \tag{10'}$$

Thus, depending on the evenness or oddness of N, the energy has a minimum at $e\Phi/ch = \frac{1}{2}$ or 0 (modulo 1). The three-dimensional problem at $T = 0$ is decomposable into many one-dimensional problems with varying values of N. Thus the above-discussed fluctuation leads to a cancellation for the three-dimensional problem, resulting in an $N^{-1}\ln Q$ versus Φ curve that is flat. (An $N^{-1}\ln Q$ curve that is flat applies to the case of a metal in its normal rather than superconducting state.) A similar cancellation obtains for $T \neq 0$.

In the neighborhood of $\Phi = +0$ the states with $n > 0$ have energies that increase with Φ, and those with $n < 0$ have energies that decrease with Φ. The average energy for the two states n and $-n$, however, increases with Φ like

$$\text{constant} + (\hbar^2/2mr^2)(e\Phi/ch)^2. \tag{11}$$

If there is a "pair correlation" of the kind proposed by Bardeen, Cooper, and Schrieffer[5] for the superconductor so that states n and $-n$ (or a pair of time-reversed states) are either both occupied or both unoccupied, (11) becomes the correct energy per particle for small Φ. (In such a case the fluctuation and cancellation phenomena disappear.) This is represented in Fig. 3 by the parabolas at $2e\Phi/ch = -2, 0, 2$, etc.

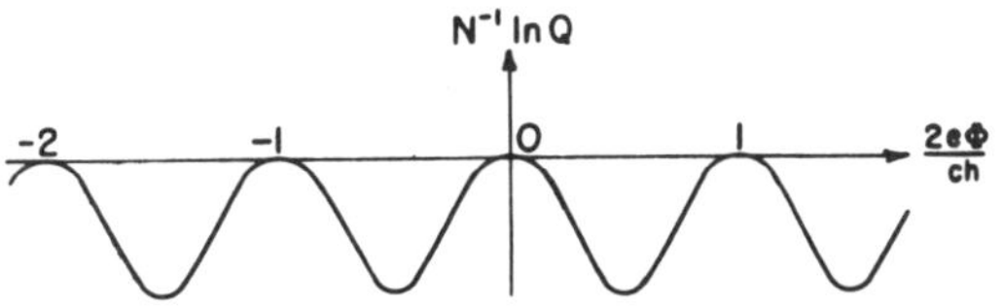

FIG. 3. A curve of $N^{-1}\ln Q$ versus $2e\Phi/ch$, showing parabolic behavior near maxima at $2e\Phi/ch$ = integer.

At $2e\Phi/ch = 1$, pairing between

$$n + \frac{e}{hc}\Phi = n + \tfrac{1}{2} = \tfrac{1}{2} \text{ and } -\tfrac{1}{2},\ \tfrac{3}{2} \text{ and } -\tfrac{3}{2},\ \text{etc.,}$$

occurs and the energy per particle remains the same as for the case $\Phi = 0$ (to order N^0). In the neighborhood of $2e\Phi/ch = 1$, the additional energy for each of these pairs is again twice

$$(\hbar^2/2mr^2)[(e\Phi/ch) - \tfrac{1}{2}]^2,$$

which give rise to the parabolas at $2e\Phi/ch = \pm 1$ in Fig. 3. In the absence of a detailed theory, we draw a smooth curve in Fig. 3 to extrapolate between the maxima.

Thus the Bardeen, Cooper, Schrieffer pairs for the superconducting state give rise to the curves such as those depicted in Fig. 3, where the parabolas are repeated at periods $\Delta(2e\Phi/ch) = 1$, and the central parabola is given by

$$N^{-1}\ln Q = -\frac{f}{kT}\frac{\hbar^2}{2m\langle r^2\rangle_{\text{av}}}\left(\frac{e\Phi}{ch}\right)^2 + \text{constant}, \tag{12}$$

where f = fraction of electrons that are paired.

It is interesting to estimate the magnitude of the body current at, say,

$$0 < 2e\Phi/ch < \tfrac{1}{2}. \tag{13}$$

It is, by (7),

$$I = -Nfc(e^2/mc^2)\Phi/(4\pi^2\langle r^2\rangle_{\text{av}}).$$

The flux induced by this current is, for a thin ring superconductor,

$$\Phi_{\text{induced}} = -f \times (\text{number of electrons in a length } e^2/mc^2 \text{ of the ring})\Phi.$$

For the experiment of reference 1, $-\Phi_{\text{induced}}/\Phi \gg 1$ if f is not too small, showing that the maxima in Fig. 3 are very pronounced.

From Fig. 3 and the argument preceding theorem 4, we conclude that the trapped flux Φ and the original flux Φ_0 are related in the following way:

$\Phi_0/(ch/2e)$	Φ
$-\frac{1}{2} \to \frac{1}{2}$	0
$\frac{1}{2} \to \frac{3}{2}$	1
$\frac{3}{2} \to \frac{5}{2}$	2
etc.	

It is interesting to notice that the existence of the variation of the energy levels of the electrons in P with the flux Φ, even when there is no magnetic field in P, is based on the same principle as the experiment proposed by Aharonov and Bohm.[6,7]

We wish to thank W. M. Fairbank and B. Deaver for informing us of the progress of their beautiful experiment and for many discussions. We also wish to thank F. Bloch for stimulating discussions. One of us (CNY) takes this opportunity to thank the members of the Physics Department of Stanford University for the hospitality extended him during his visit.

---

*Work supported in part by the U. S. Air Force through the Air Force Office of Scientific Research.

†Permanent address: Institute for Advanced Study, Princeton, New Jersey.

[1]B. S. Deaver, Jr., and W. M. Fairbank, preceding Letter [Phys. Rev. Letters 7, 43 (1961)].

[2]F. London, Superfluids (John Wiley & Sons, Inc., New York, 1950), p. 152.

[3]L. Onsager, Proceedings of the International Conference on Theoretical Physics, Kyoto and Tokyo, September, 1953 (Science Council of Japan, Tokyo, 1954), pp. 935-6.

[4]L. Onsager (private communication to W. M. Fairbank).

[5]J. Bardeen, L. N. Cooper, and J. R. Schrieffer, Phys. Rev. 108, 1175 (1957).

[6]Y. Aharonov and D. Bohm, Phys. Rev. 115, 485 (1959).

[7]R. G. Chambers, Phys. Rev. Letters 5, 3 (1960).

The Future of Physics

C. N. *Yang*

[[61f]]

Commentary begins page 51

In the last four or five years much effort and attention have been devoted by theoretical physicists to the analytic continuation from physically observable experience into unphysical regions. In particular, it has been tried by extrapolation to study properties of the singularities in the unobserved region. Such attempts have been beset from the beginning with great difficulties. But interest in them maintained. In a similar spirit, this morning we try to pursue a parallel approach: By extrapolation we want to look beyond our past experience and learn something about the so far unobserved future development of physics. We cannot hope for any real success in this pursuit. But I believe we all agree the attempt is highly interesting.

By all standards the achievements in physics in the twentieth century so far have been spectacular. Whereas at the turn of the century the atomic aspect of matter was just emerging as a new subject of study, we have today progressed dimensionwise by a refinement of a factor of a million: from atomic dimensions to subnuclear dimensions. Energywise the progress is even more impressive: from a few electron volts to multibillion volts. And the power and ingenuities of the experimental techniques have fully kept pace with the increased depth of the physicists' inquiries. The influence and impact brought about by the advances in physics upon other sciences—upon chemistry, astronomy, and even upon biology—have been important beyond description. Similar influence upon technology, upon human affairs, has been so preoccupying in the postwar years as to need no further emphasis here.

But it is not in these influences that the glory of physics and the heart of the physicist lie. It is not even in the continued enlargement of the domain of physical experimentation, important as it is, that the physicists take the greatest pride and satisfaction. What makes physics so unique as an intellectual endeavor lies in the possibility of the formulation of concepts out of which, in the words of Einstein,[1] a "comprehensive workable system of theoretical physics" can be constructed. Such a system embodies universal elementary laws, "from which the cosmos can be built up by pure deduction."

Judged by such a lofty and rigorous standard, the sixty years of the twentieth century stand out as nothing short of heroic; for besides the numerous important discoveries that *widened* our knowledge of the physical world, this period has witnessed not one, not two, but three revolutionary changes in physical concepts: the special relativity, the general relativity and the quantum theory. Out of these conceptual revolutions were constructed *deepened*, comprehensive and unified systems of theoretical physics.

Endowed with such a distinguished heritage of the recent past, what are the prospects for its future?

Surely the rapid widening of knowledge will continue, both in what Professor Peierls refers to as the foundation of physics and in what he calls behind the front line.

In the former our present knowledge is sufficient to enable us to say with some certainty that great clarification will come in the field of weak interactions in the next few years. With luck on our side we might even hope to see some integration of the various manifestations of the weak interactions.

Beyond that we are on very uncertain grounds. To be sure we can already formulate a number of questions, the answers to which seem to us at this moment to be crucial: How should one treat a system with an infinite degree of freedom? Is the continuum concept of space time extrapolatable to regions of space 10^{-14} cm to 10^{-17} cm, and to regions smaller than 10^{-17} cm? What are the bases

[1] A. Einstein, *Essays in Science*, Philosophical Library, New York, 1934.

□ Panel discussion by C. N. Yang at the MIT Centennial Celebration, April 8, 1961.

of the invariance under charge conjugation, and the invariance under isotopic spin rotation, both of which, unlike space–time symmetries, are known to be violated? What is the unifying basis of the strong, the electromagnetic and the weak interactions? What is the role of the gravitational field relative to all these? The list can go on, but we are not even sure that these questions are meaningful as we here phrased them: in fact much of the progress in physics had developed from the very recognition of the meaninglessness of some previously asked questions.

Of one thing, however, we can be sure. The accumulation of knowledge will proceed at a very rapid pace. We need only remind ourselves that not so very long ago, the time scale of discoveries in physics was measured in years, if not in decades. The Michaelson–Morely experiment, for example, was first performed in 1881, then repeated with greater precision in 1887. To explain the negative result of the experiment Fitzgerald invented a contraction hypothesis in 1982, and then Lorentz invented the Lorentz transformation in 1902, culminating in Einstein's special theory of relativity in 1905. Imagine that Michaelson's first experiment were done today!

The general awakening of mankind to the importance of science and the amazing ingenuity of the human mind at technological creativity virtually ensure for us further quickening of the pace of the experimental sciences.

But where shall we stand with respect to the "comprehensive system of theoretical physics" that was referred to a few minuts ago? Can we reasonably expect further success in the glorious tradition of the first sixty years of the twentieth century?

If it is difficult to locate singularities of functions by extrapolation, it is as difficult to predict revolutionary changes in physical concepts by forecasting. But since there seems to be too ready a tendency to have boundless faith in a "future fundamental theory," I shall sound some pessimistic notes. And in this Centennial celebration, in an atmosphere charged with excitement, with pride for past achievements and with an expansive outlook for the future, it is perhaps not entirely inappropriate to interject these somewhat discordant notes.

First, let us emphasize again that the mere accumulation of knowledge, while interesting, and beneficial to mankind, is nonetheless quite different from the aim of fundamental physics.

Second, the subject matter of subnuclear physics is already very remote from the direct sensory experience of mankind, and this remoteness is certainly going to increase as we delve into even smaller dimensions. A direct and graphic proof of this can be easily found in the growing physical size of the laboratories, the accelerators, the detectors and the computors.

What we call an experiment today consists of elaborate operations with elaborate equipments. To make any sense at all of the results of an experiment, concepts have to be formulated on all levels between our direct sensory experience and the actual experimental arrangement. The difficulty inherent in this state of affairs is as follows. Each level of concepts is connected with and in fact built upon the previous levels. When inadequacies manifest themselves, one must reach for greater depth by examining the whole complex of previous concepts. The difficulty of this task rapidly diverges with the depth of the considerations, much like in chess playing, it becomes increasingly difficult to practice always examining one more move ahead as one's skill improves.

According to Wigner's counting,[2] to reach our present investigations in field theory one must penetrate at least four levels of concepts. The details of the counting may be a subject of discussion, but there is no denying that what we envisage as a construction of a deeper and more comprehensive theoretical system represents at least one more level of penetration. Here physicists are handicapped by the fact that physical theories have their final justification in reality. Un-

[2] E. P. Wigner, *Proc. Amer. Phil. Soc.* 94, 422 (1950).

like the mathematicians, or the artists, physicists cannot create new concepts and construct new theories by free imagination.

Third, Eddington[3] has once given the example of a marine biologist who uses a net with 6 inch holes and who after careful and long studies formulates the law that all fish are larger than 6 inches. If this imaginary example sounds ridiculous, we can easily find examples in modern physics where because of the complexity and indirectness of the experiments, it has happened that one does not realize the selective nature of one's experiments, a selection that is based on concepts which may be inadequate.

Fourth, in the day-to-day work of a physicist it is very natural to implicitly believe that the power of the human intellect is limitless and the depth of natural phenomena finite. It is of course useful, or as is sometimes said, healthy, to have faith so as to have courage. But the belief that the depth of natural phenomena is finite is inconsistent and the faith that the power of the human intellect is limitless is false. Furthermore, important considerations have also to be given to the fact that psychological and social limitations on the development of the creative ability of each individual may be effectively even more stringent than natural limitations.

Having voiced these cautionary remarks, we must ask are they relevant to the development of physics, say, in the remaining forty years of this century? We cannot know the answer to this question now, but let us hope that it is in the negative.

[3] A. S. Eddington, *The Philosophy of Science,* MacMillan, New York, 1939.

[[62b]]
Commentary begins page 52

TESTS OF THE SINGLE-PION EXCHANGE MODEL

S. B. Treiman
Palmer Physical Laboratory, Princeton University, Princeton, New Jersey

and

C. N. Yang
Institute for Advanced Study, Princeton, New Jersey
(Received December 14, 1961)

The single-pion exchange model (SPEM) of high-energy particle reactions provides an attractively simple picture of seemingly complex processes and has accordingly been much discussed in recent times.[1] The purpose of this note is to call attention to the possibility of subjecting the model to certain tests precisely in the domain where the model stands the best chance of making sense.

Consider a collision between particles p and k (labelled here by their four-momenta) which results in two groups of outgoing particles, $(p_1', \ldots, p_m')$ and $(k_1', \ldots, k_n')$. We restrict ourselves to configurations where the outgoing particles, as viewed in the barycentric system, form two well-defined narrow cones and we partition the particles accordingly into the two groups $\{p_i'\}$ and $\{k_i'\}$. We suppose, in addition, that the selection rules permit the exchange of a single pion: $p+k \to \{p_i'\}+\pi+k \to \{p_i'\}+\{k_i'\}$. Define the invariant momentum transfer $\Delta = p - \sum p_i' = \sum k_i' - k$. Regarded as a function of Δ^2, the transition amplitude has a pole at $\Delta^2 = -\mu^2$ (μ=pion mass), corresponding to the diagram of Fig. 1. The residue involves a product of the amplitudes, $M(p+\pi \to \{p_i'\})$ and $M(k+\pi \to \{k_i'\})$, which describe, respectively, the indicated physical processes. The point $\Delta^2 = -\mu^2$ of course lies outside of the physical domain for the reaction $p+k \to \{p_i'\}+\{k_i'\}$. Nevertheless, in the SPEM picture one accepts the diagram of Fig. 1 as representing the dominant contribution at small enough physical Δ^2. Indeed one hopes that the only configurations which are ever very probable are those in which there is some partition of final particles corresponding to not too large Δ^2.

It is clear that, given information on the physical reactions $p+\pi \to \{p_i'\}$ and $k+\pi \to \{k_i'\}$, one is led on the basis of the diagram of Fig. 1 to quite definite, and testable, predictions. In less optimistic applications, however, one envisages allowing for at least some additional, unspecified dependence on the variable Δ^2, to correct for off-the-mass-shell effects at the vertices and in the pion propagator.[2]

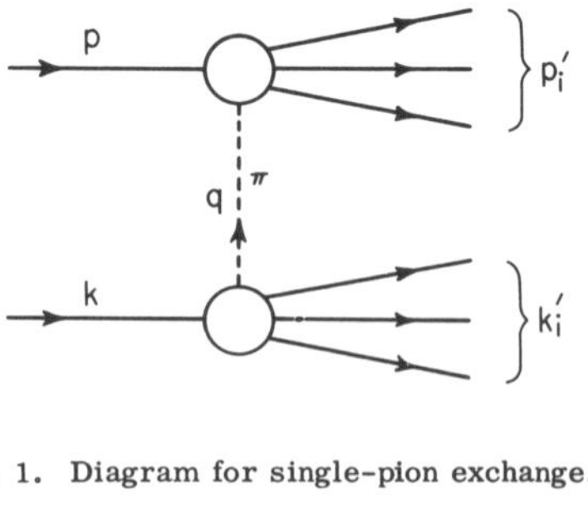

FIG. 1. Diagram for single-pion exchange.

□Reprinted from *Physical Review Letters* 8, 3 (February 1, 1962), 140–141.

Nevertheless, even if the Δ^2 dependence is left unspecified, and even if, moreover, the vertex functions of Fig. 1 are regarded as completely unknown, the general structure of the diagram leads to certain concrete and testable restrictions on the properties of the over-all reaction spectrum. This comes about because the structure of Fig. 1 implies that there is no correlation between the two groups of final particles, $\{p_i'\}$ and $\{k_i'\}$, beyond what follows from kinematics. This result, reflected in Eq. (2) below, depends in an essential way on the fact that the exchanged pion is a spinless particle.

The differential reaction cross section $d\sigma$ is given by

$$Jd\sigma = f\prod_i dp_i'\delta(p_i'^2+m_i^2) \times \prod_j dk_j'\delta(k_j'^2+\mu_j^2)\delta(p+k-\sum p_i'-\sum k_j'). \quad (1)$$

where J is the relative current of the incident particles, f is the square of the invariant transition amplitude, and all energies are positive-definite. The crucial remark is that, on the peripheral collision picture, f has the structure

$$f = G(p, p_i')H(k, k_i'). \quad (2)$$

The implications of this restriction on the structure of f are best brought out in the reference frames in which one or another of the initial particles is at rest. Thus:

1. In the system where p is at rest (the laboratory system, if p is in fact the target particle), the differential cross section should be invariant under the simultaneous rotation of all three-vectors $\vec{p}_i'$ about the momentum vector $\vec{q}$ of the virtual meson: $\vec{q} = \vec{k} - \sum_i \vec{k}_i' = \sum_i \vec{p}_i'$. This result follows from inspection of Eqs. (1) and (2).

2. Similarly, in the system where k is at rest the differential cross section should be invariant under simultaneous rotation of all three-vectors $\vec{k}_i'$ about $\vec{q} = -\sum_i \vec{k}_i' = \sum_i \vec{p}_i' - \vec{p}$.

It is easy to prove that the above two tests are exhaustive for fixed incoming energy. There are further implications of the model having to do with relations between differential cross sections at different incoming energies, but these implications are rather complicated and lend themselves less easily to experimental testing.

As an example, consider the reaction $\pi(k) + N(p) \to \pi(k_1') + \pi(k_2') + N(p')$.[3] The peripheral collision model for this process involves exchange of a single pion according to $k + p \to k + p' + \pi \to p' + k_1' + k_2'$. In the rest frame of the initial pion the implication of the model is that, for given $\vec{p}$ and $\vec{p}'$, the differential cross section should be independent of the orientation of the plane defined by $\vec{k}_1'$ and $\vec{k}_2'$ about the line $\vec{q} = -\vec{k}_1' - \vec{k}_2' = \vec{p}' - \vec{p}$. Of course for those limiting configurations in which $\vec{p}$ and $\vec{p}'$ are collinear this assertion is an empty one.[4] But where $\vec{p}$ and $\vec{p}'$ are not collinear one could, outside of the peripheral collision model, envisage a correlation between the directions defined by $\vec{k}_1' \times \vec{k}_2'$ and $\vec{p} \times \vec{p}'$. The detection of an appreciable correlation effect of this sort, especially for the subclass of events with small momentum transfer Δ^2, would weigh heavily against the single-pion exchange model.[5]

---

[1]For excellent reviews of the model, and references to the literature, see: S. D. Drell, Revs. Modern Phys. 33, 458 (1961); F. Salzman and G. Salzman, Phys. Rev. (to be published).

[2]See, for example, S. D. Drell and K. Hiida, Phys. Rev. Letters 7, 199 (1961); E. Ferrari and F. Selleri, Phys. Rev. Letters 7, 387 (1961).

[3]The single-pion exchange model was first applied here by C. Goebel, Phys. Rev. Letters 1, 337 (1958); G. F. Chew and F. E. Low, Phys. Rev. 113, 1640 (1959). Experimental results have been discussed on this model by A. R. Erwin, R. March, W. D. Walker, and E. West, Phys. Rev. Letters 6, 628 (1961); E. Pickup, D. K. Robinson, and E. O. Salant, Phys. Rev. Letters 7, 192 (1961).

[4]This is peculiar to the case where the set $\{p_i'\}$ contains only one member. For reactions in which the set $\{p_i'\}$ contains two or more particles, rotational invariance of the $\vec{k}_i'$ about $\Sigma_i \vec{k}_i'$ would be a nontrivial result even when $\vec{p}'$ and $\vec{p}$ are collinear.

[5]A second restrictive consequence of the model, as applied to the present example, is perhaps worth mentioning; namely, for unpolarized initial nucleons the outgoing nucleon beam should be unpolarized. The detection of polarization effects is, however, not an easy matter experimentally.

[[62g]]

Commentary begins page 52

Obituary for Dr. Shih-Tsun Ma

T. D. Lee and C. N. Yang

Shih-Tsun Ma died on January 27, 1962 in Sydney, Australia. The tragic news of his untimely death came as a shock to his fellow physicists and friends. To us who had known him well it brought profound sadness and regret.

Ma was born in 1913 in Peking. He received his college education at the Peking University, obtaining his B.Sc. degree in 1935. One of his teachers at college was Professor Ta You Wu who also directed his first published research work (1935) which was on a calculation of some excited states of the He atom. In 1937 he won a nationwide competition and was awarded a fellowship, enabling him to study in England with W. Heitler at Cambridge. He worked on meson theory during his four years there and received his Ph.D. degree in 1941.

He then went back to wartime China and taught at the National Southwest Associated University in Kunming. As a teacher he was extremely conscientious, and prepared very neat notes for his lectures. We had at different times (1941–1943 and 1945) taken his courses. One of us (CNY) remembers very distinctly having learned field theory from a lucid, well organized and extensive course given by Ma in the spring of 1943. Recalling today the unheated, uninsulated class rooms in which his lectures took place, with windows shattered by frequent air raids, and the floor, of tamped earth, pockmarked with depressions formed through heavy usage, we can still envisage Ma, standing in front of the blackboard, writing notes down in quick movements, young, lean and shy. We are impressed once more with the realization that quiet devotion can triumph over material shortcomings.

In 1946 Ma came to the Institute for Advanced Study in Princeton. That was at a time when Heisenberg's theory of the S-matrix was beginning to create a strong impact in physics. The famous redundant zeros of the S-matrix were discovered by Ma during his stay in Princeton. In 1947 Ma went to Ireland and joined the Institute for Advanced Study in Dublin. There he pointed out in 1949 the difficulty inherent in Fermi's method of treating quantum electrodynamics, leading a year later to the Gupta–Bleuler formalism.

Ma spent 1949–1951 at the University of Chicago, and 1951–1953 at the National Research Council, Ottawa, Canada. In 1953 he was offered positions in the U.S. He declined, even though his wife is an American, largely because he was reluctant to face the hostile and sometimes insulting attitude of the U.S. Immigration Office toward an oriental. He accepted instead a position at the University of Sydney, Australia. He was again approached, in subsequent years, by American universities with offers to join their faculty, but always declined for the same reason.

Ma continued to work in field theory, publishing in all about forty papers. Ma's writings are characterized by a simple and honest style, entirely free of pretentiousness; they reflect not only his personality, but also his life.

Theory of Charged Vector Mesons Interacting with the Electromagnetic Field

[[62i]]

Commentary begins page 52

T. D. Lee and C. N. Yang
Institute for Advanced Study, Princeton, New Jersey

(Received May 29, 1962)

It is shown that starting from the usual canonical formalism for the electromagnetic interaction of a charged vector meson with arbitrary magnetic moment one is led to a set of rules for Feynman diagrams, which appears to contain terms that are both infinite and noncovariant. These difficulties, however, can be circumvented by introducing a ξ-limiting process which depends on a dimensionless positive parameter $\xi \to 0$. Furthermore, by using the mathematical artifice of a negative metric the theory becomes renormalizable (for $\xi > 0$).

1. INTRODUCTION

The problem of a charged vector meson interacting with the electromagnetic field and other fermion fields has been discussed rather extensively in the past.[1,2] However, in the literature, there does not seem to exist any systematic study of the general case in which the charged vector meson could have an arbitrary magnetic moment. Furthermore, the question of the renormalizability of a theory of charged vector meson has not been studied in detail. The recent speculations in weak interactions[3] and the possibility that, perhaps, a vector meson $W^{\pm}$ could be produced by high-energy neutrinos[4] through its electromagnetic and weak interactions give new interest to these problems.

In this paper, an attempt is made to study these problems. We begin with a discussion of the derivation of Feynman rules for the general case of the interactions between the electromagnetic field and charged vector mesons with arbitrary magnetic moment. It turns out that by starting from the conventional canonical formalism[1] and using the Dyson-Wick procedure,[5] one obtains a set of rules for the Feynman graphs which contains terms that appear to be both infinite and noncovariant. It is then shown that this formal difficulty can be resolved by introducing a limiting process (called ξ-limiting process) which depends on a positive parameter $\xi \to 0$. The resulting rules for Feynman graphs in the ξ-limiting process become completely covariant. However, the theory continues to be divergent in a nonrenormalizable way. To remedy this, the artifice of a negative metric is introduced which makes the parameter ξ take on the role of a regulator. The final theory for $\xi > 0$ is both covariant and renormalizable. It is further shown that while the introduction of a negative metric destroys unitarity, the S matrix remains unitary as long as the total energy of the system is less than $\xi^{-\frac{1}{2}}$ times the mass of the meson.

The derivations of Feynman rules are sometimes rather complicated, because of the presence of time derivatives of field variables in the interaction Lagrangian. These detailed derivations are all given in the Appendices. As an illustration, the derivation of Feynman rules for the simple and well-known case of a charged vector meson field interacting with Fermion fields is included in Appendix A.

Strictly speaking, because of divergences there does not exist a "true" charged vector meson theory. Any theory of the charged vector meson is in this sense a separate proposal not derivable from a "true" theory. What gives the confidence that the renormalization procedure of the photon-electron interaction enjoys is, besides the impressive and accurate experimental verifications, the belief that any covariant proposal would lead essentially to the results of the usual renormalization procedure. For the charged vector meson, it is our present belief that, with the ξ-limiting process and the indefinite metric, one has a covariant theory that in some measure gives that part of the properties of the charged vector meson which is independent of specific details at very small distances.

2. CANONICAL FORMALISM

2.1 Lagrangian

We discuss a charged vector meson field φ_μ in interaction with the electromagnetic field A_μ. The charged vector mesons is assumed to possess an arbitrary magnetic moment, and is called $W^{\pm}$. The Lagrangian density of the system is[6] ($\hbar = c = 1$)

$$\mathcal{L} = -\frac{1}{2}\left(\frac{\partial A_\mu}{\partial x_\nu}\right)\left(\frac{\partial A_\mu}{\partial x_\nu}\right) - \tfrac{1}{2}G_{\mu\nu}{}^{*}G_{\mu\nu} - m^2 \varphi_\mu{}^{*}\varphi_\mu - ie\kappa F_{\mu\nu}\varphi_\mu{}^{*}\varphi_\nu, \quad (1)$$

[1] See, for example, G. Wentzel, *Quantum Theory of Fields* (Interscience Publishers, Inc., New York, 1949). In this paper we start with the formulation of vector meson field given in Wentzel's book.

[2] Feynman rules for charged vector mesons have been given by R. P. Feynman, Phys. Rev. **76**, 769 (1949) using his method of space-time approach of field theory. More detailed discussions on Feynman rules for charged vector mesons in the β formalism were given by C. N. Yang and G. Feldman, Phys. Rev. **79**, 972 (1950). See also T. Kinoshita and Y. Nambu, Progr. Theoret. Phys. **5**, 473, 749 (1950); P. T. Matthews, Phys. Rev. **76**, 1657 (1949).

[3] T. D. Lee and C. N. Yang, Phys. Rev. **119**, 1410 (1960).

[4] T. D. Lee and C. N. Yang, Phys. Rev. Letters **4**, 307 (1960). See also B. Pontecorvo and R. M. Ryndin, Dubna Report D-577 (unpublished).

[5] F. J. Dyson, Phys. Rev. **75**, 486 (1949); G. C. Wick, *ibid.* **80**, 268 (1950).

[6] Throughout this paper we use the following notations: All boldface letters such as **k**, **r**, **A**, $\boldsymbol{\varphi}$, etc., denote three-vectors. The fourth component ik_0 of the four-momentum k_μ is pure imaginary. All Greek subscripts μ, ν, $\cdots$ vary from 1 to 4 and all Roman subscripts i, j, $\cdots$ vary from 1 to 3. Repeated indices are to be summed over.

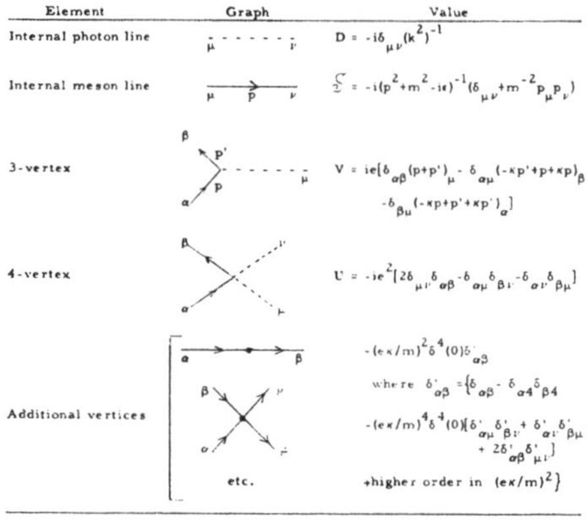

FIG. 1. Feynman diagram in momentum representation for Lagrangian (1). Each closed internal loop consisting of meson and photon lines gives rise to one integration $(2\pi)^{-4}\int[\cdots]dk_0d^3k$. A diagram includes specific assignments of momenta and polarization to all external lines, but not internal lines. The weight of each different diagram is s^{-1} where s is the symmetry number defined as follows: Label each internal line with a different integer: 1, 2, $\cdots N$. There are $N!$ different ways of labeling. Some of these labelings may lead to labeled diagrams with identical topological structure. s is simply the number of such different labelings that lead to the same labeled diagram.

where $*$ = Hermitian conjugate times $(-1)^n$, n = number of "4" subscripts,

$$
\begin{aligned}
F_{\mu\nu}&=(\partial/\partial x_\mu)A_\nu-(\partial/\partial x_\nu)A_\mu,\\
G_{\mu\nu}&=\partial_\mu\varphi_\nu-\partial_\nu\varphi_\mu,\\
G_{\mu\nu}^*&=\partial_\mu^*\varphi_\nu^*-\partial_\nu^*\varphi_\mu^*,\\
\partial_\mu&=\partial/\partial x_\mu-ieA_\mu,\\
\partial_\mu^*&=\partial/\partial x_\mu+ieA_\mu,
\end{aligned}
\tag{2}
$$

and κ is a constant. The magnetic moment $\mathfrak{M}$ and the quadrupole moment Q of W^+ is given by

$$\mathfrak{M}=(1+\kappa)(e/2m)\mathbf{S} \tag{3}$$

and

$$Q\equiv\int(3z^2-r^2)\rho d^3r=-(e\kappa/m^2), \tag{4}$$

where $\mathbf{S}$ is the spin of W and ρ is the static charge density for the state $S_z=+1$.

The equation of motion for $W^\pm$ is

$$\partial_\mu G_{\mu\nu}-m^2\varphi_\nu+ie\kappa\varphi_\mu F_{\mu\nu}=0. \tag{5}$$

2.2 Feynman Diagram

In Appendix B we carry out in detail the canonical formalism starting from the Lagrangian above: The fields[6] $\boldsymbol{\varphi}$, $\boldsymbol{\varphi}^*$, $\mathbf{A}$, and A_4 will be treated as independent canonical "coordinates." φ_4 and φ_4^* will be treated as dependent coordinates with the aid of (5). One then obtains a Hamiltonian for the system. By a unitary transformation one goes over into the interaction representation. Feynman diagrams will then be obtained through the Dyson-Wick[5] procedure.

The result of these considerations is as follows. A Feynman diagram in the present case is very much like that for the electron-photon interaction, except that there are now three kinds of vertices. The values of these vertices and the propagators, in momentum representation, are listed in Fig. 1 (proved in Appendix C).

For the purpose of easy memory we remark that the three-vertex and four-vertex functions are the matrix elements of

$$
\begin{aligned}
&-i[\mathcal{L}(e=0)-\mathcal{L}(e)]\\
&=e\kappa F_{\nu\mu}\varphi_\nu^*\varphi_\mu+\tfrac{1}{2}e\left[(A_\nu\varphi_\mu^*-A_\mu\varphi_\nu^*)\left(\frac{\partial}{\partial x_\nu}\varphi_\mu-\frac{\partial}{\partial x_\mu}\varphi_\nu\right)-\text{Herm. conj.}\right]\\
&\quad-\tfrac{1}{2}ie^2(A_\nu\varphi_\mu-A_\mu\varphi_\nu)(A_\nu\varphi_\mu^*-A_\mu\varphi_\nu^*),
\end{aligned}
\tag{6}
$$

where all operators are regarded as free fields. However, this very simple rule does not give the whole story, as the presence of the additional vertices in Fig. 1 explicitly shows.

The additional vertices are all divergent and are explicitly noncovariant. For a given process, to the lowest order in e the Feynman diagram does not contain closed loops, nor does it contain any of the additional vertices. For a higher order diagram, because of the divergent nature of the integral, the integration dk_0 gives, in addition to the usual pole contributions, contributions due to the closing of the integration contour at ∞ in the complex k_0 plane. As is discussed in Appendix E, the divergent and noncovariant vertices of Fig. 1 are the results of such extra integration contributions at infinity. Moreover, they are present only if $\kappa\neq0$. [This is because in the usual canonical formalism the components of $\boldsymbol{\varphi}$, $\boldsymbol{\varphi}^*$ are treated as coordinates, but φ_4 and φ_4^* are regarded as functions of $\boldsymbol{\varphi}$, $\boldsymbol{\varphi}^*$ and their conjugate momenta. Therefore, the interaction term $-ie\kappa F_{\mu\nu}\varphi_\mu^*\varphi_\nu$ in the Lagrangian (1) appears to contain more than one time derivative of the field variables which gives rise to these additional vertices. In this paper, Lagrangians which contain terms with more than two time derivatives of the field variables are not considered.]

3. ξ-LIMITING FORMALISM

The origin of these complications, therefore, lies in the fact that φ_4 is not treated on equal footing as the components of $\boldsymbol{\varphi}$. To circumvent this difficulty we add a term to the Lagrangian proportional to a dimensionless parameter ξ and then take the limit $\xi\to0$.

3.1 Lagrangian

Instead of (1) we thus have a Lagrangian density with an additional term:

$$\mathcal{L}_\xi = -\xi(\partial_\mu{}^*\varphi_\mu{}^*)(\partial_\nu\varphi_\nu) - \frac{1}{2}\left(\frac{\partial A_\mu}{\partial x_\nu}\right)\left(\frac{\partial A_\mu}{\partial x_\nu}\right)$$
$$-\tfrac{1}{2}G_{\mu\nu}{}^*G_{\mu\nu} - m^2\varphi_\mu{}^*\varphi_\mu - ie\kappa F_{\mu\nu}\varphi_\mu{}^*\varphi_\nu. \quad (7)$$

The equation of motion for $W^\pm$ becomes

$$\partial_\mu G_{\mu\nu} - m^2\varphi_\nu + \xi\partial_\nu(\partial_\mu\varphi_\mu) + ie\kappa\varphi_\mu F_{\mu\nu} = 0.$$

The Lagrangian density (7) can now be treated by the canonical formalism in a straightforward way. We state the result as follows.

3.2 Free Meson Field

By using (7) and setting $e=0$ one obtains the following free Hamiltonian H_0:

$$H_0 = \boldsymbol{\pi}\cdot\boldsymbol{\pi}^* + \xi^{-1}\pi_4\pi_4{}^* + m^2\varphi_\mu{}^*\varphi_\mu + (\nabla\times\boldsymbol{\varphi})\cdot(\nabla\times\boldsymbol{\varphi}^*)$$
$$+ i(\boldsymbol{\pi}\cdot\nabla\varphi_4 + \boldsymbol{\pi}^*\cdot\nabla\varphi_4{}^* - \pi_4\nabla\cdot\boldsymbol{\varphi} - \pi_4{}^*\nabla\cdot\boldsymbol{\varphi}^*), \quad (8)$$

where π_μ and $\pi_\mu{}^*$ are, respectively, the conjugate momenta of φ_μ and $\varphi_\mu{}^*$. The commutation relations at equal time are given by

$$[\pi_\mu(\mathbf{r},t), \varphi_\nu(\mathbf{r}',t)] = -i\delta_{\mu\nu}\delta^3(\mathbf{r}-\mathbf{r}'),$$
$$[\pi_\mu{}^*(\mathbf{r},t), \varphi_\nu{}^*(\mathbf{r}',t)] = -i\delta_{\mu\nu}\delta^3(\mathbf{r}-\mathbf{r}'), \quad (9)$$

and all other commutators between φ_μ, $\varphi_\mu{}^*$, π_μ, $\pi_\mu{}^*$ are zero. The free field corresponds to a system of uncoupled mesons which can be described by the annihilation operators

$a_\mathbf{k}{}^t$, $b_\mathbf{k}{}^t$ for + and − *transverse* (spin 1) mesons, $t=1, 2$
$a_\mathbf{k}{}^l$, $b_\mathbf{k}{}^l$ for + and − *longitudinal* (spin 1) mesons,
$a_\mathbf{k}{}^s$, $b_\mathbf{k}{}^s$ for + and − scalar (spin 0) mesons,

and their Hermitian conjugates, the creation operators, $a_\mathbf{k}{}^{s\dagger}$, $b_\mathbf{k}{}^{s\dagger}$, $a_\mathbf{k}{}^{l\dagger}$, etc. In terms of these operators one has the following representation:

$$\boldsymbol{\varphi} = \sum_{\mathbf{k},t}(2\Omega\omega)^{-\frac{1}{2}}[a_\mathbf{k}{}^t \exp(i\mathbf{k}\cdot\mathbf{r} - i\omega t) + b_{-\mathbf{k}}{}^{t\dagger}\exp(i\mathbf{k}\cdot\mathbf{r} + i\omega t)]\mathbf{e}_\mathbf{k}{}^t$$
$$+\sum_\mathbf{k}(2\Omega\omega)^{-\frac{1}{2}}[a_\mathbf{k}{}^l \exp(i\mathbf{k}\cdot\mathbf{r} - i\omega t) + b_{-\mathbf{k}}{}^{l\dagger}\exp(i\mathbf{k}\cdot\mathbf{r} + i\omega t)](\omega\hat{k}/m)$$
$$-\sum_\mathbf{k}(2\Omega\nu)^{-\frac{1}{2}}[a_\mathbf{k}{}^s \exp(i\mathbf{k}\cdot\mathbf{r} + i\nu t) + b_{-\mathbf{k}}{}^{s\dagger}\exp(i\mathbf{k}\cdot\mathbf{r} - i\nu t)](\mathbf{k}/m),$$

$$\varphi_4 = \sum_\mathbf{k}(2\Omega\omega)^{-\frac{1}{2}}[a_\mathbf{k}{}^l \exp(i\mathbf{k}\cdot\mathbf{r} - i\omega t) - b_{-\mathbf{k}}{}^{l\dagger}\exp(i\mathbf{k}\cdot\mathbf{r} + i\omega t)](i|\mathbf{k}|/m)$$
$$+\sum_\mathbf{k}(2\Omega\nu)^{-\frac{1}{2}}[a_\mathbf{k}{}^s \exp(i\mathbf{k}\cdot\mathbf{r} + i\nu t) - b_{-\mathbf{k}}{}^{s\dagger}\exp(i\mathbf{k}\cdot\mathbf{r} - i\nu t)](i\nu/m), \quad (10)$$

$$\boldsymbol{\pi} = \sum_{\mathbf{k},t} i(2\Omega\omega)^{-\frac{1}{2}}[a_\mathbf{k}{}^{t\dagger}\exp(-i\mathbf{k}\cdot\mathbf{r} + i\omega t) - b_{-\mathbf{k}}{}^t\exp(-i\mathbf{k}\cdot\mathbf{r} - i\omega t)]\omega\mathbf{e}_\mathbf{k}{}^t$$
$$+\sum_\mathbf{k} i(2\Omega\omega)^{-\frac{1}{2}}[a_\mathbf{k}{}^{l\dagger}\exp(-i\mathbf{k}\cdot\mathbf{r} + i\omega t) - b_{-\mathbf{k}}{}^l\exp(-i\mathbf{k}\cdot\mathbf{r} - i\omega t)](m\hat{k}),$$

$$\pi_4 = \sum_\mathbf{k}(2\Omega\nu)^{-\frac{1}{2}}m[a_\mathbf{k}{}^{s\dagger}\exp(-i\mathbf{k}\cdot\mathbf{r} - i\nu t) + b_{-\mathbf{k}}{}^s\exp(-i\mathbf{k}\cdot\mathbf{r} + i\nu t)],$$

and $\varphi_\mu{}^*$, $\pi_\mu{}^*$ are related to the Hermitian conjugates $\varphi_\mu{}^\dagger$, $\pi_\mu{}^\dagger$ of φ_μ, π_μ by

$$\boldsymbol{\varphi}^* = \boldsymbol{\varphi}^\dagger, \qquad \boldsymbol{\pi} = \boldsymbol{\pi}^\dagger,$$
$$\varphi_4{}^* = -\varphi_4{}^\dagger, \quad \pi_4{}^* = -\pi_4{}^\dagger. \quad (11)$$

In these formulas, $e_\mathbf{k}{}^1$, $e_\mathbf{k}{}^2$, and $\hat{k} = |\mathbf{k}|^{-1}\mathbf{k}$ form a right-handed orthonormal set of unit vectors,

$$\omega = (\mathbf{k}^2 + m^2)^{\frac{1}{2}} > 0, \quad \nu = (\mathbf{k}^2 + \xi^{-1}m^2)^{\frac{1}{2}} > 0, \quad (12)$$

and Ω is the normalization volume. In terms of these annihilation and creation operators H_0 becomes

$$H_0 = \sum_{\mathbf{k},t}\omega(a_\mathbf{k}{}^{t\dagger}a_\mathbf{k}{}^t + \tfrac{1}{2}) + \sum_\mathbf{k}\omega(a_\mathbf{k}{}^{l\dagger}a_\mathbf{k}{}^l + \tfrac{1}{2})$$
$$-\sum_\mathbf{k}\nu(a_\mathbf{k}{}^{s\dagger}a_\mathbf{k}{}^s + \tfrac{1}{2}) + \text{same terms with } a \to b. \quad (13)$$

These formulas show that the additional ξ-dependent term in the Lagrangian introduces scalar mesons with a negative energy

$$-\nu = -(\mathbf{k}^2 + \xi^{-1}m^2)^{\frac{1}{2}},$$

which approaches $-\infty$ as $\xi \to 0$.

3.3 Hamiltonian in Interaction Representation

The indefiniteness of the Hamiltonian makes it very doubtful that after the introduction of the coupling e when different meson states are coupled, the theory can still make physical sense. We try to remedy this by introducing a negative metric in Sec. 4. For clarity of presentation, we ignore this difficulty for the time being and proceed with the canonical formalism. All the four components of φ_μ are now regarded as canonical coordinates. In the interaction representation, the space-time dependences of the operators φ_μ and A_μ are the same as that of the free ones. In terms of these operators the interaction Hamiltonian becomes[6]

$$H_{\rm int}=-ieA_\mu[g_{\mu\nu}^*\varphi_\nu-g_{\mu\nu}\varphi_\nu^*]-ie\xi A_\mu\left[\varphi_\mu\left(\frac{\partial}{\partial x_\nu}\varphi_\nu^*\right)-\varphi_\mu^*\left(\frac{\partial}{\partial x_\nu}\varphi_\nu\right)\right]+e^2[(A_jA_j)(\varphi_k^*\varphi_k)-(A_j\varphi_j)(A_k\varphi_k^*)]$$
$$+ie\kappa F_{\mu\nu}\varphi_\mu^*\varphi_\nu+\tfrac{1}{2}e^2\kappa^2[(\varphi_4^*)^2\varphi_j\varphi_j+(\varphi_4)^2\varphi_j^*\varphi_j^*-2(\varphi_4^*\varphi_4)(\varphi_j^*\varphi_j)], \quad (14)$$

where

$$g_{\mu\nu}=\left(\frac{\partial}{\partial x_\mu}\varphi_\nu\right)-\left(\frac{\partial}{\partial x_\nu}\varphi_\mu\right) \quad (15)$$

and

$$g_{\mu\nu}^*=\left(\frac{\partial}{\partial x_\mu}\varphi_\nu^*\right)-\left(\frac{\partial}{\partial x_\nu}\varphi_\mu^*\right).$$

3.4 Feynman Diagram

Using the Dyson-Wick[5] procedure one obtains the Feynman diagrams for the Lagrangian (7). The values of the propagators and vertices are listed in Fig. 2 (proved in Appendix D).

3.5 Divergenceless Current Density

The Lagrangian (7) is[7] gauge invariant. Therefore the current density is divergenceless. An explicit proof of this fact can be obtained from an examination of the vertices and propagators of Fig. 2, in the same spirit as the corresponding proof[8] for the electromagnetic field-electron interaction. In the present case, the proof is slightly more complicated because of the momentum dependence of the vertices V which generates terms canceled by the vertices U.

4. NEGATIVE METRIC

In the ξ formalism, the propagator $\mathcal{S}$ in Fig. 2 consists of two parts: a spin-one part $-i(p^2+m^2-i\epsilon)^{-1}\times(\delta_{\mu\nu}+m^{-2}p_\mu p_\nu)$ and a spin-zero part $i(p^2+\xi^{-1}m^2+i\epsilon)^{-1}\times(m^{-2}p_\mu p_\nu)$. At first sight, it might appear that the presence of the spin-zero part acts like a regulator; therefore, we might have a renormalizable theory for $\xi\neq0$. That this is not the case can easily be seen by noticing the different signs $\pm i\epsilon$ in these two parts of the propagator. More explicitly, $\mathcal{S}$ can be written as

$$\mathcal{S}=S-2\pi i(m^{-2}p_\mu p_\nu)\delta(p^2+\xi^{-1}m^2), \quad (16)$$

where

$$S=-i(p^2+m^2-i\epsilon)^{-1}(\delta_{\mu\nu}+m^{-2}p_\mu p_\nu)$$
$$+i(p^2+\xi^{-1}m^2-i\epsilon)^{-1}(m^{-2}p_\mu p_\nu). \quad (17)$$

The second term on the right-hand side of (16) makes the theory discussed in the above section divergent in an unrenormalizable way. In the ξ-limiting formalism, in order to give a meaningful discussion of the limit $\xi\to0$, finite physical results must be first obtained before taking the limit. To achieve this we introduce a negative metric in the Hilbert space.

Element	Graph	Value
Internal photon line	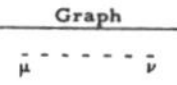	$D=-i\delta_{\mu\nu}(k^2)^{-1}$
Internal meson line	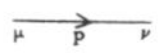	$\mathcal{S}=-i(p^2+m^2-i\epsilon)^{-1}(\delta_{\mu\nu}+m^{-2}p_\mu p_\nu)$ $+i(p^2+\xi^{-1}m^2+i\epsilon)^{-1}(m^{-2}p_\mu p_\nu)$
3-vertex	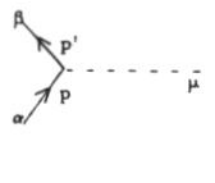	$V=ie[\delta_{\alpha\beta}(p+p')_\mu-\delta_{\alpha\mu}(-\kappa p'+p+\kappa p-\xi p')_\beta$ $-\delta_{\beta\mu}(-\kappa+p'+\kappa p'-\xi p)_\alpha]$
4-vertex	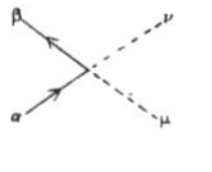	$U=-ie^2[2\delta_{\mu\nu}\delta_{\alpha\beta}-(1-\xi)\delta_{\alpha\mu}\delta_{\beta\nu}$ $-(1-\xi)\delta_{\alpha\nu}\delta_{\beta\mu}]$

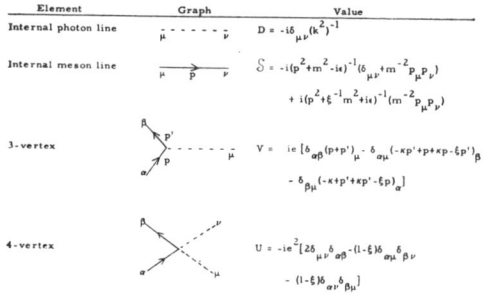

FIG. 2. Feynman diagram in momentum representation for the Lagrangian (7) (see also caption of Fig. 1).

4.1 ξ-Limiting Formalism with Negative Metric

We start with the identical Lagrangian given by (7), except that φ_μ^* and $G_{\mu\nu}^*$ are replaced by

$$\varphi_\mu^\star\equiv\eta^{-1}\varphi_\mu^*\eta,$$
$$G_{\mu\nu}^\star\equiv\eta^{-1}G_{\mu\nu}^*\eta, \quad (18)$$

respectively. Following the notation of Pauli,[9] we use η to represent the metric of the Hilbert space. It becomes clear that in order to change the sign of $(i\epsilon)$ in the spin-zero part of the propagator $\mathcal{S}$ the metric η in the *interaction representation* is chosen to be

$$\eta=(-1)^{N_s}, \quad (19)$$

where N_s is the total number of scalar mesons.

For clarity, we discuss first the free-field case ($e=0$) and then the general case in the interaction representation.

4.2 Free Meson Fields

Identical with (8) and (9) except for the replacement (18), the free Hamiltonian H_0 for the present case is

$$H_0=\boldsymbol{\pi}\cdot\boldsymbol{\pi}^\star+\xi^{-1}\pi_4\pi_4^\star+m^2\varphi_\mu\varphi_\mu^\star+(\nabla\times\boldsymbol{\varphi})\cdot(\nabla\times\boldsymbol{\varphi}^\star)$$
$$+i(\boldsymbol{\pi}\cdot\nabla\varphi_4+\boldsymbol{\pi}^\star\cdot\nabla\varphi_4^\star-\pi_4\nabla\cdot\boldsymbol{\varphi}-\pi_4^\star\nabla\cdot\boldsymbol{\varphi}^\star), \quad (20)$$

and the commutation relations are

$$[\pi_\mu(\mathbf{r},t),\varphi_\nu(\mathbf{r}',t)]=-i\delta_{\mu\nu}\delta^3(\mathbf{r}-\mathbf{r}'),$$
$$[\pi_\mu^\star(\mathbf{r},t),\varphi_\nu^\star(\mathbf{r}',t)]=-i\delta_{\mu\nu}\delta^3(\mathbf{r}-\mathbf{r}'). \quad (21)$$

All other equal-time commutators between φ_μ, $\varphi_\mu^\star$, π_μ, $\pi_\mu^\star$ are zero. We list the explicit representation of these

---

[7] Dr. T. T. Wu first pointed out the advantage of using a gauge-invariant ξ formalism.

[8] R. P. Feynman, Phys. Rev. **76**, 769 (1949).

[9] W. Pauli, Revs. Modern Phys. **15**, 175 (1945).

operators in terms of the annihilation operators $a_{\mathbf{k}}^{r}$, $b_{\mathbf{k}}^{r}$ and their Hermitian conjugates, the creation operators, $a_{\mathbf{k}}^{r\dagger}$ and $b_{\mathbf{k}}^{r\dagger}$ (where $r=t, l, s$ represent, respectively, the uncoupled transverse, longitudinal, and scalar mesons):

$$\begin{aligned}
\boldsymbol{\varphi}=&\sum_{\mathbf{k},t}(2\Omega\omega)^{-\frac{1}{2}}[a_{\mathbf{k}}{}^{t}\exp(i\mathbf{k}\cdot\mathbf{r}-i\omega t)+b_{-\mathbf{k}}{}^{t\dagger}\exp(i\mathbf{k}\cdot\mathbf{r}+i\omega t)]\mathbf{e}_{\mathbf{k}}{}^{t}\\
&+\sum_{\mathbf{k}}(2\Omega\omega)^{-\frac{1}{2}}[a_{\mathbf{k}}{}^{l}\exp(i\mathbf{k}\cdot\mathbf{r}-i\omega t)+b_{\mathbf{k}}{}^{l\dagger}\exp(i\mathbf{k}\cdot\mathbf{r}+i\omega t)](\omega\hat{k}/m)\\
&+\sum_{\mathbf{k}}(2\Omega\nu)^{-\frac{1}{2}}[a_{\mathbf{k}}{}^{s}\exp(i\mathbf{k}\cdot\mathbf{r}-i\nu t)+b_{-\mathbf{k}}{}^{s\dagger}\exp(i\mathbf{k}\cdot\mathbf{r}+i\nu t)](\mathbf{k}/m),\\
\varphi_4=&\sum_{\mathbf{k}}(2\Omega\omega)^{-\frac{1}{2}}[a_{\mathbf{k}}{}^{l}\exp(i\mathbf{k}\cdot\mathbf{r}-i\omega t)-b_{-\mathbf{k}}{}^{l\dagger}\exp(i\mathbf{k}\cdot\mathbf{r}+i\omega t)](i|\mathbf{k}|/m)\\
&+\sum_{\mathbf{k}}(2\Omega\nu)^{-\frac{1}{2}}[a_{\mathbf{k}}{}^{s}\exp(i\mathbf{k}\cdot\mathbf{r}-i\nu t)-b_{-\mathbf{k}}{}^{s\dagger}\exp(i\mathbf{k}\cdot\mathbf{r}+i\nu t)](i\nu/m), \qquad (22)\\
\boldsymbol{\pi}=&\sum_{\mathbf{k},t}i(2\Omega\omega)^{-\frac{1}{2}}[a_{\mathbf{k}}{}^{t\dagger}\exp(-i\mathbf{k}\cdot\mathbf{r}+i\omega t)-b_{-\mathbf{k}}{}^{t}\exp(-i\mathbf{k}\cdot\mathbf{r}-i\omega t)]\omega\mathbf{e}_{\mathbf{k}}{}^{t}\\
&+\sum_{\mathbf{k}}i(2\Omega\omega)^{-\frac{1}{2}}[a_{\mathbf{k}}{}^{l\dagger}\exp(-i\mathbf{k}\cdot\mathbf{r}+i\omega t)-b_{-\mathbf{k}}{}^{l}\exp(-i\mathbf{k}\cdot\mathbf{r}-i\omega t)]m\hat{k},\\
\pi_4=&\sum_{\mathbf{k}}(2\Omega\nu)^{-\frac{1}{2}}m[a_{\mathbf{k}}{}^{s\dagger}\exp(-i\mathbf{k}\cdot\mathbf{r}+i\nu t)+b_{-\mathbf{k}}{}^{s}\exp(-i\mathbf{k}\cdot\mathbf{r}-i\nu t)].
\end{aligned}$$

$\varphi_\mu{}^\star$ and $\pi_\mu{}^\star$ are related to the Hermitian conjugates $\varphi_\mu{}^\dagger$, $\pi_\mu{}^\dagger$, of φ_μ, π_μ by

$$\begin{aligned}\boldsymbol{\varphi}^\star=\eta^{-1}\boldsymbol{\varphi}^\dagger\eta, &\qquad \boldsymbol{\pi}^\star=\eta^{-1}\boldsymbol{\pi}^\dagger\eta,\\ \varphi_4{}^\star=-\eta^{-1}\varphi_4{}^\dagger\eta, &\qquad \pi_4{}^\star=-\eta^{-1}\pi_4{}^\dagger\eta,\end{aligned} \qquad (23)$$

where

$$\eta=\exp[\textstyle\sum_{\mathbf{k}} i\pi(a_{\mathbf{k}}{}^{s\dagger}a_{\mathbf{k}}{}^{s}+b_{\mathbf{k}}{}^{s\dagger}b_{\mathbf{k}}{}^{s})], \qquad (24)$$

and ω, ν, $\mathbf{e}_{\mathbf{k}}{}^{t}$, $\hat{k}$ are given by (12).

Upon substituting (22) and (23) to (20), the Hamiltonian H_0 becomes

$$\begin{aligned}H_0=&\sum_{\mathbf{k},t}\omega(a_{\mathbf{k}}{}^{t\dagger}a_{\mathbf{k}}{}^{t}+\tfrac{1}{2})+\sum_{\mathbf{k}}\omega(a_{\mathbf{k}}{}^{l\dagger}a_{\mathbf{k}}{}^{l}+\tfrac{1}{2})\\ &+\sum_{\mathbf{k}}\nu(a_{\mathbf{k}}{}^{s\dagger}a_{\mathbf{k}}{}^{s}+\tfrac{1}{2})+\text{same terms with } a\rightarrow b.\end{aligned} \qquad (25)$$

It is important to notice that the scalar mesons now have positive energy. The introduction of a negative metric is, of course, a rather drastic measure. However, we regard this only as an artifice to make possible a meaningful discussion of the limit $\xi\rightarrow 0+$. A consequence of the positive definite H_0 is that the vacuum expectation value of the time ordered product $T[\varphi_\mu(x)\varphi_\nu(0)]$ is given by

$$(2\pi)^{-4}\int S\exp(ip_\mu x_\mu)dp_0d^3p, \qquad (26)$$

where S is given by (17).

4.3 Hamiltonian in Interaction Representation

In the interaction representation, the field operators φ_μ, $\varphi_\mu{}^\star$ have the same space-time dependence as the free case. The metric η remains to be given by (24). In a similar manner to (14) the interaction Hamiltonian H_{int} is given by

$$\begin{aligned}H_{\text{int}}=&-ieA_\mu[g_{\mu\nu}{}^\star\varphi_\nu-g_{\mu\nu}\varphi_\nu{}^\star]-ie\xi A_\mu[\varphi_\mu(\partial\varphi_\nu{}^\star/\partial x_\nu)-\varphi_\mu{}^\star(\partial\varphi_\nu/\partial x_\nu)]+e^2[(A_jA_j)(\varphi_k{}^\star\varphi_k)-(A_j\varphi_j)(A_k\varphi_k{}^\star)]\\ &+ie\kappa F_{\mu\nu}\varphi_\mu{}^\star\varphi_\nu+\tfrac{1}{2}e^2\kappa^2[(\varphi_4{}^\star)^2\varphi_j\varphi_j+(\varphi_4)^2\varphi_j{}^\star\varphi_j{}^\star-2(\varphi_4{}^\star\varphi_4)(\varphi_j{}^\star\varphi_j)], \qquad (27)\end{aligned}$$

where

$$g_{\mu\nu}=(\partial\varphi_\nu/\partial x_\mu)-(\partial\varphi_\mu/\partial x_\nu)$$

and

$$g_{\mu\nu}{}^\star=(\partial\varphi_\nu{}^\star/\partial x_\mu)-(\partial\varphi_\mu{}^\star/\partial x_\nu). \qquad (28)$$

The interaction Hamiltonian is not a Hermitian matrix but one that satisfies

$$H_{\text{int}}{}^\star\equiv\eta^{-1}H_{\text{int}}{}^\dagger\eta=H_{\text{int}}. \qquad (29)$$

4.4 Feynman Diagram

In the interaction representation the S matrix for such a theory can be analyzed into sums of Feynman diagrams in exactly the same way as an ordinary theory with positive metric. The values of the propagators and vertices are listed in Fig. 3.

From the rules for Feynman diagram it is clear that the theory satisfies relativistic invariance.

4.5 Unitarity

Because of (29), the S matrix is not unitary but satisfies

$$S^\star\equiv\eta^{-1}S^\dagger\eta=S^{-1}. \qquad (30)$$

If we restrict ourselves to a system of particles with a total energy

$$E<\xi^{-\frac{1}{2}}m, \qquad (31)$$

then by using (25) it is seen that there can be *no* scalar meson in either the initial state or the final state. Thus, for the initial and final states $\eta=+1$ and the S matrix is truly unitary provided (31) holds. Consequently, if

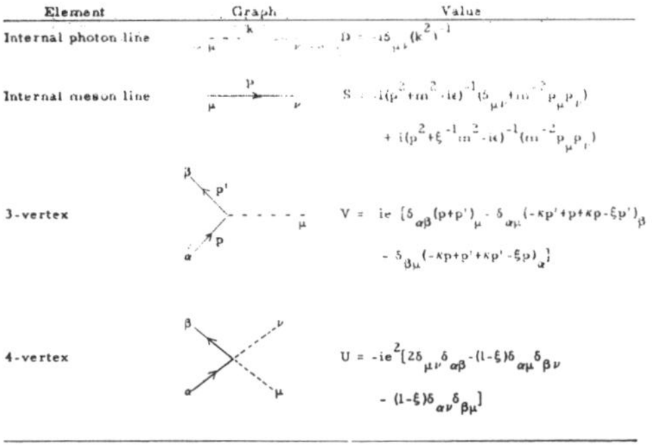

FIG. 3. Feynman diagram in momentum space for Lagrangian (7) with a negative metric (see also the caption of Fig. 1).

the limit $\xi \to 0$ exists, the limiting S matrix does become completely unitary.

4.6 Renormalization

For $\xi > 0$, the propagator of W varies asymptotically like p^{-2} at large momentum. The presence of the ξ-dependent term in the propagator acts like a regulator. Therefore, the divergencies that occur in the higher order Feynman diagrams can be eliminated by a renormalization process which is quite similar to that in the case of a charged scalar meson (except for the differences in the spin dependences).

APPENDICES

In the following appendices we give the detailed derivation of the rules for Feynman graphs for the charged vector mesons following closely the Dyson-Wick procedure.[5] These derivations are at times rather complicated. For clarity we begin with the well-known and almost trivial case of charged vector mesons interacting with the lepton fields.

APPENDIX A. CHARGED VECTOR MESON AND FERMION CURRENTS

We discuss first the derivation of Feynman graph for the simple case of charged vector mesons interacting with electrons and neutrinos.

A1. Lagrangian and Hamiltonian

The Lagrangian density for this case is given by

$$\mathcal{L} = \mathcal{L}_W + \mathcal{L}_{\text{free leptons}} + \mathcal{L}_1, \tag{A1}$$

where

$$\mathcal{L}_W = -\tfrac{1}{2} G_{\mu\lambda}{}^* G_{\mu\lambda} - m^2 \Phi_\mu{}^* \Phi_\mu \tag{A2}$$

and

$$\mathcal{L}_1 = J_\mu \Phi_\mu{}^* + J_\mu{}^* \Phi_\mu, \tag{A3}$$

in which * has the same meaning as that given in (2),

$$G_{\mu\lambda} = (\partial \Phi_\lambda / \partial x_\mu) - (\partial \Phi_\mu / \partial x_\lambda), \tag{A4}$$

$$J_\mu = i g \Psi_e{}^\dagger \gamma_4 \gamma_\lambda (1+\gamma_5) \Psi_\nu, \tag{A5}$$

and Ψ_e, Ψ_ν, Φ_λ are, respectively, the field operators for e^-, ν, and W^+. The dynamic equation for Φ is given by

$$(\partial G_{\mu\lambda}/\partial x_\mu) - m^2 \Phi_\lambda + J_\lambda = 0. \tag{A6}$$

By using $\mathcal{L}$ one obtains the following Hamiltonian:

$$H = H_W + H_{\text{free lepton}} + H_1, \tag{A7}$$

where

$$H_W = \boldsymbol{\pi}^* \cdot \boldsymbol{\pi} + m^{-2} (\boldsymbol{\nabla} \cdot \boldsymbol{\pi}^*)(\boldsymbol{\nabla} \cdot \boldsymbol{\pi}) + (\boldsymbol{\nabla} \times \boldsymbol{\Phi}^*) \cdot (\boldsymbol{\nabla} \times \boldsymbol{\Phi}) + m^2 \boldsymbol{\Phi}^* \cdot \boldsymbol{\Phi}, \tag{A8}$$

$$H_1 = -J_\mu \Phi_\mu{}^* - J_\mu{}^* \Phi_\mu + m^{-2} J_4 J_4{}^*. \tag{A9}$$

The 3-vectors $\boldsymbol{\pi}$ and $\boldsymbol{\pi}^*$ are, respectively, the conjugate momenta of $\boldsymbol{\Phi}$ and $\boldsymbol{\Phi}^*$

$$\pi_k = i G_{4k}{}^*$$

and

$$\pi_k{}^* = i G_{4k} \quad (k = 1, 2, 3). \tag{A10}$$

Because of the absence of $d\Phi_4/dt$ and $d\Phi_4{}^*/dt$ in $\mathcal{L}$, the Φ_4 and $\Phi_4{}^*$ in (A9) are not independent variables but are given by

$$\Phi_4 = m^{-2} [i \boldsymbol{\nabla} \cdot \boldsymbol{\pi}^* + J_4],$$

and

$$\Phi_4{}^* = m^{-2} [i \boldsymbol{\nabla} \cdot \boldsymbol{\pi} + J_4{}^*]. \tag{A11}$$

A2. Interaction Representation and Feynman Graphs

In the interaction representation it is convenient to introduce the following notations[10]:

$$\varphi = \boldsymbol{\Phi},$$

$$\varphi_4 = i m^{-2} \boldsymbol{\nabla} \cdot \boldsymbol{\pi}^*,$$

and

$$g_{\mu\lambda} = (\partial \varphi_\lambda / \partial x_\mu) - (\partial \varphi_\mu / \partial x_\lambda). \tag{A12}$$

Therefore, the φ_λ and $g_{\mu\lambda}$ satisfy the free-meson equation

$$\frac{\partial}{\partial x_\mu} g_{\mu\lambda} - m^2 \varphi_\lambda = 0.$$

In terms of φ_λ the interaction Hamiltonian is given by

$$H_{\text{int}} = H_1 = -j_\mu \varphi_\mu{}^* - j_\mu{}^* \varphi_\mu - m^{-2} j_4 j_4{}^*. \tag{A13}$$

Using the notation of Wick,[5] the propagator of $W^\pm$ is given by

$$\varphi_\mu{}^\cdot(x) \varphi_\lambda{}^{*\cdot}(0) \equiv \langle T[\varphi_\mu(x) \varphi_\lambda{}^*(0)] \rangle_{\text{vac}}. \tag{A14}$$

Theorem 1.

$$\varphi_\mu{}^\cdot(x) \varphi_\lambda{}^{*\cdot}(0) = \mathfrak{D}_{\mu\lambda}(x) + i m^{-2} \delta_{4\mu} \delta_{4\lambda} \delta^4(x), \tag{A15}$$

where

$$\mathfrak{D}_{\mu\lambda}(x) = [\delta_{\mu\lambda} - m^{-2} (\partial^2 / \partial x_\mu \partial x_\lambda)] \tfrac{1}{2} \Delta_F(x), \tag{A16}$$

[10] Throughout all the appendices we use capital letters to denote operators in the Heisenberg representation and small letters such as φ_λ, $g_{\mu\nu}$, a_λ, $f_{\mu\nu}$, j_μ, etc., to denote operators in the interaction representation.

$$\Delta_F(x) = -i(8\pi^4)^{-1} \times \int d^4k\, [k^2+(m-i\epsilon)^2]^{-1} \exp(ik_\lambda x_\lambda), \quad \text{(A17)}$$

$k^2 = k_\lambda k_\lambda$, $d^4k = d^3\mathbf{k}(-idk_4)$, and $\delta^4(x) = \delta^3(\mathbf{r})\delta(t)$.

Proof. To avoid possible mathematical ambiguity we shall put the whole system in a finite three-dimensional volume Ω. (The limit $\Omega \to \infty$ will be carried out only at the end.) The field operators φ_μ are then expanded into the following Fourier series:

$$\boldsymbol{\varphi}(x) = \sum_{\mathbf{k}} \sum_{t=1,2} (2\Omega\omega)^{-\frac{1}{2}}[a_{\mathbf{k}}{}^t \exp(i\mathbf{k}\cdot\mathbf{r}-i\omega t) + b_{-\mathbf{k}}{}^{t\dagger} \exp(i\mathbf{k}\cdot\mathbf{r}+i\omega t)]\mathbf{e}_{\mathbf{k}}{}^t + \sum_{\mathbf{k}} (2\Omega\omega)^{-\frac{1}{2}}[a_{\mathbf{k}}{}^l \exp(i\mathbf{k}\cdot\mathbf{r}-i\omega t) + b_{-\mathbf{k}}{}^{l\dagger} \exp(i\mathbf{k}\cdot\mathbf{r}+i\omega t)](\omega\hat{k}/m)$$

and

$$\varphi_4(x) = \sum_{\mathbf{k}} (2\omega\Omega)^{-\frac{1}{2}}[a_{\mathbf{k}}{}^l \exp(i\mathbf{k}\cdot\mathbf{r}-i\omega t) - b_{-\mathbf{k}}{}^{l\dagger} \exp(i\mathbf{k}\cdot\mathbf{r}+i\omega t)](i|\mathbf{k}|/m), \quad \text{(A18)}$$

where $\mathbf{e}_{\mathbf{k}}{}^1$, $\mathbf{e}_{\mathbf{k}}{}^2$, and $\hat{k} = |\mathbf{k}|^{-1}\mathbf{k}$ form a right-handed orthogonal set of three unit vectors, $\omega = (\mathbf{k}^2+m^2)^{\frac{1}{2}} > 0$ and $a_{\mathbf{k}}{}^r$, $b_{\mathbf{k}}{}^r$ are the annihilation operators for the transverse ($r=t=1, 2$) and the longitudinal ($r=l$) mesons. For definiteness, let us define

$$T[\varphi_\mu(x)\varphi_\nu{}^*(0)] = \varphi_\mu(x)\varphi_\nu{}^*(0) \quad \text{if} \quad t \geqq 0,$$

and

$$= \varphi_\nu{}^*(0)\varphi_\mu(x) \quad \text{if} \quad t<0. \quad \text{(A19)}$$

Therefore, the vacuum expectation value of the T product (A14) is given by (keeping Ω finite)

$$\langle T[\varphi_\mu(x)\varphi_\nu{}^*(0)]_\Omega\rangle_{\text{vac}} = \sum_{\mathbf{k}}(2\omega\Omega)^{-1}\exp(i\mathbf{k}\cdot\mathbf{r}-i\omega t) \times[\delta_{\mu\nu}-q_\mu q_\nu m^{-2}] \quad \text{if} \quad t\geqq 0,$$

and

$$= \sum_{\mathbf{k}}(2\omega\Omega)^{-1}\exp(i\mathbf{k}\cdot\mathbf{r}+i\omega t) \times[\delta_{\mu\nu}-q_\mu{}^* q_\nu{}^* m^{-2}] \quad \text{if} \quad t<0,$$

where $q_4 = i\omega$, $q_j = k_j$ ($j=1, 2, 3$), and $q_\mu{}^*$ is the complex conjugate of q_μ. Upon converting the summand on the right-hand side of the above equation into a Feynman-type integral, we find

$$\langle T[\varphi_\mu(x)\varphi_\nu{}^*(0)]_\Omega\rangle_{\text{vac}} = -i\sum_{\mathbf{k}}(2\pi\Omega)^{-1}\int_{-\infty}^{\infty} dk_0 \left[\frac{\delta_{\mu\nu}-m^{-2}k_\mu k_\nu}{k_\alpha k_\alpha+(m-i\epsilon)^2} - m^{-2}\delta_{4\mu}\delta_{4\nu}\right] \times\exp(ik_\beta x_\beta), \quad \text{(A20)}$$

where $k_4 = ik_0$. In (A20) we neglect functions that are zero if $t\neq 0$ and remain finite at $t=0$. Taking the limit $\Omega\to\infty$, we obtain (A15).

It is important to notice that

(i) An expression identical with (A20) would be obtained if instead of (A19) we define $T[\varphi_\mu(x)\varphi_\nu{}^*(0)] = \varphi_\mu(x)\varphi_\nu{}^*(0)$ for $t>0$ and $T[\varphi_\mu(x)\varphi_\nu{}^*(0)] = \varphi_\nu{}^*(0)\varphi_\mu(x)$ for $t\leqq 0$.

(ii) Because of the usual quantization procedures the limit $\lim_{\Omega\to\infty}\langle T[\varphi_\mu(x)\varphi_\nu{}^*(0)]_\Omega\rangle_{\text{vac}}$ is not covariant.

(iii) The presence of the term $\delta^4(x)$ in (A15) can also be easily seen by considering the special case $\mu=\lambda=4$. From (A14) and (A11) it follows that $\varphi_4{}^\cdot(x)\varphi_4{}^{*\cdot}(0)$ is continuous in time at $t=0$. However, $\partial\Delta_F(x)/\partial t$ approaches $-i\delta^3(\mathbf{r})$ at $t=0+$ and $+i\delta^3(\mathbf{r})$ at $t=0-$. Thus $\partial^2[\frac{1}{2}\Delta_F(x)]/\partial t^2$ contains a $\delta^4(x)$ singularity which is to be canceled by the last term on the right-hand side of (A15). Following Dyson's method[5] and by using (A13) and (A15) the S matrix can be evaluated. It can be shown quite easily that in the calculation of S matrix, after converting the appropriate T products into S products, the effects of the contact term $-m^{-2}j_4{}^*j_4$ in (A13) exactly cancel that of the term $im^{-2}\delta_{4\mu}\delta_{4\nu}\delta^4(x)$ in the propagator (A15). Therefore, one obtains the following theorem:

Theorem 2. The entire S matrix can be generated by considering an equivalent problem in which H_{int} is replaced by[11]

$$H_{\text{int}}' = -j_\mu\varphi_\mu{}'^* - j_\mu{}^*\varphi_\mu{}' \quad \text{(A21)}$$

and the propagator (A15) is replaced by

$$\varphi_\mu{}'^\cdot(x)\varphi_\lambda{}'^{*\cdot}(0) = [\delta_{\mu\lambda} - m^{-2}(\partial^2/\partial x_\mu\partial x_\lambda)]\tfrac{1}{2}\Delta_F(x). \quad \text{(A22)}$$

Theorem 2 leads to the well-known results first stated by Feynman.[2] The resulting Feynman graphs contain only one kind of vertices which connects two lepton lines and one meson line. In such a graph each internal meson line contributes only the covariant factor

$$-i(\delta_{\mu\lambda}+m^{-2}k_\mu k_\lambda)(k^2+m^2)^{-1},$$

where k_μ is the momentum carried by such a line.

---

[11] The precise meaning of Theorem 2 is as follows: Regard the S matrix as given by

$$S = \sum_{n=0}^{\infty}\frac{(-i)^n}{n!}\int T[\prod_{i=1}^{n} H_{\text{int}}'(x_i)d^3r_i dt_i].$$

In converting these T products to S products, one uses Theorems 1 and 2 of Wick's paper[8] together with the identity

$$T[\varphi_\mu'(x)\varphi_\nu'^*(y)] = :\varphi_\mu'(x)\varphi_\nu'^*(y): + [\varphi_\mu'(x)]^\cdot[\varphi_\nu'^*(y)]^\cdot,$$

where $[\varphi_\mu'(x)]^\cdot[\varphi_\nu'^*(y)]^\cdot$ is given by (A22). The resulting S matrix (expressed as a sum of S products of φ_μ' and $\varphi_\nu'^*$) is identical with the original S matrix which is obtained by using H_{int} given by (A13) and $\varphi_\mu{}^\cdot(x)\varphi_\nu{}^{*\cdot}(y)$ given by (A15).

Exactly the same meaning applies to Theorem 3, and the two lemmas.

APPENDIX B. ELECTRODYNAMICS OF CHARGED VECTOR MESONS

B1. Lagrangian and Hamiltonian

We start with the Lagrangian[10] given by (1),

$$\mathcal{L}=-\tfrac{1}{2}(\partial A_\mu/\partial x_\nu)(\partial A_\mu/\partial x_\nu)-\tfrac{1}{2}G_{\mu\nu}^*G_{\mu\nu}-m^2\Phi_\mu^*\Phi_\mu-ie\kappa F_{\mu\nu}\Phi_\mu^*\Phi_\nu,$$

and regard Φ_4, Φ_4^* as dependent variables.

Let $\boldsymbol{\pi}$, $\boldsymbol{\pi}^*$, $\mathbf{P}$, P_0 be, respectively, the conjugate momenta of $\boldsymbol{\Phi}$, $\boldsymbol{\Phi}^*$, $\mathbf{A}$, and $A_0(=-iA_4)$. We have

$$\pi_k=iG_{4k}^*,$$
$$\pi_k^*=iG_{4k},$$
$$\mathbf{P}=(d\mathbf{A}/dt)-e\kappa(\Phi_4^*\boldsymbol{\Phi}-\boldsymbol{\Phi}^*\Phi_4),$$

and

$$P_0=-dA_0/dt, \tag{A23}$$

where $k=1, 2, 3$. The Hamiltonian density H is given by[6]

$$\begin{aligned}H=&\tfrac{1}{2}(\mathbf{P}^2-P_0^2)+\tfrac{1}{2}(\partial A_\mu/\partial x_i)(\partial A_\mu/\partial x_i)+\tfrac{1}{2}G_{jk}^*G_{jk}\\&+\boldsymbol{\pi}^*\cdot\boldsymbol{\pi}+m^2(\boldsymbol{\Phi}^*\cdot\boldsymbol{\Phi}+\Phi_4^*\Phi_4)\\&+i(\boldsymbol{\pi}\cdot\nabla\Phi_4+\boldsymbol{\pi}^*\cdot\nabla\Phi_4^*)+e(\boldsymbol{\pi}^*\cdot\boldsymbol{\Phi}^*-\boldsymbol{\pi}\cdot\boldsymbol{\Phi})A_4\\&+e(\Phi_4\boldsymbol{\pi}-\Phi_4^*\boldsymbol{\pi}^*)\cdot\mathbf{A}+e\kappa(\mathbf{P}-ie\nabla A_4)(\Phi_4^*\boldsymbol{\Phi}-\Phi_4\boldsymbol{\Phi}^*)\\&+ie\kappa F_{jk}\Phi_j^*\Phi_k+\tfrac{1}{2}e^2\kappa^2(\Phi_4^*\boldsymbol{\Phi}-\Phi_4\boldsymbol{\Phi}^*)^2.\end{aligned} \tag{A24}$$

In the Hamiltonian both Φ_4 and Φ_4^* are regarded as functions of $\boldsymbol{\Phi}$, $\mathbf{A}$, A_4, $\boldsymbol{\pi}$, $\mathbf{P}$, etc. By using (5) and (A23) we obtain

$$D\Phi_4=[1-(e\kappa/m)^2\boldsymbol{\Phi}^*\cdot\boldsymbol{\Phi}]N-(e\kappa/m)^2\boldsymbol{\Phi}\cdot\boldsymbol{\Phi}N^*,$$
$$D\Phi_4^*=[1-(e\kappa/m)^2\boldsymbol{\Phi}^*\cdot\boldsymbol{\Phi}]N^*-(e\kappa/m)^2\boldsymbol{\Phi}^*\cdot\boldsymbol{\Phi}^*N,$$

where

$$m^{-2}D=[1-(e\kappa/m)^2\boldsymbol{\Phi}^*\cdot\boldsymbol{\Phi}]^2-(e\kappa/m)^4(\boldsymbol{\Phi}^*\cdot\boldsymbol{\Phi}^*)(\boldsymbol{\Phi}\cdot\boldsymbol{\Phi}),$$
$$N=i\nabla\cdot\boldsymbol{\pi}^*+e\mathbf{A}\cdot\boldsymbol{\pi}^*-e\kappa\boldsymbol{\Phi}\cdot(\mathbf{P}-i\nabla A_4),$$

and

$$N^*=i\nabla\cdot\boldsymbol{\pi}-e\mathbf{A}\cdot\boldsymbol{\pi}+e\kappa\boldsymbol{\Phi}^*\cdot(\mathbf{P}-i\nabla A_4). \tag{A25}$$

B2. Interaction Representation

In a similar manner to (A12) it is convenient to introduce the following notations in the interaction representation[10]:

$$\boldsymbol{\varphi}=\boldsymbol{\Phi}$$
$$\varphi_4=im^{-2}\nabla\cdot\boldsymbol{\pi}^*$$

and

$$a_\mu=A_\mu. \tag{A26}$$

These field operators φ_μ, a_μ, therefore, satisfy the free-meson and the free-photon equations,

$$(\partial g_{\mu\nu}/\partial x_\mu)-m^2\varphi_\nu=0$$

and

$$(\partial^2 a_\nu/\partial x_\lambda\partial x_\lambda)=0, \tag{A27}$$

where

$$g_{\mu\nu}=(\partial\varphi_\nu/\partial x_\mu)-(\partial\varphi_\mu/\partial x_\nu). \tag{A28}$$

In terms of φ_μ, a_μ the interaction Hamiltonian is given by

$$\begin{aligned}H_{\text{int}}=&-iea_\mu[g_{\mu\nu}^*\varphi_\nu-g_{\mu\nu}\varphi_\nu^*]\\&+e^2[(\mathbf{a}\times\boldsymbol{\varphi}^*)\cdot(\mathbf{a}\times\boldsymbol{\varphi})-(a_jg_{4j}^*)(a_kg_{4k})m^{-2}]\\&+ie\kappa f_{jk}\varphi_j^*\varphi_k+ie\kappa f_{4j}(\Phi_4^*\varphi_j-\varphi_j^*\Phi_4)\\&+\tfrac{1}{2}e^2\kappa^2(\Phi_4^*\boldsymbol{\varphi}-\boldsymbol{\varphi}^*\Phi_4)^2+m^2y^*y,\end{aligned} \tag{A29}$$

where

$$f_{\mu\nu}=(\partial a_\nu/\partial x_\mu)-(\partial a_\mu/\partial x_\nu), \tag{A30}$$
$$y=\Phi_4-\varphi_4-im^{-2}ea_jg_{4j},$$

and

$$y^*=\Phi_4^*-\varphi_4^*+im^{-2}ea_jg_{4j}. \tag{A31}$$

In the above, Φ_4 and Φ_4^* are regarded as functions of φ_μ, a_μ, and their derivatives. The explicit forms of the functions Φ_4 and Φ_4^* can be directly obtained by using (A25), (A26), and the following substitutions:

$$\pi_j=ig_{4j}^*,\quad \pi_j^*=ig_{4j},$$

and

$$[P_j-i(\partial a_4/\partial x_j)]=if_{4j}. \tag{A32}$$

B3. Feynman Graphs

To obtain the appropriate rules for Feynman graphs we adopt the procedures and the notations used in Wick's paper.[5] The contraction of any two operators $A(x)$ and $B(y)$ is defined to be the vacuum expectation of their T product in the interaction representation:

$$A^\cdot(x)B^\cdot(y)\equiv\langle T[A(x)B(y)]\rangle_{\text{vac}}. \tag{A33}$$

In a similar manner to (A15), many of the contractions between the operators φ_μ, $g_{\mu\nu}$, etc. cannot be expressed in terms of covariant functions. By using (A28), (A30), and (A32) we obtain the following noncovariant contractions:

$$\varphi_\mu^\cdot(x)\varphi_\nu^{*\cdot}(y)=\mathfrak{D}_{\mu\nu}(x-y)+im^{-2}\delta_{4\mu}\delta_{4\nu}\delta^4(x-y),$$

$$g_{\mu\nu}^\cdot(x)g_{\alpha\beta}^{*\cdot}(y)=-\frac{\partial^2}{\partial x_\mu\partial x_\alpha}\mathfrak{D}_{\nu\beta}-\frac{\partial^2}{\partial x_\nu\partial x_\beta}\mathfrak{D}_{\mu\alpha}+\frac{\partial^2}{\partial x_\nu\partial x_\alpha}\mathfrak{D}_{\mu\beta}$$

$$+\frac{\partial^2}{\partial x_\mu\partial x_\beta}\mathfrak{D}_{\nu\alpha}+i[\delta_{4\mu}\delta_{4\alpha}\delta_{\nu\beta}+\delta_{4\nu}\delta_{4\beta}\delta_{\mu\alpha}-\delta_{4\mu}\delta_{4\beta}\delta_{\nu\alpha}-\delta_{4\nu}\delta_{4\alpha}\delta_{\mu\beta}]\delta^4(x-y)$$

and

$$f_{\mu\nu}{}^{\cdot}(x)f_{\alpha\beta}{}^{\cdot}(y)=\left[-\delta_{\nu\beta}\frac{\partial^2}{\partial x_\mu\partial x_\alpha}-\delta_{\mu\alpha}\frac{\partial^2}{\partial x_\nu\partial x_\beta}+\delta_{\mu\beta}\frac{\partial^2}{\partial x_\nu\partial x_\alpha}+\delta_{\nu\alpha}\frac{\partial^2}{\partial x_\mu\partial x_\beta}\right]\tfrac{1}{2}D_F(x-y)$$
$$+i[\delta_{4\mu}\delta_{4\alpha}\delta_{\nu\beta}+\delta_{4\nu}\delta_{4\beta}\delta_{\mu\alpha}-\delta_{4\mu}\delta_{4\beta}\delta_{\nu\alpha}-\delta_{4\nu}\delta_{4\alpha}\delta_{\mu\beta}]\delta^4(x-y),\quad \text{(A34)}$$

where $\mathfrak{D}_{\mu\nu}(x-y)$ is given by (A16), Δ_F by (A17), and

$$D_F(x)=-i(8\pi^4)^{-1}\int d^4k(k^2)^{-1}\exp(ik_\lambda x_\lambda).\quad \text{(A35)}$$

The other relevant contractions can all be expressed in covariant forms; for example,

$$a_\mu{}^{\cdot}(x)a_\nu{}^{\cdot}(y)=\tfrac{1}{2}\delta_{\mu\nu}D_F(x-y),$$
$$g_{\mu\nu}{}^{\cdot}(x)\varphi_\lambda{}^{*\cdot}(y)=(\partial/\partial x_\mu)\mathfrak{D}_{\nu\lambda}(x-y)-(\partial/\partial x_\nu)\mathfrak{D}_{\mu\lambda}(x-y),\quad \text{(A36)}$$

etc.

It is important to notice that if $\kappa\neq 0$ the expansion of H_{int} (A29) into a power series of e actually contains an infinite number of terms. In principle, the rules for Feynman graphs can be obtained by a straight-forward application of the standard algebraic method[5] of converting the T products in the S matrix into the appropriate S products. In practice, because of the complexity of H_{int} and the presence of numerous noncovariant terms in the propagators, it is quite complicated to carry out the details. This will be done in Appendix C. It is found that, similar to the simple example of the interaction between vector mesons and lepton currents discussed in Appendix A, much of these two above mentioned complexities cancel themselves. We obtain, as a result, the following theorem (proved in Appendix C).

Theorem 3. The above S matrix can be generated by considering an equivalent problem in which H_{int} is replaced by[11]

$$H_{\text{int}}'=-iea_\mu'[g_{\mu\nu}'^*\varphi_\nu'-g_{\mu\nu}'\varphi_\nu'^*]+e^2a_\mu'a_\nu'[\delta_{\mu\nu}\varphi_\lambda'^*\varphi_\lambda'-\varphi_\mu'^*\varphi_\nu']+ie\kappa f_{\mu\nu}'\varphi_\mu'^*\varphi_\nu'+\delta H,\quad \text{(A37)}$$

where

$$\delta H=(i/2)\delta^4(0)\ln\{[1-(e\kappa/m)^2\varphi_j'^*\varphi_j']^2-(e\kappa/m)^4(\varphi_j'\varphi_j')(\varphi_k'^*\varphi_k'^*)\}.\quad \text{(A38)}$$

The contraction of the prime fields φ_μ', $g_{\mu\nu}'$, etc., are identical with that of φ_μ, $g_{\mu\nu}$, etc., *except* that all the noncovariant terms are now absent. More explicitly, (A34) is replaced by

$$\varphi_\mu'^{\cdot}(x)\varphi_\nu'^{*\cdot}(y)=\mathfrak{D}_{\mu\nu}(x-y),$$
$$g_{\mu\nu}'^{\cdot}(x)g_{\alpha\beta}'^{*\cdot}(y)=-\frac{\partial^2}{\partial x_\mu\partial x_\alpha}\mathfrak{D}_{\nu\beta}-\frac{\partial^2}{\partial x_\nu\partial x_\beta}\mathfrak{D}_{\mu\alpha}+\frac{\partial^2}{\partial x_\nu\partial x_\alpha}\mathfrak{D}_{\mu\beta}+\frac{\partial^2}{\partial x_\mu\partial x_\beta}\mathfrak{D}_{\nu\alpha},\quad \text{(A39)}$$

etc., and (A36) remains the same as before; i.e.,

$$a_\mu'^{\cdot}(x)a_\nu'^{\cdot}(y)=\tfrac{1}{2}\delta_{\mu\nu}D_F(x-y),\quad \text{(A40)}$$

etc.

Theorem 3 states that except for the term δH in (A37) the effects of the noncovariant terms in the original propagators (A34) completely cancel those that are generated by the difference between H_{int} and $-\mathfrak{L}_{\text{int}}$. If $\kappa=0$, $\delta H=0$; therefore, by using Theorem 3, the rules of deriving Feynman diagrams becomes almost trivial. However, if $\kappa\neq 0$, (δH) gives rise to additional vertices which are both divergent and noncovariant.

These results are summarized in Fig. 1.

APPENDIX C. PROOF OF THEOREM 3

C1. A Simple System of Harmonic Oscillators

Let us consider a problem of N harmonic oscillators whose coordinates are Q_1, $Q_2\cdots$, Q_n and frequencies $\omega_1=\omega_2=\cdots=1$. The Lagrangian for this system is given by

$$L=L_0+L_1,\quad \text{(A41)}$$

where

$$L_0=\frac{1}{2}\frac{d\tilde{Q}}{dt}\frac{dQ}{dt}-\tfrac{1}{2}\tilde{Q}Q,$$
$$L_1=\frac{1}{2}\frac{d\tilde{Q}}{dt}A\frac{dQ}{dt}+\frac{1}{2}\frac{d\tilde{Q}}{dt}B+\tfrac{1}{2}\tilde{B}\frac{dQ}{dt}+C,$$
$$Q=\begin{bmatrix}Q_1\\Q_2\\\vdots\\Q_N\end{bmatrix},\quad \text{(A42)}$$

$\tilde{Q}$ is the transpose of Q, A is a symmetric $(N\times N)$ matrix and B, C are, respectively, matrices of dimension $(N\times 1)$ and (1×1). All three matrices A, B, C are *functions* of Q (but do not explicitly depend on dQ/dt). The conjugate momenta P and the Hamiltonian H are given by

$$P=(1+A)(dQ/dt)+B$$

and

$$H=\tfrac{1}{2}\tilde{P}(1+A)^{-1}P+\tfrac{1}{2}\tilde{Q}Q-\tfrac{1}{2}\tilde{P}(1+A)^{-1}B-\tfrac{1}{2}\tilde{B}(1+A)^{-1}P-C+\tfrac{1}{2}\tilde{B}(1+A)^{-1}B,\quad \text{(A43)}$$

where 1 is the $(N\times N)$ unit matrix.

In the following we discuss the perturbation series in which A, B, C are treated as small but arbitrary functions of Q. It is convenient to use the interaction representation, regarding

$$H_1\equiv H-\tfrac{1}{2}\tilde{P}P-\tfrac{1}{2}\tilde{Q}Q\quad \text{(A44)}$$

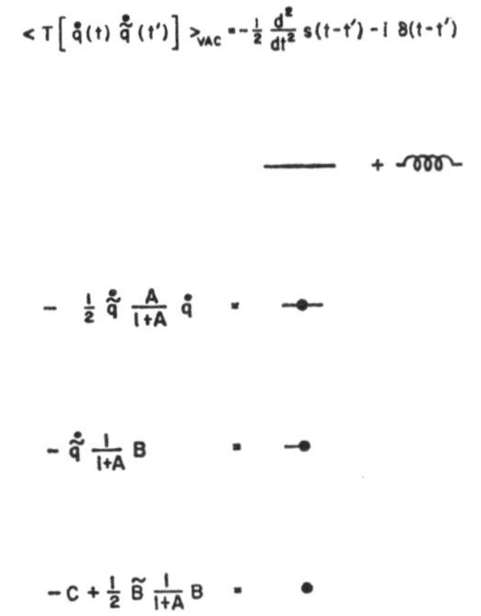

FIG. 4. Propagator and vertices for the simple system discussed in Appendix C2. The dot over $q(t)$ denotes a time derivative.

as the interaction Hamiltonian. For clarity, we introduce in the interaction representation[10]

$$q\equiv\begin{bmatrix}q_1\\ \vdots\\ q_N\end{bmatrix}\equiv Q.$$

Therefore,

$$dq/dt=P. \tag{A45}$$

The explicit time dependences of q and dq/dt are given by

$$q_n=(1/\sqrt{2})(a_ne^{-it}+a_n^{\dagger}e^{it})$$

and

$$dq_n/dt=-(i/\sqrt{2})(a_ne^{-it}-a_n^{\dagger}e^{it}), \tag{A46}$$

where a_n and $a_n^{\dagger}$ are, respectively, the annihilation and creation operators for the nth harmonic oscillator ($n=1, 2, \cdots N$). In terms of q and dq/dt the interaction Hamiltonian H_1 becomes

$$H_1=-\frac{1}{2}\frac{d\tilde{q}}{dt}A(1+A)^{-1}\frac{dq}{dt}-\frac{1}{2}\frac{d\tilde{q}}{dt}(1+A)^{-1}B$$

$$-\tfrac{1}{2}\tilde{B}(1+A)^{-1}\frac{dq}{dt}-C+\tfrac{1}{2}\tilde{B}(1+A)^{-1}B. \tag{A47}$$

We observe that the vacuum expectation values of the various T products in the interaction representation are given by

$$\langle T[q_n(t)q_m(0)]\rangle_{\rm vac}=\tfrac{1}{2}\delta_{nm}s(t),$$
$$\langle T[(dq_n/dt)(t)q_m(0)]\rangle_{\rm vac}=\tfrac{1}{2}\delta_{nm}ds/dt, \tag{A48}$$

and

$$\left\langle T\left[\frac{dq_n}{dt}(t)\frac{dq_m}{dt}(0)\right]\right\rangle_{\rm vac}=\tfrac{1}{2}\delta_{nm}s$$

$$=-\tfrac{1}{2}\delta_{nm}\frac{d^2s}{dt^2}-i\delta_{nm}\delta(t), \tag{A49}$$

where n and m vary from 1 to N and

$$s(t)=\exp(-it)\quad\text{for}\quad t\geqq 0$$
$$=\exp(it)\quad\text{for}\quad t\leqq 0. \tag{A50}$$

As in the previous case of vector mesons, (A49) states that the contraction between time derivatives of q_n and q_m differs from the time derivatives of the corresponding contraction. We now state the following lemma:

Lemma 1. The S matrix of this problem can be generated by considering an equivalent interaction Hamiltonian[11]

$$H_1'=-\frac{1}{2}\left(\frac{d}{dt}\tilde{q}'\right)A\left(\frac{d}{dt}q'\right)-\frac{1}{2}\left(\frac{d}{dt}\tilde{q}'\right)B$$

$$-\tfrac{1}{2}\tilde{B}\left(\frac{d}{dt}q'\right)-C+\delta H, \tag{A51}$$

where $\delta H=\frac{1}{2}i\delta(0)\,\text{trace}[\ln(1+A)]$ and A, B, C are the same functions as before (but replacing q by q'). The contractions between q' and dq'/dt are given by

$$[q_n'(t)]^{\cdot}[q_m'(0)]^{\cdot}=\tfrac{1}{2}\delta_{nm}s(t),$$

$$\left[\frac{dq_n'}{dt}(t)\right]^{\cdot}[q_m'(0)]^{\cdot}=\tfrac{1}{2}\delta_{nm}\frac{ds}{dt},$$

and

$$\left[\frac{dq_n'}{dt}(t)\right]^{\cdot}\left[\frac{dq_m'}{dt}(0)\right]^{\cdot}=-\tfrac{1}{2}\delta_{nm}\frac{d^2s}{dt^2}. \tag{A52}$$

It is important to notice that the term $-i\delta_{nm}\delta(t)$ in (A49) is omitted in (A52) and that H_1' is essentially the same function as $-L_1$ except for the extra term $\frac{1}{2}i\delta(0)\,\text{trace}[\ln(1+A)]$.

C2. Proof of Lemma 1

To prove the lemma we consider the usual power series expansion of the S matrix in the interaction representation

$$S=\sum_{n=0}^{\infty}S_n,$$

where

$$S_n=\frac{(-i)^n}{n!}\int T[\prod_{i=1}^{n}H_1(t_i)dt_i]. \tag{A53}$$

In converting the above S matrix from T products into S products, let us concentrate on the developments due to the conversion of $T[[dq_n(t)/dt]_t[dq_m(t)/dt]_{t'}]$:

$$T\left[\frac{dq_n}{dt}(t)\frac{dq_m}{dt}(0)\right]=:\frac{dq_n}{dt}(t)\frac{dq_m}{dt}(0):$$

$$+\left\langle T\left[\frac{dq_n}{dt}(t)\frac{dq_m}{dt}(0)\right]\right\rangle_{\rm vac}.$$

We use the graphical method that only the contraction between dq_n/dt and dq_m/dt is represented, but all other contractions such as $[q_n(t)]^{\cdot}[q_m(0)]^{\cdot}$ and $[dq_n(t)/dt]_t^{\cdot}[q_m(0)]^{\cdot}$ are suppressed (i.e., not represented explicitly in the graphs). In Fig. 4 the propagator $[dq_n(t)/dt]_t^{\cdot}[dq_m(t)/dt]_{t'}^{\cdot}$ is represented by two terms: a straight line which stands for $-\frac{1}{2}\delta_{nm}d^2s/dt^2$ and a "spring" for the second term $-i\delta_{nm}\delta(t)$ in (A49). There are three kinds of vertices which correspond, respectively, to the terms $-\frac{1}{2}(d\tilde{q}/dt)A(1+A)^{-1}(dq/dt)$, $-(d\tilde{q}/dt)(1+A)^{-1}B$, and $[-C+\frac{1}{2}\tilde{B}(1+A)^{-1}B]$. The present problem then reduces simply to one of summing over all diagrams which contain different numbers of springs but otherwise are of similar topological structures. These sums are illustrated in Fig. 5.

To understand the sum I in Fig. 5, let us define

$$\mathrm{I}_1=-\frac{1}{2}\frac{d\tilde{q}}{dt}A(1+A)^{-1}\frac{dq}{dt},$$

which contributes a term $(-i)\int \mathrm{I}_1 dt$ in S_1. In S_2 there is a corresponding term $-i\int \mathrm{I}_2 dt$ that arises from the following contraction:

$$[(-i)^2/2!]\int dtdt'\left\{-\frac{1}{2}\frac{d\tilde{q}}{dt}A(1+A)^{-1}\left[\frac{dq}{dt}(t)\right]^{\cdot}\right\}\times\left\{-\frac{1}{2}\left[\frac{d\tilde{q}}{dt}(t')\right]^{\cdot}A(1+A)^{-1}\frac{dq}{dt}\right\},$$

in which one substitutes only the $-i\delta(t-t')$ part of (A49) for $[dq(t)/dt]_t^{\cdot}[d\tilde{q}(t)/dt]_{t'}^{\cdot}$. There are altogether four such terms due to the four different ways of selecting $(dq_i/dt)(dq_j/dt)$ out of the product $[dq_n(t)/dt]_t \times[dq_m(t)/dt]_t[dq_{n'}(t)/dt]_{t'}[dq_{m'}(t)/dt]_{t'}$. Thus we find

$$\mathrm{I}_2=-\frac{1}{2}\frac{d\tilde{q}}{dt}\left(\frac{A}{1+A}\right)^2\frac{dq}{dt}.$$

Similarly, it is easy to prove that there is a corresponding term $(-i)\int \mathrm{I}_n dt$ in S_n, where

$$\mathrm{I}_n=-\frac{1}{2}\frac{d\tilde{q}}{dt}\left[\frac{A}{1+A}\right]^n\frac{dq}{dt}.$$

The total sum of all these diagrams is given by

$$\mathrm{I}=\sum_{n=1}^{\infty}\mathrm{I}_n=-\frac{1}{2}\frac{d\tilde{q}}{dt}A\frac{dq}{dt}, \qquad (A54)$$

which contributes a term $(-i)\int \mathrm{I}dt$ to the entire S matrix. Identical arguments hold for cases in which I appears only as a part of a bigger diagram. The result of eliminating $-i\delta(t)$ term in the propagators in I-type diagrams is simply to replace $-\frac{1}{2}(d\tilde{q}/dt)A(1+A)^{-1}(dq/dt)$ in the interaction Hamiltonian by $-\frac{1}{2}(d\tilde{q}/dt)A(dq/dt)$.

FIG. 5. Sums of certain diagrams discussed in Appendix C2.

To understand the sum II in Fig. 5, let us consider the term $(-i)\int \mathrm{II}_1 dt$ in S_1, where

$$\mathrm{II}_1=-\frac{d\tilde{q}}{dt}(1+A)^{-1}B.$$

By using almost identical arguments as that used in the sum I, it is easy to show that there is also a corresponding term $(-i)\int \mathrm{II}_n dt$ in S_n, where

$$\mathrm{II}_n=-(d\tilde{q}/dt)(1+A)^{-1}[A/(1+A)]^{n-1}B. \qquad (A55)$$

Thus, summing over n we obtain

$$\mathrm{II}=\sum_{n=1}^{\infty}\mathrm{II}_n=-\frac{d\tilde{q}}{dt}B. \qquad (A56)$$

In the sum III, the term III_1 is given by

$$\mathrm{III}_1=-C+\tfrac{1}{2}\tilde{B}(1+A)^{-1}B, \qquad (A57)$$

which contributes a term $(-i)\int \mathrm{III}_1 dt$ to S_1. In S_2, let us consider

$$\frac{(-i)^2}{2!}\int\left\{-\left[\frac{d\tilde{q}}{dt}(t)\right]^{\cdot}(1+A)^{-1}B\right\}\times\left\{-\left[\frac{d\tilde{q}}{dt}(t')\right]^{\cdot}(1+A)^{-1}B\right\}dtdt',$$

and again substitute only the $-i\delta(t-t')$ term for the contraction. The result gives a term $(-i)\int \mathrm{III}_2 dt$ in S_2, where

$$\mathrm{III}_2=-\tfrac{1}{2}\tilde{B}(1+A)^{-2}B. \qquad (A58)$$

Similarly, the diagram III_n contributes to S_n a term $(-i)\int \mathrm{III}_n dt$, where

$$\mathrm{III}_n=-\tfrac{1}{2}\tilde{B}(1+A)^{-n}A^{n-2}B \quad (n\geq 2). \qquad (A59)$$

Summing over n, we obtain

$$\mathrm{III}=\sum_{n=1}^{\infty}\mathrm{III}_n=-C. \qquad (A60)$$

To understand IV, let us consider in S_1 the contribution of

$$(-i)\int -\frac{1}{2}\left[\frac{d\tilde{q}}{dt}(t)\right]^{\cdot} A(1+A)^{-1}\left[\frac{dq}{dt}(t)\right]^{\cdot} dt$$

in which only $[-i\delta_{nm}\delta(t)]$ is used for the contraction. This results a term $(-i)\int \text{IV}_1 dt$ in S_1, where

$$\text{IV}_1 = +\tfrac{1}{2}i\delta(0)\,\text{trace}[A/(1+A)]. \qquad \text{(A61)}$$

Similarly, it can be shown that the diagram IV_n in Fig. 5 contributes a term $(-i)\int \text{IV}_n dt$ to S_n where

$$\text{IV}_n = +\tfrac{1}{2}i\delta(0)n^{-1}\,\text{trace}[A/(1+A)]^n. \qquad \text{(A62)}$$

The factor n^{-1} is due to the cyclic symmetry of the diagram IV_n. Summing over n, one obtains

$$\text{IV} = \sum_{n=1}^{\infty} \text{IV}_n = \tfrac{1}{2}i\delta(0)\,\text{trace}[\ln(1+A)]. \qquad \text{(A63)}$$

It is easy to see that identical sums can be performed for any part of an arbitrary diagram in which $-i\delta_{nm}\delta(t)$ occurs in the contraction $[dq_n(t)/dt]_t^{\cdot}[dq_m(t)/dt]_0^{\cdot}$. The result of such sum is Lemma 1. We recall that, since A is a function of q, IV is not a constant.

Both Theorems 2 and 3 are direct consequences of this lemma. The last term $\frac{1}{2}i\delta(0)\,\text{trace}[\ln(1+A)]$ in the lemma is the cause of the existence of (δH) in Theorem 3. This is connected with the fact that the dependence of φ_4 on $i\nabla\cdot\pi^*$ makes the extra-magnetic moment term $ie\kappa F_{\mu\nu}\varphi_\mu{}^*\varphi_\nu$ to behave like $-\frac{1}{2}(d\tilde{q}/dt)A(dq/dt)$ in the lemma. The detailed steps leading from Lemma 1 to Theorem 3 are still somewhat involved and are given in the subsequent sections.

C3. Generalization of Lemma 1

The case of vector mesons discussed in Theorem 3 differs from the problem of harmonic oscillators treated in Lemma 1 in several essential aspects. Comparison between (A26) and (A45) suggests that φ_4 and $\varphi_4{}^*$ of the vector mesons fields behave like dq/dt of the harmonic oscillators. Yet, two main differences exist:

(i) The noncovariant term $im^{-2}\delta_{4\mu}\delta_{4\nu}\delta^4(x)$ in $\varphi_\mu{}^{\cdot}(x)\times\varphi_\nu{}^{*\cdot}(0)$ [given by (A34)] does not exactly correspond to the term $-i\delta_{nm}\delta(t)$ in $[dq_n(t)/dt]_t^{\cdot}[dq_m(t)/dt]_0^{\cdot}$ which is given by (A49).

(ii) In Lemma 1, $(H_1'-\delta H)$ is the same function as $-L_1$ if one replaces dq/dt in $(-L_1)$ by dq'/dt and q by q'. The analogy between φ_4, $\varphi_4{}^*$ and dq/dt might suggest that in the case of vector mesons one could first regard $-\mathfrak{L}_{\text{int}}$ as a function of φ_μ, $\varphi_\mu{}^*$ through the relations $\Phi_4=\Phi_4(\varphi_\mu,\varphi_\mu{}^*,\cdots)$ and $\Phi_4{}^*=\Phi_4{}^*(\varphi_\mu,\varphi_\mu{}^*,\cdots)$ and then replace in $-\mathfrak{L}_{\text{int}}$ all φ_μ, $\varphi_\mu{}^*$ by φ_μ' and φ_μ'. The resulting function would, however, be completely different from $(H_{\text{int}}'-\delta H)$ given by (A37). Rather, Theorem 3 states that $(H_{\text{int}}'-\delta H)$ is the same function as $-\mathfrak{L}_{\text{int}}$ only if in $(-\mathfrak{L}_{\text{int}})$ the variables Φ_μ and $\Phi_\mu{}^*$ are replaced directly by φ_μ' and $\varphi_\mu'^*$. (The same is true also for Theorem 2 for which the term corresponds to $\delta H=0$.)

Because of the above differences between φ_4 and dq/dt, Lemma 1 has to be generalized.

Consider a problem in which the interaction Hamiltonian is given by

$$H_{\text{int}} = -\tfrac{1}{2}\tilde{\psi}A(1+A)^{-1}\psi - \tfrac{1}{2}\tilde{\psi}(1+A)^{-1}B - \tfrac{1}{2}\tilde{B}(1+A)^{-1}\psi - C + \tfrac{1}{2}\tilde{B}(1+A)^{-1}B, \qquad \text{(A64)}$$

where ψ consists of N local Hermitian operators:

$$\psi = \begin{pmatrix}\psi_1(x)\\ \psi_2(x)\\ \vdots\\ \psi_N(x)\end{pmatrix}, \qquad \text{(A65)}$$

$A(x)$ is a symmetric $(N\times N)$ matrix, B is a matrix of dimension $(N\times 1)$ and C is (1×1). The matrix elements of A, B, C are local operators. Let $M(x-y)$ be the contraction between $\psi(x)$ and $\psi(y)$,

$$\psi^{\cdot}(x)\tilde{\psi}^{\cdot}(y) = M(x-y). \qquad \text{(A66)}$$

The following lemma can then be established.[11]

Lemma 2. The S matrix of the above problem can also be generated by considering an alternative problem in which (i) the H_{int} in (A64) is replaced by

$$H_{\text{int}}' = -\tfrac{1}{2}\tilde{\psi}'A\psi' - \tfrac{1}{2}\tilde{\psi}'B - \tfrac{1}{2}\tilde{B}\psi' - C + \delta H, \qquad \text{(A67)}$$

where

$$\delta H = \tfrac{1}{2}i\delta^4(0)\,\text{trace}[\ln(1+A)], \qquad \text{(A68)}$$

and (ii) the contraction (A66) is replaced by

$$\psi'^{\cdot}(x)\tilde{\psi}'^{\cdot}(y) = M(x-y) + i\delta^4(x-y). \qquad \text{(A69)}$$

All other contractions such as that between ψ and A, B, C remain unchanged, except for the formal replacement of ψ by ψ'.

Proof. The proof of Lemma 1 can be used directly to prove Lemma 2 by simply changing dq/dt into ψ.

It is useful to observe that the functions $H_{\text{int}}(\psi)$ and $H_{\text{int}}'(\psi')$ are connected by a simple transformation similar to the usual Legendre transformation relating Lagrangian to Hamiltonian. Define

$$G(\psi') \equiv -H_{\text{int}}'(\psi') + \delta H = \tfrac{1}{2}\tilde{\psi}'A\psi' + \tilde{B}\psi' + C \qquad \text{(A70)}$$

and

$$\psi_a \equiv \psi_a' + \frac{\partial G}{\partial \psi_a'} \quad (a=1,2,\cdots N). \qquad \text{(A71)}$$

In (A70) and (A71) ψ and ψ' are considered to be c-number vectors. The function $H_{\text{int}}(\psi)$ is, then, given by

$$H_{\text{int}}(\psi) = -\tfrac{1}{2}\sum_{a=1}^{N}\left(\frac{\partial G}{\partial \psi_a'}\right)^2 - G(\psi'). \qquad \text{(A72)}$$

C4. Proof of Theorem 3

In a similar manner to (A70), let us define

$$G \equiv -H_{\mathrm{int}}' + \delta H, \tag{A73}$$

where H_{int}' is given by (A37). Let us formally regard in (A73) f_{4j}', φ_4', $\varphi_4'^*$, g_{4j}', $g_{4j}'^*$ as 11 independent c-number variables and all others such as $\boldsymbol{\varphi}'$, $\boldsymbol{\varphi}'^*$, g_{ij}', $g_{ij}'^*$, etc. as *constants*. In terms of these 11 variables the function G becomes

$$\begin{aligned} G(f_{4j}',\varphi_4',\varphi_4'^*,g_{4j}',g_{4j}'^*) &= [-(e^2\mathbf{a}\cdot\mathbf{a})\varphi_4'^*\varphi_4' + (iea_j)g_{4j}'\varphi_4'^* \\ &\quad -(iea_j)g_{4j}'^*\varphi_4' + (ie\kappa\varphi_j^*)f_{4j}'\varphi_4' - (ie\kappa\varphi_j)f_{4j}'\varphi_4'^*] \\ &\quad +[(e^2a_4\mathbf{a}\cdot\boldsymbol{\varphi}^*)\varphi_4' + (e^2a_4\mathbf{a}\cdot\boldsymbol{\varphi})\varphi_4'^* \\ &\quad -(iea_4\varphi_j^*)g_{4j}' + (iea_4\varphi_j)g_{4j}^*] + C, \end{aligned} \tag{A74}$$

where C is a constant given by

$$\begin{aligned} C &= iea_i[g_{ij}^*\varphi_j - g_{ij}\varphi_j^*] - e^2a_4^2\boldsymbol{\varphi}\cdot\boldsymbol{\varphi}^* \\ &\quad - e^2(\mathbf{a}\times\boldsymbol{\varphi})\cdot(\mathbf{a}\times\boldsymbol{\varphi}^*) - ief_{ij}\varphi_i^*\varphi_j. \end{aligned} \tag{A75}$$

In both (A74) and (A75) the values of the "constants" $\boldsymbol{\varphi}'$, $\boldsymbol{\varphi}'^*$, g_{ij}', $g_{ij}'^*$, $\mathbf{a}'$, a_4', f_{ij}' are set to be $\boldsymbol{\varphi}$, $\boldsymbol{\varphi}^*$, g_{ij}, g_{ij}^*, $\mathbf{a}$, a_4, and f_{ij}, respectively.

Similar to (A71), we define

$$\varphi_4 \equiv \varphi_4' - (\partial G/\partial\varphi_4'^*)(1/m^2),$$

$$\varphi_4^* \equiv \varphi_4'^* - (\partial G/\partial\varphi_4')(1/m^2),$$

$$g_{4j} \equiv g_{4j}' - (\partial G/\partial g_{4j}'^*),$$

$$g_{4j}^* \equiv g_{4j}'^* - (\partial G/\partial g_{4j}'),$$

and

$$f_{4j} \equiv f_{4j}' - (\partial G/\partial f_{4j}'). \tag{A76}$$

It is straightforward (though somewhat tedious) to show that the function H_{int} (A29) is related to G by

$$\begin{aligned} H_{\mathrm{int}} &= \frac{1}{m^2}\left(\frac{\partial G}{\partial\varphi_4'}\right)\left(\frac{\partial G}{\partial\varphi_4'^*}\right) \\ &\quad + \left(\frac{\partial G}{\partial g_{4j}'}\right)\left(\frac{\partial G}{\partial g_{4j}'^*}\right) + \frac{1}{2}\left(\frac{\partial G}{\partial f_{4j}'}\right)\left(\frac{\partial G}{\partial f_{4j}'}\right) - G, \end{aligned} \tag{A77}$$

where the 11 primed field variables are regarded as functions of the unprimed variables by using (A76).

In order to use Lemma 2, we define 11 Hermitian variables $\psi_1, \cdots \psi_{11}$ by

$$\psi_j' = if_{4j}',$$

$$\psi_{3+j}' = (1/\sqrt{2})[g_{4j}' - g_{4j}'^*],$$

$$\psi_{6+j}' = -i(1/\sqrt{2})[g_{4j}' + g_{4j}'^*],$$

$$\psi_{10}' = (m/\sqrt{2})[\varphi_4' - \varphi_4'^*],$$

and

$$\psi_{11}' = -i(m/\sqrt{2})[\varphi_4' + \varphi_4'^*], \tag{A78}$$

where $j=1, 2, 3$. Regarding G as a function of ψ_i', we find that equalities identical with (A78) hold between $\psi_1, \cdots \psi_{11}$ [which are defined by (A71)] and f_{4j}, g_{4j}, g_{4j}^*, φ_4 and φ_4^* [which are defined by (A76)]. Therefore, (A77) implies the validity of (A72). Furthermore, we notice that comparison between (A39) and (A34) shows that (A69) is satisfied.

Theorem 3, thus, becomes a special case of Lemma 2 provided one can show that the δH given by (A68) is, indeed, equal to (A38).

By using (A70) and (A74), the symmetric matrix A is found to be

$$A = \begin{bmatrix} 0 & 0 & 0 & 0 & 0 & 0 & 0 & 0 & 0 & R_1 & I_1 \\ & 0 & 0 & 0 & 0 & 0 & 0 & 0 & 0 & R_2 & I_2 \\ & & 0 & 0 & 0 & 0 & 0 & 0 & 0 & R_3 & I_3 \\ & & & 0 & 0 & 0 & 0 & 0 & 0 & 0 & -(ea_1/m) \\ & & & & 0 & 0 & 0 & 0 & 0 & 0 & -(ea_2/m) \\ & & & & & 0 & 0 & 0 & 0 & 0 & -(ea_3/m) \\ & & & & & & 0 & 0 & 0 & (ea_1/m) & 0 \\ & & & & & & & 0 & 0 & (ea_2/m) & 0 \\ & & & & & & & & 0 & (ea_3/m) & 0 \\ & & & & & & & & & (e^2\mathbf{a}\cdot\mathbf{a}/m^2) & 0 \\ & & & & & & & & & & (e^2\mathbf{a}\cdot\mathbf{a}/m^2) \end{bmatrix}, \tag{A79}$$

where

$$R_j = (e\kappa/m)(1/\sqrt{2})(\varphi_j^* + \varphi_j),$$

and

$$I_j = i(e\kappa/m)(1/\sqrt{2})(\varphi_j^* - \varphi_j) \quad (j=1, 2, 3).$$

Utilizing the identity,

$$\mathrm{trace}[\ln(1+A)] = \ln(\det|1+A|),$$

one finds that (A38) is true. Theorem 3 is, therefore, proved.

APPENDIX D. DERIVATIONS OF FEYNMAN RULES IN ξ-LIMITING FORMALISM

In the ξ-limiting formalism the interaction Lagrangian contains only a single time derivative of the electromagnetic field. Therefore, the results given in Figs. 2 and 3 can be directly obtained by using Lemma 1 and setting the matrix $A=0$.

APPENDIX E. REMARKS ON THE ORIGIN OF $\delta^4(0)$ TERM IN FIGURE 1

It is clear by comparing Figs. 1 and 2 (or Figs. 1 and 3) that for a given process, to the lowest order in e, the

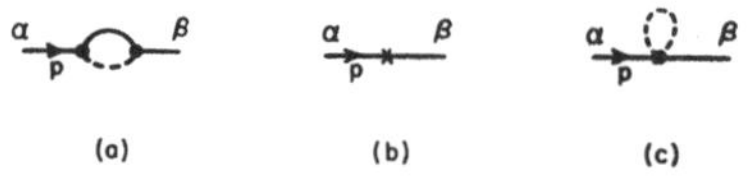

FIG. 6. Feynman diagrams for self-energy of mesons (discussed in Appendix E).

Feynman diagram does not contain any closed loops and therefore has the same value in the ξ-limiting process when $\xi \to 0$ as in the usual canonical formalism.

For higher order Feynman diagrams, the additional vertices of Fig. 1 carrying factors $\delta^4(0)$ which are infinite must be included. They are explicitly noncovariant under Lorentz transformations. The origin of these noncovariant terms is the nonidentical treatment of the space and time components of the meson field in the usual canonical formalism, as we now illustrate in the following example.

Consider the self-energy of a meson to the order e^2. There are three Feynman diagrams that contribute, as illustrated in Fig. 6 where the cross in (b) stands for the additional vertices of Fig. 1. The most divergent terms come from (a) and (b). They are, respectively, per unit volume,

$$A_a = \frac{i\kappa^2 e^2}{(2\pi)^4 m^2} \varphi_\beta^* \varphi_\alpha \int \frac{k^2\delta_{\alpha\beta} - k_\alpha k_\beta}{(p+k)^2 + m^2} d^3k dk_0, \quad \text{(A80)}$$

and

$$A_b = -\frac{i\kappa^2 e^2 \delta^4(0)}{m^2} \boldsymbol{\varphi}^* \cdot \boldsymbol{\varphi}. \quad \text{(A81)}$$

We can make the same calculation using the ξ-limiting formalism. Only diagrams (a) and (c) contribute, and the most divergent term comes from (a): This most divergent term can be written as

$$B_a = A_a - \frac{i\kappa^2 e^2}{(2\pi)^4 m^2} \varphi_\beta^* \varphi_\alpha \int \Delta d^3k dk_0, \quad \text{(A82)}$$

where

$$\Delta = \frac{k^2\delta_{\alpha\beta} - k_\alpha k_\beta}{(p+k)^2 + m^2} \frac{k^2 + m^2}{k^2 + \xi^{-1} m^2}, \quad \text{(A83)}$$

which is covariant.

Let us now evaluate the integral in (A82) by first integrating over k_0, then making $\xi \to 0$. Now,

$$\Delta \sim -\mathbf{k}^2/k_0^2 \quad \text{as} \quad k_0 \to \infty \quad \text{for} \quad \alpha = \beta = 4$$

and

$$\Delta \sim 1 \quad \text{as} \quad k_0 \to \infty \quad \text{for} \quad \alpha = \beta = 1, 2, 3.$$

Thus, as $\xi \to 0$

$$\int \Delta dk_0 \sim \int dk_0 \quad \text{for} \quad \alpha = \beta = 1, 2, 3$$

$$\sim 0 \quad \text{otherwise.}$$

It is clear that if we evaluate the integral in (A82) by first integrating over k_0, then taking $\xi \to 0$, then integrating over $\mathbf{k}$, we obtain

$$\text{(A82)} = \text{(A80)} + \text{(A81)}.$$

This example illustrates the fact that the ξ-limiting formalism is an explicitly covariant method which is more convenient than the canonical formalism.

[[62j]]
Commentary begins page 54

Concept of Off-Diagonal Long-Range Order and the Quantum Phases of Liquid He and of Superconductors

C. N. Yang
Institute for Advanced Study, Princeton, New Jersey

I. INTRODUCTION

We consider a many-particle system with fixed number of particles, with a density matrix ρ. We define the reduced density matrices $\rho_1, \rho_2, \cdots$ by

$$\mathrm{Sp}\,\rho = 1\,, \qquad (1)$$

$$\langle j|\rho_1|i\rangle = \mathrm{Sp}\; a_j\rho a_i^\dagger$$

$$\langle kl|\rho_2|ij\rangle = \mathrm{Sp}\; a_k a_l \rho a_j^\dagger a_i^\dagger \qquad (2)$$

etc.,

where $i, j, \cdots$ represent single particle states and a_i, a_j the annihilation operators for these states. In all our discussions, unless explicitly stated otherwise, we consider a collection of identical particles, either fermions or bosons.

1. This paper is concerned with the concept that in a many-body system of bosons or fermions, it is possible to have an off-diagonal long-range order (ODLRO) of the reduced density matrices in the coordinate space representation. The onset of such an order leads to a new thermodynamic phase of the system. It is reasonable to assume that superfluid He II and the superconductors are phases characterized by the existence of such an order.

2. The general characteristics of the gaseous, the liquid, and the solid phases are well known and are describable in classical mechanical terms. In particular, the solid phase is characterized by the existence of a long-range correlation. However, the long-range correlation in the solid is exhibited in quantum mechanics in the diagonal element of ρ_2 in coordinate space and is quite *different* from the off-diagonal-long-range-order that we shall discuss in this paper. Since off-diagonal elements have no classical analog, the off-diagonal long-range order discussed in this paper is a quantum phenomenon not describable in classical mechanical terms.

3. The long-range correlation in a solid is the basis of essentially all approximate calculations of the properties of a solid. If ODLRO is the characteristic of the phases He II and superconductors, it seems that a reasonable calculation of their properties can only be made with ODLRO explicitly built into the physical picture.

4. We shall show that the existence of ODLRO in ρ_n implies its existence in reduced density matrices ρ_m with $m > n$. [In fact for $m \geqq 2n$, the ODLRO occurs in a more intensified form.] The *smallest* n for which ODLRO occurs gives the collection of n particles that, in a sense, forms a basic group [hereafter called the basic group] exhibiting the long-range correlation. Of course, the system of particles that we consider may be a collection of particles of different kinds, such as nuclei and electrons. We shall give reasons to believe that the basic group must be composed of bosons and an *even* number of fermions. The phenomena of ODLRO is therefore fundamentally related to that of Bose-Einstein condensation. Or, more precisely, Bose-Einstein condensation is the simplest form of an ODLRO.

5. For a system of bosons the possible existence of ODLRO in ρ_1 was discussed in a paper by Penrose[1] and later in a paper by Penrose and Onsager.[2]

For the fermions, the ideas discussed in this paper are clearly related to the ideas of "long-range order of the average momentum," "macroscopic quantum state," etc., of London.[3] They are also clearly related to the ideas based on quasi-boson condensation in the papers of Schafroth, Butler, and Blatt.[4] Furthermore, since the wave functions assumed by Bardeen, Cooper, and Schrieffer[5] and by Bogoliubov[6] (as an *ansatz*) do have the ODLRO, the contents of

[1] O. Penrose, Phil. Mag. **42**, 1373 (1951).
[2] O. Penrose and L. Onsager, Phys. Rev. **104**, 576 (1956).
[3] F. London, *Superfluids* (John Wiley & Sons, Inc., Vol. 1, 1950, Vol. 2, 1954).
[4] M. R. Schafroth, Phys. Rev. **96**, 1442 (1954); M. R. Schafroth, S. T. Butler, and J. M. Blatt, Helv. Phys. Acta **30**, 93 (1957).
[5] J. Bardeen, L. N. Cooper, and J. R. Schrieffer, Phys. Rev. **108**, 1175 (1957).
[6] N. N. Bogoliubov, Nuovo cimento **7**, 794 (1958).

□ Reprinted from *Reviews of Modern Physics* 34, 4 (October 1962), 694–704.

the present paper are clearly also related to their work. However, it seems to us that in none of the previous works has the question of the detailed mathematical characterization of the superconducting state been raised. [Within the context of this question, the pairing idea of Bardeen, Cooper, and Schrieffer seems to be the closest in its implications to the ideas discussed in this paper.] Nor has there been an explicit understanding of exactly in what sense are the characterization of superfluidity and that of superconductivity similar, a similarity that London had emphasized.

6. It will be shown in Sec. IV that the existence of ODLRO gives rise to the phenomena of quantized magnetic flux. Furthermore, for cases where the basic group is two electrons, the unit of magnetic flux is $hc/2e$, as it was experimentally found.[7]

7. There is no discussion in this paper of the properties of the Hamiltonian that is needed to ensure the existence of ODLRO at low temperatures.

In the solid phase the existence of long-range correlation makes it necessary to introduce additional macroscopic variables, (namely, the strain) to describe the thermodynamics of the system. It is important to recognize that similarly the onset of ODLRO necessitates the introduction of additional macroscopic variables. What these variables are, however, is not discussed in this paper, except for a speculation about the fraction of superfluid in Sec. 10 and one about the penetration depth in Sec. 40.

II. PROPERTIES OF ρ_n

8. The reduced density matrices of Sec. 1 have the following properties:

$$\rho_n = \text{positive definite or semidefinite}\,, \tag{3}$$

$$\begin{aligned} \text{Sp}\,\rho_1 &= N\,, \\ \text{Sp}\,\rho_2 &= N(N-1)\,, \\ \text{Sp}\,\rho_3 &= N(N-1)(N-2)\,, \quad \text{etc.}\,, \end{aligned} \tag{4}$$

where N = total number of particles.

It is obvious that if we perform a unitary transformation on the operators a_i, the reduced density matrices ρ_n undergo a similar transformation. In fact, the transformation from, e.g., the coordinate to the momentum space representation of the ρ's follows the same law as the usual operators.

The following formulas are easy to prove:

$$\begin{aligned} \textstyle\sum_i \langle ij|\rho_2|ik\rangle &= (N-1)\langle j|\rho_1|k\rangle\,, \\ \textstyle\sum_i \langle ijk|\rho_3|ilm\rangle &= (N-2)\langle jk|\rho_2|lm\rangle\,, \quad \text{etc.} \end{aligned} \tag{5}$$

[7] B. S. Deaver and W. M. Fairbank, Phys. Rev. Letters **7**, 43 (1961); R. Doll and M. Näbauer, *ibid.* **7**, 51 (1961).

In defining the reduced density matrix ρ_2 in (2), we allow the indices i and j to run freely over all states. Clearly, there is a symmetry or antisymmetry when we switch i and j. There is, of course, a natural way to reject the superfluous elements of $\rho_2, \rho_3 \cdots$ due to these symmetries by considering $\rho_2, \rho_3, \cdots$ to operate only on states of the correct symmetry. A whole mathematical formalism can be neatly worked out for this process. We shall, however, not go into it, as it does not really add to the clarity of the physics of the problem.

9. We define λ_n as the largest eigenvalue of ρ_n. From (3) and (4) it is obvious that all eigenvalues of ρ_n are $\geqq 0$, and,

$$\begin{aligned} \lambda_1 &\leqq N\,, \\ \lambda_2 &\leqq N(N-1)\,, \\ \lambda_3 &\leqq N(N-1)(N-2)\,, \\ &\text{etc.} \end{aligned} \tag{6}$$

10. *Theorem 1.* $\lambda_2 \geqq \lambda_1^2 - \lambda_1$ for a system of bosons.

Proof. Let f_i be the normalized eigenvector for $\langle i'|\rho_1|i\rangle$ with eigenvalue λ_1.

$$\text{Define} \quad F = \textstyle\sum f_i^* a_i\,.$$

$$\text{Then} \quad \text{Sp}\,F^\dagger F\rho = \lambda_1\,.$$

Use $f_i f_j$ as a trial wave function for $\langle i'j'|\rho_2|ij\rangle$. Clearly,

$$\lambda_2 \geqq \text{trial expectation value of } \rho_2 = \text{Sp}\,F^\dagger F^\dagger FF\rho\,. \tag{7}$$

But $\quad F^\dagger F = FF^\dagger - 1\,.$

Thus $\quad \lambda_2 \geqq \text{Sp}\,F^\dagger FF^\dagger F\rho - \lambda_1\,.$

Using $\quad 0 \leqq \text{Sp}\,(F^\dagger F - \lambda_1)^2\rho = \text{Sp}\,F^\dagger FF^\dagger F\rho - \lambda_1^2 \quad (8)$

we obtain $\quad \lambda_2 \geqq \lambda_1^2 - \lambda_1\,, \quad$ Q.E.D.

Theorem 2. $\lambda_3 \geqq \lambda_1^3 - 2\lambda_1^2 - \lambda_2$ for a system of bosons.

Proof. Use the same notation as the proof of Theorem 1.

$$\begin{aligned} \lambda_3 &\geqq \text{Sp}\,F^\dagger F^\dagger F^\dagger FFF\rho \\ &= \text{Sp}\,F^\dagger F^\dagger(FF^\dagger - 1)FF\rho \\ &\geqq \text{Sp}\,F^\dagger F^\dagger FF^\dagger FF\rho - \lambda_2\,. \end{aligned}$$

$$\begin{aligned} \text{But,} \quad 0 &\leqq \text{Sp}\,F^\dagger(F^\dagger F - \lambda_1)^2 F\rho = \text{Sp}\,F^\dagger F^\dagger FF^\dagger FF\rho \\ &\quad - 2\lambda_1\,\text{Sp}\,F^\dagger F^\dagger FF\rho + \lambda_1^3 = \text{Sp}\,F^\dagger F^\dagger FF^\dagger FF\rho \\ &\quad - 2\lambda_1\,\text{Sp}\,F^\dagger FF^\dagger F\rho + 2\lambda_1\,\text{Sp}\,F^\dagger F\rho + \lambda_1^3 \end{aligned}$$

By (8), therefore,

$$\text{Sp}\, F^\dagger F^\dagger F F^\dagger F F \rho \geq 2\lambda_1 \,\text{Sp}\, F^\dagger F F^\dagger F \rho - 2\lambda_1^2 - \lambda_1^3$$
$$\geqq 2\lambda_1^3 - 2\lambda_1^2 - \lambda_1^3 = \lambda_1^3 - 2\lambda_1^2 .$$

Thus $\lambda_3 \geqq \lambda_1^3 - 2\lambda_1^2 - \lambda_2$, Q.E.D.

Theorem 3. $\lambda_4 \geqq \lambda_2^2 - 4\lambda_3 - 2\lambda_2$ for a system of bosons.

Proof. Let f_{12} be the normalized eigenvector of $\langle 1'2'|\rho_2|12\rangle$ with eigenvalue λ_2. Define

$$F = \textstyle\sum_{1,2} f_{12}^* a_1 a_2 .$$

The proof follows essentially the same lines as that of Theorem 1. In place of (8) one uses

$$0 \leqq \text{Sp}\,(F^\dagger F - \lambda_2)^2 \rho = \text{Sp}\,(F^\dagger F)^2 \rho - \lambda_2^2 .$$

In place of (7) one uses

$$\lambda_4 \geqq \text{Sp}\, F^\dagger F^\dagger F F \rho ,$$

obtained by taking a product trial wave function for ρ_4. By carrying out the detailed computation of

$$F^\dagger [F^\dagger F - F F^\dagger] F$$

one easily obtains

$$\lambda_4 \geqq \lambda_2^2 - 4\lambda_3 - 2\lambda_2 , \quad \text{Q.E.D.}$$

Theorem 4. $\lambda_4 \geqq \lambda_2^2 - 2\lambda_2$ for a system of fermions. This theorem can be proved in the same way as Theorem 3.

Notice that for fermions if we follow the reasoning that led to Theorem 1, we do not obtain any useful results.

Theorem 1 has been stated before[8] without proof. It is clear that theorems establishing lower bounds for $\lambda_5, \lambda_6, \cdots$ can be obtained in a similar fashion. These theorems presumably will show that if ρ_n has[9] an eigenvalue of the order of N, $\rho_{n+1}, \rho_{n+2}, \cdots$ have also large eigenvalues. Furthermore for the boson case, if λ_1 is of the order of N, $(\lambda_2)^{1/2}$, $(\lambda_3)^{1/3} \cdots$ form a monotonically increasing series $\leqq N$. Thus they approach a limit and it is tempting[8] to identify this limit with the size of the superfluid component of the system, if ρ is the density matrix for thermal equilibrium.

11. *Theorem 5.* For fermions, $\lambda_1 \leqq 1$.

Proof. This follows from the fact that the expectation value of $a_1^\dagger a_1$ where a_1 is the annihilation operator for any states is $\leqq 1$. Q.E.D.

Theorem 6.

$$\lambda_2 \leqq N(M - N + 2)/M \tag{9}$$

for a system of N fermions in M states. We assume both M and N to be even.

This theorem is proved in Appendix A. The proof will also show that the upper limit for λ_2 given by (9) can be reached, and can be reached in essentially only one way.

Notice that for any value of M, for fermions,

$$\lambda_2 \leqq N .$$

Theorems 5 and 6 suggest the following generalization:

Conjecture. There exists numerical constants $\beta_3, \beta_4, \cdots$ so that

$$\lambda_n \leqq (N)^{n/2} \beta_n \quad \text{for} \quad n = \text{even} ,$$
$$\lambda_n \leqq (N)^{(n-1)/2} \beta_n \quad \text{for} \quad n = \text{odd} ,$$

for a system of identical fermions.

These theorems demonstrate that large eigenvalues in the reduced density matrices for fermions essentially originate from pairs of fermions forming Bose-Einstein degeneracy.

12. *Theorem 7.* For a system of N_b bosons and N_f fermions, consider

$$\langle b'f'|\rho_2|bf\rangle , \tag{10}$$

where b and f label boson and fermion states, respectively. Its largest eigenvalue is $\leqq 1 + \lambda_1$ where λ_1 is the largest eigenvalue of $\langle b'|\rho_1|b\rangle$.

Proof. Consider the normalized eigenfunction f_{bf} for (10) with the largest eigenvalue λ. By a unitary transformation on the states of the fermion and one on the states of the boson, f_{bf} can be reduced to a paired form:

$$F \equiv \textstyle\sum_{b,f} f_{bf} a_b \alpha_f = \xi_1 a_1 \alpha_1 + \xi_2 a_2 \alpha_2 + \cdots , \tag{11}$$

where α are the annihilation operators for the fermion states and a those of the boson states. In (11) all ξ's are $\geqq 0$. Now

$$F^\dagger F + F F^\dagger = \xi_1^2 (a_1^\dagger a_1 + \alpha_1 \alpha_1^\dagger) + \xi_2^2 (a_2^\dagger a_2 + \alpha_2 \alpha_2^\dagger) + \cdots .$$

But $\lambda = \text{Sp}\, F^\dagger F \rho \leqq \text{Sp}\,(F^\dagger F + F F^\dagger)\rho$.

Furthermore, $\text{Sp}\, \alpha_i \alpha_i^\dagger \rho \leqq 1$,

$$\textstyle\sum \xi_i^2 = 1 ,$$
$$\text{Sp}\,(a_i^\dagger a_i)\rho \leqq \lambda_1 .$$

Thus $\lambda \leqq \lambda_1 + 1$, Q.E.D.

---

[8] C. N. Yang, Physica **26**, S49 (1960).

[9] We use the term "of the order of N" to apply loosely to a quantity $\geqq \alpha N$ where α is a fixed number independent of N.

This theorem suggests the following generalization:

Conjecture. In a mixture of particles, consider

$$\langle a'b',c'\cdots|\rho_n|a,b,c,\cdots\rangle \tag{12}$$

where a, b, c are states of bosons or fermions. If the collection of particles in $a, b, c,\cdots >$ contains an odd number of fermions, then the largest eigenvalue of (12) is $\leqq$ a function, independent of N, of the largest eigenvalues of the reduced density matrices

$$\langle a''',b''',\cdots|\rho_m|a'',b'',c'',\cdots\rangle\,,$$

where a'', b'', c'',$\cdots$ is a subgroup of particles in (12).

It follows from this conjectured theorem that if all lower order ρ's have no eigenvalues as large as of the order of N, then (12) also does not have such a large eigenvalue, if $a, b, c,\cdots\rangle$ contains an odd number of fermions. In the next chapter we shall demonstrate the equivalence of the existence for ρ_n of large eigenvalues and that of an off-diagonal long-range order. The above conjecture forms, then, the basis of the discussion in Sec. 4 about the basic group.

III. OFF-DIAGONAL LONG-RANGE ORDER (ODLRO)

13. We shall now discuss the equivalence of the existence for ρ_n of eigenvalues of the order of N and that of an off-diagonal long-range order. To illustrate the concept consider a system of N free fermions or bosons in a periodic box of volume Ω in thermal equilibrium. ρ commutes with the total momentum. Therefore in the momentum representation ρ_1 is diagonal:

$$\langle \mathbf{p}'|\rho_1|\mathbf{p}\rangle = \delta_{\mathbf{pp}'}n_{\mathbf{p}}\,,$$

where the diagonal element $n_{\mathbf{p}}$ is the average occupation number of the single particle state $\mathbf{p}$.

$$n = \frac{\mu e^{-p^2/T}}{1 \pm \mu e^{-p^2/T}}\,. \tag{13}$$

In coordinate representation

$$\langle \mathbf{x}'|\rho_1|\mathbf{x}\rangle = (1/\Omega)\sum n_{\mathbf{p}} \exp i\mathbf{p}(\mathbf{x}'-\mathbf{x}) = g(\mathbf{x}'-\mathbf{x})\,. \tag{14}$$

For fermions, or for free bosons at high temperatures, all n are finite and

$$\langle \mathbf{x}'|\rho_1|\mathbf{x}\rangle \to 0 \quad \text{as} \quad |\mathbf{x}-\mathbf{x}'| \to \infty\,. \tag{15}$$

But for free bosons below the Bose-Einstein transition temperature, $n_0 = N\alpha$, where α is a finite fraction. Therefore,

$$\langle \mathbf{x}'|\rho_1|\mathbf{x}\rangle \to N\alpha/\Omega \quad \text{as} \quad |\mathbf{x}-\mathbf{x}'| \to \infty\,. \tag{16}$$

The existence of a Bose-Einstein condensation is thus characterized by the nonvanishing behavior of $\langle \mathbf{x}'|\rho_1|\mathbf{x}\rangle$ as $|\mathbf{x}-\mathbf{x}'| \to \infty$.

14. If the condensation is in a state with $\mathbf{p} \neq 0$, it is clear that

$$\langle \mathbf{x}'|\rho_1|\mathbf{x}\rangle \to (N\alpha/\Omega) \exp i\mathbf{p}(\mathbf{x}'-\mathbf{x})\,, \quad \text{as} \quad |\mathbf{x}'-\mathbf{x}| \to \infty\,. \tag{17}$$

It seems that the general criterion for Bose-Einstein condensation is

$$\int_\Omega \langle \mathbf{x}'|\rho_1|\mathbf{x}\rangle d\mathbf{x}\langle \mathbf{x}|\rho_1|\mathbf{x}'\rangle = \text{order of } \Omega\,. \tag{18}$$

15. We now consider the case of N particles with any boundary condition and any density matrix ρ. By taking the trial wave function $\psi = 1/(\Omega)^{1/2}$ it is obvious that (16) implies the existence of a large eigenvalue of the order of $N\alpha$ for ρ_1. Conversely, if ρ_1 has a large eigenvalue $N\alpha$ with an eigenfunction $\phi(\mathbf{x})$, we can make a spectral resolution of ρ_1:

$$\langle x'|\rho_1|x\rangle = N\alpha\phi(\mathbf{x}')\phi^*(\mathbf{x}) + \rho_1' \tag{19}$$

where ρ_1' is a positive operator. It is reasonable to assume $\phi(\mathbf{x})$ to contain the normalization factor $1/(\Omega)^{1/2}$. Equation (19) shows that

$$\langle \mathbf{x}'|\rho_1|x\rangle \nrightarrow 0 \quad \text{as} \quad |\mathbf{x}-\mathbf{x}'| \to \infty\,. \tag{20}$$

16. We shall take (20) or (18) as the definition of the existence of an off-diagonal long-range order (ODLRO) in ρ_1. Its existence is equivalent to that of the existence of a large eigenvalue for ρ_1 of the order of N.

Proposition 1. The phase He II of liquid He is characterized by the existence of ODLRO in ρ_1 for the equilibrium density matrix of the interacting He atoms. This ODLRO is defined either by (20), or equivalently, by the condition that ρ_1 has an eigenvalue of the order of N.

17. For a system of bosons, if ODLRO exists in ρ_1, $\langle \mathbf{x}'|\rho_1|\mathbf{x}\rangle$ remains, in general, nonvanishing for all values of $\mathbf{x}$ and $\mathbf{x}'$. What is the characteristic of $\langle \mathbf{x}_1'\mathbf{x}_2'|\rho_2|\mathbf{x}_1\mathbf{x}_2\rangle$ in such a case? It is clear that a large contribution to ρ_2 comes from

$$\langle \mathbf{x}_1'|\rho_1|\mathbf{x}_1\rangle\langle \mathbf{x}_2'|\rho_1|\mathbf{x}_2\rangle$$

and therefore, in general, ρ_2 remains nonvanishing for all values of $\mathbf{x}_1$, $\mathbf{x}_2$, $\mathbf{x}_1'$, and $\mathbf{x}_2'$.

The above statement is well illustrated by the example of the free Bose gas in equilibrium. In that case, it is simplest to treat the equilibrium grand canonical ensemble. It is clear that

$$\langle \mathbf{p}_1'\mathbf{p}_2'|\rho_2|\mathbf{p}_1\mathbf{p}_2\rangle = \delta_{\mathbf{p}_1\mathbf{p}_1'}\delta_{\mathbf{p}_2\mathbf{p}_2'}n_{\mathbf{p}_1}n_{\mathbf{p}_2} + \delta_{p_1p_2'}\delta_{p_2p_1'}n_{\mathbf{p}_1}n_{\mathbf{p}_2} + \delta_{p_1p_1'}\delta_{p_2p_2'}\delta_{p_1p_2}m_{p_1} , \quad (21)$$

where $m_\mathbf{p}$ is the average of $M^2 - M - 2n_\mathbf{p}^2$ for the state $\mathbf{p}$, M being the occupation number of the state. A simple calculation shows that $m_p = 0$ for the equilibrium distribution. Thus

$$\langle \mathbf{x}_1'\mathbf{x}_2'|\rho_2|\mathbf{x}_1\mathbf{x}_2\rangle = g(\mathbf{x}_1' - \mathbf{x}_1)g(\mathbf{x}_2' - \mathbf{x}_2) + g(\mathbf{x}_1' - \mathbf{x}_2)g(\mathbf{x}_2' - \mathbf{x}_1) . \quad (22)$$

Therefore, for a free Bose gas, for the grand canonical ensemble,

with ODLRO in ρ_1, $\rho_2 \neq 0$ for all x_1, x_1', x_2, x_2' ; (23)

without ODLRO in ρ_1, $\rho_2 \approx 0$ except in the neighborhood of

(a) $\mathbf{x}_1 = \mathbf{x}_1'$, $\mathbf{x}_2 = \mathbf{x}_2'$;

and (b) $\mathbf{x}_1 = \mathbf{x}_2'$, $\mathbf{x}_2 = \mathbf{x}_1'$. (24)

18. The two cases (23) and (24) are also characterized by the fact that the largest eigenvalue of ρ_2 is of the order of N^2 and is finite, respectively. What happens if it is of the order of N? This case cannot obtain for a *free* Bose gas in equilibrium, but may obtain for other systems. Following the argument in Sec. 15 for the ODLRO in ρ_1 we make a spectral separation of ρ_2, separating the largest eigenvalue

$$\langle \mathbf{x}_1'\mathbf{x}_2'|\rho_2|\mathbf{x}_1\mathbf{x}_2/ = N\alpha\varphi(\mathbf{x}_1'\mathbf{x}_2')\phi^*(\mathbf{x}_1\mathbf{x}_2) + \rho_2' , \quad (25)$$

where ρ_2' is positive. The eigenfunction $\phi(\mathbf{x}_1,\mathbf{x}_2)$, one can expect, is zero for large separations $|\mathbf{x}_1 - \mathbf{x}_2|$ and is $\sim 1/(\Omega)^{1/2}$ for microscopic separations for $\mathbf{x}_1$ and $\mathbf{x}_2$. Thus, we have a type of behavior intermediate between (23) and (24):

With ODLRO in ρ_2, but not in ρ_1:
$\rho_2 \approx 0$ except in the neighborhood of

(a) $\mathbf{x}_1 = \mathbf{x}_1'$, $\mathbf{x}_2 = \mathbf{x}_2'$;

(b) $\mathbf{x}_1 = \mathbf{x}_2'$, $\mathbf{x}_2 = \mathbf{x}_1'$;

and (c) $\mathbf{x}_1 = \mathbf{x}_2$, $\mathbf{x}_1' = \mathbf{x}_2'$ (but $|\mathbf{x}_1 - \mathbf{x}_2'|$ may be ∞) . (26)

It is clear that if (26) obtains, ρ_2 has an eigenvalue of the order of N.

19. For a system of fermions, Theorem 5 shows that ODLRO cannot obtain for ρ_1. Theorem 6 shows, however, that ρ_2 may have eigenvalues of the order of N. Thus, for fermions ODLRO may occur in ρ_2 in the sense of (26).

Two examples are illuminating in this connection. For a system of free fermions in equilibrium, all eigenvalues of ρ_2 are finite (i.e., not of the order of N). Thus, neither in ρ_1 nor ρ_2 is there ODLRO. For a system in which the pair occupation hypothesis of Bardeen, Cooper, and Schrieffer[5] (BCS) is legitimate, it is easy to show that ρ_2 has an eigenvalue of the order of N. In fact in the proof of Theorem 6, in order to find a system with a maximum eigenvalue for ρ_2, one is forced to have pair occupation of single particle states exactly in the manner of the BCS ansatz.

20. *Proposition* 2. The superconducting state is characterized by the existence of ODLRO in

$$\langle e_1'e_2'|\rho_2|e_1e_2\rangle , \quad (27)$$

where e_1, e_2, e_1', e_2' represent electron states, for the ensemble in thermal equilibrium. This ODLRO is defined either by (26), or equivalently, by the condition that (27) has an eigenvalue of the order of N, the number of electrons in the system.

21. Actually the two propositions above are more restrictive than they need be. Take the case of He. To describe liquid He as a collection of He atoms is only an approximation. A much better description is a collection of electrons and He nuclei. A general characterization of a new phase exhibiting ODLRO should apply both to liquid He as a collection of He atoms and to liquid He as a collection of electrons and He nuclei. It is evident that in the latter description ODLRO first occurs in

$$\langle \mathrm{He}',e_1',e_2'|\rho_3|\mathrm{He},e_1,e_2\rangle$$

because any reduced density matrix of lower order would mostly describe only the internal structure of the He atom.

It thus seems that in a macroscopic system, ODLRO can set in at ρ_n. The theorems of II indicate that the reduced density matrix, to be called ρ_m, of lowest order which has ODLRO must operate on a basic group that consists of an even number of fermions and any number of bosons. For liquid He II the basic group is the He atom; for superconductors, the basic group is a set of two electrons. For ρ_m the largest eigenvalue is of the order of N. It has an ODLRO in the sense that in coordinate representation, when the unprimed coordinates are microscopically close to a point $\mathbf{x}$, and the primed coordinates are microscopically close to another point $\mathbf{x}'$, with $\mathbf{x}$ and $\mathbf{x}'$ *macroscopically* apart, ρ_m remains nonvanishing. For fixed unprimed coordinates microscopically close together, the region of the primed coordinates where ρ_m remains nonvanishing is thus a "tube" with one 3-space dimension extending macroscopically. The volume of the region is

$$\Omega \times (\text{microscopic dimension})^{m-1}$$

For higher order reduced density matrices with particle groups containing one or more basic groups, the corresponding region would have one or more 3-space dimensions extending macroscopically.

Physically the concepts of ODLRO and of the basic group are therefore directly related to the *dimensionality* of the macroscopic regions in space where the matrix elements of ρ_1, $\rho_2, \cdots$ are not vanishingly small.

22. It is easy to believe that the onset of ODLRO in an equilibrium system would lead to a phase transition. Consider, for example, a system of Bose (or Fermi) particles in thermal equilibrium. The thermodynamical function of the system can be obtained[10] as the maximum of a functional of ρ_1. This variational principle also determines ρ_1. The functional is expressed as a series of terms each of which involves integrations over products of matrix elements of ρ_1. It has been shown[11] that if an eigenvalue of ρ_1 attains the order of N, the series contains progressively larger terms and a rearrangement is necessary. Such a rearrangement is, of course, what is required by every phase transition.

The formalism of reference 10 has been generalized by De Dominicis[12] to the case where ρ_2 is also explicitly used in the argument of the functional. It is not difficult to find the successive terms in his formalism that become progressively larger when ρ_2 has a large eigenvalue of the order of N. Thus there is to be expected also a phase transition when ODLRO first sets in in ρ_2.

23. The existence of ODLRO in ρ_1 [or ρ_2] implies the possible separation (19) [or (25)]. If there is only one large eigenvalue of the order of N, then ρ_1' [or ρ_2'] vanishes as $\mathbf{x}$ becomes far separated from $\mathbf{x}'$ [or as $\mathbf{x}_1$, $\mathbf{x}_2$ become far separated from $\mathbf{x}_1'$, $\mathbf{x}_2'$]. Thus, at large spatial separations ρ_1 [or ρ_2] assumes a product form. ρ_1 thus behaves, in some respects, like a single (double) particle system in a *pure state*. It is worth noticing that the hypothesis of a product form for ρ_2 underlies many[13] discussions on superconductivity.

24. For nonequilibrium systems the existence of ODLRO requires a reformulation of transport properties. However, it is doubtful that much real progress can be made without a first understanding of the microscopic basis of the additional macroscopic equilibrium variables required by ODLRO. [Cf. Sec. 7.]

25. It is obvious that the basic group may form a bound state, as in liquid He II; or it may not form a bound state, as in superconductors.

It is also evident from these examples that ODLRO may occur in a liquid, and it may also occur in a solid. But in a solid the basic group cannot contain particles that are localized, such as the nuclei.

26. In an insulator, the electrons, because of energy considerations, have no usable empty states. Effectively, in the notation of Theorem 6, $M = N$. Thus, by that theorem, ρ_2 cannot have an eigenvalue of the order of N and consequently it cannot have an ODLRO. Thus, an insulator cannot satisfy the characterization of a superconductor as given in Sec. 20.

IV. MAGNETIC FLUX QUANTIZATION

27. To discuss the question of magnetic flux quantization we recall that[14] for a superconducting ring P with a magnetic field in the hole O, but no magnetic field in P, the vector potential can be transformed away by a gauge transformation (See Fig. 1). The Schrödinger equation for the electrons in P is then the same as that for the case where there is no magnetic field in O, but the boundary condition is that the wave function changes by a phase factor $\exp\,[i(e/c\hbar)\Phi]$ every time an electron is brought around the ring. The symbol Φ stands for the total magnetic flux through O. The Schrödinger equation together with the boundary condition determine the energy levels, and consequently the free energy $N^{-1} \ln Q$ of the system. The arguments of reference 14 show that if

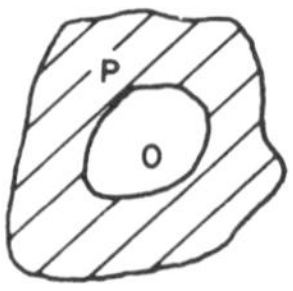

FIG. 1. Superconducting ring.

$$L^2 N^{-1} \ln Q \tag{28}$$

varies as Φ is changed, then the system would have magnetic flux quantization (because body currents would be generated if Φ is not quantized). The quantity L is the circumference of the ring.

[10] T. D. Lee and C. N. Yang, Phys. Rev. **117**, 22 (1960).
[11] T. D. Lee and C. N. Yang, Phys. Rev. **117**, 897 (1960).
[12] C. De Dominicis (to be published).
[13] See e.g., L. P. Gor'kov, J. Exptl. Theoret. Phys. U.S.S.R. **34**, 735 (1958) [translation: Soviet Phys.—JETP **7**, 505 (1958)].
[14] N. Byers and C. N. Yang, Phys. Rev. Letters **7**, 46 (1961.)

28. It is convenient for the present discussion to change slightly the geometry of the mathematical problem formulated above. Instead of a ring-shaped body P we consider a periodic box, i.e., a box of dimension $L \times L' \times L''$ with strict periodicity condition in the y and z directions:

$$\psi(y + L') = \psi(y)\,, \quad \psi(z + L'') = \psi(z)\,, \quad (29)$$

and periodicity with a phase factor in the x direction:

$$\psi(x + L) = \exp\,[i(e/c\hbar)\,\Phi]\psi(x)\,, \quad (30)$$

where Φ is a parameter. If the quantity (28) shows a variation with Φ, it is reasonable to assume that the same obtains for the ring geometry, and the physical system would show a quantization of flux, of a unit that is equal to the period in Φ of (28).

29. The periodicity conditions (29) and (30) assume a particularly simple form in the momentum representation: The lattice in momentum space is displaced from the origin by $e\Phi/cL$ in the x direction. The quantization of flux therefore depends on whether the free energy of the system changes with this displacement.

For a free Fermi gas it is not difficult to demonstrate that the free energy is independent of this displacement, as stated in reference 14. Thus, a free Fermi gas shows no magnetic flux quantization.

For a free Bose gas the same obtains for temperatures above the Bose-Einstein transition temperature. But below the Bose-Einstein transition temperature, the momentum state closest to the origin (in momentum space) is degenerate to a degree proportional to N and (28) varies with Φ quadratically for small Φ. Thus, a free Bose gas below the transition temperature should exhibit the phenomena of quantized flux. The period in Φ is clearly ch/e, which is therefore the unit of quantization.

30. While for a free particle system it is convenient to examine in momentum space the large occupation numbers, hence the large eigenvalues of ρ_1, for an interacting system of particles, it is convenient to examine the problem in coordinate space. Also, we shall use an equilibrium density matrix R with a different normalization from that of ρ as given in (1).

$$R \equiv \exp\,(-H/kT)\,. \quad (31)$$

Defining the normalization constant as Q, one has

$$R = Q\rho\,, \quad (32)$$

where $\quad Q = \mathrm{Sp}\,R = \text{partition function}\,. \quad (33)$

The contracted density matrices R_n are defined as

$$R_n = Q\rho_n\,. \quad [Q = \text{a number, not a matrix.}] \quad (34)$$

Through (4) and (34) we easily obtain

$$\begin{aligned} Q &= N^{-1}\,\mathrm{Sp}\,R_1\,, \\ Q &= [N(N-1)]^{-1}\,\mathrm{Sp}\,R_2\,, \quad (35) \\ &\text{etc.} \end{aligned}$$

Let us consider the matrix elements of R_1:

$$\langle \mathbf{x}'|R_1|\mathbf{x}\rangle\,.$$

The periodicity conditions (29) and (30) imply

$$\begin{aligned} \langle x' + L|R_1|x\rangle &= \langle x'|R_1|x - L\rangle \\ &= \exp\,[i(e/c\hbar)\Phi]\langle x'|R_1|x\rangle\,, \quad (36) \end{aligned}$$

$$\langle y' + L|R_1|y\rangle = \langle y'|R_1|y - L\rangle = \langle y'|R_1|y\rangle\,, \text{etc.} \quad (37)$$

31. We shall symbolically represent a 3-space $\mathbf{x}$ by one dimension. In what region of $\mathbf{x}$ and $\mathbf{x}'$ is R_1 nonvanishingly small, relative to its value near $\mathbf{x} = \mathbf{x}'$? The free Bose gas example of Secs. 13 and 14 shows the following:

(A) Without ODLRO, the region consists of narrow parallel strips running along

$$\begin{aligned} x &= x'\,, \\ x &= x' \pm L\,, \\ x &= x' \pm 2L\,, \quad (38) \\ &\text{etc.} \end{aligned}$$

(Cf. Fig. 2.) The values of R_1 between the strips

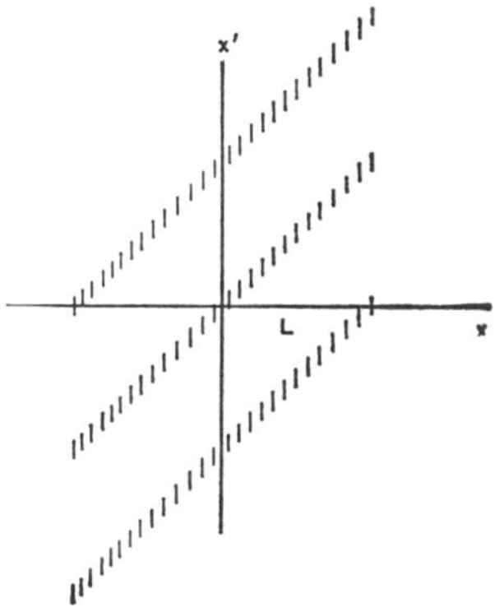

FIG. 2. Region of relatively nonvanishing values of R_1, in absence of ODLRO.

are vanishingly small. The values of R_1 in two neighboring strips, by (36), are different only by a phase factor $\exp\,[i(e/c\hbar)\Phi]$. The width of the strips is microscopic, but the distance between the strips is macroscopic.

(B) With ODLRO, the strips merge into each other, and R_1 is nonvanishing everywhere. The phase change by a factor exp $[i(e/c\hbar)\Phi]$ at distances L, however, remains.

The behavior of R_1 along a cut in the $\mathbf{x}\,\mathbf{x}'$ plane at $\mathbf{x} + \mathbf{x}' = 0$ is *schematically* illustrated in Fig. 3.

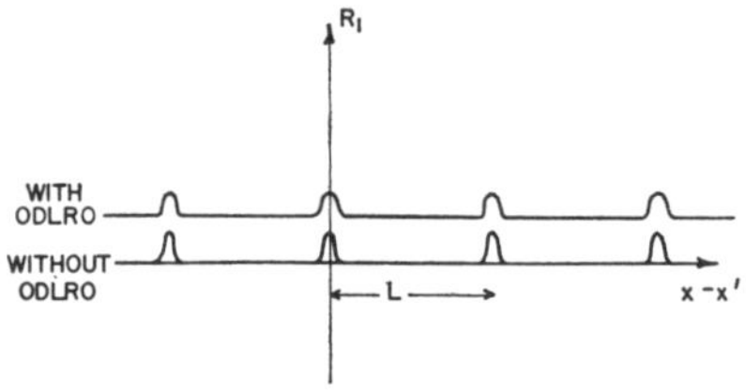

FIG. 3. Schematic plot of R_1 against $\mathbf{x} - \mathbf{x}'$. Notice that with $\Phi \neq 0$, R_1 is, in general, complex.

Notice that the merging of the strips occurs not as a consequence of the broadening of the strips, but as that of the sudden lifting of the value of R_1 between the strips.

32. The above discussion of the region of (relatively) nonvanishing values for R_1 for the cases with and without ODLRO is obviously valid for interacting particles as well.

33. When the parameter Φ is changed, it is clear from Fig. 3 that with ODLRO, the whole dependence of R_1 on $\mathbf{x} - \mathbf{x}'$ must change. Consequently, by (35) the partition function Q changes with Φ and quantization of flux follows.

If on the other hand ODLRO is not present in ρ_1, the different strips are separated from each other and the phase change at distances L can be effected by a simple multiplicative factor, as in the free Bose gas discussed in Sec. 31. Thus quantization of flux need not be present.

The difference in the behavior of R under changes in Φ for the cases with and without ODLRO is quite similar to the Bloch eigenvalue problem[15] in a periodic potential. Bloch showed that wave functions should be sought that changes by a phase factor $e^{i\Phi}$ for each lattice displacement. How does the wave function depend on Φ? If the wave function remains finite between lattice points, the energy value and the wave function would be dependent on Φ. If, however, the wave function becomes very small in a region between lattice points, caused by, e.g., a potential barrier, then the energy and the wave function would not be very much dependent on Φ. For an infinite potential barrier in between the atoms, the wave function vanishes in the barrier, and the energy levels would be independent of Φ while the wave function only picks up phase factors $e^{i\Phi}$ from one atom to the next.

The physical meaning of the effect of the presence of ODLRO on the phase condition (36) is that ODLRO preserves the memory of phases over macroscopic distances. Also in this sense, one can interpret for the Bloch problem, the effect of the small nonvanishing interatomic value of the wave function on the Φ dependence of the energy: The nonvanishing interatomic value of the wave function preserves the memory of phases from one atom to the next.

34. In the absence of an ODLRO in ρ_1 it becomes necessary to examine ρ_2. The region of relatively nonvanishing values of R_2 can be obtained from (24) if ODLRO is absent in ρ_2, and from (26) if ODLRO is present in ρ_2. To simplify matters we suppress as before y and z dimensions and consider the element

$$\langle x_1' x_2' | R_2 | x_1 x_2 \rangle$$

as a function of

$$\xi = x_1 - x_2 \,, \quad \eta = x_1' - x_2' \,, \quad \text{and} \quad \zeta = x_1 + x_2 - x_1' - x_2' \,. \tag{39}$$

Invariance under uniform displacement ensures that, for equilibrium, R_2 is independent of the fourth coordinate

$$x_1 + x_2 + x_1' + x_2' \,.$$

The periodicity conditions (29) and (30) imply

$$\begin{aligned} \langle x_1' + L, x_2' | R_2 | x_1, x_2 \rangle &= \langle x_1', x_2' + L | R_2 | x_1, x_2 \rangle \\ &= \langle x_1', x_2' | R_2 | x_1 - L, x_2 \rangle = \langle x_1', x_2' | R_2 | x_1, x_2 - L \rangle \\ &= \exp\,[i(e/c\hbar)\Phi] \langle x_1', x_2' | R_1 | x_1, x_2 \rangle \,. \end{aligned} \tag{40}$$

Repeated application of these conditions shows that in the (ξ,η,ζ) space a face-centered cubic lattice of displacements can be formed.

Displacements by ($n_1,n_2,n_3 = \pm$ integers, $n_1 + n_2 + n_3 =$ even)

$$(\xi,\eta,\zeta) \to (\xi + n_1 L, \eta + n_2 L, \zeta + n_3 L)$$

changes the value of R by a factor exp $[-\,i n_3 (e/c\hbar)\Phi]$. (41)

35. If ODLRO is not present in ρ_2, the region D of (relatively) nonvanishing values of R_2 is given by (24) in case $L = \infty$. It consists of the two lines

$$\xi = \eta \,, \quad \zeta = 0 \quad \text{and} \quad \xi = -\eta \,, \quad \zeta = 0 \,,$$
$$\text{and their microscopic neighborhoods} \,. \tag{42}$$

For finite L all displacements of (42) by the lattice

[15] F. Bloch, Z. Physik **52**, 555 (1928).

(41) should also be included in the region D. Thus, D consists of

$$\xi - \eta = m_1 L\,, \quad \zeta = m_3 L\,;$$
$$\text{and} \quad \xi + \eta = m_1 L\,, \quad \zeta = m_3 L\,;$$
and their microscopic neighborhoods,

$$(\text{where } m_1, m_3 = \pm \text{ integers}, m_1 + m_3 = \text{even})\,. \quad (43)$$

Geometrically, D consists of parallel plane square nets. Those in the even planes $\zeta = m_3 L$, $m_3 =$ even, are plotted in Fig. 4 in horizontal shading and those in the odd planes $\zeta = m_3 L$, $m_3 =$ odd, are plotted in vertical shading.

36. The values of R not on the nets are (relatively) vanishing. The dependence condition on Φ is contained in (41). Now (41) says (i) that nets in different planes should have a relative phase factor $\exp[i(\Delta m_3)(e/c\hbar)\Phi]$, and (ii) that the value of R on each net is periodic *with no phase factor* under the displacements $(\xi,\eta,\zeta) \to (\xi + n_1 L, \eta + n_2 L, \zeta)$ $n_1 + n_2 =$ even.

Since different nets are not connected to each other, (41) can be satisfied by a mere phase change from net to net, and (35) demonstrates that the free energy need not vary with Φ. Thus, there need not be a quantization of flux.

37. In case ODLRO is present in ρ_2 but not in ρ_1, the discussions of the last two sections have to be modified. The region D is now to be generated from (26). It consists of the nets (43) plus the rods

$$\xi = l_1 L\,, \quad \eta = l_2 L\,, \quad l_1, l_2 = \pm \text{ integers} \quad (44)$$
and their microscopic neighborhoods.

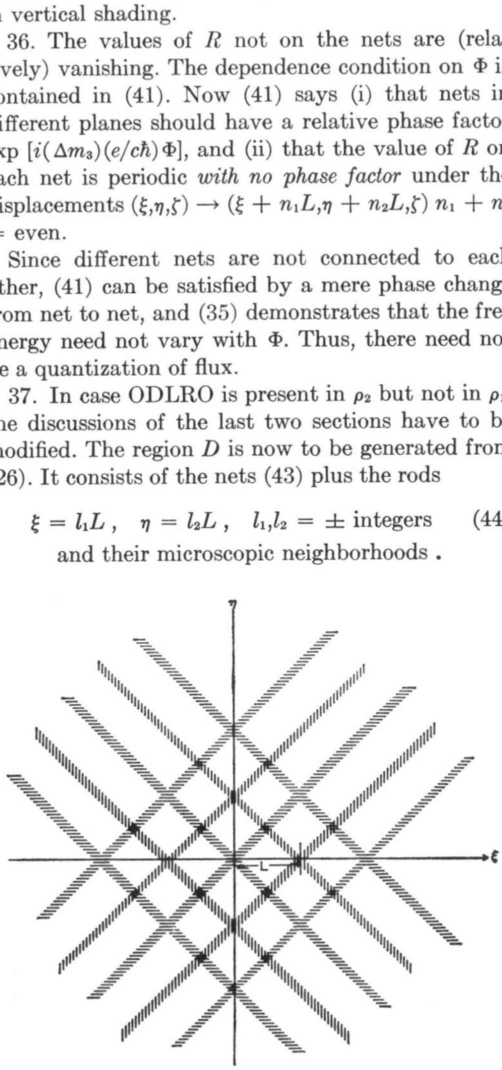

FIG. 4. Projection in $\zeta = 0$ plane of region of relatively nonvanishing values of R_2, in absence of ODLRO.

Now the rods (44) are perpendicular to the planes of the nets and connect all the nets in the even planes together, and also connect all the nets in the odd planes together. But nets in different even (odd) planes have phase-factor differences which are powers of

$$\exp 2i(e/c\hbar)\Phi\,. \quad (45)$$

Thus, R_2 changes when Φ varies, and we have the phenomena of flux quantization. The unit of quantization is, from (45),

$$ch/2e\,.$$

38. The discussions above can be generalized to the case where the basic group (Secs. 4 and 21) is of any size. If the sum of charges of the particles in the basic group is

$$\sum e \neq 0\,,$$

quantization of flux should take place with a unit of quantization $ch/(\sum e)$.

39. It is clear that the discussions above are in many respects similar to the discussions of reference 14 which was in terms of the BCS pairs. It is also similar to the discussions of Onsager in terms of a boson picture[16] and of Bardeen[17] in terms of the Ginsburg-Landau equation of a doubly charged single-particle system. Furthermore, the discussions of the present paper are based on a series of propositions and guesses. However, we believe these propositions and guesses, in fact, give the common general physical basis of the phenomena of a type of quantum phase in a many-body system.

40. We conclude this paper with a speculation. For superconductors, an important experimental quantity is the penetration depth, defined, for example, in London's book. Is it related to the function (28)? We have some arguments to indicate that it is. In fact,

$$\text{penetration depth} = \left[-4\pi L^2 \frac{\partial^2}{\Omega \partial \Theta^2} (kT \ln Q) \right]^{-1/2}$$
$$\text{at } \Phi = 0\,, \quad (46)$$

where Ω is the volume of the box.

ACKNOWLEDGMENTS

It is a pleasure to recall and to acknowledge the many discussions on various aspects of this paper that the author had in the last two or three years with F. Bloch, N. Byers, C. De Dominicis, B. Jacob-

[16] L. Onsager, Phys. Rev. Letters **7**, 50 (1961).
[17] J. Bardeen, Phys. Rev. Letters **7**, 162 (1961).

sohn, T. D. Lee, J. G. Valatin, and B. Zumino. He is much indebted to B. S. Deaver and W. M. Fairbank for informing him of their beautiful experiment and for many discussions. This paper was largely written while the author was a visitor at CERN. The financial support of CERN and of the Guggenheim Foundation is gratefully acknowledged.

APPENDIX A

To prove Theorem 6, the following lemma is useful:

Lemma. Let X be an antisymmetrical matrix

$$X = -\tilde{X} .$$

There exists a unitary matrix U so that $UX\tilde{U}$ is zero everywhere except for 2×2 diagonal blocks of the form

$$\begin{pmatrix} 0 & a \\ -a & 0 \end{pmatrix},$$

where all a's are real and positive.

Proof. Let ψ be a normalized eigenvector of $X^\dagger X$: ($\dagger$ = Hermitian conjugate)

$$X^\dagger X\psi = a^2\psi , \quad a > 0 \tag{A1}$$

Then
$$X^* X\psi = -a^2\psi . \tag{A2}$$

Define
$$\phi = -a^{-1}X^*\psi^* . \tag{A3}$$

By (A1)
$$X\phi = a\psi^* . \tag{A4}$$

By (A3)
$$X\psi = -a\phi^* . \tag{A5}$$

From (A4) and (A5),

$$\tilde{\psi}\phi^* = 0 ,$$

and
$$\tilde{\phi}\phi^* = \tilde{\psi}\psi^* = 1 .$$

Thus ψ and ϕ are orthogonal unit vectors. Taking them as the first two columns of $\tilde{U}$, one easily proves the lemma by induction.

Proof of Theorem 6. (a) To prove the theorem it is necessary and sufficient to prove it for the case that ρ is the density matrix of an N particle pure state Ψ.

(b) Consider any normalized antisymmetrical trial function f_{ij} for ρ_2: Let

$$F = \textstyle\sum_{ij} f_{ij}^* a_i a_j . \tag{A6}$$

Then the expectation value of ρ_2 is

$$\mathrm{Sp}\, F\rho F^\dagger = \Psi^\dagger F^\dagger F\Psi . \tag{A7}$$

Under a unitary transformation U on a_i, the matrix f_{ij} is transformed like the matrix X of the lemma. Thus, we can take f_{ij} to be of the diagonal 2×2 block form of the lemma. In other words, we can take without loss of generality,

$$F = \alpha_1(a_1a_2 - a_2a_1) + \alpha_2(a_3a_4 - a_4a_3) + \cdots , \tag{A8}$$

where
$$\alpha_i \geqq 0 \quad \text{and} \quad 2\textstyle\sum \alpha_i^2 = 1 , \tag{A9}$$

and Theorem 6 is equivalent to the assertion that

$$\Psi^\dagger F^\dagger F\Psi \le N(M - N + 2)/M . \tag{A10}$$

(c) Consider that Ψ and F which are of the form of (A8), and maximize $\Psi^\dagger F^\dagger F\Psi$. Not all α are 0. Without loss of generality we can assume

$$\alpha_1 > 0$$

We write

$$G = F - 2\alpha_1 a_1 a_2, K = \sum_{3,4,\ldots} a_i^\dagger a_i \tag{A11}$$

so that G and K only operate on states 3, 4, 5, $\cdots$. We write Ψ in the form of

$$\Psi = \begin{Vmatrix} \phi_{00} \\ \phi_{01} \\ \phi_{10} \\ \phi_{11} \end{Vmatrix} \tag{A12}$$

where the subscripts of ϕ represent the occupation numbers of states 1 and 2. In this notation

$$F = \begin{Vmatrix} G & & & 2\alpha_1 \\ & G & & \\ & & G & \\ & & & G \end{Vmatrix} . \tag{A13}$$

The condition on Ψ is

$$K\Psi = \begin{Vmatrix} N & & & \\ & N\text{-}1 & & \\ & & N\text{-}1 & \\ & & & N\text{-}2 \end{Vmatrix} \Psi . \tag{A14}$$

Thus both the operator $F^\dagger F$ and the condition on Ψ do not mix the subspace spanned by ϕ_{00} and ϕ_{11} with that spanned by ϕ_{01} and ϕ_{10}. Hence for maximum $\Psi^\dagger F^\dagger F\Psi$, either $\phi_{00} = \phi_{11} = 0$ or $\phi_{10} = \phi_{01} = 0$. But in the former case we can always increase $\Psi^\dagger F^\dagger F\Psi$ by proportionally increasing $\alpha_2, \alpha_3, \cdots$, and simultaneously decreasing α_1 to keep (A9) satisfied [so that G is proportionally increased]. Hence we can put

$$\phi_{10} = \phi_{01} = 0 \tag{A15}$$

which means that the states 1 and 2 are either both empty or both occupied. [Cf. Sec. 19.]

(d) Equations (A13), (A15), and Schwartz's inequality lead to

$$\begin{aligned}\Psi^\dagger F^\dagger F\Psi &= \phi_{00}^\dagger G^\dagger G\phi_{00} + 2\alpha_1[\phi_{00}^\dagger G^\dagger \phi_{11} + \text{c.c.}] \\ &\quad + 4\alpha_1^2\phi_{11}^\dagger\phi_{11} + \phi_{11}^\dagger G^\dagger G\phi_{11} \\ &\leqq \phi_{00}^\dagger G^\dagger G\phi_{00} + 4\alpha_1^2\phi_{11}^\dagger\phi_{11} + \phi_{11}^\dagger G^\dagger G\phi_{11} \\ &\quad + 4\alpha_1[(\phi_{11}^\dagger\phi_{11})(\phi_{00}^\dagger G^\dagger G\phi_{00})]^{1/2} .\end{aligned} \tag{A16}$$

(e) Let $B(M,N)$ be the maximum for $\Psi^\dagger F^\dagger F \Psi$. (A16) implies

$$\begin{aligned}\Psi^\dagger F^\dagger F \Psi \leqq\ & X^2\beta^2 B(M-2,N) + 4\alpha_1^2(1-X^2) \\ & + \beta^2(1-X^2)B(M-2,N-2) \\ & + 4\alpha_1(1-X^2)^{1/2}X\beta[B(M-2,N)]^{1/2}\,, \qquad \text{(A17)}\end{aligned}$$

where

$$\beta = (1-2\alpha_1^2)^{1/2} \geqq 0\,, \quad X = (\phi_{00}^\dagger \phi_{00})^{1/2} \geqq 0\,.$$

(f) We can now prove by induction that for even M and N

$$B(M,N) = N(M-N+2)/M \qquad \text{(A18)}$$

as follows: Substitute (A18) into the right-hand side of (A17) and maximize the resultant expression with respect to α_1 and X. After some straightforward algebra, one finds the only maximum of the right-hand side of (A17) at

$$X^2 = (M-N)/M\,, \quad \alpha_1^2 = 1/M\,,$$

where it assumes the value of $B(M,N)$ in (A18). The induction is then easily completed. Q.E.D.

It is clear from the above that the only maximum of $\Psi^\dagger F^\dagger F \Psi$ with F having the form (A8) is obtained when

$$\alpha_1 = \alpha_2 = \cdots = M^{-1/2}\,. \qquad \text{(A19)}$$

Furthermore, each pair of states (1,2), (3,4), $\cdots$ is never occupied singly. For such a problem it is easy to see that we can define $M/2$ sets of Pauli spin matrices so that

$$F = M^{-1/2} \sum\nolimits_{M/2} (\sigma_i^x + i\sigma_i^y)\,.$$

Thus, $$F^\dagger F = M^{-1}[(\sum \boldsymbol{\sigma})^2 - (\sum \sigma^z)^2 - 2\sum \sigma^z]\,. \qquad \text{(A20)}$$

The condition that the total number of particles is N is

$$\sum \tfrac{1}{2}(1-\sigma^z) = N/2\,. \qquad \text{(A21)}$$

Equations (A20) and (A21) show that there is only one largest eigenvalue for $F^\dagger F$ consistent with (A21).

[63e]
Commentary
begins
page 55

THE MASS FORMULA OF SU_3

C. N. Yang

INSTITUTE FOR ADVANCED STUDY, PRINCETON, NEW JERSEY

This paper consists of an expansion and elaboration of some points raised by R. J. Oakes and the author in a letter that has already appeared in the *Physical Review Letters*.

In the past few years there has been much discussion of the possibility of the existence of a higher symmetry SU_3. In particular, the energy levels of the resonant states $N_{3/2}{}^*$, $Y_1{}^*$, and $\Xi_{1/2}{}^*$ are experimentally found to be equidistant. A most interesting mass formula due to Gell-Mann and to Okubo has been derived from SU_3 symmetry, yielding precisely this equidistant rule. Since this is an agreement involving no parameters of adjustment, it is indeed most remarkable. Let us, however, analyze the assumptions that went into the derivation of this mass formula from SU_3 symmetry.

One assumes

$$H = H_0 + \lambda H_1 \qquad (1)$$

as the strong interaction Hamiltonian of the system, where H_0 is SU_3 symmetrical and H_1 is *assumed* to belong to an eightfold representation of the SU_3 group. λ is a parameter that is equal to 1 in the physical case. We emphasize that the separation of the Hamiltonian into the two terms in Eq. (1) is, if possible, unique.

One then makes a perturbation calculation of the energy split due to the nonsymmetrical term λH_1 to the lowest order of λ. This calculation, due to Okubo, is mathematically very nice and parallels the calculation of the Zeeman split of an atomic level in the presence of a magnetic field.

□ Reprinted from *Some Recent Advances in Basic Sciences,* Vol. 1, A. Gelbart, ed. New York: Academic Press, 1966, 157–162 (Belfer Graduate School of Science Annual Science Proceedings.) By permission of Yeshiva University.

The result is

$$M = a + bY - c\{2I(I+1) - \tfrac{1}{2}Y^2 + \tfrac{4}{3}(p-q)Y - \tfrac{1}{3}p(p+2) - \tfrac{1}{3}q(q+2) + (1/9)(p-q)^2\} \quad (2)$$

where a is independent of λ, and b and c are proportional to λ. In this formula p and q are nonnegative integers that together designate the representation of the multiplet that one is studying.

One conclusion that can be drawn from the mass formula (2) is that, for the eightfold representation, i.e., $(p, q) = (1, 1)$, one has

$$(M_{1/2} + M_{1/2}')/2 = (3M_0 + M_1)/4 \quad (3)$$

where $M_{1/2}$, $M_{1/2}'$, M_0, and M_1 are the masses of the two isospin doublets, the singlet and the triplet. On the other hand, for the decuplet, i.e., $(p, q) = (3, 0)$, one has the equidistance rule.

Now let us inquire into an underlying assumption that went into the derivation of the mass formula (2). The states in question must be represented approximately by wave functions belonging to the representation (p, q). Take, for instance, the doubly charged component of the $N_{3/2}^*$ resonance. To say that it belongs to a pure tenfold representation means that its wave function is of the form

$$(\pi N/2^{1/2}) - (K\Sigma/2^{1/2}). \quad (4)$$

Now the mass of $N_{3/2}^*$ is such that πN is an open channel and $K\Sigma$ is a closed one. This means that at large separations r of the meson and the baryon, the wave function of $N_{3/2}^*$ is certainly not of the form (4), since the two channels πN and $K\Sigma$ are then totally different. One way to give sense to assigning to $N_{3/2}^*$ a pure (3, 0) representation is to assume that at close distances the transformation property of the wave function becomes pure (at least approximately) and belongs to the (3, 0) representation. But the wave function at close distances is related to that at large distances. Whether it is consistent to assume purity for the former in face of large impurity in the latter needs detailed analysis.

The analysis cannot be made without further assumptions concerning the system. Oakes and I chose a model that gives, in a sense, maximum possibility for symmetry at small distances. We assumed that there exists a distant r_0 such that: 1) For distances less than r_0, the interaction is strong, and the system respects SU_3 symmetry. The symmetry-breaking interaction is neglected for this region. 2) For distances bigger than r_0, the interaction is neglected, and the system exists as two spatially separated particles having the experimental masses of the mesons and baryons. In this model, the symmetry-respecting and symmetry-breaking parts of the system are spatially separated. In the R matrix formulation of the scattering process, this spatial separation means that the matrix R for describing the system $r < r_0$ commutes with the symmetry group. The question to be investigated, then, is whether one can match the external wave function of the system, for example, the $N_{3/2}{}^*$, determined by the experimental kinematics of the resonance, to a relatively pure tenfold wave function for $r < r_0$.

The conclusion we reached was that it is not possible to achieve such a match. (There is no difficulty in matching *either* the wave function *or* the derivative of the wave function at $r = r_0$, but it is impossible to match *both*. It is not meaningful to match only the wave function or its derivative.)

The possibility of matching the inside pure wave function with the outside impure wave function improves if one chooses smaller and smaller values of r_0. This comes about because the system considered is in a p state. The centrifugal barrier creates an effective insulation between the external and internal wave functions. The insulation becomes very strong for values of r_0 such that $kr \ll 1$. However, in the present problem, the value of k is $\gtrsim 130$ MeV/c, and to use the centrifugal barrier to produce appreciable insulation one must choose $r_0 < [400 \text{ MeV/c}]^{-1}$. Such a small value of r_0 can hardly be accepted, because clearly the mass of the system in reality is influenced by what happens outside of such a small distance by very substantial amounts. The origin of the difficulty of marching the inside with the outside wave function is quite obvious. For *strongly* coupled channels (e.g., the channels we are discussing), the inside wave function is sensitively dependent on the wave numbers outside. For the $Y_1{}^*$, for example, the two most strongly involved channels (by Clebsh-Gordon coefficients) are the $\pi\Lambda$ and $\eta\Sigma$ (each with $\frac{1}{4}$ of the probability). The wave numbers for these two channels are $(209 - 25i)$ MeV/c and $(-12 + 467i)$ MeV/c, respectively. Such large differences in the wave numbers make it extremely difficult to produce symmetry between the two channels inside.

Another point that Oakes and I discussed was the following. The mass formula (2) yields, in addition to the well-known interval rules mentioned above, also the value of the mass of all states of a multiplet in the limit $\lambda \to 0$. This limit can be obtained from the mass value at any value of λ by the following formula.

For the octet, i.e., (1, 1),

$$\text{all masses} \to (M_0 + M_1)/2.$$

For the decuplet (3, 0), one has

$$\text{all masses} \to M_1 .$$

In other words, the mass formula (2) predicts that in the limit of no symmetry breaking one has

$$\begin{aligned} \text{all 8 baryon masses} &= 1154 \text{ MeV} \\ \text{all 8 meson masses} &= 412 \text{ MeV} \\ \text{all 10 decuplet masses} &= 1385 \text{ MeV}. \end{aligned}$$

Thus, the decuplets become bound states of the meson-baryon system with a binding energy of 181 MeV.

If the mass formula is successful when applied to the physical case $\lambda = 1$, it should apply more successfully to all values of λ between 0 and 1. We are then faced with the difficulty of stating exactly what the mass formula is to be applied to. If we apply it to the real part of the energy of the S matrix pole, we are faced with the difficulty that the $Y_1{}^*$ pole is on the $--+++$ sheet for $\lambda = 1$ and on the $+++++$ sheet for $\lambda = 0$. Now, it can be proved that no pole can cross the real axis in between the two lowest thresholds in general. It would therefore be very difficult for the $Y_1{}^*$ pole to change sheets $--+++ \to +++++$ from $\lambda = 1$ to $\lambda = 0$.

A number of authors have argued that there are other poles in "nearby" positions on other sheets. Oakes and I did consider these other poles in our paper and dropped them because one is interested not so much in poles but in the meaning of the existence of symmetry. For example, in the limit $\lambda = 0$, *where symmetry should be perfect*, poles on other sheets may exist, but they represent *impure* wave functions and are clearly irrelevant to symmetry considerations. If one claims that the mass formula is not to be applied to these impure poles for small λ, but is to be applied to them for $\lambda \simeq 1$, one will have to justify the use of the perturbation approach which led to (2) to start with. Any numerical model used to discuss this question should be carried out for all values of λ between 0 and 1, in order to bring out the supposedly increasing applicability of (2) as $\lambda \to 0$.

Other authors seem to want to apply the mass formula to some quantities other than the real part of the S matrix pole. These other quantities are some kind of averages between the pole position discussed above and the pole position on other sheets. These quantities are more analytically dependent on the parameter λ and are less subject to disturbances due to the crossing of a pole and a threshold. However, the large values of the wave numbers, especially of the closed channels, make it difficult to produce the experimental equidistant rule. Even if one manages to fit such a rule, the mass formula must then be applied not to the physically observed resonance peaks but to some quantity shifted from them by large amounts. Much room for maneuvering is left in the calculation of these shifts. The equidistant rule predicted by the mass formula (2) does not then lead to any parameterless explanation of the experimentally observed equidistant rule. For example, if the intervals between $N_{3/2}^*$, Y_1^*, and $\Xi_{1/2}^*$ are not equidistant, but in the ratio of 1 : 2, one presumably could also produce a calculation for the shifts which would yield "agreement" with the equidistant formula (2).

In conclusion, let me make a few additional remarks.

1) If there is some mixing of states, say of ψ_{27} with ψ_{10}, the effect would be additional terms in the mass formula proportional to ψ_{27}/ψ_{10}, not $(\psi_{27}/\psi_{10})^2$. Such correction terms do not yield the equidistance rule.

2) Treiman has pointed out that, taking the mass formula seriously, one has

$$\lambda = 0, \qquad m_\pi = 412 \text{ MeV}$$
$$\lambda = 1, \qquad m_\pi = 138 \text{ MeV}.$$

The same formula predicts for

$$\lambda = 1.2, \qquad m_\pi = \text{imaginary}.$$

Thus, even if the mass formula is applicable for $\lambda = 1$, it is very close to the breakdown point of its validity.

3) The only good experimental check of the mass formula so far is in its application to the eight baryons and to the decuplet. However, the check in the baryon case is to a large extent an input rather than an experimental verification. Until a reason is found for the assumption that H_1 belongs to the eight fold representation, it is hardly correct to give the baryon mass check of the mass formula full credit.

4) If there is SU_3 symmetry that is broked by large unsymmetrical interactions, the multiplet structure would presumably be more durable than mass formulas and branding ratio calculations. Unfortunately, quantitive discussions of these points promise to be messy and uncertain.

5) Empirically, the states $N_{3/2}^*$, Y_1^* and $\Xi_{1/2}^*$ present, indeed, a very striking regularity. Their masses, their total isotopic spins, and

their hypercharged numbers Y all satisfy equidistant rules. Extrapolating these rules purely empirically, one would be led to the Ω^- particle without any involved argument whatsoever. If, therefore, the Ω^- particle is indeed found satisfying this emperical prediction, it would seem that, rather than considering the fit as proving the validity of SU_3 and of the mass formula, it actually opens the question of what is the underlying principle that causes such remarkable regularity in face of such large asymmetrical interactions.

[[64e]]

Commentary begins page 55

CRITICAL POINT IN LIQUID-GAS TRANSITIONS

C. N. Yang
Institute for Advanced Study, Princeton, New Jersey

and

C. P. Yang
Ohio State University, Columbus, Ohio

(Received 31 July 1964)

In two recent experiments,[1] the specific heats $C_V(T)$ of Ar and of O_2 near the critical temperature T_C have been measured at the critical volume. The measured specific heats display sharp peaks at T_C, suggesting logarithmic infinities. Rough fits of the experimental data yield, for Ar,

$$C_V/R = 1.8L + 13, \quad T < T_C,$$

$$C_V/R = 1.8L + 3, \quad T > T_C; \tag{1a}$$

for O_2,

$$C_V/R = 2.4L + 16, \quad T < T_C;$$

$$C_V/R = 2.4L + 4, \quad T > T_C, \tag{1b}$$

where

$$L = -\ln|T - T_C|, \quad T \text{ in } ^\circ\text{C}. \tag{2}$$

The singular behavior (1) is in sharp contrast with traditional views[2] about the behavior of thermodynamical functions near T_C. We make the following remarks on these very interesting data.

(A) Thermodynamical discussion of the two-phase region.–If μ is the chemical potential, it follows from

$$SdT = -Nd\mu + Vdp$$

that

$$C_V = -NT(d^2\mu/dT^2)_V + VT(d^2p/dT^2)_V. \tag{3}$$

Now (3) is applicable to the two-phase region as well as the one-phase region. In the two-phase region $d^2\mu/dT^2$ and d^2p/dT^2 are both functions of one variable only, T. Thus for a sample with fixed N (molar weight), $C_V(T, V)$ is linear in V and its measurement in the two-phase region can yield both $d^2\mu/dT^2$ and d^2p/dT^2 along the vapor-pressure curve. We urge that this be done.

(1) and (3) show that either $d^2\mu/dT^2$ or d^2p/dT^2 (at $T = T_C-$ along the vapor-pressure curve), or both, are ∞. It is important to find out which is the actual case. [For the lattice gas to be discussed under (D), $d^2\mu/dT^2 = 0$ along the vapor-pressure curve. For real gases, it is more reasonable to expect that both $d^2\mu/dT^2$ and d^2p/dT^2 become ∞.]

If d^2p/dT^2 along the vapor-pressure curve at $T = T_C-$ for He^4 is ∞ [see (C) below], corrections must be made in the vapor-pressure table[3] of He which serves as the international standard of temperature measurement at low temperatures. A logarithmic singularity of C_V would necessitate the addition of a term $K(T - T_C)^2 \ln|T - T_C|$ in the vapor pressure. (Unless the peak in C_V for He^4 is very large, this correction to p is expected to

□ Reprinted from *Physical Review Letters* 13, 9 (August 31, 1964), 303-305.

be small, since the singularity occurs in the second derivative d^2p/dT^2. The correction to dp/dT is more important.)

(B) Thermodynamical discussion of the one-phase region.—Approaching the critical point from the one-phase region one obtains, from $C_V = T(E_{SS})^{-1}$ and (1),

$$E_{SS} \to \text{constant}(-\ln|T-T_C|)^{-1} \to 0, \tag{4}$$

where E_{SS}, E_{SV}, and E_{VV} are the second derivatives of $E(S, V)$. Now along the vapor-pressure curve

$$(dp/dT) \to \alpha = \text{constant} \tag{5}$$

at the critical point. We now assume, for $V = V_C$,

$$(\partial p/\partial T)_V = \text{continuous in } T \text{ at } T = T_C, \tag{6}$$

which is experimentally[4] true [and is satisfied by the two-dimensional lattice gas to be discussed in (D)]. Similarly we assume, for $S = S_C$,

$$(\partial p/\partial T)_S = \text{continuous in } T \text{ at } T = T_C. \tag{7}$$

If the second derivatives E_{SS}, E_{SV}, and E_{VV} approach limits as one approaches the critical point from the one-phase region, (4)-(7) require that they all approach zero in the following proportion;

$$E_{SS}:E_{SV}:E_{VV} = 1:(-\alpha):\alpha^2.$$

The velocity of sound is proportional to $(E_{VV})^{1/2}$. It therefore goes to zero as $(-\ln|T-T_C|)^{-1/2}$ near the critical point.

(C) Law of corresponding states.—de Boer[5] has pointed out that if the interatomic forces in noble gases follow a law with only two parameters, the equation of state of all noble gases in classical statistical mechanics would be the same when expressed in proper units. (The quantum correction is contained in the kinetic-energy term, which he showed was small except for He^4.) The good agreement of his results with the critical data of noble gases shows that (in noble gases) the critical phenomenon in liquid-gas transitions is not a quantum effect. Accepting this conclusion one finds that all noble gases should have the same $C_V(T)$ singularity, of the same magnitude, as that exhibited in (1) for argon, except for He^4.

Now the quantum effect is contained in the kinetic-energy term, which gives rise to a quantum dispersion of the positions of the molecules. If the peak in C_V is attributable [as will be argued in (D)] to a rapidly changing balance in the competition of holes and occupied volume, the quantum dispersion (in smearing out the difference between holes and occupied volume) should be expected to diminish the peak. Thus we expect that for He^4 the C_V peak would have a magnitude smaller than that for argon.

(D) Model of lattice gas.—The model of a lattice gas[6] was discussed more than ten years ago. It is a classical "gas" moving on discrete lattice points, with an infinite repulsive force preventing two molecules from occupying the same site, and other interactions outside of this infinite core. For the special case in which all these other interactions are attractive, the system[6] can have at most one phase transition. It was shown in reference 6 that the thermodynamics of the model can be obtained from that of an Ising model in an external magnetic field H. In this correspondence, one has the $H = 0$ region of the Ising model transformed into the two-phase region plus the $V = V_C$ line of the lattice gas, as exhibited in Table I.

Now the Ising model with nearest-neighbor interaction, with $H = 0$, has been solved in two dimensions[7] for various lattices.[8] In three dimensions many approximate methods have been devised[8,9] for locating the singularities, and for finding the thermodynamical functions. Applying the transformation of reference 6 to these results we obtained various quantities for various lattice gases. These quantities are tabulated in Table II. The results for the two-dimensional lattice gases are exact. Those for the three-dimensional ones are computed from the Ising-model approximate computations of Baker.[9] The rest of the Table lists the experimental data of references 1, 2, and 4, the values of the parameter β given by Fisher,[10] and of standard critical measurements.

It is seen that qualitatively the three-dimensional lattice gas with nearest-neighbor interaction gives a fair description of the critical phenomenon. Now the critical phenomenon of a lattice gas originates from a rapidly changing balance in the competition between the occupied and unoccupied sites. One is thus led to the suggestion that in real gases, the critical phenomenon originates from a rapidly changing balance in the competition between holes and occupied volume.[11]

Table I. Correspondence between lattice gas and Ising model.

Lattice gas	Ising model
Two-phase region	$H = 0$, $T < T_C$
$V = V_C$, $T > T_C$	$H = 0$, $T > T_C$

Table II. Critical data of lattice gases and of some real gases. L is defined in (2). $\beta = d\ln(V_G - V_L)/d\ln(T_C - T)$ at $T = T_C$.[a] V_{LO} = liquid volume at low T in equilibrium with vapor.

Lattice	C_V/RL $(T = T_C - 0)$	C_V/RL $(T = T_C + 0)$	$\left(\frac{pV}{RT}\right)$	$\left(\frac{d\ln p}{d\ln T}\right)$	V_C/V_{LO}	β
Hexagonal	0.9562	0.9562	0.076	7.0	2	$\frac{1}{8}$[b]
Square	0.9890	0.9890	0.096	6.3	2	$\frac{1}{8}$
Triangular	0.9982	0.9982	0.110	6.0	2	$\frac{1}{8}$
Simple cubic	1.18	~0.4	0.226	4.8	2	
Bcc	0.98	~0.3	0.252	4.65	2	~0.31[a]
Fcc	0.90	~0.3	0.268	4.4	2	
He			0.305	3.83	1.8	
Ne			0.296		2.3	
Ar	1.8	1.8	0.292	5.82	2.6	~0.33[a]
Kr			0.333		2.8	
Xe			0.288		2.6	
O_2	2.4	2.4	0.293		3.3	

[a]See reference 10.
[b]See reference 6.

We are indebted to W. Little for informing us of the Russian experiments and for many discussions. We also want to thank J. M. Luttinger for a conversation that eliminated a numerical error in the fit (1) to the experiments. The main points of this paper have been reported previously.[12]

---

[1]M. I. Bagatskii, A. V. Voronel, and V. G. Gusak, Zh. Eksperim. i Teor. Fiz. 43, 728 (1962) [translation: Soviet Phys.–JETP 16, 517 (1963)]; A. V. Voronel, Yu. R. Chashkin, V. A. Popov, and V. G. Simkin, Zh. Eksperim. i Teor. Fiz. 45, 828 (1963) [translation: Soviet Phys.–JETP 18, 568 (1964).

[2]See, e.g., L. D. Landau and E. M. Lifshitz, Statistical Physics (Addison-Wesley Press, Reading, Massachusetts, 1958), pp. 21, 79, 80, 81. See also p. 262, last paragraph.

[3]See, e.g., F. G. Brickwedde, H. van Dijk, M. Durieux, J. R. Clement, and J. K. Logan, The "1958 He[4] Scale of Temperatures," National Bureau of Standards Monograph 10 (U. S. Dept. of Commerce, Washington, D. C., 1960).

[4]See, e.g., A. Michels, J. M. Levelt, and W. de Graaff, Physica 24, 659 (1958).

[5]J. de Boer, Physica 14, 139, 149 (1948).

[6]T. D. Lee and C. N. Yang, Phys. Rev. 87, 410 (1952).

[7]L. Onsager, Phys. Rev. 65, 117 (1944).

[8]See, e.g., the review article of C. Domb, Advan. Phys. 9, 149 (1960).

[9]G. A. Baker, Phys. Rev. 129, 99 (1963).

[10]M. E. Fisher, to be published.

[11]For other recent discussions of the critical phenomena, see P. C. Hemmer, M. Kac, and G. E. Uhlenbeck, J. Math. Phys. 5, 60 (1964), and reference 10.

[12]C. N. Yang, Bull. Am. Phys. Soc. 9, 276 (1964).

PHENOMENOLOGICAL ANALYSIS OF VIOLATION OF CP INVARIANCE IN DECAY OF K^0 AND $\bar{K}^0$†

[64f]

Commentary begins page 58

Tai Tsun Wu* and C. N. Yang‡
Brookhaven National Laboratory, Upton, New York

(Received 18 August 1964)

1. It was recently discovered[1] that the long-lived component $K_L{}^0$ of K^0-$\bar{K}^0$ decays into the $\pi^+\pi^-$ mode. Now if CP invariance holds, the $CP = +1$ and $CP = -1$ components of K^0-$\bar{K}^0$ decay independently. The $\pi^+\pi^-$ mode in the S-wave state has $CP=1$. Hence either the short-lived component $K_S{}^0$, or $K_L{}^0$, does not decay into $\pi^+ + \pi^-$, in contradiction to the new discovery.

Accepting the experimental result of reference 1, one is thus forced to the conclusion that CP invariance is violated in K^0-$\bar{K}^0$ decay, as explicitly stated in reference 1. Notice that this conclusion is independent of the details of the Weisskopf-Wigner formulation[2] of decay amplitudes, as applied to the K_0-$\bar{K}_0$ case by Lee, Oehme, and Yang,[3] whose notation we shall follow.[4] (In particular, small corrections to the exponential decay rule of the formalism cannot alter the conclusion that CP invariance is violated.)

In the present note we shall analyze the decay properties of K^0-$\bar{K}^0$, mostly from the phenomenological viewpoint. Possible further experiments will be discussed for their theoretical significance.

We shall assume CPT invariance, the validity of the Weisskopf-Wigner formulation,[2,3] and that for the strong and electromagnetic interactions, separate C, P, and T invariance hold.

□ Reprinted from *Physical Review Letters* 13, 12 (September 21, 1964), 380–385.

In the next five sections, we shall also assume that electromagnetic interactions can be neglected, and that isotopic spin is conserved for the strong interactions. We shall come back to the electromagnetic effects in Sec. 7.

2. The experimental decay rates are tabulated in Table I. (We thank P. Franzini, J. Steinberger, and W. Willis for supplying the entries.)

To analyze the decay of K^0-$\bar{K}^0$ we consider the decay matrix

$$\Gamma = \Gamma_0 + \Gamma_2 + \Gamma_l + \Gamma_{3\pi}, \tag{1}$$

and the mass matrix

$$M = M_0 + M_2 + M_l + M_{3\pi} + \cdots, \tag{2}$$

as sums of contributions from the $\pi\pi(I=0)$, $\pi\pi(I=2)$, leptonic, and 3π modes. One has, in the notation of reference 3,

$$\Gamma_0 = \begin{pmatrix} A_0^2 & A_0^2 \\ A_0^2 & A_0^2 \end{pmatrix}, \tag{3}$$

$$\Gamma_2 = \begin{pmatrix} A_2A_2^* & A_2^2 \\ A_2^{*2} & A_2A_2^* \end{pmatrix}, \tag{4}$$

$$\Gamma_l = \begin{pmatrix} \alpha_l & x_l + iy_l \\ x_l - iy_l & \alpha_l \end{pmatrix}, \tag{5}$$

and

$$\Gamma_{3\pi} = \begin{pmatrix} \alpha_{3\pi} & x_{3\pi} + iy_{3\pi} \\ x_{3\pi} - iy_{3\pi} & \alpha_{3\pi} \end{pmatrix}, \tag{6}$$

Table I. Experimental decay rates in 10^6 sec^{-1}.

Mode	K_S^0	K_L^0
$\pi^+ + \pi^-$	$\frac{2}{3} \times 1.1 \times 10^4$	2.6×10^{-2}
$\pi^0 + \pi^0$	$\frac{1}{3} \times 1.1 \times 10^4$	Not known
Leptons	~11	~11
$\pi^+ + \pi^- + \pi^0$	$\lesssim 2$	~ 2
$3\pi^0$	$\lesssim 4$[a]	~ 4
All modes	1.1×10^4	~18

[a]No available experimental information; the number given is based on the assumption that the $|\Delta| = \frac{1}{2}$ rule is approximately valid.

where A_0 and A_2 are the decay amplitudes of

$$K^0 \to \pi + \pi \ (I = 0 \text{ standing wave})$$

and

$$K^0 \to \pi + \pi \ (I = 2 \text{ standing wave}),$$

respectively. We have chosen the phase of K^0 so that

$$A_0 = \text{real} > 0. \tag{7}$$

We emphasize that this choice, which is always possible, serves to define the phase of K and $\bar{K}$. The quantities p and q are given by

$$p^2 = A_0^2 + A_2^2 + x_l + iy_l + x_{3\pi} + iy_{3\pi} + iM_r - M_i,$$

and

$$q^2 = A_0^2 + A_2^{*2} + x_l - iy_l + x_{3\pi} - iy_{3\pi} + iM_r + M_i, \tag{8}$$

with the real parts of p and q chosen $\geqslant 0$. In Eq. (8) $M_r + iM_i = M_{12}$ is an off-diagonal element of M. The eigenstates and the eigenvalues of $\Gamma + iM$ were given by Eqs. (28) and (29) of reference 3. The decay amplitudes of K_S and K_L into π-π states are easily constructed, and are tabulated in Table II. The quantity F is

$$F = \exp[i(\delta_2 - \delta_0)], \tag{9}$$

where δ_2 and δ_0 are the π-π S-wave scattering phase shifts for the $I = 2$ and $I = 0$ states at the energy of the rest mass of K^0. Obviously,

$$R[K_S^0 \to \pi + \pi] + R[K_L^0 \to \pi + \pi]$$

$$= 2A_0^2 + 2A_2A_2^* \sim 1.1 \times 10^4,$$

$$R[K_S^0 \to \text{lep}] + R[K_L^0 \to \text{lep}] = 2\alpha_l \sim 22,$$

and

$$R[K_S^0 \to 3\pi] + R[K_L^0 \to 3\pi] = 2\alpha_{3\pi} \sim 12. \tag{10}$$

The following quantities are of intrinsic experimental interest:

$$\eta_{+-} = a_{+-}^{L}/a_{+-}^{S}, \quad \eta_{00} = a_{00}^{L}/a_{00}^{S}. \tag{11}$$

We shall also use

$$\epsilon = (p - q)/p. \tag{12}$$

These quantities are useful because they are small parameters, as we shall see later.

3. The following remarks serve to orient further analysis:

(a) If $\text{Im}A_2 = 0$, $y_l = 0$, $y_{3\pi} = 0$, and $M_i = 0$, then $p = q$, and Table II shows that $K_L \not\to \pi + \pi$, in con-

Table II. Decay amplitudes of K_S and K_L into π-π states.

State	Amplitudes[a]
$I=0$ (standing wave)	$a_0 = A_0(pp^* + qq^*)^{-1/2}(p \pm q)$
$I=2$ (standing wave)	$a_2 = (pp^* + qq^*)^{-1/2}(A_2 p \pm A_2^* q)$
$\pi^+ + \pi^-$ (outgoing wave)	$a_{+-} = (pp^* + qq^*)^{-1/2}\{[(\frac{2}{3})^{1/2}A_0 + (\frac{1}{3})^{1/2}A_2 F]p \pm [(\frac{2}{3})^{1/2}A_0 + (\frac{1}{3})^{1/2}A_2^* F]q\}$
$\pi^0 + \pi^0$ (outgoing wave)	$a_{00} = (pp^* + qq^*)^{-1/2}\{[(\frac{1}{3})^{1/2}A_0 - (\frac{2}{3})^{1/2}A_2 F]p \pm [(\frac{1}{3})^{1/2}A_0 - (\frac{2}{3})^{1/2}A_2^* F]q\}$

[a] Upper sign for K_S, lower sign for K_L.

tradiction with the experimental result of reference 1.

(b) If $\Delta Q = \Delta S$ for the leptonic decay modes of K^0-$\bar{K}^0$, then Γ_l and M_l of Eqs. (1) and (2) are both multiples of the unit matrix. Therefore, $y_l = 0$. The leptonic mode does not in this case contribute to *CP* violation as observed in reference 1, even though the lepton mode itself could violate *CP* invariance. We shall, however, not make the assumption that $\Delta Q = \Delta S$ in this paper.

(c) Phenomenologically, it is not possible to distinguish between the four M's on the right-hand side of Eq. (2). In other words, measurable quantities can only depend on M, but not on M_0, M_2, M_l, or $M_{3\pi}$ separately.

(d) If $\mathrm{Im}A_2 = 0$, then Table II gives directly

$$a_{+-}{}^{\mathrm{L}}/a_{+-}{}^{\mathrm{S}} = a_{00}{}^{\mathrm{L}}/a_{00}{}^{\mathrm{S}}. \tag{13}$$

(e) If $M_i = 0$, and $y_l + y_{3\pi} = 0$, then $p^2 - q^2 = A_2^2 - A_2^{*2}$. It will be clear in Sec. 4 that

$$|a_0{}^{\mathrm{L}}/a_2{}^{\mathrm{L}}| = O(\mathrm{Re}A_2/A_0), \tag{14}$$

provided that $|A_2|/A_0$ is small.

(f) If $\mathrm{Im}A_2 = 0$ and $M_i = 0$, then, given the experimental decay rates of K^0-$\bar{K}^0 \to$ leptons and 3π, and of $K_S \to \pi + \pi$, the rate $K_L \to \pi^+ + \pi^-$ is at most $1.75[1 + (M_\gamma/A_0{}^2)^2]^{-1} \times 10^{-2}$. This is too low to account for the experimental result of reference 1. We shall discuss this in more detail in Sec. 5.

Accordingly, roughly there are four ways to violate *CP* invariance in the decay of K^0-$\bar{K}^0$, namely, $\mathrm{Im}A_2 \neq 0$, $y_l \neq 0$, $y_{3\pi} \neq 0$, and/or $M_i \neq 0$. They correspond to *CP* (or *T*) noninvariance due to the interference between the dominant $\pi\pi(I=0)$ mode and (α) the $\pi\pi(I=2)$ mode ($\mathrm{Im}A_2/A_0 \neq 0$); ($\beta$) the lepton mode ($y_l \neq 0$), ($\gamma$) the 3π mode ($y_{3\pi} \neq 0$), and/or (δ) the off-energy shell contributions to the $K \to \bar{K}$ elements of the mass operator $M(M_i \neq 0)$. This possibility has been discussed by Sachs and Treiman.[5]

According to (f) above, (β) and (γ) together by themselves are too small to account for the magnitude of the observed effect. Thus the more important contribution to the observed *CP* violation has to come from (α) and/or (δ). [Theoretically it is, of course, to be expected that if any interference of the type (α), (β), or (γ) is present, then an interference of type (δ) is also present, in general.]

4. The $|\Delta I| = \frac{1}{2}$ rule is well verified in general, and for the $K \to \pi + \pi$ decay in particular. Thus, $|A_2/A_0| \ll 1$. Dropping A_2/A_0 in Table II, we obtain

$$\eta_{+-} \sim (p-q)/(p+q) = \epsilon/(2+\epsilon). \tag{15}$$

Thus the experimental small value of $|\eta_{+-}|$ shows that $|\epsilon| \ll 1$.

We proceed to expand various quantities to the lowest nonvanishing order of ϵ and A_2/A_0:

$$\eta_{+-} = \tfrac{1}{2}[\epsilon + (2)^{1/2} iF\, \mathrm{Im}A_2/A_0], \tag{16}$$

$$\eta_{00} = \tfrac{1}{2}[\epsilon - 2(2)^{1/2} iF\, \mathrm{Im}A_2/A_0], \tag{17}$$

and

$$R[K_S{}^0 \to \pi^+ + \pi^-] - 2R[K_S{}^0 \to \pi^0 + \pi^0]$$
$$= 2\sqrt{2}[\mathrm{Re}A_2/A_0]\cos(\delta_2 - \delta_0)$$
$$\times R[K_S{}^0 \to \pi + \pi(I=0)]. \tag{18}$$

Furthermore, since Γ_l is positive definite, $\alpha_l \geq |x_l|$, $\alpha_l \geq |y_l|$; or by Eq. (10),

$$|x_l| \lesssim 11, \quad |y_l| \lesssim 11. \tag{19}$$

Similarly,

$$|x_{3\pi}| \lesssim 6, \quad |y_{3\pi}| \lesssim 6. \tag{20}$$

Thus these elements are negligible compared with $A_0{}^2$. {In more general cases, we can establish a somewhat better bound for y. For any mode C,

$$y_C^2 \leq R[K_S^0 \to C]R[K_L^0 \to C].\}$$

Using Eq. (8) one sees that $|\epsilon| \ll 1$ implies $M_i \ll A_0^2$. Thus one has an approximate expression for the difference of the two eigenvalues of $\Gamma + iM$:

$$\lambda_+ - \lambda_- = 2A_0^2 + 2iM_r. \tag{21}$$

Therefore,

$$M_r = m_S - m_L. \tag{22}$$

Using these one obtains from Eq. (8)

$$\epsilon = \frac{-M_i + i(y_l + y_{3\pi})}{A_0^2 + i(m_S - m_L)}. \tag{23}$$

Equations (16)–(23) form the basis of a phenomenological analysis.

5. If $\mathrm{Im}A_2 = 0$ and $M_i = 0$, then it follows from Eqs. (16) and (23) that

$$R[K_L^0 \to \pi^+ + \pi^-] = 6(y_l + y_{3\pi})^2 \times 10^{-5}[1 + (M_r/A_0^2)^2]^{-1}. \tag{24}$$

Statement 3(f) then follows from Eqs. (19) and (20). This is too small by a factor of 3 or 15 for[6] $|m_S - m_L| = 1/2\tau_1$ or $3/2\tau_1$. {If, moreover, we believe that the $\Delta Q = -\Delta S$ matrix element is at most 50% of that for $\Delta Q = \Delta S$ as a result of the Paris experiment,[7] then $(x_l^2 + y_l^2)^{1/2}/\alpha_l < 2(\frac{1}{2})/[1 + (\frac{1}{2})^2] = \frac{4}{5}$. Hence, $|y_l| < 9$. Thus we can strengthen the argument by about 20%.}

6. The amplitude ratios η_{+-} and η_{00} are experimentally measurable quantities. The experiment of reference 1 gives

$$|\eta_{+-}| = 2.1 \times 10^{-3}(1 \pm 0.1). \tag{25}$$

A measurement[8] of $R[K_L^0 \to \pi^0 + \pi^0]$ would yield $|\eta_{00}|$. If one introduces further assumptions, this rate can be predicted. For example, according to Sec. 3, (d) and (e),

if $\mathrm{Im}A_2 = 0$, then $R[K_L^0 \to \pi^0 + \pi^0] = 1.3 \times 10^{-2}$;

if $M_i = y_l + y_{3\pi} = 0$, then

$$R[K_L^0 \to \pi^0 + \pi^0] = 5.2 \times 10^{-2}. \tag{26}$$

Existing experiments yield only a rough upper bound:

$$R[K_L^0 \to \pi^0 + \pi^0] < (\sim 1), \quad |\eta_{00}| < (\sim 2 \times 10^{-2}).$$

Without additional assumptions, there are no theoretical arguments that $R[K_L^0 \to \pi^0 + \pi^0]$ cannot be as large as this experimental upper bound.

It is clear that in order to measure the phases of the amplitude ratios η_{+-} and η_{00}, interference between K_L and K_S decays into these modes must be studied. To obtain greater sensitivity, the intensity ratio of K_L and K_S must be such that their decay amplitudes into these modes are about comparable.[9] The relative phase θ between the K_L and K_S beams must also be known in order to determine the phase of η_{+-} (or of η_{00}). But to determine the difference of the phases of η_{+-} and η_{00} it is not necessary to know θ.

It is convenient to construct a diagram of the complex numbers η_{+-}, η_{00}, ϵ, and iF related through Eqs. (16) and (17), as shown in Fig. 1.

(a) If η_{+-} and η_{00} are completely measured, the quantities ϵ and $F\,\mathrm{Im}A_2/A_0$ are known. Thus $\delta_2 - \delta_0$ is measured up to $\pm n\pi$, and $\mathrm{Im}A_2/A_0$ is known up to a sign. If further $m_S - m_L$ is known, then through (23), M_i and $y_l + y_{3\pi}$ are determined.

(b) If $|\eta_{+-}|$, $|\eta_{00}|$, and the phase difference

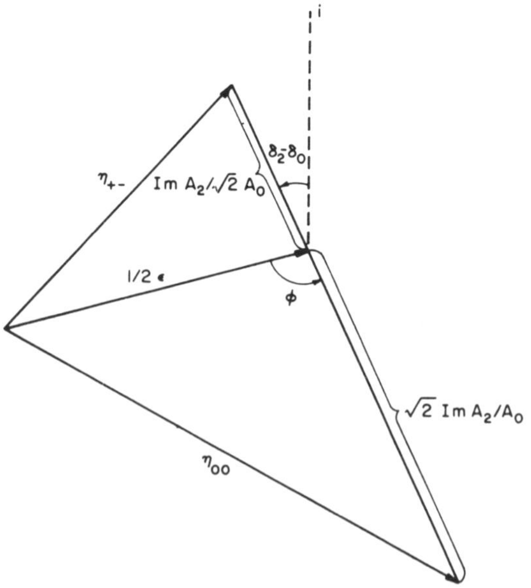

FIG. 1. Geometrical relation between η_{+-}, η_{00}, and other quantities.

of η_{+-} and η_{00} are known, then the triangle of Fig. 1 can be constructed, but its orientation relative to the real axis is known only if $\delta_2-\delta_0$ is independently obtained.

These experiments, however, do not yield any information on $\mathrm{Re}A_2/A_0$. It seems that the only experimental method of determining this ratio is to measure the rate difference on the left-hand side of Eq. (18). Existing experiment[10] gives

$$\mathrm{Re}A_2 = (0 \pm 1)\sec(\delta_2-\delta_0). \tag{27}$$

7. In the above discussion, electromagnetic effects are completely neglected. Inclusion of these effects introduces (a) mass splits between $\pi^+\pi^-$ and $\pi^0\pi^0$ states (and related effects), and (b) additional channels like $\pi\pi\gamma$. To account for (a) one introduces two eigenstates of the S matrix for the strong and electromagnetic interactions of the $\pi\pi$ S-wave state at the K^0 mass. The resultant change comprises only small real corrections to the coefficients $(\frac{2}{3})^{1/2}$ and $(\frac{1}{3})^{1/2}$ and the phase shifts δ_2 and δ_0 in Table II.

As to (b), electromagnetic effects do not introduce CP noninvariance. Thus it is reasonable to expect additional channels, such as $\pi\pi\gamma$, not to introduce matrix elements which are imaginary in phase relative to A_0. In any case, experimentally the rates $K_{S,L} \to \pi+\pi+\gamma$ are limited:

$$R[K_S \to \pi^+ + \pi^- + \gamma] < (\sim 1)$$

(from Kirsch et al.[11]) and

$$R[K_L \to \pi + \pi + \gamma] < (\sim 1)$$

(from total rate). It is reasonable to assume

$$R[K_S \to \pi^0 + \pi^0 + \gamma] < (\sim 1).$$

By an argument similar to that leading to Eq. (19), we have then

$$|y_{\pi\pi\gamma}| < (\sim 1.5).$$

Thus electromagnetic effects are expected to be negligible in the discussion of CP noninvariances of the preceding sections.

8. We now make two supplementary remarks.

One may raise the question of the evidence of T invariance in neutron[12] and Λ[13] decays and perhaps other decays. To reconcile this evidence with the CP noninvariance of reference 1, it may be that the weak interaction consists of two terms:

$$H = H_{\mathrm{strong}} + H_{\mathrm{el}} + H_{W1} + H_{W2},$$

where (a) $H_{\mathrm{strong}} + H_{\mathrm{el}} + H_{W1}$ satisfies CP invariance, (b) $H_{\mathrm{strong}} + H_{W1}$ obeys the $|\Delta I| = \frac{1}{2}$ rule, and (c) H_{W2} is weaker than H_{W1}, and violates the $|\Delta I| = \frac{1}{2}$ rule.

Another question concerns the $\pi\pi(I=2)$ decay rates of K^+ and of K^0. From Eqs. (16) and (17),

$$|\mathrm{Im}A_2| = \tfrac{1}{3}\sqrt{2}A_0|\eta_{+-} - \eta_{00}| < (\sim\tfrac{2}{3}). \tag{28}$$

If the decay amplitude of $K^+ \to \pi + \pi$ $(I=2)$ is $A_2{}^+$, the experimental rate of this process gives

$$|A_2{}^+| = 3.9. \tag{29}$$

Comparing Eqs. (27), (28), and (29) one concludes that more accurate measurements of $R[K_S \to \pi^+ + \pi^-] - 2[K_S \to \pi^0 + \pi^0]$, and information about $|\eta_{00}|$ and $\delta_2-\delta_0$ can be used to analyze the $K \to \pi + \pi$ $(I=2)$ amplitude into $|\Delta I| = \frac{3}{2}$ and $|\Delta I| = \frac{5}{2}$ components.

It is a pleasure to acknowledge the hospitality we enjoyed at Brookhaven National Laboratory where this work was done. We thank J. W. Cronin, V. L. Fitch, P. Franzini, J. Steinberger, and W. Willis for many stimulating discussions.

For other recent discussions of CP noninvariance in K^0-$\bar{K}^0$ decay, see a recent article by Sachs.[14]

---

[1]J. H. Christenson, J. W. Cronin, V. L. Fitch, and R. Turlay, Phys. Rev. Letters 13, 138 (1964).

[2]V. F. Weisskopf and E. P. Wigner, Z. Physik 63, 54 (1930); 65, 18 (1930).

[3]T. D. Lee, R. Oehme, and C. N. Yang, Phys. Rev. 106, 340 (1957).

[4]To avoid confusion we shall call the long- and short-lived components K_L and K_S. If their wave functions are called Ψ_L and Ψ_S, then, in the notation of reference 3, $\Psi_L = \Psi_-$ and $\Psi_S = \Psi_+$. Throughout this paper we use the unit 10^6 sec^{-1} for decay rates.

[5]R. G. Sachs and S. B. Treiman, Phys. Rev. Letters 8, 137 (1962).

[6]W. F. Fry, International Conference on Fundamental Aspects of Weak Interactions (Brookhaven National Laboratory, Upton, New York, 1963), p. 3; J. H. Christenson, J. W. Cronin, V. L. Fitch, and R. Turlay, ibid., p. 74.

[7]B. Aubert, L. Behr, J. P. Lowys, P. Mittner, and C. Pascaud, Phys. Letters 10, 215 (1964). The large ratio of e^+/e^- in the first few lifetimes of K_S is a strong argument that $|x|$ is quite a bit less than 1.

[8]W. Willis has been considering, independently of us, the possibility of measuring this rate.

[9]J. Steinberger has reached this same conclusion independently of us.

[10]J. L. Brown, J. A. Kadyk, G. H. Trilling, B. P. Roe, D. Sinclair, and J. C. Van der Velde, Phys. Rev. 130, 769 (1963).
[11]L. Kirsch, R. J. Plano, J. Steinberger, and P. Franzini, Phys. Rev. Letters 13, 35 (1964).
[12]M. T. Burgy, V. E. Krohn, T. B. Novey, G. R. Ringo, and V. L. Telegdi, Phys. Rev. 120, 1829 (1960).
[13]J. W. Cronin and O. E. Overseth, Phys. Rev. 129, 1795 (1963).
[14]R. G. Sachs, Phys. Rev. Letters 13, 286 (1964).

Some Speculations Concerning High-Energy Large Momentum Transfer Processes*

[65a]

Commentary begins page 59

TAI TSUN WU† AND C. N. YANG‡
Brookhaven National Laboratory, Upton, New York
(Received 23 September 1964)

It is speculated that the sharp decrease with increasing energy of differential cross sections at large angles is due to a mechanism independent of the method of excitation. Some consequences of such a possibility are discussed.

IT has been known for some time that (i) the total pp cross section remains essentially constant at high energies, and that (ii) above 300 MeV of excitation energy the nucleon has many excited states. More recently, experiments[1] have shown that the *large-angle* elastic pp cross section drops down spectacularly with energy. For example, when the center-of-mass momentum of each proton is 3.8 BeV/c, the differential cross section at 90° is only about 10^{-36} cm^2/sr.

These facts together suggest that the nucleon is an extended object with an internal structure having a "rigidity" characterized by an excitation energy of the order of a few hundred MeV. For hard collisions where the available energy is much larger than this, many degrees of freedom are excited in the nucleons, resulting in general in the emission of many particles.

Such a picture is more or less common to various statistical discussions[2] of high-energy collisions.

The spectacular drop mentioned above has been put in a more quantitative form by Orear.[3] His result is that (iii) the elastic pp differential cross section for large θ in the center-of-mass coordinate system is given by

$$(d\sigma/d\Omega)(\theta, pp \to pp) \sim Ae^{-p_\perp/0.15}, \tag{1}$$

where $p_\perp$ is the transverse momentum transfer in units of BeV/c.

Guided by these facts (i)–(iii), we attempt to speculate about the high-energy behavior of other processes. We observe that in picturing the nucleon as an extended object the difficulty in making large transverse mo-

* Work performed under the auspices of U. S. Atomic Energy Commission.
† Permanent address: Harvard University, Cambridge, Massachusetts.
‡ Permanent address: Institute for Advanced Study, Princeton, New Jersey.

[1] G. Cocconi, V. T. Cocconi, A. D. Krisch, J. Orear, R. Rubinstein *et al.*, Phys. Rev. Letters **11**, 499 (1963) and W. F. Baker, E. W. Jenkins, A. L. Read, G. Cocconi, V. T. Cocconi *et al.*, *ibid.* **12**, 132 (1964).

[2] H. W. Lewis, J. R. Oppenheimer, and S. A. Wouthuysen, Phys. Rev. **73**, 127 (1948); E. Fermi, Progr. Theoret. Phys. (Kyoto) **5**, 570 (1950); G. Fast and R. Hagedorn, Nuovo Cimento **27**, 208 (1963); L. van Hove, Rev. Mod. Phys. **36**, 655 (1964); G. Cocconi, Nuovo Cimento **33**, 643 (1964); A. Bialas and V. F. Weisskopf, CERN (to be published); and many other papers. Notice that for very small angle elastic scattering, the many modes of excitation contribute in phase so that one has an enormous "diffraction peak." See R. Serber, Rev. Mod. Phys. **36**, 649 (1964) and earlier papers.

[3] J. Orear, Phys. Rev. Letters **12**, 112 (1964). See also A. D. Krisch, *ibid.* **11**, 217 (1963), and D. S. Narayan and K. V. L. Sarma, Phys. Letters **5**, 365 (1963).

□ Reprinted from *The Physical Review* 137, 3B (February 8, 1965), B708–B711.

mentum transfers could naturally be due to the *difficulty in accelerating the various parts of a nucleon without breaking it up*. If this is the case, such a difficulty is presumably present in all high-energy collisions. Furthermore, the dominant effect of such a difficulty is to contribute a rapidly decreasing factor to the appropriate differential cross sections *independent of the specific process*. (The specific process, i.e., method of excitation, may give rise to more slowly varying factors such as polynomials of the energy.) In particular, we speculate as follows:

1. In a nucleon-nucleon collision,

$$N+N \to A+B\,, \tag{2}$$

with large θ, so many degrees of freedom are excited in the collision that the emergent particles A and B assume various excited forms of the nucleons with *relative* probabilities that do not have such a precipitous dependence on energy as (1). In other words, with a suitable change of the factor in front of the exponential, (1) holds not only for pp collision but also for any process of the form (2), where A and B are any nucleonic states satisfying all conservation laws. For example,[3a] in the high-energy limit, with fixed $\theta \neq 0, \pi$,

$$\ln\frac{d\sigma}{d\Omega}(\theta, pn \to pn) \Big/ \ln\frac{d\sigma}{d\Omega}(\theta, pp \to pp) \to 1\,, \tag{3a}$$

$$\ln\frac{d\sigma}{d\Omega}(\pi-\theta, pn \to pn) \Big/ \ln\frac{d\sigma}{d\Omega}(\theta, pn \to pn) \to 1\,, \tag{3b}$$

and

$$\ln\frac{d\sigma}{d\Omega}(\theta, pp \to pp^*) \Big/ \ln\frac{d\sigma}{d\Omega}(\theta, pp \to pp) \to 1\,. \tag{3c}$$

For some recent experimental data relating to (3c), see Ref. 4.

2. We make the same speculation for the process $p+p \to \pi^+ + D$. Similar to (3), for $\theta \neq 0, \pi$,

$$\ln\frac{d\sigma}{d\Omega}(\theta, pp \to \pi D) \Big/ \ln\frac{d\sigma}{d\Omega}(\theta, pp \to pp) \to 1\,, \tag{4}$$

in the high-energy limit. Ulrich[5] has recently extended the Orear fit to this process.

3. Again, similar statements may be made for the πp processes. For example, in the same limit,

$$\ln\frac{d\sigma}{d\Omega}(\theta, \pi^+ p \to \pi^+ p) \Big/ \ln\frac{d\sigma}{d\Omega}(\theta, \pi^- p \to \pi^- p) \to 1\,, \tag{5a}$$

$$\ln\frac{d\sigma}{d\Omega}(\theta, \pi^- p \to \pi^0 n) \Big/ \ln\frac{d\sigma}{d\Omega}(\theta, \pi^- p \to \pi^- p) \to 1\,, \tag{5b}$$

$$\ln\frac{d\sigma}{d\Omega}(\theta, \pi p \to \pi p^*) \Big/ \ln\frac{d\sigma}{d\Omega}(\theta, \pi p \to \pi p) \to 1\,, \tag{5c}$$

and

$$\ln\frac{d\sigma}{d\Omega}(\theta, \pi p \to K\Lambda) \Big/ \ln\frac{d\sigma}{d\Omega}(\theta, \pi p \to \pi p) \to 1\,. \tag{5d}$$

4. The possible validity of (3)–(5) is not dependent on the strict exponential form exhibited in (1). What is truly relevant is that existing experimental data *suggest* that at high energies, at a fixed angle θ in the center-of-mass system, *the differential cross sections all* $\to$ *0 faster than any power law*. We speculate that this fast approach to zero is independent of the specific process.

5. We observe that, in pp elastic collision, *both* of the two final protons receive large $p_\perp$ without breaking up. Consider next electron-nucleon elastic scattering at high energies. Here, it is also difficult to transfer large momentum to the nucleon without the emission of many pions. On the other hand, no such difficulty is to be associated with the electron. Therefore, it is perhaps not unreasonable to relate the electron-nucleon differential cross section at large angles to the *square root* of that of pp scattering. In other words, one would try to relate the form factors G in ep scattering to the fourth root of the elastic pp scattering differential cross section. To pursue this line of speculation one must identify the variable q^2 of $G(q^2)$ with the proper variables in pp scattering.

In order to do this, it is necessary to form a picture describing the remarkable fact implicit in (1) that large longitudinal momentum transfers in pp scattering are *not* costly, as large $p_\perp$ is. We argue that this fact is "understandable" for two reasons: (1) Since each proton is an extended object, pieces of the two may be exchanged, leading to large longitudinal momentum transfers. (2) Different parts of a proton possess instantaneous momenta relative to each other. In the laboratory system, these momenta acquire large longitudinal components.

Now, for an ep collision, the above reasons do not apply (since no exchange of "pieces" can take place between e and p), and "longitudinal" momentum transfer must be treated on a similar footing as "transverse" momentum transfer. Thus we argue that we should

---

[3a] *Note added in proof*. Equations (3)–(5) are valid in any unit chosen for $d\sigma/d\Omega$. For practical purposes, it is more convenient to write e.g., instead of (3a), as $E \to \infty$

$$\frac{\left[\ln\frac{d\sigma}{d\Omega}(\theta, pn \to pn)\right]_E - \left[\ln\frac{d\sigma}{d\Omega}(\theta, pn \to pn)\right]_{E_0}}{\left[\ln\frac{d\sigma}{d\Omega}(\theta, pp \to pp)\right]_E - \left[\ln\frac{d\sigma}{d\Omega}(\theta, pp \to pp)\right]_{E_0}} \to 1\,, \tag{3a'}$$

where E is the energy of the incoming system, and E_0 is an arbitrary fixed value for E.

[4] C. M. Ankenbrandt, A. R. Clyde, B. Cork, D. Keefe, L. T. Kerth, W. M. Layson, and W. A. Wenzel, University of California, Radiation Laboratory Report No. UCRL-11423, 1964 (to be published).

[5] B. T. Ulrich (to be published).

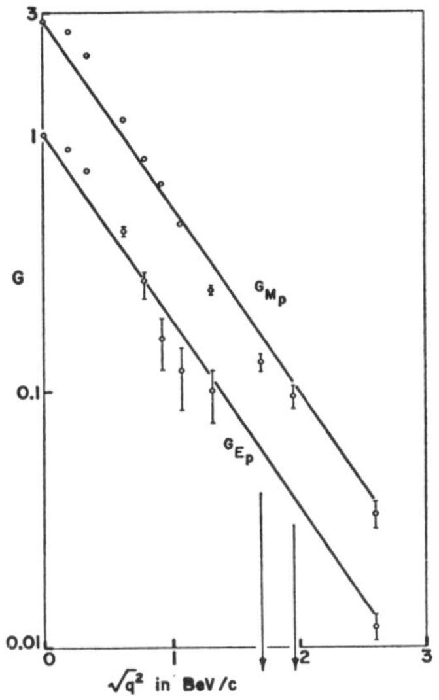

FIG. 1. Electromagnetic form factors of the proton. The straight line represents Eq. (6) with B=constant. The data are from references in footnote 7.

replace the $p_\perp$ of (1) by $(q^2)^{1/2}$, obtaining[6]

$$G(q^2) \sim B \exp[-(q^2)^{1/2}/0.6], \quad (6)$$

where q^2 is measured in $(\text{BeV}/c)^2$. Equation (6) is applicable only for sufficiently large q^2. Whether it should be applied to both form factors, or only to the one that contributes dominantly to the cross section at large q^2, is unclear to us. In Fig. 1 we plot the experimental form factors[7] G_E and G_M and compare them with (6). We notice that if the measurements are extended to higher q^2, (6) yields form factors very different from any power law.

Since the mechanism of interaction in *ep* and *pp* collisions are quite different, the factor B in (6) may vary slowly with q^2, say like a power of q^2. Our speculation, more precisely, is then

$$\frac{\ln G(q^2)}{\ln(d\sigma/d\Omega)(90°, pp \to pp)} \to \tfrac{1}{4} \quad \text{as} \quad q^2 \to \infty, \quad (7)$$

where $(d\sigma/d\Omega)(90°, pp \to pp)$ is taken at a center-of-mass momentum $= (q^2)^{1/2}$ for each proton. Again (7) may be valid even if (1) is not strictly true.

6. Similar considerations apply to processes

$$eN \to eN^*,$$
$$\nu n \to \mu p,$$
$$\nu N \to \mu N^*,$$

all of which involve cross sections falling off with energy like that of $ep \to ep$. Notice, however, that these rapid falloffs do not occur in processes involving only leptons, e.g.,

$$\nu e \to \nu e \quad \text{and} \quad \mu e \to \mu e.$$

7. We have no cogent arguments for the exponential form (1). We observe, however, that it is consistent with the following idea: Different regions of an extended object (the proton) contribute *independently* to the factor that describes the probability for the object not to break up. Such a line of reasoning also suggests that for a process such as $A+B \to C+D+E+F$ where all particles are strongly interacting particles, the rapid falloff factor at high energies is

$$\exp[-(\sum |p_\perp|)/0.3],$$

where the sum extends to all final particles.

8. The above considerations are only concerned with the rapid falloff factors for various differential cross sections at large transverse momentum transfers. Now, for high-energy collisions where many degrees of freedom are excited, it is tempting to speculate about the possibility of further statistical properties for these processes. For example, the elastic differential cross section in various isotopic spin channels may have on the average[8] the same absolute amplitude with random relative phases. This assumption is similar to the one used by Fermi[9] in his discussion of the charge distribution of multiple pion production (and is quite independent of our discussions in the previous sections). Consequences of this assumption on the large angle elastic and charge exchange differential cross sections will be discussed in the Appendix. Also discussed there are similar considerations concerning spin correlations in such scatterings.

The above arguments are of course highly speculative. It is quite possible, nonetheless, that the main point is correct, viz., that the dominant fall-off factor (at high energies) of the large-angle elastic differential cross sections is independent of the excitation process. If so, it would be rather *difficult to extract*, unfortunately, from

---

[6] Equation (6) is not inconsistent with the required analyticity of $G(q^2)$ in the cut plane. If, furthermore, $G(q^2)$ is bounded by a polynomial in q^2 in this cut plane, then (6) implies that the discontinuity of $G(q^2)$ across the cut oscillates an infinite number of times about zero.

[7] Experiments performed at Stanford, Cornell, Paris, and Cambridge, Massachusetts have been fully reported in the literature. See review article by L. N. Hand, D. G. Miller, and R. Wilson, Rev. Mod. Phys. **35**, 335 (1963). For recent data, see K. W. Chen, A. A. Cone, J. R. Dunning, Jr., S. G. F. Frank, N. F. Ramsey, J. K. Walker, and R. Wilson, Phys. Rev. Letters **11**, 561 (1963); also R. Wilson and K. W. Chen, private communication. For an earlier discussion of possible connections between *ep* and *pp* scatterings, see R. Wilson (unpublished).

[8] Here "average" means average over small energy and angular intervals.

[9] E. Fermi, Phys. Rev. **92**, 452 (1953).

high-energy large momentum transfer processes, *intrinsic information pertaining to very small distance interactions.*

We wish to thank the Brookhaven National Laboratory for the hospitality we enjoyed during our visit. We also wish to thank R. Adair, K. W. Chen, J. Orear, B. Ulrich and R. Wilson for fruitful discussions and communications.

APPENDIX

1. Consider the process

$$A+B\to C+D.$$

Let the matrix element for the process with a total isotopic spin I be denoted by a_I. The statistical hypothesis means that

$$\langle a_I a_{I'}{}^*\rangle=\delta_{II'}a, \tag{8}$$

where the average $\langle\cdots\rangle$ is defined in footnote 8. This is essentially the assumption made by Fermi.[9] Using (8), it is clear that the differential cross section on the average is proportional to

$$\textstyle\sum_I |\langle I_3(A),I_3(B)|I\rangle\langle I|I_3(C),I_3(D)\rangle|^2, \tag{9}$$

where $I_3(A)$ etc., are the I_3 component of the isotopic spin of A etc., I is the total isotopic spin, and the $\langle|\rangle$ symbols are the appropriate Clebsch-Gordan coefficients. Application of (9) to pp and πp large-angle scattering yields

$$\begin{aligned}\frac{d\sigma}{d\Omega}(\theta,pp\to pp)&=2\frac{d\sigma}{d\Omega}(\theta,pn\to pn)\\&=2\frac{d\sigma}{d\Omega}(\pi-\theta,pn\to pn),\end{aligned}\tag{10}$$

$$\begin{aligned}\frac{d\sigma}{d\Omega}(\theta,\pi^+p\to\pi^+p)&=\frac{9}{5}\frac{d\sigma}{d\Omega}(\theta,\pi^+n\to\pi^+n)\\&=\frac{9}{4}\frac{d\sigma}{d\Omega}(\theta,\pi^+n\to\pi^0p),\\&=\frac{9}{5}\frac{d\sigma}{d\Omega}(\theta,\pi^0p\to\pi^0p).\end{aligned}\tag{11}$$

2. One could discuss the spin dependence in a similar way. For example, consider large-angle $p+n\to p+n$. Denote the spin components of a particle in a direction perpendicular to the scattering plane by u (for up) and d (for down). There are the following possibilities[10] of spin arrangements:

$$uu\to uu,\quad uu\to dd,\quad ud\to ud,\quad ud\to du,$$
$$du\to ud,\quad du\to du,\quad dd\to uu,\quad dd\to dd.$$

The statistical hypothesis requires that they all have on the average[8] the same amplitude, and random phase differences.

[10] A. Bohr, Nucl. Phys. **10**, 486 (1959).

πp Charge-Exchange Scattering and a "Coherent Droplet" Model of High-Energy Exchange Processes

[66b]

Commentary begins page 61

N. Byers* and C. N. Yang
Institute for Advanced Study, Princeton, New Jersey
(Received 23 August 1965)

A model calculation is presented following a simple viewpoint to approach high-energy two-body processes $A+B \to C+D$. πp charge-exchange scattering is discussed in detail. A number of further experiments are suggested.

RECENT experiments reveal the following characteristics of high-energy collisions.

(i) The angular distribution of elastic scattering[1] is approximately

$$d\sigma/dt \cong Ae^{\gamma t}, \quad -t<0.6(\text{BeV}/c)^2, \tag{1}$$

where $-t$ is the square of the momentum transfer in the center-of-mass system. The constant γ is not drastically different for various types of collisions ($pp,\bar{p}p,\pi p,Kp$) or for different energies (5 BeV/c lab momentum and up). It ranges in value from 6 to 13 $(\text{BeV}/c)^{-2}$. For pp and πp collisions, the range[2] is even narrower;

$$8.5<\gamma<10.5\,, \quad \text{in } (\text{BeV}/c)^{-2}. \tag{2}$$

(The constancy of γ with energy strongly suggests a "size" of the interaction volume. This size has remarkably similar values in the different types of collisions.)

(ii) Recent measurements[3] of the charge-exchange scattering $\pi^- p \to \pi^0 n$ indicates an angular distribution similar to Eq. (1) for an important range of momentum transfer

$$0.1<-t<0.6.$$

(iii) Data[4] on backward elastic π^+p and π^-p scattering indicate an angular distribution similar to Eq. (1) if $-t$ is taken to be the invariant square of the 4-momentum difference between the incoming π and the outgoing p. Again, the constant γ seems to be in the range indicated by Eq. (2).

It is remarkable that in exchange processes

(a) γ is not very much dependent upon the quantum numbers exchanged. [Traditionally, it is expected that the exchange of a heavy (light) particle entails a close (distant) collision, yielding a wide (narrow) angular distribution.]

(b) The value of γ is similar to that for elastic

---

* On leave of absence from the University of California, Los Angeles, California.

[1] See, e.g., K. J. Foley, S. J. Lindenbaum, W. A. Love, S. Ozaki, J. J. Russell, and L. C. L. Yuan, Phys. Rev. Letters **11**, 425 (1963); **11**, 503 (1963); O. Czyzewski, B. Escoubes, Y. Goldschmidt-Clermont, M. Guinea-Moorhead, D. R. O. Morrison, and S. DeUnamuno-Escoubes, Phys. Letters **15**, 188 (1965), and the work of many other groups quoted in these papers.

[2] All momenta are in units of (BeV/c) in this paper, and length in $\hbar(\text{BeV}/c)^{-1}$.

[3] I. Mannelli, A. Bigi, R. Carrara, M. Wahlig, and L. Sodickson, Phys. Rev. Letters **14**, 408 (1965); A. V. Stirling, P. Sonderegger, J. Kirz, P. Falk-Vairant, O. Guisan, C. Bruneton, and P. Borgeaud, *ibid.* **14**, 763 (1965).

[4] W. R. Frisken, A. L. Read, H. Ruderman, A. D. Krisch, J. Orear, R. Rubinstein, D. B. Scarl, and D. H. White, Phys. Rev. Letters **15**, 313 (1965); C. T. Coffin, N. Dikmen, L. Ettlinger, D. Meyer, A. Saulys, K. Terwilliger, and D. Williams (unpublished report).

□ Reprinted from *The Physical Review* 142, 4 (February 25, 1966), 976–981.

scattering. (Notice that the peak in the latter is mainly shadow scattering which is absent in the exchange processes. In other words, the partial-wave amplitude in elastic processes is proportional to $e^{2i\delta}-1$, where the -1 contributes predominantly to the forward peak. In exchange processes this term is absent.[5])

(c) The forward peak rises above the differential cross sections at $-t\sim 0.6$ by factors of 100 to 1000. Furthermore the shape of the peak in $d\sigma/dt$ versus t is largely independent of the incoming energy. Both of these features are usually characteristics of elastic "diffraction" peaks, but not of inelastic processes.

There have been a number of attempts[6,7] to understand narrow peaks like (1) in exchange processes. Here we make a new attempt which is based on a very simple picture of high-energy two-body processes. Our discussions *are confined* to the forward and backward peaks where the momentum transfer $-t$ (or $-u$) is <0.6. Collisions with larger momentum transfers are, we believe, due to a physically different mechanism, and other complementary pictures must be used to describe them.

The essential independence of (1) of the incoming energy is indicative of an impact parameter picture of elastic processes. We therefore take the "eikonal" viewpoint of elastic high-energy collisions,[8] and extend it to exchange processes. We argue that, for exchange processes, the existence at small angles of enormous peaks rising above the small value of the large-angle differential cross sections, irrespective of the quantum numbers exchanged and with a shape independent of energy, is indicative of the *great difficulty in transferring large momenta, but relative ease, to varying degrees, in coherently transferring quantum numbers:* charge, spin, strangeness, *nucleon number*, etc. Because of the similarity of these peaks to the elastic one, we choose an impact parameter, or eikonal, description of such coherent transfers. Elastic and exchange processes are thus pictured as very much similar to the passage of a particle through an absorptive medium with or without coherent excitation of the medium.

In more precise terms, we express for πp scattering the spin-nonflip and spin-flip amplitudes in the usual partial-wave expansion:

$$A(\theta)=\tfrac{1}{2}\lambda\!\!\!^{-} i\sum_{l=0}^{\infty}(2l+1)\alpha_l P_l(\cos\theta)\,, \tag{3}$$

$$B(\theta)=\lambda\!\!\!^{-} i\sin\theta\sum_{l=0}^{\infty}\beta_l\frac{d}{d\cos\theta}P_l(\cos\theta)\,, \tag{4}$$

$$\alpha_l=\frac{l+1}{2l+1}a_l^{+}+\frac{l}{2l+1}a_l^{-}\,,\quad \beta_l=-(a_l^{+}-a_l^{-})/2\,,$$

$$\frac{d\sigma}{dt}=\left(\frac{d\sigma}{dt}\right)_{\text{nonflip}}+\left(\frac{d\sigma}{dt}\right)_{\text{flip}}\,,$$

$$\left(\frac{d\sigma}{dt}\right)_{\text{nonflip}}=\pi\lambda\!\!\!^{-2}|A(\theta)|^2\,, \tag{5}$$

$$\left(\frac{d\sigma}{dt}\right)_{\text{flip}}=\pi\lambda\!\!\!^{-2}|B(\theta)|^2\,, \tag{6}$$

$$-t=2k^2(1-\cos\theta),\quad k=1/\lambda\!\!\!^{-}=\text{c.m. momentum}. \tag{7}$$

The amplitudes a_l^{+} and a_l^{-} are related to (complex) phase shifts $\delta^{\pm}$ by

$$a^{\pm}=1-\exp(2i\delta^{\pm})\,. \tag{8}$$

It is clear from experimental evidence (Ref. 5; also remark H below) that all exchange processes (charge exchange or spin exchange, etc.) are small compared with the elastic forward peak at the same t. Thus $a_l^{+}\cong a_l^{-}$, and we shall write

$$\alpha=1-\exp(2i\delta)\,, \tag{9}$$

where $\delta\cong\delta^{+}\cong\delta^{-}$.

Although not necessary, we shall use in this paper, consistent with the eikonal picture, the approximations, valid for small angles,[9] of replacing $P_l(\cos\theta)$ by $J_0(b\sqrt{(-t)})$, $[b=\lambda\!\!\!^{-}(l+\tfrac{1}{2})=\text{impact parameter}]$ and $\sin\theta dP_l(\cos\theta)/d\cos\theta$ by $(l+\tfrac{1}{2})J_1(b\sqrt{(-t)})$, and the sums in (3) and (4) by integrals:

$$A(\theta)=ik\int_0^{\infty}\alpha(b)J_0(b\sqrt{(-t)})b\,db\,, \tag{3'}$$

$$B(\theta)=ik\int_0^{\infty}\beta(b)J_1(b\sqrt{(-t)})b\,db\,. \tag{4'}$$

[5] The difference of the amplitudes for $I=\frac{1}{2}$ and $I=\frac{3}{2}$ scattering is the charge-exchange amplitude. Pure I-spin $\frac{1}{2}$ and $\frac{3}{2}$ πp elastic scattering differential cross sections are essentially the same since the charge-exchange process is $\sim 1\%$ of elastic cross sections at the same energy and angle. The experimental data indicate that the t dependence of the difference of the amplitudes ($I=\frac{1}{2}$ and $\frac{3}{2}$) is similar to that of each amplitude. This is remarkable, since whatever causes the *difference* has no *a priori* reason to give an amplitude similar to that for elastic scattering. Notice also that the t dependence is so sharp that the charge-exchange differential cross section at $t=0$ is larger than the elastic differential cross section at $-t>1$.

[6] N. J. Sopkovich, Nuovo Cimento **26**, 186 (1962); A. Dar, M. Kuyler, Y. Dothan, and S. Nussinov, Phys. Rev. Letters **12**, 82 (1964); A. Dar and W. Tobocman, *ibid.* **12**, 511 (1964); A. Dar, *ibid.* **13**, 91 (1964); L. Durand and Y. T. Chiu, *ibid.* **12**, 399 (1964); **13**, 45 (1964); Phys. Rev. **137**, B1530 (1965); M. H. Ross and G. L. Shaw, Phys. Rev. Letters **12**, 627 (1964); R. C. Arnold, Phys. Rev. **136**, B1388 (1964); K. Gottfried and J. D. Jackson, Nuovo Cimento **34**, 735 (1964); J. D. Jackson, J. T. Donohue, K. Gottfried, R. Keyser, and B. E. Y. Svensson, Phys. Rev. **139**, B428 (1965).

[7] See the review article of J. D. Jackson, Rev. Mod. Phys. **37**, 484 (1965).

[8] S. Fernbach, R. Serber, and T. B. Taylor, Phys. Rev. **75**, 1352 (1949). See also R. Serber, Rev. Mod. Phys. **36**, 649 (1964); R. J. Glauber, *High Energy Collision Theory Lectures in Theoretical Physics* (Interscience Publishers, Inc., New York, 1959).

[9] G. N. Watson, *Theory of Bessel Functions* (Cambridge University Press, New York, 1944), Sec. 5.72.

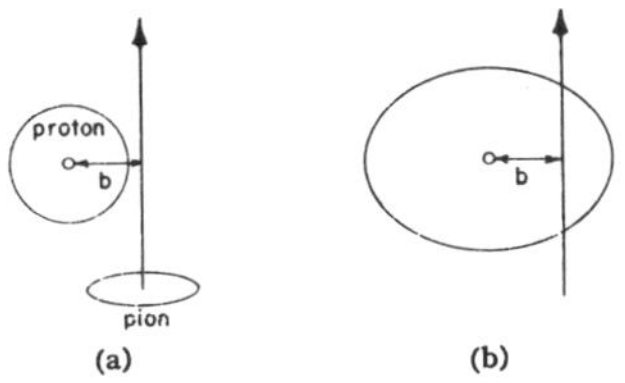

FIG. 1. (a) Schematic representation of π (ellipse) and p (circle) in rest frame of proton. (b) Eikonal description collision.

For charge-exchange scattering these formulas hold with α and β replaced by

$$\alpha^{ce} = \text{charge-exchange spin-nonflip amplitude},$$
$$\beta^{ce} = \text{charge-exchange spin-flip amplitude}.$$

Our extension of the eikonal picture to charge-exchange scattering depicts α^{ce} as being the total coherent amplitude of a charge exchange $\pi^- p \to \pi^0 n$ during a passage of a Lorentz-contracted extended object (π) through another extended object (p) with impact parameter b. [See Fig. 1(a).] For simplicity, we picture the process as a coherent excitation in the passage of a point particle in Fig. 1(b) through an extended object, and use the language one would customarily use for describing such a process (see remarks D and I below).

$$\alpha^{ce} \cong \text{(absorption factor) (probability amplitude of exchange per g/cm}^2\text{ of path) (total material thickness of path in g/cm}^2\text{)}. \quad (10)$$

Now the absorption factor is the product of that before the charge exchange with that after the charge exchange. Both factors are essentially the same for each g/cm² of path since the value of $e^{2i\delta}$ is essentially the same in $\pi^- p$ and $\pi^0 n$ scattering. The material thickness of the path is approximately proportional to δ in the eikonal picture. Thus α^{ce} is proportional to $e^{2i\delta}\delta$, or

$$\alpha^{ce} \cong -K^{ce}(1-\alpha)\ln(1-\alpha), \quad (11)$$

where K^{ce} is a numerical constant (*in general complex*) independent of the impact parameter b, but dependent on the incoming momentum.

Equation (11) above was suggested by a "geometrical picture" of high-energy exchange processes. If one adopts a "potential model" of such processes, one could also arrive at Eq. (11) in the following way. Let

$$V = \mathcal{V}\rho(r), \quad U = \mathcal{U}\rho(r), \quad (12)$$

be the elastic and charge-exchange scattering potentials, respectively. Assume $\mathcal{V}$ and $\mathcal{U}$ to be constants independent of r. Since U is off diagonal, $\mathcal{U}$ is in general complex. To lowest order in U, taking V into account in all orders, one has

$$\alpha_l^{ce} = -e^{2i\delta_l}\int_0^\infty dr\,|g_l(r)|^2 U.$$

In the WKB approximation, when $k \gg V$,

$$2i\delta_l \cong -i\mathcal{V}\int_0^\infty dz\,\rho((z^2+b^2)^{1/2}),$$

and

$$g_l(r) \cong \frac{1}{(1-b^2/r^2)^{1/4}}\cos\left\{k\int_b^r (1-b^2/r^2)^{1/2}dr - \tfrac{1}{4}\pi\right\}$$

for $r > b$. For $r < b$,

$$g_l(r) \cong \frac{1}{(|1-b^2/r^2|)^{1/4}}\exp -k\int_r^b (b^2/r^2-1)^{1/2}\,dr,$$

so as $k \to \infty$, if ρ is a smooth function, one obtains (11). [The above expression for δ may be inverted and a $V(r)$ obtained from experiment; see remark I below. Using Eqs. (15) and (23), one has $V(0) = -i0.13$ BeV.]

For the spin-flip amplitudes β and β^{ce}, we adopt the standard idea that spin flip is proportional to $\boldsymbol{\sigma}\cdot\mathbf{L}$, so that β contains a factor proportional to the orbital angular momentum, or to b. (The assumption is made here that whatever factor $\boldsymbol{\sigma}\cdot\mathbf{L}$ multiplies into has no singularity at $b=0$. This assumption is in agreement with the general picture of the nucleon and pion as structures without a central localized singularity. See remark J below.) We find, however, that experimental data[3] indicate further dominance of high b in spin flip[10]; therefore, we take

$$\beta^{ce} \cong -K_f^{ce}(1-\alpha)b^2\ln(1-\alpha). \quad (13)$$

The constant K_f^{ce} here has similar properties as K^{ce}.

To test these ideas against experiments, we take the elastic πp differential cross section as

$$(d\sigma/dt)_{el} = 41e^{10t}\,\text{mb}(\text{BeV}/c)^{-2}. \quad (14)$$

Neglecting the spin flip-term in the elastic cross sections, and neglecting the real part of $A(\theta)$, one obtains from (14)

$$\alpha(b) = 0.58e^{-b^2/20}, \quad b \text{ in } (\text{BeV}/c)^{-1}, \quad (15)$$

which can be verified by substitution into (3′) and (5). Substitution of (15) into (11) and (13), and then into

[10] To be precise, if one takes $\beta^{ce} = -K_f^{ce}(1-\alpha)b\ln(1-\alpha)$ the decrease of the theoretical $[\ln(d\sigma/dt)]_{flip}$ value from $-t=0.2$ to $-t=0.4$ is then not enough to give as good a fit with experiments as that exhibited in Fig. 2, although a passable fit can be produced.

In this connection if one takes a β^{ce} that does not vanish in the limit $b \to 0$, as, e.g., $\beta^{ce} = (\text{const})e^{-b^2c}$, (i.e., a Gaussian), the spin-flip charge-exchange amplitude $B(\theta)$ would be $\sim 1/t$ for large $-t$, which is in violent disagreement with the principal general features of the experimental data.

In nuclear physics, the dependence of the spin-flip term on the impact parameter heavily emphasizes the edge of the nucleus. See, e.g., recent data and analysis in P. G. McManigal, R. D. Eandi, S. N. Kaplan, and B. J. Moyer, Phys. Rev. **137**, B620 (1965).

(3′) and (4′) leads to a calculation[11] of the charge-exchange spin-flip and spin-nonflip amplitudes. The corresponding differential cross sections are plotted in Fig. 2, where we have chosen the constants K to fit the experimental points:

$$|K^{ce}|=0.31(p_L)^{-1/2}, \tag{16}$$

$$|K_f^{ce}|=0.18(p_L)^{-1/2}\gamma^{-1}, \tag{17}$$

$$p_L=\text{lab pion momentum in (BeV/}c\text{)},$$

$$\gamma=10\ (\text{BeV}/c)^{-2}.$$

We shall now make a few remarks:

(A) The fit with experiments exhibited in Fig. 2 has besides one normalizing constant [which is energy-dependent, as shown in Eqs. (16) and (17)] one adjustable constant, namely K_f^{ce}/K^{ce}. The energy independence[3] of the shape of the $d\sigma/dt$ versus t curve is reflected in the energy independence of K_f^{ce}/K^{ce}.

(B) For $-t>0.6$, the experimental points[3] for $\pi^-p\to\pi^0n$ show a wide peak. The total cross section under this peak is, however, very small. Its existence depends on the exact details of α^{ce} and β^{ce}. [Many crude assumptions are implicit in Eqs. (11) and (13); see, e.g., remark I below.] We have made no attempt to fit the data for $-t\gtrsim 0.6$, although assumptions (11) and (13) do lead to secondary peaks. It is interesting to speculate as to whether the "shoulder" in the elastic scattering data[4] for $-t\sim 1$ is due to a spin-flip term like (13a). (See remark E below.)

(C) For spin zero-zero elastic scattering, one has

$$\frac{d}{d(-t)}\left(\frac{d\sigma}{dt}\right)\bigg|_{t=0}=-\tfrac{1}{2}\lambda\!\!\!^{-2}\left(\frac{d\sigma}{dt}\right)_{t=0}$$
$$\times\text{Re}\{[\textstyle\sum l(l+1)(l+\frac{1}{2})\alpha_l][\sum(l+\frac{1}{2})\alpha_l]^{-1}\}. \tag{18}$$

If α is real, this is always negative. For spin $\frac{1}{2}$-spin 0 scattering one has, for the spin-nonflip part, similarly

$$\left[\frac{d}{d(-t)}\left(\frac{d\sigma}{dt}\right)_{\text{nonflip}}\right]_{t=0}=-\tfrac{1}{2}\lambda\!\!\!^{-2}\left(\frac{d\sigma}{dt}\right)_{t=0}$$
$$\times\text{Re}\{[\textstyle\sum l(l+1)(l+\frac{1}{2})\alpha_l][\sum(l+\frac{1}{2})\alpha_l]^{-1}\}. \tag{19}$$

These two formulas hold for charge-exchange processes as well if we replace α by α^{ce}. If $\arg(\alpha^{ce})$ does not depend on l, (19) is always negative. Thus, the observed positive slope of the charge-exchange differential cross section near $-t=0$ means either that the spin-flip part contributes importantly in this region, or $\arg(\alpha^{ce})$ has large variation as l increases. The former alternative is the one chosen in our model.

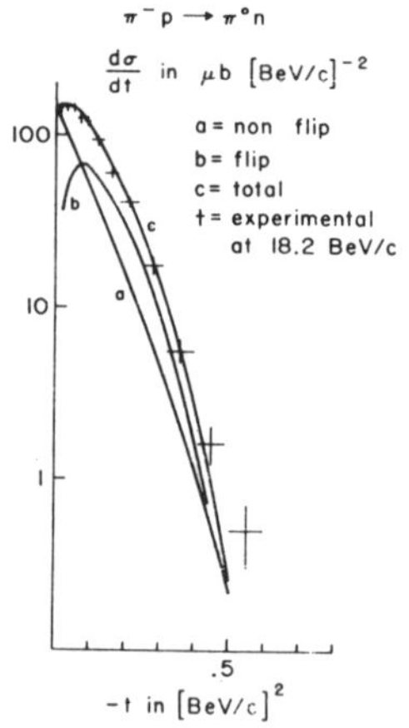

FIG. 2. $d\sigma/dt$ versus $-t$ for $\pi^-p\to\pi^0n$. The smooth curves are calculated values. Experimental data of Stirling *et al.* in Ref. 3 are indicated. Note that the data in Ref. 3 give very nearly parallel curves for $d\sigma/dt$ versus t, $0\leq -t\leq 0.4$, for lab momenta from 5.9 to 18.2 BeV/c. For $-t>0.4$, the data show faster decrease of $d\sigma/dt$ (with increasing $-t$) for higher energies. This trend suggests that our fit at $-t\approx 0.5$ will be better at higher energies.

Equation (19) for elastic scattering can be cast into the following form:

$$2\frac{d}{d(-t)}\ln\left(\frac{d\sigma}{dt}\right)_{\text{nonflip}}=-\text{Re}\langle b^2\rangle,$$

where $\langle b^2\rangle$ is the average of b^2 with weights $(2l+1)\alpha_l$. Thus, the logarithmic derivative with respect to t at $t=0$ of $d\sigma/dt$ for elastic scattering *gives a measure of the size of the interaction volume.* Using (1), one obtains

$$\langle b^2\rangle\approx 2\gamma.$$

(D) Equation (10) is applicable if the wave number difference before and after the charge exchange Δk is small. (If Δk is too large, it will destroy the coherence of phase over the interaction volume.) In the center-of-mass system, the value of

$$\Delta k \text{ is } \cong(m_1^2+m_2^2-m_3^2-m_4^2)/(4k).$$

With the Lorentz contraction taken into consideration one obtains, for high energies, the following condition for the preservation of coherence:

$$|m_1^2+m_2^2-m_3^2-m_4^2|Rp_L^{-1}\ll\pi,$$

where R is the radius of the interaction volume [$\sim 4(\text{BeV}/c)^{-1}$] and p_L is the laboratory momentum of the incident particle. This condition is well satisfied for $\pi^-p\to\pi^0n$ even at relatively low energies.

(E) In our model, the polarization perpendicular to the scattering plane of the neutron from an unpolarized target is easily computed:

$$P=\frac{2\,\text{Im}(A^*B)}{AA^*+BB^*}=\eta\xi(t), \tag{20}$$

[11] These calculations are easily effected numerically by first expanding (11) and (13) in powers of α. Each power α^n leads to an integral in (1′) and (2′) of the form

$$\int_0^\infty J_\nu(xb)\exp(-p^2b^2)b^{\mu-1}db=\frac{(x/2p)^\nu\Gamma[(\mu+\nu)/2]}{2p^\mu\Gamma(\nu+1)}\,{}_1F_1\left(\frac{\mu+\nu}{2},\nu+1,\frac{-x^2}{4p^2}\right)$$

and, since ${}_1F_1(\alpha,\gamma-z)=e^{-z}\,{}_1F_1(\gamma-\alpha,\gamma,z)$, it is easily evaluated. [See G. N. Watson, Ref. 9, Sec. 13.3.]

where

$$\eta = \sin[\arg(K_f^{ce}/K^{ce})],$$

$$\xi(t) = 2\left[\left(\frac{d\sigma}{dt}\right)_{\text{flip}}\left(\frac{d\sigma}{dt}\right)_{\text{nonflip}}\right]^{1/2}\left(\frac{d\sigma}{dt}\right)^{-1}. \tag{21}$$

Thus, if K_f^{ce} and K^{ce} are not relatively real, the polarization is proportional to $\xi(t)$ which, for our model, is quite large for $0.02 < -t < 0.6$. Its values are tabulated below:

$-t$	0.02	0.06	0.1	0.2	0.3	0.4	0.5
ξ	0.86	1.0	0.99	0.94	0.93	0.98	0.72.

It should be emphasized that while the diffraction peak in elastic scattering has predominantly an imaginary amplitude, the charge-exchange amplitudes, in our model, have complex phases equal to those of K^{ce} and K_f^{ce}. In general, these are neither real nor imaginary. For K^{ce}, experiments[3] indicate an appreciable phase since

$$\text{Re}A(0)/\text{Im}A(0) \cong 1.$$

There seems to be no *a priori* reason why η of Eq. (21) should be small. *A measurement of the neutron polarization in $\pi^- p \to \pi^0 n$, or of the right-left asymmetry in this process using a polarized target, is thus very interesting.*

In our picture, the spin-flip amplitude β in elastic scattering is small compared with α (this is generally assumed to be true); its value is given by an equation like (13); e.g.,

$$\beta = -K(1-\alpha)b^2 \ln(1-\alpha). \tag{13a}$$

If the spin-flip differential cross section is about 1% of the non-spin flip, our model could give a polarization as large as 20% when $0.02 < -t < 0.6$ [if $\arg K \approx \frac{1}{2}\pi$ (modπ)]. *It seems worthwhile to measure this polarization.*

Similarly, measurements of the polarization of the scattered proton in the backward peak in $\pi^+ p$ and $\pi^- p$ scattering would be interesting. The $\pi^- p$ backward peak[4] has a flattened top, resembling that of the forward $\pi^- p \to \pi^0 n$ process. In our model, this behavior indicates a large spin-flip contribution (see remark C above) possibly giving rise to a large polarization for the recoil proton.

(F) One can compare (11) with the assumption of the absorptive peripheral model[6,7] which is

$$\alpha^{ce} = (1-\alpha)\ \text{(one-particle exchange amplitude in lowest order perturbation calculation)}.$$

The absorption factor $1-\alpha = e^{2i\delta}$ is essential here since the one-particle exchange amplitude always gives too large values at small impact parameters. In our model, the factor $1-\alpha$ is not as crucial since "the material thickness" factor $\delta \propto \ln(1-\alpha)$ already quite effectively limits the value of α^{ce} for small impact parameters.

For extremely large energies, the collision time is short because of Lorentz contraction [see Fig. 1(a)]. During such short times the exchange of energy, momentum, and quantum numbers is an instantaneous process, and therefore not much related to the lowest mass state with those quantum numbers. In other words, it takes time for the low-mass effects to dominate, and there is not sufficient time for that in very high-energy collisions. This may account for the difficulties encountered by the absorptive peripheral model.

(G) It would be interesting to know whether for large angles, say 60°~90° in the center-of-mass system, the ratio of charge-exchange to elastic scattering becomes of the order of unity.[12] If so, one has a clear indication of the difference of the physical mechanisms for large- and for small-angle scattering.

(H) In our view, the amplitudes $a_l^{\pm}$ with specific quantum numbers dominantly follow a smooth function of b independent of parity, isotopic spin, and energy. However at energies $<\infty$, there are finite deviations from this smooth function. One can say that the difference between the spin-orbit alignment (parallel and antiparallel) gives rise to the small spin-flip amplitude β, the difference between $I=\frac{3}{2}$ and $I=\frac{1}{2}$ gives rise to the small charge-exchange scattering, and the difference between even and odd l gives rise to the small backward peaks in πp scattering (small compared to the dominant term which gives rise to the elastic forward peak). The differences all seem to $\to 0$ as $E \to \infty$. But it is perhaps more illuminating to view the situation the other way around: A spin-flip mechanism causes a small spin-flip amplitude β which can always be written as $a_l^+ - a_l^-$; a neutron-exchange mechanism causes a small backward scattering in $\pi^- p$ collisions which can be described as an even l–odd l difference; etc.

From this viewpoint, it is not surprising that the difference of a difference, for example, the charge-exchange spin-flip process, is not a second-order process which is much smaller than the charge exchange without spin flip. It would be interesting in this connection to *measure the backward $\pi^- p \to \pi^0 p$ cross section* and compare it with the forward $\pi^- p \to \pi^0 n$ and backward elastic $\pi^- p$ cross sections.

(I) Our model is essentially a droplet model of elementary particles, which are pictured as very much similar to nuclei. (It is interesting to compare the angular distributions discussed here with those in "high-energy" $\alpha-\alpha$ and α-nuclei scattering.[13]) Corresponding to the concept of the density of nuclear matter we now have a "density distribution" $\rho(r)$ which is related to α in the eikonal picture through

$$\ln(1-\alpha) = 2i\delta(b) = (\text{const})\int_{-\infty}^{\infty} \rho((b^2+x^2)^{1/2})\, dx. \tag{22}$$

[12] See, e.g., T. T. Wu and C. N. Yang, Phys. Rev. **137**, B708 (1965).

[13] For recent data, see P. Darriulat, G. Igo, H. G. Pugh, J. M. Meriwether, and S. Yambe, Phys. Rev. **134**, B42 (1964); P. Darriulat, G. Igo, H. G. Pugh, and H. D. Holmgren, *ibid.* **137**, B315 (1965) and footnote 10. We are indebted to G. E. Brown for informing us of these papers. Notice that the presence of a sharp edge in the nucleus accounts for the existence of many diffraction minima in α-Fe scattering.

If α is Gaussian as in (15), we can solve for ρ from (22), obtaining

$$\rho(r) = (\text{const})g_{1/2}(\alpha), \tag{23}$$

where $g_{1/2}$ is a familiar function in the theory of Bose-Einstein condensation of free particles:

$$g_{1/2}(\alpha)=\alpha+\alpha^2/\sqrt{2}+\alpha^3/\sqrt{3}+\cdots. \tag{24}$$

In presenting our model, we are suggesting that the same density function ρ determines the elastic and the charge-exchange scattering. [In other words, we assume the colliding particles to have no variation of "composition"; or in the language of potentials, in Eq. (12) we assume the same radial dependence for U and V.] This is a simple possibility, but at best only a crude approximation.

(J) It seems reasonable to us to describe the nucleon and pion as extended structures *without a localized central singularity*. Earlier fits to higher t scattering seem to demand such a singularity.[14] It seems to us that large t processes are due to a different physical mechanism and probably should not be fitted with a "coherent" model.

(K) Equation (15) indicates that in $\pi^- p$ collisions the interaction is not represented by a totally "black" region. If the spatial extension of higher meson resonances (say ρ) is not larger than that of the pion, there should be more interaction "per unit path length" in ρp scattering than in πp scattering. Consequently, ρp interaction should be "blacker" and should also contain more charge exchange. (We have implicitly assumed here that the "composition" of ρ is roughly the same as that of π.) We wonder whether it is possible to check these conclusions in reactions such as $\pi p \to \rho p$.

The question of the *spatial extensions* of resonances is an important one and has not, to our knowledge, been discussed in the literature. It seems to us unlikely that the resonances are much larger in size than the pion or the proton. Larger sizes lead usually to higher density of states. Experimentally, there seems to be very little increase in the density of resonant states per unit energy interval as one goes to higher resonance energies.

(L) One can ask, of course, what is ρ the density of? In other words, if p and π are extended structures, what are they made of? Clearly they are made of "stuff" which when isolated (if possible) and observed for long times would separate into mesons and hyperons, particles with extended structures themselves. In this respect, nucleons and pions are very different from a liquid drop or a nucleus.

The concept of ρ perhaps resembles more that of the probability density of the electrons in an atom. The charge-exchange scattering then resembles the scattering of x rays by an atom (except for the fact that a single atom is quite transparent to x rays).

Now, in the case of atomic physics, if one wants to study the interaction of x rays with the constituent parts of the atom (i.e., the electron), one does not study the *elastic* scattering of x rays by an atom. Instead, one studies the scattering of x rays by atoms with large momentum transfers to a single electron and very little recoil for the atom. One wonders whether similar but less clean-cut considerations should be applied to high-energy scatterings.

Is the concept of ρ useful only insofar as it enters into a coherent contribution to the amplitude of some process [with $\rho(r)d^3r$ not representing the density of any "stuff"]? In which way is it related to the possibility of "measuring" structures in spatial dimensions much less than one fermi by strongly interacting particles, which themselves have spatial extensions of $\sim 0.7 \to 1$ F? In what way is ρ related to the electromagnetic form factors? We do not know the answers to these questions.

(M) We have mentioned before the usefulness of experiments on large-angle exchange processes [remark (G)]. We also mentioned the interest in backward cross sections [remark (H)] and polarizations [remark (E)] in forward and backward directions. Many experiments of such types can be envisaged, involving $\pi p \to \Lambda K$, $pp \to pp^*$, $\pi p \to \rho p$, $pp \to p^* p^*$, etc. In those cases where the outgoing particles undergo decay processes, as, e.g., in $\pi p \to \Lambda K$, polarization measurements are, of course, relatively simple.

We have benefitted from many discussions with Professor J. R. Oppenheimer. One of us (N. B.) would like to thank him and the Institute for Advanced Study for their hospitality, and to acknowledge with gratitude the financial support of the National Science Foundation and the J. S. Guggenheim Foundation. She would also like to thank P. Falk-Vairant and the members of the Service de Physique Theorique (CEN Saclay) for the pleasure of a visit with them.

[14] R. Serber, Rev. Mod. Phys. 36, 649 (1964).

Remarks at the Dedication of the Einstein Stamp, March 14, 1966

C. N. Yang

[[66c]]

Commentary begins page 61

It is a great honor for me to have the opportunity to speak at this dedication to the memory of Einstein. The Institute for Advanced Study, which I am representing here today, is proud to have had Einstein as a Professor since its inception. I, myself, had the good fortune to attend some of his lectures and to have had a number of discussions with him when I first came to Princeton as a young physicist. But it is not of these associations that I am going to speak now. I speak as an admirer of Einstein, the greatest physicist of our time, and with Newton the two greatest physicists of all time.

The latter half of the nineteenth century brought to a successful conclusion the physical theories of matter in the bulk. The two crowning achievements of nineteenth century physics—electromagnetism, and thermodynamics–statistical mechanics—had laid down the complete framework within which the bulk properties of matter could be described. And yet there was something missing in such a description: How was one to understand the basic structural units for bulk matter? Already at the turn of the century physicists were uncovering strange phenomena totally unfamiliar to our experiences with matter in the bulk. There was the Michelson–Morley experiment on the velocity of light, there was the phenomenon of radioactivity, there was Planck's radiation law and there were many others. In short, physics was entering a new phase where properties of minute amounts of matter or energy were to be the central topic of investigation.

Into this new phase Einstein brought his unique genius and gave to us two revolutions in physical thinking: special relativity and general relativity, and contributed to and shaped a third—quantum mechanics. In the process of doing these, he not only made possible penetrating theories of atomic phenomena, he freed mankind from the concept of absolute time, the concept of Euclidean space, and brought forth a complete re-evaluation of what is meant by man's understanding of the physical world.

Einstein's work conveys his overpoweringly profound physical insight. He had a strong sense of beauty and form. He was bold and original, and yet he was also steady and stubborn. Let me quote for you in this connection the opening paragraph of what he said in 1935 at Madame Curie's death:

> At a time when a towering personality like Mme. Curie has come to the end of her life, let us not merely rest content with recalling what she has given to mankind in the fruits of her work. It is the moral qualities of its leading personalities that are perhaps of even greater significance for a generation and for the course of history than purely intellectural accomplishments. Even these latter are, to a far greater degree than is commonly credited, dependent on the stature of character.

Einstein was, himself, the embodiment of strength and endurance in the pursuit of science, and his work remains a source of inspiration and encouragement for the scientists that follow him.

[66e]
Commentary begins page 62

One-Dimensional Chain of Anisotropic Spin-Spin Interactions. I. Proof of Bethe's Hypothesis for Ground State in a Finite System

C. N. YANG
Institute for Advanced Study, Princeton, New Jersey
and
State University of New York, Stony Brook, New York

AND

C. P. YANG
Physics Department, Ohio State University, Columbus, Ohio
(Received 8 April 1966)

Bethe's hypothesis is proved for the ground state of a one-dimensional cyclic chain of anisotropic nearest-neighbor spin-spin interactions. The proof holds for any fixed number of down spins.

I. INTRODUCTION

THE eigenvalue spectrum of the Hamiltonian

$$H=-\tfrac{1}{2}\sum \{\sigma_x\sigma_x'+\sigma_y\sigma_y'+\Delta\sigma_z\sigma_z'\} \qquad (1)$$

is of current interest. In (1) $\boldsymbol{\sigma}$ are the Pauli spin matrices at a particular site ($\sigma_x^2=\sigma_y^2=\sigma_z^2=1$), $\boldsymbol{\sigma}'$ are the Pauli spin matrices at a neighboring site. Δ is a real numerical constant ($\Delta=1$ corresponds to the isotropic ferromagnetic problem, $\Delta=-1$ the isotropic antiferromagnetic problem[1]). The sum extends over all nearest neighbors in a 1-dimensional linear, 2-dimensional square, or 3-dimensional simple cubic lattices with *cyclic* boundaries.

The significance of (1) in the theory of ferromagnetism and the theory of antiferromagnetism is well known. (1) is also the problem to consider for the quantum lattice gas.[2] [In particular, the ground-state energy and the thermodynamical properties of a system with the Hamiltonian (1) can be transformed to give the ground-state energy and the thermodynamical properties of a quantum lattice gas. This quantum lattice gas is a Bose gas moving on a lattice with (a) a quantum kinetic energy, not in the form of an operator $(-\hbar^2/2\mathfrak{M})\nabla^2$, but in the form of a double difference, (b) a hard core preventing two atoms from occupying the same site, and (c) an energy of interaction equal to -2Δ for nearest neighbors. See Table I.]

Let y be the magnetization per site,

$$y=\text{eigenvalue of } (1/\mathfrak{N})\sum \sigma_z\,, \qquad (2)$$

where $\mathfrak{N}$ = total number of sites in the lattice. One is particularly interested in the function

$$f(\Delta,y)=\lim_{\mathfrak{N}\to\infty}\frac{1}{\mathfrak{N}z}\,(\text{lowest eigenvalue of } H \text{ for fixed } y)\,, \qquad (3)$$

which is half of the ground-state energy per bond for a fixed y. Here z is the number of nearest neighbors at each site. The existence of the limiting function $f(\Delta,y)$ was proved in Ref. 1. A number of general properties of f was also established there. In particular, inequalities were given between the f for one-, two-, and three-dimensional lattices.

The purpose of this and subsequent papers is to study properties of the Hamiltonian (1) for the one-dimensional linear cyclic chain.

This problem was studied by approximate methods by Bloch.[3] Bethe[4] then proposed that the eigenfunctions are of a certain specific form (to be called Bethe's hypothesis). The particular case $\Delta=-1$ (antiferro-

[1] C. N. Yang and C. P. Yang, Phys. Rev. **147**, 303 (1966).
[2] T. Matsubara and H. Matsuda, Progr. Theoret. Phys. (Kyoto) **16**, 569 (1956); **17**, 19 (1957); R. T. Whitlock and P. R. Zilsel, Phys. Rev. **131**, 2409 (1963); P. R. Zilsel, Phys. Rev. Letters **15**, 476 (1965).
[3] F. Bloch, Z. Physik **61**, 206 (1930); **74**, 295 (1932).
[4] H. A. Bethe, Z. Physik **71**, 205 (1931).

□ Reprinted from *The Physical Review* 150, 1 (October 7, 1966), 321–327.

TABLE I. Physical problems for different values of Δ.

	Quantum lattice gas with	Hamiltonian (1) is equivalent to
$\Delta>0$	Attractive interaction outside of hard core	Anisotropic ferromagnetic Hamiltonian ($\Delta=1$ corresponds to isotropic case)
$\Delta<0$	Repulsive interaction outside of hard core	Anisotropic antiferromagnetic Hamiltonian (Ref. 1) ($\Delta=-1$ corresponds to isotropic case)

magnetic isotropic case) was considered in detail by Hulthén,[5] who gave an evaluation of $f(-1,0)$ using Bethe's hypothesis. Later, Orbach[6] extended these considerations and obtained an integral equation which he numerically solved to evaluate $f(\Delta,0)$ for $\Delta\leqq-1$, again using Bethe's hypothesis. The integral equation was later solved by series expansion by Walker,[7] who obtained $f(\Delta,0)$ for $\Delta\leqq-1$ as a series. Griffiths[8] investigated the problem of $f(-1,y)$ and des Cloizeaux and Pearson[9] the excited states at $\Delta=-1$, $y=0$. (See Fig. 1.) Lieb, Schultz, and Mattis[10] and Katsura[10] studied the case $\Delta=0$.

In this series of papers we study the problem for general values of Δ and y. In the process we also establish *rigorously* the validity of Bethe's hypothesis for the ground state. These papers will use the same notation as Ref. 1 and will form a self-contained series.

2. BETHE'S HYPOTHESIS

We generalize in this section Bethe's hypothesis to the general case of $\Delta<1$.

Consider an eigenfunction ψ of H with m down spins and $\mathfrak{N}-m$ up spins. Clearly,

$$y=1-2(m/\mathfrak{N}). \tag{4}$$

We assume

$$2m\leqq\mathfrak{N}, \quad \text{or} \quad y\geqq0. \tag{5}$$

Let $x_1, x_2, \cdots, x_m$ (in ascending order) be the sites with down spins. ($1\leqq x_j\leqq\mathfrak{N}$). Bethe's hypothesis says that

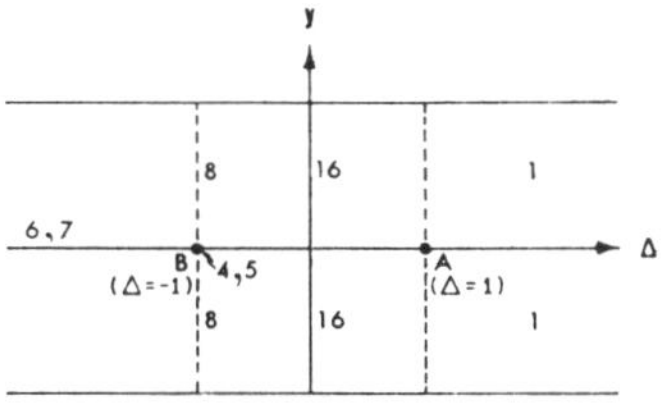

FIG. 1. Δ and y values where $f(\Delta,y)$ has been discussed in the literature. The numbers are the reference numbers quoted in this paper. The dotted line through A represents the isotropic ferromagnetic case. That through B represents the isotropic antiferromagnetic case.

[5] L. Hulthén, Arkiv. Mat. Astron. Fysik **26A**, No. 11 (1938).
[6] R. Orbach, Phys. Rev. **112**, 309 (1958).
[7] L. R. Walker, Phys. Rev. **116**, 1089 (1959).
[8] R. B. Griffiths, Phys. Rev. **133**, A768 (1964).
[9] J. des Cloizeaux and J. J. Pearson, Phys. Rev. **128**, 2131 (1962).
[10] E. Lieb, T. Schultz and D. Mattis, Ann. Phys. (N.Y.) **16**, 407 (1961); S. Katsura, Phys. Rev. **127**, 1508 (1962).

there are m *unequal real* numbers $p_1\cdots p_m$ such that the wave function ψ is a sum of $m!$ terms each of which is of the exponential form

$$(\text{constant})\ \exp i[p_{P1}x_1+p_{P2}x_2+\cdots], \tag{6}$$

where $(P1,P2,P3,\cdots Pm)$ is a permutation of 1, 2, 3, $\cdots m$. In other words,

$$\psi=\sum_P A_P \exp i[\sum_j p_{Pj}x_j]. \tag{7}$$

It will be further assumed[11] that the p's are within the following range:

$$-\pi<p_j<\pi, \quad \text{for} \quad \Delta\leqq-1; \tag{8}$$

$$-(\pi-\mu)<p_j<\pi-\mu, \quad \text{for} \quad -1\leqq\Delta<1; \tag{9}$$

where

$$0\leqq\mu<\pi, \quad \cos\mu=-\Delta. \tag{10}$$

Clearly, $\cos p_j>\Delta$. We plot the range of p_j in Fig. 2.

For large $\mathfrak{N}$ and m, but with $m/\mathfrak{N}=$fixed, the number of A_P's is larger than the number of spin arrangements. (7) is therefore not in general a hypothesis without further conditions on the A_P's. These conditions are stated below in (16) and (17) and form an integral part of Bethe's hypothesis.

We now examine the following points:

(a) Consider the equation $H\psi=E\psi$ at a configuration in which no down spins are nearest neighbors of each other. Write

$$H=-(\Delta/2)\mathfrak{N}-\tfrac{1}{2}\sum[\sigma_x\sigma_x'+\sigma_y\sigma_y'+\Delta\sigma_z\sigma_z'-\Delta].$$

The square bracket operating on any state for which the two spins in question are both up or both down gives zero. It is then easy to see that $H\psi=E\psi$ is satisfied for the configuration studied if

$$E=-(\Delta/2)\mathfrak{N}+\sum_j(2\Delta-2\cos p_j). \tag{11}$$

(One can see this most easily by taking $m=2$, then $m=3$, etc.)

(b) Consider the equation $H\psi=E\psi$ at a configuration in which among the down spins there is exactly one pair of nearest neighbors. Using (11) one sees that $H\psi=E\psi$ is satisfied if

$$\frac{A_P}{A_{P'}}=-\frac{2\Delta e^{ip}-1-e^{ip+iq}}{2\Delta e^{iq}-1-e^{ip+iq}},$$

[11] The original Bethe hypothesis was broader than that stated here. Our more restrictive form makes it easier to prove the validity of the hypothesis for the ground state.

where P and P' are any two permutations so that

$$p_{P1}, p_{P2} \cdots = \cdots p, q \cdots \tag{12}$$

and

$p_{P'1}, p_{P'2} \cdots$ = same as above except with p and q switched.

(These points are again easily proved first for $m=2$, then for $m=3$, etc.) Define[12]

$$\Theta(p,q) = +2 \tan^{-1} \times \left[\frac{\Delta \sin[(p-q)/2]}{\cos[(p+q)/2] - \Delta \cos[(p-q)/2]}\right]. \tag{13}$$

Notice

$$\Theta(p,q) = -\Theta(q,p). \tag{14}$$

Then

$$A_P/A_{P'} = -e^{-i\Theta(p,q)}. \tag{15}$$

Equations (14) and (15) lead to a solution of A_P in terms of A_0 (i.e., the A_P for P=identity):

$$A_P/A_0 = \pm \exp\{-i \sum \Theta(p_j,p_l)\}, \tag{16}$$

where the sign is + for P = even and − for odd and the summation extends over all pairs p_j, p_l for which $j>l$ and j stands to the left of l in the sequence $P1, P2, P3, \cdots$. (j and l need not be consecutive.)

(c) Consider the equation $H\psi = E\psi$ at other configurations. It is easy to prove that (15) ensures that $H\psi = E\psi$ is satisfied.

(d) The cyclic boundary condition must be imposed on (7). Using (7) the condition is fulfilled if for all P

$$A_P = A_{P''} \exp(p_{P1}\mathfrak{N}),$$

where

$$P''1, P''2, \cdots = P2, P3, \cdots Pm, P1.$$

Because of (16) this condition is in turn fulfilled if

$$\exp(ip_j\mathfrak{N}) = (-1)^{m-1} \exp[-i \textstyle\sum_l \Theta(p_j,p_l)], \quad j=1 \to m. \tag{17}$$

One of the possible sets of solutions[13] of this equation, upon taking the logarithm, is

$$\mathfrak{N} p_j = 2\pi I_j - \sum_{l=1}^{m} \Theta(p_j,p_l), \tag{18}$$

where

$$I_1, I_2, \cdots I_m = \left(-\frac{m-1}{2}\right), \left(-\frac{m-1}{2}+1\right), \cdots \left(\frac{m-1}{2}\right). \tag{19}$$

Notice[14] that for

$$\begin{aligned} m &= \text{even}, \quad I_j = \text{half-odd integer}, \\ m &= \text{odd}, \quad I_j = \text{integer}. \end{aligned} \tag{20}$$

Thus for every set of p_j satisfying (18), (8), *and* (9), *we can construct an eigenfunction* (7) *for the Hamiltonian* (1), by taking[15] $A_0 \neq 0$ and substituting (16) into (7).

3. SOME PROPERTIES OF THE FUNCTION Θ

It is convenient, to study $\Theta(p,q)$ in the interval (8) and (9), to apply the following transformations[16] $p \leftrightarrow \alpha$:

$$\Delta < -1: \ \Delta = -\cosh\lambda, \quad \lambda > 0, \tag{21a}$$

$$e^{ip} = \frac{e^{\lambda} - e^{-i\alpha}}{e^{\lambda - i\alpha} - 1}, \tag{21b}$$

$$-\pi < p < \pi \leftrightarrow -\pi < \alpha < \pi,$$

$$p(-\alpha) = -p(\alpha),$$

$$\cos p = -\cosh\lambda + \frac{\sinh^2\lambda}{\cosh\lambda - \cos\alpha},$$

$$\sin p = \frac{\sinh\lambda \sin\alpha}{\cosh\lambda - \cos\alpha},$$

$$\frac{dp}{d\alpha} = \frac{\sin p}{\sin\alpha} = \frac{\sinh\lambda}{\cosh\lambda - \cos\alpha} > 0, \tag{21c}$$

$$\Theta(p,q) = 2 \tan^{-1}\left[(\coth\lambda) \tan\frac{\beta-\alpha}{2}\right] \equiv \theta(\alpha,\beta). \tag{21d}$$

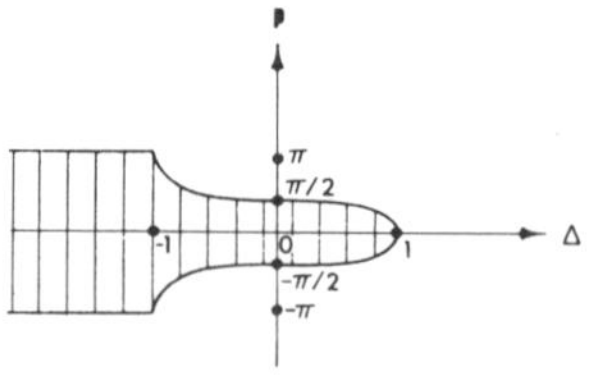

FIG. 2. Interval in which p lies.

[12] Θ is a single-valued real analytic function of Δ, p and q if the latter two are in the open interval given for p_j in (8) and (9). Θ(0,0) = 0. These conditions define uniquely the branch of $\tan^{-1}$ to take in (13). The function Θ becomes more visualizable after the transformation (21) to be discussed later. The range of values of Θ will also be given there.

[13] (18) is the same as the solution chosen by Bethe (Ref. 4), Hulthén (Ref. 5), and Orbach (Ref. 6) in their special cases. The notation here is, however, different from that in their papers. The main points in the difference are (a) we use $\tan^{-1}$ instead of $\cot^{-1}$ in (13). This difference results in our $\Delta I=1$ in (19), while in Orbach, the corresponding $\Delta\lambda=2$. (b) Our range of p as given in (8) is shifted by π from the previous convention. This is because our Hamiltonian (1) at $\Delta \leq -1$ is related to Orbach's by a unitary transformation. (See Ref. 1.) Our definition (13) and the range (8) and (9) are chosen to facilitate continuity arguments with respect to Δ which we shall need later on for proving Bethe's hypothesis.

[14] By making I_j half-odd integral for the case m = even, one can treat all values of m together. Notice that, however, this method works in the case of the quantum lattice gas only for bosons.

[15] Provided (7) is not identically zero for all x_j where $1 \leq x_j < x_2 \cdots < x_m \leq \mathfrak{N}$. This provision is probably satisfied for all $\Delta < 1$, $m \leq \mathfrak{N}/2$. We have so far, however, only succeeded in proving it, for each fixed $\mathfrak{N}$, for sufficiently small m. However, by a round-about argument in Sec. 5 we circumvent the necessity of an explicit proof.

[16] The transformation for the case $\Delta < -1$ was used by Walker (Ref. 7), and for $\Delta = -1$ by Hulthén (Ref. 5).

$-2\pi<\Theta<2\pi$, Θ = continuous in p and q.

$$\frac{\partial\theta}{\partial\beta}=-\frac{\partial\theta}{\partial\alpha}=\frac{\sinh 2\lambda}{\cosh 2\lambda-\cos(\alpha-\beta)}>0. \tag{21e}$$

$-1<\Delta<1$: $\Delta=-\cos\mu$, $0<\mu<\pi$, (21f)

$$e^{ip}=\frac{e^{i\mu}-e^{\alpha}}{e^{i\mu+\alpha}-1}, \tag{21g}$$

$-(\pi-\mu)<p<(\pi-\mu)\leftrightarrow-\infty<\alpha<+\infty$,

$p(-\alpha)=-p(\alpha)$,

$$\cos p=-\cos\mu+\frac{\sin^2\mu}{\cosh\alpha-\cos\mu},$$

$$\sin p=\frac{\sin\mu\sinh\alpha}{\cosh\alpha-\cos\mu},$$

$$\frac{dp}{d\alpha}=\frac{\sin p}{\sinh\alpha}=\frac{\sin\mu}{\cosh\alpha-\cos\mu}>0, \tag{21h}$$

$$\Theta(p,q)=2\tan^{-1}\left[(\cot\mu)\tanh\frac{\beta-\alpha}{2}\right]\equiv\theta(\alpha,\beta), \tag{21i}$$

$-|\pi-2\mu|<\theta<|\pi-2\mu|$,

$$\frac{\partial\theta}{\partial\beta}=-\frac{\partial\theta}{\partial\alpha}=\frac{\sin 2\mu}{\cosh(\alpha-\beta)-\cos 2\mu}. \tag{21j}$$

$\Delta=-1$: $\alpha=\frac{1}{2}\tan p/2$, (21k)

$-\pi<p<\pi\leftrightarrow-\infty<\alpha<+\infty$,

$$\frac{dp}{d\alpha}=4\cos^2\frac{p}{2}=\frac{4}{1+4\alpha^2}>0, \tag{21l}$$

$$\Theta(p,q)=2\tan^{-1}(\beta-\alpha)\equiv\theta(\alpha,\beta), \tag{21m}$$

$-\pi<\theta<\pi$,

$$\frac{\partial\theta}{\partial\beta}=-\frac{\partial\theta}{\partial\alpha}=\frac{2}{1+(\alpha-\beta)^2}. \tag{21n}$$

We notice that for all cases

$$\cos p=\Delta+\frac{2\pi}{C}\frac{dp}{d\alpha}, \tag{21o}$$

where

$$C=\frac{2\pi}{\sinh\lambda},\ \frac{2\pi}{\sin\mu}\quad\text{or}\quad 4\pi \tag{21p}$$

for the three cases, respectively.

Using these, and also the original form of Θ in (13), it is easy to see the following:

(a) Reference 12 is correct.

(b) Θ can be extended to the boundary of the open square (8) and (9) for p and q.

For $\Delta<-1$, there are no singularities of Θ in the closed square $-\pi\leqq p\leqq\pi$, $-\pi\leqq q\leqq\pi$. For $-1\leqq\Delta<1$, the only singularities of Θ in the closed square $-(\pi-\mu)\leqq p\leqq(\pi-\mu)$, $-(\pi-\mu)\leqq q\leqq\pi-\mu$ are at

$$p=q=\pi-\mu\quad\text{and}\quad p=q=-(\pi-\mu), \tag{22}$$

at which Θ is discontinuous.

(c) $\Theta(\pi-\mu,q)=2\mu-\pi$, (for $-1\leqq\Delta<1$), (23)

except at $q=\pi-\mu$, where Θ is discontinuous.

(d) $\Theta(-p,-q)=-\Theta(p,q)=\Theta(q,p)$. (24)

It is useful, for discussing Δ dependence, to make a further transformation (for all $\Delta<1$):

$$p\leftrightarrow\alpha\leftrightarrow a,$$

where

$$a=C\alpha/(2\pi)=\int_0^p\frac{dp}{\cos p-\Delta}. \tag{25}$$

The intervals (8) and (9) become

$$\frac{-\pi}{\sinh\lambda}<a<\frac{\pi}{\sinh\lambda}\qquad\Delta<-1, \tag{26a}$$

$$-\infty<a<\infty\qquad -1\leqq\Delta<1. \tag{26b}$$

Within this range a is analytic in p and Δ.

4. PROOF OF EXISTENCE OF SOLUTION FOR (18)

Consider the function [$\Delta<1$, p_j satisfying (8) and (9)]:

$$Z(p_1\cdots p_m,\Delta)=\sum_j r(a_j)-2\pi(\mathfrak{N}^{-1})\sum_j I_j a_j$$
$$+\tfrac{1}{2}\sum_{i,j}\mathfrak{N}^{-1}\Omega(a_i-a_j), \tag{27}$$

where C was defined in (21p), a in (25), and

$$r(x)=\int_0^x p\,da,\quad \Omega(x)=\int_0^x\theta\left(\frac{2\pi a}{C},0\right)da. \tag{28}$$

Clearly

$$\Omega(a_i-a_j)=\int_{a_j}^{a_i}\theta(\alpha,\alpha_j)da=\int_{p_i}^{p_j}\Theta(p,p_j)\frac{da}{dp}dp, \tag{29}$$

and

$$\Omega(a_i-a_j)=\Omega(a_j-a_i).$$

Thus, *Z is analytic in all p_j and Δ* for $\Delta<1$ and p_j in (8) and (9).

One has also by straightforward differentiation

$$\frac{\partial Z}{\partial p_j}=\frac{da_j}{dp_j}\left[p_j-2\pi(\mathfrak{N}^{-1})I_j+\sum_l\mathfrak{N}^{-1}\Theta(p_j,p_l)\right]. \tag{30}$$

Thus (18) is the condition for an extremum of Z at fixed Δ.

We are now in a position to prove

Theorem 1: For $m \leqq \mathfrak{N}/2$, $0 \leqq \Delta < 1$, (18) has a unique solution S so that each p_j is in (9). Each p_j is an analytic function of Δ. For any Δ, $p_i \neq p_j$ unless $i=j$.

Proof: (a) At $\Delta=0$, $\Theta=0$. (18) has then a unique solution satisfying this theorem.

(b) For $0 \leqq \Delta < 1$, Z as a function of $a_1, a_2, \cdots a_m$ has a positive-definite second-derivative matrix [*Proof:*

$$r''(x) > 0, \quad \Omega''(x) \geqq 0, \tag{31}$$

as is easily verified from (28), and (21j). Each term $\Omega(a_i - a_j)$ gives therefore a contribution to the second-derivative matrix that is positive (but not definite).] Z can thus have only one stationary point. To prove that it does have a minimum (for each Δ) at finite values of a, consider successively larger *closed* cubes C_i in p_j space approaching the open cube (9). We shall show that the position P_i of the minimum of Z in these closed cubes C_i cannot always lie on the boundary of C_i: If they always do, there would be an accumulation point P [on the boundary of the open cube (9)] of these minima P_i.

(α) Now suppose P is on the "surface" of the closed cube of (9). In other words at P, there is one p, say, p_j which is $=\pi-\mu$, all other $p<\pi-\mu$. We can approach P through a series of minima P_i at each of which $\partial Z/\partial p_j \leqq 0$, or

$$p_j - 2\pi(\mathfrak{N}^{-1})I_j + \sum_l \mathfrak{N}^{-1}\Theta(p_j, p_l) \leqq 0. \tag{32}$$

Approaching P we obtain, by (23) and the continuity of Θ,

$$(\pi-\mu) - 2\pi(\mathfrak{N}^{-1})I_j + \mathfrak{N}^{-1}(2\mu-\pi)(m-1) \leqq 0.$$

This is a contradiction since

$$I_j \leqq \tfrac{1}{2}(m-1), \quad \mu<\pi, \quad 2m \leqq \mathfrak{N}.$$

Similarly, P cannot be such that one $p=-(\pi-\mu)$.

(β) Suppose P is on an "edge" of the closed cube of (9). For example, at P, $p_j = p_l = \pi-\mu$, all other $p<\pi-\mu$. In this case we use the fact that at each P_i, since P_i is a minimum in a closed cube,

$$\frac{\partial Z}{\partial a_j} + \frac{\partial Z}{\partial a_l} \leqq 0.$$

That is,

$$p_j + p_l - 2\pi\mathfrak{N}^{-1}(I_j + I_l) + \mathfrak{N}^{-1} \times \sum_n [\Theta(p_j, p_n) + \Theta(p_l, p_n)] \leqq 0. \tag{33}$$

Using (24) and approaching P we obtain

$$2(\pi-\mu) - 2\pi\mathfrak{N}^{-1}(I_j + I_l) + \mathfrak{N}^{-1}\, 2(m-2)(2\mu-\pi) \leqq 0. \tag{34}$$

This is again a contradiction since $I_j + I_l \leqq \frac{1}{2}(m-1) + [\frac{1}{2}(m-1)-1]$.

(γ) Similarly we can prove that P is not on a "super-edge" of the closed cube of (9), etc.

(c) Thus some of the P_i are not on the boundary of the closed cube C_i. Such a P_i must give an absolute minimum of Z. Hence, Z has a unique minimum in the open cube (9), for each value of Δ. That minimum gives the unique solution of (18).

(d) Since the second-derivative matrix of Z has an inverse, one can evaluate $dp_j/d\Delta$ for every Δ in the interval $0 \leqq \Delta < 1$. This evaluation is also possible for complex values of Δ in the neighborhood of the interval. Thus p_j is an analytic function of Δ.

(e) (18) shows directly that if $p_i = p_j$, $I_i = I_j$, hence $i=j$.

Theorem 2: The solution discussed in Theorem 1 satisfies

$$p_j = -p_{m-j+1}, \quad j = 1 \to m. \tag{35}$$

Proof: For $m=$ even, consider p_j ($j=1\to m/2$) as dependent on p_j [$j=(m/2)+1 \to m$] through (35). For $m=$ odd, put $p_{(m+1)/2}=0$, and use (35) to eliminate half of the p's. Z as a function of the independent p's clearly has a positive-definite second-derivative matrix. We can prove that the minimum of Z does not lie at infinite values of a, just as in Theorem 1. Thus, Z has a minimum with respect to the independent p's satisfying (35). Using (24) one sees that (18) is satisfied at this minimum. But by Theorem 1 (18) has only one solution. Hence, Theorem 2.

Theorem 3: For $m \leqq \mathfrak{N}/2$, $\Delta \leqq 0$, (18) has solutions S forming a continuous curve in the real k_j ($j=1\to m$)$\times\Delta$ space with k_j satisfying (8) and (9). The curve extends from $\Delta=0$ down to all $\Delta<0$. At each point S on the curve

$$p_i \neq p_j \quad \text{unless} \quad i=j.$$

Furthermore,

$$p_j = -p_{m-j+1}, \quad j=1\to m. \tag{36}$$

Proof: (a) Consider the case $m=$ even. The case $m=$ odd can be treated similarly. Consider the cube $\mathfrak{C}$: $j=(m/2)+1 \to m$.

$$0 \leqq p_j \leqq (\pi-\mu)(1-\mathfrak{N}^{-1}) \qquad -1 \leqq \Delta \leqq 0, \tag{37a}$$

$$0 \leqq p_j \leqq \pi(1-\mathfrak{N}^{-1}) \qquad \Delta < -1. \tag{37b}$$

For every point in the cube $\mathfrak{C}$, we can construct a full set of p's satisfying (36). Clearly this full set lies in (8) and (9). Thus, Z is an analytic function of p and Δ in (cube $\mathfrak{C}$)$\times\Delta$. For

$$j=(m/2)+1 \to m,$$

$$\frac{\partial Z}{\partial p_j} = 2\frac{da_j}{dp_j}[p_j - 2\pi(\mathfrak{N}^{-1})I_j + \sum_l \mathfrak{N}^{-1}\Theta(p_j, p_l)]. \tag{38}$$

(b) At every point P in the cube $\mathfrak{C}$ there is a vector $v_j = -\mathfrak{N}[\quad]$ of (38). A stationary point of Z is a point where $v=0$. v is continuous in both P and Δ. Now on the boundary of $\mathfrak{C}$, the vector v is $\neq 0$ and always points *inward*. To prove this we discuss three points:

(α) For $p_j=0$, $v_j=[2\pi I_j]>0$.

(β) If $-1\leqq\Delta\leqq 0$ and $p_j=(\pi-\mu)(1-\mathfrak{N}^{-1})$, then by (21.10), (21.14), and (23),

$$\Theta(p_j,p_l)\geqq\Theta(\pi-\mu,\,p_l)=2\mu-\pi.$$

Thus,

$$\begin{aligned}v_j&\leqq-(\pi-\mu)(\mathfrak{N}-1)+2\pi I_j-(m-1)(2\mu-\pi)\\&\leqq-(\pi-\mu)(\mathfrak{N}-1)+\pi(m-1)-(m-1)(2\mu-\pi)\\&<0.\end{aligned}$$

(γ) If $\Delta<-1$, $0\leqq p<\pi$,

$$\Theta(p,p_l)+\Theta(p,\,-p_l)\geqq 2\Theta(p,0)=-4\tan^{-1}[(\coth\lambda)\tan(\alpha/2)]>-2\pi,$$

where the $\geqq$ sign can be verified by using (21.5) to calculate the derivative of its left-hand side with respect to p_l. Thus, if $p_j=\pi(1-\mathfrak{N}^{-1})$,

$$v_j<-\pi(\mathfrak{N}-1)+2\pi I_j+\pi m\leqq 0.$$

(c) Thus with respect to the vector $\mathbf{v}(P)$, the boundary of the cube $\mathfrak{C}$ has an *index* of 1. It follows from a theorem in topology[17] that there are solutions of $v=0$ which form a continuous curve in the product space of $\mathfrak{C}$ with Δ. We can then use (36) to construct a continuous curve in the product space of p_j with Δ. By (24) one easily verifies that (18) is satisfied on the curve.

(d) Obviously $p_i=p_j$ implies $I_i=I_j$, hence $i=j$.

5. PROOF THAT BETHE'S HYPOTHESIS IS VALID FOR THE GROUND STATE

We shall now use continuity arguments with respect to Δ to study the ground state. To do this we need

Theorem 4: The ground state of the Hamiltonian (1) for finite $\mathfrak{N}$ and m(m=no. of spins down) is nondegenerate for any real Δ. The ground-state energy is analytic in Δ for all real Δ.

Proof: The Hamiltonian is a matrix operator between the $\mathfrak{N}![m!(\mathfrak{N}-m)!]^{-1}$ spin arrangements. The off-diagonal elements of this matrix are -1 or 0. The diagonal elements can be all made negative if we subtract a large constant from H, i.e., there is a large number A so that $A-H$ has all elements $\geqq 0$, and all diagonal elements >0. A nonvanishing off-diagonal element connects every two spin arrangements with one pair of neighboring spins ↑↓ switched. Clearly, for large enough powers of $A-H$ all elements will be >0. Consider one such *odd* power: $(A-H)^n$. The largest eigenvalue of $(A-H)^n$ cannot be degenerate, since any corresponding wave function can be normalized so that all its elements are >0.

Now the eigenvalues are solutions of a polynomial equation with coefficients which are polynomials in Δ. Any nondegenerate solution must be analytic. Thus, Theorem 4 is proved.

Theorem 5: At $\Delta=0$, the solution of (18) is unique and gives through (20) the ground state of H.

Proof: At $\Delta=0$, $\Theta=0$. Thus (18) gives $\mathfrak{N}p_j=2\pi I_j$. (16) gives $A_P/A_0=\pm1$, + for even and − for odd permutations P. Thus, (7) becomes a determinantal wave function.

Now at $\Delta=0$, all eigenstates of H are known.[10] It is easily seen that the solution above is the ground state.

If the provision referred to in Ref. 15 is satisfied always along the solution S of Theorems 1 and 3, we have a wave function ψ for every point on S, with an eigenvalue E given by (11). The (E,Δ) plot forms a continuous curve extending over every real $\Delta<1$. Continuity in Δ and Theorems 4 and 5 would then lead to

Theorem 6: For any real $\Delta<1$ and for $2m\leqq\mathfrak{N}$, the ground state is given by Bethe's hypothesis as stated in (20). Furthermore,

$$p_j=-p_{m+1-j}\quad j=1\to m.$$

Proof: We need only examine the provision of Ref. 15. The main idea is to show that the point where the provision is not valid is discrete and therefore could be rendered harmless. This is done by showing that each element of the wave function (7) is algebraic in Δ:

(a) Put
$$u_j=\exp(ip_j). \tag{39}$$

Then

$$\exp[-i\Theta(p_j,p_l)]=\frac{2\Delta u_j-1-u_ju_l}{2\Delta u_l-1-u_ju_l}. \tag{40}$$

Now (18) implies (17) which becomes

$$u_j^{\mathfrak{N}}=(-1)^{m-1}\prod_l\frac{2\Delta u_j-1-u_ju_l}{2\Delta u_l-1-u_ju_l}. \tag{41}$$

[Notice, however, that solutions u_j of (41) may not satisfy (18).] One can eliminate all u's but one from (41), obtaining an equation

$$\mathcal{P}(u,\Delta)=0, \tag{42}$$

which is satisfied by u_1, by u_2, $\cdots$ by u_m. $\mathcal{P}$ is a polynomial of u and Δ.

Thus, each u is an algebraic function of Δ. u has no more than a finite number of cuts and poles. Furthermore, it has only a finite number of Riemann sheets.

(b) Now (16) and (40) show that A_P/A_0 is a rational function of Δ and the u's. Thus, after (16) is substituted into (7) and A_0 put $=1$, we obtain a ψ every element of which is a rational function of Δ and the u's. Define

$$\psi'=\psi\prod_{j,l}(2\Delta u_l-1-u_ju_l). \tag{43}$$

Every element of ψ' is *a polynomial of* Δ *and the* u's. At $\Delta=0$, $\Theta=0$ and all the u's are, by (18) and (39), on the unit circle and have positive real parts. Thus, at $\Delta=0$ the product in (43) is not zero and ψ' is the genuine (i.e., nonvanishing) ground-state wave function.

We have thus ψ' and E, both as polynomials in Δ and the u's so that

$$H\psi'=E\psi', \tag{44}$$

[17] P. Alexandroff and H. Hopf, *Topologie* (Springer-Verlag, Berlin, 1935).

if the u's satisfy (41). (41) defines the u's as algebraic functions of Δ. Thus, in complex Δ space except at the poles of the u's and at points where $\psi'=0$, ψ' is an eigenstate of H. These exceptional points are finite in number. We can obtain a correct eigenfunction ψ'' at these points too by properly normalizing ψ' and approaching these exceptional points. Hence, Theorem 6. (In fact the above proves a generalization of Theorem 6 to complex Δ.)

We can also prove the following theorem, which clarifies but is not essential for later discussions.

Theorem 7: The p's are analytic in Δ in an open strip containing the semi-infinite real axis $\Delta<1$.

Proof[18]*:* (a) Starting from $\Delta=0$, and moving along the real axis towards $\Delta=-\infty$, let $\Delta=\Delta_1$ be the first singularity of the u's, if any is in the way. We can form a simple closed path that loops around Δ_1 and return to $\Delta=0$, which does not pass through and does not contain, inside of it, any other singularities of any u. Now $E(\Delta)$ is analytic along the real axis, by Theorem 4. Furthermore, it is a polynomial in u. Thus, E has no singularity on or in the path and it returns to the original value when Δ goes around the path back to $\Delta=0$. Thus, ψ' returns also to the ground-state wave function at $\Delta=0$, except for a possible multiplicative factor. This wave function is a determinant. Consider its values when $x_1=1$, $x_2=2$, $\cdots x_{m-1}=m-1$, but successively $x_m=m$, $m+1$, $m+2$, $\cdots$. Its values are in the ratio of 1, $\sum u$, $\sum u^2+\sum_{j>l} u_j u_l$, $\cdots$. Thus, all symmetrical polynomials of the u's return to their original values around the loop. Hence, the u's are merely permuted in going completely around the loop. Call that permutation $P(\Delta_1)$.

(b) For $0\leqq\Delta<1$, u_j is on the unit circle. By analytic continuation, it must remain so for $\Delta_1<\Delta<0$. Thus, $p_j=-i\ln u_j$ is analytic for $\Delta_1<\Delta<1$. For $0\leqq\Delta<1$, Theorem 1 shows that (18) is satisfied. Continuing all p's to values of $\Delta<0$, (18) remains satisfied until either we reach the point Δ_1, or the p's go outside of the limits defined in (8) and (9). The latter alternative, however, does not obtain, since before the p's reach the boundary, the corresponding point must go out of the surface of the cube (37). Part (b) of the proof of Theorem 3 demonstrates that that is not possible. Thus, (18) is satisfied for all $\Delta_1<\Delta<1$.

(c) Δ_1 is not a pole for the u's, since $|u|=1$ for $\Delta=\Delta_1+0$. Since each u_j is algebraic in Δ, it has a definite value at $\Delta=\Delta_1$. (18) shows that at $\Delta=\Delta_1$, all p's are unequal. Hence, all u's are unequal.

(d) Now tighten the loop of (a) around Δ_1. Since all u's are unequal at Δ_1, the permutation $P(\Delta_1)$ must be the identity. Thus, Δ_1 is not a branch point of any u. Contradiction.

[18] One can rearrange the theorems so that the topological theorem is not needed: After Theorems 1 and 2, 4, and 5 the concept of u of (39) is introduced, together with the ψ' of (43), leading to $H\psi'=E\psi'$ for complex Δ. One then proves Theorem 7, using in part (b) of the proof the discussions following Eq. (38). This proof of Theorem 7 then automatically establishes (18) for all $\Delta<1$, with all p's within the bounds (8) and (9).

ACKNOWLEDGMENTS

It is a pleasure to acknowledge many extremely helpful discussions with Hasler Whitney and Andrew Lenard.

Treatment of Overlapping Divergences in the Photon Self-Energy Function*)

⟦66h⟧

Commentary begins page 64

R. L. MILLS†) and C. N. YANG

State University of New York at Stony Brook
Stony Brook, New York, U. S. A.

(Received August 29, 1966)

An analysis is made of the generalization of Ward's identity to the problem of renormalizing the photon self-energy function in quantum electrodynamics. It is shown that it is nontrivial to select a path of differentiation in a consistent way through the different self-energy graphs, but that a suitable set of simple rules can be given which completely resolves the overlapping divergence difficulty in the photon self-energy function.

Introduction

In the renormalization program for quantum electrodynamics, the straigtforward procedure of renormalizing divergent subgraphs in a systematic way is possible only if no two divergent proper subgraphs**) overlap, that is, if for any two such divergent parts, either one is entirely within the other, in which case the former is to be renormalized first, or else it is entirely outside, in which case they can be renormalized independently. The problem of handling overlapping divergences, which in quantum electrodynamics occur only within proper self-energy graphs, has long been a source of confusion in efforts to present the subject of renormalization in the clear and elegant manner it deserves. While the problem itself has been treated quite generally,[1)–5)] an elegant treatment[6)] exists only in the case of the electron self-energy; here Ward's identity[7)] provides a simple relation between the derivative of the electron self-energy function and the vertex function, and the latter can be renormalized straightforwardly without overlap problems. A similar relation[5),6)] does exist for the photon self-energy function, which renders the renormalization procedure equally straightforward; nevertheless, the manner of differentiation is not immediately so clear as in the electron case because of the difficulty in specifying a path of differentiation in a consistent way through every possible proper self-energy graph. It is the purpose of this note to clarify

*) Supported by the New York State Foundation of Science and Technology.

†) Permanent address: The Ohio State University, Columbus, Ohio, U. S. A.

**) A 'subgraph' will refer to a portion of a graph (a set of vertices and those lines which interconnect them) which does not include the whole graph. A 'proper' subgraph is one which cannot be disconnected by the removal of just one line.

□ **Reprinted from *Supplement of the Progress of Theoretical Physics* 37 and 38 (1966), 507–511.**

this procedure.

In the interest of clarity we ignore here the matter of infrared divergence. Our concern is solely with the overlap problem for ultraviolet divergences. We shall also ignore here the overlap problem which arises in four-photon graphs.

Let us first consider briefly the case*) of the electron self-energy function $S'(p)$, whose terms, in the Feynman-Dyson perturbation expansion, contain overlapping divergent subintegrals. This difficulty can easily be handled[6),7)] by solving for the derivatives $(\partial/\partial p^\alpha)S'^{-1}(p)$ and deducing $S'(p)$ from these, subject to the proper boundary conditions on the renormalized self-energy function. It was shown by Ward[6),7)] that the vertex function Γ is related to this derivative by

$$\frac{\partial}{\partial p^\alpha}S'^{-1}(p)=\Gamma_\alpha(p,p). \tag{1}$$

Now the graphs for Γ do not contain overlaps. So we can write

$\Gamma_\alpha(p,p)=\gamma_\alpha+\Sigma$ contributions from all skeleton graphs with insertions Γ, S', D' for all vertices and propagators. (2)

Difficulty with $\dfrac{\partial\pi}{\partial k}$

It appears, at first sight, that by defining the derivative of the proper photon self-energy function π as a new type of "vertex function" Δ, one could follow exactly the same procedure as above. Upon closer examination, however, we found that one has to use caution in this procedure, since the differentiation of any particular π graph and the subsequent reduction to skeleton graphs depends on the path chosen for the external momentum through the graph. An explicit example of the danger associated with a wrong choice of path was exhibited in reference 5).

We found (cf. reference 5), p. 1439) that to resolve the difficulty one way is to introduce a separation of each π graph into a sequence of core parts. If the core part contribution C is introduced explicitly, it turns out that the overlaps can be explicitly disentangled and the choice of path problem handled systematically by writing down equations similar to (2), which when iterated give correctly all the terms of π. This procedure will be described now.

Core part C

A π graph in which all interior self-energy parts are shrunk out and given the value S' or D' is called partially reduced. A partially reduced π

*) For simplicity, we write D for D_F and S for S_F in this paper.

graph can be uniquely separated into a sequence of core graphs in the manner shown in Fig. 1. Each core graph has one incoming electron line and one outgoing electron line on the left and similarly on the right. It must also be such that by breaking any two internal electron lines, one cannot separate it into two disjoint parts, one on the left, the other on the right. We shall refer to the electron line that enters a core part from the left or goes out of a core part towards the right as the *upper* electron line. If we denote by $C(k_1, k_2, k_3)$ the contribution from all core graphs with the same external momenta k_1, k_2, k_3 and $-k_1-k_2-k_3$ we have, as illustrated in Fig. 2,

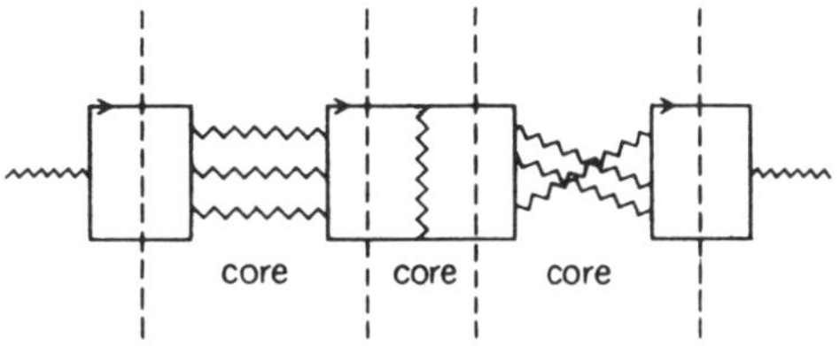

Fig. 1. Separation of a partially reduced π graph into a sequence of core graphs. In this example the π graph is separated into three core graphs as marked.

$$\pi(k) = \gamma S'S'\gamma + \gamma S'S'CS'S'\gamma + \gamma S'S'CS'S'CS'S'\gamma + \cdots. \tag{3}$$

The spin indices are suppressed in this formula in an obvious way.

This formula is not suitable for renormalization; the presence of γ, which prevents proper cancellation of the renormalization factors Z_1 and Z_3, is related to the occurrence of overlapping divergences. To resolve this difficulty we differentiate (3) after assigning in Fig. 2 the external momentum k to the *upper* electron lines S'.

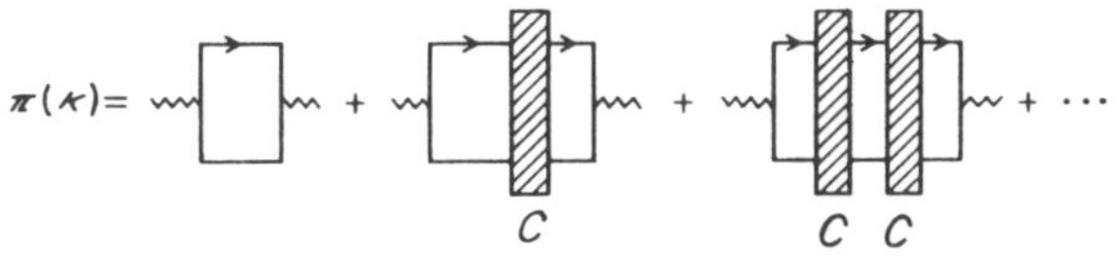

Fig. 2. π as a sum of terms involving C. All electron lines are S', not S.

$$\frac{\partial\pi}{\partial k} = \gamma\frac{\partial S'}{\partial k}S'\gamma + \gamma\frac{\partial S'}{\partial k}S'CS'S'\gamma + \gamma S'S'\frac{\partial C}{\partial k}S'S'\gamma + \gamma S'S'C\frac{\partial S'}{\partial k}S'\gamma + \cdots. \tag{4}$$

Using

$$\Gamma = \gamma + \gamma S'S'C + \gamma S'S'CS'S' + \cdots \tag{5}$$

it is obvious that (4) becomes

$$\frac{\partial\pi}{\partial k} = \Gamma\frac{\partial S'}{\partial k}S'\Gamma + \Gamma S'S'\frac{\partial C}{\partial k}S'S'\Gamma, \tag{6}$$

a formula illustrated in Fig. 3.

This removes the overlapping divergences completely since C can be reduced to skeletons without overlaps. A skeleton C graph is one in which all vertex parts are shrunk out and given the value Γ. Therefore,

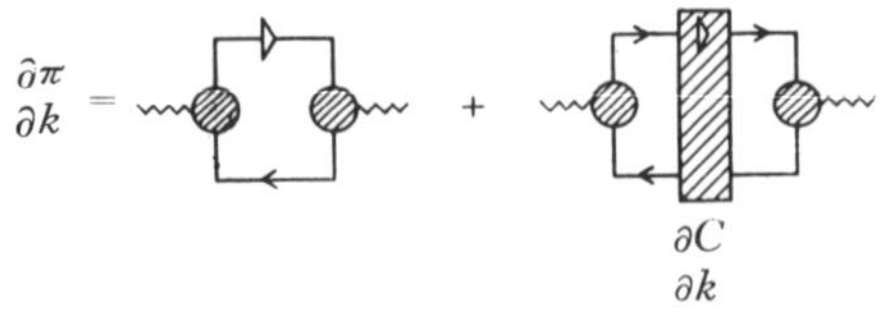

Fig. 3. Graph corresponding to Equation (6). A triangle denotes derivative with respect to k. All electron lines are S', not S.

$$C = \Sigma \text{ contribution from all skeleton } C \text{ graphs with insertions } \Gamma,\ S',\ D' \text{ for all vertices and propagators.} \tag{7}$$

as illustrated in Fig. 4.

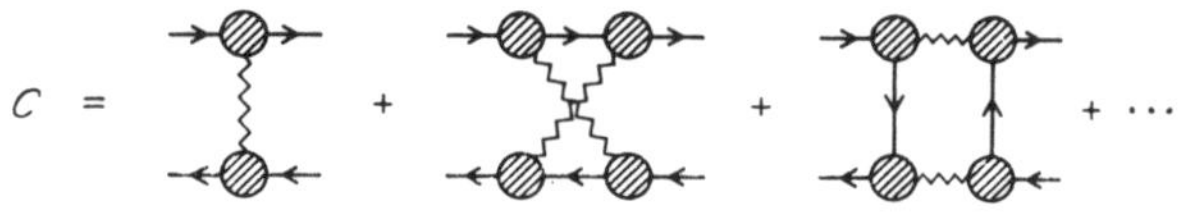

Fig. 4. C as a sum over skeletons. All internal lines are D' and S'. The blobs are Γ.

Renormalization

Equations (1) (2), (6) and (7) are equations involving S', D', Γ and C. Iteration of these equations gives the correct Feynman diagrams for S', D', Γ and C.

Furthermore, these equations are of the correct form for renormalization, since *the overlap problem*, as exhibited by the appearance of factors γ in (3) *is eliminated* through the differentiation process which leads to the replacement of (3) by (6).

To complete the renormalization program one would still have to show that subtractions from (6) and (2) actually render the integrals convergent. This is a problem that we do not propose to discuss here. But we do want to sketch a demonstration that this problem is not more involved in (6) than in the skeleton Γ diagrams. To do this we shall show that $\partial\pi/\partial k$ can be written as a sum over skeleton graphs with all possible self-energy and vertex insertions (including lower-order $\partial\pi/\partial k$ subgraphs). It is not necessary that all possible skeleton $\partial\pi/\partial k$ graphs appear, as indeed they do not, but it is necessary that all graphs with the same skeleton be included in order that the removal of renormalization factors should correspond to subtractions in the various subgraphs.

To accomplish this, we use Eq. (6). In the first term on the right we use Ward's identity (Eq. (1)), and, for the second, we give rules for differentiating C. In each C graph the path of differentiation must enter and leave by the upper external lines, as noted above. For each skeleton C graph we arbitrarily specify a standard path; then in any C graph we follow the standard path through the skeleton, use (1) and (6) for self-energy subgraphs, and an analogous prescription for Γ subgraphs. That is, for each skeleton Γ graph we choose an arbitrary standard path (one for each way of entering and leaving), and for reducible Γ graphs we use the standard path for the skeleton, and the above procedures at lower orders for self-energy and vertex subgraphs. It is convenient, though not essential, to follow electron lines wherever possible, and to follow the electron direction along each electron line.

One must then prove that the graphs generated by the above procedure satisfy the requirement discussed above. This can be done by induction. (Note in particular that a skeleton $\partial C/\partial k$ graph, for example, may arise from the differentiation of a nonskeleton C graph.)

Acknowledgement

One of the authors (RLM) wishes to thank the members of the Physics Department and the Institute for Theoretical Physics at Stony Brook for the hospitality he enjoyed during his visit.

References

1) A. Salam, Phys. Rev. **82** (1951), 217; **84** (1951), 426.
2) S. Weinberg, Phys. Rev. **118** (1960); 838.
3) J. D. Bjorken and S. D. Drell, *Relativistic Quantum Fields* (McGraw Hill Book Company, Inc., New York, 1965), Chap. 19.
4) N. N. Bogoliubov and D. V. Shirkov, *Introduction to the Theory of Quantized Fields* (Interscience Publishers, Inc., New York, 1959), Sec. 26.
Klaus Hepp, Communications in Math. Phys. (in print).
5) T. T. Wu, Phys. Rev **125** (1962), 1436.
6) J. C. Ward, Proc. Phys. Soc. (London) **A64** (1951), 54.
7) J. C. Ward, Phys. Rev. **78** (1950), 182(L).

[[67b]]

Commentary begins page 65

SOME REMARKS CONCERNING HIGH ENERGY SCATTERING

T. T. CHOU and C. N. YANG

Institute for Theoretical Physics. State University of New York, Stony Brook, New York, USA

1. INTRODUCTION

We wish to report some speculative considerations assuming the usefulness of formulating small angle high energy scattering in terms of the collision of two "droplets" with spatial structure. We have heard here in the past few days as well as this morning descriptions of nuclear scattering through a density function for the nucleus. The density function in such discussions refers of course to the density of matter or of nucleons. If the nucleon itself is to be described as an extended structure with a density function [1] of some "stuff", what this "stuff" is is unclear. However some features of the scattering process are independent of the detailed nature of the "stuff". In fact comparing small angle nucleon-nucleus scattering at high energies with small angle pion-nucleon or nucleon-nucleon scattering at high energies one could not fail to see their similarity. Such similarity has led to extensive discussions based on various views of describing pion-nucleon and nucleon-nucleon scattering as "diffraction" phenomena. We shall not go into any of these discussions in detail. We shall here only first emphasize three particular features which seem cogent to us in demonstrating the usefulness of describing high energy pion-nucleon and nucleon-nucleon scattering in terms of the interaction of extended structures. Then we shall report some speculative considerations (based on the idea that nucleons and pions are extended structures) concerning the form factors of nucleons and pions and concerning pion-pion scattering at high energies.

2. EVIDENCE FOR EXTENDED STRUCTURE

A. The nucleon as a source in its interaction with the electromagnetic field is well known to have an extended structure measurable in high energy electron nucleon scattering.

B. The high energy large-angle elastic scattering between two nuclei is extremely small. In a dramatic experiment [2] a few years ago it was demonstrated that also in proton-proton scattering the cross section at large angles is extremely small compared with that at small angles. (At 30 GeV/c incident energy the ratio is a factor of $\sim 10^{10}$). Now the smallness of the

□ Reprinted from *High Energy Physics and Nuclear Structure*. Amsterdam: North–Holland, 1967, 384–359.

large angle elastic scattering in nuclear collisions is easy to understand: extended structures shatter and break up if large momentum transfers take place between the colliding particles. It seems natural to assume [3] that this is also the mechanism for the smallness of the large angle high energy elastic p-p cross section.

C. There has been extensive analysis of the phase shifts in pion-nucleon scattering below 2 GeV/c. The transmission factor,

$$S = \exp(2i\delta) \; ,$$

which is a complex number with an absolute value $\leqslant 1$, can be plotted as a point in the unit circle.

Figs. 1, 2 and 3 exhibit some typical plots at three different energies. One observes that at these energies the points are scattered over the whole unit circle. However at higher energies the points line up along a relatively smooth line with S being essentially only a function of the impact parameter

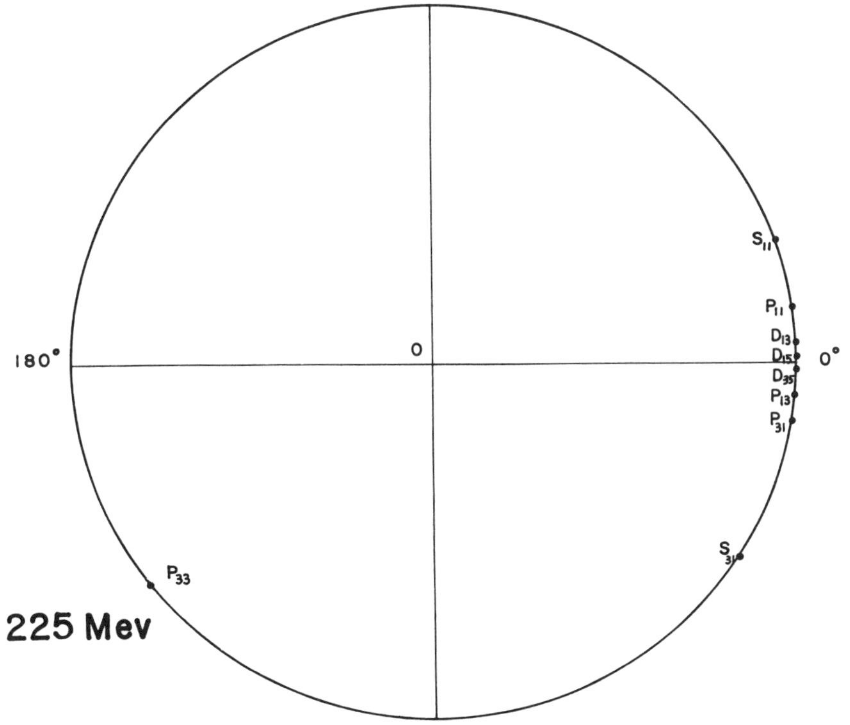

Fig. 1. Plot of the transmission factor $S = \exp(2i\delta)$ for various pion-nucleon phase shifts at 225 MeV energy. The numbers used in figs. 1, 2 and 3 are based on a compilation by A. Donnachie, R. Kirsopp, A. Lea and C. Lovelace.

$$b = (l + \tfrac{1}{2})\lambda\!\!\!^{-} .$$

The evidence for this remarkable regularity are: (i) The elastic differential cross section seems to approach at high energies a function only of the momentum transfer:

$$\frac{d\sigma}{dt} \rightarrow f(t) , \tag{1}$$

where

$$-t = 2k^2 (1 - \cos\theta)$$

where k = centre of mass momentum, and θ = centre of mass angle. (ii) The extreme smallness of the differential cross section at any but the forward angles makes it impossible for the value of S (for any specific angular momentum and parity) to deviate significantly from the smooth average behavior, since any such deviation would result in cross sections at large

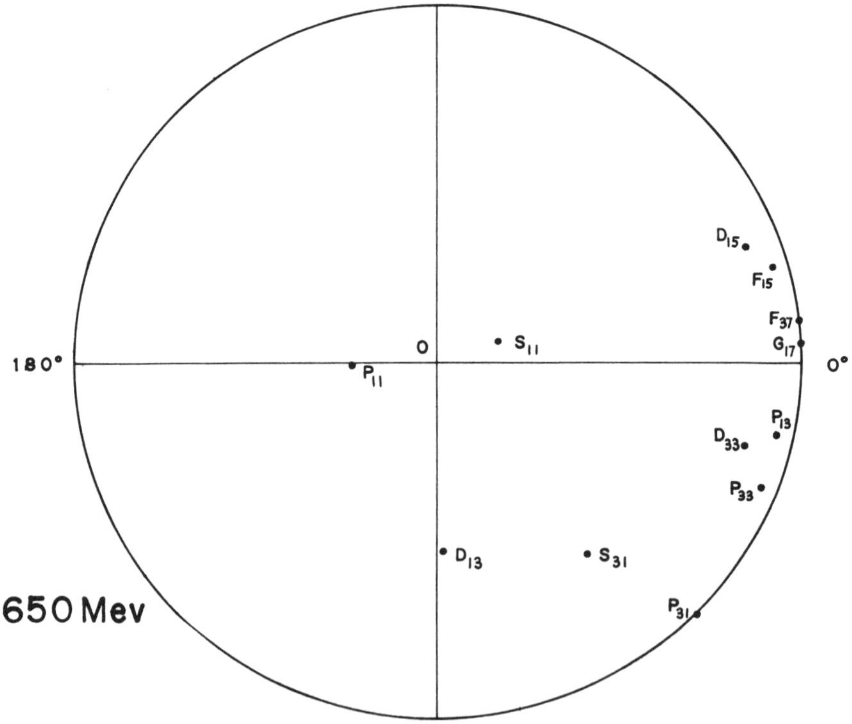

Fig. 2. Plot of the transmission factor $S = \exp(2i\delta)$ for various pion-nucleon phase shifts at 650 MeV energy.

angles many orders of magnitudes larger than the observed cross sections. To add to the strength of this argument a recent experiment [4] has shown that the large angle differential cross section is remarkably free from fluctuations with respect to angle and with respect to the incoming energy.

(iii) The backward πp peak is very small compared with the forward peak. The ratio is $\sim (3000)^{-1}$ at 3.6 GeV/c and falls rapidly with higher incoming energy. Thus the even l and odd l contributions f_{even} and f_{odd} must be approximately the same:

$$\left|\frac{f_{\text{even}} - f_{\text{odd}}}{f_{\text{even}} + f_{\text{odd}}}\right| \cong \frac{1}{\sqrt{3000}} < 2\% \ .$$

(iv) Similarly the charge exchange $\pi^- p \to \pi^0 n$ peak is small compared with the elastic forward peak and the ratio decreases rapidly with higher incoming energy. At 12 GeV/c the difference between the $I = 3/2$ and $I = 1/2$ amplitudes $f_{3/2}$ and $f_{1/2}$ is

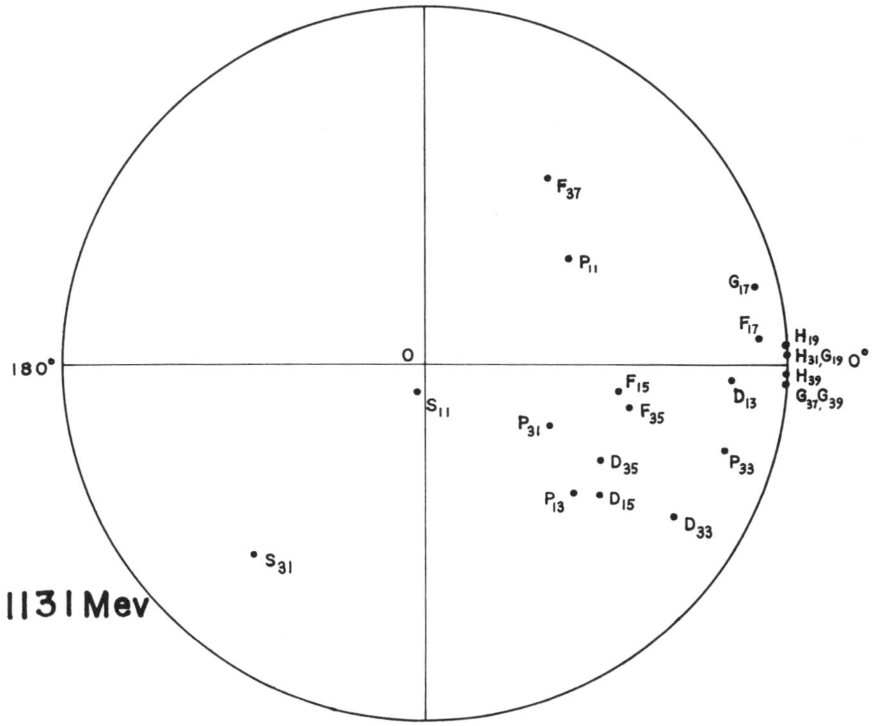

Fig. 3. Plot of the transmission factor $S = \exp(2i\delta)$ for various pion-nucleon phase shifts at 1131 MeV energy.

$$\left|\frac{f_{3/2} - f_{1/2}}{f_{3/2}}\right| \sim 15\% \ .$$

If it is true that as the incoming energy approaches infinity, eq. (1) holds and that all spin dependence and isospin dependence disappear so that S is a function only of the impact parameter $b = \lambda(l + \frac{1}{2})$, one should ask *what is the meaning of the function* $f(t)$ in eq. (1). The simplest answer that one could give to such a question is to regard the scattering as due to interactions taking place in a finite spatial region. To simplify further considerations we shall also regard the interaction as purely absorptive at infinite energies so that S is real.

3. FORM FACTORS OF PROTON AND PION

Under the assumptions stated above it is tempting to regard very high energy scattering as due to two absorptive spheres going through each other. To each sphere (proton or pion) the other sphere effectively appears as a disc with "blackness" given by

$$D(x, y) = \int_{-\infty}^{\infty} \rho(x, y, z)\mathrm{d}z \ , \tag{2}$$

where $\rho(x, y, z)$ = density of blackness. $\rho(x, y, z)$ is of course spherically symmetrical while $D(x, y)$ is cylindrically symmetrical.

For the collision between a proton and a pion the resultant blackness at an impact parameter b is therefore given by

$$F(b) = \iint_{-\infty}^{\infty} D_\pi(x, y) D_\mathrm{p}(b - x, y)\mathrm{d}x\mathrm{d}y \ . \tag{3}$$

Now we assume the transmission factor at impact parameter b to be

$$S(b) = \exp[-(\text{constant})\ F(b)] \ . \tag{4}$$

This assumption is in the spirit of the usual description of the absorption of a wave propagating through a medium. It is also the spirit behind the assumption eq. (10) in ref. [1] on the coherent droplet model.

Thus

$$\ln S_{\pi\mathrm{p}}(b) = -\ K_{\pi\mathrm{p}} \iint D_\pi(x, y) D_\mathrm{p}(b - x, y)\mathrm{d}x\mathrm{d}y \ . \tag{5}$$

Similarly

$$\ln S_{\mathrm{pp}}(b) = -\ K_{\mathrm{pp}} \iint D_\mathrm{p}(x, y) D_\mathrm{p}(b - x, y)\mathrm{d}x\mathrm{d}y \ , \tag{6}$$

$$\ln S_{\pi\pi}(b) = -K_{\pi\pi} \iint D_{\pi}(x, y) D_{\pi}(b - x, y) dx dy . \tag{7}$$

By taking two dimensional Fourier transforms (5) and (6) can be readily solved for D_p and D_{π}. Substitution into eq. (2) then leads to the density functions ρ_p and ρ_{π}. Notice that the *radial variation of these functions are dependent on the constants K only in normalization.*

We take the limiting πp and pp scattering cross sections [5] to be

$$(d\sigma/dt)_{pp} = 79.04\ e^{-10.3t}\ \mathrm{mb}\,(\mathrm{GeV}/c)^{-2} , \tag{8}$$

$$(d\sigma/dt)_{\pi p} = 31.86\ e^{-10t}\ \mathrm{mb}\,(\mathrm{GeV}/c)^{-2} . \tag{9}$$

It follows that [b in units $(\mathrm{GeV}/c)^{-1}$],

$$S_{pp}(b) = 1 - 0.78\ e^{-0.0485\ b^2} , \tag{10}$$

$$S_{\pi p}(b) = 1 - 0.51\ e^{-0.05\ b^2} . \tag{11}$$

We obtain from these the densities ρ_p and ρ_{π} and their three dimensional Fourier transforms $\tilde{\rho}_p$ and $\tilde{\rho}_{\pi}$ as exhibited in figs. 4 and 5. Numerical values for these functions are listed in table 1 and 2. The electric charge form factor F_1 for the proton is plotted in fig. 5 for comparison with $\tilde{\rho}_p$.

The root mean square radii of ρ_p and ρ_{π} are respectively 0.717 fm and 0.749 fm. The difference between these two is not large as can be seen from the similar slopes of the Fourier transforms at k = 0 in fig. 5. However it is

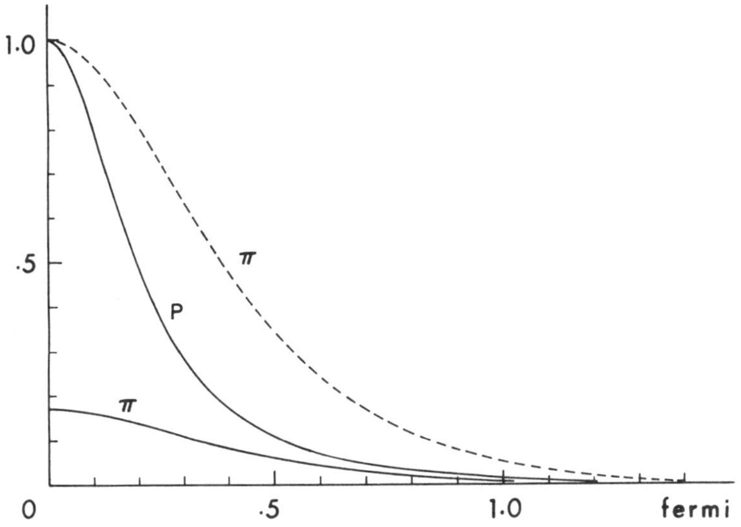

Fig. 4. The density functions ρ_p and ρ_{π} of proton and pion verus radius (solid curves). ρ_p is normalized so that $\rho_p(0)$ = 1 and $\rho_{\pi}(0)/\rho_p(0)$ is fixed by setting $K_{pp} = K_{\pi p}$. The dotted curve shows ρ_{π} normalized by $\rho_{\pi}(0)$ = 1.

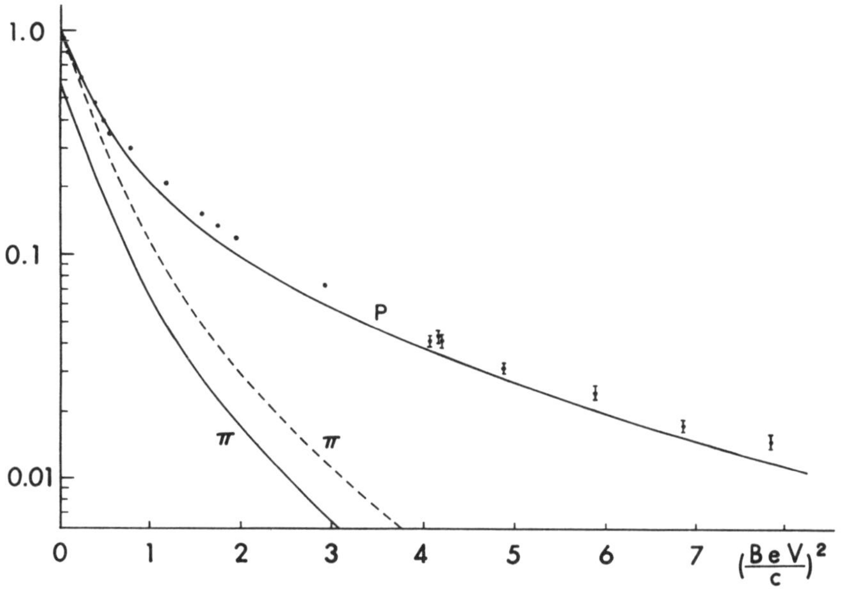

Fig. 5. The Fourier transform functions $\tilde{\rho}_p(k)/\tilde{\rho}_p(0)$ and $\tilde{\rho}_\pi(k)/\tilde{\rho}_p(0)$ versus k^2. The solid curves are obtained by putting $K_{pp} = K_{\pi p}$. The dotted curve is $\tilde{\rho}_\pi(k)/\tilde{\rho}_\pi(0)$. The experimental points are those of the form factor F_1 for the proton taken from Albrecht et al., Phys. Rev. Letters 17 (1966) 1192 for $k^2 > 3.5$ (GeV/c)^2, Bartel et al., Phys. Rev. Letters 17 (1966) 608 for $0.6 < k^2 < 3.5$ (GeV/c)^2 and Hand et al., Rev. Mod. Phys. 35 (1963) 335 for $k^2 < 0.6$ (GeV/c)^2.

noteworthy that in the p but not the π there is a very dense inside region that contributes little to the root mean square radius (cf. the two solid lines in fig. 4). This difference between p and π is further brought out if one compares their form factors normalized to 1 at the origin (see fig. 4). For the same value of $\rho(r)/\rho(0)$ between 1 and 0.1, the pion is $\sim 70\%$ larger than the proton. In momentum space this difference means that $\tilde{\rho}_\pi$ falls with higher k^2 much faster than $\tilde{\rho}_p$, as is clear from fig. 5. This conclusion is presumably testable experimentally in pion charge form factor measurements, if we assume the charge distribution in π to be proportional to ρ_π.

4. π-π SCATTERING AT HIGH ENERGIES

Eq. (7) yields directly the $\pi\pi$ elastic scattering cross section at high energies exhibited in fig. 6. These curves are very well represented by pure exponentials

$$(\mathrm{d}\sigma/\mathrm{d}t)_{\pi\pi} = A\mathrm{e}^{Bt} , \qquad (12)$$

where A and B are tabulated in table 3, together with the elastic and total cross sections.

Table 1
The density functions $\rho_p(r)$ and $\rho_\pi(r)$ of proton and pion with $\rho_p(0)/\rho_\pi(0) = 5.68$.

r (fm)	$\rho_p(r)/\rho_p(0)$	$\rho_\pi(r)/\rho_\pi(0)$
0.0	1.000	1.000
0.1	0.801	0.947
0.2	0.488	0.811
0.3	0.278	0.644
0.4	0.169	0.485
0.5	0.105	0.353
0.6	0.068	0.251
0.7	0.044	0.174
0.8	0.029	0.119
0.9	0.019	0.079
1.0	0.013	0.051
1.1	0.008	0.032
1.2	0.005	0.019
1.3	0.003	0.010
1.4	0.002	0.005

Table 2
The Fourier transform functions $\tilde{\rho}_p(k)$ and $\tilde{\rho}_\pi(k)$ of proton and pion with $\tilde{\rho}_p(0)/\tilde{\rho}_\pi(0) = 1.79$.

k^2 $(\mathrm{GeV}/c)^2$	$\tilde{\rho}_p(k)/\tilde{\rho}_p(0)$	$\tilde{\rho}_\pi(k)/\tilde{\rho}_\pi(0)$
0	1.000	1.000
0.5	0.385	0.307
1.0	0.206	0.111
2.0	0.974×10^{-1}	0.292×10^{-1}
3.0	0.579×10^{-1}	0.112×10^{-1}
4.0	0.382×10^{-1}	0.508×10^{-2}
5.0	0.268×10^{-1}	0.257×10^{-2}
6.0	0.196×10^{-1}	0.139×10^{-2}
7.0	0.148×10^{-1}	0.798×10^{-3}
8.0	0.114×10^{-1}	0.477×10^{-3}
9.0	0.897×10^{-2}	0.295×10^{-3}
10.0	0.717×10^{-2}	0.188×10^{-3}

5. REMARKS

A. The above discussion which amounts to a very simple way to factorize the collision pp (and πp and $\pi\pi$) into two factors is of course quite crude. In particular no variation of the "composition" has been allowed (i.e. K is assumed to be independent of b). However some general qualitative features of the results seem quite believable to us: (i) The πp system is considerably more transparent than the pp system. (This is evident from its smaller cross section and smaller $\sigma_{el}/\sigma_{total}$.) Thus the $\pi\pi$ system is presumably

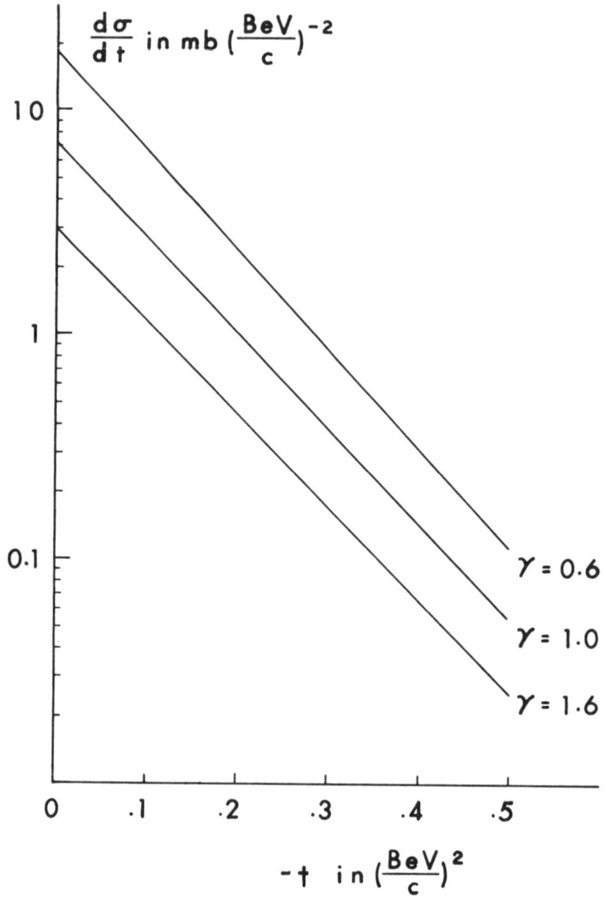

Fig. 6. $d\sigma/dt$ versus $-t$ for $\pi\pi$ elastic scattering where γ is defined as $K_{\pi p}^2/K_{\pi\pi}K_{pp}$.

Table 3
Constants in $\pi\pi$ scattering at high energies. It seems likely that the constant γ is approximately 1.

$\gamma = \frac{K_{\pi p}^2}{K_{\pi\pi}K_{pp}}$	A (mb(GeV/c)$^{-2}$)	B (GeV/c)$^{-2}$	$\sigma_{\pi\pi}$(total) (mb)	$\sigma_{\pi\pi}$(el) (mb)	$\sigma_{\pi\pi}$(el)/$\sigma_{\pi\pi}$(total)
0.6	18.3	10.3	18.9	1.78	0.094
1.0	7.16	9.78	11.8	0.73	0.062
1.6	2.94	9.52	7.59	0.31	0.041

even more so. Taking the most natural assumption $\gamma \cong 1$ we read from table 3:

$$\sigma_{\pi\pi}(\text{total}) \cong 12 \text{ mb} , \qquad \sigma_{\pi\pi}(\text{el}) \cong 0.7 \text{ mb} . \tag{13}$$

(ii) The slope B in (12) for $\pi\pi$ scattering is $\sim$ 10 $(\text{GeV}/c)^{-2}$. (iii) Neither p nor π has any sharp edges so that their form factors do not show pronounced minima like that of heavy nuclei. They also do not [1] show singularities at the origin. (iv) The pion form factor in momentum space falls much more rapidly with increasing k^2 than the proton form factor (i.e. the proton has a stiffer core than the pion). This is a consequence of the experimental fact of higher cross section for pp compared with πp, but almost identical B value as exhibited in eqs. (8) and (9).

B. At the highest energies studied $\sim$ 26 GeV/c, the forward scattering amplitude in πp seem to have a negative real part [5]. This means that S must have on the average a negative imaginary part. We guess that the points S would lie in the shaded region of fig. 7 for high but finite energies, with the left end of the region corresponding to small l and the right end to large l. At a fixed high energy one could ask how does $S \to 1$ and $l \to \infty$. The answer is that

$$S - 1 \to \text{pure imaginary} . \tag{14}$$

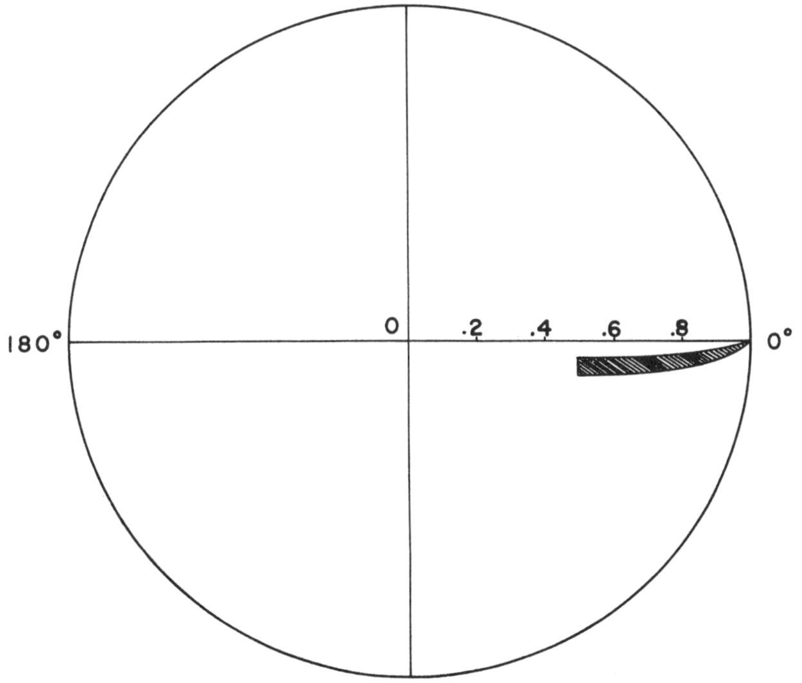

Fig. 7. Schematic drawing of the transmission factor at high energies (see text).

This conclusion comes from the fact that as $l \to \infty$ the centrifugal barrier makes $S \to 1$. Unitarity then requires eq. (14) unless the πp channel is increasingly disfavored at large l compared with other channels. But this latter situation would not arise since the centrifugal barrier works against all channels.

C. The S matrix element that we have called the transmission factor is of course the diagonal element for a given spin and parity. How do the other diagonal elements for the same spin and parity behave, such as

$$S_{K\Lambda \to K\Lambda}$$

or off diagonal elements like

$$S_{\pi p \to K\Lambda} \text{ or } S_{\pi p \to \pi\pi p}$$

or $S_{np \to np}$ in initial and final states with the same total J and parity but orthogonal spin-angular wave functions? We guess the answer is that as the energy $\to \infty$, the off diagonal elements $\to 0$ and the diagonal elements approach finite limits which are functions of b and of the nature of the incoming and outgoing particles. This guess is consistent with the view that at very high energies the interaction is an absorptive process between colliding droplets, and any specific change of nature of the colliding particles has a relative probability of zero compared with the total cross section. It is also consistent with the view that different particles such as π, K or p can have different spatial structures so that the differential cross section of $K\Lambda$ may not be the same as that of πp.

D. In a process AB $\to$ BA or AB $\to$ CD at high energies, if one looks away from the forward elastic peak, the angular distribution is experimentally known to be such that each of the two outgoing particles prefers to have approximately the same 4-momentum as one of the incoming particles. It is customary to write

$$AB \to CD \text{ (or } AB \to DC)$$

to denote the cases where C (or D) has approximately the same 4-momentum as A. It is natural of course to describe the peak in the angular distribution in AB $\to$ CD as caused by an exchange of something (X) with the right quantum numbers to bring A into C and B into D.

Experimentally while many peaks of this type have been found some peaks have been conspicuously absent, e.g. $K^-p \nrightarrow pK^-$, $\pi^-p \nrightarrow K^+\Sigma^-$, $p\bar{p} \nrightarrow \bar{p}p$. The upper limit on these peaks is, however, so far not lower than ~ 1 μb. Since it seems likely that near 90° the cross sections fall with increasing energy E like $\exp[-\alpha E^\beta]$ while the peaks fall like $E^{-\gamma}$, at high enough energies the peaks would emerge above the background of large angle cross sections. It seems worthwhile to increase the detection sensitivity and the incoming energy to test this speculation for those cases where a peak has so far not been found.

ACKNOWLEDGEMENT

We would like to thank M. C. Li for discussions more than one year ago of factorizations problems and R. Feynman, who independently pursued the factorization idea.

REFERENCES

[1] N. Byers and C. N. Yang, Phys. Rev. 142 (1966) 976.
[2] G. Cocconi, V. T. Cocconi, A. D. Krisch, J. Orear, R. Rubinstein et al., Phys. Rev. Letters 11 (1963) 499;
W. F. Baker, E. W. Jenkins, A. L. Read, G. Cocconi, V. T. Cocconi et al., Phys. Rev. Letters 12 (1964) 132.
[3] T. T. Wu and C. N. Yang, Phys. Rev. 137 (1965) B 708.
[4] J. V. Allaby, G. Bellettini, G. Cocconi, A. N. Diddens, M. L. Good, G. Matthiae, E. J. Sacharidis, A. Silverman and A. M. Wetherell, Phys. Letters 23 (1966) 389.
[5] S. J. Lindenbaum and S. Ozaki et al. (to be published), private communication.

⟦67d⟧

Commentary begins page 66

Commentary begins page 66

Some Solutions of the Classical Isotopic Gauge Field Equations

T. T. WU*

Harvard University, Cambridge, Massachusetts

and

C. N. YANG

Institute for Theoretical Physics, State University of New York, Stony Brook, New York

I. Introduction

It was pointed out[1] a number of years ago that an *isotopic gauge* can be defined in analogy with the usual electromagnetic gauge, and that the concept of local isotopic gauge invariance leads to a gauge field $\mathbf{b}_\mu$. The equations describing $\mathbf{b}_\mu$ in interaction with any source of isotopic spin are essentially uniquely determined, much like the equations describing the electromagnetic field A_μ in interaction with the electric charge. In the absence of any external sources of isotopic spin, the $\mathbf{b}_\mu$ field interacts with itself, since the $\mathbf{b}_\mu$ field possesses an isotopic spin and hence is self-generating. In this latter characteristic, the $\mathbf{b}_\mu$ field is different from the electromagnetic field, which is described by linear equations in the absence of other fields. (The nonlinear equations describing the self-generating $\mathbf{b}_\mu$ field is in some respects[2] similar to the equations of general relativity.)

We seek in this paper to find a solution of the (unquantized) $\mathbf{b}_\mu$ field in the absence of other interacting fields. Our aim is then similar to that of Born and Infeld,[3] except that they started with equations which were written down on a more or less *ad hoc* basis.

II. A Special Type of Solution

The equations for the $\mathbf{b}_\mu$ field are[1]

$$\mathbf{f}_{\mu\nu} = \mathbf{b}_{\mu,\nu} - \mathbf{b}_{\nu,\mu} - \mathbf{b}_\mu \times \mathbf{b}_\nu \tag{1}$$

$$\mathbf{f}_{\mu\nu,\nu} + \mathbf{b}_\nu \times \mathbf{f}_{\mu\nu} = 0 \tag{2}$$

$$\mathbf{b}_{\mu,\mu} = 0 \tag{3}$$

* National Science Foundation Senior Postdoctorate Fellow.

□ Reprinted from *Properties of Matter Under Unusual Conditions,* H. Mark and S. Fernbach, eds. New York: Wiley-Interscience, 1969, 349–354.

We have chosen units for $\mathbf{b}_\mu$ so that the coupling constant ε is equal to $\frac{1}{2}$. We further adopt the convention that $x_4 = ict$, and that $_{,\mu}$ means differentiation with respect to x_μ. Also subscripts μ or ν run 1 to 4, while others run 1 to 3.

Solutions in which the $\mathbf{b}_\mu$ field lies in one isotopic direction are easily found, since for them the nonlinear terms vanish, and the $\mathbf{b}_\mu$ field equations reduce to that for the free electromagnetic field. But such solutions are of no interest to us here.

To find some special solutions of Eqs. (1)–(3) we look for a static case, so that

$$\mathbf{b}_4 = 0, \qquad \mathbf{b}_{i,4} = 0 \tag{4}$$

We write the components of $\mathbf{b}_\mu$ as $b_{\mu\alpha}$, where $\alpha = 1, 2, 3$ designates the isotopic spin index.

We shall seek for a solution of the following form

$$b_{11} = b_{22} = b_{33} = 0, \qquad b_{12} = -b_{21} = x_3 f(r)/r, \text{ etc.}$$

i.e.,

$$b_{i\alpha} = \varepsilon_{i\alpha\tau} x_\tau f(r)/r \tag{5}$$

where r is the length of (x_1, x_2, x_3). Equation (3) is then automatically satisfied and (1) and (2) reduce to

$$f'' + \frac{2}{r} f' - (1 + rf)\left(\frac{2f}{r^2} + \frac{f^2}{r}\right) = 0 \tag{6}$$

where the prime means d/dr.

Writing

$$\Phi(r) = 1 + rf(r) \tag{7}$$

one has

$$r^2\Phi'' - \Phi(\Phi^2 - 1) = 0 \tag{8}$$

To study this equation we observe that putting

$$r = e^\xi \tag{9}$$

one has

$$\frac{d^2\Phi}{d\xi^2} - \frac{d\Phi}{d\xi} = \Phi(\Phi^2 - 1) \tag{10}$$

or

$$\frac{d\Phi}{d\xi} = \psi \tag{11a}$$

$$\frac{d\psi}{d\xi} = \psi + \Phi(\Phi^2 - 1) \tag{11b}$$

In the Φ–ψ plane (phase plane), $(d\Phi/d\xi,\ d\psi/d\xi)$ defines a vector field given by the right-hand side of (11a, b). The curves that are tangent to this vector field are the solutions we desire. The vector vanishes at exactly three points: $(\Phi, \psi) = (0, 0), (1, 0)$, and $(-1, 0)$.

It is not difficult to study the vector field. One finds that there are only five solutions of Eqs. (11a, b) which are finite for all $0 < r < \infty$:

(*a*) $\Phi = 0$ (12a)

(*b*) $\Phi = +1$ (12b)

(*c*) $\Phi = -1$ (12c)

(*d*) $\Phi = 1 - \frac{c}{r} + O\left(\frac{1}{r^2}\right)$ as $r \to \infty$, $(c > 0)$; $\Phi \to 0$ as $r \to 0$; (12d)

(*e*) The same as (*d*) except Φ changes sign. (12e)

To discuss the meaning of these solutions, we split the 6-vector $\mathbf{f}_{\mu\nu}$ into "electric" and "magnetic" components:

$$\mathbf{E}_j = i\mathbf{f}_{j4}, \qquad -\varepsilon_{ijk}\mathbf{H}_k = \mathbf{f}_{ij}$$

For the special type of solution satisfying (5),

$$\begin{aligned} \mathbf{E}_j &= 0, \\ -H_{j\alpha} &= -\delta_{j\alpha}\Phi'/r + x_j x_\alpha r^{-3}[\Phi' - r^{-1}\Phi^2 + r^{-1}] \\ &= -\delta_{j\alpha}\psi/r^2 + x_j x_\alpha\, r^{-4}\,[\psi - \Phi^2 + 1] \end{aligned} \tag{13}$$

It is clear that for (12b) and (12c), $\mathbf{H}_j = 0$ and the solutions are merely complicated ways of describing the vacuum $\mathbf{f}_{\mu\nu} = 0$. The non-trivial solutions are thus the three tabulated below.

1. Solution (12a)

$$\begin{aligned} \Phi &= 0, & f &= -r^{-1} \\ E_{j\alpha} &= 0, & H_{j\alpha} &= -x_j x_\alpha r^{-4} \end{aligned} \tag{14}$$

2. *Solution (12d)*

This solution has the following asymptotic behavior:

$r \to \infty$,

$$\Phi = 1 - \frac{c}{r} + O\left(\frac{1}{r^2}\right), \qquad f = -\frac{c}{r^2} + O\left(\frac{1}{r^3}\right)$$

$$E_{j\alpha} = 0, \qquad H_{j\alpha} = \frac{-c}{r^3}\left[-\delta_{j\alpha} + \frac{3x_j x_\alpha}{r^2}\right] + O\left(\frac{1}{r^4}\right) \tag{15}$$

$r \to 0$,

$\Phi \to 0$, $\psi \to 0$, as oscillatory functions of r with minima and maxima $= O(r^{1/2})$.

$$E_{j\alpha} = 0, \qquad H_{j\alpha} = -x_j x_\alpha r^{-4} + O(r^{-3/2}) \tag{16}$$

This solution is actually a one-parameter family of solutions with the parameter c of the dimension of a length. Numerical results are given in Table I.

TABLE I. Φ as a Function of r for Solution (12d),[a] with $c = 1$

r	Φ
∞	1
9.880×10	9.898×10^{-1}
1.095×10	9.136×10^{-1}
3.297	7.510×10^{-1}
1.098	4.584×10^{-1}
6.141×10^{-2}	-9.229×10^{-2} = min.
1.617×10^{-3}	1.498×10^{-2} = max.
4.296×10^{-5}	-2.441×10^{-3} = min.
1.142×10^{-6}	3.980×10^{-4} = max.
3.036×10^{-8}	-6.489×10^{-5} = min.
$8.070 \quad 10^{-10}$	1.058×10^{-5} = max.
2.15×10^{-11}	-1.725×10^{-6} = min.

[a] This table is obtained by numerical integration. As $r \to 0$, Φ oscillates with damped amplitude. The first four minima and first three maxima are tabulated.

3. Solution (12e)

This solution has the following asymptotic behavior:

$r \to \infty$,

$$\Phi = -1 + \frac{c}{r} + O\left(\frac{1}{r^2}\right), \qquad f = \frac{-2}{r} + \frac{c}{r^2} + O\left(\frac{1}{r^3}\right)$$

$$E_{j\alpha} = 0, \qquad H_{j\alpha} = \frac{-c}{r^3}\left[\delta_{j\alpha} + \frac{x_j x_\alpha}{r^2}\right] + O\left(\frac{1}{r^4}\right) \tag{17}$$

$r \to 0$,

Φ, $\psi \to 0$ as oscillatory functions of r with maxima and minima $= O(r^{1/2})$.

$$E_{j\alpha} = 0, \qquad H_{j\alpha} = -x_j x_\alpha r^{-4} + O(r^{-3/2}) \tag{18}$$

This solution is again a one-parameter family of solutions with the parameter c of the dimension of a length.

Notice that the three types of solutions (12a), (12d), and (12e) share the same dominant asymptotic form as $r \to 0$.

III. Energy

The Hamiltonian of the **b** field was given in reference 1. For the present cases,

$$H = \frac{1}{4}\int H_{j\alpha}H_{j\alpha}\,d^3x = \pi\int_0^\infty [2\Phi'^2 + (\Phi^2 - 1)^2 r^{-2}]\,dr \tag{19}$$

This integral is divergent at $r \simeq 0$.

If one replaces $\Phi \to \Phi + \delta\Phi$ in (19) for variations $\delta\Phi$ which are zero in an interval $r = 0 \to r_0$, (19) gives a stationary value at the solutions Φ discussed in Sections II-1, II-2, and II-3, above [cf. Eq. (8)]. The second variation gives

$$\delta^2 H = \pi\int_{r_0}^\infty [2(\delta\Phi')^2 + (\delta\Phi)^2(6\Phi^2 - 2)r^{-2}]\,dr \tag{20}$$

For sufficiently small r_0, $6\Phi^2 - 2 \sim -2$ near $r = r_0$ and (20) is not positive definite.

IV. Source

Are the solutions above really sourceless? The answer is clearly yes except at $r = 0$. For $r = 0$, however, the solutions are singular and this question will have to be examined in greater detail.

Another way of asking the same question is whether the solutions exhibited above do satisfy the field equation (2). To discuss this we define

$$-\mathbf{J}_\mu = \mathbf{f}_{\mu\nu,\nu} + \mathbf{b}_\nu \times \mathbf{f}_{\mu\nu} \tag{21}$$

Equation (5) leads to

$$J_{4\alpha} = 0, \qquad J_{i\alpha} = \varepsilon_{i\alpha\tau}x_\tau r^{-1} J \tag{22}$$

where

$$J = -f'' - \frac{2}{r}f' + (1 + rf)\left(\frac{2f}{r^2} + \frac{f^2}{r}\right) \tag{23}$$

At $r \neq 0$, this is clearly zero, by Eq. (6). At $r \simeq 0$, the dominant term in f is $-r^{-1}$ for all three types of solutions (12a), (12d), and (12e). Thus

$$J = -4\pi\delta^3(x) \tag{24}$$

i.e.,

$$J_{4\alpha} = 0, \qquad J_{i\alpha} = -4\pi\varepsilon_{i\alpha\tau}x_\tau r^{-1}\delta^3(x) \tag{25}$$

The source function (25) is, in the sense of Dirac's definition of δ functions or in the sense of the theory of distributions, equal to zero. We thus conclude that the solutions indeed represent classical sourceless gauge fields.

V. Total Isotopic Spin

The total isotopic spin was given in reference 1:

$$\mathbf{T} = \int \mathbf{b}_\nu \times \mathbf{f}_{4\nu} \, d^3x$$

which is equal to zero for the present solutions.

Generalizations of the above solutions are in progress.

Acknowledgment

One of the authors (T. T. Wu) would like to take this opportunity to thank Professor A. Pais and the Rockefeller University for the hospitality extended him during his visit when this work was carried out.

References

1. C. N. Yang and R. L. Mills, *Phys. Rev.*, **96**, 191 (1954).
2. R. Utiyama, *Phys. Rev.*, **101**, 1597 (1956).
3. M. Born and L. Infeld, *Proc. Roy. Soc.* (*London*), *Ser. A*, **143**, 410; **144**, 425; **147**, 522 (1934); **150**, 141 (1935).

⟦67e⟧

Commentary begins page 68

SOME EXACT RESULTS FOR THE MANY-BODY PROBLEM IN ONE DIMENSION WITH REPULSIVE DELTA-FUNCTION INTERACTION*

C. N. Yang

Institute for Theoretical Physics, State University of New York, Stony Brook, New York

(Received 2 November 1967)

The repulsive δ interaction problem in one dimension for N particles is reduced, through the use of Bethe's hypothesis, to an eigenvalue problem of matrices of the same sizes as the irreducible representations R of the permutation group S_N. For some R's this eigenvalue problem itself is solved by a second use of Bethe's hypothesis, in a generalized form. In particular, the ground-state problem of spin-$\frac{1}{2}$ fermions is reduced to a generalized Fredholm equation.

(1) Consider the one-dimensional N-body problem

$$H=-\sum_1^N \partial^2/\partial x_i^{\,2}+2c\sum_{i<j}\delta(x_i-x_j),\quad c>0, \tag{1}$$

with no limitation on the symmetry of the wave function ψ. For a given irreducible representation R_ψ of the permutation group S_N of the N coordinates x_i, we want to determine the wave function ψ. Assume Bethe's hypothesis[1] to be valid: Let $p_1,\cdots,p_N$ = a set of unequal numbers. For $0<x_{Q1}<x_{Q2}<\cdots<x_{QN}<L$,

$$\psi=\sum_P[Q,P]\exp i[p_{P1}x_{Q1}+\cdots+p_{PN}x_{QN}], \tag{2}$$

where $P=[P1,P2,\cdots,PN]$ and $Q=[Q1,Q2,\cdots,QN]$ are two permutations of the integers $1,2,\cdots,N$. $[Q,P]$ can be arranged as a $N!\times N!$ matrix. Denote the columns of this matrix by ξ_P. To satisfy the continuity of ψ and the proper discontinuity of its derivative as required by (1) at $x_{Q3}=x_{Q4}$, it is sufficient to have

$$\xi_{\dots ij\dots}=Y_{ji}^{\,34}\xi_{\dots ji\dots}, \tag{3}$$

where the subscripts for ξ on the two sides represent any two permutation P and P' so that $P1=P'1$, $P2=P'2$, $P3=i=P'4$, $P4=j=P'3$, etc. The operator Y is defined by

$$Y_{ij}^{\,34}=(y_{ij}^{\,-1}-1)+y_{ij}^{\,-1}P_{34}=Y_{ij}^{\,43}, \tag{4}$$

where

$$y_{ij}=1+x_{ij}, \tag{5}$$

$$x_{jk}=ic(p_j-p_k)^{-1}=-x_{kj}, \tag{6}$$

and P_{34} = the permutation operator on ξ so that it interchanges $Q3$ and $Q4$. Altogether there are $N!(N-1)$ equations of the form (3). Are they mutually consistent? The answer is yes for any set of unequal p's. This can be seen with the aid of the following identities:

$$Y_{ij}^{\,ab}Y_{ji}^{\,ab}=1, \tag{7}$$

and

$$Y_{jk}^{\,ab}Y_{ik}^{\,bc}Y_{ij}^{\,ab}=Y_{ij}^{\,bc}Y_{ik}^{\,ab}Y_{jk}^{\,bc}, \tag{8}$$

which are easily verified. Thus given a set of unequal p's, and $\xi_0=\xi_P$ for P = identity, all ξ_P's are determined.

(2) The imposition of the periodic boundary conditions leads to equations which, upon expressing ξ_P in terms of ξ_0, become

$$\lambda_j\xi_0=X_{(j+1)j}\times X_{(j+2)j}\cdots X_{Nj}X_{1j}X_{2j}\cdots X_{(j-1)j}\xi_0,\quad j=1,\cdots,N, \tag{9}$$

where

$$\lambda_j=\exp(ip_jL), \tag{10}$$

and

$$X_{ij}=P_{ij}Y_{ij}^{\,ij}=(1-P_{ij}x_{ij})(1+x_{ij})^{-1}. \tag{11}$$

The N Eqs. (9) say that ξ_0 is simultaneously an eigenvector of N operators. These N operators can be shown to commute with each other, using

$$X_{ij}X_{ji}=1,\quad X_{jk}X_{ik}X_{ij}X_{kj}X_{ki}X_{ji}=1,$$
$$X_{ij}X_{kl}=X_{kl}X_{ij};\quad i,j,k,\text{ and } l \text{ all unequal}. \tag{12}$$

(3) The operators P_{ij} on ξ form a $N!\times N!$ representation of S_N. To find the eigenfunctions ξ_0 in (9) we can first reduce this representation to irreducible ones. Choosing one specific irreducible representation R reduces the

□Reprinted from *Physical Review Letters* 19, 23 (December 4, 1967), 1312–1315.

eigenvalue problem (9) to one of smaller dimensions. It can be shown that the resultant wave function (2) would have a permutation symmetry R_ψ which is the same as R. For example, if R = identity representation = $[N]$, then $P_{ij}=1$, and (9) becomes 1×1 matrix equations and the result is precisely the well-known boson result.[2] If R = antisymmetric representation = $[1^N]$, then $P_{ij}=-1$, and $X_{ij}=1$, so that (9) and (10) reduce to $\exp(ip_j L)=1$, showing there is no interaction, a result to be expected for the antisymmetrical wave function.

(4) The λ_j's are functions of the p's, c, and R. It is easily seen (that R and $\tilde{R}$ being conjugate representations)

$$\lambda_j(p;c;R)=\prod_{i\neq j}\left(\frac{1-x_{ij}}{1+x_{ij}}\right)\lambda_j(p;-c;\tilde{R}). \tag{13}$$

(5) Define $\mu_j(p;c;R)$ by

$$\mu_j\Phi=X_{(j+1)j}'X_{(j+2)j}'\cdots$$
$$\times X_{Nj}'X_{1j}'X_{2j}'\cdots X_{(j-1)j}'\Phi, \tag{14}$$

where

$$X_{ij}'=(1+P_{ij}x_{ij})(1+x_{ij})^{-1}. \tag{15}$$

Clearly

$$\mu_j(p;c;\tilde{R})=\lambda_j(p;c;R). \tag{16}$$

(6) We now evaluate λ_j for $R_\psi=R=[2^M 1^{N-2M}]$. By (16) we need to find $\mu_j(p;c;[N-M,M])$. To do this we first define a convenient representation for P_{ij} of (15):

Consider N spin-$\frac{1}{2}$ particles, and consider the spin wave functions Φ for total z spin $=\frac{1}{2}(N-2M)$. These spin wave functions transform under S_N according to a sum of irreducible representations,

$$[N]+[N-1,1]+[N-2,2]+\cdots+[N-M,M]. \tag{17}$$

We consider the P_{ij}'s of (15) as operating on these spin wave functions Φ. The eigenvalue equations (14) for μ_j are then to be solved for a Φ that belongs to the symmetry $[N-M,M]$.

(7) Consider the N spins as forming a cyclic chain. The wave function Φ has $C_M{}^N$ components [$N-M$ spins up, M spins down]. The eigenvalue problem (14) can be solved with a generalized Bethe's hypothesis:

$$\Phi=\sum_P A_P F(\Lambda_{P1},y_1)$$
$$\times F(\Lambda_{P2},y_2)\cdots F(\Lambda_{PM},y_M), \tag{18}$$

where $y_1<y_2<\cdots<y_M$ are the "coordinates," along the chain, of the M down spins, and Λ_1, $\Lambda_2,\cdots,\Lambda_M$ are a set of unequal numbers. With this hypothesis, one finds

$$F(\Lambda,y)=\prod_{j=1}^{y-1}\frac{ip_j-i\Lambda-c'}{ip_{j+1}-i\Lambda+c'}\quad (c'=\tfrac{1}{2}c); \tag{19}$$

$$-\prod_j\frac{ip_j-i\Lambda_\alpha-c'}{ip_j-i\Lambda_\alpha+c'}=\prod_\beta\frac{-i\Lambda_\beta+i\Lambda_\alpha+c}{-i\Lambda_\beta+i\Lambda_\alpha-c}; \tag{20}$$

and

$$\mu_j(p;c;[N-M,M])=\prod_\beta\frac{ip_j-i\Lambda_\beta-c'}{ip_j-i\Lambda_\beta+c'}. \tag{21}$$

(8) Thus for the $R_\psi=[2^M 1^{N-2M}]$ symmetry, we need to solve

$$\exp(ip_jL)=\text{right-hand side of (21)}, \tag{22}$$

together with (20). In taking the logarithm of (20) and (22) care must be taken to add terms $2\pi i$(integer). The value of the integer can be determined by going to the limit $c\to+\infty$. One obtains, for the ground state with the symmetry $R_\psi=[2^M 1^{N-2M}]$, for the case N = even, M = odd,

$$-\sum_P\theta(2\Lambda-2p)=2\pi J_\Lambda-\sum_{\Lambda'}\theta(\Lambda-\Lambda'), \tag{23a}$$

$$Lp=2\pi I_p+\sum_\Lambda\theta(2p-2\Lambda), \tag{23b}$$

where the p's are a set of N ascending real numbers, the Λ's a set of M ascending real numbers,

$$\theta(p)=-2\tan^{-1}(p/c)\quad(-\pi\leq\theta<\pi), \tag{24}$$

and

J_Λ = successive integers from

$$-\tfrac{1}{2}(M-1)\text{ to }+\tfrac{1}{2}(M-1), \tag{24a}$$

$\frac{1}{2}+I_p$ = successive integers from

$$1-\tfrac{1}{2}N\text{ to }\tfrac{1}{2}N. \tag{24b}$$

Equation (23a) differs from that given in a recent paper,[3] in the definition of θ and our introduction of J_Λ. The present equation allows for a natural discussion of the limit $c \to +\infty$ (not $c \to 0$!) and hence the values of J_Λ.

(9) We can now approach the limit $N \to \infty$, $M = \infty$, $L \to \infty$ proportionally, obtaining

$$-\int_{-Q}^{Q} \theta(2\Lambda - 2p)\rho(p)dp = 2\pi g - \int_{-B}^{B} \theta(\Lambda - \Lambda')\sigma(\Lambda')d\Lambda', \tag{25a}$$

$$p = 2\pi f + \int_{-B}^{B} \theta(2p - 2\Lambda)\sigma(\Lambda)d\Lambda, \tag{25b}$$

$$dg/d\Lambda = \sigma, \quad df/dp = \rho. \tag{25c}$$

Or, after differentiation,

$$2\pi\sigma = -\int_{-B}^{B} \frac{2c\sigma(\Lambda')d\Lambda'}{c^2 + (\Lambda - \Lambda')^2} + \int_{-Q}^{Q} \frac{4c\rho dp}{c^2 + 4(p - \Lambda)^2}, \tag{26a}$$

$$2\pi\rho = 1 + \int_{-B}^{B} \frac{4c\sigma d\Lambda}{c^2 + 4(p - \Lambda)^2}, \tag{26b}$$

$$N/L = \int_{-Q}^{Q} \rho dp, \quad M/L = \int_{-B}^{B} \sigma d\Lambda, \tag{27a}$$

and

$$E/L = \int_{-Q}^{Q} p^2 \rho(p) dp. \tag{27b}$$

(10) Equations (26) are generalized Fredholm equations with a symmetrical kernel. It is easy to show that the equations are nonsingular by first studying the eigenvalues of the kernel in the limit $B = Q = \infty$.

(11) Equations (26) and (27) yield the ground-state energy per particle for spatial wave functions with the symmetry $[2^M 1^{N-2M}]$, at a given density N/L. For N fermions with spin $\frac{1}{2}$ interacting through the Hamiltonian (1), this spatial wave function is coupled to a spin wave function of conjugate symmetry $[N-M, M]$, i.e., the total spin of the system is $\frac{1}{2}N - M$.

(12) For $B = \infty$, integration of (26a) over all Λ yields $N = 2M$. Thus for the fermion problem with spin $\frac{1}{2}$, $B = \infty$ gives the ground state for states with total spin $= 0$. This state is also the absolute ground state for the problem, by a theorem due to Lieb and Mattis.[4]

(13) For the case $B \cong 0$, M/L is proportional to B. One can readily expand all quantities in powers of B, obtaining, for fixed $r = N/L$,

$$\frac{E}{L} = \text{const.} + \frac{M}{L}\left[cr - \left(\frac{c^2}{2\pi} + 2\pi r^2\right)\tan^{-1}\frac{2\pi r}{c}\right] + \cdots. \tag{28}$$

This result is in agreement with results already obtained by McGuire[5] for the case $M = 1$ and by Flicker and Lieb[6] for the case $M = 2$.

(14) For each symmetry R_ψ of spatial wave function ψ, the excited states near the ground state can be obtained in a similar way as in the boson case.[7] More quantum numbers are, however, necessary to designate the excitations than in the boson case, because of the existence of the integers J_Λ (which are in fact quantum numbers). Details will be published elsewhere.

(15) For the boson problem the thermodynamics and excitations for finite T were treated by Yang and Yang.[8] Extension to the present problem presents no difficulty. Details will be published elsewhere.

(16) Using (13) one could generalize all the considerations above to the case of $R_\psi = [N-M, M]$. Details will be published elsewhere. The main change is that while all Eqs. (26) and (27) remain the same, (26b) is replaced by

$$2\pi\rho = 1 - \int_{-B}^{B} \frac{4c\sigma d\Lambda}{c^2 + 4(p - \Lambda)^2} + \int_{-Q}^{Q} \frac{2c\rho(p')dp'}{c^2 + (p - p')^2}. \tag{26b'}$$

It is a pleasure to acknowledge useful discussions with J. B. McGuire in 1963 at University of California, Los Angeles, with T. T. Wu in 1964 at Brookhaven National Laboratory, with C. P. Yang in 1966 in Princeton, and with B. Sutherland in 1967 in Stony Brook.

---

*Research partly supported by U. S. Atomic Energy Commission under Contract No. AT(30-1)-3668B.

[1] H. A. Bethe, Z. Physik 71, 205 (1931), first used the hypothesis for the spin-wave problem. E. Lieb and W. Linger, Phys. Rev. 130, 1605 (1963), and J. B. McGuire, J. Math. Phys. 5, 622 (1964), first used the same hypothesis for the δ-function interaction problem. The present author believes that for all states (excited as well as the ground state) with periodic boundary condition the hypothesis is valid. Justification of this belief for some special cases is found in C. N. Yang and C. P. Yang, Phys. Rev. 150, 321 (1966), and to be published.

[2] Lieb and Linger, Ref. 1.

[3] M. Gaudin, Phys. Letters 24A, 55 (1967).

[4] E. Lieb and D. Mattis, Phys. Rev. 125, 164 (1962).

[5] J. B. McGuire, J. Math. Phys. 6, 432 (1965), and 7,

123 (1966).
[6]M. Flicker and E. H. Lieb, Phys. Rev. 161, 179 (1967).
[7]E. Lieb, Phys. Rev. 130, 1616 (1963).
[8]C. N. Yang and C. P. Yang, to be published. See also C. N. Yang, in Proceedings of Eastern Theoretical Conference, November, 1966, Brown University (W. A. Benjamin, Inc., New York, 1967), p. 215.

[[69a]]

Commentary begins page 68

Thermodynamics of a One-Dimensional System of Bosons with Repulsive Delta-Function Interaction

C. N. YANG
Institute for Theoretical Physics, State University of New York, Stony Brook, New York

AND

C. P. YANG*
Ohio State University, Columbus, Ohio

(Received 10 October 1968)

The equilibrium thermodynamics of a one-dimensional system of bosons with repulsive delta-function interaction is shown to be derivable from the solution of a simple integral equation. The excitation spectrum at any temperature T is also found.

I. INTRODUCTION

The ground-state energy of a system of N bosons with repulsive delta-function interaction in one dimension with periodic boundary condition was calculated by Lieb and Liniger.[1] The Hamiltonian for the system is

$$H = -\sum_1^N \frac{\partial^2}{\partial x_i^2} + 2c\sum_{i>j}\delta(x_i - x_j), \quad c > 0, \quad (1)$$

and the periodic box has length L. Using Bethe's hypothesis[2] they showed that the k's in the hypothesis satisfy

$$(-1)^{N-1}\exp(-ikL) = \exp\left[i\sum_{k'}\theta(k'-k)\right], \quad (2)$$

where

$$\theta(k) = -2\tan^{-1}(k/c), \quad -\pi < \theta < \pi. \quad (3)$$

Taking the logarithm of (2) is a somewhat subtle process. In this paper we shall first discuss this point and show that *all* states of (1) are given by Bethe's hypothesis with real k's. The main purpose of the paper is to then evaluate the thermodynamical properties of the system at a finite temperature T.

While we try to maintain mathematical rigor in the rest of the paper, it is to be emphasized that Secs. III and IV are far from rigorous.

II. PROOF OF BETHE'S HYPOTHESIS FOR ALL STATES

We first take the logarithm of (2):

$$kL = 2\pi I_k + \sum_{k'}\theta(k-k'), \quad (4)$$

where

$$I_k = \text{integer}, \quad \text{if } N = \text{odd},$$
$$I_k + \tfrac{1}{2} = \text{integer}, \quad \text{if } N = \text{even}. \quad (5)$$

* Partially supported by NSF Grant GP8731.

[1] E. Lieb and W. Liniger, Phys. Rev. **130**, 1605 (1963).

[2] H. A. Bethe, Z. Physik **71**, 205 (1931).

Now, for any set of real I's, $I_1, I_2, \cdots, I_N$, Eq. (4) has a unique real solution for the k's, $k_1, k_2, \cdots, k_N$. The proof of this statement (similar to but simpler than the proof of a corresponding statement[3] for the Heisenberg–Ising problem) follows. Let

$$\theta_1(k) = \int_0^k \theta(k)\,dk.$$

Define

$$B(k_1, \cdots, k_N) = \tfrac{1}{2}L\sum_1^N k_j^2 - 2\pi\sum_1^N I_j k_j - \tfrac{1}{2}\sum_{j,S}\theta_1(k_j - k_S). \quad (6)$$

Equation (4) is the condition for the extrema of B. Now the second-derivative matrix B_2 of B is positive-definite. [The first sum in (6) contributes a positive-definite part to B_2. The second sum contributes nothing. Each term in the third sum is negative-semidefinite, since $\theta_1''(k) = \theta'(k) < 0$.] Furthermore for large values of $\sum k^2$, $B \to \tfrac{1}{2}L(\sum k^2)$. Thus, B has *one* and *only one* extremum, namely, a minimum.

It is further clear from this argument that the solution above represents a point S in k space which moves continuously as c^{-1} is changed. [In fact, $dk_j/d(c^{-1})$ can be computed.] Now when $c^{-1} = 0$, $\theta_1 = 0$ and the minimum of B occurs at

$$k_j = 2\pi I_j/L. \quad (7)$$

Now the problem with $c^{-1} = 0$ is the problem of free particles with the condition that $\psi = 0$ whenever $x_i = x_j$ (any $i \neq j$). All eigenfunctions of H for this problem are easily seen to be the same as that of free fermions in the segment $0 \le x_1 \le x_2 \le x_3 \le \cdots \le x_N \le L$. Thus, when $c^{-1} = 0$, all eigenfunctions are of Bethe's form, with the k's given by (7) and with all the I's different.

[3] C. N. Yang and C. P. Yang, Phys. Rev. **150**, 321 (1966).

□Reprinted from *Journal of Mathematical Physics* 10, 7 (July 1969), 1115–1122.

By a continuity argument with respect to c^{-1} we obtain the following:

Theorem: For any set of I's satisfying (5), no two of which are identical, there is a unique set of real k's satisfying (4), with no two k's being identical. With this set of k's, one eigenfunction of H, of Bethe's form, can be constructed. The totality of such eigenfunctions form a complete set for the boson system.

The numbers I are quantum numbers for the problem.

III. ENERGY AND ENTROPY FOR A SYSTEM WITH $N = \infty$

We now consider the problem for $N = \infty$ and $L = \infty$ at a fixed density $D = N/L$. For the ground state, the quantum numbers I/L form[1] a uniform lattice between $-D/2$ and $D/2$. The k's then form[1] a non-uniform distribution between a maximum k and a minimum k. For an excited state, (5) shows that the quantum numbers I/L are still on the same lattice, but not all lattice sites are taken, and the limits $-D/2$ and $D/2$ are no longer respected. We shall call the omitted lattice sites J_j/L. We would want to define corresponding "omitted k values" to be called holes. This can be easily done: Given the I's, Eq. (4) defines the set of k's as proved in the last section. Now,

$$Lh(p) \equiv pL - \sum_{k'} \theta(p - k') \tag{8}$$

is a continuous monotonic function of p. At $p = \pm\infty$, it is equal to $\pm\infty$. Those values of p where $Lh(p) = 2\pi I$ are k's. Those values of p where $Lh(p) = 2\pi J$ will be defined as holes.

For a large system, there is thus a density distribution of holes as well as one of k's:

$$L\rho(k)\,dk = \text{No. of } k\text{'s in } dk,$$
$$L\rho_h(k)\,dk = \text{No. of holes in } dk. \tag{9}$$

By definition, the number of k's and holes in the interval dk is the number of times $Lh(k)$ ranges over values $2\pi I$ and $2\pi J$ in this interval.

Thus,

$$\frac{dh(k)}{dk} = 2\pi(\rho + \rho_h) \equiv 2\pi f(k). \tag{10a}$$

Equation (8) gives

$$h(k) = k - \int_{-\infty}^{\infty} \theta(k - k')\rho(k')\,dk'. \tag{10b}$$

Differentiation with respect to k gives

$$2\pi f = 2\pi(\rho + \rho_h) = 1 + 2c\int_{-\infty}^{\infty} \frac{\rho(k')\,dk}{c^2 + (k - k')^2}. \tag{11}$$

The energy per particle for the state is

$$E/N = D^{-1}\int_{-\infty}^{\infty} \rho(k)k^2\,dk, \tag{12}$$

where

$$D = N/L = \int_{-\infty}^{\infty} \rho(k)\,dk. \tag{13}$$

The entropy of the "state" is not zero since the existence of the omitted quantum numbers J_j allows many wavefunctions of approximately the same energy to be described by the same ρ and ρ_h. In fact, for given ρ and ρ_h, the total number of k's and holes in dk is $L(\rho + \rho_h)\,dk$, of which $L\rho\,dk$ are k's and $L\rho_h\,dk$ are holes. Thus the number of possible choices of states in dk consistent with given ρ and ρ_h is

$$\frac{[L(\rho + \rho_h)\,dk]!}{[L\rho\,dk]!\,[L\rho_h\,dk]!}.$$

The logarithm of this gives the contribution to the entropy from dk. Thus, the total entropy is, putting the Boltzman constant equal to 1,

$$S = \sum \{(L\rho\,dk + L\rho_h\,dk)\ln(\rho + \rho_h) - L\rho\,dk\ln\rho - L\rho_h\,dk\ln\rho_h\}$$

or

$$S/N = D^{-1}\int_{-\infty}^{\infty} [(\rho + \rho_h)\ln(\rho + \rho_h) - \rho\ln\rho - \rho_h\ln\rho_h]\,dk. \tag{14}$$

IV. THERMAL EQUILIBRIUM

At temperature T, we should maximize the contribution to the partition function from the states described by ρ and ρ_h. In other words, given ρ, ρ_h is defined by (11). One then computes the contribution to the partition function

$$\exp(S - ET^{-1}), \tag{14'}$$

where S and E are given by (14) and (12). The equilibrium ρ is then obtained by maximizing this contribution when ρ is varied subject to the condition (13).

The above described procedure leads in a straightforward manner to the following condition on the equilibrium ρ:

$$-A + k^2 + T\ln\frac{\rho}{\rho_h} - \frac{Tc}{\pi}\int_{-\infty}^{\infty} \frac{dq}{c^2 + (k - q)^2}\ln\left(1 + \frac{\rho}{\rho_h}\right) = 0,$$

where A is a Lagrange multiplier for the condition (13). Writing

$$\rho_h/\rho = \exp[\epsilon(k)/T], \tag{15}$$

we have

$$\epsilon(k) = -A + k^2 - \frac{Tc}{\pi}\int_{-\infty}^{\infty} \frac{dq}{c^2 + (k-q)^2} \times \ln\{1 + \exp[-\epsilon(q)/T]\}. \quad (16)$$

Equation (11) becomes

$$2\pi f(k) = 2\pi\rho(k)\{1 + \exp[\epsilon(k)/T]\} = 1 + 2c\int_{-\infty}^{\infty} \frac{\rho(q)\,dq}{c^2 + (k-q)^2}. \quad (17)$$

It will be shown in Appendix A that (16) can be solved for ϵ by iteration. Equation (17) is then a Fredholm equation for ρ. It will be shown in Appendix B that ρ can be obtained by iteration of (17). The energy, density D, and entropy can then be obtained from (12)–(14).

In Appendix C it will be shown that the maximization procedure that led to (16) can be more rigorously treated and that the conclusion of the next section can then be obtained without much algebra.

V. A = CHEMICAL POTENTIAL

We shall now show that A is the chemical potential. Multiply (16) with ρD^{-1} and integrate over k to obtain

$$A = D^{-1}\int_{-\infty}^{\infty} \rho(k^2 - \epsilon)\,dk + TD^{-1}\int_{-\infty}^{\infty} dq[(2\pi)^{-1} - f(q)] \times \ln\left(1 + \exp\left\{\frac{-\epsilon(q)}{T}\right\}\right). \quad (18)$$

In this formula, the square bracket is obtained from (17). Now use (15) to rewrite (14) as

$$S/N = D^{-1}\int_{-\infty}^{\infty} (\rho + \rho_h)\ln(1 + \exp\{-\epsilon/T\})\,dk + (DT)^{-1}\int_{-\infty}^{\infty} \rho\epsilon\,dk. \quad (19)$$

Thus, the free energy per particle is

$$FN^{-1} = (E - TS)N^{-1} = D^{-1}\int_{-\infty}^{\infty} (k^2 - \epsilon)\rho\,dk - TD^{-1}\int_{-\infty}^{\infty} (\rho + \rho_h)\ln[1 + \exp(-\epsilon/T)]\,dk. \quad (20)$$

Comparison of (18) and (20) gives, using $f = \rho + \rho_h$,

$$FN^{-1} = A - T(2\pi D)^{-1}\int_{-\infty}^{\infty} \ln[1 + \exp(-\epsilon/T)]\,dk. \quad (21)$$

If we now prove that the last term is $-PD^{-1}$ (where P is the pressure), then this formula demonstrates that A is the chemical potential, since by thermodynamics

$$F = -PL + N \times (\text{chemical potential}).$$

Now, by (21),

$$P = -\left(\frac{\partial F}{\partial L}\right)_T = -N\frac{\partial A}{\partial L} + \frac{TN}{2\pi D}\int_{-\infty}^{\infty} dq\,\frac{1}{1 + e^{\epsilon/T}}\left(-\frac{1}{T}\right)\frac{\partial\epsilon}{\partial A}\frac{\partial A}{\partial L} + \frac{T}{2\pi}\int_{-\infty}^{\infty} dk\ln(1 + e^{-\epsilon/T}), \quad (22)$$

where ϵ is considered a function of A defined by (16). Differentiating (16) with respect to A, we obtain

$$1 = -\left(\frac{\partial\epsilon}{\partial A}\right) + \frac{c}{\pi}\int_{-\infty}^{\infty} \frac{dq}{c^2 + (k-q)^2}\frac{(\partial\epsilon/\partial A)}{1 + e^{\epsilon(q)/T}}. \quad (23)$$

Comparing this equation with (17) we conclude, by the uniqueness of the solution of (17) (see Appendix B),

$$-\frac{\partial\epsilon}{\partial A} = 2\pi f(k) = 2\pi\rho(k)(1 + e^{\epsilon(k)/T}). \quad (24)$$

The first two terms in the expression (22) for P now cancel each other by (24) and (13). Thus,

$$P = \frac{T}{2\pi}\int_{-\infty}^{\infty} dk\ln(1 + e^{-\epsilon(k)/T}). \quad (25)$$

This proves the assertion that A is the chemical potential.

We shall prove in Appendix D that $P(A, T)$ is analytic in A and T. To recapitulate: ϵ is defined by (16) once A and T are given. Equation (25), then, gives P as a function of A and T. The other thermodynamical quantities are obtainable from the thermodynamical relation

$$dP = (S/L)\,dT + (N/L)\,dA. \quad (26)$$

If one wants to compute ρ, one uses either (17) or (24).

VI. SPECIAL CASES

A. $c = \infty$

The integrals in (16) and (17) do not contribute. Thus,

$$\begin{aligned} \epsilon &= -A + k^2, \\ 2\pi\rho &= z\exp(-k^2/T)[1 + z\exp(-k^2/T)]^{-1}, \\ 2\pi\rho_h &= [1 + z\exp(-k^2/T)]^{-1}, \quad (27) \\ P &= T(2\pi)^{-1}\int_{-\infty}^{\infty} dk\ln[1 + z\exp(-k^2/T)], \end{aligned}$$

where

$$z = \text{fugacity} = \exp(A/T).$$

These equations are those for a free Fermi gas, a result that is anticipated, as discussed in Sec. II.

B. $c = 0$

As $c \to 0$,

$$c(c^2 + x^2)^{-1} \to \pi\delta(x). \tag{28}$$

Thus, (16) gives

$$\epsilon = -A + k^2 - T \ln [1 + \exp(-\epsilon/T)]$$

or

$$\exp(-\epsilon/T) = [z^{-1} \exp(k^2/T) - 1]^{-1},$$

where we have used the fugacity defined in (27). Equation (25) now becomes

$$P = -T(2\pi)^{-1} \int_{-\infty}^{\infty} dk \ln [1 - z \exp(-k^2/T)]. \tag{29}$$

Equations (28) and (17) give

$$2\pi\rho_h = 1,$$
$$2\pi\rho = \exp(-\epsilon/T) = [z^{-1} \exp(k^2/T) - 1]^{-1}. \tag{30}$$

Equations (29) and (30) are precisely the corresponding expressions for a free Bose gas, as they should be.

C. $T = 0$

This is the case solved[1] by Lieb and Liniger.

It will be shown in Appendix A that $\epsilon(k)$ is a monotonically increasing function of k^2. At $T = 0$ assume the function to have a zero at $k^2 = q_0^2$ so that

$$\begin{aligned} &\epsilon(k) < 0, \quad k^2 < q_0^2, \\ &\epsilon(k) > 0, \quad k^2 > q_0^2, \\ &\epsilon(q_0) = 0. \end{aligned} \tag{31}$$

Equation (15) gives

$$\begin{aligned} &\rho = 0, \quad \text{for} \quad k^2 > q_0^2, \\ &\rho_h = 0, \quad \text{for} \quad k^2 < q_0^2. \end{aligned} \tag{32}$$

Equations (16) and (17) become

$$\epsilon(k) = -A + k^2 + \frac{c}{\pi} \int_{-q_0}^{q_0} \frac{\epsilon(q)\, dg}{c^2 + (k-q)^2}, \tag{33}$$

$$2\pi\rho = 1 + 2c \int_{-q_0}^{q_0} \frac{\rho(q)\, dq}{c^2 + (k-q)^2}, \quad \text{for} \quad k^2 < q_0^2. \tag{34}$$

Equation (34) is the equation[1] of Lieb and Liniger. Equation (33) will be useful in the next section.

VII. EXCITATION

Consider a state S, with I's and k's satisfying

$$k_j L = 2\pi I_j + \sum_i \theta(k_j - k_i), \tag{35}$$

and a state S', with primed I's and k's satisfying

$$k_j' L = 2\pi I_j' + \sum_i \theta(k_j' - k_i'). \tag{36}$$

We consider the case where

$$I_j' = I_j, \quad \text{except when} \quad j = \alpha. \tag{37}$$

[Notice that $I_1', I_2', \cdots, I_N'$ may *not* be a monotonically increasing series, since I_α' may be any integer for N = odd and any integer $+\frac{1}{2}$ for N = even.]

Subtract (35) from (36) to obtain

$$(k_j' - k_j)L = \sum_i [\theta(k_j' - k_i') - \theta(k_j - k_i)]. \tag{38}$$

We now assume that, for all $j \neq \alpha$, k_j and k_j' are approximately the same. This is the same assumption as used by Lieb[4] for the excitations near the ground state (i.e., $T = 0$). We write

$$(k_j' - k_j)L = \chi(k_j), \quad j \neq \alpha.$$

Thus, we expand those terms in (38) for which $i \neq \alpha$:

$$\chi(k_j) = \sum_{i \neq \alpha} \theta'(k_j - k_i)[\chi(k_j) - \chi(k_i)]L^{-1} + \theta(k_j - k_\alpha') - \theta(k_j - k_\alpha) \tag{39}$$

or

$$\chi(k) = \int_{-\infty}^{\infty} \theta'(k - q)[\chi(k) - \chi(q)]\rho(q)\, dq + \theta(k - k_\alpha') - \theta(k - k_\alpha). \tag{40}$$

Now we differentiate (10b) and use it to evaluate the coefficient of $\chi(k)$ in (40). Writing

$$f(k)\chi(k) = g(k), \tag{41}$$

we thus obtain

$$2\pi g(k) = -\int_{-\infty}^{\infty} \theta'(k - q) g(q) \times [1 + \exp\{+\epsilon(q)/T\}]^{-1}\, dq + \theta(k - k_\alpha') - \theta(k - k_\alpha) \tag{42}$$

or, explicitly,

$$g(k) = \frac{c}{\pi} \int_{-\infty}^{\infty} \frac{g(q)\, dq}{[c^2 + (k-q)^2][1 + \exp \epsilon(q)/T]} + \frac{1}{\pi} \tan^{-1} (k_\alpha' - k)c^{-1} - \frac{1}{\pi} \tan^{-1} (k_\alpha - k)c^{-1}. \tag{43}$$

This is a Fredholm integral equation which we shall write in operator form

$$g = \mathbf{K}g + G. \tag{44}$$

The momentum difference and energy difference

[4] E. Lieb, Phys. Rev. **130**, 1616 (1963).

between the two states are

$$\Delta K = \sum_j (k_j' - k_j) = k_\alpha' - k_\alpha + \int_{-\infty}^{\infty} \chi(k)\rho(k)\, dk \quad (45)$$

and

$$\Delta E = \sum_j (k_j'^2 - k_j^2) = k_\alpha'^2 - k_\alpha^2 + \int_{-\infty}^{\infty} \chi(k) 2k\rho(k)\, dk. \quad (46)$$

We shall prove in Appendix E the following:

Theorem: The momentum difference and energy difference[5] between the two states are

$$\Delta K = h(k_\alpha') - h(k_\alpha) \quad (47)$$

and

$$\Delta E = \epsilon_0(k_\alpha') - \epsilon_0(k_\alpha), \quad (48)$$

where

$$\epsilon_0 = \epsilon + A \quad (49)$$

and h is an odd function of k defined by (10a). These equations are accurate to the order N^0, not just N^1. (Notice that in evaluating the thermodynamical quantities, such as the energy, we only maintain accuracy up to the order N^1.)

VIII. DISCUSSIONS

(A) It is easy to prove that, for a finite number of simultaneous excitations,

$$\Delta K = \sum_\alpha h(k_\alpha') - \sum_\alpha h(k_\alpha), \quad (50)$$

$$\Delta E = \sum_\alpha \epsilon_0(k_\alpha') - \sum_\alpha \epsilon_0(k_\alpha). \quad (51)$$

Thus it is tempting to regard $h(k_\alpha)$ and $\epsilon_0(k_\alpha)$ as the momentum and energy of an elementary excitation.

To be more precise, we consider a system of noninteracting fermions with its single-particle states labeled by k. The momentum and energy of a single-particle state k are taken to be $h(k)$ and $\epsilon_0(k)$, respectively. The number of single-particle states in the k interval dk is $f(k)\, dk$. Such a system of particles will be called a model system M. At a fixed fugacity z, the model system has an average number of particles in the state k given by

$$ze^{-\epsilon_0/T}[1 + ze^{-\epsilon_0/T}]^{-1}, \quad (52)$$

so that the number of particles in the interval dk is $f\, dk$ times (52),which is also the same quantity in the true system. The model system M and the true system then have the same excitation spectra at T, *provided* only a finite number of excitations are made from thermal equilibrium. (Notice that the definition of the system M depends on h, ϵ_0, and f.)

(B) The excitation $k_\alpha \to k_\alpha'$ discussed in Sec. VII occurs with an excitation function which is proportional to a factor dependent on the method of excitation. But, in addition, it is also proportional to the number of I's in the interval dI near I_α and the number of vacancies in the interval dI' near I_α'. Thus, to excite

from (k in dk) to (k' in dk')

there is an *intrinsic* excitation factor equal to

$$\rho(k)\rho_h(k')\, dk\, dk' = \rho(k)\rho(k')e^{\epsilon(k')/T}\, dk\, dk'. \quad (53)$$

[5] In the limit $T \to 0$, the energy and momentum spectra are reducible to very simple expressions, using (31)–(34). These spectra have been obtained by Lieb in Ref. 4. Reduction to such simple equations as (33) is new.

APPENDIX A

We want to prove that (16) can be solved by iteration. Define the right-hand side of (16) as 0ϵ. Define further

$$\epsilon_1 = -A + k^2, \quad \epsilon_2 = 0\epsilon_1, \quad \epsilon_3 = 0\epsilon_2, \quad \text{etc.} \quad (A1)$$

It is easily seen that

$$\epsilon_1(k) > \epsilon_2(k) > \epsilon_3(k), \quad \text{etc.}$$

Next one can show that $\epsilon_n(k)$ is bounded from below. To do this, one proves first by induction that $\epsilon_n - k^2$ is a nondecreasing function of k^2. One then has

$$\epsilon_{n+1}(0) \geq -A - \frac{Tc}{\pi}\int_{-\infty}^{\infty} \frac{dq}{c^2 + q^2} \times \ln\,[1 + \exp\{-\epsilon_n(0)T^{-1} - q^2T^{-1}\}]. \quad (A2)$$

Now define the right-hand side of (A2) as $f[\epsilon_n(0)]$. That is,

$$f(x) = -A + x - \frac{Tc}{\pi}\int_{-\infty}^{\infty} \frac{dq}{c^2 + q^2} \ln\,(e^{x/T} + e^{-q^2/T}). \quad (A3)$$

It is clear from (A3) that $f(x) - x$ is monotonically decreasing. It has one and only one zero. Call the zero x_0 so that

$$f(x_0) = x_0. \quad (A4)$$

The right-hand side of (A2) shows that

$$f(x) \text{ is monotonically increasing} \quad (A5)$$

and that $f(x) < -A$. Thus (A4) gives

$$-A > f(x_0) = x_0. \quad (A6)$$

Equations (A4), (A5), and (A6) show that

$$\epsilon_1(0) = -A > x_0,$$
$$\epsilon_2(0) \geq f[\epsilon_1(0)] > f(x_0) = x_0,$$
$$\epsilon_3(0) \geq f[\epsilon_2(0)] > f(x_0) = x_0, \quad \text{etc.}$$

Thus,

$$\epsilon_n(k) \geq \epsilon_n(0) > x_0. \tag{A7}$$

Having shown that

$$\lim_{n\to\infty} \epsilon_n(k) = \epsilon_L(k)$$

exists, one can next prove that the limit $\epsilon_L(k)$ does indeed satisfy (16). The main point is to show successively that (i), for $\epsilon > x_0$,

$$\frac{d}{d\epsilon}\ln(1 + e^{-\epsilon/T}) > -\frac{C}{T}, \quad \text{where} \quad 0 < C < 1,$$

and (ii) $\epsilon_n \to \epsilon_L$ uniformly in k.

APPENDIX B

To show that (17) can be solved by iteration we construct the symmetrized kernel

$$\mathbf{K}' = \frac{\pi^{-1}c}{c^2 + (k-q)^2}(1 + e^{\epsilon(q)/T})^{-\frac{1}{2}}(1 + e^{\epsilon(k)/T})^{-\frac{1}{2}}. \tag{B1}$$

If ψ is any normalized function and

$$\Phi = [1 + e^{\epsilon/T}]^{-\frac{1}{2}}\psi,$$

then

$$\psi^+\mathbf{K}'\psi = \Phi^+ \frac{\pi^{-1}c}{c^2 + (k-q)^2}\Phi$$
$$\leq \Phi^+\Phi \leq [1 + e^{x_0/T}]^{-1}\psi^+\psi,$$

where x_0 was defined in Appendix A. Thus, the eigenvalues of $\mathbf{K}'$ are less than unity and iteration of (17) converges.

The solution of (17) so obtained evidently satisfies

$$\rho > 0, \quad \rho_h = \rho \exp[\epsilon/T] > 0. \tag{B2}$$

APPENDIX C

(A) We treat the maximization procedure leading to (16) and (17) more rigorously here, showing that the solution of (16) and (17) indeed leads to a minimum of the free energy, i.e., a maximum of the partition function (14′).

Consider any $\rho(k)$. If $\rho(k) \geq 0$ and the $\rho_h(k)$ defined by (11) is everywhere ≥ 0, we say that ρ is in R_0. It is clear that if ρ_1 and ρ_2 are both in R_0, then $x\rho_1 + (1-x)\rho_2$ for $0 \leq x \leq 1$ is also in R_0. Thus R_0 is convex.

We define $X(L, T, A, \rho)$ by

$$X = L\int_{-\infty}^{\infty} k^2\rho\, dk + LT\int_{-\infty}^{\infty}[\rho\ln\rho + \rho_h\ln\rho_h - (\rho + \rho_h)\ln(\rho + \rho_h)]\,dk - LA\int_{-\infty}^{\infty}\rho\, dk. \tag{C1}$$

Consider $\rho = \rho_0 + x\rho_1$ where ρ_0 and ρ_1 are independent of x. Assume ρ to be in R_0 for a real segment of x. We can take the derivatives of X with respect to x in this segment. A straightforward calculation yields

$$\frac{dX}{dx} = L\int \rho_1\, dk\Big[k^2 - A - \epsilon(k) - T \times \int B(k,q)\ln(1 + e^{-\epsilon(q)/T})\,dq\Big], \tag{C2}$$

where ϵ is defined by

$$\exp(\epsilon/T) = \rho_h/\rho \tag{C3}$$

and

$$B(k,q) = \frac{\pi^{-1}c}{c^2 + (k-q)^2} = B(q,k). \tag{C4}$$

It is easy to show that

$$-T^{-1}\frac{\partial\epsilon(k)}{\partial x} = \rho^{-1}[1 + e^{-\epsilon/T}] \times \Big\{\rho_1 - [1 + e^{\epsilon(k)/T}]^{-1}\int B(k,q)\rho_1(q)\,dq\Big\}. \tag{C5}$$

Now,

$$\frac{d^2X}{dx^2} = L\int \rho_1\, dk \times \Big\{-\frac{\partial\epsilon(k)}{\partial x} + \int B(k,q)(1 + e^{\epsilon(q)/T})^{-1}\frac{\partial\epsilon(q)}{\partial x}\,dq\Big\}. \tag{C6}$$

The double integral in (C6), after the switch $k \leftrightarrow q$, can be reduced through the use of (C5), giving

$$\frac{d^2X}{dx^2} = T^{-1}L\int\Big[\frac{\partial\epsilon(k)}{\partial x}\Big]^2\rho(k)[1 + e^{-\epsilon(k)/T}]^{-1}\,dk > 0. \tag{C7}$$

By (B2), the solution of (16) and (17) gives a ρ in R_0. By (C2), at that ρ, $dX/dx = 0$.

We conclude further, from (C7), that X has a unique minimum in R_0 at the ρ given by (16) and (17).

(B) For given L, T, and A, we denote the minimum of X discussed above by $Y = Y(L, T, A)$. Clearly,

$$\frac{\partial Y}{\partial A} = -L\int\rho\,dk = -N,$$

$$\frac{\partial Y}{\partial T} = -S,$$

by (13) and (14). Further, since Y is proportional to L,

$$dY = -N\,dA - S\,dT + (Y/L)\,dL.$$

Thus,

$$d(Y + NA) = -S\,dT + (Y/L)\,dL + A\,dN.$$

But $Y + NA$ is the free energy. Thus,

$$A = \text{chemical potential}$$

and

$$Y/L = -\text{pressure}.$$

APPENDIX D

Write (16) symbolically in the form

$$\epsilon = W[A, T, \epsilon]. \tag{D1}$$

Consider two real numbers A_0, $T_0 > 0$ and let ϵ_1 be the solution of

$$\epsilon_1 = W[A_0, T_0, \epsilon_1]. \tag{D2}$$

The existence of ϵ_1 was proved in Appendix A. Now for complex values of A and T in the neighborhood of A_0, and T_0, we shall solve (D1) by iteration:

$$\epsilon_2 = W[A, T, \epsilon_1],$$
$$\epsilon_3 = W[A, T, \epsilon_2], \quad \text{etc.}$$

It can be shown that in a sufficiently small complex neighborhood R_1 of (A_0, T_0), $\epsilon_n \to \epsilon_\infty$ as $n \to \infty$, *uniformly* in k, T, and A. Since ϵ_n is analytic in A and T within R_1, so is ϵ_∞. It then easily follows that P as computed from (25) is analytic in A and T within R_1.

APPENDIX E

To prove (47) and (48) we define the kernel of (43):

$$K(k, q) = \frac{\pi^{-1}c}{[c^2 + (k - q)^2][1 + \exp \epsilon(q)/T]}. \tag{E1}$$

Equation (43) is then equivalent to

$$(\mathbf{1} - \mathbf{K})g = \int_{k_\alpha}^{k_\alpha'} K(k, q)[1 + \exp \epsilon(q)/T]\,dq. \tag{E2}$$

Let

$$(\mathbf{1} + \mathbf{L})(\mathbf{1} - \mathbf{K}) = \mathbf{1} \tag{E3}$$

or

$$\mathbf{L} - \mathbf{K} = \mathbf{LK}. \tag{E4}$$

Equation (E2) gives

$$g(k) = \int_{k_\alpha}^{k_\alpha'} L(k, q)[1 + \exp \epsilon(q)/T]\,dq. \tag{E5}$$

Now, the $\mathbf{K}'$ of (B1) is a symmetrical kernel with a symmetrical inverse kernel. From that fact we easily obtain

$$L(k, q)[1 + \exp \epsilon(q)/T] = L(q, k)[1 + \exp \epsilon(k)/T]. \tag{E6}$$

Two other useful formulas can be obtained as follows. Equation (17) can be rewritten as

$$(\mathbf{1} - \mathbf{K})f = (2\pi)^{-1}.$$

Operating with $\mathbf{1} + \mathbf{L}$ on both sides we obtain

$$f = (2\pi)^{-1} + (2\pi)^{-1}\int_{-\infty}^{\infty} L(k, q)\,dq. \tag{E7}$$

Similarly, differentiation of (16) with respect to k yields

$$\frac{d\epsilon}{dk} = 2k + \mathbf{K}\frac{d\epsilon}{dk}.$$

Thus,

$$\frac{d\epsilon}{dk} = 2k + \int_{-\infty}^{\infty} L(k, q)2q\,dq. \tag{E8}$$

Now, by (41),

$$\chi\rho = g/[1 + \exp(\epsilon/T)].$$

Thus, (45) becomes

$$\begin{aligned}
\Delta K &= k_\alpha' - k_\alpha + \int_{-\infty}^{\infty} \frac{g(k)\,dk}{1 + \exp \epsilon(k)/T} \\
&= k_\alpha' - k_\alpha + \int_{-\infty}^{\infty} dk \int_{k_\alpha}^{k_\alpha'} dq L(k, q) \\
&\qquad \times [1 + \exp \epsilon(q)/T][1 + \exp \epsilon(k)/T]^{-1} \\
&= k_\alpha' - k_\alpha + \int_{-\infty}^{\infty} dk \int_{k_\alpha}^{k_\alpha'} dq L(q, k) \\
&= k_\alpha' - k_\alpha + \int_{k_\alpha}^{k_\alpha'} dq[2\pi f(q) - 1],
\end{aligned}$$

yielding (47). Similarly we derive (48).

APPENDIX F

We shall prove here rigorously that, for the ground state, the k's of Sec. II approach a distribution $L\rho(k)\,dk$ as $L \to \infty$, $N \to \infty$ proportionally. By continuing with respect to c^{-1} and the theorem of Sec. II, we know that for the ground state the I's form a close-packed set of integers or half-odd integers. We now define, as in (8),

$$h(k) = k - L^{-1} \sum_{k'} \theta(k - k'). \tag{F1}$$

Clearly,

$$\frac{dh}{dk} = 1 + \frac{2c}{L} \sum_{k'} \frac{1}{c^2 + (k - k')^2} > 1. \tag{F2}$$

Equation (F1) defines $h(k)$ for *all* real values of k. At the successive values of k_j, the successive values of $h(k_j)$ are $2\pi I_j/L$, by (4). Thus, the successive values of h form a lattice, with a lattice constant of $2\pi/L$, extending between $\pm(N - 1)\pi L^{-1}$. It is sometimes convenient to use h as the variable rather than k.

$k(h)$ is then a monotonically increasing odd function, defined for all real h, and approaches ∞ as $h \to \infty$.

Differentiation of (F1) gives, writing $dh/dk = h_1$,

$$h_1(k) = 1 + \frac{2c}{L} \sum_{k'} \frac{1}{c^2 + (k - k')^2}$$
$$= 1 + \frac{c}{\pi} \int_{-N\pi/L}^{N\pi/L} \frac{1}{c^2 + (k - k')^2} dh' + \text{residue}. \tag{F3}$$

The residue is in absolute value less than $A_1 L^{-1}$, since the integrand has a bounded derivative. Thus,

$$h_1(k) = 1 + \frac{c}{\pi} \times \int_{-Q}^{Q} \frac{1}{c^2 + (k - k')^2} h_1(k')\, dk' + O(L^{-1}), \tag{F4}$$

$$2N\pi/L = \int_{-Q}^{Q} h_1(k')\, dk'. \tag{F5}$$

It is now possible to complete the proof. We first use $h_1 > 1$ to obtain, from (F5),

$$Q < N\pi/L. \tag{F6}$$

With this fixed bound for Q, the inverse kernel for the integral equation (F4) is also absolutely bounded and we obtain

$$h_1(k) = 1(k, Q) + O(L^{-1}), \tag{F7}$$

where $1(k, Q)$ is the solution of (F4) when $O(L^{-1})$ is deleted. Integration of (F7) gives

$$2N\pi/L = \int_{-Q}^{Q} 1(k, Q)\, dk + O(L^{-1}). \tag{F8}$$

Thus, for fixed N/L, as $L \to \infty$, Q approaches a limit Q_0 given by

$$2\pi N/L = \int_{-Q_0}^{Q} 1(k, Q_0)\, dk. \tag{F9}$$

The rest is easy.

⟦69c⟧

Commentary begins page 69

Hypothesis of Limiting Fragmentation in High-Energy Collisions

J. Benecke, T. T. Chou, C. N. Yang, and E. Yen

Institute for Theoretical Physics, State University of New York at Stony Brook, Stony Brook, New York 11790*

(Received 28 August 1969)

A hypothesis of limiting fragmentation of the target and of the projectile in a high-energy lepton-hadron or hadron-hadron collision is defined. Arguments are given for the hypothesis. Comparisons with various models and concepts are made. Further speculations are made, including the absence of pionization processes in high-energy collisions and the dependence of multiplicity on the momentum transfer. Experiments are suggested.

INTRODUCTION

In recent years many experiments[1] have been performed on inelastic hadron-hadron collisions and inelastic *ep* collisions. These experiments, taken together with the droplet interpretation[2,3] of elastic collisions, and with cosmic-ray information, suggest a specific framework in which very-high-energy *ep* and hadron-hadron collisions could be usefully described. The framework is based on the *hypothesis of limiting fragmentation of each of the two colliding particles* in a high-energy collision which will be defined and discussed in this paper. If the hypothesis is correct, experimentally it suggests that one should measure the limiting fragment distributions. Theoretically, it suggests that the fragmentation process should be a principal subject of study for any model of high-energy collisions.

Our discussion is very much related to the traditional two-fireball model[4] used in cosmic-ray physics. We shall explicitly discuss the points of agreement between the present hypothesis and the two-fireball model, and also the specific points where the two pictures differ. Our discussion is also closely related to the concept of "diffraction dissociation" introduced by Good and Walker.[5]

A dominant feature of multiparticle processes in high-energy collisions is the small value of the transverse momenta of the outgoing particles, a property that is especially difficult to accommodate in the traditional statistical model.[6] On the other hand, longitudinal momenta are usually quite large for some of the emitted particles. The momentum distribution of the outgoing particles, especially when the multiplicity is low, strongly suggests the "persistence" of the longitudinal momentum that resides in the incoming projectile which breaks apart in the collision process. Similarly, the target under the influence of the fast-moving projectile seems, in general, to break up into many pieces. This intuitive picture of a high-energy collision process as two extended objects going through each other, breaking into fragments in the process, is defined precisely in the next section as the hypothesis of limiting fragmentation. Arguments for this hypothesis are then presented and the picture is compared with previously proposed models. Additional speculations, remarks, and suggested experiments are discussed near the end of the paper.

HYPOTHESIS OF LIMITING FRAGMENTATION

1. It has been customary to describe a collision in the c.m. system. We believe that for very-high-energy collisions, the lab system (L) and the projectile system (P, where the incoming projectile is at rest) are to be preferred, because in these systems, some of the outgoing particles *approach limiting distributions*. (Because of the large number of possible kinematic variables involved in a multiparticle process, it is important to use those variables in terms of which the process is most simply described.)

Let us consider the lab system L. In a typical high-energy collision, with incoming energy E, many particles are produced. Some of these particles tend to have increasingly high lab velocity v as E increases. Others have values of $\gamma=(1-v^2)^{-1/2}$ that remain finite as E increases. These latter particles, we propose, approach a limiting distribution as $E\to\infty$.

To be more precise, given a d^3p region for the laboratory $\mathbf{p}$ of a specific outgoing particle of mass m (say, a proton, or a pion), the probability that in a high-energy hadron-hadron or electron-hadron collision a particle of that mass will be found in that region

* Partially supported by the U. S. Atomic Energy Commission, under Contract No. AT(30-1)-3668B.

[1] See *Proceedings of the Fourteenth International Conference on High-Energy Physics, Vienna, 1968*, edited by J. Prentki and J. Steinberger (CERN, Geneva, 1968); *Proceedings of the Topical Conference on High-Energy Collisions of Hadrons, Geneva, 1968* (Scientific Information Service, Geneva, 1968).

[2] T. T. Wu and C. N. Yang, Phys. Rev. **137**, B708 (1965).

[3] N. Byers and C. N. Yang, Phys. Rev. **142**, 976 (1966); T. T. Chou and C. N. Yang, in *Proceedings of the Second International Conference on High-Energy Physics and Nuclear Structure, Rehovoth, Israel, 1967*, edited by G. Alexander (North-Holland Publishing Co., Amsterdam, 1967), pp. 348–359; Phys. Rev. **170**, 1591 (1968); Phys. Rev. Letters **20**, 1213 (1968); Phys. Rev. **175**, 1832 (1968).

[4] G. Cocconi, Phys. Rev. **111**, 1699 (1958); K. Niu, Nuovo Cimento **10**, 994 (1958); P. Coik, T. Coghen, J. Gierula, R. Holynski, A. Jurak, M. Miesowicz, T. Saniewska, and J. Pernegr, *ibid.* **10**, 741 (1958). See also R. K. Adair, Phys. Rev. **172**, 1370 (1968).

[5] M. L. Good and W. D. Walker, Phys. Rev. **120**, 1857 (1960).

[6] E. Fermi, Progr. Theoret. Phys. (Kyoto) **5**, 570 (1950); Phys. Rev. **81**, 683 (1951); **92**, 452 (1953); **93**, 1434 (1954).

 □ Reprinted from *The Physical Review* 188, 5 (December 25, 1969), 2159–2169.

approaches a limit as $E \to \infty$. In other words, we hypothesize the existence of

$$\lim_{E\to\infty} (\text{partial cross section that a particle of mass } m \text{ is emitted with lab momentum } \mathbf{p}, \text{ other emitted particles being ignored}) = \rho_1(\mathbf{p})d^3p\,, \quad (1)$$

where

$$\rho_1 > 0\,.$$

2. While in (1) we consider the limiting distribution of one particle, we believe a similar limit exists for any configuration in the lab momentum space for n particles, for any fixed $n=1, 2, \ldots$; i.e., the following limit exists:

$$\lim_{E\to\infty} (\text{partial cross section that a particle of mass } m_1 \text{ and momentum } \mathbf{p}_1\text{, and a particle of mass } m_2 \text{ and momentum } \mathbf{p}_2\text{, are emitted, together with any number of other particles}) = \rho_2(\mathbf{p}_1,\mathbf{p}_2)d^3p_1d^3p_2\,, \text{ etc.,} \quad (2)$$

where

$$\rho_2 > 0\,, \text{ etc.}$$

These limits, of course, are in general different for different collisions (ep, pp, πp, etc.).

3. In (1) and (2) above, no reference is made to the rate at which the limit is approached. Experimental data suggest that for high values of n and/or high values of lab momentum, the approach to a limiting distribution is slow.

4. The limiting distributions (1) and (2) discussed above represent *distributions of broken-up fragments of the target.* The fragments from the projectile, on the other hand, move with increasing velocity in the lab system as $E\to\infty$, and do not contribute to any limiting distribution. To study these fragments from the projectile, one must go into the projectile system P. If there should exist additional outgoing particles which are not fragments (as *defined* above) of either the projectile or the target (for example, pions in the "pionization" process which are slow in the c.m. system), they will not contribute to the limiting distribution in either the lab or the projectile system. The possible existence of a pionization process is not inconsistent with the hypothesis we are discussing here. However, for reasons to be explained later, we shall speculate in the beginning of Sec. 16 that at very high energies there are no pionization processes.

5. If limiting distributions (1) and (2) do exist, their integrals are related to average multiplicities for general or restricted types of collisions. For example,

$$\int \rho_1 d^3p = \sigma_0 \times (\text{average multiplicity per collision of particles of mass } m \text{ emitted from the target})\,, \quad (3a)$$

where σ_0 = total collision cross section. Similarly,

$$\int \rho_2 d^3p_2 = \rho_1(\mathbf{p}_1) \times (\text{average multiplicity per collision, where a particle of mass } m_1 \text{ and momentum } \mathbf{p}_1 \text{ is emitted from the target, of particles of mass } m_2 \text{ emitted from the target}). \quad (3b)$$

While the values of the integrals in (3a) and (3b) are not specified in the hypothesis under discussion here, we believe them to be divergent. This belief derives from the fact that at infinite incoming energies all multiplicities seem to become infinity. If, as we believe, there is no pionization, then the average multiplicity from the projectile plus that from the target must add up to infinity. Hence we speculate that each of them is divergent.

It may appear at first sight that a limiting fragmentation of the two colliding particles is inconsistent with increasing multiplicities at higher and higher energies. It is to be emphasized that this is not so: Increasing multiplicities derive from the fact that more and more of the *divergent* integrals (3a), (3b), etc., are made accessible at higher and higher incident energies. It seems that this view *reconciles* rather satisfactorily, for higher and higher incident energy, *the increasing average multiplicity on the one hand, with the observed persistence of cross sections for events with small multiplicities* (cf. Sec. 8.3) *on the other.*

6. We discuss now the kinematic region in which the distribution ρ_1 of (1) is defined. For those processes such as forward elastic scattering $\pi p \to \pi p$, or diffraction excitation $\pi p \to \pi^* p$, i.e., for those cases where the target remains unchanged, the contribution to ρ_1 is confined to the parabola in momentum space,

$$e - p_{\parallel} = M_t\,, \quad (4)$$

where e and $p_{\parallel}$ are, respectively, the lab energy and longitudinal momentum of the particle in question, and M_t is the mass of the target. If the target breaks into more than one particle, the region of momentum space where such processes contribute to ρ_1 is given by

$$e - p_{\parallel} < M_t\,. \quad (5)$$

Equations (4) and (5) will be proved in Appendix A.

Thus ρ_1 is a delta function σ_1 on the parabola (4) plus a distribution function τ_1 in the region (5). The region (5) is bounded by the parabola (4). A similar breakdown of ρ_2, ρ_3, etc., yields

$$\rho_n = \sigma_n + \tau_n\,, \quad (6)$$

where

$$\tau_n = \text{function defined in } R_n\,, \quad (7)$$

$$\sigma_n = \text{delta function defined on } S_n\,. \quad (8)$$

Here,

$$R_n = \text{the region } \sum_{i=1}^{n} (e - p_{\parallel})_i < M_t \qquad (9)$$

and

$$S_n = \text{boundary of } R_n$$
$$= \text{the surface defined by } \sum_{i=1}^{n} (e - p_{\parallel})_i = M_t. \qquad (10)$$

These formulas will be proved in Appendix A, with some properties of (10).

Physically, σ_n is the cross section for processes in which the target breaks up into exactly the n particles specified in its argument; τ_n, for those in which the target breaks up into *more* than n particles. Obviously,

$\sigma_1, \sigma_2, \ldots$ = limiting partial cross section of various possible fragmentations of the target under the impact of a projectile at infinite energies. [The particles in σ_n have momenta satisfying (10).] (11)

[If the target can be broken into two fragments in different ways, e.g., $p \to \pi^0 p$ and $p \to K^0\Sigma^+$, then several σ_2's must be included in the left-hand side of (11).]

Because of (11), *the distributions τ_n can be obtained from a superposition of $\sigma_{n+1}, \sigma_{n+2}, \ldots$,* etc., *after the redundant coordinates are integrated out.*

7. The discussion above about the fragmentation of the target in the laboratory system L has, of course, its counterpart for the fragmentation process of the projectile in the projectile system P. We must replace M_t by the mass of the projectile M_{inc}. (The case of ep collisions is an exception since the electron as the projectile does not break up.)

It is important to notice that a particle which has a finite energy in the projectile system P has a very large energy in the laboratory system. In fact, its laboratory energy is proportional to the energy of the incoming particle.

Viewed in the c.m. system, the fragments of the target form a jet in the backward hemisphere. It is important to recognize that if $\mathbf{p}^*$ is the center-of-mass momentum for the collision, then a particle with a finite momentum in the laboratory system would have a c.m. energy proportional to $\mathbf{p}^*$ when $\mathbf{p}^* \to \infty$. In other words, at very high energies, *almost all of the backward hemisphere in momentum space in the c.m. system represent finite momenta in the laboratory system.* A similar statement holds for the forward hemisphere when one replaces the laboratory by the projectile system. These statements can be put in a concise form for the limit $E \to \infty$, as exhibited in Table I.

TABLE I. Order of magnitude of momenta as incident energy $E \to \infty$. M_t = mass of target.

	Lab momenta	c.m. momenta
(1) Fragments of target	$O(1)$	$-O(E^{1/2})$
(2) Pionization products	$O(E^{1/2})$	$O(1)$
(3) Fragments of projectile	ξE $(\xi<1)$	$\xi(\frac{1}{2}M_t E)^{1/2}$ $(\xi<1)$
(4) Intact projectile, with excitation or fragmentation of target	$E-O(1)$	$(\frac{1}{2}M_t E)^{1/2} - O(E^{-1/2})$
(5) Elastically scattered projectile	E	$(\frac{1}{2}M_t/E)^{1/2}$

ARGUMENTS FOR HYPOTHESIS OF LIMITING FRAGMENTATION

8. We now present arguments in support of the hypothesis discussed above.

8.1. The differential cross section of inelastic ep scattering processes is written in the usual notation[7] as

$$\frac{d^2\sigma}{dq^2 d\nu} = \frac{4\pi\alpha^2}{q^4}\frac{E'}{E}\left[\cos^2(\tfrac{1}{2}\theta)\, W_2(q^2,\nu) + 2\sin^2(\tfrac{1}{2}\theta)\, W_1(q^2,\nu)\right]. \qquad (12)$$

When $E, E' \to \infty$ and $\theta \to 0$ such that q^2, ν remain finite, this reduces to

$$\lim \frac{d^2\sigma}{dq^2 d\nu} = \frac{4\pi\alpha^2}{q^4} W_2(q^2,\nu). \qquad (13)$$

The recoil particles have a total invariant mass M given by

$$M^2 = M_p^2 + 2M_p\nu - q^2. \qquad (14)$$

Thus (13) shows that

$$\lim \frac{d^2\sigma}{dq^2 dM^2} = \frac{2\pi\alpha^2}{M_p q^4} W_2(q^2,\nu). \qquad (15)$$

In other words, the cross section for the excitation of the target to any fixed M with a four-momentum transfer q^2 approaches a limit. In fact, such an excitation leads to a fixed density matrix of the recoil particles[8] in the lab system; i.e., for ep collisions the limiting distribution (11) for the fragmentations of the target proton exists. If follows that the limits (1) and (2) also exist. Thus for ep scattering, to the order (fine-structure constant)^2, the hypothesis discussed in Secs. 1–7 is proved.

8.2. High-energy elastic scattering experiments seem to indicate that as the incoming energy approaches infinity, the differential cross section $d\sigma/dt$ approaches a limit. Thus for any elastic scattering experiment, the momentum distribution of the recoil target particle approaches a limit on the parabola (4).

[7] W. K. H. Panofsky, in *Proceedings of the Fourteenth International Conference on High-Energy Physics, Vienna, 1968,* edited by J. Prentki and J. Steinberger (CERN, Geneva, 1968), pp. 23–39.

[8] The situation is entirely similar to the corresponding problem in neutrino-nucleon collisions. See T. D. Lee and C. N. Yang, Phys. Rev. **126**, 2239 (1962).

8.3. It was observed experimentally[9] that there exists a class of processes $AB \to CD$ with finite limiting cross sections at high energies, such as $pp \to pp^*$, where the p^* is a low-lying resonance having the same quantum numbers as the proton. Two consequences are the following:

(a) The recoil target particle in $pp \to p^*p$ where the target is not excited would have a limiting distribution on the parabola (4) in laboratory momentum space. (The excitation of the projectile to p^* at any finite excitation energy does not affect the allowed laboratory momenta values of the recoil target particle, in the high-energy limit under consideration.) The limiting distribution σ_1, defined in (8), is thus a sum of the contributions from this process, from the elastic scattering process of Sec. 8.2, and from all other processes where the target does not break up.

(b) In the process $pp \to pp^*$, where the target is excited to p^*, the disintegration of p^* would lead to a limiting distribution for the decay products in the lab system.

8.4. Experiments at Brookhaven and at CERN[10] studied the momentum distribution of single medium fast particles (p, π, or K) produced in pp collisions at 30 and 19.2 BeV/c. [These particles belong to category (3) of Table I.] By performing a Lorentz transformation to the projectile system P, these particles become slow. Owing to the symmetry of the pp system, this is equivalent to the study of the slow particles in the lab system L.

In Fig. 1 we plot the distribution of slow protons in L, obtained in this fashion. We observe that they are not too different at 19 and 30 BeV/c incident energy, indicating that the limit is approached at $\sim$30 BeV/c.

We have also plotted the pion and kaon distributions in the lab system, obtained in a similar fashion, in Figs. 2 and 3. Where there are data for both energies, the π^- distribution seems also to have approached a limit.

In a recent paper[11] the momentum distribution of medium-high-energy secondary π, K, and p were reported up to 70-BeV incident energy. The authors showed that for a fixed p/p_{inc} in the lab system, the

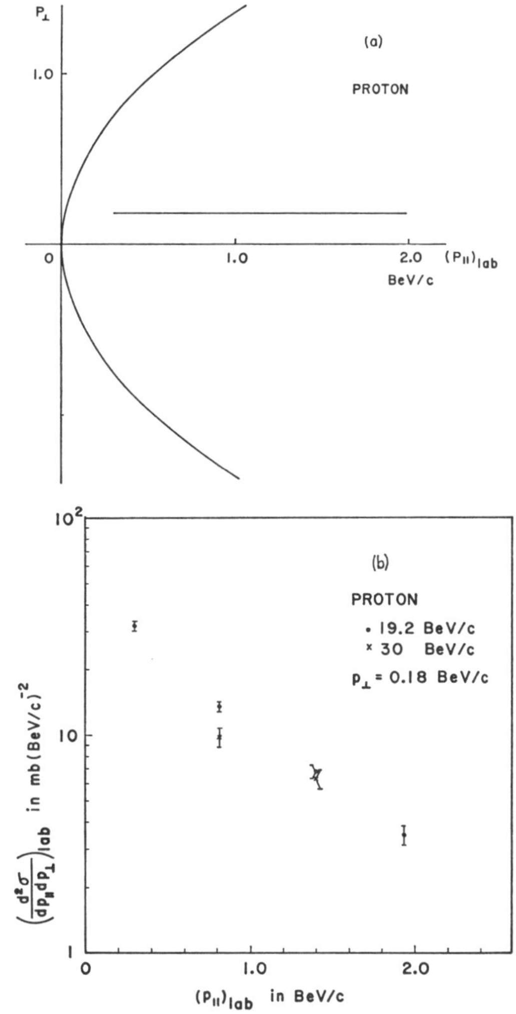

FIG. 1. (a) Kinematic boundary [a parabola, i.e., Eq. (4)] and allowed momentum space [region to the right of parabola, i.e., Eq. (5)] for single proton distribution in pp collisions at infinite energy. Line segment parallel to the $p_{\parallel}$ axis represents the region where experimental data are available at present energies with $p_\perp = 0.18$ BeV/c. The distribution of proton along this line is plotted in Fig. 1(b). All momenta are in the lab system L. (b) Distribution of proton as fragment of target in the lab system with $p_\perp = 0.18$ BeV/c at 19.2 and 30 BeV/c obtained from Ref. 10 by a procedure described in Sec. 8.4 of the text. The proton distribution at 30 BeV/c can be fitted by $d^2\sigma/dp_{\parallel}^* dp_\perp = 610\, p_\perp^2 e^{-p_\perp/0.166}$ mb (BeV/c)$^{-2}$ in c.m. system or $d^2\sigma/dp_{\parallel} dp_\perp = 2480\, p_\perp^2 e^{-p_\perp/0.166}(1-0.969 p_{\parallel}/E_{lab})$ mb (BeV/c)$^{-2}$ in the lab system, where $p_\perp$ is in BeV/c. For very small values of $(p_{\parallel})_{lab}$, the curve should dip down to zero. Compare Sec. 9.

---

[9] G. Cocconi, A. N. Diddens, E. Lillethun, G. Manning, A. E. Taylor, T. G. Walker, and A. M. Wetherell, Phys. Rev. Letters **7**, 450 (1961); E. W. Anderson, E. J. Bleser, G. B. Collins, T. Fujii, J. Menes, F. Turkot, R. A. Carrigan, Jr., R. M. Edelstein, N. C. Hien, T. J. McMahon, and I. Nadelhaft, *ibid.* **16**, 855 (1966); K. J. Foley, R. S. Jones, S. J. Lindenbaum, W. A. Love, S. Ozaki, E. D. Platner, C. A. Quarles, and E. H. Willen, *ibid.* **19**, 397 (1967).

[10] E. W. Anderson, E. J. Bleser, G. B. Collins, T. Fujii, J. Menes, F. Turkot, R. A. Carrigan, Jr., R. M. Edelstein, N. C. Hien, T. J. McMahon, and I. Nadelhaft, Phys. Rev. Letters **19**, 198 (1967); J. V. Allaby, F. Binon, A. N. Diddens, P. Duteil, A. Klovning, R. Meunier, J. P. Peigneux, E. J. Sacharidis, K. Schlüpmann, M. Spighel, J. P. Stroot, A. M. Thorndike, and A. M. Wetherell, in *Proceedings of the Fourteenth International Conference on High-Energy Physics, Vienna, 1968*, edited by J. Prentki and J. Steinberger (CERN, Geneva, 1968).

[11] Yu. B. Bushnin, S. P. Denisov, S. V. Donskov, A: F. Dunaitsev, Yu. P. Gorin, V. A. Kachanov, Yu. S. Khodirev, V. I. Kotov, V. M. Kutyin, A. I. Petrukhin, Yu. D. Prokoshkin, E. A. Razuvaev, R. S. Shuvalov, D. A. Stoyanova, J. V. Allaby, F. Binon, A. N. Diddens, P. Duteil, G. Giacomelli, R. Meunier, J.-P. Peigneux, K. Schlüpmann, M. Spighel, C. A. Stahlbrandt, J.-P. Stroot, and A. M. Wetherell, Phys. Letters **29B**, 48 (1969), especially Fig. 3(b).

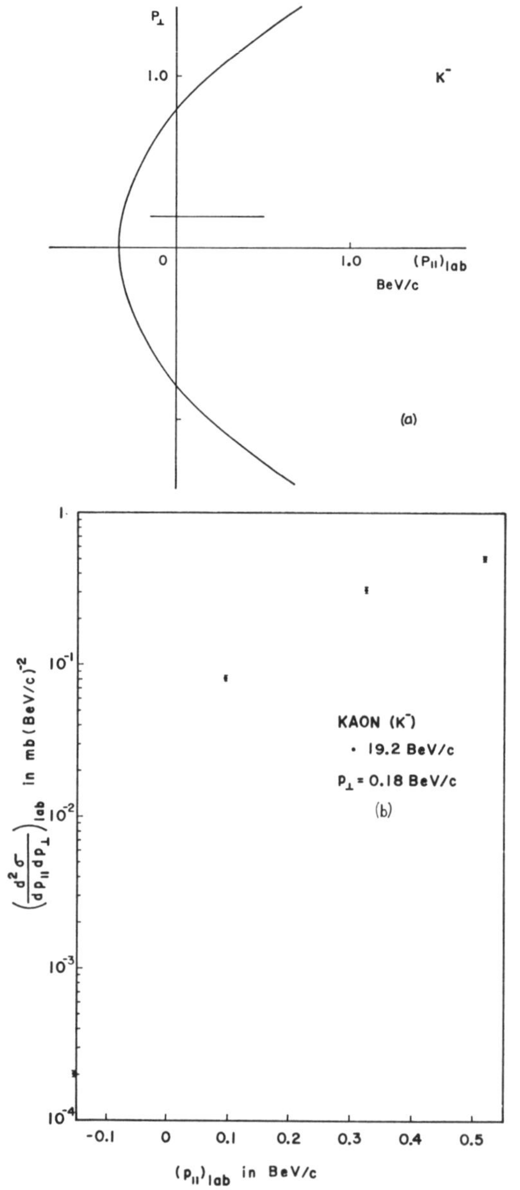

FIG. 2. (a) Kinematic boundary [a parabola, i.e., Eq. (4)] and allowed momentum space [region to the right of parabola, i.e., Eq. (5)] for single kaon (K^-) distribution in pp collisions at infinite energy. Line segment parallel to $p_{\parallel}$ axis represents the region where experimental data are available at present energies with $p_{\perp}=0.18$ BeV/c. The distribution of kaon along this line is plotted in Fig. 2(b). All momenta are in the lab system L. (b) Distribution of kaon as fragment of target in the lab system with $p_{\perp}=0.18$ BeV/c at 19.2 BeV/c obtained from Ref. 10 by a procedure described in Sec. 8.4 of the text.

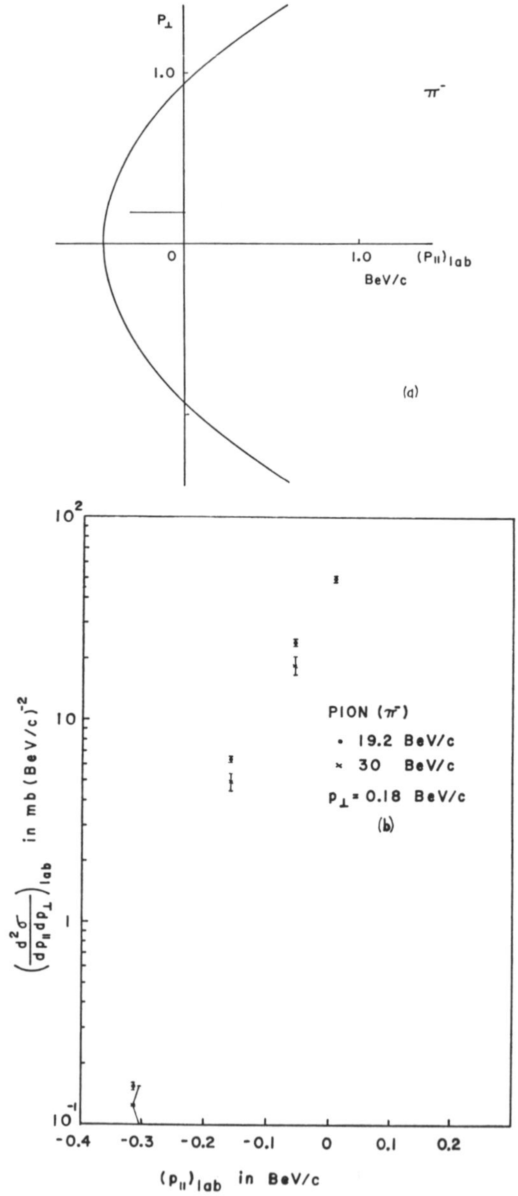

FIG. 3. (a) Kinematic boundary [parabola, i.e., Eq. (4)] and allowed momentum space [region to the right of parabola, i.e., Eq. (5)] for single pion (π^-) distribution in pp collisions at infinite energy. Line segment parallel to $p_{\parallel}$ axis represents the region where experimental data are available at present energies with $p_{\perp}=0.18$ BeV/c. The distribution of pions along this line is plotted in Fig. 3 (b). All momenta are in the lab system L. (b) Distribution of pion as fragment of target in the lab system with $p_{\perp}=0.18$ BeV/c at 19.2 and 30 BeV/c obtained from Ref. 10 by a procedure described in Sec. 8.4 of the text.

particle ratios K^-/π^- and $\bar{p}/\pi^-$ are independent of the incident energy. This independence is easily seen as a *natural consequence* of the hypothesis of a limiting distribution in the projectile system. (Fixed $x=p/p_{\text{inc}}$ in the forward direction means a fixed momentum p^{**} in the rest system of the incoming particle, for high incident energies:

$$p^{**}=\tfrac{1}{2}xM_{\text{inc}}-\tfrac{1}{2}\mu^2/xM_{\text{inc}},$$

where M_{inc}, and μ are the masses of the incoming and the secondary particles.)

8.5. At cosmic-ray energies, an incoming projectile proton emerges in a high-energy collision as an outgoing proton with an "inelasticity" (i.e., fractional energy loss) that seems[12] to approach a limiting distribution. In the projectile system P, the momenta of the outgoing protons would approach a limiting distribution.

8.6. If limiting distributions for the fragmentation on the target and the projectile exist, and if there are no pionization processes, then the transverse momentum of any outgoing particle would approach a limiting distribution, consistent with a dominant feature well known in all high-energy accelerator and cosmic-ray experiments. (See Sec. 18 for a discussion of high-multiplicity events.)

8.7. Elastic scattering at high energies have been described[2,3] in terms of a droplet picture where the target serves as an absorbing medium through which an incoming particle propagates as a wave. In such a picture, the incoming particle in the lab system shrinks into a thin disk by Lorentz contraction at high energies. The target proton has a geometrical extension of the order of 0.7×10^{-13} cm. Passage of the thin disk through the target takes about 2×10^{-24} sec, and the target is excited during this time. The excitation may cause a breakup of the target. What is the effect of higher and higher projectile momentum? The time of passage is essentially fixed, but the disk is further and further compressed (see Fig. 4). The constancy of the total cross section and of the elastic scattering cross section suggest that the momentum and quantum-number transfer process between the "stuff" in the projectile and the "stuff" in the target does not appreciably change when the projectile is further and further compressed. Thus one expects that the excitation and breakup of the target approaches a limiting distribution, which is precisely what (11) asserts.

8.8. Recently Cheng and Wu[13] showed that for certain couplings in field theory, large classes of Feynman diagrams yield, for elastic scatterings $AB\to AB$, limiting angular distributions $d\sigma/dt$ which can be expressed in terms of impact factors.[14] We believe that extension of their work to inelastic processes $AB\to CDE\cdots$ would lead to the limiting distributions (1), (2), and (11) discussed above.

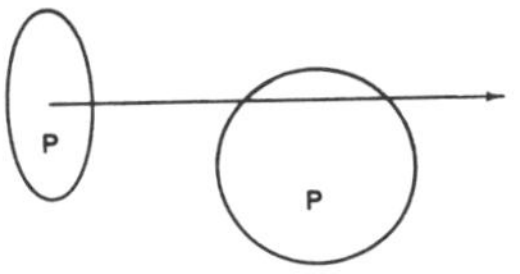

FIG. 4. Passage of Lorentz-contracted projectile through an extended target in the lab system.

DISCUSSIONS

9. Assuming that the hypothesis of limiting fragmentation of the colliding particles is correct, what do we know about the distribution of the fragments? For the single secondary proton distribution, the experiments of Ref. 10 give a rough distribution function for pp collisions as exhibited in the caption of Fig. 1. For single secondary pion and kaon distributions in pp collisions, we have only very rough information, as sketched in Figs. 2 and 3.

The distribution function τ_n satisfies

$$\tau_n=0 \text{ on boundary } S_n \text{ of } R_n. \qquad (16)$$

To see this, we observe that near S_n,

$$\sum_{i=1}^{n}(e-p_{\parallel})_i=M_t-\Delta,$$

where Δ is small. The fragmentation must, besides the n particles in τ_n, yield additional particles for which $\sum(e-p_{\parallel})=\Delta$. Since $e-p_{\parallel}$ is always positive, it must be small for each of the additional particles; i.e., these additional particles must be fast in the lab system. As $\Delta\to0$, such a fragmentation becomes increasingly unlikely.

Figures 2 and 3 demonstrate (16) very clearly. For the proton distribution in Fig. 1, (16) implies that for smaller $(p_{\parallel})_{\text{lab}}$, the limiting curve should dip to zero.

---

[12] M. Koshiba, in *Proceedings of the Tenth International Cosmic Rays Conference, Calgary, Canada, 1967* (to be published).

[13] H. Cheng and T. T. Wu, Phys. Rev. Letters **22**, 666 (1969); Phys. Rev. **182**, 1852 (1969); **182**, 1868 (1969); **182**, 1873 (1969); **182**, 1899 (1969).

[14] Cheng and Wu emphasized that their result is not consistent with a straightforward interpretation of the droplet model. We disagree with this emphasis. What seems to us to be most remarkable is, in fact, the general consistency of their results with the spirit of the droplet model. In particular, we cite the following features of their work: (i) the natural formulation in terms of two-dimensional momentum space, (ii) the factorization into two impact factors in a convolution integral, and (iii) the exponentiation, in impact-parameter space, for the transmission coefficient. All of these are characteristic features of the droplet model (Refs. 2 and 3).

Note added in proof. It has recently been shown by B. W. Lee (to be published) that, in fact, the droplet model in q-number formalism proposed by us in Phys. Rev. **175**, 1832 (1968) gives precisely the results of Cheng and Wu. That Cheng and Wu concluded otherwise in their Paper I [*ibid.* **182**, 1852 (1969)] was due to a wrong identification they made.

The main feature of the q-number formalism of the droplet model is the proposal that the S matrix should be given by an exponentiation of a convolution integral in coordinate space of q-number densities. [See Sec. 1 of Phys. Rev. **175**, 1832 (1968).] That this feature seems to be essential is also recently demonstrated by quantum-electrodynamics calculations of S. J. Chang and S. K. Ma, Phys. Rev. Letters **22**, 1334 (1969); S. J. Chang, University of Illinois Report (unpublished); Y. P. Yao, University of Michigan Reports (unpublished).

Experimental information is not at present sufficient to establish the existence of this dip.[15]

The distribution functions τ_n have mesa-like superstructures piled one on top of another. Each such superstructure derives from a process

$$\text{target} \to a^* \qquad (a^* = \text{a resonance}),$$
$$a^* \to (n+1) \text{ particles}.$$

This will be discussed in some detail in Appendix B.

Information about the fragmentation distributions $\sigma_1, \sigma_2, \ldots,$ of (11) are scanty. If we assume no pionization process, i.e., if we assume all final particles to be fragments of either of the two colliding particles, we can divide the pp collision process into four types:

$$pp \to pp, \qquad (17a)$$
$$pp \to pp^\dagger, \qquad (17b)$$
$$pp \to p^\dagger p, \qquad (17c)$$
$$pp \to p^\dagger p^\dagger, \qquad (17d)$$

where

$$p^\dagger = \text{excited state } p^*, \text{ or } p\pi, \text{ or any fragmentation of } p. \qquad (18)$$

Process (17b) represents those processes in which the projectile remains a proton and the target is excited and/or dissociated into a $p^\dagger$. It would be most interesting to know the relative probabilities of the four processes (17). One knows that (17a) has a cross section of approximately 10 mb. The experiment of Anderson *et al.*[9] shows that $pp \to pp^*$ has a total cross section of the order of 1 mb. Since the total cross section is about 40 mb, we conclude that

$$1 \text{ mb} < \sigma(pp^\dagger) < 15 \text{ mb}, \qquad (19)$$
$$0 \leq \sigma(p^\dagger p^\dagger) < 28 \text{ mb}. \qquad (20)$$

Lacking detailed information, we find it difficult to estimate $\sigma(pp^\dagger)$ and $\sigma(p^\dagger p^\dagger)$ more precisely. Let us only add that process (17b) leads to a final fast outgoing proton with a finite energy difference compared with the incoming projectile [Table I, category (4)]. Process (17b) would therefore, in cosmic-ray events, be classified as one with an inelasticity of 0.

10. The fragment distributions (11) can, of course, be reclassified according to the total mass M^* of all the fragments of the target

$$(M^*)^2 = (\sum E)^2 - (\sum \mathbf{p})^2. \qquad (21)$$

Also, it can be reclassified according to the total four-momentum transfer t to the target:

$$t = (\sum \mathbf{p})^2 - (\sum E - M_t)^2 = (\sum \mathbf{p}_\perp)^2. \qquad (22)$$

The summations in (21) and (22) extend over all n fragments of the target in the distributions σ_n of (11). $t = (\sum \mathbf{p}_\perp)^2$ follows from (10). The last identity can be restated as follows: *The total transverse momentum transfer is equal in magnitude to the total four-momentum transfer.*

Keeping M^* and t fixed, one can integrate over all redundant variables in σ_n and sum over n. The resultant cross section

$$\sigma(M^{*2},t)d(M^{*2})dt \qquad (23)$$

is then the partial cross section for fragmentation of the target, at infinite incident energy, into fragments with given values of M^*, at the momentum transfer t. For ep collisions, (15) gives

$$\sigma(M^{*2},t) = (2\pi\alpha^2/M_p t^2)W_2(t,\nu), \qquad (24)$$

where

$$2M_p\nu = M^{*2} - M_p^2 + t. \qquad (25)$$

For pp collisions, Anderson *et al.*[9] have given graphs of $d^2\sigma/dtdM^*$ ($M^* = W$ in their notation) for incident momentum 15.1 BeV/c up to $M^* \sim 2$ BeV. They measured M^* by taking the fast outgoing proton as the projectile after the collision, thereby obtaining the energy and momentum loss of the projectile. For very high energies, their procedure would give the contribution of (17b) to $\sigma(M^{*2},t)$ of (23). [At very high energies, an event where the projectile breaks up would yield another fragment X of the projectile with very high lab energy. In their procedure, X would be included in computation of M^* as a fragment of the target. Thus, M^* would be very large and the event does not contribute to $d^2\sigma/dtdM^*$ for any finite M^*. In the language of Table I, they explored the protons of category (4).]

11. It is well known[7] that in ep scattering, the function $W_2(t,\nu)$ for small t is related to the total γp cross section for an incident photon energy ν. Taking the γp total cross section to be a constant at infinite energy ν, one obtains for small t in ep scattering that

$$\sigma(M^{*2},t) \propto (M^*)^{-2}. \qquad (26)$$

It is not likely that (26) is true for hadron-hadron collision, for which one must have

$$\int\int \sigma(M^{*2},t)d(M^{*2})dt = \text{total cross section} \neq \infty. \qquad (27)$$

12. A number of very interesting experiments[16] of the type

$$\pi(\text{nucleus}) \to (\pi\pi\pi)(\text{nucleus}) \qquad (28)$$

have been performed (or are in progress) in connection

[15] Figure 1(a) of Phys. Rev. Letters **19**, 198 (1967) may be interpreted as indicating such a dip at 14±2 mrad lab angle, ~25 BeV/c lab momentum. It is to be emphasized that at present accelerator energies, the study of such a dip is necessarily difficult because background due to the breakup of the projectile may fill the dip region. In the language of Table I, the dip occurs between categories 3 and 4.

[16] Berkeley-Milan-Orsay-Saclay Collaboration; in *Proceedings of the Topical Conference on High-Energy Collisions of Hadrons, Geneva, 1968* (Scientific Information Service, Geneva, 1968), pp. 537–555.

with the concept of diffraction dissociation.[5] Viewed in the rest system of the incoming pion, these experiments, under the hypothesis of limiting fragmentation, supply information on the *fragmentation of the pion* into three pions, etc. The mass and momentum-transfer distribution for such fragmentation would therefore directly yield information on the distributions (23) and (11).

13. The hypothesis of limiting fragmentation gives emphasis to the lab and projectile systems. In this it is very different from the statistical[6] model. In the latter model, the two incoming particles collide and *arrest* each other in the c.m. system, the final product of the collision being emitted from this arrested amalgamation of the original particles. Thus the c.m. system is the important reference system in the statistical model.

We now know, through experimental observation of the apparent tendency for the longitudinal momentum to persist, that the two incoming particles *do not, in general, arrest each other*—certainly not completely. Instead, they have a tendency to go through each other and in the process to break into fragments. This observation leads to the hypothesis of limiting fragmentation in which the fragments have finite momenta in either the lab or the projectile system.

Another way to compare the statistical model with the hypothesis of limiting fragmentation may be instructive. If the statistical model is correct, there will be *no* particles of any finite energy in the lab system. Thus all the limiting cross sections $\sigma_1, \sigma_2, \ldots,$ of (11) are zero. Also $\rho_1 = \rho_2 = \cdots = 0$.

14. The spirit of the hypothesis of limiting fragmentation is very much the same as that of the two-fireball model,[4] with or without pionization. The differences are as follows:

(a) The hypothesis as discussed in Secs. 1–7 is precisely defined, while the two-fireball model is not. Perhaps because of its lack of precise definition, the two-fireball model has not served as a useful guide for experiments at accelerator energies.

(b) An essential feature of the fireball model is that each fireball is assumed to decay more or less spherically symmetrically. This is not likely to be correct in the hypothesis of limiting fragmentation, as can be seen from the following argument: In ep collisions, only the proton can break up, i.e., there is only one fireball. For the case with a small q^2 [see (14)], the proton is essentially hit by a real photon of lab energy ν. For large ν one knows from high-energy γp experiments that the angular distribution in the c.m. system of γp is very much forward-backward, and bears no resemblance to spherical symmetry. If this is fitted to a fireball model, the fireball rest system being necessarily the c.m. system of γp, the fireball decay could not give rise to anything close to spherical symmetry in its rest system.

It is useful to observe that the fragments of the target are *more clearly separated* from pionization products and from fragments of the projectile in lab momentum space than they are in the c.m. momentum space. For example, in Fig. 1, the proton fragment exhibits a peaked differential cross section which drops off as one goes to high values of lab momenta. In contrast, this same curve in the c.m. system is flat versus the c.m. longitudinal momentum[10] and exhibits no tendency of separating the fragments of the target from other particles. In the light of this observation, it seems to us that it is better to think of *the two fireballs as limiting fragment distributions in the lab and projectile system, rather than as separated concentrations of particles in the c.m. momentum space.*

It is obvious that the hypothesis of limiting fragmentation is also very much similar in spirit to the isobar model.[17] The main difference is that under our hypothesis while the fragments may be the decay product of an isobar, they also may not be. For example, in pp collisions the target proton may become p^*, but it may also become a nonresonant "background" πp. In fact, for large momentum transfers, the latter dominates over the former (cf. Sec. 18).

15. The spirit of the hypothesis of limiting fragmentation is also very much the same as that of diffraction dissociation.[5] In fact, in the hypothesis of limiting fragmentation as formulated in (11), if one assumes that

$$\text{the total } G, I^2, I_z, N, \text{ and charge of the particles in } \sigma_n = \text{that of the target}, \quad (29)$$

one would have a more restricted hypothesis which can be considered as a precise statement of diffraction dissociation. We believe this restricted hypothesis is likely to be correct.[18] [The above discussion refers to hadron-hadron collisions. For lepton-hadron collisions, one must exercise caution in drawing specific conclusions. For example, in the collision $\nu n \to \mu^-$ plus hadrons, charge is transferred from the lepton to the hadron. See a discussion on pp. 515–516 in *High Energy Collisions*, edited by C. N. Yang *et al.* (Gordon and Breach, Science Publishers, Inc., New York, 1969).]

ADDITIONAL SPECULATIONS AND REMARKS

16. In cosmic-ray experiment, one often discusses the process of pionization[12] in which slow pions are supposed to be emitted more or less isotropically in the c.m. system. The need for the pionization process arises mostly from the increasing multiplicity observed at higher energies. If the hypothesis of limiting fragmentation is correct, we have already discussed before in Sec. 5 how to accommodate phenomena of increasing multiplicity. No pionization process is needed. If pioni-

[17] S. J. Lindenbaum and R. M. Sternheimer, Phys. Rev. **105**, 1874 (1957).

[18] For some recent experiments, see W. E. Ellis, T. W. Morris, R. S. Panvini, and A. M. Thorndike, in Proceedings of the Lund International Conference on Elementary Particles, Lund, Sweden, 1969 (unpublished).

zation processes are absent, then

$$\sum_{n=1}^{\infty}\int \sigma_n d^3p_1\cdots d^3p_n = \text{total cross section.}$$

In the following two paragraphs, we give additional arguments against the pionization process.

The pionization process implies, in the c.m. system, the arresting of the colliding particles with subsequent evaporation of slow pions. Such a picture is very far removed from a model of two extended objects *going through* each other as semitransparent bodies, a model that underlies the droplet interpretation of high-energy elastic scattering.[2,3]

The c.m. system in pp or πp collisions has no intrinsic significance once one emphasizes the hadrons as extended objects with many internal degrees of freedom. To illustrate the point, let us consider π-Pb collision. The c.m. system of such a collision is of great physical importance at low energies, when, for example, one wants to know the threshold energy for the production of another π meson. However, at high energies it is well known that the π-Pb c.m. system has no great physical importance. For the same reason, once we accept, as we must, the thesis that hadrons are extended objects with many internal degrees of freedom, the c.m. system in πp collisions loses its particular physical significance. (To be more concrete, e.g., in the quark model, if the pion is supposed to be made up of two quarks and the proton of three, then the πp c.m. system is *not* the same as the quark-quark c.m. system.)

17. In (23) we discussed the fragmentation of a particle into total mass M^* at a momentum transfer t. If there is no pionization process, there would be a corresponding fragmentation of the other particle into fragments at the same momentum transfer t. The cross section will be defined to be

$$\sigma(M_1^{*2},M_2^{*2},t)d(M_1^{*2})d(M_2^{*2})dt. \tag{30}$$

This combined distribution may or may not factorize. If it does,

$$\sigma(M_1^{*2},M_2^{*2},t) = (\text{const})\sigma_{\text{target}}(M_1^{*2},t)\sigma_{\text{projectile}}(M_2^{*2},t). \tag{31}$$

We rather believe that factorization (31) is not quantitatively valid because different processes (projectile $\longrightarrow M_2^*$) should probably in general imply different excitations of the target. For example, in

$$ep\longrightarrow ep^\dagger, \tag{32}$$

$$pp\longrightarrow pp^\dagger, \tag{33}$$

$$pp\longrightarrow p^*p^\dagger, \tag{34}$$

the distribution of the different states $p^\dagger$ is probably qualitatively similar but quantitatively different for the three processes. In particular, for fixed t and M^*, the average number of hadrons that is contained in the fragment $p^\dagger$ in (32) and (33) are approximately the *same. This is an experimentally testable conclusion.* [On the other hand, the t dependence of (32) and (33) are expected to be quite *different*, as will be discussed in Sec. 18.]

18. In the fragmentation concept, the rapid decrease of elastic cross sections for large $t=q^2$ is a consequence of the idea that for large momentum transfers t, the hadron[2] breaks up in general into fragments. Consistent with this idea, it is to be expected that *for larger values of the momentum transfer t, the breakup process favors larger multiplicities of hadrons* (at fixed M^*). This particular point can be qualitatively tested in ep, μp, or hadron-hadron collisions, although exactly how the average multiplicity of the fragmentation process depends on t cannot be quantitatively predicted without a detailed model. (A great difficulty in formulating such a model lies in the following fact: Consider, say, ep collisions. *What absorbs the momentum transfer from the electron does not, in general, come out simply as one of the outgoing fragments of the proton.* Instead, it rapidly dissipates its energy-momentum to neighboring space-time points in the proton before the final fragmentation takes place. It seems that various models can be proposed to describe such a dissipation process, and one must look for guidance from future experiments.)

In fact, experimental data both from ep collisions[7] and hadron-hadron collisions[9] already give support to the speculation that the average multiplicity increases with increasing $t=q^2$ at a fixed M^*: For example, in ep collisions, the cross section (12) shows ridgelike structures when plotted against q^2 and M^*. The ridges are due to

$$ep\longrightarrow ep^*, \tag{35}$$

and the background under the ridges is due to

$$ep\longrightarrow ep^\dagger, \text{ where } p^\dagger=p\pi \text{ or } p\pi\pi, \text{ etc.}, \neq p^*. \tag{36}$$

In (35), the hadronic matter remains one piece (i.e., multiplicity of hadrons = 1), while in (36), the hadronic matter breaks up into two or more pieces (i.e., multiplicity of hadrons $\geqq 2$). Experimentally at a fixed M^*, with increasing q^2, (35) becomes rapidly insignificant[7] compared with (36). The same is true in pp and πp collisions.[9] *We regard this behavior as one of the most striking features of the inelastic data, a feature confirming the fragmentation concept.*

The concept of breaking up under large momentum transfers t suggests that the *transverse momentum for each outgoing particle may not be large even though t is large.* This is a testable proposition in hadron-hadron and lepton-hadron collisions. One can draw additional qualitative conclusions from the breakup concept. Consider an ep collision at a fixed M^* of excitation of the proton [cf. (14)] and very large t. The multiplicity is limited by the value of M^*. Thus the individual transverse momenta cannot be all small. The net result is that the cross section would be small.

Speculations along this line cannot be made more precise lacking a complete theory of the fragmentation process. [Simple heuristic arguments lead, for ep and μp collisions, to

$$W_2 = A \exp(-\alpha q^2/\langle n \rangle) . \quad (37)$$

where A and α are constants, and $\langle n \rangle$ is the average number of hadrons for the given q^2 and ν.]

The concept of fragmentation leads to a number of further testable qualitative features which we shall discuss in the rest of this section.

It is interesting to compare $\sigma(M^{*2},t)$ for ep collisions and pp or πp collisions. For hadron-hadron collisions, the experimental quantity easy to measure is not $\sigma(M^{*2},t)$, but a part of it, $\sigma_{pp\to pp^\dagger}(M^{*2},t)$, which represents those parts of σ where the incoming particle does not break up. This quantity, which we shall call σ in the rest of this paragraph, was measured by Anderson *et al.*[9] For the case that the target does not break up, i.e., $M^* = M_p$, σ represents elastic scattering and decreases with increasing t extremely fast. Our interpretation of this follows that of Ref. 2, namely, high-t elastic pp scattering is rare, because it is difficult to keep *two* protons intact: σ falls like F^4, the fourth power of the proton form factor. The same applies for the ridges in σ at M^* = resonance masses. For values of M^* in between resonances or beyond the resonance region, σ is expected to fall with increasing t like the product of F^2 (= the square of the form factor, since only one proton needs to be kept intact in this case) and a factor like (37). On the other hand, $\sigma_{ep\to ep^\dagger}(M^{*2},t)$ falls with t simply like (37).

19. Although throughout this paper we speak of limits at infinite incident energy, that is only to clarify, for an idealized case, the precise concepts under discussion. In practice, infinite energy is, of course, unattainable. Furthermore, it is quite possible that all cross sections, total or partial, have some dependence on the incoming energy through factors such as $(\ln E)^\beta$ with positive or negative values of β, or E^β with small fractional value for β. If that should turn out to be the case, the discussion of this paper may be taken to cover, at high incident energies, wide energy ranges in which the energy dependence is negligible.

SOME SUGGESTED EXPERIMENTS

20. It is very desirable to have more complete lab momentum distributions of various slow particles, p, π, K, $\bar{p}$, etc., in πp, pp, Kp, etc., collisions at high energies. In each case one wants to test whether the partial cross sections approach limits as defined in (1), (2), and (11).

One could study the same problem by measuring the momentum distribution of fast particles in the lab system and then transform to the projectile system (see Sec. 8.4). Especially interesting are the experiments of Ref. 16 mentioned in Sec. 12.

In this connection it is perhaps useful to point out some obvious kinematic facts: (a) Fast laboratory forward π and K in pp collisions are fragments of the projectile emitted *backwards* against the direction in which the projectile is hit by the target. (b) If the projectile does not break up, it will lose in the laboratory, in general, only a few BeV or less, to cause the excitation or breakup of the target [category (4), Table I]. However, if the projectile does break up, it will lose in the lab system a large amount of energy [category (3), Table I]. For example, consider a pp collision at 70 BeV. If the projectile breaks up into $p\pi$, or $p\pi\pi$, the proton will most likely lose in the lab system an energy of the order of $M_\pi/(M_\pi+M_p)$ (70 BeV) = 9 BeV or more.

21. Of particular interest, among the measurements mentioned above, are those relating to the dip (Sec. 9) and the values of $\sigma(pp^\dagger)$, $\sigma(p^\dagger p^\dagger)$, etc.

22. For ep and μp collisions, especially interesting experiments are (a) to measure the average multiplicity versus the four-momentum transfer t from the lepton to the hadrons, and (b) to measure the transverse momentum (i.e., perpendicular to incident momentum) of individual outgoing hadrons for the case of large t. The significance of these experiments was discussed at the end of Sec. 17 and in Sec. 18.

ACKNOWLEDGMENTS

The authors appreciate many interesting discussions with G. Chadwick, G. Collins, K. Lai, R. Panvini, and F. Turkot.

APPENDIX A

We prove here formulas (9) and (10). [(4) and (5) are special cases of these formulas.]

By conservation of energy and momentum,

$$\sum (e - p_{\parallel}) = M_t + (E - p_{\text{inc}}) . \quad (A1)$$

At high energies, this becomes[19]

$$\sum (e - p_{\parallel}) = M_t . \quad (A2)$$

Consider a high-energy collision where the projectile and target undergo fragmentation, with or without additional pionization process yielding slow evaporation pions in the c.m. system. At very high energies, only the fragments of the target have finite energies in the lab system (compare Table I). Thus, only these particles contribute to the sum in (A2); hence we have (10). Since $e - p_{\parallel}$ is always positive but can be arbitrarily small, omitting some fragments of the target would immediately give (9).

---

[19] Formula (A2) has been used in cosmic-ray experiments in connection with a concept called effective target mass. See N. G. Birger and Yu. A. Smorodin, Zh. Eksperim. i Teor. Fiz. **36**, 1159 (1959); **37**, 1355 (1959) [English transls.: Soviet Phys.—JETP **9**, 823 (1959); **10**, 964 (1960)].

Next we shall state the following simple theorem which gives a restatement of (10):

Theorem: In the breakup process of the target

$$\text{target} \to abcd\ \ldots\,,$$

the quantity $\sum(e-p_{\shortparallel})$ is conserved:

$$\sum(e-p_{\shortparallel})=\text{const}. \tag{A3}$$

Furthermore, this conservation law is invariant in any Lorentz frame that moves with a velocity $<c$ along the longitudinal direction.

According to (22), the total transverse momentum of the fragments in (A3) has a magnitude of $\sqrt{t}$.

For the process

$$p \to p\pi\pi\ \ldots\,, \tag{A4}$$

the quantity $M_p^{-1}(e-p_{\shortparallel})$ for the outgoing proton in the rest system of the original proton is the "elasticity" of the fragmenting projectile proton.

APPENDIX B

We prove here the statement in Sec. 9 about mesa-like superstructures for τ_n. Take $n=1$. The physical process is

$$\text{target} \to a^* \quad (a^*=\text{a resonance})\,, \tag{B1}$$

$$a^* \to bc\,. \tag{B2}$$

The components of lab momenta of a^* will be denoted by $p_{a\perp}$, $p_{a\shortparallel}$; its mass, by M^*. Then

$$M^{*2}=M_t^2+2M_tp_{a\shortparallel}-p_{a\perp}^2. \tag{B3}$$

[This is proved like (A2).] Now assume a^* to have zero width. For each $\mathbf{p}_a$, the lab momentum $\mathbf{p}_b$ of b from the decay (B2) lies *on* an ellipsoid. As $\mathbf{p}_a$ ranges over the paraboloid (B3), these ellipsoidal surfaces sweep over a region in $\mathbf{p}_b$ space (cf. Fig. 5). This region

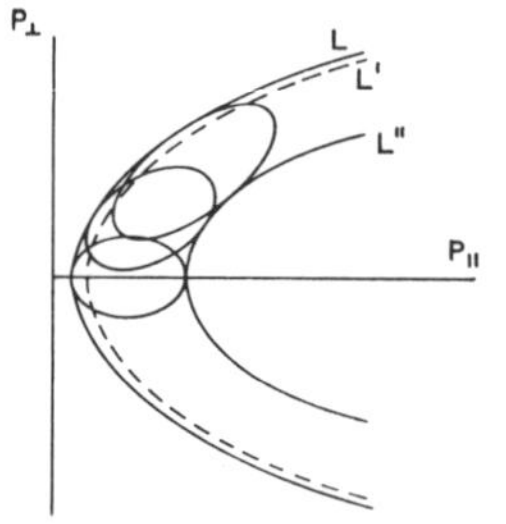

FIG. 5. Allowed values of lab momenta for particle b in processes (B1) and (B2).

becomes a mesa-like structure in the plot of $\tau_1(\mathbf{p}_b)$ versus $p_{b\shortparallel}$ and $p_{b\perp}$. To see this, consider that portion X of the ellipsoids between the envelope L and the surface L' to its right at a distance $dp_{\shortparallel}$ from L. The fractional area of each ellipsoid included in X is evidently proportional to $dp_{\shortparallel}$. Hence the partial cross section for $\mathbf{p}_b$ to lie between L and L' is $Ydp_{\shortparallel}$, where $Y\neq 0$. Thus between L and L'', (B1)-(B2) contributes a finite value to $\tau_1(\mathbf{p}_b)$, and this contribution does not vanish as one approaches L or L''; i.e., (B1) and (B2) contribute a mesa-like superstructure to $\tau_1(\mathbf{p}_b)$.

If a^* has a finite width, the bluffs of the mesa-like structure are rounded off.

Existence of mesa-like superstructures for τ_2, τ_3, ..., can be shown in a similar way.

Contributions to τ_1 due to

$$\text{target} \to a^*\,,\quad a^* \to bcd$$

or

$$\text{target} \to a^*f\,,\quad a^* \to bc$$

can be discussed similarly.

Charge Quantization, Compactness of the Gauge Group, and Flux Quantization*

C. N. YANG
Institute for Theoretical Physics, State University of New York, Stony Brook, New York 11790
(Received 5 September 1969)

The relationship between charge quantization and the compactness of the gauge group is discussed. Also, remarks are made about charge quantization and the observation of flux quantization in superconductors.

I. INTRODUCTION

A QUESTION that has been frequently raised is the equality of the absolute charges of the proton and the electron. Experimentally[1] they are equal to a high degree of accuracy. To understand the equality, Dirac has made the ingenious theoretical[2] proposal of the existence of magnetic monopoles, which have not yet been detected.

In Sec. II of this paper, we discuss the logical relationship[3] between the quantization of the electric charge and the mathematical concept of the compactness of the gauge group. We shall see that they are intimately related. Let us here emphasize that in mathematics the concept of compactness for a group[4] is of primary importance. For Lie groups, compactness is a property of the global structure of the group, which has a determining influence on the nature of the representations of the group. It is, in fact, through its influence on the representations that the compactness of the gauge group has a bearing on the quantization of charges.

In Sec. III we remark on the unit of flux quantization and the question of charge quantization.

II. CHARGE QUANTIZATION AND COMPACTNESS OF GAUGE GROUP

We consider a space-time-independent gauge transformation on charged fields ψ_j of charge e_j:

$$\psi_j \to \psi_j' = \psi_j \exp(ie_j\alpha) . \qquad (1)$$

In usual discussions one confines oneself to infinitesimal values of α. In the present discussion we consider instead *finite* values of α.

If the different e_j's $(=e_1, e_2, \ldots)$ of different fields are not commensurate with each other, the transformation (1) is different for all real values of α, and the gauge group must be defined so as to include all real values of α. Hence, the group is not compact.

If, on the other hand, all different e_j's are integral multiples of e, a universal unit of charge, then for *two values* of α different by an integral multiple of $2\pi/e$, the transformations (1) for any fields ψ_j are the *same*. In other words, two transformations (1) are indistinguishable if their α's are the same modulo $2\pi/e$. Hence the gauge group as defined by (1) is compact.

III. REMARKS ON FLUX QUANTIZATION AND CHARGE QUANTIZATION

In the experiment of flux quantization,[5] one finds that magnetic flux trapped in superconducting rings are in whole units of $2\pi ch/2e$. What should e be if the electron and the proton do not have the same charge? The answer is that e should be the electron charge, since electron pairs, not the protons, are the "basic group"[6] that possess off-diagonal long-range order in a superconductor. To illustrate this point further, let us assume that spin-up electrons and spin-down electrons have charges $-e$ and $-e'$, respectively, and that the basic group is a pair of electrons with opposite spins. The flux unit would then be $2\pi ch/(e+e')$. If, on the other hand, there are two kinds of spin-up electrons with charges e and e' which are incommensurate, and two kinds of spin-down electrons with similar charges, and, furthermore, if pairs of electrons $e-e$, $e-e'$, $e'-e$, and $e'-e'$ of opposite spins all have off-diagonal long-range order in a superconductor, then the flux unit is an integral multiple of $2\pi ch/2e$, $2\pi ch/(e+e')$, and $2\pi ch/2e'$. Hence it is ∞. To summarize, the existence of a finite flux quantization unit merely reflects on the quantized nature of the charge of the basic groups in a superconductor, and does not necessarily imply that electric charge is always quantized. But if the flux quantization unit were found to be ∞, one would have concluded that the electric charge is not quantized.

* Partially supported by U. S. Atomic Energy Commission Contract No. AT(30-1)-3668B.

[1] G. Feinberg and M. Goldhaber, Proc. Natl. Acad. Sci. (U. S.) **45**, 1301 (1959).

[2] P. A. M. Dirac, Proc. Roy. Soc. (London) **A133**, 60 (1931); Phys. Rev. **74**, 817 (1948).

[3] It has been brought to the author's attention that this relationship had been referred to in J. Schwinger, Phys. Rev. **125**, 1047 (1962); cf. also E. Progovecki, J. Math. Phys. **5**, 442 (1964); and S. Doplicher, R. Haag, and J. E. Roberts, Commun. Math. Phys. **13**, 1 (1969).

[4] See, e.g., H. Weyl, *The Classical Groups* (Princeton U. P., Princeton, N. J., 1946).

[5] B. S. Deaver and W. H. Fairbank, Phys. Rev. Letters **7**, 43 (1961); R. Doll and M. Näbauer, *ibid.* **7**, 51 (1961).

[6] C. N. Yang, Rev. Mod. Phys. **34**, 694 (1962).

□ Reprinted from *Physical Review D* 1, 8 (April 1970), 2360.

[70b]
Commentary begins page 70

Some Exactly Soluble Problems in Statistical Mechanics

C. N. Yang
Institute for Theoretical Physics, State University of New York, Stony Brook

1. INTRODUCTION

In the last few years, a number of two-dimensional classical and one-dimensional quantum mechanical problems in statistical mechanics have been exactly solved. Although these problems range over models of diverse physical interest, their solutions were obtained using very similar mathematical methods. In these lectures, I will discuss the main points of the method. I shall first, in this introductory lecture, give an overall survey of all these problems without going into the detailed method of solution. In later lectures, we shall concentrate on one particular problem: the delta function interaction in one dimension, and go into the details of that problem.

Spin–Spin Interaction in One Dimension. This problem was first discussed by Heisenberg[1] in his celebrated theory of ferromagnets. The Hamiltonian of the system is

$$H = -\frac{1}{2}\Sigma(\sigma_x\sigma_x' + \sigma_y\sigma_y' + \sigma_z\sigma_z') \ . \qquad (1)$$

The sum is over nearest neighbor spins σ and σ'. This mathematical problem was treated later by Bloch[2], who introduced the famous concept of spin waves. Bloch's solution was exact when the spin waves represent disturbances at large distances but was only approximate when the spin waves interact. Later Bethe[3] showed that in one dimension the Hamiltonian (1) can be exactly diagonalized by assuming the wave function to be a finite sum of plane waves with coefficients determined by some transcendental equations. This assumption is now known as Bethe's hypothesis. In similar form, it forms the basis of the exact solution of all the problems that I shall discuss today.

The solution of the transcendental equation resulting from Bethe's hypothesis presents some problems: Do there exist real solutions? Is the solution for the ground state always real? etc. These questions were discussed incompletely in various places. Great clarification was brought into this problem in 1966[4] when the Hamiltonian

$$H = -\frac{1}{2}\Sigma(\sigma_x\sigma_x' + \sigma_y\sigma_y' + \Delta\sigma_z\sigma_z') \qquad (2)$$

was introduced and continuity properties with respect to Δ were used. In this fashion the calculation of the ground state energy for the Hamiltonian system (2) as a function of Δ, and y, the magnetization, is reduced to that of solving Fredholm integral equations. Analytic properties of the ground state energy per particle as a function of Δ and y were fully explored.

Quantum Lattice Gas. The quantum lattice gas problem was introduced in 1956[5] as a generalization of the classical lattice gas.[6] For nearest neighbor interactions, it was shown[5] that the quantum lattice gas is equivalent to the Hamiltonian problem (2). The solution of the ground state energy for the spin–spin interaction problem (2) therefore gave the ground state solution for the quantum lattice gas problem in one dimension with nearest neighbor interaction.[7]

Two-Dimensional Ice Problem. The ice problem in two dimensions was solved by Lieb[8] in 1967 in an extremely interesting paper. It turns out that this problem, when formulated in terms of the transfer matrix, gives a transfer matrix which is diagonalized with the same wave functions as that which diagonalizes the Hamiltonian (2) for the case $\Delta = \frac{1}{2}$.

□ Lectures given at the Karpacz Winter School of Physics, February 1970. Reprinted from *Proceedings of the VII Winter School of Theoretical Physics in Karpacz.* Breslau: University of Wroclaw, 1970.

Two-Dimensional Ferroelectric Models. The two-dimension ice problem is basically a combinational problem satisfying a so-called "ice condition." A number of models of ferroelectrics, such as the *KDP* model and the *F* model, also satisfy the ice condition. After the ice problem was solved, the solution was generalized[9] to these ferroelectric models, resulting in quite general solutions possessing both horizontal and vertical electric fields.

From the viewpoint of mathematics, these solutions are especially interesting because the transcendental equation resulting from Bethe's hypothesis in this case has complex rather than real roots. The Fredholm integral equation therefore follows a path of integration which is not on the real axis.

One-Dimensional δ-Function Interaction. The quantum mechanical one-dimensional δ-function interaction problem was first discussed in 1963 by Lieb and Liniger[10] and by McGuire.[10] They treated the boson case, both for a finite number of particles and for the ground state properties in the infinite "volume" limit at a fixed density. Lieb also treated[11] the excitation spectrum near the ground state in the latter limit.

More recently generalizations were made resulting in exact solutions for:

S-matrix for any finite number of particles,[12]
Ground state energy at a finite density in infinite volume for fermions with repulsive[13] and attractive[14] forces,
Thermodynamics for bosons,[15] and
Excitation spectrum for bosons at finite temperatures.[15]

A number of further problems in the one-dimensional δ-function interaction model remain to be solved. I am of the opinion that they present no essential new mathematical difficulties.

2. *δ-FUNCTION INTERACTION*
(One Dimension)

In this and the next section we shall develop in some detail the mathematics of the δ-function interaction problem. The Hamiltonian of the system is

$$H = -\sum_i \frac{\partial^2}{\partial x_i^2} + 2c\sum_{i>j} \delta(x_i - x_j) \ , \quad i, j = 1, 2, \ldots N \ . \tag{3}$$

2.1 Two Particles (Bosons)

Bethe's hypothesis in this case asserts:

$$\psi = \alpha e^{i(k_1x_1+k_2x_2)} + \beta e^{i(k_2x_1+k_1x_2)} \quad (x_2 \geq x_1), \quad \psi(x_2, x_1) = \psi(x_1, x_2) \ . \tag{4}$$

To understand these we introduce the centre of mass coordinate X and the relative coordinate y:

$$y = x_2 - x_1, \quad X = \frac{1}{2}(x_1 + x_2), \quad K = k_1 + k_2 \ .$$

Thus

$$\psi = e^{iKX}[\alpha e^{i\frac{1}{2}(k_2-k_1)|y|} + \beta e^{i\frac{1}{2}(k_1-k_2)|y|}] \ . \tag{5}$$

The y-part of ψ, i.e., the factor in the square bracket, should satisfy

$$\left[-2\frac{\partial^2}{\partial y^2} + 2c\delta(y)\right]\psi = E\psi \ . \tag{6}$$

I.e., $\psi(y)$ is a free particle wave function for $y \neq 0$. At $y = 0$ there is a discontinuity of the derivative, but not of ψ:

$$\left.\frac{\partial\psi}{\partial y}\right|_{y=0+} - \left.\frac{\partial\psi}{\partial y}\right|_{y=0-} = c\psi|_{y=0} \ . \tag{7}$$

Substitution of (5) into (7) leads to

$$\frac{\beta}{\alpha} = -\frac{c - i(k_2 - k_1)}{c + i(k_2 - k_1)} = -e^{i\theta(k_2-k_1)} \tag{8}$$

where

$$\theta(x) = -2\tan^{-1}\frac{x}{c} \quad |\theta(x)| < \pi \ . \tag{9}$$

2.2 Many Bosons

The generalization of Bethe's hypothesis (4) reads:

$$\psi = \sum_P \alpha_P e^{i(k_{P1}X_1 + k_{P2}X_2 + \ldots + k_{PN}X_N)} \tag{10}$$

if

$$X_1 < X_2 \ldots < X_N \ . \tag{11}$$

Here P_1, P_2, ... PN is a permutation P of 1, 2, ... N. For other orderings of X_i, the wave function is obtained from (10) by permutation. The generalization of (8) becomes

$$\frac{\alpha_{P'}}{\alpha_P} = -e^{i\theta(k'-k)} \tag{12}$$

for every pair (P and P') of permutations for which

$$P'1\ P'2 \ldots P'N$$

$$P1\ P2\ \ldots PN$$

are identical except for two neighboring columns, looking like

$$\ldots k'k \ldots$$

$$\ldots kk' \ldots$$

To satisfy (12), no two k's can be identical.

2.3 Consistency of (12)

There are $(N-1)N!$ equations (12) for the $N!$ coefficients α_P. Are they consistent? The answer is[10] yes. In fact for any set of k's, all nonidentical, (12) give a unique set of α's up to a normalization factor.

2.4 Cyclic Boundary Condition

The two regions of space

$$0 < X_1 < X_2 \ldots < X_N < L \tag{13}$$

$$0 < X_2 \ldots < X_N < X_1 < L \tag{14}$$

are *next* to each other because of cyclic boundary conditions. Wave function (10) in region (13) must match correctly to the wave function in region (14) obtained from (10) by symmetrization (for the boson system we are discussing in this section). This would obtain if the term

$$\alpha_P e^{i(k_{P1}X_1+\ldots+k_{PN}X_N)} = \alpha_P e^{i(k_{P2}X_2+k_{P3}X_3+\ldots+k_{PN}X_N+k_{P1}X_1)}$$

in (13) matches smoothly into the term

$$\alpha_{P''} e^{i(k_{P2}X_2+k_{P3}X_3+\ldots+k_{PN}X_N+k_{P1}X_1)} \tag{15}$$

in (14). Here P'' stands for the permutation

$$P'' = (P2, P3, \ldots PN, P1) \ . \tag{16}$$

This smooth matching implies

$$\alpha_P = \alpha_{P''} e^{ik_{P1}L} \ . \tag{17}$$

But using (12) we find

$$\begin{aligned} \alpha_P &= [-e^{i\theta(k_{P1}-k_{P2})}]\ \alpha_{P2,P1,P3,\ldots} \\ &= [\quad \ldots \quad][-e^{i\theta(k_{P1}-k_{P3})}]\ \alpha_{P2,P3,P1,P4,\ldots} \\ &= \ldots \\ &= (-1)^{N-1} e^{i\Sigma\,\theta(k_{P1}-k_Q)} \alpha_{P''} \end{aligned} \tag{18}$$

where the summation extends over all Q. (18) and (17) together require

$$e^{ikL} = (-1)^{N-1} e^{i\sum_{k'} \theta(k-k')} \tag{19}$$

for every k.

For a set of k's, all unequal and N in number, satisfying (19), one can construct a solution of the Schrodinger equation satisfying cyclic boundary conditions.

2.5 Quantum Numbers I, Continuity with Respect to c^{-1} for $c^{-1} \geqq 0$

Taking the logarithm of (19) yields

$$kL = 2\pi I_k + \sum \theta(k - k') \tag{20}$$

where

$$\begin{aligned} I_k &= \text{integer} \qquad (\text{if } N = \text{odd}) \\ I_k &= \text{integer} + \frac{1}{2} \quad (\text{if } N = \text{even}) \ . \end{aligned} \tag{21}$$

For $c > 0$ it is not difficult to prove[14] the following:

Theorem 1

For any set of I's satisfying (21), no two of which are identical, there is a unique set of real k's satisfying (20), with no two k's being identical. With this set of k's, one eigenfunction of the Hamiltonian, of Bethe's form, can be constructed. The totality of such eigenfunctions forms a complete system. For fixed I's, such a wave function is continuous with respect to c^{-1} for $c^{-1} \geqq 0$. The numbers I are quantum numbers of the system.

2.6 Limit for the Ground State When $N \to \infty$, N/L = Constant (Bosons, $c > 0$)

When $c = +\infty$ the wave function must vanish when any two particles are at the same point x. [E.g., (7) shows that $\psi = 0$ at $y = 0$.] But this condition is satisfied also for a *free* fermion system. One thus sees that

[wave function for bosons with $c = +\infty$]
= absolute value of wave function for free fermions. (22)

Such a wave function clearly satisfies Bethe's hypothesis (10). The k's are the momenta of the fermions and satisfy (20) and (21) in which $\theta = 0$ since $c = \infty$. For the ground state, the I's form a set of close packed integers (or integers $+\frac{1}{2}$) (so that the k's form a one dimensional fermi sea):

$$I = -\frac{N-1}{2}, -\frac{N-1}{2} + 1, \ldots + \frac{N-1}{2} . \quad (23)$$

By continuity arguments, such a set of I's also gives the ground state for all $c > 0$.

We now concentrate on the ground state. Divide (20) by L:

$$k = 2\pi \frac{I_k}{L} + \frac{1}{L} \sum_{k'} \theta(k - k') . \quad (24)$$

In the limit that $N \to \infty$, $L \to \infty$ proportionally, and it can be proved that the k's approach a distribution with $L\rho(k)dk$ k's in dk. The distribution extends between $k = \pm Q$. The sum in (24) then approaches an integral and one obtains

$$k = h(k) + \int_{-Q}^{Q} \theta(k - k') \, \rho(k')dk' \quad (25)$$

where $h(k)$ is the limit of $2\pi I_k/L$. Thus

$$h(k + dk) - h(k) = \left(\frac{2\pi}{L}\right)(\text{number of } k\text{'s in } dk) = 2\pi\rho(k)dk .$$

Or

$$\frac{dh}{dk} = 2\pi\rho . \quad (26)$$

Differentiation of (25) leads to

$$1 = 2\pi\rho(k) - 2c \int_{-Q}^{Q} \frac{\rho(k')dk'}{c^2 + (k - k')^2} . \quad (27)$$

The density of particles is

$$\frac{N}{L} = \int_{-Q}^{Q} \rho dk \quad (28)$$

and the energy per unit length of space is

$$\frac{E}{L} = \int_{-Q}^{Q} \rho k^2 dk . \quad (29)$$

These equations were first derived by Lieb and Liniger.[10]

Although the arguments above for (27)–(29) are not rigorous, they can be easily made completely rigorous as follows. [The arguments below can be adapted also to the case of the anisotropic spin–spin interactions. See reference 4, second paper, p. 000.]

Define $h(p)$ by

$$h(p) = p - \frac{1}{L} \sum_{k'} \theta(p - k') . \quad (30)$$

Clearly

$$\frac{dh}{dp} = 1 + \frac{1}{L} \sum_{k} \frac{2c}{c^2 + (p - k)^2} . \quad (31)$$

Thus h is monotonic in p. Furthermore as p assumes the successive values of k, h assumes the successive values of $2\pi I_k/L$. We can thus invert the function $h(p)$ and regard p as a funciton of h. The values of p at the evenly spaced values $2\pi I_k/L$ [I_k given by (23)] of h are the values of k. Thus

(31) becomes

$$\frac{dh}{dp} = 1 + \frac{1}{L}\sum_h \frac{2c}{c^2 + [p - p(h)]^2} \qquad (32)$$

where the h's in the summation range over evenly spaced values at equal intervals of $\Delta h = 2\pi/L$. Replace the sum by an integral and a residue

$$\frac{dh}{dp} = 1 + \frac{1}{2\pi}\int \frac{2c}{c^2 + [p - p(h)]^2}\,dh + R(p) \qquad (33)$$

$$= 1 + \frac{c}{\pi}\int_{-\xi}^{\xi} \frac{dp'}{c^2 + (p - p')^2}\,\frac{dh(p')}{dp'} + R(p) \qquad (34)$$

where $\xi = \xi(L)$ is the maximum value of k. We shall now try first to bound $R(p)$. This is easy since the summand in (32) as a function of h has exactly two sections in each of which it is monotonic in h. Now for a monotonic integrand the integral is approximated by the sum with a residue not exceeding $(\Delta h)(4/c)$ in absolute value. Thus

$$|R(p)| < (\Delta h)\frac{10}{c} \ .$$

Next we write the Fredholm equation (34) in operator form

$$\frac{dh}{dp} = 1 + O_\xi \frac{dh}{dp} + R \ .$$

Or

$$\frac{dh}{dp} = \frac{1}{1 - O_\xi}[1] + \frac{1}{1 - O_\xi} R \ . \qquad (35)$$

This is meaningful: (31) implies that $dh/dp > 1$. Thus the maximum k (i.e., p) is less than the maximum h. (When $p = 0$, $h = 0$.) Thus ξ is bounded by a fixed constant A:

$$\xi \leqq A \ .$$

Now O_∞ has eigenvalues $\geqq 0$ and $\leqq 1$. Thus O_A has eigenvalues $\geqq 0$ and $\leqq B < 1$. The same is true of O_ξ. Hence $1/(1 - O_\xi)$ is a meaningful operator.

The last term of (35) is of the order of $\Delta h = 2\pi/L$. I.e.,

$$\frac{dh}{dp} = \frac{1}{1 - O_\xi}[1] + O\left(\frac{1}{L}\right) \ . \qquad (36)$$

Integrate from $-\xi$ to ξ. The left side becomes $2\pi/L(N - 1)$. The first term on the righthand side gives clearly a monotonically increasing function of ξ: $F(\xi)$, since

$$\frac{1}{1 - O_\xi}[1] = 1 + O_\xi \cdot 1 + O_\xi^2 \cdot 1 + O_\xi^3 \cdot 1 + \dots \qquad (37)$$

and every term integrated gives an increasing function of ξ. Thus

$$\frac{2\pi}{L}(N - 1) = F(\xi) + O\left(\frac{1}{L}\right) \ . \qquad (38)$$

Now approach the limit $n \to \infty$, $L \to \infty$ proportionally. (38) shows that ξ must approach a limit Q so that

$$2\pi N/L = F(Q) \ . \qquad (39)$$

This equation has one and only one solution Q for any fixed positive N/L since $F(Q)$ is monotonic and $F(0) = 0$ while $F(\infty) = \infty$. Putting the limit $\xi \to Q$ into (36) then gives

$$\text{Lim}\frac{dh}{dp} = \frac{1}{1 - O_Q}[1] \ . \qquad (40)$$

(39) and (40) are respectively (28) and (27) where

$$\text{Lim}\frac{dh}{dp} = 2\pi\rho(p) \ .$$

3. FERMION CASE WITH δ-FUNCTION INTERACTION (One Dimension)

The results of Section 2 on bosons can be generalized to the fermion case for both[13] $c > 0$ and[14] $c < 0$. The Hamiltonian is again (3) but the particles are fermions with an internal "spin" coordinate m which can range over $2J + 1$ values. [The Hamiltonian is "spin" independent.]

One starts by examining the space wave functions for (3) with arbitrary symmetry properties with respect to permutations of the identical particles. Bethe's hypothesis (10) then must be made for each section Q of phase space in which

$$x_{Q1} < x_{Q2} < x_{Q3} \ldots < x_{QN}$$

where Q stands for a permutation. Thus, instead of $N!$ coefficients α_p as in (10), one must now deal with $(N!)^2$ coefficients $\alpha_p(Q)$.

The subsequent developments will not be elaborated here. We only quote some results below.

3.1 Case[13] $c > 0$

Take the case of $J = \frac{1}{2}$. The ground state energy is then given by the following equations [which replace (27)–(29) for the boson case], which are expressed in operator language:

$$2\pi\begin{pmatrix}\rho\\ \sigma\end{pmatrix} = \begin{pmatrix}1\\ 0\end{pmatrix} + \begin{pmatrix}0 & -K_1\\ -K_1 & K_2\end{pmatrix}\begin{pmatrix}B_0 & \\ & B_1\end{pmatrix}\begin{pmatrix}\rho\\ \sigma\end{pmatrix} \quad (41a)$$

where $\rho(k)$, $\sigma(k)$ are functions of k, K_n is the integral operator

$$\langle k|K_n|k'\rangle = \frac{-nc}{(k-k')^2 + \frac{n^2c^2}{4}} \quad (42)$$

and B_0 and B_1 are projection operators: e.g.,

$$\begin{aligned} \langle k|B_0\rho\rangle &= 0 \quad \text{for } |k| > b_0 \\ \langle k|B_0\rho\rangle &= \rho(k) \text{ for } |k| \le b_0 \,. \end{aligned} \quad (43)$$

The energy E and number N of particles are given by

$$E/L = \int_{-b_0}^{b_0} k^2 \rho\, dk \quad (41b)$$

$$N/L = \int_{-b_0}^{b_0} \rho\, dk \,. \quad (41c)$$

Furthermore the wave function is constructed from space wave functions and "spin" wave functions belonging to the Young tableaus:

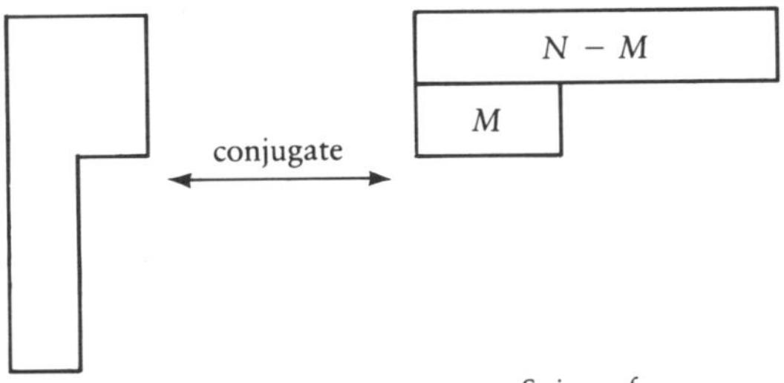

Space w.f. Spin w.f.

where

$$M/L = \int_{-b_1}^{b_1} \sigma\, dk \,. \quad (41d)$$

The generalization to higher "spins" gives similar equations as (41a)–(41d): e.g., for $J = 3/2$,

$$2\pi\begin{pmatrix}\rho\\ \sigma_1\\ \sigma_2\\ \sigma_3\end{pmatrix} = \begin{pmatrix}1\\0\\0\\0\end{pmatrix} + \begin{pmatrix}0 & -K_1 & 0 & 0\\ -K_1 & K_2 & -K_1 & 0\\ 0 & -K_2 & K_2 & -K_1\\ 0 & 0 & -K_1 & K_2\end{pmatrix} \times \begin{pmatrix}B_0 & & & \\ & B_1 & & \\ & & B_2 & \\ & & & B_3\end{pmatrix}\begin{pmatrix}\rho\\ \sigma_1\\ \sigma_2\\ \sigma_3\end{pmatrix}. \quad (44a)$$

(44b) and (44c) are the same as (41b) and (41c). The irreducible representations are then:

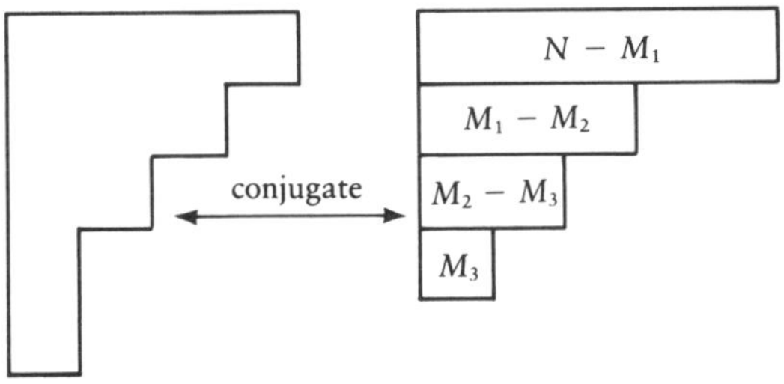

Space w.f. Spin w.f.

where

$$N/L = \int_{-b_0}^{b_0} \rho\, dk, \qquad M_i/L = \int_{-b_i}^{b_i} \sigma_i\, dk \,. \quad (i = 1, 2, 3) \quad (44d)$$

3.2 Case[14] $c < 0$.

Take the case $J = \frac{3}{2}$. The equations that replace (41a)–(41d) are

$$2\pi\begin{pmatrix}\rho_4\\ \rho_3\\ \rho_2\\ \rho_1\end{pmatrix} = \begin{pmatrix}4\\3\\2\\1\end{pmatrix} - \begin{pmatrix}K_6+K_4+K_2 & K_5+K_3+K_1 & K_4+K_2 & K_3\\ K_5+K_3+K_1 & K_4+K_2+K_0 & K_3+K_1 & K_3\\ K_4+K_2 & K_3+K_1 & K_2+K_0 & K_1\\ K_3 & K_2 & K_1 & K_0\end{pmatrix}$$

$$\times \begin{pmatrix} B_4 & & & \\ & B_3 & & \\ & & B_2 & \\ & & & B_1 \end{pmatrix} \begin{pmatrix} \rho_4 \\ \rho_3 \\ \rho_2 \\ \rho_1 \end{pmatrix} \qquad (45a)$$

$$E/L = \sum_\alpha \int_{-b_\alpha}^{b_\alpha} \left[\alpha k^2 - \frac{\alpha}{12}(\alpha^2 - 1)c^2 \right] \rho_\alpha dk \qquad (45b)$$

$$M_\alpha/L = \int_{-b_\alpha}^{b_\alpha} \rho_\alpha dk \qquad (\alpha = 1, 2, 3, 4) \qquad (45c)$$

and the symmetries are

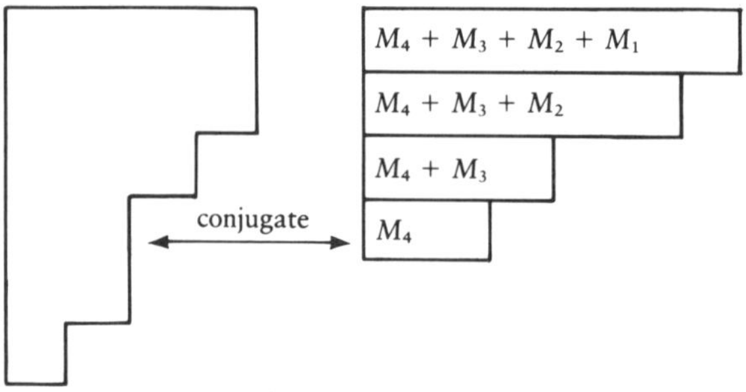

Space w.f. Spin w.f.

Generalization to other "spins" J is obvious. [The integrand in (45b) remain unchanged.]

It is interesting to consider the case that b_1, b_2, b_3, b_4 all are $\cong 0$. In that case

$$2\pi\rho_\alpha \cong \alpha$$

$$M_\alpha/L \cong \rho_\alpha 2b_\alpha = \text{small}$$

$$E/L \cong -\sum_\alpha \frac{\alpha}{12}(\alpha^2 - 1)c^2 \rho_\alpha 2b_\alpha$$

$$= -\sum_\alpha \frac{\alpha}{12}(\alpha^2 - 1)c^2 \frac{M_\alpha}{L} \,.$$

That is,

$$E \cong \sum_{\alpha=1}^{4} M_\alpha \left[-\frac{\alpha}{12}(\alpha^2 - 1)c^2 \right] . \qquad (46)$$

The energy of this dilute system is then given by that of a system of M_α bound systems of α particles each ($\alpha = 1, 2, 3, 4$). The energy of each system is $-(\alpha/12)(\alpha^2 - 1)c^2$. This last expression is in agreement with that given by McGuire[10] who gave it as the energy of an isolated bound system of α particles.

REFERENCES

1. W. Heisenberg, *Z. Physik* 49, 619 (1928).
2. F. Bloch, *Z. Physik* 61, 206 (1930); 74, 295 (1932).
3. H. A. Bethe, *Z. Physik* 71, 205 (1931).
4. C. N. Yang and C. P. Yang, *Phys. Rev.* 150, 321, 327.
5. T. Matsubara and H. Matsuda, *Progr. Theoret. Phys.* (Kyoto) 16, 569 (1956); 17, 19 (1957).
6. T. D. Lee and C. N. Yang, *Phys. Rev.* 87, 410 (1952).
7. C. N. Yang and C. P. Yang, *Phys. Rev.* 151, 258 (1966).
8. E. H. Lieb, *Phys. Rev. Letters* 18, 692 (1967).
9. E. H. Lieb, *Phys. Rev. Letters* 18, 1046 (1967); 19, 108 (1967); B. Sutherland, *Phys. Rev. Letters* 19, 103, (1967); C. P. Yang, *Phys. Rev. Letters* 19, 586 (1967); B. Sutherland, C. P. Yang, and C. N. Yang, *Phys. Rev. Letters* 19, 588 (1967).
10. E. H. Lieb and W. Liniger, *Phys. Rev.* 130, 1605 (1963); J. B. McGuire, *J. Math. Phys.* 5, 622 (1964).
11. E. H. Lieb, *Phys. Rev.* 130, 1616 (1963).
12. C. N. Yang, *Phys. Rev.* 168, 1920 (1968).
13. C. N. Yang, *Phys. Rev. Letters* 19, 1312 (1967); B. Sutherland, *Phys. Rev. Letters* 20, 98 (1968).
14. M. Gaudin, *Phys. Letters* 24A, 55 (1967) and M. Gaudin, thesis, Faculté des Sciences d'Orsay, University of Paris (Nov. 1967). The equations obtained in these papers were correct. Arguments for their correctness were, however, incomplete. See C. K. Lai and C. N. Yang, to be published.
15. C. N. Yang and C. P. Yang, *J. Math. Phys.* 10, 1115 (1969).

High-Energy Hadron–Hadron Collisions

C. N. Yang
Institute for Theoretical Physics, State University of New York, Stony Brook

[70f]

Commentary begins page 71

While high energy collision experiments yield a wealth of complicated patterns, there are a few general and very striking features that stand out. Because of the universality of these features, and because of the dominating influence they have on high energy phenomena, it is my opinion that a physical picture of high energy collisions must address itself first of all to these features before going into specific details. I shall in this short talk first state these general and striking features and proceed to describe a physical picture developed in the last few years to specifically accommodate these features. The picture was originally discussed for elastic scattering. But it leads naturally, indeed inevitably as we shall discuss, to conclusions about inelastic processes, resulting in an idea[1] called the hypothesis of limiting fragmentation. Some further speculative remarks will be made at the end of my talk.

1. *GENERAL FEATURES OF HIGH ENERGY COLLISIONS*

Three of the most striking general features of high energy collisions are:

(a) Lack of any outgoing particle with large transverse momentum. This fact has been known for many years in cosmic ray physics.

(b) The extreme smallness of elastic large angle cross sections. In fact, at a fixed angle the differential cross section seems to go down exponentially with some power of the incoming energy. This fact, first remarkably revealed in the early *pp* scattering experiments[2], is naturally accommodated in the "shattering" hypothesis[3] for high energy collisions. In this view, the hadrons are regarded as extended objects with many internal degrees of freedom. Large transverse momentum transfers, such as are required for large angle elastic scattering, in general would shatter the hadrons, leading to inelastic collisions. This view accommodates also very naturally feature (a) mentioned above.

(c) The apparent existence of the limit for very large energies of

$$\mathrm{Lim}\left(\frac{d\sigma}{dt}\right)_{\text{elastic}} = f(t) \tag{1}$$

Since for large energies $t \cong k\theta$ (k is the center of mass number and θ the center of mass scattering angle), the existence of the limit (1) means that as the wave number k increases the scattering angle θ decreases proportionally. This is, of course, precisely diffraction phenomena. In other words, (1) is natural if high energy scattering could be viewed as a diffraction from a scattering region of finite extension.

Features (b) and (c), together with their physical interpretation outlined above, could be naturally amalgamated. This has been done[4] and led to an

[1] J. Benecke, T.T. Chou, C. N. Yang, and E. Yen, *Phys. Rev.* 188, 2159 (1969).

[2] G. Cocconi et al. *Phys. Rev. Letters* 11, 499 (1963).

[3] T.T. Wu and C. N. Yang, *Phys. Rev.* 137, B708 (1965).

[4] T.T. Chou and C. N. Yang, *Phys. Rev.* 170, 1591 (1968); *175*, 1832 (1968); *Phys. Rev. Letters* 20, 1213 (1968). The physical picture was already discussed in N. Byers and C. N. Yang, *Phys. Rev.* 142, 976 (1966).

□Talk presented at the Kiev Conference, August 1970. Reprinted from *Proceedings of the Kiev Conference—Fundamental Problems of the Elementary Particle Theory*, Academy of Sciences of the Ukranian SSR, 1970, 131–133.

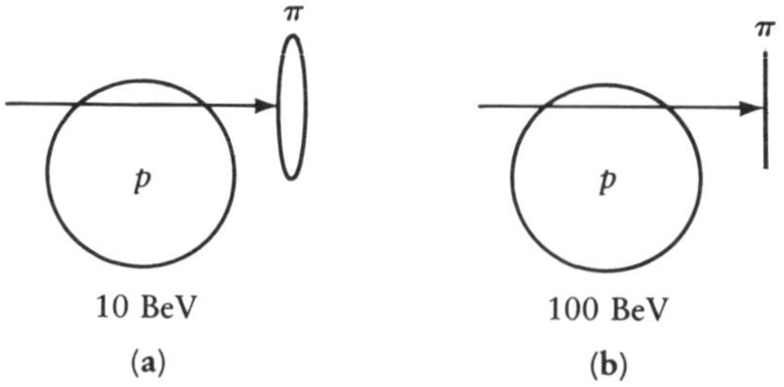

Figure 1 Schematic illustration in the laboratory frame of reference of πp collisions at 10 BeV and 100 BeV incoming energy. If the pion passes through the proton without either one being internally excited, one has an elastic scattering with the angular distribution (1) described[4] by Huygen's principle.

interesting relationship between the geometrical shapes of hadrons and elastic scattering which is in good agreement with experimental data.

2. *PHYSICAL PICTURE OF ELASTIC HIGH ENERGY COLLISIONS*

The physical picture that underlies these considerations can be illustrated by considering a πp collision in the laboratory system as shown in Figure 1a. The incoming pion, contracted by Lorentz contraction, sweeps through the target proton suffering an infinitesimal angular deflection. The target proton in the meantime takes up the recoil. Now let us consider the same collision at ten times the incoming momentum, as illustrated in Figure 1b. The velocity of the incoming particle does not appreciably change, though it is further contracted by a factor of 10. The existence of the limit (1) then states that the transmission coefficient of the pion through the proton does not change very much between Figure 1a and Figure 1b. In other words, the further contraction does not influence the transmission coefficient, presumably because the same amount of "stuff" still has to be transmitted whether further contracted or not.

3. *PHYSICAL PICTURE OF INELASTIC PROCESSES AT VERY HIGH ENERGIES*

Remembering this interpretation of (1), let us look into inelastic processes. In general, when the pion sweeps through the proton, the latter would not be left in its original internal state. It usually would be expected to vibrate or get excited (in addition to absorbing the recoil momentum). The interpretation described above of the limit (1) for elastic scattering would then lead naturally to the hypothesis that the mode of excitation or vibration of the target should also not change very much in going from Figure 2a to Figure 2b. Thus, when the target eventually breaks up, or fragments, the fragment distribution would also be practically the same for these two cases. This hypothesis was[1] called the hypothesis of limiting fragmentation.

Indeed, one could ask whether it is possible to find a theoretical model that will exhibit the limit (1) for elastic scattering, but will not lead to a limit for fragmentation processes. The answer is no: Any model in which the target has many internal degrees of freedom is essentially bound to yield either *limits for both elastic and fragmentation processes* or *no limit for either*. The reason for this is simply the fact that the unexcited and excited states of recoil for the target are, of course, coupled. As the projectile sweeps through the target, the unexcited and excited states of the target transform into each other. It is very unnatural to concoct a mathematical model which leads to a limit for unexcited recoil (i.e. elastic processes) but no limit for excited recoil (i.e. fragmentation processes).

The above discussion is also applicable to the rest system of the projectile in which the projectile exhibits limiting fragmentation.

While the discussion above assumes the strict existenceof the limit (1), for physical applicability we believe the discussion is still valid if the limit (1) holds when one neglects slow dependences on energy (such as when one replaces ln (E/M_p) by a constant). Cf. §19 of Reference 1.

In the last two years, Cheng and Wu[5] in their extensive work on the limit of Feynman diagrams

---

[5] H. Cheng and T.T. Wu, *Phys. Rev.* 1D, 2775 (1970) and papers cited therein.

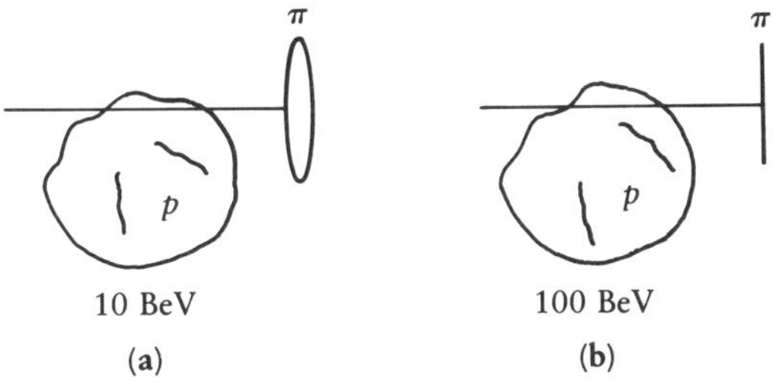

Figure 2 Schematic illustration in the laboratory frame of reference of πp collisions at 10 BeV and 100 BeV incoming energy with the target proton excited. The fact that Figures 1a and 1b lead to essentially the same transmission amplitude suggests that in Figures 2a and 2b the proton is excited in essentially the same way. This is the hypothesis of limiting fragmentation.

have obtained interesting results confirming very amazingly the general validity of the physical picture described above. I believe Professor Wu will discuss their work later today.

Recently, Chou and I[6] discussed a number of concepts natural in the framework of the hypothesis of limiting fragmentation. In particular, the idea of fragmentation fraction and of favored and disfavored fragmentation seems especially interesting.

4. SOME SPECULATIVE CONSIDERATIONS CONCERNING THE GEOMETRICAL SIZE OF HADRONS

The physical picture above envisages a hadron as an extended object with many degrees of freedom, i.e., a droplet. Elastic scattering is described as two droplets *going through* each other, scattering each other by a diffractive process. In this process the droplets are sometimes excited and then are separately fragmented.

However, it is important to recognize that in one essential aspect the hadrons behave very differently from any droplets that we have had experience with, such as a water droplet, or a droplet of nuclear mater: For these latter droplets, fragmentation of a droplet leads to *smaller* droplets, such as in nuclear fission. In the fragmentation of a hadron, however, the fragments are each of approximately the same size as the original hadron, since all hadrons seem to be approximately of the same radius of $r_0 \sim .7 \times \cdot 10^{-13}$ cm.

Could it be that this approximate universal size r_0 is due to a "distortion" of space, or "quantization" of space [rather than due to the strong interactions that give rise to the hadronic states] ? I think this is unlikely to be the case: Our extrapolation of space time concepts from macroscopic dimensions to atomic dimensions, to nuclear dimensions, and down to $R \sim .1 \times 10^{-13}$ cm. seems to agree very well with the most accurate recent measurements on the electrodynamics of leptons.[7] Since R is an order of magnitude smaller than r_0, it seems unlikely that r_0 is determined by any "distortion" of space.

If we accept this admittedly not-so-cogent argument, then we must seek for a model of hadrons as droplet states prescribed by strong interactions, which give an approximate optimum droplet size r_0. Larger droplets would of course break up into those of optimum size. But smaller droplets (i.e., droplets smaller in linear dimension than r_0) would also "break up" into those of optimum size r_0 in a way unfamiliar to us in previous physical pictures.This last point could be understood, I believe, only in a probability amplitude picture (i.e., wave picture) of droplets. How to give an example of a model of such droplets is a very speculative problem I hope to work on.

[6] T.T.Chou and C. N. Yang, *Phys. Rev. Letters* 25, 1072 (1970).

[7] See a recent review by S. J. Brodsky in *Proceedings of the Daresbury Conference*, September 1969, Daresbury Laboratories, Liverpool, England.

⟦71b⟧
Commentary begins page 71

1. Introductory Note on Phase Transitions and Critical Phenomena

C. N. Yang

Institute for Theoretical Physics,
State University of New York,
Stony Brook, New York, U.S.A.

One of the crowning achievements of nineteenth century physics was the development of the statistical (i.e. microscopic) basis of thermodynamics. While much of the ideas of this development originated with Maxwell and Boltzmann, it was Gibbs' work that more directly influenced our present formulation of equilibrium statistical mechanics. The evolvement in the last seventy years of our understanding of phase transitions and critical phenomena represents, in essence, the application of Gibbs' formulation to a wide variety of physical problems. Some of these are directly and closely related to experimental facts, others represent idealized problems formulated with a view to extracting an understanding of some essential features of various physical phenomena. If a single conclusion is to be reached summarizing the development of these seventy years in this field, it is that the statistical basis of equilibrium thermodynamics brilliantly triumphed.

It is important, in this connection, to remind ourselves that Gibbs himself did not, indeed could not, have as sanguine a view about his work as we now have. When he published his book "Elementary Principles of Statistical Mechanics" in 1901, the world of physics was witnessing, on the one hand, the great experimental discoveries of the electron, of radioactivity, of X-rays, etc. that ushered in the age of atomic (i.e. microscopic) physics, and on the other hand, the deeply perturbing contradictions inherent in classical physics when applied to the atom. He said, for example, in the preface to his book:

> "In the present state of science, it seems hardly possible to frame a dynamic theory of molecular action which shall embrace the phenomena of thermodynamics, of radiation, and of the electrical manifestations which accompany the union of atoms. Yet any theory is obviously inadequate

□Reprinted from *Phase Transitions and Critical Phenomena,* Vol. 1, C. Domb and M. S. Green, eds. London: Academic Press, 1971, 1–5.

> which does not take account of all of these phenomena. Even if we confine our attention to the phenomena distinctively thermodynamic, we do not escape difficulties in as simple a matter as the number of degrees of freedom of a diatomic gas. It is well known that while theory would assign to the gas six degrees of freedom per molecule, in our experiments on specific heat we cannot account for more than five. Certainly, one is building on an insecure foundation, who rests his work on hypotheses concerning the constitution of matter."

It is truly remarkable that "building on an insecure foundation" Gibbs arrived at such a "rational foundation of thermodynamics". (Perhaps physicists today who are frustrated by the complexities of elementary particle physics could take heart again in reflecting upon such previous difficult times in the history of physics.)

The introduction of quantum mechanics, of course, removed the chief stumbling block in physicists' confidence in equilibrium statistical mechanics. However, full realization that a *single* mathematical expression can describe *both* phases in a phase transition came much later. In 1937, J. E. Mayer published his important work (Mayer, 1937) which was discussed at the Van der Waals Centenary Congress which took place on November 26, 1937 in Amsterdam. There arose (Born and Fuchs, 1938)*:

> "a vigorous discussion on the question as to whether Mayer's explanation of the phenomena of condensation is correct. Doubts about this point were raised by the referee, because it is difficult to comprehend how a method of approximation such as that of Mayer, starting from the gaseous state, can lead to the discontinuity of the density on an isothermal curve which corresponds to condensation. The usual methods for treating the equilibrium of two phases introduce the equation of state of both phases and derive the condition for their co-existence. Mayer's theory does nothing of this kind, but treats all possible molecular arrangements with their proper weight, as if there were only one phase. How can the gas molecules 'know' when they have to coagulate to form a liquid or solid? Mayer's mathematical method is too involved to make this point quite clear".

The essential points of this confusion were rapidly clarified. Complete mathematical rigor followed some ten years later (Van Hove, 1951; Yang and Lee, 1952), which in turn led to many of the more recent rigorous results.

It is, of course, one thing to prove the existence of a thermodynamical limit for the partition function describing both phases of a phase transition, but

* A very lucid discussion can be found in Kahn and Uhlenbeck (1938). See especially pp. 400–401.

quite another thing to find out the exact nature of the discontinuity at the phase transition. For a long time, thinking on this subject was dominated by early ideas classifying transitions into first-order, second-order, etc. and subsequent discussions of critical phenomena (Ehrenfest, 1933; Landau and Lifschitz, 1958). The reasons that such ideas were widely accepted were (a) lack of experimental evidence to the contrary, (b) the empirical usefulness of Van der Waals' theory, (c) consistency with the physically appealing mean field theories (e.g. Weiss theory and Bragg–Williams theory), (d) existence of metastable phases such as the super-saturated vapour, which suggested the concept of the continuation of thermodynamical functions into these phases. In recent years, there has been an important revision in our understanding on this subject. This came about partly because of rigorous, or semi-rigorous, theoretical work such as Onsager's solution (1944) and the series expansion-extrapolation method. Partly this came about because of experimental measurements which yielded thermodynamic quantities that become singular at transition points. The whole development is especially remarkable in the amalgamation of different transition phenomena, such as gas–liquid transition and magnetic transitions, into one unified treatment. It is clear that much remains to be learned in these studies.

The beautiful solution of Onsager (1944) for the two-dimensional Ising model is mathematically very remarkable. There was, however, a time in the 1940's and 1950's when Onsager's solution was regarded as a mathematical curiosity with no real physical relevance. One heard in those days of references to "contracting the Ising disease". This feeling disappeared during the 1960's when it became clear that the lattice gas description of liquid gas transitions does capture much of the essential features of the singularities.

In a different way, the series expansion-extrapolation method (Domb, 1949; Wakefield, 1951; Domb and Sykes, 1956) for locating singularities and singular structures is a remarkable development. I must admit that in the 1950's I was extremely sceptical of such methods since it appeared to me that to predict the singularity from the first few terms of a power series was hopeless. It turns out that this method is today the most powerful one at our command and yields results, for example for the transition temperature, with fantastic accuracy. Presumably, the reason for the success lies in the rapid convergence toward a "regular" sequence in the expansions involved in these methods. It seems deeper insights in this direction are yet to come.

The wide variety of topics covered in these volumes testifies to the great vigour of the field. It is not my intention to comment here on each of these developments. Clearly exciting progress will continue to take place in many of these topics since new techniques and new materials will continue to broaden and sharpen our experience with various forms of matter. Perhaps, I may be allowed here to make a few speculative remarks about some possible future developments with the understanding that they would be strongly coloured by my personal viewpoint.

It seems to me that the prospects of finding an exactly soluble two-dimensional quantum mechanical problem or a three-dimensional classical problem of a nontrivial type are rather remote. Perhaps the combinatorial problem of paths in three dimensions is just too complicated.

On the positive side, it seems to me hopeful that some enormously important development could take place in the microscopic treatment of macroscopic quantum states. It is now known that macroscopic quantum states are a common phenomenon occurring not only in laser optics but very generally at low temperatures. While a number of important physical ideas are already well known concerning the nature of these macroscopic quantum states, a complete microscopic basis of the macroscopic qualities involved in such phenomena (on a par with the Maxwell–Boltzmann–Gibbs' microscopic theory of thermodynamics) is still lacking.

Another interesting field where fruitful results might be expected concerns the nature of the excitations in a quantum mechanical system of infinite degrees of freedom. I am convinced that the basic difficulty confronting our understanding of elementary particle physics derives from our lack of experience with systems where many degrees of freedom play important roles all at once. Now, a hadron is nothing but an excitation of a rather complex nature in the infinite degrees of freedom system which we call the vacuum. The experience gained in quantum statistical mechanics should go a long way toward clarifying the problems involved in hadron interactions. One could hope for increasing cross fertilization of ideas and methods in these two fields.

One of the great intellectual challenges for the next few decades is the question of brain organization. What is the basic mechanism for storage of memory? What are the processes that serve as the interphase between the basically chemical processes of the body and the very specific and non-statistical operations in the brain? Above all, how is concept formation achieved in the human brain? I wonder whether the spirit of the physics that will be involved in these studies will not be akin to that which moved the founders of the "rational foundation of thermodynamics".

References

Born, M. and Fuchs, K. (1938). *Proc. Roy. Soc.* **A166,** 391.
Domb, C. (1949). *Proc. Roy. Soc.* **A199,** 199 (1949).
Domb, C. and Sykes, M. F. (1956). *Proc. Roy. Soc.* **A235,** 247.
Ehrenfest, P. (1933). Leiden Comm. Supplement 75b.
Kahn, B. and Uhlenbeck, G. E. (1938). *Physica* **5,** 399.

Landau, L. D. and Lifschitz, E. M. (1958). "Statistical Physics." Pergamon Press, London.
Mayer, J. E. (1937). *J. Chem. Phys.* **5,** 67.
Onsager, L. (1944). *Phys. Rev.* **65,** 117.
Van Hove, L. (1949). *Physica* **15,** 951.
Yang, C. N. and Lee T. D. (1952). **87,** 404, 410.
Wakefield, A. J. (1951). *Proc. Camb. Phil. Soc.* **47,** 419, 799.

Some Concepts in Current Elementary Particle Physics

[[72c]]
Commentary begins page 71

Chen Ning Yang

Many years ago, when I was still a schoolboy in China, I had the unforgettable experience of reading translations of the books by Eddington and by Jeans about the new developments in physics. They described the various conceptual revolutions in 20th century physics, starting from the special theory of relativity, leading to the general theory and to quantum mechanics. I cannot say that I understood the meaning and the necessity of the Fitzgerald contraction, the Bohr atom, or the uncertainty principle, but it was impossible for me not to catch the excitement and the enthusiasm so vividly overflowing from the pages. It was clear that here was a new gate opening onto the mystery of the structure of the universe, at once full of light and darkness, enlightenment and puzzles. How much this fascination influenced my later choice of a career I could not say, because I do not know. But even today as I recall my experience so many years ago I could still feel the mysterious excitement that had then overwhelmed me.

I tell you this because I want to say how much it was a privilege for me to hear recounted last evening the personal experiences, throughout these revolutionary periods in the development of our concept of nature, of one of the prime architects of the whole enterprise. I am grateful to him and to the organizers of this conference who have made this possible.

In his talk on the physicist's conception of nature, Dirac urged us, on more than one occasion, to abandon past prejudices. Was he urging us to abandon our traditional concepts of space-time? Was he urging us to change our views about the concept of the field? Or was he urging us to introduce some new symmetries or to abandon some old ones? He did not provide us with any definite answers to these questions. But we know that in order to search for clues for new conceptual developments of physics there are two important guides. On the one hand, we must be always rooted in new experimental findings. Detached from this root physics runs into the danger of degenerating into mathematical exercises. On the other hand, we must not be shackled all the time by the desire to fit what at each moment is accepted as experimental reality. Extrapolations based on pure logic and form are essential ingredients in many great conceptual advances in our field. Perhaps in no other physicist's work is this point more clearly exhibited than that of Dirac. In all of his work, the less important as well as the more important ones, there is that insistence on elegance of form and beauty of logic that gives his papers a unique creative flavour. Dirac himself has said that

□ Talk given at the Trieste Conference in honor of P. A. M. Dirac, September 1972. Reprinted from *The Physicist's Conception of Nature,* J. Mehra, ed. Dordrecht, Holland: Reidel, 1972, 447–453.

beauty is the only requirement. If experiments contradict a beautiful idea, let us forget about experiments.

Surely that was the feeling of Einstein when he wrote after the creation of the general theory of relativity that the theory did not need experimental proof. Surely that was the feeling of Dirac when, confronted with the necessity to explain away the negative energy states, he dreamed of the crazy idea[1] of the infinite sea with holes in it.

The dichotomy of the two not quite consistent guides for conceptual development of physics discussed above is, so to say, graphically illustrated for me in a comparison of this symposium with another meeting that I just came from: the High Energy Physics conference at Batavia, near Chicago. On the surface, there is a minimum overlap of participants in the two conferences. There is a minimum overlap in the languages used. There is a minimum overlap of those topics that seem to generate the most interest. However, if one sits back and takes a long-range view and forgets about the hustle and bustle that provides the ambient atmosphere in the new laboratory in Batavia (near Chicago), if one looks away from the corridor discussions of mysterious new high transverse momentum events, if one forgets about the new preliminary data on the decay of kaons into leptons, if one takes a longer range assessment, it will become clear that above the noise level of detailed developments there is, in fact, quite a bit of overlap in interests between the two conferences. It is in this spirit of looking at the general concepts that I should like now to discuss some developments in physics that have been under extensive discussion in the last ten years or so.

In my opinion the greatest excitement in physics during the past decade was the discovery[2] of a very weak violation of *CP* invariance in 1954 by Christenson, Cronin, Fitch and Turley. The precise meaning of this discovery could be understood from the following description[3] of the meaning of the operators *C* (charge conjugation), *P* (parity, or inversion), and *T* (time reversal operator). Consider reaction *R* depicted in Fig. 1 for the process

$$A + B \rightarrow C + D.$$

Given this reaction one can conceive of the reaction $(P)R$, the reaction obtained from *R* by reflection in the plane of the paper. If *R* and $(P)R$ take place at equal rates, then we say that the laws governing *R* and $(P)R$ obey right-left symmetry, or *P* invariance holds for these reactions. Similarly for *C* invariance and *T* invariance.

In 1956 it was discovered that for weak interactions *P* invariance is violated and *C* invariance is violated. It was then thought that the combined invariance *CP* would hold for all weak interactions. Great excitement came in 1964 with the discovery that this is not true in the decay of K^0, and this discovery led to beautiful experiments.

The symmetries *C*, *P* and *T* are related by a theorem[4] discovered by Lüders, Pauli and Schwinger in the early fifties. The theorem states that under very weak general assumptions, which are accepted by essentially every physicist as valid, the combined *CPT* invariance holds, whether or not the individual invariances *C*, *P* or *T* holds or not. Because of this theorem, violation of *CP* invariance implies violation of time reversal invariance.

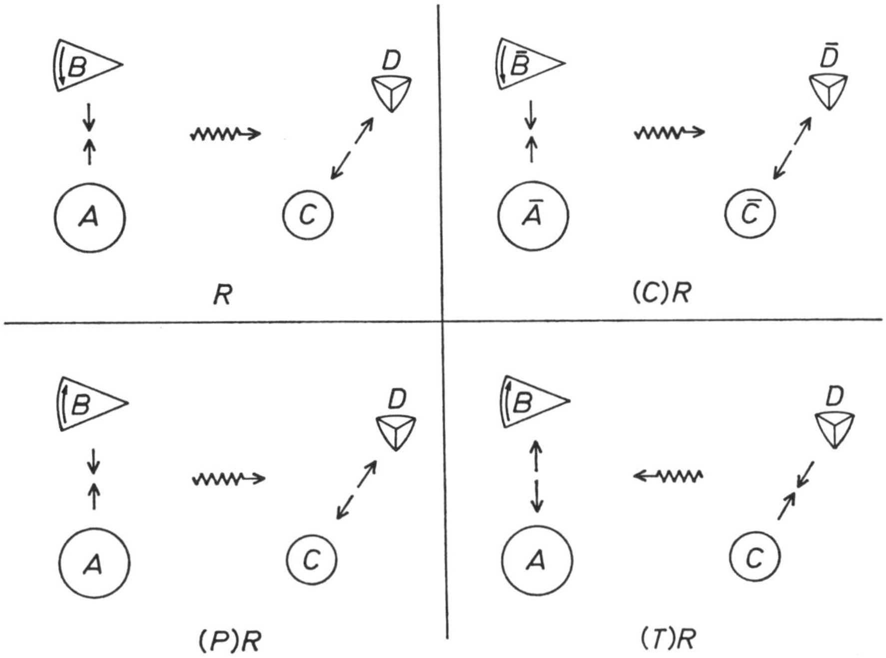

Fig. 1. The operations C, P, and T. The reaction R describes $A+B\rightarrow C+D$ where A and C are spheres, B a spinning cone, and D a tetrahedron. The charge conjugate reaction $(C)R$ has the same kinematics as R, but with all particles replaced by antiparticles. $(P)R$ is the reaction R reflected in the plane of the drawing. Notice the reversed spin of B and the reflected tetrahedron D. $(T)R$ is the time-reversed reaction of R, describing $C+D\rightarrow A+$ B. Notice the spin direction of B.

The concepts of P invariance and T invariance were rooted originally in geometrical considerations. Such was not the case for C invariance. It was Dirac's idea of particles and holes that gave rise to C invariance. As was evident from his original paper, the hole idea is very much related to the mathematical concept of complex conjugation. For this reason, C invariance is referred to as being an algebraic concept. The introduction of complex numbers as an essential element of the algebra of physical laws is one of the new concepts brought into physics by quantum mechanics. Since that led to the invariance C, we could ask: Are there other algebras that should be brought into physics. Are there other symmetries that may be due to more complicated algebras.

It is more than likely that the answers to both these questions are in the affirmative. On the one hand, from the side of mathematical concepts, there are exactly three associative division algebras (i.e. algebras in which addition, subtraction, multiplication and division rules exist satisfying the usual distributive and associative laws, which is commutative in addition and subtraction, and for which $ab=0$ implies $a=0$ or $b=0$) that contain real numbers as a subalgebra. These are real numbers, complex numbers and quaternions. (Of these the quaternions have not yet been introduced in any essential way into the laws of physics.) We note here that the generalization of the concept of complex conjugation (i.e. $i\leftrightarrow -i$ in complex number algebra) to quaternion algebra is the operation of a rotation between the three imaginary axes i, j and k. There are many such operations forming a group SO_3.

On the other hand, from the side of physical concepts there is the hitherto only phenomenologically based SU_2 symmetry which is very accurately observed by strong interactions. Fundamentally, this symmetry describes the striking similarity between the proton and the neutron. It has the mathematical structure of a three-dimensional rotational symmetry. That turns out to be exactly the structure of the group mentioned above which is the mathematical generalization of the concept of complex conjugation. All these suggest the possibility illustrated below of:

algebra	*mathematical operation leaving algebra unchanged*	*physical symmetry*
complex numbers	complex conjugation	C
quaternions	rotation of i, j, k	SU_2

SU_2 being an algebraic symmetry resulting from the incorporation of quaternions into the laws of physics. There have been discussions of this possibility[3,5] before, but all efforts in this direction have so far remained fruitless. Nevertheless, I venture to guess that nature has indeed utilized this possibility, only physicists have not yet found the right key to the incorporation of quaternions into fundamental physics so as to explain the similarity between the proton and the neutron.

Another important symmetry, discussed extensively in the sixties, is SU_3. It was found that there are amazing systematics of particle multiplets that make SU_3 a very fruitful concept even though it is badly broken. My personal prejudice is that if we finally should have an understanding of SU_3, it will turn out to be an understanding on quite a different footing from that of SU_2.

It is interesting to ask whether SU_3 could also originate from an algebra. If we investigate the next well-structured algebra beyond the quaternions it is the octonians. (In octonian algebra one has seven square roots of -1, independent of each other by real linear transformations. It is the only division algebra besides the real numbers, complex numbers and quaternion that is an 'alternate' algebra. An alternate algebra is one in which $(xy)z-x(yz)$ is antisymmetrical in x, y and z. It follows from this definition that in an alternate algebra x^n is independent of the order of association. That is, an alternate algebra is always associative for the powers of any element. A division algebra is one in which $xy=0$ implies $x=0$ or $y=0$). The group of real linear transformations that leave the octonian algebra invariant is $G2$, the first exceptional Lie group. Unfortunately $G2$ is not SU_3.

If we could turn our attention away from symmetries for a while we shall see that in the last five or six years hadron physics has been undergoing very interesting developments. The tempo of activities in this field has greatly increased, thanks to the large number of experiments on elastic and inelastic reactions studied at the high energy accelerators. While a description of a high energy hadron–hadron collision is necessarily complicated because of the large number of particles frequently produced and because of the large fluctuations in such quantities like the multiplicity of outgoing particles, the amazing thing is that the experimental results also exhibited very striking general regularities, regularities that, in my opinion, point to a direction of approach that may provide a physical description of hadronic structure. In this description hadrons go *through* each other in a high energy collision with little exchanges of quantum numbers or longitudinal momentum or energy in the centre of mass reference system. They are however excited coherently into dynamical excited states which then fragment into various outgoing particles, forming two jets. (A dynamic excited state is a superposition of states with various invariant mass values. The part consisting of the lowest invariant mass values approaches a limiting state at relatively low incoming energies. The part consisting of high invariant mass values approaches a limiting state at higher incoming energies.) In this process of going through each other, the geometrical size of the incoming particles plays an essential role, and we believe that all hadrons are approximately of the size 1.4×10^{-13} cm in diameter.

The general description outlined above is in agreement with all main features of high energy processes in both elastic and inelastic collisions. For elastic collisions one obtains thus a parameter-less relationship[6] between elastic *e–p* scattering and elastic *p–p* scattering which led, among other things, to the prediction of dips in elastic *p–p* angular distribution, a prediction recently confirmed experimentally at CERN. For inelastic collisions the general description led to the hypothesis of limiting fragmentation[7] which has found extensive and accurate experimental support in recent years.

What determines the basic size of the hadrons? Why are they of the same order of magnitude? Why can a hadron not be divided into geometrically smaller hadrons like droplets that we have been familiar with in physics: water droplets and nuclear matter? These are among the questions that we do not know how to answer, but it seems rather clear that we are basically dealing with a system of infinite degrees of freedom in which there is a vacuum, and above the vacuum, energy-wise, there are various excited states (of this system of infinite degrees of freedom) which we call hadrons. It would seem to me that an urgently needed task for us is to understand the physics of such a strongly interacting system of infinite degrees of freedom.

This is a difficult task but also a very concrete and challenging one. Fortunately, our experience with other branches of physics that deal with systems of infinite degrees of freedom would help us in this task.

Could it be that the above-quoted size of hadrons is related to some fundamental changes of space-time structure at such distances? I believe the answer to this question is no. Electromagnetic interactions of leptons have been studied to much smaller spatial dimensions than 10^{-13} cm and no deviations of the geometrical structure of space-time have been found.

I come lastly to a concept called gauge fields. This is a concept which is basically derived, on the side of physics, from the idea that some fundamental symmetries of the physical world should be related to invariance concepts at *every* space-time point. In particular, let us take isotopic spin invariance which states that the proton and the neutron are similar. If the electromagnetic field is 'switched off', there would be two entirely similar states of the nucleon, and which we choose to call the proton and which the neutron would be an arbitrary convention. Now if we adopt the view that this arbitrary convention should be independently chosen at every space-time point, then we would be naturally led to the concept of gauge fields. (Another way of putting this is that if I adopt a convention it should not bind my colleague in the next laboratory to adopt any specific convention whatsoever.)

On the mathematical side, the concept of gauge fields apparently is related to fibre bundles. But I do not know really what a fibre bundle is.

The electromagnetic field is a gauge field. Einstein's gravitational theory is intimately related to the concept of gauge fields, although to *identify* the gravitational field as a gauge field is not an absolutely straightforward matter.

During the last 15 years there have been repeated efforts to introduce a gauge field as a source, or the source, of strong interactions. These efforts have not been entirely successful, but the idea is fundamentally attractive. In my opinion it is likely to play important roles in the future.

In the last five years, efforts originating from the work of S. Weinberg (and earlier work of Gürsey, Schwinger, Salam and Ward and others) have led to great excitement about the possibility of amalgamating the concepts of gauge fields, electromagnetic fields and weak interactions. (And, in some versions, also strong interactions.) My colleague at Stony Brook, B. W. Lee, has just given at the Chicago Conference a rapporteur's talk[8] summarizing the intensive activities in this field during the past two years. In addition to the idea of gauge fields, an important new idea added is the concept of quantization at non-zero expectation values of field quantities. Personally, I think that these recent developments are along an important direction, but perhaps some further new ideas are still missing so that the current efforts end in highly non-unique and non-beautiful theories.

REFERENCES

1. The more 'crazy' an idea is, the more profound it becomes when it turns out to be relevant in the description of natural phenomena. My own feeling about Dirac's introduction of the negative sea was summarized in the following sentences in a talk at the 75th anniversary celebration of Bryn Mawr College, November 6, 1959:

'The concept of charge conjugate symmetry is a purely quantum mechanical concept and is not related to any geometrical concepts such as rotational invariance. It derives its origin from the Dirac theory of the electron, which in turn is, viewed today, a logical consequence of the fusion of the quantum theory with the requirement of relativistic invariance. To first postulate the charge conjugation concept, as Dirac did about thirty years ago, was, however, a most daring and profound step, not unlike the first introduction of the negative numbers. The later experimental verification of the existence of the antiparticles constituted not only one of the most beautiful and forceful demonstrations of the practical consequences of the symmetry principles, it represented actually one of the most gratifying and far-reaching triumphs of theoretical reasoning.'

2. J. H. Christenson, J. W. Cronin, V. L. Fitch and R. Turlay, *Phys. Rev. Letters* **13**, 138 (1964).
3. *Vistas in Research*, Vol. 3, Gordon and Breach (1968), Lecture by C. N. Yang, October 13, 1965.
4. J. Schwinger, *Phys. Rev.* **91**, 720, 723 (1953); G. Lüders, *Kgl. Danske Vidensk. Selsk. Mat-fys.*

Medd. **28** (1954); W. Pauli's article in *Niels Bohr and the Development of Physics*, Pergamon, London (1955).
5. See my comments in the discussion period after Tiomno's talk, Session 9, *Proceedings of the 7th Rochester Conference*, 1957 (the Interscience Publishers).
6. T. T. Chou and Chen Ning Yang, *Phys. Rev.* **170**, 1591 (1968).
7. J. Benecke, T. T. Chou, C. N. Yang, and E. Yen, *Phys. Rev.* **188**, 2159 (1969).
8. B. W. Lee in *Proceedings of the International High Energy Conference at Chicago*, September 1972, to be published.

[[73b]]
Commentary begins page 72

Opaqueness of pp Collisions from 30 to 1500 GeV/c *

Alexander Wu Chao and Chen Ning Yang
Institute for Theoretical Physics, State University of New York, Stony Brook, New York 11790
(Received 9 May 1973)

Assuming only the eikonal approximation and the approximate reality of the S matrix for elastic scattering we evaluate from experimental data the opaqueness of pp scattering from $p_L = 30$ to 1500 GeV/c. Parameters X and Y which characterize the shape of the function $1 - S(b)$ are defined and discussed.

INTRODUCTION

Recent CERN Intersecting Storage Rings (ISR) experiments[1] have refocused attention on the behavior of the total cross section and elastic differential cross section at very high energies. These results are especially interesting because of the earlier conjecture of Cheng and Wu[2] which seems to be remarkably confirmed. A phenomenological analysis of various data based on Cheng and Wu's picture has been made.[3] In this paper we make a less extensive phenomenological analysis, with emphasis on a description of high-energy pp collisions with as little theoretical prejudice as possible. Since the eikonal approximation and the nearly purely imaginary character of the scattering amplitudes both seem to be quite accurate, we shall adopt these assumptions but shall use no additional ones.

In the eikonal approximation[4]

$$\left(\frac{d\sigma}{dt}\right)_{\rm el} = \pi |a|^2, \tag{1}$$

$$a = \langle 1 - S(b)\rangle, \tag{2}$$

where $\langle\ \rangle$ designates the Fourier transform from the two-dimensional space of the impact parameter $\vec{b}$ to the two-dimensional space of the momentum transfer $\vec{K}$ ($K^2 = |t|$):

$$\langle X\rangle = \frac{1}{2\pi}\iint X(\vec{b})\exp(i\vec{K}\cdot\vec{b})d^2b\,. \tag{3}$$

We shall also use the same symbol to designate the inverse Fourier transform. We neglect all spin-correlation effects.

The S matrix $S(b)$ will be written as

$$S = e^{-\Omega(b)}, \tag{4}$$

where Ω, the opaqueness (or blackness), will be assumed to be real. It is, of course, dependent on the incoming energy. (Ω must be almost purely imaginary for sufficiently large b, see Ref. 5, p. 357–358. But we neglect such contributions which are probably very small.)

In Sec. I we discuss the magnitude and shape of the opaqueness $\Omega(b)$ for the 10.8-on-10.8-GeV/c and the 26.8-on-26.8-GeV/c pp collisions. In Sec. II we discuss the mathematical range of the elasticity parameter and the slope parameter.

I. OPAQUENESS AT HIGH ENERGIES

It is easy to obtain[4] the opaqueness Ω from $(d\sigma/dt)_{\rm el}$ by using

$$\langle\Omega\rangle = a + \tfrac{1}{2}a\otimes a + \tfrac{1}{3}a\otimes a\otimes a + \cdots, \tag{5}$$

where $\otimes$ is the folding integral. The inverse of (5) is

$$a = \langle\Omega\rangle - \frac{1}{2!}\langle\Omega\rangle\otimes\langle\Omega\rangle + \frac{1}{3!}\langle\Omega\rangle\otimes\langle\Omega\rangle\otimes\langle\Omega\rangle - \cdots. \tag{6}$$

For the 10.8-on-10.8-GeV/c pp collision we use the unnormalized elastic data of Barbiellini *et al.*[6] and normalize with the total cross section $\sigma_T = 39.1 \pm 0.4$ mb estimated for this collision from the data in Ref. 1. This gives the following fit:

$$a(t=0) = 7.98 \pm 0.08\ ({\rm GeV}/c)^{-2},$$

$$\frac{a}{a(t=0)} = (0.685 \pm 0.005)\exp[-(4.7 \pm 0.05)|t|] + [1 - (0.685 \pm 0.005)]\exp[-(9.0 \pm 0.5)|t|] \quad [|t| < 0.25\ ({\rm GeV}/c)^2]. \tag{7}$$

Substitution into (5) gives the value of $\langle\Omega\rangle/\langle\Omega\rangle_{t=0}$ presented in Fig. 1. Also

$$\langle\Omega\rangle_{t=0} = 10.15 \pm 0.15\ ({\rm GeV}/c)^{-2}. \tag{8}$$

We present the result this way because the error in $\langle\Omega\rangle_{t=0}$ is quite separate from that of $\langle\Omega\rangle/\langle\Omega\rangle_{t=0}$.

For the 26.8-on-26.8-GeV/c pp collisions a similar procedure yields $a(t=0) = 8.83 \pm 0.12$ (GeV/c)$^{-2}$,

$$\frac{a}{a(t=0)} = (0.82 \pm 0.01)\exp[-(5.22 \pm 0.01)|t|] + [1 - (0.82 \pm 0.01)]\exp[-(12.15 \pm 0.15)|t|] \quad [|t| < 0.4\ ({\rm GeV}/c)^2], \tag{9}$$

and the plot of $\langle\Omega\rangle/\langle\Omega\rangle_{t=0}$ which is also exhibited in Fig. 1. For this energy

 □ Reprinted from *Physical Review D* 8, 7 (October 1, 1973), 2063–2067.

$$\langle\Omega\rangle_{t=0} = 11.4 \pm 0.2\ (\mathrm{GeV}/c)^{-2}. \qquad (10)$$

The same analysis as above for pp collision at 29.7 GeV/c was already made in Ref. 4. The result is plotted also in Fig. 1. The value of $\langle\Omega\rangle_{t=0}$ is tabulated in Table I.

We notice the following facts:

(a) The opaqueness probably *decreases* slightly from $p_L = 29.7$ to 245 GeV/c, but *increases* from $p_L = 245$ to 1480 GeV/c. The over-all opaqueness in coordinate space,

$$\iint \Omega(b) d^2 b = 2\pi \langle\Omega\rangle_{t=0},$$

decreases by $(1.5 \pm 2)\%$ first and then increases by $(12.3 \pm 2.5)\%$.

(b) The shape of the opaqueness $\Omega(b)$ as a function of b *expands* from $p_L = 29.7$ to 245 GeV/c, but then does not change appreciably from $p_L = 245$ to 1480 GeV/c. (There are indications perhaps of a very slight expansion in this second energy region.)

It appears at first surprising that the increase of the slope parameter between $p_L = 245$ and 1480 GeV/c does not lead to any appreciable spatial expansion in $\Omega(b)$. Upon closer examination one finds that there is an opposite effect leading to a cancellation. With increasing energy the increasing total cross section leads to increasing importance of the higher order terms on the right-hand side of (5), which have smaller absolute values of the slope in K space.

We believe conclusions (a) and (b) to be quite firm, once one accepts the experimental data, since few assumptions have been made other than the validity of experimental data. We do not have any compelling reasons for this behavior of the magnitude and shape of the opaqueness $\Omega(b)$. We are further investigating this matter, especially considering the possibility of relinquishing the assumption that $S(b)$ is real.

In Fig. 2 we sketch the opaqueness $\Omega(b)$ vs b for all three energies. Since the errors on them are quite sensitively dependent on large K data we do not put error bars on these curves.

(c) It has been suggested[4,5] that

$$\langle\Omega\rangle = (\mathrm{constant})[F_1(K)]^2,$$

TABLE I. The opaqueness in momentum space evaluated at $t=0$ as a function of p_L.

p_L (GeV/c)	$\langle\Omega\rangle_{t=0}$ [(GeV/c)$^{-2}$]
29.7	10.3 ± 0.15
245	10.15 ± 0.15
1480	11.4 ± 0.2

where F_1 is the electric charge form factor of the proton. Figures 1 and 3 show that this suggestion is no longer good for the ISR data. Instead the formula

$$\langle\Omega\rangle = (\mathrm{constant})[G_E(K)]^2 \qquad (11)$$

seems to be relatively good.

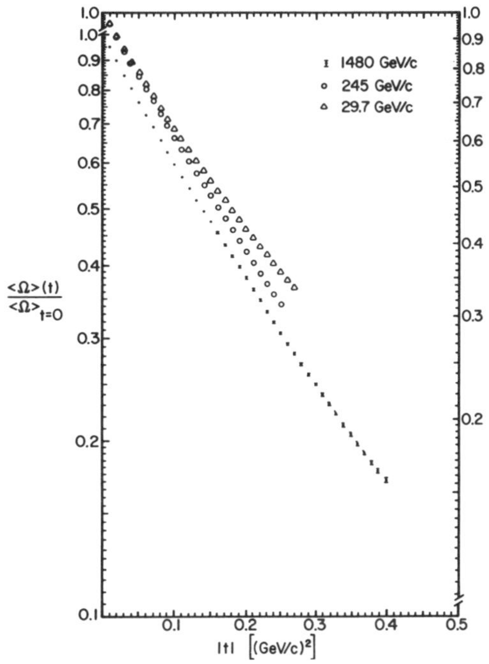

FIG. 1. The normalized opaqueness $\langle\Omega\rangle$ in momentum space for three different momenta: $p_L = 29.7$, 245, and 1480 GeV/c. They are normalized to unity at the origin. The triangles, open circles, and black dots represent the opaqueness for 29.7, 245, and 1480 GeV/c, respectively. Error bars are shown only for the last curve. The magnitude of the omitted error bars are comparable to those shown. The logarithmic scale on the left side is for the 1480-GeV/c case while that on the right-hand side is for the two lower energies. The fit used for the 29.7-GeV/c data is

$$a(t) = [7.89 \pm 0.08] \times \{(0.71 \pm 0.01)\exp[-(7.36 \pm 0.1)|t|] + [(1-(0.71 \pm 0.01)] \times \exp[-(2.36 \pm 0.1)|t|]\}\ (\mathrm{GeV}/c)^{-2}$$

which is very good for $|t| \le 0.3$ (GeV/c)^2. The experimental data for $(d\sigma/dt)/(d\sigma/dt)_{t=0}$ at 29.7 GeV/c are those of Edelstein *et al.*, Phys. Rev. D 5, 1073 (1972), normalized by $\sigma_{\mathrm{tot}} = 38.6 \pm 0.4$ mb. The experimental data for $(d\sigma/dt)/(d\sigma/dt)_{t=0}$ at 245 and 1480 GeV/c are those of Barbiellini *et al.* (Ref. 6) normalized by $\sigma_{\mathrm{tot}} = 39.1 \pm 0.4$ mb and 43.2 ± 0.6 mb, respectively.

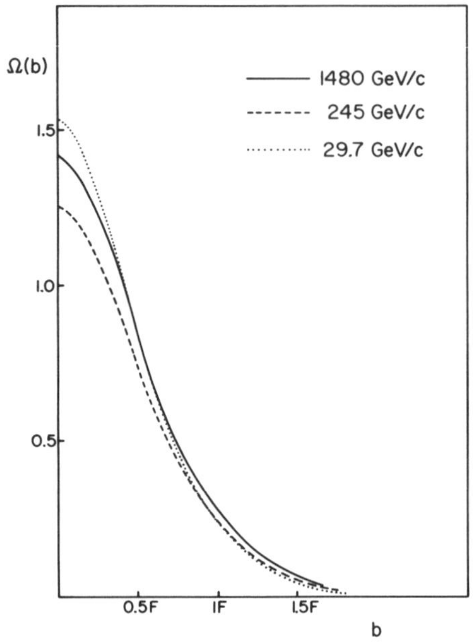

FIG. 2. The opaqueness [in units of $(\mathrm{GeV}/c)^{-2}$] in coordinate space for three different momenta: $p_L=29.7$, 245, 1480 GeV/c. The value of $\Omega(b=0)$ is quite sensitive to $a(t)$ at large $|t|$, therefore these curves are not accurate.

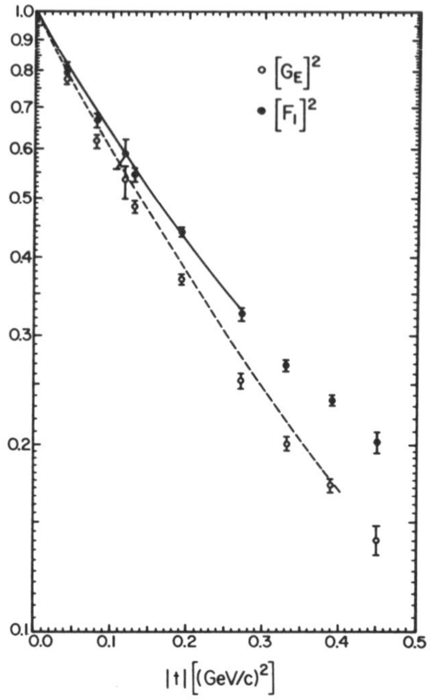

FIG. 3. The proton form factors $[F_1(t)]^2$ and $[G_E(t)]^2$ normalized to unity at $t=0$. Also drawn for comparison are two curves copied from Fig. 1. The solid curve is $\langle\Omega\rangle/\langle\Omega\rangle_{t=0}$ for 29.7 GeV/c and the dashed curve the same for 1480 GeV/c. Data for the form factors are taken from L. E. Price *et al.*, Phys. Rev. D 4, 45 (1971).

II. RANGE OF SOME PARAMETERS

Three of the most important experimental parameters are the total cross section σ_T, the total elastic cross section σ_{el}, and the slope parameter:

$$B=\left|\frac{d}{dt}\ln\left(\frac{d\sigma}{dt}\right)\right| \quad \text{at } t=0\,. \tag{12}$$

All three are of the dimension (length)^2. We define the dimensionless ratios

$$X=\frac{\sigma_{el}}{\sigma_T}, \quad Y=\frac{\sigma_T}{16\pi B}\,.$$

Experimental values of X and Y are listed in Table II for various energies. The close equality of X and Y is a reflection of the empirical fact that $\ln d\sigma/dt$ has almost a linear dependence on t. (A strict linear dependence means that $1-S$ is Gaussian in b. See the Gaussian model in Table III.)

Under the assumption that $\Omega(b)=\text{real}\geqslant 0$ one has

$$\sigma_T=4\pi\int_0^\infty (1-S)b\,db\,, \tag{13}$$

$$\sigma_{el}=2\pi\int_0^\infty (1-S)^2 b\,db\,, \tag{14}$$

$$B=\int_0^\infty (1-S)b^3db\left[2\int_0^\infty (1-S)b\,db\right]^{-1} \tag{15}$$

Thus

$$B=\tfrac{1}{2}[\text{average of } b^2 \text{ with weight } (1-S)]\,. \tag{16}$$

Table III lists for some models the values of these parameters. Notice that X and Y are range-independent. I.e., they are not changed by the transformation $S(b)\to S(cb)$ where $c=\text{constant}$. X and Y therefore are parameters characteristic of the *shape* of the function $1-S(b)$ vs b. If one increases $1-S(b)$ by a uniform factor α, X and Y both increase by the same factor α. Thus, qualitatively, transparent scatterings are indicated by small values of X and Y and opaque scatterings are indicated by large values of X and Y. Also "compact" scatterings, as in a gray-disk model, or a Gaussian model, are associated with large ratios Y/X while "noncompact" scatterings, as in a two-tiered-platform model with a large and low lower tier, are associated with a small ratio Y/X.

These qualitative features are indicated in Fig. 4. We notice that pp scattering in the $p_L\cong 5$–1500-

TABLE II. X and Y as functions of p_L. The data are taken from (and interpolated): *NN* and *ND* Interactions-Berkeley Compilation (1970), Ref. 1; S. P. Denisov *et al.*, Phys. Lett. 36B, 415 (1971); V. Bartenev *et al.*, report, 1972 (unpublished); and G. G. Beznogikh *et al.*, Phys. Lett. 39B, 411 (1972).

p_L (GeV/c)	X	Y
2.8	0.42	0.39
4.0	0.318	0.312
5.0	0.302	0.283
5.5	0.297	0.270
6.0	0.294	0.258
7.0	0.281	0.254
8.0	0.278	0.246
9.0	0.275	0.240
11.0	0.270	0.230
15	0.254	0.210
20	0.238	0.198
25	0.226	0.191
30	0.220	0.187
55	0.196	0.174
100	0.178	0.172
200	0.175	0.170
300	0.173	0.170
500	0.172	0.169
1200	0.176	0.170
1500	0.176	0.169

GeV/c region is relatively transparent. In fact the transmission coefficient (of the amplitude) for even a head-on collision is still sizable at p_L = 1500 GeV/c:

$$S^{-\Omega(0)} = e^{-1.4} = 0.25\,.$$

It is also quite compact, resembling very much a Gaussian, as Fig. 1 indicates.

Equations (13) to (15) and the condition $0 \leq S \leq 1$

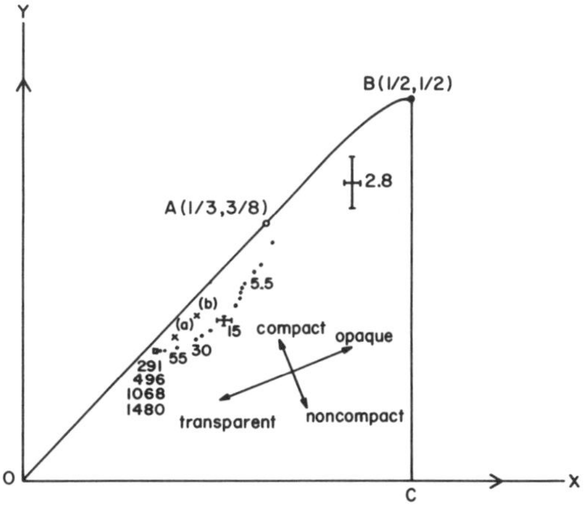

FIG. 4. The allowed region of the dimensionless variables X and Y defined in the text, OA, OC, BC are straight lines, and AB has a parametric form given in the text. Qualitative features "opaque," "transparent," "compact," and "noncompact" refer to the shape of $1-S(b)$. Also drawn in this figure are experimental values from Table II. Representative error bars have been drawn for $p_L = 2.8$ and 15 GeV/c. The square labeled by $p_L = 291$, 496, 1068, 1480 GeV/c is the point corresponding to the four ISR data in Ref. 1. The crosses labeled by a and b are calculated for collisions in which $\sigma_T = 50$ and 60 mb, respectively, under the assumption that the *shape* of $\Omega(b)$ remains the same as that for $p_L = 1480$ GeV/c, at which $\sigma_T = 43$ mb.

show that not all of the X-Y plane is allowed. We shall assume $S(b)$ to be piecewise differentiable. The allowed region is shown in Fig. 4.

The boundaries BC and OC are obvious.

To prove that OA is a boundary we use the follow-

TABLE III. X and Y for various models.

Model	$1-S$	σ_T	σ_{el}	$16\pi B$	X	Y
gray disk	α if $b<R$, 0 if $b>R$ $(0\leq\alpha\leq 1)$	$2\pi\alpha R^2$	$\pi\alpha^2 R^2$	$4\pi R^2$	$\frac{1}{2}\alpha$	$\frac{1}{2}\alpha$
Gaussian	$\alpha e^{-b^2/R^2}$ $(0\leq\alpha\leq 1)$	$2\pi\alpha R^2$	$\frac{1}{2}\pi\alpha^2 R^2$	$8\pi R^2$	$\frac{1}{4}\alpha$	$\frac{1}{4}\alpha$
two-tiered platform	α if $b<R$, $\alpha\beta$ if $R<b<(1+\gamma)^{1/2}R$, 0 if $b>(1+\gamma)^{1/2}R$ $(0\leq\alpha\leq 1,\ 0\leq\gamma,\ 0\leq\beta\leq 1)$	$2\pi\alpha(1+\gamma\beta)R^2$	$\pi\alpha^2(1+\gamma\beta^2)R^2$	$4\pi\frac{1+2\beta\gamma+\beta\gamma^2}{1+\beta\gamma}R^2$	$\frac{1}{2}\alpha\frac{1+\gamma\beta^2}{1+\gamma\beta}$	$\frac{1}{2}\alpha\frac{(1+\beta\gamma)^2}{1+2\beta\gamma+\beta\gamma^2}$
truncated parabola	α if $b<\gamma R$, $\alpha(b^2/R^2-1)(\gamma^2-1)^{-1}$ if $\gamma R<b<R$, 0 if $b>R$ $(0\leq\alpha\leq 1,\ 0\leq\gamma\leq 1)$	$\pi\alpha(1+\gamma^2)R^2$	$\pi\alpha^2(\frac{1}{3}+\frac{2}{3}\gamma^2)R^2$	$\frac{8}{3}\pi\frac{1+\gamma^2+\gamma^4}{1+\gamma^2}R^2$	$\frac{1}{3}\alpha\frac{1+2\gamma^2}{1+\gamma^2}$	$\frac{3}{8}\alpha\frac{(1+\gamma^2)^2}{1+\gamma^2+\gamma^4}$

ing transformations:

$$b^2 = \frac{\sigma_T}{2\pi} z, \quad \int_0^{b^2} (1-S)d(b^2) = \frac{\sigma_T}{2\pi}\xi . \tag{17}$$

Then

$$X = \tfrac{1}{2}\int_0^1 \frac{d\xi}{(dz/d\xi)},$$

$$Y = \frac{1}{4\int_0^1 z\,d\xi} = \frac{1}{4\int_0^1 (1-\xi)(dz/d\xi)d\xi}, \tag{18}$$

$$1 - S = \frac{d\xi}{dz}. \tag{19}$$

Using Schwarz's inequality one obtains from (18)

$$XY^{-1} \geq 2\left[\int_0^1 (1-\xi)^{1/2} d\xi\right]^2 = \tfrac{8}{9}. \tag{20}$$

This shows that all points are to the right of the line OA. This is the same restriction as one obtains at high energies from the MacDowell-Martin bound,[7] if one neglects the real part of the near forward amplitude and the spin dependence of B.

That AB is a boundary can be obtained by a variational calculation for minimizing (XY^{-n}), where $n \geq 1$. One obtains the minimum

$$XY^{-n} \geq (\tfrac{1}{2} - \tfrac{1}{6}y_0)(2 + \tfrac{2}{3}y_0^2)^n, \tag{21}$$

where y_0 is related to n by

$$n = -\frac{1}{2} + \frac{1}{2y_0} + \frac{2}{3 - y_0} \quad (0 < y_0 \leq 1). \tag{22}$$

The minimum (21) is realized by the truncated parabola model of Table II at

$$\gamma = \frac{1 - y_0}{1 + y_0} \quad \text{and } \alpha = 1 .$$

The envelope of the boundary curves (21) is AB, which is parametrically

$$X = \tfrac{1}{2} - \tfrac{1}{6}y_0, \quad Y = (2 + \tfrac{2}{3}y_0^2)^{-1}, \quad 0 \leq y_0 \leq 1 . \tag{23}$$

Are all points inside and on the curve $OABCO$ realized by some model for which $0 \leq S \leq 1$?

The answer is yes. But the line OC can only be realized if the integral for B in (15) is divergent. Also the line BC can only be realized by a model where $S = 0$ or 1 everywhere. (I.e., for a collection of black rings with or without a central black disk.) If B is assumed to be convergent and $1 - S(b)$ is assumed to be nonincreasing with increasing b, then the lines OC and BC cannot be realized except for the point B. In such a case only the open region $OABCO$, the open curve OAB and the point B can be realized.

We shall not give the detailed proof of these statements. Suffice it to mention the following two observations which are helpful: (a) Starting from any given model, replacing $1 - S$ by $\alpha(1 - S)$ leads to a new model at αX, αY. (b) Starting from any model one can add a very large but very transparent wing. If the area of the wing is A and the value of $1 - S$ on the wing is ϵ, then fixing ϵA^2 but making $A \to \infty$ would lead to no change in σ_T and σ_{el} but possible finite addition to B.

ACKNOWLEDGMENT

We would like to thank Dr. Chris Quigg for discussions and for information about experimental data.

*Supported in part by NSF Grant No. GP32998X.

[1]U. Amaldi, R. Biancastelli, C. Bosio, G. Matthiae, J. V. Allaby, W. Bartel, G. Cocconi, A. N. Diddens, R. W. Dobinson, and A. M. Wetherell, Phys. Lett. 44B, 112 (1973); S. R. Amendolia, G. Bellettini, P. L. Braccini, C. Bradaschia, R. Castaldi, V. Cavasinni, C. Cerri, T. Del Prete, L. Foa, P. Giromini, P. Laurelli, A. Menzione, L. Ristori, G. Sanguinetti, M. Valdata, G. Finocchiaro, P. Grannis, D. Green, R. Mustard, and R. Thun, Phys. Lett. 44B, 119 (1973).

[2]Hung Cheng and Tai Tsun Wu, Phys. Rev. Lett. 24, 1456 (1970), and references quoted therein.

[3]Hung Cheng, James K. Walker, and Tai Tsun Wu, report (unpublished).

[4]T. T. Chou and Chen Ning Yang, Phys. Rev. 170, 1591 (1968).

[5]T. T. Chou and Chen Ning Yang, in *Proceedings of the Second International Conference on High Energy Physics and Nuclear Structure, Rehovoth, Israel, 1967*, edited by G. Alexander (North-Holland, Amsterdam, The Netherlands, 1967).

[6]G. Barbiellini, M. Bozzo, P. Darriulat, G. Diambrini Palazzi, G. DeZorzi, A. Fainberg, M. I. Ferrero, M. Holder, A. McFarland, G. Maderni, S. Orito, J. Pilcher, C. Rubbia, A. Santroni, G. Sette, A. Staude, P. Strolin, and K. Tittel, Phys. Lett. 39B, 663 (1972).

[7]S. W. MacDowell and A. Martin, Physics (N.Y.) 135B, 960 (1964).

Integral Formalism for Gauge Fields

[74c] Commentary begins page 73

C. N. Yang
Institute of Theoretical Physics, University of Wrocław, Wrocław, Poland, and
Institute of Theoretical Physics, State University of New York, Stony Brook, New York 11990
(Received 10 June 1974)

A new integral formalism for gauge fields is described. Further developments are presented, including gravitation equations related to, but not identical with, Einstein's equations.

It was pointed out by Weyl many years ago that the electromagnetic field can be formulated in terms of an Abelian gauge transformation. This idea was extended[1] in 1954 to the concept of gauge fields for non-Abelian groups. That formulation, like the Weyl formulation for electromagnetism, was based on the replacement of ∂_μ by $\partial_\mu - ieB_\mu$. One might call such formulations differential formulations. It is the purpose of the present paper to reformulate the concept of gauge fields in an *integral formalism.* The new formalism is conceptually superior to the differential formalism and allows for natural developments of additional concepts. It further allows a mathematical and physical discussion of the gravitational field *as a gauge field,* resulting in equations related, but not identical, to Einstein's.

The basic point is the fact that *electromagnetism is a nonintegrable phase factor,* a fact discussed many years ago by Dirac, Peierls, and others, and more recently by many authors.[2] This fact is now generalized as follows:

Definition of a gauge field.—Consider a manifold with points on it labeled by x^μ ($\mu = 1, 2, \ldots, n$) and consider a gauge G which is a Lie group with generators X_k ($k = 1, 2, \ldots, m$). [For $G = U(1)$ we have electromagnetism; for G non-Abelian we have non-Abelian gauge fields.] Define a path-dependent (i.e., nonintegrable) phase factor φ_{AB} as an element of the group G associated with path AB between two points A and B on the manifold. The association is to have the group property: $\varphi_{ABC} = \varphi_{AB}\varphi_{BC}$, where the paths AB and BC are segments of ABC. Furthermore for an infinitiesimal path A to $A + dx^\mu$ the phase factor is close to the identity I of G, so that[3]

$$\varphi_{A(A+dx)} = I + b_\mu{}^k(x) X_k\, dx^\mu . \tag{1}$$

The function $b_\mu{}^k(x)$ defined on the manifold will be called a *gauge potential;* φ_{AB} will be called a *gauge phase factor.*

With this definition additional concepts and theorems are naturally developed. We summarize some of these below. Details will be published elsewhere.

Gauge field strength.—Consider a path $ABCDA$ forming the border of an infinitesimal parallelogram with sides dx and dx'. φ_{ABCDA} can be computed by multiplying four phase factors like (1) together, resulting in

$$\varphi_{ABCDA} = I + f_{\mu\nu}{}^k X_k\, dx^\mu dx^{\nu\prime} , \tag{2}$$

where

$$f_{\mu\nu}{}^k = \frac{\partial b_\mu{}^k}{\partial x^\nu} - \frac{\partial b_\nu{}^k}{\partial x^\mu} - b_\mu{}^i b_\nu{}^j C_{ij}{}^k = -f_{\nu\mu}{}^k \tag{3}$$

in which $C_{ij}{}^k$ is the structure constant of G:

$$X_k X_j - X_j X_k = C_{kj}{}^i X_i . \tag{4}$$

$f_{\mu\nu}{}^k$ will be called a *gauge field,* or gauge field strength. They are the Faraday-Maxwell fields when $G = U(1)$.

Gauge transformation.—A gauge transformation in the integral formalism is defined by a transformation

$$\varphi_{AB} \to \varphi_{AB}' = \xi_A \varphi_{AB} \xi_B{}^{-1} , \tag{5}$$

where ξ_A is an element of G which depends on the point A. It is clear that under (5)

$$\varphi_{ABCDA} \to \varphi_{ABCDA}' = \xi_A \varphi_{ABCDA} \xi_A{}^{-1} . \tag{6}$$

Thus

$$f_{\mu\nu}{}^{k\prime} = \langle k | R_{\mathrm{adj}} | j \rangle f_{\mu\nu}{}^j , \tag{7}$$

where R_{adj} is the adjoint representation for the element ξ_A. The simple transformation property (7) is the definition for the concept that $f_{\mu\nu}{}^k$ is *gauge covariant.* Generalization to other representations R of G for a gauge-covariant quantity $\psi_{\alpha\beta\gamma}{}^K$ is immediate[3]:

$$\psi_{\alpha\beta\gamma}{}^{K\prime} = \langle K | R(\xi_A) | J \rangle \psi_{\alpha\beta\gamma}{}^J . \tag{8}$$

$b_\mu{}^k$ is not gauge covariant; $f_{\mu\nu}{}^k$ is.

Gauge-covariant differentiation.—To *retain*

gauge covariance in differentiation we define

$$\psi^K|_\mu = \frac{\partial \psi^K}{\partial x^\mu} + b_\mu{}^k \langle K|Z_k|J\rangle \psi^J , \tag{9}$$

where Z_k is the matrix representation of X_k. Generalization to other cases is obvious. An interesting theorem is that

$$f_{\mu\nu}|_\lambda{}^k + f_{\nu\lambda}|_\mu{}^k + f_{\lambda\mu}|_\nu{}^k = 0 , \tag{10}$$

which is the gauge-Bianchi identity.

Introduction of a Riemannian metric.—So far we need no metric for the manifold. Now we introduce a metric for it and discuss arbitrary coordinate transformations. We come then naturally to *Riemannian covariant* quantities and *doubly covariant derivatives*. $b_\mu{}^k$ is Riemannian covariant, since φ_{AB} is coordinate-system independent. $f_{\mu\nu}{}^k$ is doubly covariant. We have

$$\begin{aligned} \psi^K{}_{\parallel\mu} &= \psi^K|_\mu , \\ \psi^{K\nu}{}_{\parallel\mu} &= \psi^{K\nu}|_\mu + \left\{ {\nu \atop \mu\alpha} \right\} \psi^{K\alpha} , \\ f_{\mu\nu\parallel\lambda}{}^k &= f_{\mu\nu}|_\lambda{}^k - \left\{ {\alpha \atop \mu\lambda} \right\} f_{\alpha\nu}{}^k - \left\{ {\alpha \atop \nu\lambda} \right\} f_{\mu\alpha}{}^k , \end{aligned} \tag{11}$$

etc. It is easily shown that

$$f_{\mu\nu\parallel\lambda}{}^k + f_{\nu\lambda\parallel\mu}{}^k + f_{\lambda\mu\parallel\nu}{}^k = 0 \tag{12}$$

which is satisfied by *all* gauge fields on *all* Riemannian manifolds.

Source of gauge fields.— We *define*, in analogy with electromagnetism, a source four-vector $J_\mu{}^k$ for a gauge field:

$$J_\mu{}^k = g^{\nu\lambda} f_{\mu\nu\parallel\lambda}{}^k = f_{\mu\nu}{}^{k\parallel\nu} . \tag{13}$$

After some computation one derives a theorem:

$$g^{\mu\lambda} J_{\mu\parallel\lambda}{}^k = 0 \quad \text{(conserved current)} , \tag{14}$$

which in electromagnetism states charge conservation. In Ref. 1 this was Eq. (14). One can also generalize Eqs. (15) and (16) of Ref. 1, leading to the concept of "total charge."

Parallel-displacement gauge field.—For any Riemannian manifold, the important concept of parallel displacement defines, along any path AB, a *linear* relationship between any vector V_A at A and its parallel vector V_B at B. Thus parallel displacement is defined by an $n\times n$ matrix M_{AB} which gives this linear relationship. M_{AB} is a representation of an element of GL(n). Thus we have the following:

Theorem.—Parallel displacement defines a gauge field with G being GL(n). The index k has n^2 values and we write $k = (\alpha\beta)$. The gauge potential and gauge fields are respectively

$$b_\mu{}^{(\alpha\beta)} = \left\{ {\alpha \atop \beta\mu} \right\}, \quad f^{(\alpha\beta)}{}_{\mu\nu} = -R^\alpha{}_{\beta\mu\nu} . \tag{15}$$

It is important to recognize that in this definition we have chosen a fixed coordinate system. A coordinate transformation would generate a linear transformation in the vector spaces V_A and V_B. In other words $M_{AB} \to N_A M_{AB} N_B{}^{-1}$. Comparison with (5) shows thus that a coordinate transformation generates a simultaneous gauge transformation of the parallel-displacement gauge potential. In fact, the usual nonlinear term in the transformation of $\{ {\alpha \atop \alpha\gamma} \}$ is precisely the nonlinear term needed in the gauge transformation of the gauge noncovariant quantity $b_\mu{}^{(\alpha\beta)}$. In this connection we observe that for GL(n),

$$C^{(\alpha\beta)}_{(\lambda\mu)(\eta\zeta)} = \delta_{\mu\eta}\varepsilon_{\alpha\lambda}\delta_{\beta\zeta} - \delta_{\lambda\zeta}\delta_{\alpha\eta}\delta_{\beta\mu} . \tag{16}$$

Thus by definitions (9) and (11)

$$\begin{aligned} \psi^{(\alpha\beta)}{}_{\parallel\mu} &= \frac{\partial \psi^{(\alpha\beta)}}{\partial x^\mu} + b_\mu{}^{(\lambda\nu)} C^{(\alpha\beta)}_{(\lambda\nu)(\eta\zeta)} \psi^{(\eta\zeta)} \\ &= \psi^\alpha{}_{\beta;\mu} , \end{aligned} \tag{17}$$

where the semicolon represents the usual Riemannian covariant differentiation with α and β treated as usual contravariant and covariant indices. The rule works also in general. E.g.,

$$f^{(\alpha\beta)}{}_{\mu\nu\parallel\lambda} = -R^\alpha{}_{\beta\mu\nu;\lambda} . \tag{18}$$

Nontrivial sourceless gauge fields.—Gauge fields for which $f_{\mu\nu}{}^k \neq 0$ and $J_\mu{}^k = 0$ are of physical interest. So far only nonanalytic examples are known.[4]

We now can construct two general types of general types of examples.

(a) Consider the natural Riemannian geometry of a semisimple Lie group. Its parallel-displacement gauge field is sourceless and analytic.

(b) Consider the same Riemannian manifold of a group G as above in (a). Define φ_{AB} as that for an infinitesimal path AB, $\varphi_{AB} = (A^{-1}B)^{1/2}$. This gauge phase factor which is itself an element of G gives a gauge field which is analytic and sourceless.

Pure spaces.—A Riemannian manifold for which the parallel-displacement gauge field is sourceless will be called a pure space. A necessary and sufficient condition for a pure space is

$$R_{\mu\alpha;\beta} = R_{\mu\beta;\alpha} \tag{19}$$

A four-dimensional Einstein space, i.e., one for

which $R_{\alpha\beta} = 0$, is a pure space.

Gravitational field as a gauge field.—The electromagnetic field and the usual gauge fields are special cases of gauge fields, satisfying (12) and (13). A natural question is whether one should identify these *same equations* for the parallel-displacement gauge field as the equations for the gravitational field. There are advantages in this identification and we shall come back to this topic in a later communication. If one adopts this identification then gravitational equations are third-order differential equations[5] for $g_{\mu\nu}$. A pure gravitational field is then described by a pure space as defined above.

Variational principles.—Equation (13) with $J_\mu^{\ k} = 0$ follows from a variational principle $\delta\int \sqrt{-g}\, d^n x = 0$, where

$$L = f_{\mu\nu}^{\ \ k} f_{\alpha\beta}^{\ \ j} g^{\mu\alpha} g^{\nu\beta} C_{ka}^{\ \ b} C_{jb}^{\ \ a}. \quad (20)$$

In the variation $g_{\mu\nu}$ is kept fixed and $b_\mu^{\ k}$ is varied, and $f_{\mu\nu}^{\ \ k}$ is given by (3); $C_{ka}^{\ \ b}$ are not varied. One could also find a variational principle which is satisfied by a pure space (19). Choose $C_{ka}^{\ \ b}$ to be the structure constants for GL(n), given by (16). Write the L of (20) as a functional of $b_\mu^{\ (\alpha\beta)}$ and $g^{\lambda\nu}$:

$$L = L(b_\mu^{\ (\alpha\beta)}, g^{\lambda\nu}), \quad (21)$$

which of course also contains derivatives of $b_\mu^{\ k}$ and $g^{\lambda\nu}$. Now form the variation

$$\delta\int\left[L(b_\mu^{\ (\alpha\beta)}, g^{\lambda\nu}) - L\left(\left\{ {\alpha \atop \beta\mu} \right\}, g^{\lambda\nu}\right)\right] \times \sqrt{-g}\, d^n x = 0, \quad (22)$$

in which $b_\mu^{\ (\alpha\beta)}$ and $g^{\lambda\nu}$ are independently varied. The resultant equations are satisfied by (15) and (19).

It is a great pleasure to acknowledge the warm hospitality extended to me during my visit to the Institute of Theoretical Physics at Wrocław where this paper was written.

[1]C. N. Yang and R. L. Mills, Phys. Rev. 96, 191 (1954).

[2]S. Mandelstam, Ann. Phys. (New York) 19 1, 25 (1962); I. Białynicki-Birula, Bull. Acad. Pol. Sci., Ser. Sci. Math. Astron. Phys. 11, 135 (1963).

[3]We use the summation convention for repeated indices. Greek indices run from 1 to n. Lower case Latin indices run from 1 to m. m of course is also the dimension of the adjoint representation of G. Upper case Latin indices run from 1 to M, where M is the dimension of a representation of G.

[4]T. T. Wu and C. N. Yang, in *Properties of Matter under Unusual Conditions*, edited by H. Mark and S. Fernbach (Wiley, New York, 1969), p. 349.

[5]R. Utiyama, Phys. Rev. 101, 1957 (1956), had concluded that Einstein's equations are gauge-field equations. We believe that was an unnatural interpretation of gauge fields.

[75c]
Commentary begins page 75

Concept of nonintegrable phase factors and global formulation of gauge fields

Tai Tsun Wu*
Gordon McKay Laboratory, Harvard University, Cambridge, Massachusetts 02138

Chen Ning Yang†
Institute for Theoretical Physics, State University of New York, Stony Brook, New York 11794
(Received 8 September 1975)

Through an examination of the Bohm-Aharonov experiment an intrinsic and complete description of electromagnetism in a space-time region is formulated in terms of a nonintegrable phase factor. This concept, in its global ramifications, is studied through an examination of Dirac's magnetic monopole field. Generalizations to non-Abelian groups are carried out, and result in identification with the mathematical concept of connections on principal fiber bundles.

I. MOTIVATION AND INTRODUCTION

The concept of the electromagnetic field was conceived by Faraday and Maxwell to describe electromagnetic effects in a space-time region. According to this concept, the field strenght $f_{\mu\nu}$ describes electromagnetism. It was later realized,[1] however, that $f_{\mu\nu}$ by itself does not, in quantum theory, completely describe all electromagnetic effects on the wave function of the electron. The famous Bohm-Aharonov experiment, first beautifully performed by Chambers,[2] showed that in a multiply connected region where $f_{\mu\nu}=0$ everywhere there are physical experiments for which the outcome depends on the loop integral

$$\frac{e}{\hbar c}\oint A_\mu dx^\mu \qquad (1)$$

around an unshrinkable loop. This raises the question of what constitutes *an intrinsic and complete description* of electromagnetism. In the present paper we wish to discuss this question and also its generalization to non-Abelian gauge fields.

An examination of the Bohm-Aharonov experiment indicates that in fact only *the phase factor*

$$\exp\left(\frac{ie}{\hbar c}\oint A_\mu dx^\mu\right), \qquad (2)$$

and *not the phase* (1), is physically meaningful. In other words, the phase (1) contains more information than the phase factor (2). But the additional information is not measurable. This simple point, probably implicitly recognized by many authors, is discussed in Sec. II. It leads to the concept of nonintegrable (i.e., path-dependent) phase factor as the basis of a description of electromagnetism.

This concept has been taken[3] as the basis of the definition of a gauge field. The discussions in Ref. 3, however, centered only on the local properties of gauge fields. To extend the concept to global problems we analyze in Sec. III the field produced by a magnetic monopole. We demonstrate how the quantization of the pole strength, a striking result due to Dirac,[4] is understood in this concept of electromagnetism. The demonstration is closely related to that in the original Dirac paper. Dirac discussed the phase factor of the wave function of an electron (which, among other things, depends on the electron energy). Our emphasis is on the nonintegrable electromagnetic phase factor (which does not depend on such quantities as the energy of the electron).

The monopole discussion leads to the recognition that in general the phase factor (and indeed the vector potential A_μ) can only be properly defined in each of many overlapping regions of space-time. In the overlap of any two regions there exists a gauge transformation relating the phase factors defined for the two regions. This discussion is made more precise in Sec. IV. It leads to the definition of global gauges and global gauge transformations.

In Sec. V generalizations to non-Abelian gauge groups are made. The special cases of SU_2 and SO_3 gauge fields are discussed in Secs. VI and VII. A surprising result is that the monopole types are quite different for SU_2 and SO_3 gauge fields and for electromagnetism.

The mathematics of these results is in fact well known to the mathematicians in *fiber bundle theory*. An identification table of terminologies is given in Sec. V. We should emphasize that our interest in this paper does not lie in the beautiful, deep, and general mathematical development in fiber bundle theory. Rather we are concerned with the necessary *concepts to describe the physics of gauge theories*. It is remarkable that these concepts have already been intensively studied as mathematical constructs.

Section VII discusses a "*gedanken*" generalized Bohm-Aharonov experiment for SU_2 gauge fields.

□ Reprinted from *Physical Review D* 12, 12 (December 15, 1975), 3845–3857.

Unfortunately, the experiment is not feasible unless the mass of the gauge particle vanishes. In the last section we make several remarks.

II. DESCRIPTION OF ELECTROMAGNETISM

The Bohm-Aharonov experiment explores the electromagnetic effect on an electron beam (Fig. 1) in a doubly connected region where the electromagnetic field is zero. As predicted[1] by Aharonov and Bohm, the fringe shift is dependent on the phase factor (2), which is equal to

$$\exp\left(\frac{-ie}{\hbar c}\Omega\right),$$

where Ω is the magnetic flux in the cylinder. Thus two cases a and b for which

$$\Omega_a - \Omega_b = \text{integer} \times (\hbar c/e) \quad (3)$$

give the same interference fringes in the experiment. This we shall state and prove as follows.

Theorem 1: If (3) is satisfied, no experiment outside of the cylinder can differentiate between cases a and b.

Consider first an electron outside of the cylinder. We look for a gauge transformation on the electron wave function ψ_a and the vector potential $(A_\mu)_a$ for case a, which changes them into the corresponding quantities for case b, i.e. we try to find $S = e^{-i\alpha}$ such that

$$S = S_{ab} = (S_{ba})^{-1},$$

$$\psi_b = S^{-1}\psi_a, \quad \text{or } \psi_b = e^{i\alpha}\psi_a, \quad (4)$$

$$(A_\mu)_b = (A_\mu)_a - \frac{i\hbar c}{e} S \frac{\partial S^{-1}}{\partial x^\mu}, \quad \text{or } (A_\mu)_b = (A_\mu)_a + \frac{\hbar c}{e}\frac{\partial \alpha}{\partial x^\mu}. \quad (5)$$

For this gauge transformation to be definable, S must be *single-valued*, but α itself need not be. Now $(A_\mu)_b - (A_\mu)_a$ is curlless; hence (5) can always be solved for α. But it is multiple-valued with an increment of

$$\Delta\alpha = \frac{e}{\hbar c}\oint [(A_\mu)_b - (A_\mu)_a]\, dx^\mu$$

$$= \frac{e}{\hbar c}(\Omega_b - \Omega_a) \quad (6)$$

every time one goes around the cylinder. If (3) is satisfied, $\Delta\alpha = 2\pi \times$ integer and S is single-valued. Case a and case b outside of the cylinder are then gauge-transformable into each other, and no physically observable effects would differentiate them. The same argument obviously holds if one studies the wave function of an interacting system of particles provided the charges of the particles are all integral multiples of e. Thus we have shown the validity of Theorem 1.

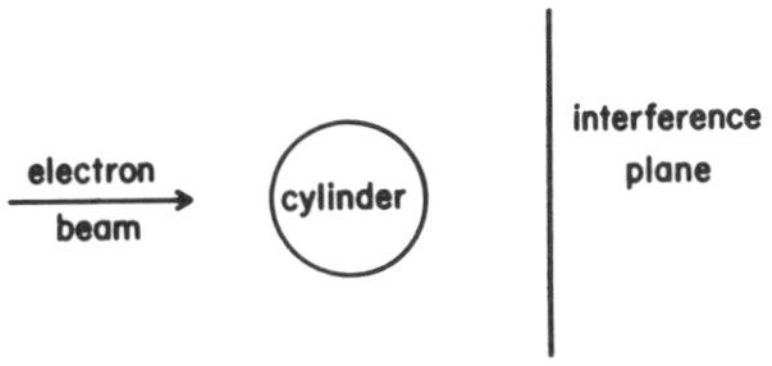

FIG. 1. Bohm-Aharonov experiment (Refs. 1, 2). A magnetic flux is in the cylinder. Outside of the cylinder the field strength $f_{\mu\nu} = 0$.

We conclude: (a) The field strength $f_{\mu\nu}$ underdescribes electromagnetism, i.e., different physical situations in a region may have the same $f_{\mu\nu}$. (b) The phase (1) overdescribes electromagnetism, i.e., different phases in a region may describe the same physical situation. What provides a complete description that is neither too much nor too little is the phase factor (2).

Expression (2) is less easy to use (especially when one makes generalizations to non-Abelian groups) as a fundamental concept than the concept of a phase factor for any path from P to Q

$$\Phi_{QP} = \exp\left(\frac{ie}{\hbar c}\int_P^Q A_\mu dx^\mu\right) \quad (7)$$

provided that an arbitrary gauge transformation

$$\exp\left(\frac{ie}{\hbar c}\int_P^Q A_\mu dx^\mu\right)$$

$$\to \exp\left(\frac{ie}{\hbar c}a(Q)\right)\exp\left(\frac{ie}{\hbar c}\int_P^Q A_\mu dx^\mu\right)\exp\left(\frac{-ie}{\hbar c}a(P)\right) \quad (8)$$

does not change the prediction of the outcome of any physical measurements. Following Ref. 3, we shall call the phase factor (7) a nonintegrable (i.e., path-dependent) phase factor.

Electromagnetism is thus the gauge-invariant manifestation of a nonintegrable phase factor. We shall develop this theme further in the next section.

III. FIELD DUE TO A MAGNETIC MONOPOLE

The definition of a nonintegrable phase factor (7) in a general case may present problems. To illustrate the problem, let us study the magnetic monopole field of Dirac.[4] Consider a static magnetic monopole of strength $g \neq 0$ at the origin $\vec{r} = 0$ and take the region R of space-time under consideration to be all space-time minus the origin $\vec{r} = 0$. We shall now show the following:

Theorem 2: There does not exist a singularity-free A_μ over all R.

If a singularity-free A_μ does exist throughout R, consider the loop integral $\oint A_\mu dx^\mu$ for time $t=0$ around a circle at fixed spherical coordinates r and θ with azimuthal angle $\phi = 0 \to 2\pi$. This integral, denoted by $\Omega(r,\theta)$ for $r>0$, is equal to the magnetic flux through a cap bounded by the loop, or more explicitly $\Omega(r,\theta) = 2\pi g(1-\cos\theta)$. At $\theta=0$, $\Omega(r,0)=0$. Increasing θ leads to a continuous increase in Ω till one approaches $\theta=\pi$, at which

$$\Omega(r,\pi) = 4\pi g\,. \tag{9}$$

But at $\theta=\pi$ the loop shrinks to a point. Therefore $\Omega(r,\pi)=0$ since A_μ has no singularity. We have thus reached a contradiction and Theorem 2 is proved.

With an A_μ which has singularities, the nonintegrable phase factor becomes undefined if the path goes through a singularity. This difficulty *must* be resolved in order to use a nonintegrable phase factor as a fundamental concept to describe electromagnetism. It can be resolved in the following way. Let us seek to divide R into two overlapping regions R_a and R_b and to define $(A_\mu)_a$ and $(A_\mu)_b$, each singularity-free in their respective regions, so that (i) their curls are equal to the magnetic field and (ii) in the overlapping region $(A_\mu)_a$ and $(A_\mu)_b$ are related by a gauge transformation. One possible choice is to take the regions to be

$$\begin{aligned} R_a:&\ 0\le\theta<\pi/2+\delta \quad 0<r,\quad 0\le\phi<2\pi,\ \text{all } t\\ R_b:&\ \pi/2-\delta<\theta\le\pi \quad 0<r,\quad 0\le\phi<2\pi,\ \text{all } t \end{aligned} \tag{10}$$

with an overlap extending throughout $\pi/2-\delta<\theta<\pi/2+\delta$. (We assume $0<\delta\le\pi/2$.) Take

$$\begin{aligned} (A_t)_a &= (A_r)_a = (A_\theta)_a = 0\,,\quad (A_\phi)_a = \frac{g}{r\sin\theta}(1-\cos\theta),\\ (A_t)_b &= (A_r)_b = (A_\theta)_b = 0\,,\quad (A_\phi)_b = \frac{-g}{r\sin\theta}(1+\cos\theta)\,. \end{aligned} \tag{11}$$

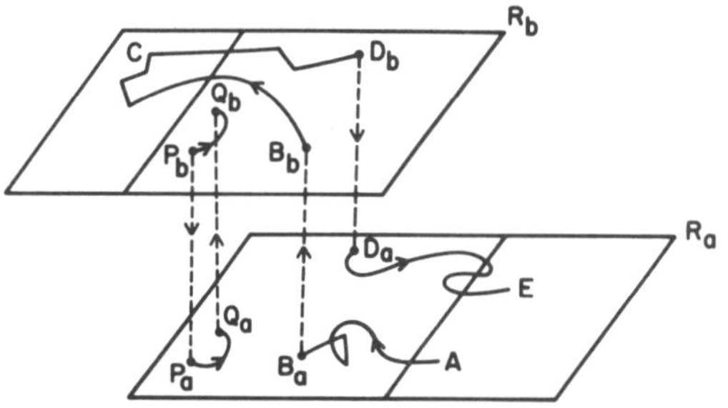

FIG. 2. Schematic diagram illustrating the relationship between R_a and R_b.

The gauge transformation in the overlap of the two regions is

$$S = S_{ab} = \exp(-i\alpha) = \exp\left(\frac{2ige}{\hbar c}\phi\right)\,. \tag{12}$$

This is an allowed gauge transformation if and only if S is single-valued, i.e.,

$$\frac{2ge}{\hbar c} = \text{integer} = D\,, \tag{13}$$

which is Dirac's quantization. With (13) we have

$$S_{ab} = \exp(iD\phi)\,. \tag{12'}$$

To define the phase factor for a path we refer to Fig. 2, where a point in the overlapping region, such as point P, is regarded as two points P_a and P_b. If a path is entirely within region a or b, we define Φ along the path by (7) with $(A_\mu)_a$ or $(A_\mu)_b$ in the integrand in the exponent. If the path $Q\leftarrow P$ is entirely within the overlapping region we have then two possible phase factors $\Phi_{Q_aP_a}$ and $\Phi_{Q_bP_b}$. It is easy to prove that

$$\Phi_{Q_bP_b} = S^{-1}(Q)\Phi_{Q_aP_a}S(P)\,, \tag{14}$$

i.e.,

$$\Phi_{Q_aP_a}S(P) = S(Q)\Phi_{Q_bP_b}\,, \tag{14'}$$

which merely states that $(A_\mu)_a$ and $(A_\mu)_b$ are related by a gauge transformation with the transformation factor (12).

For a path that crisscrosses in and out of the overlapping region, such as $A\to B\to C\to D\to E$ in Fig. 2, the definition of Φ is

$$\Phi_{EDCBA} = \Phi_{ED_a}S_{ab}(D)\Phi_{D_bCB_b}S_{ba}(B)\Phi_{B_aA}\,. \tag{15}$$

Notice that fixing the path but sliding the points B and D along it does not change Φ_{EDCBA} [because of formulas like (14')] so long as B and D remain in the overlapping region.

The phase factor so defined satisfies the group property, e.g.,

$$\begin{aligned} \Phi_{EDCBA} &= \Phi_{ED_a}\Phi_{D_aCBA}\\ &= \Phi_{ED_b}\Phi_{D_bCBA}\\ &= \Phi_{EDC}\Phi_{CBA}\,,\ \text{etc.} \end{aligned} \tag{16}$$

The relationship between the electromagnetic field and the phase factor around a loop is the same as usual. One only has to be careful that if the starting and terminating point A is in the overlapping region, the phase factor is taken to be $\Phi_{A_aBA_a} = \Phi_{A_bBA_b}$, and not $\Phi_{A_aBA_b}$ or $\Phi_{A_bBA_a}$. The phase factor around the loop is then equal to

$$\exp\left(\frac{ie}{\hbar c}\right)\Omega\,,$$

where Ω is the magnetic flux through a cap bordered by the loop. Notice that because of Dirac's quantization condition, the phase factor is the same whichever way one chooses the cap provided it does not pass through the point $\vec{r}=0$ (any t).

We have satisfactorily resolved the difficulty mentioned at the beginning of this section, provided Dirac's quantization condition (13) is satisfied. We shall now prove the following.

Theorem 3: If (13) is not satisfied (the above method of resolving the difficulty would not work since) there exists no division of R into overlapping regions $R_a, R_b, R_c, \ldots$ so that condition (i) and (ii) stated above, properly generalized to the case of more than two regions, would hold.

To prove this statement, observe that if such a division is possible, one could generalize (15) and arrive at a satisfactory definition of the phase factor. The phase factor around a loop is then a continuous function of the loop. Take the loop to be a parallel on the sphere r fixed, $t=0$, θ fixed, $\phi=0\to 2\pi$. The phase factor defined by the generalization of (15) is equal to

$$\exp\left[\frac{ie}{\hbar c}\Omega(r,\theta)\right]=\exp\left[\frac{ie}{\hbar c}2\pi g(1-\cos\theta)\right]. \qquad (17)$$

This is not equal to unity when $\theta=\pi$, since (13) is assumed to be invalid. Thus we have a contradiction.

Theorem 3 shows that if Dirac's quantization condition (13) is not satisfied, then the field of a magnetic monopole of strength g cannot be taken as a realizable physical situation in R. (Of course, if one excludes the half-line $x=y=0$, $z<0$, or any half-line starting from $\vec{r}=0$ leading to infinity, then it is possible to have any value for g.) This conclusion is the same as Dirac's, but viewed from a somewhat different point of emphasis.

IV. GENERAL DEFINITION OF GAUGE AND GLOBAL GAUGE TRANSFORMATION

Assuming that (13) holds, to round out our concept of a nonintegrable phase factor the question of the flexibility in the choice of the overlapping regions and the flexibility in the choice of A_μ in the regions must be faced. Both of these questions are related to gauge transformations.

Consider a gauge transformation ξ in R_b (ξ will be assumed to be many times differentiable, but not necessarily analytic), resulting in a new potential $(A_\mu)'_b$. We shall illustrate schematically the transformation by "elevating" the region b in Figure 3(a).

One could extend the region b. One could also contract it, provided the whole R remain covered.

One could create a new region by considering a subregion of b as an additional region R_c [Figure 3(b)], and define the gauge transformation connecting them as the identity transformation so that $(A_\mu)_c=(A_\mu)_b$. One can then "elevate" R_c and contract R_b, which results in Fig. 3(c).

Through operations of the kind mentioned in the last three paragraphs, which we shall call *distortions*, we arrive at a large number of possibilities, each with a particular choice of overlapping regions and with a particular choice of gauge transformation from the original $(A_\mu)_a$ or $(A_\mu)_b$ to the new A_μ in each region. Each of such possibilities will be called a *gauge* (or *global gauge*). This definition is a natural generalization of the usual concept, extended to deal with the intricacies of the field of a magnetic monopole.

For each choice of gauge there is a definition of a nonintegrable phase factor for every path. The group condition $\Phi_{C_cBA_a}=\Phi_{C_cB_b}\Phi_{B_bA_a}$ is always satisfied.

Notice that the original gauge we started with was characterized by (a) specifying [in (10)] the regions [R_a and R_b] and (b) specifying the gauge transformation factor (12′) in the overlap (between R_a and R_b). *It does not refer to any specific A_μ.* [A distortion may of course lead to no changes in characterizations (a) and (b). Thus two different gauges may share the same characterizations (a) and (b).] In the case of the monopole field, we had chosen the vector potential to be given by (11). But, in fact, we can attach to this gauge any $(A_\mu)_a$ and $(A_\mu)_b$ provided they are gauge-transformed into each other by (12′) in the region of overlap. (The resultant $f_{\mu\nu}$ is, of course, not a monopole field in general.) *Thus a gauge is a concept not tied to any specific vector potential.* We shall call the process of distortion leading from one gauge to another a *global gauge transformation.* It is also a concept not tied to any specific vector potential. It is a natural generalization of the usual gauge transformation.

The collection of gauges that can be globally gauge-transformed into each other will be said to

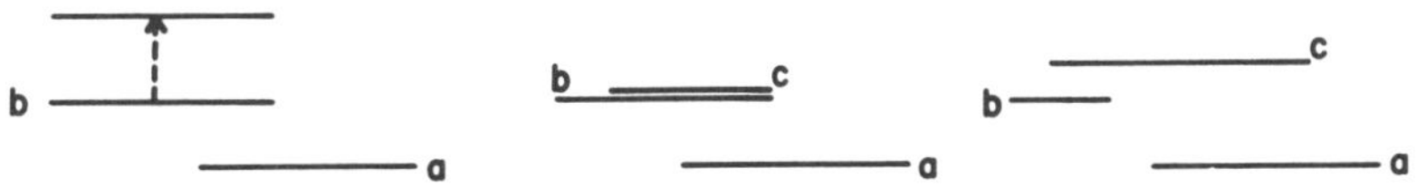

FIG. 3. Distortions allowed in gauge transformation.

belong to the same *gauge type*.

The phase factor around a loop starts and ends at the same point in the same region. Thus it does not change under any global gauge transformation, i.e. we have, for Abelian gauge fields, the following.

Theorem 4a: The phase factor around any loop is invariant under a global gauge transformation.

It follows trivially from this, by taking an infinitesimal loop, that

Theorem 5a: The field strength $f_{\mu\nu}$ is invariant under a global gauge transformation.

For a given value of D, the gauge defined by (10) and (12) will be denoted by $\mathcal{G}_D$. For $D \neq D'$, the relationship, or rather the lack of relationship, between $\mathcal{G}_D$ and $\mathcal{G}_{D'}$ is shown by Theorem 6.

Theorem 6: For $D \neq D'$, $\mathcal{G}_D$ and $\mathcal{G}_{D'}$ are not related by a global gauge transformation, i.e., they are not of the same gauge type.

To prove this theorem we use Theorem 7.

Theorem 7: Between two gauge fields defined on the same gauge there exists a continuous interpolating gauge field defined on the same gauge.

To prove Theorem 7, we simply make a linear interpolation between the two original gauge fields which we shall denote by $(A_\mu)^{(\alpha)}$ and $(A_\mu)^{(\beta)}$:

$$A^{(\gamma)} = t(A_\mu)^{(\alpha)} + (1-t)(A_\mu)^{(\beta)}, \quad 0 \le t \le 1. \tag{18}$$

In an overlap between regions a and b this interpolating vector potential assumes values $(A_\mu)_a^{(\gamma)}$ and $(A_\mu)_b^{(\gamma)}$ which are related by the proper gauge transformation belonging to this overlap. Thus we have proved Theorem 7.

Now go back to Theorem 6 and assume it to be invalid. Then we can gauge-transform the vector potential belonging to the monopole of strength $D'\hbar c/2e$ to the gauge $\mathcal{G}_D$. For this gauge we have then two monopole fields of different pole strengths. Using Theorem 7 we interpolate between them and obtain unquantized magnetic monopoles, which contradict Theorem 3.

Notice that although in this proof of Theorem 6 we have used two specific gauge fields, the theorem itself does not refer to any specific gauge fields at all.

By the same argument as used in the proof of Theorem 7, any gauge field defined on $\mathcal{G}_D$ must have a magnetic monopole of strength $D\hbar c/2e$ at the excluded point $\vec{r}=0$, in addition to possible fields produced by electric charges and currents. Thus the total magnetic flux around the origin $\vec{r}=0$ is equal to $(2\pi\hbar c/e)D$ for any gauge field defined on $\mathcal{G}_D$. We shall state this as a theorem and give another proof of it.

Theorem 8: Consider gauge $\mathcal{G}_D$ and define any gauge field on it. The total magnetic flux through a sphere around the origin $\vec{r}=0$ is *independent of the gauge field* and only depends on the gauge:

$$\oiint f_{\mu\nu}\, dx^\mu dx^\nu = \frac{-i\hbar c}{e} \oint \frac{\partial}{\partial x^\mu}(\ln S_{ab})\, dx^\mu, \tag{19}$$

where S is the gauge transformation defined by (12) for the gauge $\mathcal{G}_D$ in question, and the integral is taken around any loop around the origin $\vec{r}=0$ in the overlap between R_a and R_b, such as the equator on a sphere $r=1$.

To prove this theorem we observe that the flux through the upper half of the sphere $r=1$ is equal to the following integral around the equator:

$$\oint (A_\mu)_a dx^\mu. \tag{20a}$$

The flux through the lower half is equal to a similar integral around the equator:

$$-\oint (A_\mu)_b dx^\mu. \tag{20b}$$

Hence

$$\text{total flux} = \oint [(A_\mu)_a - (A_\mu)_b]\, dx^\mu$$
$$= \frac{-i\hbar c}{e} \oint \frac{\partial}{\partial x^\mu}(\ln S_{ab})\, dx^\mu, \tag{21}$$

which completes the proof. Using (13) and (12), the right-hand side of (21) is equal to $4\pi g$, as expected.

If one starts with any gauge which is of the same gauge type as $\mathcal{G}_D$, and makes a global gauge transformation on it, the total flux is not changed by Theorem 5a. Thus (19), which depends only on the gauge, is in fact the same for all gauges of the same type. Notice that if there are more regions in a gauge than two, (19) should be replaced by a sum of line integrals along paths that are in the various overlaps between the regions. For a case of three regions there are three paths, which are illustrated in Fig. 4. Along each path the integral is of the form (19) with S denoting the gauge transformation factor, such as (12), between the two regions containing the path. To prove Theorem 8 in this case one need only add three loop integrals to-

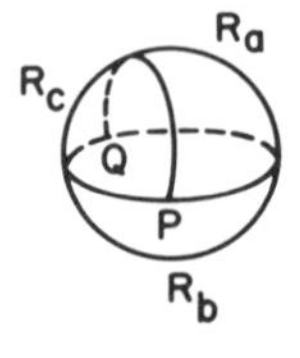

FIG. 4. Case of three regions for Theorem 8. The three paths from P to Q are in the three overlapping regions between (R_a, R_b), (R_b, R_c), and (R_c, R_a).

gether, each of the form of (20a) and (20b), and notice that along each path the integrand is always the difference of the vector potential A_μ between two regions, very much as in (21).

The first proof we gave above of Theorem 8 is easy and is "obvious" to a physicist. The second proof is more involved but is more intrinsic. The theorem is a special case of the Chern-Well theorem which evolved from the famous Gauss-Bonnet-Allendoerfer-Weil-Chern theorem, a seminal development in contemporary mathematics.[5] We want to emphasize two consequences of the theorem. (i) The right-hand side of (19) is independent of the gauge field, and only depends on the gauge type. (ii) The right-hand side of (19) has as integrand the gradient of ln$S$. Since S is single-valued, the integral must be equal to an *integral multiple* of a constant (in this case $2\pi i$). A remarkable fact is that these consequences remain valid in the general mathematical theorem, which is very deep.

V. GENERALIZATION TO NON-ABELIAN GAUGE FIELD

So far we have only considered electromagnetism and described it in terms of an Abelian gauge field that corresponds to the group U_1, or equivalently SO_2. On the basis of the discussions in the preceding section, the generalization to the non-Abelian case can be carried out without much difficulty. For a local region this has been done in Ref. 3. Extension to global considerations is our present focus of interest.

A gauge is defined by (a) a particular choice of overlapping regions and (b) a particular choice of *single-valued* gauge transformations S_{ab} in the overlapping regions. The choice of gauge transformations clearly must satisfy the following two conditions.

(1) In the overlapping region $R_a \cap R_b$, the gauge transformations S_{ba} from a to b and S_{ab} from b to a are related by

$$S_{ab} S_{ba} = 1 ,$$

where 1 is the identity element of the gauge group.

(2) If three regions R_a, R_b, and R_c overlap, then there are gauge transformations $S_{ab}, S_{ba}, S_{ac}, S_{ca}, S_{bc}, S_{cb}$ so that

$$S_{ab} S_{bc} S_{ca} = 1 , \text{ etc.}$$

in $R_a \cap R_b \cap R_c$.

As in the case of electromagnetism, both the concept of a gauge and the concept of a global gauge transformation are not tied to any specific gauge potentials, denoted in general by b^k_μ.

The *nonintegrable phase factor* for a given path is now an element of the gauge group. We shall still call it a phase factor. Since these phase factors do not in general commute with each other, Theorems 4a and 5a for the Abelian case need to be modified as follows.

Theorem 4: Under a global gauge transformation, the phase factor around any loop remains in the same class. The class does not depend on which point is taken as the starting point around the loop.

Theorem 5: The field strength $f^k_{\mu\nu}$ is covariant under a global gauge transformation.

Only theorem 4 is not immediately transparent. For a loop $ABCA$, under a gauge transformation[3]

$$\Phi_{ABCA} \to \Phi'_{ABCA} = \xi(A)\Phi_{ABCA}\xi^{-1}(A) .$$

Thus Φ'_{ABCA} and Φ_{ABCA} are in the same class. Also around the same loop if we change the starting point from A to C,

$$\Phi_{CABC} = \Phi_{CA}\Phi_{ABCA}\Phi_{AC} .$$

Hence changing the starting point does not change the class.

Theorem 4 defines *the class of a loop*. This concept is the generalization of the phase factor for electromagnetism around a loop with the magnetic flux as the exponent. It is a gauge-invariant concept.

These concepts have been extensively studied by the mathematicians in the framework of more general[6] mathematical constructs. A translation of terminology is given in Table I.

VI. CASE OF SU_2 GAUGE FIELD

For the SU_2 case we take the infinitesimal generators X_k to satisfy

$$X_1X_2 - X_2X_1 = X_3 , \text{ etc.} \tag{22}$$

and define the phase factor, as a generalization of (7), by[7]

$$\Phi_{QP} = \left[\exp\left(\int_P^Q \frac{-e}{\hbar c} b^k_\mu X_k dx^\mu\right)\right]_{\text{ordered}} , \tag{23}$$

i.e., we make the replacement

$$ieA_\mu \to -eb^k_\mu X_k , \tag{24}$$

or

$$A_\mu \to ib^k_\mu X_k . \tag{25}$$

[The subscript "ordered" means that, in the definition of the exponential in terms of a power series, the factors $b^k_\mu X_k$ are ordered along the path from P to Q with the factor $b^k_\mu(P)X_k$ at the right end of the product.] The algebraic operators X_k can be thought of as the collection of all irreducible representations of (22). The eigenvalue of iX_k with the

TABLE I. Translation of terminology.

Gauge field terminology	Bundle terminology
gauge (or global gauge)	principal coordinate bundle
gauge type	principal fiber bundle
gauge potential b^k_μ	connection on a principal fiber bundle
S_{ba} (see Sec. V)	transition function
phase factor Φ_{QP}	parallel displacement
field strength $f^k_{\mu\nu}$	curvature
source[a] J^K_μ	?
electromagnetism	connection on a $U_1(1)$ bundle
isotopic spin gauge field	connection on a SU_2 bundle
Dirac's monopole quantization	classification of $U_1(1)$ bundle according to first Chern class
electromagnetism without monopole	connection on a trivial $U_1(1)$ bundle
electromagnetism with monopole	connection on a nontrivial $U_1(1)$ bundle

[a] I.e., electric source. This is the generalization (see Ref. 3) of the concept of electric charges and currents.

minimum absolute value is $\pm\frac{1}{2}$. Therefore the minimum "charge" of all physical states can be read off from (24) by taking the 2×2 irreducible representation of X_k:

$$X_k = -\frac{i\sigma_k}{2}, \qquad (26)$$

where σ_k are the Pauli matrices. Thus

$$\text{minimum "charge"} = \frac{e}{2}. \qquad (27)$$

The particle of the gauge field belongs to the adjoint representation. Its "charges" are e, 0, and $-e$. Thus

$$\frac{\text{"charge" of gauge particle}}{\text{minimum "charge"}} = 2 \quad \text{for } SU_2. \qquad (28)$$

We shall now try to define a Dirac monopole field as a special SU_2 field along only one isospin direction $k=3$, i.e., we define

$$b^1_\mu = b^2_\mu = 0, \quad b^3_\mu = A_\mu, \qquad (29)$$

where A_μ is given in the two regions (10) by (11). In the overlapping region, transformation factor S of (12) and (14) now becomes

$$S_{ab} = \exp\left(-\frac{2ge}{\hbar c}\phi X_3\right) \qquad (30)$$

by replacement (25). This is single-valued if and only if the quantization condition

$$\frac{eg}{\hbar c} = \text{integer} = D \qquad (31)$$

is satisfied because for SU_2

$$\exp(4\pi X_3) = 1, \quad \exp(2\pi X_3) \neq 1,$$

which follows from the existence of half-integral representations such as (26).

The phase factor (30) describes a great circle, wound D times, on the manifold of SU_2 when ϕ varies from $0 \to 2\pi$. Such a circle can be continuously shrunk to the identity element, in contrast with the situation for electromagnetism. Thus, by a global gauge transformation S may be changed to $S'=1$, and the two regions a and b after the global gauge transformation can be *fused into one single region*. The gauge potential b^k_μ is then defined *everywhere* in R as a single region. Thus we have the following theorem.

Theorem 9: For the SU_2 gauge group, the gauges $\mathcal{G}_D$ for different D can be transformed into each other by global gauge transformations. The different monopole fields are therefore of the same type.

We shall only exhibit the global transformation for the case $\mathcal{G}_{-1}$ for which

$$S_{ba} = \exp(-2\phi X_3), \qquad (32)$$

$$\frac{e}{\hbar c} = \frac{-1}{g}. \qquad (33)$$

The gauge transformations we shall seek are illustrated in Fig. 5. We shall choose

$$\xi = \exp[\theta(X_1 \sin\phi - X_2 \cos\phi)], \qquad (34)$$

$$\eta = \exp[(\pi - \theta)(X_1 \sin\phi - X_2\cos\phi)]\exp(\pi X_2). \qquad (35)$$

It is easy to see that ξ is analytic in the coordinates x^μ at all points in R_a. (One only has to verify this statement at $\theta=0$, which is easily done.) Similarly η is analytic in R_b. ξ and η are therefore allowed gauge transformations in, respectively, R_a and R_b.

Now one can prove after some algebra that[8]

$$S'_{ba} = \eta S_{ba} \xi = 1 .$$

Thus after the gauge transformations ξ and η, which together form a global gauge transformation, regions R_a and R_b are related by the identity gauge transformation in their overlap, i.e., the two regions can be fused into one. To calculate the gauge potentials $b^{k'}_\mu$ after the global gauge transformation we use

$$\xi(Q)\left[1+\frac{1}{g}(b^k_\mu)'_a X_k dx^\mu\right]\xi^{-1}(P) = 1+\frac{1}{g}(b^k_\mu)_a X_k dx^\mu$$
$$= 1+\frac{1}{g}(A_\mu)_a X_3\, dx^\mu , \tag{36}$$

where A_μ is given by (11) and $Q = P + dx$. By choosing dx^μ to be along the t and r directions, one obtains $b^{k'}_t = b^{k'}_r = 0$. By choosing dx^μ to be along the θ direction, one obtains

$$b^{1'}_\theta = \frac{-g}{r}\sin\phi , \quad b^{2'}_\theta = \frac{g}{r}\cos\phi , \quad b^{3'}_\theta = 0 . \tag{37}$$

Now take dx^μ to be along the ϕ direction. We obtain, to order $d\phi$,

$$1+\frac{1}{g}(b^k_\phi)' X_k r \sin\theta\, d\phi$$
$$= \xi^{-1}(Q)\xi(P) + \frac{1}{g}(A_\phi)_a r\sin\theta\, d\phi\, \xi^{-1}(P) X_3 \xi(P) . \tag{38}$$

The first term on the right-hand side can be[8] computed in a straightforward manner:

$$\xi^{-1}(Q)\xi(P) = 1 - \sin\theta\, d\phi[(X_1\cos\phi + X_2\sin\phi) - X_3\tan\tfrac{1}{2}\theta] .$$

The second term also can be easily computed since

$$\xi^{-1}(P) X_3 \xi(P) = \sin\theta(X_1\cos\phi + X_2\sin\phi) + X_3\cos\theta$$

and $(A_\phi)_a$ was given by (11). Finally one arrives at

$$b^{1'}_\phi = \frac{-g}{r}\cos\theta\cos\phi ,$$
$$b^{2'}_\phi = \frac{-g}{r}\cos\theta\sin\phi , \tag{39}$$
$$b^{3'}_\phi = \frac{g}{r}\sin\theta .$$

Combining these results and remembering (33), we obtain

$$\frac{e}{\hbar c} b^{k'}_\mu X_k dx^\mu = \frac{-1}{r^2}\epsilon_{ikj} x^j dx^i X_k ,$$

i.e.,

$$b^{k'}_4 = 0 , \quad \frac{e}{\hbar c} b^{k'}_i = -\frac{1}{r^2}\epsilon_{ikj} x^j . \tag{40}$$

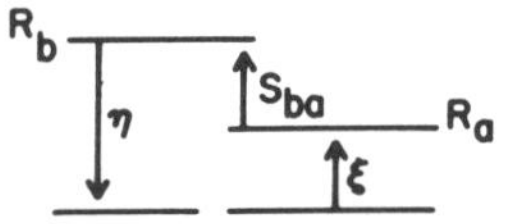

FIG. 5. A global transformation after which R_a and R_b can be fused.

Thus the new potential $b^{k'}_\mu$ is analytic in R_a. Because $\eta S_{ba}\xi = 1$ the new potential (in the overlapping region) for R_b must be the same as (40). By analyticity (40) is seen to be valid throughout R. Notice that (40) is the same potential as one of the solutions [solution (12a)], for a sourceless gauge field, in Ref. 9.

The global gauge transformation that transforms $\mathcal{G}_D$ into $\mathcal{G}_0$ for $D \neq -1$ can be obtained by slightly modifying (34) and (35).

We shall discuss Theorem 9 further in the next section.

VII. CASE OF SO_3 GAUGE FIELD

We turn to SO_3, which is locally the same as SU_2, but for which

$$e^{2\pi X_k} = 1 . \tag{41}$$

Equations (22) to (25) remain unaltered. The minimum "charge" of all physical states is now

$$\text{minimum "charge"} = e , \tag{42}$$

giving

$$\frac{\text{"charge" of gauge particle}}{\text{minimum "charge"}} = 1 \quad \text{for } SO_3 . \tag{43}$$

This last formula differentiates physically the SO_3 case from the SU_2 case.

We emphasize here a point already made in the literature[10] for electromagnetism: The local character of the gauge group is of course determined by the interactions (which determine the conservation laws). We want to ask what determines the global character. The global character (compact or noncompact in the case of electromagnetism, SU_2 or SO_3 in the isospin case) *is determined by* the representations for all states which *physically* exist. For example, in electromagnetism, if all charges are integral multiples of a single unit, the gauge group is compact,[10] because the group is *physically defined* as the simultaneous local phase factor change of *all* charge fields. There is then no physically definable meaning to the noncompact group. In the case of SU_2 or SO_3, if (43) is satisfied, then all representations of X_k physically realizable are integral representations.

Thus the simultaneous local changes of isospin phase factor of all physical systems *cannot differentiate* the group element $e^{2\pi X_k}$ from the identity. *Therefore*, the physical definition of $e^{2\pi X_k}$ is unity and the group must be SO_3.

Turning now to the monopole field for SO_3 we find that (30) is still correct. Equation (41) then leads to the quantization condition

$$\frac{2eg}{\hbar c} = \text{integer} = D \tag{44}$$

in order that S (as an element of SO_3) be a single-valued function of the coordinates x^μ in R.

As ϕ increases from 0 to 2π, the phase factor (30) describes a closed circuit in the group space of SO_3, starting from the identity element and returning to it. If one continuously traced the corresponding element of the group SU_2, one would have started from the identity and ended with the element that corresponds to

$$\begin{pmatrix} -1 & \\ & -1 \end{pmatrix}$$

in the 2×2 representation of SU_2 when D is odd. In such a case, no distortion of the closed circuit in SO_3 described by the phase factor (30) can shrink it to the identity element. This means that the gauge type for even D is not the same as that for odd D. By constructing explicit gauge transformations like (34) and (35) one can then complete the proof of the following theorem.

Theorem 10: For SO_3, all gauges $\mathcal{G}_D$ for D = even are of one type, and all gauges $\mathcal{G}_D$ for D = odd are of one type. These two types are different.

Summarizing the situation for U_1, SU_2, and SO_3 we find that in each case the "magnetic" monopole fields have quantized strengths. They belong to, respectively, infinitely many types for U_1 gauge group (electromagnetism), one type for the SU_2 gauge group, and two types for SO_3 gauge groups.

The physical meaning of these statements are as follows. In the SU_2 case, all magnetic monopole fields can be continuously changed into each other by the process of continuous changes[19] of "electric" sources. For example, starting with the "magnetic monopole" field for $\mathcal{G}_{-1}$ of Theorem 9 we can, by a gauge transformation, obtain the potentials b' [on $\mathcal{G}_0$] given in (40). We can then consider the potential (on $\mathcal{G}_0$): $b'' = \alpha b$, where $0 \le \alpha \le 1$. The gauge field for b'' is no longer electrically sourceless outside of the origin, but is magnetically sourceless except at the orgin, where it is not sourceless either magnetically or electrically. As α changes from 1 to 0 we thus have a continuous change of the original magnetic monopole field to empty space through a process during which there are continuous changes of electric charge-current distributions. Such a process is not possible for electromagnetism, by Theorem 6. (In the SO_3 case it is also not possible, although it is possible to change the magnetic monopole strength by two units by a similar process.) Thus the meaning of a magnetic monopole field in the non-Abelian case is quite different from that in electromagnetism.

It is not really surprising that in the case of electromagnetism one cannot change the magnetic monopole strength by changing electric sources: In the region R there are no magnetic monopoles. The continuity of magnetic lines of forces in R is guaranteed by the equation $\nabla\cdot\vec{H} = 0$. No continuous movement of magnetic lines of force could therefore increase or decrease the net total flux around the origin. That this state of affairs does not obtain for SU_2 and SO_3 is due to the fact that in general $\nabla\cdot\vec{H}^k \neq 0$ in the non-Abelian case, so that one cannot define the magnetic flux through a loop. However, we had seen before (Theorem 4) that in the case of a non-Abelian gauge field what takes the place of the magnetic flux is the *phase factor of a loop*. One may then ask what takes the place of the total magnetic flux outwards from a sphere around the origin $\vec{r} = 0$. To answer this question consider the loop

$$r = 1\,,\quad \theta = \text{fixed}\,,\quad \phi = 0 \to 2\pi\,. \tag{45}$$

As θ changes from 0 to π the phase factor of the loop changes and it describes a continuous circuit (in the space of the group) starting from and ending at the identity element. Clearly any other way of "looping" over the sphere only leads to a distortion of this circuit, without changing the starting and ending point. We shall call this circuit the *total circuit* for the gauge field around the origin $\vec{r} = 0$. It is a concept that replaces the total magnetic flux around $\vec{r} = 0$ in electromagnetism.

We can now prove the following generalization of Theorem 8.

Theorem 11: Consider region R and the group SU_2 or SO_3. Consider a gauge $\mathcal{G}$ and define any gauge field on it. The total circuit for the gauge field around the origin $\vec{r} = 0$ is independent of the gauge field and only depends on the gauge type of $\mathcal{G}$. For the case of $\mathcal{G}_D$,

$$\text{total circuit of the gauge field} \simeq [S_{ba}(\phi) \text{ for } \phi = 2\pi \to 0]\,, \tag{46}$$

where $\simeq$ means "can be continuously distorted into."

This last formula is the generalization of (19).

To prove Theorem 11, consider first the loop (45). The phase factor in R_a and R_b will be denoted by $\Phi^a(\theta)$ and $\Phi^b(\theta)$. They are related in the overlap by

$$\Phi^b(\theta) = \Phi^a(\theta) \tag{47}$$

since $S(\phi = 0) = I$. [One uses a generalization of (14).] Next consider the loop $L(\theta)$ which lies on the sphere $r = 1$ with its projection onto the x-y plane given in Fig. 6. It consists of a first part $(BA)_1$ around the equator and a second part $(AB)_2$ not on the equator except for points A and B. It is clear that

$$[\text{loop}(45) \text{ for } \theta = 0 \to \pi/2] \simeq [L(\theta) \text{ for } \theta = 0 \to \pi/2]$$

because both sides "loop over" the upper hemisphere. Thus

$$[\Phi^a(\theta) \text{ for } \theta = 0 \to \pi/2] \simeq [\Phi^a_{L(\theta)} \text{ for } \theta = 0 \to \pi/2]$$
$$= [\Phi^a{}_{(AB)_2} \Phi^a_{(BA)_1} \text{ for } \theta = 0 \to \pi/2].$$

$\Phi^a_{(AB)_2}$ is continuous in θ, and

$$[\Phi^a_{(AB)_2} \text{ for } \theta = 0 \to \pi/2] \simeq \text{identity element}.$$

Thus

$$[\Phi^a(\theta) \text{ for } \theta = 0 \to \pi/2] \simeq [\Phi^a_{(BA)_1} \text{ for } \theta = 0 \to \pi/2]. \tag{48}$$

Similarly

$$[\Phi^b(\theta) \text{ for } \theta = \pi/2 \to \pi] \simeq [\Phi^b_{(BA)_1} \text{ for } \theta = \pi/2 \to \pi]. \tag{49}$$

At $\theta = \pi/2$, the left-hand sides of (48) and (49) match because of (47). Also the right-hand sides match. Thus we can take (48) and (49) in tandem, obtaining

$$\text{total circuit of gauge field} \simeq [\Phi^a_{(BA)_1} \text{ for } \theta = 0 \to \pi/2 \text{ followed by } S_B \Phi^a_{(BA)_1} \text{ for } \theta = \pi/2 \to \pi], \tag{50}$$

where we have used

$$\Phi^b_{(BA)_1} = S_B \Phi^a_{(BA)_1} S^{-1}{}_A = S_B \Phi^a_{(BA)_1}. \tag{51}$$

Now

$$[\Phi^a_{(BA)_1} \text{ for } \theta = 0 \to \pi/2 \text{ followed by } \Phi^a_{(BA)_1} \text{ for } \theta = \pi/2 \to \pi] \tag{52}$$

is a loop *that doubles back* on itself, i.e., (52) can be distorted to the identity element. Applying this fact to (50) one obtains

$$\text{total circuit of gauge field} \simeq [I \text{ for } \theta = 0 \to \pi/2 \text{ followed by } S_B \text{ for } \theta = \pi/2 \to \pi]. \tag{53}$$

Now $S_B = S_{ba}(\phi = 4\pi - 4\theta)$. As $\theta = \pi/2 \to \pi$, $S_B = S_{ba}(\phi)$ for $\phi = 2\pi \to 0$. Substitution into (53) leads to (46).

To complete the proof of Theorem 11 we need the generalization of (46) to gauges that contain more than two regions. This can be done without much difficulty, e.g., for the case that region b is further divided into regions c and d, as schematically illustrated in Fig. 7(a), (46) should be replaced by

$$\begin{aligned}\text{total circuit of gauge field} \simeq [&S_{dc}(B) S_{ca}(x) \text{ for } x = A \to A \text{ along direction of arrow},\\ &\text{followed by } S_{dc}(y) S_{ca}(A) \text{ for } y = B \to B \text{ along direction of arrow}].\end{aligned} \tag{54}$$

For the case that $\mathcal{G}$ has four regions a, b, c, d as illustrated in Fig. 7(b), (46) should be replaced by

$$\begin{aligned}\text{total circuit of gauge field} \simeq [&S_{da}(x) \text{ for } x = A \to B,\\ &\text{followed by } S_{db}(y) S_{ba}(x) \text{ for } x = B \to C,\ y = B \to D,\\ &\text{followed by } S_{dc}(D) S_{cb}(x) S_{ba}(C) \text{ for } x = D \to C,\\ &\text{followed by } S_{dc}(y) S_{ca}(x) \text{ for } x = C \to E,\ y = D \to E,\\ &\text{followed by } S_{da}(x) \text{ for } x = E \to A].\end{aligned} \tag{55}$$

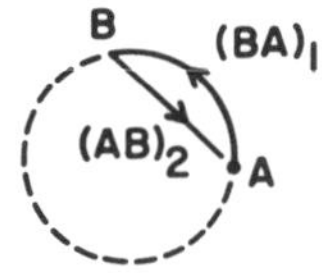

FIG. 6. Projection onto x-y plane of loop $L(\theta)$. The loop lies entirely on sphere $r=1$, and is in the upper (lower) hemisphere for $0 \le \theta \le \pi/2$ ($\pi/2 < \theta \le \pi$). The portion $(BA)_1$ lies on the equator. Coordinates for $A: r=1$, $\theta=\pi/2$, $\phi=0$. Coordinates for $B: r=1$, $\theta=\pi/2$, $\phi=h(\theta)$, where $h(\theta)=4\theta$ for $0 \le \theta \le \pi/2$ and $h(\theta)=4\pi-4\theta$ for $\pi/2 < \theta \le \pi$.

Notice that the right-hand sides of (54) and (55) are dependent only on the gauge type, and not on the specific gauge field.

VII. GENERALIZED BOHM-AHARONOV EXPERIMENT

The concept of an SU_2 gauge field was first discussed in 1954. In recent years many theorists, perhaps a majority, believe that SU_2 gauge fields do exist. However, so far there is *no experimental proof* of this theoretical idea, since conservation of isotopic spin only suggests, and does not require, the existence of an isotopic spin gauge field. What kind of experiment would be a definitive test of the existence of an isotopic spin gauge field? A generalized Bohm-Aharonov experiment would be.

If the gauge particle for isospin group SU_2 is massless, it is possible to design a *gedanken* generalized Bohm-Aharonov experiment as illustrated in Fig. 1. One constructs the cylinder of material for which the total I_z spin is not zero, e.g., a cylinder made of heavy elements with a neutron excess. One spins the cylinder around its axis, setting up a "magnetic" flux inside the cylinder, along the I_z "direction." If one scatters a proton beam around the cylinder, the fringe shift would be in the opposite direction from the corresponding shift observed with a neutron beam. To be more specific, imagine that one spins the cylinder clockwise. The magnetic flux would be emerging from the diagram towards the reader, since the cylinder has a net negative value for I_z. This means that for a proton (neutron) beam, the flux produces an increment (decrement) of path length counterclockwise around the cylinder. This increment (decrement) produces a net downward (upward) shift of the fringes, i.e., a shift toward the bottom (top) of the diagram.

If one scatters a coherent mixture of neutron and proton in a pure state, in the interference plane one would observe not only fluctuations of nucleon intensity, but also fluctuations of the neutron-proton mixing ratio. A variation of this phenomenon

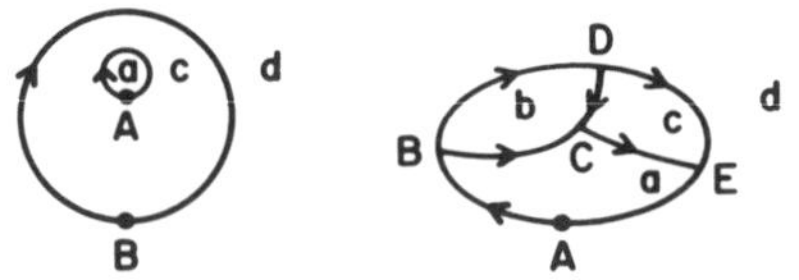

FIG. 7. Schematic diagrams for division lines in overlaps between three or more regions. The drawings are projections from the sphere $r=1$. The projection is from the south pole of the sphere onto the tangent plane at the north pole. The south pole is underneath the plane of the paper.

obtains if one imagines rotating a cylinder which has an average $\langle \vec{I} \rangle$ which is not zero, and is not in the I_z direction. A magnetic flux would then be set up which is in a "direction" other than I_z. Scattering a beam of protons would then produce some neutrons as well as protons in the interference plane. This implies, of course, that there is electric charge transfer between the beam and the cylinder together with the gauge field around it.

If the gauge particle has a finite mass $m>0$, then the experiment becomes difficult because the return flux would hug the outside surface of the cylinder, to a distance $\sim \hbar/mc$. Unless the fringe plane lies within this distance of the cylinder, the effect of the flux will be negligible.

IX. REMARKS

(a) From the viewpoint of the present paper, the electric charge and the magnetic charge play completely unsymmetrical roles. This matter deserves further comments. In the non-Abelian case it was in fact already pointed out[11] that the dual of an unquantized sourceless gauge field is not necessarily a gauge field. Thus the asymmetry between electric and magnetic phenomena is not due to the formalism, but is of an intrinsic nature in the non-Abelian case. In contrast, in the Abelian case the asymmetry is only formal since the electric and magnetic charges interact with the electromagnetic field in entirely symmetrical ways. In other words, one can use the phase factor Φ^M associated with the magnetic charges to describe the electromagnetic field, rather than the phase factor Φ discussed in the present paper. The mathematical relationship between these two kinds of phase factors (or between the associated vector potentials A_μ^M and A_μ) remains to be explored. So does the corresponding question in any second-quantized theory[12] of all the fields.

(b) In the proof of Theorem 9 we had shown explicitly how a magnetic monopole field for the SU_2 gauge group can be gauge-transformed into the solution (12a) of Ref. 9. Now a magnetic monopole field is not a gauge field at the origin $\vec{r}=0$ since it does not satisfy the Bianchi identity[3] at the origin. Thus, although solution (12a) of Ref. 9 is (electrically) sourceless at all points, including the origin, it is not a proper gauge field at the origin, a fact we did not realize before. All three solutions, (12a), (12d), and (12e), are, of course, of the same gauge type.

(c) In Sec. II it was emphasized that $f_{\mu\nu}$ underdescribes electromagnetism because of the Bohm-Aharonov experiment which involves a doubly connected space region. For non-Abelian cases, the field strength $f^k_{\mu\nu}$ underdescribes the gauge field even in a singly connected region. An example of this underdescription was given in Ref. 13.

(d) For the region of space-time outside of the cylinder of Fig. 1 there is only one gauge type. All electromagnetic fields in the region can be continuously distorted into each other by the movement of electric charges and currents inside and outside the cylinder.

(e) The phase factor for the group U_1 is the phase factor of the algebra of complex numbers. It is perhaps not accidental that such a phase factor provides the basis for the description of a physically realized gauge field—electromagnetism. Now the only possible more complicated division algebra is the *algebra of quaternions*. The phase factors of the quaternions form the group SO_3. It is tempting to speculate that such a phase factor provides the basis for the description of a physically realized gauge field—the SU_2 gauge field. Speculation about the possible relationship between quaternions and isospin has been made before.[14] Such speculations were, however, not made with reference to gauge fields. If one believes that gauge fields give the underlying basis for strong and/or weak interactions, then the fact that gauge fields are fundamentally *phase factors* adds weight to the speculation that quaternion algebra is the real basis of isospin invariance.

(f) It is a widely held view among mathematicians that the fiber bundle is a natural geometrical concept.[15] Since gauge fields, including in particular the electromagnetic field, are fiber bundles, *all gauge fields are thus based on geometry*.[16] To us it is remarkable that a geometrical concept formulated without reference to physics should turn out to be exactly the basis of one, and indeed maybe all, of the fundamental interactions of the physical world.

ACKNOWLEDGMENTS

It is a pleasure to thank Professor Shiing-shen Chern for correspondence and discussions. We are especially indebted to Professor J. Simons, whose lectures and patient explanations have revealed to us glimpses of the beauty of the mathematics of fiber bundles.

While we were making corrections on the draft of this paper, a report on the experimental discovery of a magnetic monopole[17] reached us.

Additional references to fiber bundles, monopoles and quaternions are given in footnote 18.

---

*Work supported in part by the U. S. ERDA under Contract No. AT(11-1)-3227.

†Work supported in part by the National Science Foundation under Grant No. MPS74-13208 A01.

[1]Y. Aharonov and D. Bohm, Phys. Rev. 115, 485 (1959). See also W. Ehrenberg and R. E. Siday, Proc. Phys. Soc. London B62, 8 (1949).

[2]R. G. Chambers, Phys. Rev. Lett. 5, 3 (1960).

[3]Chen Ning Yang, Phys. Rev. Lett. 33, 445 (1974). This paper introduced the formulation of gauge fields in terms of the concept of nonintegrable phase factors. The differential formulation of gauge fields for Abelian groups was first discussed by H. Weyl, Z. Phys. 56, 330 (1929); for non-Abelian groups it was first discussed by Chen Ning Yang and Robert L. Mills, Phys. Rev. 96, 191 (1954). See also S. Mandelstam, Ann. Phys. (N.Y.) 19, 1 (1962); 19, 25 (1962); I. Białynicki-Birula, Bull. Acad. Pol. Sci., Ser. Sci. Math. Astron. Phys. 11, 135 (1963); N. Cabibbo and E. Ferrari, Nuovo Cimento 23, 1146 (1962); R. J. Finkelstein, Rev. Mod. Phys. 36, 632 (1964); N. Christ, Phys. Rev. Lett. 34, 355 (1975); and A. Trautman, in *The Physicist's Conception of Nature*, edited by J. Mehra (Reidel, Boston, 1973), p. 179.

[4]P. A. M. Dirac, Proc. R. Soc. London A133, 60 (1931). Since this brilliant work of Dirac, there have been several hundred papers on the magnetic monopole. For a listing of papers until 1970, see the bibliography by D. M. Stevens, Virginia Polytechnic Institute Report No. VPI-EPP-70-6, 1970 (unpublished).

[5]See J. Milnor and J. Stasheff, *Characteristic Classes* (Princeton Univ. Press, Princeton, N.J., 1974); C. B. Allendoerfer and A. Weil, Trans. Am. Math. Soc. 53, 101 (1943); Shiing-shen Chern, Ann. Math. 45, 747 (1944). See also H. Weyl, Amer. J. Math. 61, 461 (1939), and Ref. 6 below.

[6]There are many books on fiber bundles. See e.g., N. Steenrod, *The Topology of Fibre Bundles* (Princeton Univ. Press, Princeton, N. J., 1951). For connection, see, e.g., S. Kobayashi and K. Nomizu, *Foundations of Differential Geometry* (Interscience, New York, Vol. I-1963, Vol. II-1969).

[7]The notation here is the same as that in Ref. 3, except for the normalization factor $e/\hbar c$ which was absorbed into b in Ref. 3. To avoid confusion with the azimuthal angle, we write Φ for the ϕ of Ref. 3. Notice that

$$\Phi_{(A+dx)A} = I - \frac{e}{\hbar c} b^k_\mu(x) X_k dx^\mu .$$

All formulas are the same as in Ref. 3, but the name for Φ_{QP} will now be "the phase factor from P to Q." (In Ref. 3 the same name applied to Φ_{PQ}.) The new name is in accordance with the usual convention of time ordering.

[8]Since the 2×2 representation is faithful, it is sufficient for computational purposes to use the representation (26) for X_k. This makes the algebra quite simple since one can apply the formula $e^{i\theta\sigma_k} = \cos\theta + i\,\sigma_k \sin\theta$.

[9]Tai Tsun Wu and Chen Ning Yang, in *Properties of Matter under Unusual Conditions*, edited by H. Mark and S. Fernbach (Wiley, New York, 1969), p. 349.

[10]Chen Ning Yang, Phys. Rev. D 1, 2360 (1970).

[11]Gu Chao-hao and Chen Ning Yang, Sci. Sin. 18, 483 (1975).

[12]P. A. M. Dirac, Phys. Rev. 74, 817 (1948); J. Schwinger, *Particles, Sources and Fields* (Addison-Wesley, Reading, Mass., Vol. 1-1970, Vol. 2-1973).

[13]Tai Tsun Wu and Chen Ning Yang, preceding paper, Phys. Rev. D 12, 3843 (1975).

[14]Cheng Ning Yang, comments after J. Tiomno's talk, session 9, *Proceedings of the Seventh Annual Rochester Conference on High-Energy Nuclear Physics, 1957* (Interscience, New York, 1957); Chen Ning Yang, in *The Physicist's Conception of Nature*, edited by J. Mehra (Reidel, Boston, 1973), p. 447.

[15]See, e.g., Shiing-shen Chern, Geometry of Characteristic Classes, Proceedings of the 13th Biennial Seminar, Canadian Mathematics Congress, 1972, p. 1.

[16]This is in sharp contrast with an interaction (if it exists), which is not related to gauge concepts.

[17]P. B. Price, E. K. Shirk, W. Z. Osborne, and L. S. Pinsky, Phys. Rev. Lett. 35, 487 (1975).

[18]There have been many papers on fiber bundles, monopoles, and quaternions in the physics literature. The following is only a partial list: Elihu Lubkin, Ann. Phys. (N.Y.) 23, 233 (1963); J. Math. Phys. 5, 1603 (1964); D. Finkelstein, J. M. Jauch, S. Schiminovich, and D. Speiser, J. Math. Phys. 4, 788 (1963); articles by J. A. Wheeler, B. S. DeWitt, A. Lichnerowicz, and C. W. Misner, in *Relativity, Groups and Topology*, edited by C. DeWitt and B. S. DeWitt (Gordon and Breach, New York, 1963); also C. Misner, K. Thorne, and J. A. Wheeler, *Gravitation* (Freeman, San Francisco, 1973); A. Trautman, Rep. Math. Phys. 1, 29 (1970); G. 't Hooft, Nucl. Phys. B79, 276 (1974); Hendricus G. Loos, Phys. Rev. D 10, 4032 (1974); J. Arafune, P. G. O. Freund, and C. J. Goebel, J. Math. Phys. 16, 433 (1975); B. Julia and A. Zee, Phys. Rev. D 11, 2227 (1975); M. K. Prasad and Charles M. Sommerfield, Phys. Rev. Lett. 35, 760 (1975).

[19]*Footnote added in proof.* Professor A. Lenard has raised an interesting question in this connection: In the continuous changes of "electric" sources, are sources quantized according to (44)? The answer to this question is "no," and requires explanations. The "electric" charges play two separate roles. They act as sources of gauge fields, and they also act as responders to gauge fields. In a physical situation the two roles are of course interrelated. In the discussion here, however, we separate the two roles. We therefore do not require quantization of electric charges as sources, but require quantization of electric charges as responders.

HADRONIC MATTER CURRENT DISTRIBUTION INSIDE A POLARIZED NUCLEUS AND A POLARIZED HADRON

⟦76a⟧
Commentary begins page 75

T.T. CHOU
Physics Department, University of Georgia, Athens, Georgia 30602

Chen Ning YANG *
Institute for Theoretical Physics, State University of New York, Stony Brook, New York 11794

Received 26 January 1976

The concept of *nucleonic current density* (or velocity profile) inside a polarized nucleus and the concept of *hadronic matter current density* inside a polarized hadron are introduced. Utilizing the increasing opaqueness of hadrons relative to each other at increasing relative velocities, these current densities can be obtained from a measurement of the $R(t)$ parameter in elastic hadron-nucleus and elastic hadron-hadron scattering. For very high energies, the spin dependence in elastic hadron-hadron scattering will be solely due to this non-vanishing $R(t)$ parameter. Measurements of $R(t)$ and therefore the nucleonic current or hadronic matter current thus provide powerful probes for the structure of polarized nuclei and hadrons. The results can be used to check nuclear theory and models of hadron structure. Estimates of the magnitude of R are given by assuming the proportionality between the hadronic matter density-current distribution and the electric charge-current distribution. This proportionality hypothesis which is heuristic is shown to be dependent on the concept of homogenization and the principle of minimum electromagnetic interactions for the basic constituents of hadrons. In the appendix it is shown that the spin of a Dirac particle does involve motion (currents).

1. Introduction

A number of years ago we proposed [1] a heuristic theory relating very high energy elastic pp and ep scattering. The present paper extends the heuristic basis of that theory to include spin effects. Some of the ideas of the present paper were already briefly reported [2] in Conferences in 1973. We now give details of the considerations and present numerical results for estimated values of the rotation parameter $R(t)$ in Kp and πp scattering at Fermilab energies and pp scattering at higher energies.

To describe the basic approach of the present paper, let us first analyze the theory of ref. [1]. There are two essential elements to that theory:

(A) The eikonal formulation (= formulation in terms of two dimensional impact

* Work supported in part by the National Science Foundation under Grant MPS74-13208 A01.

□ Reprinted from *Nuclear Physics* B107 (1976), 1–20.

parameter $\vec{b}$), and the formula

$$\frac{d\sigma}{dt} = \pi |\widetilde{e^{-\Omega(b)} - 1}|^2 , \tag{1}$$

where $\Omega(b)$ is the opaqueness at $\vec{b}$ and $\sim$ denotes Fourier transform.

(B) The proportionality of the opaqueness Ω and the convolution of the compressed hadronic matter density functions for the two colliding particles, e.g., in elastic Kp scattering,

$$\Omega = \text{const}\, D_K \otimes D_p \, . \tag{2}$$

(2) is based on the heuristic argument that at a fixed $\vec{b}$ each part of the K goes through the target proton along one ray. It therefore sees only the local blackness D_p of the target along the ray (fig. 1.) The overall opaqueness of the collision is a weighted average of D_p. The weight is D_K. One obtains thus eq. (2).

Central to this heuristic argument, indeed, central to all thinking about extended hadrons is the concept of a hadronic matter density distribution inside a hadron:

$$\text{hadronic matter density} \equiv \boldsymbol{\rho}(b_x, b_y, b_z) \, . \tag{3}$$

In ref. [1] the possible motion of hadronic matter inside a hadron was not considered. Since we know through experimental data [3]* that the opaqueness of a hadron to a projectile increases with their relative velocity, the motion of hadronic matter within a hadron would affect the opaqueness distribution and would necessitate changes in (2). We thus introduce in this paper the concept of a

$$\text{hadronic matter current density} \equiv \vec{j}\,(b_x, b_y, b_z) \, , \tag{4}$$

and discuss the change in (2) caused by the presence of a nonvanishing $\vec{j}$. This discussion leads to the prediction that at high energies in Kp, πp and pp elastic cross

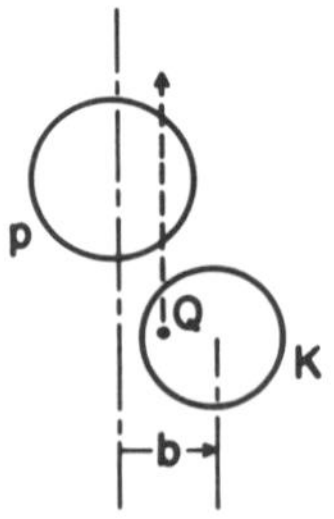

Fig. 1. Schematic drawing for Kp scattering. Each point Q of K sees only the hadronic matter in p along the dotted ray.

* The increase of the total cross section had been predicted by Hung Cheng and Tai Tsun Wu [3a] and was remarkably confirmed three years later.

sections there is a non-vanishing rotation parameter $R(t)$ [4a]*.

That the concept of $\vec{j}$ complements that of $\boldsymbol{\rho}$, forming a four-vector with it, is evident if we *examine the structure of a polarized nucleus.* It is clear that for a closed shell plus one nucleus which is polarized with $J_z = +J$ the outermost nucleon has a non-vanishing orbital angular momentum along the z axis unless it is in an S-state which we do not consider. It therefore contributes to a *nucleonic current density* as well as to a nucleon density. To a high energy projectile such as a K coming along the $+y$ axis the nucleus would appear more opaque on the $-x$ axis side than the $+x$ side (cf. fig. 2). The result of this difference of opaqueness is a non-vanishing rotation parameter $R(t)$. This parameter is measurable and its value determines, through eq. (37) of sect. 4 the nucleonic current density in the nucleus.

The result of such a measurement would provide a check on nuclear theory. (Experimentally the center of a nucleus is quite opaque and a nucleonic current near the center could not greatly affect the absorption of the projectile through that region of the target. Thus the measurement would provide meaningful information for the outer nucleons for any but the smallest nuclei.) See remark (vii) of sect. 8 for further discussions of this topic.

Can we extend the above described concepts for the structure of a nucleus to that of a hadron? It seems to us inevitable that once we accept the concept of an extended hadron with a hadronic matter density $\boldsymbol{\rho}$ in it, we must also accept the existence of a hadronic current $\vec{j}$ in a polarized hadron as an extension of the concept of a nucleonic current inside a polarized nucleus.

The discussion in the present paper will be expressed in terms of the structure of hadrons. But *all discussions* of hadron-hadron and electron-hadron collisions *are also valid for hadron-nucleus and electron-nucleus collisions* if we replace the target hadron by the target nucleus and the concept of hadronic matter current by nucleonic current (cf. sect. 8, remark (vii)).

Whereas in the case of the nucleus, there exists fairly reliable theories of its structure so that one can predict the value of $R(t)$, in the case of the hadron our understanding of its structure is still in a preliminary exploratory stage. A reasonable procedure is therefore to measure $R(t)$ and obtain from it, through eq. (37) and its

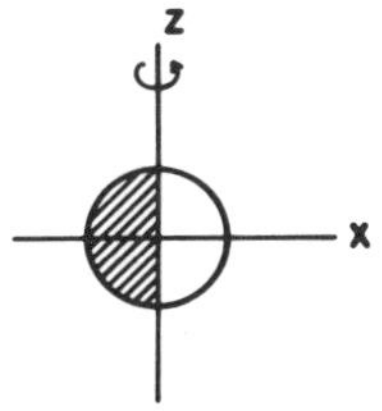

Fig. 2. A target polarized along z-axis as viewed by incoming beam A.

* In this paper we use the convention in De Lesquen et al. [4b].

generalizations to cases of higher spins, the hadronic current $\vec{j}$ inside the hadron. It seems to us that $\boldsymbol{\rho}$ and $\vec{j}$ must be fundamental quantities in hadronic structure.

Lacking a fundamental theory of the structure of hadrons, is there any estimate of $\boldsymbol{\rho}$ and $\vec{j}$? We shall make such a rough estimate by using the assumption that for a polarized target,

$$\boldsymbol{\rho} = C\boldsymbol{\rho}^{\mathrm{e}} , \tag{5}$$

$$\vec{j} = C\vec{j}^{\,\mathrm{e}} , \tag{6}$$

where $\boldsymbol{\rho}^{\mathrm{e}}$ and $\vec{j}^{\,\mathrm{e}}$ are the electric charge-current density inside the target. (5) is a heuristic assumption made for the purpose of estimation in ref. [1]. (6) is a natural extension of this assumption. Discussion of these assumptions are given in sect. 8 of the present paper.

We emphasize that even without assumptions (5) and (6), the presence of a hadronic matter current inside a polarized hadron and the phenomena of increasing absorption at increasing energies imply, in our view, that at very high energies there is a non-vanishing $R(t)$ which dominates the spin dependence of cross sections.

In the appendix we demonstrate that the spin of a Dirac particle involves motion (currents).

2. A special model

To isolate the different heuristic elements in our reasoning we shall consider first a special model of AB elastic scattering. We assume

(a) B is infinitely heavy , (7)

(b) A is without structure , (8)

(c) B is a bound state of hadronic constituents . (9)

Let us take for an elastic AB scattering the following convention:

$$\begin{aligned} &\text{incoming momentum of A} = k_\mu = (0, k, 0, ik) , \\ &\text{outgoing momentum of A} = k'_\mu = (q_x, k, 0, ik) , \\ &\text{target B polarized with fixed } J_z = m , \end{aligned} \tag{10}$$

where we have taken the limit $k \gg |q_x|$, $k \gg M_{\mathrm{A}}$. In other words, the incoming beam is along the $+y$ axis, and scatters in the xy plane. Since B is infinitely heavy, the wave function of B in coordinate space does not change in elastic scattering. The hadronic matter density-current distributions $\boldsymbol{\rho}$ and $\vec{j}$ inside of B presents an opaqueness to the projectile. For the kinematics specified in (10) there is no momentum transfer in the y direction, thus only the *compressed* charge-current densities along each ray at fixed b_x, b_z: are relevant:

$$\vec{j}(b_x, b_z) = \int_{-\infty}^{\infty} \vec{j}(b_x, b_y, b_z)\,\mathrm{d}b_y\,, \tag{11}$$

$$\rho(b_x, b_z) = \int_{-\infty}^{\infty} \boldsymbol{\rho}(b_x, b_y, b_z)\,\mathrm{d}b_y\,. \tag{12}$$

$\rho(b_x, b_z)$ was called D in ref. [1] and in eq. (2) of the present paper.

If the motion of hadronic matter in a region of the target B is toward the projectile, that region would have greater opaqueness than another region where the motion is away from the projectile. To make an estimate of this effect we assume* that

$$\text{opaqueness} \propto (P_{\text{eff}})^{\alpha}\boldsymbol{\rho}\,, \tag{13}$$

$$P_{\text{eff}} = \text{momentum of projectile in rest system of region.} \tag{14}$$

This assumption is in good agreement with experimental data on the increase of total cross section with increasing incoming momentum. Now a simple calculation shows that

$$P_{\text{eff}}^{\alpha} \simeq P_{\text{in}}^{\alpha}(1 - \alpha v_y)\,, \tag{15}$$

if quadratic terms in v are neglected. Thus the local opaqueness is

$$2\pi K(\boldsymbol{\rho} - \alpha \boldsymbol{j}_y)\,, \tag{16}$$

since

$$\boldsymbol{j}_y = \boldsymbol{\rho}\, v_y\,.$$

The factor 2π is inserted here for convenience later. K depends on P_{in}.

The sign of the term $-\alpha\boldsymbol{j}_y$ in (16) is easy to appreciate by examining fig. 3. At a fixed $\vec{b}$, if $\boldsymbol{j}_y > 0$ (<0), the target recedes (advances) against the projectile. Hence the opaqueness decreases (increases), requiring the negative sign in $-\alpha\boldsymbol{j}_y$.

Integrating (16) along a ray from $b_y = -\infty$ to ∞ yields

$$\Omega(b_x, b_z) = 2\pi K(\rho - \alpha j_y)\,. \tag{17}$$

The elastic AB scattering amplitude is obtained from this in the usual procedure [1] which is basically Huygen's principle:

$$\left.\frac{\mathrm{d}^2\sigma}{k_x\,\mathrm{d}k_x\,\mathrm{d}\phi}\right|_{\phi=0} = |\widetilde{-1 + \mathrm{e}^{-\Omega}}|^2\,, \tag{18}$$

where $\sim$ denotes a two-dimensional Fourier transform, and ϕ is the azimuthal angle

* This assumption or one similar in spirit was used or discussed in refs. [5]. Hayot and Sukhatme's paper gave curves of the shape of $\mathrm{d}\sigma/\mathrm{d}t$ as the total energy in the centre-of-mass increases from 23 GeV/c to 53 GeV/c, cf. recent data in ref. [5a].

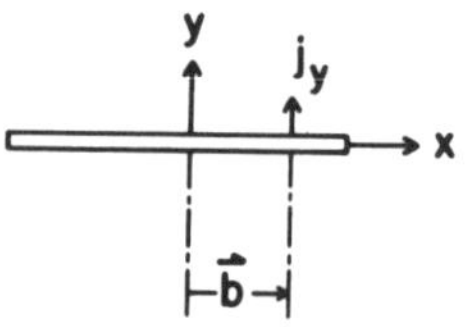

Fig. 3. The effect of j_y on the local opaqueness of a polarized target.

around the incoming beam: $\tan\phi = k_z/k_x$. We focus here on a scattering in the xy plane described by (10) for which the momentum transfer $k_z = 0$ so that $\phi = 0$.

3. Special case when spin of B is $\frac{1}{2}$

Since B has a definite parity, the density $\boldsymbol{\rho}$ must be even under parity:

$$\boldsymbol{\rho}(-b_x, -b_y, -b_z) = \boldsymbol{\rho}(b_x, b_y, b_z) . \tag{19}$$

If B has spin $\frac{1}{2}$, $\boldsymbol{\rho}$ is the product of ψ and ψ^*, both with angular momentum $\frac{1}{2}$. Thus $\boldsymbol{\rho}$ is a scalar plus a vector. Using (19) we find that $\boldsymbol{\rho}$ must be spherically symmetrical. Integrating $\int_{-\infty}^{\infty} \mathrm{d}b_y$ gives cylindrical symmetry for ρ:

$$\rho = \rho(\sqrt{b_x^2 + b_z^2}) .$$

To discuss the symmetry of $\vec{j}$, we consider not only the diagonal element $m = \frac{1}{2} \to m = \frac{1}{2}$ for the current that we have focused on so far, but also the other three elements, so that each component of $\vec{j}: j_x, j_y, j_z$ becomes a 2×2 matrix. Since B has a fixed parity, $\vec{j}$ as matrices which are functions of b_x, b_y, b_z must be odd under $\vec{b} \to -\vec{b}$. Thus

$$\vec{j} = f\vec{b}\begin{pmatrix} 1 & \\ & 1 \end{pmatrix} - g\vec{b} \times \vec{\sigma} , \tag{20}$$

where f and g are functions of $\vec{b}^2$. Now current is conserved. Thus the integral of $\vec{j} \cdot \mathrm{d}\vec{\sigma}$ over the surface of a sphere is zero. I.e. $f = 0$. For the diagonal elements $m = \pm\frac{1}{2} \to m = \pm\frac{1}{2}$,

$$j_x = \mp g b_y , \qquad j_y = \pm g b_x , \qquad j_z = 0 . \tag{21}$$

Thus the compressed current densities are

$$j_y = \pm b_x g_1 , \qquad j_x = j_z = 0 , \tag{22}$$

where g_1 is cylindrically symmetrical. The Fourier transforms are thus

$$\tilde{\rho} \equiv \frac{1}{2\pi} \iint \mathrm{d}^2 b\, \rho \mathrm{e}^{-iq \cdot b} = \text{cylindrically symmetrical} ,$$

$$\tilde{j}_x = \tilde{j}_z = 0 ,$$

$$\tilde{j}_y = \mp iq_x \cdot (\text{cylindrically symmetrical function}) . \tag{23}$$

We thus conclude that $\tilde{\rho}$ and $\tilde{j}_y$ are given by

$$\tilde{\rho} = \frac{1}{2\pi} G_1^{\mathrm{h}}(t) , \qquad \tilde{j}_y = \frac{\mp iq_x}{2\pi} G_2^{\mathrm{h}}(t) , \qquad -t = q_x^2 + q_z^2 , \tag{24}$$

($\mp$ for target spin $\pm\frac{1}{2}$ along the z axis), where G_1^{h} and G_2^{h} are "hadronic form factors", which are defined by (24).

4. Rotation parameter $R(t)$ for case B has spin = $\frac{1}{2}$

We shall now assume all quantities in (17) to be real. The scattering amplitude at high energies for a scattering in the xy plane described by (10) is proportional to

$$a(q_x, q_z) \equiv \widetilde{1 - \mathrm{e}^{-\Omega}} \tag{25}$$

evaluated at $q_z = 0$. Now express (17) as

$$\Omega = \Omega_0 + \Omega_1 \qquad \text{where} \qquad \Omega_0 = 2\pi K \rho , \qquad \Omega_1 = -2\pi\alpha K j_y . \tag{26}$$

(25) becomes, for any q_x, q_z,

$$\begin{aligned} a &= \widetilde{1 - \mathrm{e}^{-\Omega_0}} + \widetilde{\mathrm{e}^{-\Omega_0}\Omega_1} + \mathrm{O}(\Omega_1^2) \\ &= a_0 + a_1 + \mathrm{O}(\Omega_1^2) , \end{aligned} \tag{27}$$

where

$$a_0 = \widetilde{1 - \mathrm{e}^{-\Omega_0}} , \qquad a_1 = \widetilde{\mathrm{e}^{-\Omega_0}\Omega_1} . \tag{28}$$

Now by (24), $\tilde{\Omega}_0 = KG_1^{\mathrm{h}}$, $\tilde{\Omega}_1 = \pm i\alpha K q_x G_2^{\mathrm{h}}$, (29)

($\pm$ for target spin = $\pm\frac{1}{2}$ along z-axis).

Define the phase angle of a as ϕ:

$$\mathrm{e}^{i\phi} = \frac{a_0 + a_1}{|a_0|} + \mathrm{O}(a_1^2) . \tag{30}$$

By (29),

$$a_1 = iq_x \text{ (cylindrically symmetrical real function of } q_x, q_z) .$$

Thus a measurement of a_1 at $q_z = 0$, all $q_x > 0$, allows us to construct the value of a_1 for all q_x and q_z.

Since a_1 changes sign when $q_x \to -q_x$, there is no right-left asymmetry in the scattering. In other words, the polarization parameter [4]

$$P = 0 . \tag{31}$$

The phase angle ϕ is related to the rotation parameter R and the parameter A:

$$R = -\sin 2\phi, \qquad A = \cos 2\phi . \tag{32}$$

Squaring (27) we find that to order $O(\Omega_1)$,

$$\frac{d\sigma}{dt} = \pi \, |a_0|^2 = \pi \, |\widetilde{1 - e^{-\Omega_0}}|^2 + O(v^2) , \tag{33}$$

i.e. to the first order of velocity effects, the differential elastic cross section is independent of the velocity effects. a_0 can therefore be determined from the differential cross section in the usual way.

To the order $O(v)$ one can determine the compressed current j_y from a measurement of a_0 and ϕ at $q_z = 0$ in the following way. (30) allows one to determine a_1 at $q_z = 0$:

$$a_1 = |a_0| \, e^{i\phi} - a_0 . \tag{34}$$

As mentioned above, this allows one to construct $\tilde{a}_1$ at all b_x, b_z. One now uses:

$$\tilde{a}_1 = \Omega_1 \, e^{-\Omega_0} , \tag{35}$$

i.e. $$\Omega_1 = e^{\Omega_0} \tilde{a}_1 , \tag{36}$$

i.e. $$-2\pi\alpha K j_y = e^{\Omega_0} \tilde{a}_1 \tag{37}$$

to determine j_y.

Eq. (35) states that the first-order scattering amplitude $\tilde{a}_1$, in two-dimensional $\vec{b}$ space, is equal to Ω_1 times the absorption factor $e^{-\Omega_0}$. One can thus envisage Ω_1 *as a source function for first-order scattering.* Using (21) and (22) one can obtain from j_y the vector hadronic matter current $\vec{j}$. (Usage of (37) has to be made with judicious care since the operator e^{Ω_0} is very large at the center and high order terms have been neglected.)

It is instructive to notice that with the experimental situation defined in (10) (where one scatters in the xy plane a beam along the y axis on a target polarized in the z direction) there is no direct way in which one can measure R or ϕ, since ϕ is the phase of the scattering amplitude. To measure R we analyse the experimental situation (10) as follows. Take polarization $m = \frac{1}{2}$ in (10) and write the scattering amplitude a of (25) as

$$a = \alpha + i\beta . \tag{38}$$

Then

$$e^{i\phi} = (\alpha + i\beta)/(\alpha^2 + \beta^2)^{1/2}.$$

If we take $m = -\frac{1}{2}$ in (10), but keep the same q_x, then the scattering amplitude is $\alpha - i\beta$, because by (29), Ω_1 changes sign with the sign change of m. Furthermore, in the kinematics of (10), where the scattering takes place in the xy plane, the target spin component along the z-axis, i.e. m, does not change because of the scattering. (This is a general theorem due to parity conservation.) We may thus wirte the change of the spin wave function of the target in a scattering with momentum transfer q_x along the x axis in the following way:

$$\begin{pmatrix}1\\0\end{pmatrix} \to \begin{pmatrix}\alpha + i\beta\\0\end{pmatrix}, \qquad \begin{pmatrix}0\\1\end{pmatrix} \to \begin{pmatrix}0\\\alpha - i\beta\end{pmatrix}.$$

Hence

$$\frac{1}{\sqrt{2}}\begin{pmatrix}1\\1\end{pmatrix} \to \frac{1}{\sqrt{2}}\begin{pmatrix}\alpha + i\beta\\\alpha - i\beta\end{pmatrix} = \frac{|a_0|}{\sqrt{2}}\begin{pmatrix}e^{i\phi}\\e^{-i\phi}\end{pmatrix}. \tag{39}$$

This final wave function describes a recoil target with a polarization vector

$$(\cos 2\phi, -\sin 2\phi, 0)\,.$$

Thus starting with a new kinematics (10) with target spin initially polarized along the $+x$ axis, $R = -\sin 2\phi$ can be measured as the y component of the polarization of the recoil target.

As a matter of fact, one can easily show the following *theorem*: For the kinematics of (10) but with a target B having any incomplete or complete polarization vector $\vec{P}$, ($\vec{P}^2 < 1$ or $= 1$) in any direction, the recoil target has a spin density matrix with a polarization vector $\vec{P}'$ so that $\vec{P}'$ is a rotation of $\vec{P}$ through angle -2ϕ around the z-axis. where

$$\tan\phi = \frac{-ia_1^\dagger}{a_0}, \qquad \tilde{a}_1^\dagger = i\alpha K\, \widetilde{e^{-\Omega_0} q_x G_2^h}\,. \tag{40}$$

One may ask: for a scattering with a target polarized along the x-axis and a momentum transfer along the x-axis could one pursue a different line of reasoning from the argument above which considers the target as a coherent superposition of two polarization states along the z-axis? The answer is yes. For example, one could directly compute the opaqueness for a target polarized along the x-axis. But care must be taken to consider not only diagonal elements of the current but also off-diagonal elements where the spin component along the x-axis flips. Such elements can be obtained from (20). In the consideration *via* (39), on the other hand, there are no flips of spin along the z-axis, so the calculation is simpler.

5. Influence of finite target mass and projectile structure

We have introduced the concept of hadronic matter current and studied its influence on elastic scattering under assumptions (7), (8) and (9). Assumption (7) was used to prevent a recoil velocity to develop. In actual fact, the recoil mass is finite and a recoil velocity is unavoidable. However, the recoil velocity is transverse to the direction of collision and we believe its effect on the scattering to be small. (In any case, in all scatterings a similar problem exists. For example, the interpretation of the usual electric form factor of the proton as the Fourier transform of the charge distribution is saddled with the same problem. The Breit frame is undoubtedly the best frame in which to approach the problem, but that does not make the recoil velocity problem disappear.)

Assumption (8) can be dropped if we adopt the same convolution idea [1] as outlined above in fig. 1. For example take the case of Kp scattering with a compressed hadronic matter density distribution ρ_{K} inside of K. (Since K is spinless, it does not have a hadronic matter current. Notice that ρ_{K} was denoted by D_{K} in eq. (2).) The natural extension of (2) is

$$\Omega_{\mathrm{Kp}}(b_x, b_z) = (\mathrm{const.})\, \rho_{\mathrm{K}}(b_x, b_z) \otimes [\rho_{\mathrm{p}}(b_x, b_z) - \alpha j_{\mathrm{p}y}(b_x, b_z)]\,, \tag{41}$$

where we have substituted for D_{p} in (2) the right-hand side of (17).

It is now straightforward to use (41) instead of (17) to compute the amplitude for Kp scattering. We have instead of (29)

$$\widetilde{\Omega}_0 = KG^{\mathrm{h}}_{\mathrm{K}}\, G^{\mathrm{h}}_{1\mathrm{p}}, \qquad \widetilde{\Omega}_1 = \pm i\alpha KG^{\mathrm{h}}_{\mathrm{K}}\, q_x G^{\mathrm{h}}_{2\mathrm{p}}\,. \tag{29$'$}$$

K and α are numerical constants which are specific to the Kp collision (α was defined in (13)). Eqs. (30)–(36) and (28) remain unchanged. (37) becomes

$$-4\pi^2 \alpha K \rho_{\mathrm{K}} \otimes j_{y\mathrm{p}} = \mathrm{e}^{\Omega_0} \widetilde{a}_1\,. \tag{37$'$}$$

Thus for Kp scattering, to the lowest order of velocity effects we can draw the following conclusions:

(a) (33) and (29′) give the relationship between $\mathrm{d}\sigma/\mathrm{d}t$ and the density functions $\widetilde{\rho}_{\mathrm{K}}, \widetilde{\rho}_{\mathrm{p}} = (1/2\pi)G^{\mathrm{h}}_{1\mathrm{p}}$. This relationship is the same as that of ref. [1], and is independent of target polarization. (42)

(b) The polarization parameter [4] $P = 0$. (43)

(c) The rotation parameter R is given by

$$R = -\sin 2\phi\,,$$

$$\tan\phi = -ia_1^{\dagger}/a_0\,, \qquad \widetilde{a}_1^{\dagger} = i\alpha K \mathrm{e}^{-\Omega_0} \widetilde{G}^{\mathrm{h}}_{\mathrm{K}} \otimes \widetilde{[q_x G^{\mathrm{h}}_{2\mathrm{p}}]}\,, \tag{44}$$

$$\widetilde{\Omega}_0 = KG^{\mathrm{h}}_{\mathrm{K}}\, G^{\mathrm{h}}_{1\mathrm{p}}\,. \tag{45}$$

In a scattering with the kinematics of (10) but with a target proton with any polarization vector $\vec{P}$, the recoil proton has a polarization vector $\vec{P}'$ which is the vector obtained from $\vec{P}$ by a rotation through angle -2ϕ around the z-axis with ϕ given by (44) and (45).

6. Case of pp scattering

For pp scattering both the projectile and the target have (internal) hadronic matter currents. Instead of (41) we now have, to the lowest order of j_y,

$$\Omega(b_x, b_z) = (\text{const})\,[\rho(b_x, b_z) - \alpha j_y(b_x, b_z)]_{\text{proj}} \otimes [\rho(b_x, b_z) - \alpha j_y(b_x, b_z)]_{\text{target}}. \tag{46}$$

Here we have adopted the definition of the convolution operator $\otimes$ given in ref. [1] which satisfies

$$X \otimes Y = Y \otimes X\,. \tag{47}$$

The reasoning that leads to (46) is illustrated in fig. 4. At a fixed $\vec{b}$ the local opaqueness is proportional to

$$(\boldsymbol{\rho} + \alpha \boldsymbol{j}_y)_{\text{proj}}\,(\boldsymbol{\rho} - \alpha \boldsymbol{j}_y)_{\text{target}}\,. \tag{48}$$

The coordinates $\vec{b}_{\text{proj}}$, $\vec{b}_{\text{target}}$ and $\vec{b}$ are related by

$$\vec{b} = (-\vec{b}_{\text{proj}}) + \vec{b}_{\text{target}}\,. \tag{49}$$

Comparison of (49) with the definition of the convolution operator $\otimes$ in ref. [1] leads to (46) because j_y is odd in its arguments. Thus to the lowest order of j_y,

$$\tilde{\Omega} = \tilde{\Omega}_0 + \tilde{\Omega}_1\,,$$

$$\tilde{\Omega}_0 = K(G^{\text{h}}_{1\text{p}})^2, \qquad \tilde{\Omega}_1 = i\alpha K q_x\, G^{\text{h}}_{1\text{p}} G^{\text{h}}_{2\text{p}}(\sigma_{\text{proj}} + \sigma_{\text{target}})\,, \tag{29''}$$

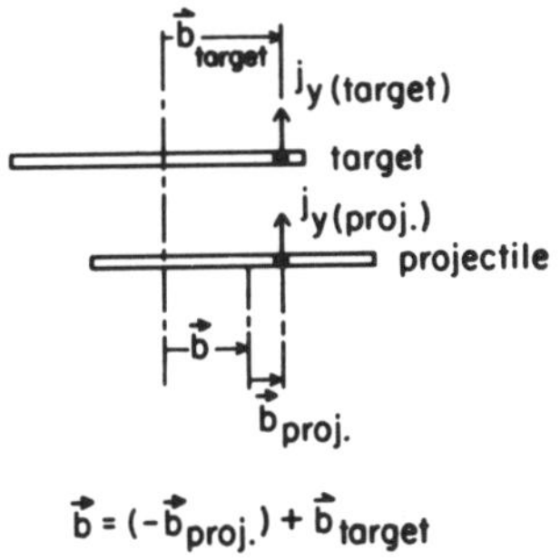

Fig. 4. Collision of a complex projectile with a complex target. $\vec{b}$ is the impact parameter of the collision.

where $\sigma_{\rm proj}$ ($\sigma_{\rm target}$) = ±1 if the projectile (target) is polarized with z spin = $\pm\frac{1}{2}$. The scattering amplitude is, as in (27),

$$a = \widetilde{1 - e^{-\Omega}} = a_0 + a_1 + O(\Omega_1^2) ,$$

where

$$a_0 = \widetilde{1 - e^{-\Omega_0}}, \qquad a_1 = \widetilde{\Omega}_1 - \widetilde{\Omega}_1 \otimes a_0 ,$$

$$\widetilde{a}_1 = \Omega_1 e^{-\Omega_0} . \tag{50}$$

One can draw the following conclusions.

(a) To the first order of the effects of $\vec{j}$, the scattering in the xy plane does not flip the z-spin of either proton. This can be shown without difficulty. There are thus four different scattering amplitudes for the four different polarizations ↑↑, ↑↓, ↓↑, ↓↓. One has

$$a_0^{\uparrow\uparrow} = a_0^{\uparrow\downarrow} = a_0^{\downarrow\uparrow} = a_0^{\downarrow\downarrow} = \text{real and even in } q_x ,$$

$$a_1^{\uparrow\downarrow} = a_1^{\downarrow\uparrow} = 0 ,$$

$$a_1^{\uparrow\uparrow} = -a_1^{\downarrow\downarrow} = \text{pure imaginary and odd in } q_x . \tag{51}$$

To the order considered, a_0 is independent of $\vec{j}$ effects and a_1 is zero without such effects. The differential cross sections are the same for all four polarizations.

(b) There are no right-left asymmetry for any of the four polarizations.

(c) If the initial spin state is

$$\psi = \begin{pmatrix} \psi_{\uparrow\uparrow} \\ \psi_{\uparrow\downarrow} \\ \psi_{\downarrow\uparrow} \\ \psi_{\downarrow\downarrow} \end{pmatrix} ,$$

the outgoing spin state is, to the order considered,

$$[a_0 + a_1^{\uparrow\uparrow} \tfrac{1}{2}(\sigma_{z\rm p} + \sigma_{z\rm t})]\psi = [a_0 e^{i\phi(\sigma_{z\rm p}+\sigma_{z\rm t})}]\psi , \tag{52}$$

where

$$i \tan\phi = \tfrac{1}{2} a_0^{-1} a_1^{\uparrow\uparrow} , \tag{53}$$

$$\widetilde{a}_0 = 1 - e^{-\Omega_0} , \qquad \widetilde{\Omega}_0 = K(G_{1\rm p}^{\rm h})^2 ,$$

$$\widetilde{a}_1^{\uparrow\uparrow} = 2 i\alpha K e^{-\Omega_0} \widetilde{[q_x G_{1\rm p}^{\rm h} G_{2\rm p}^{\rm h}]} . \tag{54}$$

(52) shows that the final spin state is a rotation of the original spin state by a rotation, through an angle -2ϕ *around* the z-axis, *of both polarization vectors of the two particles,* where ϕ is given by (53) and (54). In particular consider the case where the incoming projectile and target are *independently* polarized. Let the initial target be polarized with an incomplete polarization vector $\vec{P}$. Then the recoil target is polarized with a polarization vector $\vec{P}'$ which is the rotation of $\vec{P}$ through -2ϕ around the z-axis, irrespective of the density matrix of the initial projectile.

7. Numerical estimates

As mentioned in the introduction we shall now estimate the magnitude of the rotation parameter R, or rather -2ϕ, by using the assumptions (5) and (6), i.e. the assumption that the electric charge-current density distribution is proportional to the hadronic matter density-current distribution. Now the former is the hadronic contribution to the matrix element of elastic electron-hadron scattering.

Assumptions (5) and (6) can be expressed in another form by taking their Fourier transforms. For the case that the target is a proton the Fourier transform of the left-hand side, $\tilde{\mathbf{p}}$ and $\tilde{\vec{j}}$, are related to G_1^{h} and G_2^{h} through (24). The Fourier transforms of the right-hand side are related in a similar way to the electric and magnetic form factors G_{E} and G_{M}, as we shall see in the next paragraph. Thus (5) and (6) are equivalent to

$$G_1^{\mathrm{h}} = C_1 G_{\mathrm{E}} , \qquad G_2^{\mathrm{h}} = C_1 G_{\mathrm{M}} (2M_{\mathrm{p}})^{-1} . \tag{55}$$

where M_{p} is the mass of the proton.

The relationship between the form factors $G_{\mathrm{E}}, G_{\mathrm{M}}$ and the Fourier transform of the charge-current density was first discovered by Sachs [6]. The relationship is absolutely clear only in the case that the recoil velocity in elastic scattering is zero, i.e. in the limit that $M_{\mathrm{p}} \to \infty$ but with the static magnetic moment remaining fixed. (Examples of such systems are worked out in the appendix.) In such a limit, the target does not move and the matrix elements J^0 and $\vec{J}$ in elastic scattering are Fourier transforms of the spatial distributions:

$$J^0 = \tilde{\mathbf{p}}^{\mathrm{e}}(2\pi) , \qquad \vec{J} = \tilde{\vec{j}}^{\mathrm{e}}(2\pi) .$$

The symbol $\sim$ was defined in (23). But the matrix elements J^0 and $\vec{J}$ are expressible in terms of the usual G_{E} and $(G_{\mathrm{M}}/M_{\mathrm{p}})$. (See e.g. eqs. (9) and (10) of ref. [7]*. $(G_{\mathrm{M}}/M_{\mathrm{p}})$ is finite in the limit $M_{\mathrm{p}} \to \infty$.) Thus

$$2\pi\, \tilde{\mathbf{p}}^{\mathrm{e}} = eG_{\mathrm{E}} , \qquad 2\pi \tilde{\vec{j}}^{\mathrm{e}}_y = \mp ie(q_x)(G_{\mathrm{M}}/2M_{\mathrm{p}}) . \tag{24'}$$

Comparing this with (24) gives (55).

Let us recall that in ref. [1] it was explicitly stated that there was no principle to

* Notice that our q differs in sign with that in this paper. In the limit $M_{\mathrm{p}} \to \infty$, the Breit frame is the rest frame.

Table 1
Parameter α for various scatterings

	α	Energy region
K^+p	0.09	~ 200 GeV/c
K^-p	0.05	~ 200 GeV/c
π^+p	0.04	~ 200 GeV/c
π^-p	0.03	~ 200 GeV/c
pp	0.04	~1500 GeV/c

guide the identification of the hadronic density distribution with F_1, G_M or G_E. Assumption (55) provides now such a guidance. Although it is only an assumption, it is based on the quite natural concept that the hadronic matter density-current distribution forms a four-vector as well as the electric charge-current distribution. (Cf. remark (ii), sect. 8).

Substituting (55) into (44), (45), (53) and (54) we can now evaluate the rotation parameter R for $K^\pm p$, $\pi^\pm p$ and pp scatterings. In this procedure a_0 is determined from $d\sigma/dt$ in elastic scattering data [8], since we have seen to the order considered $d\sigma/dt$ depends only on a_0.

The parameter α in each case is determined by fitting the total cross section with the formula

$$\sigma_T \propto P_{in}^{\alpha} . \tag{56}$$

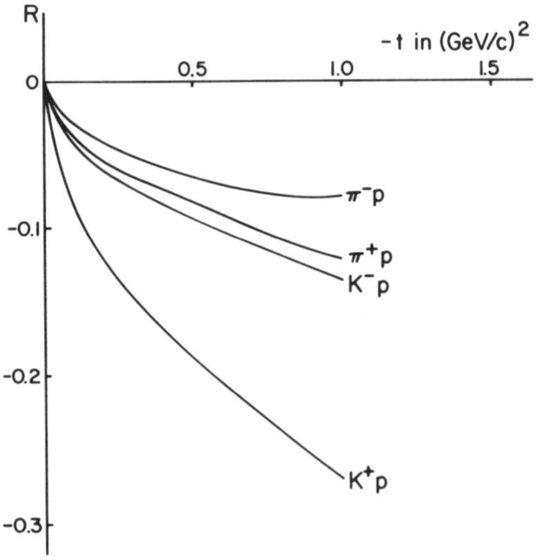

Fig. 5. Estimated value of rotation parameter $R = -\sin 2\phi$ in $K^\pm p$ and $\pi^\pm p$ scattering at 200 GeV.

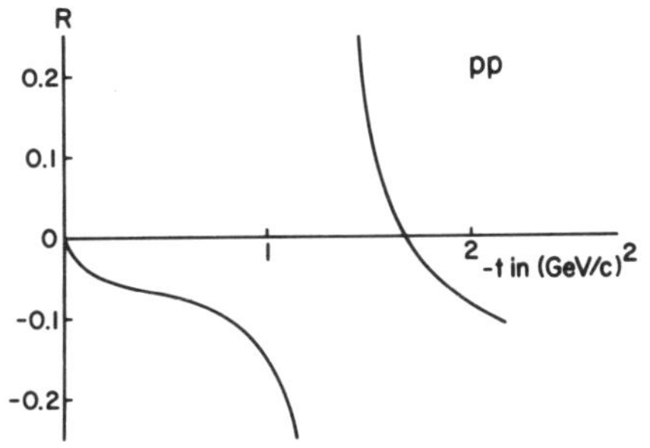

Fig. 6. Estimated value of rotation parameter $R = -\sin 2\phi$ in pp scattering at 1500 GeV. (See text for definition of ϕ). In the region $-t = 1.2$ to 1.5 $(\text{GeV}/c)^2$, R rapidly decreases to -1 and then precipitously rises to $+1$ and then decreases again. At the point $a_0 = 0$, i.e. at the dip of $d\sigma/dt$, R vanishes.

The results are given in table 1. The spin rotation angles -2ϕ ($R = -\sin 2\phi$) calculated are shown in figs. 5 and 6. Measurements of R in πp and Kp scatterings have been performed at lower energies [9] indicating that experimentally it is feasible to measure ϕ to the order of 0.1 radian and less.

These calculations show that for small t, R is negative, i.e. $\phi > 0$. This sign has a classical interpretation which was reported before [2].

8. Remarks

(i) If (55) is valid and furthermore $G_E = G_M/\mu = e^{+\gamma t}$ then it can be shown that $j_y = (\text{const})\, b_x \rho$. Thus the velocity in the y-direction is proportional to b_x. In other words, the hadronic matter motion is approximately a rigid rotation.

(ii) To discuss the validity of assumptions (5) and (6) consider a hypothetical nucleus consisting of nucleons *without* extra magnetic moments, and compare the nucleonic density-current distribution $(\vec{j}^{\mathrm{n}}, i\boldsymbol{\rho}^{\mathrm{n}})$ with the electric charge-current distribution. The Hamiltonian of the nucleus is

$$H_0 = \sum_i \{\vec{\alpha}^{(i)} \cdot (\vec{p}^{(i)} - e^{(i)}\vec{A}^{(i)}) + \beta^{(i)} m\} + V\,. \qquad (57)$$

One has

$$\boldsymbol{\rho}^{\mathrm{e}}(x) = \langle \sum_i e^{(i)}\, \delta(x - x^{(i)}) \rangle\,,$$

$$\vec{j}^{\mathrm{e}}(x) = \langle \sum_i e^{(i)} \vec{\alpha}^{(i)}\, \delta(x - x^{(i)}) \rangle\,,$$

$$\boldsymbol{\rho}^{\mathrm{n}}(x) = \langle \sum_i \delta(x - x^{(i)}) \rangle\,,$$

$$\vec{j}^{\,n}(x) = \langle \sum_i \vec{\alpha}^{(i)} \delta(x - x^{(i)}) \rangle . \tag{58}$$

Only the protons contribute to $\boldsymbol{\rho}^{e}$ and $\vec{j}^{\,e}$. Both protons and neutrons contribute to $\boldsymbol{\rho}^{n}$ and $\vec{j}^{\,n}$. For a nuclear with equal numbers of protons and neutrons, because of the *symmetry* of nucleat forces, the neutron-proton ratio does not change from point to point, i.e., they are "*homogenized*". Even if the number of neutrons is larger than that of the proton, they may still be homogenized. When homogenization obtains, it is clear that assumptions (5) and (6) are valid.

In this hypothetical model, the nucleonic density $\boldsymbol{\rho}^{n}$ is proportional to $\boldsymbol{\rho}^{e}$, and therefore is proportional to the form factor G_E *and not* F_1. (See ref. [7], eq. (9).) This provides a heuristic justification for the first equation of (55).

In reality neutrons and protons do have extra magnetic moments. But it is likely that the principle of minimum electromagnetic interaction still holds, and the nucleonic extra magnetic moments are manifestations of their complex structures. Accepting this, the assumption of homogenization would lead to (5) and (6).

In a real nucleus, because of the shell structure, homogenization is not likely to be accurately correct except for nuclei with equal number of protons and neutrons. The discussion above is therefore meant not so much for real nuclei as for an illustration of the two ingredient concepts that underlie assumptions (5) and (6): minimum electromagnetic interaction for the basic constituents and homogenization.

(iii) One must apply (55) with care. As an example, take the case of the neutron. One has $G_E(0) = 0$ but $\tilde{\boldsymbol{\rho}}^{h}(t=0)$ = total hadronic matter in the neutron $\neq 0$. (This last inequality follows because otherwise the neutron would be without hadronic matter and would be transparent to the incoming projectile.) Thus (55) cannot be valid. We believe delicate cancellation obtains in the electric charge-current distribution for a neutral hadron and assumptions (5) and (6) which are only approximate are subject to large errors for neutral hadrons.

(iv) One can compute the static magnetic moment of a hadron from its $\vec{j}^{\,e}$:

$$\iiint j_y^{e} b_x \, d^3 b = \mathcal{M} . \tag{59}$$

To prove this let us consider the example (57). Consider the static electromagnetic field $A_1 = A_3 = 0$, $A_2 = \mathcal{H} x$. The expectation value of the additional energy due to this static field is

$$-\langle \sum_i e^{(i)} \alpha_y^{(i)} (\mathcal{H} x^{(i)}) \rangle = -\mathcal{H} \iiint j_y^{e} b_x \, d^3 b ,$$

by (58). Identifying this with $-\mathcal{H}\mathcal{M}$ gives (59), which should be valid if the principle of minimum electromagnetic interaction holds. If we use, for spin-$\frac{1}{2}$ particles, $2\pi \tilde{j}_y^{e} = -i q_x G_M / 2M_p$, (59) becomes $G_M(0) = 2M_p \mathcal{M}$ which is satisfied by convention.

It may be tempting to write down formulae similar to (59):

$$\iiint (j_y b_x - j_x b_y) \, d^3 b = (\text{const})\,(\text{angular momentum})_z , \tag{60}$$

$$\iiint \boldsymbol{\rho}\, d^3b = (\text{const})(\text{mass}) . \tag{61}$$

Apply (60), (61) to a proton with z-spin equal to $\frac{1}{2}$:

$$\frac{2\iint b_x j_y \, d^2b}{\iint \rho \, d^2b} = \frac{1}{2M_p} . \tag{62}$$

If (5) and (6) are assumed, one obtains (55) which enables one to compute the left hand side of (62) in terms of G_E and G_M of the proton, resulting in

$$\frac{G_M(t=0)}{M_p G_E(t=0)} = \frac{1}{M_p}(2.79) , \tag{63}$$

which is larger than the right-hand side of (62) by $2(2.79) \approx 5.6$. We believe the reason for this "discrepancy" lies in the incorrectness of formulae (60) and (61): There are in fact three different distributions inside a proton:

electric charge-current distribution , (64)

hadronic matter density-current distribution, (65)

energy-momentum distribution . (66)

(55) expresses the proportionality between (64) and (65) while (60) and (61) expresses the proportionality between (65) and (66). We believe that there is little ground for the validity of the latter proportionality.

(v) There has been a number of theoretical papers [10] discussing polarization effects in hadron-hadron collisions. These discussions usually introduce some spin dependent interaction and is quite different in spirit from the present paper.

(vi) Throughout this paper we have treated the opaqueness as real. This is undoubtedly largely correct, but there may be some small imaginary part to the opaqueness. If such imaginary parts exist then the polarization parameter P would no longer be zero.

(vii) For the case of the *nucleonic current* inside a polarized nucleus, we emphasize that one could use as *experimental input* the projectile-nucleon scattering amplitudes near the forward direction over an energy region. This procedure has the following advantages: (i) The projectile-nucleon scattering amplitude can be measured as a complex quantity. Thus it is not necessary to assume $\Omega(b)$ to be real. (ii) The spin dependence of the projectile-nucleon scattering amplitude can be measured. (iii) Because one uses the projectile-nucleon scattering as an *input* amplitude, it is necessary to consider only the nucleus as an extended structure, while the incoming projectile is treated as a point, all the structural complexities of the latter having already been taken into consideration in the input amplitude. Thus the *convolution assumptions* such as (41) and (46) *become unnecessary.* The analysis requires, for high energies, only the Glauber type [11] of approximation. The fluctuation effects in the coherent part of the process has been discussed before [12].

Appendix

Does the spin of a Dirac particle involve motion?

Consider a "hydrogen atom" in which the proton is infinitely heavy, spinless and chargeless, and binding the electron with a potential $V(r)$. We want to compute its form factors. The Hamiltonian is

$$H = \alpha \cdot p + \beta m + V(r) \,. \tag{67}$$

We take the representation where

$$\beta = \begin{Vmatrix} 1 & & & \\ & 1 & & \\ & & -1 & \\ & & & -1 \end{Vmatrix} , \tag{68}$$

and separate the wave function ψ into

$$\psi = \begin{Vmatrix} \psi_\ell \\ \psi_s \end{Vmatrix} ,$$

where the large and small components ψ_ℓ and ψ_s both have two components. We can solve ψ_s in terms of ψ_ℓ:

$$\psi_s = \frac{-i}{m + E - V} (\sigma \cdot \nabla) \psi_\ell \simeq \frac{-i}{2m} (\sigma \cdot \nabla) \psi_\ell \,. \tag{69}$$

Thus by (58) we have

$$\vec{j}^{\mathrm{e}} = -e\psi^\dagger \vec{\alpha} \psi \simeq \frac{ie}{2m} \psi_\ell^\dagger \nabla \psi_\ell + \text{comp. conj.} \; - \frac{e}{2m} \vec{\nabla} \times (\psi_\ell^\dagger \vec{\sigma} \psi_\ell) \,. \tag{70}$$

One can evaluate this for a $s_{1/2}$ or $p_{1/2}$ state with z component of total spin $= \frac{1}{2}$. The result is

$$j_x^{\mathrm{e}} \simeq \frac{1}{2m} \frac{\partial}{\partial y} \rho^{\mathrm{e}} , \qquad j_y^{\mathrm{e}} \simeq -\frac{1}{2m} \frac{\partial}{\partial x} \rho^{\mathrm{e}} , \qquad j_z^{\mathrm{e}} = 0; \qquad (s_{1/2} \text{ state}) , \tag{71}$$

$$j_x^{\mathrm{e}} \simeq -\frac{1}{2m} \frac{\partial}{\partial y} \rho^{\mathrm{e}} - \frac{2y}{mr^2} \rho^{\mathrm{e}} , \qquad j_y^{\mathrm{e}} \simeq \frac{1}{2m} \frac{\partial}{\partial x} \rho^{\mathrm{e}} + \frac{2x}{mr^2} \rho^{\mathrm{e}} ,$$

$$j_z^e = 0\,, \qquad (p_{1/2}\ \text{state})\,. \tag{72}$$

Taking the Fourier transform of (71) and (72) and using (24′) of sect. 7 we obtain

$$(2M)^{-1}G_M = G_E(2m)^{-1}\,, \qquad (\text{for } s_{1/2}\ \text{state})\,; \tag{73}$$

$$(2M)^{-1}G_M = -G_E(2m)^{-1} + 2(mq^3)^{-1}\int_0^q G_E(K^2)K^2\,\mathrm{d}K\,, \qquad (\text{for } p_{1/2}\ \text{state}). \tag{74}$$

Notice that M and G_M are both ∞ in this model.

These formula do give the correct magnetic moments of the $s_{1/2}$ and $p_{1/2}$ states

$$G_E(0) = -e\,,$$

$$(2M)^{-1}G_M(0) = -e(2m)^{-1}\,, \qquad (\text{for } s_{1/2}\ \text{state})\,,$$

$$(2M)^{-1}G_M(0) = -e(6m)^{-1}\,, \qquad (\text{for } p_{1/2}\ \text{state})\,.$$

For three independent quarks (each with its natural magnetic moment) in the same orbital s-state in the quark model of the proton, the sum of the three eqs. (71) for the three quarks give*

$$G_M/G_E = \text{independent of } q^2\,. \tag{75}$$

This is in agreement with experimental data up to $q^2 \sim 1.2\ (\text{GeV}/c)^2$.

Eq. (71) (which is equivalent to (73)) states that for a Dirac particle in a polarized $s_{1/2}$ state, the electric current is in general *not* zero. It is zero only in regions of space where the charge density has no gradient. Where the charge density has a gradient, as at point A in fig. 7, there is a current which can be understood as due to the difference of the spinning current on the two sides of A. Thus a consistent picture is to answer

Fig. 7. Rotating particle with a density distribution that decreases with increasing distance from the origin. A net upwards current is produced at point A.

* Eq. (75) is correct if the particle is non-relativistic. The relativistic correction is model dependent and is of order $(m - E)m^{-1}$. For earlier discussion of this question in the quark model, see ref. [13].

the title question of this appendix in the affirmative: a polarized Dirac electron is a *rotating* particle. Such a rotating particle can produce an electric current even in the s-state where the "orbital" angular momentum is zero.

The discussion above for the form factors G_E and G_M of an atomic system seems to us to be of interest in general. We have, however, not been able to find in the literature such discussions for atomic systems.

This appendix originated from a question asked by K. Gottfried, T. Kinoshita, K. Lane and D.R. Yennie when one of us discussed the present paper at Cornell. It is a pleasure to acknowledge our indebtedness to these colleagues.

References

[1] T.T. Chou and C.N. Yang, Phys. Rev. 170 (1968) 1591.
[2] T.T. Chou, High-energy collisions – 1973, ed. C. Quigg (AIP Conf. Proc. No. 15, New York, 1973) pp. 118–123;
C.N. Yang, Proc. Int. Symposium on high-energy physics, ed. Y. Hara et al. (University of Tokyo, Japan, 1973) 629–634.
[3] U. Amaldi et al. Phys. Letters 44B (1973) 112;
S.R. Amendolia et al., Phys. Letters 44B (1973) 119;
A.S. Carroll et al., Phys. Rev. Letters 33 (1974) 932.
[3a] Hung Cheng and Tai Tsun Wu, Phys. Rev. Letters 24 (1970) 1456.
[4a] L. Wolfenstein, Phys. Rev. 96 (1954) 1654.
[4b] A. De Lesquen et al., Phys. Letters 40B (1972) 277.
[5] M. Kac, Nucl. Phys. B62 (1973) 402;
N. Byers, Proc. 11th Int. School "Ettore Majorara", 1973;
H. Cheng, J.K. Walker and T.T. Wu, Phys. Letters B44 (1973) 97;
Alexander Wu Chao and Chen Ning Yang, Phys. Rev. D8 (1973) 2063;
F. Hayot and U.P. Sukhatme, Phys. Rev. D10 (1974) 2183;
R. Henzi, B. Margolis and P. Valin, Phys. Rev. Letters 32 (1974) 1077.
[5a] N. Kwak et al., Phys. Letters 58B (1975) 233.
[6] R.G. Sachs, Phys. Rev. 126 (1962) 2256.
[7] L.N. Hand, D.G. Miller and Richard Wilson, Rev. Mod. Phys. 35 (1963) 335.
[8] T.T. Chou, Phys. Rev. D11 (1975) 3145;
V. Franco, Phys. Rev. D11 (1975) 1837.
[9] J. Pierrard et al., Phys. Letters 57B (1975) 393.
[10] C. Bourrely, J. Soffer and D. Wray, Nucl. Phys. B89 (1975) 32;
L. Durand, Wisconsin preprint.
[11] R.J. Glauber, Lectures in Theoretical Physics, ed. W.E. Brittin et al., (Interscience, New York, 1959) vol. 1.
[12] T.T. Chou and Chen Ning Yang, Phys. Rev. 175 (1968) 1832.
[13] G. Morpurgo, Phys. Letters 27B (1968) 378;
R.P. Feynman, M. Kislinger and F. Ravndal, Phys. Rev. D3 (1971) 2706.

DIRAC MONOPOLE WITHOUT STRINGS: MONOPOLE HARMONICS

[[76c]]

Commentary begins page 76

Tai Tsun WU *
Gordon McKay Laboratory, Harvard University, Cambridge, Massachusetts 02138

and

Chen Ning YANG **
Institute for Theoretical Physics, State University of New York, Stony Brook, New York 11794

Received 17 February 1976

Using the ideas developed in a previous paper which are borrowed from the mathematics of fibre bundles, it is shown that the wave function ψ of a particle of charge Ze around a Dirac monopole of strength g should be regarded as a *section*. The section is without discontinuities. Thus the monopole *does not* possess strings of singularities in the field around it. The eigensections of the angular momentum operators are monopole harmonics which are explicitly exhibited.

1. Introduction

In this paper, and a later on classical Lagrangian dynamics, we study the formulation of Dirac's magnetic monopoles without strings. The two papers are, however, logically and technically independent, and may be read separately.

Very soon after Dirac's original paper [1], Tamm [2] studied the wave function of an electrically charged particle around a magnetic monopole. He introduced "generalized spherical harmonics" for such wave functions. These harmonics possess discontinuities or cusps. Later Fierz [3] discussed these harmonics from a different point of view†.

Since the space around a monopole is spherically symmetrical and without singularities, the wave function of a positon or electron around the monopole *should* have no singularities. An examination of this question using the concepts developed

* Work supported in part by the US ERDA under contract no. E(11-1)-3227.
** Work supported in part by the National Science Foundation under grant MPS74-13208 A01.
† There are many papers after the publication of refs. [1–3] dealing with magnetic monopoles and generalized spherical harmonics which are more or less similar to the essence of these papers. We shall not refer to these later papers here.

□ Reprinted from *Nuclear Physics* B107 (1976), 365–380.

in a recent paper [4] shows that this is indeed the case. By a conceptual change, we shall look at the generalized spherical harmonics from a new view point and shall call them *monopole harmonics*. The monopole harmonics are *everywhere analytic* and possess no discontinuities or cusps at all. They form a complete orthonormal set and can be used as the basis of expansion of any wave function around the monopole.

In this new view point, the wave function of an electrically charged particle of charge Ze around a monopole of strength g *should not be thought of as an ordinary function*. It should instead be considered as a "*section*" characterized by a number q defined by

$$q = \tfrac{1}{2}DZ \, , \tag{1}$$

where $D = 2eg$ = monopole strength in Dirac's unit which is $(2e)^{-1}$. We have put

$$c = \hbar = 1 \, .$$

D is an integer which may be positive, negative or zero. So is $2q$.

The concept of a section is familiar in the mathematics of fibre bundles. For the case in question, the wave function is mathematically [5] a section on a C^1 vector bundle, or a line bundle.

2. Wave function as a section

The basic new point is understandable as follows. The cusps and discontinuities arise because any choice of the vector potential $\boldsymbol{A}$ around the monopole must [4] have singularities. The situation is similar to that encountered in the choice of a coordinate system on the surface of a sphere, such as the longitude and latitude system. No choice is possible which does not have some singularities. Yet the geometry of the sphere is clearly without intrinsic singularities. To avoid introducing singularities in the coordinate system one divides the sphere into more than one overlapping region and defines a singularity-free coordinate system in each region. In the overlap one has singularity-free coordinate transformations between the different coordinate systems.

Imitating this method, we divide [4] the space outside of a magnetic monopole into two regions, R_a and R_b, and define a vector potential $(A_\mu)_\mathrm{a}$ in R_a and a vector potential $(A_\mu)_\mathrm{b}$ in R_b. Using spherical coordinates r, θ, ϕ with the monopole at the origin we choose

$$R_\mathrm{a}: \quad 0 \leqslant \theta < \tfrac{1}{2}\pi + \delta \, , \quad 0 < r \, , \quad 0 \leqslant \phi < 2\pi \, , \tag{2}$$

$$R_\mathrm{b}: \quad \tfrac{1}{2}\pi - \delta < \theta \leqslant \pi \, , \quad 0 < r \, , \quad 0 \leqslant \phi < 2\pi \, , \tag{3}$$

$$R_\mathrm{ab}: \quad \tfrac{1}{2}\pi - \delta < \theta < \tfrac{1}{2}\pi + \delta \, , \quad 0 < r \, , \quad 0 \leqslant \phi < 2\pi \quad \text{(overlap)}, \tag{4}$$

where we choose δ such that $0 < \delta \leqslant \frac{1}{2}\pi$.

The vector potentials are chosen to be

$$(A_r)_a = (A_\theta)_a = 0\,, \qquad (A_\phi)_a = \frac{g}{r\sin\theta}(1 - \cos\theta)\,,$$

$$(A_r)_b = (A_\theta)_b = 0\,, \qquad (A_\phi)_b = \frac{-g}{r\sin\theta}(1 + \cos\theta)\,, \tag{5}$$

where A_r, A_θ, A_ϕ are the projections of $\boldsymbol{A}$ in the three local orthogonal directions. One has

$$(A_\mu)_a = (A_\mu)_b + \frac{i}{Ze} S_{ab} \frac{\partial S_{ab}^{-1}}{\partial x^\mu}\,, \tag{6}$$

where

$$S = S_{ab} = e^{2iq\phi} = \text{transition function}\,. \tag{7}$$

S is the gauge transformation phase factor for changing from $(A_\mu)_b$ to $(A_\mu)_a$ in the overlap R_{ab},

$$\psi_a = S_{ab}\,\psi_b\,, \tag{8}$$

where ψ_a and ψ_b are the wave function of a particle of charge Ze in R_a and R_b, respectively. A function ξ which assumes values ξ_a and ξ_b in R_a and R_b and satisfies

$$\xi_a = S_{ab}\,\xi_b = e^{2iq\phi}\,\xi_b \tag{9}$$

in the overlap R_{ab} is called a *section. ψ is thus a section.*

Let the charged particle interact with the monopole and with a potential $V(r)$ which is spherically symmetrical. We assume $V(r)$ to be without singularities for $r > 0$. Then

$$\frac{1}{2m}(\boldsymbol{p} - Ze\boldsymbol{A})^2\,\psi + V\psi = E\psi\,, \tag{10}$$

meaning

$$\frac{1}{2m}(\boldsymbol{p} - Ze\boldsymbol{A}_a)^2\,\psi_a + V\psi_a = E\psi_a\,, \qquad \text{in } R_a\,, \tag{10a}$$

$$\frac{1}{2m}(\boldsymbol{p} - Ze\boldsymbol{A}_b)^2\,\psi_b + V\psi_b = E\psi_b\,, \qquad \text{in } R_b\,. \tag{10b}$$

It is obvious that these equations are compatible with (8) because of (6).

3. Hilbert space of sections

It is clear that if ξ is a section, $x\xi$ is also a section. Also $(p - ZeA)_x\xi$ is a section. Thus $\boldsymbol{r}$ and $\boldsymbol{p} - ZeA$ are operators on the Hilbert space of sections. We define the scalar product of two sections ξ, η as

$$(\eta, \xi) \equiv \int \eta^* \xi \, \mathrm{d}^3 r \, . \tag{11}$$

(The question of convergence at $r = \infty$ and $r = 0$ is here ignored.) This integral is well defined because in the overlap R_{ab}

$$\eta_{\mathrm{b}}^* \xi_{\mathrm{b}} = \eta_{\mathrm{a}}^* \xi_{\mathrm{a}} \, .$$

It is clear that $\boldsymbol{r}$ and $\boldsymbol{p} - ZeA$ are Hermitian operators. Following Fierz [3] we shall now try to construct angular momentum operators.

Define

$$\boldsymbol{L} = \boldsymbol{r} \times (\boldsymbol{p} - Ze\boldsymbol{A}) - \frac{q\boldsymbol{r}}{r} \, . \tag{12}$$

It is clear that L_x, L_y, L_z are Hermitian operators on the Hilbert space of sections. The following commutation rules can be easily verified:

$$[L_x, x] = 0 \, , \qquad [L_x, y] = iz \, , \qquad [L_x, z] = -iy \, ,$$
$$[L_x, p_x - ZeA_x] = 0 \, , \qquad [L_x, p_y - ZeA_y] = i(p_z - ZeA_z) \, ,$$
$$[L_x, p_z - ZeA_z] = -i(p_y - ZeA_y) \, . \tag{13}$$

It follows from these that

$$[L_x, L_y] = iL_z \, , \qquad \text{etc} \, . \tag{14}$$

Eq. (13), together with its consequence (14), show that L_x, L_y, L_z are *the angular momentum operators* [3]. We emphasize that neither the Hilbert space, nor these operators, possess any "singularities". (The singularities of A_{a} and A_{b} are not real singularities because they occur outside of R_{a} and R_{b}, respectively.)

4. Monopole harmonics $Y_{q,l,m}$

Since $[r^2, \boldsymbol{L}] = 0$, we can diagonalize r^2 and study operators $\boldsymbol{L}$ for fixed r^2. I.e. we shall study sections of the form

$$\delta(r^2 - r_0^2)\xi \, ,$$

where ξ is a section dependent only on angular coordinates θ and ϕ.

$\boldsymbol{L}$ operates then on "angular sections". In the rest of this paper except sect. 11, we shall be dealing with angular sections only.

Eq. (14) shows that $[\boldsymbol{L}^2, L_z] = 0$. Simultaneous diagonalization produces the familiar multiplets with eigenvalues $l(l+1)$ and m,

$$\boldsymbol{L}^2 Y_{q,l,m} = l(l+1)\, Y_{q,l,m}\,;\; L_z Y_{q,l,m} = m Y_{q,l,m}\,, \tag{15}$$

where $l = 0, \frac{1}{2}, 1, \ldots$ and, for each value of l, m ranges from $-l$ to $+l$ in integral steps of increment. The $Y_{q,l,m}$ are the eigensections which we shall call *monopole harmonics.* We shall show later that the allowed values of l and m are

$$l = |q|, |q|+1, |q|+2, \ldots, \qquad m = -l, -l+1, \ldots, l\,, \tag{16}$$

and that each of these l, m combinations occur exactly once. We shall choose each Y normalized so that

$$\int_0^{\pi} \sin\theta\, d\theta \int_0^{2\pi} |Y_{q,l,m}|^2\, d\phi = 1\,. \tag{17}$$

(Notice that in R_{ab}, $|(Y_{q,l,m})_{\text{a}}|^2 = |(Y_{q,l,m})_{\text{b}}|^2$.) Different $Y_{q,l,m}$ (for a fixed q) are orthogonal, a fact one easily proves in the usual way from (15). We shall choose the phases of $Y_{q,l,m}$ such that the matrix elements of L_z, L_y, L_z between the Y's conform to the convention adopted in ch. 2 of Edmonds' book [6]. In particular

$$(L_x + iL_y) Y_{q,l,m} = (l-m)^{1/2}(l+m+1)^{1/2}\, Y_{q,l,m+1}\,. \tag{18}$$

These monopole harmonics will be explicitly exhibited. Each is analytic. That is, $(Y_{q,l,m})_{\text{a}}$ is analytic in R_{a} and $(Y_{q,l,m})_{\text{b}}$ is analytic in R_{b}. The set of all monopole harmonics for a fixed q forms a complete set of sections, as we shall see.

5. Explicit expressions for $Y_{q,l,m}$

Stating from (12) one easily verifies

$$\boldsymbol{L}^2 = [\boldsymbol{r} \times (\boldsymbol{p} - Ze\boldsymbol{A})]^2 + q^2\,, \tag{19}$$

$$mY_{q,l,m} = L_z Y_{q,l,m} = (-i\partial_\phi - q) Y_{q,l,m}\,, \qquad \text{in } R_{\text{a}}\,,$$

$$mY_{q,l,m} = L_z Y_{q,l,m} = (-i\partial_\phi + q) Y_{q,l,m}\,, \qquad \text{in } R_{\text{b}}\,. \tag{20}$$

Eq. (20) shows that

$$Y_{q,l,m} = \Theta_{q,l,m}(\theta)\, e^{i(m+q)\phi} \text{ in } R_{\text{a}}\,,$$

$$Y_{q,l,m} = \Theta_{q,l,m}(\theta)\, e^{i(m-q)\phi} \text{ in } R_{\text{b}}\,. \tag{21}$$

The condition for a section, (9), shows that $[\Theta_{q,l,m}(\theta)]_a = [\Theta_{q,l,m}(\theta)]_b$ in the overlap. They are, in fact, the same function. Apply (19) to $Y_{q,l,m}$. An explicit evaluation of the operator $[\boldsymbol{r} \times (\boldsymbol{p} - Ze\boldsymbol{A})]^2$ acting on $Y_{q,l,m}$ gives

$$[l(l+1) - q^2]\,\Theta_{q,l,m} = \left[-\frac{1}{\sin\theta}\frac{\partial}{\partial\theta}\sin\theta\frac{\partial}{\partial\theta} + \frac{1}{\sin^2\theta}(m + q\cos\theta)^2\right]\Theta_{q,l,m}\,. \tag{22}$$

Writing $\cos\theta = x$, this gives

$$[l(l+1) - q^2]\,\Theta = -(1-x^2)\,\Theta'' + 2x\Theta' + \frac{1}{1-x^2}(m+qx)^2\,\Theta\ ,$$

$$-1 \leqslant x \leqslant 1\ , \tag{23}$$

where prime means differentiation with respect to x. This equation can be treated in the usual way, through analyzing the indical equations at $x = \pm 1$. We shall, however, pursue a different method which yields the normalization constant and phase factor automatically.

Before proceeding we note that since Y is single valued in each region, (21) shows that

$$m - q = \text{integer}\ .$$

Thus

$$l - q = \text{integer}\ . \tag{24}$$

Now (19) shows that

$$l(l+1) \geqslant q^2\ . \tag{25}$$

Eqs. (24) and (25) show that the allowed values of l are among those given in (16).

We shall now show that each value of l in (16) is allowed, by constructing, for each of them, the explicit function $\Theta_{q,l,m}$:

$$\Theta_{q,l,-l} = N_{q,l}\sqrt{1-x}^{\,l-q}\sqrt{1+x}^{\,l+q}\,, \qquad l - |q| = \text{integer} \geqslant 0\ , \tag{26}$$

where

$$N_{q,l} = \left[\frac{(2l+1)!}{4\pi(2^{2l})(l-q)!\,(l+q)!}\right]^{1/2} > 0\ . \tag{27}$$

To show this one substitutes (26) into (23) and verifies that the latter is satisfied. The factor $N_{q,l}$ is inserted so that $Y_{q,l,-l}$ is normalized in the sense of (17).

Repeated application of (18) onto the monopole harmonics $Y_{q,l,-l}$ (given by (21) and (26)) leads to, (for l, m satisfying (26)), the explicit expression for $Y_{q,l,m}$

given below. (As stated above, this method leads to automatically normalized $Y_{q,l,m}$ starting from normalized $Y_{q,l,-l}$.)

$$(Y_{q,l,m})_a = M_{q,l,m}(1-x)^{\alpha/2}(1+x)^{\beta/2} P_n^{\alpha,\beta}(x)\, e^{i(m+q)\phi} ,$$

$$(Y_{q,l,m})_b = (Y_{q,l,m})_a e^{-2iq\phi} , \qquad (28)$$

where

$$\alpha = -q-m, \qquad \beta = q-m, \qquad n = l+m, \qquad x = \cos\theta , \qquad (29)$$

$$M_{q,l,m} = 2^m \left[\frac{2l+1}{4\pi} \frac{(l-m)!\,(l+m)!}{(l-q)!\,(l+q)!}\right]^{1/2} , \qquad (30)$$

and $P_n^{\alpha,\beta}(x)$ are [7] the Jacobi polynomials,

$$P_n^{\alpha,\beta}(x) = \frac{(-1)^n}{2^n n!}(1-x)^{-\alpha}(1+x)^{-\beta} \frac{d^n}{dx^n}[(1-x)^{\alpha+n}(1+x)^{\beta+n}] , \qquad (31)$$

which are defined if

$$n,\ n+\alpha,\ n+\beta \text{ and } n+\alpha+\beta \text{ are all integers} \geqslant 0 . \qquad (32)$$

Eq. (28) will be proved in appendix A, and some properties of the Jacobi polynomials will be discussed in appendix B.

6. Completeness of monopole harmonics

For a given q (q may be negative) the set of $Y_{q,l,m}$ with l, m satisfying (16) form a complete set of orthonormal sections. I.e. every continuous section (i.e. a section satisfying (9), with ξ_a and ξ_b being continuous in R_a and R_b) can be expanded as a series

$$\sum_{l,m} a_{l,m} Y_{q,l,m} .$$

Proof: According to appendix C, $Y_{q,l,m}$ can be expressed in terms of $P_\nu^{|\alpha|,|\beta|}(x)$. Now for fixed q = integer or half-integer, and $q+m$ = integer, there are four possible cases:

$$\alpha \geqslant 0,\ \beta \geqslant 0, \text{ so that } -m \geqslant |q| \quad \text{and} \quad \nu = l+m , \qquad (33)$$

$$\alpha \geqslant 0,\ \beta \leqslant 0, \text{ so that } |m| \leqslant -q, \quad q \leqslant 0 \quad \text{and } \nu = l+q , \qquad (34)$$

$$\alpha \leqslant 0,\ \beta \geqslant 0, \text{ so that } |m| \leqslant q,\ q \geqslant 0 \text{ and } \nu = l-q , \qquad (35)$$

$$\alpha \leqslant 0,\ \beta \leqslant 0, \text{ so that } m \geqslant |q| \quad \text{and } \nu = l-m . \qquad (36)$$

In case (33), the allowed values of l, according to (16), are $l = |m|, |m| + 1, \ldots$ which are precisely

$$\nu = 0, 1, 2, \ldots . \tag{37}$$

In case (34), the allowed values of l according to (16) are $l = -q, -q + 1, \ldots$ which are also precisely (37). Continuing this way we conclude that given q = integer or half-integer, $q + m$ = integer, the allowed values of l according to (16) are always precisely those given by (37).

Now for fixed $|\alpha|, |\beta|$, the Jacobi polynomials $P_\nu^{|\alpha|, |\beta|}$, ($\nu = 0, 1, 2, \ldots$) form [7] a complete set. The exponential functions $e^{i\phi(m+q)}$, ($m + q$ = all integers) also form a complete set. It can be proved from these results that $Y_{q,l,m}$ forms a complete set of sections for fixed q.

7. Examples and analyticity of $Y_{q,l,m}$

For the case $q = 0$, $\alpha = \beta$, and (31) shows that

$$P_{l+m}^{-m, -m} = \frac{(-1)^m}{2^m} \frac{l!}{(l+m)!} (1 - x^2)^{m/2} P_l^m , \tag{38}$$

Table 1
Examples of $\sqrt{4\pi} Y_{q,l,m}$ in region a

q	l	m	$(\sqrt{4\pi} Y_{q,l,m})_a$
$\frac{1}{2}$	$\frac{1}{2}$	$\frac{1}{2}$	$-e^{i\phi}\sqrt{1-x}$
	$-\frac{1}{2}$	$-\frac{1}{2}$	$e^0\sqrt{1+x}$
	$\frac{3}{2}$	$\frac{3}{2}$	$\sqrt{3/2}\, e^{2i\phi}\sqrt{1+x}(1-x)$
	$\frac{3}{2}$	$\frac{1}{2}$	$-\sqrt{1/2}\, e^{i\phi}\sqrt{1-x}(1+3x)$
	$\frac{3}{2}$	$-\frac{1}{2}$	$-\sqrt{1/2}\, e^0\sqrt{1+x}(1-3x)$
	$\frac{3}{2}$	$-\frac{3}{2}$	$\sqrt{3/2}\, e^{-i\phi}\sqrt{1-x}(1+x)$
1	1	1	$\sqrt{3/4}\, e^{2i\phi}(1-x)$
	1	0	$-\sqrt{3/2}\, e^{i\phi}\sqrt{1-x^2}$
	1	-1	$\sqrt{3/4}\, e^0 (1+x)$

$x = \cos\theta$. To obtain $Y_{q,l,m}$ in R_b apply (9).

where P^l_m is the associated Legendre function. Substitution of (38) into (28) shows that

$$Y_{0,l,m} = \text{usual spherical harmonics } Y_{l,m} \;.$$

We tabulate in table 1 a few of the monopole harmonics for $q = \frac{1}{2}$, 1, These examples illustrate the fact that $Y_{q,l,m}$ *is analytic everywhere.* I.e., $(Y_{q,l,m})_a$ is analytic in R_a and $(Y_{q,l,m})_b$ is analytic in R_b. For example, $(Y_{\frac{1}{2}\frac{1}{2}\frac{1}{2}})_a$ is clearly analytic in R_a, which includes the point $\theta = 0$, and

$$(Y_{\frac{1}{2}\frac{1}{2}\frac{1}{2}})_{b\,1/2})_b = \sqrt{1-\cos\theta}/\sqrt{4\pi} \tag{39}$$

is clearly analytic in R_b which includes the point $\theta = \pi$.

8. Zeros of $Y_{q,l,m}$

Table 1 shows that each of the $Y_{q,l,m}$ exhibited has at least one zero. This is in fact a special case of a general topological theorem that for $q \neq 0$, any continuous section must have at least one zero. This theorem can be proved as follows. If ξ is a continuous section and has no zeros, trace the value of $\xi_a|\xi_a|^{-1}$ in *the complex plane* as one goes along the parallel $r = 1, \theta = \theta_0$, from $\phi = 0 \to 2\pi$. $\xi_a|\xi_a|^{-1}$ (= the phase of ψ_a) describes a loop which is confined to the unit circle. As θ_0 changes, the loop is *continuously* distorted, remaining always on the unit circle. As $\theta_0 \to 0$, the loop shrinks to a point. Thus for any θ_0 satisfying $0 < \theta_0 < \frac{1}{2}\pi + \delta$ (see (2)), the loop is always shrinkable, along the circle, to a point. The same is true for the loop described by $\xi_b|\xi_b|^{-1}$ for $\frac{1}{2}\pi - \delta < \theta_0 < \pi$. Now take $\theta_0 = \frac{1}{2}\pi$. These two last statements together contradict (9) if $q \neq 0$.

9. Global gauge transformation on $Y_{q,l,m}$

The monopole harmonics exhibited above are for a special gauge [4] in which the regions R_a, R_b and the vector potential A_μ were chosen to be that given in (2) → (5). One can make global gauge transformations [4] which change the regions, the vector potential A_μ, and the value of $Y_{q,l,m}$ in a coordinated manner.

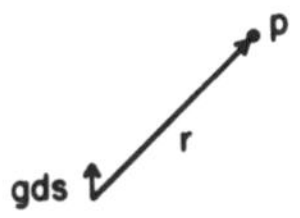

Fig. 1. Pseudomagnetic field produced by current segment ds. It is equal to $\boldsymbol{gds} \times \boldsymbol{r}r^{-3}$ where g = current.

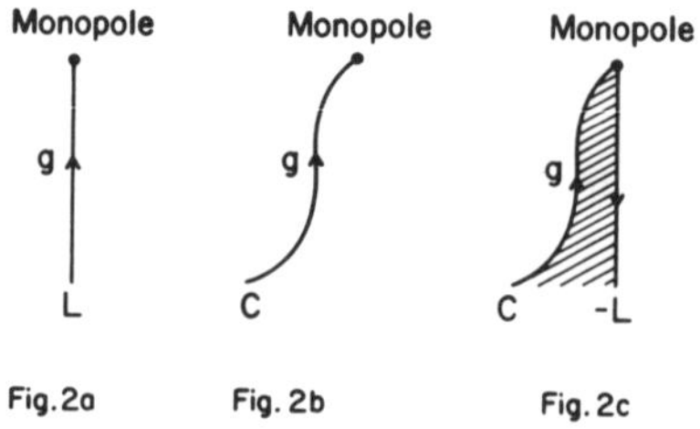

Fig. 2. Half line L and half curve C carrying current g, and complete circuit $C + (-L)$.

For example, one could make a global gauge transformation by merely contracting and expanding the regions R_a and R_b without changing either A_μ or $Y_{q,l,m}$, provided R_a and R_b together always fill the whole space outside of the origin, and provided R_a does not include the line $\theta = \pi$ while R_b does not include the line $\theta = 0$.

To discuss a more interesting global gauge transformation we shall first try to find some other possible vector potential $\boldsymbol{A}'_a$ in R_a. To this end, define as in fig. 1 the pseudomagnetic field produced, according to Biot–Savart's law, from a segment of electric current g. (The field is called a pseudomagnetic field and not a magnetic field, because of two related facts: (i) The current segment itself does not give conserved current. (ii) The pseudomagnetic field is not curlless outside of the current segment. The total pseudomagnetic field produced by a *complete* electric circuit is the magnetic field.) It is easy by straightforward integration to find that (5) satisfies

$$(\boldsymbol{A})_a = \text{total pseudomagnetic field generated by half line } L\,, \quad \text{carrying current } g \text{ (fig. 2a)}\,. \tag{40}$$

Now define

$$(\boldsymbol{A}')_a = \text{total pseudomagnetic field generated by any half curve } C\,, \quad \text{carrying current } g \text{ (fig. 2b)}\,. \tag{41}$$

Then

$$(\boldsymbol{A}')_a - (\boldsymbol{A})_a = \text{total pseudomagnetic field generated by } C + (-L)\,, \quad \text{carrying current } g \text{ (fig. 2c)}\,. \quad = \text{magnetic field generated by same}\,. \tag{42}$$

Thus

$$\nabla \times [(\boldsymbol{A}')_a - (\boldsymbol{A})_a] = 0 \text{ outside of } L \text{ and } C\,. \tag{43}$$

If C is chosen completely outside of R_a, then (43) asserts that $\nabla \times (\boldsymbol{A}')_a$ is the magnetic field of the monopole in R_a and thus we may use $(A')_a$ as the vector potential in R_a.

Using $(A'_\mu)_a$ and $(A_\mu)_b$ as the vector potentials requires a global gauge transformation from $(A_\mu)_a$ and $(A_\mu)_b$. In other words, we can find a transformation phase factor $T_{a'a}$ such that

$$(A'_\mu)_a = (A_\mu)_a + \frac{i}{Ze} T_{a'a} \frac{\partial T_{a'a}^{-1}}{\partial x^\mu} \quad \text{in } R_a \,. \tag{44}$$

For any section ξ,

$$\xi_{a'} = T_{a'a} \xi_a \,. \tag{45}$$

Comparison of (44) and (42) shows that

$$T_{a'a} = \exp(-i\, Zeg\, \Omega) = \exp(-i\, q\Omega) \,, \tag{46}$$

where Ω is the solid angle subtended by circuit $C + (-L)$ in fig. 2c at the point where $T_{a'a}$ is evaluated. Actually Ω is defined and is continuous not just in R_a but in all space outside of a surface bordered by the closed circuit $C + (-L)$. Take the surface to be the shaded area in fig. 2c. On the surface, but outside of the border $C + (-L)$, Ω increases discontinuously by 4π in going from above the diagram to underneath the diagram. Since D is an integer, $T_{a'a}$ is single valued.

The transition function in the overlap region R_{ab} is now

$$S_{a'b} = T_{a'a} S_{ab} = \exp(2iq\phi - iq\Omega) \,. \tag{47}$$

We have thus defined completely the new gauge: regions R_a, R_b and transition function $S_{a'b}$. We have also defined the gauge field in this new gauge: $(A'_\mu)_a$, $(A_\mu)_b$. We have, in addition, exhibited the global gauge transformation $T_{a'a}$ between the old gauge and the new.

Notice that the continuation of $(A'_\mu)_a$ into R_b yields singularities not on L, but on C.

If we had taken another half curve C' outside of R_a, we would have gotten another new vector potential $(A''_\mu)_a$. Any linear combination

$$(A'''_\mu)_a = \alpha (A_\mu)_a + \alpha' (A'_\mu)_a + \alpha'' (A''_\mu)_a \,, \tag{48}$$

with

$$\alpha + \alpha' + \alpha'' = 1 \,,$$

where α, α' and α'' are positive or negative real numbers, is also a possible vector potential in R_a. Notice that the continuation of $(A'''_\mu)_a$ into R_b has, in general, singularities on L, C and C'. Thus the position of the singularities of $(A_\mu)_a$, when continued into space not covered by R_a, is in general quite arbitrary.

10. Rotation of coordinate axes

A rotation of coordinate axes generates [6] a linear combination of the usual spherical harmonics,

$$Y_{0,l,m}(\theta', \phi') = \sum_{m'=-l}^{l} Y_{0,l,m'}(\theta, \phi) \mathcal{D}^{(l)}_{m'm} , \tag{49}$$

where $\mathcal{D}$ depends on the rotation. Does this hold also for the case $q \neq 0$? Define

$$Z_{q,l,m}(\theta', \phi') = \sum_{m'=-l}^{l} Y_{q,l,m'}(\theta, \phi) \mathcal{D}^{(l)}_{m'm} , \tag{50}$$

for $q \neq 0$. $Z_{q,l,m}$ is $Y_{q,l,m}(\theta' \phi')$ but in *a different gauge*, because the vector potential A_μ has not yet been changed to that which conforms with convention (5) for the new coordinate system. If one performs a global gauge transformation on $Z_{q,l,m}$ by first changing $(A_\mu)_a$ so that its singularities after continuation become the new negative z-axis, and then changing $(A_\mu)_b$ so that its singularities after continuation become the new positive z-axis then $Z_{q,l,m} \to Y_{q,l,m}(\theta', \phi') \times$ (phase factor which is independent of m).

11. Schrödinger equation

It is simple to show by explicit evaluation, and with the aid of (19) that

$$(\boldsymbol{p} - Ze\boldsymbol{A})^2 = -\frac{1}{r^2}\frac{\partial}{\partial r}\left(r^2 \frac{\partial}{\partial r}\right) + \frac{1}{r^2}\left[\boldsymbol{r} \times (\boldsymbol{p} - Ze\boldsymbol{A})\right]^2$$

$$= -\frac{1}{r^2}\frac{\partial}{\partial r}\left(r^2 \frac{\partial}{\partial r}\right) + \frac{1}{r^2}\left[\boldsymbol{L}^2 - q^2\right] . \tag{51}$$

The Hamiltonian in (10) thus commutes with $\boldsymbol{L}^2$ and L_z. Hence in solving (10) we can choose specific eigenvalues for $\boldsymbol{L}^2$ and L_z. I.e. we take

$$\psi = R(r) Y_{q,l,m} , \tag{52}$$

obtaining

$$\left[-\frac{1}{2mr^2}\frac{\partial}{\partial r}\left(r^2 \frac{\partial}{\partial r}\right) + \frac{l(l+1) - q^2}{2mr^2} + V - E\right] R = 0 . \tag{53}$$

For the case that $V = 0$ this equation was solved by Tamm [2] who found that R is a Bessel function, if $E > 0$,

$$R = \frac{1}{\sqrt{kr}} J_\mu(kr) , \tag{54}$$

where

$$\mu = \sqrt{l(l+1) - q^2 + \tfrac{1}{4}} = \sqrt{(l+\tfrac{1}{2})^2 - q^2} > 0\,, \tag{55}$$

$$k = \sqrt{2mE}\,.$$

If $E \leqslant 0$, (53) has no meaningful solution.

It is a pleasure to thank Professor Shiing-shen Chern for enlightening us on the mathematical concepts of fibre bundles and sections.

Appendix A

Proof of (28)

A straight forward computation shows that in R_a

$$L_x + iL_y = e^{i\phi}\left[-\sqrt{1-x^2}\,\frac{\partial}{\partial x} + i\frac{x}{\sqrt{1-x^2}}\frac{\partial}{\partial\phi} - q\sqrt{\frac{1-x}{1+x}}\right], \tag{A.1}$$

where

$$x = \cos\theta\,.$$

Substitute this into (18) and use (21). One obtains, in R_a,

$$\Theta_{q,l,m+1} = \frac{1}{\sqrt{(l-m)(l+m+1)}}\left[-\sqrt{1-x^2}\,\frac{d}{dx} - \frac{mx}{\sqrt{1-x^2}} - q\frac{1}{\sqrt{1-x^2}}\right]\Theta_{q,l,m}$$

$$= \frac{-\sqrt{1-x^2}^{\,m+1}}{\sqrt{(l-m)(l+m+1)}}\left[\frac{d}{dx} + q\,\frac{1}{1-x^2}\right]\sqrt{1-x^2}^{\,-m}\,\Theta_{q,l,m}\,. \tag{A.2}$$

Repeated application of (A.2) gives

$$\Theta_{q,l,m} = (\text{const})\,(1-x^2)^{m/2}\left[\frac{d}{dx} + q\,\frac{1}{1-x^2}\right]^{l+m}(1-x^2)^{l/2}\,\Theta_{q,l,-l}\,. \tag{A.3}$$

Now use

$$\frac{d}{dx} + q\,\frac{1}{1-x^2} = \sqrt{\frac{1-x^q}{1+x}}\;\frac{d}{dx}\sqrt{\frac{1+x^q}{1-x}}\,. \tag{A.4}$$

One obtains eq. (28) from (A.3). (The constant in (A.3) can be evaluated explicitly in the repeated application of (A.2).)

Appendix B

Some properties of $P_n^{\alpha,\beta}$

$P_n^{\alpha,\beta}(x)$ as defined by (31) and (32) is a polynomial of degree n. It satisfies

$$P_n^{\alpha,\beta}(-x) = (-1)^n P_n^{\beta,\alpha}(x)\,, \tag{B.1}$$

$$P_n^{\alpha,\beta}(x) = 2^{-n} \sum_{\lambda=0}^{n} \frac{(n+\alpha)!}{\lambda!\,(n+\alpha-\lambda)!}\, \frac{(n+\beta)!}{(n-\lambda)!\,(\beta+\lambda)!} (x-1)^{n-\lambda}(x+1)^{\lambda}, \tag{B.2}$$

in which $m!$ is defined to be ∞ when $m < 0$ and $0! = 1$. To prove (B.2) we arrange the square bracket in (31) into a product of $2n + \alpha + \beta$ factors, each being $(1-x)$ or $(1+x)$, and choose λ factors $(1-x)$ and $n - \lambda$ factors $(1+x)$ for differentiation. (B.2) then follows

We shall now show

$$P_{n+\alpha}^{-\alpha,\beta} = 2^{-\alpha}(x-1)^{\alpha} \frac{n!\,(n+\alpha+\beta)!}{(n+\beta)!\,(n+\alpha)!} P_n^{\alpha,\beta}\,, \tag{B.3}$$

$$P_{n+\beta}^{\alpha,-\beta} = 2^{-\beta}(x+1)^{\beta} \frac{n!\,(n+\alpha+\beta)!}{(n+\beta)!\,(n+\alpha)!} P_n^{\alpha,\beta}\,, \tag{B.4}$$

$$P_{n+\alpha+\beta}^{-\alpha,-\beta} = 2^{-\alpha-\beta}(x-1)^{\alpha}(x+1)^{\beta} P_n^{\alpha,\beta}\,. \tag{B.5}$$

To show (B.3) we use (B.2),

$$2^{n+\alpha} P_{n+\alpha}^{-\alpha,\beta} = \sum_{\lambda=0}^{n+\alpha} \frac{n!}{\lambda!\,(n-\lambda)!}\, \frac{(n+\alpha+\beta)!}{(n+\alpha-\lambda)!\,(\beta+\lambda)!} (x-1)^{n+\alpha-\lambda}(x+1)^{\lambda}$$

$$= \sum_{\lambda=0}^{n} \text{same} = (x-1)^{\alpha} \frac{n!(n+\alpha+\beta)!}{(n+\alpha)!\,(n+\beta)!}\, 2^n P_n^{\alpha,\beta}\,,$$

which leads to (B.3). Eq. (B.4) can be proved similarly. Eq. (B.5) can be proved by using (B.3) and (B.4) in succession.

Define

$$R_n^{\alpha,\beta} \equiv (1-x)^{\alpha/2}(x+1)^{\beta/2}\, 2^{-(\alpha+\beta)/2} P_n^{\alpha,\beta}\,. \tag{B.6}$$

Then (B.3), (B.4) and (B.5) together show that

$$R_n^{\alpha,\beta} = (-1)^{(\alpha-|\alpha|)/2} R_\nu^{|\alpha|,|\beta|}\,, \tag{B.7}$$

where

$$\nu = n + \tfrac{1}{2}(\alpha+\beta-|\alpha|-|\beta|)\,. \tag{B.8}$$

Using (28) as the definition of $Y_{q,l,m}$, we obtain by utilizing (B.5)

$$Y^*_{q,l,m} = (-1)^{q+m}\, Y_{-q,l,-m}\ , \tag{B.9}$$

which is a useful formula. It is correct in both regions R_a and R_b. If we take $q = 0$, (B.6) reduces to the usual formula for the complex conjugate of $Y_{l,m}$.

Appendix C

Alternative expression for $Y_{q,l,m}$

Using (28) and (B.6) we obtain

$$(Y_{q,l,m})_a = M_{q,l,m}\, 2^{(\alpha+\beta)/2} R_n^{\alpha,\beta}\, e^{i(m+q)\phi}\ . \tag{C.1}$$

Now use (B.7) to obtain

$$(Y_{q,l,m})_a = (\text{const})\,(1-x)^{|\alpha|/2}\,(1+x)^{|\beta|/2}\, P_\nu^{|\alpha|,|\beta|}\, e^{i(m+q)\phi}\ , \tag{C.2}$$

where ν is given by (B.8), and n, α and β are given by (29).

Appendix D

Clebsch-Gordan coefficients

We shall define the usual Clebsch-Gordan coefficients

$$\langle lml'm' \,|\, ll'jm_j \rangle \tag{D.1}$$

as in ref. [6]. Some usage of these coefficients for combining sections will be discussed below:

(a) Consider the product of two sections $Y_{q,l,m}(\theta,\phi)\, Y_{q',l',m'}(\theta,\phi)$ *of the same argument* θ, ϕ. The result is clearly a section with $q'' = q + q'$. The usual vector addition theorem applies and we have

$$\sum_{mm'} Y_{q,l,m}\, Y_{q',l',m'} \langle lml'm' \,|\, ll'jm_j \rangle = K\, Y_{q+q',j,m_j}\ , \tag{D.2}$$

where K depends on q, l, q', l', and j but not on m. Notice that sometimes K is zero. For example, for $Y_{q,l,m}\, Y_{q,l,m'}$, it is well known that the CG coefficients are symmetrical (with respect to $m \leftrightarrow m'$) for $j = 2l -$ even integer and antisymmetrical for $j = 2l -$ odd integer. For the latter case clearly $K = 0$. Notice also that if $j < |q + q'|$, then the right-hand side of (D.2) must vanish, since Y_{q+q',j,m_j} does not then exist.

For example, for the case $Y_{1,1,m} Y_{0,1,m'}$ the final j value is, *a priori*, 2, 1 or 0. But the case $j = 0$ vanishes since $Y_{1,0,m_j}$ does not exist. This can indeed be checked with the aid of tables 1, and the appropriate values of $Y_{0,1,m'}$ and the Clebsch-Gordan coefficients

$$Y_{1,1,1}(\sqrt{1-x^2}\,e^{-i\phi}) - Y_{1,1,0}(\sqrt{2}x) + Y_{1,1,-1}(-\sqrt{1-x^2}\,e^{i\phi}) = 0 \,. \quad \text{(D.3)}$$

Similarly, $Y_{1,1,m} Y_{1,1,m'}$ can be linearly combined, *a priori*, to give $Y_{2,2,m_j}$, $Y_{2,1,m}$ and $Y_{2,0,m}$. But the latter two do not exist, giving rise to the following identity which can be checked with table 1

$$Y_{1,1,1}Y_{1,1,-1} - Y_{1,1,0}Y_{1,1,0} + Y_{1,1,-1}Y_{1,1,1} = 0 \,. \quad \text{(D.4)}$$

(b) For a problem with two particles of different charges Ze, $Z'e$ moving in the field of a magnetic monopole, the wave function is a "double" section with respect to both $\boldsymbol{r}$ and $\boldsymbol{r}'$. Then

$$\sum_{mm'} Y_{q,l,m}(\theta,\phi)\, Y_{q',l',m'}(\theta',\phi')\langle ll'jm_j | lml'm'\rangle = F_{q,q',j,m_j} \,, \quad \text{(D.5)}$$

is a double section that transforms under a simultaneous rotation of θ and ϕ like Y_{j,m_j} does. One has to, however, remember that after the rotation one is using a different gauge, as discussed before in sect. 10.

(c) For a particle with spin $\boldsymbol{S}$, the total angular momentum is

$$\boldsymbol{J} = \boldsymbol{L} + \boldsymbol{S} = \boldsymbol{r} \times (\boldsymbol{p} - Ze\boldsymbol{A}) - \frac{q\boldsymbol{r}}{r} + \boldsymbol{S} \,. \quad \text{(D.6)}$$

The addition of $\boldsymbol{L}$ and $\boldsymbol{S}$ is achieved with the Clebsch-Gordan coefficients in the usual way with no difficulty.

References

[1] P.A.M. Dirac, Proc. Roy. Soc. A133 (1931) 60.
[2] Ig. Tamm, Z. Phys. 71 (1931) 141.
[3] M. Fierz, Helv. Phys. Acta 17 (1944) 27.
[4] Tai Tsun Wu and Chen Ning Yang, Phys. Rev. D12 (1975) 3845.
[5] S.S. Chern, Complex manifolds without potential theory (Van Nostrand, 1967).
[6] A.R. Edmonds, Angular momentum in quantum mechanics (Princeton University Press, 1960).
[7] Bateman Manuscript Project, Higher transcendental functions, ed. A. Erdelyi (McGraw-Hill, 1953) vol. 2, p. 168; ref. [6], p. 57–58.

Dirac's monopole without strings: Classical Lagrangian theory

[[76d]]

Commentary begins page 76

Tai Tsun Wu*

Gordon McKay Laboratory, Harvard University, Cambridge, Massachusetts 02138

Chen Ning Yang†

Institute for Theoretical Physics, State University of New York, Stony Brook, New York 11794

(Received 16 March 1976)

The non-quantum-mechanical interaction of a Dirac magnetic monopole and a point charge through the electromagnetic field is studied. A classical action integral which is multiple-valued is found. Stability of this action integral against variations of the world lines of the point charge and the monopole, and against variations of the electromagnetic potentials, yields the correct Lorentz equations of motion of the particles and the Maxwell equations for the field. No strings are introduced in the formalism.

I. ELECTROMAGNETIC FIELD IN INTERACTION WITH CHARGED PARTICLES AND MAGNETIC MONOPOLES

In this paper and an earlier one[9] on monopole harmonics, we study some properties of the Dirac magnetic monopole without the introduction of the concept of strings. The two papers are, however, logically and technically independent, and may be read separately.

Consider first the familiar case of one positron with world line x^μ. Then the electric current density j^μ is

$$j^\mu(\xi) = 4\pi e \int ds \frac{dx^\mu}{ds} \delta^4(\xi - x(s)), \tag{1}$$

where ξ designates a space-time point. This current density leads to an electromagnetic field described by the tensor $f^{\mu\nu} = -f^{\nu\mu}$. Let $\tilde{f}^{\mu\nu}$ be its dual such that

$$E_1 = -f^{01} = \tilde{f}^{23}, \quad H_1 = -f^{23} = -\tilde{f}^{01}, \quad \text{etc.} \tag{2}$$

[cf. (6) below], where the metric used is

$$dt^2 + dx^2 + dy^2 + dz^2 = -(dx^0)^2 + (dx^1)^2 + (dx^2)^2 + (dx^3)^2 .$$

Along a world line we define

$$ds = (dt^2 - dx^2 - dy^2 - dz^2)^{1/2} = dt(1 - v^2)^{1/2} .$$

The usual Maxwell equations for the electromagnetic field are

$$f^{\mu\nu}{}_{,\nu}(\xi) = -4\pi e \int ds \frac{dx^\mu}{ds} \delta^4(\xi - x(s)), \tag{3}$$

and

$$f_{\mu\nu,\lambda}(\xi) + f_{\nu\lambda,\mu}(\xi) + f_{\lambda\mu,\nu}(\xi) = 0$$

or (4)

$$\tilde{f}^{\mu\nu}{}_{,\nu}(\xi) = 0 .$$

Simultaneously, through the Lorentz force, the electromagnetic field also acts on the positron to determine its motion[1]:

$$\frac{dp^\mu}{ds} = -e f^{\mu\nu} \frac{dx_\nu}{ds} \qquad \left(p^\mu = m \frac{dx^\mu}{ds}\right). \tag{5}$$

The general definition of the dual field strength $\tilde{f}^{\mu\nu}$ is

$$\tilde{f}^{\mu\nu} = -\tfrac{1}{2}\epsilon^{\mu\nu\alpha\beta} f_{\alpha\beta} , \tag{6}$$

where

$$\epsilon^{0123} = -1, \quad \epsilon^{\mu\nu\alpha\beta} = \text{antisymmetrical tensor}. \tag{7}$$

Next consider the more general problem of the electromagnetic field in interaction with one positron with world line x^μ and one Dirac magnetic monopole[2] with world line X^μ. (Generalization to a finite number of positrons and a finite number of Dirac magnetic monopoles is straightforward.) The coupled equations of motion must be

$$f^{\mu\nu}{}_{,\nu}(\xi) = -4\pi e \int ds \frac{dx^\mu}{ds} \delta^4(\xi - x), \tag{Me}$$

$$\tilde{f}^{\mu\nu}{}_{,\nu}(\xi) = -4\pi g \int ds \frac{dX^\mu}{ds} \delta^4(\xi - X), \tag{Mg}$$

$$\frac{dp^\mu}{ds} = -e f^{\mu\nu} \frac{dx_\nu}{ds} , \tag{Le}$$

and

$$\frac{dP^\mu}{ds} = -g \tilde{f}^{\mu\nu} \frac{dX_\nu}{ds} \qquad \left(P^\mu = M \frac{dX^\mu}{ds}\right), \tag{Lg}$$

where p^μ and P^μ are the four-momenta of the positron and the monopole, respectively, and g is the magnetic charge of the monopole, positive if the monopole is a north (i.e., north-seeking) pole. We shall refer to these equations as the Maxwell [(Me), (Mg)] and the Lorentz [(Le), (Lg)] equations.

It has been well known since the beginning of

this century that classical electrodynamics (without magnetic monopoles), as described by (3), (4), and (5), can be formulated compactly in an action principle. It is the purpose of the present paper to generalize this formulation so that the Maxwell equations (Me) and (Mg) and the Lorentz equations (Le) and (Lg) for the interaction between electromagnetism, charged particles, and magnetic monopoles are formulated compactly in an action principle.

After Dirac introduced[2] the concept of magnetic monopoles, he came back in 1948 to the question[3] of the classical action principle. The aim of the present work is exactly to rediscuss this question of Dirac's 1948 paper. However, we shall avoid the introduction of the concept of strings attached to monopoles, which was necessary in Dirac's formulation.

Dirac originally introduced the string because his vector potential for a static magnetic monopole is singular along a semi-infinite line in three-space. Subsequently, the string has caused a number of problems, including the following two:

(i) The Dirac veto. The wave functions for all charged particles vanish on the string. This problem has been discussed most explicitly by Wentzel.[4]

(ii) Dynamic variables for the string. In studying the moving magnetic monopole, Dirac[3] used an infinite number of dynamic variables for the string. The resultant formalism becomes extremely complicated.

In this paper we circumvent all these complexities by introducing ideas[5] borrowed from the mathematics of fiber bundles.

Before going into the details, let us emphasize that throughout the present paper, except for remarks in Sec. VI, *we do not introduce Planck's constant* $\hbar$, so that the monopole strength need not satisfy Dirac's quantization condition. That is, it is possible that

$$\frac{2eg}{\hbar c} \neq \text{integer}. \tag{8}$$

Because $\hbar$ is not introduced, the concept that electromagnetism is a nonintegrable phase factor[5] is not used in the present paper.

II. EQUATION (Mg) AS KINEMATICS

The electromagnetic potential around a magnetic monopole cannot be chosen without singularities. This fact was proved in Ref. 5 for a monopole at rest where, in order to circumvent the singularity problem, the space-time outside of the monopole was divided into two overlapping regions R_a and R_b and singularity-free electromagnetic potentials $(A_\mu)_a$ and $(A_\mu)_b$ were found in R_a and R_b, respectively. We shall now use this same idea[6] to describe the electromagnetic potential outside of the world line X^μ of a magnetic monopole of strength g. The choice of R_a and R_b is very flexible. For definiteness we shall choose to define R_a and R_b in one Lorentz frame: For each t, R_a and R_b are respectively the regions defined by

$$\begin{aligned} R_a:&\ 0 \leq \theta < \tfrac{1}{2}\pi + \delta, \quad \text{all } \phi,\ r > 0 \\ R_b:&\ \tfrac{1}{2}\pi - \delta < \theta \leq \pi, \quad \text{all } \phi,\ r > 0 \end{aligned} \tag{9}$$

where r, θ, and ϕ are spherical coordinates with the monopole position at that t taken as the origin. δ is a smooth function of t satisfying $0 < \delta \leq \frac{1}{2}\pi$.

Given an electromagnetic field satisfying (Mg) we can find vector potentials $(A_\mu)_a$ and $(A_\mu)_b$ in regions R_a and R_b, respectively, so that for $i = a, b$,

$$f_{\mu\nu} = (A_{\mu,\nu})_i - (A_{\nu,\mu})_i \tag{10}$$

in R_i. In the overlap $R_{ab} = R_a \cap R_b$, the two vector potentials are related by the gauge transformation

$$(A_\mu)_a - (A_\mu)_b = \alpha_\mu \text{ in } R_{ab}, \tag{11}$$

where by (10) a_μ must satisfy

$$\alpha_{\mu,\nu} - \alpha_{\nu,\mu} = 0 \text{ in } R_{ab}. \tag{12}$$

R_{ab} is a four-dimensional region where loops around the monopole cannot be shrunk within R_{ab}. Equation (12) asserts that

$$\oint \alpha_\mu(\xi)\, d\xi^\mu = K, \tag{13}$$

where K is independent of any distortions of the loop within R_{ab}, and the integral is defined in the direction of increasing azimuthal angle ϕ (cf. Fig. 1).

To determine K we consider, at a fixed t, a spherical surface S around the monopole. The upper hemisphere S_a, where $\theta \leq \frac{1}{2}\pi$, is entirely in R_a. The lower hemisphere S_b is entirely in R_b. Hence

$$\begin{aligned} &\text{outward magnetic flux through } S_a = \oint (A_\mu)_a d\xi^\mu, \\ &\text{outward magnetic flux through } S_b = -\oint (A_\mu)_b d\xi^\mu. \end{aligned} \tag{14}$$

Thus

$$\text{outward magnetic flux through } S = \oint \alpha_\mu\, d\xi^\mu \text{ around equator}. \tag{15}$$

In (14) we have used the sign convention (2). The total outward flux is, of course, $4\pi g$. Thus

FIG. 1. Regions R_a and R_b at a given time t. $\vec{X}$ is the position of the monopole. R_a is the region above the lower cone. R_b is the region under the upper cone. The loop is in the overlap $R_{ab} = R_a \cap R_b$.

$$\oint \alpha_\mu(\xi) d\xi^\mu = 4\pi g \tag{16}$$

around *any* loop in R_{ab} which circles the monopole world line X^μ in the direction of increasing azimuthal angle ϕ.

Thus *any electromagnetic field* $f_{\mu\nu}$ *satisfying (Mg) is describable by* $(A_\mu)_a$ *and* $(A_\mu)_b$ *satisfying (10), (11), (12), and (16). Conversely, any* $(A_\mu)_a$ *and* $(A_\mu)_b$ *satisfying (10), (11), (12), and (16) gives an electromagnetic field* $f_{\mu\nu}$ *satisfying (Mg).* To prove this last statement, we observe that at a point ξ^μ not on the world line of the monopole, (10) implies the homogeneous Maxwell equation (4). Thus $\tilde{f}^{\mu\nu}{}_{,\nu}(\xi)$ is only nonvanishing on the world line of the monopole. That is,

$$\tilde{f}^{\mu\nu}{}_{,\nu}(\xi) = \int a^\mu(s) ds\, \delta^4(\xi - X(s)), \tag{17}$$

where $a^\mu(s)$ is a four-vector defined on the world line. *In any specific Lorentz frame* consider the $\mu = 0$ component of (17). Using convention (2), this component reduces to

$$-\nabla \cdot H|_\xi = a^0 \delta^3(\vec{\xi} - \vec{X}(s)) \left(\frac{dX^0}{ds}\right)^{-1}, \tag{18}$$

where on the right-hand side s is taken to be that point on the world line where $X^0(s) = \xi^0$. At this fixed ξ^0 we now integrate (18) over the interior of a sphere around the monopole at $\vec{X}(s)$. The right-hand side yields

$$a^0 \left(\frac{dX^0}{ds}\right)^{-1}.$$

The left-hand side becomes a surface integral which is (−1) times the total magnetic flux outward from the surface, which one can evaluate in the same way as in an earlier discussion that led to (15). If (16) is satisfied, this is $-4\pi g$. Thus

$$-4\pi g = a^0 \left(\frac{dX^0}{ds}\right)^{-1}, \quad \text{or } a^0 = -4\pi g \frac{dX^0}{ds}. \tag{19}$$

Since (19) holds in any Lorentz frame, we have

$$a^\mu = -4\pi g \frac{dX^\mu}{ds}.$$

Substitution of this into (17) leads to (Mg).

We have thus shown that (Mg), which describes the generation of electromagnetism by a monopole, is *equivalent* to the condition that the electromagnetic field be described by $(A_\mu)_a$ and $(A_\mu)_b$ satisfying (10), (11), (12), and (16). We shall now consider $(A_\mu)_a$ and $(A_\mu)_b$ as independent variables subject to conditions (11), (12), and (16). The field strength $f_{\mu\nu}$ described by such electromagnetic potentials *automatically* satisfies (Mg). In other words, in this approach (Mg) becomes a kinematic equation.

III. THE ACTION INTEGRAL

For the electrodynamics of positrons and electromagnetic fields described by (3), (4), and (5), the action integral is

$$\mathcal{A}(x, A) = -m\left[\int ds\right]_{\text{positron}} - (16\pi)^{-1} \int d^4\xi\, f_{\mu\nu}(\xi) f^{\mu\nu}(\xi) + e\int_{-\infty}^{\infty} A_\mu(x) \frac{dx^\mu}{ds} ds, \tag{20}$$

where the first and the last integrals are defined along the world line of the positron and ds is real and >0. Equation (4) is a kinematic condition. Equations (3) and (5) are dynamical equations which result from the condition of stability of $\mathcal{A}$ against variations respectively of $A_\mu(\xi)$ and of the world line $x^\mu(s)$.

For the electromagnetism of positrons and monopoles described by (Me), (Mg), (Le), and (Lg) we seek to find an action integral $\mathcal{A}(x, X, A)$, where x, X represent the world lines of the positron and the monopole, and A is the electromagnetic potential defined in two regions R_a and R_b and satisfies (10), (11), (12), and (16). As proved in the last section, (Mg) is a kinematic equation. We expect (Me), (Le), and (Lg) to result from the stability condition of $\mathcal{A}(x, X, A)$ against variations of A, x, and X.

Equation (20) suggests

$$\mathcal{A}(x, X, A) = -m\left[\int ds\right]_{\text{positron}} - M\left[\int ds\right]_{\text{monopole}} - (16\pi)^{-1}\int d^4\xi\, f_{\mu\nu}(\xi) f^{\mu\nu}(\xi) + \mathcal{A}_1, \tag{21}$$

where

$$\mathcal{A}_1 = e \,⨍\, A_\mu(x) dx^\mu \tag{22}$$

along the world line of the electron. We use an unusual symbol for this last integral because its definition requires careful examination. In particular, one has to define which A_μ $[(A_\mu)_a$ or $(A_\mu)_b]$ to use as the integrand. We proceed as follows:

(a) If the positron world line is entirely in one region, R_a or R_b, then the $\fint$ is defined to be the usual integral with $(A_\mu)_a$ or $(A_\mu)_b$ as the integrand. If the positron world line crosses from R_b through R_{ab} into R_a [Fig. 2(a)], the definition of $\mathcal{A}_1$ is

$$\mathcal{A}_1 = e \fint A_\mu(x)dx^\mu$$
$$= e\int_Q^\infty (A_\mu)_a dx^\mu + e\beta(Q) + e\int_{-\infty}^Q (A_\mu)_b dx^\mu\,, \tag{23}$$

where β is defined in R_{ab} by

$$\beta_{,\mu} = \alpha_\mu\,. \tag{24}$$

Given $(A_\mu)_a$ and $(A_\mu)_b$ such a β exists, because of Eq. (12), as a multiple-valued function plus an arbitrary constant independent of space-time. In view of (16), the multiple values are different from each other by $4\pi g$ times an integer.

The necessity of the term $e\beta(Q)$ in (23) is demonstrated by the fact that (23) is independent of changing the point Q to another point Q' along the positron world line in R_{ab} (Fig. 2). To see this we use (11) and (24). This demonstration is entirely similar to, but not quite the same as, a corresponding one in Ref. 5.

(b) We can rewrite the three terms on the right-hand side of (23) in a convenient notation as follows:

$$\mathcal{A}_1 = \mathcal{A}_1(\infty_a, Q_a) + \mathcal{L}(Q_a, Q_b) + \mathcal{A}_1(Q_b, -\infty_b)\,. \tag{25}$$

We write ∞_a for ∞ to emphasize that at $t=\infty$ the world line of the positron (in this case) is in R_a. A point Q in R_{ab} is treated as two points Q_a and Q_b, which can be schematically thought of as on different "floors," as illustrated in Fig. 2 of Ref. 5. The definition of $\mathcal{L}$ is

$$\mathcal{L}(Q_a, Q_b) \equiv e\beta(Q) \equiv -\mathcal{L}(Q_b, Q_a)\,. \tag{26}$$

For a world line of the positron that goes in and out of R_a and R_b as along the path (b) of Fig. 2, we have many equivalent definitions of $\mathcal{A}_1$, such as

$$\begin{aligned}\mathcal{A}_1 &= \mathcal{A}_1(\infty_a, B_a) + \mathcal{L}(B_a, B_b) + \mathcal{A}_1(B_b, A_b)\\ &\quad + \mathcal{L}(A_b, A_a) + \mathcal{A}_1(A_a, -\infty_a)\\ &= \mathcal{A}_1(\infty_a, C_a) + \mathcal{L}(C_a, C_b) + \mathcal{A}_1(C_b, A_b)\\ &\quad + \mathcal{L}(A_b, A_a) + \mathcal{A}_1(A_a, -\infty_a)\,.\end{aligned} \tag{27}$$

Notice the following identity:

$$\begin{aligned}\mathcal{A}_1(\infty_a, B_a) &= \mathcal{A}_1(\infty_a, D_a) + \mathcal{L}(D_a, D_b) + \mathcal{A}_1(D_b, C_b)\\ &\quad + \mathcal{L}(C_b, C_a) + \mathcal{A}_1(C_a, B_a)\,.\end{aligned} \tag{28}$$

(c) How does an additive constant γ (which is independent of space-time) to β affect the definition of $\mathcal{A}_1$ given above? To answer this question, we notice that for a path that begins and ends in the same region, R_a or R_b, the number of $\mathcal{L}$ terms in the definition of $\mathcal{A}_1$, such as in (27), is even and the change $\beta \to \beta + \gamma$ does not change the value of $\mathcal{A}_1$. For a path that begins and ends in different regions, the number of $\mathcal{L}$ terms is odd and the change $\beta \to \beta + \gamma$ produces a change in $\mathcal{A}_1$ by $\pm e\gamma$. In the variational principle, where one keeps the world line fixed at $t = \pm\infty$, this additive constant of $\pm e\gamma$ does not produce any changes in the final result.

(d) Since β is multiple-valued, which of its values should be chosen in the definition of $\mathcal{A}_1$ such as (23) and (26)? The answer is: Any choice is undesirable and *we just consider* $\mathcal{A}_1$ *as definable only modulo* $4\pi eg$. To see this let us consider the world line (Fig. 3) and choose a specific value of $\beta(Q)$ among its multiple values to evaluate $\mathcal{A}_1$ in (23). For simplicity we consider the case that the monopole is fixed in space at the origin so that $\vec{X} = 0$ for all times. Now continuously distort the portions of the world line of the positron between E and F so that it loops around the origin (i.e., the monopole) and return to the original position, in the direction of increasing azimuthal angle. The point Q describes a loop as shown in Fig. 3. If we use the azimuthal coordinate ϕ of Q to label the action integral $\mathcal{A}(\phi)$ for this one-parameter family of positron world lines, then a comparison between the two cases $\theta = 0$ and $\theta = 2\pi$ shows that β at Q is the only quantity that is different. More pre-

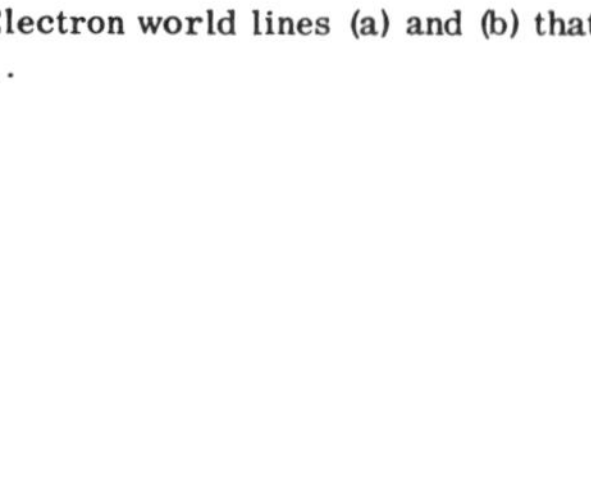

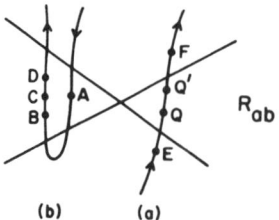

FIG. 2. Electron world lines (a) and (b) that go through overlap R_{ab}.

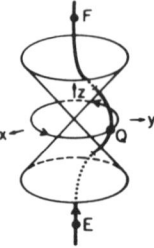

FIG. 3. Distortion of world line EF of the electron.

cisely

$$\mathcal{A}(2\pi) - \mathcal{A}(0) = \mathcal{A}_1(2\pi) - \mathcal{A}_1(0) = e\oint \alpha_\mu dx^\mu = 4\pi eg. \tag{29}$$

Thus if we want the value of $\mathcal{A}$ to be continuous with respect to the distortion of world lines, we must define its value only up to modulo $4\pi eg$.

IV. CONSTRAINTS ON x, X, A AND STABILITY OF $\mathcal{A}$ AGAINST δx^μ AND δA_μ

We have defined above the action integral $\mathcal{A}(x, X, A)$ in terms of the dynamical variables x, X, and A_μ. To be more precise, $\vec{x}(t)$ and $\vec{X}(t)$ designate the world lines of the positron and the monopole. They represent $2\infty^3$ *independent dynamical variables* subject to the constraints discussed in the next paragraph.

The world lines x and X are constrained to be timelike. Furthermore, they must not cross. That is, $\vec{x}(t) - \vec{X}(t) \neq 0$ for all t. Without this condition it would not be possible to properly define the action integral $\mathcal{A}$. For example, the one-parameter family of positron world lines labeled by ϕ in the preceding section gives rise to an action integral $\mathcal{A}(\phi)$ which varies with ϕ, with a total variation from $\phi = 0$ to $\phi = 2\pi$ given by (29). If we now distort these world lines so as to make the loop described by Q shrink toward the monopole position, we can only guarantee the continuity of $\mathcal{A}$ against such distortions if the condition is imposed that the positron and monopole world lines do not cross. The necessity of this condition had been discussed before by Rosenbaum.[7]

The dynamical variables $A_\mu(\xi)$ consist of $(A_\mu)_a$ and $(A_\mu)_b$ subject to the conditions (11), (12), and (16). As proved in Sec. II, these conditions lead to (Mg) as a kinematic equation.

We have to demonstrate now that the stability of $\mathcal{A}$ against variations of these dynamical variables, keeping the conditions discussed above satisfied, yields the remaining Maxwell and Lorentz equations.

It is easy to show that the stability of $\mathcal{A}$ against variations $\delta\vec{x}(t)$ gives rise to (Le), exactly as in the usual case without monopoles. For variations of A_μ, first consider variations $(\delta A_\mu)_a$ and $(\delta A_\mu)_b$ in the subregions of R_a and R_b outside of the overlap R_{ab}. Stability of $\mathcal{A}$ leads to equation (Me) in these subregions. Next consider variations $(\delta A_\mu)_a$ and $(\delta A_\mu)_b$ in R_{ab}, but with

$$\delta((A_\mu)_a - (A_\mu)_b) = 0. \tag{30}$$

Stability of $\mathcal{A}$ against such variations leads to equation (Me) in R_{ab}. Lastly we have to consider variations $(\delta A_\mu)_a$ and $(\delta A_\mu)_b$ in R_{ab} that violate (30). It is sufficient to consider the case $(\delta A_\mu)_b = 0$. Then conditions (11) and (12) require

$$(\delta A_\mu)_a = \delta\alpha_\mu = \delta\beta_{,\mu}, \tag{31}$$

where we have used (24). Thus (21) and (23) give

$$\delta\mathcal{A} = \delta\mathcal{A}_1 = -e\delta\beta(Q) + e\delta\beta(Q) = 0. \tag{32}$$

Clearly this conclusion also holds for more twisted world lines such as (b) of Fig. 2.

To summarize, we have demonstrated that stability of $\mathcal{A}$ against δx^μ [i.e., $\delta\vec{x}(t)$] and δA_μ, subject to the proper conditions, yields (Me) and (Le), respectively. We emphasize that the multivaluedness of $\mathcal{A}$ (which is defined modulo $4\pi eg$) does not influence the validity of this statement.

Variation of the world line of the monopole is more cumbersome to study since it necessitates a change of regions R_a and R_b. We shall circumvent this complication by investigating the dual action integral.

V. DUAL ACTION INTEGRAL

The dual field, designated by an overbar, has already been used in Sec. I. Under the dual operation, ξ^μ does not change, but

$$\bar{e} = g, \quad \bar{g} = -e, \quad \bar{f}_{\mu\nu} = -\tfrac{1}{2}\epsilon_{\mu\nu\sigma\rho} f^{\sigma\rho}, \tag{33}$$

$$\bar{E}_k = H_k, \text{ and } \bar{H}_k = -E_k. \tag{34}$$

Also

$$\bar{\bar{e}} = -e, \quad \bar{\bar{g}} = -g, \text{ and } \bar{\bar{f}}_{\mu\nu} = -f_{\mu\nu}. \tag{35}$$

Furthermore,

$$\bar{f}_{\mu\nu}\bar{f}^{\mu\nu} = -f_{\mu\nu}f^{\mu\nu}. \tag{36}$$

In particular, this dual operation can be applied to the action $\mathcal{A}$ of (21) to give

$$\bar{\mathcal{A}}(x, X, \bar{A}) = -m\left[\int ds\right]_{\text{positron}} - M\left[\int ds\right]_{\text{monopole}} - (16\pi)^{-1}\int d^4\xi\, \bar{f}_{\mu\nu}(\xi)\bar{f}^{\mu\nu}(\xi) + g\oint \bar{A}_\mu(X)dX^\mu. \tag{37}$$

In this formula we have introduced the dual potential $\bar{A}_\mu$ defined by

$$\bar{A}_{\mu,\nu} - \bar{A}_{\nu,\mu} = \bar{f}_{\mu\nu}. \tag{38}$$

Because of the presence of the positron, $\bar{A}_\mu$ is defined separately in two regions $\bar{R}_a$ and $\bar{R}_b$. At any given time t, $\bar{R}_a$ and $\bar{R}_b$ are defined exactly as R_a and R_b were defined in (9), except that $\bar{r}, \bar{\theta}, \bar{\phi}$ now refer to spherical coordinates with the *positron* position at that time as the origin. The overlap size $2\bar{\delta}$ need not be related to 2δ. The definition of the integral $⨍$ in (37) is the same as the corresponding one in (21), except that the regions are now $\bar{R}_a$ and $\bar{R}_b$. Clearly,

$$\bar{\bar{\mathfrak{a}}} = \mathfrak{a}\,, \tag{39}$$

and $\bar{\mathfrak{a}}$ is defined modulo $4\pi ge$.

Notice that the $\int d^4\xi$ terms in (37) and (21) are equal in magnitude but opposite in sign because of (35).

The discussions of Secs. II, III, and IV can of course be duplicated with appropriate changes of all quantities into their respective duals. For example, (Me) is now kinematic, while (Mg) is dynamic resulting from the condition of stability of $\bar{\mathfrak{a}}(x,X,\bar{A})$ against $\delta\bar{A}$.

Consider for *fixed* nonintersecting world lines of the positron and the monopole x^μ and X^μ (both of which are timelike) the quantity

$$\bar{\mathfrak{a}}_0(x,X) = \text{extremum of } \bar{\mathfrak{a}}(x,X,\bar{A}) \text{ with respect to } \delta\bar{A}\,. \tag{40}$$

The field strength $f_{\mu\nu}$ generated from the extremizing $\bar{A}_\mu$ satisfies both Maxwell equations (Mg) and (Me). Since it satisfies (Mg), according to Sec. II, there exists an A_μ satisfying (10), (11), (12), and (16). At such an A_μ $\mathfrak{a}(x,X,A)$ attains an extremum with respect to δA since (Me) is satisfied. We define

$$\mathfrak{a}_0(x,X) = \text{extremum of } \mathfrak{a}(x,X,A) \text{ with respect to } \delta A\,. \tag{41}$$

Notice that the extrema (40) and (41) are attained with potentials $\bar{A}$ and A that give the *same* field strengths $f_{\mu\nu}$. We shall later prove the following:

Lemma. $\bar{\mathfrak{a}}_0(x,X) - \mathfrak{a}_0(x,X) = \text{integrals at infinity.}$ (42)

Since integrals at infinity play no role in the action principle, it follows from this lemma that the stability of $\mathfrak{a}(x,X,A)$ against all variations gives all four Maxwell and Lorentz equations. In particular (Lg) follows from the stability of $\mathfrak{a}$ against δX because (42) implies $\delta\mathfrak{a}_0 = \delta\bar{\mathfrak{a}}_0$, and $\delta\bar{\mathfrak{a}}_0 = 0$ against δX implies (Lg).

Proof of lemma. (a) First consider the special case when there is a positron but no magnetic monopole. In this special case

$$\mathfrak{a}_0 - \bar{\mathfrak{a}}_0 = e\int_{-\infty}^{\infty} A_\mu\, dx^\mu - (8\pi)^{-1}\int d^4\xi\, f_{\mu\nu}(\xi) f^{\mu\nu}(\xi)\,. \tag{43}$$

Since the Maxwell equations are satisfied, (3) can be used to rewrite the first term on the right-hand side of (43):

$$e\int_{-\infty}^{\infty} A_\mu\, dx^\mu = e\int d^4\xi \int_{-\infty}^{\infty} ds \frac{dx^\mu}{ds} A_\mu(\xi)\delta^4(\xi - x) = -(4\pi)^{-1}\int d^4\xi\, A_\mu(\xi) f^{\mu\nu}{}_{,\nu}(\xi)\,. \tag{44}$$

Substitution into (43) gives

$$\mathfrak{a}_0 - \bar{\mathfrak{a}}_0 = -(4\pi)^{-1}\int d^4\xi\, [A_\mu(\xi) f^{\mu\nu}(\xi)]_{,\nu}\,, \tag{45}$$

which is equal to a surface integral at infinity. The lemma is thus proved in this special case.

(b) Next consider the case where the world lines are such that for all t, $Z(t) > z(t)$. In this case we can construct a three-dimensional surface S which at each fixed t is the plane

$$\xi^3 = \tfrac{1}{2}[Z(t) + z(t)]\,, \tag{46}$$

as illustrated in Fig. 4. S separates space-time into two regions, an upper region G containing the world line of the monopole and a lower region E containing the world line of the positron. Furthermore, E is completely in R_b and G is completely in $\bar{R}_a$. We write

$$\mathfrak{a}_0 - \bar{\mathfrak{a}}_0 = e⨍ A_\mu\, dx^\mu - g⨍ \bar{A}_\mu\, dX^\mu - (8\pi)^{-1}\int d^4\xi\, f_{\mu\nu} f^{\mu\nu} = B_E - B_G\,, \tag{47}$$

where

$$B_E = e⨍ A_\mu\, dx^\mu - (8\pi)^{-1}\int_E d^4\xi\, f_{\mu\nu} f^{\mu\nu}\,, \tag{48}$$

$$B_G = g⨍ A_\mu\, dX^\mu - (8\pi)^{-1}\int_G d^4\xi\, \bar{f}_{\mu\nu} \bar{f}^{\mu\nu}\,. \tag{49}$$

Since the positron world line is entirely in E which is in R_b, we replace the A_μ in (48) by $(A_\mu)_b$. We can then process the right-hand side of (48) in exactly the same way that we processed (43)–(45), with the space-time region E replacing the whole space-time. This leads to

$$B_E = -(4\pi)^{-1}\int_S (A_\mu)_b f^{\mu\nu}\, d\sigma_\nu + \text{terms at infinity}\,, \tag{50}$$

where $d\sigma_\nu$ is the three-dimensional surface area on S. Similarly,

$$B_G = (4\pi)^{-1}\int_S (\bar{A}_\mu)_{\bar{a}}\, \bar{f}^{\mu\nu}\, d\sigma_\nu + \text{terms at infinity}\,. \tag{51}$$

The sign difference is due to the fact that G is above S while E is below. Expressing $\bar{f}^{\mu\nu}$ in (51)

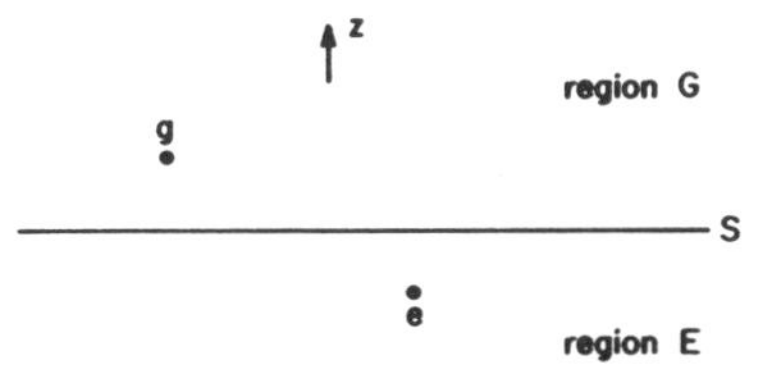

FIG. 4. Division of space-time into regions G and E. The figure shows the division at one instant of time.

in terms of $f_{\alpha\beta}$ and integrating by parts yield

$$B_G = -(8\pi)^{-1}\epsilon^{\mu\nu\alpha\beta}\int (\bar{A}_\mu)_{\bar{a}} f_{\alpha\beta}\, d\sigma_\nu + \text{terms at infinity}$$

$$= (8\pi)^{-1}\epsilon^{\mu\nu\alpha\beta}\int \bar{f}_{\mu\beta}(A_{\alpha_b})\, d\sigma_\nu + \text{terms at infinity}$$

$$= (4\pi)^{-1}\int f^{\nu\alpha}(A_\alpha)_b\, d\sigma_\nu + \text{terms at infinity}$$

$$= B_E + \text{terms at infinity}. \tag{52}$$

The lemma follows.

(c) Next consider the general case. Since the world lines of the positron and the monopole are assumed not to intersect, there always exists a three-dimensional surface S that separates all space-time into an "upper" region G containing the monopole world line and a "lower" region E containing the positron world line. But now E (and S) in general contain both points in R_a and points in R_b. Equations (47), (48), and (49) remain valid. But (50) should be replaced by

$$B_E = -(4\pi)^{-1}\bar{\int}_S A_\mu f^{\mu\nu}\, d\sigma_\nu + \text{terms at infinity}, \tag{53}$$

where we have introduced a new three-dimensional integral $\bar{\int}_S$, which is a generalization of the corresponding one-dimensional integral $\bar{\int}$. The precise definition of this new integral and its properties are given in the Appendix. Equations (51) and (52) then remain valid with the $\int$'s replaced by $\bar{\int}$'s. The lemma follows.

VI. REMARKS

1. We have assumed throughout that $m>0$, $M>0$. That is, the positron and the monopole both have nonvanishing masses.

2. Clearly the definitions of R_a and R_b are quite flexible. Given the monopole world line X^μ, continuous distortions of the boundaries of regions R_a and R_b starting from (9) are allowed, provided R_a, R_b together always cover the whole of space-time outside of the monopole world line. Amalgamation of boundaries is not allowed in this process.

3. Under a gauge transformation $A_\mu \to A'_\mu$, provided $f_{\mu\nu}$ is unchanged and (16) remains valid, $\mathcal{A}(x, X, A)$ changes by terms evaluated at infinity.

4. The condition $A^\mu{}_{,\mu}=0$ is not used at all in this paper.

5. A key point of this paper is the reduction of the Maxwell equation (Mg) to kinematics in Sec. II. While the method used for this reduction is very much in the spirit[5] of the Chern-Weil theorem in fiber-bundle theory, there is considerable difference also. The present considerations focus on the "exponent" of the phase factor of Ref. 5. The phase factor itself cannot be given meaning without the introduction of the Planck constant h.

6. If one formulates, after Feynman,[8] the quantization procedure by path integration, one would be dealing with integrals of $\exp(i\mathcal{A}/\hbar)$. Since $\mathcal{A}$ is only definable modulo $4\pi eg$, this process is meaningful only if $4\pi eg/\hbar = 2\pi$(integer). That is Dirac's quantization rule

$$2eg/\hbar = \text{integer}$$

must be satisfied. We are currently working on this problem.

7. When there are two or more neighborhoods (e.g., R_a and R_b), the barred integrals such as $\bar{\int} A_\mu dx^\mu$ are the natural ones while the corresponding ordinary integrals such as $\int A_\mu dx^\mu$ are not even definable. Thus the bar is really superfluous. We retain the bar in this paper only to draw attention to the special nature of the integral when it spans more than one region.

APPENDIX

We assume throughout this appendix that (Me) and (Mg) are satisfied, but not necessarily (Le) and (Lg).

Before defining the three-dimensional integral $\bar{\int}$ and proving (53) and (52) we shall need two preliminaries.

1. We shall rewrite the integral in (50) as follows:

$$\int_S A_\mu f^{\mu\nu}\, d\sigma_\nu = -\tfrac{1}{6}\int A_\mu f^{\mu\nu}\epsilon_{\nu\xi\eta\zeta}\, dx^{\xi\eta\zeta}$$

$$= -\tfrac{1}{2}\int A_\xi \bar{f}_{\eta\zeta}\, dx^{\xi\eta\zeta}, \tag{A1}$$

where we use the usual notation

$$dx^{\xi\eta\zeta} = -dx^{\eta\xi\zeta} = -dx^{\xi\zeta\eta}.$$

We have assumed that S is in only one region R_a or R_b.

It is easy to prove that

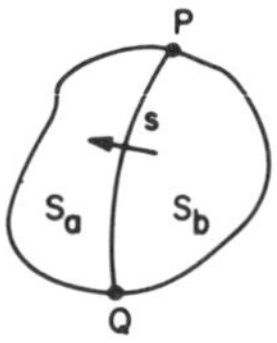

FIG. 5. Three-dimensional surface S divided into S_a and S_b. S_a is in R_a and S_b in R_b. s is the two-dimensional surface between S_a and S_b. PQ is the boundary of s and is a one-dimensional loop.

$$\oint_S A_\lambda \bar{f}_{\mu\nu} dx^{\lambda\mu\nu} = -\int_R f_{\mu\nu} f^{\mu\nu} d^4x \tag{A2}$$

if S is the boundary of a space-time region R entirely in R_a or R_b, which contains no parts of the world line of either the positron or the monopole.

2. The one-dimensional integral $⨍$ defined in Sec. III has the property that if a loop L is the border of a two-dimensional region Y, then

$$⨍_L A_\mu dx^\mu = -\tfrac{1}{2}\int_Y f_{\mu\nu} dx^\mu dx^\nu. \tag{A3}$$

This formula is obvious if Y is all in R_a or all in R_b. If Y is partly in R_a and partly in R_b, (A3) is correct because of the terms $\mathcal{L}$ in the definition of $⨍$, as in (25) or (27). The negative sign in (A3) derives from the usual convention of a right-handed loop for the sense of L.

We proceed now as follows:

3. Consider a three-dimensional surface S that is divided into two parts S_a and S_b with the boundary s between them, as illustrated in Fig. 5. We define

$$⨍_S A_\lambda \bar{f}_{\mu\nu} dx^{\lambda\mu\nu} = \int_{S_a} (A_\lambda)_a \bar{f}_{\mu\nu} dx^{\lambda\mu\nu} + \int (A_\lambda)_b \bar{f}_{\mu\nu} dx^{\lambda\mu\nu} + \int_s \beta \bar{f}_{\mu\nu} dx^{\mu\nu}, \tag{A4}$$

where s is taken to have the normal to it in the direction from S_b to S_a as indicated by the arrow in Fig. 5. The surface s is, of course, in R_{ab}. If we fix its boundary PQ, which is a one-dimensional loop, and move s, within R_{ab}, to s', it is easy to show that according to (A4), the integral $⨍$ is unchanged, provided S does not intersect either the positron or the monopole world line.

Thus the three-dimensional integral on the left-hand side of (A4) is dependent only on S and on the loop PQ which divides the boundary of S. Changing PQ on the boundary of S *does* lead to a change in the integral, just as in the one-dimensional integral case where

$$\mathfrak{A}_1(B_a, D_b) \neq \mathfrak{A}_1(B_b, D_b).$$

Now if S itself is a boundary of a four-dimensional region R, then S has no boundary of its own. Thus there is no PQ and the integral defined by (A4) is well defined. We have assumed that the world line of either the positron or the monopole does not enter the region enclosed by S. In fact, in this case,

$$⨍_S A_\lambda \bar{f}_{\mu\nu} dx^{\lambda\mu\nu} = -\int_R f_{\mu\nu} f^{\mu\nu} d^4x, \tag{A5}$$

which is the generalization of (A2), and is analogous to (A3).

4. Equation (53) is now easily proved if we take R to be the lower region E which contains the electron world line. One has instead of (A5)

$$-\int_E f_{\mu\nu} f^{\mu\nu} d^4x = ⨍_S A_\lambda \bar{f}_{\mu\nu} dx^{\lambda\mu\nu} - 8\pi e ⨍ A_\mu dx^\mu + \text{terms at infinity}. \tag{A6}$$

The first integral on the right-hand side is equal to, by the reasoning that led to (A1),

$$-2 ⨍_S A_\mu f^{\mu\nu} d\sigma_\nu .$$

Equation (53) follows immediately.

5. It remains to prove the generalization of (52), when S, the boundary of E, is not entirely in R_b, and not entirely in $\bar{R}_a$. We shall only demonstrate the proof in the case that S is entirely in $\bar{R}_a$, but not entirely in R_b. Divide S into regions S_a and S_b as in Fig. 5. Then (53) states that, omitting terms at infinity,

$$8\pi B_E = ⨍_S A_\lambda \bar{f}_{\mu\nu} dx^{\lambda\mu\nu} = \int_{S_a} A_\lambda \bar{f}_{\mu\nu} dx^{\lambda\mu\nu} + \int_{S_b} A_\lambda \bar{f}_{\mu\nu} dx^{\lambda\mu\nu} + \int_s \beta \bar{f}_{\mu\nu} dx^{\mu\nu}$$

$$-8\pi B_G = \int_S \bar{A}_\mu f_{\lambda\nu} dx^{\lambda\mu\nu},$$

where we have used $\bar{\bar{f}} = -f$. Thus

$$8\pi(B_E - B_G) = 2\int_{S_a}(A_\lambda \bar{A}_\mu)_{,\nu}dx^{\lambda\mu\nu} + 2\int_{S_b}(A_\lambda \bar{A}_\mu)_{,\nu}dx^{\lambda\mu\nu} + \int_s \beta \bar{f}_{\mu\nu}dx^{\mu\nu}$$

$$= -2\int_s (A_\lambda)_a \bar{A}_\mu dx^{\lambda\mu} + 2\int_s (A_\lambda)_b \bar{A}_\mu dx^{\lambda\mu} + \int_s \beta \bar{f}_{\lambda\mu}dx^{\lambda\mu}$$

$$= \int_s (-2\beta_{,\lambda}\bar{A}_\mu + \beta \bar{f}_{\lambda\mu})dx^{\lambda\mu} = -2\int_s (\beta \bar{A}_\mu)_{,\lambda}dx^{\lambda\mu} .$$

Since the boundary of s is at infinity we have

$B_E - B_G =$ terms at infinity.

---

*Work supported in part by the U. S. ERDA under Contract No. E(11-1)-3227.

†Work supported in part by the National Science Foundation under Grant MPS74-13208 A01.

[1]We use the notation $x^0 = -x_0 = t$, $x^i = x_i$ $(i = 1, 2, 3)$. Comma means derivative. It is well known that, because of the point nature of the positron, such a coupled system of equations does not possess any finite solution. We are here not concerned with such questions of infinities.

[2]P. A. M. Dirac, Proc. R. Soc. London A133, 60 (1931).

[3]P. A. M. Dirac, Phys. Rev. 74, 817 (1948). See also J. Schwinger, *Particles, Sources and Fields* (Addison-Wesley, Reading, Mass., 1970 and 1973), Vols. 1 and 2.

[4]Gregor Wentzel, Prog. Theor. Phys. Suppl. 37–38, 163 (1966).

[5]Tai Tsun Wu and Chen Ning Yang, Phys. Rev. D 12, 3845 (1975).

[6]The idea that one divides space into two regions is familiar in the usual problem of choosing a singularity-free coordinate system on the surface of a globe. It is well known that a single coordinate system (such as the latitudes and longitudes) must have some singularities (such as the north and south poles). To avoid singularities, one can divide the globe into two overlapping regions R_a and R_b. For example, one can choose R_a to contain more than the northern hemisphere but not the south pole, and R_b to contain more than the southern hemisphere but not the north pole. In each region a singularity-free coordinate system can be chosen. In the overlap, the two systems of coordinates are transformable into each other.

[7]D. Rosenbaum, Phys. Rev. 147, 891 (1966).

[8]R. P. Feynman, Rev. Mod. Phys. 20, 367 (1948).

[9]T. T. Wu and C. N. Yang, Nucl. Phys. B (to be published).

[[76g]]
Commentary begins page 77

What Visits Mean to China's Scientists

C. N. Yang
Institute for Theoretical Physics, State University of New York, Stony Brook

Any understanding of the People's Republic is incomplete without the appreciation that China today makes large efforts to convince the Chinese people in every field of activity of the meaning and the usefulness of Chinese policy. Communication with the United States, being one aspect of Chinese foreign policy, is no exception in this general scheme. The majority of the Chinese population are in agreement with the policy of increasing contact with the United States. As for the scientists and engineers, this general agreement with government policy seems bolstered by the observation that communciation with the United States is beneficial to Chinese development in science and technology.

The most important effect on Chinese scientific and technological development which results from scholarly communiction with the United States derives from the faster communication of new trends in various fields to the Chinese specialists. In the field of high energy physics, for example, there is in 1976 much more up-to-date information in China of developments in America and in Europe than there existed in 1973 or 1974. While the reason for this change is necessarily complex, Chinese scientists agree that increased contact with scientists abroad, especially with scientists in the United States, is a major factor.

The better up-to-date knowledge of developments abroad has a stimulating effect on Chinese research directions. Comparing the vigor with which scientific research is pursued in China in 1976 and in 1973 or 1974, it is clear that, at least in the fields of physics and mathematics, this stimulation has produced profound changes of outlook. Chinese scientists appreciate that scholarly communication with the United States has played an important role in this development. This observation applies to other fields of research activity as well.

The scholarly exchange between the two countries has also brought about the dissemination of information in the United States concerning Chinese developments in scientific and technological research, educational policies, and social priorities.This increase of understanding of China is an important fruit of the exchange efforts of the last few years and is certainly regarded as such by the Chinese scientists and engineers.

Chinese scientists feel that their American counterparts, as hosts and as guests, are generally open, hospitable, and friendly. American scientists are compared favorably in these attitudes with USSR, West European, and Japanese scientists. As host institutions, American universities score higher in openness and hospitality compared with American industry.

The question has oftentimes been raised whether in the exchange program criticisms of Chinese developments by American observers should or should not be made known to the Chinese scientists. Chinese scientists respond that they welcome and appreciate hearing matter-of-fact criticisms of developments in China. This attitude derives perhaps from two reasons. First, in the People's Republic of China one is conditioned to learn from criticism. Second, both the long history of China and the development of the last twenty-some years are reassuring enough to the Chinese scientists and engineers so that there is no cultural inferiority complex in China. Chinese research workers are thus quite capable of embracing criticism and using it in a positive way.

Generally speaking, the Chinese people share what I believe to be the world view of the Chinese government: that in the next twenty years there will be increasing parallelism in the foreign policies of the United States and of China. Chinese scientists thus look forward to increasing contact and exchanges with the American scientific community, especially after the normalization of diplomatic relations.

□ Reprinted from *Reflections on Scholarly Exchanges with the People's Republic of China, 1972–1976*, A. Keatley, ed., Committee on Scholarly Communication with the People's Republic of China, 1976.

MAGNETIC MONOPOLES, FIBER BUNDLES, AND GAUGE FIELDS

[[77e]]
Commentary begins page 78

Chen Ning Yang

Institute for Theoretical Physics
State University of New York at Stony Brook
Stony Brook, New York 11794

The reports in this monograph have shown great enthusiasm and exuberance for the unification of various interactions through the concept of gauge fields. I would like to emphasize a point that has not yet been explicitly stated by any of the other authors: gauge fields are deeply related to some profoundly beautiful ideas of contemporary mathematics, ideas that are the driving forces of part of the mathematics of the last 40 years. Recalling the relationship between physics and mathematics in earlier periods, general relativity and Riemannian geometry, quantum mechanics and Hilbert space, it is all too obvious that physicists may again be zeroing in on a fundamental new secret of nature.

The mathematical development referred to above is the theory of fiber bundles. It may appear, a priori, that this theory is quite abstract and is unrelated to the structure of the physical world. To show that this is not true, we will start with a simple demonstration that electromagnetism and quantum mechanics together lead naturally to "nontrivial fiber bundles." We will then trace the early history of the gauge field concept and its generalization, emphasizing three related but different conceptual motivations, each of which leads to a general formulation of gauge fields.

Magnetic Monopoles and Nontrivial Bundles

The magnetic monopole is the magnetic charge. Though the idea of magnetic monopoles probably was discussed in classic physics early in the history of electricity and magnetism, modern discussions of this concept date back only to 1931, when the important paper of Dirac[1] pointed out that magnetic monopoles in quantum mechanics exhibit some extra and subtle features. In particular, with the existence of a magnetic monopole of strength g, electric charges and magnetic charges must necessarily be quantized, in quantum mechanics. We will give a new derivation of this result below.

If one wants to describe the wave function of an electron in the field of a magnetic monopole, it is necessary to find the vector potential $\mathbf{A}$ around the monopole. Dirac chose a vector potential that has a string of singularities. The necessity of such a string of singularities is obvious if we prove the following theorem[2]:

Theorem: Consider a magnetic monopole of strength $g \neq 0$ at the origin, and consider a sphere of radius R around the origin. There does not exist a vector potential $\mathbf{A}$ for the monopole magnetic field that is singularity free on the sphere.

This theorem can be proved easily in the following way. If there were a singularity-free $\mathbf{A}$, we would consider the loop integral

$$\oint A_\mu \mathrm{d}x^\mu$$

□ Reprinted from *Annals of New York Academy of Sciences* 294 (November 8, 1977), 86–97.

around a parallel on the sphere, as indicated in FIGURE 1. According to Stoke's theorem, this loop integral is equal to the total magnetic flux through the cap α:

$$\oint A_\mu dx^\mu = \Omega_\alpha. \tag{1}$$

Similarly, we can apply Stoke's theorem to cap β, obtaining

$$\oint A_\mu dx^\mu = \Omega_\beta. \tag{2}$$

Here, Ω_α and Ω_β are the total upward magnetic fluxes through caps α and β, both of which are bordered by the parallel. Subtracting these two equations, we obtain

$$0 = \Omega_A - \Omega_B, \tag{3}$$

which is equal to the total flux *out* of the sphere, which, in turn, is equal to $4\pi g \neq 0$. We have thus reached a contradiction.

Having proved this theorem, we observe that R is arbitrary. Thus, one concludes that there must be a string(s) of singularities in the vector potential to describe the monopole field. Yet, we know that the magnetic field around the monopole is singularity free. This fact suggests that the string of singularities is not a real physical difficulty. Indeed, the situation is reminiscent of the problem that one faces when one wants to find a parametrization of the surface of the globe. The coordinate system that we usually use, latitude and longitude, is not singularity free. It has singularities at the north pole and at the south pole. Yet, the surface of the globe is evidently devoid of singularities. We deal with this situation usually in the manner illustrated in FIGURE 2. We consider a rubber sheet with nicely defined coordinates and stretch and wrap it downward onto the globe, so that it covers more than the northern hemisphere. Similarly, we consider another rubber sheet with nicely defined coordinates and stretch and wrap it upward, so it covers more than

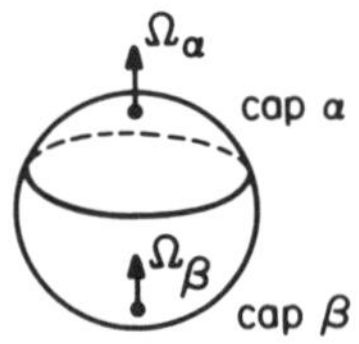

FIGURE 1. A sphere of radius R with a magnetic monopole at its center. The parallel divides the sphere into two caps α and β.

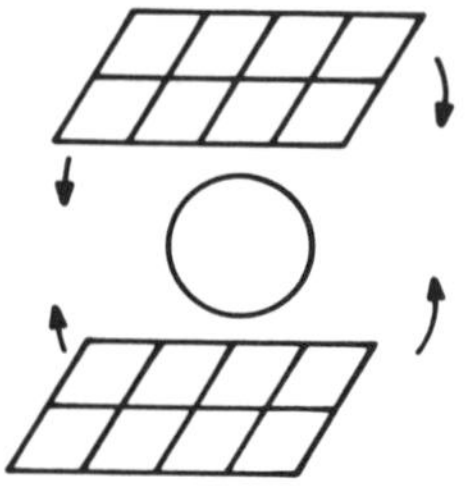

FIGURE 2. Method of parametrizing the globe.

the southern hemisphere. We now have a double system of coordinates to describe the points on the globe. The description is analytic in the domain covered by each sheet, if the globe had experienced no violence in the stretching and wrapping. In the overlapping region covered by both sheets, one has two coordinate systems that are transformable into each other by an analytic nonvanishing Jacobian. This double coordinate system is an entirely satisfactory way to parametrize the globe.

Following this idea, we will now attempt to exorcise the string of singularities in the monopole problem by dividing space into two regions. We will call the points outside of the origin, above the lower cone in FIGURE 3, region R_a. Similarly, we will call the points outside of the origin, under the upper cone, R_b. The union of these two regions gives all points outside of the origin. In R_a, we will choose a vector potential for which there is only one nonvanishing component of A, the azimuthal component:

$$(A_r)_a = (A_\theta)_a = 0, \qquad (A_\phi)_a = \frac{g}{r\sin\theta}(1-\cos\theta), \tag{4}$$

It is important to notice that this vector potential has no singularities anywhere in R_a. Similarly, in R_b, we choose the vector potential

$$(A_r)_b = (A_\theta)_a = 0 \qquad (A_\phi)_b = \frac{-g}{r\sin\theta}(1\times\cos\theta), \tag{5}$$

which has no singularities in R_b. It is simple to prove that the curl of either of these two potentials gives correctly the magnetic field of the monopole.

In the region of overlap, because both of the two sets of vector potentials share the same curl, the difference between them must be curlless and therefore must be a gradient. Indeed, a simple calculation shows

$$(A_\mu)_a - (A_\mu)_b = \partial_\mu\alpha, \text{ where } \alpha = 2g\phi, \tag{6}$$

where ϕ is the azimuthal angle. The Schrödinger equation for an electron in the monopole field is thus

$$\frac{1}{2m}(p-eA_a)^2\psi_a + V\psi_a = E\psi_a, \text{ in } R_a,$$

$$\frac{1}{2m}(p-eA_b)^2\psi_b + V\psi_b = E\psi_b, \text{ in } R_b,$$

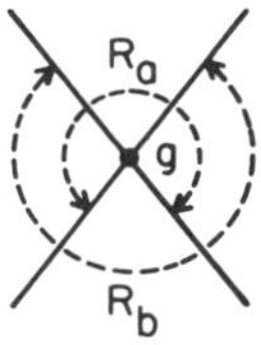

FIGURE 3. Division of space outside of monopole g into overlapping regions R_a and R_b.

where ψ_a and ψ_b are, respectively, the wave functions in the two regions. The fact that the two vector potentials in these two equations are different by a gradient tells us, by the well-known gauge principle, that ψ_a and ψ_b are related by a phase factor transformation

$$\psi_a = S\psi_b, \quad S = \exp(ie\alpha), \tag{7}$$

or

$$\psi_a = [\exp(2iq\phi)]\psi_b, \, q = eg. \tag{8}$$

Around the equator, which is entirely in R_a, ψ_a is single valued. Similarly, because the equator is also entirely in R_b, ψ_b is single valued around the equator. Therefore, S must return to its original value when one goes around the equator. That fact implies Dirac's quantization condition:

$$2q = \text{integer}. \tag{9}$$

Hilbert Space of Sections

Two ψs, ψ_a and ψ_b, in R_a and R_b, respectively, that satisfy the condition of transition (Equation **8**) in the overlap region are called a *section* by the mathematicians. We see that around a monopole, the electron wave function is a *section* and *not an ordinary function*. We will call these functions wave sections.

Different wave sections (which belong to different energies, for example) clearly satisfy the same condition of transition (Equation **8**) with the same q. Thus, we need to develop [3] the concept of a Hilbert space of sections. To develop this concept, we define the scalar product of two sections ξ, η (for the *same q*) by

$$(\eta, \xi) = \int \eta^* \xi d^3 r. \tag{10}$$

(The question of convergence at $r=0$ and $r=\infty$ is ignored here.) Notice that in the overlap

$$(\eta_a)^* \xi_a = (\eta_b)^* \xi_b, \tag{11}$$

so that Equation **10** is well defined.

It is clear that if ξ is a section, $x\xi$ is also a section, because

$$x\xi_a = S(x\xi_b).$$

Thus, x is an *operator* in the Hilbert space of sections. Similarly, we prove that the components of $(\mathbf{p} - e\mathbf{A})$ are operators, but those of $\mathbf{p}$ are not. Furthermore, $\mathbf{x}$ and $\mathbf{p} - e\mathbf{A}$ are both Hermitian.

Following Fierz,[4] we will now attempt to construct angular momentum operators. Define

$$\mathbf{L} = \mathbf{r} \times (\mathbf{p} - e\mathbf{A}) - \frac{q\mathbf{r}}{r}. \quad \textbf{(12)}$$

It is clear that L_x, L_y, and L_z are Hermitian operators on the Hilbert space of sections. The following commutation rules can be easily verified:

$$\begin{aligned} &[L_x,x] = 0, \quad [L_x,y] = iz, \quad [L_x,z] = -iy, \\ &[L_x,p_x - eA_x] = 0, \quad [L_x,p_y - eA_y] = i(p_z - eA_z), \\ &[L_x,p_z - eA_z] = -i(p_y - eA_y). \end{aligned} \quad \textbf{(13)}$$

It follows from these commutation rules that

$$[L_x,L_y] = iL_z, \text{ etc.} \quad \textbf{(14)}$$

Equation **13**, together with its consequence (Equation **14**), show that L_x, L_y, and L_z are the ***angular momentum operators***.[4] We emphasize that neither the Hilbert space nor these operators possess any "singularities." (The singularities of A_a and A_b are not real singularities, because they occur outside of R_a and R_b, respectively.)

MONOPOLE HARMONICS $Y_{q,l,m}$

Because $[r^2, \mathbf{L}] = 0$, we can diagonalize r^2 and study operators $\mathbf{L}$ for fixed r^2. That is, we will study sections of the form

$$\delta(r^2 - r_0^2)\xi,$$

where ξ is a section dependent only on angular coordinates θ and ϕ. $\mathbf{L}$ operates, then, on "angular sections."

Equation **14** shows that $[L^2,L_z] = 0$. Simultaneous diagonalization produces the familiar multiplets with eigenvalues $l(l+1)$ and m,

$$L^2 Y_{q,l,m} = l(l+1) Y_{q,l,m}; \; L_z Y_{q,l,m} = m Y_{q,l,m}, \quad \textbf{(15)}$$

where $l = 0, 1/2, 1, \ldots$, and for each value of l,m ranges from $-l$ to $+l$ in integral steps of increment. $Y_{q,l,m}$ are eigensections, which are called[3] monopole harmonics. The allowed values of l and m are

$$l = |q|, \; |q|+1, \; |q|+2, \ldots, \qquad m = -l, \; -l+1, \ldots, l. \quad \textbf{(16)}$$

Each of these l,m combinations occurs exactly once. One can choose each Y normalized, so that

$$\int_0^\pi \sin\theta d\theta \int_0^{2\pi} |Y_{q,l,m}|^2 d\phi = 1. \quad \textbf{(17)}$$

Different $Y_{q,l,m}$ (for fixed q) are orthogonal, a fact one easily proves in the usual way from Equation **15**.

The explicit values of $Y_{q,l,m}$ in terms of Jacobi polynomials were given in Reference 3. They were obtained from Equation **15**, in exactly the same way one usually obtains the spherical harmonics $Y_{l,m}$. Indeed,

$$Y_{l,m} = Y_{0,l,m}.$$

The collection of $Y_{q,l,m}$ for fixed q and values of l,m given by Equation **16** form[3] a complete orthonormal set of angular sections.

Each $(Y_{q,l,m})_a$ is analytic in R_a; so is $(Y_{q,l,m})_b$ in R_b. Thus, all of the discontinuities, cusps, and singularities in **A** and in ψ are removed in a very smooth way.

Remarks: (A) It is important to realize that the above-described way of using $(A)_a$ and $(A)_b$ together to describe the magnetic field of a monopole has an additional advantage: It gives the magnetic field **H** correctly *everywhere*. In older papers, one often used a single **A** with a string of singularities. Because, by definition,

$$\nabla \cdot (\nabla \times A) = 0,$$

the magnetic field described by $\nabla \times A$ must have *continuous* flux lines. Thus, its flux lines consist of the dotted lines of FIGURE 4, plus the bundle of lines described by the solid line, so as to make the net flux at the origin zero. Thus, $\nabla \times A$ does not correctly describe the magnetic field of the monopole, a point already emphasized by Wentzel.[5]

(B) For ordinary spherical harmonics, there are many important theorems, such as the spherical harmonics addition theorem and the decomposition of products of spherical harmonics by use of Clebsch-Gordon coefficients. These theorems can be generalized to monopole harmonics.[6]

(C) In the approximately 40 years since Dirac's first paper on monopoles, the subject has been beset with difficulties due to singularities. Now that we have removed the difficulty of string singularities through the introduction of the concept of sections, it is revealed that there is yet another difficulty, which we will call the Lipkin-Weisberger-Peshkin[7] difficulty. This difficulty occurs[8] in studying the radial wave function of a Dirac electron around a monopole (TABLE 1). It can be removed through the introduction of a small extra magnetic moment for the Dirac electron.

(D) It is instructive to go back to the reasoning represented in FIGURE 1 and attempt to repeat the steps for the combined A_a, A_b description of the magnetic field. Choose the parallel to be the equator. Then,

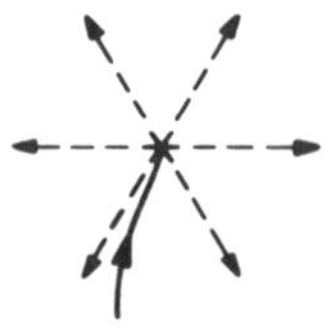

FIGURE 4. Magnetic flux lines due to **A**. Because $\nabla \cdot (\nabla \times A) = 0$, flux lines are everywhere continuous. Therefore, there is "return flux" along the solid line.

TABLE 1
DIFFICULTIES AND METHODS OF SOLUTION FOR STUDYING THE MOTION OF A DIRAC ELECTRON IN THE FIELD OF A MAGNETIC MONOPOLE

Angular Wave Function	Radial Wave Function
Difficulty of string singularity, solved by introducing sections	Lipkin-Weisberger-Peshkin difficulty, solved by introducing extra magnetic moment

$$\oint (A_\mu)_a dx^\mu = \Omega_\alpha,$$

$$\oint (A_\mu)_b dx^b = \Omega_\beta.$$

Thus,
$$4\pi g = \Omega_\alpha - \Omega_\beta = \oint [(A_\mu)_\alpha - (A_\mu)_\beta] dx^\mu,$$

which is, by Equation **6**, equal to the increment of α around the equation, that is, $2g(2\pi) = 4\pi g$.

We have arrived at an identity. I have provided this simple argument because it is exactly the gist of the proof of the famous Gauss-Bonnet-Allendoerfer-Weil-Chern theorem and the later Chern-Weil theorem, which play seminal roles in contemporary mathematics.

In fact, gauge fields, of which electromagnetism is the simplest example, are conceptually identical to some mathematical concepts in fiber bundle theory. TABLE 2 gives [2] translations for the terminologies used by physicists, on the one hand, and mathematicians, on the other. We notice that, in particular, Dirac's monopole quantization (Equation **9**) is identical to the mathematical concept of classification of U(1) bundles according to the first Chern class.

The last two entries of TABLE 2 identify electromagnetism with and without magnetic monopoles with connections to trivial and nontrivial U(1) bundles. Why is electromagnetism without monopoles "trivial"? We can gain some understanding by looking at a paper loop and a Moebius strip (FIGURE 5). If they are cut along the dotted lines, each would break into two pieces. Looking at the resultant pieces, we cannot differentiate between the two. The paper loop and the Moebius strip are different only in the way the resultant pieces are put together. For the latter, a twist of one of the resultant pieces is necessary. The difference between a trivial and a nontrivial bundle resides only in the processes of *joining:* for the nontrivial bundle, a twist is needed in the joining process. In the case of electromagnetism, the joining process is given by Equation **7** or **8**. If there is no monopole, $S = 1$, and the bundle is trivial. If there is a monopole, $S \neq 1$, and the bundle is nontrivial. (We may describe the nontrivial nature by saying that a *twist of phase* is necessary.)

EARLY HISTORY OF THE CONCEPT OF GAUGE FIELDS

Einstein's discovery of the relationship between gravitation and the geometry of space-time stimulated work by many great geometers: Levi-Civita, Cartan, Weyl, and others. In his book, ***Raum, Zeit und Materie*** (space, time and matter), Weyl[9]

TABLE 2
TRANSLATION OF TERMINOLOGIES

Gauge Field Terminology	Bundle Terminology
Gauge (or global gauge)	principal coordinate bundle
Gauge type	principal fiber bundle
Gauge potential b_μ^k	connection on a principal fiber bundle
S (Equation **8**)	transition function
Phase factor Φ_{QP}	parallel displacement
Field strength $f_{\mu\nu}^k$	curvature
Source (electric) J_μ^k	?
Electromagnetism	connection on a U_1 bundle
Isotopic spin gauge field	connection on a SU_2 bundle
Dirac's monopole quantization	classification of U_1 bundle according to first Chern class
Electromagnetism without monopole	connection on a trivial U_1 bundle
Electromagnetism with monopole	connection on a nontrivial U_1 bundle

FIGURE 5. Examples of trivial (*left*) and nontrivial (or Moebius strips, *right*) fiber bundles.

attempted to unify gravity and electromagnetism through the use of the geometric concept of a space-time-dependent scale change. The basic idea is summarized below.

		$dx\mu$ (•——→•)
scale	1	$1+S_\mu dx^\mu$
f	f	$f+(\partial f/\partial x^\mu)dx^\mu$
scale change	f	$f+(\partial/\partial x^\mu+S_\mu)f dx^\mu$

In the summary above, the first line indicates how the scale changes in going from a point x^μ to a neighboring point $x^\mu+dx^\mu$ of space-time. The second line shows how a function of space-time changes as a result of the change in argument from x^μ to $x^\mu+dx^\mu$. Finally, if the scale change is applied to the function f, one obtains at $x^\mu+dx^\mu$ the product

$$(f+\partial f/\partial x^\mu dx^\mu)\quad(1+S_\mu dx^\mu).$$

Expanding to first order in the small displacement gives the last line in the summary. The increment in f is, then,

$$(\partial/\partial x^\mu+S_\mu)f dx^\mu. \qquad \textbf{(18)}$$

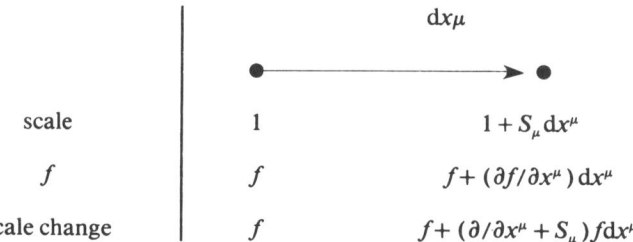

Weyl tried to incorporate electromagnetism into a geometric theory by identifying the vector potential A_μ with a space-time-dependent S_μ, generating scale changes as described. This attempt proved, however, unsuccessful.

In 1925, the concepts of quantum mechanics emerged. A key concept in quantum mechanics is the replacement of the momentum p_μ in the classic Hamiltonian by an operator:

$$p_\mu \to -ih(\partial/\partial x^\mu).$$

For a charged particle, the replacement is

$$p_\mu - (e/c)A_\mu \to -ih[\partial/\partial x^\mu - i(e/hc)A_\mu]. \tag{19}$$

In 1927, Fock[10] observed that one could base quantum electrodynamics on this operator. London[11] pointed out the similarity of Fock's to Weyl's earlier work. Comparing Equations **18** and **19**, Weyl's identification would be correct if one makes the replacement

$$S_\mu \to -i(e/hc)A_\mu.$$

In other words, instead of a *scale change*

$$(1 + S_\mu \mathrm{d}x^\mu),$$

one considers a *phase change*

$$[1 - i(e/hc)A_\mu \mathrm{d}x^\mu] \simeq \exp[-i(e/hc)A_\mu \mathrm{d}x^\mu], \tag{20}$$

which can be thought of as an *imaginary scale change*. Weyl put all of these expressions together[12] in a remarkable paper (which also first discussed the two-component theory of a spin-1/2 particle) in which the transformation of the electromagnetic potential

$$A_\mu \to A'_\mu = A_\mu + \partial_\mu \alpha \text{ (second-type transformation)}, \tag{21}$$

and the associated phase transformation

$$\psi \to \psi' = \psi \exp(ie\alpha/hc) \text{ (first-type transformation)}, \tag{22}$$

of the wave function of a charged particle were explicitly discussed.[13]

Although the phase change factor (Equation **20**) is no longer a scale factor, Weyl kept the earlier terminology*† that he used in 1918–20 and called both the transformation (Equation **20**) and the associated phase change of wave functions "gauge" transformations.

Generalization: With the discovery of many new particles after World War II, physicists explored various couplings between the "elementary particles." Many possible couplings can be written down, and the desire to find ***a principle to choose among the many possibilities*** was one of the motivations[17,18] for an attempt to generalize Weyl's gauge principle for electromagnetism. The point here is that for electromagnetism, the gauge principle determines, all at once, the way in which ***any*** particle of charge qe, a ***conserved*** quantity, serves as a ***source*** of the electromagnetic field. Because the isotopic spin **I** is also conserved, a natural question was, "Does there exist a generalized gauge principle that determines the way in which **I** serves as the source of a new field?"

Another motivation for an attempt at generalization is the observation that the conservation of **I** implies that the proton and the neutron are similar. Which to call a proton or, indeed, which superposition of the two to call a proton, is a convention that one can select arbitrarily (if the electromagnetic interaction is switched off). If one requires this freedom of choice to be independent for observers at different space-time points, that is, if one requires *localized* freedom of choice, one is led to a generalization of the gauge principle.

These two motivations were, of course, intertwined and led quite naturally to the formulation[18] of non-Abelian gauge fields.

A third approach[19] to a generalized gauge principle came later and is the "integral formalism" of gauge fields. It starts from the observation that the gauge principle of Weyl deals with a phase factor (Equation **20**) between two neighboring points. Along a path from space-time point A to space time point B, the resultant phase factor is

$$\Phi_{\mathrm{BA}} = \exp[-i(e/hc)\int_{\mathrm{A}}^{\mathrm{B}} A_\mu \mathrm{d}x^\mu], \qquad \textbf{(23)}$$

which is path dependent, that is, nonintegrable. (Dirac[1] had already discussed, in 1931, "non-integrable phases for wave functions.") If one analyzes the meaning of electromagnetism in quantum mechanics, especially through a discussion of the Bohm-Aharonov experiment,[20‡] one reaches the conclusion[2] that "electromagnetism is the gauge invariant manifestation of a non-integrable phase factor."

Once this conclusion is reached, a natural generalization is to replace a

* The idea of scale invariance, discussed in Reference 9, was developed earlier, in 1918–19, in three papers by Weyl (submitted on May 2 and June 8, 1918 and on January 7, 1919). In the first two of them, he used the term *Massstab Invarianz* (see Reference 14); in the third paper, he settled on the term *Eich Invarianz*.

The English translation of *Eich Invarianz* was "calibration invariance" in Henry Brose's 1921 translation of the fourth edition of Weyl's book *Space, Time and Matter*[15] (republished by Dover). The translation "gauge invariance" was not used, I suspect, until after Weyl's 1929 article.[12] It appeared (probably not for the first time) in Dirac's article[1] of 1931.

† The transformation (Equation **21**) that leaves field strengths unchanged must have been known in the nineteenth century. It did not, however, seem to have a specific name. In the many editions of Foppl-Abraham-Becker-Sauter on electricity and magnetism, which started in 1894, *Eich* or "gauge" was not used until the 1964 English translation *Electromagnetic Fields and Interactions,*[16] in which the term "Lorentz gauge" was inserted in a footnote.

‡ The experiment was performed by Chambers.[20]

"nonintegrable phase factor" by a "nonintegrable element of a Lie group." One thus obtains naturally an integral formalism of gauge fields.

We illustrate in FIGURE 6 the three approaches to the general concept of gauge fields. The three approaches are, of course, deeply interrelated, because phases, symmetry, and conservation laws are themselves related.

It is my opinion that, conceptually, the integral formalism of gauge fields is to be preferred to the earlier differential approach. The integral formalism has more structure and more meaning. It brings to the fore problems of global topology not easily formulated in terms of the differential approach. For example, in our earlier discussion of the field around the magnetic monopole, we did not introduce the concept of nonintegrable phase factors. We did not run into any conceptual difficulties, only because we had not raised such questions as a rotation of the coordinate axes. As soon as such questions are raised, it becomes apparent that the integral formalism is more superior, because it specifies that intrinsic meaning is unrelated to the choice of coordinate axes and of regions R_a and R_b.

Differential formalism, however, is used in computing. (The relationship between differential and integral formalisms is quite similar to that between Lie algebras and Lie groups.) In fact, a gauge-Riemannian calculus has been developed.[21]

Electromagnetism is, as we have seen, a gauge field. That gravitation is a gauge field is universally accepted, although exactly how it is a gauge field is a matter still to be clarified.[19,22] Whether weak and strong interactions are also due to gauge fields is a matter that has been intensively studied in recent years,[23] together with the question of the renormalizability of non-Abelian gauge fields.[24§] If one may borrow a term used by the biologists, one would say that there is gradually forming a "dogma" that all interactions are due to gauge fields. Because of the mathematical difficulties involved in the solution of quantized gauge fields,

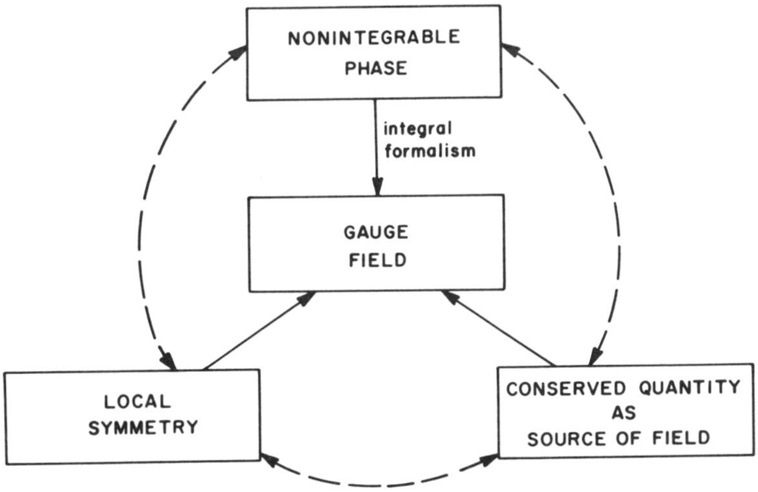

FIGURE 6. Three motivations that led to the concept of gauge fields.

§ Abers and Lee[23] also contains a review of earlier works of R. P. Feynman, L. D. Faddeev, V. N. Popov, and M. T. Veltman.

however, I believe it will be a long time before the question can be definitively answered as to exactly how strong and weak interactions are due to gauge fields.

Reflecting on how the concepts basic to gauge fields were formulated by physicists, we see that at every step, the development was tied to the problem of the conceptual description of the physical world. Firstly, Maxwell equations originated with the four fundamental experimental laws of electricity and magnetism and with Faraday's introduction of the concepts of field and flux. Maxwell's equations and the principles of quantum mechanics led to the idea of gauge invariance. Attempts to generalize this idea, motivated by physical concepts of phases, symmetry, and conservation laws, led to the theory of non-Abelian gauge fields. That non-Abelian gauge fields are conceptually identical to ideas in the beautiful theory of fiber bundles, developed by mathematicians *without reference to the physical world,* was a great marvel to me. In 1975, I discussed my feelings with Chern, and said, "This is both thrilling and puzzling, since you mathematicians dreamed up these concepts out of nowhere." He immediately protested, "No, no, these concepts were not dreamed up. They were natural and real."

REFERENCES

1. DIRAC, P.A.M. 1931. Proc. Roy. Soc. **A133:** 60.
2. WU, T. T. & C. N. YANG. 1975. Phys. Rev. D **12:** 3845.
3. WU, T. T. & C. N. YANG. 1976. Nucl. Phys. B **107:** 365.
4. FIERZ, M. 1944. Helv. Phys. Acta **17:** 27.
5. WENTZEL, G. 1966. Progr. Theor. Phys. Suppl. **37–38:** 163.
6. WU, T. T. & C. N. YANG. 1977. To be published.
7. LIPKIN, H. J., W. I. WEISBERGER & M. PESHKIN. 1969. Ann. Phys. **53:** 203.
8. KAZAMA, Y., C. N. YANG & A. S. GOLDHABER. 1977. Phys. Rev. D. In press.
9. WEYL, H. 1920. Raum, Zeit und Materie. 3rd edit. Springer Verlag. Berlin-Heidelberg. New York.
10. FOCK, V. 1927. Z. Phys. **39:** 226.
11. LONDON, F. 1927. Z. Phys. **42:** 375.
12. WEYL, H. 1929. Z. Phys. **56:** 330.
13. PAULI, W. 1933. Handbuch der Physik. 2nd edit. Vol. 24(1): 83. Geiger and Scheel.; PAULI, W. 1941. Rev. Mod. Phys. **13:** 203.
14. WEYL, H. 1918. Sitzber. Preuss Akad. Wiss.: 465; WEYL, H. 1918. Math. Z. **2:** 384; WEYL, H. 1919. Ann. Phys. **59:** 101.
15. WEYL, H. 1921. Space, Time and Matter. Dover Publications, Inc. New York, N.Y.
16. 1964. Electromagnetic Fields and Interactions. Blaisdell Publishing Co. Waltham, Mass.
17. YANG, C. N. & R. MILLS. 1954. Phys. Rev. **95:** 631.
18. YANG, C. N. & R. MILLS. 1954. Phys. Rev. **96:** 191.
19. YANG, C. N. 1974. Phys. Rev. Lett. **33:** 445.
20. AHARONOV, Y. & D. BOHM. 1959. Phys. Rev. **115:** 485; CHAMBERS, R. G. 1960. Phys. Rev. Lett. **5:** 3.
21. YANG, C. N. 1975. Proc. Sixth Hawaii Topical Conf. Particle Phys.
22. UTIYAMA, R. 1956. Phys. Rev. **101:** 1957.
23. WEINBERG, S. 1967. Phys. Rev. Lett. **19:** 1264; SALAM, A. 1968. *In* Elementary Particle Theory. N. Svartholm, Ed. Almquist and Forlag. Stockholm, Sweden.
24. 'THOOFT, G. 1971. Nucl. Phys. B **35:** 167; ABERS, E. S. & B. W. LEE. 1973. Phys. Rep. **9C:** 1.

SOME PROBLEMS ON THE GAUGE FIELD THEORIES, II*

[77h]
Commentary begins page 78

Gu Chao-hao (谷超豪) and Yang Chen-ning (杨振宁)

Received August 9, 1976.

Abstract

We prove that a gauge field can be determined locally by the field strength together with its gauge derivatives up to a certain order. It is shown, however, that in the non-Abelian case, two non-analytic gauge fields may be non-equivalent globally even when they are locally equivalent everywhere, no matter whatever the base manifold may be.

Introduction

In the present paper, we consider some general properties of the gauge fields. The gauge derivatives of the field strength are interpreted as the phase factors for certain infinitesimal loops and hence the important identities, such as the Bianchi identities, are derived in a geometrical way. Then we consider the question whether a gauge field can be determined by the field strength and its gauge derivatives. It is proved that any gauge field can be determined locally by the field strength together with its gauge derivatives up to a certain order. For the SU_3 gauge fields the gauge derivatives up to the 2nd order are sufficient. Subsequently, the relation between the local equivalence and the global equivalence is considered. If the manifold M_n is analytic and simply connected, any two analytic gauge fields equivalent in an arbitrary open set must be also equivalent globally. For the non-analytic case, we prove that two non-Abelian gauge fields on M_n may be non-equivalent globally, even when they are locally equivalent everywhere. Consequently, even for the simply connected case, the field strength and its gauge derivatives up to any order cannot determine the gauge field globally. The results obtained above are useful for understanding the structure of classical gauge fields.

I. Gauge Derivatives of the Field Strength

Let $\mathscr{F}$ be a gauge field on a differential manifold M_n[1,2] with r-dimensional Lie group G as its gauge group. In a coordinate region the field is determined by the gauge potential $b_\mu = b^i_\mu(x)X_i$ $(i = 1, 2, \cdots, r;\ \mu = 1, 2, \cdots, n)$, where x is a point

---

* This work is a result of the joint discussions and research conducted by C. N. Yang, Professor of the State University of New York at Stony Brook, U. S. A., and some teachers of Futan University during Yang's visit to Shanghai in March and April, 1976. Other members of Futan University who took part in the work were Hu He-sheng (胡和生), Li Da-qian (李大潜), Shen Chun-li (沈纯理), Si Chun-lin (司春林). Su Ru-keng (苏汝铿), Sun Xin (孙鑫), Xia Dao-xing (夏道行), and Yan Shao-zong (严绍宗).

with x^λ as its coordinates, and X_i are the bases of the Lie algebra g of the Lie group G. The field strength and its gauge derivatives are given respectively by

$$f_{\lambda\mu} = b_{\lambda,\mu} - b_{\mu,\lambda} - [b_\lambda, b_\mu], \tag{1.1}$$

$$f_{\lambda\mu|\nu} = f_{\lambda\mu,\nu} + [b_\nu, f_{\lambda\mu}], \tag{1.2}$$

etc., where the comma denotes the partial differentiation[1].

Gauge field is also determined by the phase factors for arcs in M_n. In particular, the phase factor for a differential arc $x\,x + dx$ is a differential element of G, described by the g-valued differential form $b = b_\mu dx^\mu$. We have, formally,

$$\Phi_{x\,x+dx} = I + b = I + b_\mu dx^\mu. \tag{1.3}$$

Let $ABCD$ be a differential quadrangle and x, $x + \underset{1}{dx}$, $x + \underset{1}{dx} + \underset{2}{dx}$, $x + \underset{2}{dx}$ its four vertices. It is well known that[1]

$$\Phi_{ABCDA} = I - f_{\lambda\mu} \underset{1}{dx}^\lambda \underset{2}{dx}^\mu, \tag{1.4}$$

which expresses the meaning of the field strength in phase factors for differential loops.

In the present section we shall point out at first that the gauge derivatives of field strength can be interpreted also as loop phase factors. For this purpose we construct differential parallelpiped with $ABCD$ as the lower base and $EFGH$ as the upper base where the parallelogram $EFGH$ is obtained from $ABCD$ by an infinitesimal translation $\underset{3}{dx}$. Let L be the loop $AEFGHEADCBA$ (see Fig. 1). Then

$$\Phi_L = (I + b_\lambda \underset{3}{dx}^\lambda)(I - f_{\mu\nu} \underset{1}{dx}^\mu \underset{2}{dx}^\nu - f_{\mu\nu,\rho} \underset{1}{dx}^\mu \underset{2}{dx}^\nu \underset{3}{dx}^\rho)$$
$$(I - b_\sigma \underset{3}{dx}^\sigma)(I + f_{\alpha\beta} \underset{1}{dx}^\alpha \underset{2}{dx}^\beta) \tag{1.5}$$

It follows that[2]

$$\Phi_L = I - f_{\mu\nu|\sigma} \underset{1}{dx}^\mu \underset{2}{dx}^\nu \underset{3}{dx}^\sigma. \tag{1.6}$$

For the gauge derivatives of higher order, similar results can also be obtained. Hence we have

Theorem 1. *The gauge derivatives of various orders of the field strength can be described as the phase factors for suitable differential loops.*

From the theorem 1 we see the covariance of the gauge derivatives immediately.

The important identities in gauge theories, such as the Bianchi identities, may be derived by using phase factors for differential loops. In fact, the phase factor for differential loop $ABCGHDA$ (see Fig. 2) is

$$\Phi_1 = I - (f^k_{\mu\nu} \underset{1}{dx}^\mu \underset{2}{dx}^\nu - f^k_{\mu\nu} \underset{3}{dx}^\mu \underset{1}{dx}^\nu) - f^k_{\mu\nu|\sigma} \underset{1}{dx}^\mu \underset{2}{dx}^\sigma \underset{3}{dx}^\nu.$$

---

1) All functions are assumed to be sufficiently smooth except for some specified cases.

2) Since the $\underset{1}{dx}$, $\underset{2}{dx}$ and $\underset{3}{dx}$ are independent differentials, only linear terms for each differential are to be preserved in the calculation.

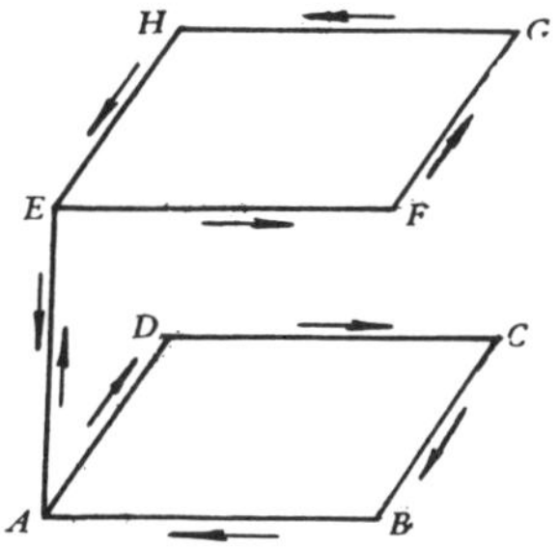

Fig. 1

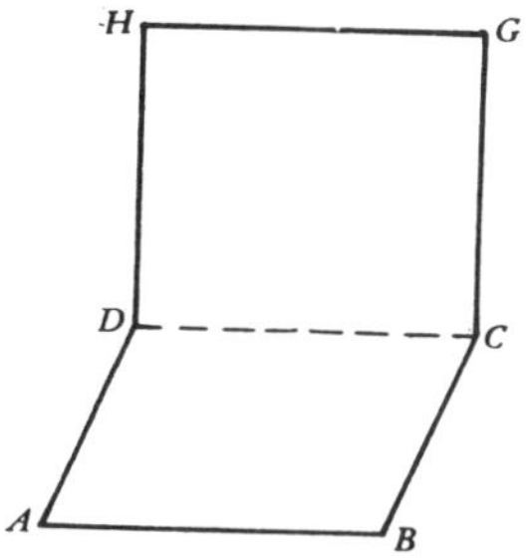

Fig. 2

Permutating the differentials $\underset{1}{dx}, \underset{2}{dx}, \underset{3}{dx}$ cyclicly we obtain the phase factors Φ_2, Φ_3 for differential loops $ADHGFEA$, $AEFGCBA$ respectively. We have $\Phi_1\Phi_2\Phi_3 = I$, since the corresponding three loops cancel each other. The Bianchi identities

$$f_{\mu\nu|\sigma} + f_{\nu\sigma|\mu} + f_{\sigma\mu|\nu} = 0 \tag{1.7}$$

are derived immediately. Similarly, if we take four differentials the Ricci identities

$$f_{\mu\nu|\sigma\tau} - f_{\mu\nu|\tau\sigma} = [f_{\sigma\tau}, f_{\mu\nu}] \tag{1.8}$$

can be obtained. Thus, using the concept of loop phase factors we have obtained some interpretation for gauge derivatives of the field strength and for the important identities between them.

II. The Determination of a Local Gauge Field Through the Field Strength and Its Gauge Derivatives

It is known that a local U_1 gauge field is determined by its field strength. For the SU_2 field, the field strength does not determine the field even locally[3]. In many cases (such as in synchro-spherically symmetrical case[4]), the field can be determined by the field strength and the source. In not a few cases, fields cannot be determined by the field strength and the source[5] or by the field strength and its first gauge derivatives[6]. However, it has been shown that an SU_2 field can be determined locally by the field strength together with its gauge derivatives up to the 2nd order[6].

Now turn to the general case. Let $\mathscr{F}$ and $\mathscr{F}_1$ be two G-gauge fields with b_μ and $\underset{1}{b}_\mu$ as gauge potentials. At first, we note that the local equivalence of $\mathscr{F}$ and $\mathscr{F}_1$ at point x_0 means that there exists a G-value function $a(x)$ in some neighbourhood of x_0 such that the equations

$$\underset{1}{b}{}^k_\mu(x) = A^k_l(a)a^l_{,\mu} + B^k_l(a)b^l_\mu \tag{2.1}$$

hold true, where a^l are the parameters of the group G, $B^k_l(a)$ the matrix elements of the adjoint representation ada and $A^k_l(a)$ the coefficients of the invariant form of group G, i.e. the element $a(a+da)^{-1}$ corresponds to the g-valued form $A^k_l(a)da^l X_k$.

We have the following theorem

Theorem 2. *For a given group G there is an integer p such that every G gauge field can be determined by the field strength together with its gauge derivatives up to the pth order.*

Proof. Let $\mathscr{F}$ and $\mathscr{F}_1$ be two G gauge fields. The integrability condition of (2.1) are

$$\underset{1}{f}^i_{\lambda\mu} = B^i_j(a) f^j_{\lambda\mu}. \tag{2.2$_0$}$$

Differentiating $(2.2)_0$ successively, we obtain

$$\underset{1}{f}^i_{\lambda\mu|\nu_1} = B^i_j(a) f^j_{\lambda\mu|\nu_1}, \tag{2.2$_1$}$$

$$\underset{1}{f}^i_{\lambda\mu|\nu_1\nu_2} = B^i_j(a) f^j_{\lambda\mu|\nu_1\nu_2}, \tag{2.2$_2$}$$

....

etc. Suppose that the field strengths $f_{\lambda\mu}$, $\underset{1}{f}_{\lambda\mu}$ of $\mathscr{F}$ and $\mathscr{F}_1$ and their corresponding gauge derivatives up to the rth order are all equal one another respectively, where r is the dimension of G. Then $(2.2)_0$, $(2.2)_1$, $\cdots$ may be written in the form

$$f^i_{\lambda\mu} = B^i_j(a) f^j_{\lambda\mu}, \tag{2.3$_0$}$$

and

$$f^i_{\lambda\mu|\nu_1\cdots\nu_s} = B^i_j(a) f^j_{\lambda\mu|\nu_1\cdots\nu_s} \quad (s = 1, 2, \cdots, r). \tag{2.3$_s$}$$

Eqs. $(2.3)_0$—$(2.3)_r$ hold true for $a = I$. Hence this set of equations is consistent with respect to the unknown a. Since G is of dimension r, there is an integer $s \leqslant r$ such that the Eqs. $(2.3)_s$ are the algebraic consequence of the Eqs. $(2.3)_0$, $(2.3)_1$, $\cdots$, $(2.3)_{s-1}$. For otherwise there would be at least $(r+1)$ independent equations in $(2.3)_0$, $(2.3)_1$, $\cdots$, $(2.3)_r$, and this is impossible. According to a theorem about partial differential equations[7], we know that the solution of Eqs. (2.1) exists and hence $\mathscr{F}_1$ is equivalent to $\mathscr{F}$. Q. E. D.

Let p be the minimal integer such that theorem 2 holds true. The proof of theorem 2 gives that $p \leqslant r$. This inequality may be improved.

At first we notice that for the Abelian group $B^i_j(a) = \delta^i_j$ for arbitrary $a \in G$. Hence $(2.3)_0$ holds true identically and (2.1) is completely integrable. Consequently, in the Abelian case the field is determined locally by the field strength. Further, if G is decomposed into a direct product $G_1 \times G_2$, then (2.1) consists of two sets of independent equations, i. e. the equations of equivalence for a G_1 gauge field and those for a G_2 gauge field respectively. Hence $p = \max(p_1, p_2)$, where p_1 and p_2 are, respectively, the corresponding integers p's for the G_1 gauge field and the G_2 gauge field.

Now we suppose that G is not a direct product, in particular, G does not contain a nondiscrete centre.

Evidently, $(2.3)_0$ means that the elements $f_{\lambda\mu} = f^i_{\lambda\mu}X_i$ in the Lie algebra g remain unchanged under the adjoint transformations ada. Hence the group elements a's satisfying the Eq. $(2.3)_0$ constitute a subgroup H_0 and the Lie algebra of H_0 consists of all y's which satisfy the condition

$$[y, f_{\lambda\mu}] = 0.$$

Similarly, Eqs. $(2.3)_0$ and $(2.3)_1$ mean that a's belong to a subgroup H_1 of H_0 and the Lie algebra of H_1 consists of all y's which satisfy the condition

$$[y, f_{\lambda\mu}] = 0, \quad [y, f_{\lambda\mu|\nu}] = 0$$

etc. The fact that $(2.3)_s$ is the consequence of $(2.3)_0$, $(2.3)_1$, $\cdots$, $(2.3)_{s-1}$ implies that $h_s = h_{s-1}$.

Consequently, in order to make an estimation of p we may list all possible sequence of proper subalgebras

$$g \supset h_0 \supset \cdots \supset h_{q-1} = h_q, \tag{2.4}$$

where h_0 is the set of elements which commute with all elements of some subspace z_0, h_1 is the set of elements which commute with all elements of some subspace $z_1 \supset z_0$ and so on. Let $q_{\max} = \max q$. Evidently, $p \leqslant q_{\max}$.

However, in some case we have $p < q_{\max}$, since z_0 should be spanned by $f_{\lambda\mu}$ and z_1 by $f_{\lambda\mu}$, $f_{\lambda\mu|\nu}$, etc. Moreover, $f_{\lambda\mu}$, $f_{\lambda\mu|\nu_1}$ $\cdots$ are certain expressions of b_μ and $\underset{1}{b_\mu}$. Considering these facts, we can obtain the exact value of p.

For SU_2 gauge field, it is easily seen that $q_{\max} = 2$ and we know also $p = 2$[6].

For SU_3 gauge field we have the following

Theorem 3. *An SU_3 gauge field is determined by the field strength together with its gauge derivatives up to the 2nd order.*

Proof. We use the Gell-Mann matrices $\lambda_i (i = 1, 2, \cdots, 8)$[8] for the bases of the Lie algebra SU_3' of SU_3.

If the field $\mathscr{F}$ is reducible, then it is equivalent to a gauge field of $U_1' \dot{+} SU_2'$ or SU_2' and hence from the argument above we see that $\mathscr{F}$ is determined by the field strength together with its gauge derivatives up to the 2nd order.

Now consider the more general case. We notice that no 3-dimensional subalgebra can be h_0. In fact, a 3-dimensional subalgebra can be transformed into $\{\lambda_1, \lambda_2, \lambda_3\}$[1)] or $\{\lambda_2, \lambda_5, \lambda_7\}$ by adjoint transformations. If $h_0 = \{\lambda_1, \lambda_2, \lambda_3\}$, then z_0 must be $\{\lambda_8\}$, while the h_0 corresponding to $\{\lambda_8\}$ would be $\{\lambda_1, \lambda_2, \lambda_3, \lambda_8\}$. If $h_0 = \{\lambda_2, \lambda_5, \lambda_7\}$, then z_0 must be $\{0\}$, and h_0 would be SU_3'. Consequently, the possible h_0's and corresponding z_0 (except for an adjoint transformation) are the following:

h_0	z_0
$\{\lambda_1, \lambda_2, \lambda_3, \lambda_8\}$	$\{\lambda_8\}$
$\{\lambda_3, \lambda_8\}$	$\{\lambda_3, \lambda_8\}$ or $\{\alpha\lambda_3 \pm \beta\lambda_8\}$ $\quad (\alpha \pm \sqrt{3}\beta \neq 0)$
$\{\alpha\lambda_3 + \beta\lambda_8\}$	
$\{0\}$	

1) $\{\lambda_1, \lambda_2, \lambda_3\}$ denotes the subspace spanned by λ_1, λ_2 and λ_3.

From Eq. (2.4) it follows immediately that in the last two cases $q = 2$ or 1, and hence we do not write down the z_0's.

When $h_0 = \{\lambda_3, \lambda_8\}$, z_0 is $\{\lambda_3, \lambda_8\}$ or $\{\alpha\lambda_3 \pm \beta\lambda_8\}$ $(\alpha \neq \pm\sqrt{3}\,\beta)$ and the possible h_1 contained in h_0 is $\{\gamma\lambda_3 + \delta\lambda_8\}$ or $\{0\}$. According to the formula $f_{\lambda\mu|\nu} = f_{\lambda\mu,\nu} + [b_\nu, f_{\lambda\mu}]$ and the definitions of z_0 and z_1, it is easily seen that z_1 must be spanned by z_0 and $[b_\nu, z_0]$. When $h_1 = \{\gamma\lambda_3 + \delta\lambda_8\}$, the corresponding z_1 exists only if the values of b_ν belong to a proper subalgebra of SU_3'. Hence the field is reducible. When $\{h_1\} = \{0\}$, we have $q = 2$.

Now consider the case $h_0 = \{\lambda_1, \lambda_2, \lambda_3, \lambda_8\}$. If $h_1 = h_0$, then $q = 1$. It is easily seen that the dimension of h_1 cannot be 3. When $h_1 = \{\lambda_3, \lambda_8\}$, we have $z_0 = \{\lambda_8\}$, $z_1 = \{\lambda_3, \lambda_8\}$. Moreover, $b_\nu \in \{\lambda_1, \lambda_2, \lambda_3, \lambda_8\}$, since z_1 is spanned by z_0 and $[b_\nu, z_0]$. Consequently, $\mathscr{F}$ is reducible. When $h_1 = \{\alpha\lambda_3 + \beta\lambda_8\}$, $(\alpha \neq \pm\sqrt{3}\,\beta)$, we can prove the reducibility of $\mathscr{F}$ in a similar way. When $h_1 = \{0\}$ we have $q = 2$.

It remains to consider the cases $h_0 = \{\lambda_1, \lambda_2, \lambda_3, \lambda_8\}$ and $h_1 = \left\{\lambda_3 \pm \frac{1}{\sqrt{3}}\lambda_8\right\}$. If $h_2 = h_1$, then $q = 2$. Hence only the case $h_2 = \{0\}$ needs to be considered. We are to prove that this case does not occur, if the field $\mathscr{F}$ is irreducible. Let $h_1 = \left\{\lambda_3 + \frac{1}{\sqrt{3}}\lambda_8\right\}$. Then $z_0 = \{\lambda_8\}$, z_1 is $\{\lambda_3, \lambda_6, \lambda_7, \lambda_8\}$ or its subspace which contains an element in the form $\kappa\lambda_3 + \sigma\lambda_6 + \tau\lambda_7$ (σ, τ not all zero). Since z_1 is spanned by z_0 and $[b_\mu, z_0]$, we have $b^4_\mu = b^5_\mu = 0$. From the Eqs. $\underset{1}{f}_{\lambda\mu|\nu} = f_{\lambda\mu|\nu}$, $\underset{1}{f}_{\lambda\mu|\nu_1\nu_2} = f_{\lambda\mu|\nu_1\nu_2}$ it follows that

$$[\underset{1}{b}_\mu - b_\mu, \lambda_8] = 0, \ [\underset{1}{b}_\mu - b_\mu, \kappa\lambda_3 + \sigma\lambda_6 + \tau\lambda_7] = 0.$$

Consequently,

$$\underset{1}{b}_\mu - b_\mu = l_\lambda\left(\lambda_3 + \frac{1}{\sqrt{3}}\lambda_8\right),$$

and hence $\underset{1}{b}{}^a_\mu = b^a_\mu$ $(a = 1, 2, 4, 5, 6, 7)$. From $\underset{1}{f}{}^1_{\lambda\mu} = f^1_{\lambda\mu}$, $\underset{1}{f}{}^2_{\lambda\mu} = f^2_{\lambda\mu}$, we have

$$b^1_\lambda = \sigma^1 l_\lambda, \ b^2_\lambda = \sigma^2 l_\lambda.$$

If b^1_λ, b^2_λ are all equal to zero, then b_λ belong to the subalgebra $\{\lambda_3, \lambda_6, \lambda_7, \lambda_8\}$ and the field $\mathscr{F}$ is reducible. In other case σ^1, σ^2 are not all equal to zero and l_λ are not all zero also. From $f^4_{\lambda\mu} = f^5_{\lambda\mu} = 0$, it follows that

$$l_\lambda(\sigma^1 b^7_\mu + \sigma^2 b^6_\mu) - l_\mu(\sigma^1 b^7_\lambda + \sigma^2 b^6_\lambda) = 0,$$
$$l_\lambda(\sigma^2 b^7_\mu - \sigma^1 b^6_\mu) - l_\mu(\sigma^2 b^7_\lambda - \sigma^1 b^6_\lambda) = 0.$$

and hence

$$b^6_\mu = \tau^1 l_\mu, \ b^7_\mu = \tau^2 l_\mu.$$

Now we see that $f^1_{\lambda\mu}$, $f^2_{\lambda\mu}$, $f^3_{\lambda\mu}$ depend on b^1_ν, b^2_ν, b^3_ν only, and hence the latter can be reduced to zero through a gauge transformation. Consequently, b_λ belong to the subalgebra $\{\lambda_3, \lambda_6, \lambda_7, \lambda_8\}$ and the field is reducible. The same fact for the case $h_1 = \left\{\lambda_3 - \frac{1}{\sqrt{3}}\lambda_8\right\}$ may be proved similarly.

Since all possible cases have been considered, we may conclude that $p=2$ for SU_3 gauge field.

III. The Relation Between the Local Equivalence and the Global Equivalence

So far, the problem of the local equivalence has been solved through the field strength and its gauge derivatives. Now, the relation between the local equivalence and the global equivalence is to be considered with the loop phase factor method[9].

Theorem 4. *Let $\mathscr{F}_1$ and $\mathscr{F}_2$ be analytic gauge fields on a connected analytic manifold M_n. Suppose that $\mathscr{F}_1$ and $\mathscr{F}_2$ are equivalent in an arbitrary neighbourhood V_0 of a point O. Then $\mathscr{F}_1$ and $\mathscr{F}_2$ are globally equivalent if M_n is simply connected or, more generally, if M_n is multi-connected and the corresponding phase factors of $\mathscr{F}_1$ and $\mathscr{F}_2$ for a complete set of independent fundamental loops are all equal in a suitable gauge.*

Proof. M_n can be covered by a system of division regions $\{V_a\}$ which are common for both $\mathscr{F}_1$ and $\mathscr{F}_2$[10]. Let l be any loop with a fixed point O as the initial and terminal point, l is called an O-loop. Let $\underset{\alpha}{\Phi_0}(l)$ $(\alpha=1, 2)$ be the phase factors for l in $\mathscr{F}_\alpha$.

According to the assumption of the theorem we have

$$\underset{1}{\Phi_0}(l)=\underset{2}{\Phi_0}(l) \tag{3.1}$$

in a suitable gauge, where $l\subset V_0$. Now suppose that l is an arbitrary O-loop which is homotopic to O. Evidently, there exists a region $\Omega\subset M_n$, homeomorphic analytically to the Euclidean space E_n, such that $l\subset\Omega$. Let the restriction of $\mathscr{F}_1$ and $\mathscr{F}_2$ on Ω be $\mathscr{F}'_1$ and $\mathscr{F}'_2$ respectively. Construct a system of analytic paths Ox as the standard paths[12], where $x\in\Omega$. Let $I+\underset{\alpha}{k}(x, dx)$ be the phase factors for the differential triangle $Ox\ x+dx\ O$ in fields $\mathscr{F}_\alpha$, where $\underset{\alpha}{k}$ are g-valued differential forms with analytic coefficients $\underset{\alpha}{k^i_\mu}(x)$. We have $\underset{1}{k}=\underset{2}{k}$ for $x\in\Omega$, since the two forms are both analytic and equal to one another for $x\in V_0\cap\Omega$. Since $\underset{\alpha}{\Phi_0}(l)$ is determined by $\underset{\alpha}{k}$, (3.1) holds true for each O-loop which is homotopic to O. If M_n is simply connected, then the conclusion of the theorem follows from the theorem 8 of [9]. If M_n is multi-connected, we may arrive at the same conclusion also, since every O-loop may be represented as a product of some fundamental loops and a loop homotopic to O.

The theorem is thus proved and it is readily seen that the supplementary condition in the multi-connected case can not be removed[10].

In the non-analytic case things are quite different.

Theorem 5. *Let G be a non-Abelian group. For any connected manifold M_n there exist gauge fields $\mathscr{F}_1$ and $\mathscr{F}_2$ such that $\mathscr{F}_1$ and $\mathscr{F}_2$ are equivalent in a certain neighbourhood of each point of M_n and are not globally equivalent.*

Proof. Since M_n contains a neighbourhood homeomorphic to E_n, it needs only to prove the above statement for the case E_n.

Construct a gauge field $\mathscr{F}_1$ such that

(1) the gauge potentials $\underset{1}{b}{}^i_\mu$ are C^∞ functions satisfying $\underset{1}{b}{}^i_\mu(x)x^\mu = 0$, $\underset{1}{b}{}^i_\mu(0) = 0$,

(2) when $r < a$ and $r > c$ the values of $\underset{1}{b}{}_\mu(x)$ generate the whole Lie algebra g, where a and c are given constants satisfying $c > a > 0$ and $r = \sqrt{(x^1)^2 + \cdots + (x^n)^2}$,

(3) $\underset{1}{b}{}^i_\mu(x) = 0$ when $a \leqslant r \leqslant c$.

Let $\mathscr{F}_2$ be a gauge field with potentials

$$\underset{2}{b}{}_\mu = \begin{cases} \underset{1}{b}{}_\mu & \text{when } r \leqslant c, \\ (ad\alpha)\underset{1}{b}{}_\mu & \text{when } r \geqslant c, \end{cases} \tag{3.2}$$

where $\alpha \in G$ and $ad\alpha \neq ade$. Evidently, $\mathscr{F}_2$ is locally equivalent to $\mathscr{F}_1$ (see Fig. 3).

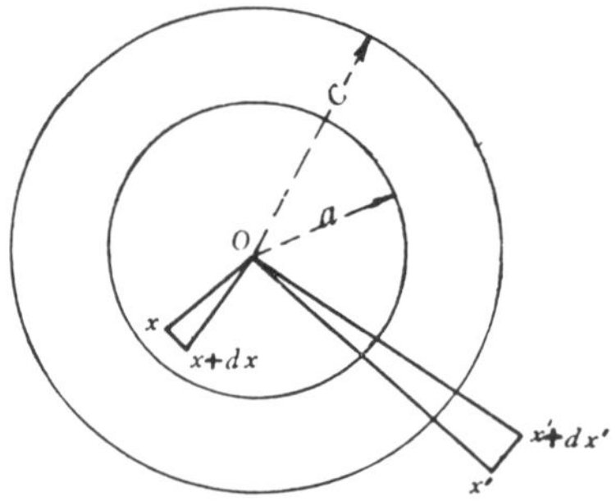

Fig. 3

$$\underset{2}{k}(x, dx) = \underset{1}{k}(x, dx)$$
$$\underset{2}{k}(x', dx') = (ad\alpha)\, \underset{1}{k}(x', dx')$$

From the condition (1) it follows that $\underset{\alpha}{\Phi}_{OP} = I$ $(\alpha = 1, 2)$ for an arbitrary P, where OP is a straight line segment. Hence the phase factors $I + \underset{\alpha}{k}$ for the differential triangle OP $P + dP$ O is equal to $I + \underset{\alpha}{b}$. Consequently, we have

$$\begin{aligned} \underset{2}{k} &= \underset{1}{k}, & (r \leqslant c), \\ \underset{2}{k} &= (ad\alpha)\underset{1}{k}, & (r \geqslant c). \end{aligned} \tag{3.3}$$

Since in both regions $r < a$ and $r > c$ the values of $\underset{1}{k}$ generate the whole Lie algebra g and $ad\alpha \neq ade$, there does not exist an element β such that the relation

$$\underset{2}{k} = (ad\beta)\underset{1}{k}$$

holds everywhere. Consequently, $\mathscr{F}_1$ is not equivalent to $\mathscr{F}_2$.

However, in the Abelian case the local equivalence at each point implies the global equivalence if the space-time region is simple connected[10].

References

[1] Yang C. N.: *Phys. Rev. Lett.*, **33** (1974), 445.
[2] Gu Chao-hao & Yang Chen-ning, *Scientia Sinica*, **18** (1975), 483.
[3] Wu T. T. & Yang C. N.: *Phys. Rev.*, **D12** (1975), 3843.
[4] Hu He-sheng: *Fudan Jour. (Natural Science)*, (1976), No. (1), 72.
[5] Xia Dao-xing: *Ibid* (1976), No. (1), 82.
[6] Shen Chun-li: *Ibid* (1976) No. (2), 61.
[7] Eisenhart, L. P.: *An Introduction to Differential Geometry*, (1940), Princeton Univ. Press.
[8] Gell-mann, M.: *Phys. Rev.*, **125** (1962), 1067.
[9] Gu Chao-hao: *Fudan Jour. (Natural Science)*, (1976), (2), 51.
[10] ————: *Ibid*, (1975), (4), 83. *Scientia Sinica* (Chinese edition) (1976), 320.

[[77i]]

Commentary begins page 79

Speech at the Benjamin W. Lee Memorial Session

C. N. Yang

Institute for Theoretical Physics, State University of New York, Stony Brook

Benjamin W. Lee was born in 1935 and died earlier this year at the age of 42. He had published more than one hundred research papers in theoretical physics in his lifetime.

The discipline of theoretical physics has as its principal aim the formulation of theoretical descriptions of the physical world which are concise and comprehensive. Its history has taught us that it is a glorious enterprise. It has produced, for understanding the subtle, complex and often confusing manifestations of nature, theoretical descriptions of unimagined accuracy. It utilizes, and helps to originate and to develop, mathematical concepts of the greatest beauty and depth.

Because nature is subtle and complex, the pursuit of theoretical physics requires bold and enthusiastic ventures into the muddy waters at the frontiers of newly discovered phenomena. Because the concepts used are beautiful and deep, the pursuit of theoretical physics requires appreciation of and insight into the structural aspects of the theoretical apparatus.

Ben Lee's work was characterized by his ability to excel in both of these requirements. His paper with Gaillard and Rosner in the *Reviews of Modern Physics*, completed before the discovery of J/ψ, was remarkable to read at the time and even more remarkable to read now, after the experimental discovery of charm. His work on the renormalizability of gauge theories is among the very important works on the fundamental structure of theoretical physics in recent years. We know that few contributions in theoretical physics remain noticeable after ten years, fewer after twenty. I venture to guess that the renormalizability of gauge fields will remain important fifty years from now.

We at Stony Brook were fortunate to have had Ben in our midst for many years. He was stimulating as a physicist, cooperative and generous as a colleague and friend. He had the admirable quality of always balancing enthusiasm with good judgment and restraint. Many of us, and I in particular, had benefited from many, many enjoyable discussions with him.

Ben's death occurred at the peak of his career as a physicist. He was exuberant and full of ideas about what is in store for our field. Let me quote from a speech he made last January at the Chicago Meeting of the American Physical Society, when Steve Weinberg was awarded the Heineman prize:

> Do we understand, or hope to understand, weak interactions as well as, say electrodynamics, in the present framework? Perhaps. We have yet to come to grips with *CP* violation and ultrahigh energy behaviors of weak interactions, on which subjects I have a few remarks to make. But I am more optimistic than ever that we are on the right track, and I can say that Steve has earned the honor bestowed on him today.

Ben's untimely death was a great loss to his family and his friends. It was a great loss to the Fermilab. It was a great loss to the science of physics.

□ Reprinted from *Unification of Elementary Forces and Gauge Theories*, D. B. Cline and F. E. Mills, eds. New York: Harwood Academic Publishers, 1977, xiii–xiv.

INTERACTION OF ELECTRONS, MAGNETIC MONOPOLES, AND PHOTONS (I)

[[78e]]
Commentary begins page 79

TU TUNG-SHENG (杜东生),
WU TAI-TSUN (吴大峻) AND YANG CHEN-NING (杨振宁)

Received February 18, 1978.

ABSTRACT

Using a method developed for treating monopole harmonics, we present in this series of papers a second quantized theory of interacting electrons, magnetic monopoles, and photons. In this first paper we discuss a system of interacting electrons, monopoles, and the electromagnetic field. The electrons and monopoles are treated as spin 1/2 Dirac particles with the negative energy states not filled.

I. INTRODUCTION

In a brilliant paper, Dirac[1] found in 1931 that in quantum mechanics, the simultaneous existence of an electron and a magnetic monopole implies that $eg = (\frac{1}{2}\hbar c)\times(\text{integer})$ where e and g are the electric and magnetic charges. Seventeen years later he went beyond this and studied[2] the second quantized theory of electrons, monopoles, and photons. Recently[3] the monopole problem with a static monopole has been reexamined with the bothersome strings of singularities eliminated. The present series of papers presents a generalization of this work to a second quantized theory of interacting electrons, monopoles, and photons, again without bothersome strings of singularities.

The key mathematical tools that we shall use are the concepts of nontrivial bundles and sections. These were already introduced in [3]. To read that reference and the present series of papers, no mathematical background on bundles and sections is necessary.

A classical Lagrangian theory has been formulated[4] for the interacting system of electrons, monopoles, and the electromagnetic field. We shall, however, not proceed by quantizing this Lagrangian system[1]. Instead, we aim at a direct Hamiltonian

1) We emphasize again here a point already discussed in [4]: with the strings of singularities, but without the regional structure, the magnetic field $\nabla\times A$ is not the correct monopole field. It contains a ''return'' field of infinite strength, but finite flux, along the singularity. This was emphasized first in G. Wentzel, *Prog. Theor. Phys. Suppl.*, **37—38**(1966), 163.

formalism, by first reviewing the usual electron-photon system without monopoles in the Fermi formulation[5], where the longitudinal field has already been eliminated. We then suitably combine this formulation with that of [3], resulting in the Hamiltonian (3.1) of Section III. The vector potential $\boldsymbol{A}$ in this Hamiltonian contains the usual transverse part, but also contains a "longitudinal magnetic" part suggested by the considerations of [3]. That this Hamiltonian does describe the interacting system correctly is then demonstrated in the remaining sections.

II. *e*-Photon Interaction

Before going into an *e*-*g*-photon interacting system, we shall first review the well-known Hamiltonian of Fermi[5] for an interacting system of electrons and photons, with the longitudinal fields already eliminated through the supplementary conditions,

$$\mathscr{H} = \sum_i \{\boldsymbol{\alpha}_i \cdot [\boldsymbol{p}_i - \varepsilon \boldsymbol{A}(\boldsymbol{x}_i)] + \beta_i m_i\} + \sum_{i>j} \varepsilon^2/r_{ij} + \sum_\gamma \omega a^\dagger a, \tag{2.1}$$

where

$$\sum_\gamma = \text{sum over all transverse photon momentum } \boldsymbol{k} \text{ and polarization.} \tag{2.2}$$

The quantity ε is the electric charge of the electron and is negative. $\omega = |\boldsymbol{k}|$ is the energy of the photon, and V the volume of normalization. The vector potential $\boldsymbol{A}(\boldsymbol{\xi})$ is ***entirely transverse.*** It is given by

$$\mathbf{A}(\boldsymbol{\xi}) = \sum_\gamma (2\pi/\omega V)^{\frac{1}{2}} \boldsymbol{e}[a e^{i\boldsymbol{k}\cdot\boldsymbol{\xi}} + \text{h.c.}], \tag{2.3}$$

where h.c. stands for Hermitian conjugate and $\boldsymbol{e}$ is the transverse polarization vector. $\boldsymbol{\xi}$ is a vector in 3 dimensional coordinate space. $a^\dagger$ and a are creation and annihilation operators. $\boldsymbol{x}_i$ is the position of the ith electron.

The commutation rules of the dynamical variable $\boldsymbol{x}_i$, $\boldsymbol{p}_i$, α_i, β_i, $\mathbf{a}$, $\mathbf{a}^\dagger$ are as usual. Eq. (2.1) defines a Hamiltonian which is conveniently studied in the Schrödinger representation. To appreciate the meaning of the dynamical system, however, one has to go into the Heisenberg representation.

We shall now use the Heisenberg representation. The scalar potentials and the fields are defined by

$$\phi(\boldsymbol{\xi}) = \sum_i \varepsilon |\boldsymbol{\xi} - \boldsymbol{x}_i|^{-1}, \tag{2.4}$$

$$\boldsymbol{E}(\boldsymbol{\xi}) = -\nabla\phi - \dot{\boldsymbol{A}},$$

$$\boldsymbol{H}(\boldsymbol{\xi}) = \nabla \times \boldsymbol{A}. \tag{2.5}$$

The time dependence of any operator B is given, as usual, by

$$i\dot{B} = B\mathscr{H} - \mathscr{H}B. \tag{2.6}$$

Using this formula we can evaluate the time derivatives of all dynamical variables, obtaining in particular

$$\dot{\boldsymbol{x}}_i = \boldsymbol{a}_i, \tag{2.7}$$

$$(\dot{p}_j)_x = \varepsilon \boldsymbol{a}_j \cdot \frac{\partial}{\partial x_j} \boldsymbol{A}(\boldsymbol{x}_j) - \varepsilon \frac{\partial}{\partial x_j} \phi(\boldsymbol{x}_j), \quad (j \text{ not summed}) \tag{2.8}$$

$$\dot{a} = -i\omega a + i\varepsilon \sum_i (\boldsymbol{a}_i \cdot \boldsymbol{e}) (2\pi/\omega V)^{\frac{1}{2}} \exp(-i\boldsymbol{k} \cdot \boldsymbol{x}_i),$$

$$\dot{a}^\dagger = i\omega a^\dagger - i\varepsilon \sum_i (\boldsymbol{a}_i \cdot \boldsymbol{e}) (2\pi/\omega V)^{\frac{1}{2}} \exp(i\boldsymbol{k} \cdot \boldsymbol{x}_i). \tag{2.9}$$

We can now evaluate the time derivative of $\boldsymbol{A}(\boldsymbol{\xi})$ using (2.9):

$$\dot{\boldsymbol{A}}(\boldsymbol{\xi}) = \sum_\gamma (2\pi/\omega V)^{\frac{1}{2}} (-i\omega)\boldsymbol{e}[a e^{i\boldsymbol{k}\cdot\boldsymbol{\xi}} - \text{h.c.}]. \tag{2.10}$$

In deriving this formula, use has been made of

$$\sum_{\text{Po.}} (\boldsymbol{a}_i \cdot \boldsymbol{e})\boldsymbol{e} = \boldsymbol{a}_i - \omega^{-2} (\boldsymbol{a}_i \cdot \boldsymbol{k}), \boldsymbol{k}, \tag{2.11}$$

where $\sum_{\text{Po.}}$ means sum over the two transverse polarizations for a fixed photon momentum $\boldsymbol{k}$. Now use (2.7) to obtain

$$\frac{d}{dt}[\boldsymbol{A}(\boldsymbol{x}_i)] = \dot{\boldsymbol{A}}(\boldsymbol{x}_i) + (\dot{\boldsymbol{x}}_i \cdot \boldsymbol{\nabla})\boldsymbol{A}(\boldsymbol{x}_i) = \dot{\boldsymbol{A}}(\boldsymbol{x}_i) + (\boldsymbol{a}_i \cdot \boldsymbol{\nabla})\boldsymbol{A}(\boldsymbol{x}_i). \tag{2.12}$$

Thus

$$\begin{aligned} \frac{d}{dt}[\boldsymbol{p}_j - \varepsilon \boldsymbol{A}(\boldsymbol{x}_j)]_x &= \varepsilon \boldsymbol{a}_j \cdot \frac{\partial}{\partial x_j} \boldsymbol{A}(\boldsymbol{x}_j) - \varepsilon \frac{\partial}{\partial x_j} \phi(\boldsymbol{x}_j) \\ &- \varepsilon \dot{\boldsymbol{A}}_x(\boldsymbol{x}_i) - \varepsilon(\boldsymbol{a}_j \cdot \boldsymbol{\nabla})\boldsymbol{A}_x(\boldsymbol{x}_j) = \varepsilon[\boldsymbol{E}(\boldsymbol{x}_j) + \dot{\boldsymbol{x}}_j \times \boldsymbol{H}(\boldsymbol{x}_j)]_x, \end{aligned}$$

i. e.

$$\frac{d}{dt}[\boldsymbol{p}_j - \varepsilon \boldsymbol{A}(\boldsymbol{x}_j)] = \varepsilon[\boldsymbol{E}(\boldsymbol{x}_j) + \dot{\boldsymbol{x}}_j \times \boldsymbol{H}(\boldsymbol{x}_j)]. \tag{2.13}$$

Eqs. (2.7) and (2.13) will be called the Lorentz equations. They exhibit the force of the electromagnetic field on the particles.

We now check whether the four Maxwell equations are satisfied. To start with, (2.3) implies

$$\boldsymbol{\nabla} \cdot \boldsymbol{A} = 0. \tag{2.14}$$

Eq. (2.5) then leads to

$$\boldsymbol{\nabla} \cdot \boldsymbol{H} = 0, \quad \boldsymbol{\nabla} \cdot \boldsymbol{E}(\boldsymbol{\xi}) = 4\pi\varepsilon \sum_i \delta^3(\boldsymbol{\xi} - \boldsymbol{x}_i). \tag{2.15}$$

Eq. (2.5) also leads to, in a straightforward manner,

$$\dot{\boldsymbol{H}} = -\boldsymbol{\nabla} \times \boldsymbol{E}. \tag{2.16}$$

To check the fourth Maxwell equation (Ampere's law), we first evaluate $\ddot{\boldsymbol{A}}$, using

(2.10) and (2.9):

$$\ddot{\boldsymbol{A}}(\boldsymbol{\xi}) = \sum_{\gamma} (2\pi/\omega V)^{\frac{1}{2}} (-i\omega)\boldsymbol{e}(-i\omega)[ae^{i\boldsymbol{k}\cdot\boldsymbol{\xi}} + \text{h. c.}]$$

$$+ \sum_{\gamma} (2\pi/\omega V)^{\frac{1}{2}} (-i\omega)\boldsymbol{e} \sum_{i} i\varepsilon(\boldsymbol{a}_i \cdot \boldsymbol{e})(2\pi/\omega V)^{\frac{1}{2}} [e^{i\boldsymbol{k}\cdot(\boldsymbol{\xi}-\boldsymbol{x}_i)} + \text{h.c.}]. \quad (2.17)$$

The first sum is equal to $-\boldsymbol{\Delta}\times\boldsymbol{H}$, on account of (2.14). The second sum is equal to, using (2.11),

$$\varepsilon \sum_{\gamma} \sum_{i} (2\pi/\omega V)\omega[\boldsymbol{a}_i - \omega^{-2}(\boldsymbol{a}_i \cdot \boldsymbol{k})\boldsymbol{k}][e^{i\boldsymbol{k}\cdot(\boldsymbol{\xi}-\boldsymbol{x}_i)} + \text{h. c.}]/2$$

$$= \varepsilon \sum_{i\gamma} \boldsymbol{a}_i (2\pi/V)[e^{i\boldsymbol{k}\cdot(\boldsymbol{\xi}-\boldsymbol{x}_i)} + \text{h. c.}]/2$$

$$+ \varepsilon \sum_{i\gamma} (2\pi/V\omega^2)(\boldsymbol{a}_i \nabla_{\boldsymbol{\xi}})\nabla_{\boldsymbol{\xi}} [e^{i\boldsymbol{k}\cdot(\boldsymbol{\xi}-\boldsymbol{x}_i)} + \text{h. c.}]/2$$

$$= 4\pi\varepsilon \sum_{i} \boldsymbol{a}_i \delta^3(\boldsymbol{\xi} - \boldsymbol{x}_i) + \varepsilon \sum_{i} (\boldsymbol{a}_i \cdot \nabla_{\boldsymbol{\xi}})\nabla_{\boldsymbol{\xi}} |\boldsymbol{\xi} - \boldsymbol{x}_i|^{-1}$$

$$= 4\pi\varepsilon \sum_{i} \dot{\boldsymbol{x}}_i \delta^3(\boldsymbol{\xi} - \boldsymbol{x}_i) - \sum_{i} (\dot{\boldsymbol{x}}_i \cdot \nabla_i)\nabla_{\boldsymbol{\xi}} \phi(\boldsymbol{\xi}). \quad (2.18)$$

Thus

$$\ddot{\boldsymbol{A}}(\boldsymbol{\xi}) = -\nabla \times \boldsymbol{H} - \sum_{i} (\dot{\boldsymbol{x}}_i \cdot \nabla_i)[\nabla_{\boldsymbol{\xi}} \phi(\boldsymbol{\xi})] + 4\pi\varepsilon \sum_{i} \dot{\boldsymbol{x}}_i \delta^3(\boldsymbol{\xi} - x_i). \quad (2.19)$$

Now

$$\dot{\boldsymbol{E}}(\boldsymbol{\xi}) = -\sum_{i} (\dot{\boldsymbol{x}}_i \cdot \nabla_i)[\nabla_{\boldsymbol{\xi}} \phi(\boldsymbol{\xi})] - \ddot{\boldsymbol{A}}. \quad (2.20)$$

Thus we have correctly the fourth Maxwell equation:

$$\nabla \times \boldsymbol{H} - \dot{\boldsymbol{E}} = 4\pi\varepsilon \sum_{i} \dot{\boldsymbol{x}}_i \delta^3(\boldsymbol{\xi} - \boldsymbol{x}_i). \quad (2.21)$$

To summarize, we have shown that Fermi's Hamiltonian (2.1), with electron and transverse photon variables as the only dynamical variables, does give a dynamical system that satisfies Maxwell and Lorentz equations in the Heisenberg representation.

III. *e*-*g*-Photon Interaction in Schrödinger Representation

We want to generalize the dynamical system of the last section to include magnetic monopoles. The generalization proceeds by comparison with the simpler problem discussed in [3] and [6] where it was explained why it was necessary to divide space into regions and why the Hilbert space of the quantum system is a Hilbert space of sections[1].

---

1) The motivation for the division into regions, explored in [3], was explained more fully in [6].

Consider an interacting[1)] system of n electrons at space coordinates $\boldsymbol{x}_i (i=1, 2, \cdots, n)$ and N magnetic monopoles at space coordinates $\boldsymbol{X}_I, (I=1, \cdots, N)$. Both interact with the electromagnetic field. Both electrons and monopoles will be taken to be Dirac spin ½ particles, with momenta $\boldsymbol{p}_j$, $\boldsymbol{P}_I$ and $\boldsymbol{\alpha}$, β matrices $\boldsymbol{\alpha}_i$, β_i, $\boldsymbol{\alpha}_I$, β_I. The Hamiltonian is

$$\mathscr{H} = \sum_i \{\boldsymbol{\alpha}_i \cdot (\boldsymbol{p}_i - \varepsilon_i \boldsymbol{A}(\boldsymbol{x}_i)) + \beta_i M_i\} + \sum_I \{\boldsymbol{\alpha}_I \cdot (\boldsymbol{P}_I - g_I \boldsymbol{B}(\boldsymbol{X}_I)) + \beta_I M_I\}$$

$$+ \sum_{i>j} \varepsilon_i \varepsilon_j r_{ij}^{-1} + \sum_{I>J} g_I g_J r_{IJ}^{-1} + \sum_\gamma \omega a^\dagger a, \tag{3.1}$$

where $\sum_\gamma$ again means summation over *transverse* photon modes,

$$\boldsymbol{B}(\boldsymbol{X}_I) = - \sum_i \varepsilon_i \boldsymbol{A}_{iI} + \sum_\gamma (2\pi/\omega V)^{\frac{1}{2}} \omega^{-1} (\boldsymbol{k} \times \boldsymbol{e})[a e^{i\boldsymbol{k}\cdot\boldsymbol{X}_I} + \text{h. c.}], \tag{3.2}$$

$$\boldsymbol{A}(\boldsymbol{x}_i) = \sum_I g_I \boldsymbol{A}_{iI} + \sum_\gamma (2\pi/\omega V)^{\frac{1}{2}} \boldsymbol{e}[a e^{i\boldsymbol{k}\cdot\boldsymbol{x}_i} + \text{h. c.}], \tag{3.3}$$

and $\boldsymbol{A}_{\xi I}$ is the instantaneous vector potential at $\boldsymbol{\xi}$ generated by a magnetic monopole at $\boldsymbol{X}_I$:

$$\boldsymbol{A}_{\xi I} = \frac{1 - \cos\theta_{\xi I}}{r_{\xi I}^2 \sin^2\theta_{\xi I}} \hat{z} \times \boldsymbol{x}_{\xi I}, \quad \text{if } \boldsymbol{x}_{\xi I} \equiv \boldsymbol{\xi} - \boldsymbol{x}_I \text{ is in } R_a, \tag{3.4}$$

$$\boldsymbol{A}_{\xi I} = - \frac{1 + \cos\theta_{\xi I}}{r_{\xi I}^2 \sin^2\theta_{\xi I}} \hat{z} \times \boldsymbol{x}_{\xi I}, \quad \text{if } \boldsymbol{x}_{\xi I} \text{ is in } R_b. \tag{3.5}$$

$\boldsymbol{A}_{iI}$ is the value of $\boldsymbol{A}_{\xi I}$ at $\boldsymbol{\xi} = \boldsymbol{x}_i$. The two regions R_a and R_b are defined in Eqs. (2) and (3) of [3]. $\hat{z}$ is a unit vector along the $+z$ axis. $\boldsymbol{A}_{\xi I}$ is exactly the vector potential at $\boldsymbol{\xi}$ chosen in [3] for a stationary magnetic monopole at $\boldsymbol{X}_I$. That we can still use it when the monopoles are not stationary, can only be fully justified after we prove in the next sections that in the Heisenberg representation all Maxwell and Lorentz equations are satisfied.

The Hamiltonian (3.1) contains dynamical variables $\boldsymbol{x}_i$, $\boldsymbol{X}_I$, $\boldsymbol{p}_i$, $\boldsymbol{P}_I$, $\boldsymbol{\alpha}_i$, β_i, $\boldsymbol{\alpha}_I$, β_I, a and $a^\dagger$. These dynamical variables satisfy the usual commutation rules. The Hilbert space on which it operates is a Hilbert space of sections[3,6]. In other words, if we diagonalize $\boldsymbol{x}_i$, $\boldsymbol{X}_I$, $a^\dagger a$, β_i, β_I, $\boldsymbol{\alpha}_{ix}\boldsymbol{\alpha}_{iy}$, $\boldsymbol{\alpha}_{Ix}\boldsymbol{\alpha}_{Iy}$ simultaneously, the wave section

$$\Phi(\boldsymbol{x}_i, \boldsymbol{X}_I, a^\dagger a, \beta_i, \beta_I, \boldsymbol{\alpha}_{ix} \boldsymbol{\alpha}_{iy}, \boldsymbol{\alpha}_{Ix} \boldsymbol{\alpha}_{Iy})$$

is still dependent on the region we choose. The total number of regions is 2^{nN}, each region being defined according to the magnitude of θ_{iI}. In an overlapping region around $\theta_{jJ} = \pi/2$, there is a difference of $\boldsymbol{A}(\boldsymbol{x}_j)$ [and of $\boldsymbol{B}(\boldsymbol{X}_J)$] for the two regions:

$$\delta \boldsymbol{A}(\boldsymbol{x}_j) = g_J \delta \boldsymbol{A}_{jJ} = -2 g_J \boldsymbol{\nabla}_j \phi_{jJ}, \tag{3.6}$$

$$\delta \boldsymbol{B}(\boldsymbol{X}_J) = -\varepsilon_j \delta \boldsymbol{A}_{jJ} = -2\varepsilon_j \boldsymbol{\nabla}_J \phi_{jJ}, \tag{3.7}$$

---

1) We have only one time coordinate t, which is shared by all four-variable coordinates. Thus $\boldsymbol{\xi}$, $\boldsymbol{x}_i$ and $\boldsymbol{x}_J$ all share the same t. This is the same situation as in § II. $\boldsymbol{\xi}$ is a coordinate point at which one evaluates field variables. The dot $(\cdot)$ means $\partial/\partial t$.

where ϕ_{jJ} is the azimuthal angle of $\boldsymbol{x}_{jJ} \equiv \boldsymbol{x}_j - \boldsymbol{X}_J$. These equations are of course straightforward generalizations of Eq. (6) of [3]. They lead to the transition function S that forms the gauge transformation between the two overlapping regions around $\theta_{jJ} = \pi/2$:

$$\psi^{(a)} = S\psi^{(b)}, \tag{3.8}$$

where

$$S = \exp[2i\varepsilon_j g_J \phi_{jJ}]. \tag{3.9}$$

As in [3], for the Hilbert space to be meaningful we need Dirac's condition:

$$2\varepsilon_j g_J = \text{integer}. \tag{3.10}$$

The transition function makes the operators $\boldsymbol{p}_i - \varepsilon_i \boldsymbol{A}(\boldsymbol{x}_i)$ and $\boldsymbol{P}_I - g_I \boldsymbol{B}(\boldsymbol{X}_I)$ valid operators (for all i and all I) on the Hilbert space, again as in the case discussed in [3]. Thus the Hamiltonian (3.1) is a valid operator.

To summarize, $\mathscr{H}$ of (3.1) is an operator on a Hilbert space of sections with transition functions (3.9). It defines a dynamical system in the Schrödinger representation. It contains the dynamical variables for the electrons, monopoles, and transverse photons.

IV. e-g-Photon Interaction in Heisenberg Representation

As in Section II, the meaning of the dynamical system of Section III can only become clear if we go into the Heisenberg representation, in which (2.6) governs the time derivative of any operator. We find thus

$$\dot{\boldsymbol{x}}_i = \boldsymbol{a}_i, \qquad \dot{\boldsymbol{X}}_I = \boldsymbol{a}_I, \tag{4.1}$$

$$\begin{aligned} \dot{a} &= -i\omega a + i\sum_i \varepsilon_i (\boldsymbol{a}_i \cdot \boldsymbol{e})(2\pi/\omega V)^{\frac{1}{2}} \exp(-i\boldsymbol{k}\cdot\boldsymbol{x}_i) \\ &\quad + i\sum_I g_I \omega^{-1} \boldsymbol{a}_I \cdot (\boldsymbol{k}\times\boldsymbol{e})(2\pi/\omega V)^{\frac{1}{2}} \exp(-i\boldsymbol{k}\cdot\boldsymbol{X}_I), \\ \dot{a}^\dagger &= \text{h. c. of above.} \end{aligned} \tag{4.2}$$

The calculation of the commutators of $\boldsymbol{p}_i - \varepsilon_i \boldsymbol{A}(\boldsymbol{x}_i)$ with $\mathscr{H}$ is a little long. To avoid confusion, we define transverse potentials $\boldsymbol{A}^T$ and $\boldsymbol{B}^T$ by

$$\begin{aligned} \boldsymbol{A}(\boldsymbol{\xi}) &= \sum_I g_I \boldsymbol{A}_{\boldsymbol{\xi} I} + \boldsymbol{A}^T(\boldsymbol{\xi}), \\ \boldsymbol{B}(\boldsymbol{\xi}) &= -\sum_i \varepsilon_i \boldsymbol{A}_{i\boldsymbol{\xi}} + \boldsymbol{B}^T(\boldsymbol{\xi}). \end{aligned} \tag{4.3}$$

We can now evaluate the time derivatives of $\boldsymbol{A}^T(\boldsymbol{\xi})$ and $\boldsymbol{B}^T(\boldsymbol{\xi})$ by using (4.2):

$$\begin{aligned} \dot{\boldsymbol{A}}^T(\boldsymbol{\xi}) &= \sum_\gamma (2\pi/\omega V)^{\frac{1}{2}}(-i\omega)\boldsymbol{e}[ae^{i\boldsymbol{k}\cdot\boldsymbol{\xi}} - \text{h. c.}] \\ &\quad + \sum_{\gamma, I} (2\pi/\omega V) i g_I (\boldsymbol{a}_I \times \boldsymbol{k})\omega^{-1}[ae^{i\boldsymbol{k}\cdot(\boldsymbol{\xi}-\boldsymbol{X}_I)} - \text{h. c.}](1/2), \end{aligned} \tag{4.4}$$

where the first sum is the same as (2.10) above, and the second sum originates from the second sum in (4.2). We have used in the second sum the identity:

$$\sum_{\text{Po.}} \boldsymbol{e}[\boldsymbol{a}_I \cdot \boldsymbol{k} \times \boldsymbol{e}] = \boldsymbol{a}_I \times \boldsymbol{k}. \tag{4.5}$$

The first sum in (4.4) is easily seen to be

$$\nabla_\xi \times \boldsymbol{B}^T(\boldsymbol{\xi}).$$

The second sum is equal to, remembering there are two polarizations for each k, $\sum_I g_I(\boldsymbol{a}_I \times \nabla_{\xi})|\boldsymbol{\xi} - \boldsymbol{x}_I|^{-1}$. Thus

$$\dot{\boldsymbol{A}}^T(\boldsymbol{\xi}) = \nabla \times \boldsymbol{B}^T + \sum_I g_I \boldsymbol{a}_I \times \nabla_{\xi} |\boldsymbol{\xi} - \boldsymbol{X}_I|^{-1}. \tag{4.6}$$

Similarly,

$$\dot{\boldsymbol{B}}^T(\boldsymbol{\xi}) = -\nabla \times \boldsymbol{A}^T - \sum_j \varepsilon_j \boldsymbol{a}_j \times \nabla_{\xi} |\boldsymbol{\xi} - \boldsymbol{x}_j|^{-1}. \tag{4.7}$$

We now define $\boldsymbol{E}$ and $\boldsymbol{H}$ by

$$\phi(\boldsymbol{\xi}) = \sum_i \varepsilon_i |\boldsymbol{\xi} - \boldsymbol{x}_i|^{-1} + \sum_I g_I \boldsymbol{a}_I \cdot \boldsymbol{A}_{\xi I}, \tag{4.8}$$

$$\boldsymbol{E}(\boldsymbol{\xi}) = -\nabla_{\xi} \phi - \dot{\boldsymbol{A}}, \tag{4.9}$$

$$\boldsymbol{H}(\boldsymbol{\xi}) = \nabla_{\xi} \times \boldsymbol{A}(\boldsymbol{\xi}). \tag{4.10}$$

It follows from these and (4.3) that

$$\boldsymbol{E}(\boldsymbol{\xi}) = -\nabla_{\xi} \sum_i \varepsilon_i |\boldsymbol{\xi} - \boldsymbol{x}_i|^{-1} - \dot{\boldsymbol{A}}^T(\boldsymbol{\xi}) - \sum_I g_I \boldsymbol{a}_I \times (\nabla_{\xi} \times \boldsymbol{A}_{\xi I}). \tag{4.11}$$

Since one has

$$(\nabla_{\xi} + \nabla_I) \ \text{(any component of } \boldsymbol{A}_{\xi I}) = 0, \tag{4.12}$$

Eq. (4.11) can be rewritten as

$$\boldsymbol{E}(\boldsymbol{\xi}) = -\dot{\boldsymbol{A}}^T + \sum_i \boldsymbol{E}_i^c - \sum_I \boldsymbol{a}_I \times \boldsymbol{H}_I^g, \tag{4.13a}$$

where

$$\boldsymbol{E}_i^c = \text{electric field at } \boldsymbol{\xi} \text{ generated by a } \textit{static} \text{ electron at } \boldsymbol{x}_i, \tag{4.14}$$

$$\boldsymbol{H}_I^g = \text{magnetic field at } \boldsymbol{\xi} \text{ generated by a } \textit{static} \text{ monopole at } \boldsymbol{X}_I. \tag{4.15}$$

The magnetic field can be written as, by (4.10) and (4.3)

$$\boldsymbol{H}(\boldsymbol{\xi}) = \nabla \times \boldsymbol{A}^T + \sum_I \boldsymbol{H}_I^g. \tag{4.13b}$$

Eqs. (4.13) express $\boldsymbol{E}$ and $\boldsymbol{H}$ in terms of $\boldsymbol{A}^T$ and $\boldsymbol{x}_i$, $\boldsymbol{X}_I$. It is important to emphasize that they are *independent of the regional structures* of the Hilbert space, just like in [3] where the magnetic field $\boldsymbol{H}$ is independent of the regional structure.

V. Symmetry Under $E \to H$, $H \to -E$

Using (4.6) and (4.7) we can rewrite Eqs. (4.13) as:

$$\boldsymbol{E}(\boldsymbol{\xi}) = -\boldsymbol{\nabla} \times \boldsymbol{B}^T + \sum_i \boldsymbol{E}_i^e, \tag{5.1a}$$

$$\boldsymbol{H}(\boldsymbol{\xi}) = -\dot{\boldsymbol{B}}^T + \sum_I \boldsymbol{H}_I^g + \sum_i \boldsymbol{\alpha}_i \times \boldsymbol{E}_i^e. \tag{5.1b}$$

These equations and (4.13) show that there is symmetry under

$$\boldsymbol{E} \to \boldsymbol{H}, \boldsymbol{H} \to -\boldsymbol{E}, \boldsymbol{A}^T \to \boldsymbol{B}^T, \boldsymbol{B}^T \to -\boldsymbol{A}^T,$$
$$\boldsymbol{x}_i \longleftrightarrow \boldsymbol{X}_I, \ \varepsilon_i \to g_I, \ g_I \to -\varepsilon_i. \tag{5.2}$$

Eqs. (5.1) express $\boldsymbol{E}$ and $\boldsymbol{H}$ in terms of $\boldsymbol{B}^T$, $\boldsymbol{x}_i$, and $\boldsymbol{X}_I$.

VI. e-g-Photon Interaction in Heisenberg Representation (Continued)

We shall prove now that the Lorentz equations and Maxwell equations are satisfied in the Heisenberg representation. To start with, we find by commuting $\boldsymbol{A}^T(\boldsymbol{x}_i)$ with $\mathscr{H}$,

$$\frac{d}{dt}\boldsymbol{A}^T(\boldsymbol{x}_i) = \dot{\boldsymbol{A}}^T(\boldsymbol{x}_i) + (\boldsymbol{\alpha}_i \cdot \boldsymbol{\nabla}_i)A^T(x_i), \tag{6.1}$$

which is the same as (2.12). Similarly,

$$\frac{d}{dt}\boldsymbol{B}^T(\boldsymbol{X}_I) = \dot{\boldsymbol{B}}^T(\boldsymbol{X}_I) + (\boldsymbol{\alpha}_I \cdot \boldsymbol{\nabla}_I)\boldsymbol{B}^T(\boldsymbol{X}_I). \tag{6.2}$$

To calculate $\frac{d}{dt}\left[\boldsymbol{p}_i - \varepsilon_i \sum_I g_I \boldsymbol{A}_{iI}\right]$ we first evaluate:

$$\left[\left\{\boldsymbol{p}_j - \varepsilon_j \sum_I g_I \boldsymbol{A}_{jI}\right\}_x, \left\{\boldsymbol{p}_l - \varepsilon_l \sum_I g_I \boldsymbol{A}_{lI}\right\}_y\right] = i\delta_{jl}\varepsilon_l \sum_I g_I \{\boldsymbol{\nabla}_l \times \boldsymbol{A}_{lI}\}_z, \tag{6.3}$$

$$\left[\left\{\boldsymbol{p}_j - \varepsilon_j \sum_I g_I \boldsymbol{A}_{jI}\right\}_x, \left\{\boldsymbol{P}_L + g_L \sum_i \varepsilon_i \boldsymbol{A}_{iL}\right\}_x\right] = 0, \tag{6.4}$$

$$\left[\left\{\boldsymbol{p}_j - \varepsilon_j \sum_I g_I \boldsymbol{A}_{jI}\right\}_x, \left\{\boldsymbol{P}_L + g_L \sum_i \varepsilon_i \boldsymbol{A}_{iL}\right\}_y\right] = \{-ig_L \varepsilon_j \boldsymbol{\nabla}_j \times \boldsymbol{A}_{jL}\}_z, \tag{6.5}$$

$$\left[\left\{\boldsymbol{p}_j - \varepsilon_j \sum_I g_I \boldsymbol{A}_{jI}\right\}_x, \boldsymbol{A}^T(\boldsymbol{x}_j)\right] = -i(\partial/\partial \boldsymbol{x}_j)\boldsymbol{A}^T(\boldsymbol{x}_j). \tag{6.6}$$

Thus

$$\frac{d}{dt}\left[\boldsymbol{p}_j - \varepsilon_j \sum_I g_I \boldsymbol{A}_{jI}\right] = \varepsilon_j \sum_I g_I (\boldsymbol{\alpha}_j - \boldsymbol{\alpha}_I) \times (\boldsymbol{\nabla}_j \times \boldsymbol{A}_{jI})$$
$$+ \varepsilon_j \boldsymbol{\nabla}[\boldsymbol{\alpha}_j \cdot \boldsymbol{A}^T(\boldsymbol{x}_i)] - \varepsilon_j \boldsymbol{\nabla}_j \sum_i \varepsilon_i r_{ij}^{-1}. \tag{6.7}$$

Eqs. (6.1) and (6.7) together give

$$\frac{1}{\varepsilon_j}\frac{d}{dt}[\boldsymbol{p}_j - \varepsilon_j \boldsymbol{A}(\boldsymbol{x}_j)] = -\dot{\boldsymbol{A}}^T(\boldsymbol{x}_j) + \boldsymbol{a}_j \times [\boldsymbol{\nabla} \times \boldsymbol{A}^T(\boldsymbol{x}_j)]$$

$$+ \sum_I g_I (\boldsymbol{a}_j - \boldsymbol{a}_I) \times (\boldsymbol{\nabla}_j \times \boldsymbol{A}_{jI}) - \boldsymbol{\nabla}_j \sum_i \varepsilon_i r_{ij}^{-1}, \tag{6.8}$$

where we have used

$$-(\boldsymbol{a}_j \cdot \boldsymbol{\nabla}_j)\boldsymbol{A}^T(\boldsymbol{x}_j) + \boldsymbol{\nabla}[\boldsymbol{a}_j \cdot \boldsymbol{A}^T(\boldsymbol{x}_j)] = \boldsymbol{a}_j \times [\boldsymbol{\nabla} \times \boldsymbol{A}^T(\boldsymbol{x}_j)]. \tag{6.9}$$

Using (4.13a) and (4.13b), we may rewrite (6.8) as

$$\frac{d}{dt}\{\boldsymbol{p}_j - \varepsilon_j \boldsymbol{A}(\boldsymbol{x}_j)\} = \varepsilon_j \boldsymbol{E}(\boldsymbol{x}_j) + \varepsilon_j \dot{\boldsymbol{x}}_j \times \boldsymbol{H}(\boldsymbol{x}_j), \tag{6.10}$$

which is one of the Lorentz equations. The other Lorentz equation is obtained similarly by commuting $\boldsymbol{P}_J - \boldsymbol{g}_J\boldsymbol{B}(\boldsymbol{X}_J)$ with the Hamiltonian, obtaining in place of (6.8),

$$\frac{1}{g_J}\frac{d}{dt}[\boldsymbol{p}_J - g_J \boldsymbol{B}(\boldsymbol{X}_J)] = -\dot{\boldsymbol{B}}^T(\boldsymbol{X}_J) + \boldsymbol{a}_J \times [\boldsymbol{\nabla} \times \boldsymbol{B}^T(\boldsymbol{X}_J)]$$

$$- \sum_i \varepsilon_i (\boldsymbol{a}_J - \boldsymbol{a}_i) \times (\boldsymbol{\nabla}_J \times \boldsymbol{A}_{iJ}) - \boldsymbol{\nabla}_J \sum_I g_I r_{IJ}^{-1}. \tag{6.11}$$

Using (5.1a) and (5.1b), we may change this equation into

$$\frac{d}{dt}[\boldsymbol{p}_J - g_J \boldsymbol{B}(\boldsymbol{X}_J)] = g_J \boldsymbol{H}(\boldsymbol{X}_J) - g_J \boldsymbol{a}_J \times \boldsymbol{E}(\boldsymbol{X}_J), \tag{6.12}$$

which is the Lorentz equation for the monopoles.

The two Lorentz equations (6.10) and (6.12) are related by the symmetry transformation (5.2).

To prove the Maxwell equations we start from the simple identities:

$$\begin{aligned}
&\boldsymbol{\nabla} \cdot \boldsymbol{A}^T = 0,\\
&\boldsymbol{\nabla} \cdot \boldsymbol{E}_i^c = 4\pi\varepsilon_i \,\delta^3(\boldsymbol{\xi} - \boldsymbol{x}_i),\\
&\boldsymbol{\nabla} \cdot (\boldsymbol{a}_I \times \boldsymbol{H}_I^g) = 0,\\
&\boldsymbol{\nabla} \cdot \boldsymbol{H}_I^g = 4\pi g_I \,\delta^3(\boldsymbol{\xi} - \boldsymbol{X}_I).
\end{aligned} \tag{6.13}$$

These equations combined with (4.13) give immediately the following two of the Maxwell equations:

$$\begin{aligned}
&\boldsymbol{\nabla} \cdot \boldsymbol{E} = \sum_i 4\pi\varepsilon_i \,\delta^3(\boldsymbol{\xi} - \boldsymbol{x}_i),\\
&\boldsymbol{\nabla} \cdot \boldsymbol{H} = \sum_I 4\pi g_I \,\delta^3(\boldsymbol{\xi} - \boldsymbol{X}_I).
\end{aligned} \tag{6.14}$$

Now

$$\boldsymbol{\nabla} \times \boldsymbol{E}_i^c = 0.$$

It follows from (4.13) then that

$$\nabla \times \boldsymbol{E} + \dot{\boldsymbol{H}} = -\sum_l \nabla_\xi \times (\boldsymbol{a}_l \times \boldsymbol{H}_l^g) + \sum_l \dot{\boldsymbol{H}}_l^g$$

$$= -\sum_l \boldsymbol{a}_l (\nabla \cdot \boldsymbol{H}_l^g) + \sum_l (\boldsymbol{a}_l \cdot \nabla_\xi) \boldsymbol{H}_l^g + \sum_l \dot{\boldsymbol{H}}_l^g .$$

The last two terms cancel out since $\boldsymbol{a}_l = \dot{\boldsymbol{X}}_l$, and $\boldsymbol{H}_l^g$ is a function of $\boldsymbol{\xi} - \boldsymbol{X}_l$. Thus

$$\nabla \times \boldsymbol{E} + \dot{\boldsymbol{H}} = -\sum 4\pi g_l \dot{\boldsymbol{X}}_l \delta^3 (\boldsymbol{\xi} - \boldsymbol{X}_l), \tag{6.15}$$

which is one of the Maxwell equations. In a similar way, but starting from (5.1) instead of (4.13), we arrive at

$$\nabla \times \boldsymbol{H} - \dot{\boldsymbol{E}} = \sum_i 4\pi \varepsilon_i \dot{\boldsymbol{x}}_i \delta^3 (\boldsymbol{\xi} - \boldsymbol{x}_i), \tag{6.16}$$

which is the fourth Maxwell equation. Eq. (6.16) is also the result of applying the symmetry operation (5.2) to (6.15), as expected.

Thus we have shown that Hamiltonian (3.1) does give the correct dynamical equations that describe the interaction between electrons, magnetic monopoles, and photons.

VII. Remarks

(i) We have in this paper discussed a system of a finite number of electrons and monopoles interacting with the electromagnetic field. The negative energy states are not filled.

(ii) Strictly speaking, the operators $\boldsymbol{E}(\boldsymbol{x}_i)$ are infinite, since the self-electric field is infinite. This problem is the same as in the usual theory without magnetic monopoles. The renormalization program, which we shall come to in later papers, will partly address this question.

(iii) The symmetry operation (5.2) which we shall denote by J has the property that

$$J^4 = 1. \tag{7.1}$$

It thus resembles the operation of multiplication by $i = \sqrt{-1}$.

References

[1] Dirac, P. A. M.: *Proc. Roy. Soc.*, **A133**(1931), 60.
[2] Dirac, P. A. M.: *Phys. Rev.*, **74**(1948), 817. See also J. Schwinger, *Particles, Sources and Fields* (Addison-Wesley, Reading, Mass., 1970 and 1973).
[3] Wu Tai-tsun & Yang Chen-ning: *Nuclear Physics*, **B107** (1976), 365. See also Wu Tai-tsun & Yang Chen-ning: *Phys. Rev.*, **D16**(1977), 1018.
[4] Wu Tai-tsun & Yang Chen-ning: *Phys. Rev.*, **D14**(1976), 437. For a Feynman path integral method of second quantization, see Xia Dao-xing, Futan Univ. Preprint (1977).
[5] Fermi, E.: *Rev. Mod. Phys.*, **4**(1932), 125.
[6] Yang Chen-ning: *Annals of N. Y. Acad. Science*, **294**(1977), 86.

POINTWISE SO_4 SYMMETRY OF THE BPST PSEUDOPARTICLE SOLUTION

[78k]
Commentary begins page 80

by Chen Ning Yang

ABSTRACT

The BPST pseudoparticle solution is shown to be everywhere pointwise SO_4 symmetrical. It is further shown that on flat Euclidean space R_4, the only SU_2 gauge field that is everywhere SO_4 symmetrical is a BPST pseudoparticle solution.

I. INTRODUCTION

In a recent generalization[1] of the Dirac monopole to SU_2 gauge fields, it was found that the concept of "pointwise SO_4 symmetry" is useful. The meaning of this concept[2] can be explained in the following way:

Consider a four-dimensional manifold with a Riemannian geometry having $+ + + +$ signature. Let P be a point on the manifold. Consider an SU_2 gauge field with field strengths $(f^i_{\mu\nu})_P$ at the point P. Does $(f^i_{\mu\nu})_P$ serve to differentiate between the various directions from P? If it does not, we say the field has pointwise SO_4 symmetry at P. To be more precise, choose coordinates so that the metric at P is $g_{\mu\nu} = \delta_{\mu\nu}$. If any SO_4 rotation for the indices μ, ν in $(f^i_{\mu\nu})_P$ can be *compensated* for by a gauge transformation on the index i, the field has pointwise SO_4 symmetry at P.

In this paper I show that the pseudoparticle solution[3] of Belavin, Polyakov, Schwartz, and Tynpkin, to be called the BPST solution, is everywhere pointwise SO_4 symmetrical. I then show that the only gauge field (sourceless or not) on R_4, (i.e., on flat $+ + + +$ space), that is everywhere pointwise SO_4 symmetrical is the BPST solution.

Chen Ning Yang is Director of the Institute of Theoretical Physics and Professor of Physics at the State University of New York, Stony Brook.

□ Reprinted from *Felix Bloch and Twentieth-Century Physics,* M. Chodorow et al., eds. Houston: Rice University, 1980, 243–247. Rice University Studies 66, 3.

II. POINTWISE SO_4 SYMMETRY OF BPST SOLUTION

It was shown in the appendix of reference 1 that the following statements are identical

(a) $f^i_{\mu\nu}$ is pointwise SO_4 symmetrical at P, (also called orthogonal or regular at P),
(b) $f^i_{\mu\nu}f^{j\lambda\nu} = a^2\delta^{ij}\delta^\lambda_\mu + a\epsilon^{ijk}f^{k\cdot\lambda}_{\mu}$ at P,
(c) $f^i_{\mu\nu}f^{j\lambda\nu} + f^j_{\mu\nu}f^{i\lambda\nu} = 2a^2\delta^{ij}\delta^\lambda_\mu$ at P,

where a is a scalar function on the manifold. It is further easy to show from lemmas 1α, 1β and 4 of reference 1 that these statements are also identical to

(d) $f^i_{\mu\nu}$ is self dual or self antidual at P, and in a coordinate system for which $g_{\mu\nu} = \delta_{\mu\nu}$ at P,

$$\tilde{\mathcal{E}}\,\mathcal{E} = \tilde{\mathcal{H}}\,\mathcal{H} = a^2$$

Theorem 1. The BPST solution is everywhere pointwise SO_4 symmetrical.

Proof: By a straightforward evaluation of the field strengths $\mathcal{E}$ and $\mathcal{H}$ for the BPST solution, we easily verify property (d) above. Hence the theorem is proved.

The square of the field strength, a^2, is easily computed to be

$$a^2 = \frac{16K^2}{[x^2 + K]^4} \quad (K > 0).$$

where $x^2 = x_1^2 + x_2^2 + x_3^2 + x_4^2$ and x_μ are the Cartesian coordinates. This function peaks at $x = 0$. Around a point P where $x \neq 0$, a SO_4 rotation of the whole field changes the magnitude of the field strength squared at most points. So the field is not SO_4 symmetrical at P. But it is pointwise SO_4 symmetrical at P in that if one considers only the value of $f^i_{\mu\nu}$ at P, then the rotation of the field strengths is equivalent to a gauge transformation of the original field strengths. Thus the value of $f^i_{\mu\nu}$ at P does not serve to choose an SO_4 frame around P.

Theorem 2. The BPST solution is the only SU_2 gauge field that is everywhere pointwise SO_4 symmetrical on R_4, the flat 4-dimensional Euclidean space.

Proof: (1) According to Appendix A of reference 1, a field that is pointwise SO_4 symmetrical can be gauge transformed to the standard form, with only one parameter, a, its amplitude. For a field that is everywhere pointwise SO_4 symmetrical, the field strengths in the proper gauge are of the standard form (IA10) or (IA11) everywhere. The amplitude a is a function of the x's. We shall write these equations in the following form

$$f^i_{\mu\nu} = a\eta^i_{\mu\nu} \quad , \tag{1}$$

or

$$f^i_{\mu\nu} = a\bar{\eta}^i_{\mu\nu} \quad , \tag{2}$$

where η and $\bar{\eta}$ are the symbols introduced by 't Hooft[4]:

$$\eta^i_{\mu\nu} = \epsilon_{i\mu\nu 4} + \delta_{\mu i}\delta_{\nu 4} - \delta_{\nu i}\delta_{\mu 4} \quad , \tag{3}$$

$$\bar{\eta}^i_{\mu\nu} = \epsilon_{i\mu\nu 4} - \delta_{\mu i}\delta_{\nu 4} + \delta_{\nu i}\delta_{\mu 4} \quad . \tag{4}$$

For the self-dual case (1), the Bianchi identity becomes[5]

$$\eta^{\xi\mu\nu\lambda}\,(\eta^i_{\mu\nu}a_{,\lambda} - aC^i_{jk}\eta^j_{\mu\nu}b^k_\lambda) = 0 \quad . \tag{5}$$

Since $\eta^i_{\mu\nu}$ is self dual, this becomes

$$\eta^{i\xi\lambda}a_{,\lambda} - aC^i_{jk}\,\eta^{j\xi\lambda}b^k_\lambda = 0 \quad . \tag{6}$$

(2) Now define a matrix M and columns Δ and b by

$$<i\xi|M|k\lambda> = C^i_{jk}\,\eta^{j\xi\lambda} \quad , \tag{7}$$

$$<i\xi|\Delta> = \eta^{i\xi\lambda}a_{,\lambda} \quad , \tag{8}$$

$$<k\lambda|b> = b^k_\lambda \quad . \tag{9}$$

Eq. (6) becomes

$$\Delta - aMb = 0 \quad . \tag{10}$$

Using[4,1]

$$\eta^i_{\alpha\lambda}\eta^j_{\beta\lambda} = \epsilon^{ijk}\,\eta_{\alpha\beta} + \delta^{ij}\delta_{\alpha\beta} \tag{11}$$

we can prove

$$(M + 1)M = 2 \quad . \tag{12}$$

Thus

$$b = a^{-1}(M + 1)\Delta/2 \quad .$$

Or

$$b^i_\xi = -a_{,\mu}\eta_{\xi\mu}(2a)^{-1} \quad . \tag{13}$$

(3) Substituting this into the equation for $f^i_{\mu\nu}$ in terms of b^i_ξ and its derivatives, we obtain as necessary and sufficient conditions for (1):

$$A_{,\mu\nu} = \delta_{\mu\nu}B + A_{,\mu}A_{,\nu} \quad , \tag{14}$$

$$2a = -A_{,\mu\mu} - (A_{,\mu})^2 \quad , \tag{15}$$

where

$$A = \frac{1}{2} ln|a| \quad , \tag{16}$$

and B is a scalar function of the coordinates. Putting

$$G = \exp(-A) \quad , \tag{17}$$

(14) becomes

$$G_{,ij} = 0,(i \neq j) \quad ,$$

and

$$G_{,11} = G_{,22} = G_{,33} = G_{,44} \quad .$$

These equations can be integrated, giving

$$G = \alpha(x - c)^2 + \beta \tag{18}$$

where c_μ is a point in R_4, and α and β are numbers. Substitution into (10), (16), and (17) gives

$$a = \frac{\pm 1}{(\alpha(x - c)^2 + \beta)^2} \quad .$$

To avoid singularities in $f^i_{\mu\nu}$, a must remain finite. Thus $\alpha\beta \nless 0$. Eq. (15) gives then $4\alpha\beta = 1$, and we obtain, with $K = \beta\alpha^{-1} > 0$,

$$a = \frac{4K}{[(x-c)^2 + K]^2} \quad , \tag{19}$$

and

$$b^i_\xi = -a_{,\mu}\,\eta^i_{\xi\mu}(2a)^{-1} \quad . \tag{20}$$

These two equations give exactly the BPST solution.[3] (20) is precisely in the form of the Corrigan-Fairlie-Wilczek-'t Hooft Ansatz.[6]

(4) The proof for the antiself dual case is entirely similar.

ACKNOWLEDGMENT

This work is supported in part by the National Science Foundation under Grant No. PHY-76-15328.

REFERENCES

1. Chen Ning Yang, J. Math. Phys. **19**, 320 (1978). Formulas in this paper will be referred to as (I2), (IA11), etc.
2. In reference 1, this concept was applied only to points P where the underlying geometry is SO_4 symmetrical at the point P. The concept is actually applicable to any point on any Riemannian geometry of signature + + + +.
3. Belavin, Polyakov, Schwartz, and Tyupkin, Phys. Lett. **59B**, 85 (1975).
4. G. 't Hooft, Phys. Rev. **D14**, 3432 (1976).
5. We use the notation that $\eta^{\xi\mu\nu\lambda} = \sqrt{g}\epsilon^{\xi\mu\nu\lambda}$ where $\epsilon = \pm 1$ is the antisymmetrical tensor. All notations follow that of reference 1.
6. E. Corrigan and D. B. Fairlie, Phys. Lett. **67B**, 69 (1977); F. Wilczek, in *Quark Confinement and Field Theory*, ed. by D. Stump and D. Weingarten (Wiley, New York, 1977); 't Hooft, unpublished.

⟦79b⟧

Commentary begins page 80

Phase shift in a rotating neutron or optical interferometer

Max Dresden and Chen Ning Yang

Institute for Theoretical Physics, State University of New York, Stony Brook, New York 11794

(Received 12 June 1979)

The phase shift caused by rotating a neutron or optical interferometer is derived as the Doppler effect due to the moving source and moving reflecting crystals.

Using the interferometer developed for x rays by Bonse and Hart,[1] Werner, Staudenmann, and Colella[2] recently did an experiment measuring the effect of the rotation of the earth on their neutron interferometer. This is the latest of a series[3,4] of beautiful experiments performed in several laboratories using the neutron interferometer. The phase shift observed in Ref. 2 due to the rotation of the earth is

$$\begin{aligned}\delta &= (\text{phase change of path } ACD - \text{that of } ABD)\\ &= (2k/v)\vec{\omega}\cdot\vec{A},\end{aligned} \tag{1}$$

where k and v are the wave number and the velocity of the neutron, $\vec{\omega}$ is the angular velocity of the spinning earth, and $\vec{A}$ is the normal area of the parallelogram $ACDB$ (Fig. 1).

Equation (1) was theoretically derived in Ref. 2 and also in earlier papers.[5] The corresponding phase shift, due to a rotation of the equipment, observed in an *optical* interferometer, had been studied[6] since 1904. Its value is given by, instead of (1),

$$\frac{2k}{c}\vec{\omega}\cdot\vec{A}, \tag{2}$$

where c is the velocity of light. Both (1) and (2) are correct only to the first order of $L\omega/v$ (or $L\omega/c$) where L is the linear dimension of the interferometer.

We give here a new derivation of (1) and (2) which is clearer than previous derivations. We stay throughout in the inertia system at rest.

DOPPLER EFFECT DUE TO MOVING SOURCE

We consider a rotation, with small angular velocity $\vec{\omega}$ of the neutron spectrometer around any fixed point 0. $\vec{\omega}$ does not have to be perpendicular to the plane of the interferring paths ACD and ABD. In fact, we shall consider a more general path illustrated in Fig. 2 which is not necessarily planar. The moving source creates a Doppler shift of the wave number along $\vec{x}_{AC_1}$, of amount

$$\delta k_1 = k_0(\vec{\omega}\times\vec{x}_{0A})\cdot\vec{x}_{AC_1}/v = (k_0/v)\vec{\omega}\cdot(\vec{x}_{0A}\times\hat{x}_{AC_1}), \tag{3}$$

where $\hat{x}_{AC_1}$ is the *unit* vector along $\vec{x}_{AC_1}$.

DOPPLER EFFECT DUE TO MOTION OF REFLECTING CRYSTALS

At the points $C_1, C_2, \ldots$ the beam is reflected by crystals that are moving in the inertia frame. Each reflection by a moving crystal causes an additional shift of the wave number of the beam. We shall now show that the cumulative effects of these shifts are to produce a total shift of the wave number of beam $\vec{x}_{C_{m-1}C_m}$ by the following amount:

$$\delta k_m = (k_0/v)\vec{\omega}\cdot(\vec{x}_{0C_{m-1}}\times\hat{x}_{C_{m-1}C_m}). \tag{4}$$

To prove this statement consider the crystal at C_1 which moves with the velocity $\vec{V}=\vec{\omega}\times\vec{x}_{0C_1}$. Let $\vec{k}_1$ be the vector wave number of the beam $\vec{x}_{AC_1}$ in the inertia frame. In the inertia frame moving with the crystal velocity $\vec{V}$, $\vec{x}_{AC_1}$ has the wave number

$$\vec{k}' = \vec{k}_1 - k_0v^{-1}\vec{V}. \tag{5}$$

In this frame the crystal is at rest and the reflected beam has the wave number

$$\vec{k}' - 2\hat{n}(\hat{n}\cdot\vec{k}'), \tag{6}$$

where $\hat{n}$ is the normal to the reflecting surface at C_1. Transforming back to the inertia frame at rest we obtain

$$\vec{k}_2 = [\vec{k}_1 - 2\hat{n}(\hat{n}\cdot\vec{k}_1)] + 2\hat{n}(\hat{n}\cdot\vec{V})k_0v^{-1}. \tag{7}$$

The term in the square brackets has the same length as $\vec{k}_1$ and is in the direction $\vec{x}_{C_1C_2}$. Thus,

$$\vec{k}_2 = |k_1|\hat{x}_{C_1C_2} + 2\hat{n}(\hat{n}\cdot\vec{V})k_0v^{-1}.$$

Thus,

$$\delta k_2 - \delta k_1 = k_2 - k_1 = 2(\hat{n}\cdot\hat{x}_{C_1C_2})(\hat{n}\cdot\vec{V})k_0v^{-1}.$$

Now

$$\hat{x}_{C_1C_2} - \hat{x}_{AC_1} = 2\hat{n}(\hat{n}\cdot\hat{x}_{C_1C_2}).$$

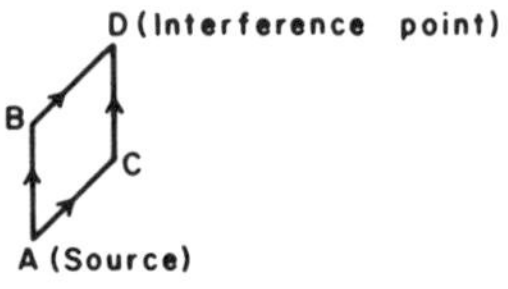

FIG. 1. Paths of interfering beams.

□Reprinted from *Physical Review D* 20, 8 (October 15, 1979), 1846–1848.

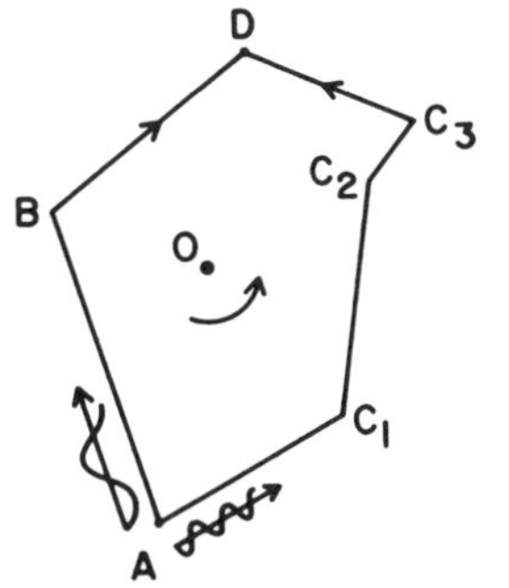

FIG. 2. Paths of intefering beams. 0 is center of rotation. The different wavelengths symbolize Doppler effects of the source A.

Thus,

$$\delta k_2 - \delta k_1 = (k_0/v)\vec{V}\cdot(\hat{x}_{C_1C_2} - \hat{x}_{AC_1}) \quad (8)$$
$$= (k_0/v)\vec{\omega}\cdot(\vec{x}_{0C_1}\times\hat{x}_{C_1C_2} - \vec{x}_{0C_1}\times\hat{x}_{AC_1})$$
$$= (k_0/v)\vec{\omega}\cdot(\vec{x}_{0C_1}\times\hat{x}_{C_1C_2} - \vec{x}_{0A}\times\hat{x}_{AC_1})\,. \quad (9)$$

Adding to (3) gives

$$\delta k_2 = (k_0/v)\vec{\omega}\cdot(\vec{x}_{0C_1}\times\hat{x}_{C_1C_2})\,. \quad (10)$$

Proceeding similarly with reflecting crystals at $C_2, C_3, \ldots$ we obtain (4). It is interesting to notice that the change of wave number δk_m in beam $\vec{x}_{C_{m-1}C_m}$, as given by (4), is only dependent on the beam position (i.e., the line $C_{m-1}C_m$) and is *independent of all previous crystals.* This is valid as long as the velocity of all crystals is given by $\vec{\omega}\times\vec{r}_m$. In an experimental arrangement where this is not true (i.e., for nonrigid rotation) δk_m would depend on the complete *history* of the beam up to $\vec{k}_m$.

DERIVATION OF PHASE SHIFT

The phase shift in the beam $\vec{x}_{C_{m-1}C_m}$ is, by (4),

$$\delta k_m(\vec{x}_{C_{m-1}C_m}) = (k_0/v)\vec{\omega}\cdot(\vec{x}_{0C_{m-1}}\times\vec{x}_{C_{m-1}C_m})$$
$$= (2k_0/v)\vec{\omega}\cdot\vec{A}_m\,, \quad (11)$$

where $\vec{A}_m$ is the vector area of the triangle $\Delta(0C_{m-1}C_m)$. Adding all contributions from the different segments of the beams we obtain (1). At this point we make the following remarks:

(a) Reflection from moving crystals also changes the law of reflection so that the path of the beam $AC_1C_2\cdots D$ is changed by first-order displacements. However, the effect of such displacements on the phase shift occurs only in the second order because of Fermat's principle for reflection.

(b) In our derivations we took the rest system, *at a given instant of time*, and considered the wave propagation in the different segments of the beam.

(c) The derivation above refers to a neutron interferometer. For an optical interferometer, all steps are valid with the replacement $v\to c$. Hence, we have (2).

(d) The earlier[6] derivations of (2) were confusing because of the apparent lack of conviction in special relativity. If we consider a neutron interferometer rather than an optical one, this confusion disappears. But another flaw remains. That is, the total phase shift was obtained in these early derivations by adding contributions from each infinitesimal element ds of the beam. That contribution was in turn evaluated from the local velocity. As was apparent from our discussion above following Eq. (10), this is in general incorrect. The contribution is not always dependent only on the local velocity if the motion of the interferometer is not a rigid rotation.

(e) A particularly simple case is an interferometer in the shape of a regular polygon of n sides where n is very large. In that case the two paths are essentially semicircles. If the whole apparatus rotates around the center 0 of the circle, there is a Doppler shift at the source of amount $\pm(r\omega/v)k$ were r is the radius of the circle. The crystals move tangentially. Hence, all scattering is forward and there is no additional Doppler shift. The total phase shift is thus

$$2\frac{r\omega}{v}k\pi r = \frac{2k}{v}\omega A\,.$$

ACKNOWLEDGMENTS

One of us (C. N. Yang) wishes to thank Professor Werner for showing him his beautiful experiment. This work was supported in part by the National Science Foundation under Grant No. PHY 76 15328A02.

[1]U. Bonse and M. Hart, Appl. Phys. Lett. 6, 155 (1965); Z. Phys. 188, 154 (1965).
[2]S. A. Werner, J. -L. Staudenmann, and R. Colella, Phys. Rev. Lett. 42, 1103 (1979).
[3]H. Rauch, W. Treimer, and U. Bonse, Phys. Lett. 47A, 369 (1974); A. W. Overhauser and R. Colella, Phys. Rev. Lett. 33, 1237 (1974); R. Colella, A. W. Overhauser, and S. A. Werner, *ibid.* 34, 1472 (1975); H. Rauch, A. Zeilinger, G. Badurek, A. Wilfing, W. Bauspiess, and U. Bonse, Phys. Lett. 54A, 425 (1975); S. A. Werner, R. Colella, A. W. Overhauser, and C. Eagen, Phys. Rev. Lett. 35, 1053 (1975).

[4]For a review see D. M. Greenberger and A. W. Overhauser, Rev. Mod. Phys. 51, 43 (1979).
[5]L. A. Page, Phys. Rev. Lett. 35, 543 (1975); J. Anandan, Phys. Rev. D 15, 1448 (1977); L. Stodolsky (unpublished).
[6]A. A. Michelson, Philos. Mag. Series 6, 8, 716 (1904); M. G. Sagnac, Compt. Rend. 152, 310 (1911); 157, 708 (1913); 157, 1410 (1913); L. Silberstein, J. Op. Soc. 5, 291 (1921); A. C. Lunn, *ibid.* 6, 112 (1922); A. A. Michelson, H. G. Gale, and F. Pearson, Astrophys. J. 61, 140 (1925).

Geometrical Model of Hadron Collisions

[[80a]]
Commentary begins page 81

T. T. Chou
Physics Department, University of Georgia
Chen Ning Yang
Institute for Theoretical Physics, State University of New York, Stony Brook

Abstract

Results obtained from the geometrical model for high energy hadron–hadron and hadron–nucleus collisions are summarized. The discussion is focused on elastic scattering, diffraction dissociation process and the concept of matter current distribution inside a polarized hadron or nucleus.

We wish to review some of the main results of the geometrical model for high energy collisions obtained since 1967. The study of this model has led to a number of specific predictions. These are:

1. For elastic pp scattering,
   a. the values for the angular distribution $d\sigma/dt$ in the region for $|t| < 1.2(\text{GeV}/c)^2$,
   b. the existence in $d\sigma/dt$ of a first minimum and a second maximum,
   c. the slow inward movement of the positions for these extrema as the incident energy increases.
2. The matter distribution inside hadrons, and the values of the rms radii for π and K mesons.
3. The dip and kink structures for the diffraction dissociation processes.
4. The hypothesis of limiting fragmentation in particle production processes.
5. The existence of a second minimum in elastic pp differential cross sections.
6. The matter current distribution inside polarized hadrons and nuclei.

While predictions (1) through (4) have all been verified, the confirmation of (5) and (6) will have to await future experiments. It should be emphasized that no adjustable parameters have been used in comparing most of these predictions with experiments.

In this talk the work on three general areas, the elastic scattering, the diffraction dissociation process, and the concept of velocity profile will be summarized. We shall present here only the main physical idea and important experimental evidence. Detailed discussions which can be found in the literature will be omitted.

I. ELASTIC SCATTERING

The geometrical description[1] of elastic hadron–hadron scattering at high energy is based on the fundamental assumption that all hadrons are extended objects with structures. This description provides a direct relationship between the hadronic matter density functions and the elastic differential cross section. The main physical concepts invoked in this picture are the following three:

1. The eikonal approximation.
2. The exponentiation for the transmission coefficient in impact-parameter space.
3. The convolution integral formalism for the opaqueness of the colliding system.

Combining the above ideas, one obtains the elastic scattering amplitude for hadrons A and B:

$$a_{AB} = \text{two dimensional Fourier transform of } \{1 - \exp[-(\text{constant})\, D_A \otimes D_B]\} \quad (1)$$

where D_A and D_B are the compressed matter densities for hadrons A and B, and $\otimes$ denotes con-

[1] T. T. Chou and C. N. Yang, *Phys. Rev.* 170, 1591 (1968).

□Reprinted from *Proceedings of the 1980 Guangzhou Conference on Theoretical Particle Physics*. Beijing: Science Press, 1980, 317–326.

volution in impact-parameter space. The quantity in the square bracket represents minus one times the opaqueness.

Some results are readily derivable from this model:

(A) The hadronic matter distributions can be computed by using the experimentally measured differential cross section. The charge form factors of hadrons may then be inferred if one makes an additional assumption that hadronic matter and charge distributions are proportional to each other. The proton form factor thus computed agrees amazingly well with the experimental G_E form factor of the proton. (See Fig. 3 of Chou and Yang, 1973.[2]) Application of the geometrical model of πp and Kp scatterings has also been used to determine the pion and kaon radii.[3] Theoretical predictions seem to be in good agreement with values obtained from direct experimental measurements.[4]

(B) With the proton form factor and the observed total cross section used as input, the model predicted[5] the pp differential cross section which shows a dip near $|t| \cong 1.4$ $(\mathrm{GeV}/c)^2$, and a maximum at a slightly higher $|t|$ value. The calculated positions of the minimum and the maximum were in good agreement with the experimental results.[6] Similar structures are also expected to exist in np elastic scattering, since it is reasonable to assume that neutron and proton have approximately the same matter form factor. This conjecture has indeed been borne out by a recent experiment.[7]

(C) The constant appearing in the opaqueness expression in equation (1) is related to the total cross section σ_T. With the discovery[8] of rising σ_T with increasing energy, this "constant" becomes dependent on the incoming energy. Because of increasing opaqueness, the model predicted[9] an inward shift of the minimum and the maximum positions with energy, and a concurrent rise of the second maximum. These characteristics are also in general agreement with the newest experiments.[10]

We now make a few remarks:

(1) We have tacitly assumed that only the spin-independent amplitude contributes predominantly to pp scatterings at high energies and that it is imaginary. The observed energy dependence of the cross section necessarily requires the existence of a real part in the scattering amplitude. This small real part is neglected in our consideration. These approximations are in accordance with the experimental fact[11] that at high energies the polarization parameter P in pp scattering is approximately zero except near the dip region.

(2) Extensive work[12] on the high energy limit of a large class of Feynman diagrams for certain couplings in field theory has been carried out by Cheng and Wu. It is remarkable that the general validity of the three basic physical concepts used in the elastic model has been confirmed, at least partly, by their studies.

(3) As the opaqueness increases with the total cross section, the geometrical model predicts[13] the existence of many dips in pp elastic scattering at very high energies. Although none of the dips other than the first one has been observed so far,

---

[2] A. W. Chao and C. N. Yang, *Phys. Rev.* D8, 2063 (1973).

[3] T. T. Chou, *Phys. Rev.* D19, 3327 (1979).

[4] E. B. Dally et al., *Phys. Rev. Lett.* 39, 1176 (1977); A. Beretvas et al., contributed paper No. 857 submitted to the XIXth International Conference on High Energy Physics, Tokyo, 1978 (unpublished).

[5] T. T. Chou and C. N. Yang, *Phys. Rev. Lett.* 20, 1213 (1968); L. Durand and R. Lipes, *Phys. Rev. Lett.* 20, 637 (1968).

[6] A. Böhm et al., *Phys. Lett.* 49B, 491 (1974).

[7] C. E. DeHaven, Jr. et al., *Phys. Rev. Lett.* 41, 669 (1978).

[8] U. Amaldi et al., *Phys. Lett.* 44B, 112 (1973); S.R. Amendolia et al., *Phys. Lett.* 44B, 119 (1973); A. S. Carroll et al., *Phys. Rev. Lett.* 33, 932 (1974). The increase of total cross section was conjectured theoretically by H. Cheng and T. T. Wu, *Phys. Rev. Lett.* 24, 1456 (1970).

[9] F. Hayot and U. P. Sukhatme, *Phys. Rev.* D10, 2183 (1974).

[10] N. Kwak et al., *Phys. Lett.* 58B, 233 (1975); E. Nagy et al., *Nucl. Phys.* B150, 221(1979).

[11] J. H. Snyder et al., *Phys. Rev. Lett.* 41, 781 (1978).

[12] H. Cheng and T. T. Wu, *Phys. Rev.* 182, 1852 (1969) and later papers.

[13] T. T. Chou and C. N. Yang, *Phys. Rev.* D17, 1889 (1978); *Phys. Rev.* D19, 3268 (1979).

the development of many dips is inevitable in view of the fact that increasing opaqueness would produce an effective black disc as the scattering center at sufficiently high energies.

(4) In describing high energy elastic scattering, the geometrical picture envisages the passage of one hadron through another with attenuation. The attenuation factor depends on the impact parameter b, and not very much on the energy. Since the elastic and inelastic channels are strongly coupled, we argued that in high energy inelastic processes the probability amplitude for the target or the projectile to be excited into any specific mode must also be a function of b alone, and would be largely independent of incident energy. This hypothesis[14] was referred to as the hypothesis of limiting fragmentation. Experimental verification[15] of the hypothesis has been carried out at the ISR at CERN.

II. DIFFRACTION DISSOCIATION

In recent years there has been experimental data[16] concerning the angular distribution of hadron–nucleus and hadron–hadron diffraction dissociation. One of the conspicuous features of all these experiments is the existence of dip or kink structures similar to that observed in pp elastic scattering. The dip in diffraction dissociation occurs generally at a smaller $|t|$ value than the dip in elastic scattering. We pointed out that the geometrical model can offer a natural explanation[17] of these dip structures.

Consider the passage of an incoming hadron through an extended target. At an impact parameter b the dissociation can take place at any point along its path during its traversal. The probability for the process to occur is approximately proportional to the thickness of the material traversed, or $\Omega(b)$. There is also absorption of the incoming wave before dissociation, and of the outgoing wave after dissociation. Assuming equal mean free path for incoming and outgoing waves, the total absorption factor can be written as $\exp[-\Omega(b)]$. Thus the source function for the outgoing hadron in diffraction dissociation may be approximated by $\Omega\exp(-\Omega)$. This approximation was first used in charge exchange scatterings and was given the name "coherent droplet model."[18]

With Ω determined from electron scattering experiments together with hadron–hadron total cross sections, numerical computations for differential cross section in diffraction dissociation processes have been made.[17] The computation contains no adjustable parameters. The calculated dip positions for both elastic scattering and diffraction dissociation process are in very good agreement with experimental values.

III. HADRONIC MATTER CURRENT DISTRIBUTION[19]

Since geometrical concepts such as sizes, matter distribution, or the opaqueness distribution of hadrons are very useful in discussing high energy collisions, one may raise the interesting question whether matter current exists inside a polarized hadron. Once we have accepted the concept of an extended hadron with a matter density in it, it seems to us inevitable that we must also accept the existence of a matter current in a polarized hadron. The concepts of matter density and matter current necessarily complement each other, resulting in a four-vector. The existence of a matter current produces observable effects. Experimental test of this idea is now possible if one utilizes the rising total cross section with increasing incoming energies.

---

[14] J. Benecke, T. T. Chou, C. N. Yang, and E. Yen, *Phys. Rev.* 188, 2159 (1969).

[15] G. Bellettini et al., *Phys. Lett.* 45B, 69 (1973).

[16] G. Goggi et al., *Phys. Lett.* 79B, 165 (1978); T. Ferbel, lecture presented at the First Workshop on Ultra-Relativistic Nuclear Collisions, LBL (1979).

[17] T. T. Chou and C. N. Yang, Dip and kink structures in hadron–nucleus and hadron–hadron diffraction dissociation (preprint).

[18] N. Byers and C. N. Yang, *Phys. Rev.* 142, 976 (1966). This approximation was later utilized in T. T. Chou, *Phys. Rev.* 176, 2041 (1968); and H. Cheng, J. K. Walker and T. T. Wu, *Phys. Rev.* D9, 749 (1974).

[19] T. T. Chou and C. N. Yang, *Nucl. Phys.* B107, 1 (1976).

Consider a Kp scattering with the target proton polarized along the x-direction. If the motion of hadronic matter in a region of the target is toward the projectile, that region would have greater opaqueness than another region in which the motion is away from the projectile. Thus, to a kaon moving parallel to the $+y$ axis, the proton would appear more opaque on the $+z$ axis side than the $-z$ side. The result of this difference in opaqueness is a non-vanishing rotation paremeter $R(t)$. I.e., the polarization vector of the recoil proton for a collision in which the K is scattered in the x-y plane toward the $+x$ axis has a y-component equal to $R(t)$. This Wolfenstein parameter R is measurable and its values for different momentum transfers t will determine the hadronic matter current density in the target.

Numerical computation showed that for small t, R is negative. This sign has a natural classical interpretation.[20] The estimated value of R at $t = -0.5(\text{GeV}/c)^2$ ranges from -0.05 to -0.2 radian for πp and Kp scatterings at 200 GeV, indicating that the matter current effect is measurable with available experimental techniques.

The given estimate for the R parameter is obtained on the assumption that the polarized target is infinitely heavy. Corrections clearly have to be made if the target has a non-vanishing recoil velocity. This effect, however, cannot be dealt with rigorously.

The above discussion is clearly applicable to elastic scattering of hadron off polarized nuclei, where the existence of nucleonic current is conceptually much easier to accept.

[20] T. T. Chou, in *High Energy Collisions—1973*, edited by C. Quigg (AIP Conf. Proc. No. 15, New York, 1973) pp. 118–123; C. N. Yang in *Proc. Int. Symposium on High Energy Physics*, edited by Y. Hara et al. (University of Tokyo, Japan, 1973) pp. 629–634.

The work of T. T. C. was supported in part by the Department of Energy under Contract No. DE-AS09-76ER00946. The work of C. N. Y. was supported in part by the National Science Foundation under Grant No. PHY7615328.

Einstein's impact on theoretical physics

⟦80b⟧
Commentary begins page 81

That symmetry dictates interactions, that geometry is at the heart of physics, and that formal beauty plays a role in describing the world are insights that have had a profound effect on current thought.

Chen-Ning Yang

There occurred in the early years of this century three conceptual revolutions that profoundly changed Man's understanding of the physical universe: the special theory of relativity (in 1905), the general theory of relativity (1915) and quantum mechanics (1925). Einstein personally was responsible for the first two of these revolutions, and influenced and helped to shape the third. But it is not about his work in these conceptual revolutions that I shall write here. Much has been written about that work already. Instead, I shall discuss, in general terms, Einstein's insights on the structure of theoretical physics and their relevance to the development of physics in the second half of this century. I shall divide the discussion into four sections which are, of course, very much related.

Symmetry dictates interaction

The first important symmetry principle discovered in fundamental physics was Lorentz invariance, which was found as a *mathematical* property of Maxwell equations, which in turn were based on the *experimental laws* of electromagnetism. In this process the invariance, or symmetry, was a secondary discovery. In his Autobiographical Notes[1] Einstein gave Hermann Minkowski credit for turning this process around. Minkowski started with Lorentz invariance, and required that field equations be covariant with respect to the invariance, as shown in the table on page 44.

Einstein himself was deeply impressed by the powerful physical consequences of symmetry principles and worked to enlarge the scope of Lorentz invariance. This idea of a more general coordinate invariance led, together with the equivalence principle, to the general theory of relativity. We might say that Einstein initiated the principle that *symmetry dictates interactions*. This principle has played an essential role in recent years in giving rise to various field theories:

▶ Coordinate-transformation invariance gives rise to general relativity
▶ Abelian gauge symmetry gives rise to electromagnetism
▶ Non-Abelian gauge symmetry gives rise to Non-Abelian gauge fields
▶ Supersymmetry gives rise to a theory with symmetry between fermions and bosons
▶ Supergravity symmetry gives rise to supergravity.

Field Theory and unification

In his articles and lectures after 1920 Einstein repeatedly emphasized the concept of the field as of central importance to fundamental physics. For example, in an article published in the Journal of the Franklin Institute in 1936 he wrote:[2]

> The escape from this unsatisfactory situation by the electric field theory of Faraday and Maxwell represents probably the most profound transformation of the foundations of physics since Newton's time.

The two field theories known around that time (1936) were Maxwell's theory and Einstein's general relativity theory. Einstein devoted the last twenty years of his life striving to unify these two theories. The necessity of doing that he explained in an article published in 1934 entitled "The problems of space, ether, and the field in physics":[3]

> ... there exist two structures of space independent of each other, the metric–gravitational and the electromagnetic ... We are prompted to the belief that both sorts of fields must correspond to a unified structure of space.

In the last editions of *The Meaning of Relativity* Einstein added an appendix in which he proposed a unified theory with a non-symmetrical metric tensor $g_{\mu\nu}$. The anti-symmetrical part was to be identified with the electro-magnetic-field tensor $F_{\mu\nu}$. This effort was not particularly successful and there has been, for some time, among some people, the impression that the idea of unification was some kind of obsession affecting Einstein in his old age. Yes, it was an obsession, but an obsession with an *insight* of what the fundamental structure of theoretical physics should be. And I would add that that insight is very much the theme of the physics of today.

In any case, Einstein's emphasis on unification produced something at once. It led many distinguished mathematicians, including Tullio Levi-Ci-

Chen-Ning Yang is the Einstein Professor and the director of the Institute of Theoretical Physics at the State University of New York at Stony Brook.

□ Reprinted from *Physics Today* 33, 6 (June 1980), 42–44, 48–49.

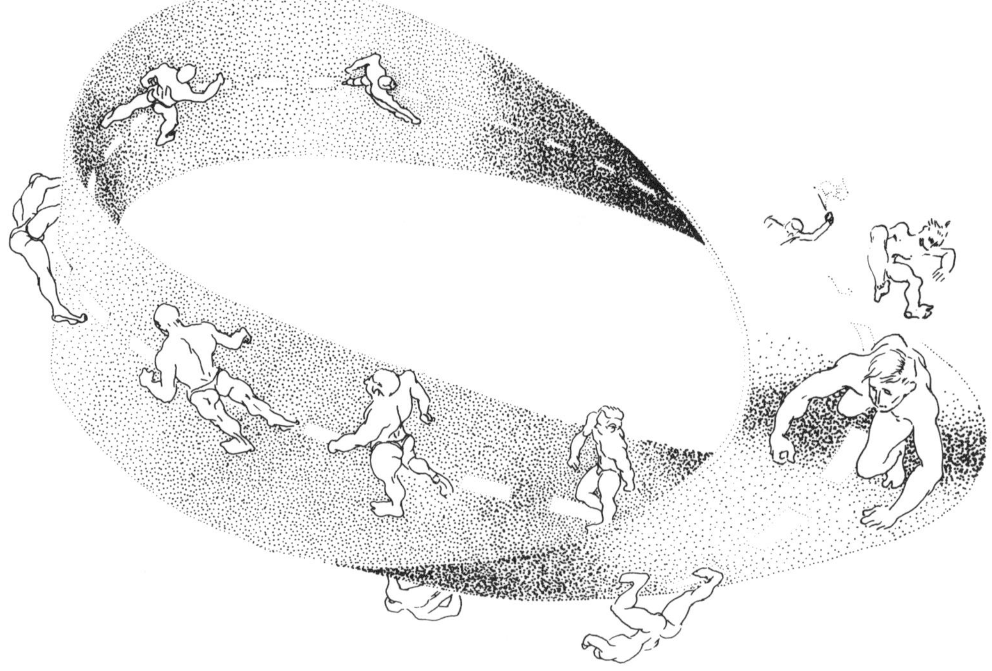

The boundary of a region has no boundary. This Möbius strip has only one surface; its boundary is a single edge, but the edge itself has no boundary. For a further explanation of this theorem, see the box on the next page. (All drawings for this article by Louis Fulgoni.)

vita, Elie Cartan and Weyl to look more deeply into possible additions to the mathematical structure of space–time.

Beginning in 1918 and 1919, Weyl made an effort to incorporate electromagnetism into gravitation. His idea led to what is called "gauge theory."[4] Since the proper treatment of coordinate invariance has produced gravity theory, Weyl thought that a new geometrical invariance could be tied to electromagnetism. His proposal was scale invariance.

If x^μ and $x^\mu + dx^\mu$ are two space-time points in the neighborhood of each other and f is some physical quantity such that it is f at x^μ and $f + (\partial f/\partial x^\mu)\,dx^\mu$ at $x^\mu + dx^\mu$, Weyl considered the space-time dependent rescaling of f, as shown in the last two rows of the table on page 48. Notice particularly the scale factor

$$1 + S_\mu dx\mu \qquad (1)$$

given in the third row.

Now Weyl observed two things about this scale factor. First, S_μ has the same number of components as the electromagnetic potential A_μ. Secondly, in a further development, he proved that when one requires the theory to be invariant under the scale change, only the curl of S_μ occurs and not S_μ itself. That is also a feature of the electromagnetic potential, A_μ. So he identified S_μ with A_μ. This idea, however, did not work. It was discussed by several people including Einstein, who demonstrated that Weyl's theory cannot possibly describe electromagnetism, and Weyl gave it up.

Then 1925 came and quantum mechanics was invented, completely independently of this development.

We all know that in classical mechanics it is not the particle momentum p_μ that occurs, but, in presence of electromagnetism, it is always the combination:

$$\pi_\mu = p_\mu - (ie/\hbar c)A_\mu \qquad (2)$$

In quantum mechanics this is to be replaced by

$$-i\hbar[\partial_\mu - (ie/\hbar c)A_\mu] \qquad (3)$$

This was pointed out[5] by Vladimir Alexandrovitch Fock in 1927. Immediately afterwards, Fritz London compared[6] expression 3 with the increment operator $(\partial_\mu + S_\mu)$ in the last expression in the table, and concluded that S_μ is to be identified not with A_μ but with the factor $-ieA_\mu/\hbar c$. The important new point is just the insertion of an imaginary unit i. This has the far-reaching consequence that expression 1 becomes:

$$1 - (ie/\hbar c)A_\mu dx_\mu \rightarrow \exp[-(ie/\hbar c)A_\mu dx^\mu] \qquad (4)$$

which is a *phase* change, not a *scale* change. Therefore, local *phase invariance* is the correct quantum mechanical characterization of electromagnetism.

Weyl himself had called his idea "*Masstab Invarianz*" at first, but later changed to "*Eich-Invarianz.*" In the early 1920's it was translated into English as "gauge invariance." If we were to rename it today, it is obvious that we should call it *phase invariance*, and gauge fields should be called *phase fields*.

Once one has understood that gauge invariance is phase invariance, one finds that the key idea is a non-integrable phase factor. The substitution for the simple phase of complex numbers with a more complicated phase, namely

A topological theorem

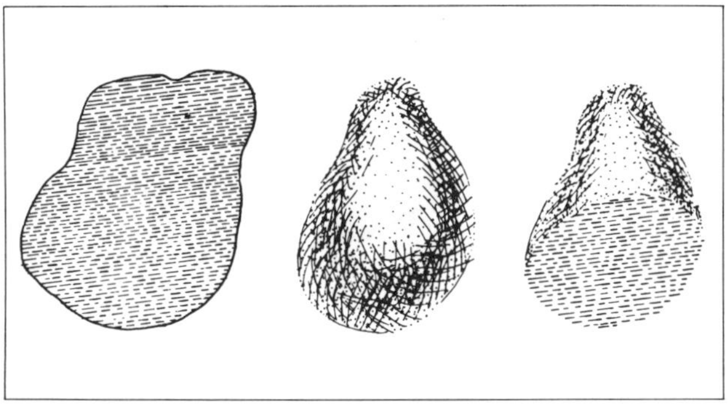

The boundary of a region has no boundary itself. In the example shown on the left, the shaded two-dimensional region has as its boundary a one-dimensional loop; the loop has no end, that is, it has no boundary itself.

The three-dimensional figure in the center is bounded by a closed two-dimensional surface; the surface again has no edges, that is no boundary.

If we cut the region to give an edge to the surface we create a second surface to complete the boundary of the smaller volume, as shown in the sketch on the right. If the edge of this cut is oriented, say, in the clockwise sense, then the edge of the part of the original surface must be oriented in the counterclockwise sense. Thus, although each of the two parts of the surface bounding the three-dimensional region has a boundary, the whole surface of the volume has no boundary; the edge is included twice, oriented in opposite directions, and therefore cancelled out.

an element of a Lie group, leads one to non-Abelian gauge theories, which were first fomulated in 1954.

We should emphasize here that the concept of phase is of great practical importance in contemporary physics. For example, the theories of superconductivity and superfluidity, the Josephson effect, holography, masers and lasers are all fundamentally based on various aspects of the concept of phase.

In 1967 Steven Weinberg and Abdus Salam independently proposed a model for a unified theory of electromagnetism with the weak interactions.The model is based on two key concepts: *Non-Abelian gauge field* and *broken symmetry*. An important further idea, due to Sheldon Glashow, was needed to eliminate contradictions with experiments. In the last six years this model has gathered amazing experimental support. The success has in turn given rise to exuberant efforts at a larger unification of strong, electromagnetic and weak interactions. I am afraid we are still some ways from a successful larger unification, and even further from a holistic unification of these interactions with general relativity. But there is little doubt that Einstein's insistence on the importance of unification was a deep insight, which he had courageously defended, against all spoken and unspoken criticism.

Geometrization of physics

Another recurrent theme in Einstein's perception about the fundamentals of theoretical physics derived from his partiality to geometrical concepts. This is not surprising since he himself created the profound concept that gravity and mechanics should be described in terms of Riemannian geometry. That he regarded electromagnetism as also geometrical was evident from the earlier quote taken from his 1934 article. He stated there that electromagnetism is a "structure" of space. If one accepts the thesis that Einstein was partial to geometrical concepts, then one might perhaps even advance the view that he liked wave mechanics because it is more geometrical and disliked matrix mechanics because it is more algebraic.

Einstein strived to find the geometrical structure that gives rise to electromagnetism. He was aware of the fact that Lorentz invariance was not enough to give Maxwell equations:[7]

> Maxwell equations imply the "Lorentz group," but the "Lorentz group" does not imply Maxwell's equations.

For example, scalar fields are seemingly simpler than Maxwell's electromagnetic field, and are consistent with Lorentz invariance, but are not the basis of electromagnetism.

Einstein was also deeply aware of the necessity to have geometrical structures that give rise to nonlinear equations:[1]

> The true laws cannot be linear nor can they derived from such.

It turns out that the structure that Einstein was seeking was the gauge field: It is a geometrical structure, as we shall presently discuss; the simplest Abelian gauge field is Maxwell's electromagnetic field and a non-Abelian gauge field is necessarily nonlinear.

We had earlier referred to the early history of gauge fields. That gauge fields are deeply related to the geometrical concept of connections on fiber bundles has been appreciated by physicists only in recent years.

To illustrate the geometrical nature of gauge fields, let us write the Gauss and Faraday laws of electromagnetism in the following well-known form:

$$\partial_\lambda f_{\mu\nu} + \partial_\mu f_{\nu\lambda} + \partial_\nu f_{\lambda\mu} = 0$$

where $f_{\mu\nu}$ is the electromagnetic field. This equation turns out to be deeply related to the theorem that the boundary of a region has no boundary itself, which is, of course, a geometrical statement (see the box on this page). Another illustration of the geometrical nature of gauge fields can be found in the fact that global considerations have become important for gauge fields through the following theoretical and experimental developments:

- ▶ Dirac's magnetic monopole (1931)
- ▶ Bohm–Aharonov experiment (1960)
- ▶ 't Hooft–Polyakov monopole (1974)
- ▶ instantons (1975)

These ideas are described in the box on page 48.

Gauge fields are also intrinsically

Symmetry and physical laws

Before Einstein and Minkowski

experiment → field equations → symmetry (invariance)

After Einstein and Minkowski

symmetry → field equations

related to general relativity which is founded on geometrical concepts. The precise relationship is, however, quite subtle and is still being explored.

On the method of theoretical physics

In his Herbert Spencer lecture of 1933, bearing the title that I have taken as the title of this section, Einstein analyzed the meaning of theoretical physics and its development. The following are some striking quotes from that lecture:[8]

> . . . the axiomatic basis of theoretical physics cannot be extracted from experience but must be freely created . . .
>
> Experience may suggest the appropriate mathematical concepts, but they most certainly cannot be deduced from it . . .
>
> But the creative principle resides in mathematics. In a certain sense, therefore, I hold it true that pure thought can grasp reality, as the ancients dreamed.

Was Einstein saying that fundamental theoretical physics is a part of mathematics? Was he saying that fundamental theoretical physics should have the tradition and style of mathematics? The answers to these questions are no. Einstein was a physicist and not a mathematician. Furthermore, he considered himself a physicist, and not a mathematician. He gave the reasons in a very penetrating way in his Autobiographical Notes:[1]

> . . . This was obviously due to the fact that my intuition was not strong enough in the field of mathematics in order to differentiate clearly the fundamentally important, that which is really basic, from the rest of the more or less dispensable erudition. Beyond this, however, my interest in the knowledge of nature was also unqualifiedly stronger; and it was not clear to me as a student that the approach to a more profound knowledge of the basic principles of physics is tied up with the most intricate mathematical methods. This dawned upon me only gradually after years of independent scientific work. True enough, physics also was divided into separate fields, each of which was capable of devouring a short lifetime of work without having satisfied the hunger for deeper knowledge. The mass of insufficiently connected experimental data was overwhelming here also. In this field, however, I soon learned to scent out that which was able to lead to fundamentals and to turn aside from everything else, from the multitude of things which clutter up the mind and divert it from the essential . . .

But he realized, from his own experience, and from the great revolutions in physics in the early years of this century, that although physics is and remains rooted in experimental laws, yet more and more, mathematical simplicity and beauty are playing a role in the formation of concepts in fundamental physics. He compared[7] theories that are "close to experience" with more mathematical ones:

> On the other hand, it must be conceded that a theory has an important advantage if its basic concepts and fundamental hypotheses are "close to experience," and greater confidence in such a theory is certainly justified. There is less danger of going completely astray, particularly since it takes so much less time and effort to disprove such theories by experience. Yet more and more, as the depth of our knowledge increases, we must give up this advantage in our quest for logical simplicity and uniformity in the foundations of physical theory . . .

As a defense against misunderstanding by his fellow physicists, he pleaded:[3]

> The theoretical scientist is compelled in an increasing degree to be guided by purely mathematical, formal considerations . . . The theorist who undertakes such a labor should not be carped at as "fanciful"; on the contrary, he should be granted the right to give free rein to his fancy, for there is no other way to the goal.

The relationship between fundamental theoretical physics and mathematics is a fascinating subject. Perhaps I

Global effects of gauge fields

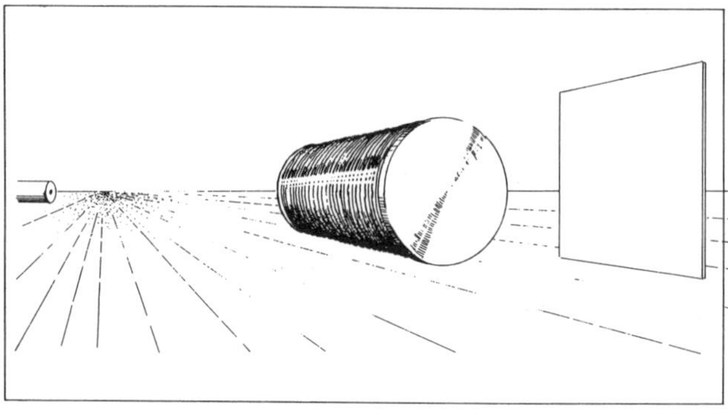

A magnetic monopole of strength g is a simple and natural idea. Dirac pointed out in 1931 that in quantum mechanics the magnitude of g must be related to the electric charge e through the condition $2eg/\hbar c =$ integer. It turns out that this condition is the simplest example of a very general and profound topological theorem: the Chern–Weil theorem. The next simplest example of the theorem is the so-called "instanton" solution of SU_2 gauge fields discovered in 1975. The 't Hooft–Polyakov monopole is a singularity-free solution for certain gauge fields, the existence of which is dependent on topological properties.

The Bohm–Aharonov experiment was proposed and performed in 1959–1960. As sketched in the figure above, electrons from the source go past a long solenoid, but are excluded from its inside. They produce an interference pattern on the screen. There is no electric or magnetic field outside of the solenoid, so the electrons suffer no local forces. Yet the interference pattern depends on the magnetic flux *inside* the solenoid, which shows that the effect of electromagnetism is not entirely local.

Scale transformations

Quantity	Value at first point	Value at neighboring point
coordinate	x^μ	$x^\mu + dx^\mu$
field	f	$f + (\partial_\mu f)dx^\mu$
scale	1	$1 + S_\mu dx^\mu$
scaled field	f	$f + (\partial_\mu + S_\mu)f dx^\mu$

We use the notation $\partial_\mu = (\partial/\partial x^\mu)$ and the summation convention.

can be allowed to tell you a story at this point.

In 1975, impressed with the fact that gauge fields are connections on fiber bundles, I drove to the house of Shiing-Shen Chern in El Cerrito, near Berkeley. (I had taken courses with him in the early 1940's when he was a young professor and I an undergraduate student at the National Southwest Associated University in Kunming, China. That was before fiber bundles had become important in differential geometry and before Chern had made history with his contributions to the generalized Gauss–Bonnet theorem and the Chern classes.) We had much to talk about: friends, relatives, China. When our conversation turned to fiber bundles, I told him that I had finally learned from Jim Simons the beauty of fiber-bundle theory and the profound Chern–Weil theorem. I said I found it amazing that gauge fields are exactly connections on fiber bundles, which the mathematicians developed *without reference to the physical world*. I added, "this is both thrilling and puzzling, since you mathematicians dreamed up these concepts out of nowhere." He immediately protested, "No, no. These concepts were not dreamed up. They were natural and real."

Deep as the relationship is between mathematics and physics, it would be wrong, however, to think that the two disciplines overlap that much. They do not. And they have their separate aims and tastes. They have distinctly different value judgments, and they have different traditions. At the fundamental conceptual level they *amazingly* share some concepts, but even there, the life force of each discipline runs along its own veins.

* * *

This article is adapted from a talk given at the Second Marcel Grossman meeting, held in Trieste, Italy, July 1979, in honor of the hundredth anniversary of the birth of Albert Einstein.

References

1. A. Einstein, "Autobiographical Notes" in *Albert Einstein, Philosopher-Scientist*, P. A. Schilpp, ed., Open Court, Evanston, Ill. (1949).
2. A. Einstein, J. Franklin Inst. **221**, 43 (1936).
3. A. Einstein in *Mein Weltbild*, Querido, Amsterdam (1934), translated in *Ideas and Opinions*, Bonanza, New York (1954).
4. For a short description of the history of gauge theory, see C.-N. Yang, Ann. N.Y. Acad. Sci. **294**, 86 (1977).
5. V. I. Fock, Z. Phys. **39**, 226 (1927).
6. F. London, Z. Phys. **42**, 375 (1927).
7. A. Einstein, Sci. Am., April 1950, page 13.
8. A. Einstein, *On the Method of Theoretical Physics*, Clarendon, Oxford (1933); reprinted in ref.3. □

List of Papers

List of Papers

(1944–1981)

(Papers in double brackets are included in this volume.)

[44a] C. N. Yang. "On the Uniqueness of Young's Differentials." *Bulletin of the American Mathematical Society* 50, 373 (1944).

[44b] C. N. Yang. "Variation of Interaction Energy with Change of Lattice Constants and Change of Degree of Order." *Chinese Journal of Physics* 5, 138 (1944).

[[45a]] C. N. Yang. "A Generalization of the Quasi-Chemical Method in the Statistical Theory of Superlattices." *The Journal of Chemical Physics* 13, 66 (1945).

[45b] C. N. Yang. "The Critical Temperature and Discontinuity of Specific Heat of a Superlattice." *Chinese Journal of Physics* 6, 59 (1945).

[47a] C. N. Yang. "On Quantized Space-Time." *The Physical Review* 72, 874 (1947).

[47b] C. N. Yang and Y. Y. Li. "General Theory of the Quasi-Chemical Method in the Statistical Theory of Superlattices." *Chinese Journal of Physics* 7, 59 (1947).

[[48a]] C. N. Yang. "On the Angular Distribution in Nuclear Reactions and Coincidence Measurements." *The Physical Review* 74, 764 (1948).

[48b] S. K. Allison, H. V. Argo, W. R. Arnold, L. del Rosario, H. A. Wilcox, and C. N. Yang. "Measurement of Short Range Nuclear Recoils from Disintegrations of the Light Elements." *The Physical Review* 74, 1233 (1948).

[[49a]] T. D. Lee, M. Rosenbluth, and C. N. Yang. "Interaction of Mesons with Nucleons and Light Particles." *The Physical Review* 75, 905 (1949).

[[49b]] E. Fermi and C. N. Yang. "Are Mesons Elementary Particles?" *The Physical Review* 76, 1739 (1949).

[[50a]] C. N. Yang. "Selection Rules for the Dematerialization of a Particle into Two Photons." *The Physical Review* 77, 242 (1950).

[50b] C. N. Yang. "Possible Experimental Determination of Whether the Neutral Meson is Scalar or Pseudoscalar." *The Physical Review* 77, 722 (1950).

[[50c]] C. N. Yang and J. Tiomno. "Reflection Properties of Spin 1/2 Fields and a Universal Fermi-Type Interaction." *The Physical Review* 79, 495 (1950).

[[50d]] C. N. Yang and David Feldman. "The S-Matrix in the Heisenberg Representation." *The Physical Review* 79, 972 (1950).

[51a] Geoffrey F. Chew, M. L. Goldberger, J. M. Steinberger, and C. N. Yang. "A Theoretical Analysis of the Process $\pi^+ + d \rightleftharpoons p + p$." *The Physical Review* 84, 581 (1951).

[51b] C. N. Yang. "Actual Path Length of Electrons in Foils." *The Physical Review* 84, 599 (1951).

⟦52a⟧ C. N. Yang. "The Spontaneous Magnetization of a Two-Dimensional Ising Model." *The Physical Review* 85, 808 (1952).

⟦52b⟧ C. N. Yang and T. D. Lee. "Statistical Theory of Equations of State and Phase Transitions. I. Theory of Condensation." *The Physical Review* 87, 404 (1952).

⟦52c⟧ T. D. Lee and C. N. Yang. "Statistical Theory of Equations of State and Phase Transitions. II. Lattice Gas and Ising Model." *The Physical Review* 87, 410 (1952).

⟦52d⟧ C. N. Yang. "Letter to E. Fermi dated May 5, 1952." Unpublished.

[52e] C. N. Yang. "Special Problems of Statistical Mechanics, Part I and II." Lectures given at University of Washington, Seattle, April–July 1952. Notes Taken by F. J. Blatt and R. L. Cooper. Mimeographed and distributed by University of Washington.

[53a] C. N. Yang. "Report on Cosmotron Experiments." In *Proceedings of the International Conference on Theoretical Physics.* Tokyo: Science Council of Japan, 1954, p. 137.

[53b] Chen Ning Yang. "Recent Experimental Results at Brookhaven." *Proceedings of the International Conference on Theoretical Physics*, p. 170. Science Council of Japan (Tokyo), 1954.

⟦54a⟧ G. A. Snow, R. M. Sternheimer, and C. N. Yang. "Polarization of Nucleons Elastically Scattered from Nuclei." *The Physical Review* 94, 1073 (1954).

⟦54b⟧ C. N. Yang and R. Mills. "Isotopic Spin Conservation and a Generalized Gauge Invariance." *The Physical Review* 95, 631 (1954).

⟦54c⟧ C. N. Yang and R. L. Mills. "Conservation of Isotopic Spin and Isotopic Gauge Invariance." *The Physical Review* 96, 191 (1954).

[55a] C. N. Yang. "Talk at 1955 Rochester Conference, Session on High Energy Pion Phenomena." In *High Energy Nuclear Physics, 1955.* New York: Wiley-Interscience, 1955, pp. 37–38.

⟦55b⟧ T. D. Lee and C. N. Yang. "Conservation of Heavy Particles and Generalized Gauge Transformations." *The Physical Review* 98, 1501 (1955).

[56a] K. M. Case, Robert Karplus, and C. N. Yang. "Strange Particles and the Conservation of Isotopic Spin." *The Physical Review* 101, 874 (1956).

[56b] K. M. Case, Robert Karplus, and C. N. Yang. "Experiments with Slow *K* Mesons in Deuterium and Hydrogen." *The Physical Review* 101, 358 (1956).

[56c] T. D. Lee and C. N. Yang. "Mass Degeneracy of the Heavy Mesons." *The Physical Review* 102, 290 (1956).

⟦56d⟧ T. D. Lee and C. N. Yang. "Charge Conjugation, a New Quantum Number *G*, and Selection Rules Concerning a Nucleon–Antinucleon System." *Il Nuovo Cimento* 10(3), 749 (1956).

⟦56e⟧ C. N. Yang. "Introductory Talk at the 1956 Rochester Conference, Session on Theoretical Interpretation of New Particles." In *High Energy Nuclear Physics, 1956*. New York: Wiley-Interscience, 1956.

[56f] C. N. Yang. "*Expanding Universes* by E. Schrödinger." *Science* 124, 370 (1956).

[56g] Kerson Huang and C. N. Yang. "Quantum Mechanical Many-Body Hard Core Interactions." *Bulletin of the American Physical Society* 2(1), 222 (1956).

⟦56h⟧ T. D. Lee and C. N. Yang. "Question of Parity Conservation in Weak Interactions." *The Physical Review* 104, 254 (1956).

[56i] T. D. Lee and C. N. Yang. "Possible Interference Phenomena Between Parity Doublets." *The Physical Review* 104, 822 (1956).

[57a] K. Huang and C. N. Yang. "Quantum Mechanical Many-Body Problem with Hard Sphere Interaction." *The Physical Review* 105, 776 (1957).

[57b] K. Huang, C. N. Yang, and J. M. Luttinger. "Imperfect Bose Gas with Hard Sphere Interaction." *The Physical Review* 105, 776 (1957).

[57c] K. M. Case, R. Karplus, and C. N. Yang. "A Reply to a Criticism by Mr. A. Gamba." *Il Nuovo Cimento* 5, 1004 (1957).

⟦57d⟧ C. N. Yang. "Present Knowledge About the New Particles." Lecture given at the Seattle International Conference on Theoretical Physics, September 1956. *Reviews of Modern Physics* 29, 231 (1957).

⟦57e⟧ T. D. Lee, Reinhard Oehme, and C. N. Yang. "Remarks on Possible Noninvariance Under Time Reversal and Charge Conjugation." *The Physical Review* 106, 340 (1957).

⟦57f⟧ T. D. Lee and C. N. Yang. "Parity Nonconservation and a Two-Component Theory of the Neutrino." *The Physical Review* 105, 1671 (1957).

[57g] T. D. Lee and C. N. Yang. "Derivative Coupling for μ Meson Decay in a Two-Component Theory of the Neutrino." Unpublished.

⟦57h⟧ T. D. Lee and C. N. Yang. "Many-Body Problem in Quantum Mechanics and Quantum Statistical Mechanics." *The Physical Review* 105, 1119 (1957).

⟦57i⟧ T. D. Lee, Kerson Huang, and C. N. Yang. "Eigenvalues and Eigenfunctions of a Bose System of Hard Spheres and Its Low-Temperature Properties." *The Physical Review* 106, 1135 (1957).

[57j] T. D. Lee, J. Steinberger, G. Feinberg, P. K. Kabir, and C. N. Yang. "Possible Detection of Parity Nonconservation in Hyperon Decay." *The Physical Review* 106, 1367 (1957).

[57k] Chen Ning Yang. "Lois de Symétrie et Particules Étranges." Lecture given at Université de Paris, May 1957. Lecture notes taken by Froissard and Mandelbrojt. Unpublished.

[57l] T. D. Lee and C. N. Yang. "Errata: Question of Parity Conservation in Weak Interactions." *The Physical Review* 106, 1371 (1957).

[57m] Kerson Huang, C. N. Yang, and T. D. Lee. "Capture of μ^- Mesons by Protons." *The Physical Review* 108, 1340 (1957).

[57n] T. D. Lee and C. N. Yang. "Possible Nonlocal Effects in μ Decay." *The Physical Review* 108, 1611 (1957).

⟦57o⟧ T. D. Lee and C. N. Yang. "General Partial Wave Analysis of the Decay of a Hyperon of Spin 1/2." *The Physical Review* 108, 1645 (1957).

[57p] T. D. Lee and C. N. Yang. *Elementary Particles and Weak Interactions*. BNL 443 (T-91). Brookhaven National Laboratory, 1957.

⟦57q⟧ Kerson Huang, T. D. Lee, and C. N. Yang. "Quantum Mechanical Many-Body Problem and the Low Temperature Properties of a Bose System of Hard Spheres." Lecture given at the Stevens Conference on the Many-Body Problem, January 1957. In *The Many-Body Problem,* ed. J. K. Percus. New York: Wiley-Interscience, 1963, p. 165.

[57r] Chen Ning Yang. "Le Problème a Plusieurs Corps en Mécanique Quantique et en Mécanique Statistique." Lecture given at Université de Paris, June 1957. Lecture notes taken by C. Bouchiat and A. Martin. Unpublished.

⟦57s⟧ C. N. Yang. "The Law of Parity Conservation and Other Symmetry Laws of Physics." In *Les Prix Nobel*. Stockholm: The Nobel Foundation, 1957, p. 95. Also in *Science* 127, 565 (1958).

⟦57t⟧ C. N. Yang. "Speech at the Nobel Banquet, December 10, 1957." In *Les Prix Nobel.* Stockholm: The Nobel Foundation, 1957, p. 53.

⟦58a⟧ T. D. Lee and C. N. Yang. "Possible Determination of the Spin of Λ^0 from Its Large Decay Angular Asymmetry." *The Physical Review* 109, 1755 (1958).

[58b] J. Bernstein, T. D. Lee, C. N. Yang, and H. Primakoff. "Effect of the Hyperfine Splitting of a μ-Mesonic Atom on Its Lifetime." *The Physical Review* 111, 313 (1958).

[58c] M. Goldhaber, T. D. Lee, and C. N. Yang. "Decay Modes of a $(\theta + \bar{\theta})$ System." *The Physical Review* 112, 1796 (1958).

⟦58d⟧ T. D. Lee and C. N. Yang. "Low-Temperature Behavior of a Dilute Bose System of Hard Spheres. I. Equilibrium Properties." *The Physical Review* 112, 1419 (1958).

[59a] T. D. Lee and C. N. Yang. "Low-Temperature Behavior of a Dilute Bose System of Hard Spheres. II. Nonequilibrium Properties." *The Physical Review* 113, 1406 (1959).

⟦59b⟧ T. D. Lee and C. N. Yang. "Many-Body Problem in Quantum Statistical Mechanics. I. General Formulation." *The Physical Review* 113, 1165 (1959).

⟦59c⟧ C. N. Yang. "Symmetry Principles in Modern Physics." Lecture given at the 75th Anniversary Celebration of Bryn Mawr College, Session on Symmetries, November 6, 1959. Unpublished.

[59d] T. D. Lee and C. N. Yang. "Many-Body Problem in Quantum Statistical Mechanics. II. Virial Expansion for Hard-Sphere Gas." *The Physical Review* 116, 25 (1959).

[60a] T. D. Lee and C. N. Yang. "Many-Body Problem in Quantum Statistical Mechanics. III. Zero-Temperature Limit for Dilute Hard Spheres." *The Physical Review* 117, 12 (1960).

[60b] T. D. Lee and C. N. Yang, "Many-Body Problem in Quantum Statistical Mechanics. IV. Formulation in Terms of Average Occupation Number in Momentum Space." *The Physical Review* 117, 22 (1960).

[60c] T. D. Lee and C. N. Yang. "Many-Body Problem in Quantum Statistical Mechanics. V. Degenerate Phase in Bose–Einstein Condensation." *The Physical Review* 117, 897 (1960).

⟦60d⟧ T. D. Lee and C. N. Yang. "Theoretical Discussions on Possible High-Energy Neutrino Experiments." *Physical Review Letters* 4, 307 (1960).

⟦60e⟧ T. D. Lee and C. N. Yang. "Implications of the Intermediate Boson Basis of the Weak Interactions: Existence of a Quartet of Intermediate Bosons and Their Dual Isotopic Spin Transformation Properties." *The Physical Review* 119, 1410 (1960).

[60f] C. N. Yang. "The Many Body Problem." Lectures given at Latin American School of Physics, Centro Brasileiro de Pesquisas Fisicás, Rio de Janeiro, June 27–August 7, 1960. Lecture notes taken by M. Bauer et al. In *Monografias de Fisica VI*. Rio de Janeiro: Centro Brasileiro de Pesquisas Fisicas, 1960.

⟦60g⟧ C. N. Yang. "Imperfect Bose System." *Physica* 26, S49 (1960).

[60h] C. N. Yang. "Some Theoretical Implications of High-Energy Neutrino Experiments." Lecture given at Berkeley Conference on High Energy Physics

Experimentation, September 12–14, 1960. Published in conference report. University of California, Berkeley, 1960.

⟦61a⟧ C. N. Yang. Introductory Notes to the Article "Are Mesons Elementary Particles?" In *The Collected Papers of Enrico Fermi*, Vol. 2. Chicago: University of Chicago Press, 1965, p. 673.

⟦61b⟧ T. D. Lee and C. N. Yang. "Some Considerations on Global Symmetry." *The Physical Review* 122, 1954 (1961).

⟦61c⟧ N. Byers and C. N. Yang. "Theoretical Considerations Concerning Quantized Magnetic Flux in Superconducting Cylinders." *Physical Review Letters* 7, 46 (1961).

[61d] T. D. Lee, R. Serber, G. C. Wick, and C. N. Yang. "Some Theoretical Considerations on the Desirability of a 300 to 1000 BeV Proton Accelerator." In *Experimental Program Requirements for a 300–1000 BeV Accelerator*, BNL 772 (T-290), p. 15. Brookhaven National Laboratory, 1961.

[61e] T. D. Lee, P. Markstein, and C. N. Yang. "Production Cross Section of Intermediate Bosons by Neutrinos in the Coulomb Field of Protons and Iron." *Physical Review Letters* 7, 429 (1961).

⟦61f⟧ C. N. Yang. "The Future of Physics." Panel Discussion at the MIT Centennial Celebration, April 8, 1961. Unpublished.

[61g] F. Gursey and C. N. Yang. "*S*-State Capture in K^-p Atoms Colliding with H Atoms." Written in May 1961. Unpublished.

[62a] C. N. Yang. "Symposium Discussion, November 4, 1961, Washington, D. C. Applied Mathematics: What Is Needed in Research and Education?" *SIAM Review* 4, 297 (1962).

⟦62b⟧ S. B. Treiman and C. N. Yang. "Tests of the Single-Pion Exchange Model." *Physical Review Letters* 8, 140 (1962).

[62c] C. N. Yang. *Elementary Particles, A Short History of Some Discoveries in Atomic Physics*. Princeton: Princeton University Press, 1962.

[62d] T. D. Lee and C. N. Yang. "High Energy Neutrino Reactions Without Production of Intermediate Bosons." *The Physical Review* 126, 2239 (1962).

[62e] M. E. Rose and C. N. Yang. "Eigenvalues and Eigenvectors of a Symmetric Matrix of 6*j* Symbols." *Journal of Mathematical Physics* 3, 106 (1962).

[62f] T. F. Hoang and C. N. Yang. *A Possible Method of Measuring the Fraction of* $\Delta Q/\Delta S = -1$ *Decay in the* K_1–K_2 *Complex*. CERN Internal Report, 4010/TH. 276, May 28, 1962.

⟦62g⟧ T. D. Lee and C. N. Yang. " Obituary for Dr. Shih-Tsun Ma." Unpublished.

[62h] C. N. Yang. "Talk at CERN, July 7, 1962." CERN preprint (1962).

[[62i]] T. D. Lee and C. N. Yang. "Theory of Charged Vector Mesons Interacting with the Electromagnetic Field." *The Physical Review* 128, 885 (1962).

[[62j]] C. N. Yang. "Concept of Off-Diagonal Long-Range Order and the Quantum Phases of Liquid He and of Superconductors." *Reviews of Modern Physics* 34, 694 (1962).

[63a] C. N. Yang. "Mathematical Deductions from Some Rules Concerning High-Energy Total Cross Sections." *Journal of Mathematical Physics* 4, 52 (1963).

[63b] C. N. Yang. "Some Properties of the Reduced Density Matrix." *Journal of Mathematical Physics* 4, 418 (1963).

[63c] C. N. Yang. "Remarks on Weak Interactions." In *Proceedings of the Eastern Theoretical Physics Conference*, ed. M. E. Rose. New York: Gordon & Breach, 1963.

[63d] R. J. Oakes and C. N. Yang. "Meson–Baryon Resonances and the Mass Formula." *Physical Review Letters* 11, 174 (1963).

[[63e]] C. N. Yang. "The Mass Formula of SU_3." In *Some Recent Advances in Basic Sciences*, Vol. 1. New York: Academic Press, 1966.

[64a] C. N. Yang. "Some Theoretical Considerations Concerning the Neutrino Experiments." In *Proceedings of The Weak Interaction Conference, Brookhaven National Laboratory*, BNL 837 (C-39), p. 249. Brookhaven National Laboratory, 1964.

[64b] C. N. Yang. "Computing Machines and High-Energy Physics." In *Proceedings of the IBM Scientific Computing Symposium on Large Scale Problems in Physics,* December 1963, p. 65. IBM, 1964.

[64c] N. Byers and C. N. Yang. "Physical Regions in Invariant Variables for n Particles and the Phase–Space Volume Element." *Reviews of Modern Physics* 36, 595 (1964).

[64d] N. Byers and C. N. Yang. "Phenomenological Analysis of Reactions Such as $K^- + p \rightarrow \Lambda + \omega$." *The Physical Review* 135, B796 (1964).

[[64e]] C. N. Yang and C. P. Yang. "Critical Point in Liquid–Gas Transitions." *Physical Review Letters* 13, 303 (1964).

[[64f]] Tai Tsun Wu and C. N. Yang. "Phenomenological Analysis of Violation of CP Invariance in Decay of K^0 and $\overline{K}^0$." *Physical Review Letters* 13, 380 (1964).

[64g] C. N. Yang. "Round-Table Discussion on High-Energy Physics. APS Washington Meeting." *Physics Today* 17, 50 (November 1964).

[64h] F. Dyson, A. Pais, B. Stromgren, and C. N. Yang. "To J. Robert Oppenheimer on His Sixtieth Birthday." *Reviews of Modern Physics* 36, 507 (1964).

⟦65a⟧ Tai Tsun Wu and C. N. Yang. "Some Speculations Concerning High-Energy Large Momentum Transfer Processes." *The Physical Review* 137, B708 (1965).

[65b] C. N. Yang. "Some Considerations Concerning Very High Energy Experiments." In *Nature of Matter: Purposes of High Energy Physics*, ed. Luke C. L. Yuan, BNL 888 (T-360), p. 74. Brookhaven National Laboratory, 1965.

[65c] C. N. Yang. "Report of the Theoretical Physics Panel to the Physics Survey Committee, February 20, 1965." In *Physics: Survey and Outlook, Reports on the Subfields of Physics,* p. 159. NAS, NRC, 1966.

[65d] C. N. Yang. "Phenomenological Description of *K* Decay."In *Proceedings of the International Conference on Weak Interactions*, p. 29. Argonne National Laboratory, 1965.

[65e] C. N. Yang. "Symmetry Principles in Physics." In *Vistas in Research*, Vol. 3. New York: Gordon & Breach, 1966. Also in *Physics Teachers* 5, 311 (October 1967).

[65f] N. Byers, S. W. MacDowell, and C. N. Yang. "*CP* Violation in *K* Decay." In *High Energy Physics and Elementary Particles.* Vienna: International Atomic Energy Agency, 1965, p. 953.

[65g] C. N. Yang. "Statement at Public Hearing, Subcommittee on Research Development, Radiation." *Congressional Record,* March 3, 1965.

[65h] C. N. Yang. "Speech on Last Day of Kyoto Conference, Commemorating the Thirtieth Anniversary of Meson Theory, September 30, 1965." Unpublished.

[66a] C. N. Yang and C. P. Yang. "One-Dimensional Chain of Anisotropic Spin–Spin Interactions." *Physics Letters* 20, 9 (1966); (Errata) *Physics Letters* 21, 719 (1966).

⟦66b⟧ N. Byers and C. N. Yang. "πp Charge-Exchange Scattering and a 'Coherent Droplet' Model of High-Energy Exchange Processes." *The Physical Review* 142, 976 (1966).

⟦66c⟧ C. N. Yang. "Remarks at the Dedication of the Einstein Stamp, March 14, 1966." Unpublished.

[66d] C. N. Yang and C. P. Yang. "Ground State Energy of a Heisenberg–Ising Lattice." *The Physical Review* 147, 303 (1966).

⟦66e⟧ C. N. Yang and C. P. Yang. "One-Dimensional Chain of Anisotropic Spin–Spin Interactions. I. Proof of Bethe's Hypothesis for Ground State in a Finite System." *The Physical Review* 150, 321 (1966).

[66f] C. N. Yang and C. P. Yang. "One-Dimensional Chain of Anisotropic Spin–Spin Interactions. II. Properties of the Ground State Energy Per Lattice Site for an Infinite System." *The Physical Review* 150, 327 (1966).

[66g] C. N. Yang and C. P. Yang. "One-Dimensional Chain of Anisotropic Spin–Spin Interactions. III. Applications." *The Physical Review* 151, 258 (1966).

⟦66h⟧ R. L. Mills and C. N. Yang. "Treatment of Overlapping Divergences in the Photon Self-Energy Function." *Supplement of the Progress of Theoretical Physics* 37 and 38, 507 (1966).

[66i] C. N. Yang. "Summary of the Conference." In *Proceedings of the Conference on High Energy Two-Body Reactions*. Stony Brook, 1966. Unpublished.

[66j] C. N. Yang. "Quantum Lattice Gas and the Heisenberg–Ising Antiferromagnetic Chain." In *Proceedings of the Eastern Theoretical Conference*. Providence, R.I.: Brown University, 1966, p. 215.

[67a] F. Abbud, B. W. Lee, and C. N. Yang. "Comments on Measuring $Re(A_2/A_0)$ in $K^0 \rightarrow \pi\pi$ Decay." *Physical Review Letters* 18, 980 (1967).

⟦67b⟧ T. T. Chou and C. N. Yang. "Some Remarks Concerning High Energy Scattering." In *High Energy Physics and Nuclear Structure*. Amsterdam: North-Holland, 1967, p. 384.

[67c] B. Sutherland, C. N. Yang, and C. P. Yang. "Exact Solution of a Model of Two-Dimensional Ferroelectrics in an Arbitrary External Electric Field." *Physical Review Letters* 19, 588 (1967).

⟦67d⟧ T. T. Wu and C. N. Yang. "Some Solutions of the Classical Isotopic Gauge Field Equations." In *Properties of Matter Under Unusual Conditions*, eds. H. Mark and S. Fernbach. New York: Wiley-Interscience, 1969, p. 349.

⟦67e⟧ C. N. Yang. "Some Exact Results for the Many-Body Problem in One Dimension with Repulsive Delta-Function Interaction." *Physical Review Letters* 19, 1312 (1967).

[68a] C. N. Yang. "S Matrix for the One-Dimensional N-Body Problem with Repulsive or Attractive Delta Function Interaction." *The Physical Review* 168, 1920 (1968).

[68b] T. T. Chou and C. N. Yang. "Model of Elastic High-Energy Scattering." *The Physical Review* 170, 1591 (1968).

[68c] T. T. Chou and C. N. Yang. "Possible Existence of Kinks in High-Energy Elastic *pp* Scattering Cross Section." *Physical Review Letters* 20, 1213 (1968).

[68d] T. T. Chou and C. N. Yang. "Model of High-Energy Elastic Scattering and Diffractive Excitation Processes in Hadron-Hadron Collisions." *The Physical Review* 175, 1832 (1968).

[68e] C. N. Yang, " General Review of Some Developments in High Energy Physics in Recent Years." Paper delivered at First Latin-American Congress, Mexico

City, July 1968. In *Primer Congreso Latino Americano de Fisica*, 1968, p. 27.

[[69a]] C. N. Yang and C. P. Yang. "Thermodynamics of a One-Dimensional System of Bosons with Repulsive Delta-Function Interaction." *Journal of Mathematical Physics* 10, 1115 (1969).

[69b] M. Goldhaber and C. N. Yang, "The K^0–$\overline{K}^0$ System in p–$\overline{p}$ Annihilation at Rest." In *Evolution of Particle Physics,* ed. M. Conversi. New York: Academic Press, 1969, p. 171.

[[69c]] J. Benecke, T. T. Chou, C. N. Yang, and E. Yen. "Hypothesis of Limiting Fragmentation in High-Energy Collisions." *The Physical Review* 188, 2159 (1969).

[69d] T. T. Chou and C. N. Yang. "Extrapolation of Elastic Differential πp Cross Section to Very High Energies and the Pion Form Factor." *The Physical Review* 188, 2469 (1969).

[69e] C. N. Yang. "Hypothesis of Limiting Fragmentation." In *Proceedings of the Third International Conference on High Energy Collisions.* New York: Gordon & Breach, 1969, p. 509.

[69f] C. N. Yang, J. A. Cole, J. Good, R. Hwa, and J. Lee-Franzini, eds. *Proceedings of the Third Interantional Conference on High Energy Collisions.* New York: Gordon & Breach, 1969.

[[70a]] C. N. Yang. "Charge Quantization, Compactness of the Gauge Group, and Flux Quantization." *Physical Review D* 1, 2360 (1970).

[[70b]] C. N. Yang. "Some Exactly Soluble Problems in Statistical Mechanics." Lectures given at the Karpacz Winter School of Physics, February 1970. In *Proceedings of the VII Winter School of Theoretical Physics in Karpacz.* University of Wroclaw, 1970.

[70c] T. T. Chou and C. N. Yang. "Remarks About the Hypothesis of Limiting Fragmentation." *Physical Review Letters* 25, 1072 (1970).

[70d] C. N. Yang. "One-Dimensional Delta Function Interaction." Lecture given at the Battelle Institute Colloquium, September 1970. In *Critical Phenomena in Alloys, Magnates, and Superconductors,* eds. R. E. Mills, E. Ascher, and R. I. Jaffee. New York: McGraw-Hill, 1971, p. 13.

[70e] C. N. Yang. "Symmetry Principles." *Encyclopedia Americana*, 1970 edition. S. v. physics.

[[70f]] C. N. Yang. "High-Energy Hadron–Hadron Collisions." Lecture given at the Kiev Conference, August 1970. In *Proceedings of the Kiev Conference—Fundamental Problems of the Elementary Particle Theory*, p. 131. Academy of Sciences of the Ukranian SSR, 1970.

[70g] C. N. Yang. "Comments After Professor Brewer's Talk." In *Proceedings of the VII Winter School of Theoretical Physics in Karpacz,* University of Wroclaw, 1970.

[71a] C. K. Lai and C. N. Yang. "Ground State Energy of a Mixture of Fermions and Bosons in One Dimension with a Repulsive δ-Function Interaction." *Physical Review A* 3, 393 (1971).

⟦71b⟧ C. N. Yang. "Introductory Note on *Phase Transitions and Critical Phenomena*." In *Phase Transitions and Critical Phenomena*, Vol. 1, eds. C. Domb and M. S. Green. New York: Academic Press, 1971, p. 1.

[71c] T. T. Chou and Chen Ning Yang. "Hadron Momentum Distribution in Deeply Inelastic *ep* Collisions." *Physical Review D* 4, 2005 (1971).

[72a] Chen Ning Yang. "Some Speculations on Colliding Beams of 100-GeV Protons and 15-GeV Electron–100-GeV Protons." In *Isabelle Physics Prospects*, BNL 17522. Brookhaven National Laboratory, 1972.

[72b] C. Quigg, Jiunn-Ming Wang, and Chen Ning Yang. "Multiplicity Fluctuation and Multiparticle Distribution Functions in High-Energy Collisions." *Physical Review Letters* 28, 1290 (1972).

⟦72c⟧ Chen Ning Yang. "Some Concepts in Current Elementary Particle Physics." Lecture given at the Trieste Conference in honor of P.A.M. Dirac, September 1972. In *The Physicist's Conception of Nature*, ed. J. Mehra. Dordrecht: D. Reidel, 1972, pp. 447–453.

[73a] T. T. Chou and Chen Ning Yang. "Charge Transfer in High-Energy Fragmentation." *Physical Review D* 7, 1425 (1973).

⟦73b⟧ Alexander Wu Chao and Chen Ning Yang. "Opaqueness of *pp* Collisions from 30 to 1500 GeV/*c*." *Physical Review D* 8, 2063 (1973).

[73c] Chen Ning Yang. "Geometrical Description of the Structure of the Hadrons." Lecture given at the International Symposium on High Energy Physics, Tokyo, July 23, 1973. In *Proceedings of the International Symposium on High Energy Physics*, University of Tokyo, 1973, pp. 629–634.

[74a] Alexander Wu Chao and Chen Ning Yang. "Possible Relationship Between the Ratio π^+/π^- and the Average Multiplicity." *Physical Review D* 9, 2505 (1974).

[74b] Alexander Wu Chao and Chen Ning Yang. "Charge Correlation Between Two Pions in a Statistical Charge Distribution Among Hadrons." *Physical Review D* 10, 2119 (1974).

⟦74c⟧ C. N. Yang. "Integral Formalism for Gauge Fields." *Physical Review Letters* 33, 445 (1974).

[74d] Chen Ning Yang. "Relationship Between Correlation Function ρ_n and Fluctuation Phenomena." Unpublished.

[75a] H. T. Nieh, Tai Tsun Wu, and Chen Ning Yang. "Possible Interactions of the *J* Particle." *Physical Review Letters* 34, 49 (1975).

[75b] Tai Tsun Wu and Chen Ning Yang. "Some Remarks About Unquantized non-Abelian Gauge Fields." *Physical Review D* 12, 3843 (1975).

⟦75c⟧ Tai Tsun Wu and Chen Ning Yang. "Concept of Nonintegrable Phase Factors and Global Formulation of Gauge Fields." *Physical Review D* 12, 3845 (1975).

[75d] Gu Chao-Hao and Yang Chen-Ning. "Some Problems on the Gauge Field Theories." *Scientia Sinica* 18, 483 (1975).

[75e] Chen Ning Yang. "Gauge Fields." In *Proceedings of Sixth Hawaiian Topical Conference in Particle Physics.* Honolulu: University of Hawaii Press, 1976, pp. 487–561.

[75f] Chen Ning Yang. "Meccanica Statistica." In *Enciclopedia del Novecento,* Vol. 4, 1979, p. 53.

⟦76a⟧ T. T. Chou and Chen Ning Yang. "Hadronic Matter Current Distribution Inside a Polarized Nucleus and a Polarized Hadron." *Nuclear Physics* B107, 1 (1976).

[76b] Tai Tsun Wu and Chen Ning Yang. "Static Sourceless Gauge Field." *Physical Review D* 13, 3233 (1976).

⟦76c⟧ Tai Tsun Wu and Chen Ning Yang. "Dirac Monopole Without Strings: Monopole Harmonics." *Nuclear Physics* B107, 365 (1976).

⟦76d⟧ Tai Tsun Wu and Chen Ning Yang. "Dirac's Monopole Without Strings: Classical Lagrangian Theory." *Physical Review D* 14, 437 (1976).

[76e] Chen Ning Yang. "Monopoles and Fiber Bundles." In *Understanding the Fundamental Constituents of Matter,* ed. A. Zichichi. New York: Plenum, 1976, p. 53.

[76f] Chen Ning Yang. "Discussions 'On Hadronic Current,' with Dr. Leader and Others." In *Understanding the Fundamental Constituents of Matter,* ed. A. Zichichi. New York: Plenum, 1976, p. 68.

⟦76g⟧ C. N. Yang. "What Visits Mean to China's Scientists." In *Reflections on Scholarly Exchanges with the People's Republic of China*, ed. A. Keatley. Committee on Scholarly Communication with the People's Republic of China, 1976.

[77a] M. L. Good, Y. Kazama, and Chen Ning Yang. "Possible Experiments to Study Incoherent-Multiple-Collision Effects in Hadron Production from Nuclei." *Physical Review D* 15, 1920 (1977).

[77b] Tai Tsun Wu and Chen Ning Yang. "Some Properties of Monopole Harmonics." *Physical Review D* 16, 1018 (1977).

[77c] Y. Kazama, Chen Ning Yang, and A. S. Goldhaber. "Scattering of a Dirac Particle with Charge *Ze* by a Fixed Magnetic Monopole." *Physical Review D* 15, 2287 (1977).

[77d] Yoichi Kazama and Chen Ning Yang. "Existence of Bound States for a Charged Spin 1/2 Particle with an Extra Magnetic Moment in the Field of a Fixed Magnetic Monopole." *Physical Review D* 15, 2300 (1977).

⟦77e⟧ Chen Ning Yang. "Magnetic Monopoles, Fiber Bundles, and Gauge Fields." *Annals of the New York Academy of Sciences* 294, 86 (1977).

[77f] Chen Ning Yang. "Conformal Mapping of Gauge Fields." *Physical Review D* 16, 330 (1977).

[77g] Chen Ning Yang. "Condition of Self-Duality for SU_2 Gauge Fields on Euclidean 4-Dimensional Space." *Physical Review Letters* 38, 1377 (1977).

⟦77h⟧ Gu Chao-hao and Yang Chen-ning. "Some Problems on the Gauge Field Theories, II." *Scientia Sinica* 20, 47 (1977).

[77i] Gu Chao-hao and Yang Chen-ning. "Some Problems on the Gauge Field Theories, III." *Scientia Sinica* 20, 177 (1977).

⟦77j⟧ C. N. Yang. "Speech at the Benjamin W. Lee Memorial Session." In *Unification of Elementary Forces and Gauge Theories*, eds. D. B. Cline and F. E. Mills, p. xiii. Harwood Academic Publishers, 1977.

[77k] Chen Ning Yang. "Symmetries in Physics." In *Unification of Elementary Forces and Gauge Theories*, eds. D. B. Cline and F. E. Mills, p. 3. Harwood Academic Publishers, 1977.

[78a] Chen Ning Yang. "Generalization of Dirac's Monopole to SU_2 Gauge Fields." *Journal of Mathematical Physics* 19, 320 (1978).

[78b] T. T. Chou and Chen Ning Yang. "Possible Existence of a Second Minimum in Elastic *pp* Scattering." *Physical Review D* 17, 1889 (1978).

[78c] Alexander W. Chao, Tai Tsun Wu, and Chen Ning Yang. "Some Inequalities in the Eikonal Approximation." In *Ta-You Wu Festschrift: Science of Matter*, ed. S. Fujita. New York: Gordon & Breach, 1978.

[78d] Chen Ning Yang. "Interaction of a Static Magnetic Monopole with a Dirac Positron." In *Proceedings of INS on New Particles and the Structure of Hadrons*, eds. K. Fujikawa et al. Institute for Nuclear Study, University of Tokyo, 1977.

⟦78e⟧ Tu Tung-sheng, Wu Tai-tsun, and Yang Chen-ning. "Interaction of Electrons, Magnetic Monopoles, and Photons (I)." *Scientia Sinica* 21, 317 (1978).

[78f] Ling-Lie Wang and Chen Ning Yang. "Classification of SU_2 Gauge Fields." *Physical Review D* 17, 2687 (1978).

[78g] Chen Ning Yang. "SU_2 Monopole Harmonics." *Journal of Mathematical Physics* 19, 2622 (1978).

[78h] Chen Ning Yang. "Developments in the Theory of Magnetic Monopoles." In *Proceedings of the 19th International Conference on High Energy Physics, Tokyo, 1978*, eds. S. Homma, M. Kawaguchi, and H. Miyazawa, p. 497. Physics Society of Japan, 1978.

[78i] Gu Chao-hao, Hu He-sheng, Shen Chun-li, and Yang Chen-ning, "A Geometrical Interpretation of Instanton Solutions in Euclidean Space." *Scientia Sinica* 21, 767 (1978).

[78j] Gu Chao-hao, Hu He-sheng, Li Da-qian, Shen Chun-li, Xin Yuan-long, and Yang Chen-ning. "Riemannian Spaces with Local Duality and Gravitational Instantons." *Scientia Sinica* 21, 475 (1978).

⟦78k⟧ Chen Ning Yang. "Pointwise SO_4 Symmetry of the BPST Pseudoparticle Solution." In *Felix Bloch and Twentieth-Century Physics,* eds. M. Chodorow et al. Houston: Rice University, 1980.

[79a] T. T. Chou and Chen Ning Yang. "Elastic Hadron–Hadron Scattering at Ultrahigh Energies and Existence of Many Dips." *Physical Review D* 19, 3268 (1979).

⟦79b⟧ Max Dresden and Chen Ning Yang. "Phase Shift in a Rotating Neutron or Optical Interferometer." *Physical Review D* 20, 1846 (1979).

[79c] Chen Ning Yang. "Fiber Bundles and the Physics of the Magnetic Monopole." In *The Chern Symposium 1979,* eds. W. Y. Hsiang et al., p. 247. Springer Verlag, 1980.

[79d] Chen Ning Yang. "Panel Discussion." In *Some Strangeness in the Proportion,* ed. H. Woolf. Reading, Mass.: Addison-Wesley, 1980, p. 500.

[79e] Chen Ning Yang. "Geometry and Physics." In *To Fulfill a Vision*, ed. Y. Ne'eman. Reading, Mass.: Addison-Wesley, 1981, pp. 3–11.

⟦80a⟧ T. T. Chou and Chen Ning Yang. "Geometrical Model of Hadron Collisions." In *Proceedings of the 1980 Guangzhou Conference on Theoretical Particle Physics.* Beijing: Science Press, 1980, p. 317.

⟦80b⟧ Chen-Ning Yang. "Einstein's Impact on Theoretical Physics." Lecture given at the Second Marcel Grossmann Meeting held in honor of the 100th Anniversary of the birth of Albert Einstein. *Physics Today* 33, 42 (June 1980). See also comments on R. Herman's letter to the editor, *Physics Today* 33, 11 (1980).

[80c] T. T. Chou and Chen Ning Yang. "Dip and Kink Structures in Hadron–Nucleus and Hadron–Hadron Diffraction Dissociation." *Physical Review D* 22, 610 (1980).

[80d] Chen Ning Yang and Chen Ping Yang. "Does Violation of Microscopic Time Reversal Invariance Lead to the Possibility of Entropy Decrease?" *Transactions of the New York Academy of Sciences* 40, 267 (1980).

[81a] T. T. Chou and Chen Ning Yang. "Dip Movement in $p\bar{p}$ and pp Elastic Collisions." *Physical Review Letters* 46, 764 (1981).

[81b] Chen Ning Yang. "After Dinner Speech." In *AGS 20th Anniversary Celebration,* Brookhaven National Laboratory, 1980, p. 85.

Index

Index

Names in Chinese Characters

A.C.T. Wu	吳期泰
C.S. Wu	吳健雄
Tai Tsun Wu	吳大峻
T.Y. Wu	吳大猷
C.P. Yang	楊振平
Chih Li Tu Yang	楊杜致禮
Eulee Yang	楊又禮
Franklin Yang	楊光諾
Gilbert Yang	楊光宇
Ko-Chuen Yang	楊克純 (楊武之)
Meng-hwa Lo Yang	楊羅孟華
E. Yan	閻愛德

Index of Names in the Commentary

Added Items for This Edition

Chapter 1

GAUGE INVARIANCE AND INTERACTIONS*

C. N. Yang
C. N. Yang Institute for Theoretical Physics
State University of New York at Stony Brook
Stony Brook, NY 11794-3840, USA

Gerardus 't Hooft wants me to write something about the early origin of non-Abelian gauge theory. I searched through my notes and found a few pages which I now contribute to this volume that he is editing.

These pages were written in March 1947 when I was a graduate student at the University of Chicago. Like graduate students of my generation, I was familiar with Pauli's description of gauge theory in his 1933 *Handbuch der Physik* article [1] and his 1941 review in the *Reviews of Modern Physics* [2], but not very much the 1929 article of Weyl [3].

I was clearly focusing on a very important problem. Unfortunately the mathematical calculations that I had carried out repeatedly in subsequent years I could not find today. They always had ended in more and more complicated formulae and total frustration. It was only in 1953–1954 when Bob Mills and I revisited the problem and tried adding quadratic terms to the field strength $F_{\mu\nu}$ that an elegant theory emerged. For Mills and me it was many years later that we realized the quadratic terms were in fact *natural* from the mathematical viewpoint.

[1] W. Pauli, *Handbuch der Physik* **24** (1933).
[2] W. Pauli, *Rev. Mod. Phys.* **13**, 203 (1941).
[3] H. Weyl, *Zeitschr. f. Phys.* **56**, 330 (1929).

*Reprinted from *50 Years of Yang–Mills Theory*, edited by G. 't Hooft (World Scientific, 2005).

March 47 (VI)

Gauge Invariance and Interact.

I. The gauge invariance in Electromagnetic theory serves for two purposes:

(i) Give rise to the definition of a charge (+ current) density.

(ii) Fix the interaction between an arbitrary field and the photon field, ~~in other~~ by the requirement of the gauge invariance.

By the second purpose, e.g. a real field cannot interact with ~~II. We can easily~~ the photon. (But notice that a complex field that does not undergo a ~~transfor~~ change under a Gauge transformat. will not interact with the photon either. e.g. the neutron field belong to this class).

II. We can easily formulate a " theory for Meson gauge transformation by requiring that $U_k \rightarrow U_k e^{i\alpha \varepsilon_k e}$, $U_k^* \rightarrow U_k^* e^{-i\alpha \varepsilon_k e}$ and get a "~~meson~~ charge & current" density when α is a constant, (ε_k = meson charge number of the particle U_k.)

In order to fix the interact. between an arbitrary field & the meson field a Gauge transformat. of the second kind for the meson field is necessary. This can be done in the case of neutral mesons by writing

(Incorrect ? see IV)

$$\mathcal{L} = -\tfrac{1}{2} F_{\nu\mu}^* F_{\nu\mu} - K^2 U_\nu^* U_\nu + 2A_\nu^* M_\nu \; ?? \; -2\tfrac{1}{K} B M_{\nu/\nu}$$

where A_ν is Stuckelberg's potential (cf. Pauli R.M.P.) & regarding A_ν & B as independent variables. Gauge transf.: $A_\nu \rightarrow A_\nu + \frac{1}{K} f_{/\nu}$, $B \rightarrow B - Kf$.

For charged mesons, the theory perhaps works too, with α = an operator not commutable with T_3 (isotopic spin).

III. Notice that for photon field, $\varepsilon_k = 0$ or 1 because all particles are either uncharged or have charge e in photon field. But for meson

V6

field $\frac{g}{e} = \varepsilon_k \cong 3$. if U_k = heavy particle field

$\frac{g}{e} = \varepsilon_k \cong 10^{-8}$ if U_k = light particle field.

IV. The gauge transformat. described in II will not lead to any fixat. of the interact. because the field part of the $\mathcal{L}$ and the particle part are respectively invariant under the transform., while in the e.m. case they both vary.

This constitutes perhaps a fundamental difference between a field with $\kappa = 0$ & one with $\kappa \neq 0$.

Masses of Particles

V. The present theory cannot give a satisfactory account of the masses of the different particles because of infinite difficulties and incompleteness of the theory, e.g. we cannot expect to be able to derive the mass ratio of the proton & the electron, or of the proton & the neutron on the present theory.

In a completely satisfactory theory there should be only three universal constants: h, c, G (gravitational const.) which in natural units are all 1. The mass or compton wave length of the particles can be derived.

We see that a theory of such kind is necessarily a merge-together of the Q. theory & general relativity (in which the masses of the particles are defined)

Point worth noticing: if electron radius = range of meson force, $M_{meson} = 137\, m$!

(V6a)

Gauge Invariance, Charge Density & Neutral Particles

* If we separate each complex wave funct. into its real & imaginary parts

$$U = v + iw,$$

the Gauge transformat. becomes

$$v \to v + \alpha w$$
$$w \to w - \alpha v.$$

$\frac{\delta L}{\delta \alpha} = 0$ represents Gauge invariance & leads to a current S_μ with divergence $= 0$,

$$S_\mu = \text{const.} \left[\frac{\partial L}{\partial v_\kappa} w - \frac{\partial L}{\partial w_\kappa} v \right].$$

* For particles of spin 0 & 1, L has terms $v.v$, $w.w$ but no vw.

∴ S_μ = term $v.w$.

∴ $S_\mu = 0$ when $v = 0$ or $w = 0$, which gives ~~Neutral particles~~ Neutral particles

" " " " " $\frac{1}{2}$, L has terms $v.w$, no $v.v$, $w.w$

∴ S_μ = term vv and ww

∴ $S_\mu \neq 0$ when $v = 0$ or $w = 0$. Thus we has no neutral particle.

If, however, we adopt Heisenberg's recipe for passing to the q-no. theory S_μ ~~becomes~~ have $v.w$ term again & we can have neutral particles. But this seems unsatisfactory because if we put $w = 0$ from the beginning we don't get a L. (see Pauli p.225, 228)

REMEMBERING ROBERT MILLS*

C. N. Yang

In 1953–1954, I was visiting Brookhaven and Bob was my office mate. We discussed many things in physics, from the experimental results pouring out of the new Cosmotron, to theoretical topics like renormalization and the Ward identity. It was in that year that we found the very elegant and unique generalization of Maxwell's equation. We were pleased by the beauty of the generalization, but neither of us had anticipated its great impact on physics 20 years later.

Bob spent one year, I think it was 1955–1956, at the Institute for Advanced Study in Princeton and we resumed our collaboration. One fruit of that was a paper on the overlapping divergence in the photon propagator which, however, was not written up for publication [1] until 1966, when he and his family visited us for the summer just after I had moved to [SUNY] Stony Brook.

Bob was an old-fashioned man. Among all the physicists whom I know, he was certainly one of the most honest and the most sincere.

Bob had a brilliant mind. He was very quick at grasping new ideas. I shall treasure the memory of our intensive collaboration and of our many discussions on diverse topics ranging from accelerator theory to the theory of computability.

[1] R. L. Mills and C. N. Yang, *Prog. Theor. Phys. Sup.* **37**, 507 (1966).

*Excerpt reprinted with permission from *Physics Today*, October 2003.